The Insects
Structure and Function

4th edition

R. F. Chapman

CAMBRIDGE
UNIVERSITY PRESS

PUBLISHED BY THE PRESS SYNDICATE OF THE UNIVERSITY OF CAMBRIDGE
The Pitt Building, Trumpington Street, Cambridge CB2 1RP, United Kingdom

CAMBRIDGE UNIVERSITY PRESS
The Edinburgh Building, Cambridge CB2 2RU, UK http://www.cup.cam.ac.uk
40 West 20th Street, New York, NY 10011-4211, USA http://www.cup.org
10 Stamford Road, Oakleigh, Melbourne 3166, Australia

First published by Edward Arnold 1969
Second edition 1971, reprinted 1972, 1974, 1975, 1978, 1980
Third edition 1982, reprinted 1983, 1985, 1988, 1991
Fourth edition published by Cambridge University Press 1998

Printed in the United Kingdom at the University Press, Cambridge

Typeset in MT Ehrhardt 9/12pt   [SE]

A catalogue record for this book is available from the British Library

Library of Congress Cataloguing in Publication data
Chapman, R.F. (Reginald Frederick)
    The insects : structure and function / R.F. Chapman. – 4th ed.
        p.   cm.
    Includes indexes.
    ISBN 0 521 57048 4 (hb). – ISBN 0 521 57890 6 (pb)
    1. Insects.   I. Title.
    QL463.C48   1998
    595.7 – dc21   97-35219 CIP

ISBN 0 521 57048 4 hardback
ISBN 0 521 57890 6 paperback

Contents

Preface

The 30 years since the first appearance of this book have seen a revolution in biology and the impact on entomology has been enormous. In writing the first edition, I tried to approach the insect as a whole organism and to combine functional morphology with physiology. That is still my aim in this rewriting, but the growth of biological science has required me continually to make decisions about where to draw the line. In particular, I felt it was impossible for me to do justice to molecular studies. I made this decision reluctantly, because molecular biology is so obviously a key to our understanding of how insects work, and I do make some reference to molecular work, especially where it causes us to rethink older ideas.

I believe it is critically important to produce a synthesis in a single volume. This is essential for the students of entomology for whom the book was, and is, primarily intended. I hope, also, to provide a useful reference for biologists in other fields who use insects as models, but I want to encourage them to think of the insect as a whole. Where does their system belong in the functioning of the organism? Systems in isolation sometimes don't make much sense.

While I have retained the major divisions that I used in the earlier editions, I have combined some chapters and rearranged them to some extent. I have adopted American spellings throughout. This was done, not without misgivings, but the principal market for the book is in North America, and the United States is my home. Nevertheless, I confess that 'esophagus' does stick in the gullet!

To facilitate further study, I have cited reviews wherever a recent review was known to me and I have considerably extended the references to primary literature. In doing this, I have tried to select recent papers, even though these may not always be the most relevant. My aim is to give an entrée to the literature, rather than to cite a particular paper. In an attempt to standardize the names of taxa, I have followed the terminology used in the 1991 edition of *The Insects of Australia*.

I have been extremely fortunate, and privileged, to have spent the last few years at the University of Arizona where the breadth of interest in insects is, perhaps, greater than anywhere else in the world. Through the Center for Insect Science, I have encountered an extraordinarily diverse array of scientists working with insects. This has greatly broadened my knowledge of many aspects of insect biology and, I believe, has contributed greatly to my ability to undertake this revision.

Throughout the several years that rewriting has taken, I have had constant support from my colleagues in the Division of Neurobiology at the University of Arizona. I thank them for their understanding and help. Various colleagues have also made significant contributions to specific chapters. They are: the late Ed Arbas, Norm Davis, John Glendinning, Rick Levine, David Morton and Leslie Tolbert of the Division of Neurobiology, Henry Hagedorn of the Department of Entomology, J.E. Baker, of the USDA, Savannah, USA, and Robin Wootton, of the University of Exeter, UK. Many colleagues around the world have graciously allowed me to see preprints of their papers and have contributed in countless other ways. I am grateful to them all. I hope they will forgive me for not referring to them all individually.

In this edition I have included a few photographs and I appreciate the willingness with which my scientific colleagues have provided them, often going to great lengths to fill my requests. They are: Dr E.A. Bernays, Dr R. Dallai, Dr A.R. Fontaine, Dr A.P. Gupta, Dr S.G.S. Gunnarson, Dr B.G.M. Jamieson, Dr J.H. Koenig, Dr M. Locke, Dr I.A. Meinertzhagen, Dr A.F. Rowley, Dr D.S. Smith, Dr R.A. Steinbrecht, Dr F. Tjallingii, and Dr L.T. Wasserthal. Chip Hedgecock has been an ever-willing helper in the preparation of these photographs for publication.

Throughout I have been supported by Elizabeth Bernays. She has read and re-read every word, checked every illustration and made innumerable suggestions to improve the content and presentation. She has also kept me fed and watered! Without her love I could never have completed the task.

Tracey Sanderson at Cambridge University Press has also been an unfailing support. She, too, has been through the whole manuscript with a fine tooth comb. Her comments on the manuscript have done much to make it more readable. I also appreciate her helpful and timely feedback to my many questions.

Some of the illustrations were done by Kristin Sonderegger and Ron Adams, and the University of Arizona's Authors' Support Fund helped defray some of the costs. Terry Villelas painstakingly typed most of the tables.

Reg Chapman

Tucson, Arizona
July, 1997

Acknowledgments

The following publishers and societies have kindly allowed me to use illustrations from previously published work:

Academic Press Ltd

Fig. 2.13a. Reprinted from *Animal Behaviour*, **34**, S.J. Simpson & R.J. Ludlow, Why locusts start to feed, pp. 480–96, 1986, by permission of the publisher Academic Press Limited London.

Fig. 3.26a. Reprinted from *Advances in Insect Physiology*, **19**, J.E. Phillips, J. Hanrahan, M. Chamberlain & B. Thomson, Mechanism and control of reabsorption in the insect hindgut, pp. 329–422, 1986, by permission of the publisher Academic Press Limited.

Fig. 5.27d. Reprinted from *Journal of Ultrastructure Research*, **64**, M. Monpeyssin & J. Baeulaton, Hemocytopoiesis in the oak silkworm, pp. 35–45, 1978, by permission of the publisher Academic Press Limited Orlando.

Fig. 10.1e,f. Reprinted from *Muscle*, D.S. Smith, 1972, by permission of the publisher Academic Press Limited Orlando.

Fig. 10.7. Reprinted from *Advances in Insect Physiology*, 1, D.S. Smith & J.E. Treherne, Functional aspects of the organization of the insect nervous system, pp. 401–84, 1963, by permission of the publisher Academic Press Limited London.

Fig. 12.18a. Reprinted from *Advances in Insect Physiology*, **24**, D.W. Stanley-Samuelson, Prostaglandins and related eicosanoids in insects, pp. 115–212, 1994, by permission of the publisher Academic Press Limited London.

Fig. 13.11b,c. Reprinted from *Developmental Biology*, **76**, G.D. Mazur, J.C. Regier, F.C. Kafatos, The silkmoth chorion: morphogenesis of surface structures, pp. 305–21, 1980, by permission of the publisher Academic Press Limited Orlando.

Fig. 15.20a. Reprinted from *Developmental Biology*, **26**, G. Schubiger, Regeneration, duplication and transdetermination in fragments of the leg disc of Drosophila, pp. 277–95, 1971, by permission of the publisher Academic Press Limited Orlando.

Fig. 17.15. Reprinted from *Advances in Insect Physiology*, **26**, L.T. Wasserthal, Interaction of circulation and tracheal ventilation in holometabolous insects, pp. 297–351, 1996, by permission of the publisher Academic Press Limited London.

Fig. 18.2. Reprinted from *International Review of Cytology*, **49**, H. Komnick, Chloride cells and chloride epithelia of aquatic insects, pp. 285–329, 1977, by permission of the publisher Academic Press Limited Orlando.

Fig. 21.4b. Reprinted from *General and Comparative Endocrinology*, W.S. Herman & L.I. Gilbert. The neuroendocrine system of Hyalophora, pp. 275–91, 1966, by permission of the publisher Academic Press Limited Orlando.

Fig. 21.10a. Reprinted from *Advances in Insect Physiology*, **18**, S.S. Tobe & B. Stay, Structure and regulation of the corpus allatum, pp. 305–432, 1985, by permission of the publisher Academic Press Limited London.

Fig. 26.12. Reprinted from *Animal Behaviour*, **29**, D.E. Cowling & B. Burnet, Courtship songs and genetic control of their acoustic characteristics in sibling species of the Drosophila melanogaster subgroup, pp. 924–35, 1981, by permission of the publisher Academic Press Limited London.

Fig. 26.20b. Reprinted from *Biological Journal of the Linnean Society*, **24**, M.F. Claridge, J. den Hollander & J.C. Morgan, Variation in courtship signals and hybridization between geographically definable populations of the rice brown planthopper, Nilaparvata, pp. 35–49, 1985, by permission of the publisher Academic Press Limited London.

Fig. 27.20a. Reprinted from *Advances in Insect Physiology*, **14**, B.W. Staddon, The scent glands of Heteroptera, pp. 351–418, 1979, by permission of the publisher Academic Press Limited London.

Addison Wesley Longman Ltd
Fig. 15.41

American Association for the Advancement of Science
Fig. 10.20 reprinted with permission from *Science*, **268**, 50–1. Copyright 1995 American Association from the Advancement of Science.

American Institute of Physics
Fig. 23.18a

American Physiological Society
Figs. 8.19, 8.27c

Annual Reviews Inc.
Fig. 9.3 with permission from the *Annual Review of Entomology*, volume **37**, copyright 1992, by Annual Reviews Inc.
Fig. 19.17c with permission from the *Annual Review of Entomology*, volume **30**, copyright 1985, by Annual Reviews Inc.

Fig. **20.8** with permission from the *Annual Review of Entomology*, volume **38**, copyright 1985, by Annual Reviews Inc.

Balaban Publishers
Fig. 12.4a

Biological Bulletin of the Marine Biological Laboratory, Woods Hole
Figs. 4.8, 15.32

Blackwell Scientific Publications Ltd
Figs. 2.8, 2.12a, 4.11, 4.12. 13.8b, 13.10, 15.2b, 24.4, 24.9b

Cambridge University Press
Figs. 10.21, 12.4b

Chapman & Hall
Figs. 2.6a,b, 2.7a, 2.13b

Churchill Livingstone
Figs. 2.16b, 2.17, 3.7a, 3.8a, 3.12b, 6.3a–c, 6.6, 24.2b

Cold Spring Harbor Laboratory Press
Figs. 15.20b–d, 15.21b,c

Company of Biologists Ltd
Figs. 2.9, 2.10b,d, 3.8b, 3.24, 4.13, 5.13, 8.16, 8.18, 9.20, 9.24, 9.25, 9.27, 9.28, 9.34b,c, 9.37, 10.4, 10.9, 13.19a,c, 15.19b, 16.23a, 17.19, 18.6, 18.11a, 18.14a, 19.1a, 19.2b, 19.3a,b, 19.8, 19.9, 19.10a, 19.11, 20.3b, 20.8b, 20.9, 21.4a, 21.7b, 22.15b, 23.12c, 23.17c, 25.11b, 26.3c, 26.10b, 26.10c, Table 10.1

Cornell University Press
Fig. **26.3a** reprinted from *Cricket Behavior and Neurobiology*, edited by Franz Huber, Thomas E. Moore, and Werner Loher; original drawing by H.C. Bennett-Clark. Copyright 1989 by Cornell University. Used by permission of the publisher, Cornell University Press.

Elsevier Science Ltd
With kind permission from Elsevier Science Ltd, The Boulevard, Langford Lane, Kidlington OX5 1GB, UK.
Fig. **3.3, 3.10b, 3.15, 3.16.** Reprinted from G.A. Kerkut & L.I. Gilbert, *Comprehensive Insect Physiology, Biochemistry and Pharmacology*, 1985, vol. 4, 165–211.
Fig. **3.5b.** Reprinted from the *International Journal of Insect Morphology & Embryology*, 7, G. Del Bene, R. Dallai & D. Marchini, Ultrastructure of the midgut and adhering tubular salivary glands in *Frankliniella*, pp. 15–24, 1991.

Fig. **3.14.** Reprinted from the *Journal of Insect Physiology*, **22**, M. Schmitz & H. Komnick, Rectale Chloridepithelium und osmoregulatorische salzaufnaume durch den Enddarm von Zygopteren und Anisopteran Libellulenlarven, pp. 875–83, 1976.
Fig. **3.22.** Reprinted from *Journal of Insect Physiology*, **10**, W.A.L. Evans & D.W. Payne, Carbohydrases in the alimentary tract of the desert locust, pp. 657–74, 1964.
Fig. **4.3.** Reprinted from G.A. Kerkut & L.I. Gilbert, *Comprehensive Insect Physiology, Biochemistry and Pharmacology*, 1985, vol. 4, pp. 313–90.
Fig. **4.7b.** Reprinted from the *Journal of Insect Physiology*, **28**, M.E. Montgomery, Life-cycle nitrogen budget for the gypsy moth, pp. 437–42, 1982.
Fig. **4.10.** Reprinted from the *Journal of Insect Physiology*, **38**, W.A. Prosser, S.J. Simpson & A.E. Douglas. How an aphid symbiosis responds to variation in dietary nitrogen, pp. 301–7, 1992.
Fig. **4.14** Reprinted from *Comparative Biochemistry and Physiology*, **98A**, M.G. Kaufman & P.A. Klug, The contribution of hindgut bacteria to dietary carbohydrate utilization by crickets, pp. 117–23, 1992.
Fig. **4.15.** Reprinted from the *Journal of Insect Physiology*, **33**, A.E. Douglas & A.F.G. Dixon, The mycetocyte symbiosis of aphids, pp. 109–13, 1987.
Fig. **4.16b.** Reprinted from the *Journal of Insect Physiology*, **38**, W.A. Prosser & A.E. Douglas, A test of the hypothesis that nitrogen is upgraded and recycled in an aphid symbiosis, pp. 93–9, 1992.
Fig. **5.6a–c.** Reprinted from the *International Journal of Insect Morphology & Embryology*, **22**, H.W. Krenn & G. Pass, Wing hearts in Mecoptera, pp. 63–76, 1993.
Fig. **5.15.** Reprinted from the *Journal of Insect Physiology*, **16**, R.R. Mills & D.L. Whitehead, Hormonal control of tanning in the American cockroach, pp. 331–40, 1970.
Fig. **5.15.** Reprinted from the *Journal of Insect Physiology*, **26**, S.W. Nicolson, Water balance and osmoregulation in Onymacris, pp. 315–20, 1980.
Fig. **5.20a,b.** Reprinted from the *Journal of Insect Physiology*, **33**, R.C. Duhamel & J.G. Kunkel, Moulting-cycle regulation of haemolymph protein clearance in cockroaches, pp. 155–8, 1987.
Fig. **5.20c.** Reprinted from *Insect Biochemistry and Molecular Biology*, **15**, L.M. Riddiford & R.H. Hice, Development profiles of the mRNAs for Manduca arylpghorin and two other storage proteins, pp. 489–502, 1985.
Fig. **5.32c,d.** Reprinted from the *Journal of Insect Physiology*, **33**, B. Guzo & D.B. Stoltz, Observations on cellular immunity and parasitism in the tussock moth, pp. 19–31, 1987.
Fig. **6.2b–d, 6.3d.** Reprinted from G.A. Kerkut & L.I. Gilbert, *Comprehensive Insect Physiology, Biochemistry and Pharmacology*, 1985, vol. 3, pp. 155–310.

Fig. 6.4. Reprinted from *Insect Biochemistry and Molecular Biology*, **20**, N.H. Haunerland, K.K. Nair & W.S. Bowers, Fat body heterogeneity during development of Heliothis zea, pp. 829–37, 1990.

Fig. 6.7. Reprinted from *Comparative Biochemistry and Physiology*, **91A**, A. Gies, T. Fromm & R. Ziegler, Energy metabolism in starving larvae of Manduca sexta, pp. 549–55, 1992.

Fig. 8.28. Reprinted from the *Journal of Insect Physiology*, **41**, D. Berrigan & D.J. Pepin, How maggots move, pp. 329–337, 1995.

Fig. 8.31. Reprinted from G.A. Kerkut & L.I. Gilbert, *Comprehensive Insect Physiology, Biochemistry and Pharmacology*, 1985, vol. 5, pp. 467–90.

Fig. 10.1d. Reprinted from *Progress in Biophysics and Molecular Biology*, **16**, D.S. Smith, The organization and function of the sarcoplasmic reticulum and T-system of muscle cells, pp. 109–42, 1966.

Fig. 10.12. Reprinted from the *Journal of Insect Physiology*, **19**, M. Anderson & L.H. Finlayson, Ultrastructural changes during growth of the flight muscles in the adult tsetse fly, pp. 1989–97, 1973.

Fig. 10.15c. Reprinted from the *Journal of Insect Physiology*, **29**, W.K. Jorgensen & M.J. Rice, Superextension and supercontraction in locust ovipositor muscles, pp. 437–48, 1983.

Fig. 12.4b. Reprinted from *Journal of Insect Physiology*, **35**, A.K. Raina, Male-induced termination of sex pheromone production and receptivity in mated females of Heliothis zea, pp. 821–26, 1989.

Fig. 13.5a–c, 13.6. Reprinted from the *International Journal of Insect Morphology & Embryology*, **22**, J. Büning, Germ cell cluster formation in insect ovaries, pp. 237–53, 1993.

Fig. 13.5d. Reprinted from *International Journal of Insect Morphology & Embryology*, **22**, E. Huebner & W. Diehl-Jones, Nurse cell–oocyte interaction in the telotrophic ovary, pp. 369–87, 1993.

Fig. 13.8a. Reprinted from the *Journal of Insect Physiology*, **33**, M.J. Klowden, Distension-mediated egg maturation in the mosquito *Aedes aegypti*, pp. 83–7, 1987.

Fig. 13.9c. Reprinted from *Insect Biochemistry and Molecular Biology*, **8**, G. Gellisen & H. Emmerich, Changes in the titer of vitellogenin and of diglyceride carrier protein in the blood of adult Locusta, pp. 403–12, 1978.

Fig. 13.10d. Reprinted from the *Journal of Insect Physiology*, **36**, C.-M. Yin, B.-X.Zou, S.-X.Yi & J.G. Stoffolano, Ecdysteroid activity during oogenesis in the black blowfly, *Phormis regina* (Meigen), pp. 375–82, 1900.

Fig. 14.26c. Reprinted from the *Journal of Insect Physiology*, **25**, M. Lagueux, C. Hetru, F. Goltzene, C. Kappler & J.A. Hoffmann, Ecdysone titre and metabolism in relation to cuticulogenesis in embryos of Locusta migratoria, pp. 709–23, 1979.

Fig. 14.30. Reprinted from *Journal of Insect Physiology*, **20**, D.C. Denlinger & W.-C. Ma, Dynamics of the pregnancy cycle in the tsetse fly, pp. 1015–26, 1974.

Fig. 15.12a. Reprinted from the *International Journal of Insect Morphology & Embryology*, **18**, R.F. Chapman & J. Fraser, The chemosensory system of the monophagus grasshopper Bootettix, pp. 111–18, 1989.

Fig. 15.35. Reprinted from the *Journal of Insect Physiology*, **36**, A. Rachinsky & K. Hartfelder, Corpora allata activity, a prime regulating element for caste-specific juvenile hormone titre in honey bee larvae, pp. 189–94, 1990.

Fig. 15.36. Reprinted from the *Journal of Insect Physiology*, **29**, D.E. Wheeler & H.F. Nijhout, Soldier determination in Pheidole, pp. 847–54, 1983.

Fig. 15.37. Reprinted from G.A. Kerkut & L.I. Gilbert, 1985, *Comprehensive Insect Physiology, Biochemistry and Pharmacology*, vol. 8, pp. 441–90.

Fig. 15.43. Reprinted from the *Journal of Insect Physiology*, **21**, R.A. Bell, D.R. Nelson, T.K. Borg & D.L. Cardwell, Wax secretion in non-diapausing and diapausing pupae of the tobacco hornworm, pp. 1725–9, 1975.

Fig. 16.19a. Reprinted from *Insect Biochemistry and Molecular Biology*, **13**, R.F. Ahmed, T.L. Hopkins & K.J. Kramer, Tyrosine and tyrosine glucoside titres in whole animals and tissues during development of the tobacco hornworm, pp. 369–74, 1983.

Fig. 17.17. Reprinted from the *Journal of Insect Physiology*, **34**, J.R.B. Lighton, Simultaneous measurement of oxygen uptake and carbon dioxide emission during discontinuous ventilation, pp. 361–7, 1988.

Fig. 17.18. Reprinted from the *Journal of Insect Physiology*, **39**, J.R.B. Lighton, Ventilation in *Cataglyphis bicolor*, pp. 687–99, 1993.

Fig. 18.11b. Reprinted from the *Journal of Insect Physiology*, **32**, A.G. Appel, D.A. Reierson & M.K. Rust, Cuticular water loss in the smokybrown cockroach, pp. 623–8, 1986.

Fig. 18.15c. Reprinted from the *Journal of Insect Physiology*, **29**, D. Rudolph, The water-vapour uptake system of the Phthiraptera, pp. 15–25, 1983.

Fig. 20.5c. Reprinted from G.A. Kerkut & L.I. Gilbert, *Comprehensive Insect Physiology, Biochemistry and Pharmacology*, 1985, vol. 5, pp. 139–79.

Fig. 21.10b. Reprinted from *Insect biochemistry and Molecular Biology*, **20**, P. Jesudason, K. Venkatesh & R.M. Roe, Haemolymph juvenile hormone esterase during the life cycle of the tobacco hornworm, pp. 593–604, 1990.

Fig. 24.4. Reprinted from the *Journal of Insect Physiology*, **39**, H. Ljunberg, P. Anderson & B.S. Hansson, Physiology and morphology of pheromone-specific sensilla on the antennae of male and female Spodoptera, pp. 253–90, 1993.

Fig. **24.10c.** Reprinted from the *Journal of Insect Physiology*, 37, L.M. Schoonhoven, M.S.J. Simmonds & W.M. Blaney, Changes in the responsiveness of the maxillary styloconic sensilla of Spodoptera, pp. 261–68, 1991.

Fig. **25.12.** Reprinted from *Insect Biochemistry and Molecular Biology*, 21, P.B. Koch, Precursors of pattern specific ommatin in red wing scales of the polyphenic butterfly Araschia levana, pp. 7853–94, 1991.

Elsevier Trends Journals
Fig. **8.20** from M. Burrows, Local circuits for the control of leg movements in an insect, *Trends in Neuroscience*, 15, 226–32, 1992.

Entomological Society of America
Figs. **3.2c, 15.11, 17.33b, 26.2**

Entomological Society of British Colombia
Fig. **16.9**

Entomological Society of Canada
Fig. **5.6h**

Evolution
Fig. **19.6**

Journal of Neuroscience
Fig. **15.12b.** Figure 1A, Pflüger *et al.*, Activity-dependent structural dynamics of insect sensory fibers. *The Journal of Neuroscience*, 14(11): 6948, (1994).

Fig. **15.23b.** Figures 1A, 1D, Truman & Reiss, Hormonal regulation of the shape of identified motor neurons in the moth Manduca sexta. *The Journal of Neoroscience*, 8(3): 766, (1988).

Kluwer Academic Publishers
Fig. **3.21b.** Reprinted from *Entomologia Experimentalis et Applicata*, 7, 1964, pp. 125–30, Studies on the secretion of digestive enzymes in *Locusta migratoria* L. II, M.A. Khan, with kind permission from Kluwer Academic Publishers.

Fig. **17.33a.** Reprinted from *Hydrobiologia*, 215, 1991, pp. 223–9, Flow patterns around cocoons and pupae of black flies, M. Eymann, with kind permission from Kluwer Academic Publishers.

Fig. **19.4.** Reprinted from *Entomologia Experimentalis et Applicata*, 9, 1966, pp. 127–78, The body temperature of the desert locust, W.J. Stower & J.F. Griffiths, with kind permission from Kluwer Academic Publishers.

Macmillan Magazines Ltd
Fig. **22.15c** reprinted with permission from *Nature*, T. Labhart, Polarization-opponent interneurons in the insect visual system. Copyright 1988 Macmillan Magazines Limited.

Plenum Press
Figs. **10.2, 27.23b**

Royal Entomological Society
Figs. **3.2, 25.18**

Royal Society of London
Fig. **9.18b–d** from J.A. Miyan & A.W. Ewing, *Philosophical Transactions of the Royal Society* of London, B 311, 1985, 271–302, Fig. 5.

Fig. **9.29b** from C.P. Ellington, *Philosophical Transactions of the Royal Society* of London, B 305, 1984, 79–113, Fig. 6b.

Fig. **17.5** from M. Burrows, *Proceedings of the Royal Society of London*, B 207, 1980, 63–78, Fig. 2.

Fig. **20.18** from L. Pearson, *Philosophical Transactions of the Royal Society of London*, B 259, 1974, 477–516, Fig. 32.

Fig. **20.23** from S.G. Matsumoto & J.G. Hildebrand, *Proceedings of the Royal Society of London*, B 213, 1981, 249–77, Fig. 26.

Fig. **24.12b** from V.O.C. Shields & B.K. Mitchell, *Philosophical Transactions of the Royal Society of London*, B 347, 1995, 459–64, Fig. 1c.

Fig. **25.3c** from R.A. Steinbrecht *et al.*, *Proceedings of the Royal Society of London*, B 226, 1985, 367–90, Fig. 6.

E. Schweizerbart'sche Verlagsbuchhandlung
Fig. **3.10c**

Society for Integrative and Comparative Biology
Figs. **3.26b, 21.12**

Oxford University Press
Fig. **9.21** From A.K. Brodsky, *The Evolution of Insect Flight*, 1994, by permission.

Springer-Verlag
Figs. **2.19, 3.21a, 5.7d,e, 5.9, 5.10a,b,d, 8.9e, 8.23d, 8.32b, 9.23, 9.30, 9.35, 12.17, 15.19c, 18.19b,c, 19.3c, 19.10b, 19.15b, 20.5a–c, 20.22, 21.6, 22.4, 22.10, 22.11, 22.12, 22.18, 22.20b, 22.22, 23.5, 23.16, 23.17b, 23.19a, 24.2a, 24.9d, 26.13, 26.15, 26.20a**

Swets & Zeitlinger Publishers
Fig. **5.19** adapted from C. Jeuniaux, G. Duchâteau-Bosson & M. Florkin (1961), *Archives Internationales de Physiologie et de Biochimie*, 69, pp. 617–27, copyright Swets & Zeitlinger Publishers. Used with permission.

Taylor & Francis
Fig. **8.8a**

Westdeutscher Verlag GmbH
also arranged with Nordrhein-Westfalische Akademie de Wissenschaften
Figs. **26.18b, 26.22b**

ACKNOWLEDGMENTS   xvii

Verlag der Zeitschrift fur Naturforschung
Fig. 15.26

John Wiley & Sons, Inc
Reprinted by Permission of Wiley–Liss, Inc., a subsidiary of John
Wiley & Sons, Inc.
Fig. 5.6e–g from Gross and fine structure of the antennal
    circulatory organ in cockroaches, G. Pass, *Journal of
    Morphology*, 1985.
Fig. 5.24 from Water compartmentalization in insects, J. Machin,
    Journal of Experimental Zoology, 1981.
Fig. 5.25a from Anthropod immune system I, A.S. Chiang, A.P.
    Gupta & S.S. Han, *Journal of Morphology*, 1988.
Fig. 10.10a from Flight muscle development in a
    hemimetabolous insect, N.E. Ready & R.K. Josephson, *Journal
    of Experimental Zoology*, 1962.
Fig. 10.10b–e from Growth and development of flight muscle in
    the locust, A.P. Mizisin & N.E. Ready, *Journal of Experimental
    Zoology*, 1986.
Fig. 10.17a from Extensive and intensive factors determining the
    performance of striated muscle, R.K. Josephson, *Journal of
    Experimental Zoology*, 1985.
Fig. 12.2a from Cytodifferentiation in the accessory glands of
    Tenebrio molitor VI, P.J. Dailey, N.M. Gadzama & G.M. Happ,
    Journal of Morphology, 1980.
Fig. 15.21a from Pattern formation in the imaginal wing disc of
    Drosophila, P.J. Bryant, *Journal of Experimental Zoology*, 1975.
Fig. 15.23c from Postembryonic neurogenesis in the CNS of the
    tobacco hornworm, R. Booker & J.W. Truman, *Journal of
    Comparative Neurology*, 1987.
Fig. 15.24c from Lysozyme in the midgut of *Manduca sexta*
    during metamorphosis, V.W. Russell & P.E. Dunn, *Archives of
    Insect Biochemistry and Physiology*, 1991.

Fig. 18.13 from Simultaneous measurements of evaporative
    water loss, oxygen consumption, and thoracic temperature
    during flight, S.W. Nicholson & G.N. Louw, *Journal of
    Experimental Zoology*, 1982.
Fig. 18.15a,b from Novel uptake systems for atmospheric water
    vapor, D. Rudolph & W. Knulle, *Journal of Experimental
    Zoology*, 1982.
Fig. 20.2a from Synaptic organization of columnar elements in
    the lamina of the wild type in Drosophila, I.A.
    Meinertzhagen & S.D. O'Neil, *Journal of Comparative
    Neurology*, 1991.
Fig. 20.2b from Distribution and morphology of synapses on
    nonspiking local interneurons in the thoracic nervous system
    of the locust, A.H.D. Watson & M. Burrows, *Journal of
    Comparative Neurology*, 1988.
Fig. 20.12 from The morphology of nonspiking interneurons in
    the metathoracic ganglion of the locust, M. Siegler & M.
    Burrows, *Journal of Comparative Neurology*, 1979.
Fig. 20.15 from Heterogeneous properties of segmentally
    homologous interneurones in the ventral nerve cords of
    locusts, K.G. Pearson, G.S. Boyan, M. Bastiani & C.S.
    Goodman, *Journal of Comparative Neurology*, 1985.
Fig. 20.20 from Neuroarchitecture of the central complex in the
    brain of the locust, U. Homberg, *Journal of Comparative
    Neurology*, 1991.
Fig. 20.24 from Correlation between the receptive fields of
    locust interneurones, their dendritic morphology and
    the central projections of mechanosensory neurones, M.
    Burrows & P.L. Newland, *Journal of Comparative Neurology*,
    1993.
Fig. 21.13 from Regulation and consequences of cellular changes
    in the prothoracic glands of Manduca sexta, W.S. Smith,
    Archives of Insect Biochemistry and Physiology, 1995.

PART I

The Head, Ingestion, Utilization and Distribution of Food

1   Head

Insects and other arthropods are built up on a segmental plan and their characteristic feature is a hard, jointed exoskeleton. The cuticle, which forms the exoskeleton, is continuous over the whole of the outside of the body and consists of a series of hard plates, the sclerites, joined to each other by flexible membranes, which are also cuticular. Sometimes the sclerites are articulated together so as to give precise movement of one on the next. Each segment of the body primitively has a dorsal sclerite, the tergum, joined to a ventral sclerite, the sternum, by lateral membranous areas, the pleura. Arising from the sternopleural region on each side is a jointed appendage.

In insects, the segments are grouped into three units, the head, thorax and abdomen, in which the various basic parts of the segments may be lost or greatly modified. Typical walking legs are only retained on the three thoracic segments. In the head, the appendages are modified for sensory and feeding purposes and in the abdomen they are lost, except that some may be modified as the genitalia and in Apterygota some pregenital appendages are retained.

1.1 HEAD

The insect head is a strongly sclerotized capsule joined to the thorax by a flexible membranous neck. It bears the mouthparts, comprising the labrum, mandibles, maxillae and labium, and also important sense organs, the antennae, compound eyes and ocelli. On the outside it is marked by grooves most of which indicate ridges on the inside, and some of these inflexions extend deep into the head, fusing with each other to form an internal skeleton. These structures serve to strengthen the head and provide attachments for muscles as well as supporting and protecting the brain and foregut.

The head is derived from the primitive pre-oral archecerebrum and a number of primitively post-oral segments. Molecular studies of *Drosophila* suggest that there are seven postoral segments: labral, ocular, antennal, intercalary, mandibular, maxillary and labial (Schmidt-Ott *et al.*, 1994). The last three segments are often called the

gnathal segments because their appendages form the mouthparts of the insect.

Reviews: Bitsch, 1973; Denis & Bitsch, 1973; DuPorte, 1957; Matsuda, 1965; Snodgrass, 1935, 1960

1.1.1 Orientation

The orientation of the head with respect to the rest of the body varies (Fig. 1.1). The hypognathous condition, with the mouthparts in a continuous series with the legs, is probably primitive. This orientation occurs most commonly in phytophagous species living in open habitats. In the prognathous condition the mouthparts point forwards and this is found in predaceous species that actively pursue their prey, and in larvae, particularly of Coleoptera, which use their mandibles in burrowing. In Hemiptera, the elongate proboscis slopes backwards between the forelegs. This is the opisthorhynchous condition.

The mouthparts enclose a cavity, the pre-oral cavity, with the mouth at its inner end (Fig. 1.2). The part of the pre-oral cavity enclosed by the proximal part of the hypopharynx and the clypeus is known as the cibarium. Between the hypopharynx and the labium is a smaller cavity known as the salivarium, into which the salivary duct opens.

1.1.2 Rigidity

The head is a continuously sclerotized capsule with no outward appearance of segmentation, but it is marked by a number of grooves. Most of these grooves are sulci (singular: sulcus), marking lines along which the cuticle is inflected to give increased rigidity. The term suture should be retained for grooves marking the line of fusion of two formerly distinct plates. The groove which ends between the points of attachment of maxillae and labium at the back of the head is generally believed to represent the line of fusion of the maxillary and labial segments and it is therefore known as the postoccipital suture.

Since the sulci are functional mechanical developments to resist the various strains imposed on the head capsule, they are variable in position in different species and any one of them may be completely absent. However, the needs

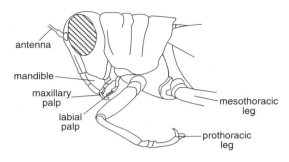

a) hypognathous

antenna

mandible

maxillary palp

labial palp

prothoracic leg

mesothoracic leg

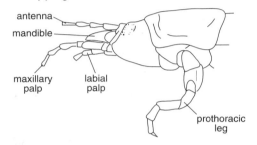

b) prognathous

antenna

mandible

maxillary palp

labial palp

prothoracic leg

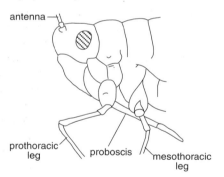

c) opisthorhynchous

antenna

prothoracic leg

proboscis

mesothoracic leg

Fig. 1.1. Orientation of the head. (a) Hypognathous – mouthparts ventral, in a continuous series with the legs (grasshopper). (b) Prognathous – mouthparts in an anterior position (beetle larva). (c) Opisthorhynchous – sucking mouthparts with the proboscis extending back between the front legs (aphid).

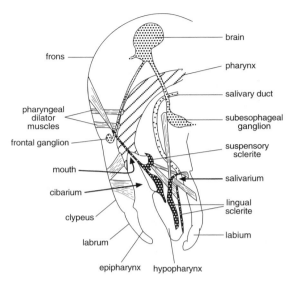

frons

brain

pharynx

salivary duct

pharyngeal dilator muscles

subesophageal ganglion

frontal ganglion

suspensory sclerite

mouth

salivarium

cibarium

lingual sclerite

clypeus

labium

labrum

epipharynx   hypopharynx

Fig. 1.2. Pre-oral cavity and some musculature. Diagrammatic vertical section through the head of an insect with biting and chewing mouthparts. Sclerites associated with the hypopharynx are black with white spots. Muscles attached to these sclerites move the hypopharynx (after Snodgrass, 1947).

between the anterior mandibular articulations. At each end of this sulcus is a pit, the anterior tentorial pit, which marks the position of a deep invagination to form the anterior arm of the tentorium. The lateral margins of the head above the mandibular articulations are strengthened by a horizontal inflexion indicated externally by the subgenal sulcus. This sulcus is generally a continuation of the epistomal sulcus to the postoccipital suture. The part of the subgenal sulcus above the mandible is called the pleurostomal sulcus, the part behind the mandible is the hypostomal sulcus. Another commonly occurring groove is the circumocular sulcus, which strengthens the rim of the eye and may develop into a deep flange protecting the inner side of the eye. Sometimes this sulcus is connected to the subgenal sulcus by a vertical subocular sulcus; the inflexions associated with these sulci act as a brace against the pull of the muscles associated with feeding. The circumantennal ridge, marked by a sulcus externally, strengthens the head at the point of insertion of the antenna, while running across the back of the head, behind the compound eyes, is the occipital sulcus.

The areas of the head defined by the sulci are given names for descriptive purposes, but they do not represent primitive sclerites. Since the sulci are variable in position, so

for strengthening the head wall are similar in the majority of insects, so some of the sulci are fairly constant in occurrence and position (Fig. 1.3). The most constant is the epistomal (frontoclypeal) sulcus, which acts as a brace

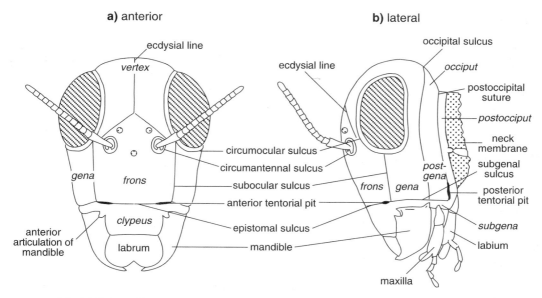

Fig. 1.3. Common lines or grooves on the insect head and the areas which they define (italicized) (modified after Snodgrass, 1960).

too are the areas which they delimit. The front of the head, the frontoclypeal area, is divided by the epistomal sulcus into the frons above and the clypeus below (Fig. 1.3). It is common to regard the arms of the ecdysial cleavage line as delimiting the frons dorsally, but this is not necessarily so (Snodgrass, 1960). From the frons, muscles run to the pharynx, the labrum and the hypopharynx; from the clypeus arise the dilators of the cibarium. The two groups of muscles are always separated by the frontal ganglion and its connectives to the brain (Fig. 1.2). Dorsally the frons continues into the vertex and posteriorly this is separated from the occiput by the occipital sulcus. The occiput is divided from the postocciput behind it by the postoccipital suture, while at the back of the head, where it joins the neck, is an opening, the occipital foramen, through which the alimentary canal, nerve cord and some muscles pass into the thorax.

The lateral area of the head beneath the eyes is called the gena, from which the subgena is cut off below by the subgenal sulcus, and the postgena behind by the occipital sulcus. The region of the subgena above the mandible is called the pleurostoma and that part behind the mandible is the hypostoma.

In hypognathous insects with a thick neck, the posterior ventral part of the head capsule is membranous. The postmentum of the labium is contiguous with this membrane, articulating with the subgena on either side. The hypostomal sulci bend upwards posteriorly and are continuous with the postoccipital suture (Fig. 1.4a). In insects with a narrow neck, permitting greater mobility of the head, and in prognathous insects, the cuticle of the head below the occipital foramen is sclerotized. This region has different origins. In Diptera, the hypostomata of the two sides meet in the midline below the occipital foramen to form a hypostomal bridge which is continuous with the postocciput (Fig. 1.4b). In other cases, Hymenoptera and the water bugs *Notonecta* and *Naucoris*, a similar bridge is formed by the postgenae, but the bridge is separated from the postocciput by the postoccipital suture (Fig. 1.4c). Where the head is held in the prognathous position, the lower ends of the postocciput fuse and extend forwards to form a median ventral plate, the gula (Fig. 1.4d), which may be a continuous sclerotization with the labium. Often the gula is reduced to a narrow strip by enlargement of the postgenae and sometimes the postgenae meet in the midline, so that the gula is obliterated. The median ventral suture which is thus formed at the point of contact of the postgenae is called the gular suture.

In all insects, the rigidity of the head is increased by four deep cuticular invaginations, known as apodemes, which usually meet internally to form a brace for the head

Fig. 1.4. Sclerotization at the back of the head. Notice the position of the bridge below the occipital foramen with reference to the posterior tentorial pit. Membranous areas stippled, compound eyes cross-hatched. The names of areas defined by sulci are italicized (after Snodgrass, 1960). **(a)** Generalized condition, no ventral sclerotization; **(b)** hypostomal bridge (*Deromyia*, Diptera); **(c)** postgenal bridge (*Vespula*, Hymenoptera); **(d)** gular bridge formed from the postoccipital sclerites (*Epicauta*, Coleoptera).

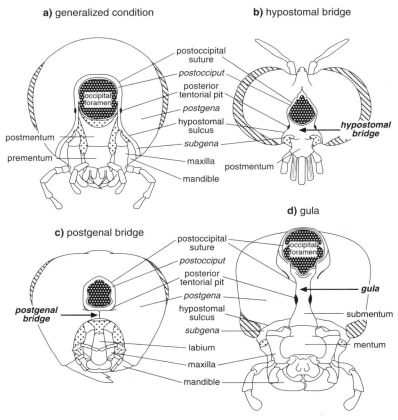

and for the attachment of muscles. The structure formed by these invaginations is called the tentorium (Fig. 1.5). Its two anterior arms arise from the anterior tentorial pits, which in Apterygota and Ephemeroptera are ventral and medial to the mandibles. In Odonata, Plecoptera and Dermaptera the pits are lateral to the mandibles, while in most higher insects they are facial at either end of the epistomal sulcus. The posterior arms arise from pits at the ventral ends of the postoccipital suture and they unite to form a bridge running across the head from one side to the other. In Pterygota the anterior arms also join up with the bridge, but the development of the tentorium as a whole is very variable. Sometimes a pair of dorsal arms arise from the anterior arms and they may be attached to the dorsal wall of the head by short muscles. In Machilidae (Archaeognatha) the posterior bridge is present, but the anterior arms do not reach it, while in Lepismatidae (Thysanura) the anterior arms unite to form a central plate near the bridge and are joined to it by very short muscles.

1.1.3 Molting

Immature insects nearly always have a line along the dorsal midline of the head dividing into two lines on the face so as to form an inverted Y (Fig. 1.3). There is no groove or ridge along this line, and it is simply a line of weakness, continuous with that on the thorax, along which the cuticle splits when the insect molts (see Fig. 16.11). It is therefore called the ecdysial cleavage line, but has commonly been termed the epicranial suture. The anterior arms of this line are very variable in their development and position and, in Apterygota, they are reduced or absent. The ecdysial cleavage line may persist in the adult insect and sometimes the cranium is inflected along this line to form a true sulcus. Other ecdysial lines may be present on the ventral surface of the head of larval insects (Hinton, 1963).

1.2 NECK

The neck or cervix is a membranous region which gives freedom of movement to the head. It extends from the postocciput at the back of the head to the prothorax, and possibly

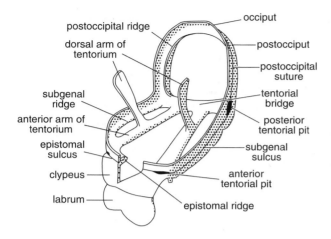

Fig. 1.5. Tentorium. Cutaway of the head capsule to show the tentorium and its relationship with the grooves and ridges of the head (after Snodgrass, 1935).

a) musculature

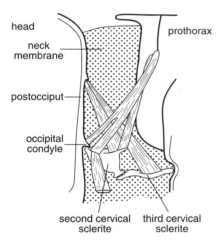

Fig. 1.6. Neck and cervical sclerites of a grasshopper. (a) Seen from the inside to show the muscles (after Imms, 1957). (b) Diagrams showing how a change in the angle between the second and third cervical sclerites retracts or protracts the head. (The first cervical sclerite is small and is not shown). Arrows indicate points of articulation.

b) head movement

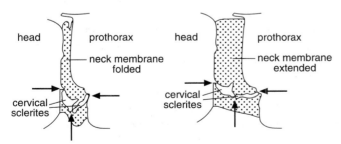

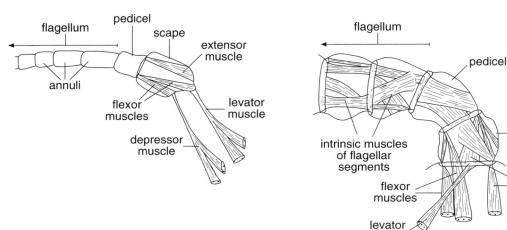

a) annulated

flagellum

pedicel

scape

extensor muscle

annuli

flexor muscles

levator muscle

depressor muscle

b) segmented

flagellum

pedicel

intrinsic muscles of flagellar segments

scape

flexor muscles

extensor muscles

levator muscle

Fig. 1.7. Antenna. Proximal region showing the musculature. (**a**) Typical insect annulated antenna. There are no muscles in the flagellum (*Locusta*, Orthoptera). (**b**) Segmented antenna of a non-insect hexapod (*Japyx*, Diplura) (after Imms, 1940).

represents the posterior part of the labial segment together with the anterior part of the prothoracic segment. Laterally in the neck membrane are the cervical sclerites. Sometimes there is only one, as in Ephemeroptera, but there may be two or three. In *Schistocerca* (Orthoptera) the first lateral cervical sclerite, which articulates with the occipital condyle at the back of the head, is very small. The second sclerite articulates with it by a ball and socket joint allowing movement in all planes. Posteriorly it meets the third (posterior) cervical sclerite and movement at this joint is restricted to the vertical plane. The third cervical sclerite connects with the prothoracic episternum, relative to which it can move in all planes.

Muscles arising from the postocciput and the pronotum are inserted on the cervical sclerites (Fig. 1.6a) and their contraction increases the angle between the sclerites so that the head is pushed forwards (Fig. 1.6b). A muscle arising ventrally and inserted on to the second cervical sclerite may aid in retraction or lateral movements of the head. Running through the neck are longitudinal muscles, dorsal muscles from the antecostal ridge of the mesothorax to the postoccipital ridge, and ventral muscles from the sternal apophyses of the prothorax to the postoccipital ridge or the tentorium. These muscles serve to retract the head on to the prothorax, while their differential contraction will cause lateral movements of the head. *Schistocerca* has 16 muscles on each side of the neck, each of which is innervated by several axons, often including an inhibitory

fiber. This polyneuronal innervation, together with the versatility of the cervical articulations and the complexity of the musculature, permits movement of the head in a highly versatile and accurately controlled manner.

1.3 ANTENNAE

All insects possess a pair of antennae, but they may be greatly reduced, especially in larval forms. Amongst the non-insectan Hexapoda, Collembola and Diplura have antennae, but Protura do not.

1.3.1 Antennal structure

The antenna consists of a basal scape, a pedicel and a flagellum. The scape is inserted into a membranous region of the head wall and pivoted on a single marginal point, the antennifer (Fig. 1.8a), so it is free to move in all directions. Frequently the flagellum is divided into a number of similar annuli joined to each other by membranes so that the flagellum as a whole is flexible. The term segmented should be avoided with reference to the flagellum of insects since the annuli are not regarded as equivalent to leg segments.

The antennae of insects are moved by levator and depressor muscles arising on the anterior tentorial arms and inserted into the scape, and by flexor and extensor muscles arising in the scape and inserted into the pedicel (Fig. 1.7a). There are no muscles in the flagellum, and the nerve which

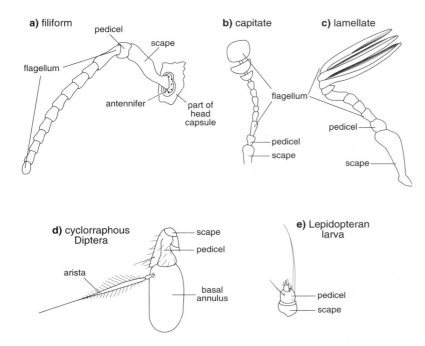

Fig. 1.8. Antennae. Different forms occurring in different insects. Not all to same scale.

traverses the flagellum is purely sensory. This is the annulated type of antenna. In Collembola and Diplura the musculature at the base of the antenna is similar to that in insects, but, in addition, there is an intrinsic musculature in each unit of the flagellum (Fig. 1.7b), and, consequently, these units are regarded as true segments.

The number of annuli is very variable between species. Adult Odonata, for example, have five or fewer annuli while adult *Periplaneta* have over 150, increasing from about 48 in the first stage larva.

The form of the antenna varies considerably depending on its precise function (Fig. 1.8). Sometimes the modification produces an increase in surface area allowing a large number of sensilla to be accommodated on the antenna (Fig. 1.9) and, in the case of the plumose antennae of some male moths, enabling them to sample a large volume of air. Sexual dimorphism in the antennae is common, those of the male often being more complex than those of the female. This often occurs where the male is attracted to or recognizes the female by her scent. Conversely, in chalcids scent plays an important part in host-finding by the female and in this case the female's antennae are more specialized than the male's.

The antennae of larval hemimetabolous insects are similar to those of the adult, but with fewer annuli. The number increases at each molt (see Fig. 15.10). In *Periplaneta*, for example, there are only 48 annuli in the first stage larva compared with over 150 in the adult. The antennae of larval holometabolous insects are usually considerably different from those of the adult. The larval antennae of Neuroptera and Megaloptera have a number of annuli, but in larval Coleoptera and Lepidoptera the antennae are reduced to three simple segments. In some larval Diptera and Hymenoptera the antennae are very small and may be no more than swellings of the head wall.

1.3.2 Sensilla on the antennae

The antennae are primarily sensory structures and they are richly endowed with sensilla in most insects. It is characteristic of insects that the pedicel contains a chordotonal organ, Johnston's organ, which responds to movement of the flagellum with respect to the pedicel (see section 23.2.3.2). In addition, the scape and pedicel often have hair plates and groups of campaniform sensilla that provide information on the positions of the basal segments with respect to the head and to each other. Scattered mechanosensory hairs are also often present on these segments.

The principal sensilla on the flagellum of most insects are olfactory, and these have a variety of forms (see section 24.1.1). It is common for contact chemoreceptors, mechanoreceptors

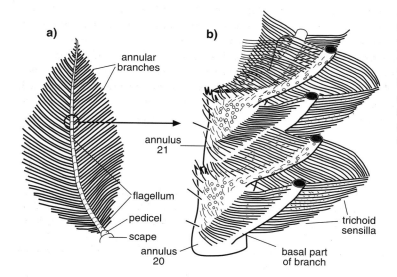

a)

annular branches

annulus 21

flagellum

pedicel

scape

annulus 20

b)

trichoid sensilla

basal part of branch

Fig. 1.9. Antenna. Plumose form providing space for large numbers of sensilla (male of the moth *Telea polyphemus*) (after Boeckh, Kaissling & Schneider, 1960). **(a)** The whole antenna seen from above. Two slender branches arise on opposite sides of each annulus. **(b)** Detail of two annuli from the side showing the bases of the branches and arrangement of long trichoid olfactory sensilla along the branches.

and thermohygroreceptors also to be present. Where the flagellum is made up of a series of similar annuli, successive annuli often have a similar arrangement of sensilla, but the sensilla are often concentrated in particular regions. In *Melanoplus* (Orthoptera), for instance, there are no basiconic or coeloconic pegs on the proximal annuli; most of these sensilla are found on the annuli in the middle of the flagellum (Fig. 1.10). In Pieridae (Lepidoptera), most of the antennal sensilla are aggregated on the terminal club. The terminal annulus often has a group of contact chemoreceptors at its tip. The total numbers of sensilla on an antenna are often very large. Adult male *Periplaneta*, for instance, have about 250 000 sensilla on each antenna and male corn borer moths, *Ostrinia*, about 8000. When the antennae are sexually dimorphic, as in many Lepidoptera, the more complex antenna bears a much larger number of sensilla. For example, male *Telea* have over 65 000 sensilla on one antenna, while the female has only about 13 000.

Review: Zacharuk, 1985

1.3.3 Functions of antennae

The antennae function primarily as sense organs and they are the primary olfactory receptors of all insects (see section 24.1.1). They also have a tactile function by virtue of the large number of mechanosensitive sensilla that are often present. Very long antennae, such as occur in the cockroach, are possibly associated with their use as feelers. Johnston's organ is important in the regulation of airspeed in flying insects (see Fig. 9.36) and in some insects, male

mosquitoes, female *Drosophila* and worker honeybees, for example, it is concerned in the perception of near-field sounds (see Fig. 23.11).

Sometimes the antennae have other functions. The adult water beetle *Hydrophilus* submerges with a film of air over its ventral surface which it renews at intervals when it comes to the surface. At the water surface the body is inclined to one side and a funnel of air, connecting the ventral air bubble to the outside air, appears between the head, the prothorax and the distal annuli of the antenna, which is held along the side of the head. The four terminal annuli of the antenna are enlarged and are clothed with hydrofuge hairs

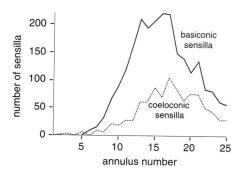

Fig. 1.10. Distribution of sensilla along the flagellum of a male grasshopper. Only olfactory sensilla are shown. 1 = most proximal annulus, adjacent to the pedicel; 25 = most distal annulus (*Melanoplus*) (data from Slifer *et al.*, 1959).

which facilitate the formation of the air funnel. In the newly hatched larva of *Hydrophilus* the antennae assist the mandibles in masticating the prey. This is facilitated by a number of sharp spines on the inside of the antennae.

In fleas and Collembola the antennae are used in mating. Male fleas use the antennae to clasp the female from below and the inner surfaces bear large numbers of adhesive discs. These discs, about 5 μm in diameter, are set on stalks above the general surface of the cuticle and within each one there is a gland, presumably secreting an adhesive material. Species with sessile or semi-sessile females lack these organs (Rothschild and Hinton, 1968). In many Collembola the males have prehensile antennae with which they hold on to the antennae of the female and, in *Sminthurides aquaticus*, the male may be carried about by the female, holding on to her antennae, for several days.

REFERENCES

Bitsch, J. (1973). Morphologie de la tête des insectes. A. Partie Générale. In *Traité de Zoologie*, vol. 8, part 1, ed. P.-P.Grassé, pp. 3–100. Paris: Masson et Cie.

Boeckh, J., Kaissling, K.-E. & Schneider, D. (1960). Sensillen und Bau der Antennengeissel von *Telea polyphemus*. *Zoologische Jahrbücher (Anatomie)*, **78**, 559–84.

Denis, J.R. & Bitsch, J. (1973). Morphologie de la tête des insectes. B. Structure céphalique dans les ordres d'insectes. In *Traité de Zoologie*, vol. 8, part 1, ed. P.-P.Grassé, pp. 101–593. Paris: Masson et Cie.

DuPorte, E.M. (1957). The comparative morphology of the insect head. *Annual Review of Entomology*, **2**, 55–70.

Hinton, H.E. (1963). The ventral ecdysial lines on the head of endopterygote larvae. *Transactions of the Royal Entomological Society of London*, **115**, 39–61.

Imms, A.D. (1940). On the antennal structure in insects and other arthropods. *Quarterly Journal of Microscopical Science*, **81**, 273–320.

Imms, A.D. (1957). *A General Textbook of Entomology*. London: Methuen.

Matsuda, R. (1965). Morphology and evolution of the insect head. *Memoirs of the American Entomological Institute*, no. 4, 334 pp.

Rothschild, M. & Hinton, H.E. (1968). Holding organs on the antennae of male fleas. *Proceedings of the Royal Entomological Society of London*, (A) **43**, 105–7.

Schmidt-Ott, U., González-Gaitán, M., Jäckle, H. & Technau, G.M. (1994). Number, identity, and sequence of the *Drosophila* head segments as revealed by neural elements and their deletion patterns in mutants. *Proceedings of the National Academy of Sciences of the United States of America*, **91**, 8363–7.

Slifer, E.H., Prestage, J.J. & Beams, H.W. (1959). The chemoreceptors and other sense organs on the antennal flagellum of the grasshopper, (Orthoptera: Acrididae). *Journal of Morphology*, **105**, 145–91.

Snodgrass, R.E. (1935). *Principles of Insect Morphology*. New York: McGraw-Hill.

Snodgrass, R.E. (1947). The insect cranium and the 'epicranial suture'. *Smithsonian Miscellaneous Collections*, **104**, no. 7, 113 pp.

Snodgrass, R.E. (1960). Facts and theories concerning the insect head. *Smithsonian Miscellaneous Collections*, **142**, 1–61.

Zacharuk, R.Y. (1985). Antennae and sensilla. In *Comprehensive Insect Physiology, Biochemistry and Pharmacology*, vol. 6, ed. G.A. Kerkut & L.I. Gilbert, pp. 1–69. Oxford: Pergamon Press.

2 Mouthparts and feeding

The mouthparts are the organs concerned with feeding, comprising the unpaired labrum in front, a median hypopharynx behind the mouth, a pair of mandibles and maxillae laterally, and a labium forming the lower lip. In Collembola, Diplura and Protura the mouthparts lie in a cavity of the head produced by the genae, which extend ventrally as oral folds and meet in the ventral midline below the mouthparts (Fig. 2.1). This is the entognathous condition. In the Insecta the mouthparts are not enclosed in this way, but are external to the head, the ectognathous condition.

2.1 ECTOGNATHOUS MOUTHPARTS

The form of the mouthparts is related to diet, but two basic types can be recognized: one adapted for biting and chewing solid food, and the other adapted for sucking up fluids.

2.1.1 Biting mouthparts

Labrum The labrum is a broad lobe suspended from the clypeus in front of the mouth and forming the upper lip (Figs. 1.2, 2.2a). On its inner side it is membranous and may be produced into a median lobe, the epipharynx, bearing some sensilla. The labrum is raised away from the mandibles by two muscles arising in the head and inserted medially into the anterior margin of the labrum. It is closed against the mandibles in part by two muscles arising in the head and inserted on the posterior lateral margins on two small sclerites, the tormae, and, at least in some insects, by a resilin spring in the cuticle at the junction of the labrum with the clypeus. Differential use of the muscles can produce a lateral rocking movement of the labrum.

Mandibles In the entognathous groups and the Archaeognatha, the mandibles are relatively long and slender and they have only a single point of articulation with the head capsule (Fig. 2.3). The mandible is rotated about its articulation by anterior and posterior muscles arising on the head capsule and on the anterior tentorial arms. The principal adductor muscles are transverse and ventral, those of the two sides uniting in a median tendon.

In Thysanura and the Pterygota, the mandibles are articulated with the cranium at two points, having a second more anterior articulation with the subgena in addition to the original posterior one (Figs. 1.3, 2.2b). These mandibles are usually short and strongly sclerotized and

a) lateral view   **b) section at AA**

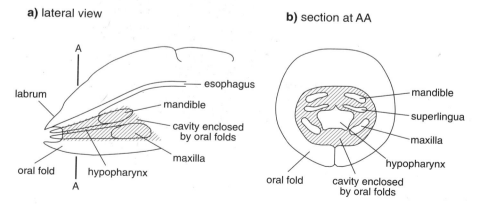

Fig. 2.1. Entognathous mouthparts (modified after Denis, 1949). **(a)** Lateral view showing the mouthparts within the cavity formed by the oral folds. The extent of the cavity is indicated by hatching. **(b)** Transverse section at AA in **(a)**.

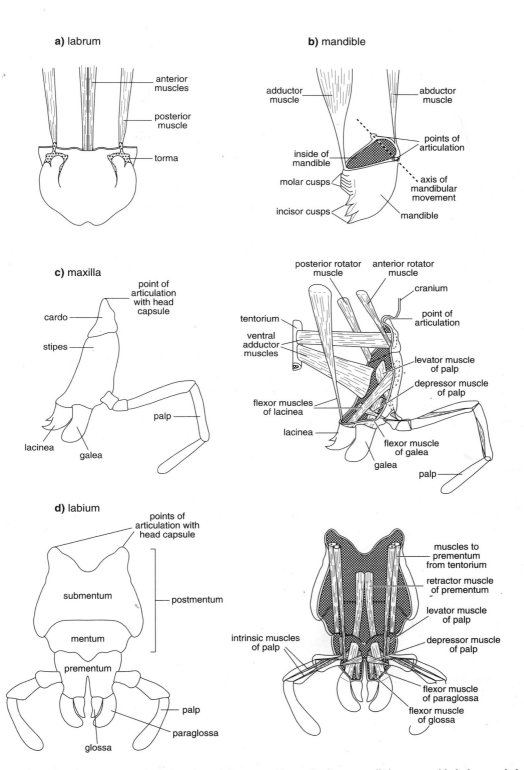

Fig. 2.2. Biting and chewing mouthparts of a pterygote insect. Surfaces normally in contact with the hemocoel, the inside of the cuticle, are shaded (after Snodgrass, 1935, 1944). (**a**) Labrum seen from the posterior, epipharyngeal, surface . (**b**) Mandible, notice the dicondylic articulation. (**c**) Maxilla from the outside (left) and inside (right). (**d**) Labium from the outside (left) and inside (right).

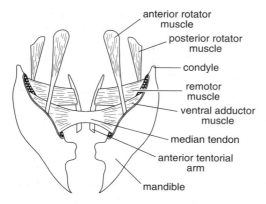

anterior rotator
muscle
posterior rotator
muscle
condyle
remotor
muscle
ventral adductor
muscle
median tendon
anterior tentorial
arm
mandible

Fig. 2.3. Monocondylic mandibles as found in Archaeognatha and non-insectan hexapods. Not all muscles are shown (after Snodgrass, 1935).

the cuticle of the cusps is often hardened by the presence of zinc or manganese (see Fig. 16.9). These cusps may become worn down during feeding, but the distribution of the harder areas of cuticle promotes self-sharpening (Chapman, 1995).

The original anterior and posterior rotator muscles of Apterygota have become abductors and adductors in the Pterygota, the adductor often becoming very powerful. The apterygote ventral adductor is retained in most orthopteroids and arises from the hypopharyngeal apophysis, but in Acrididae and the higher insects this muscle is absent, or, in insects with sucking mouthparts, may be modified as a protractor muscle of the mandible.

Maxillae The maxillae occupy a lateral position, one on each side of the head behind the mandibles. The proximal part of the maxilla consists of a basal cardo, which has a single articulation with the head, and a flat plate, the stipes, hinged to the cardo (Fig. 2.2c). Both cardo and stipes are loosely joined to the head by membrane so that they are capable of movement. Distally on the stipes are two lobes, an inner lacinea and an outer galea, one or both of which may be absent. More laterally on the stipes is a jointed, leg-like palp made up of a number of segments; in Orthoptera there are five.

Anterior and posterior rotator muscles are inserted on the cardo, and ventral adductor muscles arising on the tentorium are inserted on both cardo and stipes. Arising in the stipes are flexor muscles of lacinea and galea and another lacineal flexor arises in the cranium, but neither

lacinea nor galea has an extensor muscle. The palp has levator and depressor muscles arising in the stipes and each segment of the palp has a single muscle causing flexing of the next segment.

Labium The labium is similar in structure to the maxillae, but with the appendages of the two sides fused in the midline so that they form a median plate (Fig. 2.2d). The basal part of the labium, equivalent to the maxillary cardines and possibly including a part of the sternum of the labial segment, is called the postmentum. This may be subdivided into a proximal submentum and a distal mentum. Distal to the postmentum, and equivalent to the fused maxillary stipites, is the prementum. The prementum closes the pre-oral cavity from behind. Terminally it bears four lobes, two inner glossae and two outer paraglossae, which are collectively known as the ligula. One or both pairs of lobes may be absent or they may be fused to form a single median process. A palp arises from each side of the prementum, often being three-segmented.

The musculature corresponds with that of the maxillae, but there are no muscles to the postmentum. Muscles corresponding with the ventral adductors run from the tentorium to the front and back of the prementum; glossae and paraglossae have flexor muscles, but no extensors, and the palp has levator and depressor muscles arising in the prementum. The segments of the palp each have flexor and extensor muscles. In addition, there are other muscles with no equivalent in the maxillae. Two pairs arising in the prementum converge on to the wall of the salivarium at the junction of labium with hypopharynx (Fig. 1.2). A pair of muscles opposing these arises in the hypopharynx and the combined effect of them all may be to regulate the flow of saliva or to move the prementum. Finally, a pair of muscles arising in the postmentum and inserted into the prementum serves to retract or flex the prementum.

Hypopharynx The hypopharynx is a median lobe immediately behind the mouth (Fig. 1.2). The salivary duct usually opens behind it, between it and the labium. Most of the hypopharynx is membranous, but the adoral face is sclerotized distally, and proximally contains a pair of suspensory sclerites which extend upwards to end in the lateral wall of the stomodeum. Muscles arising on the frons are inserted into these sclerites, which distally are hinged to a pair of lingual sclerites. These, in turn, have inserted into them antagonistic pairs of muscles arising on the tentorium and labium. The various muscles serve to swing the hypopharynx forwards and back, and in the cockroach there are two more muscles running across the hypopharynx which dilate the salivary orifice and expand the salivarium.

In Apterygota, larval Ephemeroptera and Dermaptera there are two lateral lobes of the hypopharynx called the superlinguae.

2.1.2 Variation in form

The form of the mouthparts varies greatly between species. The biting surface of the mandible is often differentiated into a more distal incisor region and a proximal molar region whose development varies with diet. The mandibles of carnivorous insects are armed with strong shearing cusps; in grasshoppers feeding on vegetation other than grasses, there is a series of sharp pointed cusps, while in grass-feeding species the incisor cusps are chisel-edged and the molar area has flattened ridges for grinding. In species that do not feed as adults, the mouthparts may be greatly reduced. The mandibles of adult Ephemeroptera, for example, are vestigial or absent altogether and the maxillae and labium are also greatly reduced, being represented mainly by the palps.

The greatest divergence from the basic form occurs in the larvae of holometabolous insects. While larval Lepidoptera and Coleoptera usually have well-developed biting and chewing mouthparts, larval Diptera and Hymenoptera show some extreme modifications and reductions. Mosquito larvae, for example, have brushes of hairs on the mandibles and maxillae which are especially long in some species where they serve to filter particulate material, including food, from the water. The larvae of cyclorrhaphous flies exhibit extreme reduction of the head. The principal structures are a pair of heavily sclerotized mouthhooks with which the larva rasps at its food; sensory papillae probably represent the palps. Amongst Hymenoptera, larval Symphyta have well-developed mouthparts, similar to caterpillars, but in some parasitic species the mandibles are represented only by simple spines, and other mouthparts are not differentiated into separate sclerites.

2.1.3 Sucking mouthparts

The mouthparts of insects which feed on fluids are modified in various ways to form a tube through which liquid can be drawn into the mouth and usually another through which saliva passes. The muscles of the cibarium or pharynx are strongly developed to form a pump. In Hemiptera and many Diptera, which feed on fluids within plants or animals, some components of the mouthparts are modified for piercing, and the elongate structures are called stylets. The combined tubular structures are referred to as the proboscis, although specialized terminology is used in some groups.

In Hemiptera, mandibles, maxillae and labium are all elongate structures, while the labrum is relatively short. The food canal is formed by the opposed maxillae which are held together by a system of tongues and grooves (Figs. 2.4a, 2.8a). These allow the stylets to slide freely on each other, while maintaining the integrity of the food canal. The maxillae also contain the salivary canal. On either side of the maxillae are the mandibular stylets. These are the principal piercing structures and they are often barbed at the tip. When the insect is not feeding, the slender maxillary and mandibular stylets are held within a groove down the anterior side of the labium. The hemipteran labium is known as the rostrum. It is usually segmented, allowing it to fold as the stylets penetrate the host. There are no palps. Since the Hemiptera are hemimetabolous, the larvae and adults have similar feeding habits and both have sucking mouthparts.

Thysanoptera are also fluid feeders as larvae and adults. Their stylets are normally held in the cone-shaped lower part of the head formed by the clypeus, the labrum and labium. Only the left mandible is present. It is used to penetrate plant cells. The maxillary stylets are held together to form the food canal. There is no salivary canal; the salivary duct opens into the front of the esophagus (Chisholm & Lewis, 1984).

The adult Diptera exhibit a great variety of modifications of the mouthparts, but in all of them the food canal is formed between the apposed labrum and labium and the salivary canal runs through the hypopharynx (Fig. 2.4e,f). The mandibles and maxillae are styliform in species that suck the blood of vertebrates, but are generally lacking in other species, including the blood-sucking Cyclorrhapha. Where they are present, they are the piercing organs; in blood-sucking Cyclorrhapha tooth-like structures at the tip of the labium penetrate the host tissues by a rasping action. Similar prestomal teeth occur in other Cyclorrhapha, including the house fly, *Musca*. In many species, the tip of the labium is expanded to form a

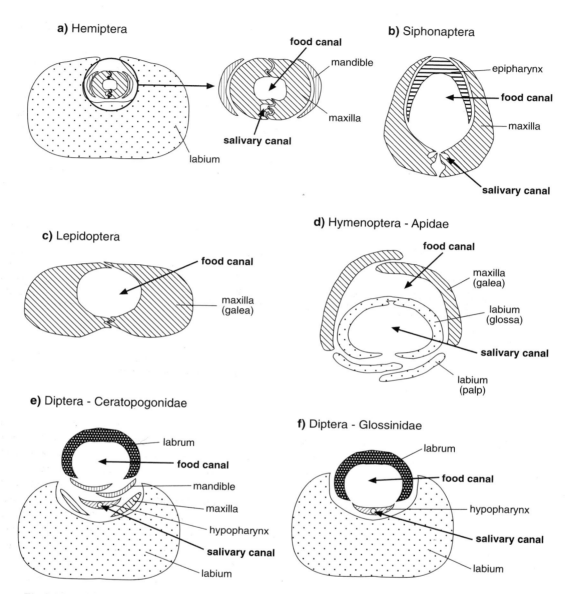

Fig. 2.4. Sucking mouthparts. Diagrammatic cross-sections of the proboscis showing the principal structures used to form tubes for delivery of saliva and intake of food. Homologous structures are indicated with the same shading in all the diagrams. In some cases the structures contain an extension of the hemocoel; this is not shown. (**a**) Hemiptera (bugs) (compare Fig. 2.8a); (**b**) Siphonaptera (fleas); (**c**) Lepidoptera (butterflies and moths); (**d**) Hymenoptera, Apidae (bees); (**e**) Diptera, Ceratopogonidae (biting midges); (**f**) Diptera, Glossinidae (tsetse flies).

a) Neuroptera   **b)** *Dytiscus*   **c)** Lampyridae

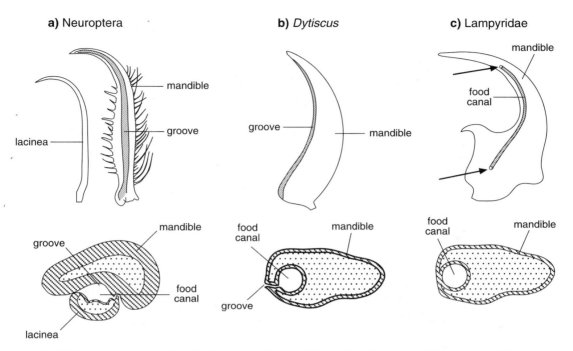

Fig. 2.5. Sucking mouthparts of larval holometabolous insects. (**a**) An antlion (Neuroptera); (**b**) *Dytiscus* larva (Coleoptera, Dytiscidae); (**c**) a firefly larva (Coleoptera, Lampyridae). Arrows show the positions at which the food canal opens to the outside (based on Cicero, 1994).

lobe, the labellum which, in Brachycera, is traversed by a series of grooves known as pseudotracheae because they are held open by cuticular ribs giving them a superficial similarity to tracheae. The pseudotracheae converge centrally on the distal end of the food canal. Diptera have maxillary palps, but no labial palps.

The food canal of fleas is formed between an extension of the epipharynx and the maxillary stylets (Fig. 2.4b). A salivary canal extends along the inside of each maxilla which also form the piercing organs. Both maxillary and labial palps are present.

The proboscis of adult Lepidoptera is formed from the galeae held together by a system of cuticular hooks ventrally and a series of plates dorsally (Fig. 2.4c). Since most Lepidoptera are nectar feeders, they do not require piercing mechanisms and the rest of the mouthparts, apart from the labial palps, are reduced or absent. There is no salivary canal although adult Lepidoptera do have salivary glands.

Adults of most Hymenoptera have biting and chewing mouthparts, but the bees are nectar-feeders and are described as lapping the nectar. This is achieved by an elongation and flattening of the galeae and labial palps which surround the fused glossae (Fig. 2.4d). The space outside the glossal tongue forms the food canal. The salivary canal is in the posterior folds of the tongue.

Larval Neuroptera and some predaceous larval Coleoptera that digest their prey extra-orally have a food canal in each of the mandibles. These function in a similar way to those of biting and chewing insects, but they are sickle-shaped. In larval Neuroptera, a groove on the inside of each mandible is converted to a tube by the juxtaposition of a slender lacinea (Fig. 2.5a). A similar groove is present in the mandibles of some larval Dytiscidae, but instead of being closed by the lacinea the lips of the groove almost join to form a tube (Fig. 2.5b). Larval Lampyridae have a tube running through each mandible and opening by a hole near the tip and another near the base within the cibarial cavity (Fig. 2.5c) (Cicero, 1994).

Associated with the production of a tube for feeding is the development of a pump for drawing up the fluids and a salivary pump for injecting saliva (see Fig. 3.15). Often the feeding pump is developed from the cibarium, which by extension of the lateral lips of the mouth becomes a closed chamber connecting with the food canal. The cibarial muscles from the clypeus enlarge so that a powerful pump is produced. In Lepidoptera and Hymenoptera the cibarial pump is combined with a pharyngeal pump which has dilators arising on the frons.

Review: Smith, 1985

2.1.4 Sensilla on the mouthparts

Most of the sensilla on the mouthparts are contact chemoreceptors, but mechanoreceptors are also common and olfactory sensilla are often present on the palps. Chordotonal organs, that probably function as pressure receptors, are present at the tips of the mandibular cusps and also in the lacinea where this is heavily sclerotized and tooth-like.

Biting and chewing insects have contact chemoreceptors on all the mouthparts except the mandibles. They also have chemoreceptors on the dorsal and ventral walls of the cibarium, often called epipharyngeal and hypopharyngeal sensilla, respectively. Orthoptera and Blattodea have large numbers of sensilla in groups (Fig. 2.6a) with especially large numbers on the tips of the maxillary and labial palps. *Gryllus bimaculatus*, for example, has over 3000 sensilla on each maxillary palp. Because each sensillum contains at least four neurons, the potential chemosensory input to the central nervous system is considerable; an adult locust has about 16 000 chemosensory neurons on the mouthparts. In the orthopteroid insects, the numbers increase each time the insect molts. By contrast, caterpillars have only about 100 neurons in mouthpart receptors and the closely associated antennae (Fig. 2.6b); the number does not increase during larval life. Fluid-feeding insects usually have chemoreceptors at the tip of the labium, on the palps when these are present, and in the walls of the cibarium (Fig. 2.6c). In addition, at least some planthoppers have an olfactory sensillum towards the tip of the rostrum. No chemoreceptors are present on the mandibular and maxillary stylets. Aphids have only mechanoreceptors at the tip of the labium; their only chemoreceptors on the mouthparts are in the cibarium.

In piercing and sucking insects (Hemiptera and Culicidae, for example) only the cibarial sensilla come directly into contact with the food as it is ingested; the labium does not enter the tissues of the host so that its sensilla can only monitor the outer surface of the food, either plant or animal. In blood-sucking species that use labial

teeth to rasp through the tissues, such as *Glossina* and *Stomoxys*, however, the labial sensilla come into direct contact with the blood. This is also true of nectar feeding insects, Lepidoptera, Apoidea and many Diptera, including female mosquitoes. Cibarial sensilla are known to be present in some of these species and are probably universal.

The axons of contact chemoreceptors and mechanoreceptors on the mandibles, maxillae and labium end in arborizations in the corresponding neuromeres of the subesophageal ganglion, but the axons from olfactory sensilla on the palps run directly to the olfactory lobes. The axons of sensilla on the labrum arborize in the tritocerebrum. Interneurons responding to chemical and mechanical stimulation of the mouthpart receptors have been demonstrated in the subesophageal ganglion of *Sarcophaga* (Mitchell & Itagaki, 1992) and must presumably occur in all insects with functional mouthparts. The cell bodies of the motor neurons regulating movements of the mouthparts are also present in this ganglion, but nothing is known of the precise pathways that connect the sensory and motor systems.

Reviews: Backus, 1988 – plant-feeding Hemiptera; Chapman, 1982 – all insect groups

2.2 MECHANICS AND CONTROL OF FEEDING

Before an insect starts to feed it exhibits a series of behavioral activities which may lead to acceptance or rejection of the food. A grasshopper first touches the surface of the plant with the sensilla at the tips of its palps. This behavior enables the insect to monitor the chemicals on the surface of the plant wax and perhaps also the odor of the plant. This may lead the insect to reject the plant without further investigation or to make an exploratory bite, presumably releasing chemicals from within the plant. This, in turn, may result in rejection, or the insect may start to feed. Essentially similar behaviors are seen in caterpillars and leaf-eating beetles.

The principal chemicals inducing feeding (phagostimulants) of leaf-eating insects are sucrose and hexose sugars, but the exploratory bite following palpation may be induced by components of the leaf wax. Insects that feed only on specific plant taxa may require the presence of a compound that is characteristic of the plant species, or group of species. For example, many caterpillars and beetles that feed on plants in the family Brassicaceae, which includes

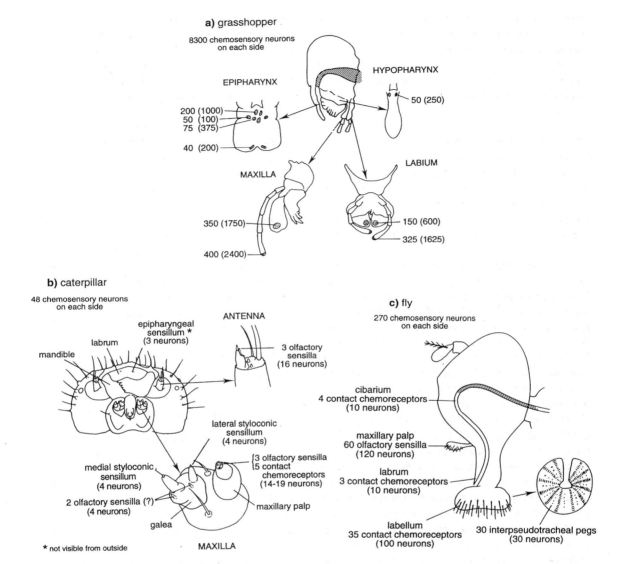

Fig. 2.6. Chemoreceptors on the mouthparts of various insects. Numbers show the number of sensilla and, in brackets, the number of chemosensitive neurons in each group of receptors. Sensilla are contact chemoreceptors unless otherwise stated. The numbers do not include sensilla scattered over the mouthparts, which may be present in addition to those in groups. **(a)** An adult locust, *Locusta*; **(b)** a caterpillar, *Pieris*; **(c)** an adult fly, *Drosophila*.

cabbage, are stimulated to feed by mustard oil glucosides (glucosinolates) that are characteristic of this plant family. Some species with restricted host ranges have chemosensory neurons in the mouthpart sensilla which respond specifically to the indicator chemicals. Other neurons are sensitive to sugars, and others to compounds that inhibit feeding, known as feeding deterrents (see section 24.2.2). Information from all the sensilla is integrated in the central nervous system. Whether or not the insect feeds depends on the balance between phagostimulants and deterrents.

Amongst nectar-feeding insects, sugars are phagostimulants. Before starting to feed, the insects exhibit a sequence of behaviors comparable with that of the leaf-chewing insects. If the tarsi contact sugar above a certain threshold concentration, the proboscis is extended. This is true of flies, bees and butterflies. Stimulation of the trichoid sensilla on the outside of the labellum of a fly causes the insect to spread the labellar lobes so that the interpseudotracheal papillae (see Fig. 2.6c) contact the food. Their stimulation initiates ingestion. When female blood-sucking insects, such as mosquitoes, feed on nectar the stylets remain enclosed in the labium and the labellar chemoreceptors are stimulated.

Blood-sucking insects fall into two classes with respect to the factors regulating ingestion: those that will gorge on saline solutions which are isotonic with vertebrate blood, and those that require the presence of an adenine nucleotide. Amongst the former are sandflies (Ceratopogonidae), anopheline mosquitoes and fleas, although fleas require the presence of a nucleotide to take a full meal. Most of the other blood-suckers require a nucleotide. ATP is generally much more stimulating than ADP, and this in turn is much more effective than AMP. ATP is normally contained within red blood cells where it would not stimulate the insect's sensory neurons. It is released by damage to these cells during probing, but is quickly degraded by the insect's own salivary apyrase.

Reviews: Bernays & Chapman, 1994 – phytophagous insects; Davis & Friend, 1995 – blood-feeding insects

2.2.1 Biting and chewing insects

Biting and chewing insects make regular opening and closing movements of the mandibles; both locusts and caterpillars commonly make up to four bites per second when feeding continuously, but the rate probably varies with temperature, the quality of the food, and the feeding state of the insect. As a result of a sequence of bites, the insect cuts off a fragment of food which is pushed back towards the mouth by the mandibles, often aided by the maxillae. Periods of continuous biting are often separated by short pauses, presumably associated with swallowing the food. The pauses get longer as a meal progresses.

The motor pattern controlling movement of the mouthparts is generated in the subesophageal ganglion. In caterpillars of *Manduca*, chewing activity occurs spontaneously in the absence of input from receptors in the wall of the thoracic segments, but in grasshoppers such spontaneous activity does not occur. The mandibular abductor muscles are stimulated to contract by mechanical stimulation of the labrum. Sensilla which detect the positions of the open mandibles then stimulate the mandibular adductor muscle motor neurons via the ganglion so that the mandibles close. This stimulates other receptors, probably those at the tips of the cusps, which presumably starts another cycle of opening.

A modulatory effect on the activity of the muscles of the mouthparts is probably exerted by serotonin. In the cockroach, *Periplaneta*, and some other insects all the nerves to these muscles have a branching network of fine serotonergic axons over their surfaces. These neurons of which the axons are a part are only active during feeding and it is presumed that the electrical activity leads to the release of serotonin from the fine branches where they locally affect the activity of the muscles moving the mouthparts (Schachtner & Bräunig, 1993). The precise effect of this modulation is not known.

Review: Chapman, 1995

2.2.2 Fluid-feeding insects

Proboscis extension by nectar-feeding flies and bees depends on the activity of muscles associated with the mouthparts (Rehder, 1989). In Lepidoptera, the proboscis is caused to unroll by increased pressure of the hemolymph. This pressure is generated in the stipes associated with each galea. A valve isolates the hemocoel of the stipes when the latter contracts. Coiling results from the elasticity of the cuticle of the galeae together with the activity of intrinsic muscles (Krenn, 1990).

Insects feeding on the internal fluids of other organisms must first penetrate the host tissues. Homoptera, which feed on plants, first secrete a blob of viscous saliva which solidifies around the labellar lobes forming a salivary flange which remains even after the insect has left (Fig. 2.7). This probably serves to prevent the stylets from slipping over the plant surface as pressure is applied. The mandibles penetrate the leaf cuticle and epidermis; subsequent deeper penetration may involve the maxillae alone, or both maxillae and mandibles. The stylets of aphids and other small Homoptera move within the walls of the plant cells aided by enzymes in the watery saliva (see section 2.5). As the stylets progress through the leaf, the insect secretes more of the viscous saliva that solidifies to form a sheath around them (Fig. 2.8a). The significance of this sheath is not known. It may serve to support the stylets and to prevent loss of plant sap and of the more fluid saliva through the wound in the epidermis. The stylets often

Fig. 2.7. Salivary flange produced from saliva by a planthopper. **(a)** Scanning electron micrograph showing the flange remaining on a leaf surface when the insect stops feeding. The saliva hardens around the mouthparts. The hole in the center was occupied by the stylets; the small holes to the right were produced by sensilla at the tip of the labium. The structures running obliquely from the top left are leaf veins (after Ribeiro, 1995). **(b)** Diagrammatic section through a flange showing its relationship to the leaf surface and the salivary sheath.

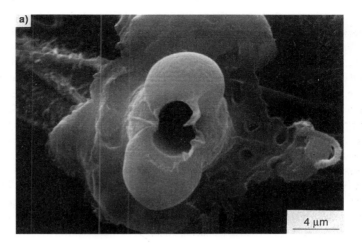

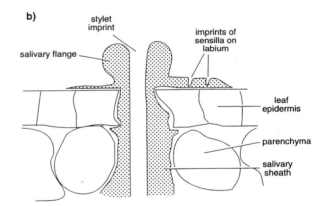

break out of the plant cell wall, and, when they do, the insect is believed to sample the chemical composition of the fluid by drawing it up to the cibarial sensilla. If the stylets are not in a sieve element of the phloem, the insect withdraws its stylets for a short distance and moves them in another direction. This process is repeated, and the path of the stylets may become very irregular (Fig. 2.8b), until the phloem is entered, then the stylets are pushed into it and ingestion begins (Fig. 2.8b) (Tjallingii & Esch, 1993). It may take anything from five minutes to three hours from the beginning of probing until a feeding site is reached. In larger Homoptera, stylet entry occurs in a similar way and a salivary sheath is formed, but in most cases the stylet pathway is through cells rather then between them.

Blood-sucking insects, like the Homoptera amongst plant-feeders, must penetrate the host tissues before starting to feed. In many species that feed on warm-blooded

vertebrates, the proboscis is moved into the feeding position in response to the warmth of the host. In mosquitoes and triatomine bugs, the stylets are pushed into the tissues and the labium folds up, but does not enter the wound. This separation of the stylets from the ensheathing labium appears to provide some additional stimulus necessary for the insects to take a full blood meal. In tsetse and stable flies, rasping movements of the prestomal (labial) teeth tear into the tissues. Blood vessels comprise less than 5% of the volume of mammalian skin so that blood is not available until a vessel is ruptured.

Salivary secretions play a critical role in blocking the hemostatic responses of the host which tend to prevent blood loss. Three different processes contribute to hemostasis: platelet aggregation, blood coagulation and vasoconstriction. The saliva of blood-sucking insects contains chemicals that inhibit all three (Table 2.1). ADP released

a)

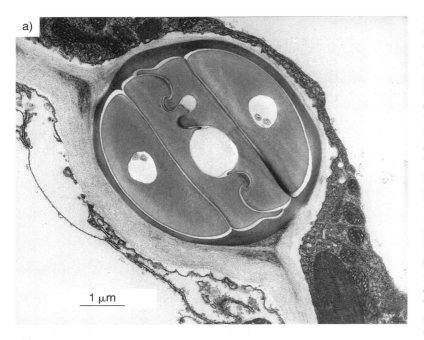

1 μm

b)

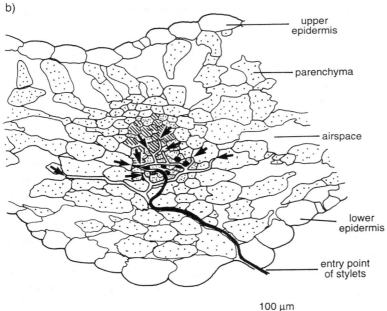

upper
epidermis

parenchyma

airspace

lower
epidermis

entry point
of stylets

100 μm

Fig. 2.8. Feeding by an aphid (after Tjallingii & Esch, 1993). **(a)** Transverse section through the stylets and salivary sheath in a leaf. The maxillary stylets interlock to form the food canal (center) and the salivary canal (above) (compare Fig. 2.4a). Each mandibular stylet has a narrow lumen, an extension of the hemocoel, containing mechanoreceptor neurons. The dark ring surrounding the stylets is the salivary sheath. Outside it, the pale fibrous material is plant cell wall. Notice that the stylets are contained within the cell wall; they do not enter the surrounding cytoplasm. **(b)** Pathways taken by the stylets of an aphid at the start of feeding. Abortive pathways are shown white with the ends of the paths indicated by arrows. The final pathway, reaching the phloem, is shown black. Phloem sieve tubes, black; xylem, cross-hatched; parenchyma, stippled.

Table 2.1. *Components of saliva of blood-sucking insects that inhibit the hemostatic responses of their vertebrate hosts*

Insect	Order	Family	Inhibiting platelet aggregation	Vasodilator	Anticoagulant
Rhodnius	Hemiptera	Reduviidae	Apyrase[a]	Nitric oxide releaser[a]	Anti-VII[c], anti-thrombin
Xenopsylla	Siphonaptera	Pulicidae	Apyrase	?	?
Aedes	Diptera	Culicidae	Apyrase	Tachykinin[b]	?
Anopheles	Diptera	Culicidae	Apyrase	Peroxidase[a]	?
Simulium	Diptera	Simuliidae	Apyrase	Marydilan	Anti-X[c], anti-thrombin
Lutzomyia	Diptera	Ceratopogonidae	Apyrase	Maxidilan[b]	?
Glossina	Diptera	Glossinidae	Apyrase	?	Anti-thrombin

Notes:

? No compound known.

[a] Enzyme.

[b] Peptide.

[c] Factor VII and factor X are substances involved in normal blood clotting.

from injured blood cells is an important signal for platelet aggregation and blood-sucking insects generally have a salivary apyrase that degrades ADP to orthophosphate and AMP. The saliva also contains a vasodilator to counteract the vasoconstriction induced by the host, but these vasodilators differ from one insect species to another. Factor X and thrombin are chemicals that regulate coagulation; peptides that counteract the effects of both chemicals are present in the saliva.

The intake of fluids depends on their viscosity and whether or not they are under pressure. Viscosity increases with the concentration of dissolved solutes so that nectar containing 40% sucrose is six times more viscous than water at the same temperature. The phloem and xylem of plants are very dilute so their viscosities are not markedly different from water. The viscosity of vertebrate blood, however, varies with the diameter of the tube through which it is being drawn. For tubes less than 100 μm in diameter, the viscosity falls as tube diameter is reduced down to about 6 μm. At smaller diameters it increases sharply because flow depends on the distortion of the red blood cells. In addition, the fluid may be under positive or negative pressure. Blood pressure in human capillaries is about 3 kPa, whereas the phloem in plants is under much higher pressure, 0.2 to 1 MPa. Xylem, on the other hand, is under strong negative pressure, as much as −2 MPa and higher. The cibarial and pharyngeal pumps which draw fluid through the proboscis

are consequently different in insects with different feeding habits. Capillarity may be important in bees, helping to move nectar up the food canal.

Once a phloem feeding insect, such as an aphid or planthopper, has reached the phloem, the pressure of the phloem is sufficient to push fluid into the insect, which appears then to play no active role in ingestion. These insects can, however, pump fluid into the gut when the food is not under pressure and do, periodically, feed from parenchyma or xylem.

The negative pressure of xylem tends to draw fluid out of the insect, and the massive cibarial pumps that are characteristic of habitual xylem-feeding insects, such as cicadas and cercopids, are necessary to overcome these pressures (Fig. 3.15). Even so, the leafhopper, *Homalodisca*, exhibits markedly reduced feeding rates at negative xylem pressures in excess of −1.5 MPa (Andersen, Brodeck & Mizell, 1992).

In nectar-feeding bees, the glossa is repeatedly extended into the nectar while the galeae and labial palps which surround it (Fig. 2.4d) remain motionless. During extension of the glossa nectar moves on to it by capillarity. The fluid moves the hairs on the glossa to a position at right angle to the surface of the glossa so that the volume of fluid that is held is increased. When the glossa is retracted into the tube formed by the galeae and labial palps the fluid is drawn into the cibarium by the pump. In *Bombus* these licking movements occur at a frequency of about 5 Hz. The intake rate of sucrose solutions at concentrations up to about 40% is around $1.75\ \mu l\ s^{-1}$. It declines at higher concentrations due to the increasing viscosity.

Blood-sucking insects produce pressure differentials well in excess of the capillary blood pressure of the host, so the latter is unlikely to play a significant role in feeding. Calculated pressure differences, which are produced by the cibarial and pharyngeal pumps, are approximately 8 kPa in the mosquito, *Aedes*, 20 kPa in the louse, *Pediculus*, 80 kPa in the bedbug, *Cimex*, and 100–200 kPa in *Rhodnius*. In the latter, the pumping rate is about 7 Hz producing an ingestion rate of about $450\ nl\ s^{-1}$. In the mosquito, *Aedes*, the intake rate is about $16\ nl\ s^{-1}$. The louse, *Pediculus*, has a food canal that is smaller than the diameter of a human red blood cell (about 7.5 μm). As a result, the feeding rate of an adult female louse is about five times lower than in the mosquito.

Little is known about the control of feeding in fluid-feeding insects, but serotonin is released into the hemolymph during feeding by *Rhodnius* probably from neurohemal organs on the abdominal nerves. It triggers plasticization of the abdominal cuticle (section 16.3.3) (Lange, Orchard & Barrett, 1989).

Review: Kingsolver & Daniel, 1995

2.2.3 Prey capture by predaceous insects

Predators catch their prey either by sitting and waiting for it to come their way, or by actively pursuing it. The mantis is an example of an insect that sits and waits. Like many predators, mantids have wide heads so that the eyes are relatively far apart. This facilitates the accurate determination of the distance of the prey from the insect (see Fig. 22.17). In addition, the head is very mobile so that movements of the prey can be followed without the whole mantis moving. The mantis strikes when the prey is of a suitable size, is at a suitable distance in front of the head, and is moving. *Sphodromantis*, a relatively large insect, strikes when the prey subtends an angle of 20–40° at the eyes and is 30 to 40 mm in front of the head. In *Tenodera* the attack zone is in an area defined by the length of the foreleg and below the body axis (Fig. 2.9). The forelegs of mantids are raptorial and armed with spines. The mantis follows the prey with its eyes (see Fig. 22.18) and moves its prothorax to orient directly at the prey. The effect of this is to make the area of attack two dimensional; the mantis has to judge the position of the prey directly in front and not to

a position on either side of the head. The attack involves the strike by the forelegs and a forwards lunge of the whole body that reduces the distance between the head of the mantid and its prey. The strike involves first the extension of the coxa and tibia and then a rapid extension of the femur and flexure of the tibia to grasp the prey. This last movement occurs in about 30 ms (Corrette, 1990; Rossel, 1991).

Dragonfly larvae also sit and wait for their prey and then capture it with the modified labium. The postmentum and prementum are both elongate and the labial palps are claw-like structures set distally on the prementum (Fig. 2.10a). At rest, the labium, often referred to as a labial mask, is folded beneath the head. If a potential prey item comes within range, the mask is extended and the prey caught by the labial palps (Fig. 2.10c). As in the mantis, the strike is very rapid, being completed in 25 ms or less, but in this case the speed of the strike depends on prior energy storage (see section 8.4.2). Before the strike, the anal sphincter is closed and the dorsoventral abdominal muscles contract (Fig. 2.10b). This results in an increase in hemolymph pressure. At the same time the flexor and extensor muscles of the prementum, which arise in the head, contract together (co-contract) so that the labial mask is held against the head and, presumably, tension builds up at the postmentum/prementum joint. Relaxation of the flexor muscle results in the sudden extension of the prementum due to the continued activity of the premental extensor muscles and release of the cuticular tension at the postmentum/prementum joint. There are no extensor muscles of the postmentum and its extension is produced by the high hemolymph pressure (Parry, 1983; Tanaka & Hisada, 1980).

Some insects that sit and wait for prey construct traps. Antlions (larval Myrmeleontidae), for instance, dig pits two to five centimeters in diameter with sloping sides in dry sand. The insect then buries itself in the sand at the bottom of the pit with only the head exposed. If an ant walks over the edge of such a pit it has difficulty regaining the top because of the instability of the sides. In addition, the larva, by sharp movements of the head, flicks sand at the ant causing it to fall to the bottom of the pit and be captured.

Larvae of the fly, *Arachnocampa* (Mycetophilidae), are luminescent and use the light to lure prey into a sticky trap consisting of vertical threads of silk studded with sticky drops.

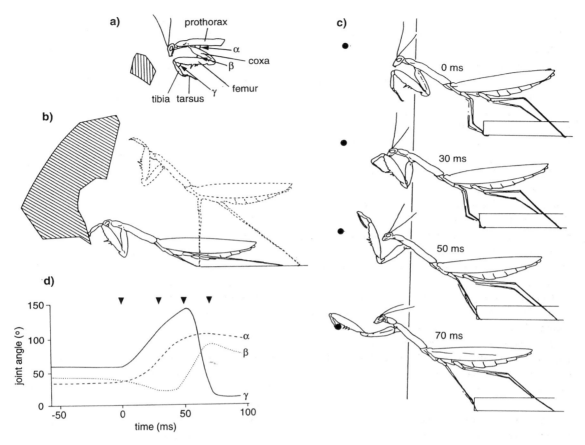

Fig. 2.9. Prey capture by a mantis, *Tenodera* (after Corrette, 1990). **(a)** Area of capture relative to the head. The vertically hatched area shows the region in which prey can be captured by striking with the forelegs without any other movement of the body. **(b)** Area of capture made possible by changes in orientation coupled with the lunge, shown by oblique hatching. Before the lunge, the insect may occupy any position between the two extremes shown in the diagram. **(c)** Diagrams of the changes in position associated with capturing prey. 0 ms, start of strike; 30 ms, start of lunge; 70 ms, capture. The vertical line is a reference showing the forward movement of the body during the lunge. The black spot represents the position of the target. **(d)** Changes in the angles between leg segments (shown in a) at times corresponding with the positions shown in **(c)**. Notice the rapid increase in β (extension of the femur) coincident with the decrease in γ (flexion of the tibia to grasp the prey).

Adult dragonflies are active aerial hunters, pursuing other insects in flight. Their thoracic segments are rotated forwards ventrally bringing the legs into an anterior position which facilitates grasping (Fig. 2.11). Tiger beetles (Coleoptera, Cicindelidae) hunt on the ground and have long legs, which enable them to move quickly, and prognathous mouthparts with large mandibles. In the trap jaw ant, *Odontomachus*, mandible closure is triggered by mechanical stimulation of a hair on the inside of the mandible. This ant hunts with its jaws wide open, and when

it encounters prey they snap shut in less than 1 ms. Such a rapid closure could not be achieved by direct muscular action, and it is probable that there is some method of developing tension at the mandibular articulation with the head analogous to that found in jumping mechanisms (section 8.4.2) (Gronenberg, Tautz & Hölldobler, 1993).

These active hunters have well-developed eyes since only vision can give a sufficiently rapid directed response to moving prey. The visual response is usually not specific and the predator will pursue any moving object of suitable

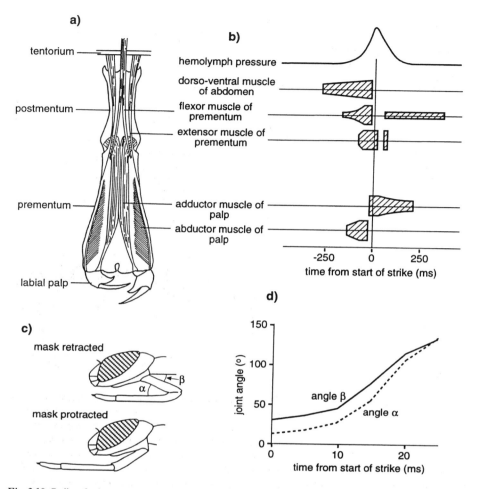

Fig. 2.10. Strike of a dragonfly larva (partly after Tanaka & Hisada, 1980). (**a**) Dorsal view of an extended labial mask with its principal muscles. (**b**) Timing of activity of the muscles relative to the strike. (**c**) Lateral view showing the labial mask retracted and extended. (**d**) Changes in the angles between the prementum and postmentum (angle α) and the postmentum and prothorax (angle β) during the strike. The positions of the angles are shown in (**c**).

size. Thus a dragonfly will turn towards a small stone thrown into the air, and the wasp, *Philanthus*, orients to a variety of moving objects, although it only catches objects having the smell of bees.

Mechanical stimulation is sometimes important in finding prey. *Notonecta* is able to locate prey trapped in the air/water interface as a result of the ripples which radiate from the struggling object. The vibrations are perceived by mechanoreceptors on the swimming legs. Coccinellid larvae preying on aphids only respond to the prey on contact.

Many Hymenoptera use a sting to inject venom into their prey (see section 27.2.7). Amongst predaceous Heteroptera, the prey is rapidly subdued, apparently by fast-acting lytic processes caused by enzymes from the salivary glands or midgut rather than by specific toxins.

2.3 REGULATION OF FEEDING

Most insects eat discrete meals separated by relatively long periods of non-feeding (Fig. 2.12, Table 2.2). A meal may weigh more than 10% of the body weight, and in some

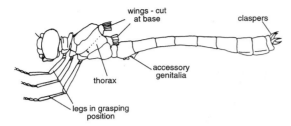

Fig. 2.11. Diagram of a male dragonfly to show the oblique development of the thorax bringing the legs into an anterior position, which facilitates grasping the prey.

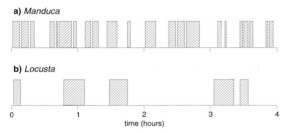

Fig. 2.12. Examples of the pattern of feeding of two phytophagous insects feeding on acceptable plants under constant conditions in the laboratory. **(a)** A caterpillar, *Manduca*, feeding on tobacco (after Reynolds, Yeomans & Timmins, 1986). **(b)** A locust, *Locusta* feeding on wheat (after Blaney, Chapman & Wilson, 1973).

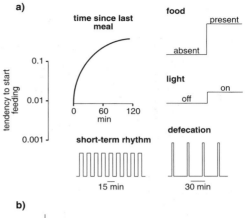

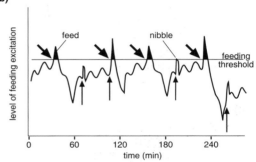

Fig. 2.13. Control of the pattern of feeding of a locust. **(a)** Effects of various factors on the probability that the insect will start to feed. The scale to the left applies to each of the diagrams (after Simpson & Ludlow, 1986). **(b)** Model, based on real data, showing how the factors in (a) interact to produce the feeding pattern. Feeding (shown in solid black) starts when the level of excitation exceeds the threshold. Small vertical arrows show increases in excitation following defecation; large, oblique arrows above the threshold show the point at which excitation is increased when the insect bites the food (after Simpson, 1995).

blood-sucking insects, such as *Rhodnius*, the quantity of blood ingested greatly exceeds the weight of the insect. Nectar-feeding insects only take very small meals relative to body weight. Those shown in the table use the nectar primarily as a flight fuel. When an ample supply is available, as might be the case with insects feeding on honeydew or some extrafloral nectaries, only a very small amount of time is spent feeding. *Lucilia* is an example of this. However, insects feeding from floral nectaries, such as *Vanessa* in Table 2.2, are limited by the small amounts of nectar found there. Moving from flower to flower and reaching the nectary of each one occupies about half of the time devoted to foraging by *Vanessa*. Homoptera, such as aphids and planthoppers, that feed on plant phloem or xylem do not have discrete meals, but feed almost continuously. Presumably they need to do so because of the very low concentrations of nutrients in their food.

The start of feeding after a pause is a probabilistic event depending on both internal and external factors which govern the level of a 'central excitatory state'. The level of central excitation increases with the time since the last meal and, in *Locusta*, a short term rhythm is superimposed on the general level of excitation (Fig. 2.13). It is further elevated by external influences such as the odor of food, and is higher in the light than in darkness. The insect is ready to feed if the central excitatory state exceeds a certain threshold. The increase in the central excitatory state is associated with the backwards movement of food in the gut and an increase in the sensitivity of the insect's receptors. If the insect now encounters suitable food it will start to feed but only for a short time in the absence of

Table 2.2. *Feeding patterns of some insects with different feeding habits. Based on laboratory observations under optimal light and temperature conditions*

Insect	Order	Food	Insect weight (mg)	Meal size (mg)	Meal size (% body weight)	Meal duration (min)	Time between meals	Time feeding %
Locusta (larva)	Orthoptera	Leaves	700	118	17	6-8	70-80 min	7-11
Nilaparvata (female)	Hemiptera	Plant fluids	2.5	—	—	Continuous	—	85
Zelus (female)	Hemiptera	Insect contents	105	~20	~20	130	>24 h	<9
Rhodnius (larva)	Hemiptera	Vertebrate blood	40	300	750	15	days	<1
Manduca (larva)	Lepidoptera	Leaves	3500	80	2	10-25	15-25 min	25-60
Vanessa (adult)	Lepidoptera	Nectar	600	30	5	21	130 min	14[a]
Lucilia (adult)	Diptera	Nectar	25	2	8	0.5-1	25-40 min	1-3.5[b]
Apis (adult, worker)	Hymenoptera	Nectar	90	30	33	30-80[c]	—	48[c]
Apis (adult, worker)	Hymenoptera	Pollen	90	20	22	10[c]	—	12[c]

Notes:

[a] Feeding on flowers in the laboratory. } Used for energy supply; in other insects listed food is for

[b] Feeding on glucose solution. } growth and/or egg development, as well as energy.

[c] Foraging on flowers in the field. Includes foraging time.

phagostimulation. Phagostimulation produces a sharp increase in excitation causing the insect to continue feeding.

If the insect loses contact with the food it exhibits a type of behavior which will increase the likelihood of relocating it. This is commonly known as 'searching' behavior. For example, once an adult coccinellid has eaten an aphid it moves less rapidly and turns more frequently. The result of this change in behavior is to keep the insect close to the point at which it encountered prey and, since aphids are commonly clumped together, to increase the likelihood of encountering more food (Nakamuta, 1985). Similar types of behavior are observed when a fly consumes a small sugar drop or when a grasshopper loses contact with its food.

On a highly acceptable food, a grasshopper feeds until the crop, and sometimes the midgut, is full. Nectar-feeding insects fill the crop and blood-sucking insects, when feeding on blood, fill the midgut. The degree of distension is monitored by stretch receptors on the wall of the alimentary canal or in the body wall. In grasshoppers, stretch receptors at the anterior end of the gut and on the ileum effect this regulation; the crop of flies is covered by a network of nerves associated with stretch receptors; in *Rhodnius* chordotonal organs in the body wall inhibit further feeding as blood in the midgut causes the abdomen to expand. The axons from these receptors pass to the central nervous system, either directly, or via the stomatogastric nervous system. The inputs from these receptors are inhibitory and are believed to function by reducing the level of the central excitatory state below the threshold for feeding.

In phytophagous insects, deterrent compounds in the food may also reduce meal size. It is believed that these, too, act via the central excitatory state.

The intervals between meals are very variable. In caterpillars they are commonly of the order of 15–30 minutes; in grasshoppers 1–2 hours (Fig. 2.12). In both caterpillars and grasshoppers, feeding on a protein rich diet results in extended intermeal intervals. Adult female mosquitoes and tsetse flies take a blood meal once every few days in relation to the cycle of oogenesis; larval *Rhodnius* only feed once in each larval stage. Consequently, the overall percentage of time spent feeding is usually very low (Table 2.2).

Variation in intermeal duration coupled with changes in meal length results in variation in the total amount of food consumed over a period. In grasshopper larvae, for

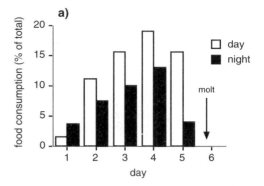

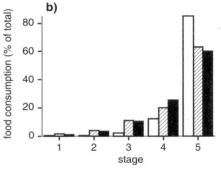

☐ *Pieris* - holometabolous, phytophagous
▨ *Schistocerca* - hemimetabolous, phytophagous
■ *Rhodnius* - hemimetabolous, blood-sucking

Fig. 2.14. Long-term variation in food intake. (a) Daily variation in the amounts of food consumed by a first stage larva of a grasshopper, *Schistocerca*. Expressed as percentages of the total amount eaten during the whole of the first stage (data from Chapman & Beerling, 1990). (b) The amount of food consumed by each larval stage of insects with different types of development and feeding habits. Expressed as percentages of the total amounts of food consumed during the whole of larval development (after Waldbauer, 1968).

example, differences may be considered on three time scales. More food is eaten during the light period than in the dark even when the temperature is constant, food intake reaches a peak in mid-stadium and ceases altogether some time before the molt, and over 50% of the food consumed over the entire developmental period of the insect is eaten during the final stadium (Fig. 2.14). Comparable changes occur in other insects. Further changes may occur during the adult period in relation to oogenesis.

Reviews: Bernays & Simpson, 1982; Chapman & de Boer, 1995

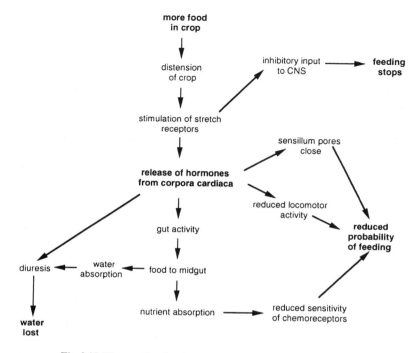

Fig. 2.15. Diagram showing the effects of foregut distension in a locust.

2.4 THE CONSEQUENCES OF FEEDING

In addition to its primary function of providing nutrients, feeding has other effects on the physiology and behavior of insects. Typically, after a full meal the insect becomes quiescent; it will not feed even in the presence of feeding stimuli, diuresis may occur and food may begin to pass backwards through the gut. In grasshoppers, all these activities are regulated, at least in part, by one or more hormones released from the corpora cardiaca as a consequence of distending the foregut and stimulating the stretch receptors (Fig. 2.15). These are the same receptors that lead to the cessation of feeding.

The reduction in readiness to feed is associated with a reduction in the sensitivity of contact chemoreceptors concerned with feeding. This results partly, in grasshoppers and caterpillars, from high levels of sugars or amino acids in the hemolymph following a meal. In grasshoppers there is also closure of the terminal pores of the contact chemoreceptors. These effects are probably not the immediate cause of the failure to feed, but have the effect of reducing sensory input during a period in which further feeding might reduce the effectiveness with which food from the previous meal is digested and absorbed. The effect of feeding on diuresis is best known in *Rhodnius* where stimulation of the abdominal stretch receptors leads to the release of diuretic hormone into the hemolymph (section 18.3.3).

2.5 HEAD GLANDS

Associated with each of the gnathal segments (mandibular, maxillary and labial) may be a pair of glands although they are not usually all present together.

2.5.1 Mandibular, hypopharyngeal and maxillary glands

Mandibular glands are found in Apterygota, Blattodea, Mantodea, Isoptera, Coleoptera and Hymenoptera and are usually sac-like structures in the head opening near the bases of the mandibles. They are large in larval Lepidoptera where they are the functional salivary glands; they are absent from adult Lepidoptera. They are especially important in social Hymenoptera where they are important sources of pheromones (see section 27.1.1 and Fig. 27.3).

Hypopharyngeal glands occur in Hymenoptera and are particularly well developed in worker honeybees. They are vestigial in queens and absent from males. There is one gland on each side of the head consisting of a long coiled duct with numerous small glandular lobes attached to it. The ducts open at the base of the hypopharynx. These glands produce an invertase as well as components of brood food.

Secretions from the hypopharyngeal and mandibular glands of worker honeybees are fed to larvae and regulate their development into queen or worker bees. Queen larvae receive mainly the secretion of the mandibular glands during their first three days of development, then, for the last two days, their food contains roughly equal amounts of secretion from the two glands. Worker larvae, by contrast receive a much greater proportion of the nutrient from the hypopharyngeal glands. The 'royal jelly' on which queen larvae are fed contains many different compounds including large amounts of 10-hydroxy-*trans*-2-decanoic acid, nucleic acids, and all the common amino acids. It also contains about 10 times more pantothenic acid (a B vitamin) and biopterin than food fed to worker larvae. Sugars, which act as phagostimulants, comprise

over 30% of royal jelly; the food of worker larvae has, at first, only about 12% sugar. Consequently, queen larvae eat much more food and grow bigger. Worker bees feed either type of larva, apparently regulating the food they give according to the type of larva they visit. The consequences of the differences in food on larval development are discussed in section 15.5.

Maxillary glands are found in Protura, Collembola, Heteroptera and some larval Neuroptera and Hymenoptera. They are usually small, opening near the bases of the maxillae, and may be concerned with lubrication of the mouthparts.

2.5.2 Labial glands

The most commonly occurring head glands are the labial glands which are present in all the major orders of insects except the Coleoptera. In most insects they function as salivary glands.

2.5.2.1 *Structure*

The labial glands of most insects are acinous glands (Fig. 2.16), but in Lepidoptera, Diptera and Siphonaptera (fleas) they are tubular with the ducts swelling to form the terminal glandular parts (Fig. 2.17). Sometimes, as in cockroaches, there is also a salivary reservoir, while in Heteroptera the gland consists of a number of separate lobes (Fig. 2.18).

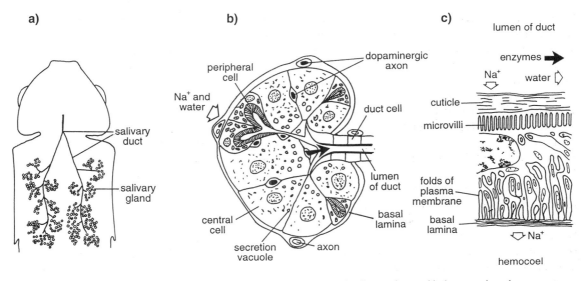

Fig. 2.16. Acinous salivary gland. Open arrows show the movement of sodium and water; black arrows show the movements of enzymes. (a) General arrangement in *Locusta*. (b) Section through an acinus of *Nauphoeta* (after Maxwell, 1978). (c) Section through a duct cell of *Schistocerca* (after Kendall, 1969).

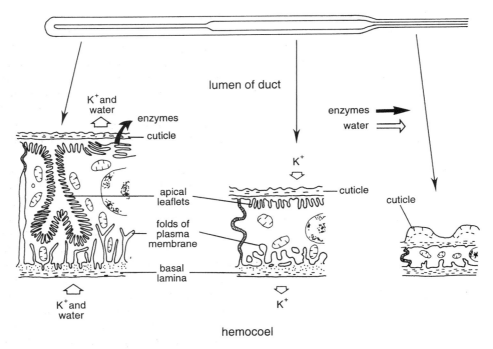

Fig. 2.17. Tubular salivary gland of *Calliphora*. At the top is a representation of the gland showing the positions of the cells depicted below. Left: secretory cell; center: cuboid cell; right: duct cell. Open arrows show the movement of potassium and water; black arrows show the movements of enzymes (after Oschman & Berridge, 1970).

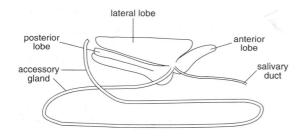

Fig. 2.18. Salivary glands of a hemipteran showing the different lobes (*Oncopeltus*) (after Miles, 1960).

The cells of the gland and duct are differentiated to perform three functions: the production and secretion of enzymes and other chemicals present in the saliva, movement of water from the hemolymph into the lumen by the active secretion of sodium or potassium, and modification of the fluid as it passes down the salivary duct by resorption of ions and, perhaps, some water. In acinous glands, the central, or zymogen, cells have extensive endoplasmic reticulum and Golgi bodies and probably produce the enzymes. The peripheral, or parietal, cells have an extensive microvillar border to the channel leading to the lumen of the duct. These cells are responsible for the movement of water into the lumen of the gland. In tubular glands both functions appear to be performed by the same cells. The movement of water results from the creation of an osmotic gradient across the cells by H^+-ATPase pumps in the microvillar plasma membrane. The pumps move protons into the lumen, and the protons are then exchanged for sodium or potassium at cation/H^+ anti-porters resulting in a high ionic concentration in the lumen. The duct cells of acinous glands, or the cuboid cells of tubular glands, remove cations from the salivary secretion, and those of the cockroach have sodium/potassium pumps in the basal plasma membrane (Just & Walz, 1994).

The ducts from the glands run forwards and unite to form a single median duct which opens just behind or on the hypopharynx. In fluid-feeding insects, muscles in the head insert on to the duct to form a salivary pump. At this point the lower wall of the duct is rigid while the upper part is flexible. It is pulled up by the dilator muscles so that fluid is drawn into the lumen of the pump. There are no compressor muscles. When the dilators relax, the upper wall springs down by virtue of the elasticity of the cuticle lining the pump and forces saliva out. In some insects, at least, valves ensure the forward flow of saliva.

2.5.2.2 *Control of secretion*

Acinous glands are innervated by axons from the sub-esophageal ganglion and from the stomatogastric system. In both cockroach and locust, one of the innervating neurons produces dopamine, another serotonin. Dopamine stimulates the secretion of fluid, while sero-tonin causes the central cells to produce and secrete the enzymes (Just & Walz, 1996). In addition, a network of branches of an octopaminergic neuron is closely associated with the salivary glands of *Locusta*.

Tubular glands are not directly innervated although in the female mosquito, *Aedes aegypti*, a plexus of nerves closely surrounds part of the gland. In both *Aedes* and *Calliphora*, serotonin, acting directly on the gland, regu-lates the production and release of saliva. In *Aedes*, the serotonin is released from the neural plexus adjacent to the gland, while in *Calliphora* it comes from neurohemal organs on the ventral nerves in the abdomen.

The production of saliva results from stimulation of chemoreceptors on the mouthparts; in *Calliphora*, stimulation of the interpseudotracheal pegs is necessary. In most insects, production and release occur together, but in insects with a salivary reservoir the two processes may be separately controlled. Production stops when sensory input is removed. In *Calliphora* it takes less than two minutes to clear the serotonin from the hemolymph when feeding stops and the interpseudotracheal pegs are no longer stimulated.

Reviews: Ali, 1997 – innervation; House & Ginsborg, 1985 – pharmacology

2.5.2.3 *Functions of saliva*

Saliva serves to lubricate the mouthparts and more is pro-duced if the food is dry. It also contains enzymes which start digestion of the food. The presence of particular enzymes is related to diet, but an amylase, converting starch to sugar, and an invertase, converting sucrose to glucose and fructose, are commonly present (Table 2.3). In blood-sucking insects, the saliva contains no digestive enzymes, but it does have components to overcome the hemostatic responses of the host (Table 2.1). In a number of insects, specific enzymes are present in the saliva that facilitate penetration and digestion of the food (Table 2.3). For example, leaf cutting ants, *Acromyrmex*, have a salivary chitinase that attacks chitin in the fungus on which the insects feed; larval warble flies, *Hypoderma*, that bore in the subcutaneous tissues of cattle, secrete a collagenase which facilitates movement of the larva through the tissues of the host.

Plant-sucking Hemiptera produce two types of saliva, a typical watery saliva carrying enzymes, and another which hardens to form the salivary sheath (Fig. 2.8). The enzymes in watery saliva are produced in the posterior lobe of the glands and water probably comes from the accessory gland. The watery saliva is produced during penetration of the plant tissues and some of its enzymes facilitate penetration through the middle lamellae of plant cell walls. Aphids, for example, have both a pectinesterase and a galacturonidase. In addition, the saliva in some species contains an amylase and a proteinase which contribute to extra-oral digestion of the plant tissue. Amino acids may be present in relatively large amounts in the saliva (Laurema & Varis, 1991) and it is possible that the aphids excrete unutilized dietary compo-nents in this way. The sheath material comes from the ante-rior and lateral lobes of the gland in the milkweed bug, *Oncopeltus*. It contains a catechol oxidase and a peroxidase which perhaps counter the effects of products produced by plants to inhibit insect feeding (Miles & Earthly, 1993).

Male scorpion flies (Mecoptera) have enlarged salivary glands and produce large quantities of saliva which are eaten by the female during copulation.

Trophallaxis Many species of wasps and ants, and some bees, exchange fluids with larvae. The larvae of these species have enlarged salivary glands (Fig. 2.19) which appear to be the source of the fluids they regurgitate. Larvae of vespid wasps produce saliva containing sugars at concentrations of 10 mg ml^{-1} or more, principally glucose and trehalose, about 1% proteins and 18–24 amino acids at total concentrations ranging from 25 to 95 μmol l^{-1} (Abe *et al.*, 1991). Larval ants produce salivary secretions from which carbohydrates are absent, but which contain high concentrations of amino acids and proteins, including a number of enzymes. These enzymes may add to the amounts present in the midguts of adult ants, although the significance of the transfer is unclear. The nutrient

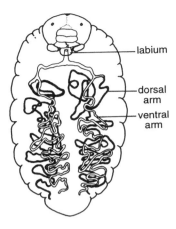

labium

dorsal
arm

ventral
arm

Fig. 2.19. Salivary glands of a larval social wasp, *Vespula*. The dorsal arm of each gland is shown black for clarity (after Maschwitz, 1966).

components of the larval saliva of wasps and ants have nutritional value for the adult insect obtaining them, and may be critically important (Abe *et al.*, 1991).

The mutual or unilateral exchange of alimentary fluid, including saliva, is called trophallaxis. Oral trophallaxis is most common, but anal trophallaxis also occurs. Oral transfer of regurgitated liquid from one adult to another is a common feature of social Hymenoptera, and can result in the rapid transfer of a chemical through a colony. In *Formica fusca*, for example, traces of honey eaten by one worker can be found in every member of the colony 24 hours later. This rapid transfer is especially important in the distribution of pheromones regulating colony structure (see section 27.1.6.2). Transfer between adult ants often involves nectar, storage of which in the crop is made possible by a valve-like proventriculus which regulates the backwards movement of fluid into the midgut. This storage reaches its most extreme in the honeypot ants, such as *Myrmecocystus*, where some workers are fed until their crops, and therefore their gasters where the crop resides, become enormously distended and their movement is greatly restricted. These 'repletes' are fed by other workers when nectar is plentiful, but serve as the source of food during dry periods.

Anal trophallaxis, or proctodeal feeding, is especially important in termites with symbiotic organisms in the rectum. In Kalotermitidae, which have symbiotic flagellates in the rectum, the flagellates are lost when the insect molts and they are regained from the anal fluid of other individuals in the colony. Production of a fluid containing the flagellates as well as wood fragments, is stimulated when a termite touches the perianal region of another with its antennae. The fluid may also have direct nutritional value.

Trophallaxis may have been of some importance in the evolution of social behavior.

Silk production In larval Lepidoptera and Trichoptera the labial glands produce silk. Silk production may begin immediately after hatching in species in which the larvae disperse on silken threads, a process known as ballooning. Other species use silk for larval shelters, and many moths spin extensive silken cocoons in which they pupate. In *Bombyx mori*, the silk moth, the glands hypertrophy at the end of the last larval stage, the increase in size resulting from 18–20 endomitotic divisions of the cells during which the nuclei become complexly branched. The posterior part of the gland, which consists only of some 500 large cells, produces the main silk protein, fibroin, as well as a polypeptide whose function is unknown. The single amino acid, glycine, comprises more than 40% of the fibroin. The central part of the gland, with about 300 cells, produces the proteins that cement the silk threads together. These are called sericins and contain large amounts of serine. The fibroin is molded to form a thread as it passes through the silk press, formed by fusion of the hypopharynx and labium around the salivarium. In Lepidoptera, the ducts from Lyonnet's gland join the ducts of the silk glands. This small gland possibly has a lubricating function as the silk passes through the press.

In Hymenoptera which spin a silken cocoon or, if they are social species, cap their cells with silk, the labial glands function first as salivary glands and change to silk production just before pupation. Weaver ants, *Oecophylla*, use silk from larvae approaching metamorphosis to bind the leaves of their nest together. Psocoptera have two pairs of labial glands, one pair produces saliva and the other, silk.

A few insects produce silk from other glands. In Neuroptera, the Malpighian tubules produce silk in the final larval stage for production of the cocoon. Embiids, which live in tunnels of silk, have silk-producing glands on the fore tarsi.

Reviews: Hölldobler & Wilson, 1990 – ants; Hunt & Nalepa, 1994 – social insects; Miles, 1972 – Hemiptera; Prudhomme *et al.*, 1985 – silk production; Ribeiro, 1995 – blood-sucking insects; Wilson, 1971 – social insects

Table 2.3. *Digestive enzymes in the saliva of insects with different feeding habits*

Insect	Order	Food	Enzymes				
			Amylase	Invertase	Proteinase	Lipase	Diet-related
Periplaneta	Blattodea	Detritus	+	+	+	+	
Locusta	Orthoptera	Leaves	+	+	−	+	
Oncopeltus	Hemiptera	Seeds	+	+	+	+	Pectinesterase[a]
Aphids	Hemiptera	Plant fluids	±	+	±	?	Polygalacturonidase[a]
Rhodnius	Hemiptera	Vertebrate blood	−	−	−	−	See Table 2.1
Zelus	Hemiptera	Insects	+	?	+	+	Hyaluronidase[b]
Platymeris	Hemiptera	Insects	?	?	+	−	
Hypoderma (larva)	Diptera	Vertebrate tissues	?	?	+	?	Collagenase[c]
Acromyrmex (adult)	Hymenoptera	Fungus	+	+	−	+	Chitinase[d]

Notes:

+, present; −, absent; ±, present in some species, not others; ?, not known.

[a] Digests components of plant cell walls.

[b] Digests insect connective tissue.

[c] Digests vertebrate collagen.

[d] Digests chitin in fungi.

REFERENCES

Abe, T., Tanaki, Y., Miyazaki, Y. & Kawasaki, Y.Y. (1991). Comparative study of the composition of hornet larval saliva, its effect on behaviour and role of trophallaxis. *Comparative Biochemistry and Physiology*, **99C**, 79–84.

Ali, D.W. (1997). The aminergic and peptidergic innervation of insect salivary glands. Journal of Experimental Biology, **200**, 1941–9.

Andersen, P.C., Brodeck, B.V. & Mizell, R.F. (1992). Feeding by the leaf hopper, *Homalodisca coagulata*, in relation to xylem fluid chemistry and tension. *Journal of Insect Physiology*, **38**, 611–22.

Backus, E.A. (1988). Sensory systems and behaviours which mediate hemipteran plant-feeding: a taxonomic overview. *Journal of Insect Physiology*, **34**, 151–65.

Bernays, E.A. & Chapman, R.F. (1994). *Host-plant Selection by Phytophagous Insects*. New York: Chapman & Hall

Bernays, E.A. & Simpson, S.J. (1982). Control of food intake. *Advances in Insect Physiology*, **16**, 59–118.

Blaney, W.M., Chapman, R.F. & Wilson, A. (1973). The pattern of feeding of *Locusta migratoria* (L.) (Orthoptera, Acrididae). *Acrida*, **2**, 119–37.

Chapman, R.F. (1982). Chemoreception: the significance of receptor numbers. *Advances in Insect Physiology*, **16**, 247–365.

Chapman, R.F. (1995). Mechanics of food handling by chewing insects. In *Regulatory Mechanisms in Insect Feeding*, ed. R.F.Chapman & G.de Boer, pp. 3–31. New York: Chapman & Hall.

Chapman, R.F. & Beerling, E.A.M. (1990). The pattern of feeding of first instar nymphs of *Schistocerca americana*. *Physiological Entomology*, **15**, 1–12.

Chapman, R.F. & de Boer, G. (ed.) (1995). *Regulatory Mechanisms in Insect Feeding*. New York: Chapman & Hall.

Chisholm, I.F. & Lewis, T. (1984). A new look at thrips (Thysanoptera) mouthparts, their action and effects of feeding on plant tissue. *Bulletin of Entomological Research*, **74**, 663–75.

Cicero, J.M. (1994). Composite, haustellate mouthparts in netwinged beetle and firefly larvae (Coleoptera, Cantharoidea: Lycidae, Lampyridae). *Journal of Morphology*, **219**, 183–92.

Corrette, B.J. (1990). Prey capture in the praying mantis *Tenodera aridifolia sinensis*: coordination of the capture sequence and strike movements. *Journal of Experimental Biology*, **148**, 147–80.

Davis, E.E. & Friend, W.G. (1995). Regulation of a meal: blood feeders. In *Regulatory Mechanisms in Insect Feeding*, ed. R.F.Chapman & G.de Boer, pp. 157–89. New York: Chapman & Hall.

Denis, R. (1949). Sous-classe des Aptérygotes. In *Traité de Zoologie*, vol. 9, ed. P.-P.Grassé, pp. 111–275. Paris: Masson et Cie.

Gronenberg, W., Tautz, J. & Hölldobler, B. (1993). Fast trap jaws and giant neurons in the ant *Odontomachus*. *Science*, **262**, 561–3.

Hölldobler, B. & Wilson, E.O. (1990). *The Ants*. Cambridge, Mass: Harvard University Press.

House, C.R. & Ginsborg, B.L. (1985). Salivary gland. In *Comprehensive Insect Physiology, Biochemistry and Pharmacology*, vol. 11, ed. G.A.Kerkut, & L.I.Gilbert, pp. 195–224. Oxford: Pergamon Press.

Hunt, J.A. & Nalepa, C.A. (ed.) (1994). *Nourishment and Evolution in Insect Societies*. Boulder, Colorado: Westview Press.

Just, F. & Walz, B. (1994). Immunocytological localization of Na^+/K^+-ATPase and $V-H^+$-ATPase in the salivary glands of the cockroach, *Periplaneta americana*. *Cell and Tissue Research*, **278**, 161–70.

Just, F. & Walz, B. (1996). The effects of serotonin and dopamine on salivary secretion by isolated cockroach salivary glands. *Journal of Experimental Biology*, **199**, 407–13.

Kendall, M.D. (1969). The fine structure of the salivary glands of the desert locust *Schistocerca gregaria* Forskål. *Zeitschrift für Zellforschung und Mikroskopische Anatomie*, **98**, 399–420.

Kingsolver, J.G. & Daniel, T.L. (1995). Mechanics of food handling by fluid-feeding insects. In *Regulatory Mechanisms in Insect Feeding*, ed. R.F.Chapman & G.de Boer, pp. 157–89. New York: Chapman & Hall.

Krenn, H.W. (1990). Functional morphology and movements of the proboscis of Lepidoptera (Insecta). *Zoomorphology*, **110**, 105–14.

Lange, A.B., Orchard, I. & Barrett, F.M. (1989). Changes in the hemolymph serotonin levels associated with feeding in the blood-sucking bug, *Rhodnius prolixus*. *Journal of Insect Physiology*, **35**, 393–9.

Laurema, S. & Varis, A.-L. (1991). Salivary amino acids in *Lygus* species (Heteroptera: Miridae). *Insect Biochemistry*, **21**, 759–65.

Maschwitz, U. (1966). Das Speichelsekret der Wespenlarven und seine biologische Bedeutung. *Zeitschrift für Vergleichende Physiologie*, **53**, 228–52.

Maxwell, D.J. (1978). Fine structure of axons associated with the salivary apparatus of the cockroach, *Nauphoeta cinerea*. *Tissue & Cell*, **10**, 699–706.

Miles, P.W. (1960). The salivary secretions of a plant-sucking bug, *Oncopeltus fasciatus* (Dall.) (Heteroptera: Lygaeidae). III. Origins in the salivary glands. *Journal of Insect Physiology*, **4**, 271–82.

Miles, P.W. (1972). The saliva of Hemiptera. *Advances in Insect Physiology*, **9**, 183–255.

Miles, P.W. & Earthly, J.J. (1993). The significance of antioxidants in the aphid-plant interaction: the redox hypothesis. *Entomologia Experimentalis et Applicata*, **67**, 275–83.

Mitchell, B.K. & Itagaki, H. (1992). Interneurons of the subesophageal ganglion of *Sarcophaga bullata* responding to gustatory and mechanosensory stimuli. *Journal of Comparative Physiology*, A **171**, 213–30.

Nakamuta, K. (1985). Mechanism of the switchover from extensive to area-concentrated search behaviour of the ladybeetle, *Coccinella septempunctata bruckii*. *Journal of Insect Physiology*, **31**, 849–56.

Oschman, J.J. & Berridge, M.J. (1970). Structural and functional aspects of salivary fluid secretion in *Calliphora*. *Tissue & Cell*, **2**, 281–310.

Parry, D.A. (1983). Labial extension in the dragonfly larva *Anax imperator*. *Journal of Experimental Biology*, **107**, 495–9.

Prudhomme, J.-C., Couble, P., Garel, J.-P. & Daillie, J. (1985). Silk synthesis. In *Comprehensive Insect Physiology, Biochemistry and Pharmacology*, vol. 10, ed. G.A.Kerkut & L.I. Gilbert, pp. 571–94. Oxford: Pergamon Press.

Rehder, V. (1989). Sensory pathways and motoneurons of the proboscis reflex in the suboesophageal ganglion of the honey bee. *Journal of Comparative Neurology*, **279**, 499–513.

Reynolds, S.E., Yeomans, M.R. & Timmins, W.A. (1986). The feeding behaviour of caterpillars (*Manduca sexta*) on tobacco and on artificial diet. *Physiological Entomology*, **11**, 39–51.

Ribeiro, J.M.C. (1995). Insect saliva: function, biochemistry and physiology. In *Regulatory Mechanisms in Insect Feeding*, ed. R.F.Chapman & G.de Boer, pp. 74–97. New York: Chapman & Hall

Rossel, S. (1991). Spatial vision in the preying mantis: is distance implicated in size detection? *Journal of Comparative Physiology*, A **169**, 101–8.

Schachtner, J. & Braunig, P. (1993). The activity patterns of identified neurosecretory cells during feeding behaviour in the locust. *Journal of Experimental Biology*, **185**, 287–303.

Simpson, S.J. (1995). Regulation of a meal: chewing insects. In *Regulatory Mechanisms in Insect Feeding*, ed. R.F.Chapman & G.de Boer, pp. 137–56. New York: Chapman & Hall.

Simpson, S.J. & Ludlow, A.R. (1986). Why locusts start to feed: a comparison of causal factors. *Animal Behaviour*, **34**, 480–96.

Smith, J.J.B. (1985). Feeding mechanisms. In *Comprehensive Insect Physiology, Biochemistry and Pharmacology*, vol. 4, ed. G.A. Kerkut & L.I. Gilbert, pp. 33–85. Oxford: Pergamon Press.

Snodgrass, R.E. (1935). *Principles of Insect Morphology*. New York: McGraw-Hill.

Snodgrass, R.E. (1944). The feeding apparatus of biting and sucking insects affecting man and animals. *Smithsonian Miscellaneous Collections*, **107**, no. 7, 113 pp.

Tanaka, Y. & Hisada, M. (1980). The hydraulic mechanism of the predatory strike in dragonfly larvae. *Journal of Experimental Biology*, **88**, 1–19.

Tjallingii, W.F. (1995). Regulation of phloem sap feeding by aphids. In *Regulatory Mechanisms in Insect Feeding*, ed. R.F. Chapman & G.de Boer, pp. 190–209. New York: Chapman & Hall.

Tjallingii, W.F. & Esch, T.H. (1993). Fine structure of aphid stylet routes in plant tissues in correlation with EPG signals. *Physiological Entomology*, **18**, 317–28.

Waldbauer, G.P. (1968). The consumption and utilisation of food by insects. *Advances in Insect Physiology*, **5**, 229–88.

Wilson, E.O. (1971). *The Insect Societies*. Cambridge, Mass: Harvard University Press.

3 Alimentary canal, digestion and absorption

3.1 ALIMENTARY CANAL

The alimentary canal of insects is divided into three main regions: the foregut, or stomodeum, which is ectodermal in origin; the midgut, or mesenteron, which is endodermal; and the hindgut, or proctodeum, which again is ectodermal (Fig. 3.1). The epithelium of all parts of the gut consists of a single layer of cells.

Since the foregut and hindgut are ectodermal in origin, the cells secrete cuticle which is continuous with that covering the outside of the body. The lining cuticle is known as the intima. It is shed and renewed at each molt. Although the midgut does not secrete cuticle, in most insects it does secrete a delicate peritrophic envelope around the food.

Usually the gut is a continuous tube running from the mouth to the anus, but in some insects that feed on a fluid diet containing little or no solid waste material the connection between the midgut and the hindgut is occluded. This is the case in some plant-sucking Heteroptera, where the occlusion is between different parts of the midgut (see Fig. 3.9), and in larval Neuroptera which digest their prey extra-orally. A similar modification occurs in the larvae of social Hymenoptera with the result that the larvae never foul the nest. In these insects a pellet of fecal matter is deposited at the larva–pupa molt.

Reviews: Chapman, 1985a; Poisson & Grassé, 1976

3.1.1 Foregut

The cells of the foregut are usually flattened and undifferentiated since they are not involved in absorption or secretion, but the cuticular lining often varies in the different regions. It is generally unsclerotized, consisting only of endocuticle and epicuticle, but in many insects sclerotized spines or teeth project from its surface. Their arrangement varies from species to species and in different parts of the foregut (Fig. 3.2a). They commonly point backwards and it is assumed that they are concerned with moving food back towards the midgut.

The foregut is commonly differentiated into the pharynx, esophagus, crop and proventriculus. The pharynx is concerned with the ingestion and backwards passage of food and has a well-developed musculature (section 3.1.4).

The esophagus usually has a simple tubular form, serving as a connection between the pharynx and the crop. It is often poorly defined in hemimetabolous insects, but, in many adult holometabolous insects, it is a long slender tube running between the flight muscles back to the abdomen. In a few insects that feed on highly resinous plants, the resin is stored in single or paired diverticula of the esophagus. Examples include caterpillars of *Myrascia* that feed on Myrtaceae (Fig. 3.3d) and sawfly larvae in the genus *Neodiprion* that feed on pines. In *Neodiprion* the diverticula have powerful circular muscles that enable the

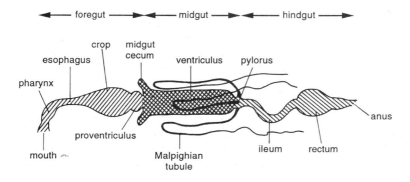

Fig. 3.1. Alimentary canal. Diagram showing the major subdivisions in a generalized insect.

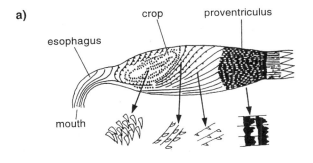

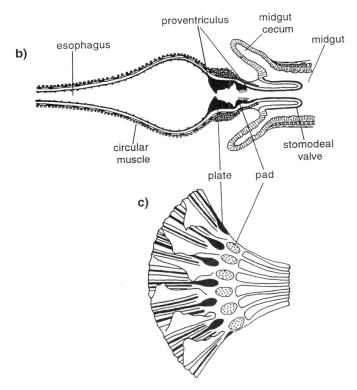

Fig. 3.2. Foregut armature. **(a)** Sagittal section through the foregut of a locust showing the pattern of cuticular spines on the intima. Enlargements show details of the spines. In the proventriculus, the spines are replaced by larger sclerotized plates with backwardly directed teeth at the posterior edges (after Williams, 1954). **(b)** Longitudinal section of the foregut of *Periplaneta* showing the development of the proventriculus to form a grinding apparatus (after Snodgrass, 1935). **(c)** Proventriculus of a cockroach slit open and laid flat showing the hexaradial symmetry (after Miller and Fisk, 1971).

larva to eject the contents through the mouth. This apparently has a defensive function.

The crop is a storage organ which in most insects is an extensible part of the gut immediately following the esophagus (Fig. 3.3a), but, in adult Diptera and Lepidoptera, it is a lateral diverticulum of the esophagus (Fig. 3.3b,c). The walls of the crop are folded longitudinally and transversely. The folds become flattened as the crop is filled, usually permitting a very large increase in volume. In *Periplaneta*, however, there is little change in volume because when the crop does not

Fig. 3.3. Storage in the gut. The different regions in which food or food components are temporarily stored are shown by hatching. **(a)–(d)** Storage in the foregut. **(a)** Is a typical caterpillar; **(d)** is a caterpillar that feeds on resinous plants and stores the resin in the esophageal diverticulum. **(e),(f)** Storage in the hind gut. **(g)** Storage in the midgut.

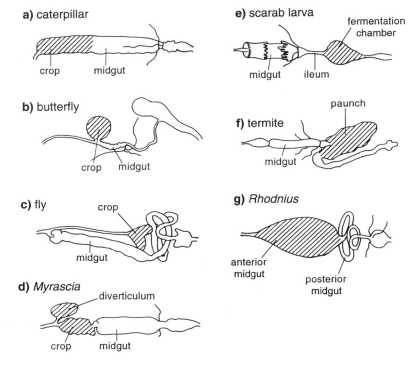

contain food it is filled with air. The effectiveness of the crop as a store, especially in fluid-feeding insects, depends on the impermeability of its cuticular lining to hydrophilic molecules (Fig. 3.13). The crop lining of *Periplaneta* is known to be permeable to free fatty acids.

The proventriculus is very variable in form. It often forms a simple valve at the origin of the midgut, projecting a short distance into the midgut lumen (as in Fig. 3.4). In other cases, as in grasshoppers (Acrididae), it forms a constriction just before the midgut and can limit the backwards movement of solid food while permitting the movement of liquids in either direction. Ants have a more specialized proventriculus varying greatly in form between species (Eisner, 1957). It allows them to separate the partly digested food in the midgut, from that in the crop which is used in trophallaxis as well as being the source of their own food. The ant's crop also functions as a filter in some species. The proventriculus of honeybees is also specialized (Fig. 3.4), allowing them to retain nectar in the crop while passing pollen grains to the midgut.

The proventriculus of many species of orthopteroid insects and some beetles, notably amongst the Adephaga, is a grinding apparatus with strong cuticular plates or teeth

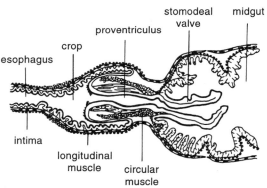

Fig. 3.4. Foregut of a worker honeybee in longitudinal section showing the development of the proventriculus. The anteriorly directed part enables the insect to extract pollen grains from nectar in the crop; the posterior part, projecting into the midgut, forms a valve (after Snodgrass, 1956).

which break up the food (Cheeseman & Pritchard, 1984). In orthopteroids it has a hexaradial symmetry with six main longitudinal folds and a series of secondary folds exhibiting varying degrees of sclerotization (Fig. 3.2b,c). Behind the teeth, in cockroaches, are six unsclerotized

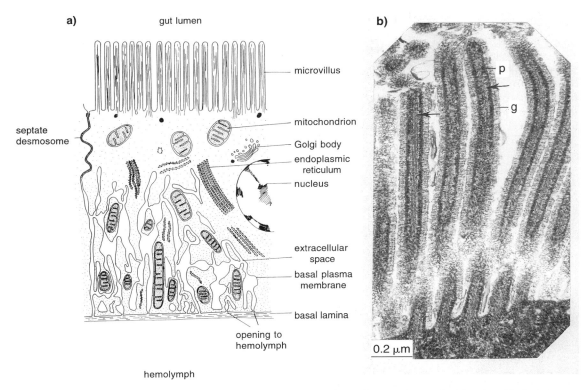

Fig. 3.5. Midgut epithelium. **(a)** Diagram of a principal midgut cell. **(b)** Electron micrograph of the microvilli of a thrips showing the glycocalyx (g) outside the plasma membrane (p) and the actin filaments (arrows) in the center of a microvillus (after Del Bene, Dallai & Marchini, 1991).

pads and, from these, folds continue on to the stomodeal valve. In beetles, the symmetry of the proventricular armature is tetraradial. Strong circular muscles surround the proventriculus.

3.1.2 Midgut

The cells of the midgut are actively involved in enzyme production and secretion, as well as in absorption of nutrients. The majority of the cells, called principal cells, are tall and columnar and the membrane on the luminal side forms microvilli (Figs. 3.5, 3.6). Each microvillus is supported by a bundle of actin filaments which arise from a layer of actin filaments beneath the apical margin of the cell.

The microvilli greatly increase the area of the cell membrane through which absorption occurs. In the grasshopper, *Schistocerca americana*, the microvilli of the cells in the anterior caeca are about 5 μm long and 0.1 μm in diameter. Each cell has some 9000 microvilli and they increase the area of the luminal surface of the cell by over two orders of magnitude. As a result, the total surface area of the midgut is about 500 cm^2 (Fig. 3.6).

The outer surface of the microvilli is covered by the glycocalyx (Fig. 3.5b) which, in most insects, is a layer of filamentous glycoproteins, but in some species its components form a regular array. In Hemiptera, the glycocalyx is replaced by lipid membranes that form a tube round each microvillus and, at least in some cases, are separated from the plasma membrane by a regular space of 9 or 10 nm. These membrane tubes extend into the lumen of the midgut when it contains food (Silva *et al.*, 1995; and see Billingsley, 1990).

When they are synthesizing enzymes, the principal cells are characterized by the presence of stacks of rough endoplasmic reticulum and Golgi bodies. In most insects, synthesis appears to occur at the time of secretion into the gut lumen so that stores of enzymes do not accumulate in the cells

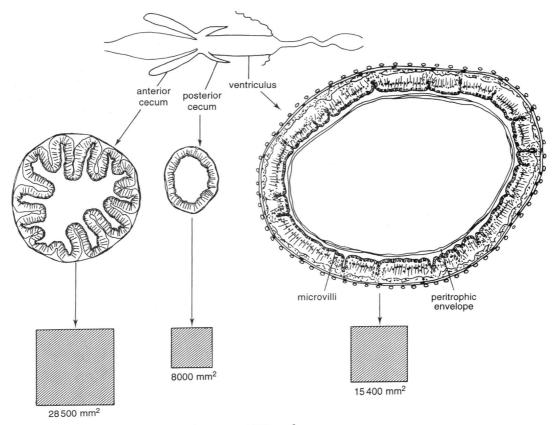

Fig. 3.6. Midgut surface area of a grasshopper. Outline of gut about twice natural size. Sections, all to same scale, through various parts of the midgut showing the extent of folding of the epithelium. The cells of all parts of the midgut have microvilli apically and so have very large surface areas. The boxes below show, to scale, the surface area of each part including the area produced by the microvilli.

(section 3.3.2.5). In a few cases, however, as in *Stomoxys*, enzymes are stored, probably as inactive precursors (zymogens), in membrane-bound vesicles in the distal parts of the cells. The enzyme is secreted within a few minutes of feeding and, in this species, the cycle of production and release occurs repeatedly while the meal is digested. In mosquitoes, however, there is only one synthetic cycle per meal.

Enzyme secretion occurs in several different ways. Membrane bound vesicles may move to the periphery of the cell, fuse with the cell membrane and release their contents into the gut lumen. This process is known as exocytosis (sometimes called eccrine secretion). In addition, extensions from the distal parts of the midgut cells may separate from the cell, carrying their contents into the gut lumen and ultimately releasing them. This is called apocrine secretion. It is also possible that vesicles containing enzymes move to the microvilli from which they bud off before releasing their contents. Different enzymes may leave a single cell by different routes, and different methods of secretion may occur in one insect (Wood & Lehane, 1991).

The principal cells of the midgut have a limited life and, in most insects, they are continually replaced from regenerative cells at the base of the midgut epithelium (Fig. 3.7a). These often occur in groups, known as nidi, and in some beetles they are found in crypts visible as small

a)

gut lumen

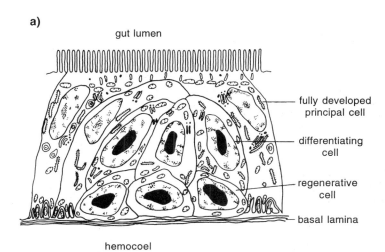

fully developed
principal cell

differentiating
cell

regenerative
cell

basal lamina

hemocoel

Fig. 3.7. Regenerative cells of the
midgut. (a) Diagram of a nidus at the
base of the midgut epithelium
showing the differentiation of
principal cells (after Fain–Maurel,
Cassier & Alibert, 1973). (b) Diagram
of a midgut crypt in a beetle extending
through the muscle layer to form a
papilla (after Snodgrass, 1935).

b)

gut lumen

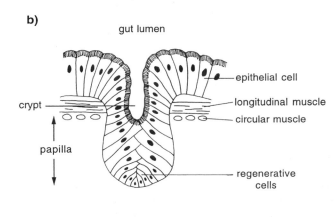

epithelial cell

crypt

longitudinal muscle

circular muscle

papilla

regenerative
cells

hemocoel

papillae on the outside of the midgut (Fig. 3.7b). In adult *Tenebrio* all the principal midgut cells are replaced over a four-day period. Regenerative cells do not occur in the midgut of cyclorrhaphous Diptera.

Endocrine cells are probably present in the midgut epithelia of all insects. They are probably involved in the regulation of enzyme production (see section 21.2.1.5).

Scattered amongst the principal cells of the midgut of caterpillars are goblet cells in which the luminal plasma membrane is invaginated to form a flask-shaped cavity (Fig. 3.8). Microvilli extend into the cavity and those at the base of the cavity contain mitochondria. Projections at the

apex of the cavity may form a valve which is capable of opening and closing. These cells create a high concentration of potassium in the gut lumen (see section 3.4). This is achieved by two processes occurring at their apical plasma membranes: a V-ATPase which pumps H^+ into the lumen and an antiporter at which H^+ is exchanged for potassium (Fig. 3.8b). The process consumes large amounts of energy and it is estimated that it uses 10% of the total ATP produced by the caterpillar.

Goblet cells also occur in Ephemeroptera, Plecoptera and Trichoptera, but it is not known if they have a similar function to those in the caterpillar.

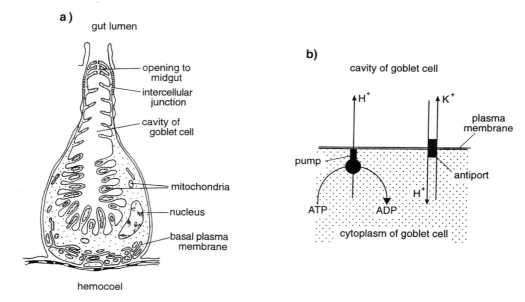

Fig. 3.8. Potassium secretion in the midgut of a caterpillar. (a) Diagram of a goblet cell (after Cioffi, 1979). (b) Diagram showing the movements of ions produced by the V-ATPase and K^+/H^+ antiporter at the apical plasma membrane (after Lepier *et al.*, 1994).

In many insects the midgut has diverticula known as ceca, usually at the anterior end (Figs. 3.1, 3.6). There are two in gryllids and some dipterous larvae; six in Acrididae (grasshoppers), eight in Blattodea and larval Culicidae, and larger numbers in some Coleoptera and Heteroptera where they are variable in position (Fig. 3.9).

The tubular part of the midgut is known as the ventriculus. It exhibits marked anatomical differentiation in the Heteroptera (Fig. 3.9), while amongst blood-sucking bugs and flies there is regional ultrastructural differentiation of the principal cells in association with storage, enzyme production and secretion, and absorption. Sometimes different cell types do not form discrete zones in the ventriculus, but are scattered through the epithelium as with endocrine cells and the goblet cells in the midgut of a caterpillar.

Reviews: Billingsley, 1990 – blood-sucking insects; Dow, 1986 – structure and function; Harvey & Nelson, 1992 – potassium secretion; Lehane & Billingsley, 1996 – structure and function; Martoja & Ballan-Dufrançais, 1984 – ultrastructure

3.1.2.1 *Peritrophic envelope*

The peritrophic envelope forms a delicate lining layer to the midgut, separating the food from the midgut epithelium (Fig. 3.6). It occurs in most insects, although it

apparently is not present in most Hemiptera. These insects have a membranous covering of the microvilli which may be analogous to a peritrophic envelope. In adult Diptera, a peritrophic envelope is absent from unfed insects, but forms within hours of the insect taking a meal. Only a blood meal induces the formation of the envelope in adult mosquitoes; a nectar meal is not followed by its production in either females or males. In adult *Aedes*, a complete envelope is produced within about 5 hours of feeding, but in other mosquitoes a longer period is required.

The envelope is usually made up of a number of separate laminae which are extracellular secretions and should not be confused with plasma membranes. Sometimes the envelope includes two different types of lamina. Each consists of a network of microfibrils, usually chitin, which may be regularly arranged to produce an open lattice structure (Fig. 3.10) or randomly oriented. The microfibrils are embedded in a matrix of proteins and glycoproteins.

Peritrophic envelopes are formed in two different ways. In larval Diptera, adult blowflies and Dermaptera, each lamina is secreted by one or more rings of cells at the anterior end of the midgut (Fig. 3.10a). In the larva of *Aedes*,

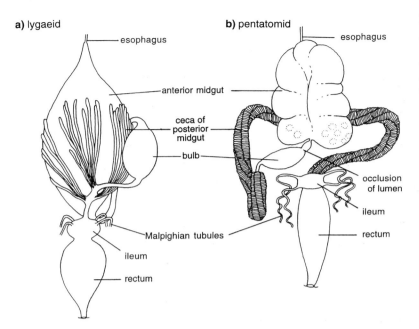

a) lygaeid

- esophagus
- anterior midgut
- ceca of posterior midgut
- bulb
- Malpighian tubules
- ileum
- rectum

b) pentatomid

- esophagus
- occlusion of lumen
- ileum
- rectum

Fig. 3.9. Midgut modifications of two plant feeding Heteroptera (after Goodchild, 1963). **(a)** *Dieuches* (family Lygaeidae). The ceca are closely applied to the anterior midgut and may be important in removing water from the food in a similar manner to the filter chamber of Homoptera shown in Fig. 3.25. **(b)** *Piezosternum* (family Pentatomidae). The midgut lumen is occluded before the bulb and possibly also at the junction of the cecal region to the ileum. The many small ceca in longitudinal rows along the posterior midgut are filled with bacteria.

for example, the most anterior 8–10 rings of cells in the midgut epithelium, a total of 300–400 cells, produce the entire peritrophic envelope. In this species, the envelope is a single lamina about 0.7 μm thick with outer and inner granular layers bounding a layer with microfibrils arranged in a grid. In contrast, in the earwig, *Forficula*, separate rings of cells in the most anterior part of the midgut produce up to four separate laminae in which the microfibrils form grids. More posterior rings of cells produce several more very thin laminae in which the microfibrils are randomly oriented. The complete peritrophic envelope thus consists of several layers of the two types of lamina. These laminae are not, apparently, produced as continuous tubes, but are formed as separate patches which join together to form the whole lamina.

In Orthoptera, Blattodea, larval Hymenoptera and Lepidoptera, and adult Nematocera, the peritrophic laminae are produced by the whole midgut epithelium. Production is not synchronized over the whole midgut but, as with laminae produced by anterior rings of cells, each lamina is formed in patches which join together. At least in some species, the microfibrils forming the laminae are assembled at the bases of the microvilli which form a template for the grid which the microfibrils form (Fig. 3.10c). This process also occurs in the anterior formation zone of *Forficula*. It is known as delamination. In an

actively feeding locust, a new lamina is produced about every 15 minutes so the peritrophic envelope becomes multilaminar. The inner layers move back with the food as more laminae are produced on the outside, with the result that, in the posterior parts of the midgut, the envelope has more laminae than it does anteriorly. The robustness of the envelope varies from species to species. For example, amongst the Acrididae (grasshoppers) *Locusta*, which is graminivorous, has a thin, delicate envelope, whereas *Schistocerca*, a polyphagous species, has a much thicker and tough envelope.

It is probable that the permeability properties of peritrophic envelopes are largely determined by proteoglycans. Experimental evidence on various Diptera suggests that molecules with an effective radius of less than 4.5 nm (with an atomic weight of about 35 kDa) pass through the envelope, but larger molecules do not. In grasshoppers and larval Lepidoptera, molecules with diameters of 25–35 nm are the largest that pass through the envelopes (Barbehenn & Martin, 1995).

The peritrophic envelope performs a number of different functions. Since it encloses the food bolus within the midgut it effectively separates the gut lumen into ecto- and endo-peritrophic spaces. These may be important in the compartmentalization of enzyme activities (section 3.3.2.4) and also make it possible, in some species, for

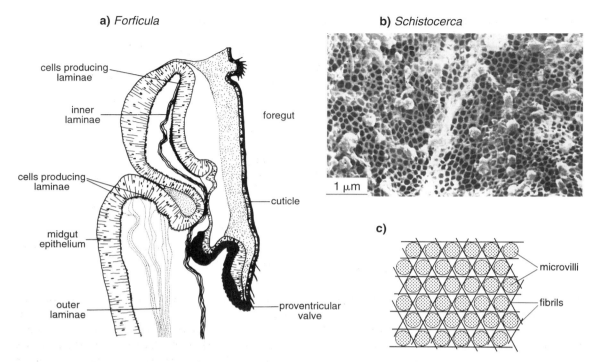

a) *Forficula*

cells producing laminae

inner laminae

foregut

cells producing laminae

cuticle

midgut epithelium

outer laminae

proventricular valve

b) *Schistocerca*

1 μm

c)

microvilli

fibrils

Fig. 3.10. Peritrophic envelope. (a) Diagram showing the origin of the peritrophic envelope in cells at the anterior end of the midgut of *Forficula* (after Peters *et al.*, 1979). (b) Scanning electron micrograph of the peritrophic envelope of *Schistocerca*. The envelope has been washed so that only the chitinous lattice remains (after Chapman, 1985a). (c) Diagram showing how the fibers which will form the lattice of the peritrophic envelope are laid down round the microvilli.

countercurrents to flow within the midgut, increasing the efficiency of absorption. In insects eating solid food it prevents the food particles from coming into contact with the microvilli of the midgut cells, perhaps avoiding damage to the cells.

The envelope confers some degree of protection against potentially harmful chemicals in the food of phytophagous insects. For example, the envelope of *Locusta* is permeable to tannic acid which causes lesions in the midgut epithelium, eventually leading to the death of the insect. In contrast, the envelope in *Schistocerca*, which often feeds on tannin-containing plants, is impermeable to tannic acid. Differences in the permeability of the peritrophic envelope to tannic acid also occur in caterpillars with different feeding habits (Barbehenn & Martin, 1992). In *Schistocerca*, and other grasshoppers with similar feeding habits, part of the epithelium of each midgut cecum forms a series of small pockets, each lined by the peritrophic envelope. Small particles and some chemicals, such as tannic

acid, collect in the pockets, apparently being swept in by the flow of water (see below). The peritrophic envelope of the pockets is pulled out intact as the whole envelope moves down the midgut and the pockets with their contents are finally voided with the feces (Bernays, 1981). Tannin molecules are much smaller than the pores in the envelopes of these species, and their failure to pass through must reflect the physicochemical properties of the matrix.

The relative impermeability of the peritrophic envelope may also confer some degree of protection against pathogenic organisms. The pore sizes are too small to permit the passage of most bacteria and, in *Schistocerca*, for example, the bacterial flora of the midgut, which is obtained adventitiously with the food, is entirely contained within the envelope. Similarly, in older honeybee larvae, *Bacillus larvae*, which causes foulbrood in bees, is unable to penetrate the envelope, but young larvae in which the envelope is not yet developed are highly susceptible to the disease. As in this instance, many pathogens of insects are able to circumvent the impermeability of the envelope by invading the tissues when the envelope is not fully developed.

The luminal surface of the peritrophic envelope of larva of *Calliphora* contains lectins. These are proteins which bind to specific carbohydrates, in this case, mannose. There is also circumstantial evidence that bacteria in the gut lumen bind to these sites, but the significance of this is not known.

Aminopeptidase is bound to the peritrophic membrane in a variety of insects. This may be an efficient way of carrying this and other enzymes into the gut lumen and avoiding their loss as fluid passes backwards from the food in the gut lumen.

Reviews: Lehane, 1997; Peters, 1992; Spence, 1991

3.1.3 Hindgut

The hindgut is usually differentiated into the pylorus, ileum and rectum (Fig. 3.1). The pylorus sometimes forms a valve between the midgut and hindgut. The Malpighian tubules often arise from it.

The ileum of most insects is a narrow tube running back to the rectum. Sometimes the posterior part is recognizably different and is called the colon. In many insects, only a single cell type is present in the ileum. The cells have extensive folding of the apical plasma membrane with abundant closely associated mitochondria. The basal plasma membrane may also be folded although the folds are less extensive than those apically (Fig. 3.11).

Amongst insects that have symbionts in their hind gut the ileum is expanded to house them. The expansion is called the paunch in termites and the fermentation chamber in larval Scarabaeidae (Fig. 3.3e,f). The cuticle lining these chambers is produced into elongate spines which probably serve for the attachment of microorganisms. Similar structures are present on the ileal cuticle of cockroaches and crickets that have a bacterial flora.

The rectum is usually an enlarged sac with a thin epithelium except for certain regions, the rectal pads, in which the epithelal cells are columnar. There are usually six rectal pads arranged radially round the rectum (Fig. 3.12a). They may extend longitudinally along the rectum or they may be papilliform as in the Diptera. The cuticle of the rectal pads is thin compared with that lining the rest of the rectum; in cockroaches it is only about 1 μm thick. It is unsclerotized except for a narrow band which forms a frame bounding each pad. Attached to the frame is a series of sheath cells

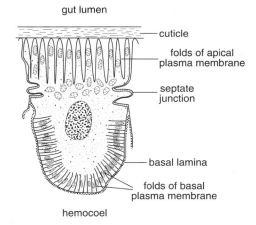

Fig. 3.11. Ileum. Diagram of an ileal cell of *Schistocerca* (based on Irvine *et al.*, 1988).

which, together with a layer of cells basally, isolate the principal cells of the pad from the hemocoel (Fig. 3.12b). In the blowfly, a pad of connective tissue, rather than a layer of cells, isolates each papilla. There is a space between the cuticle and the apical membrane of the principal cells and this membrane forms regular parallel folds containing mitochondria. The lateral plasma membranes are complexly folded and interdigitated with closely associated mitochondria (a scalariform junction) and, in termites, the development of these complexes is positively correlated with the dryness of the habitat occupied by the species. There are two layers of principal cells in the rectal pads of Neuroptera, Lepidoptera and Hymenoptera; only one layer is present in other groups. The extracellular spaces between the principal cells have limited connections with the hemocoel. The systems by which this is achieved vary from species to species. The cells have an extensive tracheal supply consistent with their high level of metabolism associated with water absorption.

The intima of the hindgut is usually no more than 10 μm thick, and differs from that of the foregut in being highly permeable. The cuticle of the locust ileum and rectum is over two orders of magnitude more permeable than that of the crop (Fig. 3.13). Permeability decreases as molecular size increases and the hindgut cuticle is virtually impermeable to polysaccharides, inulin, and other large molecules (Maddrell & Gardiner, 1980).

Insects living in freshwater possess specialized cells, called chloride cells, able to take up inorganic ions, not

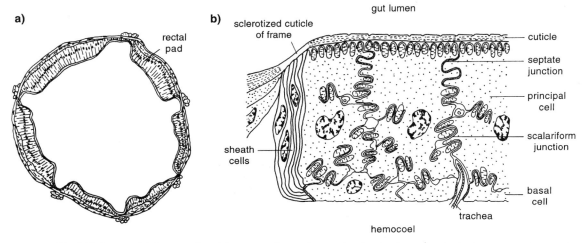

Fig. 3.12. Rectum. (a) Cross-section of the rectum of a grasshopper showing the six rectal pads. (b) Section through a rectal pad (after Noirot & Noirot-Timothée, 1976).

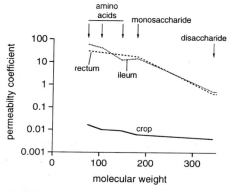

Fig. 3.13. Permeability of the intima of the fore and hindgut of *Schistocerca*. Notice that the permeability coefficient is shown with a logarithmic scale. Arrows at the top show the molecular weights of some of the compounds in the gut (based on Maddrell & Gardiner, 1980).

only chloride, from dilute solutions. In many cases these cells occur on the outer epidermis of the insects, but in larval Odonata, and some other insects, they are present in the hind gut. Such cells occur at the bases of the rectal gills of larval dragonflies (Fig. 3.14). They have microvilli at the apical margins beneath a very thin cuticle, and large numbers of mitochondria in the adjacent cytoplasm. The cells remove salts from water as it is pumped in and out of the rectum during respiration.

Review: Komnick, 1977 – chloride cells

3.1.4 Muscles of the gut

The muscles of the alimentary canal fall into two categories: extrinsic visceral muscles which arise on the body wall and are inserted into the gut, and intrinsic visceral muscles which are associated only with the gut.

Extrinsic visceral muscles are associated with the foregut and hindgut and generally function as dilators of the gut. Those in the head form pumps which suck fluids into the gut and push food back to the esophagus. Insects with biting and chewing mouthparts have weakly developed dilator muscles associated with the pharynx arising on the frons and the tentorium (Fig. 1.2). The muscles of the pump are much more strongly developed in fluid-feeding insects and, since the cibarium is tubular in these insects, it too may function as a pump. Its dilator muscles arise on the clypeus and usually pass in front of the frontal ganglion whereas those of the pharyngeal pump pass behind it. In most Hemiptera, Thysanoptera and Diptera, the cibarial pump is well-developed, while the muscles of the pharyngeal pump are often relatively weakly developed (Fig. 3.15a). Some insects, such as adult mosquitoes and larval *Dytiscus*, have both pumps well-developed (Fig. 3.15b,c) and this also seems to be true in adult Lepidoptera and bees, and in sucking lice, but the anatomical origins of the pumps are not clear (Fig. 3.15d). Cibarial pumps have no compressor muscles and compression results from the elasticity of the cuticle lining the pump, but pharyngeal pumps are compressed by intrinsic circular muscles.

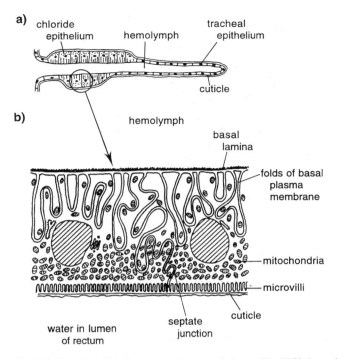

Fig. 3.14. Chloride cells in the rectum of a dragonfly larva. (Fig. 17.29 shows the position of the gills within the rectum). **(a)** Section of a rectal gill showing the position of the chloride epithelium. Note that tracheal epithelium refers to the epithelium of the gill through which gas exchange with the tracheae (not shown) occurs. **(b)** Details of the chloride cells. Notice that the luminal (water) side of the gill is at the bottom (after Schmitz and Komnick, 1976).

The extrinsic visceral muscles of the hindgut are usually present as dilators of the rectum. They are especially well-developed in larval dragonflies (Odonata) that pump water over the gills in the rectum (see Fig. 17.29a).

The intrinsic visceral muscles comprise circular muscles running round the gut and longitudinal muscles extending along parts of it. The circular muscles are not usually inserted into the gut epithelium, but are continuous all round the gut so that their contraction tends to produce longitudinal folding of the epithelium. Round the foregut, the circular muscles are external to the longitudinal muscles. They are well-developed around the pharynx and round the proventriculus where this forms a valve or has a grinding function.

Intrinsic muscles are poorly developed round the midgut. The main longitudinal muscles are outside the circular muscles, although grasshoppers have an inner layer of fine muscle fibers in connective tissue adjacent to the midgut epithelium. The longitudinal muscles appear to extend for the full length of the midgut without any insertions into it; they are inserted anteriorly into the posterior end of the foregut and posteriorly into the anterior end of the hindgut.

In caterpillars, and probably in other insects, the intrinsic musculature of the hindgut is complex. There are commonly strong muscles associated with the pyloric valve, where the Malpighian tubules join the gut. Other specific arrangements enable the insect to produce discrete fecal pellets from the cylinder of food passing along the gut. Usually the circular muscles are outside the longitudinal muscles.

Ultrastructurally, the extrinsic visceral muscles resemble typical skeletal muscles (section 10.1.2). The intrinsic muscles often differ in having 12 actin filaments round each myosin filament and the sarcoplasmic reticulum and T-system are reduced compared with skeletal muscles. Unlike vertebrate visceral muscles, however, insect visceral muscles are striated because their myofilaments are regularly arranged.

Review: Smith, 1985 – pumps

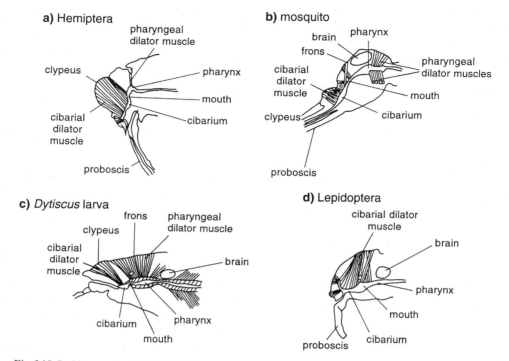

Fig. 3.15. Sucking pumps of fluid-feeding insects. The cibarial muscles are well-developed in all these insects, but the pharyngeal muscles are also important in larval *Dytiscus* and mosquito (after Chapman, 1985a).

Fig. 3.16. Stomodeal nervous system of a grasshopper seen from above (based on R. Allum, unpublished; Anderson & Cochrane, 1978).

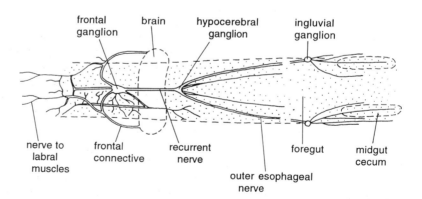

3.1.5 Innervation of the gut

The muscles of the foregut and anterior midgut are innervated by the stomodeal (or stomatogastric) nervous system. The principal ganglion of this system is the frontal ganglion which lies on the dorsal wall of the pharynx anterior to the brain (Fig. 3.16). It connects with the tritocerebrum on either side by nerves called the frontal connectives. In Orthoptera and related groups, a median nerve, the recurrent nerve, extends back from the frontal ganglion beneath the brain to join with the hypocerebral ganglion. This ganglion is closely associated with the corpora cardiaca, and from it a nerve on each side passes to the ingluvial ganglion on the side of the crop. From this ganglion, nerves extend back to the midgut. The cell bodies of most of the motor neurons controlling the muscles are in the ganglia of the stomodeal system, but a few are in the tritocerebrum.

The muscles of the hindgut are innervated from the terminal abdominal ganglion.

There is no evidence of chemosensory neurons associated with the gut, although they do occur in the cibarium (Fig. 2.6). Multipolar cells, which function as stretch receptors, are present on the outside of the gut in many insects and are involved in monitoring gut fullness. They are most abundant on the foregut, but are also present on the mid- and hindgut (see Chapman, 1985a). Trichoid and campaniform mechanoreceptors are present in the terminal region of the fermentation chamber of scarab larvae.

Review: Penzlin, 1985 – stomodeal nervous system

3.2 PASSAGE OF FOOD THROUGH THE GUT

Food is pushed back from the pharynx by the muscles of the pharyngeal pump, and subsequently passed along the gut by peristaltic movements. These movements are controlled from the stomodeal nervous system. When *Locusta* feeds after an interval of several hours without food, the solid food remains in the foregut while the insect is feeding. Backward movement of the food bolus begins shortly after the foregut becomes fully distended. This is at least partly controlled by a hormone from the corpora cardiaca.

In the females of blood-sucking flies that feed on both blood and nectar, the destination of the food is determined by its chemical qualities, detected by the receptors on the mouthparts. Stimulation with sugars causes the meal to be directed to the crop which in flies is a lateral diverticulum of the esophagus. Stimulation with ATP or ADP, but also with dilute sugar solutions, results in the meal going to the midgut (Schmidt & Friend, 1991). In the blowfly, *Phormia*, crop emptying is regulated by the osmotic pressure of the hemolymph. At high hemolymph osmotic pressures, the food is passed more slowly to the midgut than at lower pressures. Hemolymph osmotic pressure is affected by that of the food because the sugars are rapidly absorbed. Consequently, meals of concentrated sugars are retained in the crop for longer periods than meals of dilute sugars. The same thing probably occurs in *Periplaneta*.

In the midgut, the passage of food is aided by the peritrophic envelope which, as it moves down the gut, carries the enclosed food with it. Spines on the intima of the hindgut aid the backwards movement of the envelope in insects which possess them. Blood-sucking bugs, such as *Rhodnius* and *Cimex*, and flies, such as mosquitoes and stable flies, which take large, infrequent meals (Table 2.2) store food in the anterior midgut (Fig. 3.3g). No digestion occurs here, but water is absorbed. The anterior midgut also acts as a temporary food store in plant-sucking bugs (Fig. 3.9).

The movements of the hindgut are important in the elimination of undigested material. In *Schistocerca* the ileum usually forms an S-bend and at the point of inflexion the muscles constrict the gut contents. This probably breaks the peritrophic envelope and the food column so that a fecal pellet is formed. Contractions of the posterior part of the ileum and the rectum force the pellet out of the anus. In grasshoppers, the feces are enclosed in old peritrophic envelope, but in caterpillars the envelope is broken up in the hindgut.

The time taken for food to pass through the gut is very variable. In a grasshopper with continuous access to food, liquids from the food reach the midgut and are absorbed within five minutes of the start of a meal, but solid particles may take 15 minutes or more. The food from one meal has usually left the foregut in about 90 minutes, being pushed back by newly eaten food, and by this time some of the meal is present in the hindgut and may appear in the feces (Fig. 3.17). In the absence of more food, the foregut is completely empty in about 5 hours, and the midgut is empty after about 8 hours. In *Periplaneta* solid food passes through the gut in about 20 hours, but some can still be found in the crop after some days of starvation.

Termites and larval scarab beetles, which employ micro-organisms to digest cellulose, retain food fragments for long periods in the hindgut.

3.3 DIGESTION

A large part of the food ingested by insects is macromolecular, in the form of polysaccharides and proteins, while lipids are present as glycerides, phospholipids and glycolipids. Generally, only small molecules can pass into the tissues and the larger molecules must be broken down into smaller components before absorption can occur. Enzymes concerned with digestion are present in the saliva and in the secretions of the midgut. In addition, digestion may be facilitated by micro-organisms in the gut.

Review: Applebaum, 1985

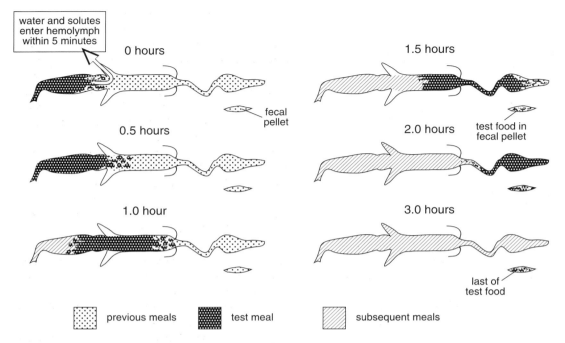

Fig. 3.17. Movement of food through the gut of a grasshopper with continuous access to food (after Baines *et al.*, 1973).

3.3.1 Extra-intestinal digestion

Some insects inject saliva into their food before starting to ingest and, because it contains enzymes, considerable digestion may occur. Such extra-intestinal (or extra-oral) digestion may constitute a major part of the total digestion. This is true in some Hemiptera at all stages of their development, in adult Asilidae and Empididae amongst the Diptera, and in the larvae of Neuroptera and some beetles such as *Dytiscus* and the Lampyridae. Extra-intestinal digestion is probably also significant in some adult carabid beetles, although the bulk of digestion occurs in the gut.

Amongst the plant-sucking Hemiptera, extra-intestinal digestion is most significant amongst seed-feeding species such as *Oncopeltus* and *Dysdercus*, although it may also be important in species feeding from the parenchyma of plants. However, it is amongst the predaceous Heteroptera that extra-intestinal digestion is most widespread. Predators occur in 30 families of Heteroptera. Belostomatidae, Reduviidae and Nabidae are families in which all the species are predaceous or feed on vertebrate blood; Lygaeidae and Pentatomidae are families in which some species are predaceous. The stylet structure of most of these insects

restricts their ability to ingest anything except fluids and very fine particles, so that food must be pre-digested. There is no evidence that these insects use a venom, in the sense of a pharmacologically acting chemical, to subdue their prey. Nevertheless, they subdue their prey very rapidly. For example, a cockroach captured by the reduviid, *Platymeris*, stops its convulsive struggling in 3–5 s, and is completely motionless after 15 s. This is apparently achieved through the rapid action of the enzymes which the bugs inject into the prey. Periods when enzymes are injected into the prey are separated by periods in which the enzymes and liquified contents of the prey are ingested. The reduviid, *Zelus*, is thus able to remove a very large proportion of the protein and glycogen content of a caterpillar in the first 45 minutes of feeding, but lipid is removed more slowly (Fig. 3.18).

Predaceous larval beetles and Neuroptera have biting mandibles with which they capture their prey (Fig. 2.5), and through which enzymes are injected into the prey and the digested contents withdrawn. Beetles have no salivary glands so it must be presumed that, in this case, the enzymes originate in the midgut. Adult carabids in the tribes Carabini and Cychrini masticate their prey before

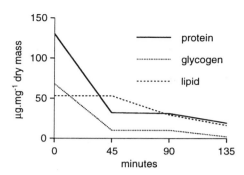

Fig. 3.18. Extra-intestinal digestion. The rate of removal of nutritional components of a caterpillar by the bug, *Zelus* (based on data of A. Cohen, unpublished).

ingesting the fluid contents. Some extra-intestinal digestion occurs and may be considerable.

Proteolytic enzymes persist in the excreta of larval blowflies and so the meat in which they live is partially liquified before it is ingested. Another instance of extra-intestinal digestion occurs in *Bombyx* where the moth, on emergence from the pupa, secretes a protease that attacks the sericin of silk, making it possible for the insect to escape from the cocoon.

Heliconiine butterflies and *Drosophila flavohirta* collect pollen but are unable to ingest it because of the form of their mouthparts. They regurgitate fluid, and nutrients may be extracted from the pollen simply by diffusion into the fluid (Nicolson, 1994).

Review: Cohen, 1995

3.3.2 Digestion in the gut lumen

Regardless of their feeding habits, most insects must digest proteins, carbohydrates and lipids and so they have a similar array of enzymes in the midgut. Nevertheless, the enzymes produced do reflect the type of food eaten by each species and stage (Table 3.1). For example, in the tsetse fly, *Glossina*, that feeds exclusively on vertebrate blood, the array of proteolytic enzymes reflects the importance of protein in the insect's diet; in adult Lepidoptera that feed only on nectar, on the other hand, proteolytic enzymes are completely lacking. In those holometabolous insects in which the larval and adult diets are different, as in the Lepidoptera, this is reflected in the enzymes present in the midgut. Where the diet is similar, as in larval and adult *Tenebrio*, the same types of enzyme are present, but their substrate specificities may differ. It is common for a particular type of enzyme to be present in several forms exhibiting specificities for different substrates even in one insect.

Even though similar enzymes are present in different species, their activities may reflect the nature of their food (Fig. 3.19). Insects feeding on stored products, rich in carbohydrates, generally have higher amylase activities than those feeding on wool or plants, but the latter usually have higher proteolytic activity.

3.3.2.1 *Digestion of proteins*

The digestion of proteins involves endopeptidases, which attack peptide bonds within the protein molecule, and exopeptidases, which remove the terminal amino acids from

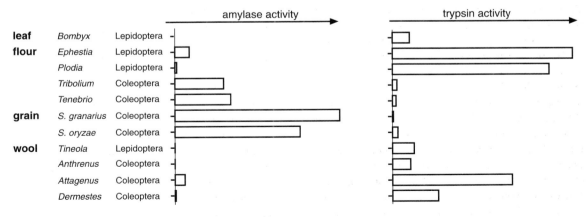

Fig. 3.19. Enzyme activity. The relative activities of amylase and trypsin in insects with different feeding habits. The grain beetles are species of *Sitophilus* (based on Baker, 1986).

Table 3.1. *Enzymes digesting proteins and carbohydrates in insects with different feeding habits. Numbers show numbers of different enzymes that are known to be present*

Food	Insects		Order	Endopeptidases				Exopeptidases		Amylase	Glucosidases		Galactosidases	
				Trypsin	Chymotrypsin	Cathepsin	Metalloproteinase	Amino-	Carboxy-		α	β	α	β
Plant	*Locusta*	A	Orthoptera	1	3	–	–	–	–	+	+	+	+	+
Plant	*Leptinotarsa*	L	Coleoptera	0	0	3	0	1	?	–	–	–	–	–
Plant	*Cylas*	L	Coleoptera	+	0	–	–	1	2	2	2	1	1	+
Plant	*Erinnyis*	L	Lepidoptera	+	–	0	0	+	–	+	+	+	–	–
Nectar	Lepidoptera	A	Lepidoptera	0	0	0	0	0	0	0	+	0	0	0
Grain	*Sitophilus*	A	Coleoptera	1	1	+	–	1	1	+	2	2	2	2
Flour	*Tenebrio*	L	Coleoptera	1	1	+	–	+	+	+	+	+	+	+
Flour	*Tenebrio*	A	Coleoptera	2	1	–	–	–	–	1	–	–	–	–
Blood	*Rhodnius*	L	Hemiptera	–	–	2	–	3	1	+*	+	+	+	–
Blood	*Glossina*	A	Diptera	2	1	–	–	2	–	–	–	–	–	–
Wool	*Tineola*	L	Lepidoptera	4	1	0	2	16	2	+	–	–	–	–
Wool	*Attagenus*	L	Coleoptera	3	4	0	–	2	2	+	–	–	–	–
Hide	*Dermestes*	L	Coleoptera	6	2	–	–	1	1	+	–	–	–	–

Notes:

+ = One or more enzymes present.

0 = Enzyme looked for, but not found.

– = No data.

* = Present, but may be from symbiont.

L, Larva; A, adult.

the molecule. Within these general categories, the enzymes are classified according to the nature of their active sites and the sites at which they cleave protein molecules.

The principal endopeptidases in the majority of insects are the serine proteases, trypsin and chymotrypsin, which have serine at the active site. Trypsin cleaves peptide linkages involving the carboxyl groups of arginine and lysine residues. Chymotrypsin is less specific, cleaving bonds involving the carboxyl groups of tyrosine, phenylalanine and tryptophan preferentially, and bonds involving other amino acid residues more slowly. Usually both types of enzyme are present in any insect with serine proteases (Table 3.1). However, in many Coleoptera and in blood-sucking Hemiptera, the main endopeptidases have cysteine or aspartic acid at their active centers (Murdock *et al.*, 1987). They are called cathepsins.

Exopeptidases fall in two categories: carboxypeptidases that attack peptides from the –COOH end, and aminopeptidases that attack the chain from the –NH$_2$ end.

Keratin Keratin is a protein occurring in wool, hair and feathers. Hard keratins contain 8–16% cystine, and disulfide linkages between cystine residues render the protein very stable. Nevertheless, a number of insects normally feed on keratinous materials. These include larvae of the clothes moth (*Tineola*), carpet beetles (Dermestidae) and numerous biting lice (Ischnocera). *Tineola* larva has a complex mixture of proteolytic enzymes (Table 3.1). In

addition to those normally present in caterpillars, it possesses a highly active cysteine desulfhydrase which produces hydrogen sulfide from cysteine. This contributes to the strong reducing conditions in the gut (see below) which promote the breaking of disulfide bonds in the keratin (Yoshimura *et al.*, 1988). Similar conditions are found in other insects that feed on wool or feathers (see Table 3.2).

3.3.2.2 *Digestion of carbohydrates*

Carbohydrates are generally absorbed as monosaccharides so disaccharides and polysaccharides in food require digestion. The polysaccharide cellulose is a major constituent of green plants, but although many insects are phytophagous, relatively few of them are able to utilize cellulose. Those that do, nearly always depend on micro-organisms to digest it (see below). However, starch and glycogen, the main storage polysaccharides of plants and insects, respectively, are digested by amylases that hydrolyse 1–4–α-glucosidic linkages. There may be separate endo- and exo-amylases, acting on starch internally or terminally.

The common disaccharides sucrose and maltose contain a glucose residue linked to another sugar by an α-linkage:

These are digested by α-glucosidases (enzymes attacking the α-linkage). As with the proteolytic enzymes, α-glucosidases may exhibit different substrate specificities. For example, a trehalase is often present, although it is not clear why this should occur in insects that feed on plants where trehalose is not found.

The naturally occurring β-glucosides (*e.g.*, salicin and arbutin) are usually of plant origin and the highest β-glucosidase activity is found in phytophagous insects. Cellobiose is a product of cellulose digestion and a cellobiase is often present even in insects where cellulose digestion is not known to occur. α- and β-galactosidases are also often present.

In the hydrolysis of carbohydrates, water is the typical acceptor for the sugar residues:

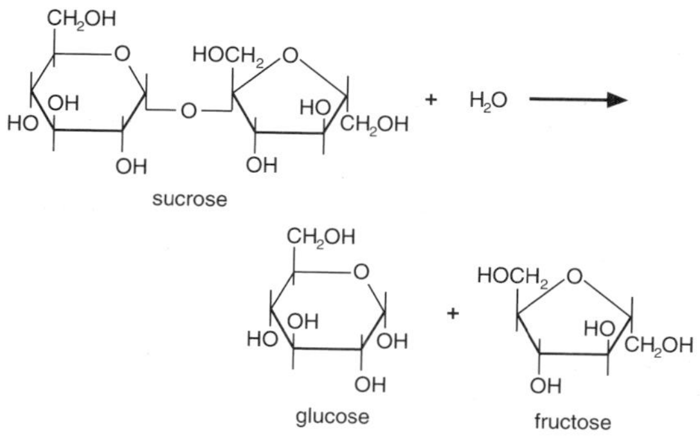

CH₂OH

HO

OH

OH

sucrose

+ H₂O ⟶

CH₂OH

HOCH₂

HO

CH₂OH

glucose

fructose

but other sugars may equally well act as acceptors with the formation of oligosaccharides. Thus, in the hydrolysis of sucrose, other sucrose molecules may act as acceptors to form the trisaccharides glucosucrose and melezitose. This process is known as transglucosylation and a similar process occurs in the hydrolysis of maltose. In some aphids glucosucrose and melezitose appear to be produced by two α–glucosidases with different acceptor specificities for the sucrose molecule. It is probably for this reason the melezitose is a common component of aphid honeydew.

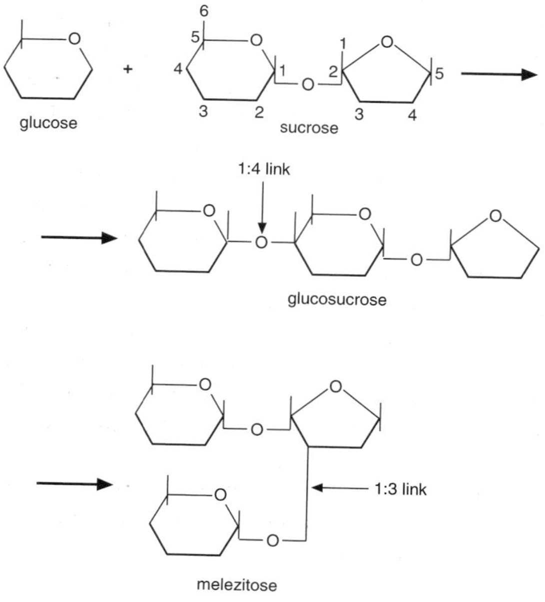

glucose   +   sucrose

⟶

1:4 link

glucosucrose

⟶

1:3 link

melezitose

Cellulose Cellulose is polymer of glucose in which the glucose molecules are joined by β–1–4 linkages. The chains of cellulose are unbranched and may be several thousand units long:

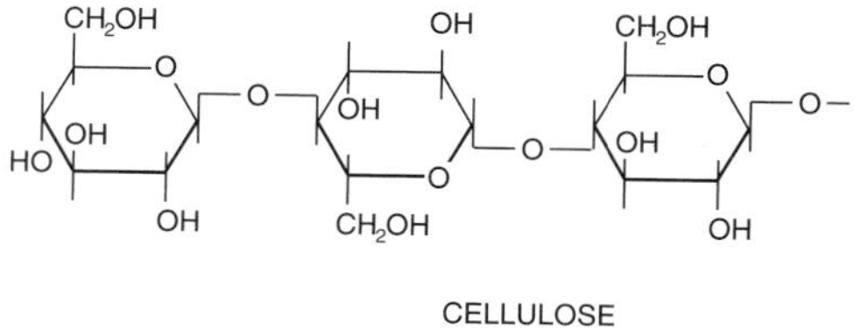

CELLULOSE

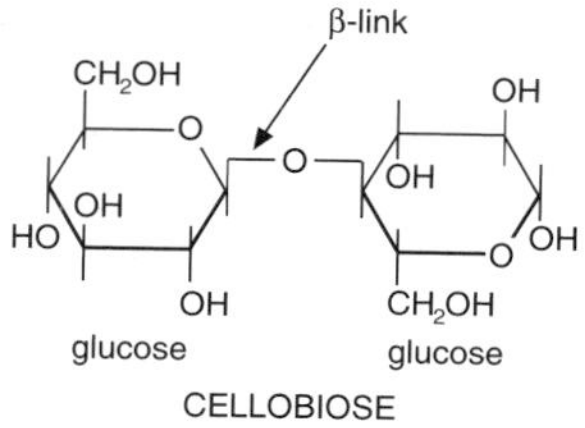

β-link

glucose   glucose

CELLOBIOSE

Hydrogen bonds occur within and between cellulose molecules, resulting in a crystalline state which contributes to the resistance of cellulose to digestion. Three classes of enzyme are involved in its hydrolysis: endoglucanases, which attack the bonds between glucose residues within the chain; exoglucanases that attack bonds near the ends of the cellulose molecule; and β-glucosidases that hydrolyze cellobiose. Exoglucanases are usually more active against crystalline cellulose than endoglucanases.

Amongst the termites, species in all the families except Termitidae have huge numbers of flagellate protozoans in the paunch (Fig. 3.3f). These organisms may constitute more than 25% of the wet weight of the insect. Many different species of flagellate may be present in one species of termite, but the species are, in general, only found in termites. The protozoans engulf fragments of plant material and ferment the cellulose, producing acetate and other organic acids, carbon dioxide and hydrogen. Fermentation is an anaerobic process and conditions in the paunch are highly reducing (see below). The organic acids are absorbed in the hindgut and provide a large proportion of the respiratory substrate used by the insect. These insects probably do produce cellulose-digesting enzymes themselves, but their activity is insufficient for the termites to sustain themselves without the aid of the symbionts. Flagellates also occur in the hindgut of the wood-eating cockroach, *Cryptocercus*.

Other termites use fungi to digest cellulose. Species of the subfamily Macrotermitinae cultivate fungi of the genus *Termitomyces* in fungus gardens. These gardens are formed from feces containing chewed, but only partially digested plant fragments. The fungus grows on this comb, producing cellulolytic enzymes, and the termites then feed on the fungus and the comb. In doing so they ingest the cellulases produced by the fungus. These may contribute to cellulose digestion in the termite gut, but probably only to small extent.

Many other insects are dependent on fungi to enable them to utilize plant material as a source of nutrients. Amongst these are the leafcutter ants, which grow fungus on fragments cut from growing plants, and species that feed on wood. Examples of wood-feeding insects that have symbiotic relationships with fungi are the ambrosia beetles in the Platypodinae, larval Cerambycidae, and wood wasps (Siricidae). These insects carry the fungus with them to inoculate a new habitat. In wood-feeding insects, the fungi may be important in concentrating nitrogen, in addition to digesting the original plant cellulose, because the nitrogen content of wood is very low, often less than 1% dry weight.

Bacteria are responsible for cellulose digestion in larval scarab beetles and in some crickets and cockroaches. The former commonly feed in decaying wood and they acquire the bacteria with the food. Digestion of the wood occurs in the fermentation chamber of the hindgut (Fig. 3.3e) where it is retained by branched spines arising from the intima. Conditions in the fermentation chamber are highly reducing (see Table 3.2).

Termites in the family Termitidae, other than Macrotermitinae, probably do produce their own cellulose digesting enzymes, but whether or not other insects do so is open to question. It is likely that in most cases, as in grasshoppers, small quantities of cellulose are digested by micro-organisms ingested with the food.

Reviews: Breznak and Brune, 1994 – termites; Martin, 1987, 1991 – cellulose digestion; Wood & Thomas, 1989 – termites

3.3.2.3 *Digestion of lipids*

Very little is known about the digestion of lipids in insects. Midgut cells produce several different esterases, which probably have specificity for different substrates. In caterpillars, galactosyl diglycerides, phosphatidylglycerols and phosphatidylcholines are hydrolysed to di- and mono-acylglycerides and free fatty acids. In the larva of *Pieris*, but not in several other species examined, some glycolipids

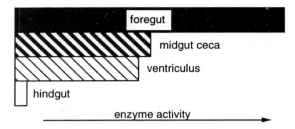

Fig. 3.20. Compartmentalization of digestion. Relative concentrations of α-glucosidase activity in the lumen of different parts of the alimentary canal of a locust (after Evans and Payne, 1964).

are produced before absorption occurs (Turunen and Chippendale, 1989).

3.3.2.4 *Compartmentalization of digestion*

Most digestion occurs in the midgut, where the enzymes are secreted, but, because of the ingestion of saliva with the food and the forwards movement of enzymes from the midgut, some digestion occurs in the foregut. In Orthoptera and some carabid beetles, enzyme activity is high in the foregut (Fig. 3.20) and much digestion occurs there.

It may sometimes be true that the permeability of the peritrophic envelope results in some degree of compartmentalization of digestion. In the larva of the moth, *Erinnyis*, amylase (molecular weight 48 kDa) and trypsin (molecular weight 55 kDa) can pass readily through the envelope. Consequently, digestion begins in the endoperitrophic space and the oligosaccharides and peptides produced can pass freely out into the ectoperitrophic space. Carboxypeptidase A is somewhat confined by its molecular size (molecular weight 102 kDa). In addition, some aminopeptidases are bound to the plasma membranes of the microvilli, and this may sometimes be true of carboxypeptidases and α-glucosidases. As a result, the final stages of digestion, especially of proteins, occur in the ectoperitrophic space and at the cell membrane.

It is not clear how general this phenomenon may be. The effective pore sizes of peritrophic membranes in many insects seem too large to act as molecular filters, but the association of certain enzymes with the plasma membranes of midgut cells is common.

Little digestion occurs in the hindgut except for the digestion of cellulose in insects with symbiotic micro-organisms in the hindgut.

Reviews: Dow, 1986; Terra, 1990

3.3.2.5 *Variation in enzyme activity*

The production and secretion of digestive enzymes is related to the feeding pattern of an insect. This is most apparent in blood-sucking insects in which meals are taken at long intervals. In a mosquito, for example, proteolytic activity in the midgut is very low before feeding. Within hours, the level of activity increases, reaching a maximum after about two days and then declining (Fig. 3.21a). In grasshoppers, which feed at relatively frequent intervals, there is no such obvious pattern, but enzyme activity is minimal at the time of a molt, rising to a maximum when the insect starts to feed and then declining again when it stops before molting (Fig. 3.21b).

Regulation of enzyme activity Enzyme production may be regulated by the chemicals in the food acting directly on the midgut epithelium. This is called a secretagogue mechanism. Secretagogue control mechanisms are known to occur in insects with a variety of feeding habits. Different enzymes are controlled independently so that only those appropriate to the food in the gut are produced. In blood-sucking insects, such as the mosquito, for example, a meal of sugar does not induce the activity of the proteolytic enzymes, but a blood meal does (Fig. 3.21a). In *Aedes*, the production of trypsin occurs in two phases after a blood meal. Some is released within three hours of feeding, its production (translation) from a pre-formed mRNA being stimulated by a protein in the ingested blood. Subsequently, the peptides resulting from the initial phase of digestion, stimulate the production of further mRNA (transcription) from which additional trypsin is produced 8–10 hours after feeding. This trypsin contains amino acids derived directly from the current blood meal (Felix *et al.*, 1991). In a number of insects, such as the caterpillar of *Spodoptera*, trypsin activity increases as the amount of protein in the food increases (Broadway and Duffey, 1986).

It is also possible that enzyme regulation is under humoral control, enzymes being produced in response to hormones released at the time of feeding. In *Tenebrio* and some other insects, the build up of midgut protease appears to be regulated by a secretion from the median neurosecretory cells of the brain. Hormonal regulation of enzyme production may occur together with secreta-gogue regulation, modulating the overall amounts of an enzyme produced under particular physiological conditions.

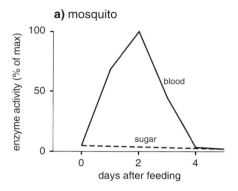

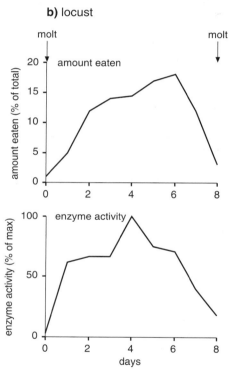

Fig. 3.21. Enzyme activity in relation to feeding. (**a**) Proteolytic activity in a mosquito after feeding on blood (full line) or sugar (broken line) (after Spiro-Kern, 1974). (**b**) Amount eaten (above) and α-glucosidase activity (below) in a locust through the final larval stage (after Khan, 1964).

Direct neural regulation of midgut enzyme production is unlikely to be important because of the time which elapses between feeding and any increase in enzyme activity. The release of salivary enzymes is, however, sometimes under direct neural control (section 2.5.1.2).

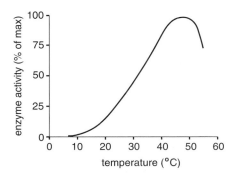

Fig. 3.22. Enzyme activity in relation to temperature. The activity of the α-glucosidase of a locust (after Evans and Payne, 1964).

The midgut environment Enzymes function efficiently only within limited ranges of temperature, pH and redox potential.

Enzyme activity increases with temperature, but at higher temperatures the enzymes are denatured. Consequently, activity rises to a maximum and then declines as the rate of denaturing becomes faster (Fig. 3.22). Maximum activity of most insect digestive enzymes is in the range 35–45 °C. Below 10 °C activity is usually slight, and above 50 °C enzymes are denatured so rapidly that very little digestion will occur. These changes are reflected in the thermal limits for survival and in thermo-regulatory behavior (Chapter 19).

Enzymes are most active only within a limited range of pH (Fig. 3.23). Most insects have enzymes with optima at pH 6–7, but in caterpillars optimal activity occurs at about pH 10. These optima correspond with the conditions found in the guts of various insects.

pH often varies along the length of the gut. In the foregut it is greatly influenced by the food and varies with diet. For example, the foregut pH of a cockroach that has eaten a diet rich in protein is about 6.3; after feeding on maltose the pH is 5.8 and, after glucose, 4.5–4.8. Although the midgut pH may also vary with the diet, it is usually buffered to maintain a relatively stable level. In most insects it is in the range pH 6.0 to 8.0, but larval Lepidoptera are marked exceptions. Here the pH is always above 8.0 and may be as high as 12.0. This is well above the pH of the food, which is usually less than 6.0. Mosquitoes have little buffering facility, and after a blood meal midgut pH rises to 7.3, the normal value for blood. Variation may occur along the length of the midgut. In caterpillars, it is lower at either end than in the middle, whereas in larval *Lucilia* the converse is true. The hindgut is usually slightly more acid than the midgut, partly due to the secretions of the Malpighian tubules.

As well as affecting the activity of the insect's own enzymes, the pH of the midgut influences the potentially harmful effects of some ingested compounds (Felton, Workman & Duffey, 1992).

Redox potential is a measure of the oxidizing or

a) proteolytic enzymes

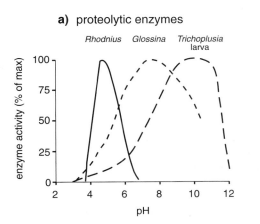

b) amylase

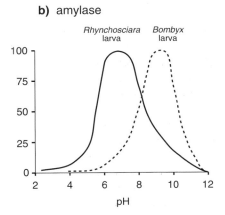

Fig. 3.23. Enzyme activity in relation to pH. **(a)** Proteolytic activity of *Rhodnius*, a blood-sucking bug; *Glossina*, a blood-sucking fly; larva of *Trichoplusia*, a leaf-eating caterpillar. The endopeptidases of *Rhodnius* are cathepsins, those of *Glossina* and *Trichoplusia* are trypsins. **(b)** Amylase activity of the larva of *Rhynchosciara*, a fly larva feeding on decaying plant material; larva of *Bombyx*, a leaf-eating caterpillar.

Table 3.2. *Redox potentials in different parts of the alimentary canals of different insects*

			Redox potential			
					Paunch/fermentation	
Food	Species	Order	Foregut	Midgut	chamber	Hindgut
Plant	*Locusta*	Orthoptera	+ + +	+ + +	×	+ + +
Plant	*Manduca* larva	Lepidoptera	+ + +	− − −	×	+ +
Plant	*Danaus* larva	Lepidoptera	+ + +	+ +	×	+ + +
Wood	*Zootermopsis*	Isoptera	+ + +	+ + +	− − −	− −
Wood	*Oryctes* larva	Coleoptera	·	+ +	− −	·
Feather	*Columbicola*	Phthiraptera	·	− − −	×	+ +
Wool	*Attagenus* larva	Coleoptera	·	− − −	×	+ + +
Wool	*Tineola* larva	Lepidoptera	·	− − −	×	+ + +
Detritus	*Ctenolepisma*	Thysanura	·	+ + +	×	+ + +
Detritus	*Periplaneta*	Blattodea	+ + +	+ + +	×	0

Notes:

+ + + Strongly oxidizing.

− − − Strongly reducing.

0 Variably oxidizing/reducing.

× Not present.

· No data.

reducing conditions in a system. A high positive potential indicates strongly oxidizing conditions, a high negative potential indicates strongly reducing conditions. It is probable that in most insects oxidizing conditions normally prevail throughout the gut (Table 3.2), but in insects feeding on substances that are intractable to digestion, such as wood, wool, or keratin, strongly reducing conditions are produced in a part of the gut. In some of these cases, as in *Zootermopsis* and the larva of *Oryctes*, microorganisms which are effective only under anaerobic conditions are believed to maintain the reducing conditions. In other cases, it is believed that the insect itself maintains the conditions. It is not known why the caterpillar of *Manduca* should maintain high reducing conditions in its midgut, while other caterpillars, such as *Danaus*, do not. It is likely that the extent of oxidizing or reducing conditions will affect the potential toxic activity of plant secondary compounds in the food of phytophagous insects (Appel and Martin, 1990).

Reviews: Chapman, 1985b – regulation of enzyme production; Dow, 1992 – caterpillar midgut pH

3.4 ABSORPTION

The products of digestion are absorbed in the midgut, but some absorption, especially of salts and water, also occurs in the hindgut. The cells in the midgut concerned with absorption are often the same cells that produce enzymes in a different phase of their cycle of activity.

Absorption may be a passive or an active process. Passive absorption depends primarily on the relative concentrations of a compound inside and outside the gut, diffusion taking place from the higher to the lower concentration. In addition, in the case of electrolytes, the tendency to maintain electrical equilibrium inside and outside the gut epithelium will interact with the tendency to diffuse down the concentration gradient. These two factors together constitute an electrochemical potential (Fig. 3.24).

Active absorption depends on some metabolic process to move a substance against its concentration gradient. In the midgut of caterpillars, and probably of other insects, the energy for these processes is derived from a V(vacuolar)-type ATPase in the apical plasma membranes. This pumps protons from the cells into the gut lumen and the protons

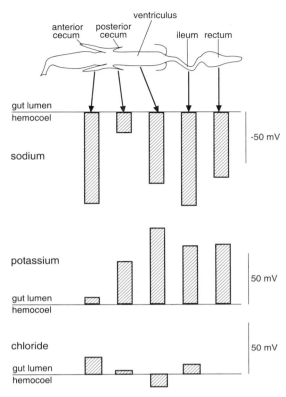

Fig. 3.24. Electrochemical gradients across the gut epithelium of a locust. The insects were recently fed. Bars extending into the lumen (upwards) indicate a positive gradient from gut to hemolymph: diffusion into the hemocoel will occur. Bars extending into the hemocoel indicate a negative gradient from gut to hemocoel: passive diffusion into the hemocoel will not occur (after Dow, 1981).

are exchanged for potassium, or perhaps sometimes, another ion (Fig. 3.8). Movement of potassium down its electrochemical gradient into the cell is coupled with that of amino acids, and perhaps other compounds, at symports. **Reviews:** Dow, 1986 – midgut absorption; Harvey & Nelson, 1992 – V-ATPases; Lepier *et al.*, 1994 – K^+/H^+ antiport in goblet cells; Turunen, 1985 – absorption

3.4.1 Absorption of water

There are two major water absorbing zones in the insect gut, one in the midgut, where water is absorbed from the food, and the other in the rectum, where water is absorbed from the feces before they exit the body. The effect of the former is to concentrate the food, enhancing both the efficiency of digestion and the maintenance of concentration gradients favorable to the absorption of nutrients across the wall of the gut. It also creates water flows which are important in nutrient absorption and perhaps enzyme conservation. Absorption in the rectum is a key component of the regulation of body water.

Water absorption is an osmotic process and depends on the establishment of an osmotic gradient across the epithelium. It is believed that potassium or sodium is actively pumped into an extra-cellular space on the hemolymph side of the epithelium. This active process necessitates the presence of large numbers of mitochondria close to the cell membrane. In order to maintain the high concentration of solute, it is pumped into a space between the epithelial cells which has only a few openings to the hemolymph. Water is thus drawn through, or between the cells from the gut lumen into the extracellular space. This creates a hydrostatic pressure which forces water out through the openings into the hemolymph (see Fig. 3.26a).

3.4.1.1 *Water absorption from the midgut*

Water absorption from the midgut often occurs in localized areas. In cockroaches, grasshoppers and in the larvae of some flies (mosquitoes and Sciaridae), it occurs in the midgut ceca, while in blood-sucking insects water is removed from the stored meal in the anterior midgut. Deep invaginations of the basal plasma membrane, which may extend more than halfway towards the apex of the cell, are closely associated with large numbers of mitochondria. These provide the energy by which potassium (sodium in the case of *Rhodnius*) is actively pumped into the intercellular spaces to create a high osmotic gradient between the gut lumen and the intercellular spaces.

The Hemiptera that feed on phloem or xylem must ingest large volumes because the concentrations of nutrients in these fluids are very low. Modifications of the gut provide for the rapid elimination of the excess water taken in. In most Homoptera, the anterior midgut forms a thin-walled bladder that wraps round the posterior midgut and the proximal ends of the Malpighian tubules. This arrangement, which is called a filter chamber, enables water to pass directly from the anterior midgut to the Malpighian tubules (Fig. 3.25). In this way the food is concentrated and dilution of the hemolymph is avoided. In this instance, the cells of the midgut are unspecialized, but an osmotic gradient is established from the Malpighian tubules to the midgut. In some bryocorine Miridae the anterior midgut makes contact

Fig. 3.25. Water absorption from the midgut of a cercopid. (**a**) General arrangement of the gut showing the filter chamber. (**b**) Postulated movement of salts and water across the wall of the gut (after Cheung and Marshall, 1973). (**c**) Transverse section of the filter chamber.

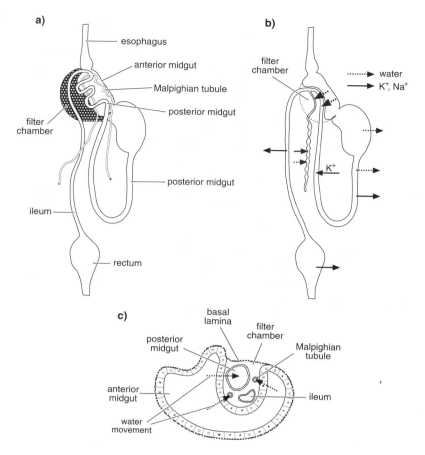

with a large accessory salivary gland. After feeding, a clear fluid is exuded from the mouthparts suggesting that water is withdrawn from the midgut directly to the salivary glands and then eliminated via the mouthparts.

Water absorption from blood stored in the anterior midgut of *Rhodnius* is dependent on the active movement of sodium into the hemocoel. It is believed that, in this case, the sodium is transported out of the cell by sodium/potassium exchange pumps in the basal plasma membrane (Farmer, Maddrell & Spring, 1981).

3.4.1.2 *Water absorption from the hindgut*

In terrestrial and saltwater insects water is absorbed from the rectum. Here, as in the midgut, V-ATPase in the apical plasma membrane probably provides the energy that drives the inward movement of ions and amino acids, principally chloride and proline in the locust (Fig. 3.26). Water moves across the epithelium due to the increased osmotic

pressure. Chloride moves passively into the lateral intercellular spaces and sodium is pumped out by sodium/potassium ATPase so that water is drawn osmotically into the spaces. Hydrostatic pressure forces the solution out to the hemolymph, but ions are resorbed as the fluid passes out from between the principal cells. This permits the recycling of the ions and the maintenance of absorption.

Reviews: Chapman, 1985a – structure; Phillips & Audsley, 1995 – physiology; Phillips *et al.*, 1986, 1988 – physiology

3.4.2 **Absorption of organic compounds**

The forward flow of water outside the peritrophic envelope produced by its absorption in the anterior parts of the midgut carries with it the products of digestion, and, where they occur, the midgut ceca are a primary area of absorption. In well-fed grasshoppers, the water is derived directly from the food, but in insects deprived of food, and in cockroaches, water is drawn forwards as it leaves the Malpighian

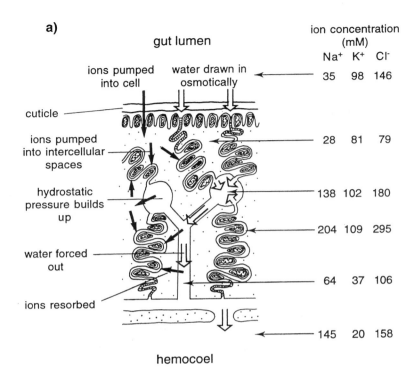

a)

gut lumen

ion concentration
(mM)
Na⁺ K⁺ Cl⁻

ions pumped into cell | water drawn in osmotically

35 98 146

cuticle

ions pumped into intercellular spaces

28 81 79

hydrostatic pressure builds up

138 102 180

204 109 295

water forced out

64 37 106

ions resorbed

145 20 158

hemocoel

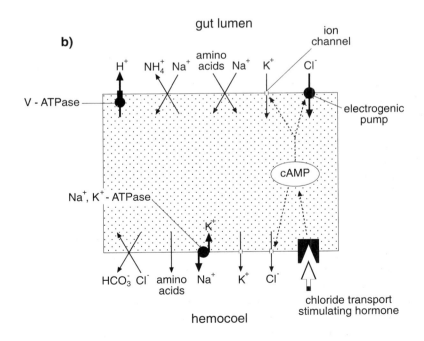

gut lumen

ion channel

b)

H⁺ NH₄⁺ Na⁺ amino acids Na⁺ K⁺ Cl⁻

V - ATPase

electrogenic pump

cAMP

Na⁺, K⁺ - ATPase

K⁺

HCO₃⁻ Cl⁻ amino acids Na⁺ K⁺ Cl⁻

chloride transport stimulating hormone

hemocoel

Fig. 3.26. Absorption of inorganic ions from the rectum of a grasshopper. **(a)** Diagram of the principal rectal cells showing the ionic concentrations measured in one experiment (after Phillips *et al.*, 1986). **(b)** diagram showing the different pumps and channels involved in the movement of ions from the rectal lumen to the hemocoel (after Phillips & Audsley, 1995).

tubules. In larval Diptera, and perhaps in some other insects, the forward movement occurs in the ectoperitrophic space, while fluid in the endoperitrophic space moves backwards. This counter current probably improves the efficiency of utilization of the digested materials in the gut.

Amino acids The absorption of amino acids is best understood in caterpillars. If amino acids are present in high concentration in the gut lumen they diffuse across the epithelium down a concentration gradient; at low concentrations, amino acid absorption occurs at symports coupled with the movement of a cation, usually potassium, into the cell. Several types of symport are present, differing in their specificity for amino acids, and cells in different parts of the midgut vary in their ability to take up different amino acids (Wolfersberger, 1996).

Different amino acids enter the hemocoel at different rates and in different amounts relative to those present in the gut lumen, partly as a result of their differential uptake, but also because they may be metabolized within the gut epithelium. In the locust, glutamate is changed to alanine and glutamine, and it is these that reach the hemocoel; little or no glutamate does so. This is important because glutamate is a neural transmitter at nerve/muscle junctions and this function is impaired if glutamate is present in the hemolymph.

Amino acids form part of the primary urine produced in the Malpighian tubules (section 18.3.1). Consequently, significant quantities reach the hind gut and are resorbed in the rectum. In the locust, glycine, serine, alanine and threonine are actively resorbed into the cells of the rectal pads by a sodium-cotransport system in the apical membrane, but uptake of proline is independent of sodium and is probably cotransported with protons (Phillips & Audsley, 1995). From these cells, amino acids probably enter the hemocoel passively, moving down a concentration gradient. Some metabolism of amino acids occurs in the rectal cells. Proline is the major metabolic substrate in the locust rectum, and, as in the midgut, glutamate is completely metabolized.

Carbohydrates Carbohydrates are absorbed mainly as monosaccharides. In some cases, at least, it is a passive process depending on diffusion from a high concentration in the gut to a low one in the hemolymph. This is facilitated by the immediate conversion of glucose to the disaccharide trehalose in the fat body surrounding the gut so that the concentration of glucose in the blood never builds up. Mannose and fructose are absorbed in a similar manner, but more slowly than glucose because their conversion to trehalose is less rapid. As a result, their concentration gradients across the gut wall are less marked. Absorption of water from the lumen of the gut also tends to maintain a relatively high concentration of sugars in the gut.

Glycogen appears in the posterior midgut cells soon after glucose is ingested by larval *Aedes*. It is possible that rapid conversion to glycogen might maintain a concentration gradient of glucose inward from the gut lumen. In *Phormia* and other dipterous larvae, however, the concentration of glucose in the hemolymph is normally high, so glucose absorption must entail another, possibly active, process.

Lipids Lipids appear to be absorbed primarily as fatty acids. In *Stomoxys*, fatty acids are absorbed in the posterior zone of the midgut and then incorporated into phospholipids and triacylglycerides. In caterpillars, the turnover of triacylglycerides is rapid and it appears that they are transported to the basal parts of the cell where they are actively exported as diacylglycerides. The turnover of phospholipids is slower and substantial amounts remain in the midgut cells 24 hours after a meal (Turunen & Chippendale, 1989). At times, the rate of export of lipids is unable to keep pace with the rate of production of intermediates and these are temporarily stored in lipid spheres. Sterols appear to be absorbed unchanged, but, in some caterpillars, sterols are esterified in the gut cells.

3.4.3 Absorption of inorganic ions

For sodium to move from the gut lumen into the hemolymph of a locust, it must move against the electrochemical gradient (Fig. 3.24). Energy for this active movement is provided by V-ATPase pumps in the apical plasma membranes of the anterior ceca, and by major sodium/potassium exchange pumps in the basal plasma membranes of the rectal cells. Potassium, on the other hand, probably moves passively down its electrochemical gradient into the hemocoel. The electrochemical gradients for chloride and calcium ions are less extreme (Dow, 1981). In the rectum, chloride ions are actively removed from the lumen by pumps in the apical membranes of the cells and pass passively from the cells to the hemolymph. Calcium is actively moved from the gut lumen to the hemocoel by cells in the midgut.

Review: Taylor, 1986 – calcium

Table 3.3. *Average values for efficiency with which foods are digested and absorbed (expressed as approximate digestibility, AD) by larval insects*

Food	Number of species examined	Principal order in sample	Average AD
Plant leaves	44	Orthoptera	35–50
	12	Coleoptera	50–83
	125	Lepidoptera	41–53
	10	Hymenoptera	26
Plant fluids	13	Hemiptera	22–60
Plant seeds	10	Coleoptera, Lepidoptera	72
	3	Hemiptera	73
Wood	10	Isoptera	54
Other insects			
parasitoids	15	Hymenoptera	68
predators	12	Various	85
Detritus	20	Coleoptera	32

3.5 EFFICIENCY OF DIGESTION AND ABSORPTION

The efficiency with which food is digested and absorbed is usually expressed as approximate digestibility [AD = (weight of food ingested − weight of feces) × 100/weight ingested]. For insects feeding on plant foliage, AD is often in the range 40–50%, but more extreme figures are recorded. Some of this variation arises from differences in plant quality; the AD for *Pieris* larvae feeding on a number of species of Brassicaceae varied from 26 to 43%. In general, it appears that insects feeding on seeds are often more efficient at utilizing their food, reflecting the fact that seeds are food stores for the potential plant (Table 3.3). Parasitoids and predators of other insects also have, on average, high levels of efficiency compared with leaf-eating insects, although there is a great deal of overlap and, on some foods, herbivorous species may be as efficient as predators.

These figures are to some extent misleading because they include components that an insect may be unable to digest. Most insects feeding on green plants, for example, cannot digest cellulose. If the more available nutrients are considered separately, soluble carbohydrates are utilized very efficiently, but proteins and fats are less effectively dealt with. For proteins, this may reflect the manner in which compounds are sequestered within the food, because when the insects are fed on artificial diets they utilize almost 100% of the protein, but this perhaps also reflects the nature of the protein.

Few data exist on the utilization of minor nutrients, but larvae of the bruchid beetle, *Bruchidius*, extract more than 90% of the copper and zinc, and 50% of iron from the seeds they eat. By contrast, utilization of nitrogen is about 40%, and of calcium, magnesium and potassium, less than 20% (Ernst, 1992).

Even though an insect has an enzyme to digest a particular compound, the form of the compound in its food may reduce the efficiency of the digestive process. For example, intact starch grains are resistant to attack by the amylase of the flour weevil, *Sitophilus*, and are only utilized efficiently if their surface is abraded. Presumably some abrasion is normally produced by the mandibles during ingestion (Baker & Woo, 1992). Similarly, pollen is largely inaccessible for digestion by bees until its coat is disrupted by osmotic shock in the midgut.

Review: Slansky and Scriber, 1985

REFERENCES

Appel, H.M. & Martin, M.M. (1990). Gut redox conditions in herbivorous lepidopteran larvae. *Journal of Chemical Ecology*, **16**, 3277–90.

Anderson, M. & Cochrane, D.G. (1978). Studies on the mid-gut of the desert locust *Schistocerca gregaria*. II. Ultrastructure of the muscle coat and its innervation. *Journal of Morphology*, **156**, 257–78.

Applebaum, S.W. (1985). Biochemistry of digestion. In *Comprehensive Insect Physiology, Biochemistry and Pharmacology*, vol. 4, ed. G.A.Kerkut & L.I.Gilbert, pp. 279–311. Oxford: Pergamon Press.

Baines, D.M., Bernays, E.A. & Leather, E.M. (1973). Movement of food through the gut of fifth-instar males of *Locusta migratoria migratorioides* (R. & F.). *Acrida*, **2**, 319–32.

Baker, J.E. (1986). Amylase/proteinase ratios in larval midguts of ten stored-product insects. *Entomologia Experimentalis et Applicata*, **40**, 41–6.

Baker, J.E. & Woo, S.M. (1992). Digestion of starch granules by α-amylases from the rice weevil, *Sitophilus oryzae*: effect of starch type, fat extraction, granule size, mechanical damage, and detergent treatment. *Insect Biochemistry and Molecular Biology*, **22**, 529–37.

Barbehenn, R.V. & Martin, M.M. (1992). The protective role of the peritrophic membrane in the tannin-tolerant larvae of *Orgyia leucostigma* (Lepidoptera). *Journal of Insect Physiology*, **38**, 973–80.

Barbehenn, R.V. & Martin, M.M. (1995). Peritrophic envelope permeability in herbivorous insects. *Journal of Insect Physiology*, **41**, 303–11.

Bernays, E.A. (1981). A specialized region of the gastric caeca in the locust, *Schistocerca gregaria*. *Physiological Entomology*, **6**, 1–6.

Billingsley, P.F. (1990). The midgut ultrastructure of hematophagous insects. *Annual Review of Entomology*, **35**, 219–48.

Breznak, J.A. & Brune, A. (1994). Role of microorganisms in the digestion of lignocellulose by termites. *Annual Review of Entomology*, **39**, 453–87.

Broadway, R.M. & Duffey, S.S. (1986). The effects of dietary protein on the growth and digestive physiology of larval *Heliothis zea* and *Spodoptera exigua*. *Journal of Insect Physiology*, **32**, 673–80.

Chapman, R.F. (1985a). Structure of the digestive system. In *Comprehensive Insect Physiology, Biochemistry and Pharmacology*, vol. 4, ed. G.A.Kerkut & L.I.Gilbert, pp. 165–211. Oxford: Pergamon Press.

Chapman, R.F. (1985b). Coordination of digestion. In *Comprehensive Insect Physiology, Biochemistry and Pharmacology*. vol. 4, ed. G.A.Kerkut & L.I.Gilbert, pp. 213–40. Oxford: Pergamon Press.

Cheeseman, M.T. & Pritchard, G. (1984). Proventricular trituration in adult carabid beetles (Coleoptera: Carabidae). *Journal of Insect Physiology*, **30**, 203–9.

Cheung, W.W.K. & Marshall, A.T. (1973). Studies on water and ion transport in homopteran insects: ultrastructure and cytochemistry of the cicadoid and cercopoid midgut. *Tissue & Cell*, **5**, 651–69.

Cioffi, M. (1979). The morphology and fine structure of the larval midgut of a moth (*Manduca sexta*) in relation to active ion transport. *Tissue & Cell*, **11**, 467–79.

Cohen, A.C. (1995). Extra-oral digestion in predaceous terrestrial Arthropoda. *Annual Review of Entomology*, **40**, 85–103.

Del Bene, G., Dallai, R. & Marchini, D. (1991). Ultrastructure of the midgut and adhering tubular salivary glands of *Frankliniella occidentalis* (Pergande) (Thysanoptera: Thripidae). *International Journal of Insect Morphology & Embryology*, **20**, 15–24.

Dow, J.A.T. (1981). Ion and water transport in locust alimentary canal: evidence from *in vivo* electrochemical gradients. *Journal of Experimental Biology*, **93**, 167–79.

Dow, J.A.T. (1986). Insect midgut function. *Advances in Insect Physiology*, **19**, 187–328.

Dow, J.A.T. (1992). pH gradients in lepidopteran midguts. *Journal of Experimental Biology*, **172**, 355–75.

Eisner, T. (1957). A comparative morphological study of the proventriculus of ants (Hymenoptera: Formicidae). *Bulletin of the Museum of Comparative Zoology, Harvard*, **116**, 439–90.

Ernst, W.H.O. (1992). Nutritional aspects in the development of *Bruchidius sahlbergi* (Coleoptera: Bruchidae) in seeds of *Acacia erioloba*. *Journal of Insect Physiology*, **38**, 831–8.

Evans, W.A.L. & Payne D.W. (1964). Carbohydrases of the alimentary tract of the desert locust, *Schistocerca gregaria*. *Journal of Insect Physiology*, **10**, 657–74.

Fain-Maurel, M.A., Cassier, P. & Alibert, J. (1973). Étude infrastructurale et cytochimique de l'intestin moyen de *Petrobius maritimus* Leach en rapport avec ses fonctions excrétrice et digestives. *Tissue & Cell*, **5**, 603–31.

Farmer, J., Maddrell, S.H.P. & Spring, J.H. (1981). Absorption of fluid by the midgut of *Rhodnius*. *Journal of Experimental Biology*, **94**, 310–6.

Felix, C.R., Betschart, B., Billingsley, P.F. & Freyvogel, T.A. (1991). Post-feeding induction of trypsin in the midgut of *Aedes aegypti* L. (Diptera: Culicidae) is separable into two cellular phases. *Insect Biochemistry*, **21**, 197–203.

Felton, G.W., Workman, J. & Duffey, S.S. (1992). Avoidance of antinutritive defense: role of midgut pH in Colorado potato beetle. *Journal of Chemical Ecology*, **18**, 571–83.

Goodchild, A.J.P. (1963). Studies on the functional anatomy of the intestines of Heteroptera. *Proceedings of the Zoological Society of London*, **141**, 851–910.

Harvey, W.R. & Nelson, N. (ed.) (1992). V-ATPases. *Journal of Experimental Biology*, **172**.

Irvine, B., Audsley, N., Lechleitner, R., Meredith, J., Thomson, B. & Phillips, J. (1988). Transport properties of locust ileum *in vitro*: effects of cyclic AMP. *Journal of Experimental Biology*, **137**, 361–85.

Khan, M.A. (1964). Studies on the secretion of digestive enzymes in *Locusta migratoria* L. II. Invertase activity. *Entomologia Experimentalis et Applicata*, **7**, 125–30.

Komnick, H. (1977). Chloride cells and chloride epithelia of aquatic insects. *International Review of Cytology*, **49**, 285–329.

Lehane, M.J. (1997). Peritrophic matrix structure and formation. *Annual Review of Entomology*, **42**, 525–50.

Lehane, M.J. & Billingsley, P.B. (1996). *The Biology of the Insect Midgut*. London: Chapman & Hall.

Lepier, A., Azuma, M., Harvey, W.R. & Wieczorek, H. (1994). K^+/H^+ antiport in the tobacco hornworm midgut: the K^+-transporting component of the K^+ pump. *Journal of Experimental Biology*, **196**, 361–73.

Maddrell, S.H.P. & Gardiner, B.O.C. (1980). The permeability of the cuticular lining of the insect alimentary canal. *Journal of Experimental Biology*, **85**, 227–37.

Martin, M.M. (1987). *Invertebrate–Microbial Interactions*. Ithaca: Cornell University Press.

Martin, M.M. (1991). The evolution of cellulose digestion in insects. *Philosophical Transactions of the Royal Society of London*, **333**, 281–8.

Martoja, R. & Ballan-Dufrançais, C. (1984). The ultrastructure of the digestive and excretory systems. In *Insect Ultrastructure*, vol. 2, ed. R.C. King & H.Akai, pp. 199–268. New York: Plenum Press.

Miller, H.K. & Fisk, F.W. (1971). Taxonomic implications of the comparative morphology of cockroach proventriculi. *Annals of the Entomological Society of America*, **64**, 671–87.

Murdock, L.L., Brookhart, G., Dunn, P.E., Foard, D.E., Kelley, S., Kitch, L., Shade, R.E., Shukle, R.H. & Wolfson, J.L. (1987). Cysteine digestive proteinases in Coleoptera. *Comparative Biochemistry and Physiology*, **87B**, 783–7.

Nicolson, S.W. (1994). Pollen feeding in the eucalypt nectar fly, *Drosophila flavohirta*. *Physiological Entomology*, **19**, 58–60.

Noirot, C. & Noirot-Timothée, C. (1976). Fine structure of the rectum in cockroaches (Dictyoptera): general organization and intercellular junctions. *Tissue & Cell*, **8**, 345–68.

Penzlin, H. (1985). Stomatogastric nervous system. In *Comprehensive Insect Physiology, Biochemistry and Pharmacology*. vol. 5, ed. G.A. Kerkut & L.I. Gilbert, pp. 371–406. Oxford: Pergamon Press.

Peters, W. (1992). *Peritrophic Membranes*. Berlin: Springer-Verlag.

Peters, W., Heitmann, S. & D'Haese, J. (1979). Formation and fine structure of peritrophic membranes in the earwig, *Forficula auricularia* (Dermaptera: Forficulidae). *Entomologia Generalis*, **5**, 241–54.

Phillips, J.E. & Audsley, N. (1995). Neuropeptide control of ion and fluid transport across locust hindgut. *American Zoologist*, **35**, 503–14.

Phillips, J.E., Audsley, N., Lechleitner, R., Thompson, B., Meredith, J. & Chamberlin, M. (1988). Some major transport mechanisms in insect absorptive epithelia. *Comparative Biochemistry and Physiology*, **90A**, 643–50.

Phillips, J.E., Hanrahan, J., Chamberlin, A. & Thompson, B. (1986). Mechanisms and control of reabsorption in insect hindgut. *Advances in Insect Physiology*, **19**, 329–422.

Poisson, R. & Grassé, P.-P. (1976). L'appareil digestif, digestion et absorption. In *Traité de Zoologie*, vol. 8, part 4, ed. P.-P.Grassé, pp. 205–353. Paris: Masson et Cie.

Schmidt, J.M. & Friend, W.G. (1991). Ingestion and diet destination in the mosquito *Culiseta inornata*: effects of carbohydrate configuration. *Journal of Insect Physiology*, **37**, 817–28.

Schmitz, M. & Komnick, H. (1976). Rectal Chloridepithelien und osmoregulatorische Salzaufnahme durch den Enddarm von Zygopteren und Anisopteren Libellenlarven. *Journal of Insect Physiology*, **22**, 875–83.

Silva, C.P., Ribeiro, A.F., Gulbenkian, S. & Terra, W.R. (1995). Organization, origin and function of the outer microvillar (perimicrovillar) membranes of *Dysdercus peruvianus* (Hemiptera) midgut cells. *Journal of Insect Physiology*, **41**, 1093–103.

Slansky, F. & Scriber, J.M. (1985). Food consumption and utilization. In *Comprehensive Insect Physiology, Biochemistry and Pharmacology*. vol. 4, ed. G.A. Kerkut & L.I. Gilbert, pp. 87–163, Oxford: Pergamon Press.

Smith, J.J.B. (1985). Feeding mechanisms. In *Comprehensive Insect Physiology, Biochemistry and Pharmacology*, vol. 4, ed. G.A. Kerkut & L.I. Gilbert, pp. 34–85. Oxford: Pergamon Press.

Snodgrass, R.E. (1935). *Principles of Insect Morphology*. New York: McGraw-Hill.

Snodgrass, R.E. (1956). *Anatomy of the Honey Bee*. London: Constable.

Spence, K.D. (1991). Structure and physiology of the peritrophic membrane. In *Physiology of the Insect Epidermis*, ed. K. Binnington & A. Retnakaran, pp. 77–93. Melbourne: CSIRO.

Spiro-Kern, A. (1974). Untersuchungen über die Proteasen bei *Culex pipiens*. *Journal of Comparative Physiology*, **90**, 53–70.

Taylor, C.W. (1986). Calcium regulation in insects. *Advances in insect Physiology*, **19**, 155–86.

Terra, W.R. (1990). Evolution of digestive systems of insects. *Annual Review of Entomology*, **35**, 181–200.

Turunen, S. (1985). Absorption. In *Comprehensive Insect Physiology, Biochemistry and Pharmacology*, vol. 4, ed. G.A. Kerkut & L.I. Gilbert, pp. 241–77. Oxford: Pergamon Press.

Turunen, S. & Chippendale, G.M. (1989). Relationship between dietary lipids, midgut lipids, and lipid absorption in eight species of Lepidoptera reared on artificial and natural diets. *Journal of Insect Physiology*, **35**, 627–33.

Williams, L.H. (1954). The feeding habits and food preferences of Acrididae and the factors which determine them. *Transactions of the Royal Entomological Society of London*, **105**, 423–54.

Wolfersberger, M.G. (1996). Localization of amino acid absorption systems in the larval midgut of the tobacco hornworm *Manduca sexta*. *Journal of Insect Physiology*, **42**, 975–82.

Wood, A.R. & Lehane, M.J. (1991). Relative contributions of apocrine and eccrine secretion to digestive enzyme release from midgut cells of *Stomoxys calcitrans* (Insecta: Diptera). *Journal of Insect Physiology*, **37**, 161–6.

Wood, T.G. & Thomas, R.J. (1989). The mutualistic association between Macrotermitinae and *Termitomyces*. *Symposium of the Royal Entomological Society of London*, **14**, 69–92.

Yoshimura, T., Tabata, H., Nishio, M., Ide, E., Yamaoka, R. & Hayashiya, K. (1988). L-cysteine lyase of the webbing clothes moth, *Tineola bisselliella*. *Insect Biochemistry*, **18**, 771–7.

4 Nutrition

Nutrition concerns the chemicals required by an organism for its growth, tissue maintenance, reproduction and the energy necessary to maintain these functions. Many of these chemicals are ingested with the food, but others are synthesized by the insect itself. In some insects, microorganisms contribute to the insect's nutrient pool.

4.1 NUTRITIONAL REQUIREMENTS

Most insects have qualitatively similar nutritional requirements since the basic chemical composition of their tissues and their metabolic processes are generally similar. Most of these requirements are normally met by the diet. Some chemicals can only be obtained in the diet: they are essential (Table 4.1). Others may be synthesized by the insect from dietary components. The dietary requirements of a species may sometimes be obscured due to chemicals having been accumulated and passed on from a previous generation.

Despite the overall similarities, major differences in nutritional requirements do occur. These may be the result of evolutionary changes associated with feeding on substrates with quantitatively, and sometimes qualitatively, different balances of nutrient chemicals.
Reviews: Dadd, 1977, 1985; Reinecke, 1985

4.1.1 Amino acids
Amino acids are required for the production of proteins which are used for structural purposes, as enzymes, for transport and storage, and as receptor molecules. In addition, some amino acids are involved in morphogenesis. Tyrosine is essential for cuticular sclerotization (section 16.5.3) and tryptophan for the synthesis of visual screening pigments (Fig. 25.6). Others, γ-aminobutyric acid and glutamate, are neurotransmitters (section 20.2.3.1), and, in some tissues and some insects, proline is an important energy source.

Amino acids are usually present in the diet as proteins and the value of any ingested protein to an insect depends on its amino acid content and the ability of the insect to digest it. Proteins vary considerably in the extent to which an insect is able to digest them. This may depend on the frequency with which appropriate points of attack occur on the protein and the extent to which these are protected by the protein's structural configuration (Broadway & Duffey, 1988). Although proteins contain some 20 different amino acids, usually only 10 of these are essential in the diet; the others can be synthesized or derived from these 10. The essential amino acids for insects are the same as those needed by rats. They are listed in Table 4.1 and Fig. 4.1. In general, the absence of any one of these amino acids prevents growth. Some insects have essential requirements for additional amino acids. Proline is the most common of these. For example, it is essential for the development of the mosquito (*Culex*), and for several other Diptera as well as the silkworm (*Bombyx*). In some other insects it is necessary for good growth and survival, although it is not absolutely essential. Aspartic acid or glutamic acid are also essential for *Phormia* and the silkworm.

Although other amino acids are not essential, they are necessary for optimal growth because their synthesis or conversion from essential amino acids is energy consuming and necessitates the disposal of surplus fragments. Consequently, alanine and glycine or serine are necessary, in addition to the essential amino acids, for optimal growth of the silkworm. These non-essential amino acids may comprise over 50% of the total amino acids necessary to produce optimal growth on an artificial diet (Fig. 4.1).

The extent to which the synthesis of non-essential amino acids can occur may be limited by the abilities of insects to make certain chemical structures. Tyrosine is a key amino acid in the process of cuticular sclerotization, but insects cannot synthesize its aromatic ring. Consequently, tyrosine can only be synthesized from compounds with the same basic structure. Phenylalanine is another aromatic amino acid from which tyrosine is often derived, but some polyphagous grasshoppers are able to use some phenolic compounds, such as protocatechuic

Table 4.1. *Qualitative dietary requirements of the larvae of insects with different feeding habits (data from Dadd, 1977)*

Food	Detritus		Leaves		Phloem	Insects		Vertebrate tissues
Insect	Blattella[a] (Blattodea)	Phormia (Diptera)	Anthonomus (Coleoptera)	Bombyx (Lepidoptera)	Myzus[a] (Hemiptera)	Chrysopa (Neuroptera)	Pseudosarcophaga (Diptera)	Cochliomyia (Diptera)
Amino acids								
arginine*	+	+	+	+	−	+	+	+
histidine*	+	+	+	+	+	+	+	+
isoleucine*	+	+	+	+	+	+	+	+
leucine*	+	+	+	+	−	+	+	+
lysine*	+	+	+	+	±	+	+	+
methionine*	−	+	+	+	+	+	+	+
phenylalanine*	−	+	+	+	−	+	+	+
threonine*	+	+	+	+	−	+	+	+
tryptophan*	+	+	+	+	−	+	+	+
valine*	+	+	+	+	−	+	+	+
alanine	−	−	−	−	−	−	−	−
aspartic acid	−	±	−	+	−	−	−	−
cystine	?	±	−	−	−	−	±	−
glutamic acid	−	±	−	±	−	−	−	±
glycine	−	−	−	−	−	−	−	+
proline	?	?	+	+	−	−	±	±
serine	?	−	−	−	−	−	−	−
tyrosine	−	−	−	−	−	−	+	−
Carbohydrates	+	+	+	.	+	.	±	.
Fatty acids								
linoleic or linolenic	+	−	+	+	−	.	−	−
others	.	.	.	.	.	.	±	.
Sterols	+	+	+	+	?	.	+	+
Fat soluble vitamins								
β-carotene	.	.	?	.	.	.	±	.
vitamin E	.	.	−	.	−	.	±	.

B vitamins							
biotin	±	±	+	+	+	+	+
folic acid	±	±	±	+	?	?	+
nicotinic acid	+	+	+	+	+	+	+
pantothenic acid	+	+	+	+	+	+	+
pyridoxine	±	+	+	+	?	?	+
riboflavin	±	+	+	+	+	+	+
thiamine	+	+	+	+	+	+	+
Lipogenic compounds							
myo-inositol	+	+	+	·	·	−	·
choline	+	+	+	·	·	+	·
Nucleic acids	?	·	−	·	±	±	+
Inorganic compounds							
sodium	·	+	+	·	·	·	·
potassium	+	+	+	·	+	+	·
calcium	+	·	+	·	·	·	·
chloride	+	·	+	·	·	·	·
iron	·	+	+	·	+	·	·
zinc	+	+	+	·	+	+	·
manganese	+	+	+	·	+	+	·

Notes:

+ Essential.

- Not needed.

· Not known.

± Not essential, but improves growth.

? Uncertain.

a Micro-organism present in these insects.

* Rat essentials.

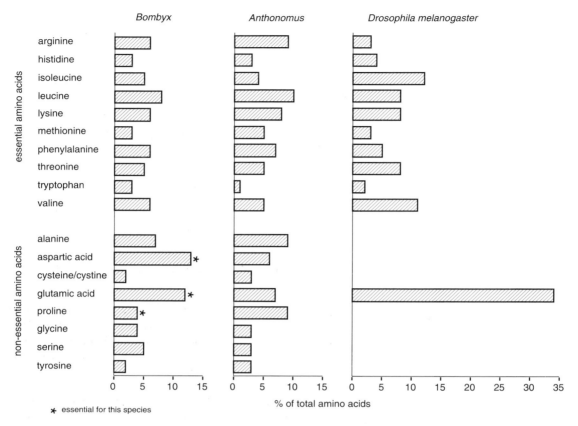

Fig. 4.1. Quantitative requirements for amino acids required in artificial diets to produce optimal growth of different insects. Expressed as a percentage of all amino acids in the diet (data from Dadd, 1985).

acid and gallic acid, in cuticle sclerotization (section 16.5.3) so conserving phenylalanine and tyrosine for protein synthesis. These compounds are non-nutrient and potentially harmful for many other insects.

Sulfur-containing amino acids are only produced from other amino acids containing sulfur. Consequently, cystine and cysteine can be synthesized from methionine and are not necessary in the diet if there is ample methionine.

Amino acid synthesis occurs primarily in the fat body although it also occurs in other tissues. The molecular skeleton may be derived from glucose or acetate being incorporated into compounds that are intermediates in glycolysis or the tricarboxylic acid cycle. From these compounds, amino acids are formed by the addition of ammonia or, more usually, by the transfer of an amino group from a pre-existing amino acid (transamination).

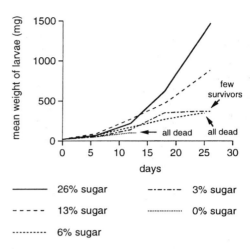

Fig. 4.2. Quantitative effect of sucrose in artificial diets on the growth from hatching of *Schistocerca* (after Dadd, 1960).

$$CH(NH_2).COOH \atop CH_2.COOH \quad + \quad CO.COOH \atop CH_2 \atop CH_2.COOH \quad \rightleftharpoons$$

aspartic acid   α - ketoglutaric acid

$$CO.COOH \atop CH_2.COOH \quad + \quad CH(NH_2).COOH \atop CH_2 \atop CH_2.COOH$$

oxaloacetic acid   **glutamic acid**

Glutamate often plays a central role in these reactions and serves both to incorporate nitrogen into the system and to distribute it among different amino acids.

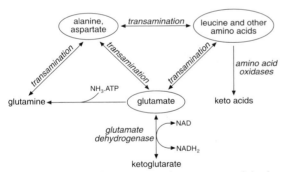

Transamination is a common phenomenon and, in the silkworm, 19 different amino acids are known to act as donors of amino groups in transamination reactions

4.1.2 Carbohydrates

Insect cuticle characteristically contains chitin, a polysaccharide (section 16.2.1.1). Carbohydrates are also used as fuels by a majority of insects. They may be converted to fats, and may contribute to the production of amino acids. They are, therefore, important components of the diets of most insects, but they are not necessarily essential because they can be synthesized from fats or amino acids. Some such conver-

sion probably occurs in most insects, and some species grow readily on artificial diets containing no carbohydrates. This is true of the larva of the screw-worm fly, which feeds on live animal tissues, and carbohydrate can be entirely replaced by wax in the diet of the wax moth, *Galleria*.

Nevertheless, most insects so far examined require some carbohydrate in the diet, and grow better as the proportion is increased. *Schistocerca* for example, needs at least 20% of digestible carbohydrates in an artificial diet for good growth (Fig. 4.2). *Tenebrio* fails to develop unless carbohydrate constitutes at least 40% of the diet, and growth is optimal with 70% carbohydrate.

The utilization of different carbohydrates depends on the insect's ability to hydrolyse polysaccharides (section 3.3.2.2), the readiness with which different compounds are absorbed, and the possession of enzyme systems capable of introducing these substances into the metabolic processes. Some insects can use a very wide range of carbohydrates, probably because they are capable of digesting the more complex structures. *Tribolium*, for example, can utilize starch, the alcohol mannitol, the trisaccharide raffinose, the disaccharides sucrose, maltose and cellobiose, as well as various monosaccharides. Other insects feeding on stored products, and some phytophagous insects, like *Schistocerca* and *Locusta*, can also use a wide range of carbohydrates, but other phytophagous insects are more restricted. The grasshopper, *Melanoplus*, is unable to utilize polysaccharides, and stem-boring larvae of the moth, *Chilo*, can only use sucrose, maltose, fructose and glucose. Most insects are unable to utilize cellulose and other plant polymers because they lack the enzymes to digest them. In some species, however, these substances are made available to the insect by the activities of micro-organisms. Pentose sugars do not require digestion since they are monosaccharides, but nevertheless do not generally support growth and may be actively toxic, perhaps because they interfere with the absorption or oxidation of other sugars.

There may be differences in the ability of larvae and adults to utilize carbohydrates. For instance, the larva of *Aedes* can use starch and glycogen, but the adult cannot.

4.1.3 Lipids

Fatty acids, phospholipids and sterols are components of cell walls as well as having other specific functions. Insects are able to synthesize many fatty acids and phospholipids so they are not usually essential dietary constituents, but many insects do require a dietary source of polyunsaturated fatty acids, and all insects require sterols.

Fatty Acids Fatty acids form an homologous series with the general formula $C_nH_{2n+1}COOH$. In the insects, they are present mainly as diacylglycerides and triacylglycerides:

```
CH2O - CO - R        CH2O - CO - R
|                    |
CH2O - CO - R'       CHO - CO - R'
                     |
                     CH2O - CO - R"

  diacylglyceride      triacylglyceride
```

R, R' and R" are different fatty acid moieties

Many different fatty acids contribute to these compounds. In *Anthonomus*, for example, 23 fatty acids have been identified, ranging in chain length from 6 to 20 carbon atoms, but palmitic and oleic acids comprise over 60% of the total. It is generally true that the major fatty acids in insect triacylglycerides and phospholipids are those with skeletons of 16 and 18 carbon atoms: palmitic (C16), palmitoleic (C16:1 double bond), stearic (C18), oleic (C18:1), linoleic (C18:2) and linolenic (C18:3) (see Fig. 4.3).

Polyunsaturated fatty acids (fatty acids with several double bonds in the chain) with 20 carbon atoms in the chain are present in the phospholipids of many insect species, and may be universal. Derivatives of polyunsaturated fatty acids,

known as eicosanoids, stimulate oviposition in crickets and may be important in the reproduction of all insects. They may also be important in thermoregulation and in lipid mobilization and there is some evidence that eicosanoids mediate the immune response of caterpillars to bacteria in the hemolymph (Stanley-Samuelson *et al.*, 1991).

Some insects, like the cockroach, *Periplaneta*, and the cricket, *Acheta*, are able to synthesize polyunsaturated fatty acids from dietary acetate, and this may also be true of some flies. Synthesis generally involves the elongation of existing components by the condensation of 2-carbon units. This may be followed by desaturation to produce compounds with double bonds; thus stearic is converted to oleic acid. Many other species, however, do have a dietary requirement for small quantities of certain polyunsaturated fatty acids. Lepidoptera, in general, appear to require linolenic or linoleic acid in the diet. A shortage of linoleic acid in the diet of *Ephestia* results in the moths emerging without scales on the wings because the scales do not separate from the pupal cuticle. Some insects are able to synthesize C20 acids from C18 acids (linoleic or linolenic acids), but others require a dietary source (Stanley-Samuelson *et al.*, 1992). Larval mosquitoes, for example, require a dietary source of C20 fatty acids. Without them, the emerging adults are weak and unable to fly. The number and position of the double bonds in the ingested fatty acids is critically important (Dadd, Kleinjan & Stanley-Samuelson, 1987). Mosquito larvae require a double bond in the ω6 position, while larval *Galleria* mainly utilize fatty acids with a double bond in the ω3 position (Fig. 4.3).

Sterols Insects are unable to synthesize sterols. As a consequence, they usually require a sterol in the diet, although some may obtain their sterols from symbiotic microorganisms. In most species, cholesterol is a necessary precursor in the synthesis of ecdysone. Insects feeding on animal tissues obtain cholesterol directly from their food and are unable to utilize plant sterols (Table 4.2). Some plants also contain a little cholesterol, but many do not, and a majority of plant-feeding insects process the common plant sterols to produce cholesterol (Fig. 4.4).

The molting hormone, ecdysone, is a sterol and sterols are also essential structural components of cell membranes. In this role, the requirement for cholesterol is less specific than it is for ecdysone synthesis and other sterols can substitute for cholesterol. They are said to have a 'sparing' role for cholesterol.

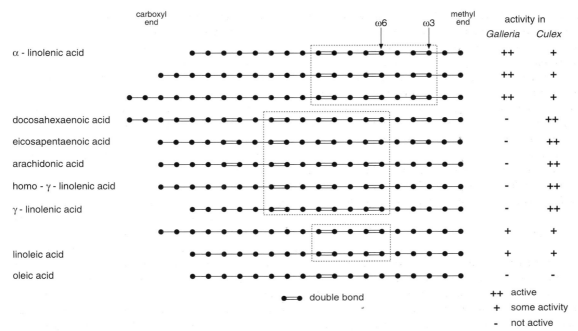

Fig. 4.3. Fatty acid utilization (shown as activity) by larvae of *Culex* and *Galleria*. The boxed sections emphasize arrangements of double bonds that are important to the two insects. *Culex* larvae utilize acids with double bonds in different positions from those utilized by *Galleria* larvae (after Dadd, 1985).

A few insects with specialized feeding habits use other sterols preferentially, reflecting the absence of cholesterol from their normal food. *Drosophila pachea* feeds on a cactus, *Lophocerus schottii*, in which the only sterols are lophenol and schottenol which it utilizes, but it is unable to use any of the usual plant sterols, including cholesterol. Another species, the ambrosia beetle, *Xyleborus*, that feeds on a symbiotic fungus (see below), can utilize cholesterol, but this compound alone is not sufficient for the beetle to complete its development; ergosterol or 7-dehydrocholesterol, typical fungal sterols, are essential.

Fat-soluble vitamins β-carotene (provitamin A) is probably essential in the diet of all insects because it is the functional component of visual pigments (section 22.2.3). It probably also has other functions. For example, the eggs of *Schistocerca* normally contain enough β-carotene to permit growth of the larvae, but in insects reared on a carotene-free diet from eggs already deficient in carotene, growth is retarded and the molt delayed. In addition, the insects are smaller and less active than usual. β-carotene is also commonly involved in the normal pigmentation of leaf-eating insects. Without it, they do not develop their normal yellow or green colors and melanization is also reduced.

Vitamin E (α-tocopherol) is necessary for reproduction in at least some insects. It improves the fecundity of some moths and beetles, and, in its absence, spermatogenesis in the house cricket is halted after spermatid formation (McFarlane, 1992).

Reviews: Bernays, 1992 – sterol requirements of phytophagous insects; Stanley-Samuelson, 1993, 1994 – eicosanoids; Svoboda & Thompson, 1985 – sterol metabolism

4.1.4 Water-soluble growth factors

B vitamins The B vitamins are organic substances, not necessarily related to each other, which are required in small amounts in the diet because they cannot be synthesized. They often function as cofactors of the enzymes catalyzing metabolic transformations. All insects require a source of seven such compounds, either in the diet or produced by associated micro-organisms. These seven are: thiamine, riboflavin, nicotinic acid, pyridoxine, pantothenic acid, folic acid and biotin.

Some of these compounds are known, in addition, to

Table 4.2. *Sterol utilization by insects with different feeding habits*[a]

Food	Detritus		Leaves			Plant sap	Decaying fruit	Cactus rot	Vertebrate tissues	Fur and hide
Insect	Blattella[b] (Blattodea)	Phormia (Diptera)	Locusta (Orthoptera)	Epilachna (Coleoptera)	Manduca (Lepidoptera)	Oncopeltus (Hemiptera)	Drosophila melanogaster (Diptera)	Drosophila pachea[a] (Diptera)	Cochliomyia (Diptera)	Dermestes (Coleoptera)
Cholesterol	+	+	+	+	+	·	+	−	+	+
7–Dehydrocholesterol	−	·	−	·	·	·	−	·	·	+
Ergosterol	−	±	−	·	·	−	+	−	−	−
β–Sitosterol	+	+	+	+	+	+	+	−	−	−
Stigmasterol	+	·	−	+	+	+	+	−	·	−
Campesterol	·	·	·	+	+	+	·	·	·	·
Spinasterol	·	·	−	+	+	−	·	·	·	·

Notes:

[a] Some of the structures are shown in Fig. 4.4.

[b] See text.

+ Utilizable.

± Probably utilizable.

− Not utilizable.

· Not known.

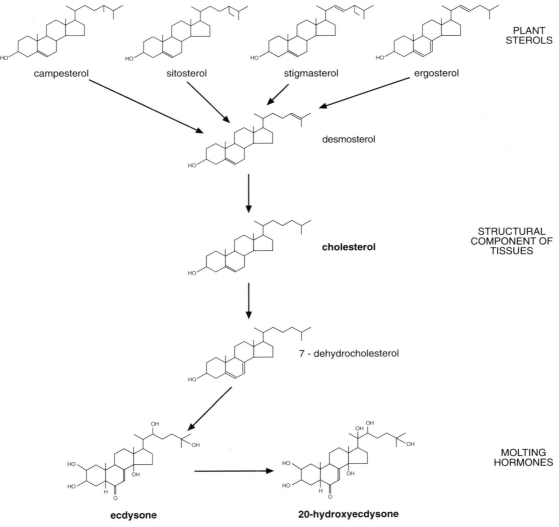

Fig. 4.4. Metabolic pathways by which plant sterols are changed to cholesterol and the molting hormones in phytophagous insects. Many intermediate steps are omitted.

have structural roles. Biotin, for example, is a component of the enzyme pyruvate carboxylase in honeybees and probably in other insect taxa (Tu & Hagedorn, 1992). Folic acid is necessary for nucleic acid biosynthesis. In these capacities, the vitamins can be spared, biotin by oleic acid and folic acid by dietary nucleic acid.

Some insects are known to have requirements for other B vitamins, in addition to the usual seven. For example, *Tenebrio* needs an external source of carnitine.

Lipogenic compounds Myo-inositol and choline are constituents of some phospholipids, the phosphatidylinositols and phosphatidylcholines (or lecithins), respectively. In this role they are required in much larger amounts than vitamins, although in small quantities compared with the main dietary constituents.

Phosphatidylcholines are the major phospholipids in all insects except Diptera, and a dietary source of choline is probably necessary for all insects. In *Drosophila*, choline has functions associated with spermatogenesis and oogenesis, in addition to its structural role in phospholipids. Choline also provides the basis for the neurotransmitter, acetylcholine.

Phosphatidylinositols are less widespread, although inositol trisphosphate is probably found as a second messenger in the nervous systems of all insects. Some insects, such as *Periplaneta*, *Schistocerca*, some Lepidoptera and some Coleoptera, are known to require a dietary source of inositol. Others are apparently able to synthesize it from glucose.

Ascorbic acid The functions of ascorbic acid are not certainly known, but its deficiency is commonly associated with abnormalities at ecdysis, suggesting that it may be concerned with some of the processes involved in cuticular sclerotization. Most, but not all, insects that feed on living plants have a dietary requirement for ascorbic acid. In contrast, insects using other types of food do not have this requirement. It is not clear whether these insects are able to synthesize ascorbic acid, or whether they do not use it.

Nucleic acids Most insects do not need a dietary source of nucleic acids, but some Diptera such as the screw-worm (*Cochliomyia*), *Drosophila* and *Culex* do. Other Diptera, though not having an absolute requirement, develop faster and with less mortality with nucleic acids in the diet. RNA or certain combinations of nucleotides are fully effective, but DNA usually is not.

Review: Kramer & Seib, 1982 – ascorbic acid

4.1.5 Inorganic compounds

Sodium, potassium, calcium, magnesium, chloride and phosphate are essential elements in the functioning of cells and are essential components of the diet of all insects. These elements are nearly always present as impurities in any artificial diet so that very little work on the precise amounts required by insects has been undertaken (McFarlane, 1991).

Iron is the central element in cytochromes and must be in the diet. Zinc is also essential, and manganese commonly so. Both metals appear to play a part in hardening the cuticle of mandibles in many insects (section 16.3.1).

4.2 BALANCE OF NUTRIENTS

Although some growth occurs on foods containing widely differing levels of nutrients, optimal growth requires the nutrient levels to be appropriately balanced. There are two

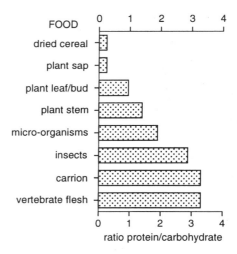

Fig. 4.5. Nutrient balance. The ratios of protein (or amino acids): carbohydrate required in artificial diets for optimal development of insects with different feeding habits (data from Dadd, 1985).

main reasons for this. First, an imbalance may require that an insect ingest and process excessive quantities of food in order to obtain enough of a particular component that is present only in low concentrations in the diet. Second, interconversions from one compound to another are metabolically costly and the rates at which they can occur are limited.

The required balance of the major constituents, amino acids or proteins and carbohydrates, is generally adapted to the natural foods of the species (Fig. 4.5). Insects that feed on other animals have high amino acid requirements relative to carbohydrates, reflecting the relatively high protein content of animal tissues. Plant-feeding species generally require approximately equal amounts of amino acids and carbohydrates. This is true for Orthoptera, Coleoptera and Lepidoptera. Insects feeding on high carbohydrate diets, such as phloem feeders and the grain beetles, have high requirements for carbohydrate.

Apart from these gross needs, an appropriate balance between specific components is also necessary. *Schistocerca gregaria* develops well on lettuce, averaging an 82% increase in mass during the final larval stadium. Supplementing the lettuce with 1 mg/day phenylalanine resulted in the insects increasing their mass by 130% relative to the controls (Fig. 4.6). These insects consumed similar amounts of lettuce to those eating lettuce without any additions, but they utilized the food more efficiently. Most of the phenylalanine was incorporated into the adult

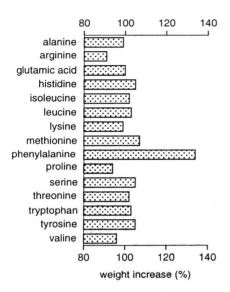

Fig. 4.6. Balance of amino acids. The weight increase of final stage larvae of *Schistocerca* feeding on lettuce supplemented daily with 1 mg quantities of single amino acids. Only phenylalanine produced an effect significantly different from insects feeding on lettuce with no supplement. Weight increase is expressed relative to the increase on lettuce without any supplement (= 100%) (after Bernays, 1982).

cuticle, presumably having been converted to tyrosine for sclerotizing the new cuticle (Bernays & Woodhead, 1984).

Another example of a single amino acid affecting the overall utilization of protein occurs in oogenesis of blood-feeding mosquitoes. Here, isoleucine is the critical amino acid. Female *Aedes* feeding on the blood of guinea pigs, which has a high isoleucine content, produce about 35 eggs per mg blood ingested. When feeding on human blood, containing similar amounts of amino acids except for a low concentration of isoleucine, they only produce about 24 eggs per mg (and see Fig. 4.20). The former use about 34% of the ingested amino acids, while those feeding on human blood utilize less than 20%; the rest is excreted (Briegel, 1985).

Amounts of minor dietary components also need to be in balance. The concentration of RNA needed for optimal development of *Drosophila* is doubled if folic acid is not also present, and an increase in dietary casein from 4% to 7% necessitates a doubling of the concentrations of nicotinic acid and pantothenic acid and a six-fold increase in folic acid for optimal growth.

Nutrients also interact with non-nutrient chemicals in the diet. For example, phenolic compounds, which are common components of leaves, may reduce the digestibility of proteins in caterpillars. However, the effects may vary in different insects, depending on their feeding habits. Whereas tannic acid is detrimental to grass-feeding grasshoppers it may serve as a nutrient for others. Both tannic acid and gallic acid can be used by the grasshopper *Anacridium* as sparing agents for phenylalanine in cuticular sclerotization (Bernays, Chamberlain & Woodhead, 1983).

4.2.1 Changes in the balance of nutrients

The nutritional requirements of an insect change with time because of the varying demands of growth, reproduction, diapause or migration. In larval insects, it is generally true that the nitrogen content of the early stages is greater than that of the later stages, at least in part due to the accumulation, in the later stages, of lipid reserves for subsequent survival, development and reproduction. Larval gypsy moths, given a choice of artificial diets with different levels of proteins and lipids, alter their choice of diets as they get older in a manner which reflects the higher lipid levels of later stage insects (Fig. 4.7a) (Stockhoff, 1993). In addition, the efficiency with which these stages utilize ingested nitrogen decreases (Fig. 4.7b). It is probable that such changes are common. Changes may also occur within a stadium. For example, the cockroach, *Supella*, has a higher carbohydrate intake relative to protein in the first half of a larval stadium than in the second (Cohen *et al.*, 1987).

Amongst those holometabolous insects that do not feed as adults, sexual differences in diet selection may already be apparent in the larvae. Female gypsy moth larvae continue to select more of a high protein diet than males in the later stages of development, and also maintain a higher level of nitrogen utilization (Fig. 4.7).

As adults, females have a higher need than males for dietary protein for egg production. This is most obvious in mosquitoes and other blood-sucking insects where the female is blood-feeding while the male feeds only on nectar, which generally contains negligible amounts of protein. Newly emerged adults of both sexes of the grasshopper, *Oedaleus*, tend to feed preferentially on the developing grains of millet rather than on the leaves. This corresponds with a period of somatic development during which the flight muscles become fully functional and it is assumed that the grain has a higher protein content than leaves. Subsequently, as oogenesis occurs, females exhibit

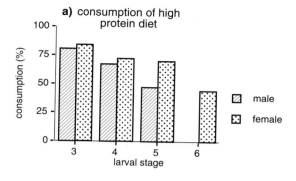

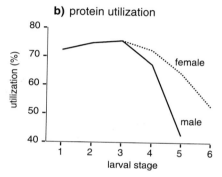

Fig. 4.7. Changes in consumption and utilization of protein in successive larval stages of the gypsy moth, *Lymantria*. (a) The percentage of the high protein diet eaten by each stage. Relatively less of this diet was eaten by the later stages. Insects were given a choice of two artificial diets. One contained high concentrations of both protein and lipid, the other contained a low protein concentration with high lipid. Females have six larval stages, males only five (data from Stockhoff, 1993). (b) The efficiency of utilization of ingested protein declines in the later stages (after Montgomery, 1982).

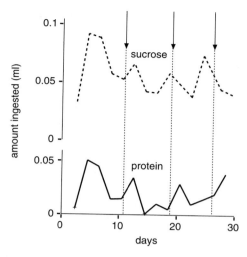

Fig. 4.8. Changes in consumption related to reproduction. An adult female *Phormia* was given a choice of 0.1 M sucrose and a brain–heart extract high in protein. Arrows show the times of oviposition. Notice that 'protein' consumption is low just before oviposition and rises immediately afterwards (after Dethier, 1961).

an even stronger preference for the grains, but males lose this preference and accept leaves and grain with equal readiness (Boys, 1978).

Some female mosquitoes do not lay eggs until they have had their first blood meal; they are said to be anautogenous. Others are autogenous and can lay their first batch of eggs without a blood meal. The protein for yolk production in autogenous mosquitoes comes from storage proteins and from the degeneration of flight muscles. Anautogenous mosquitoes obtain most of their protein from vertebrate blood. Each subsequent cycle of oogenesis is dependent on a blood meal in both autogenous and anautogenous species.

Cycles of varying nutrient intake may also occur in other species that exhibit discrete cycles of oogenesis and oviposition. In anautogenous blowflies, such as *Phormia*, the intake of protein declines in the later stages of vitellogenesis and then rises again after oviposition (Fig. 4.8). The intake of sugar may remain more or less constant, but sometimes varies inversely with protein intake (de Clerk & de Loof, 1983). Similar changes may occur in grasshoppers.

Reviews: Barton Browne, 1995 – changes during development; Simpson & Simpson, 1990 – changes due to previous feeding

4.2.2 Maintaining a balance

Many insects are known to select a diet which approximates an optimal balance of the major components. An insect can respond to a dietary imbalance in one of three ways. It can adjust the total amount ingested so that it acquires enough of the most limiting nutrient; it can move from one food to another with a different nutrient balance; or it can adjust the efficiency with which it uses nutrients. Most of our understanding of dietary regulation by insects comes from laboratory experiments using artificial diets. These experiments leave no doubt that insects do have the ability to achieve some degree of nutritional balance by regulating food intake, and there is every reason to suppose that this also occurs naturally, although the complex makeup of most natural foods and limitations of availability may restrict the degree to which an insect can achieve balance.

Several grasshoppers, caterpillars and a cockroach have been shown to increase the amount eaten if the entire nutrient composition of a diet is diluted with some inert non-nutritional substance (see Wheeler & Slansky, 1991). *Melanoplus sanguinipes*, feeding on dried wheat sprouts, was able to maintain its intake of wheat close to optimal even when the food was diluted 7:1 with cellulose. This necessitated an almost 7-fold increase in the total amount of food consumed (Fig. 4.9a). At the greatest dilution, the insects were unable to compensate. Caterpillars of *Spodoptera* compensated for dilution of the nutrients in their diet from 30 to 10% by eating three times as much (Fig. 4.9b).

Insects can also compensate for deficiencies in a class of major nutrients by adjusting the total amount eaten. Aphids feeding on artificial diets with different concentrations of amino acids increased their intake on the more dilute diets and were able to maintain their intake of amino acids at similar levels (Fig. 4.10) (Prosser, Simpson & Douglas, 1992). Locusts fed on artificial diets with either high or low levels of protein ate more when confined to the low protein diet. By contrast, they did not compensate for differences in the levels of carbohydrate, but compensation for carbohydrates has been demonstrated in adult flies (For example, Nestel, Galun & Friedman, 1985), in butterflies feeding on sugar solutions, and in cockroaches.

If they have a choice of foods with different nutrient levels, insects can regulate their nutrient intake by eating differentially from the foods available. In this way, grasshoppers and caterpillars can correct for a previous imbalance of carbohydrate or protein (Fig. 4.11), and larval *Heliothis* are able to adjust the amounts of vitamins and lipids consumed. Given a choice of foods containing different amounts of inorganic salts, *Locusta* modifies its choice so that it maintains a constant level of both salts and the major nutrients. In the absence of a choice, however, different amounts of salts are ingested because the insect maintains the optimal intake of major nutrients (Trumper & Simpson, 1993).

The importance of post-ingestive regulation of nutrient balance is demonstrated by experiments with locusts. These insects maintained a relatively constant increase in body nitrogen despite a more than 3-fold increase in the amount of nitrogen ingested. Very little of the protein was present unchanged in the feces. Most of the excess was

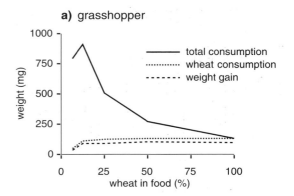

a) grasshopper

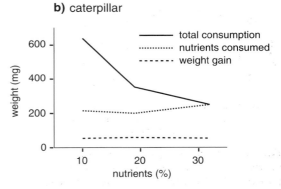

b) caterpillar

Fig. 4.9. Dietary compensation by insects with no choice of food. (**a**) A grasshopper, *Melanoplus*. Amount of food eaten in five days by final stage larvae feeding on dried, ground wheat. The wheat was mixed with indigestible cellulose in different proportions. On all but the lowest concentration, wheat consumption and weight gain were almost constant because the insect ate more (data from McGinnis & Kasting, 1967). (**b**) A caterpillar, *Spodoptera*. Amount of food eaten in the final larval stage. Nutrients were diluted with indigestible cellulose. Weight gain was similar irrespective of the percentage of nutrients in the diet (data from Wheeler & Slansky, 1991).

excreted as uric acid or some other, unknown, nitrogenous end product of catabolism (Fig. 4.12) (Zanotto, Simpson & Raubenheimer, 1993). In *Pieris*, there is an inverse relationship between the levels of particular amino acids in the diet and the metabolic utilization of each amino acid, suggesting the presence of a mechanism regulating oxidation of specific amino acids according to needs (van Loon, 1988).

The ability to adjust food intake to nutritional requirements implies some feedback of nutritional status on food selection and feeding behavior. In the locust, feedback has

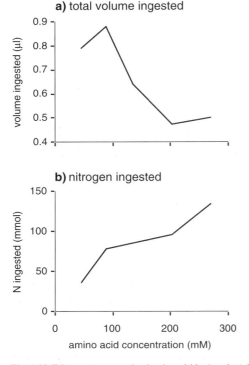

a) total volume ingested

b) nitrogen ingested

Fig. 4.10. Dietary compensation by the aphid, *Acyrthosiphon*, with no choice of food. At low amino acid concentrations, the insects ingested more food and so partially compensated for the dilution (after Prosser, Simpson & Douglas, 1992). (a) Total volume ingested in a 24-hour period. (b) Amount of nitrogen ingested.

been shown to come from the hemolymph. High levels of certain amino acids in the hemolymph, which follow from eating a diet rich in amino acids, depress the sensitivity of the peripheral contact chemoreceptors to amino acids in the diet (Fig. 4.13). The sensitivity of chemoreceptors to sucrose is, however, unaffected (Abisgold & Simpson, 1988). Conversely if the insect feeds on a diet with high levels of sucrose, the sensitivity of its receptors to sucrose is depressed. The increase in blood osmolality which follows feeding also reduces further feeding, either by extending the interval between meals or by reducing the amount eaten within a meal, depending on the state of the insects.

Learning has an important role in regulating nutrient intake. Sugars, amino acids and salts are tasted by many insects (section 24.2.2), but proteins, sterols and vitamins are not. The insect regulates the intake of these compounds by learning to associate some quality of the food with feed-back on its own nutrient status. Visual characteristics, odor and taste due to chemicals other than the nutrients may provide the stimuli and the association may be positive, resulting in feeding, or negative, leading to rejection of the food. Most work on this aspect of compensatory feeding has been on locusts and grasshoppers, but it is likely that other insects exhibit similar behavior. Locusts and grasshoppers have been shown to associate an odor with a high protein diet, and a food flavor with low protein. They have also been shown to develop an aversion to food containing unutilizable sterols. The mechanisms by which nutrient imbalance effects the insect's behavior are unknown.

Longer-term changes during larval development probably also reflect nutritional feedback. On the other hand, regulation of changes related to ovarian cycles in adult females is at least partly controlled neurally. Abdominal distension caused by the developing oocytes is important in reducing the rate of protein intake in *Phormia*, and, in *Musca*, distension of the oviducts may be important (Clifford & Woodring, 1986).

Reviews: Bernays, 1995 – learning in nutritional compensation; Simpson, Raubenheimer & Chambers, 1995; Waldbauer & Friedman, 1991

4.3 FEEDING ON NUTRITIONALLY POOR SUBSTRATES

Many insects habitually eat food that is nutritionally inadequate. Termites and many beetles feed on wood which is low in proteins and amino acids, and contains cellulose and lignin which most insects are unable to digest; many Homoptera feed on phloem with an imbalance (for insects) of amino acids; others feed on xylem which is deficient in most nutrient chemicals; some Heteroptera, sucking lice (Anoplura) and a few cyclorraphous Diptera obtain all their food from vertebrate blood which lacks some of the B vitamins and other minor nutritional components. [Note that fleas (Siphonaptera) and many nematocerous Diptera that are blood-sucking as adults have different feeding habits as larvae. This enables them to obtain the minor components of the diet that are lacking in blood].

These insects are able to use these materials through symbiotic associations with micro-organisms (Table 4.3). The nutrition of some other insects, feeding on less intractable materials, is also enhanced by symbiotic associations. This occurs in cockroaches and crickets

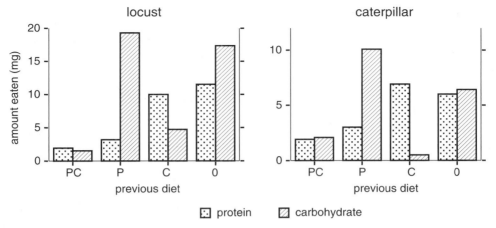

Fig. 4.11. Dietary compensation by insects with a choice of foods. Insects were given a choice of artificial diet containing either protein or carbohydrate after a period of four hours during which only one of four different diets was available. The previous diets contained protein and carbohydrate (PC), protein with no carbohydrate (P), carbohydrate with no protein (C), or neither of these components (0). After feeding on PC, little feeding occurred; after 0, carbohydrate and protein were almost equally acceptable. After feeding on P, the insects selected carbohydrate, and after feeding on C, they selected protein. The histogram for the locust shows the amounts eaten in one hour, that for the caterpillar, the amounts eaten in eight hours (after Simpson, Simmonds & Blaney, 1988).

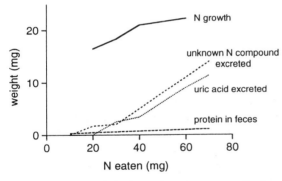

Fig. 4.12. Dietary regulation by postingestive processes. Final stage larvae of *Locusta* were given diets containing different amounts of protein. The insects maintained the amount of growth within narrow limits as the amount of protein eaten increased by excreting increasing amounts of nitrogenous materials, mostly as uric acid and other, unknown, nitrogen-containing compounds. Most of the protein was digested and absorbed, as indicated by the small amounts of unchanged protein in the feces. Only very small amounts of free amino acids were excreted (not shown) (after Zanotto, Simpson & Raubenheimer, 1993).

which are detritus feeders, in leaf-cutting ants, and in some gall-forming Cecidomyiidae (fungus gnats). The symbiotic partner is, in some cases, an ectosymbiont, but more commonly they are endosymbionts.

Review: Douglas, 1995

4.3.1 Ectosymbiotic fungi

A number of insects have ectosymbiotic relationships with fungi. The insects eat the fungus, but the association differs from that in most fungus-eating insects, because the insect manipulates the fungus, and so derives nutrients, indirectly, from substrates that would otherwise be difficult or impossible for it to utilize.

Ambrosia beetles (some Scolytinae and nearly all Platypodidae) are associated with fungi that enable them to use the xylem of woody plants. The fungi are the principal food of both larvae and adults, and their key role is probably in concentrating nitrogen, present in very low concentrations in the wood. They also provide sterols, such as ergosterol, which are essential for the beetles' development. Bark beetles (most Scolytinae) feed largely on the phloem of woody tissues which is higher in nutrients than the xylem. They also have fungal associations, but their dependence is less extreme. The beetle–fungus

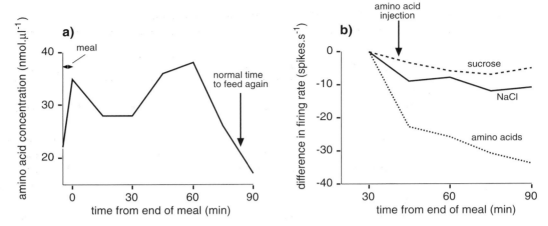

Fig. 4.13. Mechanism of diet regulation in larvae of *Locusta*. (a) changes in the total concentration of amino acids in the hemolymph following the end of a meal (time 0) on artificial diet (after Abisgold & Simpson, 1987). (b) An experiment showing the effects of increased amino acid concentration in the hemolymph on the firing rates of contact chemoreceptors on the maxillary palps. Forty-five minutes after ending a meal the insects were injected with a cocktail of amino acids to simulate the effects of feeding. The same sensilla were tested at intervals with sucrose, NaCl or a cocktail of amino acids. The responsiveness to sucrose and salt declined slightly, but the sensitivity to amino acids was greatly reduced (after Abisgold & Simpson, 1988).

associations are not species-specific. Several fungus genera are associated with ambrosia beetles. Two of the best known are *Fusarium* and *Ambrosiella*. Most of those associated with bark beetles are in the genus *Ceratocystis*.

Leaf-cutting ants (Attini) are dependent on specific fungi for larval food. Worker ants cut leaves, and other parts, from living plants and carry them to the nest. Here, the ants chew the plant fragments, removing the waxy cuticle and possibly also removing existing micro-organisms on the plant surface. Using feces, they build the chewed fragments into a garden which they inoculate with hyphae from an existing garden. The fungi are Basidiomycetes that only occur in the nests of these ants.

Macrotermitinae also cultivate fungi in gardens, called fungus combs, constructed from fresh fecal material containing wood fragments. The fungus, in the genus *Termitomyces*, is only found associated with termites. It breaks down cellulose and lignin and, when ingested by the termites, it contributes its cellulolytic enzymes to those of the insect. Nitrogen is also concentrated. In reproductive structures of the fungus, which are eaten by the termites, it reaches 8% dry weight; the wood initially ingested may have only about 0.3% dry weight. Termitidae, including the Macrotermitinae, do not have endosymbiotic protozoa, unlike all the other termites.

Reviews: Beaver, 1989 – bark and ambrosia beetles; Bissett & Borkent, 1988 – gall flies; Cherrett, Powell & Stradling, 1989 – leaf-cutting ants; Wood & Thomas, 1989 – termites

4.3.2 Endosymbionts

Many insects have micro-organisms extracellularly in the gut lumen or intracellularly in various tissues.

4.3.2.1 *Micro-organisms in the alimentary canal*
Micro-organisms are almost inevitably ingested during feeding and so an intestinal flora is present in most insects. The alimentary canal of grasshoppers, for instance, is sterile when the insect hatches from the egg, but soon acquires a bacterial flora which increases in numbers and species throughout life. In general, insects with straight alimentary canals contain fewer micro-organisms than those with complicated guts which have a range of different pH values, providing a number of different niches. The micro-organisms occurring in the gut in these cases of casual infection largely reflect what is present in the environment (Brooks, 1963). At least in the case of the locust, *Schistocerca*, and larval fruit flies, *Rhagoletis*, the gut flora does not contribute to the nutrition of the insect (Charnley, Hunt & Dillon, 1985; Howard & Bush, 1989).

Table 4.3. *Insects feeding on nutritionally poor diets and their associated symbiotic organisms*

| Type of food | Insect order/family | Micro-organisms | | Contribution to insect |
		Position in body	Type	
Wood	Blattodea			
	Cryptocercus	Hindgut	Flagellates	Cellulose digestion
	Isoptera			
	Kalotermitidae	Hindgut	Flagellates	Cellulose digestion
		Hindgut	Bacteria	Nitrogen fixation
	Macrotermitinae	Ectosymbionts	Fungi	Cellulose digestion Concentration of nitrogen
	Coleoptera			
	Anobiidae	Midgut cecal epithelium	Yeasts	Essential amino acids
	Scolytinae	Ectosymbionts	Fungus	Concentration of nitrogen
	Platypodidae	Ectosymbionts	Fungus	Concentration of nitrogen, sterols
		Hindgut	Bacteria	Nitrogen fixation
	Hymenoptera			
	Siricidae	Ectosymbionts	Fungus	Cellulose digestion
Green plants	Hymenoptera			
	Attini	Ectosymbionts	Fungus	Cellulose digestion
Phloem	Homoptera			
	Aphididae	Hemocoel	Bacteria	Amino acids
	Delphacidae	Hemocoel	Bacteria, yeasts	Amino acids, sterols
Vertebrate blood	Phthiraptera			
	Anoplura	Variable	Bacteria	B vitamins
	Hemiptera			
	Cimicidae	Hemocoel	Bacteria	B vitamins
	Diptera			
	Glossinidae	Midgut epithelium	Bacteria	B vitamins
Detritus	Blattodea	Hindgut	Bacteria	Carbohydrate digestion
		Fat body	Bacteria	Nitrogen recycling

Adult tephritid flies have a dorsal diverticulum of the esophagus in which bacteria accumulate. In general, the species of bacteria present reflect what is present on the surface of host fruit, although some species do occur consistently and in greater abundance than others (Drew & Lloyd, 1987). It is not known if these bacteria are true symbionts, or if they contribute to the nutrition of the flies.

In other cases, it is known that micro-organisms in the gut contribute to the insect's nutrition. Detritus-feeding cockroaches, such as *Periplaneta*, and crickets have bacteria in the hindgut. They may be attached to projections from the intima of the hindgut and they enhance the insects' ability to digest plant polysaccharides, such as xylans, pectins and gums, and oligosaccharides, such as raffinose.

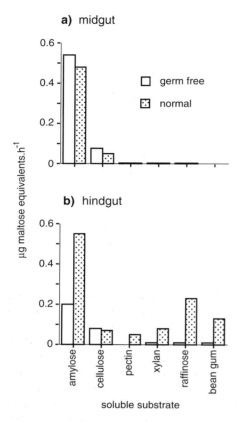

Fig. 4.14. Digestion of carbohydrates by crickets, *Acheta*, with and without their normal hindgut bacterial flora. Digestive efficiency is measured as the quantity of maltose equivalents produced in one hour (after Kaufman & Klug, 1991). **(a)** Digestion by homogenates of the midgut. The absence of hindgut bacteria has no effect. **(b)** Digestion by homogenates of the hindgut. In the presence of bacteria complex carbohydrates are digested; the germ free insect is unable to digest these compounds.

Crickets do not have enzymes capable of digesting these compounds (Fig. 4.14). The bacteria produce short-chain fatty acids which are absorbed in the hindgut (Kaufmann & Klug, 1991). These associations involve a variety of bacterial species. Whether or not some are characteristic is not known.

Scarab beetle larvae (Scarabaeidae), which feed on decaying wood, have a bacterial flora housed in an expansion of the hindgut. It is believed that the bacteria are those commonly involved in the process of wood decay and that they are ingested when the wood is eaten. In this case, the association is not truly symbiotic, although the larvae are dependent on the bacteria for digesting the food. Some termites and wood-eating cockroaches have flagellates in the hindgut which are important in the digestion of wood. These insects have a strict symbiotic relationship with their hindgut fauna, and the behavior of the insects ensures transfer from generation to generation (see below).

The bacteria which are present in the guts of higher termites are able to fix atmospheric nitrogen which is subsequently incorporated into the tissues of the insects (Bentley, 1984).

4.3.2.2 *Intracellular micro-organisms*
Intracellular micro-organisms fall into two groups: those that occur in otherwise normal cells of an insect, and those that are restricted to special cells with discrete morphology known as mycetocytes.[1] The former have been recorded from many insect orders. In general, they appear to have no effects on the biology of the host insect. However, this is not true of the bacterium *Wolbachia*, which is known to be present in ovarian tissue of some species of Orthoptera, Hemiptera, Coleoptera, Diptera, Hymenoptera and Lepidoptera (Werren, Windsor & Guo, 1995). It is transmitted cytoplasmically and causes post-zygotic incompatibility between different strains of various species, including *Tribolium confusum* and *Culex pipiens* (O'Neill *et al.*, 1992), and to cause parthenogenesis in some parasitic Hymenoptera, such as *Trichogramma*.
Review: Werren, 1997

Mycetocyte micro-organisms Mycetocyte micro-organisms are universal amongst species that feed on vertebrate blood throughout their lives: Cimicidae (bedbugs), Triatominae (kissing bugs), Anoplura (sucking lice), Glossinidae (tsetse flies), Hippoboscidae (deer flies) and Nycteribiidae (bat flies), but they are not found in fleas, mosquitoes or horseflies that have larvae which are not blood-sucking. They are almost universal amongst Homoptera, the only exceptions being those feeding on tissues other than the phloem or xylem. Many, but not all, wood-feeding beetles and Ischnocera (= Mallophaga),

---

[1]   This term has been used generically to describe cells containing any type of micro-organism. Such cells in which the micro-organisms have been positively identified as bacteria on the basis of their DNA are now known as bacteriocytes.

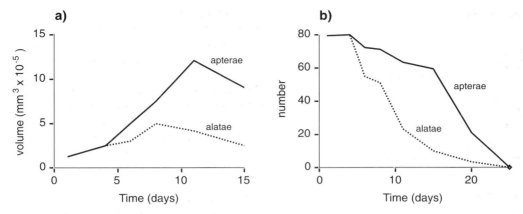

Fig. 4.15. Changes in the mycetocytes of the aphid, *Megoura* (after Douglas & Dixon, 1987). (a) Changes in the total volume of mycetocytes in an individual. The insects become adult on about day 8 and begin to produce young on day 10. (b) Changes in the numbers of mycetocytes in an individual aphid.

feeding on feathers and skin debris of birds, also have mycetocyte micro-organisms. Micro-organisms in myce-tocytes are also found in all cockroaches in the family Blattidae and ants of the tribe Camponoti. These are omnivorous insects, but it is suggested that their diet is often poor with an imbalance of amino acids.

Types of micro-organism The mycetocyte micro-organisms of aphids and weevils are bacteria (Campbell, Bragg & Turner, 1992) and this is probably also true of the myce-tocyte micro-organisms in a majority of other insects. Yeasts are present in Fulgoridae and *Laodelphax* amongst the Homoptera and in Anobiidae and Cerambycidae amongst wood-boring beetles. In Triatominae the micro-organisms are Actinomycetes.

In general, only one form of micro-organism is found in each insect species, but this is not true in many Homoptera. All Auchenorrhyncha appear to house more than one type of micro-organism and, in Fulgoridae, both yeasts and bacteria are present. Some species have as many as six different symbionts. Many aphid species have only a single bacterial symbiont, but others may have two or three different types; some also have yeasts.

Location in the insect body Mycetocytes are large poly-ploid cells. They are scattered amongst the principal cells of the midgut epithelium in *Haematopinus* (Anoplura), while in cockroaches they are scattered through the fat body. In other insects, the mycetocytes are aggregated to

form organs known as mycetomes, often in the hemocoel. Amongst holometabolous insects and some Homoptera, a mycetome is often only present in the larval stages. At metamorphosis it fragments into mycetocytes which become lodged in adult organs. The larval mycetome of the beetle *Calandra* is a U-shaped structure below the foregut, but not connected to it; in the adult, mycetomes are present in the epithelium of the midgut ceca. In Hippoboscidae and Glossinidae, the mycetome is present as a discrete zone in the midgut epithelium.

Mycetocytes generally do not divide; they increase in size, and endomitotic divisions lead to polyploidy. During the larval and early adult period of aphids, individual cells may increase in volume about 10-fold in apterous (wing-less) morphs, but to a lesser extent in alates (winged forms). The biomass of symbionts increases in parallel with the volume of the cells (Fig. 4.15). In alates, the number of mycetocytes declines sharply when the insects become adult, but, in apterous individuals, this sharp decline is delayed (Douglas & Dixon, 1987). It is not known how these changes are regulated.

Within the mycetocytes of most insects, each micro-organism is in a separate vacuole surrounded by membrane, but in *Glossina* and ants the symbionts are free in the cytoplasm.

Roles of mycetocyte micro-organisms These micro-organisms sometimes have a key role in the nitrogen economy of their host insects. In Blattidae they recycle

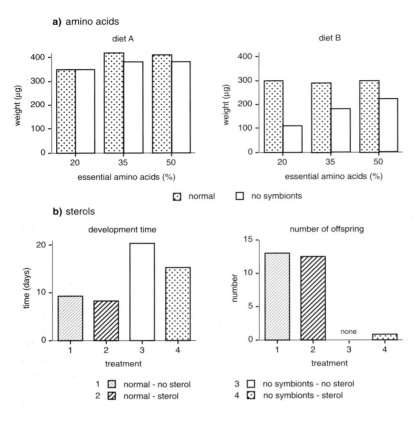

a) amino acids

b) sterols

Fig. 4.16. Contributions of symbionts to aphid survival and development on artificial diets. (a) Amino acids. Weight of *Acyrthosiphon* with and without its symbionts after feeding on artificial diet. Diet A contained a well-balanced mixture of amino acids. The aphids grew equally well irrespective of the proportion of essential amino acids and irrespective of the presence of symbionts. Diet B contained a mixture of amino acids in proportions approximating those in the phloem of *Vicia*, a normal host plant. Growth was reduced in the absence of symbionts and was most markedly affected at the lowest level of essential amino acids (after Prosser & Douglas, 1992). (b) Sterols. Performance of *Myzus* on artificial diets with or without symbionts. Aphids developed to adulthood almost equally rapidly and produced many offspring on diets with or without sterols and phospholipids when they possessed symbionts. Without symbionts, development was prolonged and few offspring were produced even when sterol was added to the diet (data from Douglas, 1988).

waste nitrogen. Cockroaches do not excrete uric acid, they store it in cells in the fat body. These stores become depleted if the insect feeds on a diet low in nitrogen. The symbionts can synthesize a range of essential amino acids, including those containing sulfur.

The symbionts of aphids and the beetle *Stegobium* are known to upgrade non-essential amino acids to essential amino acids. The symbionts of *Acyrthosiphon* contribute to the insect's ability to survive on diets containing low levels of essential amino acids, as is generally the case with phloem. Essential acids usually comprise less than 25% of the total amino acid concentration in phloem, and some essential amino acids may constitute less than 0.2 mol % of the total. The symbionts of *Acyrthosiphon* are of little importance on a high quality diet, but are most effective when the balance of amino acids approximates that occurring in the phloem of *Vicia*, a normal host of this aphid (Fig. 4.16a) (Prosser & Douglas, 1992).

There is no general agreement about the ability of the bacterial symbionts of Homoptera to produce sterols. Experiments in which the symbionts are removed from the insects, as in the experiment with aphids shown in Fig. 4.16b, may be interpreted as indicating that sterols are normally supplied by the symbionts. However, other bacteria in the same group as those found in aphids are incapable of producing sterols (see Campbell, 1989). It is, however, possible that the yeasts associated with some aphids, planthoppers and the beetle, *Lasioderma*, do produce sterols.

There is evidence that the symbionts of Anoplura and some beetles produce some of the B vitamins.

4.3.3 Transmission between generations
In view of the constant associations of specific symbionts with particular insects and their obvious importance to the insects, even though the details may not be known, it is to

be expected that mechanisms will be present to ensure their transmission from parent to offspring.

Many insects with ectosymbiotic fungi have special cuticular structures in which the fungi are transported to new feeding sites. Ambrosia beetles have cuticular pockets with associated glands. They are called mycangia and are present on the mouthparts, thorax, legs or elytra, depending on the species. They are only present in females of many species, but, in a few, males only, or both sexes, have mycangia. Entry of fungal spores into the mycangia is passive; as a result, spores of many fungal species may accumulate in them. The beetle-associated fungi grow by budding and fission in the cavity, but other fungi do not. Probably the secretions from the associated glands contribute to this selective elimination.

The queens of leaf-cutting ants and many Macrotermitinae carry fungal spores on their nuptial flight. The ants have an infrabuccal pocket, a pouch just outside the mouth that is present in workers of all ant species. When the queen has excavated her underground chamber, she regurgitates the pellet of spores, and hyphae soon develop. If the queen loses her pellet, she dies. Queens of fungus-growing termites carry a bolus of spores in the gut. It passes through the gut and, with feces, forms the beginnings of the first fungus comb in the new nest. In some Macrotermitinae, however, spores of the fungus are collected by the first workers while they are foraging away from the nest. Amongst other termites, anal trophallaxis is necessary for the renewal of the flagellate fauna after each molt (section 2.5.1.3).

In most insects with mycetocyte symbionts, the symbionts are transferred from the mycetomes to the developing oocytes and so to the next generation, a process known as transovarial transmission. Sometimes, as in *Pediculus*, the symbionts are released from the mycetocytes and migrate to the ovaries. In other cases, whole mycetocytes migrate, as in cockroaches, where they move from the fat body to the developing ovaries while the maternal cockroach is still embryonic. Bacteria are released on the exterior of the developing oocytes where they remain until vitellogenesis is complete. At this stage the bacteria are taken into the oocytes by phagocytosis (Sacchi *et al.*, 1988). In the weevil, *Sitophilus*, the association with oocytes occurs even earlier. Some of the symbionts become associated with the primordial germ cells early in embryonic development. Mycetomes are occupied by those symbionts which do not associate with the germ cells (Nardon & Grenier, 1988).

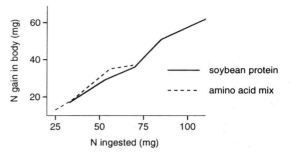

Fig. 4.17. Relationship between the amount of nitrogen ingested and growth of larval *Bombyx* on artificial diets in which the nitrogen was supplied either as a mixture of amino acids or as a protein from soybean. The similarities of the two slopes suggest that the plant protein was utilized as efficiently as the amino acids (data from Horie & Watanabe, 1983).

In other cases, as in anobiid and cerambycid beetles, the transfer does not occur until the egg is laid. For example, the mycetomes in the midgut ceca of anobiid larvae break down at metamorphosis. The yeasts from the mycetomes pass through the alimentary canal and lodge in pouches associated with the ovipositor. From here, they are smeared on the outside of the egg and are eaten when the larva hatches and eats the chorion.

In the viviparous Glossinidae and Hippoboscidae, symbionts are transferred to the developing larvae in the secretions of the milk gland (section 14.3.2.2).

Reviews: Campbell, 1989 – phytophagous insects; Douglas, 1989 – mycetocyte symbionts

4.4 NUTRITIONAL EFFECTS ON GROWTH AND DEVELOPMENT

Variations in the quantity or quality of an acceptable diet can have profound effects on insect development. The quantities of protein and amino acids ingested are important for optimal growth and reproduction. The ultimate size of the insect, reflected in the gain in body nitrogen, increases as nitrogen intake is increased (Fig. 4.17) (Horie & Watanabe, 1983). There are many examples of insects feeding on natural foods in which growth and reproduction are positively correlated with nitrogen content of the food (Ridsdill-Smith, 1991; Scriber & Slansky, 1981). Commonly, as food intake decreases, the duration of development is extended and the insect becomes smaller and lighter in weight (Fig. 4.18).

a) duration of final larval stage

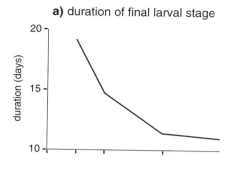

b) adult weight

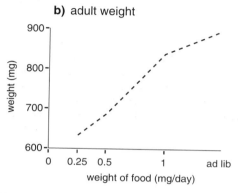

Fig. 4.18. Effects of differing amounts of food on development of a grasshopper, *Schistocerca americana*. Insects received different amounts of seedling wheat, a highly nutritious food. (a) Duration of the final larval stage. (b) Weight of newly emerged males.

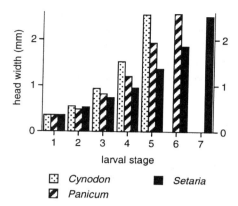

Fig. 4.19. Effects of food quality on size and number of larval stages. Larvae of *Spodoptera exempta* were fed on one of three different grasses which varied in their suitability for growth. On the two less favorable grasses the insects grew more slowly, but they developed through one or two additional stages so that their final sizes were similar. Head width is a measure of larval size (data from Yarro, 1985).

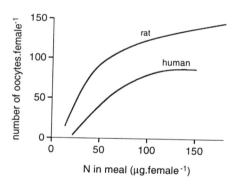

Fig. 4.20. Effects of food quantity and quality on egg production by the mosquito, *Aedes*. Insects were given different sized blood meals. As meal size increased, more eggs were produced. Human blood was utilized much less efficiently because of its relative deficiency in isoleucine (see text) (data from Briegel, 1985).

In other cases, on nutritionally poor diets, low growth rates are associated with an increase in the number of larval stages. For example, the caterpillars of *Spodoptera exempta*, which feed on grasses, grow more slowly on *Panicum* and *Setaria* than they do on *Cynodon*. When feeding on the latter grass, the insects pupate at the end of the fifth stadium, but on the other two grasses they are still small and continue to develop through one or two additional stages (Fig. 4.19). The sizes of the last stage larvae are similar irrespective of the food (Yarro, 1985).

The adequacy of larval food is reflected in the quantity of nutrients stored for subsequent egg production, but more direct effects of nutrient levels occur in insects that feed as adults. In mosquitoes, for example, egg production is proportional to the amount of nitrogen ingested with the blood meal (Fig. 4.20) (Clements, 1992).

Differences in nutrition may also produce profound differences in coloration and even in morphology.

Coloration may be affected either through the absence of a pigment or through interference with pigment metabolism. The absence of β-carotene has both effects in *Schistocerca*. Carotene is an essential constituent of the yellow carotenoid giving the background color, but in the absence of carotene melanization also is reduced.

Caterpillars of the spring brood of *Nemoria arizonaria* resemble the oak catkins on which they feed. They are yellow with a rough cuticle, and have two rows of reddish spots along the midline. Caterpillars of the summer brood resemble stems. They are green-grey and are without the rows of spots. They feed on leaves and have larger head capsules and mandibles than the spring brood. These differences result entirely from differences in the quality of food eaten by the insects (Greene, 1989).

In the honeybee, the quality of food given to larvae by the workers determines whether the larvae will become queens or workers (section 15.5).

REFERENCES

Abisgold, J.D. & Simpson, S.J. (1987). The physiology of compensation by locusts for changes in dietary protein. *Journal of Experimental Biology*, **129**, 329–46.

Abisgold, J.D. & Simpson, S.J. (1988). The effect of dietary protein levels and haemolymph composition on the sensitivity of the maxillary palp chemoreceptors of locusts. *Journal of Experimental Biology*, **135**, 215–29.

Barton Browne, L. (1995). Ontogenetic changes in feeding behavior. In *Physiological Regulation of Insect Feeding*, ed. R.F. Chapman & G.de Boer, pp. 307–42. New York: Chapman & Hall.

Beaver, R.A. (1989). Insect–fungus relationships in the bark and ambrosia beetles. *Symposium of the Royal Entomological Society of London*, **14**, 121–43.

Bentley, B.L. (1984). Nitrogen fixation in termites: fate of newly fixed nitrogen. *Journal of Insect Physiology*, **30**, 653–5.

Bernays, E.A. (1982). The insect on the plant – a closer look. In *Proceedings of the 5th International Symposium on Insect-Plant Relationships*, ed. J.H.Visser & A.K.Minks, pp. 3–17. Wageningen: Centre for Agricultural Publishing and Documentation.

Bernays, E.A. (1992). Plant sterols and host-plant affiliations of herbivores. In *Insect-Plant Interactions*, vol. 4, ed. E.A. Bernays, pp. 45–57. Boca Raton: CRC Press.

Bernays, E.A. (1995). Effects of experience on feeding. In *Physiological Regulation of Insect Feeding*, ed. R.F. Chapman & G.de Boer, pp. 279–306. New York: Chapman & Hall.

Bernays, E.A., Chamberlain, D.J. & Woodhead, S. (1983). Phenols as nutrients for a phytophagous insect *Anacridium melanorhodon*. *Journal of Insect Physiology*, **29**, 535–9.

Bernays, E.A. & Woodhead, S. (1984). The need for high levels of phenylalanine in the diet of *Schistocerca gregaria* nymphs. *Journal of Insect Physiology*, **30**, 489–93.

Bissett, J. & Borkent, A. (1988). Ambrosia galls: the significance of fungal nutrition in the evolution of the Cecidomyiidae (Diptera). In *Coevolution of fungi with plants and animals*, ed. K.A.Pirozynski & D.L.Hawksworth, pp. 203–25. London: Academic Press.

Boys, H.A. (1978). Food selection by *Oedaleus senegalensis* (Acrididae: Orthoptera) in grassland and millet fields. *Entomologia Experimentalis et Applicata*, **24**, 278–86.

Briegel, H. (1985). Mosquito reproduction: incomplete utilization of the blood meal protein for oogenesis. *Journal of Insect Physiology*, **31**, 15–21.

Broadway, R.M. & Duffey, S.S. (1988). The effect of plant protein quality on insect digestive physiology and the toxicity of plant proteinase inhibitors. *Journal of Insect Physiology*, **34**, 1111–7.

Brooks, M.A. (1963). Symbiosis and aposymbiosis in arthropods. *Symposium of the Society for General Microbiology*, **13**, 200–31.

Campbell, B.C. (1989). On the role of microbial symbiotes in herbivorous insects. In *Insect-Plant Interactions*, vol. 1, ed. E.A. Bernays, pp. 1–44. Boca Raton: CRC Press.

Campbell, B.C., Bragg, T.S. & Turner, C.E. (1992). Phylogeny of symbiotic bacteria of four weevil species (Coleoptera: Curculionidae) based on analysis of 16 S ribosomal RNA. *Insect Biochemistry and Molecular Biology*, **22**, 415–21.

Charnley, A.K., Hunt J. & Dillon, R.J. (1985). The germ-free culture of desert locusts, *Schistocerca gregaria*. *Journal of Insect Physiology*, **31**, 477–85.

Cherrett, J.M., Powell, R.J. & Stradling, D.J. (1989). The mutualism between leaf-cutting ants and their fungus. *Symposium of the Royal Entomological Society of London*, **14**, 93–120.

Clements, A.N. (1992). *The Biology of Mosquitoes*, vol. 1. London: Chapman and Hall.

Clifford, C.W. & Woodring, J.P. (1986). The effects of virginity and ovariectomy on growth, food consumption, fat body mass and oxygen consumption in the house cricket, *Acheta domesticus*. *Journal of Insect Physiology*, **32**, 425–31.

Cohen, R.W., Heydon, S.L., Waldbauer, G.P. & Friedman, S. (1987). Nutrient self-selection by the omnivorous cockroach, *Supella longipalpa*. *Journal of Insect Physiology*, **33**, 77–82.

Dadd, R.H. (1960). The nutritional requirements of locusts – III. Carbohydrate requirements and utilization. *Journal of Insect Physiology*, **5**, 301–16.

Dadd, R.H. (1977). Qualitative requirements and utilization of nutrients: insects. In *Handbook Series in Nutrition and Food*, section D, vol.1, *Nutritional Requirements*, ed. M.Rechcigl, pp. 305–46. Cleveland: C.R.C. Press.

Dadd, R.H. (1985). Nutrition: organisms. In *Comprehensive Insect Physiology, Biochemistry and Pharmacology*, vol. 4, ed. G.A. Kerkut & L.I. Gilbert, pp. 313–90. Oxford: Pergamon Press.

Dadd, R.H., Kleinjan, J.E. & Stanley-Samuelson, D.W. (1987). Polyunsaturated fatty acids of mosquitoes reared with single dietary polyunsaturates. *Insect Biochemistry*, 17, 7–16.

De Clerk, D. & De Loof, A. (1983) Effect of dietary ecdysterone on protein ingestion and copulatory behaviour of the blowfly, *Sarcophaga bullata*. *Physiological Entomology*, 8, 243–9.

Dethier, V.G. (1961). Behavioral aspects of protein ingestion by the blowfly *Phormia regina*. *Biological Bulletin of the Marine Biological Laboratory, Woods Hole*, 121, 456–70.

Douglas, A.E. (1988). On the source of sterols in the green peach aphid, *Myzus persicae*, reared on holidic diets. *Journal of Insect Physiology*, 34, 403–8.

Douglas, A.E. (1989). Mycetocyte symbiosis in insects. *Biological Reviews of the Cambridge Philosophical Society*, 64, 409–34.

Douglas, A.E. (1995). *Symbiotic Interactions*. Oxford: Oxford University Press.

Douglas, A.E. & Dixon, A.F.G. (1987). The mycetocyte symbiosis of aphids: variation with age and morph in virginoparae of *Megoura viciae* and *Acyrthosiphon pisum*. *Journal of Insect Physiology*, 33, 109–13.

Drew, R.A.I. & Lloyd, A.C. (1987). Relationship of fruit flies (Diptera: Tephritidae) and their bacteria to host plants. *Annals of the Entomological Society of America*, 80, 629–36.

Greene, E. (1989). A diet-induced developmental polymorphism in a caterpillar. *Science*, 243, 643–6.

Horie, Y. & Watanabe, K. (1983). Effect of various kinds of dietary protein and supplementation with limiting amino acids on growth, haemolymph components and uric acid excretion in the silkworm, *Bombyx mori*. *Journal of Insect Physiology*, 29, 187–99.

Howard, D.J. & Bush, G.L (1989). Influence of bacteria on larval survival and development in *Rhagoletis* (Diptera: Tephritidae). *Annals of the Entomological Society of America*, 82, 633–40.

Kaufman, M.G. & Klug, M.J. (1991). The contribution of hindgut bacteria to dietary carbohydrate utilization by crickets (Orthoptera: Gryllidae). *Comparative Biochemistry and Physiology*, 98A, 117–23.

Kramer, K.J. & Seib, P.A. (1982). Ascorbic acid and the growth and development of insects. In *Ascorbic Acid: Chemistry, Metabolism and Uses*, ed. P.A. Seib & B.M. Tolbert, pp. 275–91. American Chemical Society.

McFarlane, J.E. (1991). Dietary sodium, potassium and calcium requirements of the house cricket, *Acheta domesticus* (L.). *Comparative Biochemistry and Physiology*, 100A, 217–20.

McFarlane, J.E. (1992). Can ascorbic acid or β–carotene substitute for vitamin E in spermiogenesis in the house cricket (*Acheta domesticus*)? *Comparative Biochemistry and Physiology*, 103A, 179–81.

McGinnis, A.J. & Kasting, R. (1967). Dietary cellulose: effect on food consumption and growth of a grasshopper. *Canadian Journal of Zoology*, 45, 365–7.

Montgomery, M.E. (1982). Life-cycle nitrogen budget for the gypsy moth, *Lymantria dispar*, reared on artificial diet. *Journal of Insect Physiology*, 28, 437–42.

Nardon, P. & Grenier, A.M. (1988). Genetic and biochemical interactions between the host and its endosymbionts in the weevils *Sitophilus* (Coleoptera, Curculionidae) and other related species. In *Cell to Cell Signals in Plant, Animal and Microbial Symbioses*, vol. H17, ed. S. Scanneri, D. Smith, P. Bonfante-Fasolo & V. Gianinazzi-Pearson, pp. 255–70. Berlin: Springer-Verlag.

Nestel, D., Galun, R. & Friedman, S. (1985). Long-term regulation of sucrose intake by the adult Mediterranean fruit fly, *Ceratitis capitata* (Wiedemann). *Journal of Insect Physiology*, 31, 533–6.

O'Neill, S.L., Giordano, R., Colbert, A.M.E., Karr, T.L. & Robertson, H.M. (1992). 16 S rRNA phylogenetic analysis of the bacterial endosymbionts associated with cytoplasmic incompatibility in insects. *Proceedings of the National Academy of Sciences of the United States of America*, 89, 2699–702.

Prosser, W.A. & Douglas, A.E. (1992). A test of the hypothesis that nitrogen is upgraded and recycled in an aphid (*Acyrthosiphon pisum*) symbiosis. *Journal of Insect Physiology*, 38, 93–9.

Prosser, W.A., Simpson, S.J. & Douglas, A.E. (1992). How an aphid (*Acyrthosiphon pisum*) symbiosis responds to variation in dietary nitrogen. *Journal of Insect Physiology*, 38, 301–7.

Reinecke, J.P. (1985). Nutrition: artificial diets. In *Comprehensive Insect Physiology, Biochemistry and Pharmacology*, vol. 4, ed. G.A. Kerkut & L.I. Gilbert, pp. 391–419. Oxford: Pergamon Press.

Ridsdill-Smith, J. (1991). Competition in dung-breeding insects. In *Reproductive Behaviour of Insects*, ed. W.J.Bailey & J. Ridsdill-Smith, pp. 264–92. London: Chapman and Hall.

Sacchi, L., Grigolo, A., Mazzini, M., Bigliardi, B., Baccetti, B, & Laudani, U. (1988). Symbionts in the oocytes of *Blatella germanica* (L.) (Dictyoptera: Blattellidae): their mode of transmission. *International Journal of Insect Morphology and Embryology*, 17, 437–46.

Scriber, J.M. & Slansky, F. (1981). The nutritional ecology of immature insects. *Annual Review of Entomology*, 26, 183–211.

Simpson, S.J., Raubenheimer, D. & Chambers, P.G. (1995). The mechanisms of nutritional homeostasis. In *Physiological Regulation of Insect Feeding*, ed. R.F. Chapman & G.de Boer, pp. 251–78. New York: Chapman & Hall.

Simpson, S.J., Simmonds, M.S.J. & Blaney, W.M. (1988). A comparison of dietary selection behaviour in larval *Locusta migratoria* and *Spodoptera littoralis*. *Physiological Entomology*, **13**, 225–38.

Simpson, S.J. & Simpson, C.L. (1990). The mechanisms of nutritional compensation by phytophagous insects. In *Insect–Plant Interactions*, vol. 2, ed. E.A. Bernays, pp. 111–60. Boca Raton: CRC Press.

Stanley-Samuelson, D.W. (1993). The biological significance of prostaglandins and related eicosanoids in insects. In *Insect Lipids: Chemistry, Biochemistry and Biology*, ed. D.W. Stanley-Samuelson & D.R. Nelson, pp. 45–97. Lincoln: University of Nebraska Press.

Stanley-Samuelson, D.W. (1994). Assessing the significance of prostaglandins and other eicosanoids in insect physiology. *Journal of Insect Physiology*, **40**, 3–11.

Stanley-Samuelson, D.W., Jensen, E., Nickerson, K.W., Tiebel, K., Ogg, C.L. & Howard, R.W. (1991). Insect immune response to bacterial infection is mediated by eicosanoids. *Proceedings of the National Academy of Science*, **88**, 1064–8.

Stanley-Samuelson, D.W., O'Dell, T., Ogg, C.L. & Keens, M.A. (1992) Polyunsaturated fatty acid metabolism inferred from fatty acid compositions of the diets and tissues of the gypsy moth *Lymantria dispar*. *Comparative Biochemistry and Physiology*, **102A**, 173–8.

Stockhoff, B.A. (1993). Ontogenetic change in dietary selection for protein and lipid by gypsy moth larvae. *Journal of Insect Physiology*, **39**, 677–86.

Svoboda, J.A. & Thompson, M.J. (1985). Steroids. In *Comprehensive Insect Physiology, Biochemistry and Pharmacology*, vol. 10, ed. G.A. Kerkut & L.I. Gilbert, pp. 137–75. Oxford: Pergamon Press.

Trumper, S. & Simpson, S.J. (1993). Regulation of salt intake by nymphs of *Locusta migratoria*. *Journal of Insect Physiology*, **39**, 857–64.

Tu, Z. & Hagedorn, H.H. (1992). Purification and characterization of pyruvate carboxylase from the honey bee and some properties of related biotin-containing proteins in other insects. *Archives of Insect Biochemistry and Physiology*, **19**, 53–66.

van Loon, J.J.A. (1988). Sensory and nutritional effects of amino acids and phenolic plant compounds on the caterpillars of two *Pieris* species. Ph.D. Thesis, University of Wageningen.

Waldbauer, G.P. & Friedman, S. (1991). Self-selection of optimal diets by insects. *Annual Review of Entomology*, **36**, 43–63.

Werren, J.H. (1997). Biology of *Wolbachia*. *Annual Review of Entomology*, **42**, 587–609.

Werren, J.H., Windsor, D. & Guo, L. (1995). Distribution of *Wolbachia* among neotropical arthropods. *Proceedings of the Royal Society of London*, **262**, 197–204.

Wheeler G.S. & Slansky, F. (1991). Compensatory responses of the fall armyworm (*Spodoptera frugiperda*) when fed water- and cellulose-diluted diets. *Physiological Entomology*, **16**, 361–74.

Wood, T.G. & Thomas, R.J. (1989). The mutualistic association between Macrotermitinae and *Termitomyces*. *Symposium of the Royal Entomological Society of London*, **14**, 69–92.

Yarro, J.G. (1985). Effect of host plant on moulting in the African armyworm *Spodoptera exempta* (Walk.) (Lepidoptera: Noctuidae) at constant temperature and humidity conditions. *Insect Science and its Applications*, **6**, 171–5.

Zanotto, F.P., Simpson, S.J. & Raubenheimer, D. (1993). The regulation of growth by locusts through post-ingestive compensation for variation in the levels of dietary protein and carbohydrate. *Physiological Entomology*, **18**, 425–34.

5   Circulatory system, blood and immune systems

5.1 CIRCULATORY SYSTEM

5.1.1 Structure

Insects have an open blood system with the blood occupying the general body cavity, which is known as a hemocoel. Blood is circulated mainly by the activity of a contractile dorsal longitudinal vessel which opens into the hemocoel. The hemocoel is often divided into three major sinuses; a dorsal pericardial sinus, a perivisceral sinus, and a ventral perineural sinus (Fig. 5.1). The pericardial and perineural sinuses are separated from the visceral sinus by the dorsal and ventral diaphragms, respectively. In most insects, the visceral sinus occupies most of the body cavity, but in ichneumonids the perineural sinus is enlarged.

Reviews: Hoffman, 1976; Jones, 1977; Miller, 1985a

5.1.1.1 *Dorsal vessel*

The dorsal vessel runs along the dorsal midline, just below the terga, for almost the whole length of the body although in the thorax of adult Lepidoptera and at least some Hymenoptera, it loops down between the longitudinal flight muscles (see Fig. 19.3). It may be bound to the dorsal body wall or suspended from it by elastic filaments. Anteriorly it leaves the dorsal wall and is more closely associated with the alimentary canal, passing under the brain just above the esophagus. It is open anteriorly, ending abruptly in most insects, but as an open gutter in

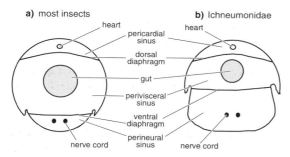

Fig. 5.1. Main sinuses of the hemocoel shown in diagrammatic cross-sections (after Richards, 1963).

orthopteroids. Posteriorly, it is closed, except in larval mayflies (Ephemeroptera) where three vessels diverge to the caudal filaments from the end of the heart. In the honeybee, *Apis*, it forms a spring-like coil in the region of the petiole. The dorsal vessel is divided into two regions: a posterior heart in which the wall of the vessel is perforated by incurrent and sometimes also by excurrent openings (ostia), and an anterior aorta which is a simple, unperforated tube (Fig. 5.2).

The heart is often restricted to the abdomen, but may extend as far forwards as the prothorax in cockroaches (Blattodea). In orthopteroids it has a chambered appearance due to the fact that it is slightly enlarged into ampullae at the points where the ostia pierce the wall. These ampullae are often more prominent in the thorax. In the larvae of dragonflies (Odonata) and the cranefly, *Tipula*, the heart is divided into chambers by valves in front of each pair of incurrent ostia and in some other cases, as in *Cloeon* (Ephemeroptera) larvae, the ostial valves themselves are so long that they meet across the lumen.

The wall of the dorsal vessel is contractile and usually consists of one or two layers of muscle cells with a circular or spiral arrangement. Longitudinal muscle strands are also present, in Heteroptera, inserting into the wall of the vessel anteriorly and posteriorly; they do not connect with other tissues. The muscles of the heart are sometimes oriented in many different directions, but this appearance may arise through the insertion of the alary muscles into the heart (see below) (Chiang, Chiang & Davey, 1990). Muscles of the dorsal vessel have short sarcomere lengths, with A-bands about 2 μm long. Each thick filament is surrounded by 9–12 thin filaments, as commonly found in visceral muscles of insects. The muscles are ensheathed inside and out by a basal lamina which also covers the nephrocytes (section 18.6).

Incurrent ostia The incurrent ostia are vertical, slit-like openings in the lateral wall of the heart. The maximum number found in any insect is 12 pairs, nine abdominal and three thoracic. All 12 pairs are present in Blattodea and

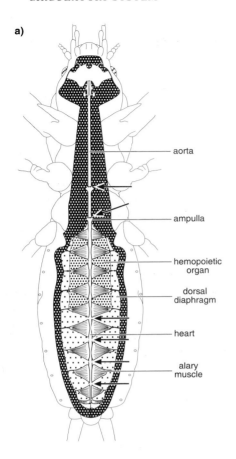

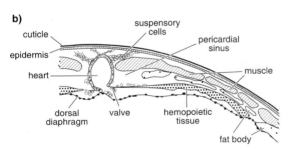

Fig. 5.2. Dorsal vessel and hemopoietic organs of the mole cricket, *Gryllotalpa* (after Nutting, 1951). **(a)** Ventral dissection. The dorsal diaphragm is continuous over the ventral wall of the heart, but is omitted from the drawing for clarity. Arrows show positions of incurrent ostia. **(b)** Transverse section through the pericardial sinus in the abdomen. Pericardial cells form supporting elements (suspensory cells) for the heart dorsally.

Orthoptera; many Lepidoptera have seven or eight. In aculeate Hymenoptera (bees, wasps and ants) there are only five pairs, while the housefly, *Musca*, has only four. Lice (Phthiraptera) and many Heteroptera have only two or three pairs and the heart is restricted to the posterior abdominal segments.

The anterior and posterior lips of each ostium are reflexed into the heart to form a valve which permits the flow of blood into the heart at diastole, but prevents its outward passage at systole (Fig. 5.3a). During diastole (expansion of the heart) the lips are forced apart by the inflowing blood. When diastole is complete the lips are forced together by the pressure of blood in the heart and they remain closed throughout systole (contraction of the heart). Towards the end of systole in the larva of the phantom midge, *Chaoborus*, the valves tend to be evaginated by the pressure, but they are prevented from completely everting by a unicellular thread attached to the inside of the heart. In the silkworm, *Bombyx*, only the hind lip of each ostium is extended as a flap within the heart (Fig.5.3b). During systole this is pressed against the wall of the heart and prevents the escape of blood. However, when the heartbeat is reversed (see below) blood flows out of the 'incurrent' ostia.

Excurrent ostia Excurrent ostia are present in the hearts of Thysanura (silverfish), Orthoptera, Plecoptera (stoneflies) and Embioptera. In the last two orders they are unpaired, but in Orthoptera they are paired ventro-lateral openings in the wall of the heart without any internal valves. Their number varies, but grasshoppers have five abdominal and two thoracic pairs. Externally, each opening is surrounded by a papilla of spongiform multinucleate cells which expands during systole, so that hemolymph is forced out, and contracts during diastole, so that entry of blood is prevented. The papillae surrounding excurrent ostia of grasshoppers penetrate the dorsal diaphragm so that the ostia discharge into the perivisceral sinus (Fig. 5.4), but in Phasmatodea (stick insects) the ostia open into the pericardial sinus and in tettigoniids the openings are between two layers of the dorsal diaphragm. This has the effect of channeling the blood laterally before it enters the general body cavity.

Segmental vessels Most Blattodea and Mantodea have no excurrent ostia, but the blood leaves the heart via segmental vessels that extend laterally (Fig. 5.5). *Periplaneta* has five pairs of abdominal segmental vessels in late stage

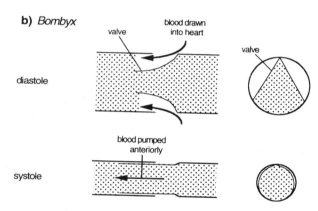

Fig. 5.3. Functioning of incurrent ostia. **(a)** In the larva of *Chaoborus*. The valves are prevented from opening outwards at systole by a unicellular thread (not shown) attached to the inside of the heart (after Wigglesworth, 1972). **(b)** In the larva of *Bombyx*. The heart is shown in horizontal (left) and transverse (right) sections.

nymphs; adults have two thoracic pairs in addition. Mantids have only abdominal vessels. At the origin of each vessel from the heart, there is a muscular valve which permits only the outward flow of blood. The walls of the vessels are non-muscular.

Innervation The dorsal vessel of some insects, such as the adult mosquito, *Anopheles*, lacks any innervation, although there are segmental nerves to the alary muscles. In most insect species, however, the heart is innervated by nerves running round the body wall from the segmental ganglia. In Odonata, Blattodea, Phasmatodea, Orthoptera, larval Lepidoptera and some adult Coleoptera, branches of the segmental nerves combine to form a lateral cardiac nerve running along each side of the heart. In *Locusta*, for example, groups of neurons with cell bodies in the midline in each abdominal ganglion send axons to the heart (Ferber & Pflüger, 1990). In addition, a pair of neurosecretory cells in the subesophageal ganglion each sends one axon forwards into the circumesophageal connective and

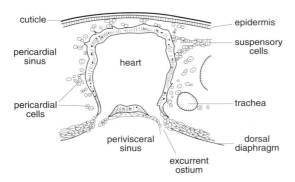

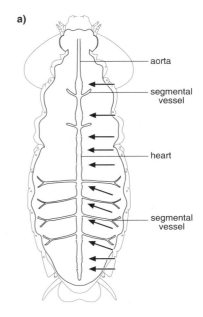

Fig. 5.4. Excurrent ostia opening directly to the perivisceral sinus. Transverse section of the heart of the grasshopper, *Taeniopoda* (after Nutting, 1951).

another back along the length of the ventral nerve cord. This axon branches in each of the abdominal ganglia sending a branch into a lateral nerve which extends dorsally to the heart. These branches contribute to the lateral cardiac nerve and have varicose terminals, typical of a neurosecretory cell, along the heart. The cells produce a FMRFamide-like peptide (Fig. 20.6) (Bräunig, 1991). The lateral cardiac nerves also receive innervation from neurons whose somata are not in the central nervous system, but lie adjacent to the heart itself. They are called cardiac neurons. *Periplaneta* has about 32 such neurons, some of which are neurosecretory.

In a majority of holometabolous insects the segmental nerves extend to the heart, but do not form lateral cardiac nerves.

5.1.1.2 *Alary muscles and dorsal diaphragm*

The dorsal diaphragm is a fenestrated connective tissue membrane. It is usually incomplete laterally, so that the pericardial sinus above it is broadly continuous with the perivisceral sinus below. The lateral limits of the diaphragm are indefinite and are determined by the presence of muscles, tracheae or the origins of the alary (or aliform) muscles which form an integral part of the diaphragm. The alary muscles stretch from one side of the body to the other just below the heart or, as in the cecropia moth, *Hyalophora*, are directly connected to the heart muscles by intercalated discs. They usually fan out from a restricted point of origin on the tergum and meet in a broad zone in the midline (Fig. 5.2a), but sometimes, as in grasshoppers, the origin of the muscles is also broad. In most orthopteroids, at least, only the part near the point of

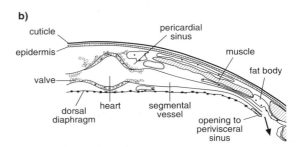

Fig. 5.5. Dorsal and segmental vessels of the cockroach, *Blaberus* (after Nutting, 1951). **(a)** Ventral dissection. The dorsal diaphragm is continuous over the ventral wall of the heart, but is omitted from the drawing for clarity. Arrows show positions of incurrent ostia. **(b)** Transverse section through the pericardial sinus showing a segmental vessel arising from the heart and discharging into the hemolymph amongst the fat body laterally.

origin is contractile, the rest, and greater part, being made up of bundles of connective tissue which branch and anastomose. Some of the connective tissue fibers form a plexus which extends to the heart wall. Orthopteroids may have as many as ten abdominal and two thoracic pairs of alary muscles, but in other insects the number is reduced. Most terrestrial Heteroptera, for instance, have from four to seven pairs. The alary muscles are visceral muscles with 10–12 thin filaments to every thick filament.

5.1.1.3 *Ventral diaphragm*

The ventral diaphragm is a horizontal septum just above the nerve cord cutting off the perineural sinus from the main perivisceral sinus (Fig. 5.1). It is present in both larvae and adults of Odonata, Blattodea, Orthoptera, Neuroptera (lacewings) and Hymenoptera, but is only found in adults of Mecoptera (scorpion flies), Lepidoptera and nematocerous Diptera. In Lepidoptera, it is unusual in having the nerve cord attached to its ventral surface by connective tissue. When present, the diaphragm is usually restricted to the abdomen, but it extends into the thorax in some grasshoppers and Hymenoptera.

The structure of the diaphragm varies. For instance, in the thorax of grasshoppers it is a delicate membrane with little or no muscle, but in the abdomen it becomes a solid muscular sheet. Laterally it is attached to the sternites, usually at one point in each segment and so there are broad gaps along the margins where perivisceral and perineural sinuses are continuous. Its structure may vary with developmental stage. For example, in *Corydalis* (Neuroptera) it forms a solid sheet in the larva, but a fenestrated membrane in the adult.

Review: Richards, 1963

5.1.1.4 *Accessory pulsatile organs*

In addition to the dorsal vessel, insects have other pulsating structures that maintain circulation through the appendages. A pulsatile organ drawing blood from the wings is present in both wing-bearing segments of most adult insects, but only in the mesothorax of Diptera and Coleoptera Polyphaga. A blood space, or reservoir, beneath the posterior part of the tergum (scutellum) which is largely or completely isolated from the remaining hemocoel of the thorax connects with the posterior veins of the wing via the axillary cord of the wing (see Fig. 5.8). The ventral wall of the reservoir forms a muscular pump. It may be derived from the heart or it may be a separate structure.

In hemimetabolous groups, other than Hemiptera, and in Coleoptera and Hymenoptera Symphyta, the wing pulsatile organ is an expansion or diverticulum of the dorsal vessel with a pair of incurrent ostia opening from the subscutellar reservoir. Odonata, for example, have, in each pterothoracic segment, an ampulla which connects with the aorta by a narrow vessel (Fig. 5.6d). It is suspended from the tergum by elastic ligaments, and its dorsal wall is muscular. When the muscles contract, the ampulla is compressed and blood is driven into the aorta. At the same time, the volume of the subscutellar reservoir is increased and blood is drawn from the wings. When the muscles relax, the elastic ligaments restore the shape of the ampulla so that blood is sucked into it from the reservoir.

Most holometabolous insects, as well as Hemiptera, have wing pulsatile organs in which a muscular diaphragm, independent of the heart, bounds the subscutellar reservoir on the ventral side. It is suspended from the scutellum by a number of filamentous strands (Fig. 5.6a–c). Contraction of the muscles, which are innervated from the ventral nerve cord, causes the diaphragm to flatten, drawing blood from the wings. Relaxation of the muscles is associated with an upward movement of the diaphragm, presumably due to the elasticity of the suspensory strands, and blood is forced into the body cavity (Fig. 5.6b,c).

A pulsatile organ is also found at the base of each antenna. It consists of an ampulla from which a fine tube extends almost to the tip of the antenna. The ampullae of Thysanura (silverfish), Archaeognatha (bristletails) and some Plecoptera have no muscles and it is presumed that in these insects the ampulla serves simply to direct the flow of hemolymph from the opening of the aorta into the antenna. In most insects, however, the ampullae have dilator muscles. Compression, which drives hemolymph into the antenna, results from the activity of elastic filaments on both sides of the wall of the ampulla. Only Dermaptera (earwigs) have compressor muscles. *Periplaneta* has a single muscle connecting the two ampullae so that when it contracts both ampullae are dilated and hemolymph flows into each one through an ostium (Fig. 5.6e–g). The contractions are myogenic in origin, but may be modulated by neural input since nerve endings do occur on the muscle. However, most of the nerve endings are concentrated in the ampulla which appears to function as a neurohemal organ. The axons originate in the subesophageal ganglion from a dorsal unpaired median neuron (section 10.3.2.4) and from a pair of somata placed laterally. Octopamine, probably from the dorsal unpaired median neuron, is present in the neurohemal area, but its function relative to the ampulla of the antenna is not known. In Lepidoptera, the aorta ends anteriorly in a sac from which the antennal vessels arise (see Fig. 5.7d).

Most insects have a longitudinal septum in the legs which divides the lumen into two sinuses and permits a bidirectional flow of blood within the leg (see below). In Hemiptera, the septum twists through 90° at the proximal

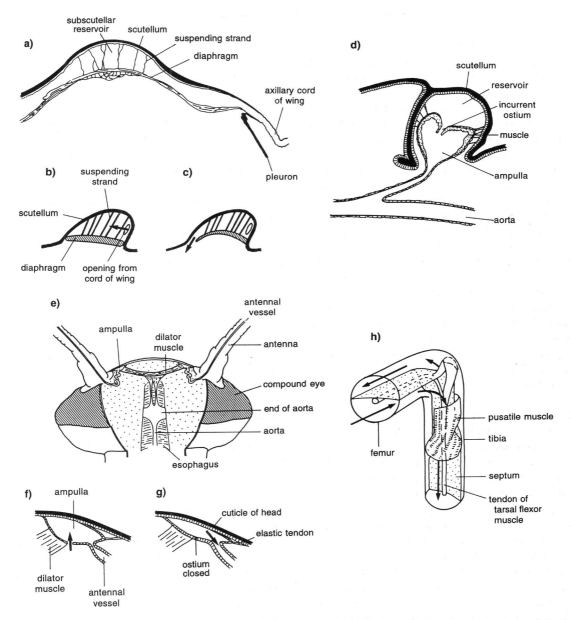

Fig. 5.6. Accessory pulsatile organs. In all diagrams, arrows indicate the direction of blood flow. **(a)–(c)** Wing heart not connected to the aorta, such as occurs in most holometabolous insects. **(a)** Transverse section through the thorax showing the connection of the subscutellar reservoir to the axillary cord of the wing. **(b)**, **(c)** Diagrammatic longitudinal section, anterior to left. When muscles of the diaphragm contract **(b)**, the diaphragm is flattened and blood is drawn in from the wing. When the muscles relax **(c)**, elastic suspensory elements draw the diaphragm up and force blood out anteriorly (after Krenn & Pass, 1993). **(d)** Wing heart connected to the aorta as in most hemimetabolous insects. Longitudinal section of the thorax. The subscutellar reservoir connects with the axillary cord of the wing on either side (modified after Whedon, 1938). **(e)–(g)** Antennal pulsatile organ of cockroach. **(e)** General arrangement as seen from above. Top of head cut away and brain removed. **(f)** Dilator muscle contracts, enlarging lumen of ampulla so that blood is drawn in from the hemocoel in the head. Lowered pressure causes constriction at the origin of the antennal vessel so that the backflow of blood from the antenna is restricted. **(g)** the muscle relaxes and the ampulla is flattened by the elasticity of its inner wall and the pull of the tendon. The ostium is closed by a valve and blood forced into the antennal vessel (after Pass, 1985). **(h)** Leg pulsatile organ of *Triatoma*. Contraction of the muscle compresses the blood sinus on one side of the septum and enlarges that on the other so that blood flows down the leg on one side of the septum and up on the other (after Kaufman & Davey, 1971).

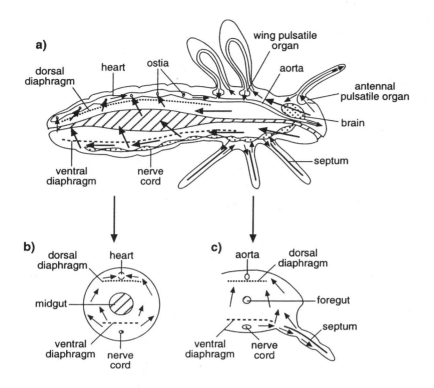

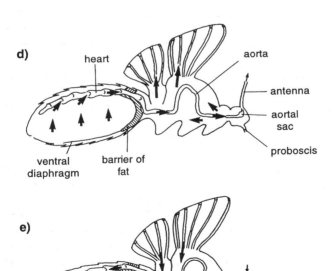

Fig. 5.7. Blood circulation. **(a)–(c)** In an insect with a fully developed circulatory system. Arrows indicate the course of the circulation. **(a)** Longitudinal section; **(b)** Transverse section of abdomen; **(c)** Transverse section of thorax. **(d),(e)** In an insect in which the blood oscillates between the thorax and abdomen. **(d)** Abdominal contraction with the heart beating forwards pushes blood into the anterior regions of the insect; **(e)** Abdominal expansion with the heart beating backwards draws blood into the abdomen (after Wasserthal, 1980).

end of the tibia, and there is a muscle at this point (Fig. 5.6h). When the muscle contracts, it compresses one sinus, forcing hemolymph into the thorax, and at the same time enlarges the other so that hemolymph is drawn into the leg. Its activity is probably myogenic although it may also be subject to neural modulation (Hantschk, 1991).

Some insects have very long cerci in which the maintenance of blood flow might require some special feature. Ephemeroptera have small, non-contractile vessels extending from the posterior end of the heart into the cerci. Plecoptera also have cercal blood vessels, but these do not connect with the heart. They open directly into the perivisceral cavity, but the remainder of the cercal lumen connects with the perivisceral cavity via the lumen of the paraproct. Changes in the volume of the paraproct produced by a small muscle draw blood from the outer lumen of the cercus and pump it into the perivisceral cavity. This flow draws blood into the cercus through the central vessel (Pass, 1987).

The muscle fibers in these different types of pulsatile organs generally have the characteristics of slow-contracting muscles.

Reviews: Krenn & Pass, 1994, 1995 – wing pulsatile organs; Pass, 1991 – antennal pulsatile organs

5.1.2 Circulation

5.1.2.1 *Movement of the blood*

During normal circulation of hemimetabolous and larval holometabolous insects, the blood is pumped forwards through the heart at systole, entering the perivisceral sinus through the anterior opening of the aorta in the head and through the excurrent ostia where these exist. The valves on the incurrent ostia prevent the escape of blood through these openings. The force of blood leaving the aorta anteriorly tends to push blood backwards in the perivisceral sinus. The backwards flow is aided by the movements of the dorsal diaphragm and by the inflow of blood into the heart, through the incurrent ostia, at diastole (Fig. 5.7a–c). Movements of the ventral diaphragm presumably help to maintain the supply of blood to the ventral nerve cord.

In adult Lepidoptera, Coleoptera and Diptera, and perhaps in some other insects, the blood is shunted backwards and forwards between the thorax and abdomen, rather than circulated. This is possible because the hemocoel in the two parts is separated by a moveable flap of fatty tissue, in Lepidoptera, or large airsacs, in cyclorrhaphous

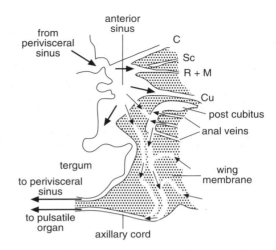

Fig. 5.8. Blood circulation in the base of the forewing of *Blattella*. Areas in which the two membranes of the wing are fused together are shaded. Well-defined channels, such as veins, are outlined by a solid line, less definite channels have no lines. Axillary sclerites are omitted (after Clare & Tauber, 1942).

Diptera, and because the heartbeat exhibits periodic reversals. As the abdomen contracts, the heart pumps blood forwards into the head (Fig. 5.7d). When the heart reverses, the abdomen actively expands, drawing blood past the barrier of fat (Fig. 5.7e). Movements of the hemolymph are coordinated with ventilatory movements (Wasserthal, 1982a). Any activity which tends to induce pressure differences between different parts of the body must affect the circulation, and, especially in adult Lepidoptera, Coleoptera and Diptera, there is a close functional link between ventilatory and hemolymph movements (see Figs. 5.10, 17.15).

Many insects appear to have a well-defined, but variable circulation through the wings, although in some, apparently, circulation only occurs in the young adult. In the absence of this wing circulation, the tracheae in the wings collapse, and the wing structure becomes brittle and dry. In most cases, blood is drawn out of the wings from the axillary cords by the thoracic pulsatile organs. In the German cockroach, *Blattella*, the anal veins connect with the axillary cord through channels between the upper and lower wing membranes. As blood is drawn out posteriorly, it is drawn in anteriorly from a blood sinus from which the major anterior veins arise (Fig. 5.8).

A different mechanism has been described in a number of Lepidoptera. Here, the blood enters the wings along all the veins (Fig. 5.9) while the heart is beating forwards and

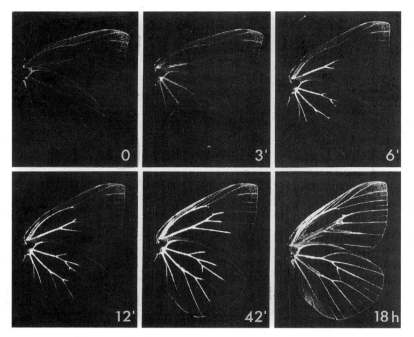

Fig. 5.9. Movement of dye into the wings of the cabbage butterfly, *Pieris*. Dye was injected into the abdomen at time 0. Notice that the dye moves out along all the veins simultaneously. Blood is pumped in and out of the veins as shown in Fig. 5.10, but the dye diffuses slowly outwards. Solutes normally present in the blood will diffuse in a similar manner (after Wasserthal, 1983).

ebbs out again during heartbeat reversal. The wing pulsatile organs are active when the heart is beating backwards (Fig. 5.10), so blood is drawn out of the wings and pushed back into the abdomen. The movement of blood out of the veins causes the tracheae inside them to expand so that air is drawn into the wing. When the pulsatile organ stops and the heart starts beating forwards again, the negative pressure on the wing tracheae is removed and their elasticity causes them to contract. This produces a negative pressure in the hemolymph space outside the trachea and blood flows in. This is only possible because of the specialized nature of the taenidia in the tracheae of the wings. The taenidial thickenings of typical tracheae function specifically to prevent such changes in volume as occur in the tracheae of the wings, but in the giant silk moth, *Attacus*, and presumably in other Lepidoptera, the taenidia of the wing tracheae are coiled along their length like a spring (Fig. 5.10c,d). Extension of the trachea induced by negative pressure outside stretches the taenidia. When the negative pressure is removed, their spring-like properties cause them to resume their relaxed position, reducing the volume of the trachea (Wasserthal, 1982a). A similar system involving the reciprocal movement of blood and air into and from the wings probably occurs in Coleoptera, and possibly in other groups.

In insects lacking leg pulsatile organs, the flow of blood through the legs is thought to be maintained by pressure differences at the base. Blood passes into the posterior compartment of the legs from the perineural sinus and out into the perivisceral sinus from the anterior compartment. In Lepidoptera, blood flow in the legs is dependent on the elasticity of the tracheae as described above for the wings (Wasserthal, 1982b).

The efficiency of this system, with respect to the rate at which substances are translocated round the body, may vary with the physiological state of the insect. Fluorescein injected into the abdominal hemocoel of *Periplaneta* can reach the pulvilli in four to eight minutes. Radiolabel from labelled sucrose was detected in the labial palps of *Locusta* within five minutes of being eaten (including digestion and absorption). Longer periods may be required for injected substances to be uniformly distributed through the hemolymph, and this is affected by temperature. At 22–25 °C, labelled material was uniformly distributed in the hemolymph of the honeybee, *Apis*, within five minutes; at 12–14 °C it required more than 15 minutes (Crailsheim, 1985). Some reports indicate that injected dyes may take more than an hour to become uniformly distributed through the hemolymph. It is possible that, in these cases, the circulation was disrupted by wounding.

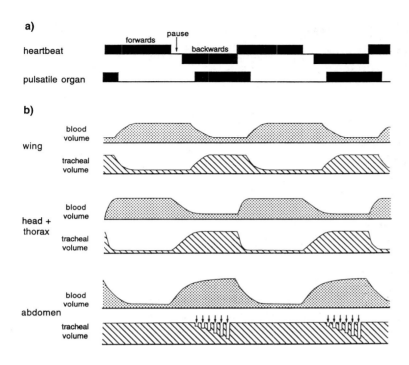

a)

heartbeat

pulsatile organ

b)

wing — blood volume / tracheal volume

head + thorax — blood volume / tracheal volume

abdomen — blood volume / tracheal volume

Fig. 5.10. Blood flow in the wings of a moth, *Attacus* (after Wasserthal, 1982a,b). **(a)** Activity of the heart and thoracic pulsatile organ, which draws blood from the wings. **(b)** Changes in the volumes of the blood spaces and tracheae in the wings, head plus thorax, and abdomen. Abdominal tracheal ventilation (indicated by arrows) is superimposed on the overall volume changes. **(c)** Oblique sections of a wing vein showing the changes in volume of the hemocoel and trachea. On the left, the trachea is of normal size with the taenidia relaxed; on the right the trachea is expanded and the taenidia stretched. **(d)** Photographs of the inside of a wing trachea showing the taenidia relaxed (left) and stretched (right). The trachea runs across the page. Notice how the extensive folding between the taenidia is stretched out when the trachea increases in diameter (right).

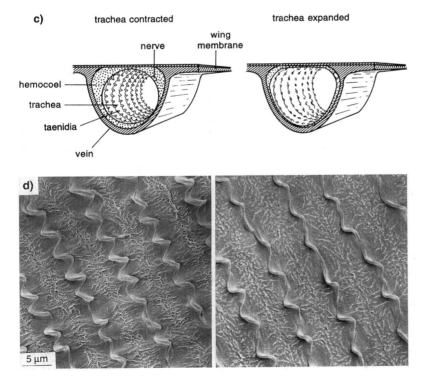

c)   trachea contracted   trachea expanded

d)

5 µm

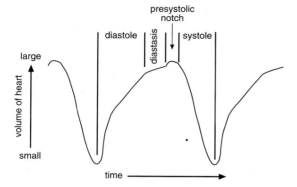

Fig. 5.11. Changes in the volume of a section of the heart during beating. The presystolic notch results from an increase in hydrostatic pressure, producing a slight increase in volume within this part of the heart due to the start of systole in the more posterior segments.

5.1.2.2 *Heartbeat*

Systole, the contraction phase of the heartbeat, results from the contractions of the intrinsic muscles of the heart wall. In a majority of insects this activity begins posteriorly and spreads forwards as a wave. In *Periplaneta* and Orthoptera, however, a synchronous contraction of the heart occurs along its whole length. This type of action is probably associated with the presence of excurrent ostia or segmental vessels. Even in these insects, peristaltic heart contractions may sometimes occur. Diastole, the expansion phase when blood enters the heart, results from relaxation of the heart muscles assisted by the elastic filaments that support the heart. In general, the alary muscles are not responsible for diastole and often contract at a lower frequency than the heart. In *Rhodnius*, however, complete diastole is dependent on the alary muscles (Chiang *et al.*, 1990). After diastole there is a third phase in the heart cycle known as diastasis in which the heart remains in the expanded state (Fig. 5.11). Increases in the frequency of the heartbeat result from reductions in the period of diastasis.

Rate of heartbeat The frequency with which the heart contracts varies considerably both within and between species. In general, the frequency of beating is higher in early than in later stage larvae. It also depends on the age within a stage of development. For example, in silkworms (*Bombyx*), the rate drops from 80 beats per minute in second stage larvae to about 50 in the fifth stage, and it drops sharply just before each molt except

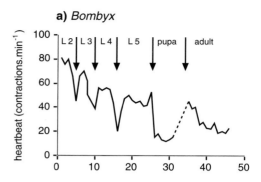

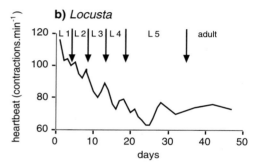

Fig. 5.12. The frequency of heartbeat at different stages of development. L1–L5 indicate larval stages; arrows indicate ecdysis. (a) In *Bombyx* (after Masera, 1933). (b) In *Locusta* (after Roussel, 1972).

the last (Fig. 5.12a). In the pupa, the heartbeat falls to 10–20 per minute and remains at this low level until shortly before adult eclosion. A similar, but slightly less marked change occurs during the development of the migratory locust, *Locusta*, but the rate of beating tends to rise before a molt (Fig. 5.12b). In Lepidoptera, a very sharp increase in the heart rate occurs at eclosion, falling to a sustained but moderate rate when eclosion is complete (Fig. 5.13).

In addition to these intrinsic changes, the rate of heartbeat is affected by environmental factors. Activity usually stops above 45–50 °C and below 1–5 °C. Within this range the rate is higher at higher temperatures. In *Locusta*, the rate of heartbeat is also higher in the light than in the dark.

It is common for the heart to stop beating, sometimes for a few seconds, but sometimes for 30 seconds or more. The heart of the young pupa of *Anopheles* sometimes stops beating altogether, and in old pupae no activity of the heart is observed.

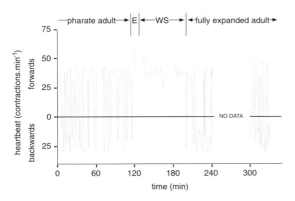

Fig. 5.13. Changes in the heartbeat of an individual *Manduca* at the time of eclosion. E = eclosion, WS = wing spreading (after Tublitz & Truman, 1985).

It is also common for the heartbeat to undergo periodic reversals, with waves of contraction starting at the front. When this occurs, blood is forced out of the 'incurrent' ostia and, at least in the mole cricket, *Gryllotalpa*, powerful currents pass out of the subterminal incurrent ostia. In female *Anopheles*, 31% of heartbeats start at the front end of the heart. Reversal of heartbeat is rare in holometabolous larvae, but begin in the pupal stage or even at the larva–pupa ecdysis. Usually the rate of heartbeat is lower when the heart is pumping backwards. In the adult blowfly, *Calliphora*, the rates are about 175 beats per minute backwards compared with about 375 forwards. In pharate adult Lepidoptera, periods of fast forward beating lasting a few minutes alternate with periods when the heart reverses and beats more slowly, but during the period of wing expansion following eclosion no reversals occur (Fig. 5.13) (Tublitz & Truman, 1985). Subsequently, periodic reversals are an essential feature of hemolymph circulation, at least in the Lepidoptera (see above).

The activity of the pulsatile organs may be different from that of the heart. The antennal ampullae of *Periplaneta* pulse at about 28 beats per minute, considerably slower than the heart rate. The wing pulsatile organs of Lepidoptera are only active during heart reversal (Fig. 5.10) when the mesothoracic organ and the heart pulsate at similar rates, although they are not necessarily in phase; the metathoracic organ pulsates more rapidly. At eclosion, the pulsatile organs pulsate more rapidly and without interruption.

Control of heartbeat The activity of the heart is basically myogenic although the myogenic pattern may be modulated neurally or hormonally. As the segmental nerves leading to the heart contain the ramifying terminals of neurosecretory axons, it might be expected that their secretions exert modulatory effects (although it is possible that the heart also functions as a neurohemal organ). In addition, hormones released into the blood at points remote from the heart are known to affect it. For example, the increase in the rate of beating at the time of eclosion in Lepidoptera is at least partly due to peptides released in the hemolymph at this time. These cardioacceleratory peptides are produced in neurosecretory cells in the ganglia of the ventral nerve cord and released into the hemolymph at the perivisceral neurohemal organs (Fig. 21.7). The same peptides, also increase the heartbeat during flight. Their effects are synergized by very low levels of octopamine which is also present in the hemolymph during wing inflation and flight (Prier, Beckman & Tublitz, 1994). A number of other neurohormones are known to affect the heart rate *in vitro*, but they are not known to be involved in regulation of heartbeat in the intact insect.

The direction of a beat, from back to front or *vice versa*, may be related to the distribution of blood pressures. If pressure at the front of the heart is so high that back pressure is set up, the heartbeat is reversed (see Wasserthal, 1981). The direction of beat after transection of the heart adds support to this suggestion. The prevalence of a reversed beat in pupal stages possibly results from blockage of excurrent ostia by histolysed tissues. In *Anopheles*, the direction of heartbeat is sometimes correlated with abdominal ventilation. If ventilation starts posteriorly, the heart beats forwards; if ventilation starts anteriorly, the heart beats from front to back. These changes might well be due to pressure.

Alternatively, or additionally, the direction of heartbeat might be related to the availability of oxygen. In the absence of a good oxygen supply, the rate of heartbeat is strongly reduced. The larva of *Bombyx* has a better tracheal supply to the posterior end of the heart than to the anterior and reversals of the heartbeat do not normally occur. If, however, the posterior spiracles are occluded, so that the oxygen supply to the posterior part of the heart is reduced, the direction of heartbeat is reversed. In the pupa, the tracheal system of the whole heart is poor and the rate of beating is low with reversals, while in the adult, the tracheal supply is well-developed both anteriorly and posteriorly and the heartbeat is rapid, again with reversals.

Review: Miller, 1985b – pharmacology

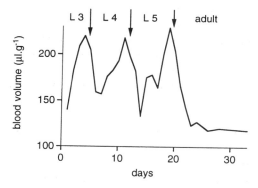

Fig. 5.14. Changes in the blood volume (expressed as volume per unit weight) during the development of *Schistocerca*. L3, L4 and L5 refer to successive larval stages. Arrows indicate the time of ecdysis (after Lee, 1961).

5.2 HEMOLYMPH

The blood, or hemolymph, circulates round the body, bathing the tissues directly. It consists of a fluid plasma in which blood cells, hemocytes, are suspended. The plasma, because of its function of maintaining the tissues throughout the body, contains many chemicals.

5.2.1 Hemolymph volume

The hemolymph volume, expressed as a percentage of the total body weight of the insect, varies with the type of insect. In the heavily sclerotized tenebrionid beetle, *Onymachus*, blood constitutes about 11% of the beetle's total mass; in mid-stadium larvae of *Locusta*, the figure is about 18%, while in mature adults it is about 12%; in cockroaches it is about 17% and in caterpillars 35–40%.

Hemolymph water comprises 20–25% of the total body water in adult insects, but in caterpillars, the figure is close to 50%. This reflects the important hydrostatic function of the hemolymph in these larval forms. This role is further evidenced in other insects at the time of the molt where an increase in hemolymph volume occurs before each ecdysis. Before each molt of the desert locust, *Schistocerca*, the relative blood volume almost doubles, and is then reduced again after the molt (Fig. 5.14). In *Periplaneta*, the pre-molt increase in volume is associated with an increase in activity of an antidiuretic hormone and the post-ecdysial fall in volume is produced by a transient rise in the hemolymph titer of diuretic hormone (Fig. 5.15). Diuresis reduces the hemolymph volume of the

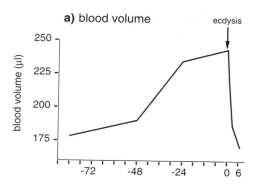

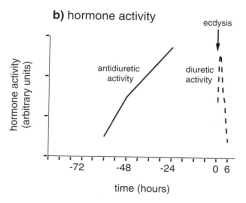

Fig. 5.15. Regulation of blood volume in relation to ecdysis in *Periplaneta* (after Mills & Whitehead, 1970). (a) Changes in blood volume. (b) Antidiuretic and diuretic activity of the hemolymph, suggesting that changes in hormonal activity regulate the changes in volume.

cabbage butterfly, *Pieris*, by about 70% in the hours following eclosion (Nicolson, 1980a).

Hemolymph volume is also affected by other factors. In *Locusta*, the hemolymph volume falls following feeding, apparently because water moves into the gut. Volume changes also result from desiccation. In the desert tenebrionid, *Onymacris*, hemolymph volume is reduced by about 60% after 12 days in desiccating conditions (Fig. 5.16c). Similar changes occur in even shorter periods under extreme environmental conditions. It is almost impossible to obtain blood samples from red locusts, *Nomadacris*, collected in the heat of the afternoon, but the insects have ample blood in the morning and late evening.

An important function of the hemolymph is to provide

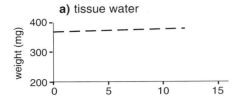

a) tissue water

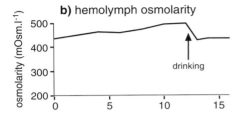

b) hemolymph osmolarity

drinking

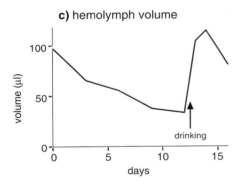

c) hemolymph volume

drinking

days

Fig. 5.16. Regulation of tissue water and hemolymph osmotic pressure in *Onymacris*. The insects were without food or water at 26 °C and 10–15% relative humidity for the first 12 days. On day 12 they were given distilled water to drink and then maintained at 50–60% relative humidity (after Nicolson, 1980b). (a) Tissue water remained almost constant over the 12 days without food or water. (b) Hemolymph osmolarity rose very slightly during the first 12 days concentration despite the marked reduction in hemolymph volume (c).

a reservoir of water to sustain the levels of water in the tissues. Thus, the tissue water of *Onymachus* does not change as a result of desiccation even though the blood volume falls drastically (Fig. 5.16). Some of the tissue water is drawn from the blood; the remainder comes from fat metabolism (Nicolson, 1980b). Body water is normally maintained from water in the food, but under extreme conditions many insects will drink (section 18.4.1.2).

Review: Reynolds, 1980 – blood volume at ecdysis

5.2.2 Constituents of the plasma

5.2.2.1 *Inorganic constituents*

Chloride is the most abundant inorganic anion in insect blood (Table 5.1). It is present in high concentrations in Apterygota and hemimetabolous insects, but is characteristically low in holometabolous insects, usually amounting to less than 10% of the total osmolar concentration (see Fig. 5.23). Other inorganic anions present are carbonate and phosphate, but these are rarely found in any quantity. Phosphates are, however, important in *Carausius* and in larval blackflies, *Simulium*.

The most abundant cation is usually sodium although the amount varies with the insect's phylogeny and its diet (Table 5.1). Most phytophagous insects have lower concentrations of sodium than insects with other feeding habits. On the other hand, potassium and magnesium levels tend to be higher in phytophagous groups reflecting the levels of these elements in plant tissues.

Amongst hemimetabolous insects, ionic concentrations appear roughly similar in larval and adult stages, although there are very few species in which both stages have been examined. In holometabolous insects, however, the patterns are very variable from species to species (Table 5.2). The blood of most adult insects has higher sodium concentrations than that of larvae, but the reverse is true for magnesium, except in Coleoptera. Calcium concentrations may be similar in both stages or lower in larvae, while potassium concentration may increase, decrease, or remain the same.

Most studies indicate changes in some components over the course of a single stage, perhaps related to feeding, but there are insufficient data to generalize. Differences in food composition affect the concentrations of some elements in the hemolymph of *Hyalophora*, but not others. Major differences in dietary potassium and calcium had no effect, but sodium and magnesium concentrations in larval hemolymph were affected by the amount in the diet. By the adult stage, however, these differences had largely disappeared. Changes also occur in relation to starvation and, in locusts, the potassium concentration increases markedly before molting. These changes may affect the behavior of the insect since neuromuscular junctions are directly exposed to the hemolymph and a low concentration of potassium raises the resting potential. Changes in hemolymph potassium are also known to cause the release of neurosecretions from neurohemal organs. In aquatic insects, the ionic composition of the hemolymph may be

Table 5.1. *Major inorganic ions in the hemolymph of different insects (concentrations in mequiv l^{-1})*

Feeding habit	Species	Order	Na	K	Ca	Mg	Cl	H₂PO₄
Phytophagous	*Locusta*	Orthoptera	60	12	17	25	98	·
	Carausius	Phasmida	9	27	16	142	93	40
	Bombyx (L)	Lepidoptera	15	46	24	101	21	3
	Leptinotarsa (L)	Coleoptera	2	55	43	147	·	·
	Pteronidea (L)	Hymenoptera	2	43	17	61	·	·
Detritivorous	*Periplaneta*	Blattodea	157	8	4	5	144	·
	Anisolabis	Dermaptera	193	5	15	11	117	3
Predaceous	*Aeschna* (L)	Odonata	145	9	7	7	110	4
	Tettigonia	Orthoptera	83	51	·	·	·	·
	Notonecta	Hemiptera	155	21	·	·	·	·
	Myrmeleon (L)	Neuroptera	143	9	7	7	·	·
	Dytiscus	Coleoptera	165	6	22	37	44	3
Blood-sucking	*Cimex*	Hemiptera	139	9	·	·	·	·
	Stomoxys	Diptera	128	11	·	·	·	·

Notes:

(L), larva.

·, No data.

affected by the composition of the environmental water (section 18.3.2).

Regular daily changes occur in the effective concentration (that is, the amount that contributes to osmotic activity) of potassium in the Madeira cockroach, *Leucophaea*, with peaks in both the light and dark periods (Fig. 5.17), but the activity of sodium does not change in a comparable way. These changes probably result from potassium being sequestered and released by hemocytes, or perhaps bound to large anionic molecules.

Various other metal elements are also found in small amounts in the blood. The most frequent are copper, iron, zinc and manganese. Iron is not free in the hemolymph, but is bound to two proteins, ferritin and transferrin (Locke & Nichol, 1992).

5.2.2.2 *Hemolymph amino acids*

Insect blood plasma is characterized by very high levels of free amino acids. The total concentration in plasma is usually more than $6\,\mathrm{mg\,ml^{-1}}$ in endopterygotes, but less than this in exopterygotes. Most of the protein amino acids are present, but their concentrations vary greatly from

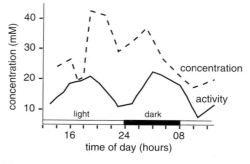

Fig. 5.17. Changes in the effective concentration (activity) of potassium (measured with a potassium-sensitive electrode) in the hemolymph of *Leucophaea* in the course of a day/night cycle and the total concentration of potassium in the blood, including that in hemocytes, which is not active as an electrolyte (after Lettau *et al.*, 1977).

insect to insect (Fig. 5.18). It is common, in endopterygotes, for glutamine and proline to be present in high concentrations relative to most other amino acids. Glutamate (glutamic acid), on the other hand, is only ever present in very small quantities, probably never exceeding

Table 5.2. *Hemolymph cation concentrations in larval and adult holometabolous insects (Concentrations are mequiv l^{-1})*

Order/species	Food		Na		K		Ca		Mg	
	Larva	Adult	Larva	Adult	Larva	Adult	Larva	Adult	Larva	Adult
Hymenoptera										
Apis	Brood food	Nectar	14	51	38	29	23	19	26	1
Vespula	Insects	Insects, fruit	21	80	45	12	15	1	19	1
Lepidoptera										
Mamestra	Leaves	Nectar	5	15	17	41	12	10	63	34
Bombyx	Leaves	None	1	8	~28	36	~11	13	~63	43
Hyalophora	Leaves	None	2	2	39	27	12	9	100	39
Coleoptera										
Oryctes	Detritus	Detritus	8	45	4	9	8	9	65	37
Timarcha	Leaves	Leaves	1	0.4	14	17	18	9	66	74
Ergates	Wood	?	13	22	19	12	7	7	53	59
Diptera										
Chironomus	Plankton	None	110	150	20	50	.	.	.	.

Note:
·, No data.

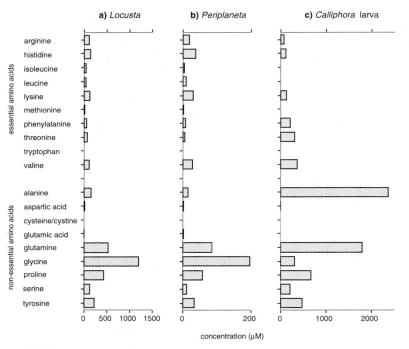

Fig. 5.18. Amino acid concentrations in the blood plasma of three insects. Note the different scale for *Periplaneta*, where concentrations are much lower (data from Irving *et al.*, 1979). (**a**) *Locusta*, a phytophagous, hemimetabolous insect. (**b**) *Periplaneta*, a detritivorous, hemimetabolous insect. (**c**) Larval *Calliphora*, a saprophagous, holometabolous insect.

10^{-5} M. This is important because glutamate is a neurotransmitter and high concentrations in the hemolymph would impair this function (Irving, Wilson & Osborne, 1979).

The concentrations of amino acids may change at different stages of the life cycle. Tyrosine, for instance, commonly accumulates before each molt and then decreases sharply as it is used in tanning and melanization of the new cuticle (see Fig. 16.19). Because free tyrosine is not very soluble, much of it is present as the more soluble glucoside. Marked changes also occur in the silkworm, *Bombyx*, when it produces its silken cocoon. Glycine is one of the major amino acids in silk. Its concentration builds up towards the end of the feeding stage and then declines during spinning (Fig. 5.19). Glutamine and asparagine are not major constituents of silk protein, but their concentrations in hemolymph fall sharply as they are taken up by the silk glands and converted to alanine, which is a major component. Other amino acids do not change in this way. After the molt to pupa, the concentrations of glycine and glutamine (included with glutamic acid in Fig. 5.19) rise sharply, possibly due to histolysis of the tissues. The amino acid concentration in the hemolymph probably rises after feeding in many insects. In *Locusta*, the concentration

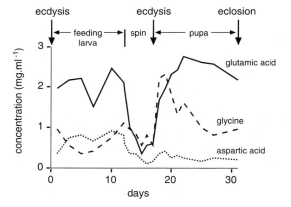

Fig. 5.19. Variations in the concentrations of some amino acids used in silk production for the cocoon of *Bombyx*. The rise in concentrations in the early pupa probably result from histolysis of the tissues. In these experiments, glutamine and glutamate were not separated, nor were asparagine and aspartate. They are shown in the figure as glutamic acid and aspartic acid, respectively (Jeuniaux *et al.*, 1961).

returns to its original level within about one hour, but, in *Rhodnius*, feeding only once in each developmental stage, it remains constant through the period of molting. It appears that the utilization of amino acids is offset by the slow, continuous digestion of the blood meal in this insect.

5.2.2.3 *Hemolymph proteins*

Insect hemolymph plasma contains many different proteins with a variety of functions. The total quantity of protein in the blood varies in the course of development, but peak concentrations in the late larval stages of Lepidoptera and Diptera may reach 100 and 200 mg ml^{-1}, respectively. These proteins are usually classified by function, although this is not always known, and this method may obscure similarities between proteins with similar structures, but different functions. Here they are grouped as follows: storage proteins, lipid transport proteins, vitellogenins (section 13.2.4.2), enzymes, proteinase inhibitors, chromoproteins, and a range of different proteins that are probably involved in the immune responses of insects (section 5.3.3). It is common for the proteins to incorporate small quantities of carbohydrate and sometimes also lipid.

Storage proteins Storage proteins have been studied primarily in Lepidoptera and Diptera, but are probably widespread in their occurrence. The proteins have six subunits and, for this reason, they are also called hexamerins. The subunits may all be the same, or be of two or three different types. Most insects studied have only one or two different storage proteins, but some Lepidoptera have three or four. One of the main classes in Lepidoptera is the arylphorins, in which the aromatic amino acids phenylalanine and tyrosine comprise 18–26% of the total, but which contain little methionine. Similar proteins occur in Coleoptera, Diptera and Hymenoptera, although in the latter the proportion of aromatic acids is less (about 12%). Lepidoptera also have methionine-rich proteins that contain 4–8% methionine, while in Hymenoptera the second protein has high levels of glutamine/glutamic acid (Wheeler & Buck, 1995). Comparable proteins are known to occur in Blattodea and Orthoptera. Coleoptera have a soluble arylphorin in the hemolymph and, in addition, have insoluble, tyrosine-rich proteins stored in the fat body (Delobel *et al.*, 1993).

Storage proteins are synthesized primarily in the fat body although in *Calpodes* arylphorin is also produced by

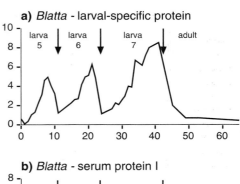

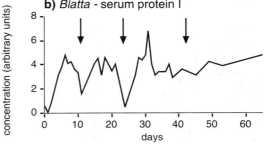

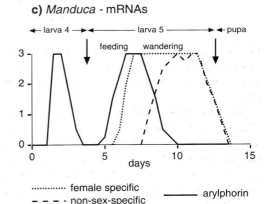

Fig. 5.20. Variations in hemolymph storage proteins. Arrows indicate the times of ecdysis. **(a),(b)** Concentration of larval specific protein **(a)** and serum protein I **(b)** in the hemolymph of *Blatta* (after Duhamel & Kunkel, 1987). **(c)** Concentration of mRNAs for three storage proteins in the fat body of female larvae of *Manduca* (after Riddiford & Hice, 1985).

the midgut epithelium and the epidermis (Palli & Locke, 1988). Their production is commonly cyclical and the hemolymph concentration builds up through each developmental stage, falling sharply during the molt (Fig. 5.20). In Diptera, synthesis only occurs in the last larval stage, and, in the late larva of the blowfly, *Calliphora*, a single storage protein comprises about 60% of the soluble protein. Arylphorins are not produced after the larva/pupa molt, but in some insects other proteins, such as serum protein I in *Blatta* (Fig. 5.20b) are still present at high concentrations in the adult stage. Females of some species produce more storage protein than males and some have a female-specific protein.

Rising ecdysone concentrations regulate the decline in arylphorin mRNA before a molt (Fig. 5.20c), but a methionine-rich protein produced only in the final larval stage of *Manduca* appears to be regulated by juvenile hormone inhibiting production of its mRNA. High concentrations of these proteins may persist in the hemolymph in the absence of synthesis if they are not removed by the fat body.

At each molt, arylphorin is taken up into the fat body where it is temporarily stored in granular form. Uptake may, in part, be unselective as there is a general turnover of proteins at metamorphosis, but selective uptake of the storage protein also occurs. The bulk of the storage protein appears to be broken down into its component amino acids in the fat body and forms the basis of tissues in the next developmental stage. Amino acids derived from larval storage protein occur in all adult tissues, but contribute especially to flight muscles and cuticle which, because of their bulk, contain large amounts of protein. It is probable that storage proteins also contribute to yolk formation, contributing the amino acid building blocks for vitellogenin synthesis, and, in autogenous mosquitoes, they disappear from females as yolk synthesis begins. In queens of many ant species, the storage proteins persist for some time after the nuptial flight and provide sufficient protein for development of the first brood of workers, which, in most species, takes place without the queen feeding (Wheeler & Buck, 1995).

Despite the apparent importance of the storage proteins, mutants of *Drosophila* can survive for many generations without any larval serum protein I although this is normally present in large amounts. Adult survival and longevity are not affected by its absence, but there is a marked reduction in fecundity. This is largely due to a failure of the flies to mature rather than any direct quantitative effect of the protein, but it appears that oocyte development may also be affected (Roberts, Turing & Loughlin, 1991).

Specific proteins also accumulate in the hemolymph of insects entering diapause. The Colorado potato beetle,

Leptinotarsa, enters diapause as an adult, while the pink bollworm, *Pectinophora*, diapauses as late stage larvae. Both species accumulate proteins that are specific to the diapause period (Koopmanschap, Lammers & de Kort, 1992; Salama & Miller, 1992), but the functions of these proteins are not known.

Lipid transport proteins Because lipids are not soluble in water their transport through the hemolymph involves combination with proteins to form lipoproteins. Insects have a single class of lipoprotein called lipophorins. The major lipid components of lipophorin are diacylglycerides and phospholipids, although smaller amounts of triacylglycerides, fatty acids and cholesterol are present. The lipid component is solubilized by the presence of two or three different proteins called apolipoproteins I, II and III. Apolipoproteins I and II are always present; apolipoprotein III is added in some species during periods of peak lipid movement through the hemolymph. The particles so produced vary in composition and density, the more lipid present, the lower the density. High-density lipophorin (HDLp) contains relatively little lipid; low-density lipophorin (LDLp) has more lipid. HDLp and LDLp are not discrete categories. Their densities can vary and particles with predominantly different densities may be present at one time. Very-high-density lipophorin (VHDL) is sometimes produced. The most common is vitellogenin.

Fatty acids are transported in the hemolymph primarily as diacylglycerides. Diacylglycerides derived from the food are taken up by HDLp at the hemolymph boundary of the midgut epithelium. From here they are transported to the tissues for utilization or, commonly, to the fat body for storage. The HDLp does not enter the cells but unloads the diacylglyceride at the cell wall and returns to circulation. Within the fat body, fatty acids are stored mainly as triacylglycerides so that their subsequent movement to other tissues requires resynthesis of diacylglycerides (Fig. 5.21b). This occurs at the cell boundary and then diacylglycerides are again loaded on to HDLp for transport to the appropriate tissues.

During normal metabolism the additional lipid added to HDLp presumably makes only small changes to the volume and density of the particle. However, when the demand for lipids is high, as in adult *Locusta* and the tobacco hornworm, *Manduca*, both of which use lipids as the main fuel for flight, so much lipid is added to the

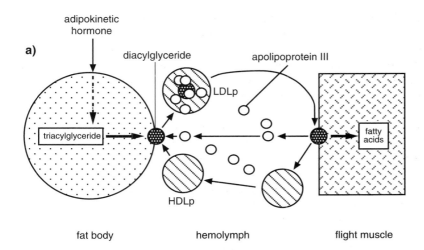

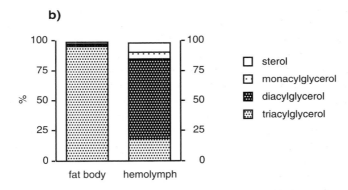

Fig. 5.21. Lipid in the hemolymph. (a) Diagrammatic representation of the movement of lipids from the fat body to flight muscles in *Manduca* by the formation of low density lipophorin (LDLp) (based on Shapiro *et al.*, 1988). (b) Proportions of major lipid components in the fat body and hemolymph of the pupa of *Hyalophora*. Most lipid is stored as triacylglycerol, but it is transported as diacylglycerol (data from Gilbert, 1967).

HDLp that its increasing size would result in a loss of solubility but for the addition of apolipoprotein III. The larger, low density particle produced by this process is LDLp (Fig. 5.21a). The mobilization of lipid in the fat body is controlled by adipokinetic hormone. At the flight muscle, the LDLp is unloaded and HDLp and apolipoprotein III return to circulation. It is possible that apolipoprotein III also becomes associated with HDLp in larval *Manduca* when lipid demands are high (Ziegler *et al.*, 1995).

About 90% of the lipid that accumulates in the developing oocytes of *Manduca* is transported as LDLp and the lipoprotein is recycled in the hemolymph. However, some of the lipid is transported by vitellogenin and by another VHDL. In this case, the lipoprotein is not recycled, but is retained in the oocyte.

Juvenile hormone is transported by lipophorins in the hemolymph. Some lipophorin has only a low affinity for the hormone, but in some insects a separate lipophorin with high affinity for juvenile hormone is also present. This serves not only to transport the hormone, but also to protect it from the action of juvenile hormone esterase (King & Tobe, 1993). Some insects have juvenile hormone binding proteins that are not lipophorins.

The apolipoproteins are synthesized in the fat body where they are combined with phospholipids and released into the hemolymph as nascent lipophorin particles. Diacylglycerides are normally added at the gut epithelium. In *Manduca* larvae, apoprotein mRNA is present during the feeding stages and total lipophorin increases, but no mRNA is available after feeding stops in each larval stage although the amount of lipophorin remains constant (Fig. 5.22). Synthesis starts again in the pupal stage about 12 hours before eclosion.

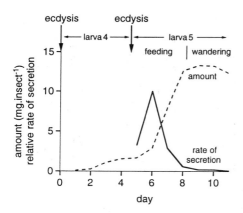

Fig. 5.22. Lipophorin secretion and accumulation in the final larval stage of *Manduca*. The amount of lipophorin in the insect, mainly in the blood, remains at a plateau even though synthesis and secretion from the fat body stops (data from Prasad *et al.*, 1987).

Enzymes The blood contains a number of enzymes, although the extent to which these occur in the plasma, rather than in the hemocytes, is not always clear. Trehalase is known to occur in the hemolymph of *Periplaneta*, *Locusta* and *Phormia* and some other insects. Its function is unknown, but it may be involved in the regulation of hemolymph levels of trehalose.

Phenoloxidase is present in the hemolymph as a proenzyme. In *Bombyx* the proenzyme is activated by peptides released from proteins by a series of enzymic reactions. Phenoloxidases catalyze the oxidation of phenols and convert catecholamines to quinones (section 16.5.3). Phenoloxidase in the hemolymph is activated by invading micro-organisms or parts of their cell walls. It probably forms part of the immune system, promoting the formation of melanin in the capsules surrounding foreign particles in the hemolymph (section 5.3.2) and perhaps forming part of the system by which foreign objects are recognized.

The hemolymph usually contains a number of esterases. One of their functions is to regulate juvenile hormone titers (section 21.4).

Protease inhibitors Protease inhibitors are present in the hemolymph of at least some insects. Their role in the insect is not known for certain, but they may be concerned with the regulation of hemolymph enzymes. In *Manduca*, for example, protease inhibitors are known to inhibit the activation of the phenoloxidase proenzyme. In *Bombyx*,

inhibitors of chymotrypsin increase markedly in the final larval stage and it is suggested that they may be important in inhibiting the activities of enzymes leaking from the tissues during histolysis (Eguchi, Matsui & Matsumoto, 1986). Other inhibitors are effective against proteases produced by pathogenic fungi. For example, the hemolymph concentration of a protease inhibitor in the hemolymph of the velvetbean caterpillar, *Anticarsia*, is enhanced by infection by the fungus, *Nomuraea*, suppressing germination of the fungal conidia (Boucias & Pendland, 1987).

Reviews: Kanost *et al.*, 1990 – hemolymph proteins; Levenbook, 1985 – storage proteins; Shapiro, Law & Wells, 1988 – lipid transport proteins; Telfer & Kunkel, 1991 – storage proteins

5.2.2.4 *Other organic constituents*

The end products of nitrogen metabolism are always present in the hemolymph, usually in very low concentrations. These commonly include uric acid and ammonia, but allantoin and urea may also be present. Various peptides and biogenic amines, acting as neurohormones or neuromodulators, are probably also always present (Chapter 21) while other hormones, such as ecdysone and juvenile hormone, occur periodically.

Trehalose, a disaccharide, is the most characteristic sugar found in insect hemolymph. Its concentration is usually in the range 4–$20\,\text{mg}\,\text{ml}^{-1}$, but it is sometimes present in greater amounts. It is not present in all insects, however. It is absent altogether in some apterygote insects and in others, like the blowfly, it is only in very low concentrations. Glucose is also often present, usually in much lower concentrations, but high concentrations occur in *Apis* and in *Phormia*. Sugar levels are normally maintained at an approximately steady level by the action of hormones (section 6.2.2).

Other carbohydrates are also sometimes present. These include the hexosamines involved in chitin synthesis and, sometimes, the sugar alcohol, inositol. Either glycerol or mannitol is probably always present, and, in insects able to tolerate freezing, the concentration may be very high (section 19.3.2.2).

The concentration of lipids in the hemolymph generally varies between about 1 and $5\,\text{mg}\,\text{ml}^{-1}$, but values approaching $15\,\text{mg}\,\text{ml}^{-1}$ are achieved in insects, such as *Locusta* and *Manduca*, that use lipids as fuels for flight. Most of the lipid is in the form of diacylglycerols (Fig. 5.21b). These components are normally carried by lipophorins.

Organic acids are present in some quantity in the plasma. The major components are acids associated with the citric acid cycle, including citrate, α-ketoglutarate, succinate and malate. Citrate is usually present in high concentration, although this varies considerably from one species to another. Organic phosphates are often present in high concentrations and in some insects tyrosine is present as a phosphate. This is an alternative to glycosylation as a means of achieving a high concentration of tyrosine.

5.2.3 Properties of the plasma

5.2.3.1 *Osmotic pressure*

The inorganic and organic solutes present in the hemolymph contribute to its osmotic pressure. Sutcliffe (1963) grouped the insects into three broad categories on the basis of the osmotic components of the plasma:

1. Sodium and chloride account for a considerable proportion of the osmolar concentration (Fig. 5.23a). This is probably the basic (in the evolutionary sense) type of blood in insects because it is similar to that in most other arthropods. It occurs in Ephemeroptera, Odonata, Plecoptera, Orthoptera and Homoptera.

2. Chloride is low relative to sodium, which constitutes 21–48% of the total osmolar concentration (Fig. 5.23b). Amino acids are also present in high concentration. This type is found in Trichoptera, Diptera, Megaloptera, Neuroptera, Mecoptera and most Coleoptera.

3. Amino acids account for 40% of the total osmolar concentration (Fig. 5.23c). There is a large category of unknown factors, but none of the other substances accounts for more than 10% of the total. Lepidoptera and Hymenoptera have this type of blood.

In many insects, the osmotic pressure of the blood is within the range 300–500 mOsmol, although this figure may vary in different stages of development, and, in overwintering insects exhibiting a high resistance to freezing, the figure may be over 500 mOsmoles. In general, insects appear able to maintain a relatively constant hemolymph osmotic pressure even when the hemolymph volume changes markedly. In the beetle, *Onymachus*, for example, hemolymph osmolality remained almost constant despite marked changes in the hemolymph volume as a result of desiccation and drinking (Fig. 5.16). The amounts of all

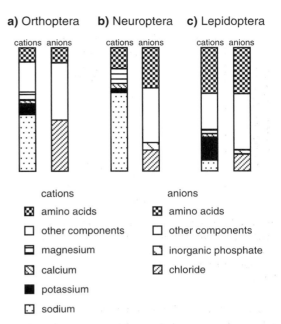

Fig. 5.23. Osmotic components of the hemolymph in different groups of insects expressed as percentages of the total osmolar concentration. Each vertical column represents 50% of the total concentration (after Sutcliffe, 1963).

the osmotic effectors varied with hemolymph volume (Fig. 5.24). It appears that they are stored in other tissues when the hemolymph volume declines and returned to the hemolymph when its volume increases. There is evidence in *Periplaneta* that sodium, and perhaps also potassium, are stored in urate granules in the fat body, but they may also remain in the hemolymph in an inactive form. In the larva of the mosquito, *Aedes*, the osmotic activity of sodium is lower than expected from its concentration, possibly because some is bound to large anionic molecules (Edwards, 1982). This may also be true in other insects. For example, in *Leucophaea* the osmotic activity of potassium is significantly lower than would be expected from its concentration in the hemolymph (Fig. 5.17). Not all insects are able to regulate hemolymph osmotic pressure within such narrow limits as *Onymachus*, but some degree of regulation appears to be a general phenomenon.

Short-term variations in osmotic pressure do occur, however. For example, in *Locusta* the osmotic pressure of the hemolymph increases by about 50 mOsmol during a meal, but returns to its original level in about an hour.

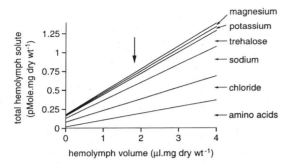

Fig. 5.24. Osmotic effectors in the hemolymph of *Onymachus* at different levels of hydration. The quantity of each component is indicated by the area between successive lines. The top line shows the total concentration of all the solutes combined. Arrow indicates the normal level of hydration (after Machin, 1981).

5.2.3.2 *pH*

The pH of insect hemolymph is usually between 6.4 and 6.8, although slightly alkaline values have been recorded in a dragonfly larva and in the larva of the midge, *Chironomus*. During normal activity there is a tendency for the blood to become more acid due to the liberation of acid metabolites, including carbon dioxide. The buffering capacity of insect blood (that is, its ability to prevent change in pH) is low in the normal physiological range, but increases sharply above and below this range. Within the normal range, bicarbonates and phosphates are the most important buffers. On the acid side of the range, carboxyl groups of organic acids are important, while on the alkaline side the amino groups of amino acids are most significant. Proteins buffer over a wide range of pH.

Review: Mullins, 1985

5.2.4 **Hemocytes**

Suspended in the blood plasma are blood cells or hemocytes. Many different types of hemocyte have been described, but a comprehensive classification is difficult because individual cells can have very different appearances under different conditions and a variety of techniques have been used in their study. Rowley & Ratcliffe (1981) and Gupta (1979a, 1979b, 1985, 1991) attempt to synonymize them across the different orders and reduce them to six main types (Fig. 5.25). They are: prohemocytes, plasmatocytes, granulocytes (which are probably the same as cystocytes or coagulocytes), spherule cells (spherulocytes), oenocytoids and adipohemocytes.

Prohemocytes are characterized by a high nuclear: cytoplasmic ratio and a general lack of organelles involved in synthesis. They rarely comprise more than 5% of the total hemocyte population. They are the stem cells from which most other hemocyte types are formed.

Plasmatocytes are very variable in shape. They contain moderate amounts of rough endoplasmic reticulum and Golgi complexes and may contain membrane-bound granules. They are amongst the most abundant hemocytes and usually account for more than 30% of the total hemocyte count. Plasmatocytes are involved in phagocytosis and encapsulation of foreign organisms invading the hemocoel.

Granulocytes contain large amounts of endoplasmic reticulum which is often extensively dilated. Golgi complexes are also abundant and the cells contain large numbers of membrane-bound granules. They comprise a considerable proportion, usually more than 30% of the hemocyte population. They discharge their contents (degranulate) on the surfaces of intruding organisms as an early part of the defense response. Granulocytes are probably derived from plasmatocytes and intermediates between the two types of cell occur (see Chain, Leyshon-Sørland & Siva-Jothy, 1992). Cystocytes are probably granulocytes in which the synthesis of granular contents is complete. They contain abundant granules, but usually contain smaller amounts of Golgi complexes and rough endoplasmic reticulum than granulocytes. They have a relatively high nucleus: cytoplasm ratio. They are often common, but have not been recognized in Diptera, Lepidoptera and Hymenoptera.

Fig. 5.25. Different types of hemocyte (**a**) after Chiang, Gupta & Han, 1988; others after Rowley and Ratcliffe, 1981): (**a**) prohemocyte of *Blattella*; (**b**) plasmatocyte of larval *Galleria*; (**c**) granulocyte of larval *Galleria*; (**d**) granulocyte (cystocyte) of *Clitumnus*. Arrowheads indicate swollen perinuclear cisterna; (**e**) spherule cell of larval *Galleria*. The large open areas, looking like vacuoles (and labelled V), are probably caused by extraction of spherules during preparation; (**f**) oenocytoid from larval *Galleria*. Inset shows size of nucleus relative to whole cell.

Abbreviations: G, granules; GO, Golgi complex; IG, developing granules; M, mitochondria; MT, microtubules; MVB, multivesicular body; N, nucleus; PE, protoplasmic extensions; PO, ribosomes; PV, pinocytotic vesicles; R, ribosomes; RER, distended cisternae of rough endoplasmic reticulum; SP, spherules; V, vacuole.

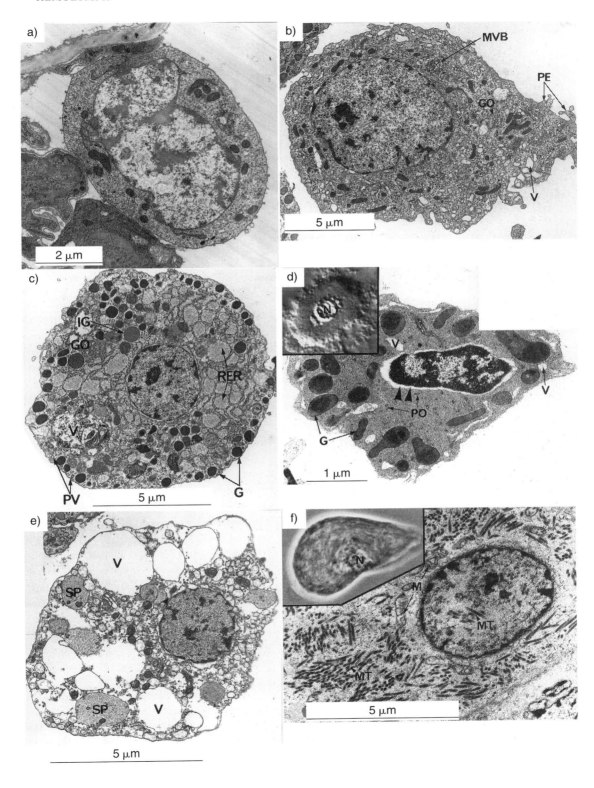

Fig. 5.26. Hemocyte production during the larval stages of *Euxoa* (data from Arnold & Hinks, 1976). **(a)** Mitotic activity in different types of hemocytes, expressed as percentage of each type. Data from a subsequent paper (Arnold & Hinks, 1983) indicates that the values obtained in the original work, and which are used in this diagram, were probably too low, but the pattern of change is probably not affected. **(b)** Hemocyte counts per microliter of blood. **(c)** Hemocyte profile – relative frequency of different types of hemocyte, expressed as percentage of total number of blood cells.

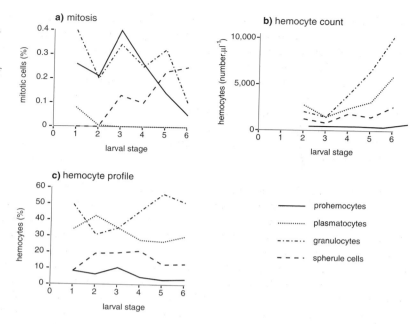

Spherule cells are characterized by the large, refractile spherules which may occupy 90% of the cytoplasm. They are not usually very common although they are found in most of the species studied. Their function is unknown.

Oenocytoids occur mainly in Lepidoptera where they are amongst the largest of the hemocytes. These cells exhibit little development of rough endoplasmic reticulum or Golgi complexes, but they have a complex array of microtubules and sometimes also crystalline inclusions. Their function is unknown.

Adipohemocytes characteristically contain lipid droplets. The nucleus: cytoplasm ratio is low, and they contain well-developed endoplasmic reticulum and Golgi complexes.
Reviews: Brehélin & Zachary, 1986; Gupta, 1979a, b, 1985, 1991; Rowley & Ratcliffe, 1981

5.2.4.1 *Origin of hemocytes*

Hemocytes are derived from the embryonic mesoderm. Subsequently, new hemocytes are produced by mitotic division of existing, circulating hemocytes, or from previously undifferentiated cells in structures known as hemopoietic organs.

Mitotic division of hemocytes The production of new hemocytes by mitosis of existing blood cells is a widespread phenomenon. In adult holometabolous insects that lack hemopoietic organs, new hemocytes can only be produced in this way. This appears also to be the case during the larval stages of the milkweed bug, *Oncopeltus*. Elsewhere, hemocyte production from existing cells appears to complement production in hemopoietic organs, but where the hemopoietic organs persist in adult insects, as in Blattodea and Orthoptera, mitotic division of existing hemocytes is relatively rare.

Not all types of cell divide and the rates of division vary even amongst those that do. Between 0.2 and 0.4% of prohemocytes were found in division in blood samples taken during the first four stages of larval development of the moth, *Euxoa*, but this level declined in the final larval stage (Fig. 5.26a). Mitotic activity was similar in granulocytes, but amongst spherule cells it increased from zero in the first two stages to about 0.25% in the final stage. Plasmatocytes only rarely divide in at least a majority of insects. Despite this, the number of plasmatocytes per unit volume of hemolymph increases throughout larval development (Fig.5.26b). They are probably derived from the prohemocytes which remain constant in relative abundance despite their high mitotic rate.

Much of the literature suggests that the mitotic rate for all the cells only rarely exceeds 1%, but some work indicates much higher rates. Arnold & Hinks (1983) suggest that in the final larval stage of *Euxoa*, the mitotic index of spherule cells may exceed 10% (see caption to Fig. 5.26) and in the final larval stage of the milkweed bug, *Oncopeltus*, a mitotic index of 4% was recorded.

On the basis of the mitotic activity of the cells, it is suggested that the whole population of granulocytes in the last larval stage of *Euxoa* turns over in about 5 days; the spherule cell population would turn over in less than one day (Arnold & Hinks, 1983). Other estimates of hemocyte longevity, in *Galleria*, suggest that plasmatocytes survive for at least nine days.

Hemopoietic organs Blood is formed in structures called hemopoietic organs. Since, in insects, only the blood cells, not the plasma, are produced in these structures, they should strictly be called hemocytopoietic organs, but the general term is more usual. Hemopoietic organs have been described in some Orthoptera, a blattid and a few larval Lepidoptera, Diptera and Coleoptera. They persist in adult Orthoptera, but not in adults of holometabolous species. No hemopoietic organs are present at any stage of the milkweed bug, *Oncopeltus*.

The positions of hemopoietic organs vary from species to species, but in most cases they are associated with, though not necessarily connected with, the heart. In the cricket, *Gryllus*, and the mole cricket, *Gryllotalpa*, they are paired, segmental structures on either side of the heart and opening into it (Fig. 5.2). In *Locusta*, *Periplaneta*, and larvae of cyclorrhaphous flies and of the beetle, *Melolontha*, they consist of irregular accumulations of cells close to the heart, but not connected with it (Fig. 5.27a,b). By contrast, in caterpillars they are groups of cells around the developing imaginal wing discs (Fig. 5.27c,d).

Only in the grylloids is the hemopoietic organ a discrete structure bounded by a cell layer and with an ill-defined lumen opening into the heart. Even here, the bounding layer of cells is incomplete. Within this boundary are irregularly shaped reticular cells apparently embedded in a connective tissue matrix. These cells undergo mitotic divisions and give rise to hemocyte stem cells. By further division, the stem cells form clusters of cells which differentiate synchronously to form hemocytes. Granulocytes and plasmatocytes are formed in this way. They separate from the cortical region and enter the circulation, presumably via the heart. The reticular cells are also phagocytic, taking up foreign material from the hemolymph. Because of this the hemopoietic organs in these insects were originally called phagocytic organs. The

process of hemopoiesis appears essentially similar in other insects although the reticular cells exist as aggregations with no bounding layer and, in Lepidoptera, reticular cells are absent.

Reviews: Feir, 1979 – mitosis; Hoffmann *et al.*, 1979 – hemopoietic organs

5.2.4.2 *Numbers of hemocytes*

Estimates of the total number of hemocytes in an insect show that small insects have many fewer hemocytes than large insects. Adult female mosquitoes have a total of less than 10 000 hemocytes, whereas adult *Periplaneta* have more than 9 000 000. Similar trends occur within a species. Second stage caterpillars of *Euxoa* have about 4000 hemocytes; sixth stage larvae have about 2 400 000. The number may also vary cyclically. For example, in the last stage larva of the wax moth, *Galleria*, the total number of cells is at first constant at about 2.2 million and then increased to almost 4 million before the insect molts (Fig. 5.28a). An even bigger relative increase occurs during the postfeeding stages of the final stage larva of *Sarcophaga*, but at the time of pupariation, when the larva becomes immobile, there is a sudden rapid decline (Fig. 5.29).

Increases in numbers of circulating cells may result from the production of new cells or, possibly, by the recruitment of cells adhering to other tissues. Reduction in hemocyte number may result from cell death or from an increase in the numbers adhering to the tissues.

Counts of the number of cells per unit volume of hemolymph (usually called the 'total hemocyte count') may not reflect the total number of hemocytes in circulation because the blood volume varies. For example, in the last stage larva of *Galleria*, when the total number of cells is constant (Fig 5.28, weight less than 200 mg) the number per unit volume decreases because the blood volume is increasing. From a functional standpoint, such as wound healing or combatting invaders, the number per unit volume may be more important than the total number.

The number of hemocytes per unit volume of blood tends to increase throughout larval development, but with additional variation within each developmental stage (Fig. 5.30). It reaches a maximum at the time of each ecdysis, except the pupa/adult ecdysis. The lack of a peak at this time may reflect the fact that major restructuring of the tissues occurs earlier in the pupal period. In hemimetabolous insects, the numbers are generally similar in larval

Fig. 5.27. Hemopoietic organs in different insects. Stippling indicates hemopoietic tissue (see also Fig. 5.2). (a) *Locusta*; (b) *Calliphora* larva; (c), (d) Caterpillar. (c) Showing the positions of the organs (arrows), (d) section through one wing disc (after Monpeyssin & Beaulaton, 1978).

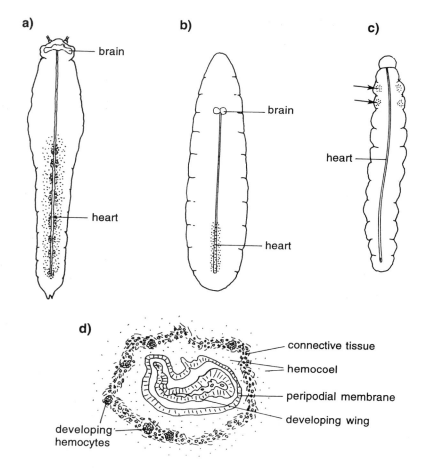

and adult insects, but in holometabolous species it is usual for larvae to have more cells per unit volume of blood than adults. In general, adult females have a higher number of hemocytes than males.

Hemocyte profile The relative abundance of different types of hemocytes (called the hemocyte profile or a differential hemocyte count) is not constant. Plasmatocytes and granulocytes are usually the most abundant, often comprising more than 80% of the total hemocyte population (Figs. 5.26c, 5.29b). The relative abundance of plasmatocytes tends to decline, and that of granulocytes to increase, through the larval period, but a sharp reversal occurs at pupariation in *Sarcophaga* when the total hemocyte count drops. The relative numbers of other cell types also change; spherule cells virtually disappear from the blood of *Sarcophaga* at pupariation. In

Rhodnius, changes in relative abundance occur in relation to feeding and molting.
Review: Shapiro, 1979

5.2.4.3 *Functions of hemocytes*
Hemocytes perform a variety of functions. Among the more obvious are wound repair and defense against parasites and pathogens (see below), but they have roles in many aspects of the normal functioning of the insect.

Granulocytes and spherule cells of larval *Calpodes* synthesize polypeptides which are secreted into the hemolymph and subsequently incorporated into the cuticle. Other peptides produced by hemocytes are probably added to the basal lamina (Sass, Kiss & Locke, 1994).

The hemocytes contain many proteases some of which appear to be involved in the breakdown of tissues at metamorphosis. For example, some hemocytes of *Sarcophaga*

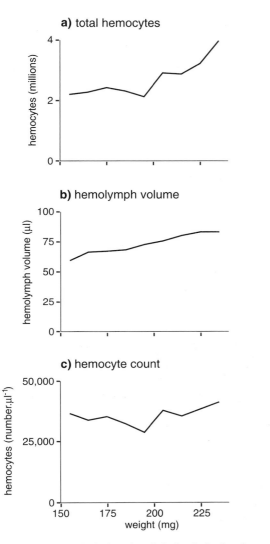

Fig. **5.28**. Changes in the hemolymph during the last larval stage of *Galleria* (data from Shapiro, 1979): (**a**) total number of hemocytes; (**b**) blood volume; (**c**) hemocyte count per microliter of blood.

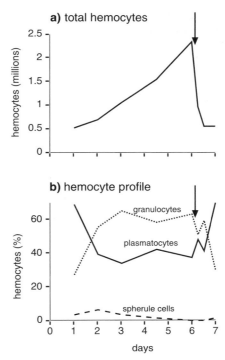

Fig. **5.29**. Changes in the hemolymph during the last larval stage of *Sarcophaga* (data from Jones, 1967): (**a**) total number of hemocytes; (**b**) hemocyte profile. The arrow shows the start of pupariation.

have a 200 kDa protein in the cell membrane. These cells increase in number at the time of pupation and the 200 kDa protein binds to sites on the basal lamina of the fat body. Here the cells release a cathepsin-type protease which dissociates the fat body (Kurata, Saito & Natori, 1993).

If the epidermis is damaged, a blood clot forms beneath the wound. Formation of the clot involves components from both the hemocytes and the plasma. Granulocytes release material which forms a gel. This gel is stabilized by plasma lipophorins and phenoloxidases from the hemocytes may also be important. It is not known what causes the cells to move to the site and degranulate, but possibly some injury factor is produced by damage to the basal lamina. On the other hand, in *Calliphora*, clotting involves the clumping and interdigitation of hemocytes without gelation.

Some time after clotting has occurred, plasmatocytes migrate to the site (Fig. 5.31). In *Rhodnius* the cells become linked to each other by zonulae adherens within 24 hours of the wound occurring and subsequently tight junctions and septate desmosomes are formed. In this way, hemocytes become bound together to form a continuous tissue. The epidermal cells migrate over the clot to repair the wound.

5.3 IMMUNITY

Insects exhibit defensive responses when their tissues are invaded by other organisms. These are now generally

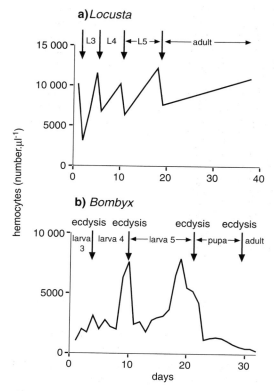

Fig. 5.30. Changes in the hemocyte count during development. Vertical arrows indicate the times of ecdysis. (**a**) *Locusta*. Blood samples were taken shortly before and after ecdysis with only a single data point in between (data from Hoffmann, 1967). (**b**) *Bombyx* (data from Nittono, 1960).

considered to constitute an immune response, although this differs in many ways from a typical vertebrate immune response. It is common to recognize two types of immune processes, cellular and humoral (hormonal), although they are probably not entirely independent.

Reviews: Dunn, 1986; Gillespie, Kanost & Trenczek, 1997; Götz & Boman, 1985; Gupta, 1986, 1991; Lackie, 1988; Ratcliffe & Rowley, 1979; Strand & Pech, 1995

5.3.1 Recognition of a pathogen or parasite

Any response by the hemocytes to an invasion of the hemocoel by an intruding organism necessitates that the cells distinguish the invader from the insect's own tissues. This is often called the recognition of non-self. It requires that the surface of an invader can be distinguished from

the surfaces of the tissues by its physical and/or chemical properties.

Insect hemolymph contains proteins that bind to carbohydrates. These are sometimes called hemagglutinins because they agglutinate mammalian erythrocytes, but the term lectin is now more generally used. Carbohydrates are incorporated into the surfaces of bacteria and fungi. For example, β–1,3–glucans are present in the walls of many fungi, and lipopolysaccharides or peptidoglycans in bacterial cell walls. These compounds stimulate phagocytosis or the activation of the prophenoloxidase cascade. It is likely that the insect's lectins interact with these compounds and this in some way makes the invading particles subject to phagocytosis (they are said to be opsonized).

In general, granulocytes appear to be involved in the initial stages of recognition of foreign tissue. Degranulation by these cells certainly occurs in the initial stages of nodule formation and encapsulation (see below). In *Bombyx*, a protein that interacts with β–1,3–glucans is present in the granules of granulocytes and in the hemolymph (Ochiai, Niki & Ashida, 1992). As a consequence of the interaction of the granulocytes with the invading organism, plasmatocytes are attracted to the site. The phagocytic effectiveness of plasmatocytes in larvae of the wax moth, *Galleria*, is greatly enhanced by the presence of granulocytes (Anggraeni & Ratcliffe, 1991; and see below). In addition, plasmatocytes of larval *Galleria* that are actively phagocytosing produce a factor which stimulates phagocytic activity in other plasmatocytes (Wiesner, Wittwer & Gotz, 1996).

5.3.2 Cellular responses

Wounding or infection cause marked changes in hemocyte counts. In larval insects and adults of hemimetabolous species, wounding produces a rapid decline in the hemocyte count, but this returns to its original level within an hour and may then become elevated above the control level for a day. If, at the time the insect is wounded, pathogenic organisms enter the hemocoel, the cell count does not recover for a longer period (Fig. 5.32a). The sudden reduction in numbers is largely a result of removing the plasmatocytes from circulation. Within five minutes of the larva of *Galleria* becoming infected by the highly pathogenic *Bacillus cereus*, the plasmatocytes almost completely disappear from the blood. They previously comprised almost 50% of the hemocyte population. Other, less pathogenic bacteria also produce a rapid decline in the number of

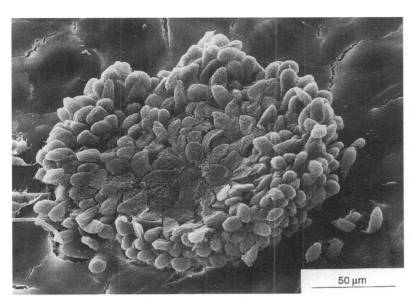

Fig. 5.31. Wound healing in *Schistocerca*. A spore of the fungus, *Metarhizium*, was placed on the outside of the cuticle 17 hours before the photograph was taken. The photograph shows the inside of the basal lamina (cracked) with an aggregation of hemocytes at the point of entry of a fungal hypha (courtesy of Dr S.G.S.Gunnarson).

50 μm

plasmatocytes, although the effect is less marked. The numbers of other types of hemocyte appear unaffected by the infection (Chain & Anderson, 1982). The reduction in plasmatocyte numbers may result from a greater tendency for the hemocytes to adhere to other objects, either the insect's own tissues or an invading organism. In the latter instance this behavior is associated with nodule or capsule formation and the sustained low hemocyte count associated with pathogenic bacteria is probably a further reflection of this as more cells adhere to the capsule. An increased tendency of hemocytes to bind to lectins following wounding or infection by the fungus, *Beauveria*, has been demonstrated in the grasshopper (Fig. 5.32b). While the effect of wounding alone is short-lived, the fungus has a more sustained effect (Miranpuri & Khachatourians, 1993). It is also probable that mortality of hemocytes is produced by toxins secreted by the invading bacteria or fungi.

The response of the hemocytes in pupal Lepidoptera is quite different. Here the number of cells increases ten-fold within an hour of wounding, and this elevated count is maintained for several days.

Phagocytosis Small numbers of bacteria, fungal spores or protozoans are phagocytosed by plasmatocytes. Non-pathogenic and weakly pathogenic organisms are dealt with effectively in this way, but the means by which they are killed are not known. Possibly bactericidal proteins produced by the hemocyte are involved. Tissue debris

produced during metamorphosis is also removed by phagocytosis.

Avoidance of phagocytosis by pathogenic organisms may depend on different mechanisms. For example, hyphal bodies of the fungal pathogen *Nomuraea*, which is capable of killing armyworms, *Spodoptera*, are not phagocytosed. *Beauveria*, another fungal pathogen which is less virulent, is phagocytosed, but nevertheless a germ tube can develop and produce hyphae (Hung, Boucias & Vey, 1993). It appears that the surface characteristics of the first are not distinguished as 'foreign' by the insect. *Beauveria* is recognized, but possibly produces immunosuppressive factors which suppress the host's immune response (section 5.3.4).

Nodule formation Large numbers of bacteria or fungal spores are attacked in a different way, by the formation of nodules (Fig. 5.33). Within a few minutes of entry, bacteria are found trapped in the coagulum produced by numbers of granulocytes. In some insect species, melanization of the necrotic granulocytes and the coagulum begins within five minutes. After about two hours, plasmatocytes arrive and begin to form a layer round the outside of the clump of cells. Little phagocytosis occurs. The intensity of the response varies, depending on the pathogenicity of the invading organisms. More strongly pathogenic organisms induce a stronger and more rapid response than non-pathogenic forms (Ratcliffe & Walters, 1983). There is evidence

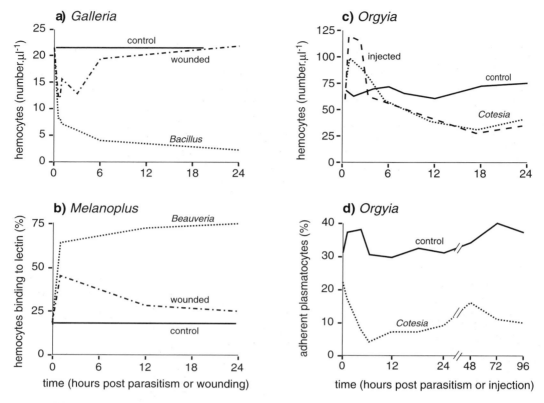

Fig. 5.32. Changes in hemocyte counts and binding properties due to wounding or infection. (a) Hemocyte count of larval *Galleria* following wounding or infection by a pathogenic *Bacillus* (data from Ratcliffe & Walters, 1983). (b) Binding properties of hemocytes of the grasshopper, *Melanoplus*, following wounding or infection by the fungus, *Beauveria* (data from Miranpuri & Khachatourians, 1983). (c) Hemocyte count of larval *Orgyia* following parasitism by the wasp *Cotesia*, or injection of a mixture of venom and the contents of the calyces of the wasp (after Guzo & Stoltz, 1987). (d) Binding properties of hemocytes of larval *Orgyia* following parasitism by *Cotesia* (after Guzo & Stoltz, 1987).

that the nodulation response is modulated by eicosanoids (Miller *et al.*, 1996).

Encapsulation Larger invaders, such as parasitoid larvae or nematodes, evoke a third type of response. They are encapsulated by large numbers of hemocytes. Granulocytes discharge their contents at the surface of the invader. This can occur within five minutes of the invasion. After about 30 minutes, plasmatocytes are attracted and start to accumulate. More plasmatocytes adhere to the outside of the clump so that the object becomes surrounded by a capsule comprising several layers of cells. The number of cell layers varies with the species and the physical and chemical nature of the surface of the object encapsulated. Recruitment of new hemocytes is usually

complete within 24 hours. The cells adjacent to the object become necrotic. Melanin is often produced in this layer as a consequence of the activation of the prophenoloxidase system. Cells in the adjacent layers become flattened and intercellular junctions, primarily tight junctions and desmosomes, develop between them. On the outside the cells are less modified.

Encapsulation normally occurs if the parasite is in an unusual host although the means by which it is killed are not understood.

5.3.3 Humoral responses
Damage to the epidermis, even without infection, induces some increase in the amounts of hemolymph proteins, but the effect is greatly enhanced by bacterial infection even if

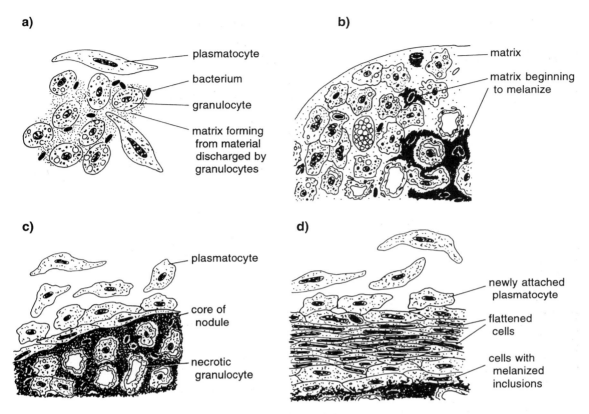

Fig. 5.33. Nodule formation (after Ratcliffe & Gagen, 1977). **(a)** One minute after injection of bacteria. Granulocytes have degranulated and bacteria are trapped in the flocculent material produced by the cells. **(b)** Thirty minutes after injection. The clumps of granulocytes and bacteria have compacted, and melanization of the matrix is beginning. **(c)** Plasmatocytes arrive at the nodule and melanization of the matrix is advanced. **(d)** Twenty-four hours after injection. Nodulation is complete. Three regions are recognizable in the layers of plasmatocytes.

the bacteria are dead. There is evidence that the response is elicited by peptidoglycans from the cell walls of the bacteria.

The hemolymph of the pupa of *Hyalophora* normally contains a small protein (48 kDa) known as hemolin. In *Manduca*, its synthesis is induced by bacterial infection. It is a member of the immunoglobulin family of proteins that are important in the immune systems of vertebrates. Peptidoglycans produce an increase in the amount of hemolin present and this appears to initiate the synthesis of a suite of proteins: a total of 15 different proteins in *Hyalophora* and 25 in *Manduca* larvae. These include two major families of proteins known as the cecropins and the attacins. The cecropins are peptides of 35–37 amino acids (4 kDa), the attacins are bigger, consisting of chains of 188 amino acids (20 kDa). They are synthesized mainly in the fat body, but also by some hemocytes and the quantity of cecropins produced increases with the number of bacteria injected into the hemolymph. These proteins are bactericidal, different proteins in each class exhibiting different specificities in their effectiveness against different species of bacteria. The induction of the different proteins is, however, the same irrespective of the nature of the bacterial infection. Comparable proteins, often belonging to different protein families, have been found in a number of other insects (For example, Chernysh *et al.*, 1996).

In addition to these proteins, lysozymes are also induced by infection. These enzymes have been found in a variety of insects, including *Locusta* and the house cricket, *Gryllus*. They appear to complement the action of cecropins and cecropin-like proteins by digesting the bacterial cell walls remaining after attack by other proteins.

Amongst the Diptera, a process analogous to encapsulation occurs, but apparently without the involvement of hemocytes. This process is known as humoral encapsulation. It is only known to occur in Chironomidae and other small Diptera with cell counts of less than 6000 hemocytes μl^{-1}. Perhaps normal cellular mechanisms are inefficient when the cell count is so low. Some 2–15 minutes after a foreign particle enters the hemolymph, strands of material begin to accumulate round it. More material is added until the invader is completely enclosed in a capsule one or more microns thick. The capsule becomes melanized, but some descriptions suggest that the protein of the capsule may become altered in some way analogous to cuticular sclerotization. Tyrosine from the hemolymph is utilized during the process. Humoral encapsulation appears to be a highly efficient means of isolating potential pathogens, and the larvae of *Chironomus* are able to withstand the injection of relatively high doses of highly pathogenic bacteria. Bacteria, fungi and nematodes can all be encapsulated by this mechanism.

Reviews: Boman & Hultmark, 1987; Götz & Vey, 1986 – humoral encapsulation

5.3.4 Overcoming the immune response

In the habitual host of a pathogen, or parasite, the invading organism usually survives. It is enabled to do this by overcoming, or by modifying, the host's normal response.

Organisms that normally are encapsulated may avoid encapsulation, or break free of the capsule. For example, as the germinating hyphae of the fungus *Metarhizium* penetrate the cuticle of *Schistocerca*, the host's hemocytes aggregate beneath the point of entry. The fungal wall contains β–1,3-glucans which presumably initiate the defensive response, but this response is suppressed by compounds consisting of five cyclically linked amino acids, known as destruxins, produced by the fungus (Huxham, Lackie & McCorkindale, 1989).

The larva of the braconid parasite, *Cotesia*, permanently suppresses encapsulation by the hemocytes of the larval tussock moth, *Orgyia*. After an initial period of elevation, the hemocyte count declines below the normal level and remains low. A similar response is produced by injecting the venom and the calyx fluid of the female parasitoid (Fig. 5.32c). The tendency of plasmatocytes to adhere to surfaces is also greatly reduced (Fig. 5.32d)

(Guzo & Stoltz, 1987). Clearly, the caterpillar is deprived of its ability to encapsulate the parasite. In the larva of the turnip moth, *Agrotis*, activation of the prophenoloxidase in the hemolymph is suppressed both by a parasitic nematode and by the nematode's symbiotic bacteria (Yokoo, Tojo & Ishibashi, 1992).

The processes involved in suppressing the host's response are best understood in parasitic Hymenoptera of the families Braconidae and Ichneumonidae. Some of these insects, such as *Cotesia*, carry a virus in the calyces of the ovaries. The virus, of a type called a polydna virus, is probably species-specific and it replicates in cells of the calyx from which it is released into the lumen of the oviduct. When the wasp oviposits in a host, she also injects a venom which probably serves to immobilize the prey, proteins secreted by the ovary, and the virus. Although it does not replicate in the host, the virus invades the host tissues and causes them to produce novel proteins. Its effectiveness may be synergized by components of the venom. In larval *Manduca* parasitized by the braconid *Cotesia*, some new proteins appear even before the parasitoid hatches and within 24 hours of oviposition they may constitute 10% of the protein in the host hemolymph. It is suggested that they bind to the surface of the newly hatched parasitoid larva and inhibit the encapsulation by the host hemocytes.

In addition to suppressing the host's immune response, some parasitoids modify the development of their hosts. A caterpillar of *Pseudaletia* parasitized by the wasp *Cotesia* produces a growth-blocking peptide, probably in response to the virus injected by the wasp. This peptide reduces the production of juvenile hormone esterase and, it is believed, in this way a high titer of juvenile hormone is maintained in the host hemolymph. This delays metamorphosis of the host. This is thought to be important because the wasp would be unable to escape from the heavily sclerotized pupa (Hayakawa, 1995).

Reviews: Beckage *et al.*, 1993; Lavine & Beckage, 1995

5.4 CONNECTIVE TISSUE

Insect connective tissue contains collagen fibers, and sometimes also elastin, glycosaminoglycans and glycoproteins. Collagen is a protein made up of three polypeptide chains with a helical structure. Glycine accounts for about 33% of the total amino acids and proline another 20%. The molecular spacing in fibrous collagen is very precise, giving rise to a characteristic banded appearance. Glycosaminoglycans are long chain polymers formed from disaccharide units. Each disaccharide consists of an amino sugar and another sugar or related molecule. Glycosaminoglycans are large molecules which form a major component of the connective tissue. Many different glycoproteins are present.

A fibrous connective tissue layer, usually 1 to 10 μm thick, surrounds many organs and may form a matrix within tissues, between muscle fibers, for example. It also forms a layer, the neural lamella, around the whole of the central nervous system (section 20.1.2). A thick layer, about 80 μm thick, surrounds the ejaculatory duct of male *Locusta*. A basal lamina, or basement membrane, is present beneath epithelia and surrounds fat body and muscles. It comprises an electron-lucent layer and an outer dense layer, but as fibrous connective tissue is produced beneath it the electron lucent layer may disappear. The basal lamina of the midgut is complex in some beetles and bugs and in the mosquito, *Aedes*. In the larva of *Oryctes*, for example, it is three layered and each layer is composed of electron dense units surrounding larger, less dense struc-tures, giving the layer an almost cellular appearance. The significance of this structure is not known.

Connective tissue is produced by the cells which it underlies and by fibroblasts within the matrix. Hemocytes also contribute to the basal lamina surrounding the hemo-coel (Sass, Kiss & Locke, 1994). In holometabolous insects, larval connective tissues are largely destroyed during the pupal period and replaced by newly formed material. These new structures are not produced until the underlying tissues are complete and they are absent from dividing tissues.

The usual functions of connective tissue are those of supporting and binding tissues together, although in insects the tracheal system also functions in this way to some extent. It is also possible that the elastic properties of some connective tissues are important. For example, the elasticity of the tunica propria (section 13.3) may help to squeeze oocytes from the ovarioles into the oviduct. It has also been suggested that the thickening of the epidermal basal lamina which occurs just before apolysis in *Rhodnius* enables it to serve as a base for the forces used in molding the new cuticle.

Review: Ashhurst, 1985

REFERENCES

Anggraeni, T. & Ratcliffe, N.A. (1991). Studies on cell-cell co-operation during phagocytosis by purified haemocyte populations of the wax moth, *Galleria mellonella*. *Journal of Insect Physiology*, **37**, 453–60.

Arnold, J.W. & Hinks, C.F. (1976). Haemopoiesis in Lepidoptera. I. The multiplication of circulating haemo-cytes. *Canadian Journal of Zoology*, **54**, 1003–12.

Arnold, J.W. & Hinks, C.F. (1983). Haemopoiesis in Lepidoptera. III. A note on the multiplication of spherule cells and granular haemocytes. *Canadian Journal of Zoology*, **61**, 275–7.

Ashhurst, D.E. (1985). Connective tissues. In *Comprehensive Insect Physiology, Biochemistry and Pharmacology*, vol. 3, ed. G.A. Kerkut & L.I. Gilbert, pp. 249–87. Oxford: Pergamon Press.

Beckage, N.E., Thompson, S.N. & Federici, B.A. (ed.) (1993). *Parasites and Pathogens of Insects*. San Diego: Academic Press.

Boman, H.G. & Hultmark, D. (1987). Cell-free immunity in insects. *Annual Review of Microbiology*, **41**, 103–26.

Boucias, D.G. & Pendland, J.C. (1987). Detection of protease inhibitors in the hemolymph of resistant *Anticarsia gem-matalis* which are inhibitory to the entomopathogenic fungus, *Nomuraea rileyi*. *Experientia*, **43**, 336–9.

Bräunig, P. (1991). A suboesophageal gan-glion cell innervates heart and retro-cerebral glandular complex in the locust. *Journal of Experimental Biology*, **156**, 567–82.

Brehélin, M. & Zachary, D. (1986). Insect hemocytes: a new classification to rule out the controversy. In *Immunity in Invertebrates*, ed. M. Brehélin, pp. 36–48. Berlin: Springer-Verlag.

Chain, B.M. & Anderson, R.S. (1982). Selective depletion of the plasmato-cytes in *Galleria mellonella* following infection of bacteria. *Journal of Insect Physiology*, **28**, 377–84.

Chain, B.M., Leyshon-Sørland, K. & Siva-Jothy, M.T. (1992). Haemocyte heterogeneity in the cockroach *Periplaneta americana* analysed using monoclonal antibodies. *Journal of Cell Science*, **103**, 1261–7.

Chernysh, S., Cociancich, S., Briand, J.-P., Hetru, C. & Bulet, P. (1996). The inducible antibacterial peptides of the Hemipteran insect *Palomena prasina*: identification of a unique family of proline-rich peptides and of a novel insect defensin. *Journal of Insect Physiology*, **42**, 81–9.

Chiang, A.S., Gupta, A.P. & Han, S.S. (1988). Arthropod immune system: I. Comparative light and electron microscopic accounts of immunocytes and other hemocytes of *Blattella germanica* (Dictyoptera: Blattellidae). *Journal of Morphology*, **198**, 257–67.

Chiang, R.G., Chiang, J.A. & Davey, K.G. (1990). Morphology of the dorsal vessel in the abdomen of the blood-feeding insect *Rhodnius prolixus*. *Journal of Morphology*, **204**, 9–23.

Clare, S. & Tauber, O.E. (1942). Circulation of haemolymph in the wings of the cockroach *Blattella germanica* L. III. Circulation in the articular membrane, the pteralia, and wing folds as directive and speed controlling mechanisms in wing circulation. *Iowa State College Journal of Science*, **16**, 349–56.

Crailsheim, K. (1985). Distribution of haemolymph in the honeybee (*Apis mellifica*) in relation to season, age and temperature. *Journal of Insect Physiology*, **31**, 707–13.

Delobel, B., Rahbé, Y., Nardon, C., Guillaud, J. & Nardon, P. (1993). Biochemical and cytological survey of tyrosine storage proteins in Coleoptera: diversity of strategies. *Insect Biochemistry and Molecular Biology*, **23**, 355–65.

Duhamel, R.C. & Kunkel, J.G. (1987). Moulting-cycle regulation of haemolymph protein clearance in cockroaches: possible size-dependent mechanism. *Journal of Insect Physiology*, **33**, 155–8.

Dunn, P.E. (1986). Biochemical aspects of insect immunology. *Annual Review of Entomology*, **31**, 321–39.

Edwards, H.A. (1982). Ion concentration and activity in the haemolymph of *Aedes aegypti* larvae. *Journal of Experimental Biology*, **101**, 143–51.

Eguchi, M., Matsui, Y & Matsumoto, T. (1986). Developmental change and hormonal control of chymotrypsin inhibitors in the haemolymph of the silkworm, *Bombyx mori*. *Comparative Biochemistry and Physiology*, **48B**, 327–32.

Feir, D. (1979). Multiplication of hemocytes. In *Insect Hemocytes*, ed. A.P.Gupta, pp. 67–82. Cambridge: Cambridge University Press.

Ferber, M. & Pflüger, H.-J. (1990). Bilaterally projecting neurones in pregenital abdominal ganglia of the locust: anatomy and peripheral targets. *Journal of Comparative Neurobiology*, **302**, 447–60.

Gilbert, L.I. (1967). Lipid metabolism and function in insects. *Advances in Insect Physiology*, **4**, 70–211.

Gillespie, J.P., Kanost, M.R. & Trenczek, T. (1997). Biological mediators of insect immunity. *Annual Review of Entomology*, **42**, 611–43.

Götz, P. & Boman, H.G. (1985). Insect immunity. In *Comprehensive Insect Physiology, Biochemistry and Pharmacology*, vol. 3, ed. G.A. Kerkut & L.I. Gilbert, pp. 453–85. Oxford: Pergamon Press.

Götz, P. & Vey, A. 1986. Humoral encapsulation in insects. In *Hemocytic and Humoral Immunity in Arthropods*, ed. A.P. Gupta, pp. 407–29. New York: Wiley.

Gupta, A.P. (1979a). Hemocyte types: their structures, synonymies, interrelationships, and taxonomic significance. In *Insect Hemocytes*, ed. A.P. Gupta, pp. 85–127. Cambridge: Cambridge University Press.

Gupta, A.P. (ed.) (1979b). *Insect Hemocytes*. Cambridge: Cambridge University Press.

Gupta, A.P. (1985). Hemocytes. In *Comprehensive Insect Physiology, Biochemistry and Pharmacology*, vol. 3, ed. G.A. Kerkut & L.I. Gilbert, pp. 453–85. Oxford: Pergamon Press.

Gupta, A.P. (ed.) 1986. *Hemocytic and Humoral Immunity in Arthropods*. New York: Wiley.

Gupta, A.P. (1991). Insect immunocytes and other hemocytes: roles in cellular and humoral immunity. In *Immunology of Insects and other Arthropods*, ed. A.P. Gupta, pp. 19–118. Boca Raton: CRC Press.

Guzo, D. & Stoltz, D.B. (1987). Observations on cellular immunity and parasitism in the tussock moth. *Journal of Insect Physiology*, **33**, 19–31.

Hantschk, A.M. (1991). Functional morphology of accessory circulatory organs in the legs of Hemiptera. *International Journal of Insect Morphology & Embryology*, **20**, 259–73.

Hayakawa, Y. (1995). Growth-blocking peptide: an insect biogenic peptide that prevents the onset of metamorphosis. *Journal of Insect Physiology*, **41**, 1–6.

Hoffman, J.A. (1967). Étude des hémocytes de *Locusta migratoria* L. (Orthoptère). *Archives de Zoologie Expérimentale et Générale*, **108**, 251–91.

Hoffmann, J.A. (1976). Appareil circulatoire et circulation. In *Traité de Zoologié*, vol. 8, part 4, ed. P.-P.Grassé, pp. 1–91. Paris, Masson et Cie.

Hoffmann, J.A., Zachary, D., Hoffmann, D. & Brehélin, M. (1979). Postembryonic development and differentiation: hemopoietic tissues and their functions in some insects. In *Insect Hemocytes*, ed. A.P. Gupta, pp. 29–66. Cambridge: Cambridge University Press.

Hung, S.-Y., Boucias, D.G. & Vey, A.J. (1993). Effect of *Beauveria bassiana* and *Candida albicans* on the cellular defense response of *Spodoptera exigua*. *Journal of Invertebrate Pathology*, **61**, 179–87.

Huxham, I.M., Lackie, A.M. & McCorkindale, N.J. (1989). Inhibitory effects of cyclodepsipeptides, destruxins, from the fungus *Metarhizium anisopliae*, on cellular immunity in insects. *Journal of Insect Physiology*, **35**, 97–105.

Irving, S.N., Wilson, R.G. & Osborne, M.P. (1979). Studies on L-glutamate in insect haemolymph. III. Amino acid analyses of the haemolymph of various arthropods. *Physiological Entomology*, **4**, 231–40.

Jeuniaux, C., Duchâteau-Bosson, G. & Florkin, M. (1961). Contributions à la biochemie du ver a soie. *Archives Internationales du Physiologie et de Biochimie*, **69**, 617–27.

Jones, J.C. (1967). Estimated changes within the haemocyte population during the last larval and early pupal stages of *Sarcophaga bullata* Parker. *Journal of Insect Physiology*, **13**, 645–6.

Jones, J.C. (1977). *The Circulatory System of Insects*. Springfield: C.C.Thomas.

Kanost, M.R., Kawooya, J.K., Law, J.H., Ryan, R.O., Van Heusden, M.C. & Ziegler, R. (1990). Insect haemolymph proteins. *Advances in Insect Physiology*, **22**, 299–396.

Kaufman, W.R. & Davey, K.G. (1971). The pulsatile organ in the tibia of *Triatoma phyllosoma pallidipennis*. *Canadian Entomologist*, **103**, 487–96.

King, L.E. & Tobe, S.S. (1993). Changes in the titre of a juvenile hormone III binding lipophorin from the hemolymph of *Diploptera punctata* during development and reproduction: functional significance. *Journal of Insect Physiology*, **39**, 241–51.

Koopmanschap, B., Lammers, H. & de Kort, S. (1992). Storage proteins are present in the hemolymph from larvae and adults of the Colorado potato beetle. *Archives of Insect Biochemistry and Physiology*, **20**, 119–33.

Krenn, H.W. & Pass, G. (1993). Wing-hearts in Mecoptera. *International Journal of Insect Morphology & Embryology*, **22**, 63–76.

Krenn, H.W. & Pass, G. (1994). Morphological diversity and phylogenetic analysis of wing circulatory organs in insects, part I: Non-Holometabola. *Zoology*, **98**, 7–22.

Krenn, H.W. & Pass, G. (1995). Morphological diversity and phylogenetic analysis of wing circulatory organs in insects, part II: Holometabola. *Zoology*, **98**, 147–64.

Kurata, S., Saito, H. & Natori, S. (1993). The 29-kDa hemocyte proteinase dissociates fat body at metamorphosis of *Sarcophaga*. *Developmental Biology*, **153**, 115–21.

Lackie, A.M. (1988). Haemocyte behaviour. *Advances in Insect Physiology*, **21**, 85–178.

Lavine, M.D. & Beckage, N.E. (1995). Polydnaviruses: potent mediators of host insect immune dysfunction. *Parasitology Today*, **11**, 368–78.

Lee, R.M. (1961). The variation of blood volume with age in the desert locust, (*Schistocerca gregaria* Forsk.). *Journal of Insect Physiology*, **6**, 36–51.

Lettau, J., Foster, W.A., Harker, J.E. & Treherne, J.E. (1977). Diel changes in potassium activity in the haemolymph of the cockroach *Leucophaea maderae*. *Journal of Experimental Biology*, **71**, 171–86.

Levenbook, L. (1985). Insect storage proteins. In *Comprehensive Insect Physiology, Biochemistry and Pharmacology*, vol. 10, ed. G.A.Kerkut & L.I.Gilbert, pp.307–46. Oxford: Pergamon Press.

Locke, M. & Nichol, H. (1992). Iron economy in insects: transport, metabolism, and storage. *Annual Review of Entomology*, **37**, 195–215.

Machin, J. (1981). Water compartmentalisation in insects. *Journal of Experimental Zoology*, **215**, 327–33.

Masera, E. (1933). Il ritmo del vaso pulsante nel *Bombyx mori*. *Rivista di Biologia*, **15**, 225–34.

Miller, J.S., Howard, R.W., Nguyen, T., Nguyen, A., Rosario, R.M.T. & Stanley-Samuelson, D.W. (1996). Eicosanoids mediate nodulation responses to bacterial infections in larvae of the tenebrionid beetle, *Zophobas atratus*. *Journal of Insect Physiology*, **42**, 3–12.

Miller, T.A. (1985a). Structure and physiology of the circulatory system. In *Comprehensive Insect Physiology, Biochemistry and Pharmacology*, vol. 3, ed. G.A.Kerkut & L.I.Gilbert, pp. 289–353. Oxford: Pergamon Press.

Miller, T.A. (1985b). Heart and diaphragms. In *Comprehensive Insect Physiology, Biochemistry and Pharmacology*, vol. 11, ed. G.A.Kerkut & L.I.Gilbert, pp. 119–30. Oxford: Pergamon Press.

Mills, R.R. & Whitehead, D.L. (1970). Hormonal control of tanning in the American cockroach: changes in blood cell permeability during ecdysis. *Journal of Insect Physiology*, **16**, 331–40.

Miranpuri, G.S. & Khachatourians, G.G. (1993). Hemocyte surface changes in the migratory grasshopper, *Melanoplus sanguinipes* in response to wounding and infection with *Beauveria bassiana*. *Entomologia Experimentalis et Applicata*, **68**, 157–64.

Monpeyssin, M. & Beaulaton, J. (1978). Hemocytopoiesis in the oak silkworm *Antheraea pernyi* and some other Lepidoptera I. Ultrastructural study of normal processes. *Journal of Ultrastructure Research*, **64**, 35–45.

Mullins, D.E. (1985). Chemistry and physiology of the hemolymph. In *Comprehensive Insect Physiology, Biochemistry and Pharmacology*, vol. 3, ed. G.A. Kerkut & L.I. Gilbert, pp. 355–400. Oxford: Pergamon Press.

Nicolson, S.W. (1980a). Diuresis and its hormonal control in butterflies. *Journal of Insect Physiology*, **26**, 841–6.

Nicolson, S.W. (1980b). Water balance and osmoregulation in *Onymacris plana*, a tenebrionid beetle from the Namib Desert. *Journal of Insect Physiology*, **26**, 315–20.

Nittono, Y. (1960). Studies on the blood cells in the silkworm, *Bombyx mori* L. *Bulletin of the Sericulture Experiment Station Tokyo*, **16**, 171–266.

Nutting, W.L. (1951). A comparative anatomical study of the heart and accessory structures of the orthopteroid insects. *Journal of Morphology*, **89**, 501–97.

Ochiai, M., Niki, T. & Ashida, M. (1992). Immunocytochemical localization of β–1,3-glucan recognition protein in the silkworm, *Bombyx mori*. *Cell and Tissue Research*, **268**, 431–7.

Palli, S.R. & Locke. M. (1988). The synthesis of hemolymph proteins by the larval fat body of an insect *Calpodes ethlius* (Lepidoptera: Hesperiidae). *Insect Biochemistry*, **18**, 405–13.

Pass, G. (1985). Gross and fine structure of the antennal circulatory organ in cockroaches (Blattodea, Insecta). *Journal of Morphology*, **185**, 255–68.

Pass, G. (1987). The 'cercus heart' in stoneflies – a new type of accessory circulatory organ in insects. *Naturwissenschaften*, **74**, 440–1.

Pass, G. (1991). Antennal circulatory organs in Onychophora, Myriapoda and Hexapoda: functional morphology and evolutionary implications. *Zoomorphology*, **110**, 145–64.

Prasad, S.V., Tsuchida, K., Cole, K.D. & Wells, M.A. (1987). Lipophorin biosynthesis during the life cycle of the tobacco hornworm, *Manduca sexta*. In *Molecular Entomology*, ed. J.H. Law, pp. 267–73. New York: Liss.

Prier, K.R., Beckman, O.H. & Tublitz, N.J. (1994). Modulating a modulator: biogenic amines at subthreshold levels potentiate peptide-mediated cardioexcitation of the heart of the tobacco hawkmoth *Manduca sexta*. *Journal of Experimental Biology*, **197**, 377–91.

Ratcliffe, N.A. & Gagen, S.J. (1977). Studies on the *in vivo* cellular reactions of insects: an ultrastructural analysis of nodule formation in *Galleria mellonella*. *Tissue & Cell*, **9**, 73–85.

Ratcliffe, N.A. & Rowley, A.F. (1979). Role of hemocytes in defense against biological agents. In *Insect Hemocytes*, ed. A.P. Gupta, pp. 331–414. Cambridge: Cambridge University Press.

Ratcliffe, N.A. & Walters, J.B. (1983). Studies on the *in vivo* cellular reactions of insects: clearance of pathogenic and non-pathogenic bacteria in *Galleria mellonella* larvae. *Journal of Insect Physiology*, **29**, 407–15.

Reynolds, S.E. (1980). Integration of behaviour and physiology in ecdysis. *Advances in Insect Physiology*, **15**, 475–595.

Richards, A.G. (1963) The ventral diaphragm of insects. *Journal of Morphology*, **113**, 17–47.

Riddiford, L.M. & Hice, R.H. (1985). Developmental profiles of the mRNAs for *Manduca* arylphorin and two other storage proteins during the final larval instar of *Manduca sexta*. *Insect Biochemistry*, **15**, 489–502.

Roberts, D.B., Turing, J.D. & Loughlin, S.A.R. (1991). The advantages that accrue to *Drosophila melanogaster* possessing larval serum protein 1. *Journal of Insect Physiology*, **37**, 391–400.

Roussel, J.-P. (1972). Rythme et régulation du coeur chez *Locusta migratoria migratorioides* L. *Acrida*, **1**, 17–39.

Rowley, A.F. & Ratcliffe, N.A. (1981). Insects. In *Invertebrate Blood Cells* vol. 2, ed. N.A. Ratcliffe & A.F. Rowley, pp. 421–88. London: Academic Press.

Salama, M.S. & Miller, T.A. (1992). A diapause associated protein of the pink bollworm *Pectinophora gossypiella* Saunders. *Archives of Insect Biochemistry and Physiology*, **21**, 1–11.

Sass, M., Kiss, A. & Locke, M. (1994). Integument and hemocyte peptides. *Journal of Insect Physiology*, **40**, 407–21.

Shapiro, J.P., Law, J.H. & Wells, M.A. (1988). Lipid transport in insects. *Annual Review of Entomology*, **33**, 297–318.

Shapiro, M. (1979). Changes in hemocyte populations. In *Insect Hemocytes*, ed. A.P.Gupta, pp. 475–523. Cambridge: Cambridge University Press.

Strand, M.R. & Pech, L.L. (1995). Immunological basis for compatibility in parasitoid-host relationships. *Annual Review of Entomology*, **40**, 31–56.

Sutcliffe, D.W. (1963). The chemical composition of haemolymph in insects and some other arthropods, in relation to their phylogeny. *Comparative Biochemistry and Physiology*, **9**, 121–35.

Telfer, W.H. & Kunkel, J.G. (1991). The function and evolution of insect storage hexamers. *Annual Review of Entomology*, **36**, 205–28.

Tublitz, N.J. & Truman, J.W. (1985). Insect cardioactive peptides II: neurohormonal control of heart activity by two cardioacceleratory peptides in the tobacco hawkmoth, *Manduca sexta*. *Journal of Experimental Biology*, **114**, 381–95.

Wasserthal, L.T. (1980). Oscillating haemolymph 'circulation' in the butterfly, *Papilio machaon* L. revealed by contact thermography and photocell measurements. *Journal of Comparative Physiology* B, **139**, 145–63.

Wasserthal, L.T. (1981). Oscillating haemolymph 'circulation' and discontinuous tracheal ventilation in the giant silk moth *Attacus atlas* L. *Journal of Comparative Physiology* B, **145**, 1–15.

Wasserthal, L.T. (1982a). Antagonism between haemolymph transport and tracheal ventilation in an insect wing (*Attacus atlas* L.). *Journal of Comparative Physiology* B, **147**, 27–40.

Wasserthal, L.T. (1982b). Reciprocal functional and structural adaptations in the circulatory and ventilatory systems in adult insects. *Verhandlungen der Deutschen zoologischen Gesellschaft*, **1982**, 105–16.

Wasserthal, L.T. (1983). Haemolymph flows in the wings of pierid butterflies visualized by vital staining (Insecta, Lepidoptera). *Zoomorphology*, **103**, 177–92.

Whedon, A.D. (1938). The aortic diverticula of the Odonata. *Journal of Morphology*, **63**, 229–61.

Wheeler, D.E. & Buck, N.A. (1995). Storage proteins in ants during development and colony founding. *Journal of Insect Physiology*, **41**, 885–94.

Wiesner, A., Wittwer, D. & Götz, P. (1996). A small phagocytosis stimulating factor is released by and acts on phagocytosing *Galleria mellonella* haemocytes *in vitro*. *Journal of Insect Physiology*, **42**, 829–35.

Wigglesworth, V.B. (1972). *The Principles of Insect Physiology*. London: Chapman & Hall.

Yokoo, S., Tojo, S. & Ishibashi, N. (1992). Suppression of the prophenoloxidase cascade in the larval haemolymph of the turnip moth, *Agrotis segetum* by an entomopathogenic nematode, *Steinernema carpocapsae* and its symbiotic bacterium. *Journal of Insect Physiology*, **38**, 915–24.

Ziegler, R., Willingham, L.A., Sanders, S.J., Tamen-Smith, L. & Tsuchida, K. (1995). Apolipophorin-III and adipokinetic hormone in lipid metabolism of larval *Manduca sexta*. *Insect Biochemistry and Molecular Biology*, **25**, 101–8.

6 Fat body

The insect fat body is the principal organ of intermediary metabolism. Most hemolymph proteins are synthesized in the fat body, and it also functions in the storage of proteins, lipids and carbohydrates. It consists of thin sheets or ribbons, usually only one or two cells thick, or of small nodules suspended in the hemocoel by connective tissue and tracheae. All the cells are consequently in immediate contact with the hemolymph, facilitating the exchange of metabolites. There is generally a peripheral, or parietal, fat body layer immediately beneath the body wall, and often a perivisceral layer surrounding the alimentary canal can also be distinguished (Fig. 6.1). The fat body is most conspicuous in the abdomen, but components extend into the thorax and head, and insinuate around the other tissues. Within a species, the arrangement is more or less constant, but there are considerable differences between insects in different orders. In hemimetabolous insects, the larval fat body persists in the adult without major changes, but in holometabolous insects the fat body is completely rebuilt at metamorphosis. In most cases this involves rebuilding a new structure from existing fat body cells following histolysis of the larval tissue, but in some adult Diptera it may be developed *de novo* (section 15.3.2.2).

Reviews: Dean, Locke & Collins, 1985; Martoja, 1976

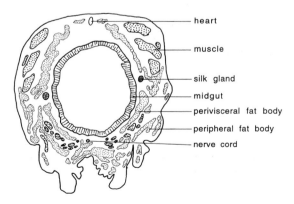

Fig. 6.1. Distribution of fat body in a caterpillar in transverse section of the abdomen.

The labels in the figure are: heart, muscle, silk gland, midgut, perivisceral fat body, peripheral fat body, nerve cord.

6.1 STRUCTURE

The principal cells of the fat body are the trophocytes, or adipocytes, and in many orders these are the only cells present. They are held together to form the sheets of tissues by desmosomes and gap junctions, and the whole tissue is clothed in a basal lamina (Fig. 6.2) attached to the cell by hemidesmosomes. In addition to the trophocytes, there may be urate cells, hemoglobin, or tracheal, cells, mycetocytes and sometimes also oenocytes.

6.1.1 Trophocytes

The form of the trophocyte varies according to the insect's developmental stage and its nutritional status. In a larva soon after ecdysis, the trophocytes are generally small with relatively little cytoplasm and little development of organelles. There are few mitochondria following the cell division preceding ecdysis (Fig. 6.3a,d), but, in a well-fed insect, there follows a preparative phase during which the trophocytes develop their capacity for synthesis. In the final larval stage of the moth, *Calpodes*, the preparative period lasts about 66 hours. During this period there is extensive replication of DNA, but no nuclear division (Fig. 6.3a,b). Most of the cells become octaploid, although some cells exhibiting 16- and 32-ploidy also occur. A similar development occurs in *Rhodnius*, while in *Calliphora*, and probably in other Diptera, polyteny occurs (the chromosomes divide, but do not separate). At the same time RNA synthesis occurs (Fig. 6.3c) as ribosomes increase in number, rough endoplasmic reticulum proliferates and the numbers of mitochondria increase by division (Fig. 6.3d). The trophocytes now have the apparatus necessary to begin synthesis. During the preparative period, the plasma membrane invaginates in a series of folds which interconnect to form the plasma membrane reticular system (Fig. 6.2). In *Calpodes*, the membranes in this system are separated from each other by 100–150 nm. The reticular system occupies the peripheral 1–1.5 μm of the cells and it presents an exceptionally large surface area to the hemolymph. However, this surface is negatively

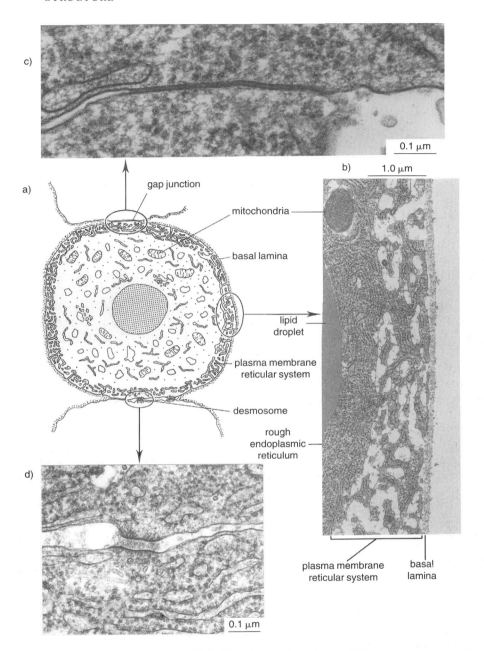

c)

0.1 μm

b) 1.0 μm

a)

gap junction

mitochondria

basal lamina

lipid
droplet

plasma membrane
reticular system

desmosome

rough
endoplasmic
reticulum

d)

0.1 μm

plasma membrane
reticular system

basal
lamina

Fig. 6.2. Structure of a mature trophocyte. (**a**) Diagram of a trophocyte. (**b**) Transmission electron micrograph of the plasma membrane reticular system of a trophocyte from the larva of *Calpodes*. (**c**) Gap junction between two trophocytes in the fat body of *Calpodes*. (**d**) Desmosome joining two trophocytes in the fat body of *Calpodes* (b, c and d after Dean *et al.*, 1985).

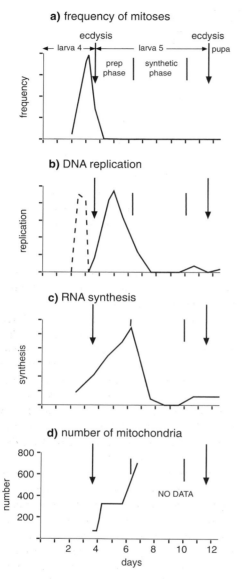

a) frequency of mitoses

ecdysis   ecdysis

← larva 4 →|← larva 5 ———→|pupa

prep | synthetic
phase | phase

frequency

b) DNA replication

replication

c) RNA synthesis

synthesis

d) number of mitochondria

number

NO DATA

2   4   6   8   10   12

days

Fig. 6.3. Changes occurring in the cells of the fat body of the caterpillar of *Calpodes* during a molt/intermolt cycle (prep phase is preparatory phase) (mainly after Locke, 1970). (**a**) The frequency of mitoses. Mitosis is limited to the period immediately before ecdysis. (**b**) DNA replication. The broken line is the presumed phase of replication associated with mitosis. The solid line shows measured replication which was not associated with cell division. (**c**) RNA synthesis in the cytoplasm. This occurs primarily in the preparatory phase. (**d**) Numbers of mitochondria in one cell (from Dean *et al*, 1985).

charged and the effect of this is to limit the access of some large charged molecules to the interior of the reticulum. It is possible that the reticulum is concerned with the reception and unloading of lipophorins (Locke, 1986; Locke & Huie, 1983). When the larva is approaching the molt to pupa, the components of the cell that have been involved in protein synthesis regress.

Immediately after eclosion, the trophocytes of adult insects commonly contain extensive lipid droplets, accumulations of glycogen, and protein granules. The trophocytes in males do not show further development and they probably play no further major role in protein synthesis. In females, however, changes occur which are comparable with those occurring in larval stages. This reflects the need, in many species, for the synthesis of vitellogenins.

Review: Locke, 1984 – vacuolar system

6.1.2 Urate cells

Urate cells, or urocytes, are present in Collembola (springtails), Thysanura (silverfish), Blattodea (cockroaches) and larval Apocrita (bees and wasps). These cells characteristically contain large crystalloid spherules of uric acid. Uric acid also accumulates as small granules in all fat body cells of larval and pupal Lepidoptera and in larval mosquitoes. It may be that in Collembola, which lack Malpighian tubules, and in larval Apocrita, which are confined to cells within the nest, the accumulation of uric acid is a means of storing the potentially toxic end-products of nitrogenous metabolism. This also appears to be the case in Lepidoptera where the uric acid starts to accumulate during the larval wandering phase preceding pupation. It continues to accumulate during the first part of the pupal period, but then is transferred to the rectum to be excreted in the meconium (see Fig. 18.4). In the cockroaches, however, the uric acid provides a store of nitrogen that can be recycled (section 18.2.2).

6.1.3 Hemoglobin cells

Hemoglobin cells have been described only in larvae of the bot fly, *Gasterophilus*, and in the backswimmers, *Anisops* and *Buenoa* (section 17.9). They are large cells, measuring about 20 × 80 μm in *Anisops* and up to 400 μm in diameter in *Gasterophilus* (Fig. 17.36). Each cell appears to be pierced by a trachea with numerous branches, but it is probable that the hemoglobin cell wraps round the tracheae, rather than being pierced by them. Because of this close association with tracheae, hemoglobin cells have been called tracheal cells, but this term is misleading. In *Anisops*, several hundred hemoglobin cells are carried on branches of a single trachea which consequently has a tree-like appearance (Miller, 1966). Hemoglobin is synthesized in these cells (Bergtrom *et al.*, 1976). Hemoglobin synthesis also occurs in the fat body of larval midges (Chironomidae), but in these insects the cells do not exhibit the anatomical specialization of hemoglobin cells.

6.1.4 Other cells

Mycetocytes are cells containing micro-organisms. In cockroaches and some Hemiptera they are scattered through the fat body (section 4.3.2.2). It is common for oenocytes, derived from the epidermis, also to be associated with the fat body (section 16.1.3).

6.2 FUNCTIONS

The fat body functions in many aspects of the storage and synthesis of proteins, lipids and carbohydrates. In a number of larval Lepidoptera and Diptera, the fat body is regionally differentiated to perform different functions, and this may be a more general phenomenon. The larva of *Helicoverpa* provides an example. During the period just before pupation, and perhaps at other times, protein synthesis occurs only in the peripheral fat body. Storage of arylphorin and a very high density lipoprotein, colored blue by noncovalently bound biliverdin, on the other hand, is restricted to the cells of the perivisceral fat body (Haunerland, Nair & Bowers, 1990). In *Drosophila*, different screening pigments for the adult eye are synthesized and sequestered in different parts of the larval fat body, and in the larva of *Chironomus*, hemoglobin synthesis appears to occur in the peripheral fat body, while storage may occur in the perivisceral fat body cells (Schin, Laufer & Carr, 1977).

Review: Haunerland & Shirk, 1995

6.2.1 Proteins and amino acids

The fat body is the principal site of synthesis of the hemolymph proteins described in section 5.2.2.3. In the larva of *Calpodes*, it synthesizes 14 out of 26 hemolymph polypeptides, amounting to about 90% of the total hemolymph protein (Palli & Locke, 1988). Figs. 5.20 and 5.22 show examples of the increase in storage proteins and lipoproteins in the hemolymph during protein synthesis by the fat

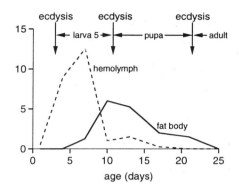

Fig. 6.4. Amounts of a blue-colored very-high-density lipoprotein present in the fat body and hemolymph in various stages of development of *Helicoverpa* (after Haunerland *et al.*, 1990).

body in larval insects. In addition, in adult females, the fat body produces vitellogenin, the protein that will form most of the yolk protein in the eggs.

Diapause proteins are also produced by the fat body. Adult Colorado potato beetles, *Leptinotarsa*, enter diapause under short day conditions. The adult beetles synthesize two vitellogenins and three different diapause proteins under all conditions. If the newly eclosed beetles experience long days, synthesis of the vitellogenins is emphasized and production of diapause proteins is low. In short days (a photoperiod of 10 hours or less), however, relatively little of the vitellogenins, but more of the diapause proteins are produced (Dortland, 1978).

As the insect prepares to pupate, protein synthesis in the fat body stops. Proteins, originally synthesized in and secreted by the fat body are now removed from the hemolymph and stored in the fat body as granules (Fig. 6.4). Some protein uptake does occur during the phase of protein synthesis, but the uptake is non-selective and proteins are lysed within the cells. Lysis stops at the end of the period of synthesis and the uptake of proteins is selective; different proteins are taken up to different extents. In both *Helicoverpa* and *Sarcophaga*, this selective uptake of specific proteins is dependent on the appearance of specific receptor proteins in the plasma membranes of the fat body. A precursor of the receptor protein is already present in the larval fat body of *Sarcophaga* and its conversion to the receptor for the uptake of storage protein is activated by molting hormone in the hemolymph before pupation (Ueno & Natori, 1984). In *Helicoverpa*, the receptor

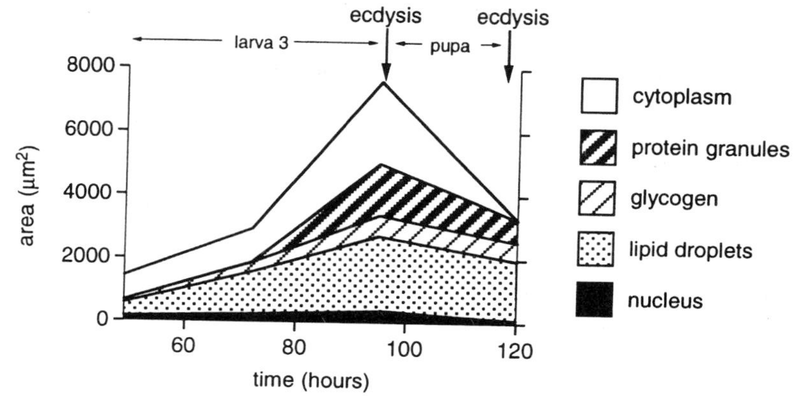

Fig. 6.5. Changes in the amounts of the major components of a trophocyte during the final larval and pupal stages of *Drosophila*. The major increases occur during the period of feeding. Amounts are expressed as the areas occupied by the components in cross-sections of the tissue (data from Butterworth *et al.*, 1965).

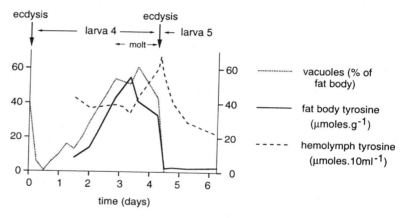

Fig. 6.6. Changes in the tyrosine content of the fat body and hemolymph of *Calpodes* larva. Tyrosine in the fat body is sequestered in vacuoles and the proportion of the fat body occupied by vacuoles is paralleled by changes in the tyrosine content. At ecdysis the vacuoles disappear as they release tyrosine into the hemolymph (after McDermid & Locke, 1983).

protein for the blue-colored protein is formed *de novo* at the time of pupation. This receptor is only present in the membranes of cells in the perivisceral fat body, not in the peripheral fat body (Wang & Haunerland, 1993). Although some increase in cell size occurs as lipid and glycogen accumulate, the uptake of proteins is associated with a considerable increase in the volumes of the fat body cells (Fig. 6.5).

While in general it appears that synthesis and storage of proteins do not occur simultaneously, in prediapause adults of *Leptinotarsa*, two diapause proteins are stored while synthesis continues.

The factors regulating protein synthesis and storage are not known with certainty although both juvenile hormone and ecdysteroids may be involved. Juvenile hormone suppresses synthesis of arylphorin in the silkworm, *Bombyx*. Synthesis starts in the final larval stage when juvenile hormone disappears from the hemolymph (Tojo, Kiguchi & Kimura, 1981). On the other hand, in adult insects of most orders, synthesis of vitellogenin is

stimulated by juvenile hormone, although in Diptera this function is performed by ecdysteroids (see Fig. 13.10). Juvenile hormone perhaps inhibits the synthesis of diapause proteins in *Leptinotarsa*, but the differences in the timing of production of the different proteins suggests that the regulation is complex (Dortland, 1978). Protein uptake into the fat body has been shown to be initiated by 20-hydroxyecdysone in a number of insects. Perhaps the hormone acts via its regulation of receptor proteins in the cell membranes of the trophocytes.

Tyrosine is a key chemical in cuticle sclerotization and insects accumulate it before a molt (section 16.5.3). In some insects, it is taken up from the hemolymph and stored in large vacuoles in the trophocytes. This has been most comprehensively studied in the fourth stage larva of *Calpodes* where the uptake of tyrosine begins about one day after ecdysis. Shortly before the next ecdysis, tyrosine is released into the hemolymph (Fig. 6.6). Accumulation in the fat body, which is an alternative to storing tyrosine as a glucoside (see Fig. 16.19) or a phosphate, may be a widespread phenomenon.

The fat body is also a major site of transamination between amino acids (section 4.1.1).

6.2.2 Carbohydrates

Carbohydrate is stored as glycogen which, in caterpillars, and probably also in other insects, is built up in the fat body during periods of active feeding. This store becomes depleted during sustained activity or over a molt, when the insect is not feeding, or if it is starved. For example, the glycogen content of the fat body of a well-fed migratory locust, *Locusta*, is about 2 mg per 100 mg fresh weight. After flying for two hours this is reduced to about 0.5 mg per 100 mg. In *Manduca* larvae, the hemolymph concentration of glucose falls within an hour of the start of starvation and remains low, but the concentration of total sugars, of which trehalose in the principal component, remains high after an initial fall (Fig. 6.7a,b). This level is maintained by conversion of glycogen to trehalose in the fat body.

The process is regulated by a hyperglycemic hormone from the corpora allata. This activates glycogen phosphorylase in the fat body (Fig. 6.7c) leading to the production of trehalose (Fig. 6.8). In the larva of *Manduca* the hormone producing this effect is the same as adipokinetic hormone in the adult moth. Release of the hormone is triggered by a low concentration of glucose in the hemolymph when the glucose is used metabolically. In the adult moth, however, adipokinetic hormone does not regulate the activity of glycogen phosphorylase; rather, activation seems to be induced by low levels of total sugars in the hemolymph (Gies, Fromm & Ziegler, 1988). In some other Lepidoptera, larval glycogen phosphorylase is believed to be activated by a hormone similar to, but distinct from the adult adipokinetic hormone.

Reviews: Friedman, 1985; Keeley, 1985

6.2.3 Lipids

The fat body is the principal store of lipid in the insect's body. Most of the lipid is present as triacylglycerol which commonly constitutes more than 70% of the dry weight of the fat body (see Fig. 5.21b). The amount stored varies with the stage of development and state of feeding of the insect. Lipid stores normally increase during periods of active feeding and decline when feeding stops (Fig. 6.5) or

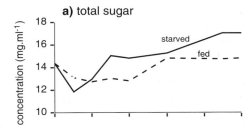

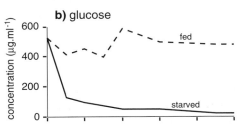

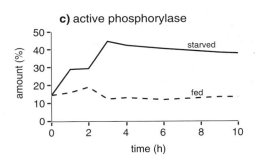

Fig. 6.7. The effects of starvation on carbohydrate metabolism in a well-fed final stage larva of *Manduca* (after Gies *et al.*, 1988). (a) Concentration of total sugars, mainly trehalose, in the hemolymph. (b) Concentration of glucose in the hemolymph. (c) Percent of active glycogen phosphorylase in the fat body (see Fig. 6.8).

when large quantities of lipid are used during oogenesis or prolonged flight.

The quantity of lipid accumulating in the fat body may greatly exceed the amount of lipid absorbed from the food (Fig. 6.9). The additional lipid is synthesized, primarily from carbohydrates. Not only is the total quantity of lipid increased by greater quantities of carbohydrate in the diet, but the relative proportions of different fatty acids may also change (Thompson, 1979a,b).

This process of lipid synthesis, known as lipogenesis, perhaps accounts for a large proportion of the lipids in most insects. Lipogenesis occurs primarily in the fat body,

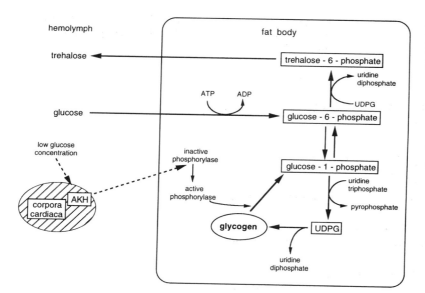

Fig. 6.8. Diagram showing the production and regulation of hemolymph trehalose from glycogen in the fat body of *Manduca* larva.

AKH = adipokinetic hormone

UDPG = uridine diphosphate glucose

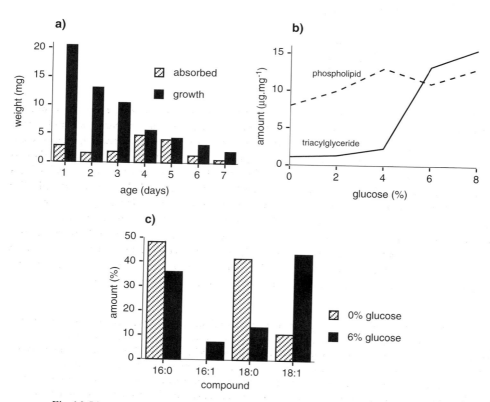

Fig. 6.9. Lipogenesis from carbohydrate. (a) The daily amount of lipid absorbed and the daily increase in lipid content during the last stage larva of *Locusta*. The increase in lipid in the first three days greatly exceeds the intake of lipid, implying that

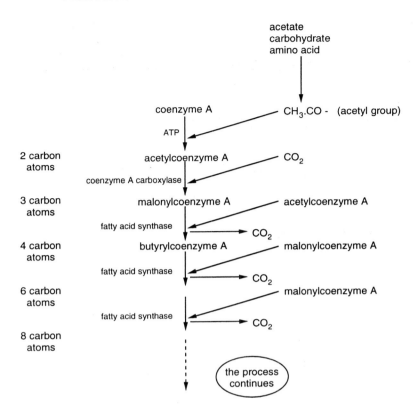

Fig. 6.10. Synthesis of fatty acids in the fat body (highly simplified).

but may also occur in the ovaries and other tissues. It involves the synthesis of fatty acids, frequently followed by their incorporation into triacylglycerol. The primary building block for fatty acids is the acetyl group ($CH_3.CO-$). In insects, this is known to be derived from acetate, glucose, or the amino acid, leucine. Combined with coenzyme A, which acts as a carrier, the active acetyl group is carboxylated to form malonyl–coenzyme A. The malonyl group contains three carbon atoms and by further condensation reactions with acetyl–coenzyme A and then with malonyl–coenzyme A, fatty acids of increasing chain length are produced (Fig. 6.10). As two carbon atoms are added to the acyl group at each of the final steps, the fatty acids produced generally have an even number of carbon atoms. In most insect species so far studied, the most commonly produced primary product of fatty acid synthesis is palmitic acid (16 carbon atoms), but myristic acid (C14) is produced by some flies and aphids, while stearic acid (C18) is the initial product in a hymenopteran parasitoid. Following this, further elongation and desaturation may occur to produce the fatty acids required by the insects (see Fig. 4.3). A large proportion of the fatty acids produced, as well as those derived directly from the diet, are subsequently combined to form triacylglycerol.

Fig. 6.9. (*cont.*)
lipogenesis from carbohydrate has occurred. Lipid growth is the increase in lipid content of the whole insect each day. Most of the lipid is in the fat body (data from Simpson, 1982). (b) Effect of dietary glucose on the lipid content of final stage larvae of *Exeristes*, a hymenopteran parasitoid. Larvae were reared on an artificial 'fat-free' diet to which various concentrations of glucose were added as the only carbohydrate. The amount of triacylglyceride increases sharply at higher glucose concentrations; relatively little increase occurs in phospholipid (data from Thompson, 1979a,b). (c) Proportions of different fatty acids in the triacylglycerides of *Exeristes* larvae reared on artificial diet containing different amounts of glucose. 16:0, palmitic acid; 16:1, palmitoleic acid; 18:0, stearic acid; 18:1, oleic acid. On the diet containing 0% glucose the total *amount* of triacylglyceride was very small (b) (data from Thompson, 1979a).

Fig. 6.11. Regulation of mobilization of diacylglycerol from triacylglycerol in the fat body.

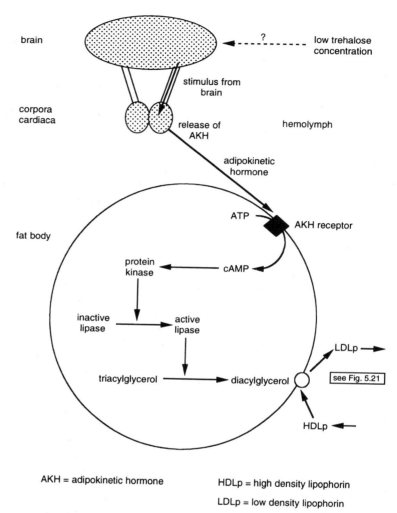

AKH = adipokinetic hormone

HDLp = high density lipophorin

LDLp = low density lipophorin

It is not known if there is any factor stimulating lipid synthesis, but there is considerable evidence that juvenile hormone inhibits synthesis. The mechanism of action, however, is not known. The mobilization of lipids from the fat body is known to be effected by the neuro-hormone, octopamine, and by adipokinetic hormone. Octopamine appears in the hemolymph when an insect is disturbed. In the desert locust, *Schistocerca*, it's titer increases fourfold within ten minutes of the start of flight while the hemolymph titer of adipokinetic hormone reaches a maximum after about 30 minutes of flight, and then is sustained at a lower level for some time. Adipokinetic hormone is released from the corpora cardiaca during flight in many insects. It is not certain how its release is regulated. Low levels of trehalose in the hemolymph produced early in the flight of the locust may stimulate release of adipokinetic hormone, but probably do not initiate it. Information from the brain is transmitted via the nerves connecting it with the corpora cardiaca and this results in release of the hormone. Both octopamine and adipokinetic hormone interact with receptor sites on the plasma membrane of fat body cells and elevate the production of cAMP. This activates a protein kinase which, in turn, activates a triacylglycerol kinase catalyzing the conversion of triacylglycerol to diacylglycerol (Fig. 6.11). Lipophorin then transports the diacylglycerol in the hemolymph (section 5.2.2.3).

Reviews: Downer, 1985; Steele, 1985; Wheeler, 1989

REFERENCES

Bergtrom, G., Gittelman, S., Laufer, H. & Ovitt, C. (1976). Haemoglobin synthesis in *Bueonoa confusa* (Hemiptera). *Insect Biochemistry*, **6**, 595–600.

Butterworth, F.M., Bodenstein, D. & King, R.C. (1965). Adipose tissue of *Drosophila melanogaster* I. An experimental study of larval fat body. *Journal of Experimental Zoology*, **158**, 141–54.

Dean, R.L., Locke, M. & Collins, J.V. (1985). Structure of the fat body. In *Comprehensive Insect Physiology, Biochemistry and Pharmacology*, vol. 3, ed. G.A. Kerkut & L.I. Gilbert, pp. 155–210. Oxford: Pergamon Press.

Dortland, J.F. (1978). Synthesis of vitellogenins and diapause proteins by the fat body of *Leptinotarsa*, as a function of photoperiod. *Physiological Entomology*, **3**, 281–8.

Downer, R.G.H. (1985). Lipid metabolism. In *Comprehensive Insect Physiology, Biochemistry and Pharmacology*, vol. 10, ed. G.A. Kerkut & L.I. Gilbert, pp. 77–113. Oxford: Pergamon Press.

Friedman, S. (1985). Carbohydrate metabolism. In *Comprehensive Insect Physiology, Biochemistry and Pharmacology*, vol. 10, ed. G.A. Kerkut & L.I. Gilbert, pp. 43–76. Oxford: Pergamon Press.

Gies, A., Fromm, T. & Ziegler, R. (1988). Energy metabolism in starving larvae of *Manduca sexta*. *Comparative Biochemistry and Physiology*, **91A**, 549–55.

Haunerland, N.H., Nair, K.N. & Bowers, W.S. (1990). Fat body heterogeneity during development of *Heliothis zea*. *Insect Biochemistry*, **20**, 829–37.

Haunerland, N.H. & Shirk, P.D. (1995). Regional and functional differentiation in the insect fat body. *Annual Review of Entomology*, **40**, 121–45.

Keeley, L.L. (1985). Physiology and biochemistry of the fat body. In *Comprehensive Insect Physiology, Biochemistry and Pharmacology*, vol. 3, ed. G.A. Kerkut & L.I. Gilbert, pp. 211–48. Oxford: Pergamon Press.

Locke, M. (1970). The molt/intermolt cycle in the epidermis and other tissues of an insect *Calpodes ethlius* (Lepidoptera, Hesperiidae). *Tissue & Cell*, **2**, 197–223.

Locke, M. (1984). The structure and development of the vacuolar system in the fat body of insects. In *Insect Ultrastructure*, vol. 2, ed. R.C. King & H. Akai, pp. 151–97. New York: Plenum Press.

Locke, M. (1986). The development of the plasma membrane reticular system in the fat body of an insect. *Tissue & Cell*, **18**, 853–67.

Locke, M. & Huie, P. (1983). A function for plasma membrane reticular systems. *Tissue & Cell*, **15**, 885–902.

Martoja, R. (1976). Le corps gras ou tissu adipeux. In *Traité de Zoologie*, vol. 8, part 4, ed. P.-P.Grassé, pp. 407–90. Paris: Masson et Cie.

McDermid, H. & Locke, M. (1983). Tyrosine storage vacuoles in insect fat body. *Tissue & Cell*, **15**, 137–58.

Miller, P.L. (1966). The function of haemoglobin in relation to the maintenance of neutral buoyancy in *Anisops pellucens* (Notonectidae, Hemiptera). *Journal of Experimental Biology*, **44**, 529–43.

Palli, S.R. & Locke, M. (1988). The synthesis of hemolymph proteins by the larval fat body of an insect *Calpodes ethlius* (Lepidoptera: Hesperiidae). *Insect Biochemistry*, **18**, 405–13.

Schin, K., Laufer, H. & Carr, E. (1977). Cytochemical and electrophoretic studies of haemoglobin synthesis in the fat body of a midge, *Chironomus thummi*. *Journal of Insect Physiology*, **23**, 1233–42.

Simpson, S.J. (1982). Changes in the efficiency of utilisation of food throughout the fifth-instar nymphs of *Locusta migratoria*. *Entomologia Experimentalis et Applicata*, **31**, 265–75.

Steele, J.E. (1985). Control of metabolic processes. In *Comprehensive Insect Physiology, Biochemistry and Pharmacology*, vol. 8, ed. G.A. Kerkut & L.I. Gilbert, pp. 99–145. Oxford: Pergamon Press.

Thompson, S.N. (1979a). Effect of dietary glucose on *in vitro* fatty acid metabolism and *in vitro* synthetase activity in the insect parasite, *Exeristes roborator* (Fabricius). *Insect Biochemistry*, **9**, 645–51.

Thompson, S.N. (1979b). The effects of dietary carbohydrate on larval development and lipogenesis in the parasite, *Exeristes roborator* (Fabricius) (Hymenoptera: Ichneumonidae). *Journal of Parasitology*, **65**, 849–54.

Tojo, S., Kiguchi, K. & Kimura, S. (1981). Hormonal control of storage protein synthesis and uptake by the fat body in the silkworm, *Bombyx mori*. *Journal of Insect Physiology*, **27**, 491–7.

Ueno, K. & Natori, S. (1984). Identification of storage protein receptor and its precursor in the fat body membrane of *Sarcophaga peregrina*. *Journal of Biological Chemistry*, **259**, 12107–11.

Wang, Z. & Haunerland, N.H. (1993). Storage protein uptake in *Helicoverpa zea*. *Journal of Biological Chemistry*, **268**, 16673–8.

Wheeler, C.H. (1989). Mobilization and transport of fuels to the flight muscles. In *Insect Flight*, ed. G.J. Goldsworthy & C.H. Wheeler, pp. 273–303. Boca Raton: CRC Press.

PART II

The Thorax and Locomotion

7 Thorax

The skeleton of the thoracic segments is modified to give efficient support for the legs and wings, and the musculature is adapted to produce the movements of these appendages.

7.1 SEGMENTATION

In larval holometabolous insects the cuticle is soft and flexible, or only partially sclerotized, and the longitudinal muscles are attached to the intersegmental folds (Fig. 7.1a). This represents a primitive condition comparable with that occurring in the annelids. Insects with this arrangement move as a result of successive changes in the shapes of the thoracic and abdominal segments (section 8.4.3), these changes of shape being permitted by the flexible cuticle. When the cuticle is sclerotized, sclerites in the intersegmental folds which have the longitudinal muscles attached to them are usually fused with the segmental sclerites behind. The large sclerite on the dorsal surface of a segment is called the tergum, or, in the thorax, the notum. Anteriorly it incorporates the intersegmental region, the original fold being marked by the antecostal sulcus where the cuticle is

inflected. The narrow rim in front of the sulcus is called the acrotergite (Figs. 7.1b, 7.2). An acrotergite never occurs at the front of the prothorax because the anterior part of this segment forms part of the neck and the muscles from the head pass directly to the acrotergite of the mesothorax.

An area at the back of each segment remains membranous, forming a new intersegmental membrane. This does not correspond with the original intersegmental groove and so a secondary segmentation is superimposed on the first, but neither corresponds precisely with the 'parasegments' defined by molecular studies (see section 14.2.5).

This basic condition with a membranous area at the posterior end of each segment occurs in the abdomen where this is sclerotized, in the meso- and meta-thoracic segments of larval insects with a sclerotized thorax, in the Apterygota and in adult Blattodea and Isoptera where the wings are not moved by indirect muscles. With this arrangement, contraction of the longitudinal muscles produces telescoping of the segments.

Review: Zrzavy & Stys, 1995

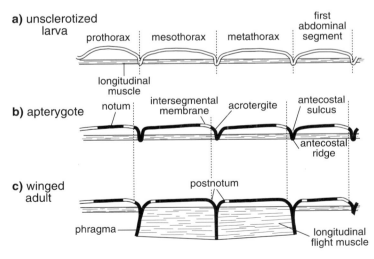

Fig. 7.1. Segmentation and the derivation of the postnotum and phragmata in pterygote insects. Sclerotized areas are indicated by a solid line, membranous areas by a double line.

7.2 THORAX

The thorax consists of three segments known as the pro-, meso- and meta-thoracic segments. In most insects all three segments bear a pair of legs, but this is not the case in larval Diptera, larval Hymenoptera Apocrita, some larval Coleoptera and a small number of adult insects which are apodous. In addition, winged insects have a pair of wings on the meso- and meta-thoracic segments and these two segments are then collectively known as the pterothorax.
Reviews: Matsuda, 1970, 1979; Snodgrass, 1935

7.2.1 Morphology of the thorax

Tergum The tergum of the prothoracic segment is known as the pronotum. It is often small serving primarily for attachment of the muscles of the first pair of legs, but in Orthoptera, Blattodea and Coleoptera it forms a large plate affording some protection to the pterothoracic segments. The meso- and meta-nota are relatively small in wingless insects and larvae, but in winged insects they become modified for the attachment of the wings. In the majority of winged insects the downward movement of the wings depends on an upwards distortion of the dorsal wall of the thorax (section 9.7.1). This is made possible by a modification of the basic segmental arrangement. The acrotergites of the metathorax and the first abdominal segment extend forwards to join the tergum of the segment in front and in many cases become secondarily separated from their original segment by a narrow membranous region. Each acrotergite and antecostal sulcus is now known as a postnotum (Fig. 7.1c). There may thus be a mesopostnotum and a metapostnotum if both pairs of wings are more or less equally important in flight, but where the hind wings provide most power, as in Orthoptera and Coleoptera, only the metapostnotum is developed. The Diptera on the other hand, using only the forewings for flight, have a well developed mesopostnotum, but no metapostnotum. To provide attachment for the large longitudinal muscles moving the wings, the antecostal ridges at the front and back of the mesothorax and the back of the metathorax usually develop into extensive internal plates, the phragmata (Figs. 7.1c and 7.6). Which of the phragmata are developed depends again on which wings are most important in flight.

Various strengthening ridges develop on the tergum of a wing-bearing segment. These are local adaptations to the mechanical stresses imposed by the wings and their muscles. The ridges appear externally as sulci which divide the notum

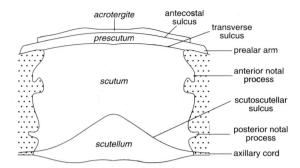

Fig. 7.2. Notum of a wing-bearing segment. Stippled areas are membrane at base of wing (axillary sclerites not shown). Names of sclerites in italics (after Snodgrass, 1935).

into areas. Often a transverse sulcus divides the notum into an anterior prescutum and a scutum, while a V-shaped sulcus posteriorly cuts off the scutellum (Fig. 7.2). These areas are commonly demarcated, but, because of their origins as functional units, plates of the same name in different insects are not necessarily homologous. In addition, the lateral regions of the scutum may be cut off by a sulcus or there may be a median longitudinal sulcus. Commonly, the prescutum connects with the pleuron by an extension, the prealar arm, in front of the wing, while behind the wing a postalar arm connects the postnotum to the epimeron (Fig. 7.5b). Laterally the scutum is produced into two processes, the anterior and posterior notal processes, which articulate with the axillary sclerites in the wing base (see Fig. 9.13). The posterior fold of the scutellum continues as the axillary cord along the trailing edge of the wing.

Sternum The primary sclerotizations on the ventral side are segmental and inter segmental plates which often remain separate in the thorax. The intersegmental sclerite is produced internally into a spine and is called the spinasternum, while the segmental sclerite is called the eusternum (Fig. 7.6). Various degrees of fusion occur so that four basic arrangements may be found:

a) all elements separate – eusternum of prothorax; first spina; eusternum of mesothorax; second spina; eusternum of metathorax (see Fig. 7.3a. Notice that in the diagram eusternum is divided into basisternum and sternellum);

b) eusternum of mesothorax and second spina fuse, the rest remaining separate;

a) *Blatta*

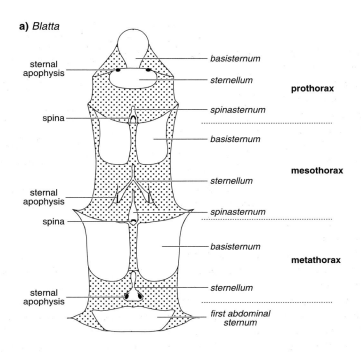

Fig. 7.3. Ventral view of the thorax. Names of sclerites in italics. The points at which the sternal apophyses and spina invaginate are slightly exaggerated in size for clarity. Membranous regions stippled. **(a)** All elements separate [*Blatta* (Blattodea)](after Snodgrass, 1935). **(b)** Complete fusion of meso- and meta-thoracic elements [*Nomadacris* (Orthoptera)](after Albrecht, 1956).

b) *Nomadacris*

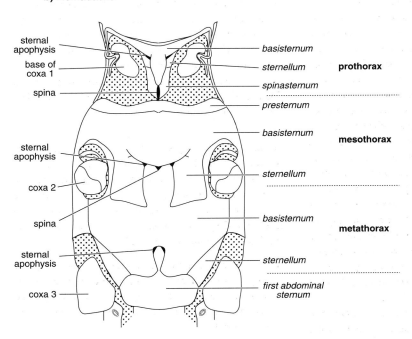

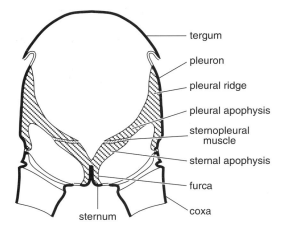

Fig. 7.4. Cross-section of a thoracic segment showing the pleural ridges and sternal apophyses (after Snodgrass, 1935).

c) eusternum of prothorax and first spina also fuse so that there are now three main elements: compound prosternum, compound mesosternum, eusternum of metathorax;

d) complete fusion of meso- and meta-thoracic elements to form a pterothoracic plate (Fig. 7.3b).

The sternum of the pterothoracic segments does not differ markedly from that of the prothorax, but usually the basisternum is bigger, providing for the attachment of the large dorsoventral flight muscles. The sternum is attached to the pleuron by pre- and post-coxal bridges.

Arising from the eusternum are a pair of apophyses, the so-called sternal apophyses (Fig. 7.6). The origins of these on the sternum are marked externally by pits joined by a sulcus (Fig. 7.3b) so that the eusternum is divided into a basisternum and sternellum, while in higher insects the two apophyses arise together in the midline and only separate internally, forming a Y-shaped furca (Fig. 7.4). Distally the sternal apophyses are associated with the inner ends of the pleural apophyses, usually being connected to them by short muscles. This adds rigidity to the thorax, while variation in the degree of contraction of the muscles makes this rigidity variable and controllable. The sternal apophyses also serve for the attachment of the bulk of the ventral longitudinal muscles, although a few fibers retain their primitive intersegmental connections with the spinasterna (Fig. 7.6).

Some insects have a longitudinal sulcus with an internal ridge running along the middle of the sternum. This is regarded by some authorities as indicating that the whole of the primitive sternum has become invaginated and that the apparent sternum in these insects is really derived from subcoxal elements (see Matsuda, 1963). The median longitudinal sulcus is known as the discrimen.

Pleuron The pleural regions are membranous in many larval insects, but typically become sclerotized in the adult. Basically there are probably three pleural sclerites, one ventral and two dorsal, which may originally have been derived from the coxa (Snodgrass, 1958). The ventral sclerite, or sternopleurite, articulates with the coxa and becomes fused with the sternum so as to become an integral part of it. The dorsal sclerites, anapleurite and coxopleurite, are present as separate sclerites in Apterygota and in the prothorax of larval Plecoptera (Fig. 7.5a). In other insects they are fused to form the pleuron, but the coxopleurite, which articulates with the coxa, remains partially separate in the lower pterygote orders forming the trochantin and making a second, more ventral articulation with the coxa (Fig. 7.5b). Above the coxa the pleuron develops a nearly vertical strengthening ridge, the pleural ridge, marked by the pleural sulcus externally. This divides the pleuron into an anterior episternum and a posterior epimeron. The pleural ridge is particularly well developed in the wing-bearing segments, where it continues dorsally into the pleural wing process which articulates with the second axillary sclerite in the wing base (Fig. 7.5b).

In front of the pleural process in the membrane at the base of the wing and only indistinctly separated from the episternum are one or two basalar sclerites, while in a comparable position behind the pleural process is a well-defined subalar sclerite. Muscles concerned with the movement of the wings are inserted into these sclerites.

Typically there are two pairs of spiracles on the thorax. These are in the pleural regions and are associated with the mesothoracic and metathoracic segments. The mesothoracic spiracle often occupies a position on the posterior edge of the propleuron, while the smaller metathoracic spiracle may similarly move on to the mesothorax. Diplura have three or four pairs of thoracic spiracles. *Heterojapyx*, for instance, has two pairs of mesothoracic and two pairs of metathoracic spiracles.

7.2.2 Muscles of the thorax
The longitudinal muscles of the thorax, as in the abdomen, run from one antecostal ridge to the next. They are

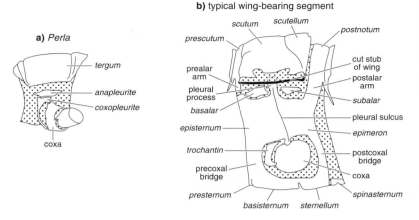

b) typical wing-bearing segment

scutum — *scutellum*

prescutum — *postnotum*

prealar arm — cut stub of wing

pleural process — *postalar arm*

basalar — *subalar*

episternum — pleural sulcus

trochantin — *epimeron*

precoxal bridge — postcoxal bridge

presternum — *coxa*

basisternum — *sternellum* — *spinasternum*

a) *Perla*

tergum

anapleurite

coxopleurite

coxa

Fig. 7.5. Lateral view of thoracic segments. Anterior to left, membranous regions stippled. Names of sclerites in italics (after Snodgrass, 1935). **(a)** The prothorax of *Perla* (Plecoptera). **(b)** A typical wing-bearing segment.

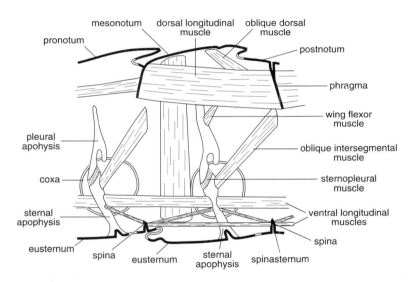

mesonotum — dorsal longitudinal muscle — oblique dorsal muscle

pronotum — postnotum

phragma

wing flexor muscle

pleural apophysis

oblique intersegmental muscle

coxa — sternopleural muscle

sternal apophysis — ventral longitudinal muscles

eusternum — spina

spina — eusternum — sternal apophysis — spinasternum

Fig. 7.6. The main muscles, other than the leg muscles, in the mesothorax of a winged insect (after Snodgrass, 1935).

relatively poorly developed in sclerotized larvae, in adult Odonata, Blattodea and Isoptera which have direct wing depressor muscles, and also in secondarily wingless groups such as Siphonaptera. In these cases they tend to telescope one segment into the next, while the more lateral muscles rotate the segments relative to each other. In unsclerotized insects, contraction of the longitudinal muscles shortens the segment.

In most winged insects, however, the dorsal longitudinal muscles are the main wing depressors and they are well developed (section 9.7.1; Fig. 7.6), running from phragma to phragma so that their contraction distorts the segments. The ventral longitudinal muscles run mainly from one sternal apophysis to the next in adult insects, producing some ventral telescoping of the thoracic segments.

Dorso-ventral muscles run from the tergum to the pleuron or sternum. They are primitively concerned with rotation or compression of the segment, but in winged insects they are important flight muscles (section 9.7.1). In larval insects an oblique intersegmental muscle runs from the sternal apophysis to the anterior edge of the following tergum or pleuron, but in adults it is usually only present between prothorax and mesothorax.

The other important muscles of the thorax are concerned with movement of the legs and wings. They are dealt with separately (sections 8.1.1, 9.6).

REFERENCES

Albrecht, F.O. (1956). The anatomy of the red locust, *Nomadacris septemfasciata* Serville. *Anti-Locust Bulletin* no. 23.

Matsuda, R. (1963). Some evolutionary aspects of the insect thorax. *Annual Review of Entomology*, **8**, 59–76.

Matsuda, R. (1970). Morphology and evolution of the insect thorax. *Memoirs of the Entomological Society of Canada*, no. 76.

Matsuda, R. (1979). Morphologie du thorax et des appendices thoraciques des insectes. In *Traité de Zoologie*, vol. 8, part 2, ed. P.-P.Grassé, pp. 1–289. Paris: Masson et Cie.

Snodgrass, R.E. (1935). *Principles of Insect Morphology*. New York: McGraw-Hill.

Snodgrass, R.E. (1958). Evolution of arthropod mechanisms. *Smithsonian Miscellaneous Collections*, **138**, no. 2.

Zrzavy, J. & Stys, P. (1995). Evolution of metamerism in Arthropoda: developmental and morphological perspectives. *Quarterly Review of Biology*, **70**, 279–95.

8   Legs and locomotion

Insects typically have three pairs of legs, one pair on each of the thoracic segments. From this, the alternative name for insects, the 'hexapods', is derived, although not all hexapods are now regarded as insects.

8.1 BASIC STRUCTURE OF THE LEGS

Each leg consists typically of six segments, articulating with each other by mono- or di-condylic articulations set in a membrane, the corium. The six basic segments are coxa, trochanter, femur, tibia, tarsus and pretarsus (Fig. 8.1a).

The coxa is often in the form of a truncated cone and articulates basally with the wall of the thorax. There may be only a single articulation with the pleuron (Fig. 8.2a), in which case movement of the coxa is very free, but frequently there is a second articulation with the trochantin (Fig. 8.2b). This restricts movement to some extent, but, because the trochantin is flexibly joined to the episternum, the coxa is still relatively mobile. In some higher forms there are rigid pleural and sternal articulations limiting movement of the coxa to rotation about these two points (Fig. 8.2c). In the Lepidoptera the coxae of the middle and hind legs are fused with the thorax and this is also true of the hind coxae in Adephaga.

The part of the coxa bearing the articulations is often strengthened by a ridge indicated externally by the basicostal sulcus which marks off the basal part of the coxa as the basicoxite (Fig. 8.3a). The basicoxite is divided into anterior and posterior parts by a ridge strengthening the articulation, the posterior part being called the meron. This is very large in Neuroptera, Mecoptera, Trichoptera and Lepidoptera (Fig. 8.3b), while in the higher Diptera it becomes separated from the coxa altogether and forms a part of the wall of the thorax.

The trochanter is a small segment with a dicondylic articulation with the coxa such that it can only move in the vertical plane (Fig. 8.1b). In Odonata there are two trochanters and this also appears to be the case in Hymenoptera, but here the apparent second trochanter is, in fact, a part of the femur.

The femur is often small in larval insects, but in most adults it is the largest and stoutest part of the leg. It is often more or less fixed to the trochanter and moves with it. In this case, there are no muscles moving the femur with respect to the trochanter, but sometimes a single muscle arising in the trochanter is able to produce a slight backward movement, or reduction, of the femur (Fig. 8.5b).

The tibia is the long shank of the leg articulating with the femur by a dicondylic joint so that it moves in a vertical plane (Fig. 8.1c,d). In most insects, the head of the tibia is bent so that the shank can flex right back against the femur (Fig. 8.1a). The tarsus, in most insects, is subdivided into from two to five tarsomeres. These are differentiated from true segments by the absence of muscles. The basal tarsomere, or basitarsus, articulates with the distal end of the tibia by a single condyle (Fig. 8.1e), but between the tarsomeres there is no articulation; they are connected by flexible membrane so that they are freely movable. In Protura, some Collembola and the larvae of most holometabolous insects, the tarsus is unsegmented (Fig. 8.4c) or, in the latter, may be fused with the tibia.

The pretarsus, in the majority of insects, consists of a membranous base supporting a median lobe, the arolium, which may be membranous or partly sclerotized, and a pair of claws which articulate with a median process of the last tarsomere known as the unguifer. Ventrally there is a basal sclerotized plate, the unguitractor, and between this and the claws are small plates called auxiliae (Fig. 8.4a). In Diptera, a membranous pulvillus arises from the base of each auxilia while a median empodium, which may be spine- or lobe-like, arises from the unguitractor (Fig. 8.4b). There is no arolium in Diptera other than Tipulidae. The development of the claws varies in different insect groups. Commonly they are more or less equally well-developed, but in Thysanoptera they are minute and the pretarsus consists largely of the bladder-like arolium. In other groups, the claws develop unequally; one may fail to develop altogether, so that in Ischnocera, for instance, there is only a single claw. In

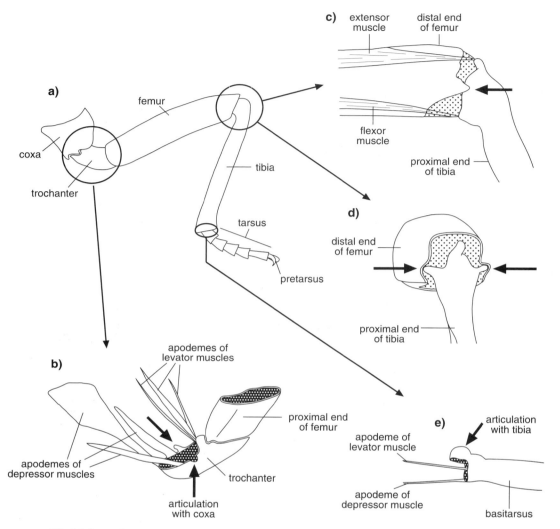

Fig. 8.1. Leg and articulations. Points of articulation shown by bold arrows (mainly after Snodgrass, 1935, 1952). **(a)** Typical insect leg. **(b)** Dicondylic articulation of trochanter with coxa and the apodemes of muscles moving the trochanter. Notice that the femur is united with the trochanter; there is no moving joint. **(c)**, **(d)** dicondylic articulation of tibia and femur, (c) side view, (d) end view. **(e)** Monocondylic, ball articulation of tarsus with tibia.

Protura, some Collembola and many holometabolous larvae the entire pretarsus consists of a single claw-like segment (Fig. 8.4c).

8.1.1 Muscles of the legs

The muscles which move the legs fall into two categories: extrinsic, arising outside the leg, and intrinsic, wholly within the leg and running from one segment to another. The coxa is moved by extrinsic muscles arising in the thorax and a fairly typical arrangement is shown in Fig. 8.5a with promotor and remotor muscles arising on the tergum, abductor and adductor muscles from the pleuron and sternum, and rotator muscles also from the sternum (see section 8.4.1.1. for definitions). The roles of the muscles vary, depending on the activities of other muscles and also on the type of articulation. In *Apis* (Hymenoptera), which has rigid pleural and sternal articulations, promotor and remotor muscles from the

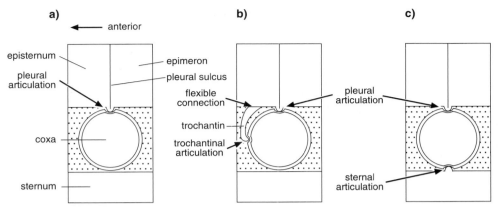

Fig. 8.2. Three types of coxal articulation with the thorax. Points of articulation shown by arrows. Membranous regions stippled (after Snodgrass, 1935).

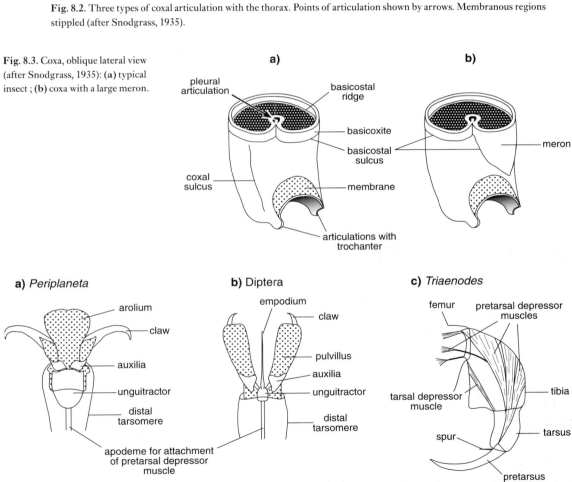

Fig. 8.3. Coxa, oblique lateral view (after Snodgrass, 1935): (a) typical insect ; (b) coxa with a large meron.

Fig. 8.4. Pretarsus. (a) Pretarsus of *Periplaneta*, ventral view (after Snodgrass, 1935). (b) Pretarsus of a dipteran, ventral view (after Snodgrass, 1935). (c) Distal part of prothoracic leg of larval *Triaenodes* (Trichoptera) showing a simple pretarsal segment (after Tindall, 1964).

a) extrinsic muscles

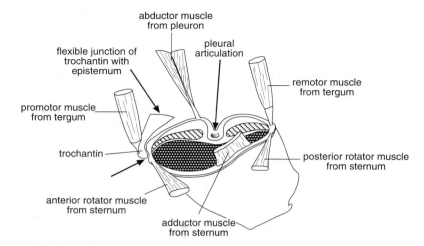

abductor muscle
from pleuron

pleural
articulation

flexible junction of
trochantin with
episternum

remotor muscle
from tergum

promotor muscle
from tergum

trochantin

posterior rotator muscle
from sternum

anterior rotator muscle
from sternum

adductor muscle
from sternum

b) intrinsic muscles

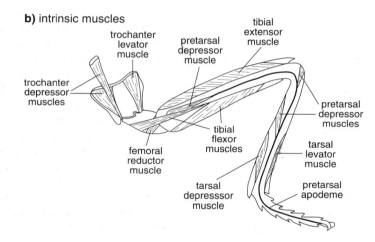

trochanter
levator
muscle

pretarsal
depressor
muscle

tibial
extensor
muscle

trochanter
depressor
muscles

pretarsal
depressor
muscles

femoral
reductor
muscle

tibial
flexor
muscles

tarsal
levator
muscle

tarsal
depresssor
muscle

pretarsal
apodeme

Fig. 8.5. Leg muscles. (a) Extrinsic muscles of coxa as seen from the midline of the insect. Muscles arising from the areas marked with diagonal hatching are omitted (from Snodgrass, 1935). (b) Intrinsic muscles. Note that one trochanter depressor muscle is extrinsic (after Snodgrass, 1927).

tergum are absent. In the pterothoracic segments, muscles (insertions marked with oblique hatching in Fig. 8.5a) run from the coxae to the basalar and subalar sclerites. They are concerned with movements of the wings as well as the legs.

The intrinsic musculature is much simpler than the coxal musculature, consisting typically only of pairs of antagonistic muscles in each segment (Fig. 8.5b). In *Periplaneta*, there are three levator muscles of the trochanter arising in the coxa and three depressor muscles, two again with origins in the coxa and a third arising on the pleural ridge and the tergum.

The femur is usually immovably attached to the trochanter, but the tibia is moved by extensor and flexor muscles arising in the femur and inserted into apodemes from the membrane at the base of the tibia (see Fig. 8.21a). Levator and depressor muscles of the tarsus arise in the tibia and are inserted into the proximal end of the basitarsus, but there are no muscles within the tarsus moving the tarsomeres.

It is characteristic of the insects that the pretarsus has a depressor muscle, but no levator muscle. The fibers of the depressor occur in small groups in the femur and tibia and are inserted into a long apodeme which arises on the unguitractor (Figs. 8.5b, 8.21a). Levation of the pretarsus results from the elasticity of its basal parts.

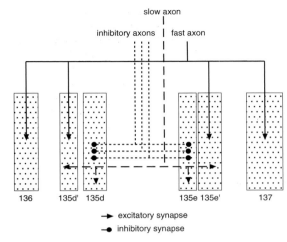

Fig. 8.6. Muscle innervation. Diagrammatic representation of the motor neurons to the depressor muscles of the trochanter in the mesothoracic coxa of *Periplaneta*. Muscle units are numbered as in the text (after Pearson & Iles, 1971).

Innervation of the muscles The innervation of the leg muscles is complex. Most muscles are innervated by both fast and slow axons and by inhibitory axons (section 10.1.5), but not all the fibers within a muscle are innervated by all three types of motor neuron. For example, in *Periplaneta*, each of the meso- and metathoracic coxae has four muscles which depress the trochanter (Fig. 8.6). Two of these, 136 and 137, are innervated only by a fast axon which also goes to parts of the other two muscles, 135′, 135e′. These parts are also innervated by a slow axon, which also supplies other parts of these muscles, 135d and e, which have no fast nerve supply. In addition, three inhibitory fibers innervate these parts of the muscles; inhibitory fibers do not run to parts of the muscles that receive input from the fast axon.

The extensor tibiae muscle in the hind leg of a grasshopper has an even more complex supply. In addition to fast, slow and inhibitory axons, it receives input from an axon which releases the neuromodulator, octopamine (section 10.3.2.4). This axon is called the dorsal unpaired median axon of the extensor tibiae muscle (DUMETi). A majority of fibers are innervated only by the fast axon, but this is probably not a general feature of leg muscles as the extensor tibiae muscle of grasshoppers is specialized for jumping.

Nearly all the muscles moving the coxa, trochanter and tibia in each leg of a locust are innervated by the same

inhibitory neuron which is consequently known as the common inhibitor (see Fig. 10.9). Two other inhibitory neurons innervate the muscles of the distal parts of the legs, both running to the same muscles.

8.1.2 Sensory system of the legs

The legs of insects have an extensive sensory system. Some of the sensory elements are proprioceptors, monitoring the positions of the leg segments and the stance of the insect. Other mechanoreceptors and chemoreceptors are involved in the perception of environmental stimuli.

Review: Seelinger & Tobin, 1981 – cockroach

8.1.2.1 *Proprioceptors*

The proprioceptors include hair plates and campaniform sensilla (section 23.1.3.2) and chordotonal organs (section 23.2.1). *Periplaneta* has hair plates at the proximal end of the coxa and also at the coxa–trochanter joint (Fig. 8.7a). There are groups of campaniform sensilla on the trochanter, a group proximally on the femur and another on the tibia, and a small number on the dorsal surface at the distal end of each tarsomere. In total, there are about 140 sensilla in hair plates and 80 campaniform sensilla on each front leg. The other legs have similar numbers. In addition, *Periplaneta* has a chordotonal organ associated with each joint of the leg, and multipolar neurons at the trochanter-femur and femur–tibia joints. There may also be strand receptors like those described in the locust (section 23.3.2). Similar arrangements of proprioceptors occur on the legs of the migratory locust, *Locusta* (Field & Pflüger, 1989; Hustert, Pflüger & Bräunig, 1981), the stick insect, *Carausius* (Bässler, 1983), and the adult tobacco hornworm, *Manduca* (Kent & Griffin, 1990).

8.1.2.2 *Exteroceptors*

Many exteroreceptive sensilla are also present on the legs. Mechanosensitive trichoid sensilla are distributed all over the legs, and the final larval stage of the grasshopper, *Schistocerca americana*, has about 140 of these sensilla on each tarsus (Fig. 8.7b); others are present on the femur and tibia. The axons from the sensilla in different areas of the leg converge to separate interneurons so that spatial information is maintained within the central nervous system (Fig. 20.24). This is true not only of the first order spiking local interneurons, but also of non-spiking interneurons and spiking intersegmental neurons (see Fig. 8.20) (Burrows, 1989). This ensures that the insect

a) proprioceptors

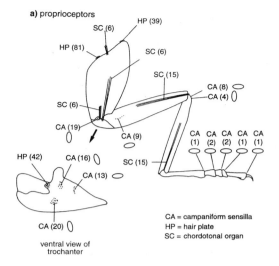

CA = campaniform sensilla
HP = hair plate
SC = chordotonal organ

ventral view of trochanter

b) exteroreceptors

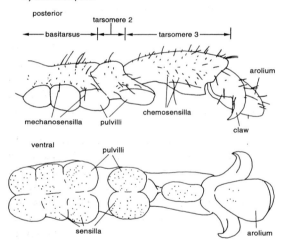

Fig. 8.7. Sensory system of the leg. **(a)** Proprioceptors on the foreleg of *Periplaneta*. Numbers in brackets show the number of sensilla in each group. Ellipses show orientations of campaniform sensilla. **(b)** Exteroreceptors on the fore tarsus of the grasshopper, *Schistocerca americana*. Posterior view: there are 140 long mechanosensitive sensilla and 200 small chemosensitive sensilla on the upper surface and sides of the tarsus. Ventral view: sensilla on the pulvilli and arolium. There are 180 sensilla. About 60% are chemoreceptors.

responds in an appropriate manner when a particular part of a leg is touched.

In addition, each leg has many contact chemoreceptors. *S.americana* has about 200 on the upper surface of the tarsus and over 100 on the pulvillar pads on which the insect normally stands (Fig. 8.7b). Although some contact chemoreceptors may be present on the femur and tibia, most are on the tarsus. Tarsal chemoreceptors are of general occurrence in hemimetabolous insects and adult holometabolous insects, but there is no evidence that they occur on the legs of holometabolous larvae. The thoracic legs of caterpillars possess only small numbers of mechanosensitive hairs. Most insects also have a subgenual organ, sensitive to substrate vibration (section 23.2.3.1).

Review: Chapman, 1982 – chemoreceptors

8.1.3 Adhesion

Many insects are able to climb and hold on to smooth surfaces. Different insects use different structures and probably different mechanisms for adhesion, but many use adhesive setae, also sometimes called tenent hairs. These setae are grouped together to form adhesive pads which occur on various parts of the legs. *Rhodnius* and some other Reduviidae have adhesive pads at the distal ends of the tibiae of the front and middle legs. Amongst the flies, the pulvilli have adhesive properties, and many beetles have pads of setae on the underside of the tarsomeres. In each of these cases the adhesive structures are areas of membranous cuticle covered by large numbers of small setae. For example in the lady beetle, *Epilachna*, there are two pads on the underside of each tarsus. Each pad carries about 800 setae which are 70–120 μm long. Many of the setae are expanded at the tip to form flattened, foot-like structures 5–10 μm in diameter (Fig. 8.8a). In the fly, *Calliphora*, the adhesive setae are also on the tarsi. They are much smaller than those in the beetle, only 9–15 μm high with a 'foot' about 1 μm in diameter. The fly has about 42 000 adhesive hairs altogether. The flexibility of the setae enables them to make contact with irregular surfaces much more efficiently than would be true of a single, larger structure. This greatly increases the power of adhesion. The males of many species of beetle have more adhesive setae than the females. These additional setae are used by the male to grasp the female during mating.

Hairless adhesive pads occur in a number of insects. The arolia of cockroaches and grasshoppers can function

a) tenent hairs

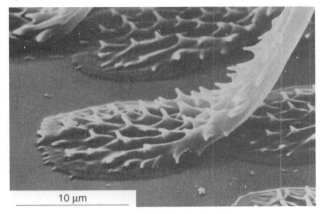

10 μm

b) suckers

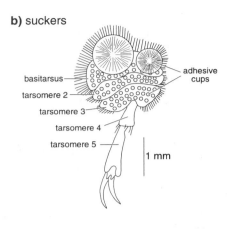

basitarsus

adhesive cups

tarsomere 2

tarsomere 3

tarsomere 4

tarsomere 5

1 mm

Fig. 8.8. Adhesive pads. **(a)** Tip of tenent hair of *Philonthus* (Coleoptera, Staphylinidae) (after Stork, 1983). **(b)** Suckers on the foreleg of male *Dytiscus* (Coleoptera, Dytiscidae) (after Miall, 1922).

as adhesive organs, and in the latter group they are bigger in habitually climbing species. Some aphids have an eversible pulvillus at the tibio–tarsal articulation and planthoppers have a pair of pulvillus-like pads on the pretarsus which function as adhesive organs.

Adhesion is the result of surface tension of a fluid at the tips of the hairs or on the pads. It contains lipoproteins and is produced by gland cells closely associated with the adhesive organs, probably reaching the surface of the cuticle through wax canals. In many beetles, however, there is no evidence of fluid and it is possible that adhesion results from molecular forces operating when the two surfaces, substrate and tips of the adhesive setae, are very closely applied together (Lees & Hardie, 1988; Stork, 1983).

The pulling force exerted by several of these insects when walking vertically up a pane of glass is often well in excess of 10 times the insect's own body mass.

In male dytiscid beetles a different mechanism of adhesion occurs. The first three tarsomeres of the foreleg of male *Dytiscus* are enlarged to form a circular disc. On the inside, this disc is set with stalked cuticular cups, most of which are only about 0.1 mm in diameter, but two of which are much larger than the rest, one being about 1 mm across (Fig. 8.8b). It seems that these cups act as true suckers, although it is not certain how the suction is created. The suckers are used by the male to grasp the female, but may also be used occasionally to grasp prey.

8.2 MODIFICATIONS OF THE BASIC LEG STRUCTURE

The basic insect walking leg may be modified in various ways to serve a number of functions. Amongst these are jumping, swimming, digging, grasping, grooming and stridulation. Modifications associated with jumping and swimming are considered in sections 8.4.2.1. and 8.5.2.2.

Digging Legs modified for digging are best known in the Scarabaeoidea and the mole cricket, *Gryllotalpa*. In *Gryllotalpa*, the forelimb is very short and broad, the tibia and tarsomeres bearing stout lobes which are used in excavation. In the scarab beetles, the femora are short, the tibiae are again strong and toothed, but the tarsi are often weakly developed. Larval cicadas are also burrowing insects. They have large, toothed fore femora, the principal digging organs, and strong tibiae which may serve to loosen the soil (Fig. 8.9a). The tarsus is inserted dorsally on the tibia and can fold back. In the first stage larva it is three-segmented, but it becomes reduced in later instars and may disappear completely.

Grasping Modifications of the legs for grasping are frequent in predatory insects. Often pincers are formed by the apposition of the tibia on the femur. This occurs in the forelegs of mantids (see Fig. 2.9) and mantispids (Neuroptera), in some Heteroptera such as Phymatidae and Nepidae, and in some Empididae and Ephydridae

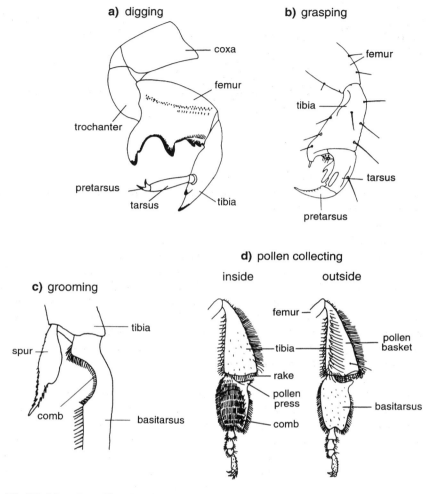

Fig. 8.9. Adaptations of legs. (**a**) Digging. Foreleg of a larval cicada (after Pesson, 1951). (**b**) Grasping. Leg of *Haematopinus* (Phthiraptera) (after Séguy, 1951). (**c**) Grooming. Foreleg of a mutillid (Hymenoptera) (after Schönitzer & Lawitzky, 1987). (**d**) Hind tibia and tarsus of a worker honeybee showing the pollen-collecting apparatus (partly after Snodgrass, 1956).

amongst the Diptera. In some Empididae the middle legs are modified in this way, while in *Bittacus* (Mecoptera) the fifth tarsomere on all the legs closes back against the fourth to form a grasping structure.

The ability to hold on is important in ectoparasitic insects. These usually have well-developed claws and the legs are frequently stout and short as in Hippoboscidae, Ischnocera and Anoplura. In the latter two groups, the tarsi are only one or two segmented and often there is only a single claw which folds against a projection of the tibia to form a grasping organ (Fig. 8.9b).

Grooming Insects commonly use the legs or mandibles to groom parts of the body, removing particles of detritus in the process. The eyes and antennae are often groomed, and so are the wings. Cockroaches clean their antennae by passing them through the mandibles, which chew lightly at the surface, but many insects use the forelegs for this purpose, then cleaning the legs with the mandibles. Neuroptera and Diptera hold an antenna between the two forelegs, which are drawn forwards together towards its tip. Mosquitoes have a comb, consisting of several rows of setae, at the distal end of the

Fig. 8.10. Silk production in the foreleg of an embiid. (**a**) Basitarsus seen in transparency to show the silk glands. (**b**) A single silk gland showing its connection to a seta.

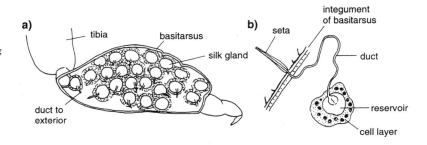

fore tibia. The combs are scraped along the proboscis or antennae in rapid strokes. In many other insects each antenna is cleaned by the ipsilateral foreleg which is often modified as a toilet organ. *Schistocerca* (Orthoptera) has a cleaning groove between the first and second pads of the first tarsomere. This is fitted over the lowered antenna and then drawn slowly along it by an upward movement of the head and extension of the leg. In *Apis* and other Hymenoptera there is a basal notch in the basitarsus lined with spinelike hairs forming a comb. A flattened spur extends down from the tip of the tibia in such a way that when the metatarsus is flexed against the tibia the spur closes off the notch to form a complete ring (Fig. 8.9c). This ring is used to clean the antenna. First it is closed round the base of the flagellum and then the antenna is drawn through it so that the comb cleans the outer surface and the spines on the spur scrape the inner surface. A similar, though less well-developed organ, occurs in Coleoptera of the families Staphylinidae and Carabidae. Lepidoptera have a mobile lobe called the strigil on the ventral surface of the fore tibia. It is often armed with a brush of hairs and is used to clean the antenna and possibly the proboscis.

The hind legs of Apoidea are modified to collect pollen from the hairs of the body and accumulate it in the pollen basket. Pollen collecting is facilitated by pectinate hairs which are characteristic of the Apoidea. In the honeybee, *Apis*, pollen collected on the head region is brushed off with the forelegs and moistened with regurgitated nectar before being passed back to the hind legs which also collect pollen from the abdomen using the comb on the basitarsus (Fig. 8.9d). The pollen on the combs of one side is then removed by the rake of the opposite hind leg and collects in the pollen press between the tibia and basitarsus. By closure of the press, pollen is forced outwards and upwards on to the outside of the tibia and is held in place by the hairs of the pollen basket. On returning to the nest, the pollen is kicked off into a cell by the middle legs.

Silk production Insects in the order Embioptera are unique in having silk glands in the basitarsus of the front legs in all stages of development of both sexes. The basal tarsomere is greatly swollen, and within it are numerous silk glands each with a single layer of cells surrounding a reservoir (Fig. 8.10). There may be as many as 200 glands within the tarsomere, each connected by a duct to its own seta with a pore at the tip through which the silk is extruded.

Reduction of legs Some reduction of the legs occurs in various groups of insects. For example, among butterflies, adults of many species have reduced anterior tarsi, and the Nymphalidae are functionally four-legged with the front legs being held alongside the thorax. In male nymphalids, the tarsus and pretarsus of the foreleg are completely lacking, while in females the fore tarsus consists only of very short tarsomeres. In the male of *Hepialus* (Lepidoptera), on the other hand, the hind leg lacks a tarsus.

More usually, reduction of the legs is associated with a sedentary or some other specialized habit, such as burrowing, in which legs would be an encumbrance. Thus female coccids are sedentary and are held in position by the stylets of the proboscis. The legs are reduced, sometimes to simple spines, and in some species they are absent altogether in the later stages of development. Female Psychidae (Lepidoptera) show varying degrees of leg reduction, some species being completely apodous. These insects never leave the bags constructed by their larvae. Legs are also completely absent from female Strepsiptera, which are parasitic in other insects.

Apart from the Diptera, all the larvae of which are apodous, legless larvae are usually associated with particular modes of life. There is a tendency for larvae of leaf-mining Lepidoptera, Coleoptera and Tenthredinoidea to be apodous (see Hering, 1951). Parasitic larvae of Hymenoptera and Strepsiptera are apodous, and in larval Meloidae (Coleoptera) the legs are greatly reduced. Finally, in the social and semisocial Hymenoptera in which the larvae are provided with food by the parent, apodous forms are also the rule.

Review: Schönitzer & Lawitzky, 1987 – antenna cleaners

8.3 MAINTENANCE OF STANCE

Even standing still requires muscular activity and the continual adjustment of this activity to compensate for small shifts in position. Some leg muscles, like the extensor tibiae muscles of the front and middle legs of *Schistocerca*, are continuously active in a stationary insect. Slow axons innervating these muscles fire at low rates, 5–30 Hz, their activity varying with leg position. The fast axons are not active in a stationary insect, and the inhibitory neurons are only sporadically active, but do fire in response to contact or vibration.

When an insect is standing still, any force tending to change the angles between the segments of the legs, such as a sudden gust of wind, is opposed by a muscular reflex, called the postural position reflex, and in most cases this is mediated by the chordotonal organs. Most studies have examined the femur–tibia joint. In this case, any tendency to reduce the angle between the femur and tibia is opposed by the tibial extensor muscle, while an increase in the angle is opposed by the tibial flexor muscle.

Similar reflexes have been described at other joints of the leg. These reflexes are produced in response to passive changes in the angles between segments whatever the initial degree of flexion or extension, but the input from the chordotonal organ, and the activity of the motor neurons to the muscles, varies according to the position. When *Locusta migratoria* is standing on a horizontal surface, the angles between the femora and tibiae of the front and middle legs are usually close to 90°. In the hind legs, the angle is more variable, but is usually less than 45°. The postural resistance reflexes contribute to the maintenance of these positions in the middle and hind legs, but not in the forelegs (Field and Coles, 1994).

Campaniform sensilla on the legs monitor strains in the cuticle and produce reflex responses in muscles tending to alleviate those strains. Some campaniform sensilla at the proximal end of the tibia of *P. americana* are oriented parallel with the long axis of the leg, others are at right angles to it (Fig. 8.7a). Axial forces, such as would be produced by the weight of the insect when it is standing still, are perceived by the most proximal, transversely oriented sensilla, although their responses to such axial forces are relatively weak. Bending the tibia, however, produces strong responses. When the insect is standing on a horizontal surface with the tibia inclined away from the body, the mass of the body will tend to cause an upward bending of the tibia, compressing the proximal sensilla (Fig. 8.11b). Twisting the tibia causes both groups of sensilla to respond. Stimulation of the proximal sensilla excites slow motor neurons to the extensor tibiae and extensor trochanteris muscles, and inhibits the slow motor neurons to the flexor tibiae and flexor trochanteris muscles (Fig. 8.11c). Stimulation of the distal sensilla has the opposite effect (Zill, Moran and Varela, 1981). A comparable reflex system compensates for stress in the trochanter of the stick insect, *Carausius morosus*, when the femur is bent anteriorly or posteriorly. Here there are two groups of campaniform sensilla on the trochanter which reflexly activate the retractor or protractor muscles of the coxa (Schmitz, 1993). The pathways between proprioceptive sense cells and motor neurons are, in some cases at least, monosynaptic (Skorupski and Hustert, 1991). Similar systems for the control of stance are almost certainly present in all insects.

The hair plates on the legs contribute to the insects' gravitational sense (section 23.1.3.2).

8.4 LOCOMOTION

Mobility at some stage of the life history is a characteristic of all animals. They must move in order to find a mate, to disperse and, in many cases, to find food. The success of insects as terrestrial animals is in part due to their high degree of mobility arising from the power of flight (see Chapter 9), but more local movements by walking or swimming are also important.

8.4.1 Walking and running

Most insects move over the surface of the ground by walking or running. The legs move in sequences which are varied at different speeds in such a way that stability is

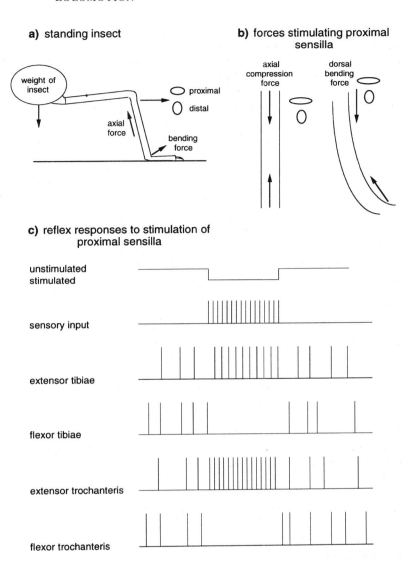

a) standing insect

b) forces stimulating proximal sensilla

c) reflex responses to stimulation of proximal sensilla

Fig. 8.11. Functioning of the proximal campaniform sensilla of a leg in a standing insect (based on Zill & Moran, 1981a; Zill *et al.*, 1981). (**a**) Cross-section of an insect. The two orientations of the sensilla in the proximal group are shown by the ellipses. (**b**) Diagram showing the effects of an axial force on dorsal bending. The transversely oriented sensilla, compressed along their short axes, are stimulated. (**c**) Activity in the motor neurons of tibial and trochanteral extensor muscles is enhanced when the sensilla are stimulated. The activity of the flexor muscles is inhibited. Vertical lines represent action potentials.

usually maintained. Co-ordination of these movements involves central mechanisms, but segmental reflexes are also important.

Reviews: Delcomyn, 1985; Hughes and Mill, 1974

8.4.1.1 *Movements of the legs*

In describing the movements of the legs the following terms are used (Hughes, 1952):

Protraction – complete movement forwards of the whole limb relative to its articulation with the body.

Promotion – the movement of the coxa resulting in protraction.

Retraction – the backward movement of the leg relative to its articulation between the time the foot is placed on the ground and the time it is raised.

Remotion – the corresponding movement of the coxa.

Adduction – the movement of the coxa towards the body.

Abduction – the movement of the coxa away from the body.

Levation – the raising of the leg or a part of the leg, part of protraction.

Depression – lowering the leg, or a part of the leg.

Extension – an increase in the angle between two segments of the leg.

Flexion – a decrease in the angle between two segments of the leg.

8.4.1.2 *Mechanism of walking*

A leg may act simply as a strut with the forces acting down it depending on its angle of inclination to the body and the weight of the insect (Fig. 8.12a). Equal and opposite forces will be exerted by the leg on the body. The force acting down the leg can be resolved into two components, horizontal and vertical, and because the leg is splayed out lateral to the body the horizontal force can be resolved into longitudinal and transverse components (Fig. 8.13). The relative sizes of the longitudinal and transverse components will vary according to the leg's position. In Fig. 8.13 it is assumed that only three legs are on the ground (see below) and it is clear that for most of its movement the strut effect of the foreleg tends to retard forward movement (longitudinal thrust forwards), while that of the middle and hind legs promotes forward movement (longitudinal thrust backwards). So long as all the opposing longitudinal (forwards, backwards) and lateral (left, right) forces balance each other there will be no movement, but if the forces are not balanced the body will be displaced due to a fall in the center of gravity.

A leg can also act as a lever, that is a bar on which external work is done so that it rotates about a fulcrum. This effect is produced by the extrinsic muscles which move the leg relative to the body and so lever the insect along (Fig. 8.12b).

The leg, however, is not a simple, rigid strut or bar. It has intrinsic muscles which can also exert forces on the body by flexing or extending the leg. If a leg is extended anteriorly, flexion of the joints will pull the body forwards (Fig. 8.12d), while, in a leg directed backwards, straightening the joints will push the body forwards (Fig. 8.12c).

When *Periplaneta* starts to move, the foreleg is fully protracted due to maximum coxal promotion and extension of all the leg segments. At this stage it exerts a strut action retarding forward movement. Retraction begins by coxal remotion which produces a lever effect drawing the animal forwards, an effect which is added to by flexion of the trochanter on the coxa and the tibia on the femur. This phase continues until the leg is at right angles to the long axis of the insect. When it has passed this position it exerts a strut effect which, aided by leg extension, tends to push the insect forwards.

During protraction the leg is lifted and flexed so that it exerts no forces on the body. The coxal promotor muscle probably starts to contract before retraction is complete, so the change over from retraction to protraction is smooth. As the foreleg swings forwards it extends again, so that in each cycle of movement the intrinsic muscles undergo two phases of contraction and relaxation, while the extrinsic muscles only contract and relax once.

The mid and hind tarsi are always placed on the ground behind their coxae, so their longitudinal strut effect always assists forward movement (Fig. 8.13). The main propulsive forces of both pairs of legs are derived from extension of the trochanter on the coxa and of the tibia on the femur pushing the insect forwards.

The longitudinal forces produced by these movements are generally such that the insect moves forwards. At the same time lateral forces are produced and when, for instance, the right foreleg is on the ground it tends to push the head to the left. This is partly balanced by the other legs, but there is some tendency for the head to swing from side to side during walking (see Hughes, 1952).

The precise functions of a muscle or a leg during locomotion vary with the orientation of the insect. For instance, when *Carausius* (Phasmatodea) is walking on a horizontal plane, the middle and hind legs provide most support and most of the propulsive force is provided by the hind legs; the forelegs have a largely sensory function. However, if the insect is suspended beneath a surface, all the legs are used to hold on and much of the power for movement comes from the middle legs.

8.4.1.3 *Patterns of leg movement*

Each step comprises a period of protraction, when the leg is swung forwards with the foot off the ground, and a period of retraction, as the leg moves back relative to the body with the foot on the ground. The pattern of stepping often varies with the speed of movement. At the lowest speeds, the insect may have most of its feet on the ground most of the time and the legs are protracted singly in the sequence R3 R2 R1 L3 L2 L1 R3 etc. (where R and L indicate right and left, and 1, 2 and 3 the fore, middle and hind legs respectively) (Fig. 8.14a) or more irregularly. At higher stepping rates, three legs, the fore and hind legs of one side and the middle leg of the other, are lifted more or

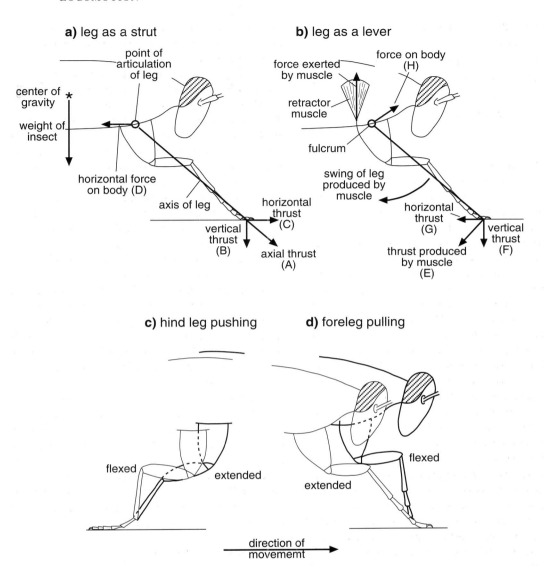

a) leg as a strut

point of articulation of leg

center of gravity

weight of insect

horizontal force on body (D)

axis of leg

vertical thrust (B)

horizontal thrust (C)

axial thrust (A)

b) leg as a lever

force exerted by muscle

force on body (H)

retractor muscle

fulcrum

swing of leg produced by muscle

horizontal thrust (G)

thrust produced by muscle (E)

vertical thrust (F)

c) hind leg pushing

flexed

extended

d) foreleg pulling

flexed

extended

direction of movememt

Fig. 8.12. Mechanical functioning of leg. (**a**) A leg acting as a strut. The axial thrust (A) is exerted down the length of the leg by virtue of the weight of the insect. The size of the axial thrust depends, among other things, on how much of the weight is borne by the other legs. It can be resolved into vertical and horizontal components (B and C), but because the foot is held by friction with the substratum it does not move. Instead, a horizontal force (D), equal and opposite to force (C) acts on the body and, in this case, tends to push it back unless balanced by other forces. (**b**) A leg acting as a lever. Contraction of the retractor muscle tends to swing the leg back so that the foot exerts a thrust (E) on the ground. This can be resolved into vertical and horizontal components (F and G), but since the foot is held still by friction an equal and opposite force (H) acts on the body pushing it forwards and upwards. (**c**) Extension of the coxotrochanteral and femoro-tibial joints of the hind leg pushes the body forwards. (**d**) Flexion of the coxotrochanteral and femoro-tibial joints of the foreleg pulls the body forwards.

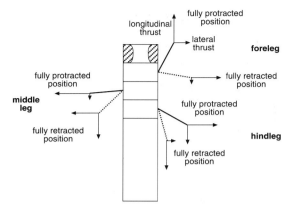

Fig. 8.13. Stability. Diagram to show the positions of the legs forming a typical triangle of support when fully protracted (solid line) and fully retracted (dotted line). The arrows show longitudinal and lateral components of the horizontal strut effect which the legs exert on the ground at these times. The forces acting on the body will be in the opposite directions (after Hughes, 1952).

Higher speeds result from increases in the frequency with which the legs are moved and in the length of each stride. In the slow-moving cockroach, *Blaberus discoidalis*, stride frequency increases to a maximum of about 13 Hz as the insect's speed increases, but then reaches a plateau (Fig. 8.16a). Further increases in speed result from increases in stride length (Full and Tu, 1990). The faster moving *P. americana* moves its legs much more quickly, but shows relatively little increase in stride frequency as it runs faster, and, in this species, increases in stride length account for most of the increase in speed (Full and Tu, 1991).

Except at very low stepping rates (below 3 steps s^{-1} in Fig. 8.17), the rate of stepping is increased primarily by shortening the period of retraction; the period of protraction also shortens but less markedly, so that the ratio protraction time/retraction time increases from about 0.3 at moderate speeds (3 steps s^{-1}) approaching 1.0 at high speeds (15 steps s^{-1}) (Fig. 8.17).

At high speeds, worker ants continue to use alternating triangles of support (Zollikofer, 1994), but *Periplaneta americana* switches to quadrupedal and even bipedal gaits, running on its hind legs and raising the front of the body at an increasing angle to the ground (Full and Tu, 1991). In both ants and *Periplaneta* there are periods when all the legs are off the ground at the same time. The worker ants appear to 'bounce' from one triangle of support to the next. Under these conditions, the stability normally provided by a triangle of support is greatly reduced; dynamic stability due to the insect's movement probably becomes increasingly important as the insect runs faster (Ting, Blickham & Full, 1994).

less simultaneously, so that the insect is supported on a tripod, or triangle, formed by the other three legs. As one set is protracted, the other is retracted, so that the insect is always supported on three legs (Fig. 8.14b). Stability is enhanced by the fact that the body is slung between the legs so the center of gravity is low (Fig. 8.15). This use of alternating triangles of support during walking and running occurs in all the terrestrial insects so far examined, although not necessarily at all speeds. A key feature of this type of locomotion is that the legs on either side of a segment are in antiphase.

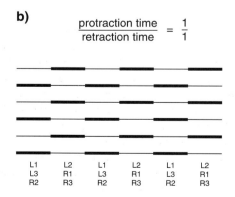

Fig. 8.14. Stepping patterns. Diagram showing the disposition of the feet with different protraction time:retraction time ratios. Thick lines indicate retraction with the foot on the ground, thin lines protraction with the foot in the air (based on Hughes, 1952)

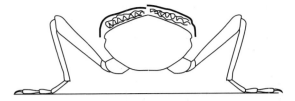

Fig. 8.15. Stability. Transverse section through the mesothorax of *Forficula* (Dermaptera) to show the body suspended between the legs (after Manton, 1953).

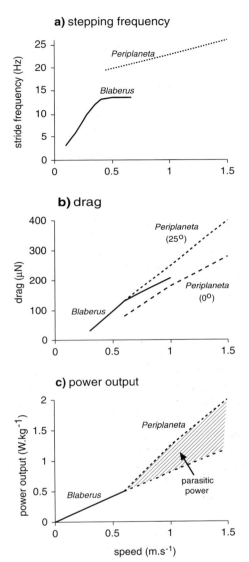

Fig. 8.16. Mechanics of locomotion in *Blaberus*, slow moving, and *Periplaneta*, fast moving (after Full & Tu, 1990; Full & Koehl, 1993). (a) Stepping (stride) frequency at different speeds. (b) Drag at different speeds and, for *Periplaneta*, different angles of attack. (c) Power output at different speeds. The hatched area shows the parasitic power exerted by *Periplaneta* in order to overcome drag.

Some insects are effectively quadrupedal due to the modification of the fore or hind legs for other purposes. It is common for grasshoppers, in which the hind leg is modified for jumping to use only the anterior two pairs of legs in walking. At high speed, *Tropidopola* moves the legs of each segment together in the sequence L1/R1 L2/R2 L1/R1 etc. It maintains stability by using the tip of the abdomen as an additional point of support. Mantids often walk with the forelegs off the ground, with the stepping sequence L3 L2 R3 R2 L3 etc., or L3/R2 L2/R3 L3/R2 etc.

Speed of movement The speed of movement varies very greatly from one insect to another, but in general it is higher at higher temperatures. At 25 °C *Periplaneta* moves at about 70 cm s^{-1} with top speeds up to 130 cm s^{-1} (see Hughes and Mill, 1974). Speed also depends to some extent on size: insects with longer legs can take longer strides, so that for the same frequency of stepping they will move further than smaller insects. Thus first stage larvae of *Blattella* (Blattodea) can move at about 3 cm s^{-1}, while adults are capable of speeds up to 20 cm s^{-1}.

8.4.1.4 *Co-ordination of leg movements*

During walking it is probably usual for all three axons that innervate each muscle (fast, slow and inhibitory, see above) to be active. For example, as *Schistocerca* lifts its fore foot off the ground at the beginning of protraction, both the fast and slow axons to the extensor tibiae begin to fire (Fig. 8.18a). This has the effect of extending the tibia as the leg swings forwards. The fast neuron stops firing before the insect puts its foot down, but the slow neuron keeps firing at a moderately high rate through the first half of retraction. The activity of the inhibitory neuron peaks during the second half of retraction, when the slow axon is silent (Burns & Usherwood, 1979; but see Wolf, 1990)

The alternating movements of protraction and retraction are the result of regular patterns of activity of antagonistic muscles in the different segments of the leg. In *Periplaneta*, for example, the levator muscle of the trochanter is active during protraction, lifting the foot off the ground (Fig. 8.18b); the trochanter depressor muscle is

a) protraction and retraction times

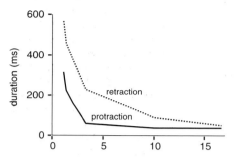

b) protraction/retraction

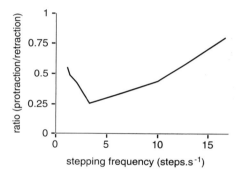

stepping frequency (steps.s^{-1})

Fig. 8.17. Mechanics of locomotion. Changes in duration of retraction and protraction times of a leg in *Periplaneta* in relation to the stepping frequency (after Delcomyn & Usherwood, 1973). (**a**) Protraction and retraction times. (**b**) Ratio of protraction/retraction.

inactive but it starts to contract before the foot is placed on the ground while the levator muscle is still active. At the same time, the contralateral depressor muscle is continuously active as this foot is on the ground. Similar patterns of activity have been recorded for other pairs of antagonistic muscles and for other insects.

These patterns are generated by a program within the central nervous system, and appropriate patterns of activity of the motor neurons can be produced by isolated ganglia (Fig. 8.19) (Ryckebusch & Laurent, 1993). Each leg is believed to be driven by its own pattern generator and integration between the legs is achieved through the central nervous system. Details of the manner in which this integration is achieved are not known, although various models are discussed by Bässler (1983) and Delcomyn (1985).

However, the pattern of leg movement is not fixed even at a constant speed. It varies with the load on each leg and

according to the nature of the terrain (Duch & Pflüger, 1995). This flexibility involves modulation of the central pattern by input from the proprioceptors and to some extent also from other mechanoreceptors on the leg. As in standing, different groups of proprioceptors may be involved in different processes. For example, as a slowly walking cockroach puts its foot down at the end of protraction, the proximal campaniform sensilla on the tibia are stimulated by the dorsal bending of the leg (see Fig. 8.11). The slow motor neuron to the extensor tibiae muscle starts to fire soon afterwards and it is probable that the input from the campaniform sensilla influences both the timing and the rate of firing of the motor neuron. This reduces the degree of dorsal bending and the proximal sensilla stop firing. If the rate of firing of the slow neuron to the extensor tibiae muscle exceeds 300 Hz, the *distal* campaniform sensilla fire. Their activity contributes to the inhibition of the extensor tibiae motor neuron and prevents excessive muscle contraction which might damage the tibia. If the leg encounters obstacles during its movement, the input from these sensilla is affected (Zill and Moran, 1981b). The timing of activity of the campaniform sensilla is altered during fast walking. The chordotonal organs in the legs are also important in regulating the stepping movements (Bässler, 1988).

The proprioceptors often connect directly, via monosynaptic pathways, with the motor neurons (Fig. 8.20). In contrast, the input from exteroceptive mechanoreceptors on the leg is integrated, primarily, by spiking local interneurons. The axon from one hair synapses with several interneurons, and each interneuron receives input from many sensilla in its receptive field (a specific part of the leg, see Fig. 20.24). The spiking interneurons connect, in turn, with a network of non-spiking interneurons. In the locust, spiking local interneurons in the midline of the ganglion make inhibitory synapses with non-spiking interneurons, whereas the interneurons in another group (anteromedial) generally make excitatory connections. Each spiking local interneuron synapses with more than one non-spiking interneuron, and each of the latter receives input from several spiking interneurons. Finally, each non-spiking interneuron connects with several motor neurons and each motor neuron receives input from several non-spiking interneurons. Some of these inputs are excitatory, but others are inhibitory. As a result of all these interconnections, each non-spiking interneuron responds to a stimulus in the context of the activity generated by many other interneurons and the output to the motor neurons is varied accordingly.

a)

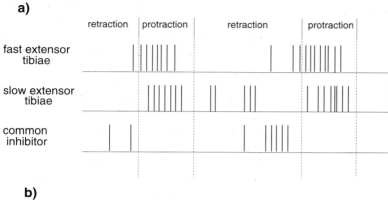

b)

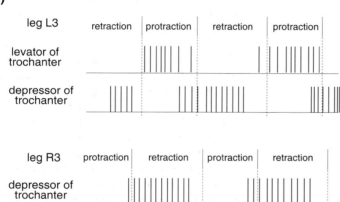

Fig. 8.18. Neuromuscular activity during walking. **(a)** Activity of the fast, slow and inhibitory axons to the extensor tibiae muscle of the prothoracic leg of *Schistocerca* (data from Burns & Usherwood, 1979). **(b)** Activity of muscles moving the trochanters of the hind legs of *Periplaneta*. Antagonistic muscles in L3 are in antiphase, and the depressor of L3 is in antiphase with the depressor of R3 (data from Delcomyn & Usherwood, 1973).

The non-spiking interneurons are known to exhibit oscillations of their membrane potentials which are synchronized with rhythmic bursts of activity in the motor neurons. The source of these oscillations remains unknown.

Reviews: Burrows, 1992, 1996

8.4.1.5 *Forces acting on the body during walking and running*

When an animal moves through air or water it generates mechanical power to move its mass horizontally and vertically. In walking, and most cases of jumping and swimming, it generates this power by moving its legs. In flight, of course, the power is generated by wing movement.

Whether the insect is in air or water, it encounters forces acting in opposition to its motion. The force acting in the opposite direction to forward movement is known as the drag force. Lift is defined as the force acting at right angles to the direction of movement, and is usually almost vertically upwards. The power that an insect generates to overcome these forces, as distinct from that which moves it forwards or vertically, is called the parasitic power.

Drag (D) is proportional to the density (ρ) of the fluid in which the insect is moving, its area at right angles to its direction of movement (its projected area, S), its drag coefficient (C_D), and the velocity of the relative wind (U)

$$D = 0.5 C_D \rho S U^2$$

The *drag coefficient* (C_D) depends on the shape of the organism (a streamlined form presents less resistance than an irregular form), its surface texture (a smooth surface presents less resistance than a rough one), and the Reynolds number.

Reynolds number (Re) is a measure of the relative importance of inertial and frictional forces acting on the moving insect

$$\mathrm{Re} = \rho V l / \mu$$

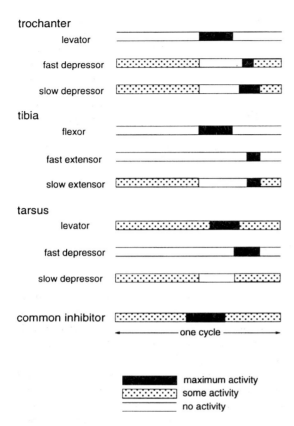

Fig. 8.19. Central nervous control of walking. Rhythmic activity in the motor neurons in an isolated metathoracic ganglion of *Schistocerca* during one complete cycle equivalent to protraction and retraction of the leg. Dark bars show periods of maximal activity; dotted bars show periods when some activity may occur; open bars, no activity. Notice that the neurons to antagonistic muscles are in antiphase, periods of maximal activity in one coinciding with minimal activity in the other. As this output is to a hindleg, flexion of the tibia occurs when the foot is off the ground due to the levation of the trochanter and tarsus (after Ryckebusch & Laurent, 1993; data for the fast extensor tibiae neuron is less well documented than for other neurons).

where ρ = the density of the medium, μ = the viscosity of the medium, V = the speed of the insect, l = a measure of size. Reynolds number increases with the size of the insect and the viscosity of the medium becomes less important.

The *relative wind* (U) is the movement of air relative to the animal. The faster the animal moves, the greater the relative wind.

The *angle of attack* (α) is the angle at which the relative wind strikes the body of the organism. If the body of an insect walking on a horizontal surface is horizontal, the angle of attack is zero; if the front end of the body is raised, the angle of attack is positive. A positive angle of attack will create lift.

Lift (L) is determined in a similar manner to drag, substituting a lift coefficient for the drag coefficient.

$$L = 0.5 C_L \rho S U^2$$

The drag exerted on the insect during locomotion is a function of its speed, or more correctly of the velocity of the relative wind, and the angle of attack. Consequently, in a fast running cockroach, which raises the front part of its body off the ground, drag is considerably greater than in a slowly moving insect (Fig. 8.16b). The insect must generate power to overcome this drag, and, in *P.americana* running at $1–1.5\ \mathrm{m\ s^{-1}}$ this parasitic power accounts for 20–35% of the power generated (Fig. 8.16c). The effect of drag is, however, negligible in the case of the more slowly moving *B.discoidalis* (Full & Koehl, 1993).

Lift acts in the opposite direction to gravity and so reduces the effective weight of the insect. It would be expected, therefore, that lift might reduce the power output needed by a running insect. Although lift forces acting on *P.americana* increase as it runs faster because of the increased tilt of its body, this lift never amounts to more than 2% of the body mass, and it is considered to have a negligible effect on the power output required (Full & Koehl, 1993).

8.4.2 Jumping

In Orthoptera, Siphonaptera, Homoptera and some beetles, jumping is produced by the hind legs, but the ant, *Harpegnathos*, uses both the middle and hind pairs of legs (Urbani *et al.*, 1994). Other jumping mechanisms, not involving the legs, occur in Collembola, Elateridae and *Piophila* (Diptera). In most cases jumping is a means of escape and landing is uncontrolled. Fleas, however, jump in order to attain their hosts, and *Harpegnathos* jumps to catch prey.

The rapid release of energy required for jumping cannot be achieved by direct muscular contraction, but depends on the storage of energy derived from the relatively slow contraction of a muscle and then its sudden release to produce the jump. Energy is stored in the muscle itself and in the cuticle; its release is controlled independently of the muscle producing the power.

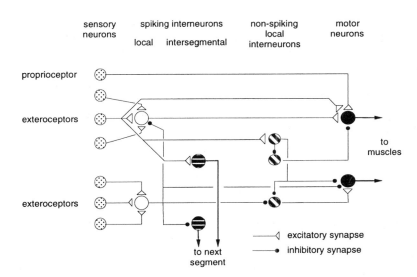

Fig. 8.20. Peripheral modulation of walking. Pathways in the metathoracic ganglion of *Schistocerca* of signals from peripheral mechanoreceptors of the hind leg. Each local and motor neuron receives input from several neurons in each of the categories to its left, and each local interneuron outputs onto several neurons in the categories to its right (partly after Burrows, 1992).

8.4.2.1 *Jumping with legs*

Orthoptera and jumping beetles In Orthoptera, Alticini (flea beetles) and *Orchestes* (a weevil) the hind femora are greatly enlarged, housing the large extensor tibiae muscles which provide the power for the jump. In all these insects, the jump results from the sudden straightening of the femoro-tibial joint extending the tibia, which is also elongate, so that the tarsus is pushed against the substratum with great force.

The structure of the hind leg, and especially of the femoro-tibial joint, is adapted to permit the development of maximum force by the extensor tibiae muscle, the storage of the energy they produce, and its rapid release resulting in the sudden extension of the tibia. The extensor tibiae muscle consists of a series of short fibers inserted obliquely into a long, flat apodeme (Fig. 8.21). In *Schistocerca*, many of the fibers of this muscle are innervated only by a fast axon (see Table 10.1). Collectively, because of their oblique arrangement, they have a large cross-sectional area and can develop a force up to 16 N compared with only 0.7 N by the flexor tibiae muscle. Just above the articulation with the tibia, the cuticle of the femur is heavily sclerotized forming a dark area known as the semilunar process (Fig. 8.22). Beneath the articulation, the cuticle of the femur is thickened internally to form a process, known as Heitler's lump, over which the apodeme of the flexor tibiae slides (Fig. 8.23a).

Before a jump, the tibia is flexed against the femur by the action of the flexor muscle. The axons to the extensor muscle remain silent (Fig. 8.24). Then, after a pause of 100–200 ms, both flexor and extensor muscles contract together. There is no movement of the tibia at this time despite the much greater power of the extensor muscle because the lever ratio between the flexor and extensor muscles (f/e in Fig. 8.23) greatly favors the flexor muscle. When the tibia is closed up against the femur, this ratio is about 21:1 (Fig. 8.23d). However, as both muscles exert parallel forces on the distal end of the femur, this becomes distorted and energy is stored in the semilunar processes (Fig. 8.22). Rapid extension of the tibia occurs when the flexor muscles suddenly relax due to the cessation of their motor input and to the activity of their inhibitory nerve supply (Fig. 8.24). The semilunar process springs back to its relaxed shape so that the femur–tibia articulation moves distally at the same time as the extensor muscle pulls the head of the tibia proximally. Because of the length of the tibia, a small movement of the head of the tibia produces a big movement of the distal end. This mechanical advantage varies with the position of the tibia, but is greatest, about 150:1, at the start of the movement. As a result of the force exerted by the distal ends of the two tibiae, the insect is hurled into the air. The insect's center of gravity is close to the line joining the insertions of the metathoracic coxae, so little torque is produced when the legs extend and the insect moves through the air without rolling. The initial thrust exerted by the tibiae is directly downwards because just before the jump the flexed femur–tibia is moved so that the tibia is parallel with the ground. In the course of extension, however, a backward component develops, pushing the insect forwards.

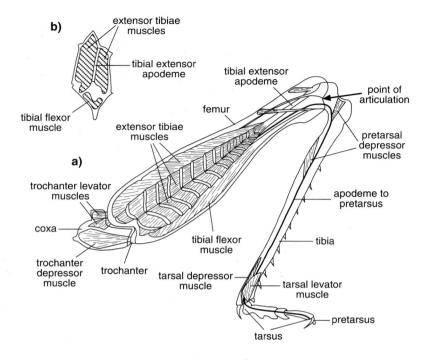

Fig. 8.21. Jumping. Hind leg of a grasshopper (after Snodgrass, 1935). (a) Leg seen in transparency to show the musculature. (b) Transverse section of the femur.

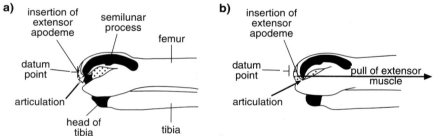

Fig. 8.22. Jumping. Specializations of the hind femoro-tibial joint of a locust. Membranous areas stippled, heavily sclerotized areas black (after Bennet-Clark, 1975). (a) The tibia is flexed, but the extensor muscle remains relaxed. (b) The extensor muscle contracts at the same time as the flexor muscle. Because the insertion of the extensor apodeme is almost in line with the articulation, contraction of the extensor muscle causes distortion of the head of the femur, straining the semilunar process. Notice that the femur has shortened slightly with reference to the datum point which shows the position of the origin of the extensor apodeme in (a), and the semilunar process is bowed.

Fifth stage larvae of *Locusta* can make long jumps of up to 70 cm, reaching a height of 30 cm. However, the distance jumped by a grasshopper varies with its size. Larger insects jump longer distances. For example, the longest jump made by third stage *Schistocerca* is about 50 cm, but adults can jump as far as 1 m. Jumping ability also varies within a developmental stage and, in the final larval stage, reaches a maximum about three days after molting and

then slowly declines until day 11 before falling sharply on the day before the next molt (Katz & Gosline, 1993; Queathem, 1991).

Homoptera In the jumping Homoptera of the families Cercopidae, Cicadellidae, Membracidae and Psyllidae, the jump is produced by a rotation of the leg on the coxo-trochanteral joint. Powerful muscles from the furca, pleuron

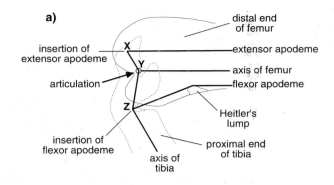

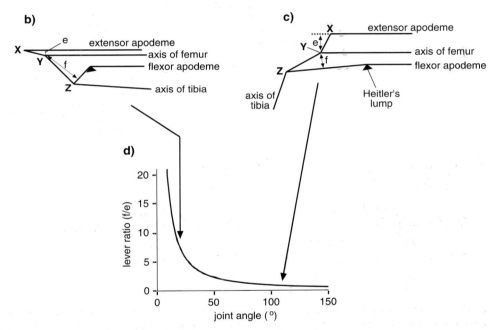

Fig. 8.23. Jumping. Mechanical relationship between the tibial extensor and flexor muscles of a locust. X shows the position at which the apodeme of the extensor muscle is attached to the head of the tibia, and Z the position at which the apodeme of the flexor muscle is attached. Y is the point of articulation of the tibia with the femur (after Heitler, 1974, 1977). **(a)** Diagram showing the position of Heitler's lump and arrangement of the apodemes. The distal end of the femur is shown in outline (solid line) and the proximal end of the tibia (dotted). **(b), (c)** Changes in positions of apodemes and their insertions with the tibia flexed **(b)** and extended **(c)**. **(d)** Changes in the lever ratio of the tibial flexor and extensor muscles as the tibia swings away from the femur. The very high lever ratio (f/e) when the tibia is fully flexed enables the flexor muscle to hold the tibia in position despite the much larger size of the extensor muscle which is contracting at the same time.

and notum are inserted into an apodeme from the trochanter and the coxa opens very widely to the thorax to permit the entry of these muscles. In psyllids, the coxa is fused with the thorax and the position of the trochanteral articulations is altered so as to bring the femora parallel with the trunk.

Siphonaptera The muscles producing the jump of a flea are depressors of the trochanter/femur which arise in the thorax. When the femur is raised by the trochanter levator muscle, the insertion of the depressor muscle relative to the point of articulation of the trochanter with the coxa is

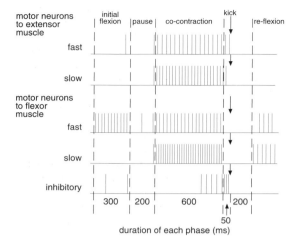

Fig. 8.24. Jumping. Activity of the motor neurons controlling the extensor and flexor tibiae muscles of the hind leg of a locust in relation to a jump. Vertical lines represent action potentials (after Heitler & Burrows, 1977).

such that it compresses a pad of resilin (section 16.3.3) above the base of the leg without rotating the femur. The energy produced is stored in the resilin until a slight forwards movement of the depressor muscle, produced by a muscle inserted into its tendon, causes the coxa to swing suddenly downwards as the tension in the system is released. This has the effect of projecting the insect into the air (Bennet-Clark & Lucey, 1967).

Reviews: Bennet-Clark, 1990 – grasshopper jumping; Burrows, 1996 – grasshoppers, neural control

8.4.2.2 *Mechanisms of jumping not involving legs*

Collembola Collembola jump using modified abdominal appendages (Fig. 8.25). Arising from the posterior end of the fourth abdominal segment is a structure called the furca, which consists of a basal manubrium bearing a pair of rami, each divided into a proximal dens and a distal mucron. The furca can be turned forwards and held flexed beneath the abdomen by a retinaculum on the posterior border of segment three. The jump is produced by the furca swinging back rapidly to the extended position so that it strikes the substratum and throws the animal into the air. Movement of the furca is produced by powerful longitudinal muscles in the abdomen coupled, in at least some species, with distortion of the cuticle so that it acts like a spring (Christian, 1978, 1979).

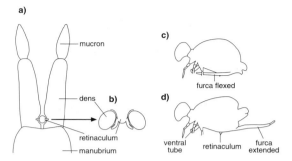

Fig. 8.25. Jumping in Collembola (partly after Denis, 1949). (a) Furca and retinaculum seen from below. (b) Diagram showing the retinaculum holding the dentes. (c), (d) Diagrams of a collembolan with the furca in the flexed and extended positions. Jumping is produced by the swing from the flexed to the extended position.

Other insects Jumping as a result of the sudden release of tension previously developed also occurs in Elateridae and the larvae of various Diptera. Elaterids (click beetles) jump if they are turned on their backs and the jump serves as a means by which they can right themselves. The insect first arches its back between the prothorax and the mesothorax so that it is supported anteriorly by the prothorax and posteriorly by the elytra with the middle of the body off the ground (Fig. 8.26). This movement is produced by the median dorsal muscle and results in the withdrawal of a median prosternal peg from the pit in which it is normally at rest. The massive prothoracic intersegmental muscle acts antagonistically to the median dorsal muscle, but when it starts to contract a small process on the upper side of the peg catches on a lip on the anterior edge of the mesosternum and no movement occurs. Tension builds up in the muscle because of its continued contraction and energy is stored within it and, possibly, in the associated cuticle. Finally, the prosternal peg slips off its catch and this energy is released as rotational energy, both prothorax and the posterior end of the body rotating upwards towards each other, and as translational energy, the center of gravity of both parts of the body being moved upwards with considerable velocity (Fig. 8.26c). This translational energy carries the insect into the air at an initial speed of about $2.5\,\mathrm{m\,s^{-1}}$, the whole jumping action being complete in about 0.5 ms. This is a relatively inefficient process and only 60% of the energy expended during the jump is used in lifting the beetle off the ground.

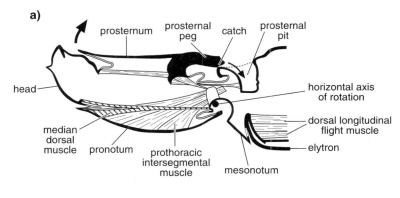

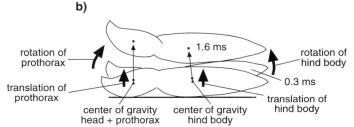

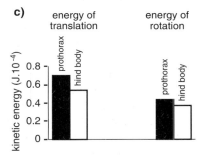

Fig. 8.26. Jumping by a click beetle (after Evans, 1972, 1973). **(a)** Longitudinal section of the head and thorax, ventral side uppermost, with the prosternal peg withdrawn from the mesosternal pit. Arrows show direction of rotation of the head and prothorax about the horizontal axis which contributes to the jump. Sclerotized cuticle black, membranous cuticle white. **(b)** Diagram showing the positions of the body in the first 1.6 ms of a jump, starting with the insect at rest on its dorsal surface. **(c)** Energy produced by the movements shown in (b).

The larva of *Piophila* (Diptera) lives in cheese and in the last stage it is able to jump. It does this by bending the head beneath its abdomen so that the mandibles engage in a transverse fold near the posterior spiracles which are at the end of the abdomen. The longitudinal muscles on the outside of the loop so formed contract and build up a tension until suddenly the mandibles are released and the larva jerks straight, striking the ground so that it is thrown into the air, sometimes as high as 20 cm. A similar phenomenon occurs in the larvae of some Clusiidae and Tephritidae. In cecidomyiid larvae, anal hooks catch in a forked prosternal projection producing a leap by building up muscular tension and then suddenly releasing it.

8.4.3 Crawling

The larvae of many holometabolous insects move by changes in the shape of the body rather than by movements of the legs as in walking or running by adult insects. This type of locomotion can be differentiated as crawling. In the majority of crawling forms, the cuticle is soft and flexible and does not, by itself, provide a suitable skeleton on which the muscles can act. Instead, the hemolymph within the body provides a hydrostatic skeleton. Muscles lining the body wall of caterpillars keep the body turgid and, because of the incompressibility of the body fluids, compression of one part of the body due to muscular contraction is compensated for by expansion of some other part. The place and form of these compensating changes is controlled by the differences in tension of the muscles throughout the body.

Caterpillars typically have, in addition to the thoracic legs, a pair of prolegs on each of abdominal segments three to six and another pair on segment ten (see Fig. 15.5). The prolegs are hollow cylindrical outgrowths of the body wall, the lumen being continuous with the hemocoel (Fig. 8.27a). An apical area, less rigid than the sides, is known as the planta and it bears one or more rows or circles of outwardly curved hooks, or crochets, with which the proleg obtains a grip. Retractor muscles from the body wall are inserted into the center of the planta so that when they contract it is drawn inwards and the crochets are disengaged. The leg is evaginated by turgor pressure when the muscles relax. On a smooth surface the prolegs can function as suckers. The crochets are turned up and the planta surface is first pressed down on to the substratum and then the center is slightly drawn up so as to create a vacuum (Hinton, 1955).

Caterpillars move by serial contractions of the longitudinal muscles coupled with leg movements starting posteriorly and continuing as a wave to the front of the body. The two legs of a segment, including those on the thoracic segments, move in synchrony. Each segment is lifted by contraction of the dorsal longitudinal muscles of the segment in front and its own dorso-ventral muscles, while at the same time the prolegs are retracted (Fig. 8.27b). Subsequently, contraction of the ventral longitudinal muscles brings the segment down again and completes the forward movement as the legs are extended and obtain a fresh grip. As the wave of contraction passes forwards along the body at least three segments are in different stages of contraction at any one time. These patterns are the product of rhythmic activity in the ventral nerve cord which occurs even in the absence of sensory input (Fig. 8.27c) although it is probable that this central control is normally modified by local reflexes involving the stretch receptors (Johnston & Levine, 1996a,b; Weevers, 1965).

Many geometrid larvae have prolegs only on abdominal segments six and ten. These insects loop along, drawing the hind end of the body up to the thorax and then extending the head and thorax to obtain a fresh grip.

In the apodous larvae of cyclorrhaphous Diptera, movement again depends on changes in the shape of the body as a result of muscles acting against the body fluids. The posterior segments of the body have raised pads usually running right across the ventral surface of a segment and armed with stiff, curved setae, which may be distributed evenly or in rows or patches. Each pad, or welt, is provided with retractor muscles. In the larva of *Musca* there are locomotory welts on the anterior edges of segments six to twelve and also on the posterior edge of segment twelve and behind the anus (see Fig. 15.7f). In movement, the anterior part of the body is lengthened and narrowed by the contraction of oblique muscles, while the posterior part maintains a grip with the welts. Hence the front of the body is pushed forwards over, or through, the substratum. It is then anchored by the mouthhooks, which are thrust against the substratum until they are held by an irregularity of the surface and the posterior part moved forwards by a wave of longitudinal shortening. As a result, the larva exhibits a regular sequence of lengthening and shortening (Fig. 8.28) (Berrigan & Pepin, 1995). In soil-dwelling larvae of Tipulidae, Bibionidae and Hepialidae (Lepidoptera), and probably in other burrowing forms, the anterior region is anchored by the broadening of the body which accompanies shortening.

8.5 LOCOMOTION IN AQUATIC INSECTS

8.5.1 Movements on the surface of water

Some insects and other hexapods are able to move on or in the film at the surface of water. Aquatic Collembola, such as *Podura aquatica*, sometimes occur on the surface film in large numbers. These animals have hydrofuge cuticles which prevent them from getting wet, but the ventral tube on the first abdominal segment is wettable and anchors the insect to the surface, while the claws, which are also wettable, enable it to obtain a purchase on the water. These animals can spring from the water surface using the caudal furca in the same way as terrestrial Collembola.

Gerris (Heteroptera) stands on the surface film and rows over the surface. All the legs possess hydrofuge properties distally, so they do not break the surface film. At the start of a power stroke, the forelegs are lifted off the surface and the long middle legs sweep backwards producing an indentation of the water surface which spreads backwards as a wave (Fig. 8.29). The mesotarsus pushes against the wave giving extra impetus to the forward movement of the insect which continues as a glide as the middle legs protract off the water surface. During retraction, muscles

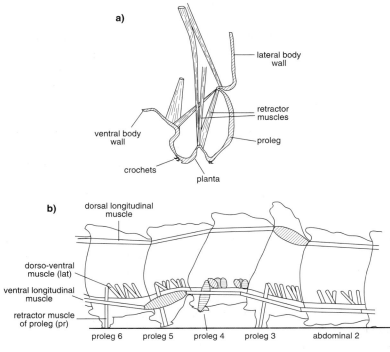

Fig. 8.27. Caterpillar crawling. **(a)** Transverse section through part of an abdominal segment of a caterpillar showing a proleg (after Hinton, 1955). **(b)** Longitudinal section through the abdomen of a caterpillar showing a wave of contraction which passes along the body from behind forwards (left to right) and produces forward movement. Contracted muscles are shown hatched. There are no prolegs on abdominal segment 2 (based on Hughes, 1965). **(c)** Central nervous control of crawling. Rhythmic activity in nerves from an isolated nerve cord of *Manduca*. Each black bar shows the periods of activity of motor neurons in a segmental nerve. The thoracic nerves shown innervate the femoral levator muscles and extensor muscles of more distal leg segments. They produce protraction of the leg. The abdominal nerves shown innervate the lateral body wall muscles (lat in b) and the proleg retractor muscles (pr in b). There are no prolegs on abdominal segments 1 and 7 (after Johnston & Levine, 1996a).

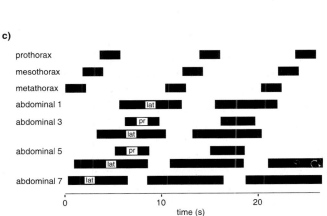

inserted into the coxa and trochanter contract simultaneously, most of the power being provided by two muscles arising in the mesothorax and inserted into the trochanter. These muscles start to contract before the leg begins to move and the rapid acceleration of the leg suggests that some temporary storage of energy in the cuticle might be occurring (Bowdan, 1978). Steering may be achieved by unequal contractions of the retractor muscles on the two sides and fast turning is produced by movement of the legs of one side while the legs of the other, towards which the insect is turning, remain still.

Stenus (Coleoptera) lives on grass stems bordering mountain streams in situations such that the beetles fall into the water quite frequently. The beetle can walk on the surface of the water, but only slowly. More rapid locomotion is produced following the secretion of chemicals from the pygidial glands opening beneath the last abdominal tergite. Five chemicals are released, of which the most

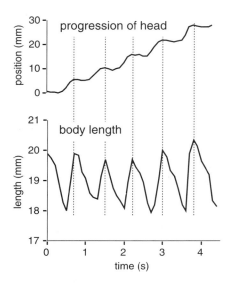

Fig. 8.28. Blowfly larva crawling. The head is pushed forwards as the larva elongates, slipping back slightly during shortening because the mouthhooks are not firmly anchored (after Berrigan & Pepin, 1995).

important is probably a terpenoid called stenusin. This substance lowers the surface tension of the water and at the same time it makes the surface of the beetle hydrophobic, so drag is reduced as the insect is drawn through the water by the higher surface tension in front. It may move at up to $70 \, \text{cm s}^{-1}$ using its abdomen as a rudder (Schildknecht, 1977).

8.5.2 Movement under water

The activity of aquatic insects is affected by their respiratory habits (see Chapter 17). Permanently submerged forms which respire by gills or a plastron have a density greater than that of the water and can move freely over the bottom of their habitat. In swimming, these insects must produce a lift force to take them off the bottom. Many other insects come to the surface to renew their air supply and submerge with a store of air, which gives them buoyancy. This buoyancy must be overcome by the forces of propulsion when the insect dives.

Review: Nachtigall, 1985

8.5.2.1 *Bottom dwellers*

Bottom-dwelling aquatic insects, such as *Aphelocheirus* (Heteroptera) and larval Odonata and Trichoptera, can walk over the substratum in the same way as terrestrial insects. The larva of *Limnephilus* (Trichoptera) basically uses an alternation of triangles of support, but because of the irregularity of the surface the stepping pattern tends to be irregular. The forelegs may step together instead of alternating and the hind legs may follow the same pattern. Normally the power for walking comes primarily from traction by the fore and middle legs and pushing by the hind legs, but under difficult conditions the hind legs may be extended far forwards outside the middle legs so that they help the other legs to pull the larva along.

The larval case of the caddis fly, *Triaenodes*, is built of plant material arranged in a spiral, the most anterior whorl of which extends dorsally beyond the rest of the case above the thorax (Fig. 8.30). This dorsal whorl provides a certain amount of lift, helping to carry the case off the bottom. This lift is controlled by the movements of the legs, which tend to produce a downward thrust.

Larval Anisoptera can walk across the substratum using their legs, but they are also able to make sudden escape movements by forcing water rapidly out of the branchial chamber so that the body is driven forwards (section 17.5.2.1). The branchial chamber is compressed by longitudinal and dorso-ventral contractions of the abdomen, the contractions being strongest in segments six to eight, in which the branchial chamber lies. Before this contraction, the anal valves close and then open slightly leaving an aperture about $0.01 \, \text{mm}^2$ in area. The contractile movement lasts about 100 ms and water is forced through the anus at a velocity of about $250 \, \text{cm s}^{-1}$ propelling the larva forwards at $30–50 \, \text{cm s}^{-1}$. As the abdomen contracts, the legs are retracted so as to lie along the sides of the body, offering minimal resistance to forward movement. Successive contractions may occur at frequencies up to 2.2 Hz, continuing for up to 15 s. Co-ordination involves giant fibers running in the ventral nerve cord.

8.5.2.2 *Free-swimming insects*

Larval and pupal Diptera, larval and adult Heteroptera and adult Coleoptera form the bulk of free-swimming insects and, apart from the Diptera, most of them use the hind legs, sometimes together with the middle legs, in swimming. The point of attachment of the hind legs is displaced posteriorly compared with terrestrial insects and in dytiscids and gyrinids, the coxae are immovably fused to the thorax. This limits the amount of movement at the base of the leg and the basal muscles are concentrated into

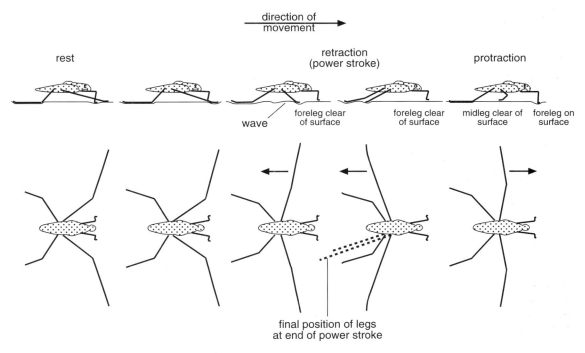

Fig. 8.29. Movement on the water surface. Positions of the legs of *Gerris*. Arrows show the directions of movement of the legs relative to the body; insect moving from left to right (after Nachtigall, 1974).

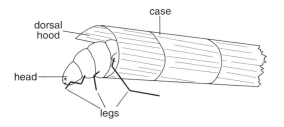

Fig. 8.30. Diagram of *Triaenodes* larva (Trichoptera) in its case (after Tindall, 1964).

two functional groups, a powerful trochanter/femur retractor group and a weaker protractor group. Intrinsic muscles of the legs tend to be reduced and the movements of the distal parts of the legs during swimming are largely passive.

The two legs of a segment move together, contrasting with the alternating movement of the legs in terrestrial insects, but *Hydrophilus* (Coleoptera) is an exception. This beetle uses the middle and hind legs in swimming, the middle leg of one side being retracted simultaneously with the hind leg of the opposite side, but out of phase with the contralateral middle leg.

The effect of drag during swimming Because of the high density of water relative to air, drag presents a much more serious problem to aquatic insects than to terrestrial species. Most aquatic insects are streamlined so that the drag imposed on the body is much less than would be the case with other shapes and is only slightly above the drag produced by the 'ideal streamlined body' (Fig. 8.31). Drag also increases markedly with the angle of attack and is large in larger insects because of their greater frontal areas. There is also a marked increase in resistance if the insect turns broadside or ventral side to the direction of movement and this facilitates turning and braking. Turns are made by producing strokes of unequal amplitude on the two sides or, in making a sharp turn, the leg on the inside may be extended and kept still while the contralateral leg paddles.

Thrust To overcome drag, and in order to move forwards, the insect produces thrust by retraction of its hind legs,

a)

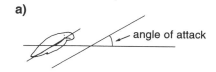

— angle of attack

b)

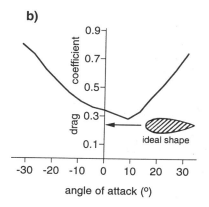

Fig. 8.31. Drag on a swimming beetle as its angle of attack changes (after Nachtigall, 1985).

but, because it is surrounded by the medium, protraction of the legs also produces forces and these tend to drive the insect backwards. If it is to move forwards, the forward thrust produced on the backstroke must exceed the backward thrust produced on the forward, recovery stroke of the legs.

The thrust which a leg exerts in water is proportional to its area and the square of the velocity with which it moves. Hence to produce the most efficient forward movement a leg should present a large surface area and move rapidly on the backstroke, while presenting only a small surface and moving relatively slowly on the recovery stroke. Higher speeds will be produced if the greatest surface area is furthest from the body.

To achieve a large surface area the hind tibiae and tarsi, and sometimes also those of the middle legs, are flattened antero-posteriorly to form a paddle, which, in *Acilius* and *Dytiscus*, is increased in area by inflexible hairs and, in *Gyrinus* (Coleoptera), by cuticular blades 1 µm thick and 30–40 µm wide (Fig. 8.32). In *Acilius* (Coleoptera) the hairs constitute 69% of the total area of the hind tibiae and 83% of the area of the tarsi. The hind legs of these insects are relatively shorter than the hind legs of related terrestrial insects, but the tarsi are relatively longer.

On the backstroke (retraction) the swimming legs of

Dytiscus are straight with the fringing hairs, which are articulated at the base, spread to expose a maximum area (Fig. 8.33a–d). On the forward stroke (protraction), however, the femorotibial joint flexes so that the tibia and tarsus trail out behind (Fig. 8.33e–h). At the same time the tibia rotates through 45° so that the previously dorsal surface becomes anterior and the fringing hairs fold back. The tarsus, which articulates with the tibia by a ball and socket joint, rotates through 100° in the opposite direction. These movements are passive, resulting from the form of the legs and the forces exerted by the water, and they ensure that the tibia and tarsus are presented edge on to the movement, producing a minimum of thrust. Subsequently, at the beginning of the backstroke the leg and hairs extend passively to expose a maximum surface area again. There is no extensor tarsi muscle and the extensor tibiae is weak. The power for the stroke comes from the muscles moving the trochanter on the fixed coxa.

Similar devices for exposing a maximum leg area during the power stroke and a minimum during the recovery stroke are employed by other insects. The swimming blades fringing the leg of *Gyrinus* are placed asymmetrically so that they open like a venetian blind, turning to overlap and produce a solid surface during the power stroke. In the recovery stroke the tarsomeres collapse like a fan and are concealed in a hollow of the tibia, which in turn is partly concealed in a hollow of the femur (Fig. 8.34a).

The relative power developed on the forward and backward strokes also depends on the relative speeds of the strokes. In *Acilius* (Coleoptera) the backstroke is faster than the forward stroke, so that for a given leg area the forward thrust on the body exceeds the backward thrust. In *Gyrinus*, on the other hand, the backstroke is slower than the forward stroke, so that for a given area the backward thrust is greater than the forward thrust. Hence if the area of the legs of *Gyrinus* remained constant the insect would tend to move backwards and it is only because the great reduction in area of the leg on the forward stroke reduces the backward thrust on the body that the net effect is to push the insect forwards.

The legs move in an arc and so lateral thrust is produced in addition to the longitudinal thrust (Fig. 8.34b). In most insects, where the legs of the two sides move in phase, the lateral forces developed on the two sides balance each other out, but in *Hydrophilus*, where the legs are used alternately, there is some deviation to either side, although the lateral thrust of the hind leg on one side is

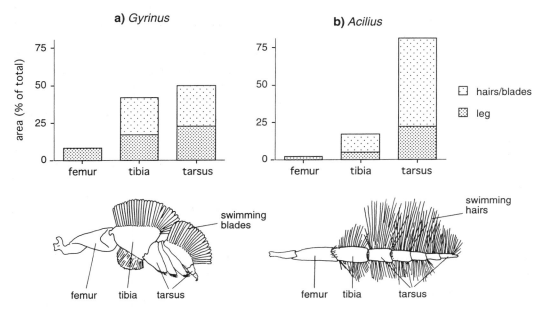

Fig. 8.32. Hind leg modifications of aquatic beetles. Structural adaptations and surface area of each part of the leg (after Nachtigall, 1962). **(a)** *Gyrinus*, **(b)** *Acilius*.

largely balanced by the opposite lateral thrust of the contralateral middle leg. Forward thrust is minimal at the beginning and end of each stroke (Fig. 8.34b, leg at A and C), but when the legs are at right angles to the body the whole of the thrust developed is longitudinal (Fig. 8.34b, leg at B). It is advantageous if the velocity of the leg is greatest at this point and is low at the beginning and end of the stroke so that the lateral forces, which are produced mainly during these phases, are kept to a minimum. This is the case, at least in *Acilius* and *Gyrinus*. In the latter the leg is moving most rapidly when it is at an angle of 90–135° to the body. After this the velocity rapidly falls to zero.

Buoyancy Many free-swimming insects are buoyant, that is, they have lift within the water, because of the air in their tracheae and in air stores. When they stop swimming they come to rest at the surface of the water in a characteristic position which results from the distribution of air stores on and in the body. Most species float head down and *Notonecta* (Heteroptera), for instance, rests at an angle of 30° to the surface. As it kicks with its swimming legs this angle is increased to 55° so that the insect is driven down, but as it loses momentum during the recovery stroke of the

legs it will tend to rise again (Fig. 8.35). If the driving movements of the legs are repeated rapidly, before the insect rises very much, the path may be straightened out and by controlling the rate of leg movement the insect can dive, move at a constant level or rise to the surface (Fig. 8.35b). The beat tends to be faster at higher temperatures and so movement becomes more uniform as the temperature rises. In *Dytiscus*, the buoyancy effect is offset at higher speeds by using the middle and hind legs alternately, while *Hydrophilus* achieves the same effect by using the legs of the two sides out of phase. Hence these insects produce a continuous driving force which offsets their buoyancy. Amongst water beetles, secretions from the pygidial glands increase the wettability of the cuticle. This may be particularly important in enabling small species to break the water surface film.

A few insects, such as larval *Chaoborus* (Diptera) and *Anisops* (Heteroptera), can control their buoyancy so that they can remain suspended in mid water (Teraguchi, 1975).

Stability Because insects swimming beneath the surface are surrounded by the water and have no contact with a solid object, they are, like flying insects, subject to instability in the rolling, pitching and yawing planes (see

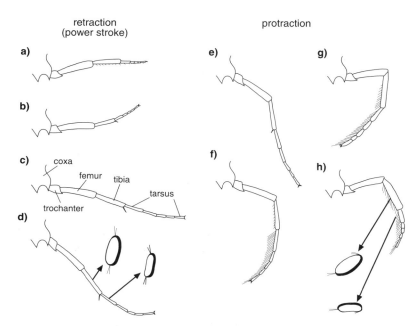

retraction
(power stroke)

protraction

Fig. 8.33. Swimming. Successive positions of the right hind leg of *Dytiscus* during swimming as seen from above (assuming the body of the insect to be transparent). Insets are cross-sections of the tibia and tarsus showing their orientation during retraction (d) and protraction (h). The thick line represents the morphologically anterior side of the leg (after Hughes, 1958). **(a)–(d)** Stages of retraction (the power stroke) with the femur swinging back relative to the coxa. **(e)–(h)** Stages of protraction (the recovery stroke) with the femur moving forwards.

Fig. 9.33). The dorso-ventral flattening of many aquatic insects provides stability in the rolling and pitching planes. The control of yawing involves the eyes, antennae and possibly also receptors on the legs, these receptors acting so that any unequal stimulation as a result of deviation from a straight course is corrected for. In *Triaenodes* the long case (Fig. 8.30) acts as a rudder, giving some stability in the pitching and yawing planes. Rolling may be controlled by the long, outstretched hind legs.

Swimming movements are maintained by the flow of water past the head as the insect moves forwards (the equivalent of the relative wind). This tends to push the antennal flagellum back, but the scape and pedicel are held at a constant angle. The water movement is probably monitored by Johnston's organ (section 23.2.3.2) which maintains swimming movements in a manner analogous to that which maintains wing movements in a flying insect (section 9.11.4; Gewecke, 1985).

Speed The speed of movement depends on the frequency with which strokes are made and the lengths and velocities of the strokes (Gewecke, 1985). *Gyrinus* can swim on the surface at up to $100\,\mathrm{cm\,s^{-1}}$ in short bursts, the hind leg making 50–$60\,\mathrm{strokes\,s^{-1}}$. Beneath the surface its speed rarely exceeds $10\,\mathrm{cm\,s^{-1}}$. In general, the maximum sustainable velocity increases with the size of

the insect. *Acilius* can reach $35\,\mathrm{cm\,s^{-1}}$, larger beetles may attain speeds of $100\,\mathrm{cm\,s^{-1}}$. The larva of *Triaenodes* only moves at about $1.7\,\mathrm{cm\,s^{-1}}$ because of the high drag produced by its case.

8.5.2.3 *Other forms of swimming*

Appendages other than the legs are sometimes used in swimming. Mosquito larvae when suspended from the surface film or browsing on the bottom can glide slowly along as a result of the rapid vibrations of the mouth brushes in feeding. In *Aedes communis* this is the normal method of progression. *Caraphractus cinctus* (Hymenoptera) parasitizes the eggs of dytiscids, which are laid under water. The parasite swims jerkily through the water by rowing with its wings, making about two strokes per second. Larval Ephemeroptera and Zygoptera move by vertical undulations of the caudal gills and the abdomen, while in the larva of *Ceratopogon* (Diptera) lateral undulations pass down the body from head to tail driving the insect through the water (Fig. 8.36a). Many other dipterous larvae flex and straighten the body alternately to either side, often increasing the thrust by a fin-like extension of the hind end. Mosquito larvae, for instance, have a fan of dense hairs on the last abdominal segment and as a result of the lateral flexing of the body move along tail first (Fig. 8.36b). The relative density of mosquito larvae is very close

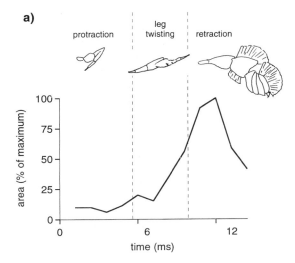

a)

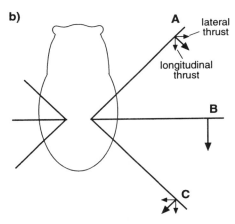

b)

Fig. 8.34. Swimming. Changes in thrust and the effective area of the hindleg. (a) Effective area of the leg of *Gyrinus* during one complete stroke (after Nachtigall, 1962). (b) Thrust exerted with the hind leg at different points of the power stroke. It is assumed that the velocity of the leg at A and C is only half its velocity at B, where, since the leg is at right angles to the body, only longitudinal thrust is produced. Equal and opposite forces act on the body (based on Nachtigall, 1965).

to that of water and its value affects their locomotion. Early stage larvae are usually less dense than the medium, so they rise to the surface when they stop swimming. This is also true of the pupae, but last stage larvae may be slightly denser and so they sink when they stop moving actively. The larvae of *Chaoborus* and other Diptera make similar movements to those of mosquito larvae.

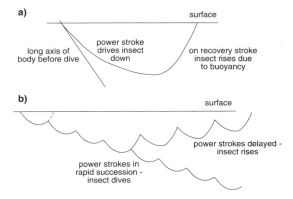

Fig. 8.35. Path through water of a buoyant insect such as *Notonecta* (after Popham, 1952): (a) the path due to a single swimming stroke by the legs; (b) different paths produced by differences in the timing of successive strokes.

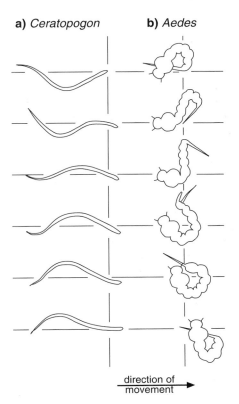

Fig. 8.36. Swimming by larval Diptera (after Nachtigall, 1965): (a) *Ceratopogon*, (b) *Aedes*.

REFERENCES

Bässler, U. (1983). *Neural Basis of Elementary Behavior in Stick Insects.* Berlin: Springer-Verlag.

Bässler, U. (1988). Functional principles of pattern generation for walking movements of stick insect forelegs: the role of the femoral chordotonal afferences. *Journal of Experimental Biology*, **136**, 125–47.

Bennet-Clark, H.C. (1975). The energetics of the jump of the locust *Schistocerca gregaria. Journal of Experimental Biology*, **63**, 53–83.

Bennet-Clark, H.C. (1990) Jumping in Orthoptera. In *Biology of Grasshoppers* ed. R.F. Chapman & A. Joern, pp. 173–203. New York: Wiley & Sons.

Bennet-Clark, H.C. & Lucey, E.C.A. (1967). The jump of the flea: A study of the energetics and a model of the mechanism. *Journal of Experimental Biology*, **47**, 59–76.

Berrigan, D. & Pepin, D.J. (1995). How maggots move: allometry and kinematics of crawling in larval Diptera. *Journal of Insect Physiology*, **41**, 329–37.

Bowdan, E. (1978). Walking and rowing in the water strider, *Gerris remigis* II. Muscle activity associated with slow and rapid mesothoracic leg movement. *Journal of Comparative Physiology*, **123**, 51–7.

Burns, M.D. & Usherwood, P.N.R. (1979). The control of walking in Orthoptera II. Motor neurone activity in normal free-walking animals. *Journal of Experimental Biology*, **79**, 69–98.

Burrows, M. (1989). Processing of mechanosensory signals in local reflex pathways of the locust. *Journal of Experimental Biology*, **146**, 209–27.

Burrows, M. (1992). Local circuits for the control of leg movements in an insect. *Trends in Neuroscience*, **15**, 226–32.

Burrows, M. (1996). *The Neurobiology of an Insect Brain.* Oxford: Oxford University Press.

Chapman, R.F. (1982). Chemoreceptors: the significance of receptor numbers. *Advances in Insect Physiology*, **16**, 247–356.

Christian, E. (1978). The jump of the springtails. *Naturwissenschaften* **65**, 495.

Christian, E. (1979). Der Sprung der Collembolen. *Zoologische Jahrbücher. Zoologie und Physiologie der Tiere*, **83**, 457–90.

Delcomyn, F. (1985). Walking and running. In *Comprehensive Insect Physiology, Biochemistry and Pharmacology*, vol. 5, ed. G.A. Kerkut & L.I. Gilbert, pp. 439–66. Oxford: Pergamon Press.

Delcomyn, F. & Usherwood, P.N.R. (1973). Motor activity during walking in the cockroach *Periplaneta americana. Journal of Experimental Biology*, **59**, 629–42.

Denis, R. (1949). Sous-classe des Apterygotes. In *Traité de Zoologie*, vol. 9, ed. P.-P. Grassé, pp. 111–275. Paris: Masson et Cie.

Duch, C. & Pflüger, H.J. (1995). Motor patterns for horizontal and upsidedown walking and vertical climbing in the locust. *Journal of Experimental Biology*, **198**, 1963–76.

Evans, M.E.G. (1972). The jump of the click beetle (Coleoptera: Elateridae) – a preliminary study. *Journal of Zoology, London*, **167**, 319–36.

Evans, M.E.G. (1973). The jump of the click beetle (Coleoptera: Elateridae) — energetics and mechanics. *Journal of Zoology, London*, **169**, 181–94.

Field, L.H. & Coles, M.M.L. (1994). The position-dependent nature of postural resistance reflexes in the locust. *Journal of Experimental Biology*, **188**, 65–88.

Field, L.H. & Pflüger, H.-J. (1989). The femoral chordotonal organ: a bifunctional orthopteran (*Locusta migratoria*) sense organ. *Comparative Biochemistry and Physiology*, **93A**, 729–43.

Full, R.J. & Koehl, M.A.R. (1993). Drag and lift on running insects. *Journal of Experimental Biology*, **176**, 89–101.

Full, R.J. & Tu, M.S. (1990). Mechanics of six-legged runners. *Journal of Experimental Biology*, **148**, 129–46.

Full, R.J. & Tu, M.S. (1991). Mechanics of a rapid running insect: two-, four- and six-legged locomotion. *Journal of Experimental Biology*, **156**, 215–31.

Gewecke, M. (1985). Swimming behaviour of the water beetle *Dytiscus marginalis* L. (Coleoptera, Dytiscidae). In *Insect Locomotion*, ed. M. Gewecke & G. Wendler, pp. 111–20. Berlin: Paul Parey.

Heitler, W.J. (1974) The locust jump: specialization of the metathoracic femoral–tibial joint. *Journal of Comparative Physiology*, **89**, 93–104.

Heitler, W.J. (1977). The locust jump III. Structural specializations of the metathoracic tibiae. *Journal of Experimental Biology*, **67**, 29–36.

Heitler, W.J. & Burrows, M. (1977). The locust jump I. The motor programme. *Journal of Experimental Biology*, **66**, 203–19.

Hering, E.M. (1951). *Biology of Leaf Miners.* 's-Gravenhage: Junk.

Hinton, H.E. (1955). On the structure, function, and distribution of the prolegs of the Panorpoidea, with a criticism of the Berlese–Imms theory. *Transactions of the Royal Entomological Society of London*, **106**, 455–545.

Hughes, G.M. (1952). The co-ordination of insect movements. I. The walking movements of insects. *Journal of Experimental Biology*, **29**, 267–84.

Hughes, G.M. (1958). The co-ordination of insect movements. III. Swimming in *Dytiscus, Hydrophilus*, and a dragonfly nymph. *Journal of Experimental Biology*, **35**, 567–83.

Hughes, G.M. (1965). Locomotion: terrestrial. In *The Physiology of Insecta*, 1st edition, vol. 3, ed. M. Rockstein, pp. 227–54. New York: Academic Press.

Hughes, G.M. & Mill, P.J. (1974). Locomotion: terrestrial. In *The Physiology of Insecta*, 2nd edition, vol. 3, ed. M. Rockstein, pp. 335–79. New York: Academic Press.

Hustert, R., Pflüger, J.H. & Bräunig, P. (1981). Distribution and specific central projections of mechanoreceptors in the thorax and proximal leg joints of locusts. *Cell and Tissue Research*, **216**, 97–111.

Johnston, R.M. & Levine, R.B. (1996a). Crawling motor patterns induced by pilocarpine in isolated larval nerve cords of *Manduca sexta*. *Journal of Neurophysiology*, **76**, 3178–95.

Johnston, R.M. & Levine, R.B. (1996b). Locomotor behavior in the hawkmoth *Manduca sexta*: kinematic and electromyographic analyses of the thoracic legs in larvae and adults. *Journal of Experimental Biology*, **199**, 759–74.

Katz, S.L. & Gosline, J.M. (1993). Ontogenetic scaling of jump performance in the African desert locust (*Schistocerca gregaria*). *Journal of Experimental Biology*, **177**, 81–111.

Kent, K.S. & Griffin, L.M. (1990). Sensory organs of the thoracic legs of the moth *Manduca sexta*. *Cell & Tissue Research*, **259**, 209–223.

Lees, A.D. & Hardie, J. (1988). The organs of adhesion in the aphid *Megoura viciae*. *Journal of Experimental Biology*, **136**, 209–28.

Manton, S.M. (1953). Locomotory habits and the evolution of the larger arthropodan groups. *Symposia of the Society for Experimental Biology*, **7**, 339–76.

Miall, L.C. (1922). *The Natural History of Aquatic Insects*. London: MacMillan.

Mill, P. J. & Pickard, R.S. (1975). Jet-propulsion in anisopteran dragonfly larvae. *Journal of Comparative Physiology*, **97**, 329–38.

Nachtigall, W. (1962). Funktionelle Morphologie, Kinematic und Hydromechanik des Ruderapparates von *Gyrinus*. *Zeitschrift für Vergleichende Physiologie*, **45**, 193–226.

Nachtigall, W. (1965). Locomotion: swimming (hydrodynamics) of aquatic insects. In *The Physiology of Insecta*, 1st edition, vol. 2, ed. M. Rockstein, pp. 255–81. New York: Academic Press.

Nachtigall, W. (1974). Locomotion: mechanics and hydrodynamics of swimming in aquatic insects. In *The Physiology of Insecta*, 2nd edition, vol. 3, ed. M. Rockstein, pp. 381–432. New York: Academic Press.

Nachtigall, W. (1985). Swimming in aquatic insects. In *Comprehensive Insect Physiology, Biochemistry and Pharmacology*, vol. 5, ed. G.A. Kerkut & L.I. Gilbert, pp. 467–90. Oxford: Pergamon Press.

Pearson, K.G. & Iles, J.F. (1971) Innervation of coxal depressor muscles in the cockroach, *Periplaneta americana*. *Journal of Experimental Biology*, **54**, 215–32.

Pesson, P. (1951). Ordre des Homoptères. In *Traité de Zoologie*, vol. 10, ed. P.-P. Grassé, pp. 1390–656. Paris: Masson et Cie.

Popham, E.J. (1952). A preliminary investigation into the locomotion of aquatic Hemiptera and Coleoptera. *Proceedings of the Royal Entomological Society of London* A, **27**, 117–9.

Queathem, E. (1991). The ontogeny of grasshopper jumping performance. *Journal of Insect Physiology*, **37**, 129–38.

Ryckebusch, S. & Laurent, G. (1993). Rhythmic patterns evoked in locust leg motor neurons by the muscarinic agonist pilocarpine. *Journal of Neurophysiology*, **69**, 1583–95.

Schildknecht, H. (1977). Protective substances of arthropods and plants. *Scripta Varia*, **41**, 59–107.

Schmitz, J. (1993). Load-compensating reactions in the proximal leg joints of stick insects during standing and walking. *Journal of Experimental Biology*, **183**, 15–33.

Schönitzer, K. & Lawitzky, G. (1987). A phylogenetic study of the antenna cleaner in Formicidae, Mutillidae, and Tipulidae (Insecta, Hymenoptera). *Zoomorphology*, **107**, 273–85.

Seelinger, G. & Tobin, T.R. (1981). Sense organs. In *The American Cockroach*, ed. W.J. Bell & K.G. Adiyodi, pp. 217–45. New York: Chapman & Hall.

Séguy, E. (1951). Ordre des Anoploures ou Poux. In *Traité de Zoologie*, vol. 10, ed. P.-P. Grassé, pp. 1365–84. Paris: Masson et Cie.

Skorupski, P. & Hustert, R. (1991). Reflex pathways responsive to depression of the locust coxotrochanteral joint. *Journal of Experimental Biology*, **158**, 599–605.

Snodgrass, R.E. (1927). Morphology and mechanism of the insect thorax. *Smithsonian Miscellaneous Collections*, **80**, no. 1.

Snodgrass, R.E. (1935). *Principles of Insect Morphology*. New York: McGraw-Hill.

Snodgrass, R.E. (1952). *A Textbook of Arthropod Anatomy*. Ithaca: Cornell University Press.

Snodgrass, R.E. (1956). *Anatomy of the Honey Bee*. London: Constable.

Stork, N.E. (1983). The adherence of beetle tarsal setae to glass. *Journal of Natural History*, **17**, 583–97.

Teraguchi, S. (1975). Correction of negative buoyancy in the phantom larva, *Chaoborus americanus*. *Journal of Insect Physiology*, **21**, 1659–70.

Tindall, A.R. (1964). The skeleton and musculature of the larval thorax of *Triaenodes bicolor* Curtis (Trichoptera: Limnephilidae). *Transactions of the Royal Entomological Society of London*, **116**, 151–210.

Ting, L.H., Blickham, R. & Full, R.J. (1994). Dynamic and static stability in hexapedal runners. *Journal of Experimental Biology*, **197**, 251–69.

Urbani, C.B., Boyan, G.S., Blarer, A., Billen, J. & Ali, T.M.M. (1994). A novel mechanism for jumping in the Indian ant *Harpegnathos saltator* (Jerdon) (Formicidae, Ponerinae). *Experientia*, **50**, 63–71.

Weevers, R.de G. (1965). Proprioceptive reflexes and the co-ordination of locomotion in the caterpillar of *Antheraea pernyi* (Lepidoptera). In *The Physiology of the Insect Central Nervous System*, ed. J.E. Treherne & J.W.L. Beament, pp. 113–24. London: Academic Press.

Wolf, H. (1990). Activity patterns of inhibitory motoneurones and their impact on leg movement in tethered walking locusts. *Journal of Experimental Biology*, **152**, 281–304.

Zill, S.N. & Moran, D.T. (1981a). The exoskeleton and insect proprioception. I. Responses of tibial campaniform sensilla to external and muscle-generated forces in the American cockroach, *Periplaneta americana*. *Journal of Experimental Biology*, **91**, 1–24.

Zill, S.N. & Moran, D.T. (1981b). The exoskeleton and insect proprioception. III. Activity of tibial campaniform sensilla during walking in the American cockroach, *Periplaneta americana*. *Journal of Experimental Biology*, **94**, 57–75.

Zill, S.N., Moran, D.T. & Varela, F.G. (1981). The exoskeleton and insect proprioception. II. Reflex effects of tibial campaniform sensilla in the American cockroach, *Periplaneta americana*. *Journal of Experimental Biology*, **94**, 43–55.

Zollikofer, C.P.E. (1994). Stepping patterns in ants II. Influence of body morphology. *Journal of Experimental Biology*, **192**, 107–18.

9   Wings and flight

9.1 OCCURRENCE AND STRUCTURE OF WINGS

Fully developed and functional wings occur only in adult insects although the developing wings are present in larvae. In hemimetabolous larvae they are visible as external pads (section 15.3.1), but they develop internally in holometabolous species (section 15.3.2.2). The Ephemeroptera are exceptional in having two fully-winged stages. The final larval stage molts to a subimago, which resembles the adult except for having fringed and slightly translucent wings and rather shorter legs. It is able to make a short flight, after which it molts and the adult stage emerges. In the course of this molt the cuticle of the wings is shed with the rest of the cuticle.

The fully developed wings of all insects appear as thin, rigid flaps arising dorsolaterally from between the pleura and nota of the meso- and meta-thoracic segments. Each wing consists of a thin membrane supported by a system of veins. The membrane is formed by two layers of integument closely apposed, while the veins are formed where the two layers remain separate and the cuticle may be thicker and more heavily sclerotized (Fig. 9.1). Within each of the major veins is a nerve and a trachea, and, since the cavities of the veins are connected with the hemocoel, hemolymph can flow into the wings (section 5.1.2.1).

9.1.1 Basic structure of the wing

The structure of the wing is determined primarily by the need to optimize the production of favorable aerodynamic forces during flight. In addition, in many insects, the structure allows the wing to fold when the insect is not flying.
Review: Wootton, 1992

9.1.1.1 *Veins and venation*

The principal support of the wing membrane is provided by a number of well-marked veins running along the length of the wing and connected to each other by a variable number of cross-veins. There is a tendency for the wings of lower orders of insects to be pleated in a fan-like

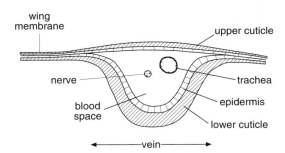

Fig. 9.1. Diagrammatic section through part of a wing including a transverse section of a vein.

manner with the longitudinal veins alternately on the crests or in the troughs of folds (Fig. 9.2). A vein on a crest is called convex (indicated by + in Fig. 9.2a), while a vein in a trough is called concave (− in Fig. 9.2a).

The basic longitudinal veins which can be distinguished in modern insects are shown in Fig. 9.2a following the recommendations of Wootton (1979). These are generally similar to the widely used Comstock-Needham system, but with some significant rationalizations. From the leading edge of the wing backwards they are:

costa (abbreviated to C) on or just behind the leading edge

subcosta (Sc)

radius (R)

radial sector (Rs)

anterior media (MA) ⎱ Media (M) where the two

posterior media (MP) ⎰ cannot be distinguished

anterior cubitus (CuA)

posterior cubitus (CuP)

anals (1A, 2A, etc.)

Any of these veins may branch, the branches then being given subscripts 1, 2, 3, etc. as in Fig. 9.2a. It is important to recognize that these branches are not necessarily

[185]

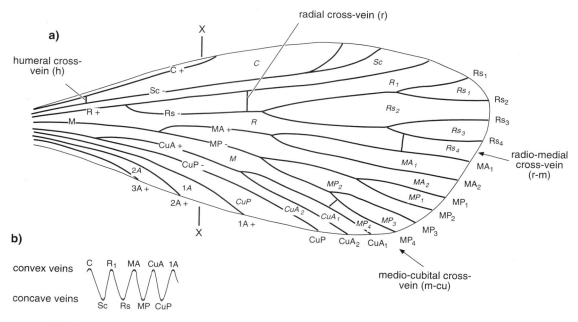

Fig. 9.2. Wing venation. **(a)** Diagram of wing venation showing the main cross veins and the names of the cells (italicized) enclosed by the veins. See text for abbreviations. **(b)** Section at X–X in (a) showing the concave and convex veins with the depth of pleating greatly exaggerated.

homologous in different groups of insects (see also Wootton & Ennos, 1989).

In some very small insects, the venation may be greatly reduced. In Chalcidoidea, for instance, only the subcosta and part of the radius are present (see Fig. 9.8f). Conversely, an increase in venation may occur by the branching of existing veins to produce accessory veins or by the development of additional, intercalary veins between the original ones, as in the wings of Orthoptera. Large numbers of cross-veins are present in some insects, and they may form a reticulum as in the wings of Odonata and at the base of the forewings of Tettigonioidea and Acridoidea.

The form of an individual vein reflects its role in the production of useful aerodynamic forces by the wing as a whole. On the leading edge of the wing, the longitudinal veins form a rigid spar supporting the wing as it moves through the air. In Lepidoptera, for example, the subcostal vein is circular in cross-section and so is equally resistant to bending in any direction (Fig. 9.3a), and, in dragonflies, the cross veins along the leading edge of the wing form angle brackets which contribute to its rigidity (Fig. 9.3d).

Behind the leading edge, the wing is often longitudinally corrugated (Fig. 9.2b). This, in itself, confers some degree of resistance to longitudinal bending, and the elliptical cross-section of some of the veins (Fig. 9.3b) confers further resistance to vertical bending. The arrangement of folds and veins in the hindwing of Orthoptera also limits vertical flexibility while facilitating folding (Fig. 9.3c). Cross veins are often circular in cross-section (Fig. 9.3e). Where flexibility is required the veins are annulated (Fig. 9.3f) or have short narrow or unsclerotized regions (Fig. 9.4b–f).

9.1.1.2 *Wing flexion*

The wings flex and twist during flight to maximize the output of aerodynamic power. As the wing moves down, bending forces act from below, and significant inertial forces act on it as it changes direction at the top and bottom of the stroke. The production of useful aerodynamic forces requires that the wing remain relatively rigid on the downstroke, but exhibits some flexibility on the upstroke. Flexion occurs at specific points of the wing and requires breaks or flexibility in the veins at these points.

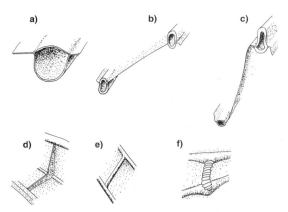

Fig. 9.3. Vein morphology (after Wootton, 1992). (**a**) Principal supporting vein, resistant to bending and twisting in any direction (Lepidoptera, *Papilio*). (**b**) Supporting veins near the leading edge of a dragonfly wing, resistant to bending up or down (Odonata, *Calopteryx*). (**c**) Veins from the pleated area of a hindwing. Ridge vein is resistant to bending up, or down, but the trough vein is compliant in all directions (Orthoptera, *Schistocerca*). (**d**) Cross-veins forming a rigid angle bracket and linking the anterior longitudinal veins to form a stout spar supporting the leading edge of the wing of a dragonfly (Odonata, *Calopteryx*). (**e**) Normal cross-vein, circular in cross-section (Odonata, *Calopteryx*). (**f**) Annulate cross-vein permitting flexibility (Diptera, *Eristalis*).

Two longitudinal flexion lines are of widespread occurrence. These are the median flexion line, which usually arises close to the media and runs just behind the radial sector for much of its length, and the claval furrow which runs close to CuP (Fig. 9.4a). This furrow allows the posterior part of the wing to flap up and down with respect to the rest of the wing. There is also commonly a transverse flexion line, such as the nodal flexion line of cicadas (Fig. 9.4b). This line permits the distal regions of the wing to bend down on the upstroke, but does not permit upward flexion during the downstroke. This is achieved, in cicadas, by lines of weakness on the ventral sides of some veins while sclerotization along the dorsal surface is continuous (Fig. 9.4c–f). In many other insects, the arched (cambered) cross-section of the whole wing seems adequate to prevent dorsal bending while allowing the wing to flex ventrally (Wootton, 1992).

9.1.1.3 *Wing folding*

When at rest, the wings are held over the back in most insects. This may involve longitudinal folding of the wing

membrane and sometimes also transverse folding. Folding may sometimes occur along the flexion lines. In addition, most Neoptera have a jugal fold just behind vein 3A on the forewings (Fig. 9.4a). It is sometimes also present on the hindwings. Where the anal area of the hindwing is large, as in Orthoptera and Blattodea, the whole of this part may be folded under the anterior part of the wing along a vannal fold a little posterior to the claval furrow (see Fig. 9.9). In addition, in Orthoptera and Blattodea, the anal area is folded like a fan along the veins, the anal veins being convex, at the crests of the folds, and the accessory veins concave. Whereas the claval furrow and jugal fold are probably homologous in different species, the vannal fold varies in position in different taxa (Wootton, 1979).

Folding is produced by a muscle arising on the pleuron and inserted into the third axillary sclerite in such a way that, when it contracts, the sclerite pivots about its points of articulation with the posterior notal process and the second axillary sclerite. As a result, the distal arm of the third axillary sclerite rotates upwards and inwards, so that finally its position is completely reversed. The anal veins are articulated with this sclerite in such a way that when it moves they are carried with it and become flexed over the back of the insect. Activity of the same muscle in flight affects the power output of the wing and so it is also important in flight control (see section 9.11).

In orthopteroid insects, the elasticity of the cuticle causes the vannal area of the wing (section 9.1.1.4) to fold along the veins. Consequently energy is expended in unfolding this region when the wings are moved to the flight position. In general, wing extension probably results from the contraction of muscles attached to the basalar sclerite or, in some insects, to the subalar sclerite.

Paper wasps (Vespidae) and some other Hymenoptera have a longitudinal fold line close to the cubital vein. The form of the cuticle where the fold line crosses the claval furrow ensures that only two positions of the fold are stable: with the wing fully extended or fully flexed. The former condition occurs when the wing is pulled into the flight position, the latter when the wing is in the rest position (Danforth & Michener, 1988).

The hindwings of Coleoptera, Dermaptera and a few Blattodea fold transversely as well as longitudinally so that they can be accommodated beneath the protective forewings. This transverse folding necessitates a modification of the venation and in Coleoptera there is a discontinuity between the proximal and distal parts of the

Fig. 9.4. Fold lines and flexion lines (mainly after Wootton, 1981). **(a)** Diagram illustrating the main flexion lines. **(b)** A transverse flexion line, shown by dots, on the forewing of a cicada. **(c)** Forewing of a cicada showing the areas enlarged in d, e and f. Notice, in d–f, that breaks in the veins are incomplete above, but complete below so that the wing will flex down, but not up. **(d)** Enlargement of part of the leading edge of the wing shown in c. Open arrow shows the complete break on the ventral surface of C+Sc. **(e)** Break in M_{2+3}. **(f)** Break in the cubital cross-vein complete on the lower surface (open arrow).

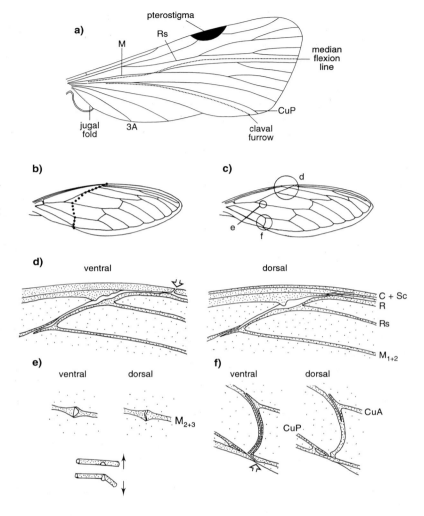

Fig. 9.5. Hindwing folding in a beetle (Coleoptera, *Melolontha*) (after Jeannel, 1949). **(a)** Wing extended in the flight position. **(b)** wing folded back over body. It would normally be concealed beneath the elytra. Parts of veins seen through folds of the wing membrane are hatched.

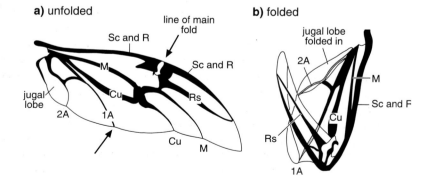

veins (Fig. 9.5). The folding results from the structure and elasticity of the cuticle of the veins (Hammond, 1979).

The wings are sometimes held in the folded position by being coupled together or fastened to the body. For instance, in Psocoptera, the costal margin of the hindwing is held by a fold on the pterostigma of the forewing. The elytra of Coleoptera are held together by a tongue and groove mechanism, but they are also held to the body by a median longitudinal groove in the metathorax which holds the reflexed inner edges of the elytra. Dermaptera have rows of spines on the inside edge of the tegmen which catch on to combs on the metathorax, while many aquatic Heteroptera have a peg on the mesothorax which fits into a pit in the margin of the hemelytron. Symphyta have specialized lobes, the cenchri, on the metanotum which engage with rough areas on the undersides of the forewings to hold them in place.

9.1.1.4 *Areas of the wing*

The flexion- and fold-lines divide the wing into different areas. The region containing the bulk of the veins in front of the claval furrow is called the remigium (Fig. 9.6a). The area behind the claval furrow is called the clavus except in hindwings in which this area is greatly expanded, when it is known as the vannus. Finally the jugum is cut off by the jugal fold where this is present (Wootton, 1979). In some Diptera there are three separate lobes in this region of the wing base, known from proximally outwards as the thoracic squama, alar squama and alula (Fig. 9.6b). There is some confusion in the terminology and homologies of these lobes, but it is probable that the thoracic squama is derived from the posterior margin of the scutellum, the alar squama represents the jugum and the alula is a part of the claval region which has become separated off from the rest. Some Coleoptera have a lobe called an alula folded beneath the elytron. It appears to be equivalent to the jugum.

The wing margins and angles are also named (Fig. 9.6a). The leading edge of the wing is called the costal margin, the trailing edge is the anal margin and the outer edge is the apical margin. The angle between the costal and apical margins is the apical angle, that between the apical and anal margins is the anal angle, while the angle at the base of the wing is called the humeral angle.

The veins divide the area of the wing into a series of cells which are most satisfactorily named after the vein forming the anterior boundary of the cell (Fig. 9.2a). A cell

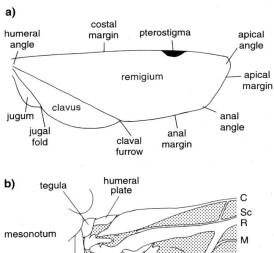

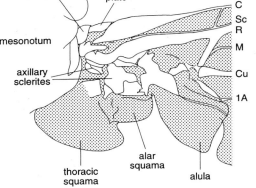

Fig. 9.6. Wing areas. (a) The terminology applied to different parts of the wing. (b) Base of the right wing of a tabanid (Diptera) showing the arrangement of the lobes at the base of the wing (after Oldroyd, 1949).

entirely surrounded by veins is said to be closed, while one which extends to the wing margins is open.

Pterostigma On the anterior margin of the wing in some groups is a pigmented spot, the pterostigma (Fig. 9.4a). This is present on both pairs of wings of Odonata and on the forewings of many Hymenoptera, Psocoptera, Megaloptera and Mecoptera. The mass of the pterostigma is frequently greater than that of an equivalent area of adjacent wing and its inertia influences the movement of the whole wing membrane. In Odonata it is believed to reduce wing flutter during gliding, thus raising the maximum speed at which gliding can occur. In smaller insects it provides some passive control of the angle of attack of the wing during flapping flight, giving enhanced efficiency at the beginning of the wing stroke without the expenditure of additional energy (Norberg, 1972).

a)

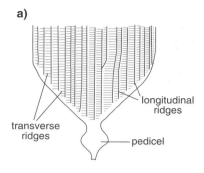

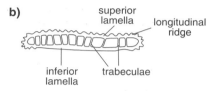

Fig. 9.7. Lepidopteran scale. **(a)** Basal half of a scale showing the pedicel that attaches to the wing membrane. **(b)** Transverse section of a scale (after Bourgogne, 1951).

9.2 MODIFICATIONS OF THE WINGS

9.2.1 The wing membrane

The wing membrane is typically semitransparent and often exhibits iridescence as a result of its structure (section 25.2.2). Sometimes, in addition, the wings are patterned by pigments contained in the epidermal cells. This is true in some Mecoptera and Tephritidae, while in many insects which have hardened forewings, such as Orthoptera and Coleoptera, the whole forewing is pigmented.

The surface of the wing membrane is often set with small non-innervated spines called microtrichia. In Trichoptera, larger macrotrichia clothe the whole of the wing membrane giving it a hairy appearance. In Lepidoptera, the wings are clothed in scales which vary in form from hair-like to flat plates. They usually cover the body as well as the wings. A flattened scale consists of two lamellae with an airspace between, the inferior lamella, that is the lamella facing the wing membrane, being smooth, the superior lamella usually having longitudinal and transverse ridges (Fig. 9.7). The two lamellae are supported by internal struts called trabeculae. The scales are set in sockets in the wing membrane and are inclined to the surface, overlapping each other to form a complete covering. In primitive Lepidoptera, the scales are randomly distributed on the wings, but in butterflies (Papilionoidea) and some other groups they are arranged in rows.

Pigments in the scales are responsible for the colors of many Lepidoptera, the pigment being in the wall or the cavity of the scale. In other instances, physical colors result from the structure of the scale (section 25.2.2). Some specialized scales are associated with glands (section 27.1.5.2), while scales may also be important in smoothing the air-flow over the wings and body (see Fig. 9.30). On the body they are also important as an insulating layer helping to maintain high thoracic temperatures (section 19.1.2).

Scales also occur on the wing veins and body of mosquitoes (Culicidae) and on the wings of some Psocoptera and a few Trichoptera and Coleoptera. Scales and hairs on the wing membrane are not innervated, but mechano- and chemosensitive hairs are often present on the veins.

9.2.2 Wing form

The shape of the wings is probably determined, primarily, by aerodynamic considerations, but other ecological factors may provide different selective pressures.

Wings with narrow, petiolate bases are found in relatively slow-flying insects, such as some damselflies (Zygoptera) and antlions (Myrmeleontidae) (Fig. 9.8a). This shape probably minimizes drag on the body due to the downwash of air from the flapping wings. Wings with broad bases, on the other hand, are associated with capacity for rapid flight. They occur in Orthoptera, many Hemiptera and Lepidoptera (Fig. 9.8b), as well as in dragonflies (Anisoptera) and ascalaphids (Neuroptera, Ascalaphidae) (Betts & Wootton, 1988; Wootton, 1992).

In Odonata, Isoptera, Mecoptera and male Embioptera the two pairs of wings are roughly elliptical in shape and similar in form, but in most other groups of insects the fore and hindwings differ from each other. Sometimes the hindwings are small, relative to the forewings, as in Ephemeroptera, Hymenoptera (Fig. 9.8c,f) and male coccids, while in some Ephemeroptera, such as *Cloeon*, and some male coccids they are absent altogether. In Diptera, the hindwings are modified to form halteres. In other insects, most of the power for flight is provided by the hindwings which have a much bigger area than the forewings (Fig. 9.8d). This is the case in Blattodeea, Mantodea, Orthoptera, Dermaptera and most Plecoptera and Coleoptera and in male Strepsiptera where the forewings are dumb-bell shaped.

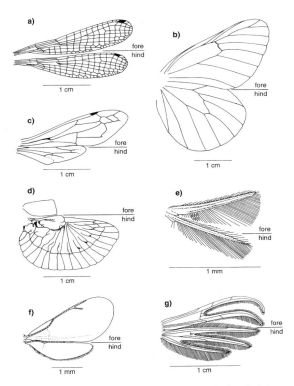

Fig. 9.8. Wing forms. **(a)** Both wings power-producing. Petiolate wings of a damsel fly. Not anatomically coupled (Zygoptera, *Ischnura*). **(b)** Both wings power-producing. Broad-based wings with amplexiform coupling (Lepidoptera, *Aporia*). **(c)** Forewing power-producing. Hindwing reduced and coupled to the forewing by hamuli in a hornet (Hymenoptera, *Vespa*). **(d)** Hindwing power-producing. Forewing reduced to a short tegmen in an earwig. Not anatomically coupled (Dermaptera, *Echinosoma*). **(e)** Fringed wings and reduced venation of a thrips. Frenate-type wing coupling (Thysanoptera, *Thrips*). Notice the small size. **(f)** Reduced venation of a chalcid wasp. Hindwing coupled to forewing by hamuli (Hymenoptera, *Eulophus*). Notice the small size. **(g)** Deeply divided wings of a plume moth. Frenate wing coupling (Lepidoptera, *Alucita*).

The wings of very small insects are often reduced to straps with one or two supporting veins and long fringes of hairs (Fig. 9.8e). These forms occur in Thysanoptera, in Trichogrammatidae and Mymaridae amongst the Hymenoptera, and in some of the small Staphylinoidea amongst the Coleoptera. The wings of plume moths, Pterophoridae and Orneodidae, are deeply cleft and fringed with scales (Fig. 9.8g). Wing fringes are common in Lepidoptera and Culicidae, and in some Tinaeoidea

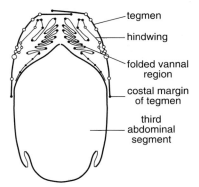

Fig. 9.9. Wing folding. Transverse section through the abdomen of a grasshopper showing the hindwings folded beneath the tegmina (after Uvarov, 1966).

they are so extensive as to greatly increase the effective area of the wing.

In other cases, particular wing forms are presumed to have some ecological significance apart from the production of aerodynamic forces, although their real significance is not always certain. The forewings of many insects are thicker than the hindwings and serve to protect the latter when they are folded at rest (Fig. 9.9). Forewings modified in this way are known as elytra or tegmina. Leathery tegmina occur in Blattodeea, Mantodea, Orthoptera and Dermaptera, while in Heteroptera only the basal part of the wing is hardened, such wings being known as hemelytra (Fig. 9.10a). The basal part of the hemelytron may be subdivided into regions by well-marked veins and, in mirids, where the development is most complete, the anterior part of the wing is cut off as a proximal embolium and distal cuneus, the center of the wing is the corium, and the anal region is cut off as the clavus. In lygaeids only the corium and the clavus are differentiated (see Wootton, 1996).

The elytra of Coleoptera are usually very heavily sclerotized and the basic wing venation is lost, although it may be indicated internally by the arrangement of tracheae. The two surfaces of the elytron are separated by a blood space (Fig. 9.10b) across which run cuticular columns, the trabeculae, arranged in longitudinal rows and marked externally by striations. There are usually nine or ten such striae, although the number may be as high as 25 in some Carabidae. The elytra of beetles do not overlap in the midline, but meet and are held together by a tongued and grooved joint, while in some Carabidae, Curculionidae and

a) hemelytron

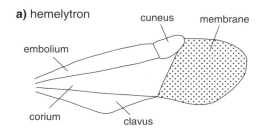

b) elytron

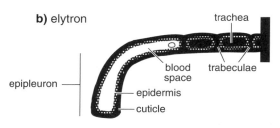

Fig. 9.10. Protective forewings. (a) Hemelytron of a mirid (Heteroptera) (after Comstock, 1918). (b) Diagrammatic transverse section through part of an elytron of a beetle.

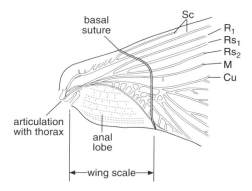

Fig. 9.11. Wing base of a termite showing the basal suture at which the distal part of the wing breaks off (after Grassé, 1949).

Ptinidae they are fused together so that they cannot open and in these species the hindwings are also reduced. At the sides, the elytra are often reflexed downwards, the vertical part being called the epipleuron and the horizontal part the disc.

In Orthoptera, the forewings are often modified for sound production and they may be retained for this function in species in which they are no longer used in flight (section 26.1.2.1).

The shape of the wings may also have significance not directly related to flight. Swallow-tailed butterflies and some Lycaenidae have a projection from the hind margin of the hindwing, while in the Nemopteridae and some Zygaenidae the hindwings are slender ribbons trailing out behind the insect. This probably tends to divert the attention of a predator away from the head and thorax, at least in some of these insects. An irregular outline to the wings, such as occurs in some butterflies, serves to break up the outline of a resting insect and presumably has a camouflage function.

Some insects have both pairs of wings reduced and they are said to be brachypterous or micropterous. This occurs, for instance, in some Orthoptera and Hemiptera. The completely wingless, or apterous, condition is also widespread. Winglessness occurs as a primitive condition

in the Apterygota, while the ectoparasitic orders Phthiraptera and Siphonaptera are secondarily wingless. Wingless species are also widespread in most other orders, but apparently do not occur in Odonata or Ephemeroptera. Sometimes both sexes are wingless, but frequently the male is winged and only the female is apterous. This is the case in coccids, Embioptera, Strepsiptera, Mutillidae and some Chalcididae. In the ants and termites, only the reproductive caste is winged and here the wings are shed after the nuptial flight, breaking off by a basal suture so that only a wing scale remains (Fig. 9.11). The break is achieved in different ways, but termites frequently rest the wing on the ground and then break it off by twisting the wing base. After loss of the wings, the flight muscles degenerate.

Commonly the development of the wings varies within a species either geographically or seasonally. The extent to which this is genetically determined is often unclear, but in many species environmental factors have a dominant effect. Such wing polyphenism occurs in various groups. Within a species of the grasshopper, *Chrotogonus*, the tegmina and hindwings may vary in length from fully developed to very short. In many other insects, however, and notably in Hemiptera, species may either be apterous or macropterous (with fully developed wings), without any intermediates (section 15.5).

Review: Séguy, 1973

9.3 WING COUPLING

The major movements of the wings during flight are produced by distortions of the thorax (see below) and, because

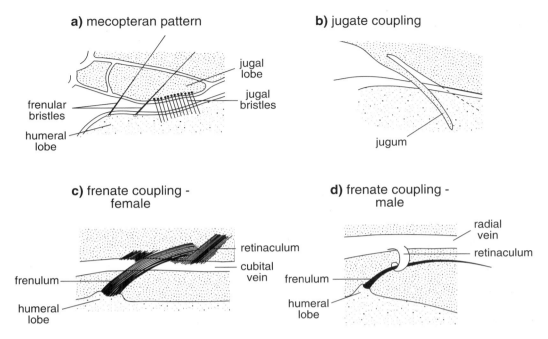

Fig. 9.12. Wing coupling mechanisms involving the jugal and humeral regions of the wings. All diagrams represent the mechanisms as seen from below with the attachment to the thorax immediately to the left. Membrane of the forewing with dark stippling, that of the hindwing with light stippling (after Tillyard, 1918). (**a**) Primitive mecopteran pattern (Mecoptera, *Taeniochorista*). (**b**) Jugate coupling in a hepialid moth (Lepidoptera, *Charagia*). (**c**) Frenate coupling in a female sphingid moth (Lepidoptera, *Hippotion*). (**d**) Frenate coupling in a male sphingid moth (Lepidoptera, *Hippotion*).

they are so closely associated, the movements of each of the thoracic segments must influence the other. Hence, it is impossible for the fore and hindwings to beat completely independently of each other, and, in Orthoptera and Odonata, where the wings are not otherwise linked, both pairs of wings vibrate with the same frequency and with the hindwing beat consistently more advanced than the forewing beat (Fig. 9.32; see Brodsky, 1994).

In the majority of insects the fore and hindwings are linked anatomically so that they move together as a single unit. This wing coupling may take various forms, but, in many species, it involves lobes or spines at the wing base. A primitive arrangement is found in some Mecoptera of the family Choristidae in which there is a jugal lobe at the base of the forewing and a humeral lobe at the base of the costal margin of the hindwing. Both lobes are set with setae known as the jugal and frenular bristles, respectively (Fig. 9.12a), and, although they do not firmly link the wings, they overlap sufficiently to prevent the wings moving out of phase.

In some Trichoptera, only the jugum is present; it lies on top of the hindwing and the coupling mechanism is not very efficient. However, the Hepialidae (Lepidoptera) have a strong jugal lobe which lies beneath the costal margin of the hindwing so that this is held between the jugum and the rest of the forewing (Fig. 9.12b). This is called jugate wing coupling. In Micropterygidae, the jugum is folded under the forewings and holds the frenular bristles. This type of coupling is jugo–frenate coupling. Many other Lepidoptera have a well-developed frenulum which engages with a catch or retinaculum on the underside of the forewing usually near the base of the subcostal vein but sometimes elsewhere. This is frenate coupling. Female noctuids, for instance, have from two to 20 frenular bristles and a retinaculum of forwardly directed hairs on the underside of the cubital vein (Fig. 9.12c); in the male, the frenular bristles are fused together to form a single stout spine and the retinaculum is a cuticular clasp on the radial vein (Fig. 9.12d). Thysanoptera have the wings coupled in a comparable way by hooked spines at

Fig. 9.13. Wing articulation with the thorax. Axillary sclerites with dark stippling for clarity (modified after Snodgrass, 1935).

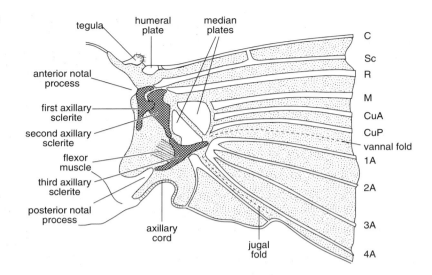

the base of the hindwing which catch a membranous fold of the forewing.

Other insects have the wings coupled by more distal modifications which hold the costal margin of the hindwing to the anal margin of the forewing. Hymenoptera have a row of hooks, the hamuli, along the costal margin of the hindwing which catch into a fold of the forewing; Psocoptera have a hook at the end of CuP of the forewing which hooks on to the hind costa; and the forewing of Heteroptera has a short gutter edged with a brush of hairs on the underside of the clavus which holds the costal margin of the hindwing. Homoptera exhibit a variety of modifications linking the anal margin of the forewing to the costal margin of the hindwing (see Pesson, 1951).

The wings of the Papilionoidea and some Bombycoidea are coupled by virtue of an extensive area of overlap between the two. This is known as amplexiform wing coupling. A similar arrangement is present in some Trichoptera, often occurring together with some other method of coupling.

9.4 ARTICULATION OF THE WINGS WITH THE THORAX

Where the wing joins the thorax, its dorsal and ventral cuticular layers are membranous and flexible. In these membranes are the axillary sclerites, which transmit movements of the thorax produced by the flight muscles to the wing. Typically there are three axillary sclerites (Fig.

9.13). The first is in the dorsal membrane and articulates proximally with the anterior notal process and distally with the subcostal vein and the second axillary sclerite. The second extends to both dorsal and ventral membranes and articulates ventrally with the pleural wing process (see Fig. 7.5b) and distally with the base of the radius. It is also connected with the third axillary sclerite, which articulates proximally with the posterior notal process and distally with the anal veins. The third axillary sclerite is Y-shaped with a muscle inserted into the crutch of the Y. In Hymenoptera and Orthoptera, there is a fourth axillary sclerite between the posterior notal process and the third axillary sclerite. The precise arrangement of the axillary sclerites relative to each other and the flexion lines of the wing is extremely complex and plays a significant role in changes in wing form during flight (see Brodsky, 1994; Wootton, 1979).

In addition to the axillary sclerites, there are other plates in the wing base. Connected with the third axillary, and perhaps representing a part of it, may be one or two median plates from which the media and the cubitus arise. At the base of the costa is a humeral plate and often, proximal to it, is another plate derived from the edge of the articular membrane and called the tegula. In *Locusta* this has been shown to be an important sensory structure which modulates the basic pattern of wing movements (Wolf, 1993, and see below) This may also be true in other insects. It is well-developed in Lepidoptera, Hymenoptera and Diptera (Fig. 9.13).

All present-day insects other than Ephemeroptera and Odonata are able to fold their wings back over the body when at rest. It might be expected that this folding would be associated with greater complexity of the sclerites at the wing base and that in Ephemeroptera and Odonata the arrangement would be simpler. Odonata have only two large plates hinged to the tergum and supported by two arms from the pleural wing process. The plates are called the humeral and axillary plates. However, the wing base of Ephemeroptera is very similar to that in other insects (see Snodgrass, 1935).

It has been shown in the tsetse fly, *Glossina*, that the base of the wing disarticulates from the pleural process when the wing is at rest (Chowdhury & Parr, 1981). This probably also occurs in other related flies, if not in other insects, since it has been shown that the radial stop, which fits into a groove on the pleural process (see below), separates from the pleural process during each wing stroke (Miyan & Ewing, 1985a) and, in some species, can effect this fit at two different positions on the pleural wing process. This 'gear changing' is apparently involved in producing different power output from the wings and may enable one wing to remain stationary while the other beats (Nalbach, 1989).

Although the movement of the wings on the thorax involves some condylic movement at the pleural process, a great deal of movement is permitted by the presence of resilin ligaments, such as the wing hinge ligament of Orthoptera (section 16.3.3). In this way, the problems of friction and lubrication which would occur at a normal articulation moving at the high frequency of the wings are avoided.

9.5 SENSILLA ON THE WINGS AND THE HALTERES

Many insects have hair sensilla along the wing veins. In general, these are probably mechanoreceptors responding to touch and possibly to the flow of air over the wings in flight, but, in many species of Diptera, contact chemo-receptors are also known to be present. Campaniform sensilla are present at the base of the wing, mainly in groups on the subcostal and radial veins (Fig. 9.14 and see Gnatzy, Grünert & Bender, 1987). These groups are often present on both upper and lower surfaces of the wing, but in Acrididae and Blattodea they are only present ventrally. The sensilla in the groups are generally oval, all those in a

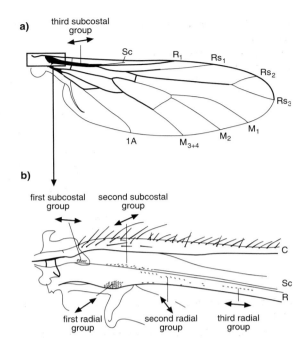

Fig. 9.14. Distribution of groups of campaniform sensilla at the base of the wing of a fly (Diptera, *Empis*). Arrows indicate the orientation of the long axes of the sensilla (after Pringle, 1957). (**a**) Whole wing showing, enclosed within the box, the position of the area enlarged in b. (**b**) Proximal parts of the anterior veins.

group being similarly oriented, so that they are sensitive to distortions of the wing base in a particular direction. More distally on the veins are other scattered campaniform sensilla, but these are large and almost circular, so that, unlike those in the basal groups, they have no directional sensitivity. The number of sensilla in each group varies, there being more in more highly maneuverable species. Thus *Apis* has about 700 campaniform sensilla at the base of each forewing, while the scorpion fly (Mecoptera, *Panorpa*) has only about 60. The sensilla at the wing base are probably concerned in the control of stability in flight. The more distal sensilla are stimulated by changes in the camber of the wing, but their functions are not understood (Dickinson, 1992). Diptera have an additional group of campaniform sensilla on the tegula.

In Orthoptera, each wing has a stretch receptor and a chordotonal organ in the thorax associated with the wing base; in the mesothorax of *Schistocerca* they arise together on the mesophragma. The stretch receptor extends to just behind the subalar sclerite, while the chordotonal organ is

Fig. 9.15. Halteres. Dorsal and ventral views of the haltere of a fly showing the groups of sensilla. The orientation of the campaniform sensilla is indicated by the arrows (Diptera, *Lucilia*) (after Pringle, 1948).

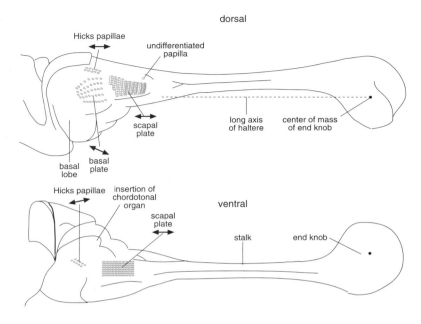

attached a little more ventrally (Gettrup, 1962). These organs have been identified in acridids, gryllids and tettigoniids, but not in a gryllotalpid or a blattid. A chordotonal organ is also associated with each wing in Odonata. These sense organs are concerned with the control of wing movements (section 9.10.3). Insects with asynchronous flight muscles do not have internal proprioceptors connected with the wings.

Halteres The hindwings of Diptera are modified to form halteres, which are sense organs concerned with the maintenance of stability in flight. Each haltere consists of a basal lobe, a stalk and an end knob which projects backwards from the end of the stalk so that its center of mass is also behind the stalk (Fig. 9.15). The whole structure is rigid except for some flexibility of the ventral surface near the base which allows some freedom of movement, while the cuticle of the end knob is thin but kept distended by the turgidity of large vacuolated cells inside it. The haltere is variable in size. In crane flies (Tipulidae) and robber flies (Asilidae) it is relatively long, exceeding 12% of forewing length, but in mosquitoes and Cyclorrhapha it is only about 6% of the forewing length.

On the basal lobe of the haltere are groups of campaniform sensilla which can be homologized with the groups at the base of a normal wing. Dorsally there are two large groups of sensilla: the basal and scapal plates (Fig. 9.15). In

Calliphora there are about 100 sensilla in each group. The sensilla of the scapal plate are parallel with the axis which passes through the main point of articulation and the center of mass of the haltere (indicated as the long axis of the haltere in Fig. 9.15); those of the basal plate are oriented with their long axes at about 30° to the axes of the longitudinal rows in which they are arranged. Near the basal plate is a further small group of campaniform sensilla known as Hicks papillae. These are set below the general surface of the cuticle and are oriented parallel with the long axis of the haltere. There is also a single round, undifferentiated papilla near the scapal plate. On the ventral surface there is another scapal plate with about 100 sensilla and a group of ten Hicks papillae. These are oriented parallel with the long axis of the haltere. Also attached to the ventral surface is a large chordotonal organ oriented at about 45° to the long axis of the haltere. A smaller chordotonal organ runs vertically across the base.

These sensilla react to the forces acting at the base of the haltere during flight. They perceive the vertical movements of the haltere and also the torque produced by lateral turning movements of the fly (section 9.11.2).

9.6 MUSCLES ASSOCIATED WITH THE WINGS

The muscles concerned with wing movement fall, functionally, into three groups: direct muscles are inserted into

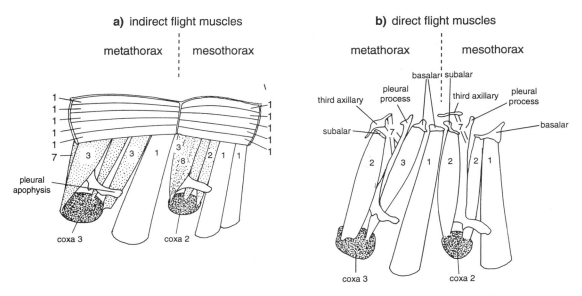

Fig. 9.16. Flight muscles of a locust and their innervation. Numbers on each muscle show the numbers of motor axons, including inhibitory axons, innervating each unit. Dorso–ventral muscles not arising from the coxae have their origins on the sternum. (**a**) Indirect flight muscles. Bifunctional muscles that move the wings and the legs are stippled. (**b**) Direct flight muscles.

the wing base and their contractions have direct effects on wing movement; indirect muscles moving the wings indirectly by causing distortions of the thorax; and accessory muscles which modulate the wing stroke by changing the shape or mechanical properties of the thorax. For details of the musculature of Odonata see Simmons, 1977; of Orthoptera: Wilson & Weis-Fogh, 1962; of Diptera: Heide, 1971; Miyan & Ewing, 1985a; and of Lepidoptera: Kondoh & Obara, 1982; Rind, 1983.

Reviews: Brodsky, 1994; Kammer, 1985

9.6.1 Direct wing muscles

The direct wing muscles are inserted on to the axillary sclerites or the basalar and subalar sclerites (see Fig. 7.5) which connect to the axillary sclerites by ligaments. One muscle, arising on the pleuron and inserted into the third axillary sclerite, flexes the wing backwards. It is also active in flight, producing some remotion of the wing during both the up and down strokes and, in this way, effecting steering (see section 9.11.3). The basalar and subalar muscles commonly consist of several units arising on different parts of the pleuron, as well as the sternum and the coxa (Fig. 9.16). Both muscles are involved in wing twisting (see below) as well as wing depression. In addition, the basalar muscle is involved in wing extension from

the flexed position. Odonata have two muscles arising from the episternum inserted into the humeral plate and two from the edge of the epimeron inserted into the axillary plate. The relatively large number of direct muscles (Table 9.1), reflects their role in changing the form of the wing during the wingbeat cycle to provide the precision necessary for efficient aerodynamics and steering.

9.6.2 Indirect wing muscles

The indirect wing muscles are usually the main power-producing muscles for flight. In all insects, wing elevation is produced by indirect dorso-ventral muscles inserted into the tergum of the wing-bearing segment. There are usually several muscles with different points of origin (Fig. 9.16). These muscles are not always homologous. In many insects they arise on the sternum or the coxae, but in Auchenorrhyncha and Psyllidae the tergosternal muscles are small and are functionally replaced as wing elevators by the oblique dorsal muscles. These arise on the postphragma and so are normally obliquely longitudinal, but in the groups mentioned the phragma extends ventrally carrying the origins of the muscles with it so that they come to exert their pull vertically instead of horizontally.

Wing depression is produced, in most insects, by dorsal longitudinal indirect muscles usually comprising five or six muscle units. These extend from the anterior to the posterior phragma of each wing-bearing segment (Fig. 9.16). The mode of action of these muscles is illustrated in the next section.

9.6.3 Accessory muscles

These are pleurosternal and tergopleural muscles. The pleurosternal muscles control the lateral elasticity of the thorax.

9.6.4 Muscle innervation

The structure and innervation of muscles reflects their functions. Most of the indirect flight muscles are dedicated to flight; they have no other function. In keeping with this, they usually comprise one or a small number of similar muscle units each innervated by a single fast motor neuron (Fig. 9.16; Table 9.1). In contrast, the direct and accessory muscles as well as bifunctional indirect muscles may have polyneuronal innervation. For example, one of the tergo-coxal muscles of *Schistocerca* is innervated by six motor neurons and the common inhibitory neuron (see Fig. 10.9). In addition, these muscles may contain units that are physiologically different from each other. For example, the third axillary muscle of *Manduca* has three units, two which are of the physiologically 'intermediate' type and one which is a 'tonic' muscle (section 10.1.2). Each intermediate unit is innervated by a separate fast neuron, while the tonic unit has a slow neuron. This permits increased versatility in the activity of the muscle, reflecting the need for versatility in the control of flight (Rheuben & Kammer, 1987; see also Trimarchi & Schneiderman, 1994).

Because the main power-producing muscles are innervated by a single motor neuron, the total number of neurons controlling power production in flight is small; the forewings of the locust are controlled by about 20, and those of a fly by about 26. A larger number of neurons is concerned with modulating the power output via the direct wing muscles.

The effects of the motor neurons may be modulated by octopamine. Dorsal unpaired median (DUM) neurons, which are known to secrete octopamine in locusts, have been described innervating the dorsal longitudinal muscles of insects from the Orthoptera, Heteroptera, Lepidoptera and Diptera. In *Bombyx*, the dorsoventral indirect flight muscles also receive inputs from an unpaired median neuron, while in *Locusta* the tergocoxal

muscles receive input from four putative octopaminergic neurons, and the subalar muscle has input from five such neurons (Kutsch & Schneider, 1987)

9.7 MECHANISMS OF WING MOVEMENT

The up and down movements of the wings are produced by direct and indirect wing muscles but they also involve the elasticity of the thorax, the wing base and the muscles themselves.

9.7.1 Movements produced by the muscles

The upward movement of the wings is produced by the indirect dorso-ventral muscles. By contracting, they pull the tergum down and hence also move down the point of articulation of the wing with the tergum. The effect of this is to move the wing membrane up, with the pleural process acting as a fulcrum (Fig. 9.17a–d).

The downward movement of the wings in Odonata and Blattodea is produced by direct muscles inserted into the basalar and subalar sclerites. These muscles pull on the wings outside the fulcrum of the pleural process and so pull the wings down (Fig. 9.17b). By contrast, in Diptera and Hymenoptera, the downward movement is produced by the dorsal longitudinal indirect muscles. Because the dorsum of the pterothorax is an uninterrupted plate, without membranous junctions (see Fig. 7.1), contraction of the dorsal longitudinal muscles does not produce a telescoping of the segments as in the abdomen. Instead, the center of the tergum becomes bowed upwards (Fig. 9.17e,f). This moves the tergal articulation of the wing up and the wing membrane flaps down (Fig. 9.17d). In Coleoptera and Orthoptera the downward movement is produced by the direct and indirect longitudinal muscles acting together. The direct muscles are also concerned in twisting the wing during the course of the stroke (see section 9.8.5).

Details of the movement may be much more complex than this simple account suggests. In cyclorrhaphous Diptera, contraction of the dorsal longitudinal muscles causes the scutellar lever to swing upwards (Fig. 9.18). The scutellar lever articulates with the first axillary sclerite and its upward movement causes the sclerite to rotate until it hits the paranotal plate, pushing it and the scutum upwards and stretching the dorsoventral muscles. The second axillary sclerite rotates in a socket on the inner face of the pleural wing process. As it swings, the wing is

Table 9.1. *Numbers of muscles associated with one mesothoracic wing and the number of neurons by which they are innervated*[a]

Order	Indirect muscles						Direct muscles	
	Depressor muscles		Unifunctional levator muscles		Bifunctional levator muscles			
	Number (units)	Neurons/unit	Number	Neurons/muscle	Number	Neurons/muscle	Number	Neurons/muscle[b]
Odonata	0	—	2	3	2	3,5	9	3–10
Orthoptera	1(5)	1	2	1	4	1–3	4	1–8
Lepidoptera	2(5)	1	2	2	3	2–3	4	1
Diptera	1(5–6)	1	5	1	2	1	17	1–2

Note:

[a] In some cases the number of units in a muscle is not known so that more than one neuron/muscle may indicate polyneuronal innervation of a single unit, or may indicate a number of separate units. These numbers do not include DUM neurons.

[b] In most cases the innervation is known for only some of the direct muscles.

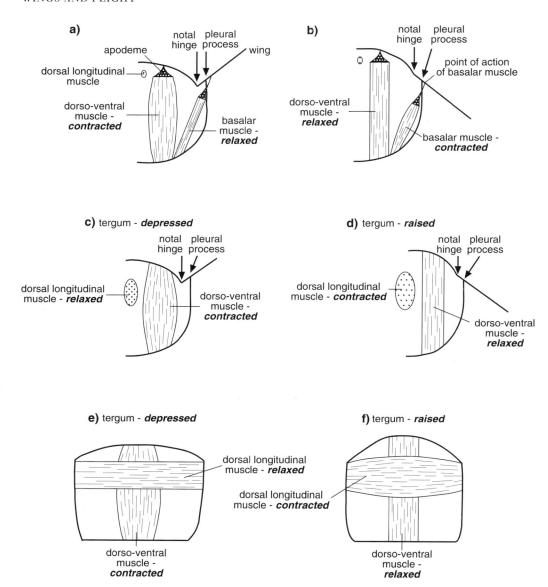

Fig. 9.17. Muscular basis of wing movements. (**a**), (**b**) In an insect, such as a dragonfly, in which the direct wing muscles cause depression of the wings. (**a**) Indirect dorso–ventral muscles cause wing elevation. (**b**) Direct dorso–ventral muscles cause depression. (**c**)–(**f**) In an insect, such as a fly, in which both up and down movements of the wing are produced by indirect muscles. (**c**), (**d**) Cross-sections of the thorax. (**e**) and (**f**) Sagittal sections of the wing-bearing segment from the inside corresponding with (**c**) and (**d**), respectively. In (**f**), contraction of the dorsal longitudinal muscles raises the tergum (as seen in cross-section in d) and the wing flaps down.

depressed until a process on the underside of the radial vein, known as the radial stop, fits into a groove on top of the pleural process. This contact now becomes the pivotal point for further wing depression which results from wing bending. On the upstroke, these changes are reversed and the dorsal longitudinal muscles are elongated as the dorsoventral muscles contract. This description assumes that there is no click mechanism (see below).

A number of the muscles moving the wings arise in the coxae, which are themselves moveable. Whether these muscles move the legs or the wings appears to be determined by the activity of other muscles and the position of the appendages: if the wings are closed the muscles move the legs, but in flight, with the legs in the flight position, the wings are moved.

9.7.2 Movement due to elasticity

The capacity to store elastic energy and subsequently to release it at high rates is an essential feature of the flight mechanism of most insects. At the beginning of a wing stroke (up or down), energy is expended to overcome the inertia of the wing, while at the end of a stroke the wing has momentum and must be stopped. Insects store the energy derived from this momentum in elastic systems which may be in the cuticle or in the flight muscles themselves. This energy is then used on the return stroke (section 10.4.4).

In *Schistocerca* (Orthoptera), and probably in most other insects, much of the energy involved in the upstroke is stored as elastic forces in the pad of resilin which forms the main wing hinge. This is possible because the aerodynamic forces produced at this time act in the same direction as the wing movement so assisting its movement. Thus the muscles have only to overcome the forces of inertia of the wing and elasticity of the wing base, and as a result some 86% of the energy they produce is stored for use in the downstroke. The elastic properties of this pad are almost perfect and all but 3% of the energy imparted to it when it is stretched during the upward movement of the wings is available for pulling the wing down. The radial stop in the wing articulation of cyclorrhaphous Diptera (see above) may also enable energy to be stored in the cuticle at the end of the downstroke. Elastic (resilin) ligaments connect the subalar sclerite to the second and third axillary sclerites of *Calliphora* which also has resilin in the apodemes to which the pleuroaxillary and tergo-

pleural muscles attach. Dragonflies have a similar elastic apodeme where the subalar muscle attaches to the subalar sclerite.

Elastic forces are also stored as a result of distortions of the thorax. It has been widely believed that the lateral stiffness of the thorax, produced by the sternopleural articulation and, to a lesser extent, the tergopleural articulation, was of major importance. As a result of this lateral stiffness the position of the wings was thought to be unstable for much of the stroke and the wings were only stable in the extreme up and down positions. This arrangement is called a 'click' mechanism, the wings 'clicking' automatically to one of the stable positions once they passed the position of maximum instability. The accompanying sudden changes in shape of the thorax were believed to be essential components of the stretch-activation system of the asynchronous muscles (section 10.4.2).

However, critical observations of flies suggest that, at least in these insects, a click mechanism does not normally occur (Ennos, 1987; Miyan & Ewing, 1985a,b, 1988; but see Pfau, 1987 for a contrary view). Energy is stored during both the up and down strokes, probably in the pleural process which is slightly deformed in both halves of the stroke, but probably also in other elements of the wing articulation. Release of this energy contributes to both parts of the stroke, but most obviously to the upstroke. Stretch activation of the synchronous muscles still occurs, as in the 'click' mechanism, but the mechanism and timing of stretching are different. The dorsoventral muscles, producing the upstroke, are activated by the raising of the scutum by the action of the second axillary sclerite (see Fig. 9.18). This occurs towards the end of the downstroke so that the muscles are activated at the appropriate time. Stretching of the dorsal longitudinal muscles, however, occurs throughout the upstroke, and especially at the beginning. It is inferred that the delay in activation, following stretching, must be sufficient for the wing to be at or near the top of the upstroke before they start to contract (Miyan & Ewing, 1985a). There is also good evidence that stretch-activation in bees does not depend on a click mechanism (see Esch & Goller, 1991).

In insects with fibrillar muscles it is probable that the muscles themselves are the principal site of energy storage. These muscles are characterized by a greater resistance to stretch compared with other muscles due to the elastic properties of the contractile system (section 10.4.2).

Fig. 9.18. Wing articulation in a fly. **(a)** Lateral view of the thorax. Membranous parts are stippled (after Pringle, 1957). **(b)–(d)** Diagrammatic representations of the movements of the axillary sclerites during wing depression. The first axillary sclerite (dotted) is pushed upwards by the scutellar lever and **(c)** makes contact with the underside of the paranotal plate (black) pushing it upwards **(d)**. This raises the height of the scutum. At the same time the radial stop hits the top of the pleural process **(c)** and the wing bends down **(d)** (after Miyan & Ewing, 1985a).

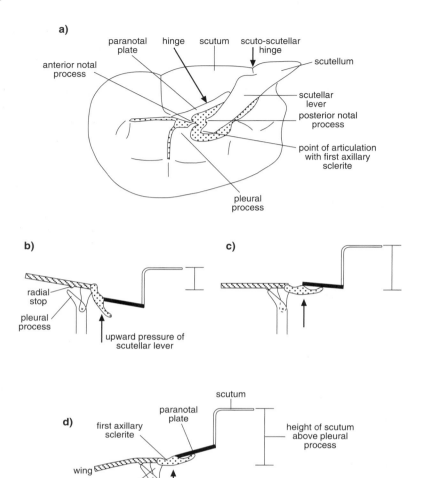

9.8 MOVEMENTS OF THE WINGS

9.8.1 Definitions (Figs. 9.19, 9.20)

Stroke plane: the average plane through which the wing moves relative to the body in one complete cycle.

Stroke plane angle: the slope of the stroke plane relative to the horizontal (Dudley & Ellington, 1990a).

Wingstroke amplitude: the angle through which the wing moves from the extreme top to the extreme bottom of the stroke, measured in the stroke plane.

Body angle: the angle of the longitudinal axis of the body relative to the horizontal.

9.8.2 Stroke plane

The stroke plane varies with the flight behavior of the insect (Fig. 9.20a). It is almost horizontal in many hovering insects but becomes oblique as the insect flies faster. The angle between the stroke plane and the long axis of the body remains more or less constant in many insects with low wingbeat frequencies such as Orthoptera and Blattodea. In these groups it is commonly between 70 and 100°. In flies and bees, the relationship is more variable and marked changes occur when the insect is maneuvering. Differences in the stroke plane of the wings on the two sides of the body produce turning movements.

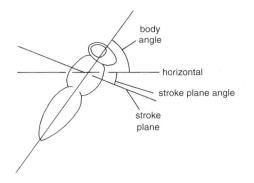

Fig. 9.19. Stroke plane angle and body angle.

The wings do not make simple up and down movements, but in the course of each cycle they also move backwards and forwards to some extent. As a result, the tip of the wing moves in an ellipse relative to the body (Fig. 9.20a), moving forward and down on the downstroke, and up and back on the upstroke. In some insects, such as honeybees and some flies, the wing tip traces a more complex figure relative to the body (Nachtigall, 1976). The upstroke is faster than the downstroke and when the insect is moving forwards the path of the wing tip through the air has a backwards component (Figs. 9.20b, 9.23).

9.8.3 Amplitude of wingstroke

The amplitude of the wingstroke is often within the range 70–130°. Greater amplitudes commonly result in the

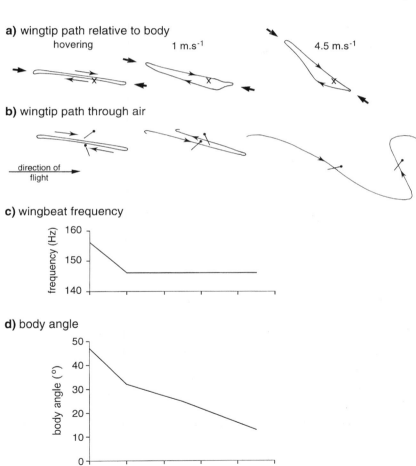

Fig. 9.20. Wingbeat of a bumblebee flying at different airspeeds (Hymenoptera, *Bombus*) (after Dudley & Ellington, 1990a,b). (a) Path of the wingtip relative to the body (anterior to the right). Arrows at either end of the paths indicate the stroke plane. X shows the position of the wing base. (b) Path of the wing-tip as the insect flies from left to right. Short bars show the orientation of the wing at midpoints on the up and down strokes. The leading edge of the wing is indicated by a dot. (c) Wingbeat frequency. (d) Body angle. Notice that at high airspeeds the body angle decreases while the stroke plane angle (shown in a) increases.

Fig. 9.21. Wingbeat frequencies. Horizontal lines show the ranges of frequencies at which insects in different orders flap their wings (after Brodsky, 1994).

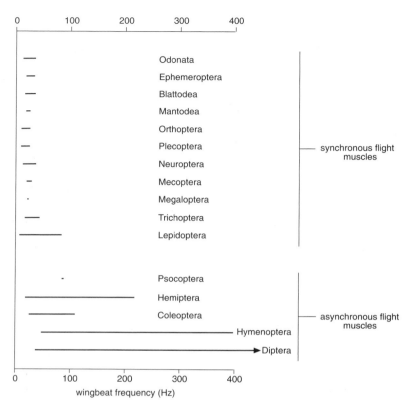

wings 'clapping' together above the body (see below) and are associated with greater power output. They occur most frequently at the beginning of flight, when high lift forces are necessary to raise the insect off the ground. In *Drosophila*, an increase in amplitude from 90° to 140° is associated with a change in the stroke plane, but in *Apis* the amplitude varies independently of stroke plane. Variation in the amplitude of wingbeat on the two sides of the body may be used in steering, the insect turning away from the side of greatest amplitude.

Wings whose primary function is protective, such as the elytra of beetles and tegmina of grasshoppers, beat with lower amplitudes since they are not the primary power producers. The forewings of *Locusta*, for instance, commonly move through 70–80° compared with 110–130° for the hindwings, and the elytra of *Oryctes* (Coleoptera) have an amplitude of only about 20°, nearly all the power coming from the hindwings.

At least in Diptera, amplitude is largely controlled by the activity of the muscle which inserts on to the third axillary sclerite. Because it inserts at a point distal to the pleural process, it functions as a wing depressor. It may be active at any time during the wing stroke, including the upstroke, but, because of the geometry of its insertion, it is most effective when the wing is almost horizontal. At this point, its action can bend the wing down from the radial stop (see Fig. 9.18) and so increase the amplitude. Amplitude may also be affected by a 'gear change' (see above).

9.8.4 Wingbeat frequency

Amongst the Odonata and orthopteroid insects, which are usually moderate to large sized, the wingbeat frequency is usually within the range 15–40 Hz (Fig. 9.21). Hemipteroid insects usually have higher wingbeat frequencies than this and, in the small whiteflies (Aleyrodidae), the frequency may exceed 200 Hz. Hymenoptera and Diptera commonly beat the wings even more rapidly, while different species of Lepidoptera have wingbeat frequencies ranging from about 4 Hz to 80 Hz. Insects with high wingbeat frequencies generally have asynchronous flight muscles.

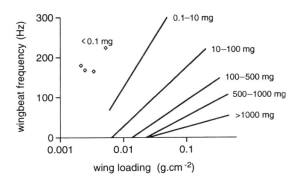

Fig. 9.22. Relationship between wing loading and wingbeat frequency for insects of different body weight. For insects weighing less than 0.1 mg, data are available for only four species, indicated by separate symbols (data from Byrne, Buchmann & Spangler, 1988).

Wing loading is the mass of the insect per unit area of the wing. Wingbeat frequency within a size class of insects is positively correlated with wing loading (Fig. 9.22), but the relationship varies with the weight of the insects. Small insects have much higher wingbeat frequencies than larger insects at any given level of wing loading.

Variation of wingbeat frequency with thoracic temperature has been recorded in some small to moderate sized insects amongst the Diptera, Lepidoptera, Hymenoptera and Coleoptera as well as in larger Odonata and Orthoptera (May, 1981; Oertli, 1989). In general, the increase is slight with Q_{10} (the change in relative rate for a 10 °C rise in temperature) varying between 1.0 and 1.4. So, for example, a rise in the thoracic temperature of the beetle, *Popillia*, from 30 to 40 °C results in an increase in wingbeat frequency from about 110 to 130 Hz ($Q_{10} = 1.18$). In some larger insects, thoracic temperature is regulated during flight over a range of ambient temperatures and large changes in wingbeat frequency would not be expected in these insects.

In insects with asynchronous flight muscles, these changes are the result of changes in the resonant properties of the thorax rather than changes in the neural output to the muscles whereas, in Orthoptera, the increase in wingbeat frequency is, at least partly, due to temperature effects on the central nervous system (Foster & Robertson, 1992).

Adult age affects wingbeat frequency in some hemimetabolous insects. In the Australian plague locust (Orthoptera, *Chortoicetes*), for instance, wingbeat frequency increases from 15–20 Hz soon after molting to 25–35 Hz about ten days later. At about the time of eclosion, the firing pattern of motor neurons necessary for flight gradually develops. Within five days, the pattern is fully established, but the frequency with which the output of each motor neuron oscillates remains low, at 15–20 Hz. Within the next few days the frequency of oscillation rises to 25–35 Hz probably because of altered inputs from the proprioceptors at the bases of the wings associated with changes in the development of the flight muscles and the cuticle (see section 9.10.3) (Altman, 1975).

An increase in wingbeat frequency gives greater power output and correlations between wingbeat frequency and lift or airspeed are known to occur in a number of insect species. This indicates that individual insects have some control over the wingbeat frequency, although in free-flying bumblebees the frequency remained almost constant at airspeeds varying from 1 to 4.5 m s^{-1}, with only a slight increase when hovering (Fig. 9.20c).

9.8.5 Wing twisting

In addition to variations in the form of the wingbeat, the wing may twist in different ways in different phases of the stroke. Such cyclical wing twisting may be produced passively by inertial and aerodynamic forces acting on the wing, or by changes in the movements of the axillary sclerites relative to each other during the wingbeat. This alters the forces which the wing exerts and twisting may itself produce aerodynamically useful power. Wootton (1993) recognizes two types of wing twisting: rotation, and internal torsion.

Rotation occurs at the top and bottom of each stroke as the wing changes from upstroke to downstroke and *vice versa*. On the downstroke, the wing is usually pronated, that is, with the leading edge down, while on the upstroke it is supinated, with the leading edge up (Fig. 9.23). This change is produced, at least in part, by differential action of the basalar and subalar muscles acting at the base of the wing. The former pulls the leading edge down, while the latter pulls the trailing edge down. In addition, the momentum of the wing assists this process and, in Diptera, is sufficient to account for much of the rotation. The axis about which the wing twists is close to the radial vein, but the center of mass of the wing lies behind the torsional axis. Consequently, as the wing decelerates and then reverses, inertia causes it to twist about the axis (Fig. 9.24) (Ennos, 1988). However, *Drosophila*, and presumably other flies, can regulate the timing of wing rotation (Dickinson,

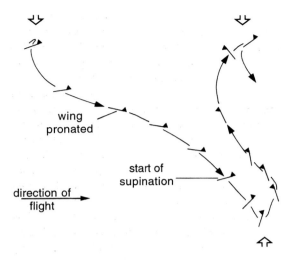

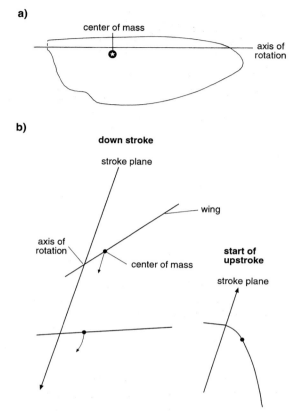

Fig. 9.23. Wing twisting in a fly. The orientation of the wing is shown at intervals of approximately 0.5 ms. The upper side of the leading edge is shown by a triangle. Notice the rapid rotations at the top and bottom of the stroke (indicated by open arrows) (after Nachtigall, 1966).

Lehmann & Götz, 1993). The extent of wing rotation may be relatively small, as in grasshoppers, but in insects with narrow wings, and especially in insects that hover, the wings may rotate through large angles at high speeds. The angular velocity of the wing of *Drosophila* at the ventral reversal exceeds $100\,000\,^{\circ}\,\mathrm{s}^{-1}$ producing significant aerodynamic power (Zanker, 1990).

Internal torsion during the wing stroke is a consequence of the aerodynamic forces acting on the wing membrane. The camber of the wing, or of the leading edge (Fig. 9.25), tends to limit distortion during the downstroke, but may facilitate supination on the upstroke (Wootton, 1993). In addition, many of the flexion lines also permit downward, but not upward, bending of the distal parts of the wing (Fig. 9.4b) so that they contribute to supination on the upstroke.

9.9 AERODYNAMICS

9.9.1 **Flapping flight**

The movements of an insect's wings in flight are complex. Not only do the wings move up and down, they twist during the wingstroke and rotate, often very rapidly, at the top and bottom of each half stroke. These movements generate the aerodynamic forces enabling the insect to

Fig. 9.24. Wing rotation in a fly. **(a)** The center of mass of the wing lies behind the axis of rotation which approximately coincides with the radial vein. **(b)** As the wing decelerates at the end of the downstroke, the inertia of the center of mass causes the wing to start rotating and, as the wing starts to move up, flex downwards about the axis of rotation (after Ennos, 1988).

remain airborne. It had commonly been accepted that, for larger insects at least, these forces could be satisfactorily accounted for by considering the wing as if it were an aerofoil, like the fixed wing of an aeroplane, in a steady airflow, but with changing angle of attack at different positions during the wing stroke. The overall force acting on the wing could then be found by integrating the forces calculated for all the positions. This was called 'quasi-steady' analysis. However, more precise measurements of tethered locusts and calculations based on high-speed film of a variety of insects hovering freely and in free flight over a wide range of speeds have indicated that the forces produced are generally well in excess of those predicted by

a)

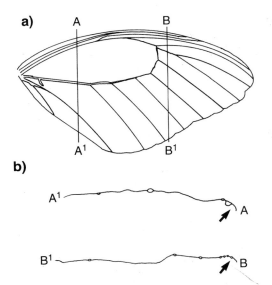

b)

Fig. 9.25. Camber on the wing of a butterfly (Lepidoptera, *Papilio*) (after Wootton, 1993). (**a**) Diagram of the wing showing the venation. (**b**) Sections through the wing at the points marked (anterior to right). Notice the camber at the leading edge of the wing, indicated by arrows.

quasi-steady theory (Cloupeau, Devillers & Devezeaux, 1979; Dudley & Ellington, 1990b; Ellington, 1984f, 1995). As a result, it is now generally accepted that insects make extensive use of 'non-steady' high-lift mechanisms resulting from the wing's accelerations during the flapping cycle.

These mechanisms all involve the generation of vortices which accelerate the air over the wings and are shed as a wake behind and below the insect, but different mechanisms are used, varying between insect groups and the type of flight involved. Some involve the rotation of the wings at the top and bottom of each stroke while others relate to the translatory movement of the wing during the downstroke and sometimes also the upstroke (Brodsky, 1991, 1994; Dickinson & Gotz, 1996; Grodnitsky & Morozov, 1993). Particularly important may be 'delayed stall' which allows the wing to develop high lift by operating at angles of attack above those at which they would normally stall (suddenly lose lift) (see Ellington, 1995).

The best studied mechanism is the 'clap and fling' which operates in the flight of the chalcid wasp, *Encarsia*. This insect has a wing span of about 1.3 mm and a wingbeat frequency of about 400 Hz. At the top of the upstroke the wings clap together and then the leading edges of the wings are separated quickly (are 'flung' apart) while the trailing edges remain in contact. Air is sucked into the increasing gap between the upper surfaces creating bound vortices round the wings (Fig. 9.26). Immediately after the 'fling', the wings separate completely, each carrying a bound vortex with it. As a consequence of the fling, an air circulation exists over the wings from the start of the down stroke, and lift equal to the body weight is produced almost from the beginning (Weis-Fogh, 1973).

A similar 'clap and fling' has been described in thrips, a whitefly, and in *Drosophila* (Ellington, 1984c). It, or the

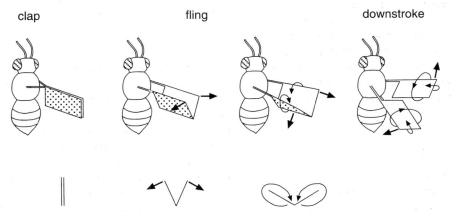

Fig. 9.26. 'Clap and fling' flight mechanism in a small parasitic wasp (Hymenoptera, *Encarsia*). Upper row shows an oblique dorsal view of the movements of the wings with the underside stippled, lower row the cross-section at the midpoint of the wings. Heavy arrows show wing movements, thin arrows air movements (after Weis-Fogh, 1973).

a) start of downstroke   **b)** late downstroke   **c)** upstroke

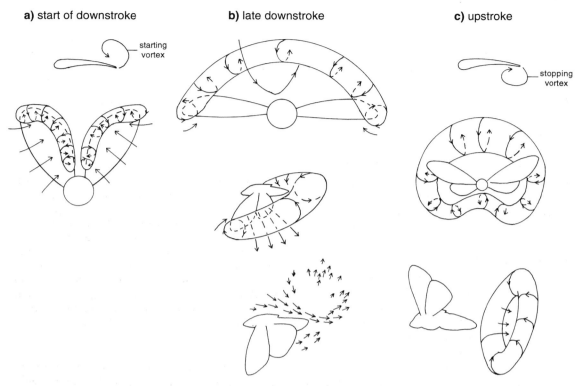

Fig. 9.27. Vortices round the moving wings (after Grodnitsky & Morozov, 1993). **(a)** Start of the downstroke. A starting vortex forms on the upper side of the trailing edge of each wing, as seen in a section of the wing (upper diagram) and from the front of the insect (lower diagram). **(b)** Mid–late downstroke. The vortices of the two sides join, remaining attached to the insect at the wing tips, seen from in front (top diagram) and the side (lower diagrams). Air (arrows) is drawn through the vortex over the dorsal surface of the insect. **(c)** At the start of the upstroke, a stopping vortex forms beneath the wing. Subsequently, the starting and stopping vortices join and are shed behind the insect.

similar 'clap and peel' in which the wings 'peel' apart, starting at the leading edge, also occurs periodically in larger insects such as locusts and butterflies. In these cases, the phenomenon may be important in generating high lift forces (Cooter and Baker, 1977).

Figure 9.27 is an example of the type of airflow developing round the wings of a larger insect in normal flight. As the wings move apart at the beginning of the downstroke, a flow of air round the trailing edge of the wing creates a vortex, called the starting vortex, on the upper side of the wing (Fig. 9.27a). The speed of rotation of this vortex is accelerated as the wing moves downwards, and the vortex tubes of the two sides join in the middle, but remain connected to the wings at their tips (Fig. 9.27b). Soon after the beginning of the upstroke, a new vortex is created on the

underside of the wing. It is called the stopping vortex; it rotates in the opposite direction to the starting vortex (Fig. 9.27c). On the upstroke it gains in velocity, and the vortices of the two sides join so that starting and stopping vortices form a ring connected to each other and to the wing tips. Finally, as the wings pronate at the top of the vortex, the vortex ring is thrown off backwards. The series of vortex rings behind and below the insect, impart a backwards and downwards flow to the column of air which they surround producing an equal and opposite movement of the insect.

It might be expected that small insects would use their wings as paddles rather than as aerofoils because of their low Reynolds number and the consequent greater importance of viscosity of the air (see section 8.4.1.5; Brodsky, 1994). There is, however, no evidence of any insects doing

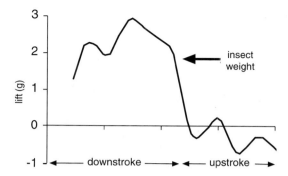

Fig. 9.28. Variation in lift in the course of a wing stroke of the locust (Orthoptera, *Schistocerca*). The movements of the hindwings are shown. The forewings begin their strokes about 0.1 of a wingbeat later (see Fig. 9.32) (after Cloupeau *et al.*, 1979).

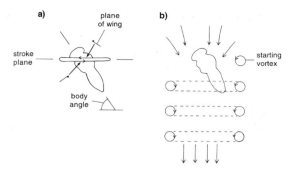

Fig. 9.29. Hovering. (**a**) Diagram showing positions of the body and the stroke plane adopted by many insects when they hover. On the anatomical downstroke, the wing is strongly pronated (leading edge of morphological upper side indicated by a triangle). On the anatomical upstroke, the wing is strongly supinated. Rapid rotations at the end of each half stroke create the vortex rings shown in (b). (**b**) Diagram of the airflow round a hovering insect. The vortex rings (shown in section) are thrown off below and behind the insect (compare Fig. 9.27). The vortices draw air downwards as indicated by the arrows. An equal and opposite reaction supports the mass of the insect (after Ellington, 1984d).

this. Even thrips, small Hymenoptera and Diptera appear to depend on circulatory lift forces (vortices) in the same way as larger insects (Ellington, 1984c,d).

Although 'quasi-steady' aerodynamic analysis seems inadequate to explain most insect flight, it may be enough to account for that of butterflies flying at air speeds in excess of $2\,\mathrm{m\,s}^{-1}$ (Dudley, 1991) and in parts of the wingstroke of large insects, like locusts, when the wing is moving steadily on the downstroke. This does not mean, however, that no non-steady mechanisms are in use.

Gliding flight (section 9.9.2) is well enough explained by orthodox steady-state aerodynamic theory.

Aerodynamic forces acting on the insect: lift, drag and thrust The principal function of the wings is to generate lift (section 8.4.1.5). Since the direction of movement of the wings is constantly changing, so is the direction of lift. It is reduced on the upstroke and may become negative (Fig. 9.28), but the average lift over the whole wingstroke of an insect is positive, inclined upwards and slightly forwards. Most of the upward component of lift is produced by the action of the wings, but a small amount, usually less than 10% of the total, is produced by the action of the relative wind on the body, known as body lift. In general, drag is small relatively to the mass of the insect so that the greater part of the power output is involved in supporting the weight rather than in overcoming drag.

Drag results partly from the friction of the air on the body and wings, and partly from the kinetic energy given to the vortices which are left behind. Some of this vorticity is an inevitable consequence of the movement of the body

and wings through the air, but some is essential in accelerating the air downwards to produce lift. The component of drag which results from lift generation is known as the induced drag.

Reviews: Brodsky, 1994; Ellington, 1995

9.9.1.1 *Hovering*

Many insects are able to hover. Sometimes this behavior is particularly associated with feeding, as in *Macroglossum* (Lepidoptera), or with mating, as in swarms of some flies, and it often occurs before landing, enabling the insect to land precisely on a particular spot.

Hovering by many larger insects is achieved with the body approaching vertical and the stroke plane almost horizontal. The wings rotate through $100°$ or more at the end of each half stroke so that their angle of attack is similar on the morphologically 'up' and 'down' strokes (Fig. 9.29). A series of vortex rings is produced beneath the insect by the rapid rotations of the wing at the end of each half stroke in the manner shown in Fig. 9.29b (see Ellington, 1984a–f). However, hovering by hover flies (Syrphidae) and dragonflies is carried out with the body nearly horizontal, and presumably involves a different mechanism to produce the aerodynamic forces.

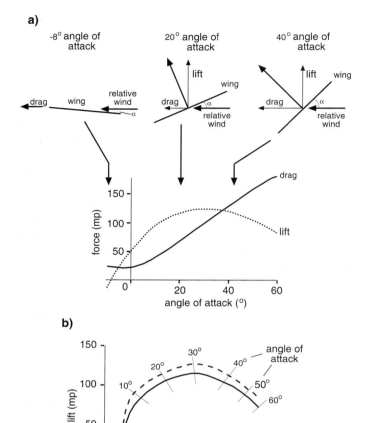

Fig. 9.30. Gliding. Experimental data from a moth which demonstrate the principles that would apply to any gliding insect (Lepidoptera, *Agrotis*) (after Nachtigall, 1967). **(a)** Variations in lift and drag with the angle of attack. At the top are shown sections of the wing with various angles of attack (α) corresponding to three points indicated by arrows on the graph. The insect is flying from left to right at a constant speed. **(b)** The relationship between lift and drag at different angles of attack showing the effect of removing the scales.

9.9.2 Gliding

Occasionally insects are seen to glide with the wings out-stretched. This behavior has been observed in Odonata, Orthoptera and Lepidoptera and ranges from a pause in wing movement lasting only a fraction of a second, to prolonged glides lasting many seconds. The ability to glide depends on the maintenance of a high lift/drag ratio produced by having the wings at a suitable angle to the relative wind. As the angle of attack is increased, both lift and drag increase, but above about 35° lift starts to decrease while drag continues to increase (Fig. 9.30). For various Lepidoptera, the lift/drag ratio is maximal at an angle of attack between 5 and 15°. The scales on the wings are believed to contribute to the lift but do not affect drag and so enable butterflies to glide for longer than would be possible without them (Nachtigall, 1976).

During a glide, the insect expends very little energy and it is suggested that the inability of dragonflies to fold their wings is a secondary adaptation to gliding. Locusts are able to lock their forewings in an outstretched position (Neville, 1965) and this may facilitate gliding. Short glides by *Locusta* are described by Baker and Cooter (1979).

9.10 CONTROL OF WINGBEAT

9.10.1 The initiation of wing movements

In most insects the wings start to beat as a result of loss of tarsal contact with the substratum and in the locust this occurs when the insect jumps into the air. When the legs are touching the ground, movement of the wings is inhibited, contact being perceived through the sensilla of the legs. In the cockroach, hair sensilla on the undersides of the tarsi as well as campaniform sensilla at all the leg joints are involved in this inhibition, but only a small subset of these sensilla is necessary to produce the effect (Krämer & Markl, 1978). This inhibition is overridden in insects that engage in preflight warmup.

Since fibrillar, asynchronous muscles only oscillate at high frequency when stretch-activated, the start of flight in insects with such muscles depends on some mechanism to initiate the process. In bees, the separate units in the dorso-ventral muscles are stimulated synchronously by neural inputs. This results in a muscle twitch of large amplitude, effecting the stretch of the dorsal longitudinal muscles which, at about the same time, are activated via their motor neurons (Esch & Goller, 1991).

At least in locusts, but probably also in other insects, octopamine appears to have an important arousal effect. Its hemolymph titer rises rapidly in the first few minutes of flight. It has a direct effect on the activity of interneurons producing the flight pattern (see Fig. 20.8a) and on the input of the forewing stretch receptor (Fig. 20.8b); it may affect the activity of the flight muscles (section 10.3.2), and it may be responsible for the early mobilization of lipid from the fat body before the release of adipokinetic hormone (see Orchard, Ramirez & Lange, 1993).

9.10.2 Maintenance of wing movements

The loss of tarsal contact with the substratum is sufficient to maintain the movement of the wings of *Drosophila* as well as initiating it, but in most other insects flight soon stops unless the insect receives further stimulation. This is provided by the movement of wind against the head. A wind speed of only $2 \, \mathrm{m \, s^{-1}}$ is sufficient to maintain the wing movements of *Schistocerca* and, since this is less than the flight speed of the insect, the relative wind produced in flight will provide sufficient stimulus. Air movement is perceived in locusts by hair beds on the face. In Diptera, and in the water beetle, *Dytiscus*, the wind is perceived by movements of the third antennal segment relative to the second, probably involving Johnston's organ (Bauer & Gewecke, 1985).

These stimuli also result in the legs being drawn up close to the body in a characteristic manner. Thus in locusts stimulation of the hair beds causes the fore legs to

assume the flight position. Diptera hold their legs in the flight position when their antennae are stimulated in flight.

9.10.3 Nervous control of wing movements

In locusts, and other insects with synchronous flight muscles, muscle contraction is regulated directly by the motor neurons: each time a neuron fires, the muscle which it innervates contracts. The basic pattern of muscular contractions involved in the flight of the locust can be produced in the complete absence of input from peripheral sensilla. This rhythm is generated by a complex of interneurons (forming a pattern generator, section 20.5.4), which drive the motor neurons (Robertson & Pearson, 1985). However, the centrally generated rhythm is slower and different in certain details from that normally required for efficient flight. Inputs from the tegulae and the stretch receptors at the base of each wing reset the rhythm and keep the pattern generator active. The tegulae of the hindwings affect the activity of both fore and hindwings; the forewing tegulae affect the activity of motor neurons to the levator muscles (Wolf & Pearson, 1988).

The axon of a forewing stretch receptor has a complex of branches in all three thoracic ganglia, while that from a hindwing stretch receptor has branches in the meso- and meta-thoracic ganglia. These branches synapse with dendrites of the flight motor neurons without any intervening interneurons so that the pathways between sensillum and effector muscles are monosynaptic. In addition, they connect with interneurons in the central pattern-generating system. The stretch receptors fire close to the time of maximum elevation of the wings, sometimes just after depression has begun. Elevation of the wing causes the stretch receptors to fire and their input inhibits the activity of motor neurons to the levator muscles and promotes the activity of motor neurons to the depressor muscles (Pearson & Ramirez, 1990). The chordotonal organs associated with the tegulae, in contrast, start to fire soon after the beginning of the downstroke and may continue to be active through most of the stroke. Their input signals the completion of the downstroke, perhaps by monitoring the velocity of the wing movement, and they elicit an immediate upstroke (Wolf, 1993). In this way, the stretch receptors and tegulae control the wingbeat frequency even though the basic rhythm is generated within the central nervous system.

Superimposed on this basic system are inputs concerned with the maintenance of steady flight and with steering. In the locust, about 20 descending interneurons,

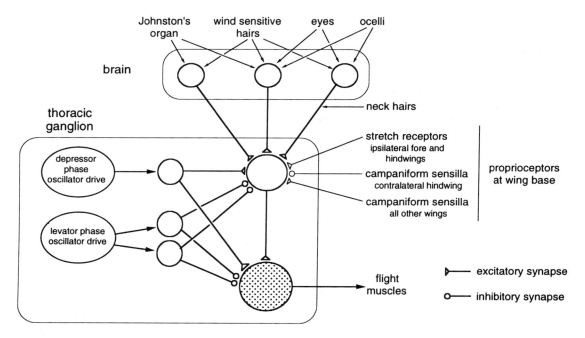

Fig. 9.31. Neural control of wingbeat in an insect with synchronous flight muscles. Sensory inputs from the sense organs of the head and neck affect the activity of interneurons in the brain (open circles), which vary in the selectivity of their responses. These interneurons have axons extending (descending) as far as the anterior abdominal neuromeres. In each ganglion (only one is shown) of the flight system, these descending neurons synapse with a premotor interneuron (large open circle) which also receives input from the proprioceptors at the wing bases. These different inputs modulate the activity of the premotor interneuron, which is driven or inhibited by the outputs from the pattern generating systems. The premotor interneuron activates the appropriate motor neuron (stippled) controlling a flight muscle (after Rowell & Reichert, 1991).

with cell bodies in the brain and axons extending as far as the fourth abdominal neuromere, are known to be involved in conveying information from the head to the motor neurons in the thoracic and abdominal segments. They convey information from the compound eyes, the wind-sensitive hairs, the ocelli, the antennae and proprioceptive hairs on the neck. Each descending interneuron tends to be most sensitive to one or two of the sensory inputs, although many also respond to other inputs. Each one also tends to respond best to deviations in a particular direction, for example, changes in the visual field resulting from rolling in one direction rather than the other (see Fig. 9.34) (Baader, Schäfer & Rowell, 1992; Rowell, 1989).

Within each of the thoracic ganglia, numerous other interneurons are involved in steering. In the mesothoracic ganglion of *Locusta*, there are at least 28 of these neurons, which are additional to those producing the basic pattern of oscillation. They receive inputs from the descending interneurons and from proprioceptors associated with the wings. They are driven and inhibited by inputs from the pattern generator, and they output, sometimes directly, to the motor neurons of the flight muscles (Fig. 9.31) (Rowell & Reichert, 1991). Basically similar arrays of interneurons occur in the anterior abdominal neuromeres to control the activity of the abdominal muscles involved in steering.

The force exerted by a muscle can be increased either by increasing the number of units which are active or by increasing the strength of the pull exerted by each unit. Although the units of the indirect flight muscles are only innervated by fast axons they contract more strongly if, instead of being stimulated by a single action potential, they are stimulated by two or more spikes following close together (section 10.4.4). This provides a means by which graded information can be transmitted to muscles despite the all-or-nothing code of the nervous system. Thus when the insect is producing only low lift forces, the second

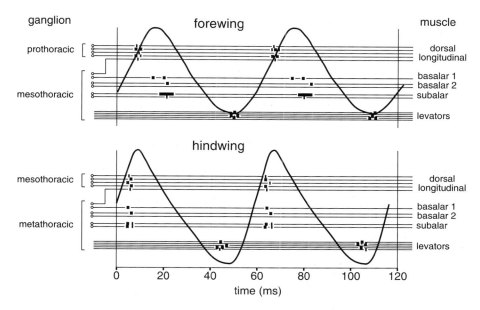

Fig. 9.32. Neural control of wingbeat in an insect with synchronous flight muscles. Diagram illustrating the timing of firing of motor neurons to the flight muscles of the fore and hindwings of a locust in relation to the wingbeat cycle. Each neuron is shown by a horizontal line with its origin in the appropriate ganglion on the left-hand side. Each dot on a line represents an action potential occurring at that time; a small dot indicates that activity in the neuron may or may not occur, a large dot that it always occurs. The heavy bar on the motor neuron to the forewing subalar muscle indicates that firing occurs within this period, but not at a precisely fixed time as with the other units. The heavy curves indicate the angular displacement of the wings (after Wilson and Weis-Fogh, 1962).

basalar muscle and some units of the dorsal longitudinal muscle of the hindwing may be inactive, whereas in producing high lift forces all the units come into action and the forces exerted by the individual units are increased by double firing of the motor neurons. In Lepidoptera, however, each motor neuron produces five to seven action potentials in each cycle and the burst length is directly correlated with the duration of muscle contraction, and so is inversely correlated with wingbeat frequency (Kammer, 1985).

The twisting of the wings by controller muscles in the locust is precisely timed by the pattern of motor impulses to the muscles (Fig. 9.32). Only in the case of one muscle, the mesothoracic subalar muscle, is the firing of the motor neuron very variable in its timing and this is the muscle which varies the twisting of the forewing to control lift. The precise co-ordination of the other neurons does not arise from a fixed pattern of connections between them as some function during walking in sequences completely different from that involved in flight.

The problem of control of the wingbeat is different in insects in which the wingbeat is produced by asynchronous muscles. Here, too, the muscles act in a precise sequence, but this sequence is not directly related to nervous input and the timing of firing of the motor neurons does not coincide with any particular phase of the wingbeat cycle. The nervous input to the flight muscles serves only as a general stimulator, maintaining the muscle contractions (section 10.3.2). In general, the motor neurons only produce single spikes, and their rate of firing varies from five to 25 spikes s^{-1} while wingbeat frequency is commonly in excess of 100 Hz.

Wingbeat frequency in these insects is controlled to some extent by the muscles that control the mechanical properties of the thorax; an increase in the lateral stiffness of the thorax produces an increased wingbeat frequency, while a decrease in stiffness leads to a reduced frequency. There is evidence, however, that the rate of firing of the motor neurons is positively correlated with the wingbeat frequency (Kammer, 1985).

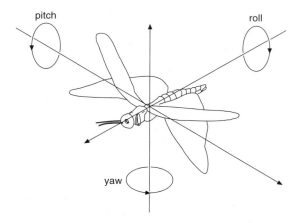

Fig. 9.33. Instability. Diagram showing the axes about which an insect may rotate when in flight (after Weis-Fogh, 1956).

9.10.4 Landing

During flight, the insect's legs are held close to the body, but before landing they must obviously be extended so that the insect lands on its feet. In *Lucilia*, leg extension results from visual stimuli. Particularly important in producing leg extension is a marked contrast in the stimulation of adjacent ommatidia and a rapid change in the illumination of successive ommatidia. Such changes might occur as the insect approaches a surface as the angular movement will increase as it gets closer, and details with contrasting shadows will become more apparent. In addition, to produce leg movements, a relatively large number of ommatidia must be stimulated and so the insect will not continually respond to small features of the environment which are visible in normal flight. The information from the eyes is integrated by interneurons that respond differentially to movements across the eyes in different directions (Borst & Bahde, 1988).

9.11 STABILITY IN FLIGHT

Because of variations in the forces acting on it during flight, there is a tendency for an insect to deviate from a steady path. This instability may involve rotation about any of the three major axes passing through the center of gravity of the body (Fig. 9.33). Rotation about the long axis of the body is called rolling, rotation about the horizontal, transverse axis is pitching, and rotation about the vertical axis is yawing. To some extent such deviations may be corrected passively through the shape of the body and form of the wingbeat, but insects also have the capacity to make active changes in the aerodynamic forces acting on the body in order to maintain a steady flight. **Review:** Kammer, 1985

9.11.1 Passive stability

Some degree of passive stability about the rolling axis results from the wings being inserted above the center of gravity. Some stability in yaw is achieved if the maximum thrust is delivered when the wings are in front of the insect's center of gravity; this appears to be the case in flies, for instance. The long abdomen of insects such as dragonflies and locusts acts as a rudder giving stability about yawing and pitching axes.

9.11.2 Active maintenance of stability

Deviations from a steady path are perceived by various sensory receptors and the nervous input from these exerts a controlling influence on the wingbeat so that the deviation is corrected. Of primary importance in this respect are the eyes, Johnston's organ in the antenna, the hair beds on the front of the head and the sensilla at the base of the wings. The halteres of Diptera are of fundamental importance in this order and they are considered separately.

Rolling Vision plays an important part in the control of rolling. Odonata and Orthoptera, and probably also other insects, have a dorsal light reaction by which they align the head so that the dorsal ommatidia receive maximal illumination. To produce this reaction, a number of ommatidia must be illuminated, but the response does not depend on stimulation of any particular part of the eye as it is still apparent if the most dorsal ommatidia, which are normally concerned in the response, are covered.

As most light normally comes from the sun or the sky overhead, the dorsal light reaction ensures that the head is usually held in a vertical position. In Odonata, where the head is loosely articulated to the thorax, the head also tends to stay in a vertical plane due to its own inertia, but this is not the case in the locust where the head and thorax are broadly attached and rolling by the thorax is immediately transmitted to the head. As a result, a locust flying in complete darkness is unable to orient in this plane and will fly upside down or at any other angle.

If insects controlled rolling exclusively by a dorsal light reaction they would sometimes have a tendency to fly at

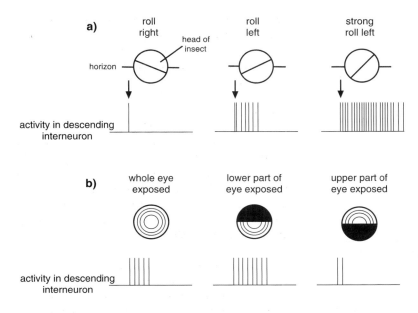

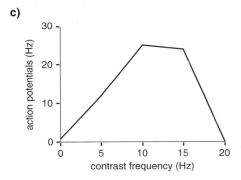

Fig. 9.34. Maintenance of stability. Examples of the activity of visually sensitive descending interneurons to perturbations of the visual field comparable with those occurring in flight in a locust (*Schistocerca*). (**a**) A wind-sensitive interneuron that responds to roll to the left. Arrows mark the onset of the wind (after Rowell & Reichert, 1985). (**b**) A neuron responding to image patterns simulating forward flight. The neuron responds most strongly when only the lower part of the eye is stimulated by the moving pattern, as would be the case in normal flight. Interneurons with these characteristics are probably involved in the optomotor response (based on Baader *et al.*, 1992). (**c**) The sensitivity of a neuron similar to that in (**b**) to different rates of pattern movement (contrast frequency) over the eye (data from Baader *et al.*, 1992).

unusual angles. This might occur, for instance, with the sun low in the sky just before sunset. The fact that this does not occur indicates that other stimuli are also important. *Schistocerca* also orients to the horizon, keeping it transversely across the eyes with the upper ommatidia more brightly illuminated than the lower ones. Orientation to the horizon is accurate and, in good light, the insect can follow slow changes and even oscillations of the horizon up to 40 Hz although under these conditions it is unable to stabilize the position of the head. Both the compound eyes and ocelli are involved in the horizon-detecting response. Although fine tuning of the response depends on the compound eyes, the ocelli decrease the latency between a change in the horizon and the insect's response so that the rate of stabilization of the image in the compound eye is

optimized. The information is relayed to the flight control system in the thoracic ganglia via interneurons some of which are sensitive to specific types of input. Fig. 9.34a shows the firing of such an interneuron in *Schistocerca*. It responds mostly when the insect rolls strongly to the left, but only if there is wind on the head. It barely responds to roll to the right.

The dorsal light reaction gives stability to the head, and deviations from the stable position cause flight steering responses to restore stability. However, the head may reach its stable position before the rest of the body is aligned with the head. Any deviation from this alignment is signalled by proprioceptors between the head and the thorax. In *Schistocerca* there are hair beds on the cervical sclerite and hairs along the anterior border of the pronotum which are involved in this orientation. Unequal stimulation of the sensilla on the two sides due to a turning of the thorax relative to the head leads to differential twisting of the wings so that the thorax is brought back into alignment again (Taylor, 1981).

Pitching When flying steadily, an insect tends to keep the body angle to the horizontal more or less constant. In bumblebees the body angle is reduced at higher speeds (Fig.9.20d), but Baker, Gewecke & Cooter (1981) did not observe any correlation between flight speed and body angle in free-flying locusts. The average value for the body angle was 7.4°. In the locust, any tendency to pitch is counteracted by changes in the twisting of the forewing during the downstroke so that the forces it exerts are modified. There is no regulation of the upstroke or of the hindwing in any phase. The twisting of the forewing is regulated by the campaniform sensilla at the bases of the wings. The mechanism by which bumblebees control body angle, and hence body lift, is not known.

In Diptera the halteres are important in controlling pitching, but it is also probable that the eyes and Johnston's organ in the antenna exert some controlling influence over the wing movements.

Yawing Vision plays a part in the control of yaw, but in locusts the sensilla in the facial hair beds have directional sensitivity. Oblique stimulation of these sensilla, such as occurs during yaw, leads to a change in the form of the wingbeat so that the original orientation is restored. The insect may also maintain stability in yaw by actively using the abdomen as a rudder.

Sensilla at the wing base In the normal vibration of a wing, a twisting-force, or torque, is produced in the cuticle at the wing base. If the wing were to move up and down in a vertical plane, only vertical torque would be produced. Thus with the wing in the up position, the cuticle at the base on the upper side would be compressed, while that on the ventral side was stretched, and vice versa with the wing in the down position; all the forces would be acting parallel with the long axis of the wing. But because of the complexity of the wing movement the torque will differ in strength and direction in different parts of the wing stroke. The torque is perceived by the sensilla, and particularly the campaniform sensilla, of the wing base. These are arranged in groups, all those within a group having a

similar orientation (Fig.9.14), so that each group responds maximally to torque in a particular direction and, if the sensitivity of the sensilla is appropriate, do so only once during a wing cycle. It is very likely that any tendency for the insect to deviate from a stable orientation would result in differential changes in the stimulation of these sensilla, which could thus exert a controlling influence on the wingbeat to correct for the deviation. This is certainly the case in the control of lift and pitching in *Schistocerca*, but the situation is best understood in the halteres of Diptera, which are specialized organs of stability.

Halteres The halteres vibrate with the same frequency as the forewings, but in antiphase. Their movement is less complex than that of the wings. Because of their structure, with the center of mass in the end knob (Fig.9.15), and because of the nature of their articulation with the thorax, they vibrate in a vertical or near vertical plane without making the complex fore and aft movements of the wings. Hence the forces acting at the base of the haltere and stimulating the campaniform sensilla are limited to a vertical plane when the haltere is oscillating with the insect in steady flight, and dorsal and ventral torques oscillate with the same frequency as the vibration of the halteres. These torques may be perceived by the sensilla of dorsal and ventral scapal plates which are oriented parallel with the long axis of the haltere.

The path of the end knob during vibration represents an arc of a circle about the long axis of the insect and the haltere may thus be regarded as a gyroscope whose axis of rotation corresponds with the long axis of the insect. In *Calliphora*, they have an angular rotation of about $50\,000°\,s^{-1}$. As in a gyroscope, the moving halteres possess inertia, tending to maintain a fixed orientation in space so that, if the insect rotates about its axes of yaw or pitch, torques will be produced at right angles to the stroke plane. These torques will vary cyclically through the course of each oscillation by the haltere with yaw producing maximum torques in both up and down strokes, while the torques produced by pitch have a single cycle during the stroke (Fig. 9.35) (Nalbach, 1993). Roll also produces torques, but these are in the plane of movement of the haltere and are small relative to the torques produced by the normal vibration of the haltere. A fly can modulate the wingbeat to correct for yaw and pitch with only a single haltere, but both are required for correction of roll. It is to be expected that, normally, information

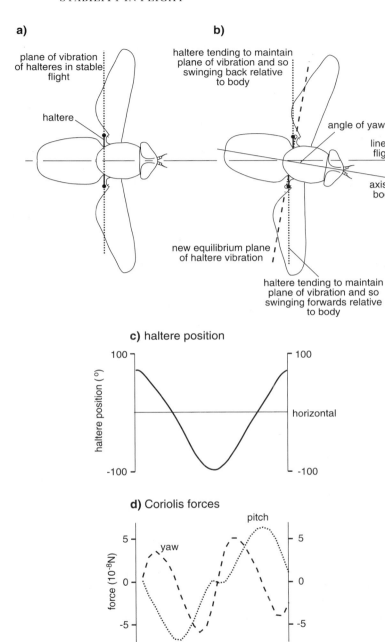

a)

plane of vibration
of halteres in stable
flight

haltere

b)

haltere tending to maintain
plane of vibration and so
swinging back relative
to body

angle of yaw

line of
flight

axis of
body

new equilibrium plane
of haltere vibration

haltere tending to maintain
plane of vibration and so
swinging forwards relative
to body

c) haltere position

100   100

haltere position (°)

horizontal

-100   -100

d) Coriolis forces

pitch

5   5

force (10⁻⁸N)

yaw

0   0

-5   -5

Fig. 9.35. Maintenance of stability. Diagrams to illustrate the action of the halteres. (a) In stable flight the halteres swing outward and vibrate in a vertical plane at right angles to the insect's body (dotted line). (b) If the insect makes a yawing movement the halteres have a tendency to continue vibrating in their original planes (dotted) and a horizontal torque is created at the base of the haltere. If the yaw is not corrected, the halteres rapidly assume the equilibrium position (dashed line). (c) Path of the end of a haltere through the air as the insect flies from left to right. Contrast the regularity of the path with that of the forewing in Fig. 9.23. (d) Calculated forces perpendicular to the plane of vibration of the haltere (Coriolis forces) produced at its base by yaw and pitch. These forces will stimulate some of the campaniform sensilla. Notice that the forces produced by yaw oscillate twice during each complete up and down movement of the haltere, with a maximum on both the down and up strokes (after Nalbach, 1993).

from the halteres of the two sides is integrated in the central nervous system.

In yawing, the haltere on the outer side of the rotation will tend to swing back relative to the insect, while that on the inside will swing forwards (Fig. 9.35a,b). The campaniform sensilla respond to compression forces along their short axes, and bending the haltere backwards compresses the campaniform sensilla of the basal plate, and also stimulates the chordotonal organ. When the insects rolls, the torques in the two halteres will be in antiphase, while in pitching both halteres are affected in the same way at the same time, by torques at right angles to the stroke plane.

9.11.3 Steering

Insects must steer continually while in flight, both to maintain stability and to maintain or change direction. In locusts, steering may involve differences in wing-stroke amplitude and the degree of wing twisting on the two sides, changes in the phase relationships of the fore and hind wings, and use of the abdomen and the hind legs as a rudder. All or only some of these activities may be employed at different times. Changes in the wingbeat are also involved in insects where the abdomen is too short to be an effective rudder. In flies, one wing may remain stationary while the other is actively moving, producing a sharp turn.

9.11.4 Control of flight speed

Flight speed is regulated with respect to movement over the ground, groundspeed, and with respect to the air, airspeed.

Some insects are known to maintain a constant groundspeed despite variations in the speed and direction of the wind. This response depends on perceiving the apparent movement of objects in the visual field as the insect moves relative to them. It is called an optomotor reaction. A flying insect appears to prefer images to pass over the eye from front to back at a certain moderate speed, the preferred retinal velocity. If an insect flies downwind, this velocity may be exceeded and the insect turns and flies into the wind. An upwind orientation is maintained only as long as it can make headway against the wind. If the wind is too strong for this to occur, the insect lands.

Amongst the descending interneurons from the brain are some that have the potential for mediating this response. In locusts, for example, two such neurons arise in each side of the brain. They are insensitive to image movements from back to front, as would occur if the insect was flying backwards, and are most sensitive to movements from front to back across the ventral region of the eye (Fig. 9.34b). They respond maximally when the contrasting pattern flickers with a frequency of 10 to 20 per second (Fig. 9.34c) (Baader *et al.*, 1992). Thus these neurons convey the type of information needed for the optomotor response to the wing-regulating organization in the thorax. Similar interneurons are present in flies (Gronenberg & Strausfeld, 1990,1991) and probably in all other flying insects. Connections within the brain are described in Chapter 20.

Air speed is measured and regulated by the antennae, at least in *Apis*, *Calliphora*, some Lepidoptera and locusts

(section 23.2.3.2). In flight, the antennae are held horizontally and directed forwards. Movement through the air tends to push the flagellum backwards relative to the insect, but it compensates for this deflection by swinging the pedicel forwards. At higher air speeds the compensation is greater and so the antennae are pointed more directly forwards (Fig. 9.36). Johnston's organ only responds to movement, and it is probable that, in flight, vibrations of the flagellum produced by the flapping wings stimulate Johnston's organ. Different scolopidia are stimulated depending on the degree of deflection of the flagellum and it is probably on this basis that airspeed is determined (Gewecke, 1974).

9.12 POWER FOR FLIGHT

Flight demands a great deal of power to lift the insect off the ground and drive it forwards. The forces exerted by flight muscles are in no way unusual, and because of chemical and mechanical inefficiencies only a small proportion of the energy expended by the muscles is effectively available. About 80% of the energy consumed by the muscles is lost as heat, and of the mechanical work performed by the muscles only about one half may be aerodynamically useful. Consequently only 5–10% of the energy consumed by the flight muscles contributes to flight. The high power output necessary for flight is achieved by their high frequencies of contraction.

The metabolic rates of insects in flight are often 50–100 times higher than their resting rates. Oxygen consumption, a measure of metabolic rate, increases linearly with the body weight of the insect (Fig. 9.37). The relationship between oxygen consumption in flight and weight is essentially the same for moths, which have synchronous flight muscles, and euglossine bees, which have asynchronous flight muscles (see Casey, 1989; Casey, May & Morgan, 1985)

The high metabolic rate depends on the availability of oxygen, a suitably high muscle temperature, and an abundant supply of fuel. The demand for oxygen is met by modifications of the tracheal system (see Fig. 17.4). Insects use various behavioral and physiological devices to achieve a body temperature at which the flight muscles can function efficiently (section 19.1).

9.12.1 Fuels for flight

The fuels providing energy for flight vary in different insects. Hymenoptera and Diptera commonly use

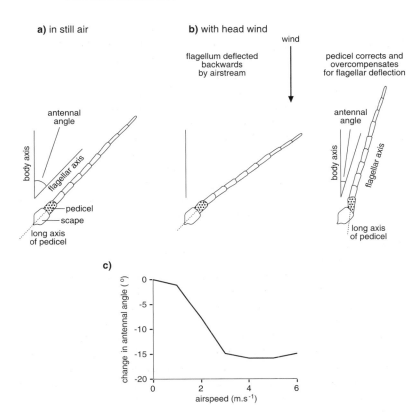

a) in still air

b) with head wind

wind

flagellum deflected
backwards
by airstream

pedicel corrects and
overcompensates
for flagellar deflection

antennal
angle

body axis

flagellar axis

pedicel
scape
long axis
of pedicel

antennal
angle

body axis

flagellar axis

long axis
of pedicel

c)

Fig. 9.36. Control of airspeed (after Gewecke, 1974). (a) Position of the antenna in still air. (b) Changes in the position of an antenna resulting from a head wind. Long axis of pedicel shown as a dotted line. (c) Changes in the antennal angle of a tethered locust associated with increasing wind speed (equivalent to increasing airspeed of a freely flying insect).

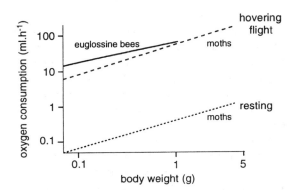

Fig. 9.37. Oxygen consumption, as a measure of metabolic activity, in freely flying insects. Data for the bees is based on observations of nine species, that for moths on 62 species from six families (data from Bartholomew & Casey, 1978; Casey *et al.*, 1985).

carbohydrates; locusts, aphids and migratory Lepidoptera depend mainly on fats, but use carbohydrates during short flights and the early stages of sustained flight. Some Diptera and possibly Coleoptera metabolize amino acids, especially proline, at the flight muscles, although the energy is ultimately derived from lipid reserves. Fat is more suitable than carbohydrate as a reserve for insects that make long flights because it produces twice as much energy per unit weight; a gram of fat yields 39 000 J, but a gram of carbohydrate yields only 17 000 J. In addition, glycogen, a common carbohydrate reserve, is strongly hydrated so that it is eight times heavier than isocaloric amounts of fat. Thus an insect can store large amounts of energy more readily as fat and 85% of the energy stored by the locust is in this form.

Initially during flight, fuel reserves within the muscles themselves are utilized, but in most insects these are adequate only for very short flights and further supplies of fuel are drawn from elsewhere. The blood-sucking bugs, *Rhodnius* and *Triatoma*, are exceptional in storing

Fig. 9.38. Fuel consumption in flight. Changes in the concentrations of fuels during the first hour of flight in the migratory locust and the blowfly. **(a)** Glycogen and glucose in flight muscle. **(b)** Trehalose and lipid in the hemolymph.

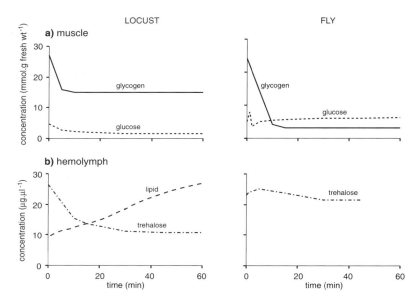

relatively large amounts of triacylglycerol in the flight muscles (Ward, Candy & Smith, 1982). Beyond the muscles themselves, the fat body is the principal store of fuel.

Reviews: Beenakkers, van der Horst & van Marrewijk, 1985; Candy, 1989

Carbohydrates Carbohydrates in the flight muscles are usually the immediate source of energy at the start of flight. In the locust, the muscle concentration of glucose starts to decline immediately after flight begins, but, in blowflies, there is a transient increase due to the rapid production of glucose from trehalose (Fig. 9.38). Subsequently, after about 30 minutes flight, the glucose concentration stabilizes in both species. These stable levels indicate that glucose is provided from other sources, initially from glycogen in the muscle itself, and then from elsewhere.

Trehalose in the hemolymph forms an important carbohydrate reserve in many insects, and is also the form in which carbohydrates are transported from other sites. As a result, the hemolymph concentration of trehalose may fall, as it does in the locust, where it becomes stable after about 30 minutes of flight. In the fly, *Calliphora*, however, the concentration first rises and then returns to its original level. The stable levels in both species result from the synthesis of trehalose from glycogen in the fat body. In some

other insects, sugars in the gut, in the crop of *Tabanus* (Diptera) and the honey stomach of *Apis*, may be converted to trehalose immediately after absorption and transported directly to the flight muscles.

The oxidation of carbohydrate in insect muscle involves the usual processes of glycolysis and the citric acid cycle (Figs. 9.39, 9.40). Dihydroxyacetone phosphate and glyceraldehyde-3-phosphate are formed in equal amounts from fructose-1,6-diphosphate. They are interchangeable and oxidation can occur either by the direct transfer of hydrogen to the electron transfer chain or via pyruvate and the citric acid cycle. The latter pathway is advantageous because it results in the conservation of a greater amount of energy as ATP. Consequently, a system favoring the conversion of glyceraldehyde-3-phosphate to dihydroxyacetone phosphate is desirable. However, the oxidation of glyceraldehyde-3-phosphate is limited by the availability of nicotinamide adenine dinucleotide (NAD^+).

Regeneration of NAD^+ from NADH is achieved in the glycerol phosphate shuttle (Fig. 9.40). This involves the transfer of hydrogen to the electron transfer system via dihydroxyacetone phosphate and glycerol-3-phosphate. However, the electron transfer system is in the mitochondria, while glycolysis occurs in the cytosol. The shuttle involves the transfer of glycerol-3-phosphate into the mitochondria and the movement of dihydroxyacetone

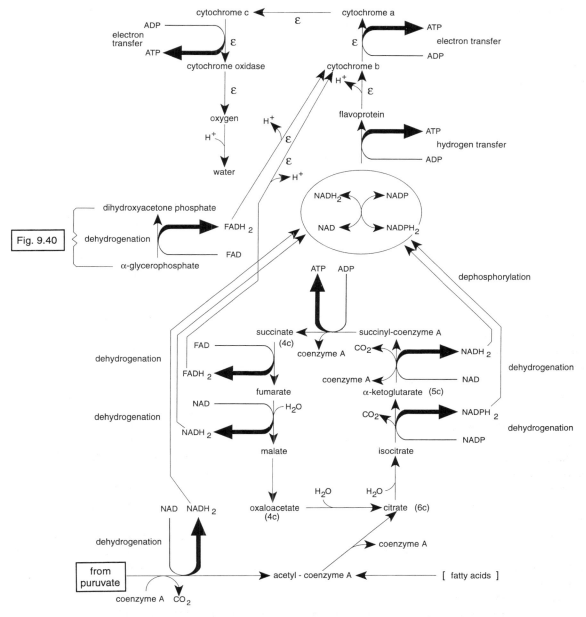

Fig. 9.39. The citric acid cycle and terminal oxidation in the mitochondria. Notice how the glycerol phosphate shuttle, which channels virtually all carbohydrate metabolism via pyruvate (see Fig. 9.40), results in the greatest production of ATP. Bold, curved arrows show energy conserving steps. Fatty acids are introduced into the citric acid cycle via acetyl–coenzyme A.

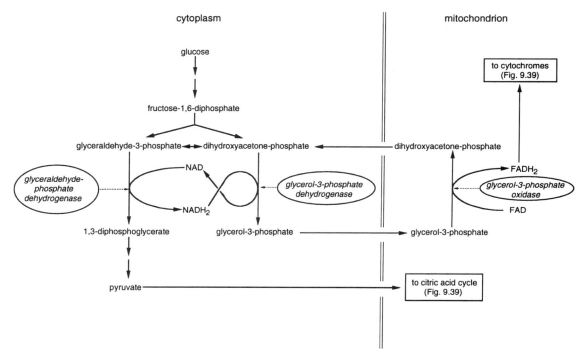

Fig. 9.40. The glycerol phosphate shuttle. Enzymes referred to in table 9.2 are shown in italics. The points at which they are active are indicated by broken arrows.

phosphate out again. The interconversion of glycerol–3–phosphate and dihydroxyacetone phosphate is catalyzed by the enzyme glycerol phosphate dehydrogenase which is present in the cytosol and in the mitochondria (shown in Fig. 9.40 and Table 9.2 as glycerol-3-phosphate dehydrogenase in the cytoplasm and glycerol-3-phosphate oxidase in the mitochondria). The level of the enzyme in the cytoplasm of flight muscle is consistently greater than that in the mitochondria. This limits the rate at which hydrogen is transferred to the electron transfer chain by this route, but it ensures that the catalytic amounts of dihydroxyacetone phosphate required to regenerate NAD^+ are available. Glyceraldehyde-phosphate dehydrogenase (producing 1,3-diphosphoglycerate) is present in large amounts, especially in those insects where carbohydrates form the principal flight fuel (*Calliphora* in Table 9.2). This system ensures that virtually all the carbon derived from the original substrate, glucose-6-phosphate, passes via pyruvate to the citric acid cycle and the enzymes of this pathway are present at high levels (Table 9.2). This ensures that the greatest amount of energy is conserved and made available for muscle contraction. The activity of the flight muscles is entirely aerobic and the enzyme lactate dehydrogenase is present at very low levels. By contrast, in leg muscles, the level is more than 30 times higher.

Lipids Insects engaging in long-range migration, such as locusts, some butterflies and planthoppers switch from using carbohydrates as the main source of fuel to using lipids. This switch occurs some 15–30 minutes after take-off and, in general, the lipids are obtained from the fat body where they are stored as triacylglycerides. They are transported through the hemolymph as diacylglycerides and their concentration increases and then stabilizes in the first 2–3 hours of flight (Fig. 9.38b).

At the flight muscles, the glycerides are degraded in a series of steps into two–carbon units. Mitochondrial membranes are impermeable to fatty acids, and transfer of fatty acids into the mitochondria is facilitated by carnitine, a water-soluble vitamin. Within the mitochondria, the fats enter the tricarboxylic acid cycle as acetyl-coenzyme A, condensing with oxaloacetate to form citrate. The

Table 9.2. *Flight muscle activity of enzymes associated with the utilization of different fuels in flight. Activity of the enzymes in the fat body is given for comparison where data are available*

| | Insect (Order) | | | | | |
| | *Locusta* (Orthoptera) | | *Philosamia* (Lepidoptera) | *Leptinotarsa* (Coleoptera) | | *Calliphora* (Diptera) |
Enzyme	Muscle	Fat	Muscle	Muscle	Fat	Muscle
Enzymes of glycolysis						
glyceraldehyde phosphate dehydrogenase	69	—	18	30	—	194
lactate dehydrogenase	0.3	—	0.6	0.1	—	<0.05
Enzymes of the glycerolphosphate shuttle						
glycerol-3–phosphate dehydrogenase	33	4	13	17	—	48
glycerol-3–phosphate oxidase	2.3	—	0.8	1.2	—	—
Enzymes of the citric acid cycle						
citrate synthase	57	1	80	11	5	45
succinate dehydrogenase	4	0.5	4	11	0.4	—
Enzyme introducing fatty acids						
3–hydroxyacyl-coenzyme A dehydrogenase	66	—	98	11	49	<0.05
Enzymes introducing amino acids						
glutamate dehydrogenase	1	1	—	58	2.4	—
alanine aminotransferase	4	9	—	80	74	—

Notes:
Units are μmol substrate converted/mg muscle protein/h.

—, No data.

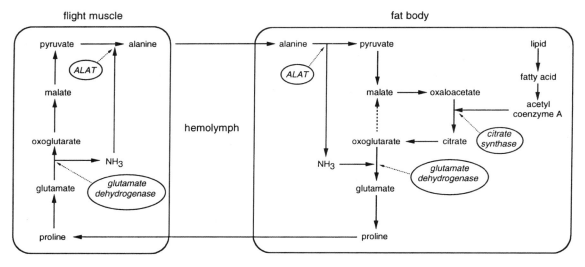

Fig. 9.41. Proline metabolism in flight muscle associated with the utilization of lipids in the fat body. Enzymes referred to in Table 9.2 are shown in italics. The points at which they are active are indicated by broken arrows. ALAT=alanine aminotransferase.

enzymes catalyzing these various reactions are present at high concentrations in the muscles of species using lipids as the primary flight fuels, but are at low levels or absent from species using mainly carbohydrates as well as from other tissues of the same species (Table 9.2).

Amino acids Oxidation of amino acids may occur to a minor extent in the flight muscle of most insects, but, in a few species, proline provides the major substrate oxidized by the flight muscles although lipids in the fat body are the ultimate source of fuel. This occurs in the tsetse fly, *Glossina*, and some beetles, such as *Leptinotarsa*. The initial reserve of proline is small, and in the tsetse fly sufficient to last only for about two minutes, and it is probable that proline is synthesized during flight. Proline is first converted to glutamate, which then undergoes transamination with pyruvate to produce alanine and oxoglutarate (Fig. 9.41). The latter enters the citric acid cycle while the former is returned to the fat body for the resynthesis of proline. In the flight muscles, the level of glutamate dehydrogenase is higher than in other insects, while the level of alanine aminotransferase is high in both flight muscle and the fat body (Table 9.2). The activity of coenzyme A dehydrogenase in the fat body is comparable with that in the flight muscles of insects oxidizing lipids as their primary flight fuel.

9.12.2 Mobilization of fuel for flight
The energy sources used by flight muscles are usually stored in a form that is not immediately metabolizable and is often at some point remote from the muscles. Consequently, the mobilization of these reserves must be coordinated with muscle activity.

The nerve impulse that initiates contraction of the flight muscles activates the fibrillar ATPase by the release of Ca^{2+} from the sarcoplasmic reticulum. This calcium also promotes the activation of the phosphorylase involved in the breakdown of glycogen to glucose-1-phosphate and probably of glycerol-3-phosphate dehydrogenase, which is involved in the glycerol phosphate shuttle. As a consequence, utilization of carbohydrate proceeds at a fast rate.

The mobilization of glycogen in the fat body of locusts and some flies is regulated hormonally. In locusts, the adipokinetic hormone (AKH, see below) also mediates glycogen metabolism by activating a glycogen phosphorylase. In the moth, *Manduca*, however, AKH does not act in this way. In this species, activation of the glycogen phosphorylase is regulated by the titer of hemolymph carbohydrates. A fall in the level of hemolymph trehalose as it is utilized by the flight muscles during flight stimulates the activation of the enzyme (Ziegler & Schulz, 1986).

In locusts and some other insects, information from the brain leads to the release of adipokinetic hormone from the corpora cardiaca, but this release is inhibited if the carbohydrate concentration in the hemolymph is high. In addition to its effect on glycogen mobilization, this hormone causes the mobilization of lipids in the fat body by activating a lipase. Diacylglycerides are released from the fat body and these are transported in the hemolymph to the muscles (see Fig. 6.11). In locusts the release of diacylglycerides into the hemolymph is apparent within 5 minutes of the start of flight and the concentration more than doubles. It is also possible that octopamine is involved in the early stages of lipid mobilization (Orchard *et al.*, 1993).

Review: Wheeler, 1989

REFERENCES

Altman, J.S. (1975). Change in the flight motor pattern during the development of the Australian plague locust, *Chortoicetes terminifera*. *Journal of Comparative Physiology*, **97**, 127–42.

Baader, A., Schäfer, M. & Rowell, C.H.F. (1992). The perception of the visual flow field by flying locusts: a behavioural and neuronal analysis. *Journal of Experimental Biology*, **165**, 137–60.

Baker, P.S. & Cooter, R.J. (1979). The natural flight of the migratory locust, *Locusta migratoria* L. II. Gliding. *Journal of Comparative Physiology*, **131**, 89–94.

Baker, P.S., Gewecke, M. & Cooter, R.J. (1981). The natural flight of the migratory locust, *Locusta migratoria* L. III. Wing-beat frequency, flight speed and attitude. *Journal of Comparative Physiology*, **141**, 233–7.

Bauer, C.K. & Gewecke, M. (1985). Flight behavior of the water beetle Dytiscus marginalis L. (Coleoptera, Dytiscidae). In *Insect Locomotion*, ed. Gewecke, M. & Wendler, G., pp. 205–14. Berlin: Verlag Paul Parey.

Bartholomew, G.A. & Casey, T.M. (1978). Oxygen consumption of moths during rest, pre-flight warm-up, and flight in relation to body size and wing morphology. *Journal of Experimental Biology*, **76**, 11–25.

Beenakkers, A.M.T., van der Horst, D.J. & van Marrewijk, W.J.A. (1985). Biochemical processes directed to flight muscle metabolism. In *Comprehensive Insect Physiology, Biochemistry and Pharmacology*, vol. 10, ed. G.A.Kerkut & L.I.Gilbert, pp. 451–86. Oxford: Pergamon Press.

Betts, C.R. & Wootton, R.J. (1988). Wing shape and flight behavior in butterflies (Lepidoptera: Papilionoidea and Hesperioidea): a preliminary analysis. *Journal of Experimental Biology*, **138**, 271–88,

Borst, A. & Bahde, S. (1988). Spatio-temporal integration of motion. A simple strategy for safe landing in flies. *Naturwissenschaften*, **75**, 265–7.

Bourgogne, J. (1951). Ordre des Lépidoptères. In *Traité de Zoologie*, vol. 10, ed. P.-P.Grassé, pp. 174–448. Paris: Masson et Cie.

Brodsky, A.K. (1991). Vortex formation in the tethered flight of the peacock butterfly *Inachis io* L. (Lepidoptera, Nymphalidae) and some aspects of insect flight evolution. *Journal of Experimental Biology*, **161**, 77–95.

Brodsky, A.K. (1994). *The Evolution of Insect Flight*. Oxford: Oxford University Press.

Byrne, D.N., Buchmann, S.L. & Spangler, H.G. (1988). Relationship between wing loading, wingbeat frequency and body mass in homopterous insects. *Journal of Experimental Biology*, **135**, 9–23.

Candy, D.J. (1989). Utilization of fuels by the flight muscles. In *Insect Flight*, ed. G.J. Goldsworthy & C.H. Wheeler, pp. 305–19. Boca Raton, Florida: CRC Press.

Casey, T.M. (1989). Oxygen consumption during flight. In *Insect Flight*, ed. G.J. Goldsworthy & C.H. Wheeler, pp. 257–72. Boca Raton, Florida: CRC Press.

Casey, T.M., May, M.L. & Morgan, K.R. (1985). Flight energetics of Euglossine bees in relation to morphology and wingstroke frequency. *Journal of Experimental Biology*, **116**, 271–89.

Chowdhuri, V. & Parr, M.J. (1981). The 'switch mechanism' and sound production in tsetse flies (Diptera: Glossinidae). *Journal of Natural History*, **15**, 87–95.

Cloupeau, M., Devillers, J.F. & Devezeaux, D. (1979). Direct measurements of instantaneous lift in desert locusts; comparison with Jensen's experiments on detached wings. *Journal of Experimental Biology*, **80**, 1–15.

Comstock, J.H. (1918). *The Wings of Insects*. New York: Comstock Publishing Co.

Cooter, R.J. & Baker, P.S. (1977). Weis–Fogh clap and fling mechanism in *Locusta. Nature, London.* **269**, 53–4.

Danforth, B.N. & Michener, C.D. (1988). Wing folding in the Hymenoptera. *Annals of the Entomological Society of America*, **81**, 342–9.

Dickinson, M.H. (1992). Directional sensitivity and mechanical coupling dynamics of campaniform sensilla during chord-wise deformations of the fly wing. *Journal of Experimental Biology*, **169**, 221–33.

Dickinson, M.H. & Gotz, K.G. (1996). The wake dynamics and flight forces of the fruit fly *Drosophila melanogaster*. *Journal of Experimental Biology*, **199**, 2085–104.

Dickinson, M.H., Lehmann, F.-O. & Götz, K.G. (1993). The active control of wing rotation by *Drosophila*. *Journal of Experimental Biology*, **182**, 173–89.

Dudley, R. (1991). Biomechanics of flight in neotropical butterflies: aerodynamics and mechanical power requirements. *Journal of Experimental Biology*, **159**, 335–57.

Dudley, R. & Ellington, C.P. (1990a). Mechanics of forward flight in bumblebees I. Kinematics and morphology. *Journal of Experimental Biology*, **148**, 19–52.

Dudley, R. & Ellington, C.P. (1990b). Mechanics of forward flight in bumblebees II. Quasi-steady lift and power requirements. *Journal of Experimental Biology*, **148**, 53–88.

Ellington, C.P. (1984a). The aerodynamics of hovering insect flight. I. The quasi-steady analysis. *Philosophical Transactions of the Royal Society of London* B, **305**, 1–15.

Ellington, C.P. (1984b). The aerodynamics of hovering insect flight. II. Morphological parameters. *Philosophical Transactions of the Royal Society of London* B, **305**, 17–40.

Ellington, C.P. (1984c). The aerodynamics of hovering insect flight. III. Kinematics. *Philosophical Transactions of the Royal Society of London* B, **305**, 41–78.

Ellington, C.P. (1984d). The aerodynamics of hovering insect flight. IV. Aerodynamic mechanisms. *Philosophical Transactions of the Royal Society of London* B, **305**, 79–113.

Ellington, C.P. (1984e). The aerodynamics of hovering insect flight. V. A vortex theory. *Philosophical Transactions of the Royal Society of London* B, **305**, 115–44.

Ellington, C.P. (1984f). The aerodynamics of hovering insect flight. VI. Lift and power requirements. *Philosophical Transactions of the Royal Society of London* B, **305**, 145–81.

Ellington, C.P. (1995). Unsteady aerodynamics of insects flight. *Symposium of the Society for Experimental Biology*, **49**, 109–29.

Ennos, A.R. (1987). A comparative study of the flight mechanism of Diptera. *Journal of Experimental Biology*, **127**, 355–72.

Ennos, A.R. (1988). The inertial cause of wing rotation in Diptera. *Journal of Experimental Biology*, **140**, 161–9.

Esch, H. & Goller, F. (1991). Neural control of fibrillar muscles in bees during shivering and flight. *Journal of Experimental Biology*, **159**, 419–31.

Foster, J.A. & Robertson, R.M. (1992). Temperature dependency of wing-beat frequency in intact and deafferented locusts. *Journal of Experimental Biology*, **162**, 295–312.

Gettrup, E. (1962). Thoracic proprioceptors in the flight systems of locusts. *Nature, London*, **193**, 498–9.

Gewecke, M. (1974). The antennae of insects as air-current sense organs and their relationship to the control of flight. In *Experimental Analysis of Insect Behaviour*, ed. L. Barton Browne, pp. 100–13. Berlin: Springer-Verlag.

Gnatzy, W., Grünert, U. & Bender, N. (1987). Campaniform sensilla of *Calliphora vicina* (Insecta, Diptera). I. Topography. *Zoomorphology*, **106**, 312–9.

Grassé, P.-P. (1949). Ordre des Isoptères ou termites. In *Traité de Zoologie*, vol. 9, ed. P.-P. Grassé, pp. 408–544. Paris: Masson et Cie.

Grodnitsky, D.L. & Morozov, P.P. (1993). Vortex formation during tethered flight of functionally and morphologically two-winged insects, including evolutionary considerations on insect flight. *Journal of Experimental Biology*, **182**, 11–40.

Gronenberg, W. & Strausfeld, N.J. (1990). Descending neurons supplying the neck and flight motor of Diptera: Physiological and anatomical characteristics. *Journal of Comparative Neurology*, **302**, 973–91.

Gronenberg, W. & Strausfeld, N.J. (1991). Descending pathways connecting the male-specific visual system of flies to the neck and flight motor. *Journal of Comparative Physiology* A, **169**, 413–26.

Hammond, P.M. (1979). Wing-folding mechanisms in beetles, with special reference to investigations of adephagan phylogeny (Coleoptera). In *Carabid Beetles, Their Evolution, Natural History and Classification*, ed. T.L. Erwin, G.E. Ball, D.R. Whitehead & A. Halpern, pp. 113–80. The Hague: Junk.

Heide, G. (1971). Die Funktion der nicht-fibrillären Flugmuskeln von *Calliphora* Teil I. Lage, Insertionsstellen und Innervierungsmuster der Muskeln. *Zoologische Jahrbucher (Physiologie)*, **76**, 87–98.

Jeannel, R. (1949). Ordre des Coleoptèröides. In *Traité de Zoologie*, vol. 9, ed. P.-P. Grassé, pp. 771–1077. Paris: Masson et Cie.

Kammer, A.E. (1985). Flying. In *Comprehensive Insect Physiology, Biochemistry and Pharmacology*, vol. 5, ed. G.A. Kerkut & L.I. Gilbert, pp. 491–552. Oxford: Pergamon Press.

Kondoh, Y. & Obara, Y. (1982). Anatomy of motoneurones innervating mesothoracic indirect flight muscles in the silkmoth, *Bombyx mori*. *Journal of Experimental Biology*, **98**, 23–37.

Krämer, K. & Markl, H. (1978). Flight inhibition on ground contact in the American cockroach, *Periplaneta americana* – I. Contact receptors and a model for their central connections. *Journal of Insect Physiology*, **24**, 577–86.

Kutsch, W. & Schneider, H. (1987). Histological characterization of neurones innervating functionally different muscles of *Locusta*. *Journal of Comparative Neurology*, **261**, 515–28.

May, M.L. (1981). Wingstroke frequency of dragonflies (Odonata: Anisoptera). in relation to temperature and body size. *Journal of Comparative Physiology*, **144**, 229–40.

Miyan, J.A. & Ewing, A.W. (1985a). How Diptera move their wings: a re-examination of the wing base articulation and muscle systems concerned with flight. *Philosophical Transactions of the Royal Society of London*, **311**, 271–302.

Miyan, J.A. & Ewing, A.W. (1985b). Is the 'click' mechanism of dipteran flight an artefact of CCl_4 anaesthesia. *Journal of Experimental Biology*, **116**, 313–22.

Miyan, J.A. & Ewing, A.W. (1988). Further observations on dipteran flight: details of the mechanism. *Journal of Experimental Biology*, **136**, 229–41.

Nachtigall, W. (1966). Die Kinematic der Schlagflügelbewegungen von Dipteren. Methodische und analytische Grundlagen zur Biophysik des Insektenfluges. *Zeitschrift für Vergleichende Physiologie*, **52**, 155–211.

Nachtigall, W. (1967). Aerodynamische Messungen am Tragflügelsystem segelnder Schmetterlinge. *Zeitschrift für Vergleichende Physiologie*, **54**, 210–31.

Nachtigall, W. (1976). Wing movements and the generation of aerodynamic forces by some medium-sized insects. *Symposia of the Royal Entomological Society of London*, **7**, 31–47.

Nalbach, G. (1989). The gear change mechanism of the blowfly (*Calliphora erythrocephala*) in tethered flight. *Journal of Comparative Physiology* A, **165**, 321–31.

Nalbach, G. (1993). The halteres of the blowfly *Calliphora* I. Kinematics and dynamics. *Journal of Comparative Physiology* A, **173**, 293–300.

Neville, A.C. (1965). Energy and economy in insect flight. *Science Progress, London*, **53**, 203–20.

Norberg, R.A. (1972). The pterostigma of insect wings an inertial regulator of wing pitch. *Journal of Comparative Physiology*, **81**, 9–22.

Oertli, J.J. (1989). Relationship of wing beat frequency and temperature during take-off flight in temperate-zone beetles. *Journal of Experimental Biology*, **145**, 321–38.

Oldroyd, H. (1949). Diptera. I. Introduction and key to families. *Handbooks for the Identification of British Insects*, 9, part 1.

Orchard, I, Ramirez, J.-M. & Lange, A.B. (1993). A multifunctional role for octopamine in locust flight. *Annual Review of Entomology*, **38**, 227–49.

Pearson, K.G. & Ramirez, J.M. (1990). Influence of input from the forewing stretch receptors on motorneurones in flying locusts. *Journal of Experimental Biology*, **151**, 317–40.

Pesson, P. (1951). Ordre des Thysanoptera. In *Traité de Zoologie*, vol. 10, ed. P.-P.Grassé, pp. 1805–69. Paris: Masson et Cie.

Pfau, H.K. (1987). Critical comments on a 'novel mechanical model of dipteran flight' (Miyan & Ewing, 1985). *Journal of Experimental Biology*, **128**, 463–8.

Pringle, J.W.S. (1948). The gyroscopic mechanism of the halteres of Diptera. *Philosophical Transactions of The Royal Society of London* B, **233**, 347–84.

Pringle, J.W.S. (1957). *Insect Flight*. Cambridge: Cambridge University Press.

Rheuben, M.B. & Kammer, A.E. (1987). Structure and innervation of the third axillary muscle of *Manduca* relative to its role in turning flight. *Journal of Experimental Biology*, **131**, 373–402.

Rind, F.C. (1983). The organization of flight motoneurones in the moth, *Manduca sexta*. *Journal of Experimental Biology*, **102**, 239–51.

Robertson, R.M. & Pearson, K.G. (1985). Neural circuits in the flight system of the locust. *Journal of Neurophysiology*, **53**, 110–28.

Rowell, C.H.F. (1989). Descending interneurones of the locust reporting deviations from flight course: what is their role in steering? *Journal of Experimental Biology*, **146**, 177–94.

Rowell, C.H.F. & Reichert, H. (1985). Compensatory steering in locusts: the integration of non-phase locked input with a rhythmic motor output. In *Insect Locomotion*, ed. M. Gewecke & G. Wendler, pp. 175–82. Berlin: Paul Parey.

Rowell, C.H.F. & Reichert, H. (1991). Mesothoracic interneurons involved in flight steering in the locust. *Tissue & Cell*, **23**, 75–139.

Séguy, E. (1973). L'aile des insectes. In *Traité de Zoologie*, vol. 8, part 1, ed. P.-P.Grassé, pp. 595–702. Paris: Masson et Cie.

Simmons, P. (1977). The neuronal control of dragonfly flight I. Anatomy. *Journal of Experimental Biology*, **71**, 123–40.

Snodgrass, R.E. (1935). *Principles of Insect Morphology*. New York: McGraw-Hill.

Taylor, C.P. (1981). Contribution of compound eyes and ocelli to steering of locusts in flight. *Journal of Experimental Biology*, **93**, 1–18.

Tillyard, R.J. (1918). The panorpoid complex. I. The wing-coupling apparatus, with special reference to the Lepidoptera. *Proceedings of the Linnean Society of New South Wales*, **43**, 286–319.

Trimarchi, J.R. & Schneiderman, A.M. (1994). The motor neurons innervating the direct flight muscles of *Drosophila melanogaster* are morphologically specialized. *Journal of Comparative Neurobiology*, **340**, 427–43.

Uvarov, B.P. (1966). *Grasshoppers and Locusts*, vol. 1. Cambridge: Cambridge University Press.

Ward, J.P., Candy, D.J. & Smith, S.N. (1982). Lipid storage and storage during flight by triatomine bugs (*Rhodnius prolixus* and *Triatoma infestans*). *Journal of Insect Physiology*, **28**, 527–34.

Weis-Fogh, T. (1956). Biology and physics of locust flight. II. Flight performance of the desert locust (*Schistocerca gregaria*). *Philosophical Transactions of the Royal Society of London* B, **239**, 459–510.

Weis-Fogh, T. (1973). Quick estimates of flight fitness in hovering animals, including novel mechanisms for lift production. *Journal of Experimental Biology*, **59**, 169–230.

Wendler, G., Müller, M. & Dombrowski, U. (1993). The activity of pleurodorsal muscles during flight and at rest in the moth *Manduca sexta* (L..). *Journal of Comparative Physiology*, **173**, 65–75.

Wheeler, C.H. (1989). Mobilization and transport of fuels to the flight muscles. In *Insect Flight*, ed G.J. Goldsworthy & C.H. Wheeler, pp. 273–303. Boca Raton, Florida: CRC Press.

Wilson, D.M. & Weis-Fogh, T. (1962). Patterned activity of co-ordinated motor units, studied in flying locusts. *Journal of Experimental Biology*, **39**, 643–67.

Wolf, H. (1993). The locust tegula: significance for flight rhythm generation, wing movement control and aerodynamic force production. *Journal of Experimental Biology*, **182**, 229–53.

Wolf, H. & Pearson, K.G. (1988). Proprioceptive input patterns elevator activity in the locust flight system. *Journal of Neurophysiology*, **59**, 1831–53.

Wootton, R.J. (1979). Function, homology and terminology in insect wings. *Systematic Entomology*, **4**, 81–93.

Wootton, R.J. (1981). Support and deformity in insect wings. *Journal of Zoology, London*, **193**, 447–68.

Wootton, R.J. (1992). Functional morphology of insect wings. *Annual Review of Entomology*, **37**, 113–40.

Wootton, R.J. (1993). Leading edge section and asymmetric twisting in the wings of flying butterflies (Insecta, Papilionoidea). *Journal of Experimental Biology*, **180**, 105–17.

Wootton, R.J. (1996). Functional wing morphology in Hemiptera. In *Hemipteran Phylogeny*, ed. C.W. Schaefer. Thomas Say Symposium of the Entomological Society of America.

Wootton, R.J. & Ennos, A.R. (1989). The implications of functions on the origin and homologies of the dipterous wing. *Systematic Entomology*, **14**, 507–20.

Zanker, J.M. (1990). The wing beat of *Drosophila melanogaster* I. Kinematics. *Philosophical Transactions of the Royal Society of London* B, **327**, 1–18.

Ziegler, R. & Schulz, M. (1986). Regulation of lipid metabolism during flight in *Manduca sexta*. *Journal of Insect Physiology*, **32**, 903–8.

10 Muscles

10.1 STRUCTURE

10.1.1 Basic muscle structure

Each muscle is made up of a number of fibers, which are long, usually multinucleate cells running the whole length of the muscle. The characteristic feature of muscle fibers is the presence of myofibrils (fibrils) which are embedded in the cytoplasm (sarcoplasm) and extend continuously from one end of the fiber to the other. The fibrils in their turn are composed of molecular filaments consisting mainly of two proteins: myosin and actin. These filaments are much shorter than the whole muscle and they are arranged in units called sarcomeres. Each fibril comprises a large number of sarcomeres stacked end to end (Fig. 10.1a,b).

The thick (myosin) filaments are stouter and are made up of numerous myosin molecules. These are elongate structures with two globular 'heads' at one end, and, in each sarcomere (see below), all the molecules in one half are aligned in one direction, while all those in the opposite half are aligned in the opposite direction (Fig. 10.1c). The myosin molecules are probably arranged round a core of another protein, paramyosin, with their heads arranged in a helix.

The thick filaments are each surrounded by a number of thin (actin) filaments which consist of two chains of actin molecules twisted round each other. The actin filaments are oriented in opposite directions at either end of the sarcomere and those of adjacent sarcomeres slightly overlap each other and are held by an amorphous material. All the filaments in a fiber are aligned with each other so that the joints between the ends of the actin filaments form a distinct line, known as the Z-line or Z-disc, running across the whole fiber.

On either side of each Z-line, actin filaments extend towards, but do not reach, the center of the sarcomere. The myosin filaments do not normally reach the Z-lines, although there is some controversy concerning the presence of a connection with the Z-lines in fibrillar muscle.

As a result of these arrangements and the alignment of the components across a whole fiber, the muscles have a banded appearance when viewed under phase contrast or in stained preparations. Each sarcomere has a darkly staining band extending for most of its length with a lightly staining band at each end; the darker region is due to the myosin filaments and the lighter regions occur where they are absent. These are known respectively as the anisotropic, A and isotropic, I bands. In the center of the A band, where actin filaments are absent, is the rather paler H band (Fig. 10.1a–c). Other bands may also be present, and changes occur when the muscle contracts.

The 'heads' of the myosin molecules form cross-bridges which provide structural and mechanical continuity along the whole length of the muscle fiber. Further proteins, tropomyosin and troponin, are also present in small quantities in the contractile elements, and another, called flightin, occurs in the asynchronous flight muscles of *Drosophila* (Vigoreaux *et al.*, 1993).

In skeletal muscles, other than flight muscles, and in visceral muscle, each thick (myosin) filament is surrounded by 12 actin filaments, with a ratio of thin to thick filaments of 6:1 (Fig. 10.1f). In flight muscles, however, whether these are synchronous or asynchronous (see below), six thin filaments surround each thick one, with a ratio of 3:1 (Fig. 10.1e) (Smith, 1984).

Each fiber is bounded by the sarcolemma comprising the plasma membrane of the cell plus the basal lamina. The cytoplasm of the fiber is called sarcoplasm and the endoplasmic reticulum, which is not connected to the plasma membrane, is known as the sarcoplasmic reticulum. The plasma membrane is deeply invaginated into the fiber, often as regular radial canals between the Z-discs and the H bands (see below). This system of invaginations is called the transverse tubular, or T, system. It is extensive; for example, in the moth, *Philosamia*, about 70% of the muscle plasma membrane is invaginated within the fiber. The T-system is associated with vesicles of the sarcoplasmic reticulum (Fig. 10.1d). When the two systems are very close, the space between their membranes is occupied by electron-dense material and the arrangement is called a dyad. The nuclei occur in different positions in the cell in

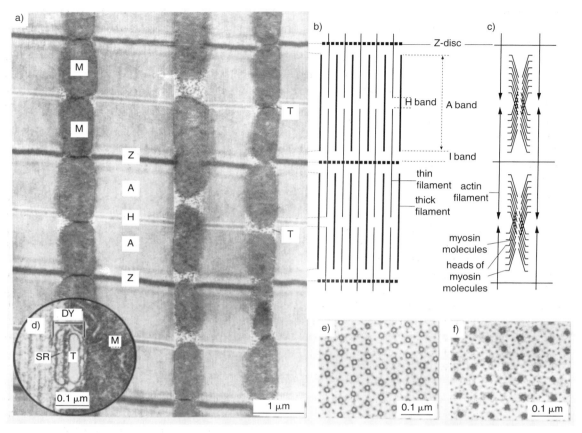

Fig. 10.1. Basic structure of a muscle. (a) Electron micrograph of a longitudinal section of part of muscle (asynchronous flight muscle of the wasp, *Polistes*). Abbreviations: A = A band, H = H band, I = I band, M = mitochondrion, T = transverse tubule, Z = Z-disc (after Smith, 1968). (b) Diagram showing the arrangement of the filaments that produces the banding pattern seen in (a). The filaments are aligned with the bands of two sarcomeres in (a). (c) Diagrammatic representation of the orientation of the actin (thin) and myosin (thick) molecules in a muscle. (d) Electron micrograph of a dyad (DY). Abbreviations: M = mitochondrion, SR = sarcoplasmic reticulum, T = transverse tubule (from the flight muscle of a dragonfly, *Celithemis*) (after Smith, 1966). (e) Electron micrograph of a transverse section of a flight muscle with a ratio of thin:thick filaments of 3:1 (after Smith, 1972). (f) Electron micrograph of a transverse section of an intersegmental muscle with a ratio of thin:thick filaments of 6:1 (after Smith, 1972).

different types of muscle. The arrangement of the fibrils within the fiber varies (see below), but they are always in close contact with the mitochondria, which are sometimes known as sarcosomes.

A single muscle is made up of a number of fibers, sometimes a very large number. For example, the metathoracic dorsal longitudinal muscle of the adult grasshopper, *Schistocerca*, is made up of over 3000 fibers, and even in a fourth stage larva there are over 500 fibers in this muscle. A tergocoxal muscle in the same insect contains about 50 fibers in the fourth stage larva and about 400 fibers in the

adult (Mizisin & Ready, 1986). Even in muscles not concerned with flight, the number of fibers may be large. For example, in the coxal depressor muscle of the cockroach, *Periplaneta*, there are about 765 fibers in the fifth (out of ten) larval stage. This number only increases to about 870 in the adult (Nüesch, 1985). In contrast to these numbers, the number of fibers in asynchronous flight muscles is small (see below).

The muscle fibers are collected into units separated from neighboring units by a tracheolated membrane, and each muscle consists of one or a few such units. For example, there are five units in the dorsal longitudinal flight muscles of grasshoppers, and three in the tergocoxal muscle referred to above. Each muscle unit may have its own independent nerve supply, and is thus the basic contracting unit of the muscle, but in other cases several muscle units have a common innervation and so function together as the motor unit.

Reviews: Aidley, 1985 – contractile machinery; Maruyama, 1985 – muscle biochemistry

10.1.2 Variations in structure

Two broad categories of muscle can be distinguished: skeletal muscles and visceral muscles. Skeletal muscles are attached at either end to the cuticle and move one part of the skeleton relative to another. Visceral muscles move the viscera and have only one or, commonly, no attachment to the body wall. Many form circular muscles around the gut and ducts of the reproductive system.

Skeletal muscles can be differentiated functionally into synchronous and asynchronous muscles. Most skeletal muscles are synchronous muscles: that is, they exhibit a direct relationship between motor neuron activity and contraction (see section 10.5.2). Asynchronous muscles, which do not have this direct relationship, only occur in the flight muscles of Thysanoptera, Psocoptera, Homoptera, Heteroptera, Hymenoptera, Coleoptera and Diptera and in the tymbal muscles of some Cicadidae.

10.1.2.1 *Synchronous skeletal muscles*

The form and arrangement of the fibrils in synchronous muscles is very variable. Tubular muscles have the myofibrils arranged radially round a central core of cytoplasm containing the nuclei. This arrangement is common in leg and trunk muscles and also occurs in the flight muscles of Odonata and Blattodea (Fig. 10.2a). In close-packed muscles, on the other hand, the fibrils are only $0.5–1.0\,\mu m$ in diameter and are packed throughout the whole fiber; the nuclei are flattened and peripheral. Fibers of this type occur in some larval insects, in Apterygota and in the flight muscles of Orthoptera, Trichoptera and Lepidoptera (Fig. 10.2b).

The abundance and arrangement of mitochondria is related to the level and type of activity of the muscles. In the tubular muscles of Odonata and the close-packed flight muscles of Orthoptera they are large and numerous, occupying about 40 % of the fiber volume (Fig. 10.3a). This is also true of the muscles which oscillate at high frequencies to produce the sounds of cicadas and bush crickets (Fig.

10.3c). In muscles which do not oscillate rapidly the mitochondria generally occupy a much smaller proportion of the fiber volume (Fig. 10.3e–h) making it possible for a larger proportion of the fiber to be occupied by the contractile elements. The mitochondria may be in pairs on either side of a Z-line or scattered irregularly between the fibers. Fibers with abundant mitochondria may be colored pink by the high cytochrome content.

The development of the sarcoplasmic reticulum is correlated with the mechanical properties of the muscle, and in particular with the rates of relaxation of fibers. In muscles which tend to maintain a sustained contraction, such as the locust extensor tibiae muscle (Fig. 10.3e) and the accessory flight muscles of Diptera, it is relatively poorly developed. On the other hand, in fast-contracting synchronous muscles, such as those associated with the sound-producing apparatus of male *Neoconocephalus* (Orthoptera) and cicadas, the sarcoplasmic reticulum comprises 15% or more of the total fiber volume (Fig. 10.3c). A characteristic of synchronous flight muscles, whether they are tubular or close-packed, is that the distance from the sarcoplasmic reticulum to the myofibrils is short, generally less than $0.5\,\mu m$ and even less in very fast-contracting muscle. In moths with a high wingbeat frequency, none of the myosin filaments is more than about 10 nm from the nearest element of the sarcoplasmic reticulum (Elder, 1975). This close proximity facilitates the rapid release and resequestration of calcium ions during cycles of contraction and relaxation (section 10.3.2).

In synchronous muscles, the invaginations of the T-system occur midway between the Z-disc and the center of the sarcomere. Sarcomere length in many synchronous muscles ranges from about $3\,\mu m$ to $9\,\mu m$, but is usually only about $3–4\,\mu m$ in flight muscles. The I band usually constitutes 30–50% of the resting length of the sarcomere, but *in situ* the extent of muscle shortening may be much less than this. For instance, the I bands of locust flight muscle constitute about 20% of the sarcomere length, but, during flight, the muscle shortens by only about 5%. Fiber diameter is commonly greater in close-packed muscle than in tubular muscle; up to $100\,\mu m$ in the former compared with $10–30\,\mu m$ in the latter.

Structurally and functionally, synchronous skeletal muscles form a continuum with slow, or tonic fibers at one extreme and fast fibers at the other. Slow fibers have little sarcoplasmic reticulum, the volume occupied by mitochondria is relatively large (Fig. 10.3g), and the ratio of thin to thick filaments is high (6:1). The filaments occupy

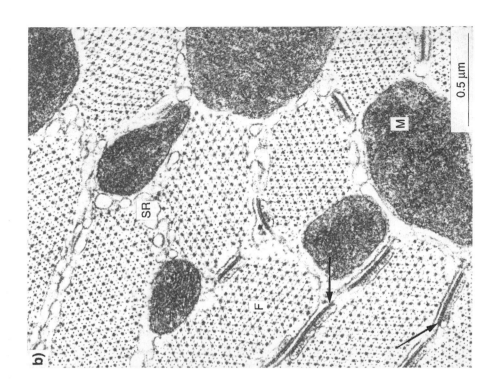

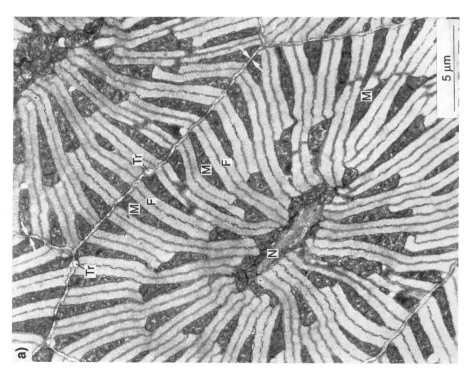

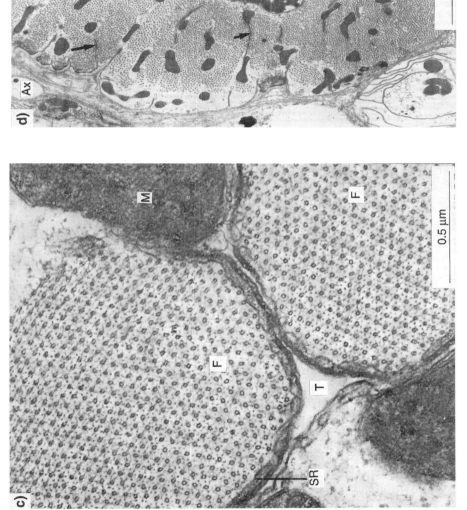

Fig. 10.2. Muscle types. Electron micrographs of transverse sections of different muscles. Abbreviations: Ax = axon, BL = basal lamina, F = myofibril, M = mitochondrion, N = nucleus, SR = sarcoplasmic reticulum, T = transverse tubule, Tr = trachea. **(a)** Part of a tubular muscle with radial arrangement of myofibrils within each fiber. Opposing white arrows indicate intercellular space between fibers (flight muscle of dragonfly, *Enallagma*) (after Smith, 1968). **(b)** Part of a close-packed muscle fiber. Arrows point to dyads (after Smith, 1984). **(c)** Fibrillar muscle showing parts of two myofibrils. Notice their large size (after Smith, 1984). **(d)** One fiber of *visceral* muscle. The myofilaments are not grouped into myofibrils. Arrows point to dyads (spermatheca of cockroach, *Periplaneta*) (after Smith, 1968).

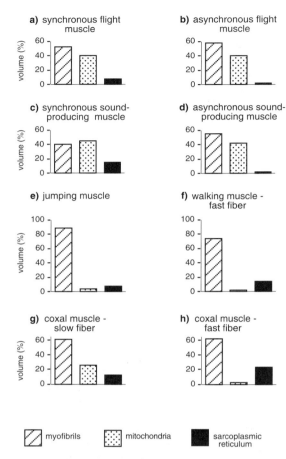

Fig. 10.3. Proportions of the muscle volume occupied by the different major components, myofibrils, mitochondria and sarcoplasmic reticulum, in different types of muscle: (a) synchronous flight muscle; (b) asynchronous flight muscle (c) synchronous sound-producing muscle; (d) asynchronous sound-producing muscle; (e) jumping muscle (extensor tibiae of locust); (f) skeletal muscle with fast fibers; (g) skeletal muscle with slow fibers; (h) skeletal muscle with fast fibers.

the greater part of the fibers and may not be grouped into discrete myofibrils, and the sarcomeres are long. Fast fibers have the opposite characteristics: extensive sarcoplasmic reticulum, a small volume of mitochondria (Fig. 10.3f,h), low ratios (3:1) of thin to thick filaments, and short sarcomeres. Intermediate fibers have intermediate characteristics (Cochrane, Elder & Usherwood, 1972; Hoyle, 1978b).

Some muscles have a uniform complement of fibers. For example, the posterior coxal depressor muscle of

Periplaneta (muscle 136 in Fig. 8.6) consists entirely of fast fibers. Others include more than one type of fiber. Muscle 135d′ in Fig. 8.6, for example, has a bundle of about 250 fast fibers ventrally and almost 700 slow fibers dorsally. The extensor tibiae muscle of the locust hind leg contains different fiber types mixed in various proportions in different parts of the muscle (Fig. 10.4, Table 10.1).

To a large extent, the anatomical characteristics of the fibers are reflected in the types of innervation they receive, with fast fibers being innervated by fast axons and slow fibers by slow axons (see below, Table 10.1) (Hoyle, 1978b; Morgan & Stokes, 1979).

10.1.2.2 *Asynchronous skeletal muscles*

Asynchronous muscles are characterized by the large size of the fibrils, up to 5 μm in diameter, with a corresponding increase in the diameters of the fibers, which range from 30 μm in carabid beetles to 1.8 mm in *Rutilia* (Diptera). Because the fibrils are so conspicuous, muscles with this characteristic are sometimes called fibrillar muscles. The fibrils, with nuclei scattered between them, are distributed through the entire cross-section of the fiber (Fig. 10.2c).

Asynchronous muscles may contain only a few fibers because these are so big. For example, the dorsal longitudinal flight muscles of Muscidae consist of only six fibers. Further, sarcomere length is short, only one or two microns in *Tenebrio*, and the I band makes up less than 10% of this. In some cases, the myosin filaments taper towards the Z-line and may be attached to it, so that there is no distinct I band.

The plasma membrane is invaginated in a T-system as in other muscles, but the positions of the invaginations are variable. In the wasp, *Polistes* (Hymenoptera), the T-tubules are aligned with the H band, but in the fibrillar muscles of *Tenebrio* (Coleoptera) and *Megoura* (Homoptera) the system is more complex and less regular. In these insects invaginations of the plasma membrane are produced by indenting tracheoles and, from these invaginations, fine tubules of plasma membrane extend in to entwine each fibril. Associated with the T-system are vesicles of the sarcoplasmic reticulum, but this differs markedly in its development from the sarcoplasmic reticulum of other muscles, consisting only of a number of unconnected vesicles scattered without reference to sarcomere pattern. It occupies only a very small proportion of the fiber volume (Fig. 10.3b,d).

Mitochondria are large, as in all flight muscles, and occupy 30–40% of the fiber volume. Almost the whole

Table 10.1. *Fiber types in the extensor tibiae muscle of the locust. The percentages of muscle fibers having different innervation and ultrastructure*

Axons innervating*	Structural type of muscle fibers	Region of location in extensor tibiae (see Fig. 10.4)	% of total number of fibers in the muscle
F	Fast	a, b, c, d, e, f	68
F D	Fast	b, c, d	6
F S	Intermediate	a, d, e, f	3
F S I	Intermediate	a, c, d, e, f	12
F D S	Fast	a, d	0.5
F I	Fast	a, d, e, f	2
S I	Slow	a, f, I35c, I35d	8
S	Slow	a	0.5

Notes:

After Hoyle (1978b).

* F, fast axon.

  S, slow axon.

  D, octopamine axon.

  I, inhibitory axon.

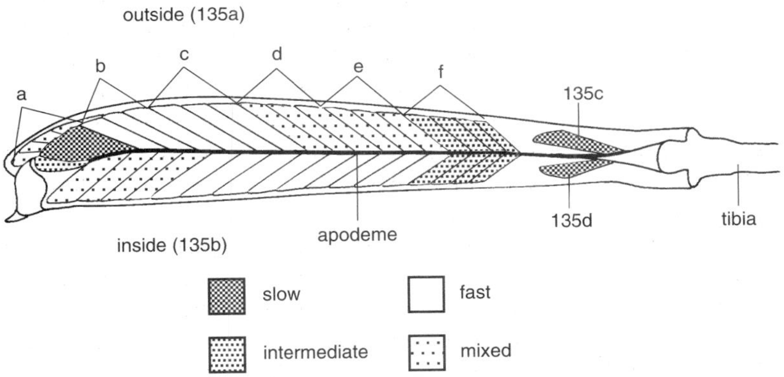

Fig. 10.4. Arrangement of fibers with different properties in the extensor tibiae muscle of a locust. Shading shows the dominant type of fiber in each region. Lettering indicates the type of innervation (see Table 10.1) and numbers indicate the anatomically distinct parts of the muscle (after Hoyle, 1978b).

surface of each myofibril can be in direct contact with mitochondria. These may be regularly arranged, as between the Z-discs and H bands in *Polistes*, or without any regular arrangement, as in *Calliphora*.

10.1.2.3 *Visceral muscles*

Visceral muscles differ in structure from skeletal muscles in several respects. Adjacent fibers are held together by desmosomes, which are absent from skeletal muscle, and in some cases the fibers may branch and anastomose. Further, each fiber is uninucleate and the contractile material is not grouped into fibrils but packs the whole fiber (Fig. 10.2d).

As in other muscles, it consists of thick and thin filaments, presumably representing myosin and actin, often with a ring of ten to twelve actin filaments round each myosin filament (Fig. 10.1f). A T-system with a regular arrangement is present in *Periplaneta*, but in *Carausius* and *Ephestia* (Lepidoptera) it is irregularly disposed. The sarcoplasmic reticulum is poorly developed; mitochondria are small and often few in number (Miller, 1975).

The muscles appear striated due to the alignment of the filaments. They therefore resemble skeletal muscle but contrast with the visceral muscle of vertebrates, which is not striated. The Z-discs and H bands are irregular and

Fig. 10.5. Attachment of a muscle fiber to the integument (after Caveney, 1969).

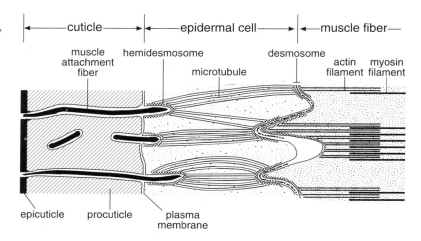

sarcomere length is very variable. In cardiac muscle, the sarcomeres are short, about 3 μm, but they may be as long as 10 μm in other visceral muscles.

Visceral muscles may be innervated from the stomodeal nervous system or from the ganglia of the ventral nerve cord, but are sometimes without innervation, as in the heart of *Anopheles* (Diptera) larvae.

Reviews: Aidley, 1985; Elder, 1975; Huddart, 1985 – visceral muscle; Smith, 1984 – ultrastructure

10.1.3 Muscle insertion

Skeletal muscles are fixed at either end to the integument, spanning a joint in the skeleton so that contraction of the muscle moves one part of the skeleton relative to the other. Typically such muscles are said to have an origin in a fixed or more proximal part of the skeleton, and an insertion into a distal, movable part, but these terms become purely relative where muscles have a dual function (section 9.7.1). In many cases, muscles are attached to invaginations of the cuticle called apodemes.

At the point of attachment to the epidermis, the plasma membranes of muscle and epidermal cells interdigitate and are held together by desmosomes (Fig. 10.5). Within the epidermal cell, microtubules run from the desmosomes to hemidesmosomes on the outer plasma membrane, and, from each hemidesmosome, a dense attachment fiber passes to the epicuticle through a pore canal. In earlier studies, microtubules and attachment fibers were not recognized as separate structures and were termed tonofibrillae. In most muscles, actin filaments reach the terminal plasma membrane of the muscle fiber,

inserting into the dense material of desmosomes. In asynchronous flight muscle, the terminal region of each myofibril consists of a dense body of microfibers, perhaps made up of extended actin filaments. Each of these regions is inserted on to an extension of an epidermal cell containing microtubules as in other muscle attachments (Smith, 1984).

The muscle attachment fibers in the cuticle are not digested by molting fluid so that during molting they retain their attachment to the old cuticle across the exuvial space between the new and old cuticles. As a result, the insect is able to continue its activities after apolysis during the development of the new cuticle. The connections to the old cuticle are broken at about the time of ecdysis (Lai-Fook, 1967).

Muscle attachment fibers extending to the epicuticle can only be produced at a molt and most muscles appear to form their attachments at this time. Muscle attachment can occur later, however, if cuticle production continues in the postecdysial period, but in this case the attachment fibers are only connected to the newly formed procuticle and do not reach the epicuticle (Hinton, 1963).

10.1.4 Oxygen supply

Muscular contraction requires metabolic energy, and the muscles have a good tracheal supply. This is particularly true of the flight muscles, where the respiratory system is specialized to maintain the supply of oxygen to the muscles during flight (section 17.1.3). In most muscles, the tracheoles are in close contact with the outside of the muscle fiber. This arrangement provides an adequate supply of oxygen to relatively small muscles or those whose oxygen demands are not high, but in the flight muscles of all insects, except perhaps those of Odonata and Blattodea, the tracheoles indent the muscle membrane becoming functionally, but not anatomically, intracellular within the muscle fiber (see Fig. 17.1). The tracheoles follow the invaginations of the T-system and so penetrate to the center of the fibers. Fine tracheoles, less that 200 nm in diameter, branch off from the tracheoles forming the secondary supply to the muscles. They come very close to the mitochondria, and probably every mitochondrion in the flight muscles is supplied by one or two tracheoles of this tertiary system so that the distance that oxygen has to pass through the tissue is reduced to a minimum (Wigglesworth & Lee, 1982). In Odonata and blattids, on the other hand, tracheoles remain superficial to the muscle fibers and these fibers, with a radius of about 10 μm, are believed to approach the limiting size for the diffusion of oxygen at sufficiently high rates. Wigglesworth & Lee (1982), however, suggest that the flight muscles of these insects may also have an invaginating system of terminal tracheoles.

Locust flight muscles use some $80 \, \text{liters} \, O_2 \, \text{kg}^{-1} \, \text{h}^{-1}$ during flight and consumption by the flight muscles of some other insects may exceed $400 \, \text{liters} \, O_2 \, \text{kg}^{-1} \, \text{h}^{-1}$. The special adaptations of the thoracic tracheal system enable these demands to be met. In locusts, pterothoracic ventilation produces a supply of oxygen well in excess of needs, and the specialized system of tracheae and tracheoles in the muscles ensures that oxygen reaches the site of consumption. The fine branches of the tertiary tracheal supply to the muscles may be fluid-filled when the insect is not flying, but, in flight, the fluid is withdrawn so that air extends to the very tips of the branches close to the mitochondria. The cuticle lining these fine branches is believed to be very permeable (Wigglesworth & Lee, 1982).

10.1.5 Innervation

Characteristically in insects, each axon innervating a muscle has many nerve endings spaced at intervals of 30–80 μm along each fiber (Fig. 10.6). This is called multiterminal innervation. The form of the nerve ending varies. The fine nerve branches in flight muscles of Diptera run longitudinally over the surface of the muscle; in the flight muscles of the flour beetle (*Tenebrio*), they are completely invaginated into the muscle. In Orthoptera, the axon divides at the surface of the muscle and the branches, with their sheathing glial cells, form a claw-like structure.

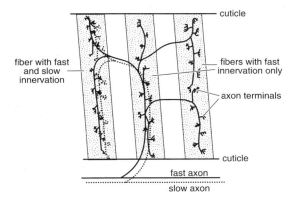

Fig. 10.6. Multiterminal and polyneuronal innervation. Diagram illustrating the innervation of three fibers of a muscle unit. All three receive branches of the fast axon, while one (on the left) also has endings from the slow axon (after Hoyle, 1974).

The terminal branches of the axons are often expanded into a series of swellings, or boutons, on the muscle surface. Different axons, with different physiological properties (see below) may have boutons of different sizes even on the same muscle fiber. The boutons contain the synapses and there are more in larger boutons (Atwood & Cooper, 1995). At the neuromuscular junction, glial cells are absent so that the plasma membranes of nerve and muscle lie close together, separated by a synaptic gap of about 30 nm (Fig. 10.7). The terminal axoplasm contains synaptic vesicles, comparable with those in the presynaptic terminal of a synapse in the central nervous system. They vary in diameter from 20 to 60 nm.

In muscles with a single, discrete function, such as some of the indirect flight muscles, the fibers are innervated by a single axon, but it is common for a single muscle fiber to be innervated by more than one motor axon. Such multiple innervation is called polyneuronal and it allows the muscle to function more variably. Some examples of flight muscles in which at least some fibers are innervated by more than one axon are given in Fig 9.16 and table 9.1. The locust extensor tibiae muscle is an example of a muscle made of a large number of fibers differing in their innervation. Because this muscle is of primary importance in jumping, 68% of the fibers are innervated only by a single axon, the 'fast' axon (see below) (Table 10.1). Most other fibers are also innervated by this axon, but also receive inputs from one or two others. A small number, about 8%, of the fibers are not innervated by the 'fast'

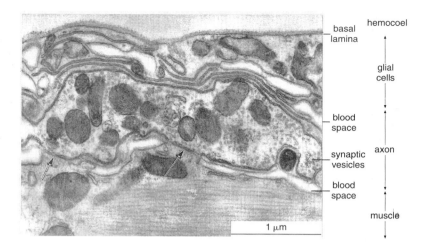

Labels on figure: hemocoel, basal lamina, glial cells, blood space, synaptic vesicles, axon, blood space, muscle, 1 μm

axon, but by a separate 'slow' axon, usually accompanied by a second axon. The additional axons may have inhibitory or more subtle modulatory effects on the activity of the muscle (see below).

While it is usual for the different neurons innervating a single fiber to have qualitatively or quantitatively different effects, there are examples where this is not so. The transverse sternal muscles in the abdomen of the bushcricket, *Decticus*, receive input from three excitatory neurons with similar properties (see below) and one of these muscles in the cricket, *Gryllus*, is innervated by four neurons with similar effects (Consoulas *et al.*, 1993).

Different units of a muscle sometimes serve different functions and in this case they have completely separate nerve supplies. Thus, the posterior part of the basalar muscle of the beetle, *Oryctes*, is concerned only with wing depression and has only a fast innervation, but the anterior part also controls wing twisting and its innervation is complex, consisting of up to four axons, one of which is inhibitory.

The cell bodies of the motor neurons controlling skeletal muscles are usually in the ganglion of the segment in which the muscle occurs. Sometimes, however, a muscle is innervated by a motor neuron with its cell body in a different ganglion. An extreme example of this occurs in the abdominal transverse sternal muscles of crickets and bushcrickets. These muscles are continuous across the midline, but the nerve supply of each side is separate. In the bushcricket, *Decticus*, each side of the muscles in abdominal segments 3 to 7 is innervated by excitatory

neurons from three ganglia. In addition, each side of each muscle receives input from an inhibitory cell arising in the ganglion of its own segment (Fig. 10.8) (Consoulas *et al.*, 1993)

Sometimes a single neuron innervates more than one muscle. This is most likely to occur where different muscles commonly act in unison. For example, each of the meso- and metathoracic coxae of *Periplaneta*, has four depressor muscles (Fig. 8.6). A single fast axon innervates all four, although it does not innervate all units in two of the muscles (135d,e). All the units of two muscles (135d,e) are innervated by a single slow axon, and some are also innervated by inhibitory axons, one of which goes to most of the fibers, the distribution of the others being more restricted. The innervation of sternal muscles in *Decticus* and in *Teleogryllus* (Fig. 10.8) is another example of one neuron innervating different muscles, in this case in different body segments. This arrangement does not mean that activity of the muscles is linked in a fixed way, because they often receive additional input from separate inhibitory neurons.

Extreme examples of a neuron innervating more than one muscle are the common inhibitors in the locust meso- and metathoracic ganglia. These neurons innervate the 12 and 13 muscles moving the middle and hind legs, respectively (Fig. 10.9). These include muscles that normally act antagonistically to each other and which have separate excitatory innervation. Two other inhibitory neurons in each segment each innervate four muscles in the femur and tibia (Hale & Burrows, 1985). Another example of a

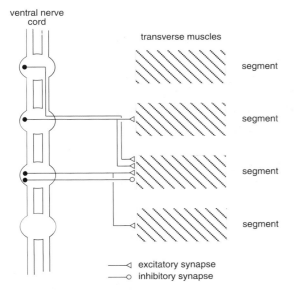

Fig. 10.8. Polyneuronal innervation. Innervation of a transverse abdominal muscle of *Teleogryllus* by neurons with cell bodies in different segmental ganglia. Muscles of one side only shown; muscles on the other side of the body are similarly innervated. The complete innervation of the muscle in segment 6 is shown. Each of the other muscles receives a similar innervation. Some neurons innervate muscles in two segments (after Consoulas *et al.*, 1993).

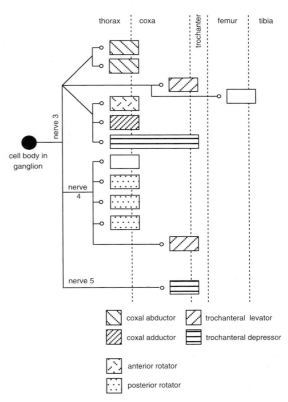

Fig. 10.9. Innervation of many muscles by a single neuron. Muscles innervated by the common inhibitor neuron of the metathoracic leg of a locust. Each rectangle represents a muscle with its origin in the segment to the left and insertion in the segment to the right. Nerves 3, 4 and 5 are nerves emanating from the metathoracic ganglion. The axon of the inhibitor neuron has a branch in each nerve. Note that the neuron innervates muscles that are antagonistic to each other. Unshaded muscles are those whose antagonists are not innervated by the neuron (after Hale & Burrows, 1985).

common inhibitor has its cell body in the brain of the cricket. It innervates six out of seven antennal muscles (Allgäuer & Honegger, 1993).

Many of the muscles of the foregut are innervated by neurons with cell bodies in the ganglia of the stomodeal nervous system.

10.2 CHANGES DURING DEVELOPMENT

Changes occur in many muscles as the insect develops. During the larval stages, muscles generally increase in size. At metamorphosis, larval muscles may be destroyed or modified, or new adult muscles may be developed. Muscles often continue development in the early days of adult life and this is reflected in the insect's inability to undertake extended flights during this period. The period during which an adult insect's ability to fly is not fully developed is called the teneral period. Flight muscles may regress in adults that have completed their flight period. Some muscles may be associated specifically with hatching

or with ecdysis, regressing or completely disappearing after a brief period of use. These changes are considered in this section, but see section 15.3.2.2 for changes occurring at metamorphosis.

10.2.1 Growth during the development of hemimetabolous insects

In hemimetabolous insects, muscles increase in size throughout larval development and the increase continues for the first few days after adult eclosion (Fig. 10.10). Growth during the larval stages results primarily from an

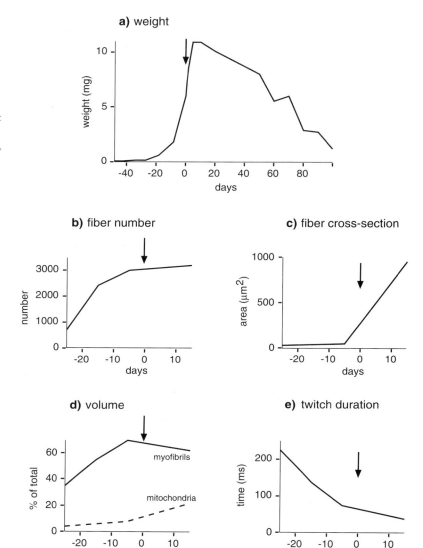

Fig. 10.10. Muscle growth in hemimetabolous insects. Changes in the dorsal longitudinal flight muscle associated with the development of flight behavior. Arrow (day 0) indicates the time of adult eclosion. The period before eclosion includes several larval stadia. (a) Change in wet weight of the muscle in the cricket, *Teleogryllus* (after Ready & Josephson, 1982). (b) Increase in fiber number in the grasshopper, *Schistocerca nitens*. (c) Increase in cross-sectional area of fibers in the grasshopper, *Schistocerca nitens*. (d) Changes in the percentage of muscle occupied by mitochondria and myofibrils in the grasshopper, *Schistocerca nitens*. (e) Decrease in twitch duration in the grasshopper, *Schistocerca nitens*.
Notice that in b–d the final two points represent the mid-point of the last larval stadium and day 15 of adult life. For this reason, they do not show unequivocally that growth occurred after the final molt. Data from other sources, however, show that a marked increase in fiber cross-section and changes in the volume densities of the components do occur in the early adult (after Mizisin & Ready, 1986).

increase in the number of fibers forming the muscles whereas growth after eclosion is produced by an increase in the size of the fibers. This difference is probably a reflection of the fact that new attachments to the cuticle are normally formed at molts (see above). The difference in the number of fibers between early stage larvae and adults is relatively small in muscles which perform the same functions throughout life, but much more marked in the case of flight muscles.

An increase in fiber size involves all the elements of the muscle: myofibrils, mitochondria and sarcoplasmic reticulum (Fig. 10.10c,d), but the relative rates of increase of these components varies from muscle to muscle. In a tergocoxal muscle of the grasshopper, *Schistocerca nitens*, all the components increase in proportion during larval development, but in the metathoracic dorsal longitudinal muscle the percentage of volume occupied by myofibrils approximately doubles during the last three larval stages. In both muscles, the proportion of muscle volume occupied by mitochondria increases during the first two weeks

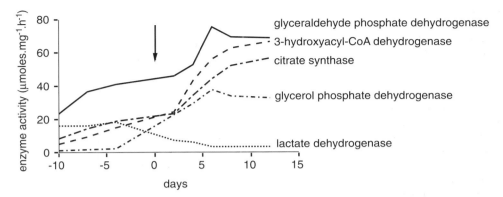

Fig. 10.11. Enzyme activity in the respiratory pathways in the dorsal longitudinal flight muscle of the locust, *Locusta*, in relation to the development of flight behavior. Arrow at day 0 indicates adult eclosion. About seven days after eclosion, the insect is capable of sustained flight. Enzyme activity is expressed as μmoles substrate digested per mg muscle protein per hour. The actual values for lactate dehydrogenase are one-tenth of those shown (after Beenakkers, van den Broek & de Ronde, 1975).

of adult life. Fiber growth in young adults involves an increase in the size of the myofibrils and, in the dorsal longitudinal muscle of *Locusta*, a doubling of the number of filaments in each fibril (Bücher, 1965; Mizisin & Ready, 1986; Ready & Najm, 1985; Ready & Josephson, 1982).

These anatomical changes are sometimes accompanied by changes in enzyme activity (Fig. 10.11). The activity of enzymes involved in flight metabolism increases through the final stages of muscle development, tending to reach a plateau about seven days after the final molt. At the same time, the activity of lactate dehydrogenase declines.

10.2.2 Post-eclosion growth of the flight muscles of holometabolous insects
At the time of eclosion, when the cuticle expands, the flight muscles of cyclorrhaphous flies rapidly increase in length by 25% or more. The change in length is correlated with a comparable increase in the lengths of the sarcomeres due to an increase in the lengths of the filaments. Synthesis of the actin and myosin necessary for this elongation appears to be stimulated by stretching resulting from air swallowing during expansion. The number of filaments within a myofibril does not increase at this time (Houlihan & Breckenridge, 1981). Similar changes in the lengths of flight muscles probably occur in all insects that exhibit a marked increase in body size at eclosion.

An increase in flight muscle volume continues over the days following eclosion in cyclorrhaphous flies and some

Hymenoptera. This growth involves an increase in the sizes of both mitochondria and myofibrils. In the tsetse fly, *Glossina*, for example, post-eclosion changes comparable with those described in Orthoptera occur. In these insects, flight muscle maturation is dependent on the insects taking several blood meals. The mass of the muscles increases considerably, mostly due to an increase in the absolute size and relative proportion of each fiber occupied by the myofibrils (Fig. 10.12). Within each myofibril the number of thick filaments more than doubles. The mitochondria also increase in size (Anderson & Finlayson, 1973). At the same time the wingbeat frequency increases, although this is at least partly due to changes in the cuticle. Comparable changes have been documented in other flies and in the honeybee, *Apis*, where the mitochondria undergo a 12-fold increase in volume during the first three weeks of adult life. In Lepidoptera, however, no post-eclosion changes occur; the flight muscles are fully functional immediately after wing expansion.

10.2.3 Regressive changes in flight muscles
The flight muscles of the reproductive castes of termites and ants regress completely after the nuptial flight, beginning when the wings are shed. It is believed that the products from the muscles contribute to oogenesis in the period before workers are available to feed the queen. Flight muscle degeneration following dealation (casting the wings) also occurs in some crickets (Tanaka, 1993; Walker, 1972).

a) volume

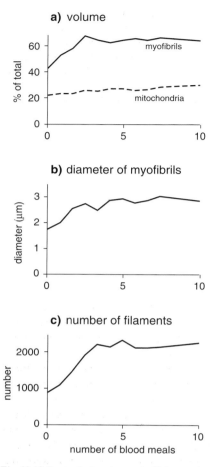

b) diameter of myofibrils

c) number of filaments

number of blood meals

Fig. 10.12. Post-eclosion changes in flight muscles of the tsetse fly, *Glossina*. Muscle development is dependent on the insect having several blood meals (after Anderson & Finlayson, 1976). (a) Percentage of the muscle volume occupied by myofibrils and mitochondria. (b) Diameter of myofibrils. (c) Number of thick filaments per myofibril.

In some other insects, flight muscle histolysis follows a dispersal flight even though the wings are not shed. This is true of some aphid species and other Hemiptera, and in some crickets and bark beetles. Amongst bark beetles, such degeneration may be followed, after a period of reproduction, by redevelopment of the muscles, and further flight (see Kammer & Heinrich, 1978). In the Colorado potato beetle, temporary regression of the flight muscles is associated with reproductive diapause.

In general, muscle histolysis is correlated with, and may be caused by, the increase in juvenile hormone titer

a) ecdysial muscles

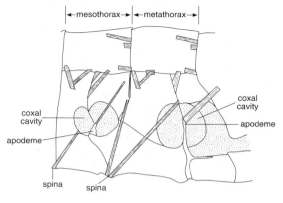

b) permanent muscles

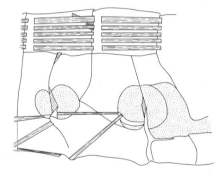

Fig. 10.13. Molting muscles. Muscles in the pterothorax of a locust in the first larval stage. Note that the flight muscles are not well-developed at this time. Membranous areas stippled (after Bernays, 1972). (a) Muscles which disappear after the final molt. (b) Muscles which remain throughout life.

that regulates reproductive development. In *Leptinotarsa*, however, the converse is true. Here, the insects enter diapause, and the flight muscles degenerate, when the juvenile hormone titer is low; they regenerate when it rises. In the aphid, *Acyrthosiphon*, and probably in other insects, breakdown of the indirect flight muscles is a genetically programmed event (Kobayashi & Ishikawa, 1994).

Regression of the flight muscles occurs in older insects of many species. This occurs, for example, in some crickets (Fig. 10.10a) and in mosquitoes (*Aedes*) and *Drosophila*. As the size of the mitochondria declines, so does flight activity (Rockstein and Bhatnagar, 1965; Kammer and Heinrich, 1978).

Review: Finlayson, 1975

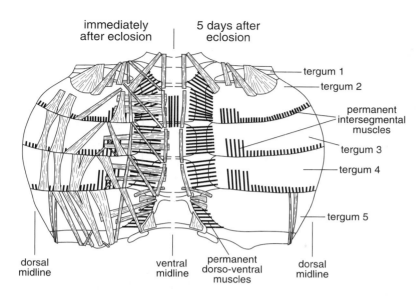

immediately after eclosion | 5 days after eclosion

tergum 1
tergum 2
permanent intersegmental muscles
tergum 3
tergum 4
tergum 5

dorsal midline

ventral midline

permanent dorso-ventral muscles

dorsal midline

Fig. 10.14. Eclosion muscles. The abdominal muscles of a fly, *Sarcophaga*, immediately after eclosion (left) and 5 days later (right). The abdomen has been split along the dorsal midline, laid out flat and viewed from inside (after Cottrell, 1962).

10.2.4 Muscles associated with hatching and molting

Muscles specifically associated with hatching are known to occur in grasshoppers and crickets. Newly hatched grasshopper larvae have a pair of ampullae in the neck membrane which are used when the insect escapes from the egg shell and makes its way to the surface of the soil. Once the insect reaches the surface, the ampullae are no longer used and their associated retractor muscles degenerate. The muscles associated with molting (see below) are probably also active at this time, but they do not degenerate after hatching. In *Rhodnius*, the muscles that develop at each molt (see below) are also present when the insect hatches, but regress shortly afterwards.

Muscles associated with molting are known from several hemimetabolous species and may commonly be present. Their function is to facilitate ecdysis and expansion of the cuticle at molting by increasing the hydraulic pressure of the hemolymph. In grasshoppers, and probably in other insects, some of these muscles are attached to membranous regions of the cuticle and they may prevent excessive ballooning of these areas so that the greatest forces are exerted on the presumptive sclerites, ensuring maximum expansion. Grasshoppers have many of these accessory muscles in the neck, thorax and abdomen which regress after the final molt (Fig. 10.13).

In the blood-sucking bug, *Rhodnius*, some ventral intersegmental muscles in the abdomen are fully differentiated only when the insect molts. Within a few days of molting they regress and lose their contractile function. They redevelop, in readiness for the next molt, when the insect has a blood meal.

Amongst holometabolous insects, muscles that function only at the time of eclosion are known in Diptera and Lepidoptera and they probably also occur in other groups. In the blowfly, *Sarcophaga*, there are special muscles in the abdomen that appear to be involved in escape from the puparium and subsequent cuticle expansion (Fig. 10.14). These muscles are internal to the definitive muscles and extend from the front of one presumptive sclerite to the front end of the next so that when they contract the sclerites buckle because they are not yet hardened, and the abdominal cavity is reduced in volume. These muscles break down when the cuticle is expanded and hardened (Cottrell, 1962).

Breakdown of these muscles involves two processes, loss of contractility and muscle degeneration. The first stage, in Lepidoptera, is triggered by the falling concentration of molting hormone. It may begin before eclosion, but the effect is delayed if the motor neurons controlling the muscles are active (as they normally are at this time). Under these circumstances, breakdown does not occur until expansion of the adult wings and cuticle is complete, and the first signs of degeneration appear about 5 hours after eclosion. In the tobacco hornworm moth, *Manduca*, the absence of molting hormone accounts for the complete disappearance of the muscles in the days after eclosion, but in the silkmoth, *Antheraea*, eclosion hormone provides the signal inducing their final degeneration. In the flesh fly, *Sarcophaga*, eclosion hormone starts a slow degeneration, but some other factor triggers the rapid degeneration of thoracic eclosion muscles. The effect of the hormones is to switch on genetically programmed cell death (Bothe & Rathmayer, 1994; Kimura & Truman, 1990).

Review: Truman, 1985

10.2.5 Effects of activity on muscles

A few examples are known in which muscle size is affected by activity. In *Glossina*, flight following eclosion results in a rapid increase in the mass of the flight muscles. This increase, which occurs more slowly in the absence of induced activity, is due to an increase in the volumes of myofibrils and mitochondria (Anderson & Finlayson, 1976).

The hardness of food has been shown to affect the size of the mandibular adductor muscles in caterpillars of the moth, *Pseudaletia*. It is not known how quickly the change occurs, but insects feeding on harder food develop larger heads and larger muscles than those eating soft foods. Similar changes in head size and, by implication, muscle development also occur in grasshoppers (Bernays, 1986).

Flight to exhaustion (which probably rarely occurs naturally) has been shown to affect mitochondrial structure in the flight muscles of a locust, a wasp and a mosquito. They become swollen and form a continuous mitochondrial 'mass' round the myofibrils sometimes to an extent that discrete mitochondria cannot be distinguished (Johnson & Rowley, 1972).

Regression of muscles that are not used also occurs. This effect is not a direct consequence of lack of use, but is controlled indirectly. For example, regression of the dorsal longitudinal flight muscles of numerous insects following loss of the wings is coincident with, and probably controlled by, an increase in the titer of juvenile hormone in the hemolymph. Grasshoppers have the ability to autotomize the hind leg, a break appearing between the trochanter and the femur. Muscles in the thorax associated with this leg subsequently atrophy. This is the result of severing the nerve to the leg even though this nerve does not innervate the muscles that degenerate (Arbas & Weidner, 1991).

10.3 MUSCLE CONTRACTION

10.3.1 Mechanics of contraction

Muscle contraction results from the thin and thick filaments sliding relative to each other so that each sarcomere is shortened (Fig. 10.15a). It is believed that actin and myosin filaments first become linked together by cross-bridges formed by the heads of myosin molecules and that movement of these links with subsequent breaking and recombination causes the actin filaments to slide further between the myosin filaments so that the sarcomeres, and hence the muscle, shortens.

In most muscles, the extent of contraction appears to be limited by the length of the I band. Shortening can continue only until the end of the thick filaments reach the Z-disc, obliterating the I band. However, some body-wall muscles of blowfly larvae and caterpillars and other larvae with a hydrostatic skeleton are able to supercontract to less than half of their relaxed length. Intrinsic muscles of the gut, heart and reproductive ducts have a similar capacity. In these muscles, the myosin filaments pass through pores in the Z-line and project into the adjacent sarcomeres (Fig. 10.15b). This may be made possible by the cross bridges on myosin filaments linking with actin filaments of the next sarcomere as they pass through the pores of the Z-disc.

In contrast, some other muscles have the capacity to superextend. This is true of the intersegmental muscles between segments 4–5, 5–6 and 6–7 in the abdomen of female grasshoppers which extend during oviposition. A single muscle between segments 5 and 6 in *Locusta* can vary from 1.2 mm in length when fully contracted to over 11 mm when fully extended. This is possible because the Z-discs are discontinuous and fragment into Z-bodies, to which the actin filaments remain attached, when the muscle extends (Fig. 10.15c). When the muscles contract after oviposition, the Z-discs may not be completely restored (Jorgensen & Rice, 1983).

10.3.2 Control of contraction

Muscle contraction results from a series of steps initiated by the arrival of an action potential at the nerve/muscle junction. At an excitatory synapse this leads, first, to excitation and then to activation of the muscle fiber.

Review: Aidley, 1985

10.3.2.1 *Excitation of the muscle*

With the exception of some visceral muscles, muscles are stimulated to contract by the arrival of an action potential

a) normal muscle

b) supercontracting muscle

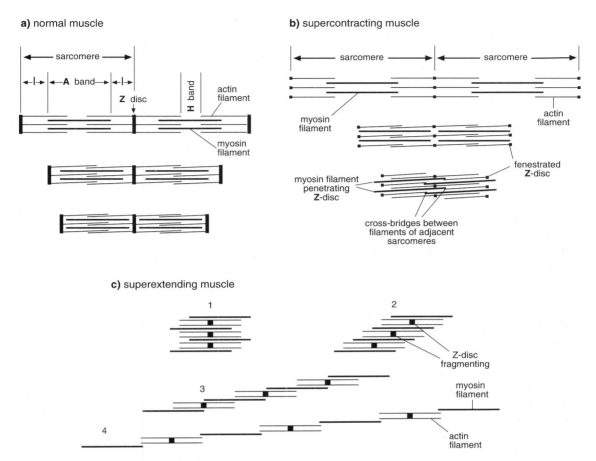

c) superextending muscle

Fig. 10.15. Mechanics of muscle contraction. (a) Normal muscle showing the sliding of the filaments producing shortening of the sarcomeres (I = I band). (b) Supercontracting muscle. When the muscle is fully contracted, the myosin filaments extend through perforations of the Z–discs (after Osborne, 1967). (c) Superextending muscle. Numbers indicate successive stages of extension. When the muscle is extended, the Z–discs become fragmented (after Jorgensen & Rice, 1983).

at the nerve/muscle junctions. Where the junction involves excitation of a skeletal muscle it is almost certain that L-glutamate is the chemical transmitter across the synaptic gap and this may also be true with visceral muscles (Miller, 1975). As is the case for acetylcholine at a central nervous synapse, the transmitter substance is present in synaptic vesicles at the nerve ending. Some spontaneous discharge of transmitter substance into the synaptic gap normally occurs, but the rate of release of the vesicles is greatly enhanced by the arrival of the action potential (Usherwood, 1974).

The unstimulated muscle fiber has a difference in electrical potential across the plasma membrane. This resting potential is within the range 30–70 mV, the inside being negative with respect to the outside. Its occurrence appears to depend largely on maintenance of an excess of potassium ions inside the membrane associated with an inflow of chloride ions. This distribution is maintained, in larval *Phormia* (Diptera), by a sodium/potassium pump and the passive movement of chloride ions, but in larval *Spodoptera* (Lepidoptera) by an H^+/K^+-ATPase together with a K^+/Cl^- cotransporter system (Fitzgerald, Djamgoz & Dunbar, 1996).

The arrival of the excitatory transmitter substance at the postsynaptic membrane on the muscle surface causes a change in permeability leading to a rise (that is, a

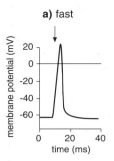

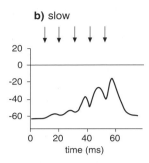

Fig. 10.16. Electrical changes at the muscle membrane following stimulation by (a) a fast axon, and (b) a slow axon. Arrows show times of neural stimulation (after Hoyle, 1974).

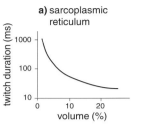

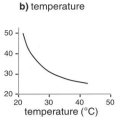

Fig. 10.17. Twitch duration. (a) Relationship with the degree of development of the sarcoplasmic reticulum in various insect muscles (after Josephson, 1975). (b) Effect of temperature in the locust dorsal longitudinal flight muscle (after Neville & Weis-Fogh, 1963).

depolarization) in the muscle membrane potential. The evidence suggests that the inward current producing the depolarization is carried by calcium ions. The short-lived increase in potential produced by these changes is called the postsynaptic potential. The postsynaptic potential spreads from the synapse but decreases rapidly; its effect is therefore localized and large numbers of nerve endings are necessary to stimulate the whole fiber (multiterminal innervation).

The invaginations of the T-system probably convey the changes in potential deep into the muscle and close to the fibrils. This is important since activation of the fibrils is dependent on calcium released from the sarcoplasmic components of dyads. If the calcium had to diffuse from the surface membrane to the central fibrils there would be a considerable delay in contraction. The T-system greatly reduces this delay by bringing the plasma membrane to within a few microns of each fibril.

The size of the muscle twitch produced by the arrival of an excitatory action potential depends on the anatomical characteristics of the muscle fiber and on whether stimulation occurs via the fast or slow axons. It must be understood that the terms 'fast' and 'slow' do not refer to the speed of conduction of the action potential, but to the size of postsynaptic potential, and hence muscle twitch, that is produced. The difference probably resides in the amount of neurotransmitter released at the nerve/muscle junction following the arrival of the action potential (Usherwood, 1974). Stimulation via the fast axon is presumed to release a large amount of neurotransmitter and produces a large postsynaptic potential of consistent size followed by a brief, powerful contraction of the muscle (Fig. 10.16a).

Contractions tend to fuse if the rate of stimulation exceeds ten per second and at 20–25 stimuli per second the muscle undergoes a smooth, maintained contraction: it is in a state of tetanus.

A single action potential from the slow axon, on the other hand, probably releases a small amount of neurotransmitter and produces only a small postsynaptic potential followed by a very small twitch. With increasing frequency of action potentials, the postsynaptic potential increases in size (Fig. 10.16b) and the velocity and force with which the muscle contracts increases progressively; the response is said to be graded. In the extensor tibiae muscles of the locust, for instance, less than five action potentials per second via the slow axon produce no response in the muscle, 15–20 Hz produces muscle tonus, and stimulation by over 70 Hz produces rapid extension of the tibia. The speed of response increases up to 150 Hz.

Review: Pichon & Ashcroft, 1985

10.3.2.2 *Activation of the muscle fiber*

Activation of the contractile mechanism involves the release of calcium from the sarcoplasmic reticulum and it is presumed that this occurs where the T-system and sarcoplasmic reticulum form dyads. In the resting muscle, the formation of cross-bridges between the actin and myosin filaments is inhibited by a protein, tropomyosin, which blocks myosin binding sites on the actin molecules. Tropomyosin is held by another protein, troponin, whose configuration is altered by binding with calcium so that tropomyosin no longer blocks the binding sites and myosin heads become linked to actin. Combination of ATP with

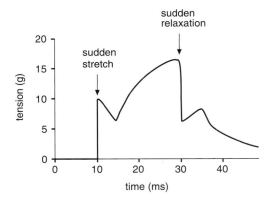

Fig. 10.18. Stretch activation. The effects of sudden small changes in length on the tension developed by asynchronous muscle (after Pringle, 1965).

the myosin cross-bridges causes a conformational change and the link to actin is broken.

A cycle of muscle contraction and relaxation is associated with calcium release and then its sequestration. The rapid removal of calcium from the system is associated with the well-developed sarcoplasmic reticulum and, in general, twitch duration is inversely correlated with the quantity of sarcoplasmic reticulum present (Fig. 10.17a).

In asynchronous muscle the picture is different. An isolated muscle contracts in the same way as synchronous muscle when stimulated at low frequencies and, at higher frequencies of stimulation, the muscle remains in a state of sustained contraction (tetanus). However, this does not happen in an oscillatory system, like the thorax, where the muscles act antagonistically to each other. Here, the muscles start to contract in response to a burst of action potentials, which presumably results in the release of calcium ions from the sarcoplasmic reticulum. However, subsequent muscle oscillations are not related directly to nervous stimulation, hence the term asynchronous. Further nervous stimulation is necessary only to maintain the level of muscle activity, perhaps by maintaining the level of calcium which remains in the sarcoplasm. The reduced development of the sarcoplasmic reticulum appears to be related to this failure to resequester calcium rapidly.

Asynchronous muscles occur only in oscillating systems such as the thorax where the contraction of one flight muscle moves the wings and at the same time lengthens the antagonistic muscles. These sudden changes in length produce corresponding changes in muscle tension,

a sudden increase in length produces a sudden increase in tension and vice versa. It is an intrinsic property of the contractile proteins of fibrillar muscle that a sudden change in tension is followed, after a delay, by a further change (Fig. 10.18). Thus a sudden increase in tension due to stretching is followed by a further rise in tension, and a sudden drop in tension is followed by a delayed fall. As the flight muscles occur in antagonistic pairs, a decrease in tension in one corresponds to an increase in tension in the other. The increased tension acting on a pliant system will produce movement and, as a result of this, the muscles alternately contract and relax, the rate of oscillation being determined by the mechanical and elastic properties of the thorax and the muscles. The process by which sudden elongation of a muscle leads to its contraction is called stretch activation.

10.3.2.3 *Inhibition of muscle contraction*
In addition to the normal excitatory innervation, some muscle fibers of some muscles are innervated by a neuron that inhibits their activation. At an inhibitory nerve/muscle junction, a neural transmitter, probably γ-aminobutyric acid (GABA), is released. It causes a change in permeability at the postsynaptic membrane, but, unlike the process occurring at an excitatory synapse, results in an influx of chloride ions. As a result, the membrane potential becomes even more negative, the membrane is hyperpolarized, and the tension exerted by the fiber decreases (Hoyle, 1974; Usherwood, 1974).

10.3.2.4 *Neuromodulation*
Many muscles, in addition to being innervated by excitatory, and perhaps also inhibitory neurons, receive input from neurons that release compounds modifying the muscle's response to normal excitation. Three chemicals have been commonly identified as such neuromodulators: octopamine, 5-hydroxytryptamine (serotonin) and proctolin.

The octopamine-producing cells are situated in the midline of a ganglion and are unpaired, their axons branching to innervate muscles on either side of the body. Because of their positions they are called dorsal, unpaired, median cells (abbreviated to DUM) with a suffix to indicate the muscles that they innervate. Thus: DUMeti innervates the extensor tibiae muscle in the hind leg of a grasshopper; DUMdl innervates the dorsal longitudinal muscle; and DUMovi the muscles of the oviduct. Octopaminergic cells are also known to supply the antennal muscles and, in

Periplaneta, the muscles of the male accessory glands (Allgäuer & Honegger, 1993; Sinakevitch *et al.*, 1994). Not all DUM cells are concerned with modulating muscular activity and, in the desert locust, *Schistocerca*, only eight, out of a total of 90 DUM cells in the metathoracic ganglion are known to terminate at muscles.

The axons from DUM cells commonly accompany motor axons to the muscles. Their terminal branches form series of swellings, called varicosities which contain vesicles varying in diameter from 60 to 230 nm and with an electron dense core (unlike synaptic vesicles which are electron lucent). They may, or may not contain octopamine, but are clearly related to the secretion of the compound (Hoyle, Colquhoun & Williams, 1980). Unlike the terminals of motor axons, there are no discrete nerve/muscle junctions and the axon terminals are separated from the muscle by thickened sarcolemma.

Octopamine may have its effect both pre- and postsynaptically (Fig. 10.19). It may cause more of the normal neurotransmitter, L-glutamate, to be released, and it may elevate the level of cAMP in the muscle (Whim & Evans, 1991). These changes have been shown to have a variety of effects on muscle activity. In the extensor tibiae muscle of Stenopelmatidae (Orthoptera) it produces a sustained tension. Intrinsic rhythms in the same muscle in grasshoppers are eliminated by octopamine although its effects are dependent on the level of activity of the slow axon to the muscles (Hoyle, 1985). In the dorsal longitudinal flight muscles, octopamine increases the force generated when the muscle contracts and it also increases the rates of contraction and relaxation. In locusts, octopamine does not induce these changes after the teneral period (Whim & Evans, 1989).

Clearly, octopamine is likely to have a very important role in modulating muscular activity during walking, jumping and flying, contributing to the behavioral versatility of insects. However, there is no certain information on its real role in the whole insect. (see Hoyle, 1985; Orchard, Ramirez & Lange, 1993; Whim & Evans, 1989 for some possible roles).

Cells which produce serotonin and which have axons to the muscles of the mouthparts have been described in locust, a cricket and a cockroach. *Periplaneta* has two or three pairs of such neurons in the subesophageal ganglion. Their axons branch to form a fine network over the surfaces of all the nerves to the mouthparts, forming a neurohemal organ (Davis, 1987). In *Locusta* the fibers extend over the surface of the mandibular adductor

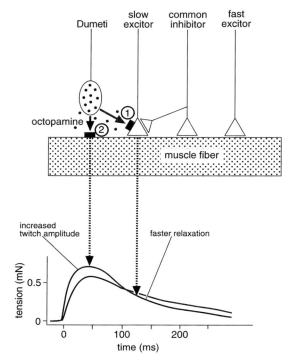

Fig. 10.19. Neuromodulation. Effect of octopamine on muscle activity. The effect may be presynaptic (1) or postsynaptic (2). The latter increases the speed and amplitude of the twitch, the former produces faster relaxation (after Evans, 1985).

muscles and some labral and antennal muscles. Serotinergic neurons also run to the muscles of the fore-, mid- and hindgut in the locust and cricket, and to the muscles of the reproductive system (Nässel, 1988).

In *Locusta*, the serotinergic neurons of the mouthpart muscles are known to be active during feeding, and the serotonin they release is believed to modulate the activity of the muscles since, *in vitro*, it increases the amplitude and rate of contraction and the rate of relaxation (Baines, Tyrer & Downer, 1990; Schachtner & Bräunig, 1993). Like octopamine, serotonin causes an increase in cAMP in the muscle. Caterpillars do not have serotinergic neurons immediately associated with the muscles, but there may be a serotinergic neurohemal organ in the head which serves a similar function (Griss, 1990).

A peptide, proctolin, is present in some slow motor neurons, apparently being released at the same time as the principal neurotransmitter. It may act presynaptically, by increasing the rate at which the transmitter is released, and postsynaptically, enhancing the tension produced by neural stimulation. It is known to be present in some axons to the antennal muscles of *Gryllus*, *Locusta* and *Periplaneta* and to an opener muscle of the ovipositor of *Locusta*. Neurons innervating a variety of visceral muscles in the reproductive system and alimentary canal of *Rhodnius* are also proctolinergic (Allgäuer & Honegger, 1993; Belanger & Orchard, 1993; Lange, 1993).

It is probable that modulation of muscle activity by these and perhaps other compounds is a widespread and possibly universal phenomenon in insects. Even where the muscles do not receive a direct neural supply of the compounds, it is likely that they are affected by their presence in the hemolymph.

Review: Rheuben, 1995

10.3.2.5 *Control of visceral muscles*

In innervated visceral muscles, the principles of muscle control are the same as in skeletal muscles. L-glutamate may be involved as a neurotransmitter, but it is conceivable that different transmitters are involved in different muscles (Usherwood, 1974).

In some insects, the heart is not innervated and the contractions of its muscles are myogenic. This does not mean that they are uncontrolled, and, for example, in *Manduca* it is known that heart muscle activity is modulated by a neurosecretion from the corpora cardiaca. It is probable that the activity of all non-innervated muscles is controlled by blood-borne factors.

Myogenic contractions are commonly slow and rhythmic, but fast myogenic contractions also occur in some muscles at some times. These are produced by action potentials generated spontaneously within some muscle fibers. Not all the fibers in a muscle appear able to produce these action potentials and electrical activity spreads from cell to cell, decreasing as the distance from the active cell increases (Kalogianni & Theophilidis, 1995).

Some muscles exhibit both neurogenic and myogenic contractions. This is the situation in muscles associated with the oviducts of orthopterans.

10.4 ENERGETICS OF MUSCLE CONTRACTION

10.4.1 Definitions

Force is an influence causing a mass to change its state of motion, so that it accelerates or decelerates. A muscle generates a force as it contracts because of the resistance of the object being moved. The unit of force is called a Newton (N). $N = kg\,m\,s^{-2}$.

Tension is produced by opposing forces pulling on an object. Tension in a muscle normally rises to a peak as the muscle begins to shorten and then falls again as shortening continues. If there is no measurable change in length of the muscle, it is said to be contracting *isometrically*. During isometric contraction, tension increases to a maximum. If the muscle decreases in length, while tension remains constant, the contraction is said to be *isotonic*.

Work is the application of a force over a distance, or a measure of the energy transferred by a force.

Energy is the ability to do work. The unit of energy is the same as the unit of work, the joule (J). $J = kg\,m^2\,s^{-2}$.

Power is the rate of doing work, or the rate at which energy is supplied. The unit of power is the watt (W). $W = kg\,m^2\,s^{-3}$

10.4.2 Tension and force

The tension exerted by insect muscles is not exceptional. For instance, the mandibular muscles of various insects exert tensions of $3.6–6.9\,g\,cm^2$, and the extensor tibiae muscle of *Decticus* (Orthoptera) $5–9\ g\ cm^2$ compared with the values of $6–10\ g\ cm^2$ in humans.

Because a muscle has intrinsic elasticity, tension does not fall to zero in the absence of stimulation. Some of this elasticity is attributed to the muscle attachments to the cuticle and some to the sarcolemma, but the greater part is due to elastic elements in the contractile system itself. Energy is stored in this elastic system when the muscle is stretched. Flight muscles, and especially fibrillar muscles, have a much higher elasticity than other muscles.

The force exerted by a muscle is proportional to its cross-sectional area and, in general, this is not very great in insects. In some muscles, however, such as the extensor tibiae of a locust, a considerable cross-sectional area is achieved by an oblique insertion of the muscle fibers into a large apodeme (see Fig. 10.4). As a result, this muscle can exert a force of up to 15 N.

10.4.3 Twitch duration

The duration of each muscle twitch, the time for it to shorten and relax, is dependent on temperature (Fig. 10.17b). This is of critical importance in flight because twitch duration limits the rate at which antagonistic muscles can operate efficiently. To produce the aerodynamic forces necessary for flight, the wings must beat at a certain minimum rate. If the muscle twitch durations of antagonistic muscles overlap to a significant extent, some of the muscle energy is wasted. Efficient flight by *Schistocerca* requires a wingbeat frequency of about 20 Hz, with a period of about 50 ms. Only above 30 °C is twitch duration short enough to avoid significant overlap of antagonistic muscle twitches (Fig. 10.17b) and it is therefore not surprising that sustained flight only occurs at relatively high temperatures.

Twitch duration also varies in different fiber types, being shorter in fast than slow fibers. Fast fibers (short twitch) have fibrils that are small in cross-sectional area with a relatively large proportion of sarcoplasmic reticulum. These factors are related to the readiness with which calcium reaches the most distant myofilaments and is resequestered (see above). Twitch duration is not correlated with fiber diameter, with sarcomere length or with the volume of the fiber occupied by mitochondria (Josephson & Young, 1987; Müller *et al.*, 1992).

10.4.4 Power output

A muscle's mechanical efficiency is defined as the ratio of mechanical power output to metabolic energy consumption. Direct measurements of the mechanical power output of flight muscles of *Manduca* and *Bombus* (insects with synchronous and asynchronous flight muscles, respectively) give maximum values of 130 and 110 W kg^{-1}, respectively (Gilmour & Ellington, 1993; Stevenson & Josephson, 1990). These values indicate mechanical efficiencies of 10% or less. Not all this power is available for mechanically useful work. In a flying insect, energy is required not only to move the wing to produce aerodynamic force, but also to start it moving from an extreme, up or down, position, and for braking at the end of a half stroke (Fig. 10.20). The wing's inertia towards the end of the half stroke stretches the antagonistic muscles, and this energy may be stored in elastic elements within the muscle. As flight muscles only shorten by very small amounts, approximately 2% of their length, it is possible that the cross-bridges between thick and thin filaments remain attached throughout the cycle of elongation and contraction and function as springs (Alexander, 1995; Dickinson & Lighton, 1995). Energy may also be stored in elastic elements of the wing hinge as this is stretched towards the end of the half stroke. This stored

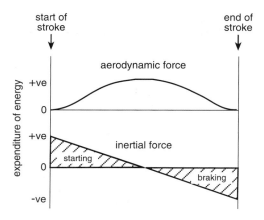

Fig. 10.20. The forces involved in moving a wing. At the start of the stroke, most energy is used overcoming the inertia of the wing. In midstroke, useful aerodynamic forces are produced. At the end of the stroke, the wing has considerable inertia (after Alexander, 1995).

energy then contributes to the beginning of the next half stroke.

Maximum efficiency is achieved only under certain conditions. Work output per cycle of contraction and relaxation rises to a maximum and then declines as the frequency of cycling increases. For the flight muscle of *Manduca* at 35 °C, maximum efficiency occurs at a cycle frequency of about 30 Hz. Work per cycle, at the optimal cycle frequency, increases with temperature, at least up to 40 °C.

The power output of a muscle may be increased by multiple stimulation via the motor axon, and such double (or multiple) firing is commonly recorded when insects appear to produce more power. This occurs, for instance, in the double firing of the axon to the second basalar muscle of *Schistocerca* (see Fig. 9.32), which may result in the muscle more than doubling the amount of work which it does. The extra force exerted varies with the timing of the second impulse relative to the first. In the basalar muscle at 40 °C, the force is maximal when the second stimulus follows about 8 ms after the first. In the tettigoniid, *Neoconocephalus*, maximum power output from a tergocoxal muscle is achieved when the motor neuron fires three times with intervals of 4 ms between spikes (Josephson, 1985). Multiple firing does not increase the power output of all muscles at their normal operating frequencies, however (Stevenson & Josephson, 1990).

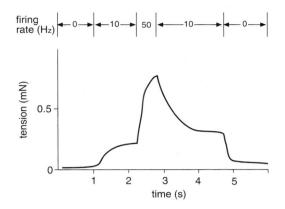

Fig. 10.21. The development of 'catch' tension in a muscle stimulated via a slow axon (mesothoracic extensor tibiae muscle of the desert locust, *Schistocerca*). Notice that the tension exerted when the muscle is stimulated at 10 Hz is higher after a brief burst of stimulation at 50 Hz (after Evans & Siegler, 1982).

A great deal of power is required to lift the insect off the ground during flight. Because of the low level of muscle efficiency, most insects are able to produce sufficient power only by a high wingbeat frequency, reflecting the oscillation frequency of the flight muscles. Energy consumption by flying insects is very high. The metabolic rates of flying insects are commonly 100 time higher than those of resting insects (see Fig. 9.37).

Review: Josephson, 1981

10.5 MUSCULAR CONTROL IN THE INTACT INSECT

Insects have only small numbers of separate motor units in their muscles compared with vertebrates. Consequently, precision and flexibility of movement is achieved not by employing different numbers of units but by changes in the strengths of contraction of individual units. This fine control is effected through the polyneuronal innervation of the muscles and through neuromodulation.

10.5.1. Muscle tonus

The muscles of a stationary insect are not completely relaxed. As in any animal, they must maintain some degree of tension if the insect is to maintain its stance and be ready to make an immediate response. Maintaining this tension or tonus may involve three different mechanisms.

In many cases it is dependent on a low level of neural input to slow (tonic) muscle fibers. The tonic fibers of some muscles, when stimulated by a high frequency burst from the slow axon, sustain a higher tension than before the burst at a low level of stimulation (Fig. 10.21). This is known as a 'catch' tension. It is eliminated by the activity of a fast or inhibitory axon (Burns & Usherwood, 1978). Second, some muscles may exhibit a steady tension in the absence of neural input. This has been demonstrated in some spiracle muscles and in the extensor tibiae muscle of the locust. Finally, some muscles are known to undergo slow rhythmic changes in tension which are myogenic in origin. These muscles also respond to neural stimulation (Hoyle, 1978a).

10.5.2 Locomotion

Most behavioral activities result from the coordinated activity of sets of muscles. This is most obvious in locomotion which involves the oscillation of an appendage such as a leg or a wing. For example, during slow walking by a cockroach, only the slow axon to the coxal depressor muscles is active and so only the muscles numbered 135 in Fig. 8.6 are involved; the strength and speed of contraction depends on the frequency of nerve firing. At walking speeds of more than 10 cycles per second, the slow axon is reinforced by the activity of the fast axon, which also activates muscles 136 and 137. It is presumed that the inhibitory axons fire at the end of contraction and so ensure complete and rapid relaxation as the antagonistic muscle contracts. The presence of three separate inhibitors, perhaps innervating different fibers in the muscles, gives increased flexibility to the system, but the situation is complicated by the fact that they also innervate other muscles.

The control of muscle activity in jumping by locusts and the control of stridulation in grasshoppers provide other examples of the interaction between fast, slow and inhibitory axons (see Figs. 8.24, 26.16).

Oscillation of antagonistic pairs of muscles Skeletal muscles usually occur in antagonistic pairs, such as the extensor and flexor muscles of the tibia, and the levator and depressor muscles of the wings. Sometimes a muscle is opposed only by the elasticity of the cuticle. Thus, depression of the pretarsus is produced by a muscle, but extension results entirely from the elasticity of the cuticle at the base of the segment. The tymbal muscle in cicadas has no

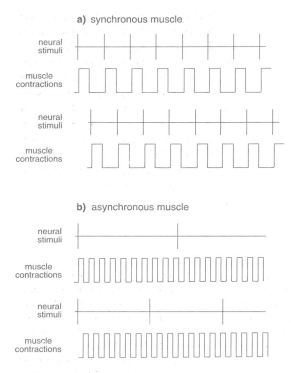

a) synchronous muscle

neural
stimuli

muscle
contractions

neural
stimuli

muscle
contractions

b) asynchronous muscle

neural
stimuli

muscle
contractions

neural
stimuli

muscle
contractions

Fig. 10.22. Contraction of antagonistic muscle pairs (as in flight muscles) in relation to neural stimulation. **(a)** Synchronous muscle. Each contraction results from the arrival of an action potential. The stimuli to antagonistic muscles are in antiphase. **(b)** Asynchronous muscle. Muscle oscillations occur independently of the arrival of action potentials which serve only to keep the muscle in an activated state. The rate of neural input to the antagonistic muscles may differ, as in this example.

antagonist; it is stretched when the tymbal buckles outwards (section 26.1.3).

When antagonistic muscles oscillate, they are driven by motor neurons firing in antiphase (except for asynchronous flight muscles, see below). The precise timing of neural activity and the interplay of fast and slow excitatory neurons and inhibitory neurons permits a wide range of modulation. Section 8.4.1.4 describes neural regulation in walking insects.

The high wingbeat frequencies necessary for flight are produced by the rapid oscillation of pairs of antagonistic muscles. This is achieved despite the relatively low rate of shortening of about $40\,\mathrm{mm\,s^{-1}}$ in *Schistocerca* and only $11\,\mathrm{mm\,s^{-1}}$ in *Sarcophaga*. Three factors combine to reduce the duration of the muscle twitch and so to make

flight possible: (i) the loading of the muscle, (ii) the temperature of the muscle, and (iii) the very slight contraction necessary to move the wing.

A maximum rate of shortening is achieved if the tension and loading of the muscle are maximal at the beginning of contraction. Loading of flight muscles involves the inertia of the wings, mechanical leverage of the wings, which changes in the course of a stroke, damping of the movement of the wings by the air, and elastic loading due to the straining of the thorax and stretching the antagonistic muscles. Inertia is highest at the beginning of the stroke when the wing may be momentarily stationary, or even moving in the opposite direction (Fig. 10.20). Mechanical leverage of the wings and loading due to the elasticity of the thorax are also greatest at this time and so will favor a high rate of shortening.

The importance of temperature in governing the duration of the muscle twitch is considered in section 10.4.3. Flight can only occur when muscle temperature is high enough to ensure rapid shortening and relaxation.

Finally, the wings are articulated with the thorax in such a way that only a very small contraction of the muscles is necessary to produce a large movement of the wing: the flight muscles of *Sarcophaga*, for instance, only shorten by 1 or 2%, in the course of a wingstroke. As a result the muscle twitch is brief despite the low rate of shortening.

In insects with synchronous flight muscles, each contraction of the flight muscles is produced by the arrival of a nerve impulse and the motor neurons to the antagonistic muscles fire approximately in antiphase (Fig. 10.22a). These muscles occur in Odonata and orthopteroid insects as well as in Neuroptera, Trichoptera and Lepidoptera amongst the holometabolous insects. Wingbeat frequencies are generally less than 50 Hz although in some Lepidoptera the frequency approaches 100 Hz.

These muscles are generally activated by one or two action potentials in the motor axons, but in some Lepidoptera with very low wingbeat frequencies (5–10 Hz) the motor burst is characteristically 3–7 action potentials. The occurrence of two or more action potentials is associated with high lift requirements by the insects and results in higher power output by the muscles. Where longer bursts occur, the longer period of excitation keeps the muscles contracting for relatively long periods, and burst length is inversely proportional to wingbeat frequency. Manoeuvers in flight are achieved by alterations in the power output of specific muscles and by changes in the times at which different direct muscles, which affect wing twisting, are activated relative to each other. Figure 9.32 illustrates some aspects of muscle coordination in flight.

In insects with asynchronous muscles, the wingbeat frequency is often in excess of 100 Hz and it is a characteristic of these muscles that several contractions follow the arrival of each nerve impulse (Fig. 10.22b). Wingbeat frequency in these insects is determined primarily by the mechanical properties of the thorax and the muscles themselves. Neural inputs serve to maintain the muscles in an active state (see section 10.3.2.2).

Each neural signal to an asynchronous muscle normally consists of a single spike; double firing does not usually occur. The frequency of the single spikes varies from 5 to about 25 Hz in different insects while wingbeat frequency varies from about 50 to over 200 Hz. The ratio of action potentials to wing cycles varies from about 1 : 5 to 1 : 40, but the rate of firing to antagonistic muscles is not necessarily the same. For example, during flight in the honeybee the dorsal longitudinal muscles are stimulated at a lower frequency (about 86%) than the dorsoventral muscles (Esch & Goller, 1991). Nevertheless, an increase in the firing rate of the motor neurons does sometimes correlate with an increase in wingbeat frequency.

Review: Kammer, 1985

REFERENCES

Aidley, D.J. (1985). Muscular contraction. In *Comprehensive Insect Physiology, Biochemistry and Pharmacology*, vol. 5, ed. G.A. Kerkut & L.I. Gilbert, pp. 407–37. Oxford: Pergamon Press.

Alexander, R.M. (1995). Springs for wings. *Science*, 268, 50–1.

Allgäuer, C. & Honegger, H.W. (1993). The antennal motor system of crickets: modulation of muscle contractions by a common inhibitor, DUM neurons, and proctolin. *Journal of Comparative Physiology* A, 173, 485–94.

Anderson, M. & Finlayson, L.H. (1973). Ultrastructural changes during growth of the flight muscles in the adult tsetse fly, *Glossina austeni*. *Journal of Insect Physiology*, 19, 1989–97.

Anderson, M. & Finlayson, L.H. (1976). The effect of exercise on the growth of mitochondria and myofibrils in the flight muscles of the tsetse fly, *Glossina morsitans*. *Journal of Morphology*, 150, 321–6.

Arbas, E.A. & Weidner, M.H. (1991). Transneuronal induction of muscle atrophy in grasshoppers. *Journal of Neurobiology*, 22, 536–46.

Atwood, H.L. & Cooper, R.L. (1995). Functional and structural parallels in Crustacean and *Drosophila* neuromuscular systems. *American Zoologist*, 35, 556–65.

Baines, R.A., Tyrer, N.M. & Downer, R.G.H. (1990). Serotinergic innervation of the locust mandibular closer muscle modulates contractions through the elevation of cyclic adenosine monophosphate. *Journal of Comparative Neurology*, 294, 623–62.

Beenakkers, A.M.T., van den Broek, A.T.M. & de Ronde, T.H.A. (1975). Development of catabolic pathways in insect flight muscles. A comparative study. *Journal of Insect Physiology*, 21, 849–59.

Belanger, J.H. & Orchard, I. (1993). The locust ovipositor opener muscle: proctolinergic central and peripheral neuromodulation in a centrally driven motor system. *Journal of Experimental Biology*, 174, 343–62.

Bernays, E.A. (1972). The muscles of newly hatched *Schistocerca gregaria* larvae and their possible functions in hatching, digging and ecdysial movements (Insecta: Acrididae). *Journal of Zoology, London*, 166, 141–58.

Bernays, E.A. (1986). Diet-induced head allometry among foliage-chewing insects and its importance for graminivores. *Science*, 231, 495–7.

Bothe, G.W.M. & Rathmayer, W. (1994). Programmed degeneration of thoracic eclosion muscle in the flesh fly, *Sarcophaga bullata*. *Journal of Insect Physiology*, 40, 983–95.

Bücher, T. (1965). Formation of the specific structural and enzymic pattern of the insect flight muscle. In *Aspects of Insect Biochemistry*, ed. T.W. Goodwin, pp. 15–28. London: Academic Press.

Burns, M.D. & Usherwood, P.N.R. (1978). Mechanical properties of locust extensor tibiae muscles. *Comparative Biochemistry and Physiology*, 61A, 85–95.

Caveney, S. (1969). Muscle attachment related to cuticle architecture in Apterygota. *Journal of Cell Science*, 4, 541–59.

Cochrane, D.G., Elder, H.Y. & Usherwood, P.N.R. (1972). Physiology and ultrastructure of phasic and tonic skeletal muscle fibers in the locust, *Schistocerca gregaria*. *Journal of Cell Science*, 10, 419–41.

Consoulas, C., Hustert, R. & Theophilidis, G. (1993). The multisegmental motor supply to transverse muscles differs in a cricket and a bushcricket. *Journal of Experimental Biology*, 185, 335–55.

Cottrell, C.B. (1962). The imaginal ecdysis of blowflies. Observations on the hydrostatic mechanisms involved in digging and expansion. *Journal of Experimental Biology*, 39, 431–48.

Davis, N.T. (1987). Neurosecretory neurons and their projections to the serotonin neurohemal system of the cockroach *Periplaneta americana* (L.), and identification of mandibular and maxillary motor neurons associated with this system. *Journal of Comparative Neurology*, **259**, 604–21.

Dickinson, M.H. & Lighton, J.R.B. (1995). Muscle efficiency and elastic storage in the flight motor of *Drosophila*. *Science*, **268**, 87–90.

Elder, H. Y. (1975). Muscle structure. In *Insect Muscle*, ed. P.N.R. Usherwood, pp. 1–74. London: Academic Press.

Esch, H. & Goller, F. (1991). Neural control of fibrillar muscles in bees during shivering and flight. *Journal of Experimental Biology*, **159**, 419–31.

Evans, P.D. (1985). Octopamine. In *Comprehensive Insect Physiology, Biochemistry and Pharmacology*, vol. 11, ed. G.A. Kerkut & L.I. Gilbert, pp. 499–530. Oxford: Pergamon Press.

Evans, P.D. & Siegler, M.V.S. (1982). Octopamine mediated relaxation of catch tension in locust skeletal muscle. *Journal of Physiology*, **324**, 93–112.

Finlayson, L.H. (1975). Development and degeneration. In *Insect Muscle*, ed. P.N.R. Usherwood, pp. 75–149. London: Academic Press.

Fitzgerald, E.M., Djamgoz, M.B.A. & Dunbar, S.J. (1996). Maintenance of the K^+ activity gradient in insect muscle compared in Diptera and Lepidoptera: contributions of metabolic and exchanger mechanisms. *Journal of Experimental Biology*, **199**, 1857–72.

Gilmour, K.M. & Ellington, C.P. (1993). Power output of glycerinated bumble-bee flight muscle. *Journal of Experimental Biology*, **183**, 77–100.

Griss, C. (1990). Mandibular motor neurons of the caterpillar of the hawk moth *Manduca sexta*. *Journal of Comparative Neurology*, **296**, 393–402.

Hale, J.P. & Burrows, M. (1985). Innervation patterns of inhibitory motor neurones in the thorax of the locust. *Journal of Experimental Biology*, **117**, 401–13.

Hinton, H. E. (1963). The origin and function of the pupal stage. *Proceedings of the Royal Entomological Society of London* A **38**, 77–85.

Houlihan, D.F. & Breckenridge, L. (1981). Stretch-induced growth of blowfly muscle. *Journal of Insect Physiology*, **27**, 521–5.

Hoyle, G. (1974). Neural control of skeletal muscle. In *The physiology of Insecta*, vol. 4, ed. M. Rockstein, pp. 176–236. New York: Academic Press.

Hoyle, G. (1978a). Intrinsic rhythm and basic tonus in insect skeletal muscle. *Journal of Experimental Biology*, **73**, 173–203.

Hoyle, G. (1978b). Distributions of nerve and muscle fibre types in locust jumping muscle. *Journal of Experimental Biology*, **73**, 205–33.

Hoyle, G. (1985). Generation of motor activity and control of behavior: the roles of neuromodulator octopamine, and the orchestration hypothesis. In *Comprehensive Insect Physiology, Biochemistry and Pharmacology*, vol. 5, ed. G.A. Kerkut & L.I. Gilbert, pp. 607–21. Oxford: Pergamon Press.

Hoyle, G., Colquhoun, W. & Williams, M. (1980). Fine structure of an octopaminergic neuron and its terminals. *Journal of Neurobiology*, **11**, 103–26.

Huddart, H. (1985). Visceral muscle. In *Comprehensive Insect Physiology, Biochemistry and Pharmacology*, vol. 11, ed. G.A. Kerkut & L.I. Gilbert, pp. 131–194. Oxford: Pergamon Press.

Johnson, B.G. & Rowley, W.A. (1972). Ultrastructural changes in *Culex tarsalis* flight muscle associated with exhaustive flight. *Journal of Insect Physiology*, **18**, 2391–9.

Jorgensen, W.K. & Rice, M.J. (1983). Superextension and supercontraction in locust ovipositor muscles. *Journal of Insect Physiology*, **29**, 437–48.

Josephson, R.K. (1975). Extensive and intensive factors determining the performance of striated muscle. *Journal of Experimental Zoology*, **194**, 135–54.

Josephson, R.K. (1981). Temperature and the mechanical performance of insect muscle. In *Insect Thermoregulation*, ed. B. Heinrich, pp. 19–44. New York: Wiley.

Josephson, R.K. (1985). Mechanical power output from striated muscle during cyclic contraction. *Journal of Experimental Biology*, **114**, 493–512.

Josephson, R.K. & Young, D. (1987). Fiber ultrastructure and contraction kinetics in insect fast muscles. *American Zoologist*, **27**, 991–1000.

Kammer, A.E. (1985). Flying. In *Comprehensive Insect Physiology, Biochemistry and Pharmacology*, vol. 5, ed. G.A. Kerkut & L.I. Gilbert, pp. 491–552. Oxford: Pergamon Press.

Kammer, A. E. and Heinrich, B. (1978). Insect flight metabolism. *Advances in Insect Physiology*, **13**, 133–228.

Kalogianni, E. & Theophilidis, G. (1995). The motor innervation of the oviducts and central generation of the oviductal contraction in two orthopteran species (*Calliptamus* sp. and *Decticus albifrons*). *Journal of Experimental Biology*, **198**, 507–20.

Kimura, K. & Truman, J.W. (1990). Postmetamorphic cell death in the nervous and muscular systems of *Drosophila melanogaster*. *Journal of Neuroscience*, **10**, 403–11.

Kobayashi, M. & Ishikawa, H. (1994). Mechanism of histolysis in indirect flight muscles of alate aphid (*Acyrthosiphon pisum*). *Journal of Insect Physiology*, **40**, 33–8.

Lai-Fook, J. (1967). The structure of developing muscle insertions in insects. *Journal of Morphology*, **123**, 503–28.

Lange, A.B. (1993). The association of proctolin with the spermatheca of the locust, *Locusta migratoria*. *Journal of Insect Physiology*, **39**, 517–22.

Maruyama, K. (1985). Biochemistry of muscle contraction. In *Comprehensive Insect Physiology, Biochemistry and Pharmacology*, vol. 10, ed. G.A. Kerkut & L.I. Gilbert, pp. 487–98. Oxford: Pergamon Press.

Miller, T. A. (1975). Insect visceral muscles. In *Insect Muscle*, ed. P.N.R. Usherwood, pp. 545–606. London: Academic Press.

Mizisin, A.P. & Ready, N.E. (1986). Growth and development of flight muscle in the locust (*Schistocerca nitens*, Thünberg). *Journal of Experimental Zoology*, **237**, 45–55.

Morgan, C.R. & Stokes, D.R. (1979). Ultrastructural heterogeneity of the mesocoxal muscles of *Periplaneta americana*. *Cell & Tissue Research*, **201**, 305–14.

Müller, A.R., Wolf, H., Galler, S. & Rathmayer, W. (1992). Correlation of electrophysiological, histochemical, and mechanical properties in fibres of the coxa rotator muscle of the locust, *Locusta migratoria*. *Journal of Comparative Physiology* B, **162**, 5–15.

Nässel, D.R. (1988). Serotonin and serotonin–immunoreactive neurons in the nervous system of insects. *Progress in Neurobiology*, **30**, 1–85.

Nüesch, H. (1985). Control of muscle development. In *Comprehensive Insect Physiology, Biochemistry and Pharmacology*, vol. 2, ed. G.A. Kerkut & L.I. Gilbert, pp. 425–52. Oxford: Pergamon Press.

Neville, A. C. & Weis-Fogh, T. (1963). The effect of temperature on locust flight muscle. *Journal of Experimental Biology*, **40**, 111–21.

Orchard, I., Ramirez, J.-M. & Lange, A.B. (1993). A multifunctional role for octopamine in locust flight. *Annual Review of Entomology*, **38**, 227–49.

Osborne, M.P. (1967). Supercontraction in the muscles of the blowfly larva: an ultrastructural study. *Journal of Insect Physiology*, **13**, 1471–82.

Pearson, K.G. & Iles, J.F. (1971). Innervation of coxal depressor muscles in the cockroach, *Periplaneta americana*. *Journal of Experimental Biology*, **54**, 215–32.

Pichon, Y. & Ashcroft, F.M. (1985). Nerve and muscle: electrical activity. In *Comprehensive Insect Physiology, Biochemistry and Pharmacology*, vol. 5, ed. G.A. Kerkut & L.I. Gilbert, pp. 85–113. Oxford: Pergamon Press.

Pringle, J.W.S. (1965). Locomotion: flight. In *The Physiology of Insecta*, vol. 2, ed. M.Rockstein, pp. 283–329. New York: Academic Press.

Ready, N.E. & Josephson, R.K. (1982). Flight muscle development in a hemimetabolous insect. *Journal of Experimental Zoology*, **220**, 49–56.

Ready, N.E. & Najm, R.E. (1985). Structural and functional development of cricket wing muscles. *Journal of Experimental Zoology*, **233**, 35–50.

Rheuben, M.B. (1995). Specific associations of neurosecretory or neuromodulatory axons with insect skeletal muscles. *American Zoologist*, **35**, 566–77.

Rockstein, M. and Bhatnagar, P. L. (1965). Age changes in size and number of the giant mitochondria in the flight muscle of the common housefly (*Musca domestica* L.). *Journal of Insect Physiology*, **11**, 481–91.

Schachtner, J. & Bräunig, P. (1993). The activity pattern of identified neurosecretory cells during feeding behavior in the locust. *Journal of Experimental Biology*, **185**, 287–303.

Sinakevitch, I.G., Geffard, M., Pelhate, M. & Lapied, B. (1994). Octopamine-like immunoreactivity in the dorsal unpaired median (DUM) neurons innervating the accessory gland of the male cockroach *Periplaneta americana*. *Cell & Tissue Research*, **276**, 15–21.

Smith, D.S. (1966). The organization and function of the sarcoplasmic reticulum and T-system of muscle cells. *Progress in Biophysics and Molecular Biology*, **16**, 109–42.

Smith, D.S. (1968). *Insect Cells: Their Structure and Function*. Edinburgh: Oliver & Boyd.

Smith, D.S. (1972). *Muscle: A Monograph*. New York: Academic Press.

Smith, D.S. (1984). The structure of insect muscles. In *Insect Ultrastructure*, vol. 2, ed. R.C. King & H. Akai, pp. 111–150. New York: Plenum Press.

Smith, D.S. & Treherne, J.E. (1963). Functional aspects of the organisation of the insect nervous system. *Advances in Insect Physiology*, **1**, 401–84.

Stevenson, R.D. & Josephson, R.K. (1990). Effects of operating frequency and temperature on mechanical power output from moth flight muscle. *Journal of Experimental Biology*, **149**, 61–78.

Tanaka, S. (1993). Allocation of resources to egg production and flight muscle development in a wing dimorphic cricket, *Modicogryllus confirmatus*. *Journal of Insect Physiology*, **39**, 493–8.

Truman, J.W. (1985). Hormonal control of ecdysis. In *Comprehensive Insect Physiology, Biochemistry and Pharmacology*, vol. 8, ed. G.A. Kerkut & L.I. Gilbert, pp. 413–40. Oxford: Pergamon Press.

Usherwood, P. N. R. (1974). Nerve-muscle transmission. In *Insect Neurobiology*, ed. J. Treherne, pp. 245–305. Amsterdam: North-Holland Publishing Co.

Vigoreaux, J.O., Saide, J.D., Valgeirsdottir, K. & Pardue, M.L. (1993). Flightin, a novel myofibrillar protein of *Drosophila* stretch-activated muscles. *Journal of Cell Biology*, **121**, 587–98.

Walker, T.J. (1972). Deciduous wings in crickets: a new basis for wing dimorphism. *Psyche* **79**, 311–3.

Whim, M.D. & Evans, P.D. (1989). Age-dependence of octopaminergic modulation of flight muscle in the locust. *Journal of Comparative Physiology* A, **165**, 125–37.

Whim, M.D. & Evans, P.D. (1991). The role of cyclic AMP in the octopaminergic modulation of flight muscle in the locust. *Journal of Experimental Biology*, **161**, 423–38.

Wigglesworth, V.B. & Lee, W.M. (1982). The supply of oxygen to the flight muscles of insects: a theory of tracheole physiology. *Tissue & Cell*, **14**, 501–18.

PART III

The Abdomen, Reproduction and Development

11 Abdomen

The insect abdomen is more obviously segmental in origin than either the head or the thorax, consisting of a series of similar segments, but with the posterior segments modified for mating and oviposition. In general, the abdominal segments of adult insects are without appendages except for those concerned with reproduction and a pair of terminal, usually sensory, cerci. Pregenital appendages are, however, present in Apterygota and in many larval insects as well as in non-insectan hexapods. Aquatic larvae often have segmental gills, while many holometabolous larvae, especially amongst the Diptera and Lepidoptera, have lobe-like abdominal legs called prolegs.

Reviews: Bitsch, 1979; Matsuda, 1976; Snodgrass, 1935

11.1 SEGMENTATION

11.1.1 Number of segments

The basic number of segments in the abdomen is eleven plus the postsegmental telson which bears the anus, although Matsuda (1976) regards the telson as a twelfth segment. Only in adult Protura and the embryos of some hemimetabolous insects is the full complement visible. In all other instances there is some degree of reduction. The telson, if it is present at all, is generally represented only by the circumanal membrane, but in larval Odonata three small sclerites surrounding the anus may represent the telson.

In general, more segments are visible in the more generalized hemimetabolous orders than in the more specialized holometabolous insects. Thus in Acrididae all eleven segments are visible (Fig. 11.1a) whereas in adult Muscidae only segments 2–5 are visible and segments 6–9 are normally telescoped within the others (Fig. 11.1b). Collembola are exceptional in having only six abdominal segments, even in the embryo.

The definitive number of segments is present at hatching in all hexapods except Protura. All the segments differentiate in the embryo and this type of development is called epimorphic. In Protura, on the other hand, the first stage larva hatches with only eight abdominal segments plus the telson; the remaining three segments are added at subsequent molts, arising behind the last abdominal segment, but in front of the telson. This type of development is called anamorphic.

In general, the abdomen is clearly marked off from the thorax, but this is not the case in Hymenoptera where the first abdominal segment is intimately fused with the thoracic segments and is known as the propodeum. The waist of Hymenoptera Apocrita is thus not between the thorax and abdomen, but between the first abdominal segment and the rest of the abdomen. Often segment 2 forms a narrow petiole connecting the two parts. The swollen part of the abdomen behind the waist is called the gaster (Fig. 11.2).

11.1.2 Structure of abdominal segments

A typical abdominal segment, such as the third, consists of a sclerotized tergum and sternum joined by membranous pleural regions which are commonly hidden beneath the sides of the tergum, as in figures 11.1 and 11.2. In many holometabolous larvae, however, there is virtually no sclerotization and the abdomen consists of a series of membranous segments. This is true in many Diptera and Hymenoptera, some Coleoptera and most lepidopterous larvae. In these, the only sclerotized areas are small plates bearing trichoid sensilla. Even where well-developed terga and sterna are present these may be divided into a number of small sclerites as in the larva of the beetle, *Calosoma* (Fig. 11.3). In contrast, the tergum, sternum and pleural elements sometimes fuse to form a complete sclerotized ring. This is true in the genital segments of many adult male insects, in segment 10 of Odonata, Ephemeroptera and Dermaptera and segment 11 of Machilidae.

Typically, the posterior part of each segment overlaps the anterior part of the segment behind (Fig. 11.4a), the two being joined by a membrane, but segments may fuse together, wholly or in part. For instance, in Acrididae the terga of segments 9 and 10 fuse together (Fig. 11.1a), while in some Coleoptera the second sternum fuses with the next two and the sutures between them are largely obliterated.

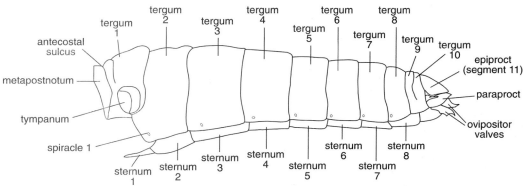

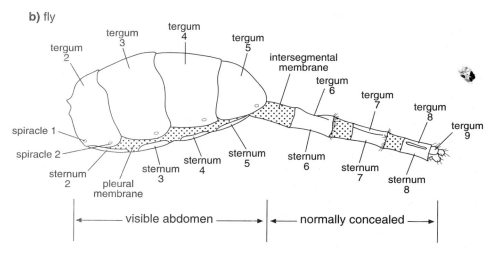

Fig. 11.1. Abdomen in lateral view. (a) An insect in which parts of all 11 segments are present in the adult (female red locust, *Nomadacris* (Orthoptera)) (after Albrecht, 1956). (b) An insect with a reduced number of segments in the adult. Segments 6–9, which form the ovipositor, are normally retracted within the anterior segments (female housefly, *Musca* (Diptera)) (after Hewitt, 1914).

The more anterior segments have a spiracle on either side. This may be set in the pleural membrane (Fig. 11.3), or in a small sclerite within the membrane, or on the side of the tergum (Fig. 11.1) or sternum. The reproductive opening in male insects is usually on segment 9, while in the majority of female insects the opening of the oviduct is on or behind segment 8 or 9. The Ephemeroptera and Dermaptera are unusual in having the opening behind segment 7. These genital segments may be highly modified, in the male to produce copulatory apparatus and in the females of some orders to form an ovipositor. This may be formed by the sclerotization and telescoping of the posterior abdominal segments, or it may involve modified abdominal appendages.

In front of these genital segments the abdominal segments are usually unmodified, although segment 1 is frequently reduced or absent. Behind the genital segments, segment 10 is usually developed, but segment 11 is often represented only by a dorsal lobe, the epiproct, and two lateroventral lobes, the paraprocts. In Plecoptera, Blattodea and Isoptera the epiproct is reduced and fused with the tergum of segment 10, while in most holometabolous insects segment 11 is lacking altogether and segment 10 is terminal.

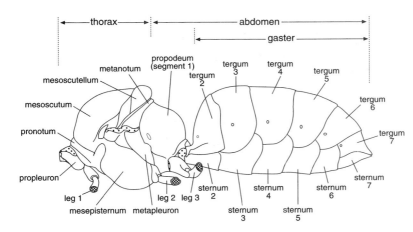

Fig. 11.2. Thorax and abdomen of a hymenopteran to show the waist between abdominal segments 1 and 2 (honeybee, *Apis*) (after Snodgrass, 1956).

Modifications of the terminal abdominal segments often occur in aquatic insects and are concerned with respiration (see Chapter 17).

11.1.3 Musculature

Where the cuticle of the abdomen is largely membranous, as in many holometabolous larvae, most longitudinal muscles run from one intersegmental fold to the next. In most insects with well-sclerotized abdominal segments, the dorsal and ventral abdominal longitudinal muscles are in two series, external and internal (Fig. 11.4). The internal muscles run from one antecostal ridge to the next and so retract the segments within each other. The external muscles are much shorter and only extend from the posterior end of one segment to the anterior end of the next and, because of the degree of overlap between the segments, the origins may be posterior to the insertions (Fig. 11.4b). Hence they may act as protractor muscles, extending the abdomen, and their efficiency is sometimes increased by the development of apodemes so that their pull is exerted longitudinally instead of obliquely. If such a protractor mechanism is absent, extension of the abdomen results from the hydrostatic pressure of blood.

There are also lateral muscles which usually extend from the tergum to the sternum, but sometimes arise on or are inserted into the pleuron. They are usually intrasegmental, but sometimes cross from one segment to the next. Their effect is to compress the abdomen dorso-ventrally (Fig. 11.4c). Dilation of the abdomen often results from its elasticity and from blood pressure, but in some insects some of the lateral muscles function as dilators. This occurs when the tergal origins of the muscles are carried ventrally by extension of the terga, while the sternal inser-

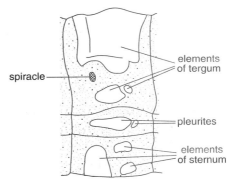

Fig. 11.3. Small sclerotized plates in an abdominal segment. Membrane stippled (larva of a beetle, *Calosoma*) (after Snodgrass, 1935).

tions may also be carried dorsally on apodemes (Fig. 11.4d).

In addition to the longitudinal and lateral muscles, others are present in connection with abdominal appendages, especially the genitalia, and the spiracles (section 17.2.2), while transverse bands of muscle form the dorsal and ventral diaphragms (sections 5.1.1.2, 5.1.1.3).

11.2 ABDOMINAL APPENDAGES AND OUTGROWTHS

Insects are generally believed to have been derived from an arthropod ancestor with a pair of appendages on each segment. Typical legs, such as are found on the thorax, never occur on the abdomen of insects, but various appendages do occur and some of these are probably derived from typical appendages. Molecular studies indicate that the

Fig. 11.4. Abdominal musculature (from Snodgrass, 1935). (a) Diagram of the dorsal longitudinal musculature in an abdominal segment. Typical arrangement of external and internal muscles, both acting as retractors. (b) Origin of external muscle (arrow) shifted posteriorly so that it acts as a protractor. (c) Transverse section of the right-hand side showing a typical arrangement with the tergosternal muscles acting only as compressors. (d) Transverse section of the right-hand side showing tergosternal muscles differentiated into compressor and dilator muscles. Notice how the insertion of the dilator muscle on to the sternum is shifted dorsally by the apodeme.

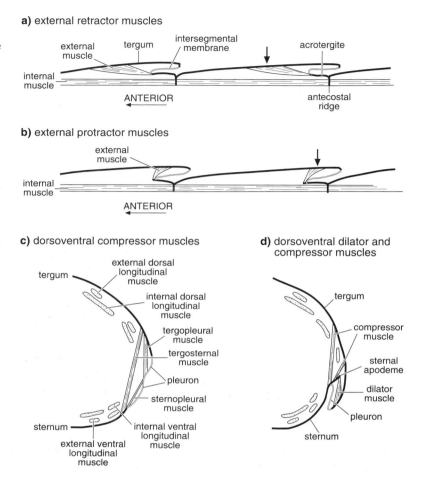

a) external retractor muscles

b) external protractor muscles

c) dorsoventral compressor muscles

d) dorsoventral dilator and compressor muscles

prolegs of caterpillars are homologous with the thoracic legs (Panganiban, Nagy & Carroll, 1994). Some other appendages are probably secondary structures which have developed quite independently of the primitive appendages.

The structure and functioning of the male and female genitalia are considered in chapters 12 and 13. Apart from the genitalia and the cerci, abdominal appendages or other outgrowths of the body wall tend to occur in larvae rather than in adults. The appendages of Apterygota and other primitive hexapods are present in all stages of development, however.

11.2.1 Abdominal appendages of primitive hexapods
Styliform appendages Styliform structures, often associated with eversible vesicles, are present on the abdomen of

Apterygota and some related non-insect hexapods. On abdominal segments 2–9 of Machilidae, 7–9 or 8–9 of Lepismatidae, 1–7 of Japygidae and 2–7 of Campodeidae there are pairs of small, unjointed styli, each inserted on a basal sclerite which is believed to represent the coxa (Fig. 11.5a). Since similar styli are present on the coxae of the thoracic legs of *Machilis* (Archaeognatha) these styli are regarded as coxal epipodites.

Associated with the styli, but occupying a more median position, are eversible vesicles. These are present on segments 1–7 of Machilidae and 2–7 of *Campodea* (Diplura), but in Lepismatidae and Japygidae there are generally fewer or none. The vesicles evert through a cleft at the posterior margin of the segment, being forced out by blood pressure (Fig. 11.5c,d). The retractor muscles of the vesicles arise close together on the anterior margin of the

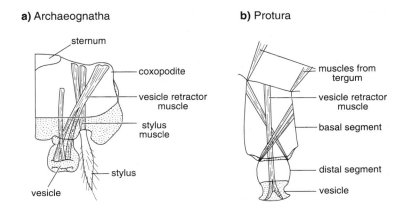

a) Archaeognatha

sternum

coxopodite

vesicle retractor muscle

stylus muscle

stylus

vesicle

b) Protura

muscles from tergum

vesicle retractor muscle

basal segment

distal segment

vesicle

Fig. 11.5. Styliform appendages and eversible vesicles. (a) Archaeognatha. Abdominal appendage of *Nesomachilis*. The sternum and coxopodite are seen from the inside (from Snodgrass, 1935). (b) Protura. Abdominal appendage of *Acerentomon* (from Snodgrass, 1935). (c) Diplura. Section through an eversible vesicle of *Campodea*, retracted (after Drummond,1953). (d) Diplura. Section through an eversible vesicle of *Campodea*, everted (after Drummond,1953).

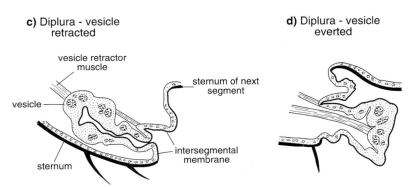

c) Diplura - vesicle retracted

vesicle retractor muscle

vesicle

sternum of next segment

intersegmental membrane

sternum

d) Diplura - vesicle everted

sternum. Like those on the ventral tube of Collembola, these vesicles can absorb water from the substratum (Drummond, 1953).

There are pairs of appendages on each of the first three segments of the abdomen of Protura. At their most fully developed they are two-segmented with an eversible vesicle at the tip (Fig. 11.5b). The appendages are moved by extrinsic and intrinsic muscles, which include a retractor muscle of the vesicle.

Abdominal appendages of Collembola The Collembola have pregenital appendages on three abdominal segments (see Fig. 8.25d). From the first segment a median lobe projects forwards and down between the last pair of legs. This is known as the ventral tube and at its tip are a pair of eversible vesicles which in many Symphypleona are long and tubular. The unpaired basal part of the ventral tube is believed to represent the fused coxae of the segmental appendages and the vesicles are thus coxal

vesicles. The vesicles are everted by blood pressure from within the body and are withdrawn by retractor muscles.

The ventral tube appears to have two functions. In some circumstances it functions as an adhesive organ enabling the insect to walk over smooth or steep surfaces. To facilitate this on a dry surface the vesicles are moistened by a secretion from cephalic glands opening on to the labium and connecting with the ventral tube by a groove in the cuticle in the ventral midline of the thorax. The ventral tube also enables Collembola to adhere to the surface film on water since it is the only part of the cuticle which is wettable; all the rest is strongly hydrofuge.

The second function of the vesicles of the ventral tube is the absorption of water from the substratum (section 18.4.1.2).

The appendages of the third and fourth segments of the abdomen of many Collembola form the retinaculum and the furca, which are used in locomotion (section 8.4.2.2).

11.2.2 Larval structures associated with locomotion and attachment

Leg-like outgrowths of the body wall, known as prolegs, are common features of the abdomen of holometabolous larvae. These appendages are expanded by blood pressure and moved mainly by the muscles of the adjacent body wall together with others inserted at the base of the proleg and a retractor muscle extending to the sole or planta surface (see Fig. 8.27).

Well-developed prolegs are a feature of lepidopterous larvae, which usually have a pair on each of abdominal segments 3–6 and 10 (see Fig. 15.5). Megalopygidae have more prolegs than other Lepidoptera with prolegs on segments 2–7 and 10. Those on segments 2 and 7 have no crochets. More frequently the number of prolegs is reduced and in Geometridae there are usually only two pairs, on segments 6 and 10. Prolegs are completely absent from some leaf-mining larvae and from the free-living Eucleidae, some of which, however, have weak ventral suckers on segments 1–7.

Distally, where it makes contact with the substratum, the proleg is flattened, forming the planta surface. This is usually armed with hook-shaped structures called crochets (see Fig. 8.27). The arrangement of the crochets varies. They may form a complete ring, or be arranged in transverse or longitudinal rows, reflecting the behavior of the larva and the nature of the substrate on which it lives (see Stehr, 1987).

Digitiform prolegs without crochets occur on the first eight abdominal segments of larval Mecoptera. They have no intrinsic musculature, but are moved by changes in blood pressure and by the action of muscles on adjacent parts of the ventral body wall. Prolegs without crochets also occur on the abdomen of larval Symphyta and particularly in the Tenthredinoidea. The number varies from six to nine pairs.

Larval Trichoptera have anal prolegs on segment 10 (see Fig. 11.8b). Their development varies, but in the Limnephilidae, where they are most fully developed, there are two basal segments and a terminal claw, having both levator and depressor muscles. These appendages, together with a dorsal and two lateral retractile papillae on the first abdominal segment, enable the larva to hold on to its case.

Sometimes, when prolegs are not developed, their position is occupied by a raised pad armed with spines. Such a pad is called a creeping welt and is functionally comparable with a proleg. Creeping welts and prolegs are present in many dipterous larvae, some of which have several prolegs

a) tabanid prolegs

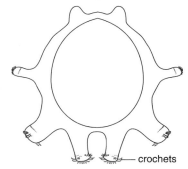

crochets

b) blepharocerid sucker

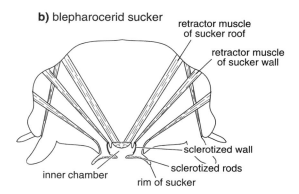

retractor muscle of sucker roof

retractor muscle of sucker wall

sclerotized wall

sclerotized rods

inner chamber

rim of sucker

Fig. 11.6. Prolegs and suckers of larval Diptera (after Hinton, 1955). (a) Cross-section of an abdominal segment of a tabanid larva showing several pairs of prolegs, some with crochets. (b) Transverse section through the sixth abdominal segment of a blepharocerid larva showing the ventral sucker.

on each segment (Fig. 11.6a) while others have creeping welts which extend all round the segment. The larvae of a number of families of Diptera have abdominal suckers which may be derived from prolegs. Thus the larva of the psychodid *Maruina* has a sucker on each of abdominal segments 1–8 and these suckers enable the larva to maintain its position along the sides of waterfalls. In another larva, of *Horaiella*, a single large sucker, bounded by a fringe of hairs, extends over the ventral surface of several segments.

Larval Blepharoceridae, which live in fast-flowing streams and waterfalls, have a sucker on each of abdominal segments 2–7. Each sucker has an outer flaccid rim with an incomplete anterior margin. The central disc of the sucker is supported by closely packed sclerotized rods and in the middle a hole leads into an inner chamber with strongly

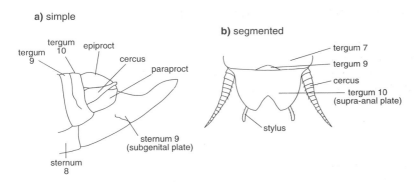

a) simple

b) segmented

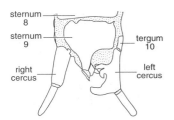

c) asymmetrical

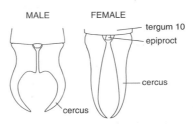

d) sexually dimorphic

Fig. 11.7. Different types of cerci. (a) Simple cercus [lateral view of tip of abdomen of male red locust, *Nomadacris* (Orthoptera)]. (b) Segmented [dorsal view of the tip of the abdomen of male *Periplaneta* (Blattodea)]. (c) Asymmetrical [ventral view of tip of abdomen of male *Idioembia* (Embioptera)]. (d) Sexually dimorphic [male and female forceps (cerci) *Forficula* (Dermaptera)].

sclerotized walls and an extensively folded roof (Fig. 11.6b). Muscles inserted into the roof and the rim of the sclerotized walls of the inner chamber increase the volume of the chamber when they contract and if at the same time the rim of the sucker is pressed down on to the substratum a partial vacuum is created so that the sucker adheres to the surface.

Even if a well-formed sucker is not present, many dipterous larvae can produce a sucker-like effect by raising the central part of the ventral surface while keeping the periphery in contact with the substratum, the sucker being sealed and made effective by a film of moisture.

11.2.3 Sensory structures

Most insects have mechanosensitive sensilla on the abdominal segments, and grasshoppers also have small contact chemoreceptors scattered amongst the mechanoreceptors (Thomas, 1965). In addition, the appendages of segment 11 often form a pair of structures called cerci which usually function as sense organs.

Cerci are present and well-developed in the Apterygota and the hemimetabolous orders other than the hemipteroids. In holometabolous insects, cerci are present in the adults of Mecoptera and some Diptera; they are not present in holometabolous larvae. The cerci may be simple, unsegmented structures as in Orthoptera (Fig. 11.7a), or multi-segmented as in Blattodea and Mantodea (Fig. 11.7b). They may be very short and barely visible or long and filamentous, as long or longer than the body as in Thysanura, Ephemeroptera and Plecoptera. Even within a group, such as the Acridoidea, the range of form of the cerci is considerable (Uvarov, 1966). In cockroaches, where the cerci are segmented, additional segments are added at each molt. The first instar cercus of *Periplaneta* has three segments, while in the adult male it has 18 or 19 and in the adult female 13 or 14. Growth results from division of the basal segment.

The cerci are usually set with large numbers of trichoid sensilla. Sometimes these sensilla are filiform and are sensitive to air movements. This is true in cockroaches and crickets where different filiform hairs are maximally sensitive to air movements from different directions (section 23.1.3.1). In the female of the sheep blowfly, *Lucilia*, there are also a small number of contact chemoreceptors and olfactory receptors on the cerci (Merritt & Rice, 1984). This may also be true in other insects.

Sometimes the cerci differ in the two sexes of a species, and they may play a role in copulation. Thus the cerci of female *Calliptamus* (Orthoptera) are simple cones, but in the male they are elongate, flattened structures with two or three lobes at the apex armed with strong inwardly directed points. There is similar dimorphism in Embioptera, where the male

Fig. 11.8. Abdominal appendages of pterygote larvae. Membrane stippled. (a) Gill (*Sialis* (Megaloptera), dorsal view). (b) Gills and anal proleg (*Hydropsyche* (Trichoptera) lateral view of terminal abdominal segments). (c) Urogomphus (*Oodes* (Coleoptera) lateral view of terminal abdominal segments). (d) Caudal filament and segmented cerci (*Heptagenia* (Ephemeroptera) dorsal view of terminal abdominal segments).

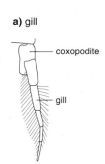

a) gill

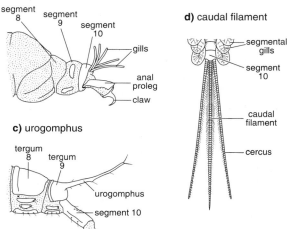

cerci are generally asymmetrical with the basal segment of the left cercus forming a clasping organ (Fig. 11.7c), and amongst the earwigs the cerci form powerful forceps which are usually straight and unarmed in the female, but incurved and toothed in the male (Fig. 11.7d). Similar forceps-like cerci in the Japygidae are used in catching prey. In some holometabolous insects they form part of the external genitalia.

The cerci of larval Zygoptera are modified to form the two lateral gills (see Fig. 17.28), while in the ephemeropteran *Prosopistoma* the long, feather-like cerci, together with the median caudal filament, can be used to drive the insect forwards by beating against the water.

11.2.4 Gills

Gills are present on the abdominal segments of the larvae of many aquatic insects. Ephemeroptera usually have six or seven pairs of plate-like or filamentous gills (see Figs. 15.4, 18.2) which are moved by muscles. They may play a direct role in gaseous exchange, but perhaps are more important in maintaining a flow of water over the body. Gill tufts may also be present on the first two or three abdominal segments, or in the anal region of larval Plecoptera. The larva of *Sialis* (Megaloptera) has seven pairs of five-segmented gills, each arising from a basal sclerite on the side of the abdomen (Fig. 11.8a), and a terminal filament of similar form arises from segment 9. Similar, but unsegmented gills are present in other larval Megaloptera and in some larval Coleoptera. Larval Trichoptera have filamentous gills in dorsal, lateral and ventral series.

Some aquatic larvae have papillae, often incorrectly called gills, surrounding the anus. They are concerned with salt regulation (section 18.3.2.2) and are found in larval mosquitoes and chironomids, where a group of four papillae surrounds the anus (see Fig. 18.3), and in some larval Trichoptera.

11.2.5 Secretory structures

Some insects have glands opening on the abdomen which probably have a defensive function in most cases (see Chapter 27). Most aphids have a pair of tubes, known as siphunculi, or cornicles, projecting from the dorsum of segment 5 or 6, or from between them. Each cornicle has a terminal opening which is normally closed by a flap of cuticle controlled by an opener muscle and the whole structure can be moved by a muscle inserted at the base so that the cornicle can be pointed in various directions, even forwards. Aphids release an alarm pheromone from the cornicles if they are attacked by parasites or predators (section 27.1.6.1). This causes a response in other aphids of the same species, but the response differs in different species. Individuals of *Schizaphis graminum* usually drop off the plant when they perceive the pheromone; *Myzus persicae* may drop off the plant, or walk away from the feeding site; other species jerk about in a manner which is presumed to discourage attack without withdrawing their stylets from the host. The effective radius of the pheromone may extend up to about 3 cm from the emitting aphid (Nault & Phelan, 1984).

The first abdominal appendages are well developed in the embryos of insects belonging to a number of groups.

They are known as pleuropodia (section 14.2.10), but they do not persist after hatching. Perhaps their primary function is the secretion of enzymes which digest the serosal cuticle prior to hatching.

11.2.6 Other abdominal structures

Apart from the segmentally arranged prolegs and gills some insects have other abdominal appendages which often appear to have a defensive role, but whose function is sometimes unknown. Some groups of insects have a median process projecting from the last segment. In Thysanura and Ephemeroptera this is in the form of a median caudal filament which resembles the two cerci (Fig. 11.8d). Larval Zygoptera have a median terminal gill on the epiproct, while in larval hawkmoths (Sphingidae) a terminal spine arises from the dorsum of segment 10.

Some larval Coleoptera have a pair of processes called urogomphi, which are outgrowths of the tergum of segment 9 (Fig. 11.8c). They may be short spines or multi-articulate filaments and they may be rigid with the tergum or arise from the membrane behind it so that they are mobile. Jeannel (1949) regards them as homologous with cerci, but see Crowson (1960).

Caterpillars of some families have branched projections of the body wall on some or all of the body segments, both thorax and abdomen. These projections are called scoli and possibly have defensive functions. Similar structures are found on some larval Mecoptera and Coleoptera (Stehr, 1987).

Sometimes the prolegs are modified for functions other than walking. In some Notodontidae the anal prolegs are modified for defensive purposes. Thus in the larva of the puss moth, *Cerura*, they are slender projections which normally point posteriorly, but, if the larva is touched, the tip of the abdomen is flexed forwards and a slender pink process is everted from the end of each projection. At the same time the larva raises its head and thorax from the ground and emits formic acid from a ventral gland in the prothorax. This reaction is presumed to be a defensive display.

In a few larval Diptera, prolegs may be used for holding prey. The larva of *Vermileo* lives in a pit in dry soil and feeds in the same way as an antlion. It lies ventral side up, and prey which fall into the pit are grasped against the thorax by a median proleg on the ventral surface of the first abdominal segment.

REFERENCES

Albrecht, F. O. (1956). The anatomy of the red locust, *Nomadacris septemfasciata* Serville. *Anti-Locust Bulletin* no. 23, 9 pp.

Bitsch, J. (1979). Morphologie abdominale des insectes. In *Traité de Zoologie*, vol. 8, part 2, ed. P.-P.Grassé, pp. 291–587. Paris: Masson et Cie.

Crowson, R. A. (1960). The phylogeny of Coleoptera. *Annual Review of Entomology*, **5**, 111–34.

Drummond, F. H. (1953). The eversible vesicles of *Campodea* (Thysanura). *Proceedings of the Royal Entomological Society of London* A, **28**, 115–8.

Hewitt, C. G. (1914). *The House-Fly, Musca domestica Linn*. Cambridge: Cambridge University Press.

Hinton, H. E. (1955). On the structure, function, and distribution of the prolegs of the Panorpoidea, with a criticism of the Berlese–Imms theory. *Transactions of the Royal Entomological Society of London*, **106**, 455–545.

Jeannel, R. (1949). Ordre des Coléoptéroïdes. In *Traité de Zoologie*, vol. 9, ed. P.-P.Grassé, pp. 771–891. Paris: Masson et Cie.

Matsuda, R. (1976). *Morphology and Evolution of the Insect Abdomen*. Oxford: Pergamon Press.

Merritt, D.J. & Rice, M.J. (1984) Innervation of the cercal sensilla on the ovipositor of the Australian sheep blowfly (*Lucilia cuprina*). *Physiological Entomology*, **9**, 39–47.

Nault, L.R. & Phelan, P.L. (1984). Alarm pheromones and presociality in pre-social insects. In *Chemical Ecology of Insects*. ed. W.J. Bell & R.T. Cardé, pp. 237–56. New York: Chapman & Hall.

Panganiban, G., Nagy, L. & Carroll, S.B. (1994). The role of the *distal-less* gene in the development and evolution of insect limbs. *Current Biology*, **4**, 671–5.

Snodgrass, R. E. (1935). *Principles of Insect Morphology*. New York: McGraw-Hill.

Snodgrass, R. E. (1956). *Anatomy of the Honey Bee*. London: Constable.

Stehr, F. W. (ed.) (1987) *Immature Insects*, vol. 1. Dubuque, Iowa: Kendall/Hunt Publishing Co.

Thomas, J.G. (1965) The abdomen of the female desert locust (*Schistocerca gregaria* Forskål) with special reference to the sense organs. *Anti-Locust Bulletin* no. 42, 20pp.

Uvarov, B. P. (1966). *Grasshoppers and Locusts*, vol 1. Cambridge: Cambridge University Press.

12   Reproductive system: male

12.1 ANATOMY OF THE INTERNAL REPRODUCTIVE ORGANS

The male reproductive organs typically consist of a pair of testes connecting with paired seminal vesicles and a median ejaculatory duct (Fig. 12.1). In most insects there are also a number of accessory glands which open into the vasa deferentia or the ejaculatory duct.

Testis The testes may lie above or below the gut in the abdomen and are often close to the midline. Usually each testis consists of a series of testis tubes or follicles ranging in number from one in Coleoptera Adephaga to over 100 in grasshoppers (Acrididae). Sometimes, as in Lepidoptera, the follicles are incompletely separated from each other (Fig. 12.2c), and the testes of Diptera consist of simple, undivided sacs, which may be regarded as single follicles. Sometimes the follicles are grouped together into several separate lobes (Fig. 12.1b). In the cerambycid, *Prionoplus*, for example, each testis comprises 12 to 15 lobes each with 15 follicles. The testes of Apterygota are often undivided sacs, but it is not certain in this case that they are strictly comparable with the gonads of other insects since the germarium occupies a lateral position in the testis instead of being terminal.

The wall of a follicle is a thin epithelium, sometimes consisting of two layers of cells, standing on a basal lamina. The follicles are bound together by a peritoneal sheath and if the two testes are close to each other they may be bound together. In some Lepidoptera, the two testes fuse completely to form a single median structure.

Vas deferens and seminal vesicle From each testis follicle, a fine, usually short, vas efferens connects with the vas deferens (plural: vasa deferentia) (Fig. 12.2b), which is a tube with a fairly thick bounding epithelium, a basal lamina and a layer of circular muscle outside it. The vasa deferentia run backwards to lead into the distal end of the ejaculatory duct[1].

---

[1] Note: distal and proximal are used throughout this text with reference to the origin of any structure. The ejaculatory duct originates as an invagination of the ectoderm. Thus, the part nearest the body wall at the point of origin is proximal; the part most remote from the body wall is distal.

At least some of the cells of the vas deferens are glandular, secreting their products into the lumen (Riemann & Giebultowicz, 1991).

The seminal vesicles, in which sperm are stored before transfer to the female, are dilations of the vasa deferentia in many insects (Fig. 12.1b), but in some Hymenoptera and nematoceran Diptera they are dilations of the ejaculatory duct. Lepidoptera have both structures: sperm are stored temporarily in expanded regions of the vasa deferentia and then are transferred to dilations in the upper part of the ejaculatory duct, known as the duplex (Fig. 12.14). In Orthopteroidea and some Odonata, Phthiraptera and Coleoptera, they are not simply expansions of the male ducts, but are separate structures (Fig. 12.1a). In some insects, the seminal vesicles are epidermal in origin and in these cases they are lined with cuticle.

The cellular lining of the seminal vesicles is glandular and probably provides nutrients for the sperm.

Ejaculatory duct The vasa deferentia join a median duct, called the ejaculatory duct, which usually opens posteriorly in the membrane between the ninth and tenth abdominal segments (gonopore in Fig. 12.7). Ephemeroptera have no ejaculatory duct and the vasa deferentia lead directly to the paired genital openings. Dermaptera, on the other hand, have paired ejaculatory ducts, although, in some species, one of the ducts remains vestigial. Thus in *Forficula* the righthand ejaculatory duct is fully functional while the lefthand duct is vestigial (Popham, 1965).

The epithelium of the ejaculatory duct is one cell thick and, as it is epidermal in origin, it is lined with cuticle. Often at least a part of the wall is muscular, although the ejaculatory duct in *Apis* is entirely without muscles. Parts of the wall of the duct may be glandular, contributing to the formation of the spermatophore.

Where a complex spermatophore is produced, the ejaculatory duct is also complex. Thus in *Locusta* (Orthoptera), the ejaculatory duct consists of upper and lower ducts connected via a funnel-like constriction (Fig. 12.13). The upper part of the duct into which the accessory glands open has a columnar epithelium and thin cuticle and the lumen is a vertical slit in cross section. In

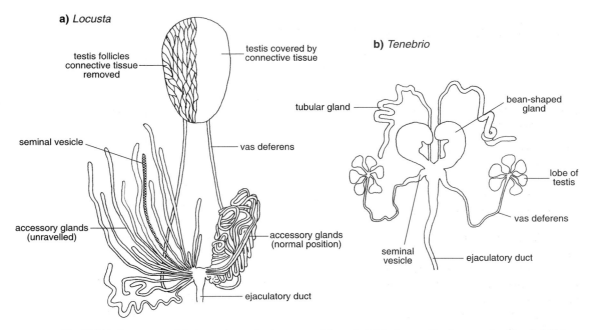

Fig. 12.1. Basic structure of the internal reproductive organs of the male. **(a)** An insect with a large number of testis follicles and accessory glands. The testes lie close together in the midline, but are distinct. The glands are of different types (but this is not shown in the diagram) (*Locusta*, Orthoptera) (from Uvarov, 1966). **(b)** An insect with several distinct testis lobes and only two pairs of accessory glands (*Tenebrio*, Coleoptera) (from Imms, 1957).

the funnel, the cuticle is thicker and forms nine (usually) ridges on either side. These curve upwards posteriorly as they run back to meet in the dorsal midline and they project so that they almost completely divide the lumen. The lumen of the lower duct is circular in cross section and leads to the ejaculatory sac and spermatophore sac. Scattered muscle fibers are present in the wall of the upper duct but are absent elsewhere. The ejaculatory duct of the milkweed bug, *Oncopeltus*, is also extremely complex, being specialized for erection of the penis (Fig. 12.10).

The ejaculatory duct in Lepidoptera is extended inwards as an unpaired duct of mesodermal origin and so not lined with cuticle. This is called the simplex. It bifurcates distally to connect with the accessory glands and vasa deferentia and this part is known as the duplex (Fig. 12.14). The simplex produces most of the spermatophore since Lepidoptera have only a single pair of accessory glands. In *Calpodes*, it is divided into seven sections partially separated from each other by constrictions, and some of these are further differentiated into zones of cells with different structures. These different sections and zones

produce a variety of secretions which contribute to the spermatophore. The constrictions make it possible for the secretions of different sections to be used separately and in sequence (Lai-Fook, 1982a,b).

Accessory glands The male accessory glands may be ectodermal or mesodermal in origin, when they are known as ectadenia or mesadenia, respectively. Ectadenia, which open into the ejaculatory duct, occur in many Coleoptera, in the Diptera Nematocera and some Homoptera. Mesadenia, which open into the vasa deferentia or the distal end of the ejaculatory duct, are found in Orthoptera and many other orders. In some species of Heteroptera and Coleoptera, both ectadenia and mesadenia are present.

The number and arrangement of accessory glands varies considerably between different groups of insects. In Lepidoptera there is a single pair of glands (Fig. 12.14); in *Tenebrio*, there are two pairs (Fig. 12.1b). In contrast, *Schistocerca* and *Locusta* have 15 pairs of accessory glands, not counting the seminal vesicles with which they are closely associated (Fig. 12.1a), and *Gryllus* has over 600. Sometimes there are no morphologically distinct accessory

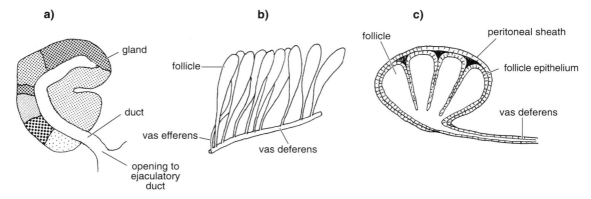

Fig. 12.2. Male reproductive organs. (**a**) Accessory gland. Diagram of the bean-shaped gland of *Tenebrio* (see Fig. 12.1b) showing that different regions contain cells producing different secretions (shown by different shading). Two additional cell types also occur, but are not in the plane of this section (after Dailey, Gadzama & Happ, 1980). (**b**) Testis. Diagram of a series of testis follicles opening independently into the vas deferens, as in Orthoptera. (**c**) Testis. Diagrammatic section through a testis in which the follicles are incompletely separated from each other and have a common opening to the vas deferens, as in Lepidoptera (from Snodgrass, 1935).

glands. This is the case in Apterygota, Ephemeroptera and Odonata, and muscoid Diptera.

Each accessory gland consists of a single layer of epithelial cells whose fine structure varies depending on their stage of development and also on the nature of the secretion produced. Where the glands are few in number, different regions within them may be functionally distinguishable. In *Tenebrio*, for example, the tubular glands produce three classes of compounds, and the bean-shaped glands have eight morphologically distinguishable cell types secreting a number of different products (Fig. 12.2a). In Lepidoptera, the accessory glands are regionally differentiated to produce two different secretions (Fig. 12.14). On the other hand, where many glands are present, several glands may produce the same proteins. This is the case in the Orthoptera. In *Schistocerca*, seven of the 15 pairs of glands appear to produce a single product, while most of the other pairs each produce a unique product. Where there are few or no accessory glands, their role is sometimes taken over by glandular cells in the ejaculatory duct. This is most obvious in the lepidopteran simplex, but also occurs in muscoid Diptera.

Outside the epithelium is a muscle layer which, in *Gryllus*, consists of a single layer of fibers wound round the gland in a tight spiral, but with a more complex arrangement round the openings of the glands to the ejaculatory duct. In this insect, both sets of muscles are innervated by

proctolinergic DUM neurons (section 10.3.2.4) with cell bodies in the terminal abdominal ganglion. At least some of the muscles are also innervated by other neurons which may be inhibitory (Kimura, Yasuyama & Yamaguchi, 1989). In *Periplaneta*, the DUM cells innervating the accessory glands are octopaminergic (Sinakevitch *et al.*, 1994).

The accessory glands become functional in the adult insect. Their secretions are involved in producing the spermatophore, where one is present, and also in transferring to the female chemicals that modify her behavior and physiology (see below).

Reviews: Chen, 1984 – structure and biochemistry; Gillott, 1988 – general; Happ, 1984 – structure and development; Happ, 1992 – structure and control of development; Snodgrass, 1935 – structure

12.2 SPERMATOZOA

12.2.1 Structure of mature spermatozoa

The mature sperm of most insects are filamentous, often about 300 μm long and less than a micron in diameter with head and tail regions of approximately the same diameter (Fig. 12.3) (but the sperm of some *Drosophila* (Diptera) species may be 15 mm long). The sperm has a typical cell membrane about 10 nm thick, coated on the outside by a

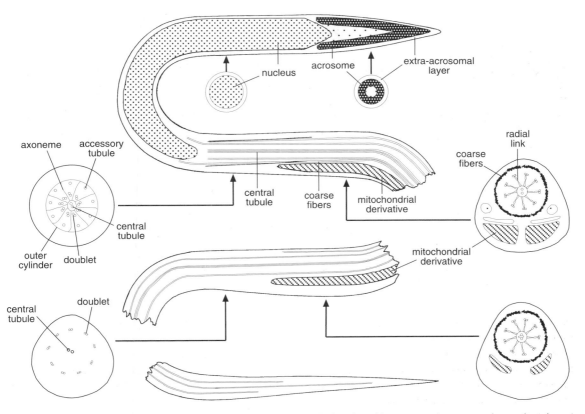

Fig. 12.3. Diagram showing the structure of a sperm in longitudinal section with representative cross-sections at the points shown.

layer of glycoprotein known as the glycocalyx. In fleas this is about 13 nm thick, and in grasshoppers about 30 nm. The glycocalyx is made up of rods at right angles to the surface of the sperm (Fig. 12.4a). Lepidopteran sperm have a series of projections called lacinate appendages running along their length. They are made up of thin laminae stacked parallel with, but external to, the surface membrane (Fig. 12.4b). These structures are no longer present when the sperm are released from the cyst (section 12.2.4), but the material of which they are composed may subsequently be used in binding the sperm together.

The greater part of the head region is occupied by the nucleus (Fig. 12.3). In mature sperm of most species, the nucleus is homogeneous in appearance, but sometimes, as in the grasshopper *Chortophaga*, it has a honeycomb appearance. The DNA is apparently arranged in strands parallel with the long axis of the sperm. In Lepidoptera, a second kind of sperm is produced in addition to the normal nucleate (eupyrene) sperm. These sperm are

without nuclei (apyrene) so that they cannot effect fertilization of the egg. Spermatids which give rise to them have numerous micronuclei instead of a single nucleus and these micronuclei subsequently break down completely. Apyrene sperm are formed in separate cysts from the eupyrene sperm.

In front of the nucleus is the acrosome. This is a membrane-bound structure of glycoprotein with, in most insects, a granular extra-acrosomal layer and an inner rod or cone. Neuropteran sperm have no acrosome and occasional species with no acrosome occur in other orders. The acrosome is probably involved with attachment of the sperm to the egg and possibly also with lysis of the egg membrane, thus permitting sperm entry.

Immediately behind the nucleus, the axial filament, or axoneme, arises. In most cases this consists of two central tubules with a ring of nine doublets and nine accessory tubules on the outside (Figs. 12.3, 12.4). The central tubules are surrounded by a sheath and are linked radially

Fig. 12.4. Sperm structure. (a) Electron micrograph of a transverse section through the tail region showing the glycocalyx (grasshopper, after Longo *et al.*, 1993). (**b**) Electron micrograph of a transverse section of a lepidopteran sperm showing the lacinate appendages (after Jamieson, 1987).

a) Orthoptera – tail region

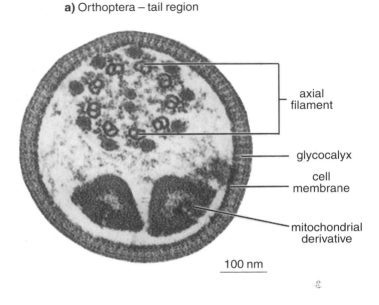

axial
filament

glycocalyx

cell
membrane

mitochondrial
derivative

100 nm

b) Lepidoptera – head region

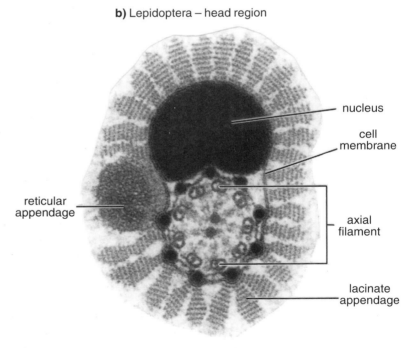

nucleus

cell
membrane

reticular
appendage

axial
filament

lacinate
appendage

100 nm

to the doublets. Additional fibers are usually present between the accessory tubules. Some unusual exceptions to this 9 + 9 + 2 arrangement occur. Accessory tubules are lacking in Collembola, Japygidae (Diplura), Mecoptera and Siphonaptera.

The sperm of some insects have two axial filaments. This occurs in Psocoptera, Phthiraptera, Thysanoptera and many bugs. In *Sciara* (Diptera), there is no well-organized axial filament, but 70–90 tubule doublets, each with an associated accessory tubule, are arranged in a spiral which encloses the mitochondrial derivative posteriorly. It is presumed that the axial filament or the equivalent structure causes the undulating movements of the tail which drive the sperm forwards.

The sperm of Pterygota have two mitochondrial derivatives which flank the axial filament. Within these the cristae become arranged as a series of lamellae projecting inwards from one side of the derivative and at right angles to its long axis. The matrix of the derivative is occupied by a paracrystalline material. Sperm of Mecoptera and Trichoptera and species of some other orders have only one mitochondrial derivative, while phasmids have none at all (but this does not mean that they are without respiratory enzymes, see below). More or less normal mitochondria persist in the sperm of Apterygota and non-insect Hexapoda except that they fuse together and become elongated. There are three such mitochondria in the sperm tail of Collembola, and two in Diplura and Machilidae.

Coccid sperm occur in bundles and lack all the typical organelles. The nucleus is represented by an electron-opaque core with no limiting membrane. Mitochondrial derivatives are absent, but the homogeneous cytoplasm of the sperm probably contains the enzymes with a respiratory function. Each sperm has 45–50 microtubules in a spiral round a central mass of chromatin. They run the whole length of the sperm and may be concerned with its motility, replacing the typical axial filament (Swiderski, 1980).

Sperm of Kalotermitidae and Rhinotermitidae have no flagellum at all. The sperm of *Reticulitermes* (Rhinotermitidae) is spherical with no acrosome, but it has a few normal mitochondria and two short axial filaments although these do not extend into a tail. It is presumed that this sperm is non-motile. Non-motile sperm also occur in the dipteran family Psychodidae and in *Eosentomon* (Protura).

Review: Jamieson, 1987 – structure and evolution

12.2.2 Sperm bundles

In a number of insects, sperm are grouped together in bundles which sometimes persist even after transfer to the female. The sperm of *Thermobia* (Thysanura) normally occur in pairs, the two individuals being twisted round each other with their membranes joined at points of contact (Dallai & Afzelius, 1984). Pairs of sperm also occur in some Coleoptera.

Coccids have much more specialized sperm bundles. In these insects, each cyst (see section 12.2.3) commonly produces 32 sperm which may become separated into two bundles of about 16. Some species have 64 sperm in a bundle. Each bundle becomes enclosed in a membranous sheath and the cyst wall degenerates. The sheath of many species, such as *Pseudococcus*, is longer than the sperm, which occupy only the middle region, and the head-end of the sheath has a corkscrew-like form. The sperm bundles of *Parlatoria*, on the other hand, are only the same length as the sperm, which are all oriented in the same direction within the bundle. Movement of the bundles does not normally occur until they enter the female. It results from the combined activity of the sperm within. Subsequently, the sheath of the bundle ruptures and the sperm are released.

In some Orthoptera and Odonata, different types of sperm bundle, known as spermatodesms, are formed. The spermatodesms of tettigoniids comprise about ten sperm anchored together by their acrosomes. These bundles are released from the testis and the sperm heads then become enclosed in a muff of mucopolysaccharide secreted by the gland cells of the vas deferens. In acridids, the spermatodesms are completely formed within the testis cyst and may include all the sperm within the cyst. The spermatids come to lie with their heads oriented towards a cyst cell and extracellular granular material round the acrosome of each coalesces to form a cap in which the heads of all the sperm are embedded. The spermatodesms of Acrididae persist until they are transferred to the female.

12.2.3 Spermatogenesis

At the distal end of each testis follicle is the germarium, in which the germ cells divide to produce spermatogonia (cells which divide mitotically to produce spermatocytes; spermatocytes divide meiotically to produce spermatids) (Fig. 12.5). In Orthoptera, Blattodea, Homoptera and Lepidoptera, the spermatogonia probably obtain nutriment

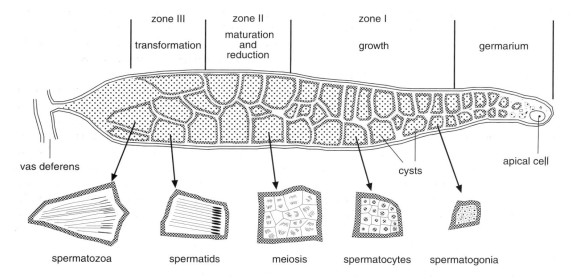

Fig. 12.5. Diagram of a testis follicle showing the sequence of stages of development of the sperm (from Wigglesworth, 1965).

from a large apical cell with which they have cytoplasmic connections, while in Diptera and Heteroptera an apical syncytium performs a similar function. Transfer of mitochondria from this syncytium to the spermatogonia has been observed in Diptera.

The apical connections are soon lost and the spermatogonia associate with other cells which form a cyst around them (Fig. 12.5). One, or sometimes more, spermatogonia are enclosed in each cyst and, in *Prionoplus*, there are initially two cyst-cells round each spermatogonium. They may supply nutriment to the developing sperm and, in *Popillia* (Coleoptera), nutrient transfer may be facilitated by the sperm at one stage having their heads embedded in the cyst-cells. In Heteroptera, large cells with irregular nuclei, called trophocytes, are scattered amongst the cysts.

As more cysts are produced at the apex of a follicle, they displace those which have developed earlier so that a range of developmental stages is present in each follicle with the earliest stages distally in the germarium and the oldest in the proximal part of the follicle adjacent to the vas deferens. Three zones of development are commonly recognized below the germarium (Fig. 12.5):

I: a zone of growth, in which the primary spermatogonia, enclosed in cysts, divide and increase in size to form spermatocytes;

II: a zone of maturation and reduction, in which each spermatocyte undergoes the two meiotic divisions to produce spermatids;

III: a zone of transformation, in which the spermatids develop into spermatozoa, a process known as spermiogenesis.

The number of sperm ultimately produced by a cyst depends on the number of spermatogonial divisions and this is fairly constant for a species. In grasshoppers (Acrididae), there are between five and eight spermatogonial divisions and *Melanoplus*, which typically has seven divisions before meiosis, usually has 512 sperm per cyst. Normally four spermatozoa are produced from each spermatocyte, but in many coccids the spermatids which possess heterochromatic chromosomes degenerate so that only two sperm are formed from each spermatocyte and 32 are present in each cyst. In *Sciara* (Diptera), only one spermatid is formed from each spermatocyte because of an unequal distribution of chromosomes and cytoplasm at the meiotic divisions.

Review: Tuzet, 1977a

12.2.3.1 *Spermiogenesis*

The spermatid produced at meiosis is typically a rounded cell containing normal cell organelles. It subsequently becomes modified to form the sperm and this process of spermiogenesis entails a complete reorganization of the cell. It is convenient to consider separately each organelle of the mature sperm.

Acrosome The acrosome is derived, at least in part, from Golgi material, which in spermatocytes is scattered through the cytoplasm in the form of dictyosomes. There may be 30 or 40 of these in the cell and they consist of several pairs of parallel membranes with characteristic vacuoles and vesicles. After the second meiotic division the dictyosomes in *Acheta* fuse to a single body called the acroblast, which consists of 6–10 membranes forming a cup with vacuoles and vesicles both inside and out. In the later spermatid, a granule, called the pro-acrosomal granule, appears in the cup of the acroblast and increases in size. The acroblast migrates so that the open side faces the nucleus, and then the granule, associated with a newly developed membrane, the interstitial membrane, moves towards the nucleus and becomes attached to it. As the cell elongates, the acroblast membranes migrate to the posterior end of the spermatid and are sloughed off together with much of the cytoplasm and various other cell inclusions. The pro-acrosomal granule then forms the acrosome, becoming cone-shaped and developing a cavity in which an inner cone is formed. In *Gelastocoris* (Heteroptera), the pro-acrosome is formed from the fusion of granules in the scattered Golgi apparatus and no acroblast is formed. This may also be the case in Acrididae.

Nucleus In the early spermatid of grasshoppers the nucleus appears to have a typical interphase structure with the fibrils which constitute the basic morphological units of the chromosomes unoriented. As the sperm develops, the nucleus becomes very long and narrow and the chromosome fibrils become aligned more or less parallel with its long axis. The nucleoplasm between the fibrils is progressively reduced until finally the whole of the nucleus appears to consist of a uniformly dense material. A similar linear arrangement of the chromosomes occurs in other groups.

Mitochondria In the spermatid, the mitochondria fuse to form a single large body, the nebenkern, consisting of an outer limiting membrane and a central pool of mitochondrial components. The nebenkern separates into two mitochondrial derivatives associated with the developing axial filament immediately behind the nucleus. They elongate

to form a pair of ribbon-like structures. At the same time, their internal structure is reorganized so that the cristae form a series of parallel lamellae along one side and the matrix is replaced by paracrystalline material.

Centriole and axial filament Young spermatids contain two centrioles oriented at right angles to each other and each composed, as in most cells, of nine triplets of tubules. One gives rise to the axial filament, but ultimately both centrioles disappear. The tubules of the axial filament grow out from the centriole and finally extend the length of the sperm's tail. The accessory tubules arise from tubule doublets, appearing first as side arms, which become C-shaped and then separate off and close up to form cylinders. The development of accessory tubules is discussed by Dallai and Afzelius (1993).

12.2.3.2 *Biochemical changes*

The repeated cell divisions during spermatogenesis entail the synthesis of large amounts of DNA and RNA, but synthesis of DNA stops before meiosis occurs, while RNA synthesis continues into the early spermatid. Subsequently, no further synthesis occurs and RNA is eliminated first from the nucleus and then from the cell as the nucleus elongates. The reduction in RNA synthesis is associated with a rise in the production of an arginine-rich histone which forms a complex with DNA stopping it from acting as a primer for RNA synthesis, and, perhaps, insulating the genetic material from enzymic attack during transit to the egg.

12.2.3.3 *Control of spermatogenesis*

The spermatocytes reach meiosis before the final molt in most insect species and, in species which do not feed as adults, the whole process of spermatogenesis may be complete before adult eclosion (Fig. 12.6). In many species, however, spermatogenesis continues for an extended period and new sperm continue to be produced throughout adult life. The time taken for the completion of spermatogenesis varies, but in *Melanoplus* the spermatogonial divisions take eight or nine days and spermiogenesis, ten. In the skipper butterfly, *Calpodes*, spermiogenesis takes about four days (Lai-Fook, 1982c).

The factors regulating spermatogenesis are not well understood. At least in Lepidoptera, ecdysteroids have some role, but this differs from species to species. Spermatogenesis in the larva proceeds as far as prophase of

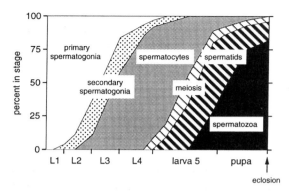

Fig. 12.6. Spermatogenesis sometimes continues through the life of the insect. Notice that, in this example, the earliest developing cells undergo meiosis during the fourth larval stage, but later developing cells are in this stage in the late pupa (*Bombyx*, Lepidoptera) (after Engelmann, 1970).

the first meiotic division apparently without any hormonal involvement, but is then delayed. In *Manduca*, further development is initiated by a peak of ecdysteroid during the wandering period of the last larval stage and then it continues through the early pupal period. High titers of ecdysteroid late in the pupal stage may inhibit further meiosis in spermatocytes developing into eupyrene sperm (Friedländer & Reynolds, 1992). In contrast, in *Heliothis*, a factor produced by the testis sheath is necessary for meiosis to occur. This factor is not an ecdysteroid although subsequently the sheath does produce ecdysteroids that regulate both the fusion of the two testes into a single structure and the development of the internal genital tract during the pupal period (Giebultowicz, Loeb & Borkovec, 1987; Loeb, 1991).

Juvenile hormone regulates maturation of the accessory glands in the newly eclosed adult insect and may affect reproductive behavior (Happ, 1992).

12.2.4 Transfer of sperm to the seminal vesicle
In some Heteroptera, in *Chortophaga* (Orthoptera), and possibly in other insects, the sperm make a complex circuit of the testis follicle before they leave the testis, moving in a spiral path to the region of the secondary spermatocytes and then turning back and passing into the vas deferens. In *Chortophaga*, the movement occurs after the spermatodesm is released from the cyst, but in the heteropteran, *Leptocoris*, the sperm are still enclosed in the cyst. In this case, the displacement starts while the spermatids are still

differentiating and is at least partly due to the elongation of the cyst which occurs during sperm development.

The fate of the cyst-cells is variable. In *Prionoplus* they break down in the testis, but in *Popillia*, although the sperm escape from the cysts as they leave the testis, the cyst-cells accompany the sperm in the seminal fluid into the bursa of the female. Here they finally break down and it has been suggested that they release glycogen used in the maintenance of the sperm.

In some Lepidoptera, the release of sperm from the testis occurs in the pharate adult. At first, this is inhibited by the high titer of ecdysteroid in the hemolymph, but release is permitted when this drops to a low level and then occurs with a circadian rhythmicity. Sperm start to move into the vas deferens towards the end of the light phase and remain there during the scotophase. Then, early in the next light phase, they are moved to the seminal vesicles. The cells of the upper vas deferens also show secretory activity which is at a maximum when the sperm move out of the testes and again when the sperm are transferred to the seminal vesicles (Riemann & Giebultowicz, 1991). The sperm are inactive in the vas deferens and are carried along by peristaltic movements of the wall of the tube. They remain immobile in the seminal vesicle, where they are often very tightly packed and in some cases, as in *Apis*, the heads of the sperm are embedded in the glandular wall of the vesicle.

12.3 TRANSFER OF SPERM TO THE FEMALE

Sperm transfer from a male to a female involves several different activities: the location of one sex by the other, courtship, pairing, copulation, and, finally, the insemination of the female. Location of one sex by the other may involve the visual, auditory and olfactory senses. These are considered in Chapters 25–27. Courtship is not considered here (see Thornhill & Alcock, 1983). This section deals with the mechanisms of copulation and female insemination.

12.3.1 External reproductive organs of the male
The external reproductive organs of the male are concerned in coupling with the female genitalia and with the intromission of sperm. They are known collectively as the genitalia.

There is considerable variation in structure and terminology of the genitalia in different orders (see Tuxen,

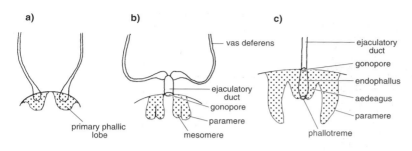

Fig. 12.7. Diagrams illustrating the origin and development of the phallic organ (after Snodgrass, 1957).

1956, for terminology) and the problems of homologizing the different structures are outlined by Scudder (1971). According to Snodgrass (1957) the basic elements are derived from a pair of primary phallic lobes which are present in the posterior ventral surface of segment 9 of the embryo (Fig. 12.7a). They are commonly regarded as representing limb buds and the structures arising from them as derived from typical appendages. Snodgrass (1957), however, believes that they may represent ancestral penes rather than appendages of segmental origin. These phallic lobes divide to form an inner pair of mesomeres and outer parameres, collectively known as the phallomeres (Fig. 12.7b). The mesomeres unite to form the aedeagus, the intromittent organ. The inner wall of the aedeagus, which is continuous with the ejaculatory duct, is called the endophallus, and the opening of the duct at the tip of the aedeagus is the phallotreme (Fig. 12.7c). The gonopore is at the outer end of the ejaculatory duct where it joins the endophallus and hence is internal, but in many insects the endophallic duct is eversible and so the gonopore assumes a terminal position during copulation. The parameres develop into claspers, which are very variable in form. They may be attached with the aedeagus on a common base, the phallobase, and in many insects these basic structures are accompanied by secondary structures on segments 8, 9 or 10. The term phallus is used by Snodgrass (1957) to mean the parameres together with the aedeagus, but is often used to mean the aedeagus alone; penis is sometimes used instead of phallus.

No intromittent organ is present in Collembola or Diplura. Male Archaeognatha and Thysanura have terminal segments similar to those in females, but with a median phallus which is bilobed in Thysanura. In these groups, sperm are not transferred directly to the female (see section 12.3.3.1). Paired penes are present in Ephemeroptera and some Dermaptera, but in the majority of pterygote insects there is a single median aedeagus. This is protected from injury in various ways. In grasshoppers and fulgorids, the sternum of the last abdominal segment extends to form a subgenital plate (see Fig. 11.7a). In many Endopterygota, protection is afforded by withdrawal of the genital segments within the preceding abdominal segments.

Other components of the male external genitalia are concerned with grasping the female. They are often called claspers. They may be derived from the parameres, from cerci, as in Dermaptera and many Orthoptera, or from the paraprocts, as in Zygoptera and some Tridactyloidea. In many Plecoptera, and occasionally in other orders, there are no claspers, the sexes being held together by the fit of the intromittent organ into the female bursa.

Many male Diptera have the terminal abdominal segments rotated so that the relative positions of the genitalia are altered. In Culicidae, some Tipulidae, Psychodidae, Mycetophilidae and some Brachycera, segment 8 and the segments behind it rotate through 180° soon after eclosion. As a result, the aedeagus comes to lie above the anus instead of below it and the hindgut is twisted over the reproductive duct (Fig. 12.8a). The rotation may occur in either a clockwise or an anticlockwise direction. In Calliphora, and probably in all Schizophora, the terminal segments are rotated through 360° in the pupa so that the genitalia are in their normal positions at eclosion, but the movement is indicated by some asymmetry of the preceding sclerites and by the ejaculatory duct looping right round the gut (Fig. 12.8b). The extent to which different segments rotate varies in different groups. Amongst the Syrphidae, a total twist of 360° is achieved by two segments rotating through 90° and one through 180° so that there is an obvious external asymmetry. Temporary rotation of the genital segments during copulation occurs in some other insects, such as Heteroptera (Fig. 12.10).

Male Odonata differ from all other insects in having the intromittent organs on abdominal segments 2 and 3.

a) 180° rotation

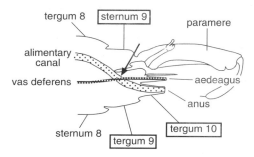

b) 360° rotation

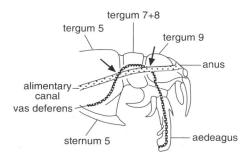

Fig. 12.8. Diagrams illustrating rotation of the terminal segments in male Diptera. Arrows show points at which the relative positions of alimentary canal and vas deferens change (from Séguy, 1951). (**a**) Rotation of the ninth and following segments through 180°. Notice that segments 9 and 10 are inverted (boxed labels) (*Aedes*). (**b**) Rotation through 360° indicated by the vas deferens twisting over the hindgut in a muscid.

Appendages which are used to clasp the female are present on segment 10, but the genital apparatus on segment 9 is rudimentary. A depression on the ventral surface of segment 2, known as the genital fossa, opens posteriorly into a vesicle derived from the anterior end of segment 3. In Anisoptera, the vesicle connects with a three-segmented penis and laterally there are various accessory lobes which guide and hold the tip of the female abdomen during intromission; the whole complex is termed the accessory genitalia (Fig. 12.9). Sperm are transferred to the vesicle from the terminal gonoduct by bending the abdomen forwards. This may occur before the male grasps the female, as in *Libellula*, or after he has grasped her, but before copulation,

as in *Aeschna*. The possible origins of the accessory genitalia are discussed by Corbet (1983). In species of dragonfly that copulate when settled, the penis serves to remove sperm of other males already in the female's spermatheca before he introduces his own sperm (Corbet, 1980).

12.3.2 Pairing and copulation

When the sexes come together, one sex commonly mounts on the back of the other. In cockroaches, and some gryllids and tettigoniids, the female climbs on to the back of the male. Another common position is with the male on the female; this occurs, for instance, in Tabanidae. Sometimes, as in Acrididae, although the male sits on top of the female, his abdomen is twisted underneath her during copulation. The abdomen of the male is also twisted under the female in insects, such as the scorpion fly (*Panorpa*) which lie side by side at the start of copulation. In other groups, the insects pair end to end and, in this case, the terminal segments of the male are often twisted through 180°. The end-to-end position is achieved with the male on his back in some tettigoniids and a few Diptera, while, in Culicidae, male and female lie with their ventral surfaces in contact.

In pairing the male usually grasps the female with his feet. In *Aedes aegypti*, for instance, the insects lie with their ventral surfaces adjacent and the male holds the female's hind legs in a hollow of the distal tarsomere by flexing back the pretarsus. His middle and hind legs push up the female abdomen until genital contact is established and then his middle legs may hook on to the wings of the female, while his hind legs hang free. Some male Hymenoptera, such as *Ammophila*, hold the female with their mandibles instead of, or, in some species, as well as, the legs.

The legs of males are sometimes modified for grasping the female. For example, the fore legs of *Dytiscus* and some other beetles bear suckers (see Fig. 8.8b); spines on the middle femora of *Hoplomerus* (Hymenoptera) fit between the veins on the female's wings; and the hind femora of male *Osphya* (Coleoptera) are modified to grip the female's abdomen and elytra. Male fleas (Siphonaptera) and some male Collembola and Ischnocera (biting lice), have modified antennae with which they hold the female.

Male Odonata are exceptional in their manner of holding the female. At first, the male grasps the thorax of the female with his second and third pairs of legs, while the first pair touch the basal segments of her antennae. He then flexes his abdomen forwards and fits two pairs of claspers on his abdominal segment 10 into position on the head or

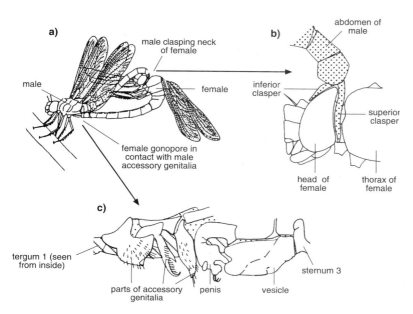

Fig. 12.9. Mating in Odonata. (a) Male and female copulating (*Aeschna*) (after Longfield, 1949). (b) Position of the male claspers round the neck of the female (*Aeschna*) (after Tillyard, 1917). (c) Male accessory genitalia, terga of left side removed (based on Chao, 1953).

thorax of the female. This completed, he lets go with his legs and the two insects fly off 'in tandem'. The claspers consist of superior and inferior pairs and, in Anisoptera, the superior claspers fit round the neck of the female while the inferior claspers press down on top of her head (Fig. 12.9b). In most Zygoptera, the claspers grip a dorsal lobe of the pronotum and, in some Coenagriidae, they appear to be cemented on by a sticky secretion (Corbet, 1983).

Copulation may occur immediately after the insects have paired or there may be a considerable interval before they copulate. In the cockroach, *Eurycotis*, where the female climbs on the male's back, the signal that the insects are in an appropriate position for copulation comes from mechanosensitive hairs on the male's first abdominal tergite. The hairs are stimulated when the female attempts to feed on a secretion produced by glands close to the hairs and this only occurs when she is appropriately placed for the male to copulate (Farine *et al.*, 1996). Once the male and female genitalia are linked, the insects may alter their positions and it is common among Orthoptera and Diptera for an end to end position to be adopted at this time.

The details of copulation vary from group to group depending on the structure of the genitalia, and only a few examples are given. In Acrididae, the tip of the male's abdomen is twisted below the female and the edges of the epiphallus (a plate on top of the genital complex) grip the sides of the female's sub-genital plate and draw it down

into the male's anal depression. The male uses his cerci to hold the female's abdomen and the aedeagus is inserted between the ventral valves of the ovipositor.

The male of *Oncopeltus* mounts the female, the genital capsule is rotated through 180°, mainly by muscular action, and the male's parameres (claspers) grasp the female's ovipositor valves. Following insertion of the aedeagus, the insects assume an end to end orientation in which they are held together mainly by the aedeagus (Fig. 12.10). An end to end position is also taken up by *Blattella* (Blattodea), but at first the female climbs on the back of the male, who engages the hook on his left phallomere on a sclerite in front of the ovipositor. Then, in the end to end position, the lateral hooks on either side of the anus and a small crescentic sclerite take a firm grip on the ovipositor (Fig. 12.11).

Copulation in Odonata involves the male flexing his abdomen so that the head of the female touches his accessory genitalia; she then brings her abdomen forwards beneath her so as to make contact with the accessory genitalia (Fig. 12.9a,c). Some species, such as *Crocothemis*, copulate and complete sperm transfer in flight and, in these insects, copulation is brief, lasting less than 20 s. Many species, however, settle before copulating and the process may last for a few minutes or an hour or more. During most of this time, the male is removing sperm already present in the female's spermatheca; sperm transfer takes only a few seconds (Corbet, 1980).

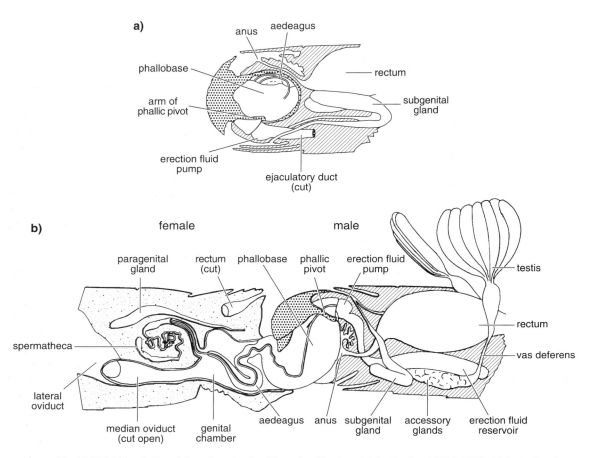

Fig. 12.10. Male genitalia and direct insemination (*Oncopeltus*, Hemiptera) (after Bonhag & Wick, 1953). **(a)** Sagittal section of the male genital capsule with the aedeagus retracted. **(b)** Sagittal section through the terminal segments of a copulating pair. Notice the inversion of the male genital capsule (with the anus now ventral to the phallic pivot) and the insertion of the aedeagus into the spermatheca.

The duration of copulation in other insects is equally variable. In mosquitoes the process is complete within a few seconds, while in *Oncopeltus* the insects may remain coupled for five hours, in *Locusta* for eight to ten hours, and in *Anacridium* (Orthoptera) for up to 60 hours. Insemination is completed much more rapidly than this; in *Locusta* sperm reach the spermatheca within two hours of the start of copulation.

12.3.3 Insemination

In the insects, the transfer of sperm to the female (insemination) is a quite separate process from fertilization of the eggs, which in some cases does not occur until months or even years later. During this interval the sperm are stored in the female's spermatheca. Sperm may be transferred in a spermatophore produced by the male, or they may be passed directly into the spermatheca without a spermatophore being produced.

12.3.3.1 *Spermatophores*

The primitive method of insemination in insects involves the production by the male of a spermatophore, a capsule in which the sperm are conveyed to the female. Spermatophores, of varying complexity, are produced by the Apterygota, Orthoptera, Blattodea, some Heteroptera, all the Neuroptera except Coniopterygidae, some Trichoptera, Lepidoptera, some Hymenoptera and Coleoptera and a few Diptera Nematocera and *Glossina*.

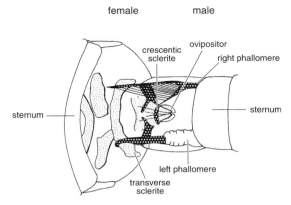

female   male

crescentic   ovipositor
sclerite
                          right phallomere

sternum   sternum

                    left phallomere

transverse
sclerite

Fig. 12.11. Male genitalia. Ventral view of the terminal segments of a male and female cockroach (*Blattella*) showing the manner in which the male genitalia clasp the female. The insects are represented in the end-to-end position with the subgenital plates and endophallus of the male removed. Female reproductive sclerites, light stippling; male reproductive sclerites, dark stippling (after Khalifa, 1950).

The spermatophore may be little more than a drop of sperm-containing fluid deposited in the environment, but more usually it is a discrete structure that may be preformed by the male and deposited on the female, or produced by the male when he encounters a female, or produced by male secretions in the female ducts during copulation.

Structure and transmission In Collembola, the male deposits spermatophores on the ground quite independently of the presence of a female. The spermatophore consists of a sperm-containing droplet, without any surrounding membrane, mounted on top of a stalk which is often about 500 μm high. Sometimes spermatophores are produced in aggregations of Collembola, so there is a good chance of a female finding one and inserting it into her reproductive opening, while, in other cases, the male grasps the female by her antennae and leads her over the spermatophore. The spermatophores of *Campodea* (Diplura) are also produced in the absence of the female and, like those of Collembola, each one consists of a globule 50–70 μm in diameter mounted on a peduncle 50–100 μm high (Fig. 12.12a). The globule has a thin wall which encloses a granular fluid, floating in which are from one to four bundles of sperm. The sperm can survive in a spermatophore for two days. A male may produce some 200 spermatophores in a week, but at least some of these

will be eaten by himself and other insects. Male *Lepisma* (Thysanura) also deposit spermatophores on the ground, but only in the presence of females. The male spins silk threads over the female, making side to side movements with his abdomen, so that her movements are restricted, and she is then guided over the spermatophore, which she inserts into her genital duct. In *Machilis* (Archaeognatha), sperm-containing droplets are deposited on a thread and the male then twists his body round the female, guiding her genitalia with his antennae and cerci into positions in which they can pick up the droplets (Schaller, 1971).

In the Pterygota, spermatophores are passed directly from the male to the female. Where the spermatophore is produced outside the female, one or two sperm sacs may be embedded, or closely associated with, a gelatinous proteinaceous mass called the spermatophylax (Fig. 12.12b). In phasmids, gryllids and tettigoniids only the neck of the spermatophore penetrates the female ducts, the body of the structure remains outside and is liable to be eaten by the female or other insects (see below). In Blattodea the body of the spermatophore, although still outside the female ducts, is protected by the female's enlarged subgenital plate. The signal to transfer a spermatophore is provided to a male cricket by mechanical stimulation of small trichoid sensilla in a cavity enclosed by his epiphallus. These hairs are stimulated by the female's copulatory papillae when the insects copulate (Sakai *et al.*, 1991).

The spermatophore is specialized in some Acrididae to form a tube which is effectively a temporary elongation of the intromittent organ (Fig. 12.12c). It consists of two basal bladders in the ejaculatory and sperm sacs of the male (Fig. 12.13) leading to a tube which extends into the female's spermathecal bulb.

In most other insect groups, many species practice direct insemination without producing a spermatophore. Where a spermatophore is retained, it is often formed in the female's bursa copulatrix.

Spermatophore production The spermatophore is produced from secretions of glands of the male's reproductive system, usually the accessory glands, but where these are absent or reduced, from glands in the ejaculatory duct (or the simplex in Lepidoptera). The secretions are molded in the ducts of the male and sometimes also the female. Secretions from different glands, or different cells within a gland, are produced in sequence to form separate parts of the spermatophore.

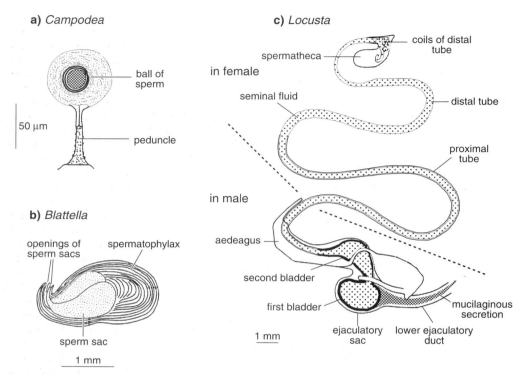

a) *Campodea*

ball of sperm

50 μm

peduncle

b) *Blattella*

openings of sperm sacs

spermatophylax

sperm sac

1 mm

c) *Locusta*

coils of distal tube

spermatheca

in female

seminal fluid

distal tube

proximal tube

in male

aedeagus

second bladder

first bladder

mucilaginous secretion

1 mm

ejaculatory sac

lower ejaculatory duct

Fig. 12.12. Spermatophores. **(a)** Produced independently of the female and left on the ground (*Campodea*, Diplura) (after Bareth, 1964). **(b)** Deposited at the opening of the female duct (*Blattella*, Blattodea) (after Khalifa, 1950). **(c)** Produced within the male and female ducts during copulation and connecting directly with the spermatheca. See Fig. 12.13(a) for details of male genitalia (*Locusta*, Orthoptera) (after Gregory, 1965).

Production of the spermatophore by *Locusta* begins within two minutes of the start of copulation. Most of it is formed in the male, although the ducts of the female serve to mold the tubular part. The first secretion builds up in the upper ejaculatory duct and is forced down through the funnel, the shape of which produces a series of folds, molding the secretion into a cylinder (Fig. 12.13). A white semi-fluid secretion is then forced into the core of the cylinder so that it becomes a tube (Fig. 12.13b). This is enlarged in the ejaculatory sac to form the first bladder, while the part remaining in the ejaculatory duct (known at this time as the reservoir tube) ultimately forms the second bladder in the spermatophore sac. At this stage seminal fluid is passed into the rudimentary spermatophore and then a separate cylinder of material is formed and pushed into the bladder, where it becomes coiled up (Fig. 12.13c). This will form the distal tube and a further series of secretions forms the proximal tube (Fig. 12.13d). As the last

part of the proximal tube enters the bladder it draws the wall of the reservoir tube with it, so that this becomes invaginated, and finally the whole of the tube except for the tip is pushed inside the bladder by a mucilaginous secretion (Fig. 12.13e).

At this time the ejaculatory sac starts to contract and squeezes the tube of the spermatophore out of the bladder, while the pressure of mucilage in the ejaculatory duct forces the tip backwards through the gonopore, which is now open, and out through the aedeagus into the duct of the female's spermatheca (Fig. 12.13f,g). This process involves the tube being turned inside out and finally the second bladder is everted and molded in the sperm sac (see Gregory, 1965 for a full account of this process). In *Gomphocerus*, the tube is not turned inside out, but elongates by expansion (Hartmann, 1970).

The spermatophore of Lepidoptera is formed wholly within the female ducts after the start of copulation. Here,

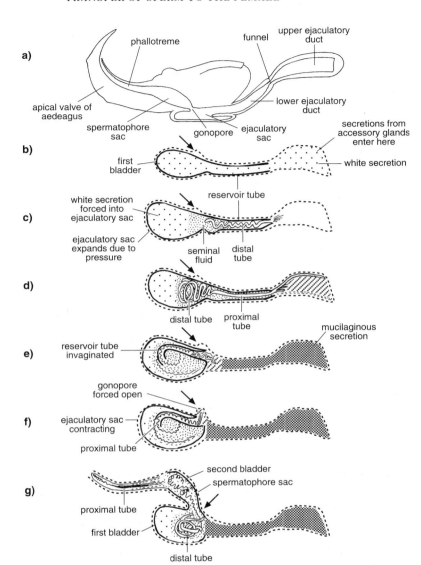

a)
phallotreme — funnel — upper ejaculatory duct
apical valve of aedeagus
spermatophore sac — gonopore — ejaculatory sac
lower ejaculatory duct
secretions from accessory glands enter here

b)
first bladder
white secretion

c)
white secretion forced into ejaculatory sac
ejaculatory sac expands due to pressure
reservoir tube
seminal fluid — distal tube

d)
distal tube — proximal tube

e)
reservoir tube invaginated
mucilaginous secretion

f)
gonopore forced open
ejaculatory sac contracting
proximal tube

g)
second bladder
spermatophore sac
proximal tube
first bladder
distal tube

Fig. 12.13. Spermatophore production in *Locusta* (after Gregory, 1965). Compare the fully developed spermatophore in Fig. 12.12(c). (a) Sagittal section through the male ejaculatory duct and genitalia. Diagrams (b)–(g) are aligned with this. Arrow in (b)–(g) indicates the position of the gonopore. (b) The first secretion is molded to form the first bladder and reservoir tube by the white secretion. Ejaculatory duct shown as broken line. (c) After movement of the seminal fluid and sperm into the bladder, further secretions form the distal tube. (d) The proximal tube of the spermatophore is formed. (e) A mucilaginous secretion invaginates the reservoir tube inside the first bladder (distal tube not shown, see d). (f) Continued pressure and contractions of the ejaculatory sac start to force the end of the proximal tube through the gonopore (previously closed) (distal tube not shown, see d). (g) Continued contraction of the ejaculatory sac everts the reservoir tube into the spermatophore sac where it forms the second bladder. Similar pressure forces the proximal and then the distal tubes into the female ducts (see Fig. 12.12c).

the different glands function in sequence, starting with the section of the simplex nearest the aedeagus (Fig. 12.14). The aedeagus projects into the female's bursa copulatrix and the secretion of the lower region of the simplex forms a pearly body (not shown in Figure) and the wall of the spermatophore. This is followed by the contents of the seminal vesicle, the sperm and seminal fluid, partly mixing with them to form the inner matrix of the spermatophore. The secretion from the lower parts of the accessory glands forms the outer matrix, and, finally, the secretions from the ends of the glands form the spermatophragma which blocks the

duct to the female's bursa copulatrix (Osanai, Kasuga & Aigaki, 1987, 1988; and see Fanger & Naumann, 1993).

For an account of the production of a spermatophore from multiple secretions of the accessory glands of *Tenebrio* see Dailey, Gadzama & Happ (1980) and references cited in Happ (1984).

Reviews: Mann, 1984; Tuzet, 1977b

Transfer of sperm to the spermatheca Immediately following the transfer of the spermatophore, the sperm migrate to the spermatheca, where they are stored.

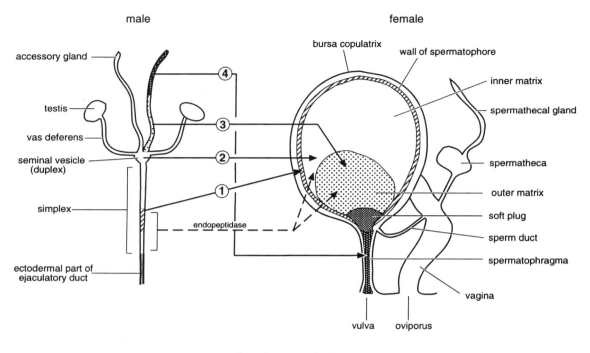

Fig. 12.14. ABOVE: male and part of the female reproductive systems of the silkmoth, *Bombyx*, showing how secretions from different parts of the male system contribute to different parts of the spermatophore (shown by corresponding shading). Note that the female system is shown at a much larger scale. Numbers show the sequence in which secretions are produced from different regions (based on Osanai, Kasuga & Aigaki, 1987, 1988). **BELOW**: chemical reactions occurring in the spermatheca which produce the respiratory substrate for sperm activity.

Sometimes they are able to escape from the sperm sac through a pore, but in other cases, where the sperm sac is completely enclosed within the spermatophore, they escape as a result of the spermatophore rupturing. In Lepidoptera and *Sialis*, the inside of the bursa copulatrix is lined with spines or bears a toothed plate, the signum dentatum, to which muscles are attached. The spermatophore is gradually abraded by movements of the spines until it is torn open. In *Rhodnius*, the first sperm reach the spermatheca within about 10 minutes of the end of mating, while in *Acheta* transfer takes about an hour, and in *Zygaena* (Lepidoptera) 12–18 hours.

The transfer of sperm to the spermatheca may be either active or passive. In *Acheta*, sperm are held in the body of the spermatophore (the ampulla), which remains external to the female, and the spermatophore is specialized to force the sperm out into the female ducts. An outer reservoir of fluid, the evacuating fluid, with a low osmotic pressure is separated by an inner layer with semipermeable properties from an inner proteinaceous mass called the pressure body, which has a high osmotic pressure (Fig. 12.15a). When the spermatophore is deposited, water passes from the evacuating fluid into the pressure body because of the difference in osmotic pressure. The

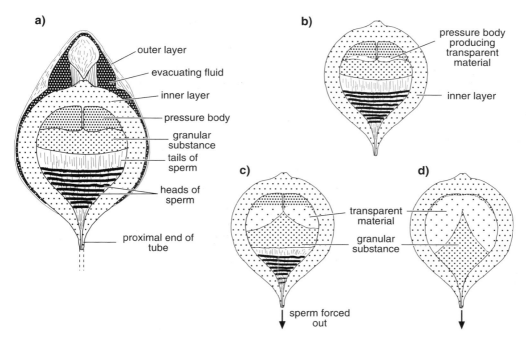

Fig. 12.15. Spermatophore of the house cricket (*Acheta*, Orthoptera) (after Khalifa, 1949). (a) Horizontal section through the ampulla which remains outside the female. (b) The pressure body starts to swell and releases a transparent material due to the absorption of evacuating fluid from outside the inner layer. In (b)–(d), the outer layer and evacuating fluid are not shown. (c) Production of the transparent material continues starting to push sperm out of the ampulla. (d) Sperm have been completely forced out of the ampulla.

pressure body produces a transparent material and swells, forcing the sperm out of the ampulla and down the tube of the spermatophore into the spermatheca (Fig. 12.15b–d). In *Locusta*, also, the sperm in the spermatophore are initially outside the female (in this case, in the first bladder of the spermatophore). From here they are pumped along the spermatophore by contractions of the ejaculatory sac, first appearing in the spermatheca about 90 minutes after the start of copulation.

In many insects the spermatophore is placed in the bursa copulatrix of the female and the transfer of sperm to the spermatheca is probably brought about by the contractions of the female ducts. An opaque secretion from the male accessory glands of *Rhodnius* injected into the bursa with the spermatophore induces rhythmic contractions of the oviducts, probably by way of a direct nervous connection from the bursa to the oviducal muscles. The contractions cause shortening of the oviduct, and, it is suggested, cause the origin of the oviduct in the bursa to make bite-like movements in the mass of semen in the bursa so that

sperm are taken into the oviduct. As this process continues, the more anterior sperm are forced forwards along the oviduct and are passed into the spermatheca. In Diptera Nematocera, it is probable that fluid is absorbed from the spermatheca creating a current which transports the sperm into the spermatheca (Linley & Simmonds, 1981).

In contrast to these examples, however, the sperm of *Bombyx* are activated before they leave the spermatophore (see below), and they may contribute actively to their transfer to the spermatheca.

Fate of the spermatophore Females of some species eject the spermatophore some time after insemination. *Blattella* and *Rhodnius*, for instance, drop the old spermatophores some 12 and 18 hours, respectively, after copulation. The female *Sialis* pulls the spermatophore out and eats it and this commonly happens in Blattodea, where specific post-copulatory behavior often keeps the female occupied for a time to ensure that the sperm have left the spermatophore before she eats it. In some grasshoppers, such as

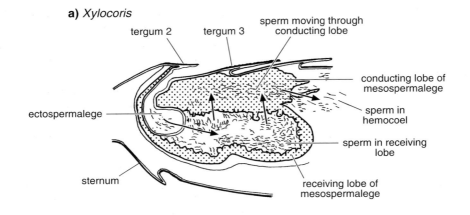

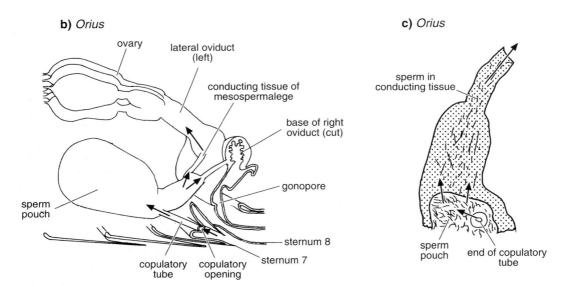

Fig. 12.16. Hemocoelic insemination in Hemiptera. Arrows show direction of sperm movement. (**a**) Longitudinal section through the ectospermalege and mesospermalege of *Xylocoris galactinus* about one hour after copulation (after Carayon, 1953a). (**b**) Diagram of the internal reproductive organs of female *Orius*. (**c**) Longitudinal section of part of the mesospermalege of *Orius* showing the sperm pouch and conducting tissue (after Carayon, 1953b).

Gomphocerus, the spermatophore is ejected by muscular contraction of the spermathecal duct.

The spermatophore is dissolved by proteolytic enzymes in other insects, such as Lepidoptera and Trichoptera, and in many Trichoptera only the sperm sac remains one or two days after copulation. In the wax moth, *Galleria*, digestion is complete in ten days, but the neck of the spermatophore persists. The spermatophore of

Locusta breaks when the two sexes separate, either where the tube fits tightly in the spermathecal duct or at its origin with the bladders in the male. The part remaining in the male is ejected within about two hours by the contractions of the copulatory organ, while in the female the distal tube disappears within a day, presumably being dissolved, but the proximal tube dissolves much more slowly, and persists for several days until it is ejected, probably by contractions of the spermathecal duct. In *Chorthippus* the activity of the enzyme-secreting cells of the spermatheca is controlled by the corpora allata.

12.3.3.2 *Direct insemination*

Various groups of insects have dispensed with a spermatophore and sperm are transferred directly to the female ducts, and often into the spermatheca, by the penis, which may be long and flagelliform for this purpose. Such direct insemination occurs in some members of the orders Heteroptera, Mecoptera, Trichoptera, Hymenoptera, Coleoptera and Diptera.

Direct insemination occurs in *Aedes aegypti*, and, in this insect, the paraprocts expand the genital orifice of the female while the aedeagus is erected by the action of muscles attached to associated apodemes. The aedeagus only penetrates just inside the female opening, where it is held by spines engaging with a valve of the spermatheca. A stream of fluid from the accessory glands is driven along the ejaculatory duct and into the female by contractions of the glands, and sperm are injected into the stream by the contractions of the seminal vesicles. Thus a mass of semen is deposited inside the atrium of the female and from here the sperm are transferred to the spermatheca. The sperm of *Drosophila* are similarly deposited in the vagina and then pass to the spermatheca.

Oncopeltus has a long penis which reaches into the spermatheca and deposits sperm directly into it (Fig. 12.10). Erection of the phallus in this insect is a specialized mechanism involving the displacement of an erection fluid into the phallus from a reservoir of the ejaculatory duct. The fluid is forced back from the reservoir by pressure exerted by the body muscles, and this pressure is maintained throughout copulation. At the end of the ejaculatory duct, the fluid is forced into a vesicle and then pumped into the phallus. In those Coleoptera and Hymenoptera with a long penis, erection is probably produced by an increase in blood pressure resulting from the sudden contraction of the abdominal walls.

12.3.3.3 *Hemocoelic insemination*

In some Cimicoidea (Hemiptera), the sperm, instead of being deposited in the female reproductive tract, are injected into the hemocoel. A good deal of variation occurs between the species practicing this method and they can be arranged in a series showing progressive specialization (Hinton, 1964). In *Alloeorhynchus flavipes* the penis enters the vagina, but a spine at its tip perforates the wall of the vagina so that the sperm are injected into the hemocoel. They are not phagocytosed immediately, but disperse beneath the integument and later collect under the peritoneal membrane surrounding the ovarioles, possibly being directed chemotactically. The sperm adjacent to the lowest follicle penetrate the follicular epithelium and fertilize the eggs via the micropyles.

Primicimex shows a further separation from the normal method of insemination. Here the left clasper of the male penetrates the dorsal surface of the abdomen of the female, usually between tergites 4 and 5 or 5 and 6. The clasper ensheathes the penis, and sperm are injected into the hemocoel. They accumulate in the heart and are distributed round the body with the blood. Many are phagocytosed by the blood cells, but those that survive are stored in two large pouches at the base of the oviducts. The holes made in the integument by the claspers become plugged with tanned cuticle.

In other species, the sperm are not injected directly into the hemocoel, but are received into a special pouch called the mesospermalege or organ of Ribaga or Berlese, which is believed to be derived from blood cells. Other genera have a cuticular pouch, called the ectospermalege, for the reception of the clasper and the penis. There may be one or two ectospermalegia and their positions vary, but in *Afrocimex* they are situated in the membrane between segments 3 and 4 and segments 4 and 5 on the left-hand side. *Xylocoris galactinus* has a mesospermalege for the reception of sperm immediately beneath the ectospermalege (Fig. 12.16a). The mesospermalege is formed from vacuolated cells surrounding a central lacuna into which the sperm are injected and from here they move down a solid core of cells, forming the conducting lobe, into the hemocoel. They finally arrive at the conceptacula seminis at the bases of the lateral oviducts, where they accumulate. In *Cimex* this migration takes about 12 hours and after the female takes her next blood meal the sperm are carried intracellularly in packets to the ovaries through special conduit cells. At the base of each ovariole they accumulate in a corpus seminalis derived from the follicular cells (Davis, 1964).

In *Anthocoris* and *Orius*, perforation of the integument is not necessary because a copulatory tube opens on the left between the sternites of segments 7 and 8 and passes to a median sperm pouch, where the sperm accumulate (Fig. 12.16b,c). From here the mesospermalege forms a column of conducting tissue along which the sperm pass to the oviducts, so that they are never free in the hemocoel.

In all these examples some sperm are digested by blood cells or by phagocytes in the mesospermalege. They probably have nutritional value, and perhaps hemocoelic insemination and its associated digestion of sperm facilitates more prolonged survival of the recipients in the absence of food (Hinton, 1964), although this has been questioned by other authorities (see Leopold, 1976).

In Strepsiptera, sperm also pass into the hemocoel to fertilize the eggs, but they do so via the genital canals of the female (see Fig. 14.31).

Review: Carayon, 1977

12.4 OTHER EFFECTS OF MATING

In the course of sperm transfer, secretions of the male accessory glands are transferred to the female and these may have various effects on her subsequent behavior and physiology.

12.4.1 Nutritive effects of mating

In insects, such as crickets and tettigoniids, where the spermatophore includes a spermatophylax, the latter is usually eaten by the female and serves two functions which, in different ways, favor the survival of the male's offspring. First, it gives time for the sperm to leave the sperm sac which is ultimately eaten along with the spermatophylax. Females of the tettigoniid, *Requena verticalis*, for example, take about five hours to eat the spermatophylax (Gwynne, Bowen & Codd, 1984). Second, it provides extra nutrients to the female. In some tettigoniids which produce spermatophores with a large spermatophylax, the spermatophore can exceed 20% of the weight of the male producing it; in *Requena verticalis* it is as much as 40%. More than 80% of the dry weight of the spermatophylax is protein, and the amino acids from the proteins are incorporated into the female tissues after she eats the spermatophore, greatly increasing the weight of oocytes in the ovaries. More than one third of the protein of each egg produced by a zaprochiline tettigoniid is derived from the male (Simmonds & Gwynne, 1993). This material permits the female to produce bigger and, in some species, more eggs (Fig. 12.17).

Even where the spermatophore is produced internally, and so is not eaten by the female, the male may contribute to the nutrient pool of the female. Amino acids, and even

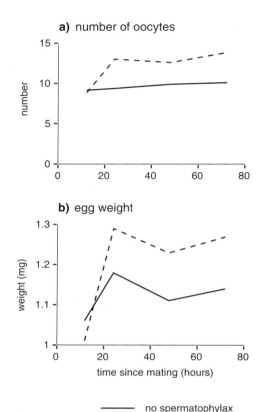

Fig. 12.17. Nutritional value of the spermatophylax in a tettigoniid. Insects were either allowed to eat the spermatophylax in the normal way, or the spermatophylax was removed so that it could not be eaten (after Simmons, 1990). (a) Total number of developing oocytes in the ovaries. (b) Weight of one chorionated egg.

some proteins, from the spermatophore of *Melanoplus* are subsequently found in the eggs. This is also known to occur in some Blattodea, Lepidoptera and Coleoptera (Boggs, 1981; Huignard, 1983; LaMunyon & Eisner, 1994; Mullins, Keil & White, 1992).

In grasshoppers belonging to the subfamilies Cyrtacanthacridinae, Melanoplinae and Pyrgomorphidae, several spermatophores are produced during a single copulation. *Melanoplus* produces about seven and *Schistocerca* six at each normal mating. These spermatophores are simple sac-like structures quite different from those of *Locusta* and related species (Pickford and

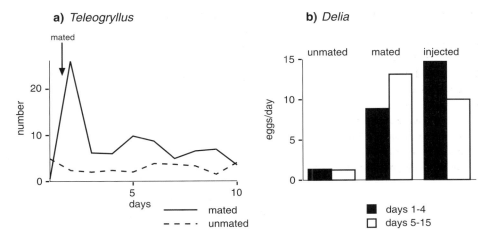

a) *Teleogryllus*   **b)** *Delia*

Fig. 12.18. Enhancement of oviposition as a result of mating and transfer of male accessory gland material to the female. **(a)** Number of eggs per day laid by females of the Australian field cricket, *Teleogryllus* after mating. Unmated females lay only a few eggs each day (after Stanley-Samuelson, 1994). **(b)** Average number of eggs per day laid by females of the onion fly, *Delia* when unmated, mated or after injection of an extract of male accessory glands into unmated females. Bars show averages for the first four days and the next 11 days (data from Spencer *et al.*, 1992).

Padgham, 1973). A single spermatophore contains sufficient sperm to fertilize several batches of eggs and it is possible that mating after the first oviposition is more important in maintaining fecundity than in maintaining fertility (Leahy, 1973). Although only one spermatophore is produced each time *Chorthippus* mates, repeated mating by the female enhances both the rate of egg production and the number of eggs per pod irrespective of whether the insects have ample food or only a limited supply (Butlin, Woodhatch & Hewitt, 1987). There is also evidence that multiple mating by *Drosophila mojavensis* contributes material for oogenesis (Markow & Ankney, 1984).

Some butterflies exhibit puddling behavior, aggregating at muddy pools and drinking. A principal function of this behavior is to accumulate sodium. In many species, only males exhibit this behavior and they may be anatomically, and presumably physiologically, adapted to accumulate sodium which they then transfer to the females during copulation. Males of the skipper butterfly, *Thymelicus lineola*, transfer more than 10% of their body weight as sodium to females in this way. The sodium increases female fertility (Pivnick & McNeil, 1987; Smedley & Eisner, 1995).

12.4.2 Enhancement of female fertility
In addition to the possible nutrient effects that accompany sperm transfer in many insects, copulation may also be a

trigger for oviposition and sometimes oogenesis. Virgin females usually do not lay eggs, or lay relatively few. Mating, or the experimental injection of a component of the male accessory glands, causes them to oviposit (Fig. 12.18).

In *Drosophila* and the beetle, *Acanthoscelides*, the fertility-enhancing compound is a peptide. That in *Drosophila* contains 36 amino acids (Kubli, 1992). In some other insects, grasshoppers, crickets and a mosquito, the active compound is known to be a protein. In addition, males of some crickets and lepidopterans transfer prostaglandins or prostaglandin-synthesizing chemicals to the female during copulation. These have also been shown to stimulate oviposition (Stanley-Samuelson, 1994).

Within the female, the peptides and proteins may act in one of two ways. Either they may stimulate the tissue of the reproductive tract to produce a hormone, or they may enter the hemocoel and affect a distant target site. In *Rhodnius*, the fertility-enhancing factor induces the tissues of the spermatheca and associated ducts to produce a hormone, known as the spermathecal factor. This causes some neurosecretory cells in the brain to release a peptide that produces contractions of the ovarian sheath leading to ovulation (Kriger & Davey, 1984).

The sex peptide of *Drosophila* is derived from a larger peptide transferred to the female and then cleaved in the bursa copulatrix. After absorption into her hemolymph, it acts directly on the brain, at least with respect to its function in inhibiting female receptivity (see below). The fecundity enhancing compound produced by *Bombyx* acts directly on the terminal abdominal ganglion.

In the mosquito, *Aedes aegypti*, a secretion from the male accessory glands also induces the female to respond to the stimuli of potential oviposition sites to which she was previously unresponsive (Yeh & Klowden, 1990).

12.4.3 Reduction of female's readiness to remate

Males of many species employ some mechanism to reduce the likelihood that a female with whom he has just mated will mate again. This may take the form of mate guarding, but often involves a male-induced effect on the female's ability or willingness to mate again. In some species, the reduction in receptivity is permanent and the female never mates again; in others, receptivity subsequently returns. Species in which the female mates only once are common amongst Hymenoptera and Diptera. The solitary bee (*Centris pallida*), the onion fly (*Delia antiqua*) and tsetse flies (*Glossina*), are examples. Similar behavior is recorded in a few species from other orders. Cyclical changes in receptivity, resulting from the resumption of mating after an interval, occur in many species in most insect orders, often in relation to the female's cycle of oogenesis. Female *Drosophila melanogaster* will remate about a week after a previous mating (Spencer *et al.*, 1992).

These changes in female behavior induced by the male may be associated with the physical blocking of the female ducts by the spermatheca or some other component of the male accessory gland secretion forming a mating plug. In other instances, components of the male accessory gland secretion affect female behavior without physically preventing further sperm transfer.

Mating plugs, distinct from spermatophores, are produced in a number of insects. Thus, in mosquitoes belonging to the genera *Anopheles*, *Aedes* and *Psorophora*, a plug, formed from the accessory gland secretions of the male, is deposited in the genital chamber of the female, even though these insects do not produce a spermatophore. A similar plug is produced by some species of *Drosophila* (Alonso-Pimentel, Tolbert & Heed, 1994) and by the honeybee, *Apis*. A comparable structure is present in some Lepidoptera where it is produced immediately after the spermatophore (Fig. 12.14). It is called the spermatophragma or sphragis and it effectively prevents

further mating by the female. It may also ensure that sperm in the bursa copulatrix are transferred to the spermatheca rather than moving back towards the vulva.

Loss of receptivity by the female is sometimes regulated directly through neural pathways. For instance, females of *Nauphoeta* and *Gomphocerus* are unreceptive while there is a spermatophore in the spermatheca; cutting the ventral nerve cord results in the return of receptivity. In *Aedes*, receptivity is also first switched off by mechanical stimulation resulting from filling the bursa copulatrix with seminal fluid. Subsequently, however, the female remains unreceptive for the rest of her life as a result of a substance, known as matrone, in the secretions of the male accessory glands. Matrone passes from the bursa copulatrix into the hemolymph and then acts directly on the central nervous system of the female. Receptor sites probably exist in the terminal abdominal ganglion (Leopold, 1976). The male accessory gland secretions also affect host-seeking by female mosquitoes (Fernandez & Klowden, 1995). A sex peptide also switches off female receptivity in *Drosophila*, but the sustained depression of sexual activity over a period of days is probably dependent on the presence of sperm in the spermatheca (Kubli, 1992).

In the housefly, *Musca*, and the stable fly, *Stomoxys*, where accessory glands are absent, a secretion from the ejaculatory duct inhibits female receptivity (see below) (Morrison, Venkatesh and Thompson, 1982). Compounds that inhibit receptivity are also known to occur in Lepidoptera.

Not only is female receptivity reduced, in *Helicoverpa* some chemical component of the spermatophore leads to an immediate reduction in the synthesis of sex attractant pheromone by the female (Fig. 12.19) (Kingan, Thomas-Laemont & Raina, 1993).

12.4.4 Transfer of other ecologically relevant compounds

Some examples are known of males transferring chemicals which have ecological relevance to the species beyond the immediate context of reproduction. For example, males of the 'Spanish fly' (*Lytta*, Coleoptera) synthesize cantharidin, presumably as a defensive substance. It is probably synthesized and stored in the accessory reproductive glands. Females do not synthesize the compound, but large quantities are transferred to the female during copulation (Sierra, Woggon & Schmid, 1976).

a) after normal mating

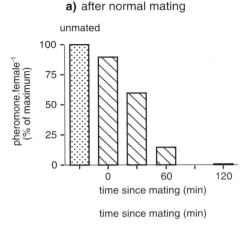

b) after injection of accessory gland extract

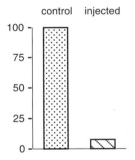

Fig. 12.19. Reduction in pheromone production as a result of mating and transfer of male accessory gland material to the female of the moth, *Helicoverpa*. **(a)** Reduction in the amount of the sex attractant pheromone component, Z-11-hexadecanal, in the female following mating (after Raina, 1989). **(b)** Effect of injecting an extract of the male accessory glands on the amount of Z-11-hexadecanal in the female (data from Kingan *et al.*, 1993).

A similar transfer of compounds that confer protection against predators is known to occur in some Lepidoptera. Larvae of the moth, *Utetheisa*, feed on plants containing pyrrolizidine alkaloids, and adult males of the milkweed butterfly, *Danaus*, obtaining the same chemicals by licking at the plant surface. In both species, males transfer large amounts of these compounds to the females during copulation. They are subsequently found in the eggs where they have been shown to have a protective function (Dussourd *et al.*, 1988). The spermatophore of *Utetheisa* is large, as much as 10% of the body weight, presumably in relation to its function in conveying nutrients and alkaloids to the female (LaMunyon & Eisner, 1994).

12.5 SPERM CAPACITATION

In a number of species, sperm undergo changes after they are transferred to the female spermatheca and in some cases, at least, these changes are essential before the sperm can fertilize an egg. This process of maturation of sperm within the female is known as capacitation. In *Musca*, the sperm lose the plasma membrane from the head and most of the granular material from the acrosome.

More obvious changes occur where the sperm are still grouped together when they leave the spermatophore. In grasshoppers, the glycoprotein binding the sperm together is dissolved in the spermatheca so that they separate. The glycocalyx is also removed. This occurs within 6 hours in one part of the spermatheca, in another part it takes more than 15 hours from the time of insemination. Sperm from this region are those that will ultimately fertilize the eggs (Longo *et al.*, 1993).

The breakup of bundles of eupyrene sperm in the spermatophore of Lepidoptera is associated with, and probably caused by the activation of apyrene sperm. A specific protease from the lower simplex digests the glycoprotein covering the flagellar membrane of the apyrene sperm. The same protease liberates arginine from proteins in the seminal fluid, and the arginine is metabolized to 2-oxoglutarate, a preferred substrate for the sperm (Fig. 12.14) (Aigaki, Kasuga & Osanai, 1987; Osanai & Kasuga, 1990). Cyclic AMP is also necessary to activate the sperm. Once the apyrene sperm of *Bombyx* are activated, they appear to break up the eupyrene sperm bundles by their physical activity. A protease liberating arginine and cAMP are also necessary for sperm activation in grasshoppers (Osanai & Baccetti, 1993; Osanai & Kasuga, 1990).

REFERENCES

Aigaki, T., Kasuga, H. & Osanai, M. (1987). A specific endopeptidase, BAEE esterase, in the glandula prostatica of the male reproductive system of the silkworm, *Bombyx mori*. *Insect Biochemistry*, 17, 323–8.

Alonso-Pimentel, H., Tolbert, L.P. & Heed, W.B. (1994). Ultrastructural examination of the insemination reaction in *Drosophila*. *Cell & Tissue Research*, 275, 467–79.

Bareth, C. (1964). Structure et dépôt des spermatophores chez *Campodea remyi*. *Compte Rendus Hebdomadaire de Séances de l'Academie des Science, Paris*, 259, 1572–5.

Boggs, C.L. (1981). Selection pressures affecting male nutrient investment at mating in heliconine butterflies. *Evolution*, 35, 931–40.

Bonhag, P.F. & Wick, J.R. (1953). The functional anatomy of the male and female reproductive systems of the milkweed bug, *Oncopeltus fasciatus* (Dallas) (Heteroptera: Lygaeidae). *Journal of Morphology*, 93, 177–283.

Butlin, R.K., Woodhatch, C.W. & Hewitt, G.M. (1987). Male spermatophore investment increases female fecundity in a grasshopper. *Evolution*, 41, 221–5.

Carayon, J. (1953a). Organe de Ribaga et fécondation hémocoelienne chez les *Xylocoris* du groupe *galactinus* (Hemipt. Anthocoridae). *Compte Rendus Hebdomadaire de Séances de l'Academie des Sciences, Paris*, 236, 1009–101.

Carayon, J. (1953b). Existence d'un double orifice genital et d'un tissu conducteur des spermatozoides chez les Anthocorinae (Hemipt. Anthocoridae). *Compte Rendus Hebdomadaire de Séances de l'Academie des Sciences, Paris*, 236, 1206–8.

Carayon, J. (1977). Insémination extra-génitale traumatique. In *Traité de Zoologie*, vol. 8, part 5A, ed. P.-P. Grassé, pp. 351–90. Paris: Masson et Cie.

Chao, H.-F. (1953). The external morphology of the dragonfly *Onychogomphus ardens* Needham. *Smithsonian Miscellaneous Collections*, 122, no.6, 1–56.

Chen, P.S. (1984). The functional morphology and biochemistry of insect male accessory glands and their secretions. *Annual Review of Entomology*, 29, 233–55.

Corbet, P.S. (1980). Biology of Odonata. *Annual Review of Entomology*, 25, 189–217.

Corbet, P.S. (1983). *A Biology of Dragonflies*. Farringdon: Classey.

Dailey, P.J., Gadzama, N.M. & Happ, G.M. (1980). Cytodifferentiation in the accessory glands of *Tenebrio molitor*. VI. A congruent map of the cells and their secretions in the layered elastic product of the male bean-shaped gland. *Journal of Morphology*, 166, 289–322.

Dallai, R. & Afzelius, B.A. (1984). Paired spermatozoa in *Thermobia* (Insects, Thysanura). *Journal of Ultrastructure Research*, 86, 67–74.

Dallai, R. & Afzelius, B. (1993). Development of the accessory tubules of insect sperm flagella. *Journal of Submicroscopic Cytology and Pathology*, 25, 494–504.

Davis, N.T. (1964). Studies on the reproductive physiology of Cimicidae (Hemiptera) – I. Fecundation and egg maturation. *Journal of Insect Physiology*, 10, 947–63.

Dussourd, D.E., Ubik, K., Harvis, C., Resch, J., Meinwold, J. & Eisner, T. (1988). Biparental defensive endowment of eggs with acquired alkaloid in the moth *Utetheisa ornatrix*. *Proceedings of the National Academy of Sciences of the United States of America*, 85, 5992–6.

Engelmann, F. (1970). *The Physiology of Insect Reproduction*. Oxford: Pergamon Press.

Fanger, H. & Naumann, C.M. (1993). Correlation between the male genital ducts and the spermatophore structure in a ditrysian moth, *Zygaena trifolii* (Esper, 1783) (Insecta, Lepidoptera, Zygaenidae). *Acta Zoologica*, 74, 239–46.

Farine, J.P., Everaerts, C., Abed, D., Ntari, M. & Brossut, R. (1996). Pheromonal emission during the mating behavior of *Eurycotis floridana* (Walker) (Dictyoptera: Blattidae). *Journal of Insect Behavior*, 9, 197–213.

Fernandez, N.M. & Klowden, M.J. (1995). Male accessory gland substances modify the host-seeking behavior of gravid *Aedes aegypti* mosquitoes. *Journal of Insect Physiology*, 41, 965–70.

Friedländer, M. & Reynolds, S.E. (1992). Intratesticular ecdysteroid titres and the arrest of sperm production during pupal diapause in the tobacco hornworm, *Manduca sexta*. *Journal of Insect Physiology*, 38, 693–703.

Giebultowicz, J.M., Loeb, M.J. & Borkovec, A.B. (1987). *In vitro* spermatogenesis in Lepidopteran larvae: role of the testis sheath. *International Journal of Invertebrate Reproduction and Development*, 11, 211–26.

Gillott, C. (1988). Arthropoda – Insecta. In *Reproductive Biology of Invertebrates. III. Accessory Sex Glands*, ed. K.G. Adiyodi & R.G. Adiyodi, pp. 319–471. Chichester: Wiley & Sons.

Gregory, G.E. (1965). The formation and fate of the spermatophore in the African migratory locust, *Locusta migratoria migratorioides* Reiche and Fairmaire. *Transactions of the Royal Entomological Society of London*, 117, 33–66.

Gwynne, D.T., Bowen, B.J. & Codd, C.G. (1984). The function of the katydid spermatophore and its role in fecundity and insemination (Orthoptera, Tettigoniidae). *Australian Journal of Zoology*, 32, 17–22.

Happ, G.M. (1984). Structure and development of male accessory glands in insects. In *Insect Ultrastructure*, vol. 2, ed. R.C. King & H. Akai, pp. 365–96. New York: Plenum Press.

Happ, G.M. (1992). Maturation of the male reproductive system and its endocrine regulation. *Annual Review of Entomology*, **37**, 303–20.

Hartmann, R. (1970). Experimentelle und histologische Untersuchungen der Spermatophorenbildung bei der Feldheuschrecke *Gomphocerus rufus* L. (Orthoptera, Acrididae). *Journal der Morphologie der Tiere*, **68**, 140–76.

Hinton, H.E. (1964). Sperm transfer in insects and the evolution of haemocoelic insemination. *Symposium of the Royal Entomological Society of London*, **2**, 95–107.

Huignard, J. (1983). Transfer and fate of male secretions deposited in the spermatophore of females of *Acanthoscelides obtectus* Say. *Journal of Insect Physiology*, **29**, 55–63.

Imms, A.D. (1957). *A general Textbook of Entomology*. 9th edition revised by O.W. Richards and R.G. Davies. London: Methuen.

Jamieson, B.G.M. (1987). *The ultrastructure and Phylogeny of Insect Spermatozoa*. Cambridge: Cambridge University Press.

Khalifa, A. (1949). The mechanism of insemination and the mode of action of the spermatophore in *Gryllus domesticus*. *Quarterly Journal of Microscopical Science*, **90**, 281–92.

Khalifa, A. (1950). Spermatophore production in *Blattella germanica* L. (Orthoptera: Blattidae). *Proceedings of the Royal Entomological Society of London* A, **25**, 53–61.

Kimura, T., Yasuyama, K. & Yamaguchi, T. (1989). Proctolinergic innervation of the accessory gland in male crickets (*Gryllus bimaculatus*): detection of proctolin and some pharmacological properties of myogenically and neurogenically evoked contractions. *Journal of Insect Physiology*, **35**, 251–64.

Kingan, T.G., Thomas-Laemont, P.A. & Raina, A.K. (1993). Male accessory gland factors elicit change from 'virgin' to 'mated' behavior in the female corn earworm moth *Helicoverpa zea*. *Journal of Experimental Biology*, **183**, 61–76.

Kriger, F.L. & Davey, K.G. (1984). Identified neurosecretory cells in the brain of female *Rhodnius prolixus* contain a myotropic peptide. *Canadian Journal of Zoology*, **62**, 1720–3.

Kubli, E. (1992). The sex-peptide. *BioEssays*, **14**, 779–84.

Lai-Fook, J. (1982a). Structure of the noncuticular simplex of the internal male reproductive tract of *Calpodes ethlius* (Lepidoptera, Hesperiidae). *Canadian Journal of Zoology*, **60**, 1184–201.

Lai-Fook, J. (1982b). Structure, function, and possible evolutionary significance of the constrictions in the male reproductive system of *Calpodes ethlius* (Hesperiidae, Lepidoptera). *Canadian Journal of Zoology*, **60**, 1828–36.

Lai-Fook, J. (1982c). Testicular development and spermatogenesis in *Calpodes ethlius* Stoll (Hesperiidae, Lepidoptera). *Canadian Journal of Zoology*, **60**, 1161–71.

LaMunyon, C.W. & Eisner, T. (1994). Spermatophore size as determinant of paternity in an arctiid moth (*Utetheisa ornatrix*). *Proceedings of the National Academy of Sciences of the United States of America*, **91**, 7081–4.

Leahy, M.G. (1973). Oviposition of virgin *Schistocerca gregaria* (Forskål) (Orthoptera: Acrididae) after implant of the male accessory gland complex. *Journal of Entomology* A, **48**, 69–78.

Leopold, R.A. (1976). The role of male accessory glands in insect reproduction. *Annual Review of Entomology*, **21**, 199–221.

Linley, J.R. & Simmons, K.R. (1981). Sperm motility and spermathecal filling in lower Diptera. *International Journal of Invertebrate Reproduction*, **4**, 137–46.

Loeb, M.J. (1991). Growth and development of spermducts in the tobacco budworm moth *Heliothis virescens, in vivo* and *in vitro*. *International Journal of Invertebrate Reproduction and Development*, **19**, 97–105.

Longfield, C. (1949). *The Dragonflies of the British Isles*. London: Warne.

Longo, G., Sottile, L., Viscuso, R., Giuffrida, A. and Provitera, R. (1993). Ultrastructural changes in sperm in *Eyprepocnemis plorans* (Charpentier) (Orthoptera: Acrididae) during storage of gametes in female genital tract. *Invertebrate Reproduction and Development*, **24**, 1–6.

Mann, T. (1984). *Spermatophores. Development, Structure, Biochemical Attributes and Role in the Transfer of Spermatozoa*. Berlin: Springer-Verlag.

Markow, T.A. & Ankney, P.F. (1984). *Drosophila* males contribute to oogenesis in a multiple mating species. *Science*, **224**, 302–3.

Morrison, P.E., Venkatesh, K. & Thompson, B. (1982). The role of male accessory-gland substance on female reproduction with some observations on spermatogenesis in the stable fly. *Journal of Insect Physiology*, **28**, 607–14.

Mullins, D.E., Keil, C.B. & White, R.H. (1992). Maternal and paternal nitrogen investment in *Blattella germanica* (L.) (Dictyoptera; Blattellidae). *Journal of Experimental Biology*, **162**, 55–72.

Osanai, M. & Baccetti, B. (1993). Two-step acquisition of motility by insect spermatozoa. *Experientia*, **49**, 593–5.

Osanai, M. & Kasuga, H. (1990). Role of endopeptidase in motility induction in apyrene silkworm spermatozoa; micropore formation in the flagellar membrane. *Experientia*, **46**, 261–4.

Osanai, M., Kasuga, H. & Aigaki, T. (1987). The spermatophore and its structural changes with time in the bursa copulatrix of the silkworm, *Bombyx mori*. *Journal of Morphology*, **193**, 1–11.

Osanai, M., Kasuga, H. & Aigaki, T. (1988). Functional morphology of the glandula prostatica, ejaculatory valve, and ductus ejaculatorius of the silkworm, *Bombyx mori*. *Journal of Morphology*, **198**, 231–41.

Pickford, R. & Padgham, D.E. (1973). Spermatophore formation and sperm transfer in the desert locust, *Schistocerca gregaria* (Orthoptera: Acrididae). *Canadian Entomologist*, **105**, 613–8.

Pivnick, K.A. & McNeil, J.N. (1987). Puddling in butterflies: sodium affects reproductive success in *Thymelicus lineola*. *Physiological Entomology*, **12**, 461–72.

Popham, E.J. (1965). The functional morphology of the reproductive organs of the common earwig (*Forficula auricularia*) and other Dermaptera with reference to the natural classification of the order. *Journal of Zoology*, **146**, 1–43.

Raina, A.K. (1989). Male-induced termination of sex pheromone production and receptivity in mated females of *Heliothis zea*. *Journal of Insect Physiology*, **35**, 821–6.

Riemann, J.G. & Giebultowicz, J.M. (1991). Secretion in the upper vas deferens of the gypsy moth correlated with the circadian rhythm of sperm release from the testes. *Journal of Insect Physiology*, **37**, 53–62.

Sakai, M., Taoda, Y., Mori, K. Fujino, M. & Ohta, C. (1991). Copulation sequence and mating termination in the male cricket *Gryllus bimaculatus* DeGeer. *Journal of Insect Physiology*, **37**, 599–615.

Schaller, F. (1971). Indirect sperm transfer by soil arthropods. *Annual Review of Entomology*, **16**, 407–46.

Scudder, G.G.E. (1971). Comparative morphology of insect genitalia. *Annual Review of Entomology*, **16**, 379–406.

Séguy, E. (1951). Ordre des Diptères. In *Traité de Zoologie*, vol. 10, ed. P.-P. Grassé, pp. 449–744. Paris: Masson et Cie.

Sierra, J.R., Woggon, W.-D. & Schmid, H. (1976). Transfer of cantharadin (1) during copulation from the adult male to the female *Lytta vesicatoria* ('Spanish flies'). *Experientia*, **32**, 142–4.

Simmons, L.W. (1990). Nuptial feeding in tettigoniids: male costs and the rates of fecundity increase. *Behavioral Ecology and Sociobiology*, **27**, 43–7.

Simmons, L.W. & Gwynne, D.T. (1993). Reproductive investment in bushcrickets: the allocation of male and female nutrients to offspring. *Proceedings of the Royal Society of London* B, **252**, 1–5.

Sinakevitch, I.G., Geffard, M., Pelhate, M. & Lapied, B. (1994). Octopamine-like immunoreactivity in the dorsal unpaired median (DUM) neurons innervating the accessory gland of the male cockroach *Periplaneta americana*. *Cell and Tissue Research*, **276**, 15–21.

Smedley, S.R. & Eisner, T. (1995). Sodium uptake by puddling in a moth. *Science*, **270**, 1816–8.

Snodgrass, R.E. (1935). *Principles of Insect Morphology*. New York: McGraw-Hill.

Snodgrass, R.E. (1957). A revised interpretation of the external reproductive organs of male insects. *Smithsonian Miscellaneous Collections*, **135**, no. 6, 60 pp.

Spencer, J.L., Bush, G.L., Keller, J.E. & Miller, J.R. (1992). Modification of female onion fly, *Delia antiqua* (Meigen), reproductive behavior by male paragonial gland extracts (Diptera: Anthomyiidae). *Journal of Insect Behavior*, **5**, 689–97.

Stanley-Samuelson, D.W. (1994). Prostaglandins and related eicosanoids in insects. *Advances in Insect Physiology*, **24**, 115–212.

Swiderski, Z. (1980). The fine structure of the sperm bundles of the coccid, *Aspidiotus perniciosus* Cumst., and sperm movement. *International Journal of Invertebrate Reproduction*, **2**, 331–9.

Thornhill, R. & Alcock, J. (1983). *The Evolution of Insect Mating Systems*. Cambridge, Mass: Harvard University Press.

Tillyard, R.J. (1917). *The Biology of Dragonflies*. Cambridge: Cambridge University Press.

Tuxen, S.L. (1956). *Taxonomist's Glossary of Genitalia in Insects*. Copenhagen: Monksgaard.

Tuzet, O. (1977a). La spermatogenèse. In *Traité de Zoologie*, vol. 8, part 5A, ed. P.-P.Grassé, pp. 139–276. Paris: Masson et Cie.

Tuzet, O. (1977b). Les spermatophores de insectes. In *Traité de Zoologie*, vol. 8, part 5A, ed. P.-P.Grassé, pp. 277–349. Paris: Masson et Cie.

Uvarov, B.P. (1966). *Grasshoppers and Locusts*, vol. 1. Cambridge: Cambridge University Press.

Wigglesworth, V.B. (1965). *The Principles of Insect Physiology*. London: Methuen.

Yeh, C. & Klowden, M.J. (1990). Effects of male accessory gland substances on the pre-oviposition behavior of *Aedes aegypti* mosquitoes. *Journal of Insect Physiology*, **36**, 799–803.

13 Reproductive system: female

13.1 ANATOMY OF THE INTERNAL REPRODUCTIVE ORGANS

The female reproductive system consists of a pair of ovaries, which connect with a pair of lateral oviducts. These join to form a median oviduct opening posteriorly into a genital chamber. Sometimes the genital chamber forms a tube, the vagina, and this is often developed to form a bursa copulatrix for reception of the penis. Opening from the genital chamber or the vagina is a spermatheca for storing sperm, and, frequently, a pair of accessory glands is also present (Fig. 13.1).

Review: Martoja, 1977

Ovary The two ovaries lie in the abdomen above or lateral to the gut. Each consists of a number of egg-tubes, or ovarioles, comparable with the testis follicles in the male. The oocytes develop in the ovarioles. The ovaries of Collembola are probably not homologous with those of insects, but are sac-like with a lateral germarium and no ovarioles.

The number of ovarioles in an ovary varies in relation to size and life style of the insect as well as its taxonomic position. In general, larger species within a group have more ovarioles than small ones; thus small grasshoppers commonly have only four ovarioles in each ovary, while larger ones may have more than 100. Similarly, in the higher Diptera, *Drosophila* has 10–30 ovarioles in each ovary, whereas *Calliphora* has about 100. By contrast, most Lepidoptera have four ovarioles on each side irrespective of their size. Variation in relation to size may also occur within a species. In the mosquito, *Aedes punctor*, the number of ovarioles in the two ovaries combined varies from about 30 to 175 as the insect's size increases, but the cockroach, *Periplaneta americana*, has eight on each side irrespective of its size.

Viviparous species exhibit extreme reduction in the numbers of ovarioles. The viviparous Diptera, *Melophagus*, *Hippobosca* and *Glossina*, have only two in each ovary and some viviparous aphids have only one functional ovary with a single ovariole. At the other extreme are the queens of some species of social insects: each ovary of the queen termite, *Eutermes*, has over 2000 ovarioles, and that of a queen honeybee 150–180. Species which disperse as first stage larvae may also have large numbers of ovarioles. Scale insects have several hundred ovarioles and the blister beetle, *Meloe*, 1000.

Other than in Diptera, there is no sheath enclosing the ovary as a whole, but each ovariole has a sheath which is made up of two layers: an outer ovariole sheath, or tunica externa, and an inner tunica propria (Fig. 13.4b,c). The tunica externa is an open network of cells, sometimes including muscle cells. The cells of this net are rich in lipids and glycogen and are metabolically active, but there is no evidence that they are directly concerned with oocyte development. Tracheoles also form part of the external sheath, but they do not penetrate it and all the oxygen utilized by the developing oocytes diffuses in from these elements. In *Periplaneta*, mycetocytes (section 4.3.2.2) are present in the sheath.

The tunica propria is a basal lamina with elastic properties possibly secreted by the cells of the terminal filament and the follicle cells. It surrounds the whole of each ovariole and the terminal filament. During the early stages of development it increases in thickness, but during the period of yolk uptake, when the oocytes enlarge rapidly, it becomes stretched and very thin. The tunica propria has a supporting function, maintaining the shape of the ovariole, and, in addition, because of its elasticity, playing a part in ovulation (section 13.3). It also has the potential to function as a molecular sieve since, in *Phormia*, it does not permit the passage of molecules larger than 500 kDa.

Each ovariole is produced, distally, into a long terminal filament consisting of a cellular core bounded by the tunica propria (Figs. 13.1a, 13.4). Usually the individual filaments from each ovary combine to form a suspensory ligament and sometimes the ligaments of the two sides merge into a median ligament. The ligaments are inserted into the body wall or the dorsal diaphragm and so suspend the developing ovaries in the hemocoel.

a) *Schistocerca*   **b)** *Rhagoletis*

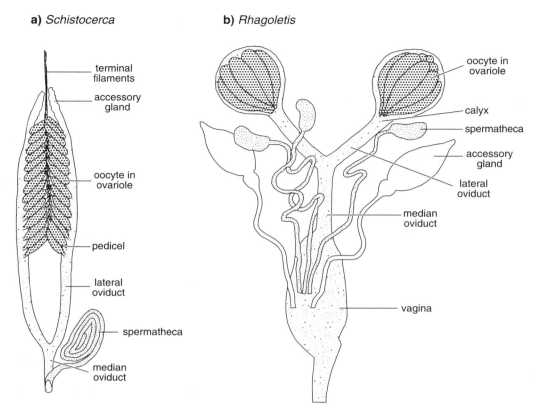

Fig. 13.1. Female reproductive system (partly after Snodgrass, 1935). **(a)** An insect with many ovarioles opening sequentially into the lateral oviducts. The oviduct and spermatheca open independently (Orthoptera: *Schistocerca*). **(b)** An insect in which the ovarioles open together into the end of the lateral oviduct which forms the calyx. The genital chamber is a continuation of the median oviduct, forming the vagina. The spermathecae arise at the junction of the vagina with the median oviduct (Diptera: *Rhagoletis*).

Proximally, the ovariole narrows to a fine duct, the pedicel, which connects with the oviduct. In an immature insect the lumen of the ovariole is cut off from the oviduct by an epithelial plug, but this is destroyed at the time of the first ovulation and subsequently is replaced by a plug of follicular tissue.

The ovarioles may connect with the oviduct in a linear sequence and, if there are only a few, as in some Apterygota and Ephemeroptera, they may appear to be segmental. This appearance is probably coincidental and there is no suggestion of a segmental arrangement in insects with a larger number of ovarioles (Fig. 13.1a). In other groups, such as the Lepidoptera and Diptera, the ovarioles open together into an expansion of the oviduct known as the calyx (Fig. 13.1b).

Oviducts The oviducts are tubes with walls consisting of a single layer of cuboid or columnar cells standing on a basal lamina and surrounded by a muscle layer. Usually, the two lateral oviducts join a median oviduct which is ectodermal in origin and hence lined with cuticle. However, the Ephemeroptera are exceptional in having the lateral oviducts opening separately, each with its own gonopore. The median oviduct is usually more muscular than the lateral ducts, with circular and longitudinal muscles. It opens at the gonopore which, in Dermaptera, is ventral on the posterior end of segment 7, but in most other groups opens into a genital chamber invaginated above the sternum of segment 8 (Fig. 13.2a). Sometimes the genital chamber becomes tubular and it is then effectively a continuation of the oviduct through segment 9. Such a continuation is

a) *Locusta*

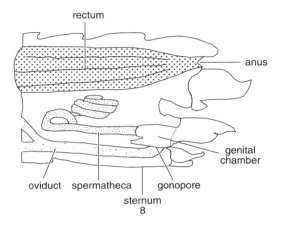

b) ditrysian lepidopteran

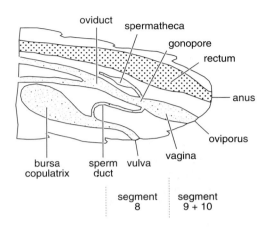

Fig. 13.2. Female reproductive system. Diagrammatic sagittal sections of the end of the abdomen. (**a**) An insect in which the genital chamber is open, not tubular (Orthoptera: *Locusta*) (after Uvarov, 1966). (**b**) A ditrysian lepidopteran in which the genital chamber forms the vagina (after Imms, 1957).

called the vagina and its opening to the exterior, the vulva (but notice the different terminology in ditrysian Lepidoptera, Fig. 13.2b). The vagina may not be distinguishable in structure from the median oviduct, but its anterior end, and the position of the true gonopore, is marked by the insertion of the spermatheca (Snodgrass, 1935).

Frequently the vagina is developed to form a pouch, the bursa copulatrix, which receives the penis, while in viviparous Diptera the anterior part of the chamber is enlarged to form the uterus, in which larval development occurs.

The females of the ditrysian Lepidoptera are unusual in having two reproductive openings (Fig. 13.2b). One, on segment 9, serves for the discharge of eggs and is known as the oviporus; the other, on segment 8, is the copulatory opening, known as the vulva. The latter leads to the bursa copulatrix which is connected with the oviduct by a sperm duct. Two openings also occur in the water beetles *Agabus*, *Ilybius* and *Hydroporus*, but here both openings are terminal with the bursa copulatrix opening immediately above the vagina.

Spermatheca A spermatheca, used for the storage of sperm from the time the female is inseminated until the eggs are fertilized, is present in most female insects. Sometimes there are two, as in *Blaps* (Coleoptera) and *Phlebotomus* (Diptera), and most of the higher flies have

three (Fig. 13.1b). In Orthoptera and other lower insects orders, the spermatheca opens into the genital chamber independently of the oviduct (Fig. 13.2a), but, where the genital chamber forms a vagina, the spermathecal opening is internal and is effectively within the oviduct (Fig. 13.2b). The spermatheca is ectodermal in origin and is lined with cuticle. Typically it consists of a storage pouch with a muscular duct leading to it. A gland is often associated with it, or the spermathecal epithelium may itself be glandular. The contents of the spermatheca, derived from the glands, are known to contain several proteins and include a carbohydrate–protein complex. The functions of these secretions are not known for certain, but they probably provide nutrients for the sperm during storage, or they may be concerned with sperm activation (Gillott, 1988).

Accessory glands Female accessory glands often arise from the genital chamber or the vagina. Where such glands are apparently absent the walls of the oviducts may themselves be glandular. This is the case in grasshoppers where the lateral oviducts also usually have a wholly glandular anterior extension (Fig. 13.1a).

Accessory glands often produce a substance for attaching the eggs to the substratum during oviposition and hence are often called colleterial (glue) glands. In a number of insects, they produce an ootheca that protects the eggs after oviposition (section 13.5.4). Glands associated with the genitalia, which are often modified to form a sting, perform a variety of functions in female Hymenoptera (see Chapter 27).

13.2 OOGENESIS

Each ovariole consists of a distal germarium in which oocytes are produced from oogonia, and a more proximal vitellarium in which yolk is deposited in the oocytes. These two regions reflect two phases of oocyte growth: the first is regulated directly by the oocyte's genome and contains species specific information (all the substances whose synthesis is regulated by the DNA of the germ line are known collectively as the euplasm); the second is mainly regulated by genes outside the oocyte, producing pools of molecules that will subsequently be involved in embryonic growth. The vitellarium in a mature insect forms by far the greater part of the ovariole.

The germarium contains prefollicular tissue and the stem line oogonia and their derivatives. The stem line oogonia are derived directly from the original germ cells (section 14.2.13), and, in *Drosophila*, there are only two or three of these in each ovariole (Spradling, 1993). When they divide, one of the daughter cells remains a functional stem line cell, while the other becomes a definitive oogonium (sometimes called a cystocyte) and develops into an oocyte. Oocytes pass back down the ovariole, enlarging as they do so, and as each oocyte leaves the germarium it is clothed by the prefollicular tissue which forms the follicular epithelium. At first this may be two- or three-layered, but ultimately consists of a single layer of cells. Oocyte growth continues, the follicular epithelium keeping pace by cell division so that its cells become cuboid or columnar. In *Drosophila*, the number of follicle cells surrounding each oocyte increases from an initial figure of about 16 to about 1200. During yolk accumulation, growth of the oocyte is very rapid, but at this time the follicle cells no longer divide and they become stretched over the oocyte as a flattened, squamous epithelium. Nuclear division may continue in follicle cells without cell division, so the cells become binucleate or endopolyploid, permitting the high levels of synthetic activity in which these cells are involved. The functions of follicle cells change during oocyte development (Fig. 13.3). At first, they produce some minor yolk proteins and perhaps some of the enzymes that will later be involved in processing the yolk.

The follicle cells also produce ecdysone, or a precursor of ecdysone, which, at least in some insects, accumulates in the oocyte (section 14.2.15 and see section 13.2.4.2). In the later stages of oogenesis, they produce the vitelline envelope and the ligands responsible for determination of the terminals of the embryo and its dorso-ventral axis (section 14.2.5). Finally, they produce the egg shell, or chorion (section 13.2.5).

Each ovariole typically contains a linear series of oocytes in successive stages of development with the most advanced in the most proximal position furthest from the germarium (Fig. 13.4). An oocyte with its surrounding follicular epithelium is termed a follicle and successive follicles are separated by interfollicular tissue derived from the prefollicular tissue. In many species, new follicles are produced as the oldest oocytes mature and are ovulated (but see below). As a result, the number of follicles in an ovariole may be approximately constant for a species, although there is variation between species. Thus *Schistocerca* commonly has about 20 follicles in each ovariole even in senile females which have oviposited several times. In *Drosophila* there are usually six follicles per ovariole, while *Melophagus* has only one follicle in each ovariole at any one time.

13.2.1 Types of ovariole
There are two broad categories of ovarioles: panoistic, in which there are no special nurse cells, and meroistic, in which nurse cells, or trophocytes, are present. Further, there are two types of meroistic ovariole: telotrophic, in which all the trophocytes remain in the germarium, and polytrophic, in which trophocytes are closely associated with each oocyte and are enclosed within the follicle. Trophocytes are sister-cells of the oocytes so that they have the same genome, they retain their connections with the oocyte because cell division is incomplete, and they supplement the oocyte in the synthesis of the euplasm. As a result, the oocytes in meroistic ovarioles have much larger amounts of euplasm than those in panoistic ovarioles. This has important consequences for the development of the embryo (14.2.4).

Rapid synthesis of euplasmic constituents can also be achieved by replication of the chromosomes in the trophocytes without further cell division (endopolyploidy, see below) and by the amplification of genes extrachromosomally in DNA bodies. This is known to occur in some species with meroistic ovarioles. An increase in the

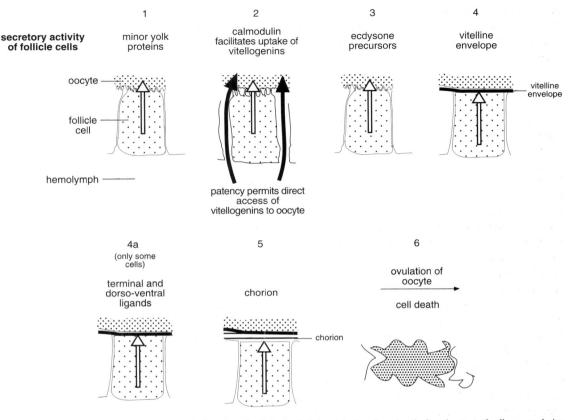

Fig. 13.3. Diagram of the changes in function of cells in the follicle epithelium from just before the start of yolk accumulation in the oocyte (1) to cell death at the time of ovulation (6). Notice that in (2) the cells are widely separated from each other allowing the hemolymph to reach the oocyte.

DNA content of these bodies occurs up to the early stages of meiosis, but then they fragment and ultimately disappear.

As the oocyte grows, its nucleus increases proportionately in size and is now known as the germinal vesicle. Transcription by the nuclear DNA appears to be suppressed soon after the beginning of yolk uptake. During yolk deposition, as the oocyte grows much more rapidly, the germinal vesicle becomes relatively smaller and, finally, the nuclear membrane breaks down.

Panoistic ovarioles Panoistic ovarioles (Fig. 13.4), which have no specialized nurse cells, are found in the more primitive orders of insects, the Thysanura, Odonata, Plecoptera, Orthoptera and Isoptera. Amongst the holometabolous insects, only Siphonaptera have ovarioles

of this type. The prefollicular tissue may be cellular, but sometimes, as in *Thermobia* (Thysanura), it consists of small scattered nuclei in a common cytoplasm.

Telotrophic ovarioles Telotrophic ovarioles are characterized by the presence of trophic tissue as well as oogonia and prefollicular tissue in the terminal regions. This arrangement is found in Hemiptera and Coleoptera Polyphaga. In Hemiptera, each ovariole has only a single stem cell that divides to form a cluster of cells which remain connected to each other by cytoplasmic bridges (Fig. 13.5). The most proximal cells are the precursors of the oocytes while more distal cells are the trophocytes. These may divide more frequently than the oocytes so that a larger number of trophocytes is produced. The cells all remain connected to a central region called the trophic

Fig. 13.4. Parts of a panoistic ovariole: (a) diagram of a whole ovariole; (b) the distal region (germarium); (c) a proximal region (part of the vitellarium).

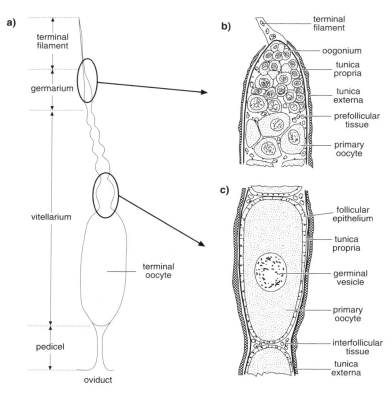

core. A basically similar phenomenon occurs in the Polyphaga except that here there are several stem cells in an ovariole and each gives rise to columns of cells, the most proximal of which is the oocyte (Fig. 13.6). In the more advanced beetle families, the intercellular bridges and cell membranes of the trophocytes break down so that the oocytes come to be connected with a common trophic core.

As in other types of ovariole, the oocytes become clothed by follicle cells as they leave the germarium, but each oocyte remains connected to the germarium by a cytoplasmic nutritive cord which extends to the trophic tissue, elongating as the oocyte passes down the ovariole. In the beetles this cord is extremely slender, less than 10 μm in diameter. At the time of yolk uptake, the nutritive cord finally breaks and the follicle cells form a complete layer round the oocyte.

Unlike panoistic and polytrophic ovarioles where the production of new oocytes is a continuous process, division of the germ cells does not continue after oocyte growth begins. As a result, the potential number of oocytes produced by an ovariole is fixed.

Polytrophic ovarioles As in telotrophic ovarioles, divisions of the cells derived from stem cells in polytrophic ovarioles are incomplete so that clusters of cells are formed. However, unlike telotrophic ovarioles, the trophocytes move down the ovariole with their associated oocyte and become enclosed in the follicle (Fig. 13.7). Polytrophic ovarioles occur in Dermaptera, Psocoptera, Phthiraptera and throughout the holometabolous orders, except for most Coleoptera and the Siphonaptera. Divisions of the trophocytes within a cluster are synchronized by cues (as yet unknown) which pass through the cytoplasmic bridges. The number of divisions, and so the number of trophocytes associated with each oocyte, is characteristic for each species, although in those species with larger numbers of trophocytes some variation may occur. *Aedes* and *Melophagus* (Diptera) have seven trophocytes associated with each oocyte as a result of three successive cell divisions, *Drosophila* and *Dytiscus* (Coleoptera) have 15 trophocytes from four cell divisions, *Habrobracon* (Hymenoptera) 31, and vespid wasps 63. No more than six successive cell divisions occur in the trophocyte clusters in any insect so the maximum

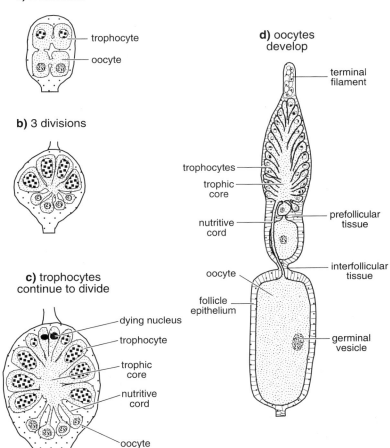

a) 2 divisions

trophocyte

oocyte

b) 3 divisions

c) trophocytes continue to divide

dying nucleus

trophocyte

trophic core

nutritive cord

oocyte

d) oocytes develop

terminal filament

trophocytes

trophic core

nutritive cord

oocyte

follicle epithelium

prefollicular tissue

interfollicular tissue

germinal vesicle

Fig. 13.5. Telotrophic ovariole of a hemipteran. **(a)**–**(c)** Successive stages in the formation of a cluster of oocytes and trophocytes (after Büning, 1993). **(d)** diagram of a later stage in the development of the oocytes which remain connected to the trophic core by the nutritive cords (after Huebner & Diehl-Jones, 1993).

number of cells in a cluster is 64 (1 oocyte + 63 trophocytes). Dermaptera are exceptional in having only one trophocyte associated with each oocyte, but this results from the separation of pairs of cells from larger interconnected groups.

As each oocyte with its trophocytes leaves the germarium, the oocyte always occupies a proximal position with respect to the base of the ovariole. All the cells become enclosed within a common follicular epithelial layer which soon becomes flattened over the trophocytes, but is thicker, with cuboid cells, round the oocyte. A fold of follicular epithelium pushes inward separating the oocyte from the trophocytes except for a median pore. In Neuroptera, Coleoptera and Hymenoptera the tropho-

cytes are pinched off in a separate, but connected, follicle from the beginning (Fig. 13.7b).

At first the trophocytes are larger than the oocyte, and the trophocyte nuclei enlarge considerably. In *Drosophila*, the trophocyte nuclei increase in volume about 2000-fold and the chromosomes undergo up to 10 doublings to produce polytene chromosomes. In the trophocytes of the moth, *Antheraea*, ploidy levels up to 2^{16} occur. Unlike earlier cell divisions, these mitoses are not synchronized in the different trophocytes of a follicle. In *Drosophila*, the trophocytes adjacent to the oocyte have larger nuclei, and their chromosomes undergo one more replication than the anterior (distal) trophocytes. These cells subsequently lose their DNA, but the anterior cells do not. In most cases the whole genome is replicated to an equal extent, but there is also evidence for selective replication of those elements that are particularly important in oocyte development (Telfer, 1975).

Reviews: Büning, 1993, 1994; King & Büning, 1985

13.2.2 Transport from trophocytes to oocyte

During the early stages of oocyte growth, mRNAs and ribonucleoproteins are transported to the oocyte. Subsequently, ribosomes and other major constituents of the euplasm are transferred. Finally, in polytrophic ovarioles, all the cytoplasmic contents are moved to the oocyte when the trophocytes collapse. In *Drosophila*, this final movement causes the oocyte to almost double in volume in less than 30 minutes. The RNAs transferred include those that determine the long axis of the embryo (section 14.2.5).

The movement of material into the oocyte from the trophocytes may be along an electrical potential gradient. In insects with polytrophic ovarioles, there is a difference in electrical charge between the trophocytes and the oocyte so that charged molecules are transported electrophoretically, contributing to the movement of material into the oocyte (Singleton & Woodruff, 1994; Telfer, 1981). Alternatively, or perhaps additionally, transport may involve special proteins associated with cytoskeletal elements running between the cells. In *Rhodnius*, it is known that microtubules extend from the nurse cells to the

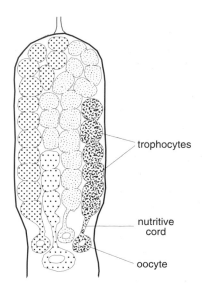

Fig. 13.6. Telotrophic ovariole of a beetle. Each oocyte is shown with its associated cluster of trophocytes (Coleoptera, Polyphaga) (after Büning, 1993).

oocytes (Huebner & Diehl-Jones, 1993) and these could form the basis for such a transport mechanism. In polytrophic ovarioles, the final collapse of the trophocytes, perhaps mediated by contractile units within the cells, might force the contents into the oocyte.

Review: Büning, 1994

Fig. 13.7. Polytrophic ovariole. (a) Diagram showing the interconnections that remain between the oocyte (stippled) and its associated trophocytes as a result of incomplete cell divisions. The oocyte always occupies the most posterior (proximal) position as it moves down the ovariole (after King, 1964). (b) Diagram of part of an ovariole in which the trophocytes are in separate follicles from the oocytes (Hymenoptera: *Bombus*) (after Hopkins & King, 1966).

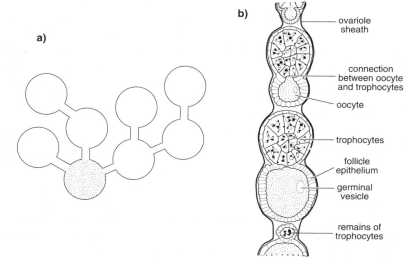

13.2.3 Meiosis

Meiosis begins early in the development of the oocyte, in most insects, but the meiotic divisions are not completed in the ovary, and oocytes usually leave the ovarioles in the metaphase of the first maturation division. In *Locusta*, meiosis does not proceed beyond the first prophase until the time of ovulation, and the resumption of meiosis coincides with sperm entry in many insects although other factors are apparently responsible for the effect. In *Locusta*, continuation of the process is stimulated by ecdysone from the follicle cells to which the oocyte becomes sensitive just before chorionogenesis (Lanot *et al.*, 1990), and, in *Drosophila*, ovulation triggers the resumption of meiosis through some hydration effect (Foe, Odell & Edgar, 1993).

In viviparous species, such as *Hemimerus* (Dermaptera), and in Heteroptera which practice hemocoelic insemination, however, maturation of the oocytes and fertilization takes place in the ovary.

13.2.4 Yolk

The euplasm in insect oocytes usually constitutes much less than 10% of the total oocyte content. The remaining 90% or more is yolk, consisting largely of lipids and proteins. Lipids often comprise about 40% of the dry weight of the terminal oocyte and most of the lipid is stored as triacylglycerol (Kawooya & Law, 1988). The protein content of the yolk is usually approximately equal to the lipid content, and 60–90% of yolk proteins are derived from vitellogenins. These are proteins manufactured outside the oocyte, only by females, and with specific uptake mechanisms at the oocyte membrane. Most insects produce only one or two vitellogenins which range in molecular weight, in different species, from 210 000 to 652 000. The vitellogenins are glycolipophosphoproteins in which lipid makes up 7–15% of the total mass and carbohydrates 1–14%. The carbohydrates in the molecule are oligosaccharides of mannose. The higher Diptera differ from other insects in producing three to five smaller yolk proteins varying in molecular weight from 44 000 to 51 000. When vitellogenins enter the oocyte they often undergo some relatively minor modifications and then are known as vitellins. Lipophorin (section 5.2.2.3) may also constitute a significant part of yolk protein. It makes up 15% of the total protein in the oocyte of *Manduca*. Other proteins are present in smaller amounts in addition to free amino acids.

Small amounts of carbohydrate are also present in the yolk. Some is associated with the vitellins; most of the rest is stored as glycogen. In the terminal oocyte of *Manduca*, free carbohydrates contribute about 2% of the dry weight.

13.2.4.1 *Patterns of accumulation in the oocyte*

The accumulation of yolk in the oocytes occurs in the lower, proximal part of each ovariole, known as the vitellarium, and results in a very rapid increase in size of the oocyte. In *Drosophila*, the oocyte volume increases about 100 000 times during the course of its development.

The process of yolk accumulation in insect oocytes is commonly called vitellogenesis (see Büning, 1994, for example). However, the same term is used to describe the synthesis of vitellogenin (see Raikhel & Dhadialla, 1992, for example) and it is also used in this sense in other animal groups. To avoid confusion, synthesis of the specific yolk protein will be referred to as 'vitellogenin synthesis', while 'yolk accumulation' will be used to describe the accumulation of yolk in the oocyte.

In those Lepidoptera, Ephemeroptera and Plecoptera which do not feed as adults, yolk accumulation is completed in the late larva or pupa. In most species, however, a period of maturation is required in the adult before the oocytes are ready for ovulation. Commonly this period is only a matter of days, but in cases of adult diapause it may be very prolonged. Hence it is necessary, in insects, to differentiate between becoming adult (that is, having the outward appearance of an adult with, in most species, wings, and being able to fly) and becoming sexually mature.

In many species, yolk accumulation is largely restricted to the oocyte nearest the oviduct in each ovariole (known as the terminal oocyte). The succeeding oocytes remain relatively small until the first is discharged from the ovariole into the oviduct, a process known as ovulation. Hence there is an interval between successive ovulations from any single ovariole. However, in many Lepidoptera and Diptera, and in some other insects, yolk accumulation occurs simultaneously in a number of oocytes in each ovariole and fully developed oocytes are stored before oviposition.

Cyclical yolk accumulation is often dependent on nutrition. This is most obvious in blood-sucking insects, like mosquitoes and triatomine bugs, which take discrete meals with relatively long intervals between them. These insects produce a batch of eggs after each meal and the

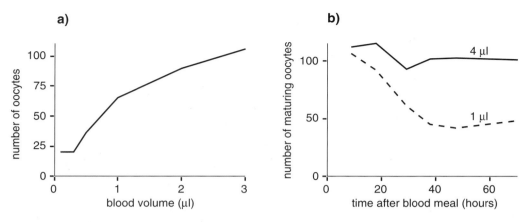

Fig.13.8. Affect of 'meal' size on egg production and resorption (Diptera: *Aedes*). Insects were given blood by enema. (**a**) Number of oocytes greater than 200 μm long in relation to the amount of blood placed in the midgut (Klowden, 1987). (**b**) Effect of amount of blood on resorption. Notice that insects given only 1 μl started to accumulate yolk in a similar number of oocytes to those given 4 μl but subsequently more than half the oocytes were resorbed. Oocytes were considered to be maturing if they were more than 100 μm long (after Lea, Briegel & Lea, 1978).

number of eggs produced is proportional to the amount of blood ingested (Fig. 13.8).

Periodicity of yolk deposition also occurs in viviparous and ovoviviparous species perhaps due to the physical constraints imposed by the developing embryos. The viviparous cockroach, *Diploptera*, retains its ootheca in a brood sac and mechanical stimulation by the ootheca is largely responsible for the inhibition of yolk accumulation in the oocytes remaining in the ovarioles (see below).
Review: Wheeler, 1996 – effects of nutrition

13.2.4.2 *Yolk proteins*
Most yolk protein is synthesized in the fat body independently of the ovaries. In addition to vitellogenins and lipophorin, some insects are known to produce smaller amounts of other yolk proteins in the fat body. The fat body of *Aedes aegypti* also produces proenzymes that are probably involved in yolk metabolism during embryogenesis. These enzymes are sequestered in the oocyte outside the yolk granules in which the vitellins are stored. Similar enzymes are produced by many insects.

Follicle cells also contribute proteins to the oocyte. In the higher Diptera, they produce the same yolk proteins as the fat body, though in smaller amounts. Follicle-specific proteins are known to be produced by the follicle cells of several Lepidoptera (van Antwerpen & Law, 1993). In the silkmoth, *Bombyx*, one of these proteins constitutes about

20% of soluble protein in the fully developed oocyte. In addition to these major components, the follicle cells may contribute other proteins in smaller amounts. Eggs of *Blattella*, and probably those of other insects, have high concentrations of the protein calmodulin in the cortical regions and outside the bounding membranes of yolk granules. Calmodulin is synthesized in the follicle cells and may enter the oocyte through gap junctions (Zhang & Kunkel, 1994). It probably binds calcium which is involved in the uptake of vitellogenin and perhaps also in the ooplasmic transport of yolk vesicles.

Control of vitellogenin synthesis Vitellogenin synthesis is initiated by juvenile hormone acting directly on the fat body in most insects, but in Diptera juvenile hormone acts indirectly. Here, it causes both the fat body and the resting stage ovaries to acquire competence to respond to other signals.

The synthesis of juvenile hormone by the corpora allata of the female is regulated by signals from the brain where relevant information from other tissues is integrated. Synthetic activity may be switched on or off. The signal to initiate juvenile hormone synthesis is generally related to nutrient or mated status. In locusts, mating and some food odors act via neurosecretory cells in the brain to initiate the synthesis and release of juvenile hormone (Fig. 13.9). Blood-feeding also leads to juvenile hormone

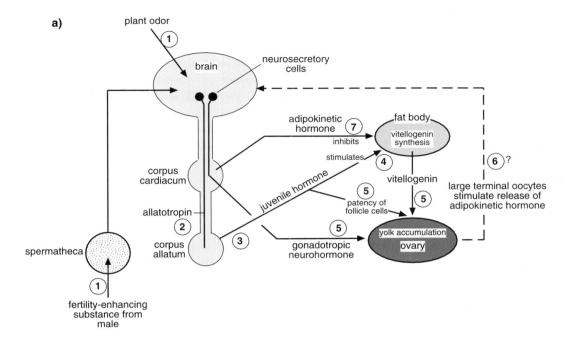

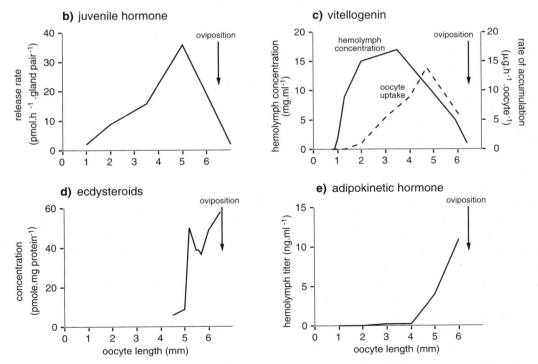

Fig. 13.9. Control of vitellogenin synthesis and yolk accumulation in *Locusta*. (**a**) Diagram showing the sequence of events that follow insemination of the female and the transfer of fertility enhancing substance by the male. Numbers show the sequence of events. Stage 6 is inferred (see text). (**b**) Release of juvenile hormone from the corpora allata during the first vitellogenic cycle (stage 3 in a) (data from Gadot & Applebaum, 1985). (**c**) Vitellogenin concentration in the hemolymph and rate of uptake by the terminal oocytes (stage 5 in a) (after Gellissen & Emmerich, 1978). (**d**) Accumulation of ecdysteroid (mainly ecdysone) in the posterior quarter of the terminal oocyte in the later stages of yolk accumulation. Ecdysone leads to a resumption of meiosis in the oocyte nucleus (data from Lanot *et al.*, 1987). (**e**) Hemolymph titer of adipokinetic hormone during yolk accumulation (stage 7 in a) (data from Moshitsky & Applebaum, 1990).

synthesis in mosquitoes and in *Rhodnius*; a high-protein meal has the same effect in *Phormia* (Fig. 13.10) (Yin et al., 1994). There is also some evidence of a direct component of regulation from the ovaries.

Juvenile hormone synthesis is regulated by allatotropins and allatostatins acting on the corpora allata directly via the neural connection to the brain or via the hemolymph (section 21.4). In the viviparous cockroach, *Diploptera*, small terminal oocytes, between 0.6 and 1.4 mm long, enhance the rate of synthesis, but large, fully developed oocytes have an inhibitory effect which acts on the brain via the hemolymph. When an ootheca is present, it stimulates mechanoreceptors in the brood sac which inhibit juvenile hormone synthesis via the neural pathway (Stay et al., 1994).

Although there is a decline in juvenile hormone synthesis at the end of the first gonotrophic cycle in *Locusta*, vitellogenin synthesis in this insect is further reduced by adipokinetic hormone acting directly on the fat body (Fig. 13.9). The hemolymph titer of this hormone increases markedly when the terminal oocytes are large. This effect is produced by a hormone titer only about one tenth of that required to initiate lipid mobilization in the fat body during flight (section 6.2.3) so that its two functions do not conflict (Moshitzky & Applebaum, 1990). Although vitellogenin synthesis by the fat body stops, the mRNA responsible for its synthesis is stored for the next cycle of synthesis (Glinka et al., 1994).

In the Diptera, synthesis of vitellogenin is regulated by ecdysone (Fig. 13.10). The prothoracic glands which are the principal source of ecdysone in the immature stages, are no longer present in adults, and ecdysone is produced by the follicle cells. In the female mosquito, *Aedes*, ecdysone synthesis is itself regulated by a hormone, egg development neurosecretory hormone, produced by cells in the brain. Release of this hormone is triggered by a blood meal, but it stimulates the follicle cells to produce ecdysone only if they have already been rendered competent to respond by juvenile hormone. At the fat body, ecdysone is changed to 20-hydroxyecdysone which is the active hormone. Ecdysone, or a precursor, is synthesized by the follicle cells in other insects such as the locust (Fig. 13.9d), but here its primary function is the regulation of molting during the embryonic period (section 14.2.10).

In some Lepidoptera, all the oocytes are fully developed at eclosion. This is the case, for example, in the cecropia moth, *Hyalophora*. In this insect, vitellogenin appears in the hemolymph when the juvenile hormone titer declines and the larva spins its cocoon. Vitellogenin concentration in the hemolymph remains high throughout the pupal period, finally declining in the pharate adult as the oocytes accumulate yolk. In *Plodia*, ovarian development early in the pupal period is correlated with high titers of ecdysone, but vitellogenin synthesis only proceeds when ecdysteroid titers are low (Shaaya et al., 1993).

Reviews: Hagedorn, 1985 – role of ecdysteroids; Hagedorn, 1994 – oogenesis in mosquitoes; Koeppe et al., 1985 – role of juvenile hormone

Accumulation of proteins in the oocyte The oocyte is surrounded by the follicle epithelium, but, when the oocyte starts to accumulate vitellogenin, extensive intercellular spaces develop between the follicle cells although the cells remain in contact with each other through gap junctions (Giorgi et al., 1993). The intercellular spaces permit direct access of the hemolymph to the surface of the oocyte and the follicle is said to be patent. Vitellogenin is taken up selectively from the hemolymph and this is made possible by specific receptors for vitellogenin on the plasma membrane of the oocyte. The receptors with their bound vitellogenin are taken into the oocyte by endocytosis (Ferenz, 1993). The protein is accumulated and stored in yolk bodies, while the receptors are recycled to the surface membrane.

The vitellogenins are changed following their accumulation in the oocyte although, in most insects, the changes are small. For example, *Rhodnius* vitellin contains the same four subunits and has the same molecular weight as the vitellogenin but has more phospholipids and triacylglycerol (Wang & Davey, 1992). Minor changes to the vitellogenins also occur in *Carausius* (Giorgi et al., 1993), but in *Blattella* the larger of two subunits comprising the vitellogenin is cleaved into two, and cleavage of the vitellogenin also occurs in the boll weevil, *Anthonomus* (Heilmann, Trewitt & Kumaran, 1993). Proteins from the follicle cells are also modified as they enter the oocyte (Sato & Yamashita, 1991).

The uptake of proteins from the hemolymph is regulated by juvenile hormone. This induces patency (Yin, Zou & Stoffolano, 1989) and also produces changes in the surface of the oocyte. It does not, however, determine the specificity of uptake, at least in *Drosophila* (Bownes et al., 1993). The fact that, in many species, only the terminal oocyte in each ovariole accumulates yolk appears to occur

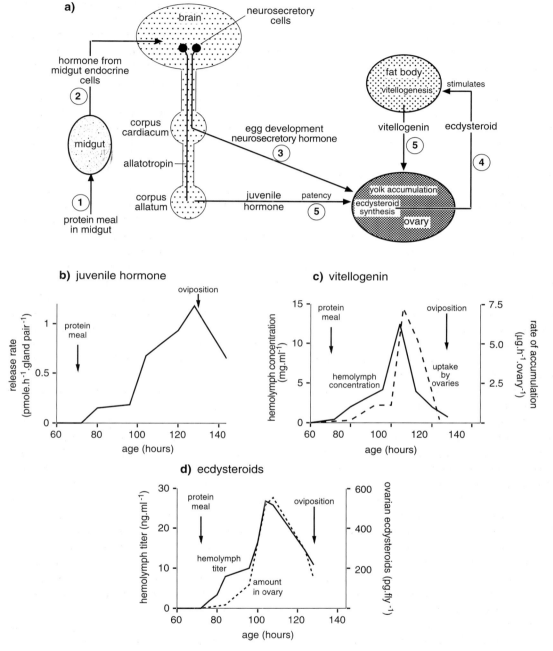

Fig. 13.10. Control of vitellogenin synthesis and yolk accumulation in *Phormia*. (**a**) Diagram showing the sequence of events that follow a high protein meal. Numbers show the sequence of events (based on Yin *et al.*, 1994). (**b**) Release of juvenile hormone from the corpora allata during the vitellogenic cycle (stage 5 in a) (after Zou *et al.*, 1989). (**c**) Vitellogenin concentration in the hemolymph and rate of uptake by the ovaries (stage 5 in a) (after Zou *et al.*, 1989). (**d**) Ecdysteroid: hemolymph titer and amount in the ovaries (after Yin *et al.*, 1990).

because it produces a factor that inhibits yolk uptake by the younger follicles. This may act via a local intercellular interaction or via the hemolymph.

In addition to the selective uptake of vitellogenins, some non-specific uptake from the hemolymph also occurs. As a result of this, substances present in the hemolymph that are not specifically associated with embryonic development may be present in the yolk.

Review: Raikhel & Dhadialla, 1992

13.2.4.3 *Yolk lipids*

Most of the lipid that accumulates in the oocyte is synthesized in the fat body; less than 1% is synthesized by the oocyte. In *Manduca*, 90% of the lipid is carried to the oocyte by low density lipophorins. At the oocyte membrane, the lipophorin unloads its lipids and is recycled. Most of the remainder of the lipid comes from high density lipophorin which is taken up by the oocyte in *Manduca* so that it also contributes to the protein content of the yolk. In *Hyalophora*, however, apolipoprotein III is recycled and only a very high density lipophorin is taken up (section 5.2.2.3). The uptake is receptor mediated. Much of the phospholipid that accumulates in the oocyte is also transported by lipophorin (Gondim, Oliveira & Masuda, 1989).

13.2.5 **Vitelline envelope and chorionogenesis**

The follicle cells start to produce the proteins of the vitelline envelope (section 14.1.1) as yolk accumulation nears its end. In *Drosophila*, it starts to form as a number of discrete plaques which subsequently coalesce, but in some other insects it is secreted uniformly over the surface of the oocyte from the start. It is possible that, in some insects, the oocyte also contributes to formation of the envelope (Zimowska *et al.*, 1995). In insects with polytrophic ovarioles, the vitelline envelope is only produced round the oocyte, not round the trophocytes. At about the same time the vitelline envelope is produced, the follicle cells also produce the ligands responsible for determining the extreme anterior and posterior ends of the embryo, and those that determine its dorso–ventral axis (section 14.2.5).

Synthesis of the vitelline envelope in mosquitoes is initiated and maintained by 20-hydroxyecdysone which reaches a peak in the hemolymph at this time (Raikhel & Lea, 1991). After fertilization, as the egg is laid, the vitelline envelope becomes compacted in a number of insects.

The chorion (section 14.1.1) is produced wholly by the follicle cells. The various layers are laid down sequentially, usually by addition to the outer surface, but, in the lamellate chorion of Lepidoptera, additional material is added to layers already present by the insertion of new sheets of fibers into the lamellae. As a result, in *Lymantria*, the lamellae increase in thickness from about 0.04 μm when first laid down to 0.2 μm in the fully developed chorion (Leclerc & Regier, 1993). Subsequently, the fibers which form the lamellae thicken and ultimately fuse together so that the lamellae now appear homogeneous rather than fibrous. During this phase there is some compaction of the lamellae and the effect of this is to increase the density of the chorion, a process known as 'densification' (Mazur, Regier & Kafatos, 1989). In many eggs, the outlines of the individual follicle cells producing the chorion are visible on the outside of the completed structure.

Proteins constitute over 90% of the endochorion and much of the exochorion when this is present. Many different proteins are involved: 19 have been recorded in the endochorion of *Drosophila* eggs, and 100 or more are present in the chorion of various Lepidoptera. In *Acheta* (Orthoptera), *Oncopeltus* (Hemiptera) and *Drosophila*, most of the proteins are large with molecular weights in excess of 20 kDa. In the Lepidoptera, by contrast, most are in the range 10–20 kDa. These proteins include structural proteins and enzymes responsible for post-secretion changes in the chorion.

Some of the proteins are produced at different times by the follicle cells. As a result, successive layers of the chorion may be chemically different (Fig. 13.11b). Where marked structural differences occur in different parts of the chorion, specific proteins may be produced by different regions of the follicle epithelium. The specific proteins are additional to a population of proteins that provide the basic material of the chorion and which are produced by all the follicle cells. For example, the cells producing the aeropyle crown of the chorion of *Antheraea* synthesize eight proteins that are not made by any other follicle cells and a further 15 proteins that are also produced by cells forming some other specific areas of the chorion (Mazur, Regier & Kafatos, 1980). These specific proteins are produced principally towards the end of chorionogenesis (Fig. 13.11c). This may be true for area-specific proteins in general since they are often concerned with surface structures of the chorion.

The endochorion of many insect eggs contains

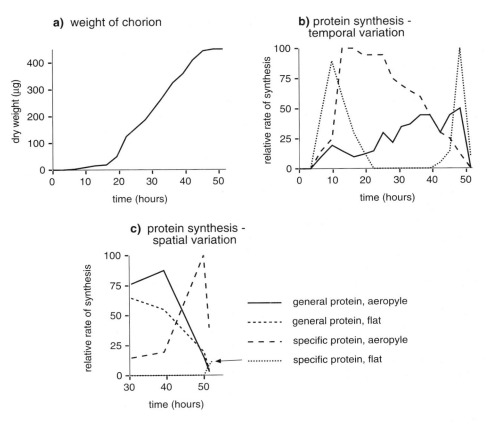

Fig. 13.11. Chorionogenesis in a moth (*Antheraea*). (a) Increase in the weight of the chorion with time (data from Paul & Kafatos, 1975). (b) Temporal variation in the rates of synthesis of three chorionic proteins (indicated by full, dashed and dotted lines). These examples were selected to show different patterns of synthesis (after Mazur, Regier & Kafatos, 1980). (c) Spatial variation in protein synthesis. The general protein is produced by all follicle cells. The specific protein is only produced in the final stages of chorionogenesis in the follicle cells producing the aeropyle region of the egg. Virtually none is produced by the other follicle cells (represented here by cells forming a 'flat' area of chorion). The specific protein is only produced in small amount relative to the general protein; in the figure amounts are multiplied by 10 compared with the general protein (after Mazur, Regier & Kafatos, 1980).

extensive spaces where the outer endochorion is supported by pillars (see Fig. 14.3). Where this occurs, the follicle cells produce a 'flocculent material' or 'filler' which occupies the space between the pillars and supports the newly formed outer endochorion which roofs it (Fig. 13.12). As the chorion dries, this filler collapses, leaving an airspace. Functionally similar material is produced at the surface of the eggs of *Antheraea* where it forms a mold for the production of crowns around the openings of aeropyles.

Aeropyles, micropyles and other pores in the chorion are formed by long cellular processes from one, two or three cells (Fig. 13.13). These processes contain bundles of microtubules and, if there is more than one, they may be held together by cell junctions or simply twist around each other, as in *Dacus* (Mouzaki, Zarani & Margaritis, 1991). Proteins of the chorion are laid down round the processes. Late in chorionogenesis the cells producing the processes degenerate, leaving pores through the chorion.

The chorion of many insect eggs hardens after it is produced though sometimes not until the egg is laid. This is the case in mosquitoes, for example, where the egg hardens and darkens during the day after oviposition. Hardening results from the formation of cross-links between the protein molecules, but these links are produced in different

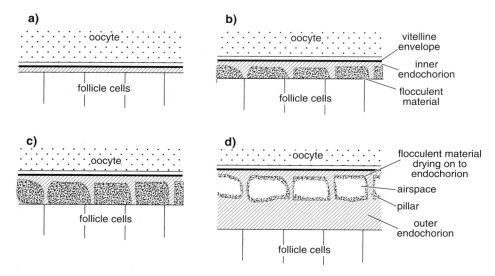

Fig. 13.12. Chorionogenesis. (a)–(d) Diagrammatic representation of the formation of air spaces in the chorion. (a) The vitelline envelope is complete and the first layers of the inner endochorion (hatched) are being produced. (b) A flocculent material is produced over most of the surface. In gaps between this material the endochorion starts to form pillars. (c) Secretion of flocculent material is complete and the outer endochorion starts to form outside it (below in diagram). (d) More outer endochorion is produced and the flocculent material shrinks, leaving air spaces.

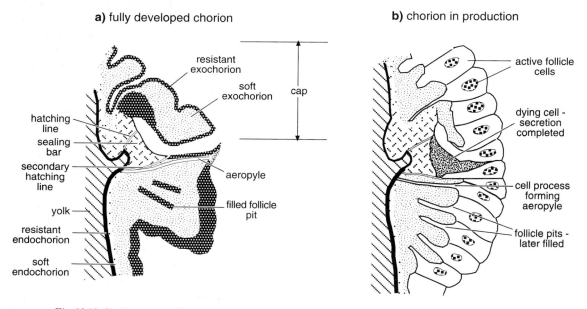

Fig. 13.13. Chorionogenesis. Formation of pores in the chorion. Sections through the chorion of *Rhodnius* at the junction of the cap with the main shell: (a) the complete chorion; (b) formation of an aeropyle by a process from one of the cells. Notice also the differential activity of other cells at the junction and lines of weakness where the chorion breaks when the larva hatches (after Beament, 1946).

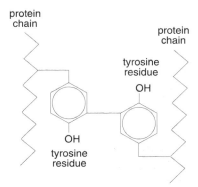

Fig. 13.14. Hardening of the chorion by forming dityrosine bonds between protein chains, as in some fruitflies.

ways in different insects. In *Aedes* and the lubber grasshopper, *Romalea*, hardening involves a process similar to the sclerotization that occurs in the cuticle (Li & Nappi, 1992). In silkmoths, disulfide bonds are formed, reflecting the fact that cysteine is present in relatively large amounts in the proteins. The fruit flies, *Drosophila* and *Ceratitis*, form direct links between tyrosine residues in the proteins producing dityrosine and trityrosine bonds (Fig. 13.14). Finally, the chorions of crickets are hardened by disulfide linkages, but, in addition, cross-links are formed by calcium.

Regulation of these processes that occur after secretion is not generally understood. Probably, enzymes that catalyze the reactions are incorporated into the chorion in inactive forms and subsequently activated. In *Drosophila*, for example, an inactive peroxidase in the chorion is activated by hydrogen peroxide produced at the plasma membrane of the follicle cells.

In grasshoppers, some beetles and in the moth, *Micropteryx*, a coating is added to the outside of the egg as it passes down the oviduct. In grasshoppers, it is called the extrachorion.

Reviews: Kafatos *et al.*, 1995 – genetic regulation in Lepidoptera; Margaritis, 1985 – general; Regier & Kafatos, 1985 – molecular

13.2.6 Resorption of oocytes

Oocytes in the ovarioles may be destroyed and their contents resorbed by the insect. Resorption, or oosorption, may occur when an oocyte is at any stage of development, but it is most commonly observed in terminal oocytes during yolk accumulation. In extreme cases, the terminal oocytes of all the ovarioles may be resorbed; in others, only some oocytes are destroyed while those in most of the ovarioles continue to develop normally.

During resorption, the yolk spheres break down and protein and lipid yolk disappear from the oocyte. It is probable that vitellogenins (or vitellins) are released from the yolk spheres and returned to the hemolymph as a consequence of an increase in permeability of the oocyte membrane. The role of the follicle cells is not clear, but they become folded on each other as the oocyte shrinks and lose their connections with the oocyte (Huebner, 1981). Finally, the whole follicle collapses to form a resorption body, which persists at the base of the ovariole. In grasshoppers, the resorption body is frequently colored orange. In *Machilis* (Thysanura) and Hymenoptera, resorption is known to occur after the chorion is formed, but the mechanism by which yolk is withdrawn through the chorion is not known.

Resorption is a widespread phenomenon that has been recorded from Orthoptera, Blattodea, Dermaptera, Heteroptera, Coleoptera, Diptera and Hymenoptera. Although it most commonly occurs under adverse conditions, some degree of resorption is evident even in insects under apparently optimal conditions. For example, females of *Locusta* resorb about 25% of the oocytes even when they have access to unlimited amounts of high quality food; autogenous females of the screwworm fly, *Chrysomya*, resorb 30% of their first batch of oocytes and still resorb 10% if given additional protein (Spradberry & Schweizer, 1981).

More extreme levels of resorption are often a consequence of lack of nutrients in the adult (Fig. 13.8b). In *Culicoides barbosai* (Diptera), the number of oocytes completing development for a second oviposition is proportional to the size of the blood meal taken by the insect, the remainder are resorbed; in *Locusta* the percentage of oocytes resorbed is inversely proportional to the quantity of food eaten; and low levels of protein in the food lead to oocyte resorption in *Cimex*. In all these cases, all the oocytes start to accumulate yolk, but then some, or even all, of them are resorbed (Fig. 13.8b). Resorption is increased if the insect is parasitized (for example, Liu, 1992) and sometimes also in relation to season and age of the insects. All these effects are probably a consequence of reduction in, or competition for, the available nutrients in the adult. Food shortage in larval stages has not been recorded as a factor influencing the extent of resorption except in Thysanura.

Fig. 13.15. Control of fertilization in a grasshopper. As the egg enters the genital chamber it stimulates mechanoreceptors which trigger contraction of muscles around the spermatheca so that sperm are expressed on to the egg (based on Okelo, 1979).

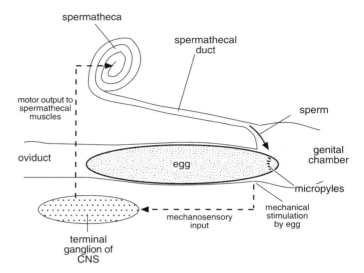

The inability to produce or lay fertile eggs also leads to resorption. Yolk accumulation begins in *Schistocerca* before the female mates, but in the absence of mating many oocytes are resorbed. In this species, mature males produce a maturation pheromone and even the presence of mature males, without mating taking place, reduces the amount of resorption. Lack of suitable oviposition sites causes mosquitoes to resorb terminal oocytes, though fully-developed eggs are retained, and in parasitic Hymenoptera resorption can occur in the absence of a suitable host.

Juvenile hormone is generally involved in the control of yolk accumulation in the oocytes and the lack of this hormone is associated with resorption, at least when all the developing oocytes are affected. The control mechanism involved when only some of the oocytes regress is not clear, but it is possible that competition between oocytes for the available nutrients is involved. In some species, at least, it is clear that resorption is not a direct response to a lack of vitellogenins in the hemolymph.

Review: Bell & Bohm, 1975

13.3 OVULATION

The passage of the oocyte into the oviduct, a process known as ovulation, involves the oocyte's escape from the follicular epithelium and the breakdown of the epithelial plug at the entrance to the pedicel. In *Periplaneta*, the elasticity of the tunica propria helps to force the oocyte

into the oviduct, where it may be stored temporarily before oviposition. In species where the external ovariole sheath contains muscle fibers, these probably assist the movements of the oocyte. The elasticity of the tunica propria causes it to fold up after the oocyte is shed and this pulls the next oocyte down into the terminal position. The empty follicle epithelium of the first oocyte usually persists, but it becomes greatly folded and compressed and forms a new plug at the entrance to the pedicel. The compressed follicle epithelium is known as a corpus luteum and sometimes the corpora lutea of two or three successive ovulations may be present together despite the fact that they break down progressively. In *Melophagus*, only a small relic of the follicle epithelium persists as much of the debris from the follicle cells and trophocytes is passed out of the ovariole when this contracts after ovulation.

Sometimes, as in Orthoptera, all the ovarioles ovulate simultaneously, but in other cases, as in viviparous Diptera, the ovarioles function alternately or in sequence. In Lepidoptera, which commonly lay large batches of eggs although possessing a total of only eight ovarioles, the oocytes may accumulate in the very long pedicels that join the ovarioles to the lateral oviducts until a large number is present. Similarly in some Diptera and parasitic Hymenoptera, such as *Apanteles*, large numbers of eggs may be stored, in this case in the lateral oviducts, thus enabling the insect to lay a large number of eggs quickly when it finds a suitable oviposition site.

It is not clear how ovulation is controlled. Where it is followed immediately by oviposition, as in *Rhodnius*, both ovulation and oviposition may be regulated by the same factors (see below).

13.4 FERTILIZATION OF THE EGG

The egg is fertilized as it passes into the genital chamber when a few sperm are released from the spermatheca. The release of sperm in *Schistocerca* is in direct response to mechanosensory input produced by an egg entering the genital chamber. This input to the terminal abdominal ganglion results in contractions of muscles which compress the spermatheca and force sperm out (Fig. 13.15). A similar mechanism of sperm release occurs in the cricket, *Teleogryllus*. In this case, the egg, on entering the genital chamber, presses on the cuticle and stimulates subcuticular mechanoreceptors. The input from these receptors inhibits the activity of the oviducal muscles so that the egg stops moving, and, at the same time, activates the muscles of the spermathecal duct which squeeze the sperm, tail first, towards the egg (Sugawara, 1993). In other species, sperm may be forced out of the spermatheca by pulsed increases in hemolymph pressure.

In the some insects, the female clearly has the capacity to release or withhold sperm according to environmental cues. Male Hymenoptera are haploid, developing from unfertilized eggs. Queen bees lay male eggs in large cells in the comb and female eggs in smaller cells. Females of some parasitic Hymenoptera tend to lay haploid, male, eggs in small hosts and diploid, female, eggs in larger hosts. For example, the ichneumonid wasp, *Coccygomimus*, lays its eggs in the pupa of the wax moth, *Galleria*. It determines whether or not to fertilize an egg from the size of the cocoon enclosing the host pupa, but it only differentiates between sizes if the cocoon exhibits some chemical characteristic of the pupa (Sandlan, 1979). The measurement of size made by the parasitoid is relative so that it tends to lay male eggs in the smaller members of the host population irrespective of their absolute size (Godfray, 1994). Amongst the Aphelinidae, which parasitize homopterans, female eggs are laid in host nymphs, but male eggs are laid if the ovipositor enters a larval parasitoid already within the host. The parasitoid may be of the same species as the egg-laying female, or of a different species (Hunter & Godfray, 1995). The distinction between parasitized and unparasitized hosts is presumably made on the basis of chemicals encountered by the ovipositor.

How these insects measure size is not generally understood. Queen honeybees apparently use the forelegs to measure the sizes of the cells in which they are ovipositing. The parasitic wasp, *Trichogramma minutum*, lays its eggs in the eggs of other insects and regulates the number of eggs it lays (but not their sex) by determining the surface area of the host egg. It does this by measuring the curvature of the egg surface, probably determined from the angle between the antennal scape and the head, and the time taken to walk across the egg from one side to the other (the transit time) (Schmidt & Smith, 1989). The mechanism by which the female permits or prevents the release of sperm is not known in any of these cases although queen bees have a valve and pump on the spermathecal duct which presumably regulates release.

Females of some insect species have the capacity to select for sperm quality from amongst the sperm in the spermatheca. Males of *Drosophila subobscura* produce long and short sperm. Male ejaculate contains a predominance of the former, but long sperm are stored preferentially in the female's spermathecae. They are also released preferentially from the spermatheca so that most eggs are fertilized by long sperm (Bressac & Hauschteck-Jungen, 1996). There is also evidence that twice-mated females of the moth, *Utetheisa ornatrix*, use sperm from the larger of two males, irrespective of whether she mated first or last with the larger insect (LaMunyon & Eisner, 1993).

Sperm entry to the oocyte is facilitated by the orientation of the egg in the genital chamber where the micropyles are aligned opposite the opening of the spermathecal duct. It is usually stated in the literature that several sperm penetrate each oocyte and that fertilization is effected by only one of the sperm while the rest degenerate. However, there is increasing evidence that monospermy is also common (Retnakaran & Percy, 1985)

In a few insects, fertilization occurs while the oocytes are still in the ovary. This is true of those Cimicoidea which practice hemocoelic insemination, and also occurs in *Aspidiotus*, a coccid, in which the sperm become attached to cells which proliferate in the common oviduct and then migrate to the pedicels.

13.5 OVIPOSITION

13.5.1 Female genitalia: the ovipositor
The gonopore of the female insect is usually situated on or behind the eighth or ninth abdominal segment, but the

Ephemeroptera and Dermaptera are exceptional in having the gonopore behind segment 7. In many orders there are no special structures associated with oviposition, although sometimes the terminal segments of the abdomen are long and telescopic forming a type of ovipositor (see Fig. 11.1). Such structures are found in some Lepidoptera, Coleoptera and Diptera. In *Musca*, the telescopic section is formed from segments six to nine and normally, when not in use, it is telescoped within segment five. The sclerites of the ovipositor are reduced to rods in this species. In others, as in tephritids, the tip of the abdomen is hardened and forms a sharp point, which enables the insect to insert its eggs in small holes and crevices.

An ovipositor of a quite different form, derived from the appendages of abdominal segments eight and nine, is present in Thysanura, some Odonata, Orthoptera, Homoptera, Heteroptera, Thysanoptera Terebrantia and Hymenoptera. Scudder (1961) has attempted to rationalize the terminology associated with the ovipositor and his terms will be used together with the more generally used terminology of Snodgrass (1935).

Scudder (1961) believes that *Lepisma* has a basic form of ovipositor from which those of other insects can be derived. At the base of the ovipositor on each side are the coxae of segments 8 and 9. These are known as the first and second gonocoxae (first and second valvifers of Snodgrass, 1935) (Fig. 13.16) and articulating with each of them is a slender process which curves posteriorly. These processes are called the first and second gonapophyses (valvulae of Snodgrass, 1935) and, together, they form the shaft of the ovipositor. In *Lepisma* (Fig. 13.16a), the second gonapophyses of the two sides are united, so that the shaft is made up of three elements fitting together to form a tube down which the eggs pass. At the base of the ovipositor there is a small sclerite, the gonangulum, which is attached to the base of the first gonapophysis and articulates with the second gonocoxa and the tergum of segment 9. The gonangulum probably represents a part of the coxa of segment 9. It is not differentiated in *Petrobius* (Archaeognatha).

In some Thysanura and in the Pterygota, an additional process is present on the second gonocoxa. This is the gonoplac (third valvula of Snodgrass, 1935). It may or may not be a separate sclerite and may form a sheath round the gonapophyses. The gonoplacs are well developed in the Orthoptera, where they form the dorsal valves of the ovipositor with the second gonapophyses enclosed within

the shaft as in tettigoniids (Fig. 13.16b) or reduced as in the gryllids. Throughout the Orthoptera the gonangulum is fused with the first gonocoxa.

The Hemiptera and Thysanoptera have the gonangulum fused with tergum 9, while the gonoplac may be present or absent. In Pentatomomorpha and Cimicomorpha the development of the ovipositor is related to oviposition habit. If the insect oviposits in plant or animal tissue the valves are sclerotized and lanceolate and the anterior strut of the gonangulum is heavily sclerotized. Species laying on leaf surfaces, however, have membranous and flap-like gonapophyses and the anterior strut of the gonangulum is membranous or absent (Scudder, 1959).

In the Hymenoptera, the first gonocoxae are usually absent (although they may be present in Chalcidoidea) and the second gonapophyses are united. The first gonapophyses slide on the first by a tongue and groove mechanism (Fig. 13.16c). In the Symphyta and parasitic groups, the ovipositor retains its original function, but in Aculeata it forms the sting. This does not involve any major modifications of the basic structure, but the eggs, instead of passing down the shaft of the ovipositor, are ejected from the opening of the genital chamber at its base. In *Apis*, the first gonapophyses are known as the lancets and the fused second gonapophyses as the stylet. This forms an inverted trough, which is enlarged into a basal bulb into which the reservoir of the poison gland discharges.

Reviews: Mickoleit, 1973 – evolution in Endopterygota; Quicke *et al.*, 1994 – Hymenoptera; Scudder, 1961 – general

Sensilla on the ovipositor   The terminal abdominal segments, and the ovipositor, where one is present, have an array of mechanoreceptors. Although chemoreception may be an important factor in oviposition, the number of chemoreceptors on the ovipositor is usually very small. Insects receive most chemical information about their oviposition substrates from tarsal or antennal sensilla.

Trichoid mechanoreceptors are generally present and these may have directional sensitivity (Kalogianni, 1995). In addition, small numbers of campaniform sensilla are also sometimes reported and may be a usual occurrence. Chemoreceptors have been recorded on the ovipositors of Orthoptera, Diptera, parasitic Hymenoptera and Lepidoptera. In most cases there are only a few contact chemoreceptors. The stemborer moth, *Chilo*, for example has four, and the flour moth, *Ephestia*, 10 (Anderson &

a) Thysanura

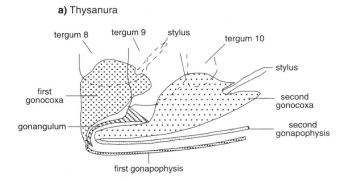

b) tettigoniid

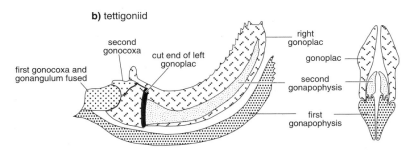

c) Hymenoptera

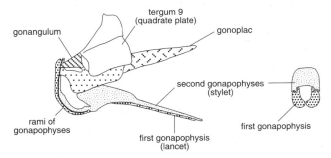

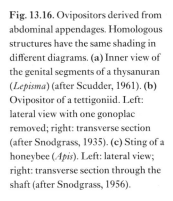

Fig. 13.16. Ovipositors derived from abdominal appendages. Homologous structures have the same shading in different diagrams. (a) Inner view of the genital segments of a thysanuran (*Lepisma*) (after Scudder, 1961). (b) Ovipositor of a tettigoniid. Left: lateral view with one gonoplac removed; right: transverse section (after Snodgrass, 1935). (c) Sting of a honeybee (*Apis*). Left: lateral view; right: transverse section through the shaft (after Snodgrass, 1956).

Hallberg, 1990). Multiporous, presumed olfactory, sensilla have been recorded around the ovipositor in *Musca* and other cyclorrhaphous flies (Merritt & Rice, 1984), and the sunflower moth, *Homoeosoma*, has 59 on each side (Faucheux, 1991). By contrast, *Manduca* has no chemoreceptors associated with the ovipositor, nor has the damsel fly, *Sympecma*, even though it oviposits in plants (Gorb, 1994).

13.5.2 Mechanisms of oviposition

In the majority of species which lack an appendicular ovipositor, eggs are simply deposited on a surface or, if the terminal segments of the abdomen are elongated or telescopic, may be inserted into crevices. In some cases, however, specialized structures are involved in the oviposition process. Many Asilidae, for instance, have spine-bearing plates called acanthophorites at the tip of the abdomen. These push the soil aside as the insect inserts the tip of its abdomen and oviposits and then, when the abdomen is withdrawn, the soil falls back and covers the eggs. The beetle *Ilybius* has two finely toothed blades forming an ovipositor. The points of these blades are pushed into the surface of a suitable plant and then worked upwards by a rapid, rhythmic, saw-like action so that a

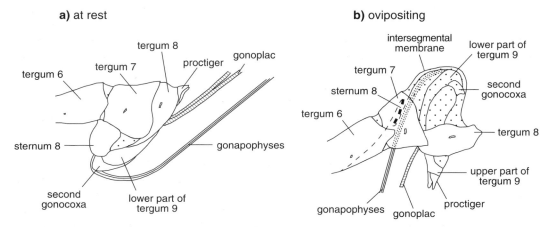

a) at rest   **b)** ovipositing

Fig. 13.17. Oviposition. Changes in the positions of the basal parts of the ovipositor and the terminal abdominal segments during oviposition by an ichneumonid wasp (*Megarhyssa*) (after Snodgrass, 1935). (a) At rest; (b) during oviposition.

tongue of plant tissue is cut away at the sides. An egg is laid in the hole beneath the blades and is covered by the tongue of plant tissue when the blades are withdrawn.

Many species with an ovipositor derived from the appendages of segments 8 and 9 penetrate tissues using a sliding movement of the valves relative to each other similar the sting mechanism of *Apis*. In an ichneumon, the tip of the abdomen is turned down at the start of oviposition so that the valves point ventrally instead of posteriorly (Fig. 13.17). The gonapophyses then work their way by rapid to and fro movements into the host tissue (or through the wood in which the host is boring in the case of *Rhyssa*). The gonoplacs do not enter the wound, but become deflected outside it. In this way *Rhyssa* can bore through 3 cm of wood in 20 minutes.

In grasshoppers (Acrididae), the action of the valves is quite different, involving an opening and closing movement of the dorsal and ventral valves rather than a sliding movement (Fig. 13.19a). These movements are produced by muscles inserted on to an apodeme at the base of the valves, and the whole structure is rhythmically protracted and retracted. The insect starts by raising the body on the first two pairs of legs and arching the tip of the abdomen downwards so that it presses more or less vertically on the ground. The opening movement of the valves scrapes particles of the substratum sideways and upwards and pressure is exerted down the abdomen so that the valves slowly dig a hole. As the hole deepens, the abdomen lengthens by the unfolding and stretching of the intersegmental

membranes between segments 4 and 5, 5 and 6, and 6 and 7. The membranes are specialized to permit stretching, having a lamellate endocuticle under a thin epicuticle which is folded at right angles to the long axis of the body. As the abdomen lengthens, these folds become smoothed out and the endocuticle stretches. The intersegmental membranes of the male do not stretch to the same extent as in the mature female, nor do those of the immature female, indicating that some change occurs during maturation. The change is controlled by the corpora allata and it is probable that hydrogen bonding between the protein chains in the cuticle is reduced by an enzyme from the epidermis (Tychsen and Vincent, 1976). The intersegmental muscles are also stretched and they are specialized to permit this (see Fig. 10.15).

The extension of the abdomen of grasshoppers may be very considerable. For instance, the abdomen of *Anacridium* stretches from 3–5 cm to 10 cm in length and *Schistocerca* can dig to a depth of 14 cm. During digging the ventral valves of the ovipositor lever the abdomen downwards while the upper valves push soil away and effect the excavation. The pull exerted by the ventral valves is transmitted to the abdomen and this results in the extension of the intersegmental membranes between segments 4 and 7 (Vincent, 1975). At intervals during digging, the female partly withdraws her abdomen and the walls of the hole are smoothed and compacted by small movements of the ovipositor valves together with twisting movements of the abdomen. Even in apparently suitable soil a female frequently abandons a hole and starts to dig again, but when a suitable hole has been constructed the process of oviposition proper begins. Just before an egg is laid, the female pumps air into her tracheal system by rapid movements of the head and then, with the thoracic spiracles closed, forces the air backwards so that the abdomen becomes turgid. It remains turgid until the egg is laid, then the head moves forwards again and the pressure is released. Eggs are passed out micropylar end first and the abdomen is slowly withdrawn as more eggs are laid. When all the eggs have been laid, the female fills the upper part of the hole with a frothy plug (Fig. 13.21b) and, finally, after withdrawing her abdomen, she scrapes soil over the top of the hole with her hind tibiae. The whole process may take about two hours, of which egg-laying occupies some 20 minutes.

13.5.3 Control of oviposition

The readiness to oviposit is influenced by mating. A female *Bombyx*, for example, lays all her eggs within 24 hours of mating whereas a virgin female retains most of her eggs for some days. An essentially similar change of behavior occurs in insects belonging to many different orders. It is induced by peptides or other substances transferred to the female by the male during sperm transfer (see section 12.4.2). Under normal conditions, this substance presumably causes the female to seek an oviposition site and then, when she receives appropriate stimulation from her mechano- and chemoreceptors, to begin ovipositing. In *Bombyx*, the eggs are normally laid close together in a single layer; if the trichoid mechanoreceptors on the anal papillae are damaged the eggs are deposited in uneven clumps. In the intact insect the information from the hairs is relayed to the brain (Fig. 13.18) and from the brain a command to motor neurons arising in the terminal abdominal ganglion leads to contractions of the oviduct so that eggs are deposited, and to movements of the abdomen which result in the eggs being spread regularly over a surface (Yamaoka and Hirao, 1971, 1977).

In grasshoppers, the activity of the muscles moving the ovipositor valves is regulated by a central pattern generator in the terminal abdominal ganglion (Fig. 13.19). In the non-ovipositing insect the activity of the pattern generator is inhibited by neural activity from the head ganglia and the metathoracic ganglion (Thompson, 1986a,b). Sensory input from mechanoreceptors on the ovipositor valves probably helps to maintain and modulate the basic rhythm produced by the pattern generator

(Kalogianni, 1995). The ventral opener muscle is innervated by a motor neuron producing proctolin, and this may also be true of other muscles associated with the ovipositor valves. It is suggested that octopamine is also necessary for the normal functioning of the muscle and to maintain the activity of the central pattern generator (Belanger & Orchard, 1993).

The movement of eggs down the oviducts appears, in *Locusta*, to result from myogenic activity of the oviducal muscles and a myotropic peptide is present in the hemolymph during oviposition. These muscles are also controlled neurally, but neural stimulation has the effect of preventing the backwards movement of eggs. A second central pattern generator and octopamine from dorsal unpaired median (DUM) cells in the subterminal abdominal ganglion are also involved in integrating these activities, but their precise functions are not clear (Facciponte & Lange, 1992; Kalogianni & Theophilidis, 1993). A myotropic hormone is also present at maximum titers during the period of oviposition of *Rhodnius* (Davey & Kriger, 1985). It is presumed to have an important role in oviposition, as in *Locusta*.

13.5.4 Role of the accessory glands

The function of the female accessory glands is generally to fix eggs in position or protect them from desiccation and predators. Many insects attach their eggs on or close to the larval food source. Lice fix their eggs to hairs; many plant-feeding Heteroptera and Lepidoptera lay eggs on the host plant. The glue holding the eggs in position comes from the female accessory glands. Within the Lepidoptera, insects that drop their eggs from the air, such as the Hepialidae, lack accessory glands. However, the glands are well-developed in species that cover their eggs with secretion (Hinton, 1981). In *Chrysopa* (Neuroptera), the eggs are laid on tall stalks formed from silk produced in the accessory glands.

The accessory glands of the parasitic wasp, *Pimpla* (Hymenoptera), produce hyaluronic acid and a lipoprotein which coats the egg as it is laid into the host. It is suggested that this is a critical factor in preventing encapsulation by the host hemocytes (Blass & Ruthmann, 1989) and this may also occur in other parasitic insects (section 5.3.4).

The accessory gland secretions of many aquatic insects form a gelatinous mass enclosing the eggs. For instance, *Chironomus dorsalis* produces a structure in which the eggs

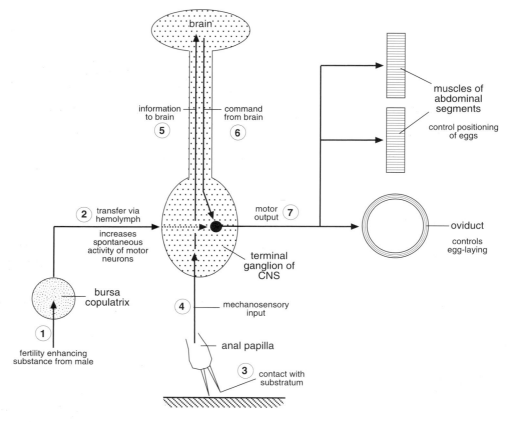

Fig. 13.18. Control of oviposition in the silk moth, (*Bombyx*). numbers indicate the sequence of events (after Yamaoka & Hirao, 1971, 1977).

loop backwards and forwards round the circumference of the matrix while a pair of fibers, anchoring the mass to the surface, run through the center (Fig. 13.20b). Other species of chironomids and Trichoptera lay their eggs in masses or strands of gelatinous material.

Female accessory gland secretions have another important function in muscid flies. Here, the secretion contains proteolytic enzymes and an esterase which have two essential roles in fertilization of the egg. They breakdown the acrosomal membrane of the sperm, and they lead to the digestion of a cap over the micropyle. Whether this also occurs in other insects is not known (Gillott, 1988).

Oothecae Species from several insect groups lay their eggs in oothecae formed by secretions of the female accessory glands. Oothecae are produced by a majority of Blattodea, Mantodea and Acridoidea, and by the tortoise beetles, Cassidinae. These structures have precise and characteristic forms which permit respiration by the eggs and the escape of the newly hatched larvae while, at the same time, protecting them from desiccation and from predators. Amongst the Blattodea, *Blatta*, for instance, lays its eggs in two rows, each of eight eggs, inside a capsule which becomes tanned as it is formed (Fig. 13.21a). A crest along the top of the capsule contains cavities connecting via small pores to the outside and via narrow tubes to the inside of the ootheca. Immediately below the opening of each tube, the chorion of an egg is produced into an irregular lobe containing airspaces which communicate with the airspace around the oocyte. Each egg is associated with an air tube in the ootheca.

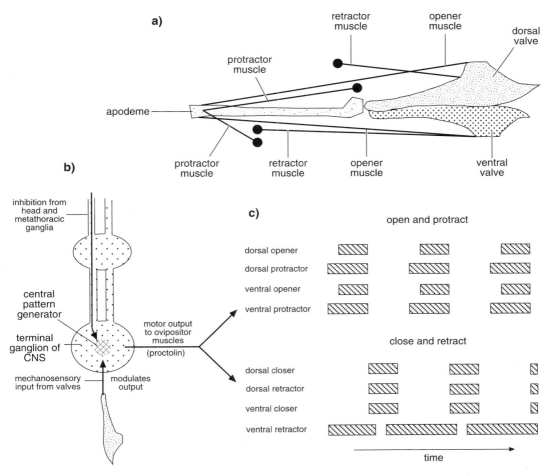

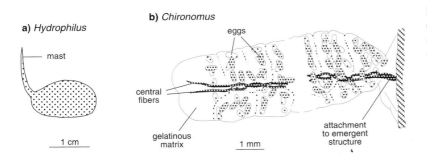

Fig. 13.19. Control of oviposition in the locust (*Locusta*) (mainly after Thompson, 1986a,b). (**a**) Diagram of the ovipositor valves and the apodeme to which the dorsal valve hinges. Black spots indicate the origins of muscles on the body wall. Closer muscles and some accessory muscles are not shown. (**b**) Diagram of the terminal abdominal ganglion showing some features of regulation. (**c**) Pattern of activity in the principal muscles produced by the central pattern generator.

Fig. 13.20. Eggs of aquatic insects: (**a**) egg cocoon of *Hydrophilus* (Coleoptera) (after Miall, 1922); (**b**) egg rope of *Chironomus* (Diptera).

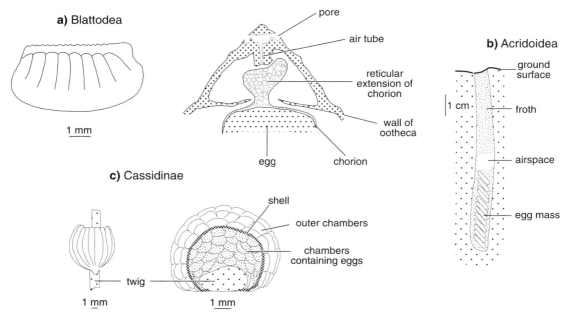

Fig. 13.21. Oothecae. **(a)** Cockroach (*Blatta*). Left: lateral view of the ootheca; right: transverse section through the crest showing the passage by which air reaches the eggs (after Wigglesworth & Beament, 1950). **(b)** Grasshopper (*Acrida*) (after Chapman & Robertson, 1958). **(c)** Tortoise beetle (*Basipta*). Left: side view; right: view from above showing the chambers which are produced by secretions of the accessory glands. Each chamber contains one egg (after Muir & Sharp, 1904).

Most Acridoidea lay their eggs in egg pods in the ground. A pod consists of a mass of eggs held together by a frothy secretion and is sometimes also enclosed by a layer of the same substance. The hole above the egg-mass is plugged by more froth (Fig. 13.21b). The number of eggs in a pod reflects the number of ovarioles in the ovaries and may vary from one to over 300. The frothy plug permits respiration and provides an easy route to the soil surface for the hatching larvae.

Tortoise beetles also produce oothecae. The form and complexity of the ootheca varies from species to species, but *Basipta*, for example, attaches its ootheca to the stem of its food plant (Fig. 13.21c). The theca is formed of a large number of lamellae produced from an accessory gland secretion which is compressed into a plate-like form as it is extruded between the terminal sclerites of the abdomen. The lamellae form an open cup, occupied by about 30 chambers also formed by the lamellae. An egg is placed in each chamber as it is formed and round the outside of the cup the lamellae are plastered firmly together to form a hard 'shell' with looser lamellae outside it.

Amongst aquatic insects, all the Hydrophilidae (Coleoptera) produce egg cocoons, or oothecae. These cocoons are made from silk synthesized in modified ovarioles. The cocoon of *Hydrophilus* is molded to the shape of the abdomen with the aid of the forelegs, then the abdomen is withdrawn and the eggs are laid. Finally, the cocoon is sealed off and remains floating on the surface of the water. It is equipped with a silken 'mast' about an inch high, which has a respiratory function (Fig. 13.20a).

In all these insects, the material from the glands, which is initially fluid, is modified as the chemicals from different glands mix and come into contact with the outside air. The process is best understood in *Periplaneta*. The two accessory glands each consist of a mass of branched tubules lined with cuticle. The gland cells differ in different parts of the glands and in the left gland, which is larger than the right, three types are recognizable. They produce the protein from which the ootheca is formed, a β-glucoside of protocatechuic acid and an oxidase. The right-hand gland has two types of secretory cell. It secretes a β-glucosidase which liberates protocatechuic acid from its β-glucoside when the secretions of the two glands mix in the genital chamber. The protocatechuic acid is oxidized to a quinone by the oxidative enzymes and this tans the protein to produce a cuticle-like structure. A similar process occurs in the production of an ootheca by the mantis, *Tenodera* (Yago *et al.*, 1990).

Some insects cover the eggs with material which is not derived from the accessory glands. The heteropteran *Plataspis* forms a type of ootheca, laying its eggs in two rows and then covering them with hard elongate pellets of a secretion produced by specialized cells in the gut.

Reviews: Gillott, 1988 – accessory glands; Hinton, 1981 – oothecae

REFERENCES

Anderson, P. & Hallberg, E. (1990). Structure and distribution of tactile and bimodal taste/tactile sensilla on the ovipositor, tarsi and antennae of the flour moth, *Ephestia kuehniella* (Zeller) (Lepidoptera: Pyralidae). *International Journal of Insect Morphology & Embryology*, **19**, 13–23.

Beament, J.W.L. (1946). The waterproofing process in eggs of *Rhodnius prolixus* Stahl. *Proceedings of the Royal Society of London*, B, **133**, 407–18.

Belanger, J.H. & Orchard, I. (1993). The locust ovipositor opener muscle: proctolinergic central and peripheral neuromodulation in a centrally driven motor system. *Journal of Experimental Biology*, **174**, 343–62.

Bell, W. J. & Bohm, M. K. (1975). Oosorption in insects. *Biological Reviews*, **50**, 373–96.

Blass, S. & Ruthmann, A. (1989). Fine structure of the accessory glands of the female genital tract of the ichneumonid *Pimpla turionellae* (Insecta, Hymenoptera). *Zoomorphology*, **108**, 367–77.

Bownes, M., Ronaldson, E., Mauchline, D. & Martinez, A. (1993). Regulation of vitellogenesis in *Drosophila*. *International Journal of Insect Morphology & Embryology*, **22**, 349–67.

Bressac, C. & Hauschteck-Jungen, E. (1996). *Drosophila subobscura* females preferentially select long sperm for storage and use. *Journal of Insect Physiology*, **42**, 323–8.

Büning, J. (1993). Germ cell cluster formation in insect ovaries. *International Journal of Insect Morphology & Embryology*, **22**, 237–53.

Büning, J. (1994). *The Insect Ovary: Ultrastructure, Previtellogenic Growth and Evolution*. London: Chapman & Hall.

Chapman, R.F. & Robertson, I.A.D. (1958). The egg pods of some tropical African grasshoppers. *Journal of the Entomological Society of Southern Africa*, **21**, 85–112.

Davey, K.G. & Kriger, F.L. (1985). Variations during the gonotrophic cycle in the titer of the myotropic ovulation hormone and the response of the ovarian muscles in *Rhodnius prolixus*. *General and Comparative Endocrinology*, **58**, 452–7.

Facciponte, G. & Lange, A.B. (1992). Characterization of a novel central pattern generator located in the VIIth abdominal ganglion of *Locusta*. *Journal of Insect Physiology*, **38**, 1011–22.

Faucheux, M.J. (1991). Morphology and distribution of sensilla on the cephalic appendages, tarsi and ovipositor of the European sunflower moth, *Homoeosoma nebulella* Den. & Schiff. (Lepidoptera: Pyralidae). *International Journal of Insect Morphology & Embryology*, **20**, 291–307.

Ferenz, H.-J. (1993). Yolk protein accumulation in *Locusta migratoria* (R. & F.) (Orthoptera: Acrididae) oocytes. *International Journal of Insect Morphology & Embryology*, **22**, 295–314.

Foe, V.E., Odell, G.M. & Edgar, B.A. (1993). Mitosis and morphogenesis in the *Drosophila* embryo: point and counterpoint. In *The Development of Drosophila melanogaster*, vol. 1, ed. M. Bate & A.M. Arias, pp. 149–300. New York: Cold Spring Harbor Laboratory Press.

Gadot, M. & Applebaum, S.W. (1985). Rapid in vitro activation of corpora allata by extracted locust brain allatotropic factor. *Archives of Insect Biochemistry and Physiology*, **2**, 117–29.

Gellisen, G. & Emmerich, H. (1978). Changes in the titer of vitellogenin and of diglyceride carrier lipoprotein in the blood of adult *Locusta migratoria*. *Insect Biochemistry*, **8**, 403–12.

Gillott, C. (1988). Arthropoda – Insecta. In *Reproductive Biology of Invertebrates. III. Accessory Sex Glands*, ed. K.G. Adiyodi & R.G. Adiyodi, pp. 319–471. Chichester: Wiley.

Giorgi, F., Masetti, M., Ignacchiti, V., Cecchettini, A. & Bradley, J.T. (1993). Postendocytic vitellin processing in ovarian follicles of the stick insect *Carausius morosus* (Br.). *Archives of Insect Biochemistry and Physiology*, **24**, 93–111.

Glinka, A.V., Rodik, O.A., Pshennikova, E.S. & Valjushok, L.N. (1994). Stored forms of vitellogenin mRNA in fat body cells of *Locusta migratoria*. *Insect Biochemistry and Molecular Biology*, **24**, 249–55.

Godfray, H.C.J. (1994). *Parasitoids. Behavioral and Evolutionary Ecology*. Princeton: Princeton University Press.

Gondim, K.C., Oliveira, P.L. & Masuda, H. (1989). Lipophorin and oögenesis in *Rhodnius prolixus*: transfer of phospholipids. *Journal of Insect Physiology*, **35**, 19–27.

Gorb, S.N. (1994). Central projections of ovipositor sense organs in the damselfly, *Sympecma annulata* (Zygoptera, Lestidae). *Journal of Morphology*, **220**, 139–46.

Hagedorn, H.H. (1985). The role of ecdysteroids in reproduction. In *Comprehensive Insect Physiology, Biochemistry and Pharmacology*, vol. 8, ed. G.A. Kerkut & L.I. Gilbert, pp. 205–62. Oxford: Pergamon Press.

Hagedorn, H.H. (1994). The endocrinology of the adult female mosquito. *Advances in Disease Vector Research*, **10**, 109–48.

Heilmann, L.J., Trewitt, P.M. & Kumaran, A.K. (1993). Proteolytic processing of the vitellogenin precursor in the boll weevil, *Anthonomus grandis*. *Archives of Insect Biochemistry and Physiology*, **23**, 125–34.

Hinton, H.E. (1981). *Biology of Insect Eggs*, vol. 1. Oxford: Pergamon Press.

Hopkins, C.R. & King, P.E. (1966). An electron-microscopical and histochemical study of the oocyte periphery in *Bombus terrestris* during vitellogenesis. *Journal of Cell Science*, **1**, 201–16.

Huebner, E. (1981). Oocyte-follicle cell interactions during normal oogenesis and atresia in an insect. *Journal of Ultrastructure Research*, **74**, 95–104.

Huebner, E. & Diehl-Jones, W. (1993). Nurse cell–oocyte interaction in the telotrophic ovary. *International Journal of Insect Morphology & Embryology*, **22**, 369–87.

Hunter, M.S. & Godfray, H.C.J. (1995). Ecological determinants of sex allocation in an autoparasitoid wasp. *Journal of Animal Ecology* **64**, 95–106.

Imms, A.D. (1957). *A General Textbook of Entomology*. 9th edition revised by O.W. Richards & R.G. Davies. London: Methuen.

Kafatos, F.C., Tzertzinis, G., Spoerel, N.A. & Nguyen, H.T. (1995). Chorion genes: an overview of their structure, function, and transcriptional regulation. In *Molecular Model Systems in the Lepidoptera*, ed. M.R. Goldsmith & A.S. Wilkins, pp. 181–215. Cambridge: Cambridge University Press.

Kalogianni, E. (1995). Physiological properties of wind-sensitive and tactile trichoid sensilla on the ovipositor and their role during oviposition in the locust. *Journal of Experimental Biology*, **198**, 1359–69.

Kalogianni, E. & Theophilidis, G. (1993). Centrally generated rhythmic activity and modulatory function of the oviductal dorsal unpaired median (DUM) neurones in two Orthopteran species (*Calliptamus* sp. and *Decticus albifrons*). *Journal of Experimental Biology*, **174**, 123–38.

Kawooya, J.K. & Law, J.H. (1988). Role of lipophorin in lipid transport to the insect egg. *Journal of Biological Chemistry*, **263**, 8748–53.

King, R.C. (1964). Studies on early stages of insect oogenesis. *Symposium of the Royal Entomological Society of London*, **2**, 13–25.

King, R.C. & Büning (1985). The origin and functioning of insect oocytes and nurse cells. In *Comprehensive Insect Physiology, Biochemistry and Pharmacology*, vol. 1, ed. G.A. Kerkut & L.I. Gilbert, pp. 37–82. Oxford: Pergamon Press.

Klowden, M.J. (1987). Distension-mediated egg maturation in the mosquito, *Aedes aegypti*. *Journal of Insect Physiology*, **33**, 83–7.

Koeppe, J.K., Fuchs, M., Chen. T.T., Hunt, L.-M., Kovalick, G.E. & Briers, T. (1985). The role of juvenile hormone in reproduction. In *Comprehensive Insect Physiology, Biochemistry and Pharmacology*, vol. 8, ed. G.A. Kerkut & L.I. Gilbert, pp. 165–203. Oxford: Pergamon Press.

LaMunyon, C.W. & Eisner, T. (1993). Postcopulatory sexual selection in an arctiid moth (*Utetheisa ornatrix*). *Proceedings of the National Academy of Sciences of the United States of America*, **90**, 4689–92.

Lanot, R., Sellam, M.O., Bucher, B. & Thiebold, J.J. (1990). Ecdysone- controlled meiotic reinitiation in oocytes of *Locusta migratoria* involves a decrease in cAMP levels. *Insect Biochemistry*, **20**, 639–44.

Lanot, R., Thiebold, J., Lagueux, M., Goltzene, F. & Hoffmann, J.A. (1987). Involvement of ecdysone in the control of meiotic reinitiation in oocytes of *Locusta migratoria* (Insecta, Orthoptera). *Developmental Biology*, **121**, 174–81.

Lea, A.O., Briegel, H. & Lea, H.N. (1978). Arrest, resorption, or maturation of oocytes in *Aedes aegypti*: dependence on the quantity of blood and the interval between blood meals. *Physiological Entomology*, **3**, 309–16.

Leclerc, R.F. & Regier, J.C. (1993). Chorionogenesis in the Lepidoptera: morphogenesis, protein synthesis, specific mRNA accumulation, and primary structure of a chorion cDNA from the gypsy moth. *Developmental Biology*, **160**, 28–38.

Li, J. & Nappi, A.J. (1992). *N*-acetyltransferase activity during ovarian development in the mosquito *Aedes aegypti* following blood feeding. *Insect Biochemistry and Molecular Biology*, **22**, 49–54.

Liu, T.P (1992). Oöcyte degeneration in the queen honey bee after infection by *Nosema apis*. *Tissue & Cell*, **24**, 131–138.

Martoja, R. (1977). Les organes génitaux femelles. In *Traité de Zoologie*, vol. 8, part 5A, ed. P.-P.Grassé, pp. 1–123. Paris: Masson et Cie.

Margaritis, K. (1985). Structure and physiology of the eggshell. In *Comprehensive Insect Physiology, Biochemistry and Pharmacology*, vol. 1, ed. G.A. Kerkut & L.I. Gilbert, pp. 153–230. Oxford: Pergamon Press.

Mazur, G.D., Regier, J.C. & Kafatos, F.C. (1980). The silkmoth chorion: morphogenesis of surface structures and its relation to synthesis of specific proteins. *Developmental Biology*, **76**, 305–21.

Mazur, G.D., Regier, J.C. & Kafatos, F.C. (1989). Morphogenesis of silkmoth chorion: sequential modification of an early helicoidal framework through expansion and densification. *Tissue & Cell*, **21**, 227–42.

Merritt, D.J. & Rice, M.J. (1984). Innervation of the cercal sensilla on the ovipositor of the Australian sheep blowfly (*Lucilia cuprina*). *Physiological Entomology*, **9**, 39–47.

Miall, L.C. (1922). *The Natural History of Aquatic Insects*. London: MacMillan.

Mickoleit, G. (1973). Uber den ovipositor der Neuropteroidea und Coleoptera und seine phylogenetische Bedeutung (Insecta, Holometabola). *Zeitschrift für Morphologie der Tiere*, **74**, 37–64.

Moshitzky, P. & Applebaum, S.W. (1990). The role of adipokinetic hormone in the control of vitellogenesis in locusts. *Insect Biochemistry*, **20**, 319–23.

Mouzaki, D.G., Zarani, F.E. & Margaritis, L.H. (1991). Structure and morphogenesis of the eggshell and micropylar apparatus in the olive fly, *Dacus oleae* (Diptera: Tephritidae). *Journal of Morphology*, **209**, 39–52.

Muir, F. & Sharp, D. (1904). On the egg-cases and early stages of some Cassididae. *Transactions of the Entomological Society of London*, (1904), 1–23.

Okelo, O. (1979). Mechanisms of sperm release from the receptaculum seminis of *Schistocerca vaga* Scudder (Orthoptera: Acrididae). *International Journal of Invertebrate Reproduction*, **1**, 121–31.

Paul, M. & Kafatos, F.C. (1975). Specific protein synthesis in cellular differentiation. II. The program of protein synthetic changes during chorion formation by silkmoth follicles, and its implementation in organ culture. *Developmental Biology*, **42**, 141–59.

Quicke, D.L.J., Fitton, M.G., Tunstead, J.R., Ingram, S.N. & Gaitens, P.V. (1994). Ovipositor structure and relationships within the Hymenoptera with special reference to the Ichneumonoidea. *Journal of Natural History*, **28**, 635–82.

Raikhel, A.S. & Dhadialla, T.S. (1992). Accumulation of yolk proteins in insect oocytes. *Annual Review of Entomology*, **37**, 217–51.

Raikhel, A.S. & Lea. A.O. (1991). Control of follicular epithelium development and vitelline envelope formation in the mosquito; role of juvenile hormone and 20-hydroxyecdysone. *Tissue & Cell*, **23**, 577–91.

Regier, J.C. & Kafatos, F.C. (1985). Molecular aspects of chorion formation. In *Comprehensive Insect Physiology, Biochemistry and Pharmacology*, vol. 1, ed. G.A. Kerkut & L.I. Gilbert, pp. 113–51. Oxford: Pergamon Press.

Retnakaran, A. & Percy, J. (1985). Fertilization and special modes of reproduction. In *Comprehensive Insect Physiology, Biochemistry and Pharmacology*, vol. 1, ed. G.A. Kerkut & L.I. Gilbert, pp. 231–93. Oxford: Pergamon Press.

Sandlan, K. (1979). Sex ratio regulation in *Coccygomimus turionella* Linnaeus (Hymenoptera: Ichneumonidae) and its ecological implications. *Ecological Entomology*, **4**, 365–78.

Sato, Y. & Yamashita, O. (1991). Synthesis and secretion of egg-specific protein from follicle cells of the silkworm, *Bombyx mori*. *Insect Biochemistry*, **21**, 233–8.

Schmidt, J.M. & Smith, J.J.B. (1989). Host examination walk and oviposition site selection of *Trichogramma minutum*: studies on spherical hosts. *Journal of Insect Behavior*, **2**, 143–71.

Scudder, G.G.E. (1959). The female genitalia of the Heteroptera: morphology and bearing on classification. *Transactions of the Royal Entomological Society of London*, **111**, 405–67.

Scudder, G.G.E. (1961). The comparative morphology of the insect ovipositor. *Transactions of the Royal Entomological Society of London*, **113**, 25–40.

Shaaya, E., Shirk, P.D., Zimowska, G., Plotkin, S., Young, N.J., Rees, H.H. & Silhacek, D.L. (1993). Declining ecdysteroid levels are temporally correlated with the initiation of vitellogenesis during pharate adult development in the Indian meal moth, *Plodia interpunctella*. *Insect Biochemistry and Molecular Biology*, **23**, 153–8.

Singleton, K. & Woodruff, R.I. (1994). The osmolarity of adult *Drosophila* hemolymph and its effect on oocyte–nurse cell electrical polarity. *Developmental Biology*, **161**, 154–67.

Snodgrass, R.E. (1935). *Principles of Insect Morphology*. New York: McGraw-Hill.

Snodgrass, R.E. (1956). *Anatomy of the Honey Bee*. London: Constable.

Spradbery, J.P. & Schweizer, G. (1981). Oosorption during ovarian development in the screw-worm fly, *Chrysomya bezziana*. *Entomologia Experimentalis et Applicata*, **30**, 209–14.

Spradling, A.C. (1993). Developmental genetics of oogenesis. In *The Development of Drosophila melanogaster*, vol. 1, ed. M. Bate & A.M. Arias, pp. 1–70. New York: Cold Spring Harbor Laboratory Press.

Stay, B., Sereg Bachmann, J.A., Stoltzman, C.A., Fairbairn, S.E., Yu, C.G. & Tobe, S.S. (1994). Factors affecting allatostatin release in a cockroach (*Diploptera punctata*): nerve section, juvenile hormone analog and ovary. *Journal of Insect Physiology*, **40**, 365–72.

Sugawara, T. (1993). Oviposition behaviour of the cricket *Teleogryllus commodus*: mechanosensory cells in the genital chamber and their role in the switch-over of steps. *Journal of Insect Physiology*, **39**, 335–46.

Telfer, W.H. (1975). Development and physiology of the oocyte nurse cell syncytium. *Advances in Insect Physiology*, **11**, 223–319.

Telfer, W.H. (1981). Electrical polarity and cellular differentiation in meroistic ovaries. *American Zoologist*, 21, 675–86.

Thompson, K.J. (1986a). Oviposition digging in the grasshopper. I. Functional anatomy and the motor program. *Journal of Experimental Biology*, 122, 387–411.

Thompson, K.J. (1986b). Oviposition digging in the grasshopper. II. Descending neural control. *Journal of Experimental Biology*, 122, 413–25.

Tychsen, P.H. & Vincent, J.F.V. (1976). Correlated changes in mechanical properties of the intersegmental membrane and bonding between proteins in the female adult locust. *Journal of Insect Physiology*, 22, 115–25.

Uvarov, B.P. (1966). *Grasshoppers and Locusts*, vol. 1. Cambridge: Cambridge University Press.

van Antwerpen, R. & Law, J.H. (1993). Immunocytochemical localization of a follicle specific protein of the hawkmoth *Manduca sexta*. *Tissue & Cell*, 25, 885–92.

Vincent, J.F.V. (1975). How does the locust dig her oviposition hole? *Journal of Entomology* A, 50, 175–81.

Wang, Z. & Davey, K.G. (1992). Characterization of yolk protein and its receptor on the oocyte membrane in *Rhodnius prolixus*. *Insect Biochemistry and Molecular Biology*, 22, 757–67.

Wheeler, D. (1996). The role of nourishment in oogenesis. *Annual Review of Entomology*, 41, 407–31.

Wigglesworth, V.B. & Beament, J.W.S. (1950). The respiratory mechanisms of some insect eggs. *Quarterly Journal of Microscopical Science*, 91, 429–52.

Yago, M., Sato, H., Oshima, S. & Kawasaki, H. (1990). Enzymic activities involved in the oothecal sclerotization of the praying mantid, *Tenodera aridifilia sinensis* Saussure. *Insect Biochemistry*, 20, 745–50.

Yamaoka, K. & Hirao, T. (1971). Role of nerves from the last abdominal ganglion in oviposition behaviour of *Bombyx mori*. *Journal of Insect Physiology*, 17, 2327–36.

Yamaoka, K. & Hirao, T. (1977). Stimulation of virginal oviposition by male factor and its effect on spontaneous nervous activity in *Bombyx mori*. *Journal of Insect Physiology*, 23, 57–63.

Yin, C.-M., Zou, B.-X., Li, M.-F. & Stoffolano, J.G. (1990). Ecdysteroid activity during oögenesis in the black blowfly, *Phormia regina* (Meigen). *Journal of Insect Physiology*, 36, 375–82.

Yin, C.-M., Zou, B.-X., Li, M.-F. & Stoffolano, J.G. (1994). Discovery of a midgut peptide hormone which activates the endocrine cascade leading to oögenesis in *Phormia regina* (Meigen). *Journal of Insect Physiology*, 40, 283–92.

Yin, C.-M., Zou, B.-X. & Stoffolano, J.G. (1989). Precocene II treatment inhibits terminal oöcyte development but not vitellogenin synthesis and release in the black blowfly, *Phormia regina* Meigen. *Journal of Insect Physiology*, 35, 465–74.

Zhang, Y. & Kunkel, J.G. (1994). Most egg calmodulin is a follicle cell contribution to the cytoplasm of the *Blattella germanica* oocyte. *Developmental Biology*, 161, 513–21.

Zimowska, G., Shirk, P.D., Silhacek, D.L. & Shaaya, E. (1995). Vitellin and formation of yolk spheres in vitellogenic follicles of the moth, *Plodia interpunctella*. *Archives of Insect Biochemistry and Physiology*, 29, 71–85.

Zou, B.-X., Yin, C.-M., Stoffolano, J.G. & Tobe, S.S. (1989). Juvenile hormone biosynthesis and release during oocyte development in *Phormia regina* Meigen. *Physiological Entomology*, 14, 233–9.

14 The egg and embryology

14.1 THE EGG

Insect eggs are typically large relative to the size of the females that produce them because they contain a great deal of yolk. It is generally believed that the eggs of Endopterygota contain less yolk and are smaller than those of Exopterygota (Anderson, 1972b). To some extent, this may reflect the types of ovariole they possess. For example, in two locust species which have panoistic ovarioles, each egg weighs about 0.5% of female weight; amongst insects with telotrophic ovarioles, the egg of *Trialeurodes vaporarium* is over 1% of the female weight, and that of *Callosobruchus maculata* (Coleoptera) 0.6%. By contrast, amongst insects with polytrophic ovarioles, comparable figures for *Apis mellifera* and *Grammia geneura* (Lepidoptera) are 0.07% and 0.11%, respectively.

Egg size is affected by factors other than the type of ovariole, however. Amongst Lepidoptera from temperate regions, species overwintering in the egg stage have larger eggs than species that overwinter in some other stage, and species feeding on woody plants have bigger eggs than those feeding on herbaceous plants (Reavey, 1992). Individual females of at least some butterflies lay smaller eggs as they grow older, and females of the cornborer moth, *Ostrinia*, lay smaller eggs if they do not receive adequate nutrition (Leahy & Andow, 1994). The parasitic Hymenoptera which lay their eggs in their hosts have very small eggs and this is also true amongst the Tachinidae (Diptera).

Insect eggs have a wide variety of forms. Commonly, as in Orthoptera and many Hymenoptera, they are sausage shaped (Fig. 14.10), but they may be conical, as in *Pieris*, or rounded, as in many moths and Heteroptera. The eggs of Nepidae and some Diptera have respiratory horns (Fig. 14.2), while the eggs of many parasitic Hymenoptera have a projection called a pedicel at one end. The eggs of *Encyrtus* (Hymenoptera) are unusual in consisting of two bladders connected by a tube. During the process of oviposition the contents of the egg pass from the proximal to the distal bladder, and the proximal bladder is lost. It is

suggested that this may facilitate the entry of the egg into a host through a relatively small hole.

Review: Hinton, 1981

14.1.1 Egg structure

The egg contains some cytoplasm and a relatively large amount of yolk, and it is enclosed by a shell formed from the vitelline envelope and the chorion. The composition and synthesis of yolk is considered in section 13.2.4.

14.1.1.1 *Cytoplasm*

At the time of oviposition, the egg cytoplasm forms a bounding layer, the periplasm, and an irregular reticulum within the yolk. The cytoplasm (euplasm) is much more extensive in insects with meroistic ovarioles than in those with panoistic ovarioles, reflecting the greater contribution of the germ line genome to the oocyte (section 13.2). The zygote nucleus usually occupies a posterior position.

14.1.1.2 *Vitelline envelope*

The vitelline envelope is a proteinaceous layer surrounding the oocyte. It is often called the vitelline membrane, but it is preferable to limit use of the term 'membrane' to cellular structures; the vitelline envelope is extracellular. It is commonly about 1–2 μm thick, but, in the parasitic Hymenoptera, it is much thinner, about 150 nm in both *Habrobracon* and *Nasonia*, while in the damselfly, *Sympetrum*, it is 9 μm thick. It may vary in thickness in different parts of the egg, and is not necessarily uniform in structure over the entire surface of the oocyte. In *Drosophila*, for example, there are spaces within it where the chorion splits at the time of hatching, and in the beetle, *Lytta*, it is perforated beneath the micropyles.

14.1.1.3 *Chorion*

The chorion is a complex structure produced by the follicle cells while the egg is in the ovary (section 13.2.5). It varies in thickness from less than one micron in eggs of parasitic Hymenoptera to over 50 μm in some orthopteran

Fig. 14.1. Airspaces in the chorion. (a) A chorion in which the air-filled inner meshwork connects with the exterior through a limited number of aeropyles [*Musca* (Diptera)](after Hinton, 1960). (b) A chorion in which extensive airspaces extend through its thickness [*Tetrix* (Orthoptera)](after Hartley, 1962).

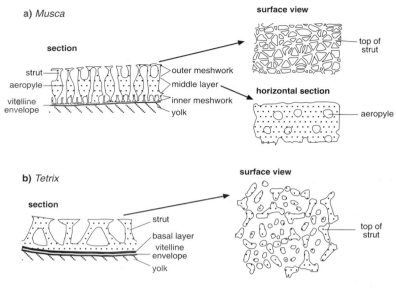

and lepidopteran eggs. Even closely related species may have chorions differing in thickness. In the silkmoths, for example, the chorion of *Bombyx* is about 25 μm thick while that of *Hyalophora* is 55 μm thick. These differences are presumed to have ecological significance.

A wax layer is sometimes present on the inside of the chorion immediately adjacent to the vitelline envelope. This has been observed in *Drosophila*, *Rhodnius*, grasshoppers and some Lepidoptera. It may be of general occurrence, at least amongst insects laying their eggs in places where they are subject to desiccation, but it is difficult to detect and may not always have been recognized.

Two anatomically distinct layers, sometimes called the endochorion and exochorion, are present in the chorion of the eggs of some species, but others lack these distinct layers. Even where both are present they are not necessarily homologous in different species (Hinton, 1969). The endochorion, which forms the greater part of the chorion in all eggs, is largely proteinaceous, whereas the exochorion, when it is present, contains some carbohydrate.

The innermost layer of the chorion (but outside any wax layer) in many insects is a crystalline layer less than one micron thick, consisting of proteins stabilized by disulfide or dityrosine bonds. In *Drosophila*, holes in the crystal lattice are 2–4 nm across, large enough to permit the free movement of oxygen and carbon dioxide. This layer is absent from the eggs of parasitic Hymenoptera.

Extensive airspaces are usually present in the chorion adjacent to the oocyte and, in species that lay their eggs in moist environments, these airspaces may extend through the entire thickness of the chorion. This is the case in the grouse locust, *Tetrix* (Fig. 14.1b). In the eggs of most species, however, openings through the chorion are restricted, and this limits water loss. For example, in *Musca*, the inner meshwork in the chorion, which provides a continuous layer of air all round the developing embryo, connects with an outer meshwork by pores (aeropyles) through an otherwise solid middle layer (Fig. 14.1a). In many other species, the connections are even more restricted. For example, in *Calliphora* (Diptera), the outer meshwork and aeropyles connecting with the inner meshwork are absent over the greater part of the egg, and are present only between the hatching lines (Fig. 14.3). Eggs of *Ocypus* (Coleoptera) have an equatorial band of aeropyles connecting with the inner airspaces, *Rhodnius* eggs have a ring of aeropyles just below the cap (see Fig. 13.13), and those of *Carausius*, have a single small pore at which the reticular inner chorion is exposed at the egg surface. The respiratory horns of some Diptera and the Nepidae (Hemiptera) serve the same function of connecting the inner layer of air with the atmosphere outside, while at the same time restricting the area through which rapid loss of water can occur (Fig. 14.2). This is also true of the micropylar processes of the Pentatomomorpha which also have a respiratory function (Fig. 14.4a).

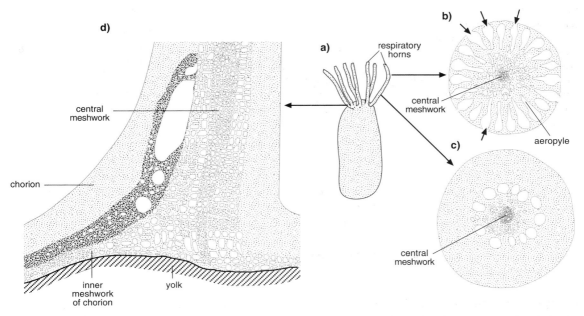

Fig. 14.2. Respiratory horns on the egg of *Nepa* (Hemiptera) (after Hinton, 1961): **(a)** whole egg; **(b)** cross-section of a distal region of a horn. Arrows indicate the openings of the airspaces to the outside; **(c)** cross-section of a proximal region of a horn; **(d)** longitudinal section of the basal region of a horn where it connects to the chorion round the egg. Notice the continuity of the air-filled central meshwork of the horn with the inner meshwork that surrounds the egg.

The chorion outside the inner layer of airspaces is sometimes thick and, in Lepidoptera, it is lamellate. The lamellae are formed from sheets of protein fibers, all the fibers within a sheet having the same orientation. The orientation of fibers in successive sheets is rotated so that they have a helicoidal arrangement through the thickness of the chorion. Rotation through 180° creates the appearance of lamellae, in exactly the same way as in chitinous cuticle (section 16.2.1.1).

The outer surface of the chorion is often sculptured, frequently with a basically hexagonal pattern reflecting the shapes of the follicle cells which secrete it. Outside this, grasshoppers have an additional layer, the extrachorion which differs from the chorion in being produced by gland cells in the oviducts rather than by the follicle cells.

The outer covering of the egg of the moth, *Micropteryx*, is unusual in consisting of a forest of knobbed projections. The stalks carrying the knobs are about 50 μm high and 2–4 μm in diameter, and the knobs themselves are about 10 μm in diameter. These are produced in the hour following oviposition by exudation from the oocyte itself. The exudate forces the viscous outer layer of the shell, produced by the female accessory glands, outward to form the projections (Chauvin & Chauvin, 1980).

Specializations of the chorion

ESCAPE FROM THE EGG. The first stage larvae of many species escape from the egg by chewing through the chorion or by using special egg bursting devices (section 15.1.1). The chorion of some species, however, has a line of weakness along which it splits when the larva exerts pressure from within. These hatching lines are usually visible from the outside, and sometimes define a cap or operculum.

A cap is present in the eggs of Phasmatodea, Embioptera, Cimicomorpha (Heteroptera) and Phthiraptera. In *Rhodnius*, the cap is joined to the rest of the chorion by a sealing bar, which is formed from a very thin layer of resistant endochorion (perhaps equivalent to the crystalline endochorion described above) and a thick amber-colored layer (Fig. 13.13). There are lines of weakness (hatching lines) in the amber layer above and below the sealing bar. Some pentatomid eggs also appear to have a cap, but this has the same structure as the rest of the chorion and is not joined to it by a sealing bar.

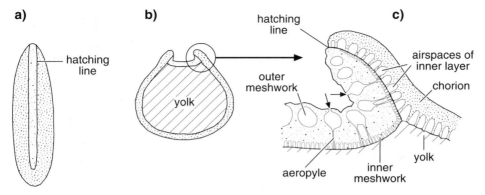

Fig. 14.3. Hatching lines in the egg of *Calliphora* (Diptera) (after Anderson, 1960; Hinton, 1960). (**a**) Dorsal view of the whole egg. (**b**) Cross-section through the middle of the egg. (**c**) Detail of a section through one of the hatching lines where the air-filled spaces are back-to-back. Arrows indicate the openings of the airspaces to the outside.

Amongst the Diptera, the eggs of Muscoidea have hatching lines in the form of two ridges running longitudinally along the length of the egg (Fig. 14.3). Along these lines, the inner layer of the chorion extends outward so that each ridge contains two inner layers which are back to back, creating a line along which the chorion is relatively weak. In *Drosophila melanogaster*, there is a discontinuity in the endochorion where an operculum meets the body of the egg, and, along this line, the two parts are held together only by the thin innermost layer of the endochorion and a layer of exochorion. The vitelline envelope is also modified along this line. It is not certain how important these devices are as many species, even some *Drosophila* species, appear to have no weak lines in the chorion.

MICROPYLES. Because the chorion is laid down in the ovary, some provision is necessary to allow the subsequent entry of sperm. This takes the form of micropyles, which are funnel shaped pores passing right through the chorion, usually near the anterior pole of the egg. The pores are 1–2 μm in diameter, often with a wider funnel at the surface of the chorion. The vitelline envelope beneath the micropyles is also modified.

The number of micropyles varies: most dipteran eggs have only a single terminal micropyle, while Acrididae (grasshoppers) commonly have 30 or 40 arranged in a ring at the posterior end of the egg (Fig. 14.4). In the Heteroptera, the number varies from 0 to 70, but is more or less constant within a species. When there is a large number, they are usually arranged in a circle around the anterior end of the egg. Most Cimicoidea have two micro-

pyles, but different species of Coreidae have between 4 and 60, and pentatomids from 10 to 70. Eggs produced by older females of *Rhodnius* have a reduced number of micropyles compared with those laid by younger females. There are no micropyles in the eggs of species of Cimicoidea which are fertilized in the ovary (section 12.3.3.3).

Sometimes the micropylar openings are associated with aeropyles and are raised above the surface of the egg. This occurs in Pentatomomorpha, and in *Oncopeltus* each consists of a cup on a stem (Fig. 14.4a). The micropylar canal passes through the middle of the process and through the chorion and is surrounded by an open reticulum enclosing airspaces.

Reviews: Cobben, 1968 – Heteroptera; Margaritis, 1985

14.1.2 Respiration

Gaseous exchange occurs via the aeropyles and the chorionic airspaces in the eggs of most terrestrial insects. The layer of air in the inner chorion has direct access to the ovum through pores in the innermost layer of the chorion and connects with the outside air via aeropyles. Where only a single aeropyle is present, as in the beetle, *Callosobruchus*, its size is critically important. Too small a pore will not permit the entry of an adequate supply of oxygen, and the eggs of two strains of *Callosobruchus maculatus* with different metabolic rates have differently shaped aeropyles (Daniel & Smith, 1994).

The eggs of some terrestrial insects which are laid in soil or detritus are subject to periodic flooding. Eggs of some species can survive this because the chorion, with its hydrophobic characteristics, maintains a layer of air

a) *Oncopeltus*

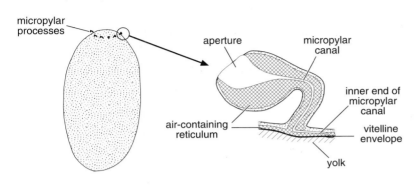

b) *Locusta*

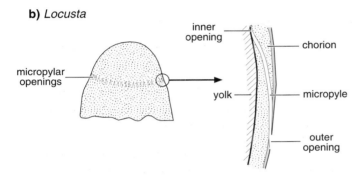

Fig. 14.4. Micropyles. **(a)** *Oncopeltus* (Hemiptera). Left: whole egg; right: longitudinal section through a micropylar process. The process also connects the air layer round the egg with the atmosphere (after Southwood, 1956). **(b)** *Locusta* (Orthoptera). Left: posterior pole of egg; right: longitudinal section along the length of a micropylar canal (after Roonwal, 1954).

around the egg into which gas from the surrounding water may diffuse. Thus the chorion acts as a plastron (section 17.5.2.2), but the effectiveness of a plastron depends on the area available for gaseous exchange, that is, on the extent of the air/water interface. In the eggs of Lepidoptera and most Heteroptera, *Rhodnius*, for example, the air/water interface is too small to be of significance since it is limited to the openings of the aeropyles. Nevertheless, these eggs may survive some degree of flooding by virtue of the fact that they can tolerate a great reduction in their metabolic rate.

In other insects, the plastron is larger and more effective. In *Calliphora*, where the plastron occurs between the hatching lines, and in *Musca*, where it extends over the whole egg, normal development continues if the egg is immersed in well-aerated water. The respiratory horns of many dipteran eggs also form an efficient plastron if the eggs are flooded and this is also true in the eggs of Nepidae, which are essentially terrestrial in their respiration as the horns normally project above the surface of the water.

The surface tension of water contaminated with organic acids and other surface active substances is lower than that of clean water, and the ease with which a plastron is wetted, and hence ceases to function, is inversely proportional to the surface tension. Thus the plastron of insect eggs which are laid in organic materials subject to flooding needs to have a high resistance to wetting if it is to continue functioning in spite of the low surface tension. It must also be able to withstand wetting by raindrops which, momentarily, may exert a pressure approaching 50 kPa. Hence, although eggs in dung and similar situations can rarely be subject to flooding by more than a few centimeters of water, they possess a plastron capable of withstanding flooding by clean water to a much greater depth, and in some cases their resistance is greater than that exhibited by the plastron of some aquatic insects (Hinton, 1969).

Eggs laid in water, such as those of dragonflies, obtain their oxygen from that dissolved in the water. The chorion in these eggs is generally without any obvious system of spaces and oxygen diffuses through the solid material. This is presumably because, in the absence of any need to restrict water loss, the chorion is relatively porous.

14.1.3 Water regulation

All insect eggs are subject to water loss, but the extent to which water loss occurs is affected by the choice of oviposition site by the female and by the permeability of the chorion. Subsequently, during embryonic development and in the pharate first stage larva, water loss may be modified by the development of new structures within the egg. The eggs of many insects also take up water, and this is often essential for continued embryonic development.

14.1.3.1 *Water loss*

The rate of water loss from newly laid insect eggs is often very low, especially in eggs of species which do not take up water during development. In *Rhodnius*, for instance, water is lost at a rate of $3\,\mu\mathrm{g\,cm^{-2}\,h^{-1}\,mmHg^{-1}}$ from the egg, but in species that lay their eggs in damp places, the rate of water loss may be much higher if the eggs are exposed to desiccating conditions. It is $60\,\mu\mathrm{g\,cm^{-2}\,h^{-1}\,mmHg^{-1}}$ in exposed eggs of the beetle, *Phyllopertha*, for example. The rate of loss may also change with time as the membranes covering the embryo alter (see below), and, in the beetle *Acanthoscelides*, fertilization causes a change in the vitelline envelope that reduces water loss (Biemont, Chauvin & Hamon, 1981).

The chorion itself limits water loss to some extent and differences in its thickness may be important in this respect. A thick chorion and a reduction in the number of respiratory horns is a characteristic of the eggs of heteropteran species that are laid in exposed situations subject to desiccation (Southwood, 1956). Some tropical grasshoppers that survive the dry season in the egg stage have thick, tough chorions, and the inner chorion of the eggs of mosquitoes such as *Aedes*, which resist desiccation, is thicker and darker than that in the non-resistant eggs of *Culex*.

However, it is generally considered that the principal factor in limiting water loss is the layer of wax on the inside of the chorion. Like the wax layer on the outside of the cuticle (section 18.4.1.1), it has a critical temperature above which water loss increases sharply (Fig. 14.5). The critical temperature for the eggs of *Rhodnius* is 42.5 °C, and for the eggs of *Lucilia* (Diptera) and *Locustana* (Orthoptera) 38 °C and 55–58 °C, respectively.

A second layer of wax is subsequently laid down in the serosal cuticle (section 14.2.10). Functionally, this secondary wax layer may supplement the primary wax layer

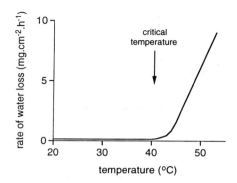

Fig. 14.5. Water loss from the egg showing the critical temperature (*Rhodnius*, Hemiptera) (after Beament, 1946).

on the inside of the chorion, or it may replace it if the latter becomes broken by the increase in size of the egg which sometimes occurs. The rate of evaporation from the exposed egg of *Locustana* at 35 °C and 60% relative humidity drops from $0.35–0.54\,\mathrm{mg\,egg^{-1}\,24h^{-1}}$ at oviposition, when only the primary wax layer is present, to $0.11–0.36\,\mathrm{mg\,egg^{-1}\,24h^{-1}}$ five days later when the serosal cuticle and secondary wax layer are completed all over the egg except for the hydropylar region (see below). When this region is also sealed off, evaporation rate falls to $0.03–0.04\,\mathrm{mg\,egg^{-1}\,24h^{-1}}$.

There is little data on the effects of embryonic cuticles on water loss from the egg.

14.1.3.2 *Absorption of water*

The eggs of *Rhodnius*, and probably of many other Heteroptera and Lepidoptera which are laid in dry, exposed situations, develop without any uptake of water, but the eggs of many insect species absorb water from the environment in the course of development. Amongst terrestrial insects, it is recorded in Orthoptera, some Hemiptera and some Coleoptera; in aquatic species it occurs in Odonata, Heteroptera, Coleoptera and some Diptera. It results in a considerable increase in volume and weight (Fig. 14.6).

Water uptake usually occurs over a limited period of development which varies from species to species. For instance, amongst the grasshoppers, it occurs before there has been any significant development in *Camnula*, during early embryonic development in *Locusta*, and after blastokinesis in *Melanoplus differentialis*. Immediately after oviposition, no water is taken up by the eggs of these

a) surface membranes

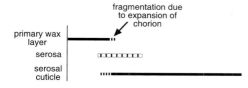

fragmentation due
to expansion of
chorion

primary wax
layer

serosa

serosal
cuticle

b) egg weight

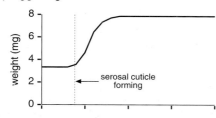

serosal cuticle
forming

c) osmotic pressure

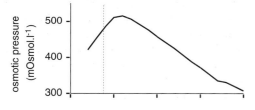

d) rate of water loss

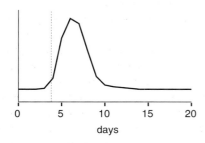

days

Fig. 14.6. Water uptake by the egg of a grasshopper
(*Chorthoicetes*) (based on Lees, 1976). (**a**) Structural changes in
the egg that affect water uptake and loss. (**b**) Change in egg
weight due to water uptake. (**c**) Changes in the osmotic pressure
of the yolk. (**d**) Changes in the rate at which water is lost if the
egg is exposed to a desiccating environment, reflecting changes in
permeability of the egg membranes.

species. Then follows a period of rapid uptake, followed by
a further period in which no marked change in water
content occurs provided the egg is not subject to desiccation (Fig. 14.6). In the cricket, *Allonemobius*, water is taken
up in an earlier stage of development at 30 °C than at 20 °C
(Tanaka, 1986).

The mechanism of water uptake is not fully understood, but it is generally considered to be passive and to
reflect changes in the permeability of the egg membranes,
in the osmotic pressure of the yolk (which normally is
greater than that of any surrounding water) and in the
hydrostatic pressure within the egg.

The initial failure of grasshopper eggs to absorb water
is due to the primary wax layer in the chorion. This forms a
barrier to the osmotic uptake of water and is equally
effective in preventing water loss. The period of water
uptake begins at the time the serosa is completed and the
serosal cuticle is first laid down (see below). The increase
in volume resulting from water uptake produces cracking
of the chorion and the primary wax layer must be disrupted. This leads to a great increase in permeability of the
chorion and at the same time initiates mobilization of the
yolk so that the osmotic pressure within the egg increases.
These two factors combined lead to a rapid uptake of water
(Fig. 14.6). Following blastokinesis in *Chortoicetes*, the
osmotic pressure of the yolk falls, contributing to, but not
wholly accounting for, the reduction in water intake. This
reduction may also be a consequence of the completion of
the serosal cuticle and the secondary wax layer or of the
reduction in stress on the egg membranes so that they
become less permeable. The eggs remain capable of taking
up water during this period if they are subject at any stage
to water loss.

The mechanism may be different in the cricket,
Teleogryllus. Here, the chorion remains permeable at all
times, but water uptake is prevented initially, despite a
high internal osmotic pressure, by the hydrostatic pressure
within the egg. Subsequently, the structure of the inner
chorion changes and its permeability increases, enabling
the egg to take up water osmotically. The resulting
increase in hydrostatic pressure may then be responsible
for preventing further water uptake.

In Heteroptera, it is suggested that water uptake by the
eggs of *Notostira* does not begin until osmotically active
substances are produced within the egg, while water
uptake by the eggs of *Phyllopertha* stops following a
modification of the chorion which makes it waterproof.

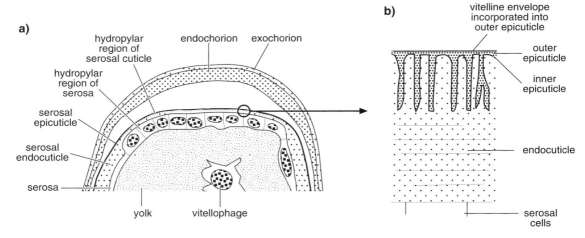

Fig. 14.7. The hydropyle of a grasshopper. **(a)** Longitudinal section through the posterior pole of the egg (based on Roonwal, 1954). **(b)** Detail of the hydropylar serosal cuticle (based on Slifer & Sekhon, 1963).

Water uptake by the eggs of many species is limited to a specialized region of the shell called the hydropyle. It may involve a modification of the serosal cuticle, as it does in grasshoppers, or a modification of the chorion itself, as in some Hemiptera and perhaps also some beetles. In some other insects, as in *Notostira* and *Gryllulus* (Orthoptera), water is absorbed over the whole of the egg and there is no hydropyle.

In grasshoppers, the hydropyle is at the posterior pole of the egg. The inner layer of the serosal epicuticle is deeply invaginated into the serosal endocuticle, and the outer epicuticle is a finely fibrous, proteinaceous material which fills the folds (Fig. 14.7). The serosal endocuticle is thinner in this region than elsewhere.

By contrast, in eggs of the water scorpion, *Nepa*, the hydropyle comprises a fine meshwork within the chorion which is isolated from the surrounding chorion. The region, which is normally in the water, provides a direct route through the chorion to the surface of the egg cell.

Whether the hydropyle is formed from the chorion or the serosal cuticle, the outer part is hydrophilic so that it readily holds water. In the grasshoppers, the outer serosal epicuticle performs this function and the thin, folded inner epicuticle is freely permeable, whereas it is impermeable over other parts of the egg. It becomes impermeable in species that have an egg diapause, and the end of diapause is associated with an increase in permeability.

In most cases, the serosal cells underlying the hydropylar region are larger or more columnar than other cells in the serosal epithelium. Even where there are no visible differences between the serosal cells, they may be functionally different, although this is not known to be so. There is no evidence, however, that these cells have an active role in water uptake.

The eggs of whiteflies (Aleyrodidae) have a pedicel at one end formed from a hollow extension of the chorion. The insects attach their eggs to leaves, either by cutting a slit with the ovipositor and inserting the pedicel into the slit or by pushing the pedicel into a stoma. The eggs take up water from the leaf via the pedicel (Byrne, Cohen & Draeger, 1990).

The eggs of the stick insect, *Extatosoma tiaratum*, are exceptional in being able to take up water from the atmosphere even at very low vapor pressures. The mechanism of uptake, which is active, is not known (Yoder & Denlinger, 1992).

Review: Hadley, 1994

14.2 EMBRYOLOGY

Immediately following fertilization, as the egg is laid, the zygote nucleus divides and the daughter nuclei migrate to the periphery of the egg to form a layer of cells, the blastoderm, surrounding the yolk. Part of this cell layer becomes thickened to form the germ band from which the embryo develops. Gastrulation follows, as a result of which an inner layer of cells is formed beneath the germ band. The details of gastrulation vary and the process is not immediately comparable with gastrulation in other animals. The embryo becomes cut off from the surface of the egg by extra-embryonic membranes, which break and disappear when the embryo undergoes more or less extensive movements in the yolk. These movements bring the embryo to its final position with the yolk now enclosed within the body wall.

The ectoderm forms the body wall, which invaginates to form the tracheal system, and the stomodeum and proctodeum; the nervous system and sense organs are also ectodermal in origin. The mesoderm may at first form coelomic sacs, but these break down to form muscles and the circulatory and reproductive systems. The germ cells from which the sex cells are ultimately derived are differentiated early in development, sometimes after only a few nuclear divisions.

Reviews: Anderson, 1972a,b – general; Haget, 1977 – general; Nagy, 1995 – Lepidoptera, early stages; Sander, Gutzeit & Jäckle, 1985 – general

14.2.1 Cleavage and the blastoderm

After oviposition, the zygote nucleus of an insect egg starts to divide, the first division occurring within about 30 minutes of zygote formation in *Dacus* (Diptera) and within 10 minutes in *Drosophila*. Nuclear division is not accompanied by cell division, but each daughter nucleus is accompanied by a halo of cytoplasm and each unit of nucleus and cytoplasm is called an energid (Fig. 14.8a). The first few divisions of the daughter nuclei are synchronized, synchrony perhaps being facilitated by the fact that they are in cytoplasmic continuity. During interphase periods, the cytoplasm of the energids increases at the expense of the cytoplasmic reticulum.

The energids move apart as they divide and become arranged in a layer within the yolk, bounding a spherical or elongate mass of yolk which roughly corresponds with the form of the egg. In the eggs of hemimetabolous insects, the nuclei at this time are more superficial than those of holometabolous insects, perhaps related to the smaller amount of cytoplasm in the former. Migration of the energids continues until they reach and enter the periplasm all round the egg (Fig. 14.8c). In the higher Diptera, their arrival at the periplasm appears to be synchronized, but in other cases this is not so.

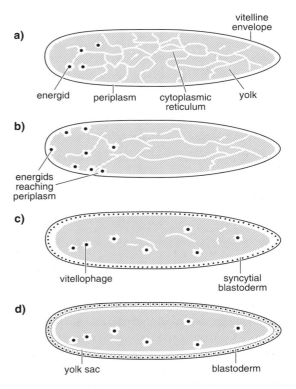

Fig. 14.8. Early development in the egg leading up to the formation of the blastoderm.

Within the periplasm, which is commonly invaded after about the eighth mitotic division, nuclear division continues, but is often no longer synchronized. Synchronous division does continue in *Dacus* and *Drosophila*, however, while in *Apis*, *Calandra* (Coleoptera) and *Calliphora* waves of mitoses pass along the egg from one end to the other. At the same time, at least in Diptera, the periplasm thickens due to the addition of cytoplasm from the reticulum.

Mitotic activity during this period of development is very high, but the time taken to complete a mitotic cycle is longer in the more primitive groups, such as Orthoptera, where it may take some hours, than in more advanced groups such as Lepidoptera and Diptera. Amongst the Lepidoptera a complete mitotic cycle normally takes less than an hour, while in *Drosophila* at 25 °C it takes only about ten minutes.

In *Drosophila*, folds of the plasma membrane develop between adjacent nuclei in the periplasm during the fourteenth cycle of mitosis which lasts much longer than the earlier nuclear divisions, marking the point at which protein synthesis dependent on zygotically transcribed mRNA becomes important. The folds spread inward and, when they are about 40 μm below the nuclei, they spread laterally to pinch off the cells. About 6000 cells are formed simultaneously, so that the undivided mass of yolk becomes surrounded by a layer of cells called the blastoderm (Fig. 14.8d). The formation of a cell layer round the periphery of the egg while the central mass of yolk remains undivided is called superficial cleavage.

In *Dacus* and *Drosophila*, the inner cell walls of the blastoderm cut off an inner undivided layer of cytoplasm. This anucleate layer is called the yolk sac. It subsequently becomes nucleated due to invasion by some of the vitellophages, but finally it is digested with the yolk in the midgut.

Review: Foe, Odell & Edgar, 1993 – *Drosophila*

14.2.2 Vitellophages

In many insects, only some of the energids migrate to the surface to form the blastoderm, the rest remain behind in the yolk to form the yolk cells, or vitellophages. Thus in *Dacus* about 38 of 128 energids remain in the yolk to form the primary vitellophages and they subsequently increase in number to about 300. Commonly, the vitellophages begin to separate after the sixth or seventh divisions and develop an enlarged nucleus, which increases in size through endomitotic division of the chromosomes. In most orders, some cells migrate back from the blastoderm to form secondary vitellophages and, in Blattodea, some Lepidoptera and Nematocera, all the vitellophages apparently have this secondary origin. There is also some evidence that vitellophages are derived from some of the pole cells (section 14.2.13) and, in *Dacus*, some tertiary vitellophages are formed from the proliferating anterior midgut rudiment.

The vitellophages have a variety of functions. They are concerned with the breakdown of yolk, engulfing it in vacuoles (Giorgi & Nordin, 1994). Later, when the yolk is enclosed in the midgut, they may form part of the midgut epithelium. They are also involved in the formation of new cytoplasm and are responsible for contractions of the yolk, producing the local liquefactions which are necessary for this.

In the eggs of Orthoptera, Lepidoptera and Coleoptera the yolk may become temporarily divided by membranes into large masses or spherules containing one or more

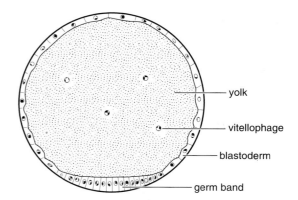

Fig. 14.9. Germ band. Diagrammatic cross-section of an egg showing the origin of the germ band as a ventral thickening of the blastoderm.

vitellophages. These yolk spherules are first formed close to the embryo and under the serosa, but ultimately extend throughout the yolk.

14.2.3 Other types of cleavage

The superficial pattern of cleavage occurring in most insects is a consequence of the large amount of yolk present, but in species with less yolk other forms of cleavage occur. Complete cleavage occurs in the small eggs of some parasitic Hymenoptera and those of Collembola. In the latter, each cell produced consists of a mass of yolk, in the center of which is an island of cytoplasm containing the nucleus. Cleavage is equal in *Isotoma*, so that cells of similar size result, but, in *Hypogastrura*, cleavage is unequal with the formation of micro- and macromeres. Complete cleavage continues to about the 64-cell stage, when the nuclei in their islands of cytoplasm migrate to the surface and become cut off from the yolk by cell boundaries. A blastoderm is thus formed and subsequent cleavage becomes superficial. Some nuclei remain in the yolk forming the vitellophages and the original boundaries within the yolk disappear leaving a single central mass.

14.2.4 Formation of the germ band

The blastoderm of most insects is initially a uniform layer of cuboid cells all over the yolk. Subsequently, it becomes thicker in the ventral region of the egg due to the aggregation of the cells so that they become columnar. This thickening is the embryonic primordium, or germ band. It develops into the future embryo, while the rest of the blastoderm remains extra-embryonic (Fig. 14.9). In some

a) *Kalotermes* **b)** *Acheta* **c)** *Notonecta* **d)** *Chrysopa* **e)** *Apis*

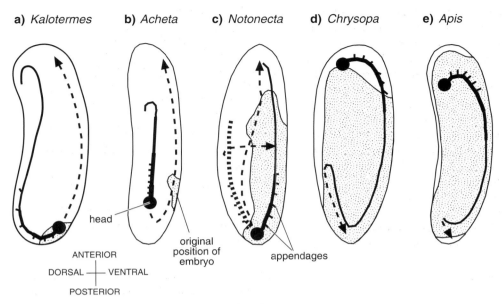

head

original
position of
embryo

appendages

ANTERIOR

DORSAL —┼— VENTRAL

POSTERIOR

Fig. 14.10. Diagram showing the position of the germ band (stippled), the position of the embryo before blastokinesis (solid black, showing the protocephalon and gnathal and thoracic appendages), and movement of the embryo at katatrepsis. In all cases the embryo finally comes to lie on the ventral surface of the egg with its head at the anterior pole. **(a)** termite (*Kalotermes*, Isoptera); **(b)** Cricket (*Acheta*, Orthoptera); **(c)** backswimmer. The embryo first develops on the dorsal side of the egg and moves to the ventral side (transverse arrow) before reversing its position in the egg (*Notonecta*, Hemiptera); **(d)** green lacewing (*Chrysopa*, Neuroptera); **(e)** honeybee (*Apis*, Hymenoptera).

Lepidoptera, the blastoderm is differentiated into germ band and extra-embryonic tissue from the time of its first appearance, and, in some other insects, such as Mallophaga and *Apis*, the whole blastoderm is thick initially but subsequently becomes thinner except for the germ band.

In eggs containing little cytoplasm, like those of most insects with panoistic ovarioles, the germ band is a small disc or streak of tissue over the posterior end of the egg (Fig. 14.10). In insects with polytrophic ovarioles it is more extensive and, in dipteran eggs, which contain a relatively small amount of yolk, the germ band is formed from most of the blastoderm and there is very little extra-embryonic tissue. Where the germ band is short relative to the whole egg, embryonic development usually involves the differentiation of additional tissue posteriorly. Where the germ band is large, all the parts of the larval body are represented from the outset.

The germ band is differentiated into a broad head region, the protocephalon, and a narrow 'tail', the protocorm (Fig. 14.11a) and, in a typical hemimetabolous insect, the midventral cells of the germ band are the pre-

sumptive mesoderm and midgut (Fig. 14.12a). The presumptive stomodeum and proctodeum lie at either end of this. In higher Diptera and some Lepidoptera and Hymenoptera, the fate of the different parts of the egg is much more precisely determined, but the same general arrangement of presumptive areas exists (Fig. 14.12b). In holometabolous insects, these presumptive areas relate to the development of larval tissues, but, where most or all of the adult tissue develops from imaginal discs (section 15.3.2.2), these areas may also be determined at a very early stage of embryonic development.

14.2.5 Determination of the body axes and segmentation

Regulation of blastoderm production and its subsequent development into the embryo is most fully understood in *Drosophila*. The features of the longitudinal axis are determined by three systems that are maternal in origin (Fig. 14.13). While the oocyte is still in the ovary, the trophocytes produce a specific mRNA that is transferred to the oocyte and accumulates at the anterior end. Soon after the egg is laid, translation of this mRNA occurs to produce a

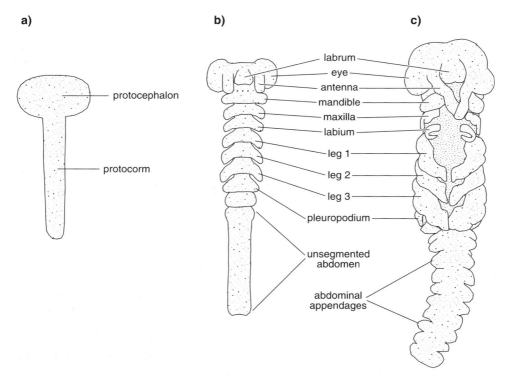

Fig. 14.11. Early stages in the development of an insect with a short germ band showing the whole embryo with the embryonic membranes removed (*Ornithacris*, Orthoptera).

Fig. 14.12. Presumptive larval areas on the blastoderm. Anterior to right. Gnathal and thoracic segments with distinctive shading to facilitate comparison. **(a)** An insect with a short germ band (*Acheta*, Orthoptera). **(b)** An insect with a long germ band (*Dacus*, Diptera).

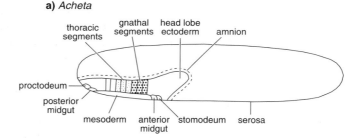

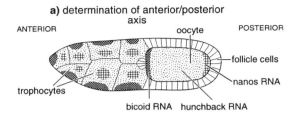

a) determination of anterior/posterior axis

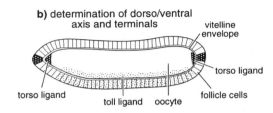

b) determination of dorso/ventral axis and terminals

Fig. 14.13. Diagram showing the maternal contributions to formation of the axes in the embryo of *Drosophila*. Different RNAs and ligands shown with distinctive shading. Cells synthesizing the products and ultimate positions in the oocyte are shown with similar shading. **(a)** The anterior/posterior axis is determined by RNA transferred to the oocyte from the trophocytes. All the trophocytes synthesize each of the RNAs. **(b)** The anterior and posterior terminals and the dorso/ventral axis are determined by ligands transferred to the perivitelline space from some groups of follicle cells. These ligands bind to receptors on the surface of the blastoderm and so produce localized signals in the embryo.

protein known as 'bicoid' which diffuses away from the anterior end of the egg. Different concentrations of the protein lead to the expression of different genes in the zygotic nuclei, either directly or indirectly through a second protein, 'hunchback'. Compounds that act in this way to produce local gene expression are called morphogens. Hunchback RNA is produced by the maternal trophocytes, but is also transcribed by the zygotic nuclei in proportion to the quantity of bicoid protein present. As a result, it exists in a gradient which is highest at the anterior pole of the egg and declines posteriorly. At the posterior end of the egg, hunchback RNA activity is disabled by the product of another gene, 'nanos'. Nanos RNA is also produced in the trophocytes and becomes localized at the posterior end of the egg.

Segmentation results from the differential effects of the concentration of the hunchback protein on a series of gap genes. The effect of these is to transform a gradient

(of hunchback protein) into a series of discrete steps that ultimately result in the formation of segments. The products of the gap genes limit the expression of a further set of genes, the pair rule genes. These genes are expressed in alternate segments and define 14 'parasegments' equating to the three gnathal segments, the three thoracic segments and the first eight abdominal segments. However, these 'parasegments' do not exactly correspond with the segments as seen in the larva or the adult fly. Each segment, as normally understood, is derived from the posterior compartment of one parasegment and the anterior compartment of the next.

The third system defining the insect's polarity regulates development at the extreme anterior and posterior ends of the embryo. Unlike bicoid and nanos, it depends on positional signals produced by the follicle cells and held in the vitelline envelope or the perivitelline space (Fig. 14.13b). In *Drosophila*, the signal is produced by two cells at the posterior end of the follicle, and a small group of cells within the follicle at the anterior end. Soon after oviposition, the ligand in the vitelline envelope binds to newly formed receptors, leading to the local production of a morphogen, known as 'torso', regulating the formation of the acron and telson.

The dorso-ventral axis of the embryo is determined in a similar way, but here the initial signal is produced by the oocyte itself while it is still in the ovary. This signal causes the follicle cells to produce a ligand, 'toll', which ultimately results in the production of a protein called 'dorsal' in the egg. This protein is more actively taken up by nuclei on one side of the blastoderm which becomes the ventral side.

Reviews: Bate & Arias, 1993; Lawrence, 1992; Patel, 1994; St Johnston & Nüsslein-Volhard, 1992; Zrzavý & Stys, 1995

14.2.6 Gastrulation

Gastrulation is the process by which the mesoderm and endoderm are invaginated within the ectoderm. In insects, there is no deep invagination as in other animals, but an inner layer is formed by the cells initially lying along the midline of the germ band. In most insects, these cells become columnar and then migrate inward so that a mid-ventral groove is formed. They are isolated progressively from the outside by more lateral cells spreading beneath them. At the same time, the invaginated cells proliferate to form an inner layer (Fig. 14.14). In some beetles, as in *Clytra*, the invagination is so marked that it is at first almost tubular, while in *Apis* a broad middle plate sinks in

a) *Clytra*

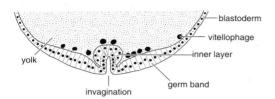

b) *Apis*

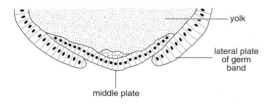

c) *Locusta*

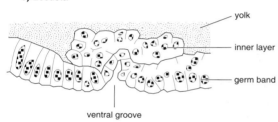

Fig. 14.14. Different types of gastrulation. **(a)** Invagination of the center of the germ band (*Clytra*, Coleoptera). **(b)** Overgrowth from the sides (*Apis*, Hymenoptera). **(c)** Proliferation of cells from the upper surface of the germ band (*Locusta*, Orthoptera).

without rolling up and the ectoderm grows across to cover it. The inner layer which is formed in this way comprises the mesoderm with a midgut rudiment anteriorly and posteriorly. Insects do not have a discrete layer comparable with the endoderm of other animals.

In Diptera, the invagination of the posterior midgut rudiment deep into the yolk during the latter part of gastrulation bears a superficial resemblance to the process that occurs in other animals. Mesodermal invagination begins along the ventral surface, but extension of the mesoderm, and the ectoderm which comes to cover it, pushes the invagination of the posterior midgut rudiment and the proctodeum anteriorly along the dorsal surface of

the embryo (Fig. 14.15a,b). Invagination of the proctodeum then carries the posterior midgut rudiment deep into the yolk (Fig. 14.15c).

In the collembolan, *Isotoma*, all the cells of the blastoderm divide tangentially to form outer and inner layers. Later, the cells of the inner layer migrate to the region of the germ band so that the extra-embryonic region comes to consist of a single layer of cells (Jura, 1972).

14.2.7 Formation of embryonic membranes
The germ band does not usually remain exposed at the surface of the yolk, but becomes covered by one or more embryonic membranes. Soon after its formation, folds appear at its periphery (Fig. 14.16a) and extend ventrally beneath the embryo until they meet and fuse in the ventral midline (Fig. 14.16b). Thus the embryo lies on the dorsal surface of a small cavity, the amniotic cavity, bounded by a thin membrane, the amnion. The membrane round the outside of the yolk is now called the serosa. Amnion and serosa may remain connected where the embryonic folds fuse or they may become completely separated, so that the embryo sinks into the yolk and yolk separates amnion and serosa (Fig. 14.16c). No further cell division occurs in the serosa, but endomitosis may occur so that the nuclei become very large. In *Gryllus* (Orthoptera), they contain four times as much DNA as the nuclei in the germ band. In the Cyclorrhapha, the embryo occupies the whole egg from the beginning of development and the amnion is vestigial and the serosa absent.

The Apterygota exhibit variations in the development of the membranes. In *Ctenolepisma*, the amnio-serosal folds fuse as they do in pterygote insects, but in *Machilis* (Archaeognatha) the embryo does not become cut off in an amniotic cavity. It is invaginated within the yolk, but remains connected to the exterior. Adjacent to the embryo is a zone of cells with small nuclei whereas the rest of the egg is covered by cells with large nuclei (Fig. 14.17). From their superficial resemblance to amnion and serosa in other insects, these zones are called proamnion and proserosa. In *Lepisma*, the amnion-like cells are restricted to the membrane lining the invagination.

In Orthoptera, Phthiraptera and at least some Heteroptera, Coleoptera and Diptera, the serosa secretes a cuticle to the outside. A non-cellular layer is produced on the inside of the serosa by some Coleoptera, while in the collembolan, *Isotoma*, two successive serosal cuticles are formed on the outside.

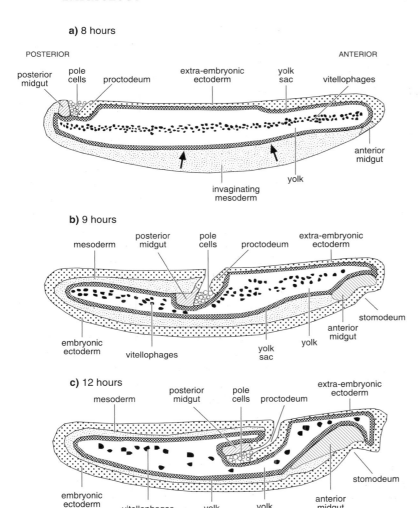

a) 8 hours

POSTERIOR   ANTERIOR

posterior midgut — pole cells — proctodeum — extra-embryonic ectoderm — yolk sac — vitellophages — anterior midgut — yolk — invaginating mesoderm

b) 9 hours

mesoderm — posterior midgut — pole cells — proctodeum — extra-embryonic ectoderm — stomodeum — anterior midgut — yolk — yolk sac — vitellophages — embryonic ectoderm

c) 12 hours

mesoderm — posterior midgut — pole cells — proctodeum — extra-embryonic ectoderm — stomodeum — anterior midgut — yolk — yolk sac — vitellophages — embryonic ectoderm

Fig. 14.15. Early stages in the development of a dipteran with a long germ band. Arrows in **(a)** indicate the movement of the mesoderm as it folds within the ectoderm (fruitfly, *Dacus*, Diptera) (after Anderson, 1962).

14.2.8 Movements of the embryo

The early embryos of most hemimetabolous orders are small relative to the size of the egg and in many of these groups the embryo makes extensive movements during development. In Blattodea, Dermaptera and Isoptera the germ band initially lies on the dorsal surface of the egg near its posterior pole. In most of these insects, as the germ band lengthens it extends onto the ventral surface and the head moves to a more posterior position (Fig. 14.10a, 14.18). In other hemimetabola, the movements are more marked and the embryo becomes immersed within the yolk (Fig. 14.10b, 14.19d). The posterior end of the germ band is flexed upwards into the yolk, so that the embryo comes to lie with its head-end towards the posterior pole of the egg. This movement is known as anatrepsis. Comparable displacements of the embryo occur in Ephemeroptera, Odonata, most Orthoptera and hemipteroid insects.

Following the period of elongation and development of the appendages, the amnion and serosa, which enclose these embryos, fuse and rupture close to the head. The embryonic membranes now pull back from the embryo leaving it exposed on the surface of the yolk (Fig. 14.19e, 14.20). In species in which the embryo is already on the ventral surface of the egg, it stays where it is, but in species where it has developed on the dorsal side of the egg with its

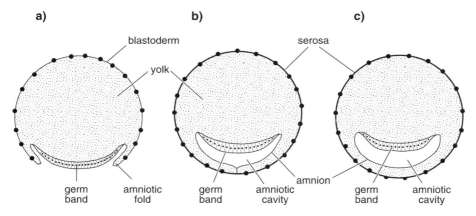

a)   **b)**   **c)**

Fig. 14.16. Development of the amniotic cavity: **(a)** lateral folds beginning to grow over the germ band; **(b)** lateral folds meet beneath the germ band; **(c)** amnion and serosa separated; embryo immersed in yolk.

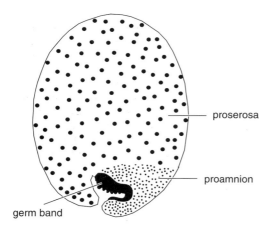

Fig. 14.17. Early stage in the invagination of the embryo of *Machilis* (Thysanura) showing the differentiation of the extra-embryonic membrane into proamnion, with small nuclei, and proserosa, with large nuclei (after Johannsen & Butt, 1941).

head towards the posterior pole, it now moves round to lie with its head towards the anterior pole (Figs. 14.10, 14.19f). This movement is known as katatrepsis. The term blastokinesis is sometimes used synonymously with katatrepsis, but it is also, and preferably, used to refer to all the movements of the embryo within the egg, including both anatrepsis and katatrepsis.

In holometabolous insects, the embryo elongates and differentiates with its head towards the anterior pole of the egg, and the extensive movements which occur in many hemimetabolous groups do not take place (Fig. 14.10d,e). In some Coleoptera, such as *Dytiscus* and *Tenebrio*, the embryonic membranes fuse and rupture, and the embryo shortens rapidly, but without changing its position. Shortening also occurs in embryos of other holometabolous insects but the fate of the embryonic membranes varies (see below).

Lepidoptera are unusual amongst holometabolous insects in making extensive movements during embryogenesis. The movements are called blastokinesis though they differ entirely from the movements occurring during blastokinesis in hemimetabolous insects (Anderson, 1972b).

14.2.9 Dorsal closure

One effect of katatrepsis in many insects is to reverse the relative positions of embryo and yolk. At first, the embryo lies on or in the yolk, but when the movements are completed the yolk is contained within the embryo. At first, the dorsal enclosure of the yolk is formed by the extra-embryonic membranes; later the provisional tissue is replaced by the embryonic ectoderm which grows upwards to form the definitive dorsal closure. As the ectoderm extends, amnion and serosa shrink and become confined to an antero-dorsal region where the serosa finally invaginates into the yolk in the form of a tube. This is known as the dorsal organ; it is ultimately digested in the midgut (Fig. 14.20a).

Where no marked katatrepsis occurs, the dorsal closure is produced by rearrangement of the embryonic

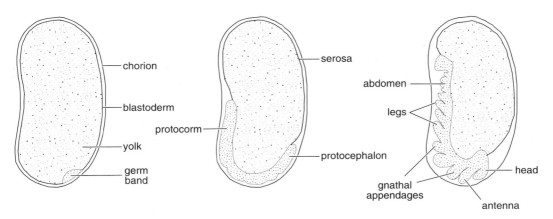

Fig. 14.18. Early stages in the development of an embryo with a short germ band seen from the side. The amnion is not shown; it covers the outer surface of the embryo (termite, *Zootermopsis*, Isoptera)

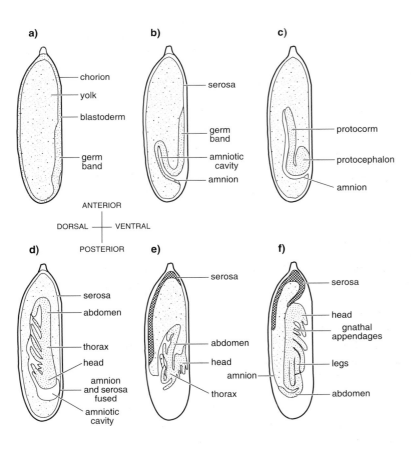

Fig. 14.19. Stages in the embryonic development of a damsel fly, *Agrion* (from Johannsen & Butt, 1941). (a) Blastoderm with germ band; (b) formation of the amniotic cavity (c) anatrepsis; (d) embryo completely immersed in yolk. Amniotic cavity complete; (e) katatrepsis; (f) katatrepsis almost complete. Amnion forming the provisional dorsal closure.

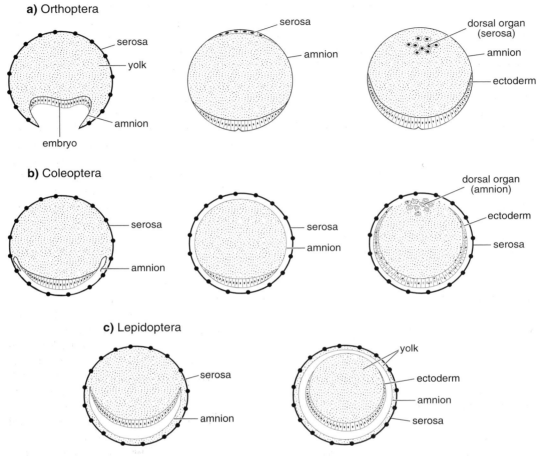

Fig. 14.20. Dorsal closure and the fate of the embryonic membranes. The earliest stages, figures on the left of the diagram, can be derived from **(b)** or **(c)** in Fig. 14.16 (from Imms, 1957). **(a)** Amnion forms the provisional dorsal closure. Serosa contracts and is destroyed (*Oecanthus*, Orthoptera). **(b)** Amnion forms the provisional dorsal closure. Serosa remains intact around the outside of the embryo (*Leptinotarsa*, Coleoptera). **(c)** Ectoderm forms the dorsal closure. Amnion and serosa remain intact, enclosing a layer of yolk outside the embryo (Lepidoptera).

membranes even though the embryo itself remains relatively static. In *Leptinotarsa* (Coleoptera) and other Chrysomelidae, the amnion breaks and grows up inside the serosa (Fig. 14.20b). Later it is replaced by the ectoderm while the serosa remains intact round the outside. In Lepidoptera, Tenthredinidae and Nematocera, amnion and ectoderm grow dorsally together so that the ectoderm forms the dorsal closure from the beginning and the amnion forms a membrane all round the outside (Fig. 14.20c). The serosa is invaginated and destroyed in Nematocera, but it persists in Lepidoptera and

Tenthredinidae so that a layer of yolk is present all round the embryo, held between the amnion and serosa. This provides the first meal for larval Lepidoptera when they hatch.

Although there is a considerable increase in the area of the epidermis during dorsal closure, it appears to be produced by changes in the shapes of the epidermal cells, and not by the addition of more cells.

14.2.10 *Ectodermal structures*

Appendages The whole outer wall of the embryo represents the ectoderm, and the appendages are formed by its outgrowth. The labrum develops in front of the sto-modeum, and the antennal rudiments grow on either side on the protocephalon (Fig. 14.11c). The protocorm becomes segmented and, in the lower orders, each segment extends laterally to form the rudiment of an appendage. The anterior appendages in grasshoppers are first apparent when about 30% of the total incubation period has elapsed (Fig. 14.11, 14.26).

Immediately behind the protocephalon are the rudiments of the mandibles, maxillae and labium. The latter arises in series with the rest as a pair of limbs that later fuse in the midline to form the definitive labium. The appendages of the next three segments form the walking legs. These grow longer and become folded and grooved where later the cuticle will become segmented (Fig. 14.11c).

The abdominal appendages subsequently disappear except that in some insects the appendages of segments 8 and 9 contribute to the ovipositor and those on segment 11 form the cerci. In Orthoptera and some other orders, the appendages of the first abdominal segment also persist for a time. They are known as the pleuropodia and have a distal area in which the cells become very large. They then degenerate, becoming torn off when the insect hatches. In Orthoptera, they probably secrete an enzyme which digests the serosal endocuticle before hatching. They probably serve the same purpose in *Belostoma* (Heteroptera), where they sink into the body so that only the tip of the appendage projects from a bowl-shaped cavity and they reach their greatest development just before hatching. However, in the cockroach, *Diploptera*, the external and lateral membranes of the cells of the pleuropodia are associated with large numbers of mitochondria, resembling fluid-transporting tissues. Possibly they are concerned with fluid regulation or osmoregulation in the embryo and the fluid surrounding it.

The pleuropodia have a specialized function in *Hesperoctenes* (Heteroptera). The egg of this insect has no yolk or chorion as it develops within the female parent. Its pleuropodia grow and fuse together to form a membrane which completely covers the embryo and makes contact with the wall of the oviduct to function as a pseudo-placenta (section 14.3.2.1).

The pleuropodia assume a variety of forms in Coleoptera, but in Dermaptera, Hymenoptera and Lepidoptera they are only ever present as small papillae which soon disappear.

Growth of the appendages, as of the embryo as a whole, occurs partly as a result of cell division and partly by changes in the shape of the appendages and rearrangement of the cells. Four phases of growth are recognizable in a grasshopper embryo (Fig. 14.26 center). In the first, which lasts up to blastokinesis, the increase in size is relatively slow and results from an increase in cell number. Just after blastokinesis, a rapid increase in length occurs (phase 2) associated with a further increase in cell numbers. After this, it is probable that no further mitosis occurs in the epidermal cells until apolysis of the second embryonic cuticle. During this period (phase 3) the embryo remains almost constant in length. Once the cuticle of the first stage larva is produced, no further mitosis occurs and the slow growth before hatching (phase 4) is due to changes in the shapes of the cells. Growth at this time is limited because the embryo occupies the whole of the egg, but at the time of hatching a very rapid extension occurs due to the flattening of the cuticle (see Fig. 16.22). This final expansion is more marked in the appendages than in the body as a whole.

In *Drosophila*, mitosis in the epidermal cells remains relatively synchronized through the sixteenth division which is completed by the time 30% of the developmental period has elapsed (for brevity, times are subsequently referred to as '*x*% development', and see section 14.2.15). No further mitosis occurs in the epidermis, and growth results from changes in cell shape and arrangement.

Amongst holometabolous insects where the imaginal appendages develop from imaginal discs, the latter may already be apparent in the embryo. In *Drosophila*, the discs invaginate beneath the surface of the embryonic epidermis at about 60% development. Initially, the imaginal cells remain an integral part of the epidermis, connecting with the surface by long, narrow extensions which collectively form a small placode in the epidermis, but they subsequently become completely separated from the surface in the peripodial cavities (section 15.3.2.2).

Nervous system The central nervous system arises from ectodermal cells on the ventral side of the embryo. Individual cells divide tangentially cutting off large cells called neuroblasts on the inside. This occurs at about 20% of development in *Drosophila*. The number of neuroblasts is constant in each segment, and is similar between segments (Fig. 14.21a) and even between species as widely different as a grasshopper and *Drosophila*. Most neuroblasts divide to form neurons, but one gives rise to glial

a) number of neuroblasts

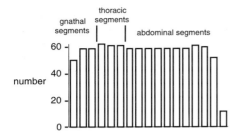

b) neurogenesis

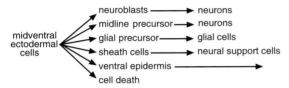

Fig. 14.21. Development of the central nervous system in an embryonic grasshopper (*Schistocerca*). **(a)** Number of neuroblasts formed in the body segments (data from Doe & Goodman, 1985). **(b)** Fate of cells from the midventral ectoderm which forms the ganglia of the ventral nerve cord (based on Doe & Goodman, 1985).

cells. Having produced the appropriate number of neurons, the neuroblasts die.

In grasshoppers, similar numbers of neurons are produced in all the segments, but differential mortality results in differing final numbers in the different ganglia. For example, 98 neurons are produced by the median neuroblast of the metathoracic segment of the grasshopper and 86 from that in the first abdominal segment. Subsequently, however, more of the latter die, so the ganglia finally have 93 and 59 midline neurons, respectively (Fig. 14.22). Cell death corresponds with the times of embryonic molts, suggesting some hormonal regulation (Doe & Technau, 1993; Thompson & Siegler, 1993). By contrast, differences in the numbers of neurons in different ganglia of *Drosophila* result primarily from differences in their postembryonic production (see Fig. 15.23).

The sheath cells which surround the central nervous system are also ectodermal in origin, being derived from ventral cells in the same way as the neuroblasts (Fig. 14.21b).

As the embryo segments, the ganglia become

differentiated. Three paired groups of neuroblasts, corresponding with the protocerebrum, deutocerebrum and tritocerebrum, develop in the protocephalon and, in addition, when the full complement of segments is present, 17 postoral ganglia may be recognizable: three in the gnathal segments, three in the thorax and eleven in the abdomen. The first three fuse to form the subesophageal ganglion in all insects and some fusion of abdominal ganglia always occurs, but the extent to which fusion of other ganglia occurs varies with taxon.

Axons grow as processes known as growth cones from which transient extensions spread out, apparently sensing guidance cues in the surrounding environment. The first axons to grow along a particular pathway are called pioneer axons. They use cues on the surfaces of, or secreted by, other cells. These cues may be attractive or repulsive. A variety of different cells may be involved, including other neurons, glial cells, and specific epithelial tissue. Once the pathway is established, the growth cones of later axons follow the pioneer axons, and are able to distinguish between bundles of axons to follow the appropriate path. Synapses start to form in the central nervous system at about 70% of development and from this time normal action potentials occur in some interneurons indicating that the nervous system is becoming functional (Fig. 14.22) (Leitch, Laurent & Shepherd, 1992).

The optic lobes, which come to be associated with the protocerebrum, contain no neuroblasts, although similar-looking large cells are present. In Orthoptera, the optic lobes are formed by delamination from the ectodermal thickening which forms the eye, but, in Hymenoptera and Coleoptera, they develop from an ectodermal invagination arising outside the eye rudiment.

The cells of the stomatogastric system are formed from the ectoderm of the stomodeum (Copenhaver & Taghert, 1991). The receptors of the peripheral nervous system are formed from epidermal cells. In *Drosophila*, the cells giving rise to larval sensilla are already distinguishable by 23% development. Each of these cells divides twice to form cells that will become a neuron and three sheath cells. If the fully developed sensillum contains more than one neuron, additional neurons are formed by division of the initial neuron mother cell. In *Drosophila*, this process is complete by about 40% development. Clusters of cells which will form sensilla in the grasshopper embryo are well-developed by 55% development.

Review: Goodman & Doe, 1993 – neurogenesis in *Drosophila*

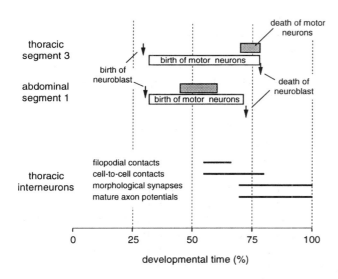

Fig. 14.22. Development of neurons in the ventral ganglia of an embryonic grasshopper. **Top**: birth and death of neurons derived from the median neuroblasts of the metathoracic and first abdominal segments (based on Thompson & Seigler, 1993). **Below**: maturation of median spiking local interneurons in the metathoracic ganglion (based on Leitch, Laurent & Shepherd, 1992).

Other ectodermal structures The tracheal system arises as bilateral segmental invaginations of the epidermis which become T-shaped. The arms of the T in adjacent segments fuse to form the longitudinal trunks and further invaginations from these develop into the finer branches of the system.

Oenocytes (section 16.1.3) are cut off from the epidermis of all the abdominal segments except possibly the last two.

The foregut and hindgut are ectodermal in origin, being formed from the stomodeum and proctodeum, respectively. Their development is considered together with the midgut.

The Malpighian tubules arise as evaginations from the tip of the proctodeum and so are ectodermal in origin. Early in their development, a large cell becomes apparent at the tip of each evagination. This cell does not divide, but regulates the continuous division of cells close to it so that the proximo–distal axis of the tubule becomes elongated. Development begins at 36% of development in *Rhodnius* (Skaer, 1992). Usually only two or three pairs of tubules develop in the embryo, but others may be produced during larval development.

Review: Manning & Krasnow, 1993 – development of the tracheal system in *Drosophila*

Embryonic cuticles The first cuticle to be produced during the embryonic development of many insects is the serosal cuticle which is secreted by the blastoderm and forms a continuous layer on the inside of the eggshell until the first stage larva hatches. Another very thin cuticle is produced by embryos of Acrididae after about 35% of the developmental period has elapsed (Fig. 14.26). A similar cuticle is present in Phasmatodea and at least some Heteroptera and Lepidoptera. Soon after blastokinesis in insects belonging to the hemimetabolous orders and at least some belonging to the Neuroptera, Trichoptera, Lepidoptera and Coleoptera, a second embryonic cuticle is formed. It is thicker than the first, about 6 μm in the grasshopper, *Melanoplus*, and comprises an epicuticle and a chitinous endocuticle. The first embryonic cuticle becomes broken up and is no longer visible in the later stages of embryogenesis. At 75–80% development, the second embryonic cuticle also separates from the epidermis and the cuticle of the first stage larva is laid down. The second embryonic cuticle, however, remains intact. In grasshoppers, it forms the outer covering of the vermiform larva (section 15.1.2), serving to hold the appendages close against the body of the insect during hatching.

The Diptera do not have embryonic cuticles, and the cuticle of the first larval stage starts to form at about 80% development as dorsal closure becomes complete.

Cuticle production and molting in the embryo appear to be regulated by ecdysteroids just as in postembryonic molts. In *Locusta* and *Bombyx*, the three major ecdysteroids in the embryo are ecdysone, 20-hydroxyecdysone and 2-deoxyecdysone. They are produced in the ovary of the female parent by the follicle cells and are already present in the newly laid egg, but in conjugated, inactive forms. Free ecdysone reaches peak levels when the serosal cuticle, the two embryonic cuticles and the first stage larval cuticle are produced (Fig. 14.26). It is not known if the prothoracic glands of the embryo produce ecdysteroids after they develop. Juvenile hormone is also present in the embryo, but its role in the embryonic molts is unclear.

Review: Hoffmann & Lagueux, 1985

14.2.11 Mesoderm and body cavities

The mesoderm is derived from the inner layer of the germ band which forms two lateral strands running the length of the body and joined across the midline by a thin sheet of cells. In the lower orders, the lateral strands become segmented and the somites separate off from each other, but in Lepidoptera and Hymenoptera the somites remain connected together. Amongst the Cyclorrhapha, there is a tendency for the mesoderm to remain unsegmented. In *Dacus*, for example, the strands of mesoderm only become segmented as they differentiate into the definitive structures. The mesoderm in the protocephalon arises *in situ* in the lower orders, but moves forwards from a postoral position in the more advanced groups.

Coelomic cavities develop as clefts in the somites in *Carausius* and *Formica* (Hymenoptera), but by the mesoderm rolling up to enclose a cavity in *Locusta* and *Sialis* (Fig. 14.23). In the Heteroptera, the coelomic sacs remain open to the epineural sinus, while in Diptera no coelomic cavities are formed.

Where they are most fully developed, a pair of coelomic cavities is present in each segment of the protocorm, while, in the protocephalon, pairs of cavities develop in association with the premandibular and antennal segments. Sometimes one or two more pairs are present in front of the antennae. Subsequently, in Orthoptera and Coleoptera, the thoracic and abdominal coelomic cavities become confluent forming a tube on either side.

At the same time as the coelom forms, the primary body cavity develops as a space between the upper surface of the embryo and the yolk. This cavity is called the epineural sinus and, in Orthoptera and *Pediculus* (Phthiraptera), it is bounded dorsally by a special layer of cells forming the yolk cell membrane (Fig. 14.23).

Soon, the walls of the coelomic sacs break down as the mesoderm which forms them differentiates to form muscles and other tissues. As a result, the coelomic cavities and the epineural sinus become confluent. Thus, although the body cavity of insects is called a hemocoel, it is strictly a mixocoel in those insects developing coelomic cavities. Some of the coelomic sacs, particularly those associated with the antennae, may be quite large and make a significant contribution to the final cavity.

After katatrepsis, when the midgut is formed, the mesoderm extends dorsally between the body wall and the gut so that the body cavity also extends until it completely surrounds the gut.

The outer walls of the coelomic sacs form the somatic muscles, the dorsal diaphragm, the pericardial cells and the subesophageal body. The latter is found in the Orthoptera, Plecoptera, Isoptera, Ischnocera, Coleoptera and Lepidoptera and consists of a number of large binucleate cells in the body cavity and closely associated with the inner end of the stomodeum. The cells become vacuolated and usually disappear at about the time of hatching, but in Isoptera they persist until the adult stage is reached. It is usually assumed that these cells are concerned with nitrogenous excretion, but they may be concerned with the breakdown of yolk.

The somatic muscles are formed from cells known as myoblasts. These fuse together to form syncytia which are progressively enlarged by the incorporation of more myoblasts. The syncytia produce processes resembling axon growth–cones that move over the inner surface of the epidermis until they reach their attachment points.

The inner walls of the coelomic sacs form the visceral muscles. The gonads, fat body and blood cells are also mesodermal in origin. The heart is formed from special cells, the cardioblasts, originating from the upper angle of the coelomic sacs, while the aorta is produced by the approximation of the median walls of the antennal coelomic sacs on either side.

Review: Bate, 1993 – muscle development in *Drosophila*

14.2.12 Alimentary canal

The foregut and hindgut arise early in development as ectodermal invaginations, the stomodeum and proctodeum (Fig. 14.24). These invaginations carry the anterior and posterior rudiments of the midgut into the embryo. These rudiments then extend towards each other forming two longitudinal strands of tissue beneath the yolk and above the visceral mesoderm. From these strands, midgut tissue spreads out over the surface of the yolk, eventually completely enclosing it.

Review: Skaer, 1993 – *Drosophila*

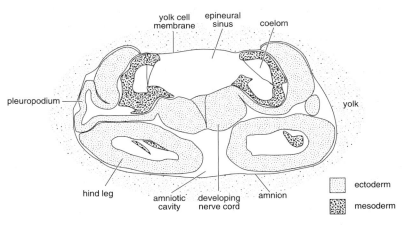

Fig. 14.23. Transverse section
through a grasshopper embryo
showing coelomic cavities,
pleuropodia and the yolk cell
membrane.

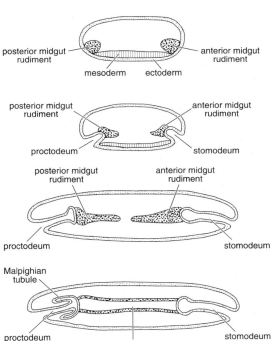

Fig. 14.24. Diagrams illustrating the development of the midgut (after Henson, 1946).

14.2.13 Reproductive system

The germ cells, from which sperm and oocytes are ultimately produced, become separated from the somatic cells very early in development. This early separation of the germ line results in a very direct cell lineage from the gametes of one generation to the gametes of the next, in -

isolation from the structural cells of the body (Fig. 14.25). This presumably helps ensure the integrity of the genetic system by reducing the possibility of DNA replication errors.

In the Diptera and some Coleoptera and Hymenoptera, the first nuclei to reach the surface of the egg are at the posterior end, and they become cut off in separate cells, called pole cells, while the rest of the blastoderm is still syncytial. The pole cells also differentiate during blastoderm formation in Dermaptera, Psocoptera and Homoptera, but in some groups of insects they are not recognizable and the germ cells appear at about the time the mesoderm differentiates. In *Locusta* they first appear in the walls of the coelomic sacs in abdominal segments 2 and 5.

In Nematocera, the number of energids which migrates into the pole plasm is constant for a species: one in *Miastor* and *Wachtliella*, two in *Sciara*, and six in *Culex*. However, in most of the Cyclorrhapha studied the number is variable, between three and eleven in *Drosophila*, for example. The invading nuclei divide to produce eight pole cells in *Miastor* and about 40 in *Drosophila*. These may be outside the blastoderm or, as in *Dacus*, in a circular polar opening of the blastoderm (Fig. 14.12b).

The region of cytoplasm in which these cells develop is already differentiated before the egg is laid; it contains polar granules rich in RNA and protein. In *Drosophila*, three proteins and five types of RNA have been identified in the pole plasm, probably in the polar granules, and there are probably even more components. As the pole cells form, the polar granules become fragmented and are incorporated in the cells. It is most likely that the various components of the polar granules are responsible for the

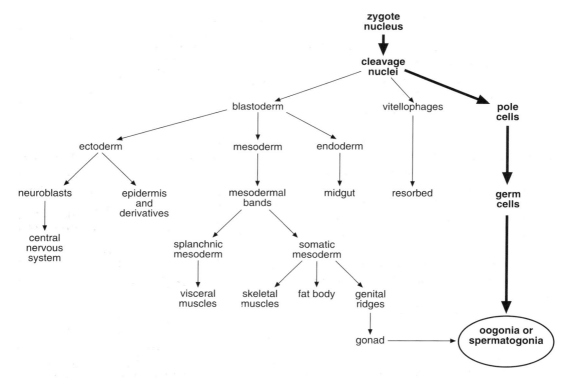

Fig. 14.25. Early segregation of the germ cells during embryonic development. The development of all other tissues involves extensive nuclear division (based on Anderson, 1962).

determination of cells entering the posterior region of the egg as pole cells.

In the Nematocera, all the pole cells migrate to form the germ cells in the gonads, but, in Cyclorrhapha, some are lost during the migration, perhaps becoming vitellophages, and only a proportion of them become germ cells. Some of these migrate through the blastoderm before gastrulation, but others do so only during or after gastrulation and are carried forwards and invaginated with the proctodeum (Fig. 14.15). The germ cells become enclosed by the mesoderm and increase in number before they become separated into columns by the ingrowth of the mesoderm. These columns form the germaria of the ovarioles or the testis follicles.

The mesoderm which forms the gonads is derived from several segments, but condenses to form a single structure on each side of the body. The rudiment thickens ventrally and gives rise to solid strands of cells in which cavities appear to form the lateral ducts. The median ducts arise from ectodermal invaginations.

Review: St. Johnston, 1993 – *Drosophila*

14.2.14 Metabolic changes

Oxygen uptake by the egg increases throughout development as the embryo increases in size, while the respiratory quotient, which at first equals one, soon falls to a low level. This suggests that, at first, the small carbohydrate reserves of the egg are used and that fat subsequently becomes the main metabolic substrate. It has been calculated that, in the egg of a grasshopper, 75% of the oxygen uptake is concerned in the oxidation of fat. The oxidation of lipid is advantageous because it produces relatively large quantities of metabolic water compared with carbohydrates and there is no nitrogenous waste which would be produced if the embryo used its protein reserves for energy metabolism.

Initially, the RNA present in the egg is derived from the female parent. Transcription of new RNA by the zygotic nuclei appears to begin in the early stages of cleavage in Coleoptera and Hymenoptera, but in other orders not until the syncytial blastoderm or even the blastoderm proper is formed. This is correlated with the fact that, in the Coleoptera and Hymenoptera, mitotic cycles are slow relative to those in the other orders, taking 25 minutes or more to complete compared with less than 10 minutes in *Drosophila*.

The rate of RNA synthesis by each nucleus is very high at first, but as new nuclei and DNA are produced, the rate declines (Fig. 14.27). Subsequently, the rate of synthesis per nucleus rises again so that, although the total amount of DNA in the embryo is stable, the total amount of RNA continues to increase.

Many different proteins are produced; over 400 have been recorded in both *Drosophila* and the chironomid, *Smittia*. The majority of these are produced throughout embryonic development, but production of some (10% in *Smittia* and 20% in *Drosophila*) varies systematically in the course of development. In general, the identities of these proteins are unknown, but there are marked changes in histones. Histones are small proteins that bind to DNA and prevent non-specific transcription. At first, the histones in the egg are of maternal origin, but once transcription by the zygotic nuclei begins, the rate of synthesis of histone RNA parallels that of DNA (Anderson & Lengyel, 1980).

14.2.15 Temporal pattern of events

In describing the timing of events in embryology, it is usual to refer to time as a percentage of the total embryonic period from egg-laying to hatching. This makes it possible to compare insects with different developmental times. The temporal sequence of the major events described above are illustrated in Fig. 14.26 for grasshoppers, as representative hemimetabolous insects with panoistic ovarioles, and for *Drosophila* (Fig. 14.27), a holometabolous insect with polytrophic ovarioles. There is some variation in the times at which the insects reach particular stages of development, but, in general, the pattern is similar in the two species. Completion of the blastoderm is slightly later in *Drosophila* than in the grasshopper, and so is the definitive dorsal closure. Neurogenesis occurs at about 25% of developmental time in both species with the first stage larval cuticle produced at about 80%.

The total duration of embryonic development varies considerably and it is generally considered that insects with a long germ band (most holometabolous insects) have shorter embryonic periods that those with short germ bands (Orthoptera and related orders). Thus at 30 °C, the complete development of *Culex* (long germ band) takes about 30 hours, that of *Ostrinia* (Lepidoptera) (long germ band) 82 hours, that of *Oncopeltus* (intermediate germ band) 5 days, that of *Schistocerca* (short germ band) 15

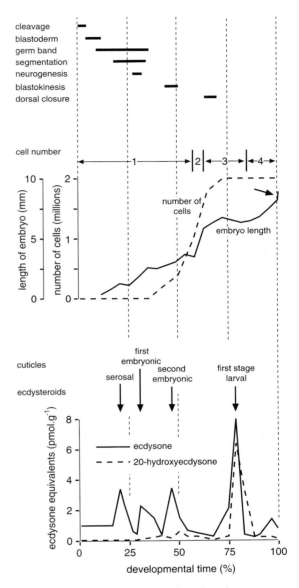

Fig. 14.26. Timing of major events in the embryonic development of an insect with a short germ band (*Locusta*, Orthoptera). **Top**: timing of major events. **Center**: total number of cells and growth of the embryo. Numbers 1–4 are the different phases of growth (partly after Salzen, 1960). **Below**: changes in ecdysteroid titer (measured in ecdysone equivalents) in relation to cuticle development (after Lagueux *et al.*, 1979).

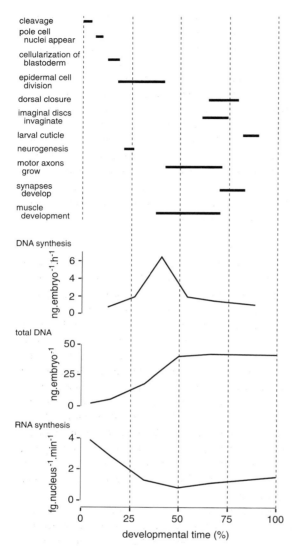

Fig. 14.27. Timing of major events in the embryonic development of an insect with a long germ band (*Drosophila*, Diptera). **Top**: timing of major events. **Below**: changes in DNA and RNA synthesis and accumulation. Notice that DNA total and synthesis are expressed per embryo; RNA synthesis per nucleus (data from Anderson & Lengyel, 1979, 1980).

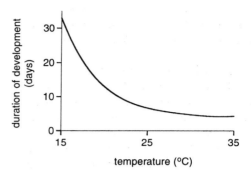

Fig. 14.28. The effect of temperature on the duration of embryonic development (milkweed bug, *Oncopeltus*, Hemiptera) (from Richards, 1957).

the temperature exceeds a certain level, often in the range 35–40 °C, nor if it falls below a certain level, which is about 14 °C in *Oncopeltus*, and 13 °C in *Cimex*. Some development does occur at lower temperatures, however, and in *Oncopeltus* some morphogenesis occurs even at 5 °C. In addition to these developmental thresholds, there is also a behavioral temperature threshold below which the fully developed embryo will not hatch. It is thus necessary to distinguish between the threshold temperature for some development, below which no differentiation occurs, the threshold for complete development, and the hatching threshold. These distinctions are not always clear in the literature.

Above the minimum temperature for full development the total heat input (temperature × time) necessary to produce full development and hatching is approximately constant whatever the temperature. Thus, in *Schistocerca gregaria*, complete development requires 225 degree days above 15 °C. For instance, at 30 °C, development takes about 15 days [(30–15 °) × 15 days = 225 degree days] and at 20 °C about 45 days [(20–15 °) × 45 days = 225 degree days]. In *Schistocerca*, this relationship also holds with fluctuating temperatures, including periods below the minimum for full development, but in *Oncopeltus* and some other insects this is not entirely the case as some development occurs below the minimum for complete development. Hence periods of low temperature do have an influence on the total number of degree days necessary for development in some insects (Hagstrum & Milliken, 1991).

Humidity also influences egg development in some species. Eggs of *Musca* only develop at high humidities

days, and that of *Ornithacris* (Orthoptera) (short germ band) 43 days. The extent to which these differences are also a reflection of the overall size of the egg is not clear.

The duration of development decreases as temperature increases (Fig. 14.28). Development is not completed if and, even at 80% relative humidity, only 15% of the eggs survive to hatching. In *Lucilia* (Diptera) there is a linear relationship between the time of development and saturation deficit.

Many eggs have to absorb water before they can complete their development. If there is sufficient moisture in the environment to prevent death through desiccation, but not enough for development to continue, the eggs may remain quiescent for some time. Under such circumstances the eggs of *Schistocerca* develop to the beginning of katatrepsis and then remain quiescent and viable for up to six weeks. At any time during this period development will proceed if more water becomes available.

In some species, the embryonic period is greatly prolonged by an egg diapause. Many insect species overwinter in embryonic diapause (section 15.6), and, as an extreme example, the diapause eggs of the brown locust, *Locustana*, may survive for over three years. Embryonic diapause occurs at different stages of development in different species even within one family. For instance, in the family Acrididae (grasshoppers), it occurs just after blastoderm formation in *Austroicetes*, just before katatrepsis in *Melanoplus femur-rubrum*, but not until 80% development in several other species of *Melanoplus*.

14.3 VIVIPARITY

14.3.1 Ovoviviparity

Many species retain the eggs in the genital tracts until the larvae are ready to hatch, hatching occurring just before or as the eggs are laid. All the nourishment for the embryo is present in the egg and no special nutritional structures are developed in the eggs or parent. Viviparity of this sort is called ovoviviparity and it differs from normal oviparity only in the retention of the eggs.

Ovoviviparity occurs spasmodically in various orders of insects: Ephemeroptera, Blattodea, Psocoptera, Homoptera, Thysanoptera, Lepidoptera, Coleoptera and Diptera, being particularly widespread in the last group from which the following examples are drawn. Sometimes a species of *Musca* that is normally oviparous retains its eggs and deposits larvae. Some other Diptera, particularly amongst the Tachinidae, are always ovoviviparous. In these species, the eggs are retained in the median oviduct which becomes enlarged to form the uterus.

Ovoviviparous tachinids produce large numbers of eggs as do many oviparous Diptera, but in other ovovivipa-

rous species, such as *Sarcophaga*, small numbers of relatively large eggs are produced at each ovulation, and in *Musca larvipara* only one large egg is produced at a time. The increased size of the eggs permits the accumulation of more nutriment so that the embryo can develop beyond the normal hatching stage and larvae are born in a late stage of development. In *Hylemya strigosa*, for instance, the larva passes through the first larval stage and molts to the second while still in the egg, casting the first stage cuticle immediately after hatching. In *Termitoxenia*, development in the uterine egg goes even further. A fully developed third stage larva hatches from the egg immediately it is laid. It pupates a few minutes afterwards and the free-living larva never feeds.

14.3.2 Viviparity

In some insects in which the eggs are retained after fertilization, the embryos receive nourishment directly from the parent in addition to, or instead of, that present in the yolk. Such insects are regarded as truly viviparous and some anatomical adaptations are present in the parent or egg which facilitate the transfer of nutriment. Viviparous species commonly produce fewer offspring than related oviparous species and this may be associated with a reduction in the number of ovarioles. Thus *Melophagus* and *Glossina* (Diptera) have only two ovarioles on each side, while in the related, oviparous, *Musca domestica* there are about 70 ovarioles in each ovary. Similarly amongst the viviparous Dermaptera, *Hemimerus* has 10–12 ovarioles on each side, but only about half are functional, and *Arixenia* has only three ovarioles on each side.

Sometimes the eggs are retained and development occurs in the ovariole, as in *Hemimerus*, the aphids and Chrysomelidae. In other insects, such as the viviparous Diptera, the vagina is enlarged to form a uterus. In Strepsiptera and a few parthenogenetic Cecidomyiidae the eggs develop in the hemocoel of the parent.

Elaphothrips tuberculatus is facultatively viviparous. Some females lay eggs that develop into females; others are viviparous and produce only males; and others exhibit both modes of reproduction. The switch to viviparity appears to be related to the quality of the leaves in which the insects are feeding (Crespi, 1989).

Hagan (1951) recognizes three main categories of viviparity and his scheme is followed here.

Review: Hagan, 1951

14.3.2.1 *Pseudoplacental viviparity*

Insects exhibiting pseudoplacental viviparity produce eggs containing little or no yolk. They are retained by the female and are presumed to receive nourishment via embryonic or maternal structures called pseudoplacentae. There is, however, no physiological evidence concerning the importance of these structures. Viviparous development continues up to the time of hatching, but the larvae are free-living.

In *Hemimerus*, the fully developed oocyte has no chorion or yolk, but is retained in the ovariole during embryonic development. It is accompanied by a single nurse cell and enclosed by a follicular epithelium one cell thick. At the beginning of embryonic development, the follicle epithelium becomes two or three cells thick and at the two ends thickens still more to form the anterior and posterior maternal pseudoplacentae (Fig. 14.29). As the embryo develops it comes to lie in a cavity (the pseudoplacental cavity) produced by the enlargement of the follicle, but it becomes connected with the follicle by cytoplasmic processes extending out from the cells of the amnion and, later, from those of the serosa (Fig. 14.29b). Furthermore, some of the embryonic cells form large trophocytes which come into contact with the anterior maternal pseudoplacenta. The follicle epithelium and pseudoplacentae show signs of breaking down and this is taken to indicate that nutriment is being drawn from them.

Later in development, the serosa spreads all round the embryo and, with the amnion, enlarges anteriorly to form the fetal pseudoplacenta (Fig. 14.29c). By this time, dorsal closure is complete except anteriorly, where the body cavity is open to an extra embryonic cavity, the cephalic vesicle. It is presumed that nutriment passes from the pseudoplacenta to the fluid in the cephalic vesicle and then circulates into and around the embryo. The embryonic heart is probably functional at this time thus aiding the circulation.

Parthenogenetic eggs of aphids also develop in the ovarioles. They are much smaller than eggs that are fertilized when these are produced by the same species of aphid, and they have very little yolk. The oocyte is only briefly surrounded by follicle cells and, as it moves down the ovariole, it separates from them. As a result, it does not become surrounded by a vitelline envelope or a chorion since these are normally produced by the follicle cells. At first, nutriment is received via the nutrient cords as aphids have telotrophic ovarioles, but later, when the nutritive cord is broken, the ovariole sheath appears to assume a trophic function. It thickens and shows obvious signs of metabolic activity. Extensive gaps develop in the sheath which permit the direct access of hemolymph, and it is probable that the embryo obtains some nutriment from this source. When the embryo subsequently develops a serosa, this also appears to have a trophic function. The oocyte is invaded by endosymbionts before the ovariole sheath thickens, while the embryo is at an early stage of development. They enter through an extensive pore in the blastoderm, and accumulate in cells which become mycetocytes. These cells are differentiated very early in development. The possible roles of the endosymbionts in embryonic nutrition are unknown (Couchman & King, 1980).

Pseudoplacental viviparity is also known to occur in a psocopteran, *Archipsocus*, in which the serosa is an important trophic organ, and in the Polyctenidae (Heteroptera) in which first the serosa and then the pleuropodia function as pseudoplacentae.

Review: Blackman, 1987 – aphids

Viviparity in Blattodea Cockroaches are anomalous with respect to viviparity, compared to other orders. Fundamentally, all cockroaches are oviparous, laying their eggs in an ootheca which is extruded from the genital ducts. In some species, like *Periplaneta*, the ootheca may be carried projecting from the genital opening, but it is finally dropped some time before the eggs hatch. Other species, such as *Blattella*, continue to carry the ootheca externally until the time of hatching. Some others extrude the ootheca, but then withdraw it into the body where it is held in a median brood sac extending beneath the rest of the reproductive system. In these species, the ootheca may be poorly developed and as the eggs increase in size they project beyond the ootheca. In most species the increase in egg size results only from the absorption of water, but in *Diploptera*, in which the eggs increase in length by five or six times during embryonic development, there is also an increase in dry weight, indicating that some nutriment is obtained from the parent after ovulation, although the mechanism by which they do so is not known.

14.3.2.2 *Adenotrophic viviparity*

In adenotrophic viviparity, fully developed eggs with chorions are produced and passed to the uterus where they are retained. Embryonic development follows normally, but when the larva hatches it remains in the uterus and is

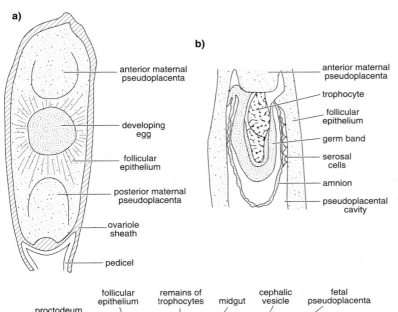

a)

anterior maternal
pseudoplacenta

developing
egg

follicular
epithelium

posterior maternal
pseudoplacenta

ovariole
sheath

pedicel

b)

anterior maternal
pseudoplacenta

trophocyte

follicular
epithelium

germ band

serosal
cells

amnion

pseudoplacental
cavity

Fig. 14.29. Pseudoplacental viviparity. Stages in the embryonic development of *Hemimerus* (Dermaptera) (from Hagan, 1951). (**a**) Early cleavage showing the maternal pseudoplacentae. (**b**) Fully developed germ band. (**c**) End of blastokinesis. The maternal pseudoplacentae have been replaced by an embryonic (fetal) pseudoplacenta.

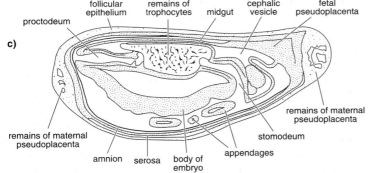

c)

proctodeum

follicular
epithelium

remains of
trophocytes

midgut

cephalic
vesicle

fetal
pseudoplacenta

remains of maternal
pseudoplacenta

remains of maternal
pseudoplacenta

amnion   serosa   body of
embryo

appendages

stomodeum

nourished by special maternal glands. Parturition occurs when the larva is fully developed and pupation follows within a short time, there being no free-living feeding phase. This type of viviparity only occurs in a few blood-sucking Diptera: Glossinidae (tsetse flies), Hippoboscidae, Streblidae and Nycteribiidae.

In *Glossina*, each ovary has only two ovarioles and these function in sequence: first one ovariole on the right produces a mature oocyte, then one on the left, then the second on the right, and so on. Only one fully chorionated oocyte is produced at a time. It is fertilized in the uterus and embryonic development proceeds rapidly, lasting three or four days at 25 °C. The larva hatches in the uterus, on the ventral wall of which is a small pad of glandular cells with a cushion of muscle beneath and other muscles running to the ventral body wall. This structure is known

as the choriothete and it is responsible for removing the chorion and the cuticle of the first stage larva. It undergoes cyclical development, degenerating during the later stages of larval development and starting to regenerate just before larviposition, so that it is fully developed by the time the next larva is ready to hatch. The choriothete adheres to the chorion and, when this is split longitudinally by the larval egg-burster, pulls it off by muscular action. The chorion becomes folded up against the ventral wall of the uterus. The cuticle of the first stage larva is pulled off in the same way. When the second stage larva molts, its cuticle is not shed immediately but it is subsequently split by the growth of the third stage larva. The cast cuticles are expelled by the female at parturition.

The larva feeds in the uterus on the secretion of specialized accessory glands, known as milk glands. The

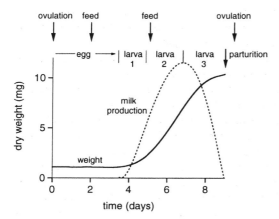

Fig. 14.30. Adenotrophic viviparity. Growth of the tsetse fly, *Glossina*, through its embryonic and larval stages in the uterus of the adult female. The larva feeds on 'milk' produced by the female accessory gland. Milk production is shown on an arbitrary scale. The female parent may take a blood meal on any day up to day six, but feeding after this is unusual (after Denlinger & Ma, 1974).

milk contains approximately equal quantities of lipids and proteins plus amino acids, and very little else. Its production is cyclical, beginning just before a larva hatches and continuing until the third larval stage is well-developed (Fig. 14.30). These changes in production are reflected in changes in the development of the milk glands.

The nutrients in the milk are derived from the blood meals taken by the parent fly. Most flies feed on the first day of pregnancy and then after an interval of three or four days. A relatively large proportion of the amino acids in the first feed are incorporated into lipids in the fat body and transferred to the milk gland during the later stages of gestation. Amino acids from the second feed are probably taken up directly by the milk glands and there is little lipid synthesis. The adult fly does not usually feed in the later stages of pregnancy when the large larva occupies most of the space in its abdomen.

The larval respiratory system opens by a pair of posterior spiracles in the first two larval stages, but the system is highly specialized in the third, enabling the larva to respire while still in the uterus, but limiting water loss. The terminal segment of the abdomen bears two heavily sclerotized lobes (polypneustic lobes) each of which is crossed by three longitudinal bands of perforations leading into the tracheal system. Each of these perforations is guarded by a

valve which permits air to be drawn into the system but does not allow it to be forced out. In addition to these openings in the polypneustic lobes, the spiracles of the second stage larva on the insides of the lobes remain open because the second instar cuticle is not shed. As a further consequence of this, the tracheal system is lined by two layers of cuticle, that of the second stage larva being broken only at the inner ends of the system.

Indirectly acting dorso-ventral muscles produce a piston-like movement in a specialized part of the tracheal system in the polypneustic lobes and it is suggested that this movement sucks air in through the valved perforations and forces it forwards between the two linings of the tracheae. An exhalent current flows through the loose second stage linings and out through the persistent spiracles of the second stage larva. The respiratory muscles contract 15–25 times per minute. By this means the larva is able to draw air in through the genital opening of the parent, but this mechanism can only function while the second stage cuticle persists, a period of four or five days at the beginning of the third larval stage. In the earlier larval stages, oxygen may be obtained, at least partly, by diffusion from the female tracheal system which invests the uterus, while in the late third stage larva the valves in the polypneustic lobes disappear and a two-way airflow through the perforations is possible.

The hindgut of the larva is occluded at its connection with the midgut and again at the anus, so that waste materials from the midgut are not voided and the hindgut forms a reservoir for nitrogenous waste. This arrangement prevents the larva from fouling the female ducts.

Following parturition, the larva moves to a suitable site for puparium formation; it does not feed during the short period for which it is active.

Maturation of the next oocyte occurs during pregnancy and ovulation follows about 30 minutes after parturition, so that maximum fecundity is achieved. For a discussion of the possible endocrine control of reproduction in *Glossina* see Tobe and Langley (1978).

As far as is known, the development of the Hippoboscidae and Nycteribiidae does not differ in essentials from that of *Glossina*, but there is no evidence for a complex air circulation similar to that which occurs in the larva of *Glossina*.

14.3.2.3 *Hemocoelous viviparity*
Hemocoelous viviparity differs from the other forms of viviparity in that development occurs in the hemocoel of

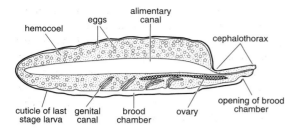

Fig. 14.31. Hemocoelous viviparity. Sagittal section through a female strepsipteran showing eggs in the hemocoel (from Clausen, 1940).

the parent female. This type of development occurs throughout the Strepsiptera and in some larval Cecidomyiidae which reproduce pedogenetically.

Female Strepsiptera have two or three ovarial strands on either side of the midgut, but there are no oviducts and mature oocytes are released into the hemocoel by the rupture of the ovarian walls. In *Stylops*, the eggs contain very little yolk, but some is present in other genera such as *Acroschismus*. Sperm enter through the genital canals which arise in the ventral midline of the female and open into the hemocoel (Fig. 14.31). Fertilization and development occur in the hemocoel with direct transfer of nutriment from the hemolymph to the embryo. The larvae hatch and find their way to the outside through the genital canals (Hagan, 1951).

In *Miastor* (Diptera) the eggs are similarly liberated into the hemocoel from simple sacs. The developing egg is nourished via nurse cells, which arise independently of the oocyte, and later via the serosa which becomes thickened and vacuolated. When the larvae hatch they feed on the tissues of the female and any unhatched eggs, finally escaping through a rupture in the wall of the parent.

14.4 POLYEMBRYONY

Sometimes an egg, instead of giving rise to a single larva, produces two or more. This is called polyembryony. It occurs occasionally in Orthoptera and perhaps in other groups, but in some endoparasitic insects it is a regular phenomenon. This is true, for instance, in *Halictoxenos* (Strepsiptera), and in a number of Hymenoptera including several genera of Encyrtidae and Ichneumonidae which parasitize the eggs and larvae of Lepidoptera, and *Aphelopus* and *Platygaster*. The eggs of all these parasites

are small and relatively free from yolk. Nutriment is derived from the host tissues in which the developing embryos are situated.

Cleavage of the egg is complete, the resulting cells become completely separated from each other, and each develops into an embryo. They become surrounded by a membrane called the trophamnion (but not in any way comparable with the amnion in normal embryos) which, in the Hymenoptera, is derived from the cytoplasm surrounding the polar bodies produced during the meiotic divisions of the oocyte. Sometimes the polar bodies fuse to form a single paranuclear mass which subsequently divides, or the polar bodies may proliferate without prior fusion. The cytoplasm containing the derivatives of the polar nuclei does not divide. The trophamnion is presumed to transfer nutrients from the host tissues to the developing embryos, although later in development it forms only a very thin membrane round the embryos.

The number of embryos produced from a single egg varies. In *Platygaster hiemalis*, each egg produces only two larvae, while each egg of *Platygaster vernalis* forms eight larvae. At the other extreme, the eggs of some encyrtids are known to give rise to 500 larvae, and in *Litomastix* the number is even greater. This is a parasite of the larva of *Plusia* (Lepidoptera). Its zygote nucleus and associated cytoplasm divides to form two blastomeres, which become surrounded by the trophamnion. By further division (Fig. 14.32b) over 200 blastomeres are produced, some of them becoming spindle-shaped and pushing between the others to form nucleated inner membranes which divide the embryonic region into 15–20 primary embryonic masses each containing up to 50 blastomeres (Fig. 14.32c). These cells continue to divide and the embryonic masses are further divided into secondary and tertiary embryonic masses by ingrowths of the inner layer and the trophamnion. Finally the tertiary masses, which may become separated from each other, divide to form embryos of which 1000 or more may be derived from one egg.

In the chalcid, *Ageniaspis*, the number of embryos produced depends on the size of the host. If an egg is laid in a host whose volume is less than $10\,mm^3$, it forms less than 20 embryos, but in a host with a volume greater than $40\,mm^3$, more than 100 embryos are produced.

Polyphenism occurs amongst the larvae of the encyrtid, *Copidosomopsis tanytmemus*. About 10% of the larvae hatch into the host hemocoel much earlier than the

Fig. 14.32. Polyembryony. Early stages in the development of the polygerm of *Litomastix* (Hymenoptera) (from Johannsen & Butt, 1941).

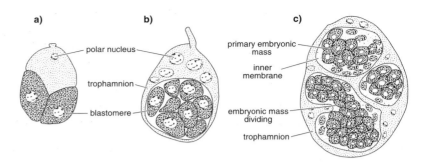

others. These larvae have well-developed mouthparts, unlike the majority, and they attack larvae of any competing species of parasite which may lay its eggs in the same host. These larvae ultimately die without reproducing (Cruz, 1986).

The offspring of one egg are usually all of one sex in polyembryonic species of Hymenoptera. There are, apparently, exceptions where both sexes are produced, but, in such cases, it is not certain that these might not have arisen from two separate eggs. The development of the parasite embryos is synchronized with specific events in the development of the host larva suggesting that it is regulated by the host's endocrine system (Strand, Goodman & Baehrecke, 1991).

In the strepsipteran, *Halictoxenos*, formation of the 'trophamnion' does not apparently involve the polar bodies and additional embryos are derived from the amnion covering the original embryo.

Review: Ivanova-Kasas, 1972

14.5 PARTHENOGENESIS

Sometimes eggs develop without being fertilized. This phenomenon is known as parthenogenesis. Occasional parthenogenesis, following from the failure of a female to find a mate, is probably widespread, but a number of insects use parthenogenesis as a normal means of reproduction. It has been recorded from all the insect orders except Odonata, Dermaptera, Neuroptera and Siphonaptera.

The sex of the offspring developing from an unfertilized egg is dependent on the sex-determining mechanism of the insect and the behavior of the chromosomes at the meiotic division of the oocyte nucleus. In the majority of insects, the female is homogametic (XX) and the male heterogametic (XY or XO), but the Lepidoptera are exceptional with the females having the heterogametic

constitution. Hence the unfertilized eggs of most insects contain only X-chromosomes as any Y-chromosome must come from the male. Whether the egg contains one or two X-chromosomes, that is, whether it is haploid or diploid, depends on the behavior of the chromosomes at meiosis. Sometimes no reduction division occurs, or reduction is followed by doubling of the chromosome number so that the diploid, XX, composition of the egg is maintained. These eggs will give rise only to females. Eggs which undergo a normal reduction division and in which no chromosome doubling occurs remain haploid and, if they develop at all, become males. Such haploid males are characteristic of some insect groups.

Parthenogenesis may be classified according to the behavior of the chromosomes at the maturation division of the oocyte:

Haplo-diploidy – A normal reduction division occurs in the oocyte, fertilized eggs developing into females, unfertilized eggs into males. This is characteristic of Hymenoptera and some smaller groups.

Apomictic (ameiotic) parthenogenesis – No reduction division occurs so that the offspring have the same genetic constitution as the mother and all are female. This is of common occurrence in blattids, aphids, tenthredinids and curculionids.

Automictic (meiotic) – A normal reduction division occurs, but is followed by the fusion of two nuclei so that the diploid number of chromosomes is restored. Often the female pronucleus fuses with the second polar nucleus, or two cleavage nuclei may fuse. In *Solenobia* (Lepidoptera, Psychidae) two pairs of nuclei fuse after the second cleavage division. *Moraba* (Orthoptera) is exceptional in having a pre-meiotic doubling of the chromosomes followed by a normal division so that the diploid number is restored. This type of parthenogenesis, in which only females are produced, also occurs in phasmids and coccids.

An alternative classification based on the sex of the offspring produced as a result of parthenogenesis is as follows:

Arrhenotoky – Only males are produced.

Thelytoky – Only females are produced.

Amphitoky – Individuals of either sex may be produced. It is possible, though not proven, that reproduction in the tailor ant, *Oecophylla*, is an example of amphitoky.

14.5.1 Arrhenotoky

Facultative arrhenotoky, in which the eggs may or may not be fertilized, is characteristic of a few groups of insects. It occurs throughout the Hymenoptera, in some Thysanoptera, the Coccidae Iceryini, some Aleyrodidae and the beetle *Micromalthus*. In all of these, the unfertilized, haploid eggs, produce males.

In the Hymenoptera, the parent female determines whether or not an egg is fertilized by controlling the release of sperm from the spermatheca as the eggs pass down the oviduct. The stimuli prompting the female to withhold sperm are largely unknown. *Aphidius* lays unfertilized eggs during the first two or three hours of oviposition, but then switches to fertilized eggs. *Microplitis* lays more unfertilized than fertilized eggs at 27 °C, but equal numbers at 20 °C. Some other factors are discussed in section 13.4.

Two species of Hymenoptera are known to produce haploid males from fertilized eggs. These are the parasitic wasps, *Encarsia pergandiella* and *Nasonia*. In both cases, the chromosomes contributed by the male are lost immediately after fertilization. In *Nasonia*, this is controlled by an extrachromosomal factor from the male (Hunter, Nur & Werren, 1993).

The coccid, *Icerya purchasi*, is of interest in that, apart from a few haploid males, the adult population consists entirely of hermaphrodites which are diploid with diploid ovaries, but which also have haploid testes. No true females occur. When the larva giving rise to an hermaphrodite hatches from the egg, all the cells are diploid, but after a time haploid nuclei appear in the gonad. They form a core from which the testis develops surrounded by the ovary. Oocytes undergo a normal reduction division, but the spermatocytes, which are already haploid, do not. The

hermaphrodites are normally self-fertilizing, but can be fertilized by the occasional males which arise from unfertilized eggs. Cross-fertilization between the hermaphrodites does not occur.

14.5.2 Thelytoky

Thelytokous parthenogenesis probably occurs occasionally in many species of insect. For example, in the Orthoptera, unmated females of *Schistocerca* live much longer than mated females, but lay about the same number of eggs. Nearly all of these start to develop, but only about 25% hatch and further heavy mortality occurs in the first larval stage. Thus the viability of unfertilized eggs is much less than that of fertilized eggs, but, nevertheless, *Schistocerca* can develop parthenogenetically for several generations. It appears that the only eggs to survive are those in which the chromosomes double after meiosis. Unmated cockroaches also live longer than mated females, but produce fewer eggs with poor viability. In *Bombyx* the tendency for sporadic parthenogenesis varies in different strains.

In some other insects, thelytoky is a regular occurrence and, in *Carausius* and some Thysanoptera, males are extremely rare, the whole population normally reproducing parthenogenetically. Sometimes, as in some Psychidae (Lepidoptera) and Coccidae, a parthenogenetic race exists together with a normal bisexual race. Thus, *Lecanium* (Homoptera) has one race consisting entirely of females which reproduce apomictically and another race which is bisexual and exhibits facultative thelytoky. In this case, fertilized eggs may become males or females, while unfertilized eggs develop automictically and so produce females. Commonly such races occur in different areas and in the weevil *Otiorrhynchus dubius*, for instance, a parthenogenetic race occurs in northern Europe and a bisexual race in central Europe. In general, parthenogenesis occurs more commonly in the higher latitudes than closer to the equator and, in the genus *Otiorrhynchus*, 78% of the species occurring in Scandinavia reproduce parthenogenetically, while only 28% of those occurring in the Austrian Alps do so.

Constant thelytoky occurs in a few Lepidoptera, such as the psychid *Solenobia*, raising the question of sex determination as the females of Lepidoptera are heterogametic. Thelytoky in these insects may result from the passage of the X chromosome to the polar body at maturation so that only the Y chromosome remains in the egg; or two polar nuclei may fuse to give a female XO or XY constitution, while the egg nucleus degenerates.

A completely different explanation supposes that the females are homozygous, YY, and so can give rise only to females (Soumalainen, 1962).

Thelytoky can result from apomixis or automixis. In the parasitic wasps of the genus *Trichogramma*, it is induced by bacteria in the insect (section 4.3.2.2). In the absence of these bacteria, the wasps reproduce sexually (Stouthamer & Kamer, 1994).

An unusual type of thelytoky, known as gynogenesis, occurs in the form *mobilis* of *Ptinus clavipes* (Coleoptera). This form exists only as triploid females which reproduce parthenogenetically, but the development of eggs is triggered by healthy sperm of *P. clavipes* or, less successfully, of *P. pusillus*.

14.5.3 Alternation of generations

A number of insects combine the advantages of parthenogenesis with those of bisexual reproduction using an alternation of generations. This occurs, for instance, in the gall wasps, Cynipidae (Hymenoptera), which are commonly bivoltine (having two generations each year), a generation of parthenogenetic females alternating with a bisexual generation. *Neuroterus lenticularis*, for example, forms galls on the underside of oak leaves in which the species overwinters. Females emerge in the spring and lay eggs. The eggs develop into haploid males, or into females in which the diploid condition is apparently restored by endomitosis at the blastoderm stage. In this way the bisexual generation arises, the insects emerging from catkin galls in early summer. After mating, the females of this generation lay eggs that are fertilized and which produce the parthenogenetically reproducing females of the following spring generation.

Aphids have a more complex alternation of generations with several parthenogenetic generations occurring during the summer (see Fig. 15.37). Sometimes, as in *Aphis fabae*, an alternation of host plants also occurs. The first generation emerges on spindle (*Euonymus*) in spring from overwintering eggs. It consists entirely of females, the fundatrices, which may or may not produce wingless generations of fundatrigeniae before a winged generation develops. These migrate to annuals such as beans and produce wingless virginopara, of which there may be many successive generations. Ultimately, the virginopara produce the sexuparae, some of which are winged and return to the spindle while others are wingless. The former produce females, the latter winged males which then join

the females; they mate and winter eggs are produced. All these generations, except for the last, reproduce parthenogenetically and consist entirely of females. In some genera, such as *Terraneura*, only one class of sexuparae is produced and these individuals give birth to both male and female sexual forms, an instance of amphitoky.

Female aphids, which have an XX constitution, are produced apomictically. Males are also diploid, but XO. In the parthenogenetic production of males, no reduction of the autosomes occurs, but one of the X chromosomes is lost. The production of males is ultimately under environmental control (see section 15.5).

Spermatogenesis in male aphids is unusual. The two spermatocytes produced from each spermatogonium at the first meiotic division are different; both have the full complement of autosomes, but one has an X chromosome, the other does not. The cell without an X chromosome degenerates so that only those with an X chromosome undergo the second meiotic division and only one type of sperm is produced. Hence the fertilized eggs can only be female (Blackman, 1987).

Aphids reproduce very rapidly, combining the advantages of parthenogenesis with viviparity and pedogenesis (section 14.6), so that successive generations are telescoped. In the tropics, where conditions are continuously favorable, parthenogenesis may continue indefinitely without the intervention of a sexual generation.

An alternation of generations may also occur in Cecidomyiidae.

Review: Sanderson, 1988 – Cynipidae

14.6 PEDOGENESIS

Sometimes immature insects mature precociously and are able to reproduce. This phenomenon is known as pedogenesis. Most insects reproducing pedogenetically are also parthenogenetic and viviparous. Development of the offspring produced pedogenetically usually begins in the larval insect.

The larvae of *Miastor* (Diptera, Cecidomyidae) and *Micromalthus* (Coleoptera) give birth to other larvae or, occasionally, lay eggs. Pedogenesis occurs in *Miastor* only under very good or poor nutritional conditions. Under intermediate nutritional conditions normal adults are produced. Young larvae are set free in the body cavity of the pedogenetic larva and they feed on the maternal tissues, eventually escaping through the body wall of the parent.

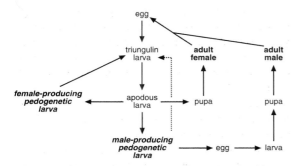

Fig. 14.33. Pedogenesis. The life history of *Micromalthus*. Bold type: reproductive forms; italics: pedogenetically reproducing larvae (based on Pringle, 1938).

Micromalthus has five reproductive forms: adult males, adult females, male-producing larvae, female-producing larvae, and larvae producing males and females. The species has a complex heteromorphosis (Fig. 14.33). The form emerging from the egg is a triungulin which molts to an apodous larva that can develop in one of three ways. It can develop through a pupa to a normal adult female, it can molt to a larval form which gives rise pedogenetically to a male, or it can molt to a pedogenetic larva which produces triungulins. Male-producing larvae lay a single egg

containing a young embryo, but the egg adheres to the parent and when the larva hatches it eats the parent larva. If, for some reason the parent larva is not eaten, it subsequently produces a small brood of female larvae.

In the cecidomyids *Tekomyia* and *Henria*, pupal forms give birth to larvae. The larvae of these insects are of two types: one produces a pupa and, ultimately, a normal adult; the other forms a hemipupa, a rounded structure with, in *Henria*, vestiges of wings and legs. A brood of larvae, commonly between 30 and 60, escapes from the hemipupa by rupturing the cuticle. Pedogenesis is the normal method of reproduction in these insects and although normal adults may be produced it is not certain that they are capable of producing viable offspring (Wyatt, 1963).

Pedogenesis also occurs in aphids. Although the young are not born until the parent aphid has reached the adult stage, their development may begin before she is born while she is still in the ducts of the grandparental generation. Development of the offspring continues throughout the larval life of the parent.

The hemipteran, *Hesperoctenes*, is an example of a pedogenetic form in which fertilization occurs. Some last stage larvae are found with sperm in the hemocoel as a result of hemocoelic insemination. These sperm fertilize the eggs which develop in the ovaries of the larva.

REFERENCES

Anderson, D.S. (1960). The respiratory system of the egg-shell of *Calliphora erythrocephala*. *Journal of Insect Physiology*, **5**, 120–8.

Anderson, D.T. (1962). The embryology of *Dacus tryoni* (Frogg.)[Diptera, Trypetidae (=Tephritidae)], the Queensland fruit fly. *Journal of Embryology and Experimental Morphology*, **10**, 248–92.

Anderson, D.T. (1972a). The development of hemimetabolous insects. In *Developmental Systems: Insects*, vol. 1, ed. S.J. Counce & C.H. Waddington, pp. 95–163. New York: Academic Press.

Anderson, D.T. (1972b). The development of holometabolous insects. In *Developmental Systems: Insects*, vol. 1, ed. S.J. Counce & C.H. Waddington, pp. 165–242. New York: Academic Press.

Anderson, K.V. & Lengyel, J.A. (1979). Rates of synthesis of major classes of RNA in *Drosophila* embryos. *Developmental Biology*, **70**, 217–31.

Anderson, K.V. & Lengyel, J.A. (1980). Changing rates of histone synthesis and turnover in *Drosophila* embryos. *Cell*, **21**, 717–27.

Bate, M. (1993). The mesoderm and its derivatives. In *The Development of Drosophila melanogaster*, ed. M. Bate & A.M. Arias, pp. 1013–90. New York: Cold Spring Harbor Laboratory Press.

Bate, M. & Arias, A.M. (eds.) (1993). *The Development of Drosophila melanogaster*, vol. 1. New York: Cold Spring Harbor Laboratory Press.

Beament, J.W.S. (1946). The waterproofing process in eggs of *Rhodnius prolixus* Stähl. *Proceedings of the Royal Society of London* B, **133**, 407–18.

Biemont, J.C., Chauvin, G. & Hamon, C. (1981). Ultrastructure and resistance to water loss in eggs of *Acanthoscelides obtectus* Say (Coleoptera: Bruchidae). *Journal of Insect Physiology*, **10**, 667–79.

Blackman, R.L. (1987). Reproduction, cytogenetics and development. In *Aphids. Their Biology, Natural Enemies and Control*, vol. 2 A, ed. A.K. Minks & P. Harrewijn, pp. 163–195. Amsterdam: Elsevier.

Byrne, D.N., Cohen, A.C. & Draeger, E.H. (1990). Water uptake from plant tissue by the egg pedicel of the greenhouse whitefly, *Trialeurodes vaporarium* (Westwood) (Homoptera: Aleyrodidae). *Canadian Journal of Zoology*, 68, 1193–5.

Chapman, R.F. (1970). Growth of the embryo of *Ornithacris turbida* (Walker) (Orthoptera: Acrididae) with special reference to the antennae. *Proceedings of the Royal Entomological Society of London* A, 45, 29–38.

Chauvin, J.T. & Chauvin, G. (1980). Formation des reliefs externes de l'oeuf de *Micropteryx calthella* L. (Lepidoptera: Micropterygidae). *Canadian Journal of Zoology*, 58, 761–6.

Clausen, C.P. (1940). *Entomophagous Insects*. New York: McGraw-Hill.

Cobben, R.H. (1968). *Evolutionary Trends in Heteroptera*. Wageningen: Centrum voor Landbouwpublikaties en Landbouwdocumentatie.

Copenhaver, P.F. & Taghert, P.H. (1991). Origins of the insect enteric nervous system: differentiation of the enteric ganglia from a neurogenic epithelium. *Development*, 113, 1115–32.

Couchman, J.R. & King, P.E. (1980). Ovariole sheath structure and its relationship with developing embryos on a parthenogenetic viviparous aphid. *Acta Zoologica (Stockholm)*, 61, 147–55.

Crespi, B.J. (1989). Facultative viviparity in a thrips. *Nature*, 337, 357–8.

Cruz, Y.P. (1986). Development of the polyembryonic parasite *Copidosomopsis tanytmemus* (Hymenoptera: Encyrtidae). *Annals of the Entomological Society of America*, 79, 121–7.

Daniel, S.H. & Smith, R.H. (1994). Functional anatomy of the egg pore in *Callosobruchus maculatus*: a trade-off between gas-exchange and protective functions? *Physiological Entomology*, 19, 30–8.

Denlinger, D.L. & Ma, W.-C. (1974). Dynamics of the pregnancy cycle in the tsetse *Glossina morsitans*. *Journal of Insect Physiology*, 20, 1015–26.

Doe, C.Q. & Goodman, C.S. (1985). Early events in insect neurogenesis I. Development and segmental differences in the pattern of neuronal precursor cells. *Developmental Biology*, 111, 193–205.

Doe, C.Q. & Technau, G.M. (1993). Identification and cell lineage of individual neural precursors in the *Drosophila* CNS. *Trends in Neuroscience*, 16, 510–4.

Foe, V.E., Odell, G.M. & Edgar, B.A. (1993). Mitosis and morphogenesis in the *Drosophila* embryo: point and counterpoint. In *The Development of* Drosophila melanogaster, ed. M. Bate & A.M. Arias, pp. 149–300. New York: Cold Spring Harbor Laboratory Press.

Giorgi, F. & Nordin, J.H. (1994). Structure of yolk granules in oocytes and eggs of *Blattella germanica* and their interaction with vitellophages and endosymbiotic bacteria during granule degradation. *Journal of Insect Physiology*, 40, 1077–92.

Goodman, C.S. & Doe, C.Q. (1993). Embryonic development of the *Drosophila* central nervous system. In *The Development of* Drosophila melanogaster, ed. M. Bate & A.M. Arias, pp. 1131–1206. New York: Cold Spring Harbor Laboratory Press.

Hadley, N.F. (1994). *Water Relations of Terrestrial Arthropods*. San Diego: Academic Press.

Hagan, H.R. (1951). *Embryology of the Viviparous Insects*. New York: Ronald Press.

Haget, A. (1977). L'embryologie des insectes. In *Traité de Zoologie*, vol 8, part 5B, ed. P.-P.Grassé, pp. 1–387. Paris: Masson et Cie.

Hagstrum, D.W. & Milliken, G.A. (1991). Modeling differences in insect developmental times between constant and fluctuating temperatures. *Annals of the Entomological Society of America*, 84, 369–79.

Hartley, J.C. (1962). The egg of *Tetrix* (Tetrigidae, Orthoptera), with a discussion on the probable significance of the anterior horn. *Quarterly Journal of Microscopical Science*, 103, 253–9.

Henson, H. (1946). The theoretical aspect of insect metamorphosis. *Biological Reviews of the Cambridge Philosophical Society*, 21, 1–14.

Hinton, H.E. (1960). Plastron respiration in the eggs of blowflies. *Journal of Insect Physiology*, 4, 176–83.

Hinton, H.E. (1961). The structure and formation of the egg-shell in the Nepidae (Hemiptera). *Journal of Insect Physiology*, 7, 224–57.

Hinton, H.E. (1969). Respiratory systems of insect egg shells. *Annual Review of Entomology*, 14, 343–68.

Hinton, H.E. (1981). *Biology of Insect Eggs*. Oxford: Pergamon Press.

Hoffmann, J.A. & Lagueux, M. (1985). Endocrine aspects of embryonic development in insects. In *Comprehensive Insect Physiology, Biochemistry and Pharmacology*, vol. 1, ed. G.A. Kerkut & L.I. Gilbert, pp. 435–460. Oxford: Pergamon Press.

Hunter, M.S., Nur, U. & Werren, J.H. (1993). Origin of males by genome loss in an autoparasitoid wasp. *Heredity*, 70, 162–71.

Imms, A.D. (1957). *A General Textbook of Entomology*. 9th edition revised by O.W. Richards & R.G. Davies. London: Methuen.

Ivanova-Kasas, O.M. (1972). Polyembryony in insects. In *Developmental Systems: Insects*, vol. 2, ed. S.J. Counce & C.H. Waddington, pp. 243–271. London: Academic Press.

Johannsen, O.A. & Butt, F.H. (1941). *Embryology of Insect and Myriapods*. New York: McGraw-Hill.

Jura, C. (1972). Development of apterygote insects. In *Developmental Systems: Insects*, vol. 2, ed. S.J. Counce & C.H. Waddington, pp. 49–94. London: Academic Press.

Lagueux, M., Hetru, C., Goltzene, F., Kappler, C. & Hoffmann, J.A. (1979). Ecdysone titre and metabolism in relation to cuticulogenesis in embryos of *Locusta migratoria*. *Journal of Insect Physiology*, **25**, 709–23.

Lawrence, P.A. (1992). *The Making of a Fly: the Genetics of Animal Design*. Oxford: Blackwell.

Leahy, T.C. & Andow, D.A. (1994). Egg weight, fecundity, and longevity are increased by adult feeding in *Ostrinia nubilalis* (Lepidoptera: Pyralidae). *Annals of the Entomological Society of America*, **87**, 342–9.

Lees, A.D. (1976). The role of pressure in controlling the entry of water into the developing eggs of the Australian plague locust *Chorthoicetes terminifera* (Walker). *Physiological Entomology*, **1**, 39–50.

Leitch, B., Laurent, G. & Shepherd, D. (1992). Embryonic development of synapses on spiking local interneurones in locust. *Journal of Comparative Neurology*, **324**, 213–36.

Manning, G. & Krasnow, M.A. (1993). Development of the *Drosophila* tracheal system. In *The Development of* Drosophila melanogaster, ed. M. Bate & A.M. Arias, pp. 609–85. New York: Cold Spring Harbor Laboratory Press.

Margaritis, L.H. (1985). Structure and physiology of the eggshell. In *Comprehensive Insect Physiology, Biochemistry and Pharmacology*, vol. 1, ed. G.A. Kerkut & L.I. Gilbert, pp. 153–230. Oxford: Pergamon Press.

Nagy, L.M. (1995). A summary of lepidopteran embryogenesis and experimental embryology. In *Molecular Model Systems in the Lepidoptera*, ed. M.R. Goldsmith & A.S. Wilkins, pp. 136–64. Cambridge: Cambridge University Press.

Patel, N.H. (1994). The evolution of arthropod segmentation: insights from comparisons of gene expression patterns. *Development*, 1994 supplement, 201–7.

Pringle, J.A. (1938). A contribution to the knowledge of *Micromalthus debilis* LeC. (Coleoptera). *Transactions of the Royal Entomological Society of London*, **87**, 271–86.

Reavey, D. (1992). Egg size, first instar behaviour and the ecology of Lepidoptera. *Journal of Zoology, London*, **227**, 277–97.

Richards, A.G. (1957). Cumulative effects of optimum and suboptimum temperatures on insect development. In *Influence of Temperature on Biological Systems*, ed. F.H.Johnson, pp. 145–62. Washington: American Physiological Society.

Roonwal, M.L. (1954). The egg-wall of the African migratory locust, *Locusta migratoria migratorioides* Reiche and Frm. (Orthoptera, Acrididae). *Proceedings of the National Institute of Sciences, India*, **20**, 361–70.

Salzen, E.A. (1960). The growth of the locust embryo. *Journal of Embryology and Experimental Morphology*, **8**, 139–62.

Sander, K., Gutzeit, H.O. & Jäckle, H. (1985). Insect embryogenesis: morphology, physiology, genetical and molecular aspects. In *Comprehensive Insect Physiology, Biochemistry and Pharmacology*, vol. 1, ed. G.A. Kerkut & L.I. Gilbert, pp. 319–85. Oxford: Pergamon Press.

Sanderson, A.R. (1988). Cytological investigations of parthenogenesis in gall wasps (Cynipidae, Hymenoptera). *Genetica*, **77**, 189–216.

Skaer, H. (1992). Cell proliferation and rearrangement in the development of the Malpighian tubules of the hemipteran, *Rhodnius prolixus*. *Developmental Biology*, **150**, 372–80.

Skaer, H. (1993). The alimentary canal. In *The Development of* Drosophila melanogaster, ed. M. Bate & A.M. Arias, pp. 941–1012. New York: Cold Spring Harbor Laboratory Press.

Slifer, E.H. & Sekhon, S.S. (1963). The fine structure of the membranes which cover the egg of the grasshopper, *Melanoplus differentialis*, with special reference to the hydropyle. *Quarterly Journal of Microscopical Science*, **104**, 321–34.

Soumalainen, E. (1962). Significance of parthenogenesis in the evolution of insects. *Annual Review of Entomology*, **7**, 349–66.

Southwood, T.R.E. (1956). The structure of the eggs of the terrestrial Heteroptera and its relationship to the classification of the group. *Transactions of the Royal Entomological Society of London*, **108**, 163–221.

St Johnston, D. (1993). Pole plasm and the posterior group genes. In *The Development of* Drosophila melanogaster, ed. M. Bate & A.M. Arias, pp. 325–63. New York: Cold Spring Harbor Laboratory Press.

St Johnston, D. & Nüsslein-Volhard, C. (1992). The origin of pattern and polarity in the Drosophila embryo. *Cell*, **68**, 201–19.

Stouthamer, R. & Kamer, D. (1994). Cytogenetics of microbe-associated parthenogenesis and its consequence for gene flow in *Trichogramma* wasps. *Heredity*, **73**, 317–27.

Strand, M.R., Goodman, W.G. & Baehrecke, E.H. (1991). The juvenile hormone titer of *Trichoplusia ni* and its potential role in embryogenesis of the polyembryonic wasp *Copidosoma floridanum*. *Insect Biochemistry*, **21**, 205–14.

Tanaka, S. (1986). Uptake and loss of water in diapause and non- diapause eggs of crickets. *Physiological Entomology*, **11**, 343–51.

Thompson, K.G. & Siegler, M.V.S. (1993). Development of segment specificity in identified lineages of the grasshopper CNS. *Journal of Neuroscience*, **13**, 3309–18.

Tobe, S.S. & Langley, P.A. (1978). Reproductive physiology of *Glossina*. *Annual Review of Entomology*, **23**, 283–307.

Wyatt, I.J. (1963). Pupal paedogenesis in the Cecidomyiidae (Diptera) – II. *Proceedings of the Royal Entomological Society of London* A, **38**, 136–44.

Yoder, J.A. & Denlinger, D.L. (1992). Water vapour uptake by diapausing eggs of a tropical walking stick. *Physiological Entomology*, **17**, 97–103.

Zrzavý, J. & Stys, P. (1995). Evolution and metamerism in Arthropoda: developmental and morphological perspectives. *Quarterly Review of Biology*, **70**, 279–95.

15   Postembryonic development

15.1 HATCHING

15.1.1 Mechanisms of hatching

Most insects force their way out of the egg by exerting pressure against the inside of the shell. The insect increases its volume by swallowing the extra-embryonic fluid and in some cases by swallowing air which diffuses through the shell. Then, waves of muscular contraction pump hemolymph forwards so that the head and thoracic regions are pressed tightly against the inside of the shell. In grasshoppers, and perhaps in other insects, these muscular waves are interrupted periodically by a simultaneous contraction of the abdominal segments which causes a sudden increase in pressure in the anterior region. The dorsal membrane of the neck in grasshoppers has a pair of lobes, the cervical ampullae, which are inflated by the increase in hemolymph pressure (Fig. 15.1a). They serve to focus the pressure on a limited area of the shell. If the shell does not split, the ampullae are withdrawn and a further series of posterior to anterior waves of contraction follows ending with another sudden abdominal contraction. One of these sudden contractions ultimately ruptures the shell.

The position of the rupture generally depends on where the insect puts pressure on the chorion. In grasshoppers, the chorion is split transversely above the ampullae; in the water beetle, *Agabus*, the split is longitudinal, while in some species it is variable in position. The chorion of some species has a line of weakness along which it splits (section 14.1.1.3). The egg of *Calliphora*, for example, has a pair of longitudinal hatching lines running along its length (see Fig. 14.3), and in Heteroptera a hatching line runs round the egg where the cap joins the body of the chorion (see Fig. 13.13). In eggs of *Aedes*, there is a line of weakness in the serosal cuticle and a split in the chorion follows this passively, perhaps because the serosal cuticle and chorion are closely bound.

In species with a thick serosal cuticle, such as Acrididae and Heteroptera, hatching is aided by an enzyme that digests the serosal endocuticle before hatching begins. The enzyme is produced by the pleuropodia, and, because they are not covered by the embryonic cuticle, it is secreted directly into the extra-embryonic space (section 14.2.10).

Cuticular structures known as egg bursters aid hatching in a number of insects. These are usually on the head of the embryonic cuticle of Odonata, some Orthoptera, Heteroptera, Neuroptera and Trichoptera, but are on the cuticle of the first stage larva in Nematocera, Carabidae and Siphonaptera. Their form varies. Pentatomids have a T- or Y-shaped central tooth, while Cimicomorpha have a row of spines running along each side of the face from near the eye to the labrum (Fig. 15.1c). Some sucking lice (Anoplura) have a pair of spines, and several pairs of lancet-shaped blades arising from the embryonic cuticle over the head. Often, as in the fleas, mosquitoes and *Glossina*, a cuticular tooth is in a membranous depression which can be erected by blood pressure (Fig. 15.1b). In *Agabus*, the egg-burster is a spine on either side of the head, but many Polyphaga have egg-bursters on the thoracic or abdominal segments of the first stage larva (van Emden, 1946). For example, *Meligethes* larvae have a tooth on each side of the mesonotum and metanotum, while larval tenebrionids have a small tooth on either side of the tergum of these segments and the first eight abdominal segments (Fig. 15.1d).

It is not clear how these various devices function and Jackson (1958) believes that, in *Agabus*, the egg of which has a soft chorion, they are no longer functional. In other cases they appear to be pushed against the inside of the shell until it is pierced and then a slit is cut by appropriate movements of the head. The larva of *Dacus* (Diptera) uses its mouth hooks in a similar way, repeatedly protruding them until they tear the chorion. The blades in lice and the spines in *Cimex* are used to tear the vitelline envelope, the chorion then being broken by pressure (Sikes and Wigglesworth, 1931).

Larval Lepidoptera gnaw their way through the chorion and, after hatching, continue to eat the shell until only the base is left. In *Pieris brassicae*, where the eggs are laid in a cluster, newly hatched larvae may also eat the tops off adjacent unhatched eggs.

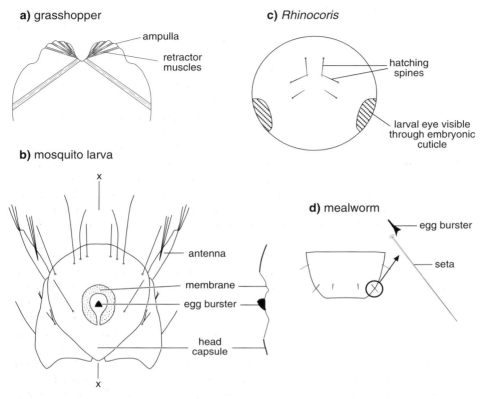

Fig. 15.1. Hatching devices. **(a)** Cross-section through the neck membrane of a grasshopper showing the ampullae (after Bernays, 1972a,b). **(b)** The head of a first instar larva of a mosquito. The position of the egg-burster is shown from above and in a diagrammatic vertical section along the line XX (after Marshall, 1938). **(c)** Embryonic cuticle over the head of a sucking bug showing the hatching spines. The cuticle is shed as soon as the larva escapes from the egg (*Rhinocoris*, Heteroptera) (after Southwood, 1946). **(d)** Position of the egg-burster spine at the base of a hair (seta) on the eighth abdominal segment of the first instar mealworm (*Tenebrio*, Coleoptera) (after van Emden, 1946).

When the egg is enclosed in an ootheca, the larva escapes from this after leaving the egg. In the Blattodea, the ootheca is split open before hatching by the swelling of the eggs as they absorb water. The first stage larvae of grasshoppers, still enclosed in the embryonic cuticle, wriggle through the frothy plug secreted above the eggs (see Fig. 13.21b). The head is thrust forwards through the froth by elongating the abdomen, the tip of which is pressed against the substrate to give a point of support, while the cervical ampullae are withdrawn. Then, when elongation is complete, the cervical ampullae are expanded to give a purchase while the abdomen is drawn up. By repeating this cycle the larva tunnels its way to the soil surface following the line of least resistance offered by the soft plug.

Special muscles which assist the hatching process are known to occur in some insects. They break down soon after hatching (section 10.2.4).

15.1.2 Intermediate molt

In those insects which possess an embryonic cuticle this separates from the underlying epidermis some time before hatching, but it is not shed, so that the insect hatches as a pharate first stage larva (section 15.2). The embryonic cuticle is shed during or immediately after hatching in a process commonly known as the intermediate molt. As the larva of *Cimex* or a louse emerges from the egg, it swallows air and, by further pumping, splits the embryonic cuticle over the head. The cuticle is shed as the larva continues to hatch because, in Heteroptera, it is attached inside the chorion at two or three places (Sikes and Wigglesworth, 1931).

The intermediate molt in acridids begins as the newly hatched larva reaches the surface of the soil. Ecdysis behavior is triggered by the lack of all-round contact with the substrate which the larva has experienced up to that time. The larva swallows air, increasing its volume by about 25%, and, as it does so, waves of shortening and lengthening pass along the body of the larva causing it to move forwards within the intact embryonic cuticle. It maintains its forward position inside the embryonic cuticle by backwardly directed spines on the abdominal sternites. As a result of these movements the embryonic cuticle is pulled taut over the head and thorax, eventually splitting mid-dorsally. The first stage larval cuticle expands so that the insect swells out of the embryonic cuticle, which is worked backwards. It is held at the tip of the abdomen by two spiny knobs, called brustia, at the bases of the cerci so that the hind legs can be withdrawn and then it is finally kicked off by the hind legs.

15.1.3 Hatching stimuli

The stimuli that promote hatching are largely unknown and, in many cases, insects appear to hatch whenever they reach the appropriate stage of development. Even in these instances, however, it is possible that some external stimulus influences hatching.

Suitable temperatures are necessary for any insect to hatch and there is a threshold temperature below which hatching does not occur. This temperature varies from species to species, but is about 8 °C for *Cimex*, 13 °C for *Oncopeltus* and 20 °C for *Schistocerca*. It is independent of the threshold temperature for full embryonic development, which may be either higher, as in *Cimex* (13 °C), or lower, as in *Schistocerca* (about 15 °C). The failure of fully formed larvae to hatch at low temperatures is related to inactivity. Newly hatched *Schistocerca* larvae, for instance, are not normally active below about 17 °C and their activity remains sluggish below 24 °C, and *Cimex* is not normally active below 11 °C. Further, temperatures must be sufficiently high for the enzyme digesting the serosal cuticle to function efficiently (see above).

If temperature is suitable and the larva is fully formed, hatching may be induced by specific environmental stimuli. For example, larvae of the stem-boring moth, *Chilo partellus*, hatch just after dawn in response to the high light intensity (Fig. 15.2a). Amongst aquatic insects, *Aedes* eggs hatch when immersed in deoxygenated water, the lower the oxygen tension the greater the percentage hatching, but responsiveness varies with age. The larvae are most sensitive soon after development is complete and will then hatch even in aerated water, but if they are not wetted for some time they will only hatch at very low oxygen tensions. Low oxygen tension is perceived by a sensory center in the head or thorax and maximum sensitivity coincides with a period of maximum activity of the central nervous system as indicated by the concentration of acetylcholine. Low oxygen tension has completely the opposite effect on the hatching of *Agabus* (Coleoptera) larvae, which only occurs in oxygenated water. Some species of *Lestes* (Odonata) lay their eggs above the water. Fully formed larvae hatch when they are wetted.

In other species, hatching at a particular time of day results from the entrainment of activity to the cycle of changing light or temperature over the previous days. Unhatched larvae of the pink bollworm moth, *Pectinophora gossypiella*, for example, are entrained to the light cycle in the days before hatching so that they hatch at dawn (Saunders, 1982). They hatch at the appropriate time of day even if they are in continuous light for the last day of development. Similarly, the hatching of larval *Schistocerca gregaria* at dawn is not an immediate response to the changing temperature of the surrounding soil, but results from an entrainment of activity to the 24-hour cycle of temperature changes during the last days of embryonic development (Fig. 15.2b).

15.2 LARVAL DEVELOPMENT

Post-embryonic development is divided into a series of stages, each separated from the next by a molt. The form that the insect assumes between molts is known as an instar, that which follows hatching or the intermediate molt being the first instar, which later molts to the second instar, and so on until at a final molt the adult or imago emerges. No further molts occur once the insect is adult except in Apterygota and non-insect hexapod groups. The periods between molts are called stages, or stadia – for example, first larval stage, second larval stage, and so on – although the term instar is commonly used to refer to this period as well as to the form of the insect during the period.

During larval development there is usually no marked change in body form, each successive instar being similar

Fig. 15.2. Hatching stimuli. **(a)** Direct response to light by unhatched caterpillars of the moth, *Chilo partellus*. Light (white) and dark (black) periods are shown by the bar above each graph. In continuous light (top left) the insects hatched over an extended period with a peak after about 100 hours of egg incubation. 'Lights on' after about 94 hours of incubation stimulated hatching (bottom left), darkness after this delayed it (right-hand graphs). In all cases, eggs were maintained at a constant temperature of 30 °C with 12 h light : 12 h dark cycles up to the time of the experiment (based on Chapman *et al.*, 1983). **(b)** Entrainment to a temperature cycle leading to hatching by larvae of the desert locust, *Schistocerca gregaria*. Insects in the upper panel experienced cycling temperatures throughout life. Insects in the lower panel experienced cycling temperatures up to the beginning of the experiment (time 0 in figure), but then were maintained at 33 °C. They hatched at the same time as insects experiencing cycling temperatures despite the fact that they were experiencing a constant temperature. The eggs were in constant darkness throughout the experiment (after Padgham, 1981)

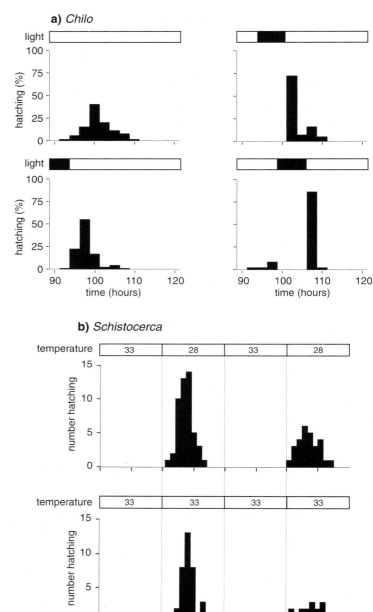

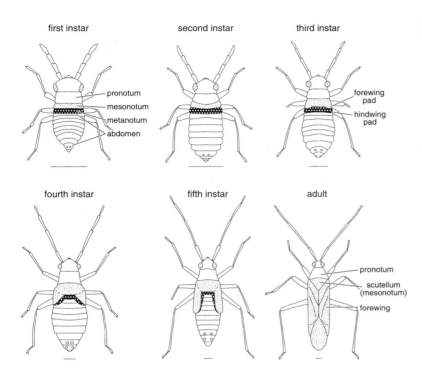

first instar

pronotum
mesonotum
metanotum
abdomen

second instar

third instar

forewing pad
hindwing pad

fourth instar

fifth instar

adult

pronotum
scutellum (mesonotum)
forewing

Fig. 15.3. Successive stages in the postembryonic development of a hemimetabolous insect. The horizontal line under each stage represents 0.5 mm (*Cyllecoris*, Heteroptera) (after Southwood & Leston, 1959).

to the one preceding it, but the degree of change from last instar larva to adult varies considerably and may be very marked. This change is called metamorphosis (Snodgrass, 1954; Wigglesworth, 1965). In morphological terms, Snodgrass (1954) relates metamorphosis to the loss of adaptive features peculiar to the larva, and the extent of change occurring as a reflection of the degree of ecological separation of the larva from the adult. It can also be defined in physiological terms as the change which accompanies a molt in the absence of juvenile hormone. The term metamorphosis is sometimes applied to all the changes occurring in the life history, from egg to adult (see e.g. Imms, 1957), but it is better not to use it in this wide sense.

The insects can be grouped in three categories, ametabolous, hemimetabolous or holometabolous, according to the extent of the change at metamorphosis. Ametabolous insects have no metamorphosis, the adult form resulting from a progressive increase in size of the larval form. This is characteristic of the Apterygota and hexapods other than insects in which the larva hatches in a form essentially like the adult apart from its small size and lack of development of genitalia. At each molt the larva

grows bigger and the genitalia develop progressively. Adults and larvae live in the same habitat.

In hemimetabolous insects, the larva hatches in a form which generally resembles the adult except for its small size and lack of wings and genitalia (Fig. 15.3), but, in addition, usually with some other features which are characteristic of the larva but which are not present in the adult. At the final molt these features are lost. The Orthoptera, Isoptera, Hemiptera are commonly regarded as hemimetabolous. Snodgrass (1954) calls these groups ametabolous or paurometabolous (that is, with a very slight metamorphosis), but a quantitative analysis of growth changes shows a gradual transformation through the larval instars and a sharp discontinuity at the molt from larva to adult. This discontinuity applies not only to typical adult features such as the wings and genitalia, but to other features that are not regarded as typically adult. In *Rhodnius*, the larval cuticle with its stellate folds and abundant plaques bearing sensilla is replaced by the adult cuticle, which has transverse folds, a few sensilla and no plaques (section 15.3.1).

The Plecoptera, Ephemeroptera and Odonata have aquatic larvae and their specific larval adaptations are

Fig. 15.4. Late larval instar and adult of an aquatic hemimetabolous insect showing conspicuous adaptive features in the larva. These features are indicated in heavy type (*Ephemera*, Ephemeroptera) (after Macan, 1961; Kimmins, 1950).

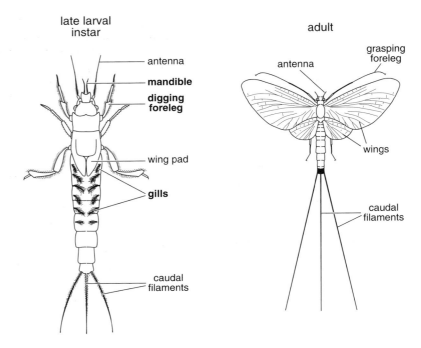

much more marked than in the previous groups. Hence these forms undergo a more conspicuous metamorphosis involving, among other things, loss of the gills (Fig. 15.4). The general body form, nevertheless, resembles that of the adult and these insects are also regarded as hemimetabolous.

In holometabolous insects, the larvae are usually quite unlike the adult and a pupal stage is present between the last larval stage and the adult (Fig. 15.5). The pupa is characteristic of holometabolous development, which occurs in all the Neuroptera, Trichoptera, Lepidoptera, Coleoptera, Hymenoptera, Diptera and Siphonaptera.

Amongst the insect groups that typically have a hemimetabolous development, a few have life histories somewhat analogous to those of holometabolous insects. Hemiptera have a typical hemimetabolous development, but in Thysanoptera, which are phylogenetically close to Hemiptera, the last two larval stages do not feed, and the final stage, which is often called a pupa, is sometimes enclosed in a cocoon. Both these stages have external wing pads, but the earlier larval stages have none; instead, some development of the wings occurs internally. Within the Hemiptera, the last two larval stages of male scale insects (Coccoidea) also do not feed. These insects clearly have a life history that can be regarded as holometabolous even though they are phylogenetically far removed from the majority of holometabolous insects.

In whiteflies (Aleyrodidae), the final larval stage feeds for a short period but then has an extended non-feeding period during which the pharate adult develops. This stage is commonly known as a pupa (Byrne & Bellows, 1991) and some texts refer to whitefly development as holometabolous, although there is really no resemblance to the holometabolous development in other groups.

The term 'molt', as commonly used, includes two distinct processes: apolysis, the separation of the epidermis from the existing cuticle; and ecdysis, the casting of the old cuticle after the production of a new one. In the interval between apolysis and ecdysis the insect is said to be pharate. During this time the cuticle of one developmental stage conceals the presence of the next. In most hemimetabolous insects and holometabolous larval stages, the pharate period is relatively short (Fig. 15.6a), but it is often extended at the larval/pupal molt and pupal/adult molt (eclosion) (Fig. 15.6b) and sometimes may be very long. It is most important that this pharate stage is appreciated, especially in relation to the physiology and behavior of insects.

Some difficulties in terminology arise because of the pharate period. The terminology usually applied to

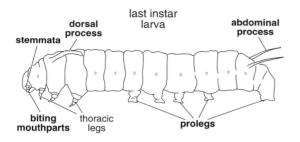

last instar larva

dorsal process

stemmata

abdominal process

biting mouthparts

thoracic legs

prolegs

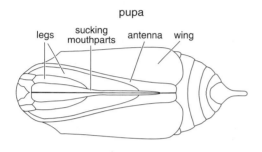

pupa

legs

sucking mouthparts

antenna

wing

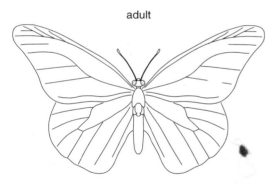

adult

Fig. 15.5. Successive stages in the postembryonic development of a holometabolous insect. Adaptive features of the larva are indicated in heavy type. Only one larval stage is shown (*Danaus*, Lepidoptera) (after Urqhart, 1960).

different stages, such as first stage larva, or pupa, refers to the outward appearance of the insect between ecdyses (Fig. 15.6). An alternative is to consider the stages as extending from apolysis to apolysis. The importance of this distinction becomes apparent when extreme examples are considered. For instance, in cyclorrhaphan Diptera the cuticle of the last larval stage forms the puparium and from this the adult emerges. There is a normal pupal stage, but it is always enclosed within the old larval cuticle (Figs. 15.6c, 15.16b). Some moths which diapause as adults do so

within the pupal cuticle, so that, although they outwardly appear to be pupae, they are, in fact, adults.

In practice, the pharate stage is not always readily recognizable and the more usual terminology, with each stage extending from ecdysis to ecdysis, has been retained in this book, but with reference to the pharate condition where it is obviously important.

Review: Sehnal, 1985

15.2.1 Types of larvae

Larvae of hemimetabolous insects essentially resemble the adults and they are sometimes called nymphs to distinguish them from the more radically different larvae of holometabolous insects. The most conspicuous difference between hemimetabolous and holometabolous larvae is in the development of the wings. In the former, the wings develop as external buds which become larger at each molt, finally enlarging to form the adult wings (Fig. 15.3). In the latter, however, the wings (and often other appendages) develop in invaginations of the epidermis beneath the larval cuticle and so are not visible externally (see Figs. 15.20, 15.21). The invaginations are finally everted so that the wings become visible externally when the larva molts to a pupa (Fig. 15.5). In this text, The term larva is used for the immature stages of both hemimetabolous and holometabolous insects to emphasize the basic similarities in physiological regulation of the different forms (see section 15.4).

Many different larval forms occur amongst the holometabolous insects. The least modified with respect to the adult is the oligopod larva which is hexapodous with a well-developed head capsule and mouthparts similar to the adult, but no compound eyes. Two forms of oligopod larvae are commonly recognized: a campodeiform larva, that is well sclerotized, dorso-ventrally flattened and is usually a long-legged predator with a prognathous head (Fig. 15.7a); and a scarabaeiform larva, that is fat with a poorly sclerotized thorax and abdomen, and which is usually short-legged and inactive, burrowing in wood or soil (Fig. 15.7b). Campodeiform larvae occur in the Neuroptera, Trichoptera, Strepsiptera and some Coleoptera, while scarabaeiform larvae are found in the Scarabaeoidea and some other Coleoptera.

A second basic form is the polypod larva which has abdominal prolegs in addition to the thoracic legs. It is generally poorly sclerotized and is a relatively inactive form living in close contact with its food (Fig. 15.7c). The

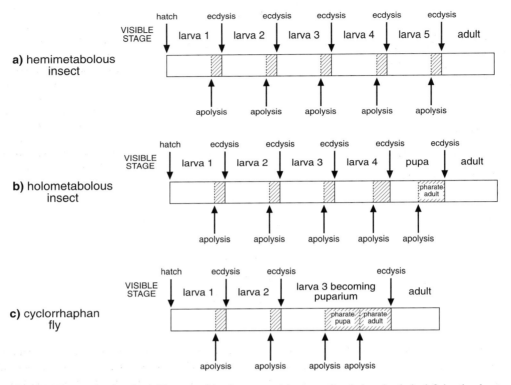

Fig. 15.6. Diagram showing the visible stages of development and the times of apolysis and ecdysis, defining the pharate periods (hatched), in: **(a)** a hemimetabolous insect; **(b)** a holometabolous insect; **(c)** a cyclorrhaphan fly in which the pupa is always concealed within the puparium.

larvae of Lepidoptera, Mecoptera and Tenthredinidae are of the polypod type.

The third basic form is the apodous larva, which has no legs and is very poorly sclerotized. Several different forms can be recognized according to the degree of sclerotization of the head capsule:

1 Eucephalous – with a well-sclerotized head capsule (Fig. 15.7d). Found in Nematocera, Buprestidae, Cerambycidae and Aculeata.
2 Hemicephalous – with a reduced head capsule which can be retracted within the thorax (Fig. 15.7e). Found in Tipulidae and orthorrhaphous Brachycera.
3 Acephalous – without a head capsule (Fig. 15.7f). Characteristic of Cyclorrhapha.

15.2.2 Heteromorphosis
In most insects, development proceeds through a series of essentially similar larval forms leading up to metamorpho-

sis, but sometimes successive larval instars have quite different forms. Development which includes such marked differences is termed heteromorphosis. [Hypermetamorphosis is sometimes used for this type of development, but this implies the use of the term metamorphosis to refer to change of form throughout the life history rather than restricting it to the larva/adult transformation.] Heteromorphosis is common in predaceous and parasitic insects in which a change in habit occurs during the course of larval development. Two types of heteromorphosis occur, one in which the eggs are laid in the open and the first stage larva searches for the host, and a second in which the eggs are laid in or on the host.

In the first type, the first stage larva is an active campodeiform larva (Fig. 15.8a). In Strepsiptera, this larva attaches itself to a host, often a bee or a sucking bug, when the latter visits a flower in which the larva is lurking. Subsequently, it becomes an internal parasite and loses all trace of legs, while developing a series of dorsal projections

a) campodeiform

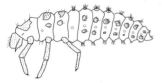

b) scarabaeiform

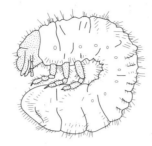

c) polypod

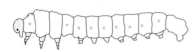

d) eucephalous

e) hemicephalous

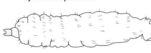

f) acephalous

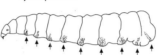

Fig. 15.7. Larval forms of holometabolous insects. Sclerotized parts stippled: **(a)** oligopodous, campodeiform (*Hippodamia*, Coleoptera); **(b)** oligopodous, scarabaeiform (*Popillia*, Coleoptera); **(c)** polypodous (*Neodiprion*, Hymenoptera); **(d)** apodous, eucephalous (*Vespula*, Hymenoptera); **(e)** apodous, hemicephalous (*Tanyptera*, Diptera); **(f)** apodous, acephalous. Arrows point to creeping welts (*Musca*, Diptera).

which increase its absorptive area. Later, in the sixth and seventh stages it develops a cephalothorax. A basically similar life history with an active first stage larva followed by inactive parasitic stages occurs in Mantispidae (Neuroptera), Meloidae and some Staphylinidae (Coleoptera), Acroceridae, Bombyliidae and Nemestrinidae (Diptera), Perilampidae and Eucharidae (Hymenoptera) and Epipyropidae (Lepidoptera). The first stage larvae of Meloidae and Strepsiptera are sometimes called triungulins because, in some species, they have three pretarsal claws (Clausen, 1940; Snodgrass, 1954).

The second type of heteromorphosis occurs in some endoparasitic Diptera and Hymenoptera. Amongst the parasitic Hymenoptera exhibiting this type of development, the first stage larva is known as a protopod larva. It has many different forms in different species and is often quite unlike a normal insect (see Fig. 15.8b,c; and Clausen, 1940). The first stage larva of the braconid, *Helorimorpha*, for instance, has a big head, a small unsegmented body and a tapering tail (Fig. 15.8b). The third stage larva, on the other hand, is a fairly typical eucephalous hymenopteran larva. In the Platygasteridae, the first stage larva is even more specialized with an anterior cephalothorax bearing rudimentary appendages, a

segmented abdomen and various tail appendages (Fig. 15.8c). These larvae hatch from eggs which contain very little yolk and some authorities regard them as embryos which hatch precociously (Chen, 1946), but others believe them to be specialized forms adapted to their environment (Snodgrass, 1954).

15.2.3 Numbers of instars

Primitive insects usually have more larval instars than advanced species. Thus Ephemeroptera molt as many as 40 times, and cockroaches often have 10 molts. In contrast, most hemipteroid insects and endopterygotes have five or fewer larval stages (Table 15.1). Even within a group of related insects there is variation. Amongst the grasshoppers, Acridoidea, the Pyrgomorphidae have five or more larval instars while the Gomphocerinae, which are also usually smaller, have four.

Despite these generalizations, the number of larval stages through which a species passes is not absolutely constant. In the Orthoptera, the female, which is often bigger than the male, commonly has an extra larval stage, and larvae hatching from small eggs grow slowly and have an additional stage. The red locust, *Nomadacris*, may have six, seven, or occasionally eight larval instars depending on its environment and that of the parents. In *Plusia* and some

Fig. 15.8. Heteromorphosis: **(a)** larval stages of *Corioxenos* (Strepsiptera) (after Clausen, 1940); **(b)** first and late larval stages of *Helorimorpha* (Hymenoptera) (after Snodgrass, 1954); **(c)** first instar larva of *Platygaster* (Hymenoptera).

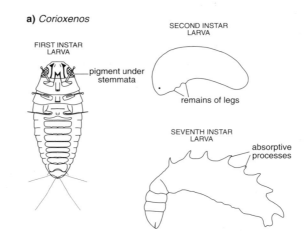

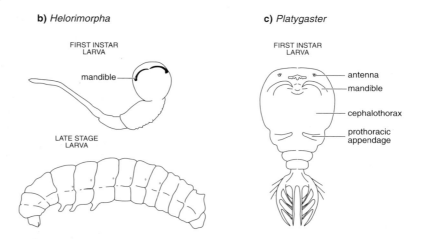

other Lepidoptera, larvae reared in isolation may pass through five, six or seven instars, while nearly all of those reared in a crowd have only five. At least amongst the Lepidoptera, much larger numbers of molts may occur if the insect has a poor food supply, and the larva of the clothes moth, *Tineola*, is recorded as molting as many as 40 times. Not all insects have this flexibility, and in most Hemiptera and cyclorrhaphous Diptera and many Hymenoptera and Coleoptera, the number of instars is constant.

15.2.4 Growth

15.2.4.1 *Weight*

There is a progressive increase in weight throughout the larval stages (Fig. 15.9). Typically, the weight increases steadily throughout a stage of development and then falls slightly at the time of molting due to the loss of the cuticle and some loss of water that is not replaced because the insect is not feeding. Following the molt, the weight rapidly increases above its previous level. Expressed in terms of the increase in absolute weight, the growth rate (increase in weight per unit time) is usually greater in the later stages, but the relative growth rate (the increase in weight relative to existing weight in a defined period of time – often expressed as $mg\,mg^{-1}\,day^{-1}$) normally decreases as the organism increases in size (Fig. 15.9a) (Slansky & Scriber, 1985). In some aquatic insects, there is no decrease in weight before a molt, but, conversely, there is a sharp increase due to the absorption of water, either

Table 15.1. *Numbers of larval stages in different orders of insects. The commonly occurring numbers are given. Numbers outside this range may occur under some conditions*

Order	Common name	Number of larval stages
Archaeognatha		10–14
Thysanura	Bristletails	9–14
Ephemeroptera	Mayflies	20–40
Blattodea	Cockroaches	6–10
Mantodea	Mantids	5–9
Grylloblattodea	Rock crawlers	8
Orthoptera	Crickets	5–11
Phasmida	Stick insects	8–12
Isoptera	Termites	5–11
Dermaptera	Earwigs	4–6
Embioptera	Webspinners	4–7
Plecoptera	Stoneflies	22–23
Hemiptera	Bugs	3–5
Thysanoptera	Thrips	5–6
Psocoptera	Book lice	6
Phthiraptera	Lice	3–4
Neuroptera	Lacewings	3–5
Mecoptera	Scorpion flies	4
Siphonaptera	Fleas	3
Diptera	Flies	3–6
Trichoptera	Caddis flies	5–7
Lepidoptera	Butterflies	5–6
Coleoptera	Beetles	3–5
Hymenoptera	Bees	3–6

through the cuticle or via the alimentary canal (Fig. 15.9c).

In blood-sucking insects, such as *Rhodnius*, which feeds only once during each larval stage, the pattern of growth is different. During the non-feeding period of each stage there is a slow, steady loss in weight due to water loss and respiration, but feeding is accompanied by a sharp increase in weight followed by a fairly rapid fall as water is eliminated. There is, of course, a net increase in weight from one instar to the next.

The final weight of the adult insect often varies according to the conditions under which the larva develops. Rapid development at high temperatures results in adults that are relatively light in weight. Crowding, possibly through its effect on the rate of development, also influences adult size, insects from crowded conditions being smaller than others reared in isolation. For example, in one experiment on *Locusta*, adult females from larvae reared in isolation weighed 1.5 g, while others from larvae reared in crowds weighed only 1.2 g. Where isolation is correlated with the production of an extra larval instar, the difference in weight may be even more marked. Finally, adult weight may be influenced by the food on which the larva is nourished. This is particularly well illustrated in phytophagous insects, such as the grasshopper *Melanoplus sanguinipes* in which the weight of females varies from 140 mg to 320 mg, depending on the food available to the larvae. However, insects in which the timing of metamorphosis is determined by size (see below) are much less variable in size as adults.

15.2.4.2 *Increase in size of the cuticle*
Fully sclerotized cuticle does not expand, so growth of sclerotized parts only occurs when an insect molts and a new, soft cuticle is produced and expanded. Consequently, hard parts increase in size in a series of steps (Fig. 15.9b). Membranous regions can expand, however, by the pulling out of folds. Thus a structure with a wholly membranous cuticle, or one, such as the abdomen of *Locusta*, in which the membranes are extensible, may grow continuously. Other regions in which there is rather less membrane show an intermediate type of growth with some extension occurring in the course of each stage together with a marked increase at each molt.

In many hemimetabolous and ametabolous insects, the number of annuli in the antennal flagellum increases during postembryonic life. Thus, the first instar larva of *Dociostaurus* (Orthoptera) has antennae with 13 annuli, while in the adult there are 25. In annulated antennae new annuli are added basally by division of the most proximal annulus of the flagellum known as the meriston. In orthopteroid insects and Odonata some of the annuli adjacent to the meriston, and derived from it, may also divide. They are known as meristal annuli. At each molt in *Periplaneta* (Blattodea) the meriston divides to produce 4–14 new annuli and each meristal annulus may divide once.

Where the cerci are segmented, as in Blattodea and Mantodea, they increase in size at successive molts by the addition of segments basally.

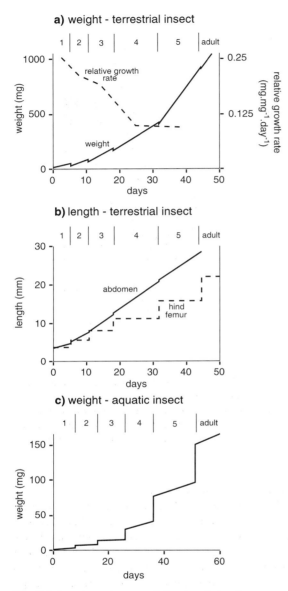

Fig. 15.9. Larval growth patterns in hemimetabolous insects. Numbers at the top indicate successive larval stages and vertical lines the times of molting. (**a**) Change in weight and relative growth rate of a terrestrial insect (*Locusta*, Orthoptera) (data from Duarte, 1939). (**b**) Changes in linear dimensions of a terrestrial insect. The abdomen increases almost linearly by unfolding of intersegmental membranes, but the hind femur only increases in length at each molt (*Locusta*, Orthoptera) (after Clarke, 1957). (**c**) Change in weight of an aquatic insect (*Notonecta*, Heteroptera) (from Wigglesworth, 1965).

It is often true that different parts of the body grow at different rates when compared with some standard such as total body length. If the rate of growth of the part, such as the head, is the same as that of the standard, growth is said to be isometric; if the rate is different from the standard, growth is said to be allometric. If the part in question grows relatively faster than the standard, it exhibits positive allometry, while slower growth is negative allometry. For example, in the parasitic dermapteran, *Hemimerus*, the meriston with the meristal annuli grows faster than the antenna as a whole, so that in the adult it contributes a greater proportion of the length than it does in the earlier larval instars (Fig. 15.10a). Conversely the five apical segments grow more slowly than the whole antenna, so their final contribution is proportionately less than originally.

The straight line relationship between two parts on a log/log plot as illustrated in Figure 15.10a occurs if growth rates are consistent over time, but this is not always the case. The mesothorax of *Dysdercus*, increases at roughly the same rate as the body as a whole at the first molt, but at later molts it exhibits positive allometry (Fig. 15.10b); the seventh abdominal segment increases at about the same relative rate as the whole body in the early stages, but also exhibits marked positive allometry at the final molt as the genitalia develop. In none of the segments is growth relative to the body as a whole uniform throughout larval life. Similar variations in relative growth rate occur in *Ectobius* and are probably usual in larvae of hemimetabolous insects.

Dyar's law suggests that a particular sclerotized part of the body (such as the head capsule or a leg segment) increases in linear dimensions from one instar to the next by a ratio which is constant throughout development (often in the range 1.2–1.4). For example, the locust's hind femur increases in length from one instar to the next by a ratio of 1.36–1.46 (Fig. 15.9). Przibam's rule suggests that the ratio by which cuticular structures increase in length at each ecdysis is constant (1.26). However, from the account given above it is obvious that considerable variation may occur in the increase in size of one body part from one stage to the next. Consequently, although these 'rules' are sometimes useful in specific instances, they cannot be generally applied (see Sehnal, 1985). In general, the sclerotized parts of holometabolous larvae tend to increase from one stage to the next by a greater proportion than do those of hemimetabolous larvae (Fig. 15.11).

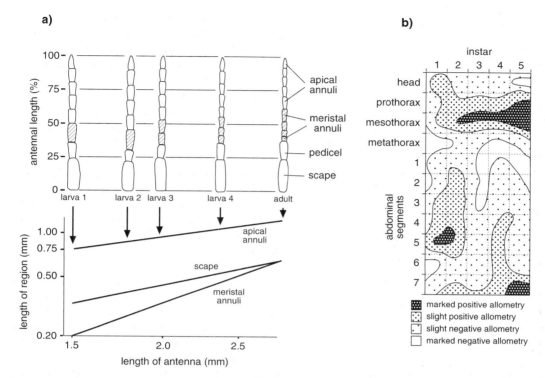

Fig. 15.10. Allometric growth. **(a)** Changes in the relative proportions of different parts of the antenna of a hemimetabolous insect in different stages of development. Above: representation of the whole antenna with lengths of different parts shown as a percentage of the whole; below: changes in lengths of different parts plotted against the length of the whole antenna (log/log scale) (*Hemimerus*, Dermaptera) (after Davies, 1966). **(b)** Diagram showing the growth rates of different parts of the body relative to the growth of the body as a whole from one larval stage to the next in a hemimetabolous insect (*Dysdercus*, Heteroptera) (after Blackith, Davies & Moy, 1963).

Regeneration Some insects are able to regenerate an appendage following its accidental loss. This occurs in Blattodea, some Orthoptera, Phasmatodea, some Hemiptera and larval holometabolous insects, but not in Acridoidea. Regeneration of cuticular structures can only occur at a molt as this is the only time at which new cuticle is produced. Consequently, it is restricted to larval stages; regeneration of appendages does not occur in adults. If the loss occurs early in a developmental stage, before the production of molting hormone, the appendage reforms at the next molt, but replacement of a limb lost after this does not occur until the following (next-but-one) molt. It grows at each successive molt, but generally remains smaller than a normal appendage.

Regeneration of muscles and parts of the nervous system also occurs.

Review: Bullière & Bullière, 1985 – regeneration

15.2.4.3 *Growth of the tissues*

The form of the cuticle is determined by the epidermis which may grow either by an increase in cell number or by an increase in cell size. Cell numbers increase just before molting in many insects (see Fig. 16.15), but, in larval Cyclorrhapha, the increase in larval size during development results entirely from an increase in the size of the epidermal cells. In this case the nuclei also increase in volume as a consequence of endomitosis and this results in an increase in the amount of DNA present in each nucleus.

Growth of the central nervous system of hemimetabolous insects does not involve the production of new neurons except in the brain. In the terminal abdominal ganglion of *Acheta*, for example, there are about 2100 neurons at all stages of development. On the other hand, the number of glial cells in the ganglion increases from about 3400 in the first stage larva to 20 000 in the adult and

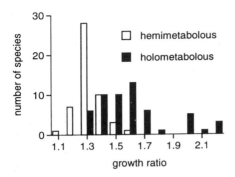

Fig. 15.11. Average growth ratios from one larval stage to the next for a range of hemimetabolous and holometabolous insects. The growth ratio is the ratio of size in one stage to the size in the previous stage averaged over all larval stages for each insect. The data are derived primarily from measurements of head width (after Cole, 1980).

the volume of the ganglion is increased 40-fold. In most holometabolous insects there is extensive reconstruction of the nervous system at metamorphosis and undifferentiated neuroblasts persist through the larval period up to this time (see below).

Marked changes occur in the sensory system during larval development of hemimetabolous insects. Additional mechanoreceptors and chemoreceptors are added to those already present at each molt (Fig. 15.12). The numbers of ommatidia forming the compound eyes also increase. By contrast, the numbers of sensilla are constant through the larval life of holometabolous insects and compound eyes are only present in adults (Chapman, 1982).

The musculature of most larval hemimetabolous insects closely resembles that of adults and probably all the adult muscles, including those that become flight muscles, are represented at hatching. In addition, there may be muscles that are operative only during molting and which disappear after the final molt (section 10.2.4). The muscles grow by an increase in fiber size between molts and by the addition of new fibers at molts (see Fig. 10.10). The musculature of larval holometabolous insects is, by contrast, usually fundamentally different from that of the adults. During larval development, muscles increase in size, but there is no basic change in their arrangement.

As with the epidermis, an increase in the size of an internal organ may result from an increase in cell size or in cell number, or sometimes both. The fat body of larval *Aedes* grows by an increase in cell number, but most other

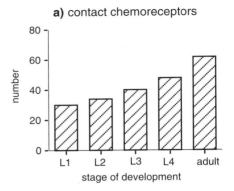

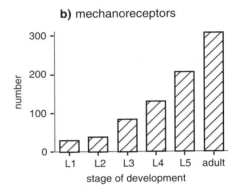

Fig. 15.12. Changes in numbers of peripheral sensilla during the development of hemimetabolous insects. (**a**) Changes in the numbers of contact chemoreceptors on the tip of the maxillary palp of a grasshopper in successive stages of development (*Bootettix*, Orthoptera) (after Chapman & Fraser, 1989). (**b**) Changes in the numbers of mechanosensitive filiform hairs on the prosternum of a locust in successive stages of development (after Pflüger *et al.*, 1994).

tissues in this insect and in *Drosophila* have a constant number of cells and grow by cell enlargement. This enlargement is accompanied by endomitosis. In the midgut both processes occur; the epithelial cells enlarge, but ultimately break down during secretion and each is replaced by two or more cells derived from the regenerative cells. In some insects the whole of the midgut epithelium is replaced at intervals by regenerative cells (section 3.1.2). In general, it appears that tissues which are destroyed at metamorphosis grow by cell enlargement while those that persist in the adult grow by cell multiplication.

The development of the Malpighian tubules varies. In

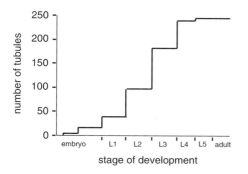

Fig. 15.13. Changes in the number of Malpighian tubules during the development of a hemimetabolous insect (*Schistocerca*, Orthoptera) (based on data in Savage, 1956).

orthopteroid orders, they increase in number throughout larval life. The primary tubules arise as diverticula from the proctodeum in the embryo (see Fig. 14.24). There are four primary tubules in *Blatta*; some other insects have six. Secondary tubules develop later, largely post-embryonically. *Schistocerca*, for example, has six primary tubules, but twelve more are added before the larva hatches and more develop in each larval stage up to the adult (Fig. 15.13). Secondary tubules appear as buds at the beginning of each larval stage, but after their initial development they increase in length without further cell division as a result of an increase in cell size. In holometabolous insects, the number of Malpighian tubules remains constant throughout larval life, although they do increase in length by increasing cell size and by cell rearrangement. New tubules only form at metamorphosis.

The increase in volume of internal structures, especially the fat body, is limited by the cuticle. In holometabolous larvae with soft, folded cuticle, considerable growth is possible. Extension of the abdomen by unfolding intersegmental membranes occurs in species with more rigid cuticles. In addition, in grasshoppers, and probably in some other insects, some growth of internal organs occurs at the expense of airsacs which become increasingly compressed during each developmental stage.

15.2.4.4 *Control of growth*

Larval growth is characterized by periodic molts and to some extent internal changes are correlated with the molting cycle. Molting is initiated by the growth and molting hormone and at larval molts the effect of this hormone is modulated by juvenile hormone so that larval

genes are activated and hence larval characters produced (section 15.4).

While hormones exert an overall controlling influence, local factors control the form of particular areas of the body. For example, epidermal cells often show distinct polarity, secreting cuticle in a form giving an obvious anterior–posterior pattern. In the first stage larva of *Schistocerca*, the cuticular plates associated with each epidermal cell on the sides of the abdominal sternites are produced into backwardly pointing spines; similarly, in *Oncopeltus*, a row of spines marks the posterior end of the area of cuticle secreted by each of the cells forming the abdominal sternites; and the scales of Lepidoptera grow out with a particular orientation. Experimental manipulation shows that the polarity of the cells within a body segment is produced by a gradient of a diffusible substance known as a morphogen. Similar gradients occur in the legs.

In addition to having a particular orientation, cuticular structures are dispersed in regular patterns characteristic of the species. For instance, the abdominal tergites of larval *Rhodnius* bear a number of evenly-spaced sensilla. At each molt these increase in number, new sensilla being formed in the biggest gaps between the existing sensilla. This is consistent with the hypothesis that a determining substance present in the epidermis is absorbed by existing sensilla, but accumulates between them if they become widely spaced due to growth of the epidermis. If this concentration exceeds a certain threshold the development of a new sensillum is initiated. The development of sensilla on the adult cuticle of *Oncopeltus* can be accounted for in a similar way.

Where two or more integumental features are present in an integrated pattern they may be controlled by the same substance. In *Rhodnius*, for instance, it is suggested that a differentiating substance (morphogen) in high concentration produces the sensilla and that the same substance in low concentration initiates the development of dermal glands, which are thus arranged round each sensillum. Where the integumental features are not arranged in an integrated manner, as with the hairs and scales on the abdomen of *Ephestia*, two determining substances might be involved.

In *Drosophila*, and almost certainly in the other insects, the boundaries of the parasegments (section 14.2.5) are sources of signals that organize the patterning and orientation of associated cellular fields (Fietz *et al.*, 1994).

There is relatively little information on the control of

growth of internal organs, but some show cyclical activity which coincides with the molt. The fat body cells of *Rhodnius*, for example, exhibit a marked increase in RNA concentration and mitochondrial number just before a molt, and the ventral abdominal intersegmental muscles become fully developed only at this time (section 10.2.4). In insects in which the Malpighian tubules increase in number, mitosis and development of new tubules are phased with respect to the molt.

Review: Caveney, 1985

15.3 METAMORPHOSIS

The changes that occur in the transformation of the larva to the adult may be more or less extensive depending on the degree of difference between the larva and the adult. In hemimetabolous insects, the changes are relatively slight; in holometabolous insects, the changes are very marked and a pupal stage is interpolated between the final larval stage and the adult.

15.3.1 Hemimetabolous insects

Epidermal mitosis and expansion only occurs at the time of a molt and, in hemimetabolous insects, progressive development of the wing buds occurs at each molt. Apart from being small, the larval wing buds differ from the adult wings in being continuous sclerotizations with the terga and pleura; the basal region of the wing is not membranous and no accessory sclerites are present. These appear at the final molt.

In general, the wings arise in such a way that the lateral margins of the wing buds become the costal margins of the adult wings (Fig. 15.3), but, in Odonata, the buds arise in an erect position, the margin nearer the midline ultimately becoming the costal margin (Fig. 15.14a). The wing buds of Acrididae originate as simple outgrowths of the terga, but at the antepenultimate molt they become twisted into the position found in the Odonata. This twisting results from the lower epidermis growing more rapidly than the upper. At the final molt the wings twist back so that the costal margins of the folded adult wings are ventro-lateral in position.

The genitalia develop progressively by modification of the terminal abdominal segments, but with a more marked alteration at the final molt (Fig. 15.14b).

Although accessory wing sclerites are not developed in the larva, the muscles that are attached to them in the adult are already attached to appropriate positions on the larval cuticle. For instance, in adult *Locusta*, the promotor-extensor muscle of the mesothoracic wing is inserted into the first basalar sclerite, but the equivalent muscle of the metathoracic wing is inserted into both basalar sclerites. In the larva, although the sclerites are not developed, the muscle in the mesothorax has one point of attachment to the pleural cuticle, while that in the metathorax has two.

In grasshoppers and dragonflies, all the flight muscles are present in the larva, although some are lacking striations and are presumably nonfunctional. These muscles increase in size throughout the larval period and further changes occur in the young adult (section 10.2.2). Many muscles disappear after the final molt. They are necessary for the process of molting (section 10.2.4) and their disappearance is accompanied by the death of the motor neurons supplying them. Other changes also occur in the nervous system. For example, changes in the arborization patterns of afferent sensory axons may occur within the central nervous system even though the peripheral sensilla remain functional (Pflüger *et al.*, 1994). The effect of this will be to alter the neural pathways of which the sensory neurons form a part.

15.3.2 Holometabolous insects

The development of adult features in holometabolous insects varies with the degree of modification of larval features involved. In Neuroptera and Coleoptera, where the larvae have some resemblance to the adults, relatively little reconstruction occurs, but, in Diptera, the tissues are almost completely rebuilt following histolysis and phagocytosis of the larval tissues. This reconstruction occurs in the pupa.

15.3.2.1 *Pupa*

All the features of the adult become recognizable in the pupa which consequently has a greater resemblance to the adult than to the larva. At the larva/pupa molt, the wings and other features that up to now have been developing internally are everted and become visible externally although not fully expanded to the adult form (Fig. 15.5). The appendages extend freely from the body in the pupae of some insects and this condition is known as exarate, but in other species the appendages are glued down by a secretion produced at the larva-pupa molt. This is the obtect condition and obtect pupae are usually more heavily scle-

a) wings

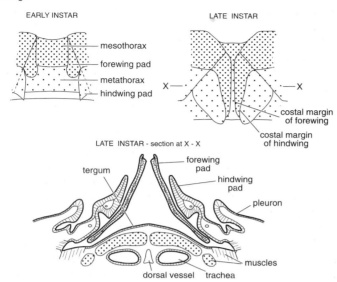

Fig. 15.14. Development of adult external structures in hemimetabolous insects.
(a) Wings. Above: dorsal views of the pterothorax of young and older larvae of a dragonfly; below: transverse section through the dorsal metathorax of a dragonfly larva at the same stage as that shown above right. The section is in the plane marked X–X above right (after Comstock, 1918). **(b)** Genitalia: ventral view of the tip of the abdomen of a female grasshopper at different stages of development (*Eyprepocnemis*, Orthoptera) (after Jago, 1963).

b) external genitalia

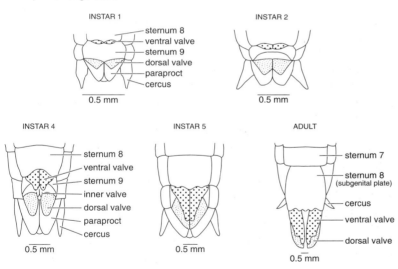

rotized than exarate pupae. A further differentiation can be made on the presence or absence of articulated mandibles in the pupa. When articulated mandibles are present (the decticous condition), they have apodemes fitting closely within the adult mandibular apodemes (Fig. 15.15) and hence can be moved by the mandibular muscles of the pharate adult. The alternative condition, with immobile mandibles, is known as adecticous.

Decticous pupae are always exarate. They occur in Megaloptera, Neuroptera, Trichoptera and some Lepidoptera. Some adecticous pupae are also exarate as in Cyclorrhapha, Siphonaptera and most Coleoptera and Hymenoptera, but others are obtect. Most Lepidoptera, Nematocera, Orthorrhapha, Staphylinidae, some Chrysomelidae and many Chalcidoidea have obtect, adecticous pupae.

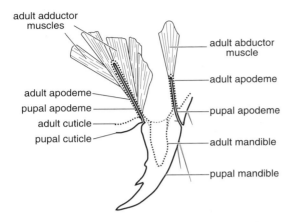

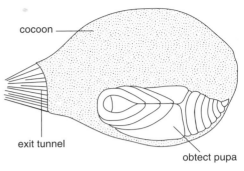

Fig. 15.15. Mandible of a decticous pupa showing the pupal apodemes inside the adult apodemes. Adult cuticle shown as dotted lines (*Rhyacophila*, Trichoptera) (after Hinton, 1946).

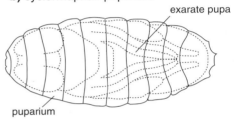

Fig. 15.16 Protection of the pupa. (**a**) A lepidopteran pupa inside a silk cocoon (*Saturnia*). (**b**) A pupa (dotted) inside the puparium of a cyclorrhaphan dipteran. The puparium is formed from the cuticle of the last stage larva.

Prepupa The last stage larva is often quiescent for two or three days before the ecdysis to a pupa, and in some cases the insect is a pharate pupa for a part of this time. During this period, the insect is sometimes referred to as a prepupa, but it does not usually represent a distinct morphological stage separated from the last stage larva by a molt. A separate morphological stage known as the prepupa does exist in Thysanoptera and male Coccidae. In these insects the prepupa is a quiescent stage following the last stage larva and it is succeeded by a pupal stage.

Protection of the pupa Most insect pupae are immobile and hence vulnerable, and most insects pupate in a cell or cocoon which affords them some protection. Many larval Lepidoptera construct an underground cell in which to pupate, cementing particles of soil with a fluid secretion. The larva of the puss moth, *Cerura* (Lepidoptera), constructs a chamber of wood fragments glued together to form a hard enclosing layer, and some beetle larvae pupate in cells in the wood in which they bore. Many larvae produce silk which may be used to hold other structures, such as leaves, together to form a chamber for the pupa, but in some other species a cocoon is produced wholly from silk (Fig. 15.16a). Silken cocoons are produced by Bombycoidea amongst the Lepidoptera and by Siphonaptera, Trichoptera and some Hymenoptera.

Cocoon formation by the silkmoth, *Antheraea*, takes about two days. It follows gut purging, in which the larva expels the gut contents by a series of waves of contraction passing along the abdomen from front to back. The larva subsequently enters an active wandering phase that ends when it finds a suitable site in which to pupate. The first phase of cocoon formation is the construction of a scaffold of silk threads between leaves of the food plant and the production of a stalk which attaches the cocoon to the leaf petiole. Subsequent behavior consists of a series of cycles in which the insect weaves loops of silk using figure-of-eight movements of the head to construct one end of the cocoon and then turns through 180° to form the other end. A complete layer of silk has been produced after a period of about 14 h, and the insect turns from one end of the cocoon to the other at much shorter intervals (5 min compared with 80 min at 23 °C) and at the same time coats the inside of the cocoon with a liquid from the anus containing crystals of calcium oxalate. This liquid accumulates in the hindgut after purging and the calcium oxalate is produced by the Malpighian tubules.

a) *Simulium*   **b)** mosquito

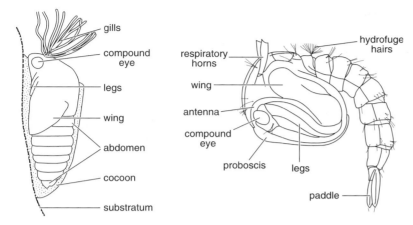

Fig. 15.17. Aquatic pupae. **(a)** Fixed pupa of a simuliid (Diptera). **(b)** Mobile pupa of a mosquito (Diptera).

When the salivary glands produce silk, they also secrete compounds that will link the silk proteins in the same way that proteins are linked in the cuticle (section 16.5.3). These tanning agents, which are phenolic compounds derived from the food, are produced as glucosides and so are inactive. The salivary gland also produces the enzymes to activate the tanning agents, but, as the silk dries very quickly, activation does not occur until it is rewetted by fluid from the caterpillar's hindgut. At this time the chains of silk protein are linked together and the wall of the cocoon becomes stiff and colored yellow-brown. This period of impregnation of the cocoon lasts for about an hour. Afterwards, more silk is added to the inside of the cocoon, but the spinning cycles are interrupted by periods of inactivity that become progressively longer until the larva becomes completely quiescent.

An exceptional protective structure is produced from the cuticle of the last stage larva by cyclorrhaphous Diptera. Procuticle is laid down throughout the last larval stage at the end of which the larva shortens and the outer part of the cuticle is tanned to form a rigid ovoid structure known as the puparium (Fig. 15.16b). The process is called pupariation. Subsequently, after apolysis, the pupal/adult head and appendages, previously concealed beneath the larval epidermis, are everted and the pupal cuticle is secreted (Fristrom *et al.*, 1991). The pupa remains inside the puparium and the adult, when it emerges, escapes from both the pupal and puparial cuticles.

A few insects form unprotected pupae. These are particularly well known in the Nymphalidae and Pieridae, where the pupae are suspended from a silk pad. These exposed pupae exhibit homochromy (section 25.5.2.2) whereas protected pupae are normally brown or very pale in color.

Pupae of aquatic insects The behavior of aquatic insects on pupation varies considerably. Some larvae, such as those of the aquatic Syrphidae and the beetle, *Hydrophilus*, leave the water and pupate on land, but many others, particularly the aquatic Diptera, pupate in the water. Sometimes the pupae are fastened to the substratum. For example, the pupae of Blepharoceridae have ventro-lateral pads on the abdomen with which they attach themselves to stones, while Simuliidae construct open cocoons attached to stones and rocks (Fig. 15.17a). The pupa projects from the open end of the cocoon, which is constructed more strongly in faster flowing water than it is in a weak current. Chironomidae pupate in the larval tubes or embedded in the mud, while *Acentropus* (Lepidoptera) forms a silken cocoon with two chambers separated by a diaphragm. The pupa lies in the lower chamber, which is air-filled.

Other aquatic pupae obtain oxygen from the air, either directly or indirectly. The pupae of most Culicidae and Ceratopogonidae are free-living and active. They are buoyant so that, when undisturbed, they rise to the surface and respire via prothoracic respiratory horns (Fig. 15.17b). If disturbed, movements of the anal paddles drive them downwards. The pupae of some Culicidae and Ephydridae have their respiratory horns embedded in the tissues of aquatic plants, obtaining their oxygen via the aerenchyma.

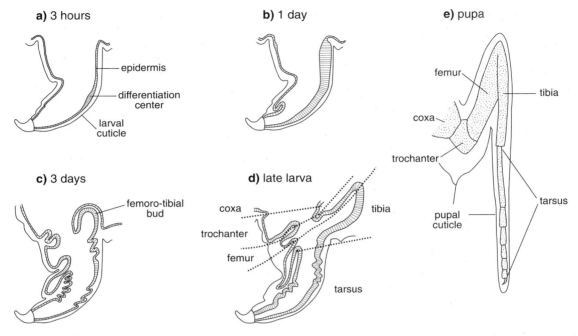

Fig. 15.18. Development of the adult leg of a lepidopteran at different times after the molt to the last larval stage and in the pupa (*Pieris*) (based on Kim, 1959). **(a)–(d)** changes in the epidermis within the larval leg. The presumptive areas of the adult leg are shown separated by dotted lines in d). **(e)** Fully developed leg (shaded) of the pharate adult within the pupal cuticle.

Significance of the pupa The occurrence of a pupa is indicative of the broad differences which occur between larval and adult forms of holometabolous insects. It is a stage during which major internal reconstruction occurs and is of particular importance in permitting the development and attachment of adult muscles to the cuticle and the full development of the wings.

Most of the adult thoracic muscles are different from those of the larva. It has been suggested that muscles will only develop in an appropriate form and length if they have a mold in which to do so. The pupa provides a mold for the adult muscles. Further, most muscles are attached to the cuticle by filamentous projections from the epicuticle (section 10.1.3) that can only develop at a molt, in this case the pupa to adult molt.

Development of the wings internally within the larva is restricted by lack of space and this problem becomes more acute as the insect approaches the adult condition and the flight muscles also increase in size. Thus development can only be completed after the wings are everted and, for this reason, two molts are necessary in the transformation from larva to adult. At the first, from larva to pupa, the wings are everted and grow to some extent. Further growth occurs and the adult cuticle is laid down at the pupa–adult molt.

The importance of the pupa in wing development and associated changes is emphasized by the absence of a pupal stage in the life histories of female Strepsiptera and Coccidae, which are wingless and larviform. The males in these groups are winged and have a pupal stage (see section 15.2).

Reviews: Hinton, 1948, 1963

15.3.2.2 *Development of adult features*

Epidermal structures If an adult appendage does not differ markedly from that of the larva it may be formed by a proliferation of the tissue within and at the base of the larval organ. This occurs in the legs of Lepidoptera, for example. Soon after the larva of the cabbage butterfly, *Pieris*, enters its final stage the epidermis of its thoracic legs becomes separated from the cuticle except at points of muscle attachment. It is now free to thicken and fold. The first thickening, well supplied with tracheae, develops at the junction of the second and third leg joints (Fig. 15.18a)

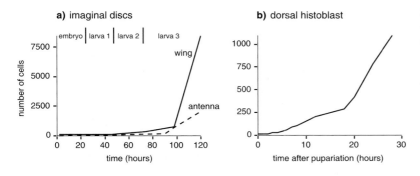

a) imaginal discs

embryo | larva 1 | larva 2 | larva 3

wing

antenna

number of cells

7500

5000

2500

0

time (hours): 0 20 40 60 80 100 120

b) dorsal histoblast

1000

750

500

250

0

time after pupariation (hours): 0 10 20 30

Fig. 15.19. Development of adult epidermis in *Drosophila*. (a) Numbers of cells in antennal and wing imaginal discs during the embryonic and larval periods (based on Oberlander, 1985). (b) Number of cells in the anterior dorsal histoblasts during pupal development (after Madhavan & Madhavan, 1980). (c) The positions of imaginal discs and histoblast nests in the larva (after Nöthiger, 1972).

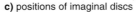

c) positions of imaginal discs

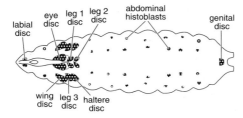

labial disc · eye disc · leg 1 disc · leg 2 disc · abdominal histoblasts · genital disc

wing disc · leg 3 disc · haltere disc

and from this differentiation center a wave of cell multiplication spreads out (Fig. 15.18b,c). As a result of the increase in area, the epidermis becomes folded and a particularly large fold develops basally. Later, when the epidermis expands to form the pupal leg, this basal fold becomes divided by a longitudinal septum to form the femur and tibia. Epidermis from the more proximal parts of the larval leg forms the coxa and trochanter, and more distal tissue forms the tarsus. Further differentiation continues in the pupa to produce the adult leg (Fig. 15.18e).

Where the difference between larval and adult organs is more marked the adult tissues develop from epidermal thickenings called imaginal buds or discs. The founder cells comprising these discs divide to form clones. Since the production of adult organs is restricted to small groups of cells the remainder of the epidermis is free to undergo larval modifications. The discs may be regarded as islands of embryonic tissue which remain undifferentiated until they give rise to the adult structures. They do not produce cuticle in the larva and the cells continue to divide between molts (Fig. 15.19a), not just at the time of the molt as occurs with larval epidermal cells.

The imaginal disc commonly becomes invaginated beneath the larval epidermis forming a cavity known as the peripodial cavity (Fig. 15.20b). It is lined with epidermis (the peripodial membrane) and as the imaginal disc

enlarges the appendage forms and pushes into the cavity (Fig. 15.20c) becoming folded as it increases in size. In Lepidoptera, the wings and antennae develop in peripodial cavities, but, in Diptera, all the main ectodermal features of the adult develop in this way (Fig. 15.19c).

At pupation, the appendage is everted and the peripodial membrane comes to form part of the epidermis of the general body wall (Fig. 15.20c,d). Evagination probably results from a rearrangement of the cells of the disc which, at the same time, also tend to increase in surface area. This process is initiated by an increase in 20-hydroxyecdysone in the hemolymph; at earlier molts, when the titer of this hormone also increases, evagination is probably inhibited by juvenile hormone. At some stage, the discs acquire the 'competence' to undergo metamorphosis. Prior to this, metamorphosis does not occur even in the appropriate humoral environment. The acquisition of competence appears to be associated with the number of divisions that the cells of the disc have undergone.

The details of development of the imaginal discs vary from one insect to another and from organ to organ. Where an appendage is present in the larva as well as the adult the imaginal disc is closely associated with the larval structure. Thus, in *Pieris*, the adult antenna is first apparent in the first stage larva as a thickening of the epidermis at the base of the larval antenna. The cells divide and, in

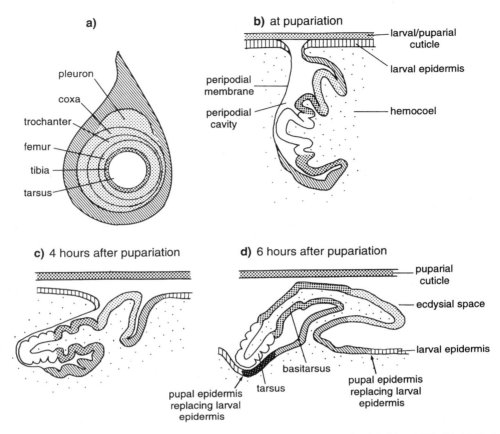

Fig. 15.20. Development of an adult leg in *Drosophila*. **(a)** Fate map of a leg disc (after Schubiger, 1971). **(b)–(d)** Sections through the disc at different times after pupariation showing the elongation and eversion of the leg (from Fristrom & Fristrom, 1993).

the succeeding stages, an invagination is produced which pushes upwards, deep into the larval head. In the fifth stage larva, the adult antennal tissue grows more quickly than the peripodial membrane, so that it is thrown into folds, and towards the end of the larval stage the larval antenna starts to degenerate and is invaded by imaginal cells. When the peripodial cavity (which opens by a slit on the front of the epidermis of the head) evaginates, the antenna is carried to the outside and the peripodial membrane now forms a part of the wall of the head. The maxilla develops in an essentially similar way, but very little development of the labium takes place until the fifth larval stage.

The wings also develop from imaginal discs (Fig. 15.21). In some Coleoptera, they form as simple evaginations of the epidermis beneath the larval cuticle, but more usually they develop in peripodial cavities. In *Pieris*, the wing imaginal discs are apparent in the embryo and invaginate in the second and third larval stages. During the fourth stage, the wing starts to develop in the peripodial cavity, finally becoming external at the larva–pupa molt. In *Drosophila*, on the other hand, invagination of the peripodial cavity is complete before the larva hatches, but the wing itself does not develop until the second larval stage, growing more extensively in the third stage.

During development, the cells of each imaginal disc become programmed for their ultimate phenotypic expression in the adult. They are said to become determined, and their subsequent development is fixed; they will produce the same structure even if they are moved experimentally to another position on the appendage. More detailed features are not determined until after the more basic characters. For example, the dorsal and ventral

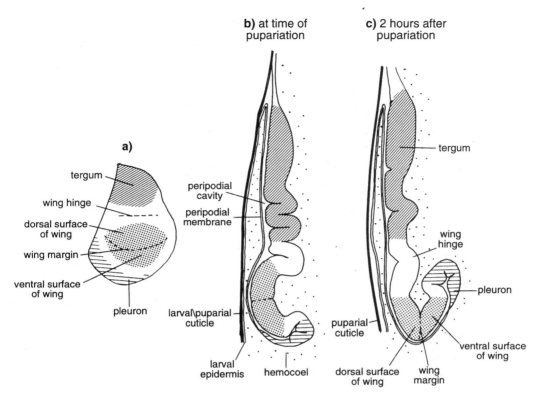

Fig. 15.21. Development of a forewing in *Drosophila*. (a) Fate map of a wing disc (after Bryant, 1975). (b) Section through the disc at the time of pupariation (from Fristrom & Fristrom, 1993). (c) Section through the disc two hours after pupariation (from Fristrom & Fristrom, 1993).

surfaces of the wing are determined before the details of the pattern, and cells become committed to a segment of a leg before a specific structure on that segment. Some degree of determination of the adult tissues may already be apparent in the embryo, and determination of wing regions is advanced at the time of pupariation (Fig. 15.21). In *Drosophila*, cells along the dorso–ventral boundary of the wing have a critical organizational role (Kim, Irvine & Carroll, 1995).

The internal development of the wings is complex, involving great expansion and the formation of the veins. The development of a *Drosophila* wing is used here as an example. When the wing evaginates at pupation, the upper and lower surfaces have already come together although they remain separated along certain lines called lacunae (Fig. 15.22a,b). Four lacunae run the length of the wing rudiment, the second dividing into two distally. A nerve and a trachea enter the second lacuna, and at about this stage, some six hours after the formation of the puparium,

the pupal cuticle is laid down. After this the upper and lower surfaces of the wing are forced apart by an increase in blood pressure (Fig. 15.22c,d). The cells at first become stretched across the gap as narrow threads connecting the two surfaces, but these connections are finally broken except at the wing margins. A less extensive inflation occurs in *Tenebrio* (Coleoptera) and *Habrobracon* (Hymenoptera). Perhaps the inflation has the effect of expanding the newly formed pupal cuticle to the greatest possible extent so that the development of the adult wing can proceed.

Following inflation, the wing flattens again. The epidermal layers on the two sides first become apposed round the edges (Fig. 15.22e) and then the contraction spreads inward so that a flat double sheet of cells is produced (Fig. 15.22f). During this process the definitive wing veins are formed along lines where the two epidermal layers remain separated (Fig. 15.22g). The veins are at first wide channels, but they become narrower as the wing continues to

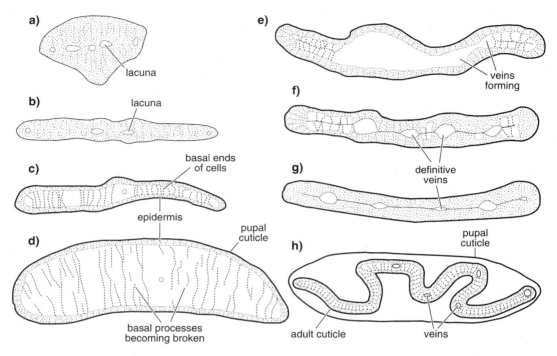

Fig. 15.22. Diagrammatic transverse sections of a developing forewing of *Drosophila* (after Waddington, 1941): **(a),(b)** successive stages in the larva before pupation; **(c)–(g)** successive stages in the pupa; **(h)** pharate adult.

expand. Cell division proceeds actively, especially above the veins, so that here the cells become crowded and columnar, while elsewhere they are flattened. The fully developed wing finally secretes the adult cuticle (Fig. 15.22h).

Even in the imaginal discs, some programmed cell death (apoptosis) occurs in the course of differentiation. This is especially obvious where normal adult structures are of reduced size. The female vaporer moth, *Orgyia*, for example, has only vestigial wings, but up to the pupal molt her wings develop normally as imaginal discs. Then, during apolysis, as the epidermis retracts from the pupal cuticle, extensive cell death occurs so that the wings become greatly reduced in size. Cell death on a much smaller scale occurs in the developing wings of the male which is fully winged (Nardi, Godfrey & Bergstrom, 1991).

When the imaginal appendages are everted from the peripodial cavities at the time of pupation, the peripodial membrane contributes to the general epidermis of the adult body wall (Figs. 15.20, 15.21). The extent to which the larval epidermis is replaced varies. In Coleoptera there is no extensive replacement, but in Hymenoptera and Diptera the epidermis is completely renewed from imaginal cells. The epidermis of the head and thorax are formed by growth from the imaginal appendage discs, while the abdominal epithelium is formed from special groups of cells called histoblasts. In *Drosophila*, the abdominal segments have pairs of dorsal, ventral and spiracular histoblasts (Fig. 15.19c). These initially comprise 5–15 cells which remain unchanged throughout larval development, but after pupariation the cells start to divide. They gradually replace the larval cells, but at no time is there any discontinuity in the epidermis. The larval cells are completely replaced and are sloughed off into the body cavity and phagocytosed.

Reviews: Campbell & Tomlinson, 1995 – axial determination in legs; Cohen, 1993 – *Drosophila* imaginal discs; Fristrom & Fristrom, 1993 – *Drosophila* epidermis; Fristrom *et al.*, 1991 – pupal cuticle in *Drosophila*; Larsen-Rapport, 1986 – determination; Oberlander, 1985 – imaginal discs; Williams & Carroll, 1993 – determination

Muscles The muscular system usually undergoes extensive modification at metamorphosis and the muscles fall into four categories:

1 Larval muscles may pass unchanged into the adult

2 Existing larval muscles are reconstructed

3 Larval muscles may be destroyed and not replaced

4 New muscles, not represented in the larva, may be formed.

The adult muscles of some flies are already present in the larva as rudimentary non-functional fibers. The dorsal longitudinal muscles in *Simulium*, for instance, are only about four microns in diameter in the first stage larva. They grow throughout the larval period and their nuclei increase in number. During the pharate pupal period the muscle rudiment becomes divided to produce the definitive number of fibers and myofibrils appear for the first time. They continue to grow until some time after the final molt.

Reconstruction of larval muscles occurs in two ways. In the Neuroptera, Coleoptera and some Lepidoptera, larval muscles contain two sets of nuclei, the functional larval nuclei and other, small, nuclei which are scattered through the cytoplasm. At metamorphosis the small nuclei multiply and, with associated cytoplasm, form myocytes. These migrate into the body of the muscle and associate in strands to form new fibers. In Diptera, and some Lepidoptera and Hymenoptera, on the other hand, myoblasts originating outside the larval muscle are concerned in the production of adult muscle, adhering to the outside or penetrating the sarcolemma in order to form new fibers. The dorsal longitudinal flight muscles of adult *Drosophila* are formed by new myoblasts, but larval muscles form a scaffold that determines the appropriate development of the adult muscles (Farrell, Fernandes & Keshishian, 1996).

Most larval muscles are completely histolyzed and disappear during the early pupal period, but the precise timing varies because some larval muscles have specific functions in pupal development or adult eclosion and are destroyed much later than others. For example, the larval leg muscles of *Manduca* start to degenerate when the larva enters its wandering phase and degeneration is complete by the time the larva pupates (Consoulas, Anezaki & Levine, 1997). Some abdominal muscles, however, do not degenerate until after pupation. In *Drosophila*, most muscles of the head and thorax start to break down before puparium formation and are fragmented before the larva pupates, but the dilator muscles of the pharynx remain unchanged until after pupation. They are apparently important in the evagination of the head region of the insect at pupation, and only degenerate after this has occurred. In addition, one pair of muscles persists in each abdominal segment for about half the pupal period. They may help to establish the segmentation of the pupal abdomen by telescoping each segment into the preceding one.

The first sign of muscle degeneration is liquefaction of the peripheral parts of the fiber. This is followed by separation of the fibrils and, in the flour moth, *Ephestia*, phagocytes penetrate the sarcolemma and assist the destruction. The sarcolemma breaks down and the muscles separate from their attachments and fragment, the remains being consumed by phagocytes.

New muscles are always formed by free myoblasts that aggregate to form multicellular groups within which cells fuse to form the multinucleate muscle fibers. Where the new muscle is associated with an existing motor neuron that becomes respecified (see below), the myoblasts may migrate along the axon to their definitive positions and the neuron may influence the rate of myoblast proliferation (Currie & Bate, 1991; Fernandes, Bate & Vijayraghavan, 1991). Many of the muscles that are produced during the pupal period have only short lives in Diptera and Lepidoptera. These are the muscles specifically associated with eclosion which break down after the adult insect has emerged from the pupal cuticle (see Fig. 10.14).

Reviews: Fernandes & Keshishian, 1995; Schwartz, 1992 – programmed cell death

Nervous system The central nervous system of holometabolous insects is extensively restructured at metamorphosis. The brain becomes considerably enlarged, mainly due to the development of the antennal and optic lobes in relation to the much larger antennae of adult insects and the appearance of compound eyes. In most holometabolous insects, particularly those that are more specialized, fusion of some ventral ganglia commonly occurs at metamorphosis. This condensation is effected by a forward movement of the more posterior ganglia resulting from the shortening of the interganglionic connectives. For instance, the larva of *Manduca* has, in addition to the head ganglia, three thoracic and seven separate abdominal ganglia. In the adult, the meso- and meta-thoracic ganglia are fused with the first two abdominal ganglia to form a compound ganglion close behind the prothoracic ganglion. The next three abdominal ganglia remain separate, but the last three fuse together to form another compound ganglion (Fig. 15.23a). In the course of these changes the perineurium is histolyzed and the neural lamella digested, the former being redeveloped from remaining glial cells. The higher Diptera are exceptional in having a more condensed central nervous system in the larva than in the adult.

Over 90% of the neurons in the nerve cord of adult *Drosophila* develop during the larval period from groups of neuroblasts that persist from the embryo. In the first stage larva of *Drosophila*, there are 47 neuroblasts in each thoracic neuromere, but only six in each abdominal neuromere. *Manduca* has over 20 neuroblasts in each of the thoracic ganglia, but only four in the abdominal neuromeres behind the first. These cells start to divide at different times during larval development. The most rapid division occurs in the final larval stages (Fig. 15.23c). Some of these cells die just before and during pupal development, but, despite this, the neuroblasts contribute over 2500 new interneurons to each of the thoracic ganglia and about 50 to each abdominal ganglion in *Manduca*. Similar numbers are produced in *Drosophila*. All these new cells become interneurons; probably no new motor neurons are formed (Booker & Truman, 1987; Truman & Bate, 1988).

The motor neurons already present in the larval ganglion have two possible fates, depending on the muscles which they innervate:

1   If the larval muscle degenerates, but a new muscle develops in a similar position, the neuron may persist and innervate the new muscle. It is said to be respecified.

2   If the larval muscle is destroyed and not replaced, the motor neuron dies. This may occur at various times during metamorphosis. About 16% of all the motor neurons in the abdominal ganglia of *Manduca* die soon after pupation following the deaths of their target muscles.

Respecification of neurons occurs in relation to leg and flight muscles as well as to the newly formed abdominal intersegmental muscles. For example, the neuron innervating a femoral flexor muscle in the larva of *Manduca* is respecified to innervate the femoral extensor muscle in the adult leg. This involves regression and then regrowth of the dendritic arborization in the ganglion and the remaking of contact with the new target muscle by the axon.

The sensory system is almost entirely renewed at metamorphosis in holometabolous insects. The compound eyes and antennae of the adult are completely new structures, and this is true of the sensilla associated with the mouthparts, at least in a majority of insects. It is known, however, that a few mechanoreceptors persist during the pupal stage of *Manduca* and some stretch receptors and chordotonal organs are still present in the adult.

Reviews: Levine, Morton & Restifo, 1995 – molecular biology; Truman, Thorn & Robinow, 1992 – programmed cell death; Weeks & Levine, 1992 – control

Alimentary canal The alimentary canal is extensively remodelled at metamorphosis in species that have different larval and adult diets. In larval Lepidoptera, for example, the gut is relatively simple and the midgut occupies most of the body cavity, but the adult has a large crop and rectal sac and only a small midgut (Fig. 15.24a,b). This change is associated with the change from leaf-feeding, with continuous access to food, to fluid-feeding, with the need to store nectar in the crop between feeds and the requirement to digest and absorb only sugars. Reconstruction of the stomodeum and proctodeum results from the renewed activity of the larval cells without any accompanying cell destruction in Coleoptera, but, in Lepidoptera and Diptera, new structures develop, at least partly, from proliferating centers known as imaginal rings at the inner ends of the foregut and hindgut. The larval cells are sloughed into the body cavity.

The midgut is probably completely renewed in all holometabolous insects, usually being reformed from the regenerative cells at the base of the epithelium. These cells proliferate and form a layer round the outside of the larval cells which thus come to lie in the lumen of the new alimentary canal. Sometimes this process occurs twice, once on the formation of the pupa and again when the adult tissues are forming. It is suggested that the special pupal midgut enables the insect to digest the sloughed remains of the larval midgut so that these can be assimilated and used in the reconstruction. In *Manduca*, the newly formed epithelium contains large vacuoles containing the enzyme, lysozyme, which is discharged into the lumen when the epithelium is complete (Fig. 15.24c). It is presumed to have an antibacterial function (Russell & Dunn, 1991).

Malpighian tubules Sometimes the larval Malpighian tubules remain unchanged in the adult, or slight

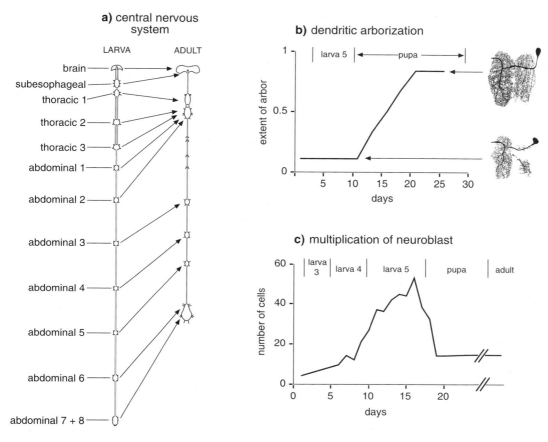

a) central nervous system

LARVA ADULT

brain
subesophageal
thoracic 1
thoracic 2
thoracic 3
abdominal 1
abdominal 2
abdominal 3
abdominal 4
abdominal 5
abdominal 6
abdominal 7 + 8

b) dendritic arborization

larva 5 | pupa

extent of arbor

days

c) multiplication of neuroblast

larva 3 | larva 4 | larva 5 | pupa | adult

number of cells

days

Fig. 15.23. Metamorphosis of the central nervous system of *Manduca*. (a) Brain and ventral nerve cord of the larva and adult showing the fusion of ganglia. (b) Changes in the form of an abdominal motor neuron from larva to adult. The extent of the arbor is expressed on an arbitrary scale (from Truman & Reiss, 1988). (c) Changes in the number of neurons produced from a neuroblast nest in an abdominal ganglion. The decline in cell number at pupation is due to the deaths of many cells (after Booker & Truman, 1987).

modifications may occur as in the Lepidoptera. Caterpillars have a cryptonephridial arrangement of the Malpighian tubules (Fig. 18.14), but at metamorphosis the parts associated with the rectum are histolyzed while the more proximal parts form the adult tubules. In Coleoptera, the tubules are rebuilt from special cells in the larval tubules, while, in Hymenoptera, the larval tubules break down completely and are replaced by new ones developing from the tip of the proctodeum.

Fat body The fate of the fat body at metamorphosis has been most fully studied in Lepidoptera and Diptera. In both groups, the cells of the fat body become disassociated at the larva–pupa molt through the activity of hemocytes. Subsequently, in *Drosophila*, the cells persist independently but are progressively histolyzed. Some may persist in the adult's head, but the bulk of the adult fat body is formed from mesenchyme cells on the inside of the imaginal discs. In Lepidoptera, the cells of the peripheral fat body which synthesize storage proteins in the caterpillar are destroyed, but those of the perivisceral fat body that accumulate the proteins reassociate to form the adult fat body.

Review: Haunerland & Shirk, 1995

Other systems In general, the tracheal system shows little change other than the development of new branches to

a) larva

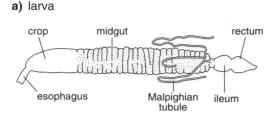

b) adult

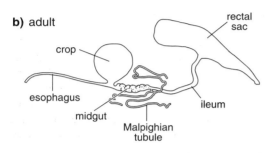

c) lysozyme

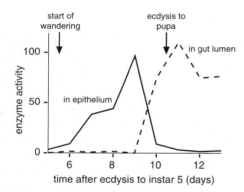

Fig. 15.24. Metamorphosis of the alimentary canal in a lepidopteran: (a) larval gut; (b) adult gut; (c) lysozyme activity in the midgut epithelium and lumen during the final larval and early pupal stages of *Manduca* (after Russell & Dunn, 1991).

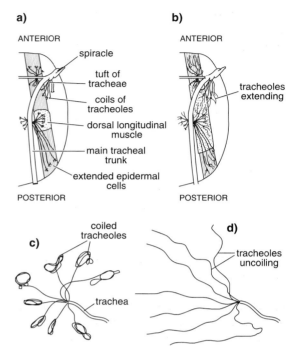

Fig. 15.25. Metamorphosis of the tracheal system in the thorax of *Calliphora* (after Houlihan & Newton, 1979). (a),(b) Diagrams of the right half of the thorax of the pharate adult three and four days after pupariation showing the tracheal supply to the developing dorsal longitudinal muscle. At first, the muscle is very short, and connects with long extensions of the epidermal cells. As it elongates, the extensions become shorter. (c),(d) A tuft of tracheoles arising from the end of a trachea. The tracheoles are coiled on day 3 (c), but have uncoiled by day 4 (d).

accommodate the particular needs of the adult, such as the supply to the flight muscles, and the elimination of some specifically larval elements. In cyclorrhaphous Diptera, however, an extensive reconstruction occurs. The pupal tracheal system consists of four main tracheal trunks which extend into the head and as far as the anterior segments of the abdomen from the thoracic spiracle. From these main trunks, tufts of fine, unbranched tracheae arise and each ends in several tracheoles which become tightly coiled about a third of the way through the puparial period (Fig. 15.25). Up to this time the developing adult flight muscles are still very short and their development is not dependent on an oxygen supply. Subsequently, however, they elongate rapidly and the tracheoles uncoil, extending with the muscle fibers whose metabolic processes at this time are at least partially dependent on aerobic respiratory processes. Subsequently, the adult tracheal supply develops from tracheoblasts largely independently of the pupal supply, but it remains fluid-filled until about three hours before eclosion. The filling of the adult tracheae with air corresponds with a sharp increase in oxygen consumption, partly associated with increased metabolic activity in the tissues and partly with an increase in movements by the fly.

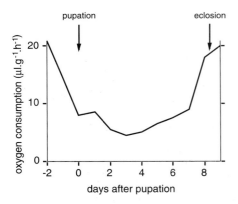

Fig. 15.26. Reduced oxygen consumption during pupal development (*Galleria*, Lepidoptera) (after Sláma, 1982).

The circulatory system undergoes little change from larva to adult.

Biochemical changes During the pupal period oxygen consumption at first falls and then rises again, following a characteristic U-shaped curve (Fig. 15.26). This is associated with changes in the enzyme systems regulating energy release. In *Calliphora*, the activity of lactate dehydrogenase is high two days after pupariation, but falls sharply about midway through the pupal period, while citrate synthase and glycerolphosphate dehydrogenase increase (Fig. 10.11). These changes indicate a switch from predominantly anaerobic catabolic processes to aerobic processes corresponding with the time at which the pupal tracheal system forms a well-developed air supply to the developing muscles. A comparable change in the relative emphasis of anaerobic and aerobic metabolic pathways occurs in *Philosamia* (Lepidoptera) after about two-thirds of the pupal period has elapsed.

The components of adult proteins are derived partly from the histolysis of larval tissues and partly from proteins built up and stored in the larval hemolymph for this specific purpose (section 5.2.2.3).

The waste products of pupal metabolism accumulate in the rectum and are discharged as the meconium when the adult emerges. Uric acid accumulates throughout the pupal period, but especially during histolysis, while in Lepidoptera and Hymenoptera allantoic acid comprises an appreciable part of the nitrogenous waste of the pupa. In *Phormia* (Diptera), urea accumulates during the development of the adult, suggesting that it is the end product of nitrogen metabolism in this insect.

15.3.3 Eclosion

The escape of the adult insect from the cuticle of the pupa or, in hemimetabolous insects, of the last larval stage is known as eclosion. Where the pupa is enclosed in a cell or cocoon the adult also has to escape from this.

15.3.3.1 *Escape from the cocoon or cell*

Sometimes the pharate adult is sufficiently mobile to make its escape from the cocoon or cell while still within the pupal cuticle. This is the case in species with decticous pupae which use the pupal mandibles, actuated by the adult muscles, to bite through the cocoon. Sometimes, as in Trichoptera, the adult mouthparts are non-functional and the sole function of the adult mandibular muscles is to work the pupal mandibles at eclosion; subsequently they degenerate. The pupa moves away from the cocoon before the adult emerges, aided by its freely moveable appendages and backwardly directed spines on the pupal cuticle.

In species with adecticous pupae, other methods are employed. Amongst the Lepidoptera, the pupa of Monotrysia and primitive Ditrysia works its way forwards with the aid of backwardly directed spines on the abdomen, forcing its way through the wall of the cocoon with a ridge or tubercle known as a cocoon cutter on the head. The pupa does not escape completely from the cocoon, but is held with the anterior part sticking out by forwardly directed spines on the ninth and tenth abdominal segments. With the pupal cuticle fixed in this way the adult is able to pull against the substratum and so drag itself free more readily. Cocoon cutters are also present in Nematocera although in this group they are usually multiple structures.

In many insects with adecticous pupae, the adult emerges from the pupa within the cocoon or cell, making its final escape later, often while its cuticle is still soft and unexpanded. This is true of many higher Ditrysia, whose escape is facilitated by the flimsiness of the cocoon or the presence of a valve at one end of the cocoon through which the insect can force its way out while the ingress of other insects is prevented. The cocoon of *Saturnia* is of this type (Fig. 15.16a), while, in Megalopygidae, a trap door is present at one end.

Some Lepidoptera produce secretions which soften the material of the cocoon. *Cerura*, for instance, produces an

Fig. 15.27. Escape from pupal protective structures. **(a)** Ptilinum of a cyclorrhaphan fly. Dorsal views showing the ptilinum expanded (left) and retracted (right). **(b)** False mandible used by the beetle, *Polydrosus*, to escape from its cocoon. The false mandible is deciduous (after Hinton, 1946). **(c)** Cocoon cutter of the flea, *Trichopsylla*. The cutter is shed after escape from the cocoon (after Hinton, 1946).

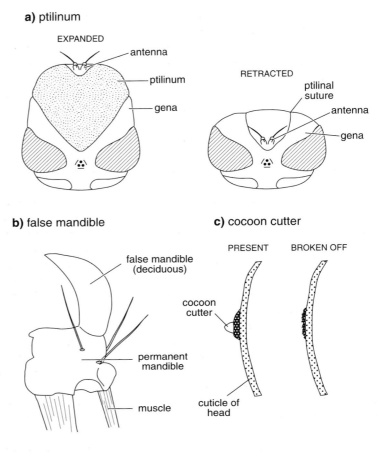

a) ptilinum

EXPANDED — antenna — ptilinum — gena

RETRACTED — ptilinal suture — antenna — gena

b) false mandible

false mandible (deciduous) — permanent mandible — muscle

c) cocoon cutter

PRESENT   BROKEN OFF — cocoon cutter — cuticle of head

oral secretion containing potassium hydroxide which softens one end of its cell of agglutinated wood chips. This enables the adult insect to push its way out protected by the remnants of the pupal cuticle. The silkmoths *Bombyx* and *Antheraea* produce a proteinase which softens the silk wall of the cocoon sufficiently for the adult to push its way out. In *Antheraea*, the proteinase is secreted on to the surface of the galeae two days before eclosion. It dries, forming a semi-crystalline encrustation. At the time of eclosion a liquid is secreted from labial glands which open by a single median pore just below the mouth. This liquid dissolves the enzyme and wets the inside of the cocoon. It contains potassium and functions as a buffer keeping the enzyme solution at about pH 8.5. The enzyme digests the sericin coating of the silk so that the fibroin threads are readily separated. A few Noctuidae also produce softening secretions.

Cyclorrhaphous Diptera have an eversible sac called the ptilinum at the front of the head which assists in their escape from the puparium (Fig. 15.27a). It can be expanded in the newly emerged fly by blood forced into the head from the thorax and abdomen and is then withdrawn again by muscles which force blood back to the thorax. Pressure of the ptilinum splits off the cap of the puparium and, if the puparium is buried in the soil, the ptilinum is also used by the fly to dig its way to the surface. Once the fly's cuticle has hardened, the ptilinum is no longer eversible and the muscles associated with it degenerate. Its position is indicated in the mature fly by the ptilinal suture.

The degree of hardening that these insects undergo before escaping from the cocoon varies. In some, most of the cuticle remains soft until after eclosion, but some parts, particularly those involved in locomotion, harden beforehand. Thus, in *Calliphora*, the legs and apodemes harden, so do the bristles which protect the soft cuticle, and such specialized parts as the halteres, antennae and genitalia. The remainder of the cuticle does not harden until after it is expanded when the insect is free. In Lepidoptera, however, the body does not expand greatly after eclosion and hardening of the cuticle is extensive before the insect emerges from the cocoon, although the wings remain limp.

Other insects emerge from the pupa and harden fully before making their escape from the cocoon and they may have specialized features to assist this. Coleoptera and Hymenoptera use their mandibles to bite their way out. Some weevils of the subfamily Otiorrhynchinae have an appendage, known as the false mandible, on the outside of the mandible (Fig. 15.27b), which is used in escaping from the cocoon and then, in most species, falls off. Amongst the Cynipidae (Hymenoptera), which do not feed as adults, the sole function of the adult mandibles is to allow the insect to escape from the host in which the larva pupated.

The cuticle of fleas also hardens before they escape from their cocoons and they may remain in the cocoon for some time after emergence. Their escape is stimulated by mechanical disturbances and in many species is facilitated by a cocoon cutter on the frons. In *Trichopsylla*, the cocoon cutter is deciduous (Fig. 15.27c). Finally, in the males of Strepsiptera, the mandibles are used to cut through the cephalothorax of the last stage larva in which pupation occurs. The larval cephalothorax is earlier extruded through the cuticle of the host so that the adult insect can easily escape (Hinton, 1946).

15.3.3.2 *Eclosion of aquatic insects*

Of the insects that pupate under water, some emerge under water and swim to the surface as adults; in others, the pupa rises to the surface before the adult emerges. In the Blepharoceridae, the adult undergoes some degree of hardening within the pupal cuticle, so that as soon as it emerges it rises to the surface and is able to fly. *Simulium* and *Acentropus* also emerge beneath the surface, but come up in a bubble of air. In *Acentropus*, this is derived from the air in the cocoon, while *Simulium* pumps air out into the gap between the pupal and adult cuticles. Thus the adult emerges into a bubble of air and is able to expand its wings before rising to the surface.

The pupae of Culicidae are buoyant, while some other insects, such as *Chironomus*, whose pupae are normally submerged, increase their buoyancy just before emergence by forcing gas out of the spiracles into the space beneath the pupal cuticle, or by increasing the volume of the tracheal system. Aided by backwardly directed spines, these pupae then escape from their cocoons or larval tubes and rise to the surface. Many Trichoptera swim to the surface as pharate adults, the middle legs of the pupae of these species being fringed to facilitate swimming, and the insect may continue to swim at the surface until it finds a suitable object to crawl out on. In other Trichoptera, the pharate adults crawl up to the surface while the last stage larvae of Odonata and Plecoptera crawl out onto emergent vegetation so that the adult emerges above the water. Larval Ephemeroptera also come to the surface, but the form that emerges is a subimago, not the imago. The subimago resembles the imago, but its legs and caudal filaments are shorter and the wings are translucent instead of transparent and are fringed with hairs. The subimago flies off as soon as it emerges, but settles a short distance away and soon molts to the imago. This is the only known example of a fully winged insect undergoing a molt.

The act of eclosion at the surface, in mosquitoes, results from the pharate adult swallowing air. Before the pupal cuticle is split, the air is probably drawn from the space between the adult and pupal cuticles and is derived from air drawn into the tracheal system through the respiratory horns of the pupa. A comparable process probably occurs in other insects with aquatic pupae.

15.3.3.3 *Timing of eclosion*

It is important that eclosion is timed so that an insect's life history is synchronized with suitable environmental conditions and so that the meeting of the two sexes is facilitated. Synchrony with the environment results from the common reaction to the environment of the members of the species. Temperature is particularly important since, to a large extent, it governs the rate of development and the activity of the insect. In long-lived species, a diapause may also be important in synchronizing eclosion so that the sexes meet (see below).

Apart from these general seasonal effects, many insects emerge at particular times of day, often at night or in the early morning. This probably gives the insect some degree of protection against predators while it is at its most vulnerable in the period before it is able to fly. Thus, in Britain, the last stage larvae of *Anax* (Odonata) which are ready to molt leave the water between 20.00 and 21.00 hours and by 23.00 hours most adults have emerged and are expanding their wings. The timing of emergence of *Sympetrum* (Odonata) and most tropical dragonflies is similar, although some temperate species emerge during

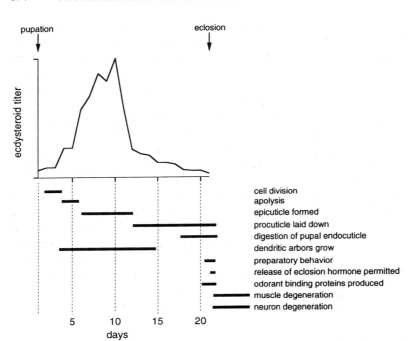

Fig. 15.28 Changes in ecdysteroid titers in the hemolymph of *Manduca* in relation to changes in different tissues during the pupal period (data from various sources).

the day, perhaps because activity is limited by low temperature at night.

Moths are also known to emerge at particular times of day, the timing being controlled by the eclosion hormone acting in response to photoperiod or temperature cycles and differing in its time of release from species to species. For example, with a cycle of 17 h light:7 h dark, *Hyalophora* escapes from the pupal cuticle in the first hours of the light period, while *Antheraea pernyi* does so towards the end of the light period. The emergence of *Manduca* is also affected by light, but is regulated more effectively by an increase in temperature of 3 °C or more. This is to be expected since the insect pupates beneath the soil surface.

It is probable that the time of emergence of many insects is determined by entrainment to light or temperature cycles. This is true of the moths described above and of *Drosophila*. In other cases, emergence may be a direct response to environmental conditions, even if these act at some previous time. For example, synchrony within a brood of *Aedes taeniorhynchus* is brought about by a tendency for the larvae to pupate at about sunset, but different broods emerge at different times depending on the temperature during the pupal period (Clements, 1992).

It is common for male insects to emerge as adults a little before the females, although the difference is not great. This probably reflects, in part, the smaller size of many male insects.

15.4 CONTROL OF POSTEMBRYONIC DEVELOPMENT

Cuticular growth and changes in form only occur when an insect molts. These processes are governed, primarily, by two classes of hormone: juvenile hormone and molting hormones (ecdysteroids) (Chapter 21). Molting is induced and regulated by ecdysteroids. As their titer in the hemolymph rises, the epidermal cells exhibit a complex pattern of DNA and RNA synthesis. This is known as the preparatory phase; it ends with division of the epidermal cells (Fig. 15.28 shows some of the processes involved at the molt to adult in *Manduca*; similar changes occur in the epidermis and cuticle at all molts). Except for cells in the imaginal discs of holometabolous insects, this is the only time when mitosis occurs in epidermal cells. At this time, with the ecdysteroid titer approaching its peak, apolysis occurs and the epidermal cells produce the new epicuticle. Then, the falling titer of ecdysteroid leads to the production of chitin and the proteins of the new procuticle. Ecdysis follows soon afterwards. (For a more complete account of the cuticular changes occurring during a molt, see section 16.4.)

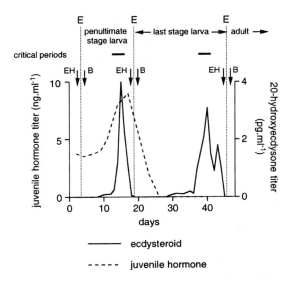

Fig. 15.29. Changes in hormone titers regulating molting and metamorphosis in a hemimetabolous insect. At the molt from larva to larva, juvenile hormone is present during the critical period; at the molt from larva to adult there is a critical period, but no juvenile hormone is present at this time. Experimental application of juvenile hormone during this sensitive period would lead to the production of another larval stage. Eclosion hormone (EH) and bursicon (B) are assumed to be produced for a brief period immediately before and after each ecdysis (E), respectively. Their production in this insect has not been demonstrated experimentally (*Nauphoeta*, Blattodea) (based on Lanzrein *et al.*, 1985).

The type of cuticle produced, whether it is larval, pupal (in the case of holometabolous insects) or adult depends on whether or not juvenile hormone is present during a critical period in each developmental stage. This critical period usually occurs when the ecdysteroid initiates the next molt. In hemimetabolous insects, if juvenile hormone is present during the critical period, the insect retains its larval characteristics; if juvenile hormone is absent, the insect becomes an adult (Fig. 15.29). It is apparent that, in the absence of juvenile hormone, a program of gene expression is initiated which differs from that occurring when juvenile hormone is present.

In holometabolous insects, similar critical periods occur (Fig. 15.30). In the absence of juvenile hormone, the pupal program is initiated. A small peak of ecdysteroid (called the commitment peak) on the fourth day of the final larval stage in *Manduca* follows the removal of juvenile

hormone from the hemolymph. After this, larval cuticular proteins are no longer produced and the insect is committed to metamorphosis. There is, however, a second critical period in the final larval stage when juvenile hormone or its immediate precursor is normally present. Juvenile hormone at this stage prevents the premature development of the imaginal discs. After the molt to pupa, the epidermal cells switch to the production of adult cuticle, but this can be prevented experimentally by the presence of juvenile hormone during a critical period. If juvenile hormone is applied at this time, the insect forms a second pupa.

Although the major changes are effected by these two hormones, many other factors are concerned in regulating molting. The sequence of events is: initiation of molting, switching on the production of ecdysteroids, controlling the process of ecdysis, and controlling sclerotization. These are described in the following sections and illustrated diagrammatically in Fig. 15.31.

Reviews: Nijhout, 1994; Riddiford, 1985; Riddiford & Truman, 1993; Steel & Davey, 1985

15.4.1 Initiation of molting and metamorphosis
The factors responsible for initiating molting are only poorly understood. In general, it is probably true that the insect responds to reaching a certain size, but how it measures size is not known in most cases. In *Rhodnius*, molting is induced by a full meal of blood resulting in abdominal distension that stimulates receptors in the ventral body wall. Abdominal stretch is also important in the milkweed bug, *Oncopeltus*, although here the stretch is achieved progressively over a prolonged period of feeding. In the larva of *Manduca*, however, the insect initiates molting when it reaches a certain body weight. It is not known how the caterpillar assesses its size, but stretching of the body wall is not involved.

The size at which molting occurs is not absolute, but depends on the insect's size at the beginning of the stage. Some species molt if they are starved and they may become smaller. This is known to occur in beetles and caterpillars living in stored products such as grain or flour. Clearly some factor apart from size is involved in initiating molting in these insects.

Metamorphosis occurs when ecdysteroids are produced in the absence of juvenile hormone. In *Manduca*, the switch to metamorphosis is related to head size at the beginning of the larval stage. If the head width is less than 5 mm at this time, juvenile hormone production continues and

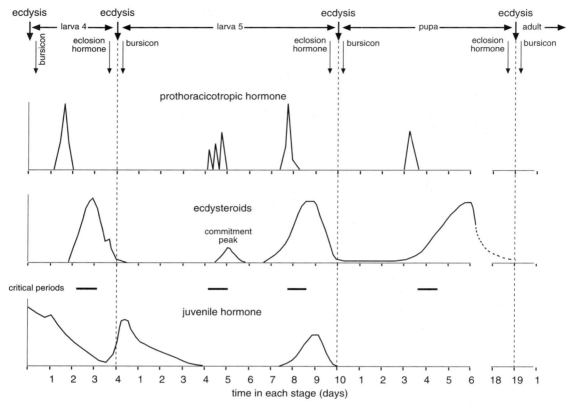

Fig. 15.30. Changes in hormone titers regulating molting and metamorphosis in a holometabolous insect. At the molt from larva to larva, juvenile hormone is present during the critical period; at the molt from larva to pupa, no juvenile hormone is present at the first critical period. The second critical period of sensitivity to juvenile hormone in the fifth stage larva regulates development of the imaginal discs. Eclosion hormone and bursicon are produced for a brief period before and after each ecdysis (based on data for *Manduca*, Lepidoptera).

the larva is committed to molting to another larva. If head width exceeds 5 mm, however, the larva is committed to become a pupa at the next molt (Fig. 15.32). Head size may not be the relevant parameter for the insect, but it seems likely that some aspect of size is relevant. It is not known how the insect is able to assess its size, or how the information is transmitted within the body. If the larva of *Manduca* is committed to pupate, juvenile hormone production is switched off when the insect reaches a weight of about 5 g.

15.4.2 Production of ecdysteroid hormones

The prothoracic gland is stimulated to produce ecdysone by the prothoracicotropic hormone (PTTH) (Chapter 21). Release of this hormone, usually via the corpora cardiaca (but from the corpora allata in Lepidoptera), results from

the presumed neural signal produced by size. Each period of ecdysteroid production is preceded by a peak of PTTH. In the final stage larva of *Manduca*, and perhaps also of other insects, the release of PTTH is inhibited by a high titer of juvenile hormone in the hemolymph although in earlier larval stages PTTH is released in the presence of juvenile hormone.

In *Manduca*, PTTH release only occurs within a certain period of the day, extending from just before until sometime after lights-out in the laboratory. Its production is said to be gated (Fig. 15.33). This probably ensures that, under natural conditions, the insect molts during the night. Photoperiodic gating of PTTH secretion is probably a common phenomenon.

Ecdysteroid hormones regulate many other activities

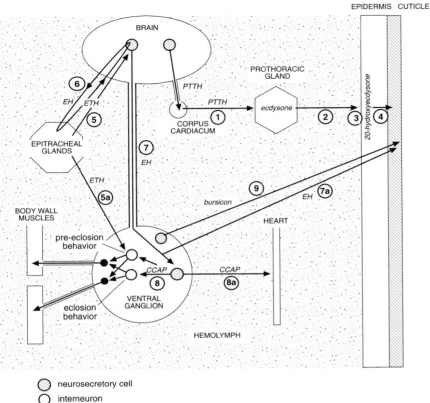

CONTROL OF APOLYSIS AND CUTICLE PRODUCTION

1 PTTH stimulates synthesis and release of ecdysone

2 ecdysone in hemolymph

3 ecdysone hydroxylated at tissues

4 20-hydroxyecdysone regulates genes producing cuticle

CONTROL OF ECDYSIS

5 ecdysis triggering hormone causes release of eclosion hormone

5a ecdysis triggering hormone switches on pre-eclosion behavior

6 positive feedback loop between ETH and EH results in massive release of EH

7 central release of EH causes release of CCAP

7a EH acting via hemolymph plasticizes cuticle

8 CCAP switches on eclosion behavior and switches off pre-eclosion behavior

CONTROL OF EXPANSION AND SCLEROTIZATION

8a CCAP acting via hemolymph increases heartbeat

9 bursicon first plasticizes cuticle, then switches on cuticular sclerotization

Fig. 15.31. The hormones involved in regulation of events at a molt. Juvenile hormone is not shown. Names of hormones are italicized. CCAP, crustacean cardioactive peptide; EH, eclosion hormone; ETH, ecdysis triggering hormone; PTTH, prothoracicotropic hormone.

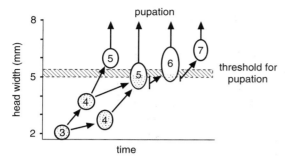

Fig. 15.32. Critical head size for the termination of larval development. The vertical axis of each ellipse indicates the range of head widths that might occur at each stage. If the head width is less than the threshold immediately following a molt (shaded areas in ellipses), the insect will subsequently molt to another larval stage. If the head width is greater than the threshold, the next molt will be to a pupa. Number in each ellipse is the larval stage. *Manduca* normally has five larval stages, but with poor food quality more stages are produced (after Nijhout, 1975).

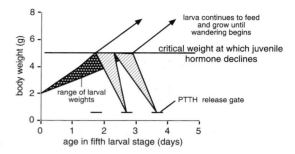

Fig. 15.33. The critical body weight at which the molting program is initiated in the final larval stage of *Manduca*. The release of prothoracicotropic hormone (PTTH) is gated. Larvae developing more slowly miss the gate for PTTH release on day 3 and are delayed until day 4 (after Nijhout, 1981).

in addition to the activity of the epidermal cells (see Fig. 15.28). The commitment peak in the final stage of caterpillars causes gut purging and wandering (see above). The prepupal peak of ecdysteroids causes the dendritic arbors of some motor neurons in *Manduca* to regress and other neurons to die. It also initiates the degeneration of muscles specifically associated with ecdysis, although this may be offset while the neurons innervating these muscles continue to be active. Changes in the ecdysteroid titer in the pupa and pharate adult are responsible for further changes in the nervous system including the regrowth of motor neuron dendrites, the development of olfactory glomeruli and of sensitivity in the adult olfactory system (Fig. 15.28) (Vogt *et al.*, 1993).

The action of ecdysteroid hormones depends on the presence of ecdysteroid receptors in the target tissues. The occurrence of these varies temporally, and there are at least three isoforms. They may be present in many or all tissues, in which case the hormone provides a general cue, or they may be present only in specific cells of specific tissues, so that the hormone exerts specific regulatory effects in addition to the general one (Fahrbach, 1992; Riddiford & Truman, 1993).

15.4.3 Control of ecdysis

Three neuropeptides are involved in the control of eclosion in *Manduca*, and the process is probably similar at other molts and in other insects. They are: ecdysis triggering hormone from the epitracheal glands, eclosion hormone from neurosecretory cells in the brain, and crustacean cardioactive peptide (CCAP) from cells in the ventral nerve cord (see Chapter 21 for details of the hormones and their origins). These neuropeptides interact to induce the behaviors that culminate in eclosion (Fig. 15.31).

The initiation of eclosion is dependent on signals indicating that the insect is physiologically in an appropriate state and that it is appropriately positioned for the process to be completed successfully. (Grasshoppers, for example, suspend themselves from their hind legs before a molt). These signals are largely unknown, but the falling ecdysteroid titer may permit release of the neuropeptides as well as promoting the sensitivity of target tissues to eclosion hormone.

Ecdysis triggering hormone is the first to be released into the hemolymph at the appropriate time. It acts directly on cells in the ventral ganglia of the nerve cord to switch on pre-ecdysis behavior (see below). It also stimulates cells in the brain to release eclosion hormone. Eclosion hormone and ecdysis triggering hormone then appear to form a positive feedback loop, each stimulating the release of the other, so their concentrations rapidly build up to a peak about an hour before eclosion occurs, declining over the next few hours. In *Manduca*, the release of eclosion hormone (and perhaps ecdysis triggering hormone) is gated so that eclosion occurs at a specific time of day, but it is not gated at earlier molts. Gated production of eclosion hormone probably accounts for the specific times of eclosion in many other insects cited in section 15.3.3.

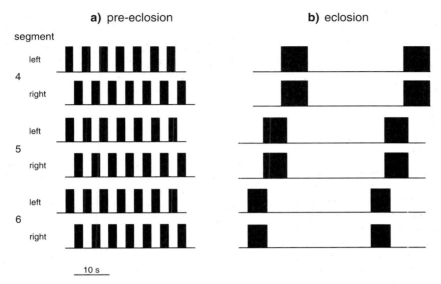

a) pre-eclosion **b)** eclosion

segment

left

4

right

left

5

right

left

6

right

10 s

Fig. 15.34. Patterns of motor activity in nerves to the abdominal muscles initiated by hormones at eclosion of an adult moth (*Manduca*) (see Fig. 15.31) (based on Truman, 1978). (**a**) Pre-eclosion behavior is switched on by ecdysis triggering hormone. (**b**) Eclosion behavior is switched on by crustacean cardioactive peptide, the release of which is stimulated by eclosion hormone.

Eclosion hormone is released locally within the central nervous system as well as into the hemolymph, and this leads to the release of CCAP in the ventral ganglia. Crustacean cardioactive peptide switches off pre-eclosion behavior and switches on eclosion behavior. Acting via the hemolymph, it also causes an accelerated heartbeat.

The programs of motor neuron activity which produce pre-eclosion and eclosion behavior are built into the abdominal ganglia and, in some insects, can be produced even in the isolated abdominal nerve cord. Pre-eclosion behavior consists of a series of abdominal rotations. These are produced by the muscles of the two sides contracting in antiphase, while those of successive segments contract together (Fig. 15.34a). This behavior, which probably serves to free the adult abdomen from the overlying pupal cuticle, lasts about 30 minutes in *Hyalophora*. It is followed by a period of quiescence of about the same duration and then by eclosion behavior. The latter consists of waves of abdominal contraction and of 'shrugging' of the wing bases that continues until eclosion is complete. These movements are produced by the synchronous activity of muscles on the two sides of the body with activity spreading forwards from the posterior abdominal segments (Fig.

15.34b). After eclosion, this behavior is inhibited by higher centers in the central nervous system.

Eclosion hormone also regulates plasticization of the cuticle of the wings, the release of bursicon, and breakdown of the eclosion muscles in silkmoths.

Reviews: Hesterlee & Morton, 1996; Horodyski, 1996; Truman, 1985, 1988; also see Ewer, Gammie & Truman, 1997

15.4.4 Controlling expansion and sclerotization

After ecdysis, wing spreading behavior is triggered via the subesophageal ganglion. Tonic contraction of the abdomen forces blood into the wings and wing spreading begins. The start of this activity triggers the release of the hormone bursicon from the neurohemal organs associated with the abdominal ganglia. Bursicon first produces a further increase in the plasticity of the wing cuticle and then initiates tanning. The release of bursicon is prevented by mechanical contact of the insect with its surroundings such as occurs while the insect is still within its old cuticle. Similar inhibition occurs in insects, such as some flies, that pupate below the surface of the ground, and in first stage grasshoppers when they hatch from their underground eggs. In both cases, bursicon release is inhibited while the

Fig. 15.35. Polyphenism. Control of development of female honeybees. The development of larvae diverges in the final (fifth) stage as the production of juvenile hormone by the corpora allata becomes markedly higher in larvae destined to become queens. This corresponds with a JH-sensitive period late in the fourth stage when the switch from workers to queens occurs. The small peak in juvenile hormone production before pupation corresponds with a second JH-sensitive period, perhaps regulating development of the imaginal discs (c, time at which cells are capped by workers) (after Rachinsky & Hartfelder, 1990, and other sources).

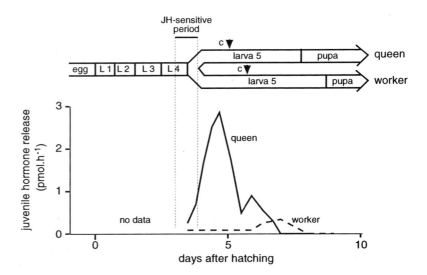

insects are digging, but begins as soon as they are free on the surface.

In *Manduca*, and probably in other insects, a cardioactive peptide accelerates the pumping activity of the heart facilitating wing expansion.

Review: Reynolds, 1985

15.5 POLYPHENISM

The term polymorphism has been used in a general sense to indicate variation in body form or color within a species. It is evident, however, that, while some instances of polymorphism are under strict genetic control, others are a product of the environment in which the insect develops. These instances are now usually referred to as polyphenism. Thus, polyphenism refers to the occurrence of different phenotypes within a species where the development of the phenotype is governed by environmental conditions.

Polyphenism is a characteristic feature of social insects, where the different castes are determined by the needs of the colony, and of aphids, where different morphs occur seasonally. In addition, many insects exhibit seasonal variations in form that can be attributed to environmental conditions. Different color forms, or differences in wing development are among the more common.

In all these cases, the environmental conditions modulate the amounts or timing of hormone secretion. Juvenile

hormone or ecdysone are often involved, but, in other cases, neurosecretions from the brain provide the link between the nervous system and the tissues.

Castes of social insects Amongst the social Hymenoptera, the queen and workers are all female. In *Apis*, the newly hatched female larva has the potential to develop into either form. Its subsequent development is determined by the quality and quantity of the food supplied to it. If given a diet based primarily on secretions of the nurse bees' mandibular glands, it will become a queen. If it receives larger proportions of secretion from the hypopharyngeal glands of the workers, it will itself become a worker. Similarly, amongst the larvae of worker ants, the differentiation of minor or major workers, or of soldiers, is dependent on feeding, which is ultimately reflected in the sizes of the larvae.

These size differences lead to different titers of juvenile hormone which controls the insects' development. The switch from worker to queen in *Apis* is made in the late fourth and early fifth larval stages. Queen larvae develop a high titer of juvenile hormone late in the fourth larval stage, whereas in workers the titer is very low (Fig. 15.35). Amongst larval ants, major worker or soldier development is apparently the result of a high titer of juvenile hormone early in the final larval stage. If the titer is above a threshold, the critical size at which pupation occurs is reset so that the larva continues to develop until

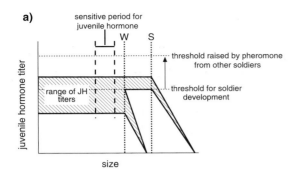

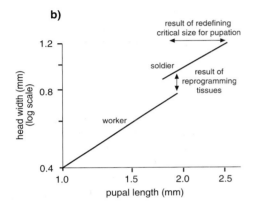

Fig. 15.36. Polyphenism. Control of soldier development in the ant *Pheidole*. (a) Soldier development is determined by the hemolymph titer of juvenile hormone. Above a threshold titer, the critical size for completion of larval development (dotted vertical lines) is increased and the larva is destined to become a soldier. Soldiers already in the colony produce a pheromone which raises the threshold for juvenile hormone so that production of more soldiers is unlikely to occur. W, size threshold for pupation as worker; S, size threshold for pupation as soldier (after Wheeler & Nijhout, 1984). (b) Reprogramming of the critical size for completion of larval development shown in (a) results in the production of adults with greater head widths. At the same time adult tissues are reprogrammed to those of soldiers, resulting in a break in the allometric relationship between head width and body size (after Wheeler & Nijhout, 1983).

it reaches a new threshold (Fig. 15.36a). The threshold level for soldier development is variable and soldiers already present in a colony can suppress the development of further soldiers, apparently via a pheromone comparable with those that regulate colony structure in termites (section 27.1.6.3). It is believed that the pheromone raises

the threshold for juvenile hormone that must be reached for the switch to soldier production to occur. Not only does the high titer of juvenile hormone reset the critical size for pupation, it may also reprogram the development of imaginal tissues so that some parts of the body are increased in size relative to the overall size of the insect (Fig. 15.36b).

Aphid morphs Aphids occur in a variety of different forms. Most generations consist entirely of parthenogenetic females, but a sexual generation usually occurs each year. Some, like *Aphis fabae*, have separate winter and summer hosts. Movement between the two hosts demands the production of winged morphs (alatae or alates), whereas at other times the aphids are commonly wingless (apterae) although winged forms may also be produced on the secondary, summer host (Fig. 15.37).

The sexual generation is usually produced in response to the short days of autumn acting on the parent aphid. Males are produced by the loss of an X chromosome during the single maturation division of the oocytes [female aphids have two X chromosomes, males only one (section 14.5.3)] mediated by a neurohormone acting directly on the oocytes (Blackman, 1980).

Under long photoperiods, the largest embryos in the ovarioles are determined for the production of more virginoparae (insects that will produce more parthenogenetic forms). Under short photoperiods, however, the embryos become determined as gynoparae (giving rise to sexual forms) or directly as oviparae (the sexual female). The production of virginoparae is stimulated by a neurosecretion from cells in the brain which are influenced by light passing directly through the cuticle of the head, not via the compound eyes. Juvenile hormone is also involved in virginipara production.

The production of winged (alate) forms is also under environmental control, and is induced by crowding, short photoperiods, and low temperatures. The production of winged gynoparae in *Aphis fabae* requires exposure of both the parent female and the postnatal insect to short days. Initially, the embryo is affected indirectly by the light regime experienced by its parent, but well-developed embryos are themselves sensitive to photoperiod (Hardie & Lees, 1983; Vaz Nunes & Hardie, 1992). It appears that juvenile hormone acting early in development suppresses wing development, but how the environmental factors regulate juvenile hormone titers is unclear.

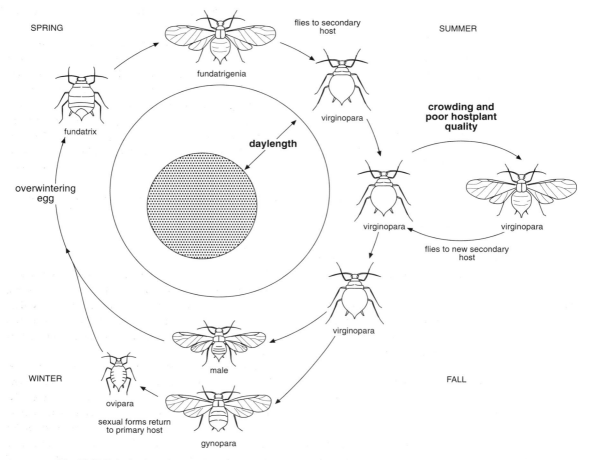

Fig. 15.37. Polyphenism. Control of wing development and the production of sexual forms in an aphid. Seasonal changes in form result largely from responses to long or short daylengths. The production of winged virginopara in summer is a response to crowding or poor hostplant quality (based on Hardie & Lees, 1985).

Brachyptery in other insects Short-winged forms are known to occur in many species of Orthoptera and Hemiptera. Long wings are often produced in response to crowding or poor food quality. The suppression of wing development is a response to high titers of juvenile hormone in final stage larvae, but experiments on crickets indicate that this effect may be very subtle (Zera & Tanaka, 1996).

Color and Form Many insects exhibit color variation in response to environmental changes. These may be regular occurrences, as in the seasonal forms of some butterflies, or more unpredictable, as in the phenomenon of homochromy (section 25.5.2.2).

Crowding the larvae of some Lepidoptera and grasshoppers causes them to become darker, or to have more dark patterning, than siblings reared in isolation. This is due to the deposition of greater quantities of melanin in the cuticle, which in caterpillars is controlled by a neuropeptide. At least in *Spodoptera littoralis*, crowded larvae have higher hemolymph titers of this neuropeptide (Altstein, Ben-Aziz & Gazit, 1994). The same peptide in the adults controls pheromone biosynthesis (section 27.1.5).

In locusts, insects reared in isolation are usually green with little black pigmentation; those reared in crowds are yellow or orange with extensive black pigmentation. The pronotum of *Locusta* reared in isolation is produced into

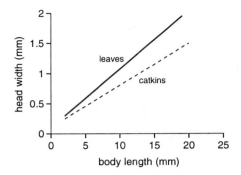

Fig. 15.38. Polyphenism. The relationship between head width and body size in relation to feeding by caterpillars of *Nemoria arizonaria*. Insects feeding on the harder food (leaves) develop progressively bigger heads (after Greene, 1989).

a prominent dorsal crest, while in crowded insects it is saddle shaped. Insects reared in crowds have fewer sensilla on the antenna than their isolated siblings (Greenwood & Chapman, 1984). Similar changes in color and anatomy may occur in non-swarming grasshoppers and differences in an insect's exposure to odors may affect the numbers of sensilla on the antennae, but whereas locusts also exhibit marked changes in behavior as a consequence of crowding, grasshoppers usually do not do so. The regulatory mechanisms underlying these differences are not well understood although neurosecretion from the brain is known to be involved. Juvenile hormone titers also differ in the final stage larvae depending on whether they are crowded or not (Pener, 1991). A behavioral change can occur within four hours of a switch from isolation to crowding (Bouaïchi, Roessingh & Simpson, 1995).

An extreme example of polyphenism which involves differences in morphology as well as in color occurs in the caterpillar of *Nemoria arizonaria*. The spring generation of this species resembles the catkins of the oak trees on which it feeds. A second generation occurs when the tree is in leaf. The insects of this generation feed on the leaves and are twig mimics rather than catkin mimics. They are greenish-grey in color, rather than yellow, and they lack many of the cuticular outgrowths which help to give insects in the spring generation their resemblance to catkins. The development of the two phenotypes is determined by the food eaten by the insects and, since the leaves of these oaks are very tough, the insects of the summer generation develop relatively larger heads (Fig. 15.38) (Greene, 1989). It is possible that food quality affects form

and color through a peptide comparable with that described above in *Spodoptera*.

Many adult butterflies also exhibit seasonal color polyphenism. Different color morphs are produced by long or short days, acting in concert with high or low temperatures. These environmental effects produce differences in the timing of ecdysteroid action sometimes, perhaps, synergized by the action of a neuropeptide from the brain (Koch, Brakefield & Kesbeke, 1996; Rountree & Nijhout, 1995). Production of the different morphs is often associated with the presence or absence of a pupal diapause (see below).

Reviews: Hardie & Lees, 1985; Nijhout, 1994; Zera & Denno, 1997

15.6 DIAPAUSE

Diapause is a delay in development evolved in response to regularly recurring periods of adverse environmental conditions. It is not referable to immediately prevailing adverse environmental conditions, and thus differs from a delay produced by currently adverse conditions such as low temperature.

In temperate regions, diapause facilitates winter survival; in the tropics, it is commonly associated with surviving regularly occurring dry seasons. Diapause also contributes to the synchronization of adult emergence so that the chances of finding a mate are considerably improved. This is particularly necessary in long lived species such as the dragonfly, *Anax*, in which larval development usually extends over two summers and during this time individual rates of development vary so that at any one time a wide variety of larval sizes is present.

15.6.1 Occurrence of diapause
Many insects living in regions where winter temperatures are too low for development enter a state of diapause at some stage of the life history, but the occurrence of diapause in insects from warmer regions depends on the severity of their environment and the conditions in their particular micro-habitat. Many insects in the tropics survive without a diapause.

Diapause can occur in any stage of development from the early embryo to reproductive adult, but in the majority of diapausing species, only one stage exhibits diapause (Fig. 15.39). Egg diapause is common in grasshoppers and Lepidoptera, occurring early in embryogenesis in some species, and much later in others. Many holometabolous

Fig. 15.39. Diapause in different insects showing the separation of the sensitive stage (dotted) from the stage of diapause (black) (based on Nijhout, 1994).

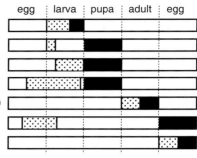

	egg	larva	pupa	adult	egg
Diatraea grandiosella (Lepidoptera)					
Sarcophaga crassipalpis (Diptera)					
Sarcophaga argyrostoma (Diptera)					
Manduca sexta (Lepidoptera)					
Leptinotarsa decemlineata (Coleoptera)					
Bombyx mori (Lepidoptera)					
Lymantria dispar (Lepidoptera)					

[⋮⋮⋮] sensitive period   [■] diapause period

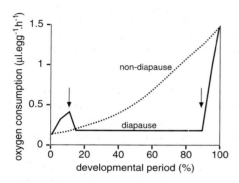

Fig. 15.40. Oxygen consumption during diapause in the egg of *Melanoplus*. The dotted line shows the progressive increase in consumption that would occur if the egg developed without diapause. Note that the developmental period is expressed as a percentage of the total time for embryonic development. In the absence of diapause this would be about 30 days; with diapause it could be 200 days. Arrows mark the beginning and end of diapause (partly from Lees, 1955).

insects diapause in the pupal stage. In eggs and pupae, morphogenesis comes to a standstill during diapause reflected in the reduced oxygen consumption (Fig. 15.40). Diapausing larvae may stop feeding altogether, but some species remain active and feed periodically. They may molt, but such molts are not associated with growth, and are called stationary molts. Adult diapause is characterized by a lack of sexual development or behavior. Feeding may occur in some species.

Some insects have an obligate diapause with every individual in every generation entering diapause. Insects with an obligate diapause usually have only one generation each year, a univoltine cycle. In other species, some generations may be completely free of diapause while in other generations some or all of the insects enter diapause. This is facultative diapause. It occurs in insects with two or more generations per year, a multivoltine cycle. Facultative diapause is well suited to regions with a long developmental season as it enables the insect to make the best use of the time available. The difference between obligate and facultative diapause is probably only one of degree. Insects with an obligate diapause apparently respond to such a wide range of environmental factors that they invariably undergo diapause, but in the laboratory it may be possible to avoid diapause by exposing them to conditions outside their normal range.

15.6.2 Initiation of diapause

The most reliable and consistent indicator of seasons is daylength, or photoperiod, and this is the most important of the stimuli initiating diapause. Other possible indicators are temperature, the state of the food, and the age of the parent. The sensitive stage at which diapause is induced of necessity occurs before that at which diapause occurs (Fig. 15.39). The embryonic diapause of *Bombyx* is an extreme example. Here diapause is primarily induced by exposure of the egg of the previous generation.

The period of sensitivity to stimulation also varies in different species. Larval *Diataraxia* (Lepidoptera) are only sensitive to photoperiod for two days, while in *Bombyx*, although the well-developed embryo is the most sensitive stage, the first three larval stages are also sensitive, although to decreasing extent. Further, a number of photoperiodic cycles are necessary in order to produce an effect. The larva of *Dendrolimus* (Lepidoptera), for instance, must be subjected to about 20 short-day impulses to induce diapause, while exposure of the larva to 15 and 11 short-day impulses are required to induce diapause in

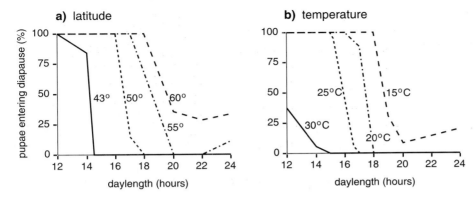

Fig. 15.41. Diapause in the moth *Acronycta* (after Danilevskii, 1965). **(a)** The effects of photoperiod. Diapause is prevented by long days, but the critical daylength, at which the switch from non-diapause to diapause development occurs, increases at higher latitudes. All insects were reared at 23 °C. **(b)** The interaction of photoperiod with temperature. The critical photoperiod for diapause is reduced in insects reared at higher temperatures. All insects were from the same population, from 50° N.

pupal *Acronycta* (Lepidoptera) and *Pieris*, respectively. This number varies, however, under the influence of temperature and nutrition.

Photoperiod Outside the tropics, long days occur in summer and short days in winter with increasing or decreasing daylength in spring or autumn. The relatively short days of autumn herald the approach of winter and for many species they act as a stimulus initiating diapause. There is a critical daylength around which small differences in photoperiod produce a complete change from non-diapause to diapause development (Fig. 15.41). As these insects develop without diapause under long-day conditions they are known as long-day insects.

Some insects react in the opposite way. *Bombyx* is an example of such a short-day insect. Here exposure of the eggs of the bivoltine race to long days, ensures that the eggs of the next generation will enter diapause (Fig. 15.42). Adults developing from eggs exposed to a photoperiod of less than 14 hours lay non-diapausing eggs in the next generation. Thus the long-day conditions of spring acting on the eggs lead to the production of diapause eggs in the autumn, but these eggs, being subjected to short days, will ensure that the eggs of the next generation, laid the following spring, develop without diapause.

Photoperiod interacts with temperature, and in long-day insects the critical day-length, below which diapause occurs, is often longer at lower temperatures. For example, the critical daylength inducing pupal diapause in

Acronycta is about 16 hours at 25 °C, but almost 19 hours at 15 °C. (Fig. 15.41b). In *Pieris*, however, the critical daylength is not influenced by temperature.

The response of many animals to photoperiod involves the perception of small changes in daylength rather than the actual duration of the light period, but in insects this is not the case. Most insects respond to the absolute length of the photoperiod. A few cases in which the insect is believed to respond to changes in daylength are recorded, but even these can probably be accounted for in terms of reactions to the absolute daylength.

Cells in the brain are directly sensitive to light in many insects, and, in these species, the compound eyes and ocelli do not mediate the photoperiodic response, and sometimes, as in the larva of *Pieris*, there is a relatively transparent area in the cuticle of the head above the sensitive area in the brain. On the other hand, in the beetle, *Pterostichus*, in which the cuticle of the head is heavily pigmented, the compound eyes mediate the response.

Light intensity during the photoperiod is not important provided that it exceeds a very low threshold value. This varies with the species, but commonly is about 170 lux or less. Hence daily fluctuations in light intensity due to clouds have no effect on photoperiod and the 'effective daylength' includes the periods of twilight. As a result of this high sensitivity, insects inside fruit and even the pupa of *Antheraea* inside its cocoon are affected by photoperiod. Some insects are so sensitive that they are stimulated by moonlight (about 5 lux) which might, therefore, contribute

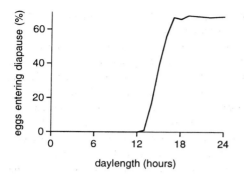

Fig. 15.42. Diapause in the egg of the silkmoth, *Bombyx*. Long days experienced by the embryo induce diapause in the eggs of the next generation (see Fig. 15.39) (after Danilevskii, 1965).

to the effective daylength. This effect, however, might be offset by the relatively low temperatures occurring at night.

In most insects, only short wavelengths are concerned in the photoperiodic reaction.

Review: Takeda & Skopik, 1997 – measurement of photoperiod

Temperature Temperature also plays a part in the induction of diapause and, in general, in temperate regions high temperature suppresses and low temperature enhances any tendency to enter diapause. Thus, it will reinforce photoperiod. With changes in latitude, however, this reinforcement does not occur as at higher latitudes, although summer days are longer, the temperatures are lower than they are nearer the equator. Hence species with extensive geographical ranges are differently adapted to photoperiod in different parts of their range. For instance, in southern Russia, *Acronycta* enters diapause only when the daylength falls below 15 hours, but with increasing latitude there is an increase in the critical daylength so that specimens from St.Petersburg (60 °N) only avoid diapause when the photoperiod exceeds 18 hours (Fig. 15.41a). These differences are inherited characteristics of the populations.

Because of these differences, the life history of *Acronycta* varies in different parts of its range. In the south (latitude 43 °N), three partly overlapping generations occur and only in the last do the pupae undergo diapause. Farther north (latitude 50 °N), two generations occur, the whole of the second being determined for diapause, while around St.Petersburg (latitude 60 °N) only a very small number of individuals avoid diapause and the species is largely univoltine. Comparable differences in the life

histories are known in other insects. *Bombyx* again differs from most other insects in that high temperature induces diapause and low temperature prevents it. In *Nasonia* (Hymenoptera), temperature acts independently of photoperiod, and chilling the female causes her to lay eggs which will give rise to diapausing larvae.

Other factors There is evidence in a few cases that the amount or quality of the food can influence diapause. A shortage of prey results in reproductive diapause in the green lacewing, *Chrysopa*, and hostplant condition, perhaps related to protein shortage, induces diapause in the larvae of a number of stem-boring moths.

Rearing conditions of the parents may sometimes be important. For example, solitary phase females of the brown locust, *Locustana*, lay 100% diapausing eggs compared with only 42% by gregarious phase females. Further, old gregarious phase females lay more diapausing eggs than young females.

15.6.3 Preparations for diapause

Before insects become inactive in diapause there is usually a build up of reserve food substances, particularly in the fat body, with a consequent reduction in the proportion of water in the body. Comparison of different forms determined for diapause or non-diapause shows that those destined for diapause build up bigger food reserves. In adult *Pyrrhocoris* (Hemiptera), for example, diapause-associated storage proteins accumulate in the hemolymph (Sula, Kodrík & Socha, 1995), and, in *Bombyx*, eggs destined for diapause receive about 10% more lipid during vitellogenesis than non-diapause eggs, and the lipid contains a higher proportion of unsaturated fatty acids (Shimuzu, 1992).

Pupal diapause in some moths and flies is associated with the presence of considerably greater quantities of hydrocarbons in the cuticular surface wax. In *Manduca*, wax continues to be secreted for about 10 days after pupation in diapause pupae, whereas in non-diapause pupae wax production is complete in four or five days (Fig. 15.43). The puparium of diapausing *Sarcophaga crassipalpis* has about double the quantity of hydrocarbons compared with non-diapause insects, most of it on the inside of the cuticle. This increase in wax is probably responsible for a reduction in water loss (Yoder *et al.*, 1992).

In many insects from temperate regions, diapause is associated with an increase in cold tolerance, although the

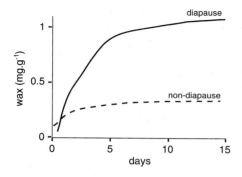

Fig. 15.43. Preparation for diapause. The laying down of cuticular wax in the first few days of pupal life in diapause and non-diapause *Manduca* (after Bell *et al.*, 1975).

two phenomena are not necessarily closely linked. Diapause is frequently associated with the accumulation of polyhydric alcohols, such as sorbitol and glycerol, which can function as cryoprotectants. However, the amounts that accumulate initially may not be sufficient to affect the supercooling temperature of the insect. Rather, they appear to be formed as a result of the general depression of metabolic activity. Further increase in these cryoprotectants may occur later as part of a thermal acclimation process which is associated with, but is not a component of, diapause (section 19.3.2) (Hodkova & Hodek, 1994; Pullin & Wolda, 1993).

15.6.4 Diapause development

Except in the adult, a delay in morphogenesis is characteristic of diapause, but, although morphogenesis is at a standstill, physiological changes do occur. This physiological development is referred to as diapause development. As with other physiological processes, diapause development occurs most rapidly under certain environmental conditions of which temperature is often of overriding importance. The range of temperatures at which diapause development occurs, unlike that for morphogenesis, varies with the geographical distribution of the species and, in temperate regions, the optimum temperature for diapause development is commonly within the range 0–10 °C, well below the temperature necessary for morphogenesis (Fig. 15.44). At higher or lower temperatures, diapause development proceeds more slowly and in the extremes stops altogether. On the other hand in tropical species, where diapause is concerned with survival of the dry season, the temperature range for

diapause development is often little, if any, lower than the range for morphogenesis.

Diapause is commonly associated with a relatively low water content of the tissues. This probably enhances the ability of the insect to survive periods of extremely low temperature, but the role of water in diapause development is not clear. In the eggs of many orthopteran species, morphogenesis is not resumed until water is available, but this water is only effective after a period of diapause development has been completed. Hence, although water is essential for the resumption of morphogenesis, as is true even of grasshopper eggs developing without diapause (section 14.1.3.2), it is not concerned in diapause development. A similar restoration of the water balance accompanying reactivation after diapause is known to occur in other insects in other stages of development.

The duration of diapause development varies considerably with conditions and from species to species. Under optimal conditions the diapause development of *Gryllulus commodus* (Orthoptera) is completed in 15 days; on the other hand *Cephus* (Hymenoptera) requires a minimum of 90–100 days. Once diapause development is complete, morphogenesis is resumed provided the environmental conditions are suitable. If they are not, the insect remains in a state of quiescence until conditions become more favorable.

In the vast majority of insects, photoperiod is of no importance once diapause has been initiated, but, in a few cases, it directly affects the duration of diapause. For example, the larva of *Dendrolimus* (Lepidoptera) resumes its growth after a delay of only two weeks if the days during diapause are long, but remains dormant for twice as long in short days.

15.6.5 Control of diapause

Diapause in the larva and pupa, and probably also in late embryonic stages, results from a deficiency of ecdysone so that growth and molting do not occur. The release of ecdysteroids is normally triggered by prothoracicotropic hormone and the lack of this hormone is ultimately responsible for the insect's failure to continue development. In the larvae of many species, high titers of juvenile hormone may suppress PTTH secretion, but this is not always true. Sometimes juvenile hormone seems to be necessary only at the very beginning of diapause, and in the wasp, *Nasonia*, it seems to play no part at all. Similarly in lepidopteran pupae, the suppression of PTTH secretion

Fig. 15.44. Diapause development. Temperature ranges for diapause development and normal morphogenesis. (**a**) Eggs of the Australian grasshopper, *Austroicetes* (based on Andrewartha, 1952). (**b**) Pupae of the moth, *Saturnia* (adapted from Lees, 1955).

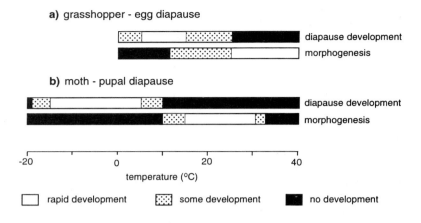

does not seem to be a consequence of high juvenile hormone titers.

Diapause early in embryogenesis is produced by a diapause hormone in the parent female. In *Bombyx*, the hormone acts only on oocytes weighing about 500 μg; smaller or larger oocytes are not affected. The hormone results in an increase in trehalase activity in the oocytes leading to the conversion of hemolymph trehalose to glycogen in the yolk (Yamashita, 1996). In addition, the hormone causes a change in pigmentation of the serosa due to the uptake of 3-hydroxykynurenine which is subsequently converted to an ommochrome.

Adult diapause is linked to the absence of juvenile hormone which is usually involved with vitellogenin synthesis or uptake (section 13.2.4.2), and, in males, with the development of reproductive behavior.

Reviews: Danilevskii, 1965; Denlinger, 1985; Tauber, Tauber & Masaki, 1986

REFERENCES

Altstein, M. Ben-Aziz, O. & Gazit, Y. (1994). Pheromone biosynthesis activating neuropeptide (PBAN) and colour polymorphism: an immunochemical study in *Spodoptera littoralis*. *Journal of Insect Physiology*, **40**, 303–9.

Andrewartha. H.G. (1952). Diapause in relation to the ecology of insects. *Biological Reviews of the Cambridge Philosophical Society*, **27**, 50–107.

Bell, R.A., Nelson, D.R., Borg, T.K. & Cardwell, D.L. (1975). Wax secretion in non-diapausing and diapausing pupae of the tobacco hornworm, *Manduca sexta*. *Journal of Insect Physiology*, **21**, 1725–9.

Bernays, E. A. (1972a). The muscles of newly hatched *Schistocerca gregaria* larvae and their possible functions in hatching, digging and ecdysial movements (Insecta: Acrididae). *Journal of Zoology, London*, **166**, 111–58.

Bernays, E. A. (1972b). The intermediate moult (first ecdysis) of *Schistocerca gregaria* (Forskål) (Insecta: Orthoptera). *Zeitschrift für Morphologie der Tiere*, **71**, 160–79.

Blackith, R. E., Davies, R. G. & Moy, E. A. (1963). A biometric analysis of development in *Dysdercus fasciatus* Sign. (Hemiptera: Pyrrhocoridae). *Growth*, **27**, 317–34.

Blackman, R.L. (1980). Chromosomes and parthenogenesis in aphids. *Symposium of the Royal Entomological Society of London*, **10**, 133–48.

Booker, R. & Truman, J.W. (1987). Postembryonic neurogenesis in the CNS of the tobacco hornworm, *Manduca sexta*. I. Neuroblast arrays and the fate of their progeny during metamorphosis. *Journal of Comparative Neurology*, **255**, 548–59.

Bouaïchi, A., Roessingh, P. & Simpson, S.J. (1995). An analysis of the behavioural effects of crowding and re-isolation on solitary-reared adult desert locusts (*Schistocerca gregaria*) and their offspring. *Physiological Entomology*, **20**, 199–208.

Bryant, P.J. (1975). Pattern formation in the imaginal wing disc of *Drosophila melanogaster*: fate map, regeneration and duplication. *Journal of Experimental Zoology*, **193**, 49–78.

Bullière, D. & Bullière, F. (1985). Regeneration. In *Comprehensive Insect Physiology, Biochemistry and Pharmacology*, vol. 2, ed G.A. Kerkut & L.I. Gilbert, pp. 372–424. Oxford: Pergamon Press.

Byrne, D.N. & Bellows, T.S. (1991). Whitefly biology. *Annual Review of Entomology*, **36**, 431–57.

Campbell, G. & Tomlinson, A. (1995). Initiation of the proximodistal axis in insect legs. *Development*, **121**, 619–28.

Caveney, S. (1985). Intercellular communication. In *Comprehensive Insect Physiology, Biochemistry and Pharmacology*, vol. 2, ed G.A. Kerkut & L.I. Gilbert, pp. 320–70. Oxford: Pergamon Press.

Chapman, R.F. (1982). Chemoreception. The significance of receptor numbers. *Advances in Insect Physiology*, **16**, 247–356.

Chapman, R.F., Bernays, E.A., Woodhead, S., Padgham, D.E. & Simpson, S.J. (1983). Control of hatching time of eggs of *Chilo partellus* (Swinhoe) (Lepidoptera: Pyralidae). *Bulletin of Entomological Research*, **73**, 667–77.

Chapman, R.F. & Fraser, J. (1989). The chemosensory system of a monophagous grasshopper, *Bootettix argentatus*. *International Journal of Insect Morphology and Embryology*, **18**, 111–8.

Chen, S. H. (1946). Evolution of the insect larva. *Transactions of the Royal Entomological Society of London*, **97**, 381–404.

Clarke, K. U. (1957). On the increase in linear size during growth in *Locusta migratoria* L. *Proceedings of the Royal Entomological Society of London* A, **32**, 35–9.

Clausen, C.P. (1940). *Entomophagous Insects*. New York: McGraw-Hill.

Clements, A.N. (1992). *The Biology of Mosquitoes*, vol. 1. London: Chapman & Hall.

Cohen, S.M. (1993). Imaginal disc development. In *The Development of* Drosophila melanogaster, ed. M. Bate & A.M. Arias, pp. 747–841. New York: Cold Spring Harbor Laboratory Press.

Cole, B.J. (1980). Growth ratios in holometabolous and hemimetabolous insects. *Annals of the Entomological Society of America*, **73**, 489–91.

Comstock, J. H. (1918). *The Wings of Insects*. New York: Comstock Publishing Company.

Consoulas, C., Anezaki, M. & Levine, R.B. (1997). Development of adult thoracic leg muscles during metamorphosis of the hawk moth *Manduca sexta*. *Cell & Tissue Research*, **287**, 393–412.

Currie, D. A. & Bate, M. (1991). The development of adult abdominal muscles in *Drosophila*: myoblasts express twist and are associated with nerves. *Development*, **113**, 91–102.

Danilevskii, A.S. (1965). *Photoperiodism and Seasonal Development of Insects*. Edinburgh: Oliver & Boyd.

Davies, R. G. (1966). The postembryonic development of *Hemimerus vicinus* Rehn & Rehn (Dermaptera: Hemimeridae). *Proceedings of the Royal Entomological Society of London* A, **41**, 67–77.

Denlinger, D.L. (1985). Hormonal control of diapause. In *Comprehensive Insect Physiology, Biochemistry and Pharmacology*, vol. 8, ed G.A. Kerkut & L.I. Gilbert, pp. 353–412. Oxford: Pergamon Press.

Duarte, A.J. (1939). Problems of growth of the African migratory locust. *Bulletin of Entomological Research*, **29**, 425–56.

Ewer, J., Gammie, S.C. & Truman, J.W. (1997). Control of insect ecdysis by a positive-feedback endocrine system: roles of eclosion hormone and ecdysis triggering hormone. *Journal of Experimental Biology*, **200**, 869–81.

Fahrbach, S.E. (1992). Developmental regulation of ecdysteroid receptors in the nervous system of *Manduca sexta*. *Journal of Experimental Zoology*, **261**, 245–53.

Farrell, E.R., Fernandes, J. & Keshishian, H. (1996). Muscle organizers in *Drosophila*: the role of persistent larval fibers in adult flight muscle development. *Developmental Biology*, **176**, 220–9.

Fernandes, J., Bate, M. & Vijayraghavan, K. (1991). Development of the indirect flight muscles of *Drosophila*. *Development*, **113**, 67–77.

Fernandes, J. & Keshishian, H. (1995) Neuromuscular development in *Drosophila*: insights from embryos and pupae. *Current Opinion in Neurobiology*, **5**, 10–8.

Fietz, M.J, Concordet, J.-P., Barbosa, R., Johnson, R., Krauss, S., McMahon, A.P., Tabin, C. & Ingham, P.W. (1994). The *Hedgehog* gene family in *Drosophila* and vertebrate development. *Development*, 1994 Supplement, 43–51.

Fristrom, D. & Fristrom, J.W. (1993). The metamorphic development of the adult epidermis. In *The Development of* Drosophila melanogaster, ed. M. Bate & A.M. Arias, pp. 843–97. New York: Cold Spring Harbor Laboratory Press.

Fristrom, J.W., Fristrom, D.K., Apple, R.T., Birr, C., Fechtel, K. & Wolfgang, W.J. (1991). Hormone-induced differentiation of the imaginal disc epidermis: pupal cuticle formation in *Drosophila*. In *Physiology of the Insect Epidermis*, ed. K. Binnington & A. Retnakaran, pp. 55–76. Canberra: CSIRO.

Greene, E. (1989). Diet-induced developmental polymorphism in a caterpillar. *Science*, **243**, 643–6.

Greenwood, W. & Chapman, R.F. (1984). Differences in numbers of sensilla on the antennae of solitarious and gregarious *Locusta migratoria* (Orthoptera: Acrididae). *International Journal of Insect Morphology and Embryology*, **13**, 295–301.

Hardie, J. & Lees, A.D. (1983). Photoperiodic regulation of the development of winged gynoparae in the aphid, *Aphis fabae. Physiological Entomology*, **8**, 385–91.

Hardie, J. & Lees, A.D. (1985). Endocrine control of polymorphism and polyphenism. In *Comprehensive Insect Physiology, Biochemistry and Pharmacology*, vol. 8, ed G.A. Kerkut & L.I. Gilbert, pp. 441–90. Oxford: Pergamon Press.

Haunerland, N.H. & Shirk, P.D. (1995). Regional and functional differentiation in the insect fat body. *Annual Review of Entomology*, **40**, 121–45.

Hesterlee, S. & Morton, D.B. (1996). Insect physiology: the emerging story of ecdysis. *Current Biology*, **6**, 648–50.

Hinton, H. E. (1946). A new classification of insect pupae. *Proceedings of the Zoological Society of London*, **116**, 282–328.

Hinton, H. E. (1948). On the origin and function of the pupal stage. *Transactions of the Royal Entomological Society of London*, **99**, 395–405.

Hinton, H. E. (1963). The origin and function of the pupal stage. *Proceedings of the Royal Entomological Society of London* A, **38**, 77–85.

Hodkova, M. & Hodek, I. (1994). Control of diapause and supercooling by the retrocerebral complex in *Pyrrhocoris apterus. Entomologia Experimentalis et Applicata*, **70**, 237–45.

Horodyski, F.M. (1996). Neuroendocrine control of insect ecdysis by eclosion hormone. *Journal of Insect Physiology*, **42**, 917–24.

Houlihan, D. F. & Newton, J. R. L. (1979). The tracheal supply and muscle metabolism during muscle growth in the puparium of *Calliphora vomitoria. Journal of Insect Physiology*, **25**, 33–44.

Imms, A. D. (1957). *A General Textbook of Entomology*. 9th edition, revised by O.W. Richards & R.G. Davies. London: Methuen.

Jackson, D. J. (1958). Egg-laying and egg-hatching in *Agabus bipustulatus* L., with notes on oviposition in other species of *Agabus* (Coleoptera: Dytiscidae). *Transactions of the Royal Entomological Society of London*, **110**, 53–80.

Jago, N. D. (1963). Some observations on the life cycle of *Eyprepocnemis plorans meridionalis* Uvarov, 1921, with a key for the separation of nymphs at any instar. *Proceedings of the Royal Entomological Society of London* A, **38**, 113–24.

Kim, C.-W. (1959). The differentiation centre inducing the development from larval to adult leg in *Pieris brassicae* (Lepidoptera). *Journal of Embryology and Experimental Morphology*, **7**, 572–82.

Kim, J., Irvine, K.D. & Carroll, S.B. (1995). Cell recognition, signal induction, and symmetrical gene activation at the dorsal-ventral boundary of the developing Drosophila wing. *Cell*, **82**, 795–802.

Kimmins, D. E. (1950). Ephemeroptera. *Handbooks for the Identification of British Insects*, **1**, part 9.

Koch, P.B., Brakefield, P.M. & Kesbeke, F. (1996). Ecdysteroids control eyespot size and wing color pattern in the polyphenic butterfly *Bicyclus anynana* (Lepidoptera: Satyridae). *Journal of Insect Physiology*, **42**, 223–30.

Lanzrein, B., Gentinetta, V., Abbegglen, H., Baker, F.C., Miller, C.A. & Schooley, D.A. (1985). Titers of ecdysone, 20-hydroxyecdysone and juvenile hormone III throughout the life cycle of a hemimetabolous insect, the ovoviviparous cockroach *Nauphoeta cinerea. Experientia*, **41**, 913–7.

Larsen-Rapport, E.W. (1986). Imaginal disc determination: molecular and cellular correlates. *Annual Review of Entomology*, **31**, 145–75.

Lees, A.D. (1955). *The Physiology of Diapause in Arthropods*. Cambridge: Cambridge University Press.

Levine, R.B., Morton, D.B. & Restifo, L.L. (1995). Remodelling of the insect nervous system. *Current Opinion in Neurobiology*, **5**, 28–35.

Macan, T. T. (1961). A key to the nymphs of the British species of Ephemeroptera. *Freshwater Biological Association Scientific Publication*, no. 20, 63 pp.

Madhavan, M.M. & Madhavan, K. (1980). Morphogenesis of the epidermis of adult abdomen of *Drosophila. Journal of Embryology and Experimental Morphology*, **60**, 1–31.

Marshall, J. F. (1938). *The British Mosquitoes*. London: British Museum.

Nardi, J.B., Godfrey, J.L. & Bergstrom, R.A. (1991). Programmed cell death in the wing of *Orgyia leucostigma* (Lepidoptera: Lymantriidae). *Journal of Morphology*, **209**, 121–31.

Nijhout, H.F. (1975). A threshold for metamorphosis in the tobacco hornworm, *Manduca sexta. Biological Bulletin*, **149**, 214–25.

Nijhout, H.F. (1981). Physiological control of molting in insects. *American Zoologist*, **21**, 631–40.

Nijhout, H.F. (1994). *Insect Hormones*. Princeton: Princeton University Press.

Nöthiger, R. (1972). The larval development of imaginal discs. In *The Biology of Imaginal Discs*, ed. H. Ursprung & R. Nöthiger, pp. 1–34. New York: Springer-Verlag.

Oberlander, H. (1985). The imaginal discs. In *Comprehensive Insect Physiology, Biochemistry and Pharmacology*, vol. 2, ed. G.A. Kerkut & L.I. Gilbert, pp. 151–82. Oxford: Pergamon Press.

Padgham, D. E. (1981). Hatching rhythms in the desert locust, *Schistocerca gregaria. Physiological Entomology*, **6**, 191–8.

Pener, M.P. (1991). Locust phase polymorphism and its endocrine relations. *Advances in Insect Physiology*, **23**, 1–79.

Pflüger, H.-J., Hurdelbrink, S., Czjzek, A. & Burrows, M. (1994). Activity-dependent structural dynamics of insect sensory fibers. *Journal of Neuroscience*, **14**, 6946–55.

Pullin, A.S. & Wolda, H. (1993). Glycerol and glucose accumulation during diapause in a tropical beetle. *Physiological Entomology*, **18**, 75–8.

Rachinsky, A. & Hartfelder, K. (1990). Corpora allata activity, a prime regulating element for caste-specific juvenile hormone titre in honey bee larvae (*Apis mellifera carnica*). *Journal of Insect Physiology*, **36**, 189–94.

Reynolds, S.E. (1985). Hormonal control of cuticle mechanical properties. In *Comprehensive Insect Physiology, Biochemistry and Pharmacology*, vol. 8, ed. G.A. Kerkut & L.I. Gilbert, pp. 335–51. Oxford: Pergamon Press.

Riddiford, L.M. (1985). Hormone action at the cellular level. In *Comprehensive Insect Physiology, Biochemistry and Pharmacology*, vol. 8, ed. G.A. Kerkut & L.I. Gilbert, pp. 37–84. Oxford: Pergamon Press.

Riddiford, L.M. & Truman, J.W. (1993). Hormone receptors and the regulation of insect metamorphosis. *American Zoologist*, **33**, 340–7.

Rountree, D.B. & Nijhout, H.F. (1995). Hormonal control of a seasonal polymorphism in *Precis coenia* (Lepidoptera: Nymphalidae). *Journal of Insect Physiology*, **41**, 987–92.

Russell, V.W. & Dunn, P.E. (1991). Lysozyme in the midgut of *Manduca sexta* during metamorphosis. *Archives of Insect Biochemistry and Physiology*, **17**, 67–80.

Saunders, D.S. (1982). *Insect Clocks*. Oxford: Pergamon Press.

Savage, A. A. (1956). The development of the Malpighian tubules of *Schistocerca gregaria* (Orthoptera). *Quarterly Journal of Microscopical Science*, **97**, 599–615.

Schubiger, G. (1971). Regeneration, duplication and transdetermination in fragments of the leg disc of *Drosophila melanogaster*. *Developmental Biology*, **26**, 277–95.

Schwartz, L.M. (1992). Insect muscle as a model for programmed cell death. *Journal of Neurobiology*, **23**, 1312–26.

Sehnal, F. (1985). Growth and life cycles. In *Comprehensive Insect Physiology, Biochemistry and Pharmacology*, vol. 2, ed. G.A. Kerkut & L.I. Gilbert, pp. 1–86. Oxford: Pergamon Press.

Shimuzu, I. (1992). Comparison of fatty acid composition in lipids of diapause and non-diapause eggs of *Bombyx mori* (Lepidoptera: Bombycidae). *Comparative Biochemistry and Physiology*, **102B**, 713–6.

Sikes, E. K. & Wigglesworth, V. B. (1931). The hatching of insects from eggs and the appearance of air in the tracheal system. *Quarterly Journal of Microscopical Science*, **74**, 165–92.

Sláma, K. (1982). Inverse relationship between ecdysteroid titers and total body metabolism in insects. *Zeitschrift für Naturforschung*, **37c**, 839–44.

Slansky, F. & Scriber, J.M. (1985). Food consumption and utilization. In *Comprehensive Insect Physiology, Biochemistry and Pharmacology*, vol. 4, ed. G.A. Kerkut & L.I. Gilbert, pp. 87–163. Oxford: Pergamon Press.

Snodgrass, R. E. (1954). Insect metamorphosis. *Smithsonian Miscellaneous Collections*, **122**, no. 9, 124 pp.

Southwood, T.R.E. (1946). The structure of the eggs of the terrestrial Heteroptera and its relationship to the classification of the group. *Transactions of the Royal Entomological Society of London*, **108**, 163–221.

Southwood, T. R. E. & Leston, D. (1959). *Land and Water Bugs of the British Isles*. London: Warne.

Steel, C.G.H. & Davey, K.G. (1985). Integration in the insect endocrine system. In *Comprehensive Insect Physiology, Biochemistry and Pharmacology*, vol. 8, ed G.A.Kerkut & L.I.Gilbert, pp. 1–35. Oxford: Pergamon Press.

Sula, J., Kodrík, D. & Socha, R. (1995). Hexameric haemolymph protein related to adult diapause in the red firebug, *Pyrrhocoris apterus* (L.) (Heteroptera). *Journal of Insect Physiology*, **41**, 793–800.

Takeda, M. & Skopik, S.D. (1997). Photoperiodic time measurement and related physiological mechanisms in insects and mites. *Annual Review of Entomology*, **42**, 323–49.

Tauber, M.J., Tauber, C.A. & Masaki, S. (1986). *Seasonal Adaptations in Insects*. Oxford: Oxford University Press.

Truman, J.W. (1978). Hormonal release of stereotyped motor programmes from the isolated nervous system of the cecropia silkmoth. *Journal of Experimental Biology*, **74**, 151–73.

Truman, J.W. (1985). Hormonal control of ecdysis. In *Comprehensive Insect Physiology, Biochemistry and Pharmacology*, vol. 8, ed. G.A. Kerkut & L.I. Gilbert, pp. 413–40. Oxford: Pergamon Press.

Truman, J.W. (1988). Hormonal approaches for studying the nervous system development in insects. *Advances in Insect Physiology*, **21**, 1–34.

Truman, J.W. & Bate, M. (1988). Spatial and temporal patterns of neurogenesis in the central nervous system of *Drosophila melanogaster*. *Developmental Biology*, **125**, 145–57.

Truman, J.W. & Reiss, S.E. (1988). Hormonal regulation of the shape of identified motorneurons in the moth *Manduca sexta*. *Journal of Neuroscience*, **8**, 765–75.

Truman, J.W., Thorn, R.S. & Robinow, S. (1992). Programmed neuronal death in insect development. *Journal of Neurobiology*, **23**, 1295–311.

Urquhart, F. A. (1960). *The Monarch Butterfly*. Toronto: University of Toronto Press.

van Emden, F. I. (1946). Egg-bursters in some more families of polyphagous beetles and some general remarks on egg-bursters. *Proceedings of the Royal Entomological Society of London*, **21**, 89–97.

Vaz Nunes, M. & Hardie, J, (1992). Photoperiodic induction of winged females in the black bean aphid, *Aphis fabae*. *Physiological Entomology*, **17**, 391–6.

Vogt, R.G., Rybczynski, R., Cruz, M. & Lerner, M.R. (1993). Ecdysteroid regulation of olfactory protein expression in the developing antenna of the tobacco hawk moth, *Manduca sexta*. *Journal of Neurobiology*, **24**, 581–97.

Waddington, C. H. (1941). The genetic control of wing development in *Drosophila*. *Journal of Genetics*, **41**, 75–139.

Weeks, J.C. & Levine, R.B. (1992). Endocrine influences on the post-embryonic fates of identified neurons during insect metamorphosis. In *Determinants of Neuronal Identity*, ed. M. Shankland & E.R. Macagno, pp. 293–322. San Diego: Academic Press.

Wheeler, D.E. & Nijhout, H.F. (1983). Soldier determination in *Pheidole bicarinata*: effect of methoprene on caste and size within castes. *Journal of Insect Physiology*, **29**, 847–54.

Wheeler, D.E. & Nijhout, H.F. (1984). Soldier determination in *Pheidole bicarinata*: inhibition by adult soldiers. *Journal of Insect Physiology*, **30**, 127–35.

Wigglesworth, V. B. (1965). *The Principles of Insect Physiology*. London: Methuen.

Williams, J.A. & Carroll, S.B. (1993). The origin, patterning and evolution of insect appendages. *BioEssays*, **15**, 567–77.

Yamashita, O. (1996). Diapause hormone in the silkworm, *Bombyx mori*: structure, gene expression and function. *Journal of Insect Physiology*, **42**, 669–79.

Yoder, J.A., Denlinger, D.L., Dennis, M.W. & Kolattukudy, E. (1992). Enhancement of diapausing flesh fly puparia with additional hydrocarbons and evidence for alkane biosynthesis by a decarboxylation mechanism. *Insect Biochemistry and Molecular Biology*, **22**, 237–43.

Zera, A.J. & Denno, R.F. (1997). Physiology and ecology of dispersal polymorphism in insects. *Annual Review of Entomology*, **42**, 207–31.

Zera, A.J. & Tanaka, S. (1996). The role of juvenile hormone and juvenile hormone esterase in wing morph determination in *Modicogryllus confirmatus*. *Journal of Insect Physiology*, **42**, 909–15.

PART IV

The Integument, Gas Exchange and Homeostasis

16 Integument

The integument is the outer layer of the insect, comprising the epidermis and the cuticle. The cuticle is a characteristic feature of arthropods and is, to a large extent, responsible for the success of insects as terrestrial animals.

16.1 EPIDERMIS

16.1.1 Epidermal cells

The epidermis is the outer cell layer of the insect. It is one cell thick with cell densities ranging from about $3000\,mm^{-2}$ in a trachea of *Rhodnius* to $11\,000\,mm^{-2}$ in the sternal area of larval *Tenebrio*, but cell density and cell depth at right angles to the surface of the epidermis change during development (see below).

The apical plasma membrane of an epidermal cell forms a series of short projections or ridges, flattened at the tips where the membrane is electron-dense on the inside (Figs. 16.1a, 16.18). These specialized regions of membrane are known as plasma membrane plaques and they are the sites of secretion of the outer epicuticle and of chitin fibers. During and just after a molt, the epidermal cells also have cytoplasmic processes on the outside extending into the pore canals of the cuticle (see Fig. 16.2), but these processes may be withdrawn as the cuticle matures.

The epidermal cells are held together near their apices by zonulae adhaerens and lower down by septate junctions (Fig. 16.1a). At greater distances from the cuticle, adjacent cells are not tightly bound to each other, and the spaces between them (lateral lymph spaces) are, to some extent, isolated from the hemolymph by desmosomes close to the basal lamina (Locke, 1991). Gap junctions between epidermal cells probably provide a pathway for the movement of low molecular weight substances, such as hormone second messengers, and perhaps morphogens. This enhances coordination between cells, reducing any minor differences that might occur due to slight differences in the timing or amounts of hormonal signals carried by the hemolymph.

The basal plasma membrane is commonly flat and attached to the basal lamina by hemidesmosomes, but in some specialized wax-secreting epithelia it forms a plasma membrane reticular system (section 6.1.1), and in gland cells it is folded inwards (Fig. 16.1b).

Epidermal cells have extensive rough endoplasmic reticulum and Golgi complexes, and they often contain membrane-bound pigment granules.

All the epidermal cells are glandular in the sense that they secrete cuticle and the enzymes concerned in its production and its digestion at the time of molting. Some of the cells have an additional specialized glandular function, and these fall into three classes (Noirot & Quennedy, 1974). Class 1 gland cells have the outer plasma membrane produced as microvilli or parallel lamellae which may abut directly on to the cuticle, but are often separated from it by a space in which it is presumed their secretion accumulates (Fig. 16.1b). The cuticle above the cells is usually unmodified and it is presumed that the secretion reaches the external surface of the cuticle via the pore canals and epicuticular filaments. Class 1 cells are often involved in pheromone production (section 27.1.1).

Class 2 gland cells are derived from epidermal cells, but have no direct contact with the cuticle, nor do they have a duct. They are only known from the sternal glands of termites (see Noirot & Quennedy, 1974).

Class 3 gland cells are also below the epidermis, but connect with the exterior by a duct (Fig. 16.1c). The distal surface of the gland cell is developed into microvilli which surround a cavity. This cavity is extracellular and therefore is not a vacuole, although it is often referred to as such. The lining of the duct leading to the outside is continuous with the epicuticle on the outside of the body. Internally it projects into the cavity of the gland cell and it is perforated, forming a structure called the end apparatus. The duct is produced by a separate cell.

Gland cells of this type produce the cement on the outer surface of the cuticle (see below) and exhibit cycles of development that are synchronized with the molting cycle (Horwath & Riddiford, 1991). In all

[415]

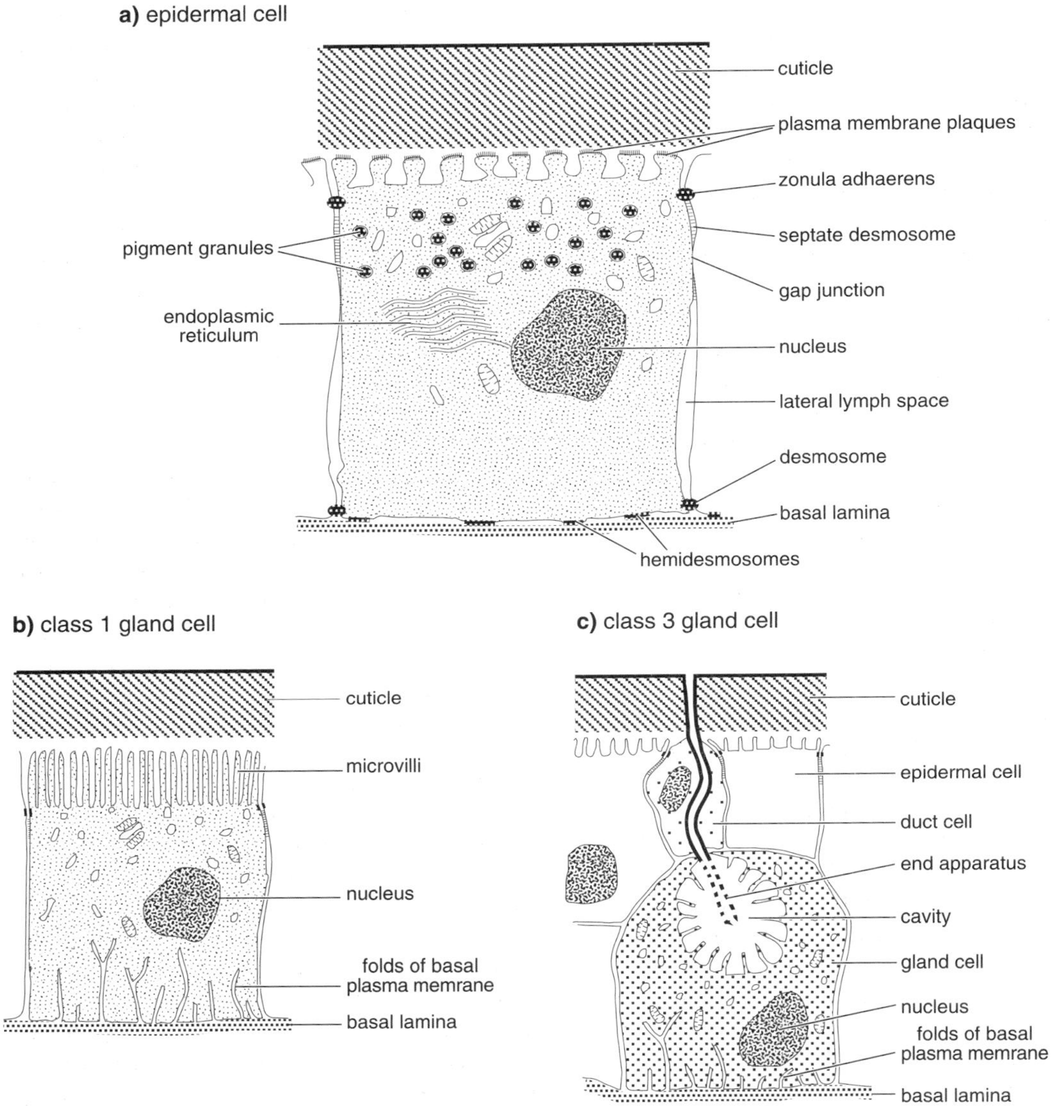

a) epidermal cell

- cuticle
- plasma membrane plaques
- zonula adhaerens
- septate desmosome
- gap junction
- pigment granules
- endoplasmic reticulum
- nucleus
- lateral lymph space
- desmosome
- basal lamina
- hemidesmosomes

b) class 1 gland cell

- cuticle
- microvilli
- nucleus
- folds of basal plasma memrane
- basal lamina

c) class 3 gland cell

- cuticle
- epidermal cell
- duct cell
- end apparatus
- cavity
- gland cell
- nucleus
- folds of basal plasma memrane
- basal lamina

Fig. 16.1. Epidermal cells. **(a)** Principal features of an epidermal cell during the intermolt period. **(b)** Class 1 gland cell with no duct to the exterior (based on Noirot & Quennedey, 1974). **(c)** Class 3 glandular unit (based on Noirot & Quennedey, 1974).

dipterous larvae they form peristigmatic glands. These are composed of large cells with ducts opening to the outside near the edges of the spiracles. Their secretion, which is produced continuously, is responsible for the hydrofuge properties of the cuticle surrounding the spiracles which prevent the entry of water into the tracheal system. In addition, class 3 gland cells are often involved in the production of defensive secretions and pheromones (Chapter 27).

The neurons and support cells of mechano- and chemoreceptors associated with the cuticle are also derived from epidermal cells.

Reviews: Caveney, 1985, 1991 – cell coupling; Locke, 1984, 1985a, 1991 – ultrastructure; Noirot & Quennedy, 1974 – glands

16.1.2 Basal lamina

The epidermal cells stand on a basal lamina, or basement membrane. The chemistry of the lamina is not well understood in insects, but the primary components are the fibrous protein, collagen, glycoproteins and glycosaminoglycans. The latter are polymers of disaccharides consisting of amino sugar and uronic acid moieties. The molecules are charged and they probably contribute to the functioning of the basal lamina as a molecular sieve (Locke, 1991). The lamina forms a continuous sheet beneath the epidermis and, at points where muscles are attached, it is continuous with the sarcolemma. In *Rhodnius*, the basal lamina is about 0.15 μm thick in the fourth stage larva, but six days after feeding it thickens to about 0.5 μm. The thickening probably serves to strengthen the basal lamina allowing it to serve as a stable platform for the forces used in molding the new epicuticle which is produced soon afterwards. The epidermal cells are anchored to the basement membrane by hemidesmosomes.

The basal lamina may be produced by the epidermal cells, but plasmatocytes also contribute to it.

Review: Ashhurst, 1985

16.1.3 Oenocytes

Oenocytes are derived from epidermal cells. They are often large cells, more than 100 μm in diameter, with, at some stages of development, an extensive plasma membrane reticular system (Jackson & Locke, 1989) (section 6.1.1). The nucleus is large and there is an extensive tubular endoplasmic reticulum, but few mitochondria. In Ephemeroptera, Odonata and Heteroptera, the oenocytes remain in the epidermis between the bases of the epidermal cells and the basal lamina. In other groups they move away from the epidermis. In Lepidoptera and Orthoptera they form clusters in the body cavity, while in Homoptera, Hymenoptera and some Diptera they are dispersed and embedded in the fat body. When in the fat body, they often form close associations with the fat body trophocytes.

Oenocytes may be formed continuously, or a new generation of cells may be produced at each molt, or, in holometabolous insects, there may be separate larval and adult generations. They show cycles of development that, in immature insects, are associated with the molting cycle.

The oenocytes synthesize the hydrocarbons, and perhaps other lipids, that contribute to the epicuticle. The hydrocarbons are transported in the hemolymph from the oenocytes to all parts of the epidermis by lipophorins (Chino, 1985; Gu *et al.*, 1995) and the plasma membrane reticular system possibly provides chambers in which the lipids are loaded on to the lipophorins. Where the oenocytes are in direct contact with the epidermis, lipids are transferred directly to the epidermal cells via cytoplasmic strands (Wigglesworth, 1988).

Some of these cuticular hydrocarbons function as sex pheromones in some species (section 27.1.2). In *Tenebrio*, and perhaps in *Gryllus* (Romer & Bressel, 1994), the oenocytes also synthesize ecdysone, but this may not be a general phenomenon (Rees, 1985).

16.2 BASIC STRUCTURE OF CUTICLE

The cuticle is a secretion of the epidermis and covers the whole of the outside of the body as well as lining ectodermal invaginations such as the stomodeum and proctodeum and the tracheae. It is differentiated into two major regions: an inner region, up to 200 μm thick, characterized by the presence of chitin and forming the bulk of the cuticle, and the thin outer epicuticle, 1–4 μm thick, which contains no chitin.

Reviews: Hepburn, 1985; Neville, 1975

16.2.1 Chitinous cuticle

Chitin is a characteristic constituent of insect procuticle, commonly comprising 20–50% of its dry weight. In insect cuticle it is always associated with protein, perhaps being bound to it by covalent bonds (Andersen, 1979). Chitinous cuticle as it is first secreted is known as procuticle. Subsequently, the outer part often becomes hard and rigid (tanned or sclerotized) to form exocuticle while the inner, undifferentiated part is called endocuticle (Fig. 16.2a). Between the two there may be a region of hardened, but not fully darkened cuticle which is fuchsinophil (staining with the red dye, acid fuchsin; fully sclerotized cuticle does not stain readily) and such a layer is called mesocuticle.

16.2.1.1 *Chitin*

Chitin is a polysaccharide made up largely of N-acetylglucosamine residues, but it also probably contains some glucosamine. The sugar residues are linked by 1–4 β-linkages so that they form a chain in which all the residues are oriented in the same direction (for example, in the following diagram, C1 is always at the left-hand end of the sugar residue).

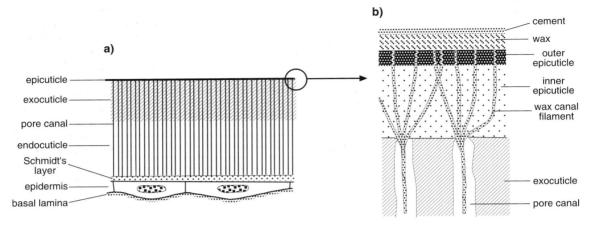

Fig. 16.2. Basic structure of the integument: (**a**) section through mature integument; (**b**) section through the epicuticle at greater magnification.

acetylglucosamine   acetylglucosamine   glucosamine

Adjacent chitin chains are held together by hydrogen bonds to form microfibrils. Neighboring chains run in opposite directions, with the C1 atoms at opposite ends of the residues. The suggested linkages are shown in Fig. 16.3.

The chitin microfibrils are about 2.5–3.0 nm in diameter and are embedded in a protein matrix. They lie parallel to each other in the plane of the cuticle, but their orientation is often different in successive levels through the thickness of the cuticle. In most insects, the microfibrils in the outer parts of the procuticle, which subsequently becomes the exocuticle, rotate anticlockwise through a fixed angle in successive levels so that their arrangement is helicoidal and a series of thin lamellae is produced (Fig. 16.4a). This is called lamellate cuticle. The inner procuticle may also be lamellate throughout, or layers with helicoidally arranged microfibrils may alternate with layers in which the microfibrils are uniformly oriented. Wholly lamellate cuticle is found in Apterygota and in larval and pupal Lepidoptera, Diptera and Coleoptera. Where helicoidal and unidirectional layers alternate, all

the unidirectional layers may have the same orientation (Fig. 16.4b), as in locusts and cockroaches, or they may have different orientations (Fig. 16.4c), as in beetles and bugs. In some such cases the intervening layers of helicoidal cuticle are very thin, so that the orientation appears to change suddenly from one layer to the next. This is called a pseudo–orthogonal arrangement. The endocuticle contains a higher proportion of lamellate cuticle in larval insects than in adults of the same species.

The alternate production of helicoidally arranged and unidirectional layers of microfibrils has a circadian periodicity in most species. In locusts, cuticle laid down at night has microfibrils with a helicoidal arrangement whereas that produced in the daytime has them uniformly oriented. Such daily growth layers are found in larval and adult hemimetabolous insects and in adults of many holometabolous species. In some cases, however, shortage of food may limit cuticle production so that growth layers are not produced regularly. **Reviews:** Kramer & Koga, 1986 – chitin; Neville, 1983 – growth layers

16.2.1.2 *Proteins*

Proteins are the major constituents of insect cuticle. A hundred or more are present in the cuticles of most insects studied. Cuticle from any one part of an insect contains several different proteins which may differ from the proteins in the cuticle of other parts of the same insect. Proteins produced by one family of genes are characteristically associated with membranous regions of the

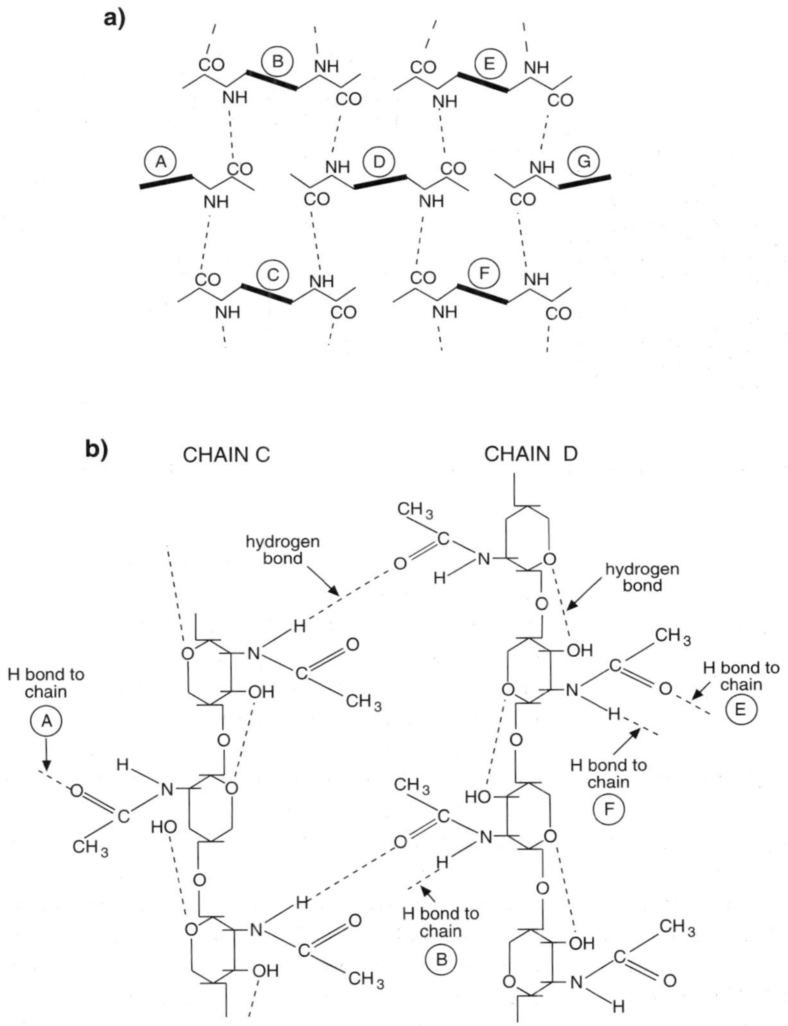

a)

b)

CHAIN C   CHAIN D

Fig. 16.3. Diagrammatic representation of part of a chitin microfibril showing the cross-links between the chitin chains. Dotted lines represent hydrogen bonds. **(a)** Cross-section through part of a microfibril. Thick lines represent a chain of acetylglucosamine residues oriented at right angles to the plane of the page and seen end on. Chains A, D and G run in one direction, chains B, C, E and F in the opposite direction. **(b)** Chains C and D seen in face view. Linkages are distorted to show the two chains in the same plane.

cuticle, while proteins produced by another gene family characterize hard cuticle. These differences are apparent even before the cuticle is hardened, and the differing physical properties of different parts of the cuticle appear to be at least partly a consequence of the different proteins they contain, independent of the hardening process. The proteins of soft cuticle generally contain more aspartic acid, glutamic acid, histidine, lysine and tyrosine and are more hydrophilic than those of hard cuticles. The same proteins are present in cuticle with similar properties from different stages of an insect: the proteins appear to characterize the type of cuticle rather than the stage of development.

Hardening of cuticle is primarily a consequence of cross-links between protein molecules so that they form a rigid matrix. The process of cross-linking is called tanning or sclerotization and the cuticle is then said to be sclerotized (see below, section 16.5.3).

Reviews: Andersen, Højrup & Roepstorff, 1995; Willis, 1987, 1991

16.2.1.3 *Lipid*

Lipids are present in the procuticle and, in *Rhodnius*, they impregnate the walls of the pore canals and are present in layers at intervals of 0.5–1.0 μm. These lipid layers become

SEEN FROM ABOVE   SEEN IN SECTION

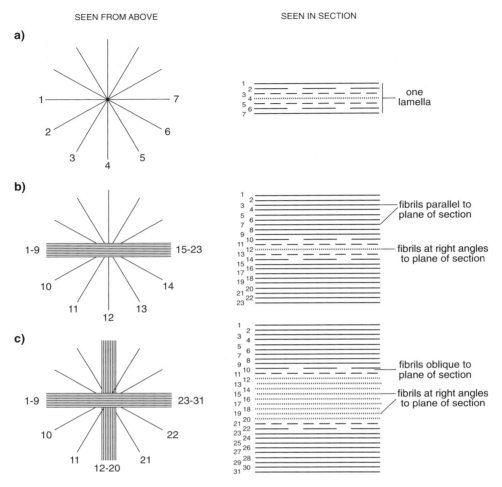

Fig. 16.4. Diagrams showing the arrangement of chitin microfibrils in cuticle. Numbers indicate successive layers in the cuticle with (**left**) the orientation of microfibrils in successive layers as seen in transparency through the cuticle, and (**right**) the appearance of the fibrils in sections of the cuticle running parallel with the fibrils in layer 1. (**a**) A helicoidal arrangement producing a single lamella. (**b**) Two layers with uniformly oriented microfibrils (1–9 and 15–23) separated by a single lamella with a helicoidal arrangement (10–14). (**c**) Layers of uniformly directed fibrils at right angles to each other (1–9, 12–20 and 23–31) separated by layers with a helicoidal arrangement (10–11 and 21–22).

dispersed when the insect feeds and the cuticle stretches, but more layers are laid down with the new cuticle. Wigglesworth (1988) believes that lipid, together with the sclerotized proteins, plays a significant role in cuticle hardening.

16.2.2 Epicuticle
The epicuticle is made up of several layers. The thickest layer, 0.5 to 2.0 μm thick, is the inner epicuticle immediately outside the procuticle. Outside it, is a very thin outer

epicuticle, only about 15 nm thick, and outside this again is a wax layer of variable thickness. Some insects have a thin 'cement' layer outside the wax (Fig. 16.2b).

Inner epicuticle This layer is chemically complex and is known to consist primarily of tanned lipoproteins. During its production, phenolic substances and phenoloxidase are also present. These are probably concerned with tanning the proteins. Phenoloxidase persists as an extracellular

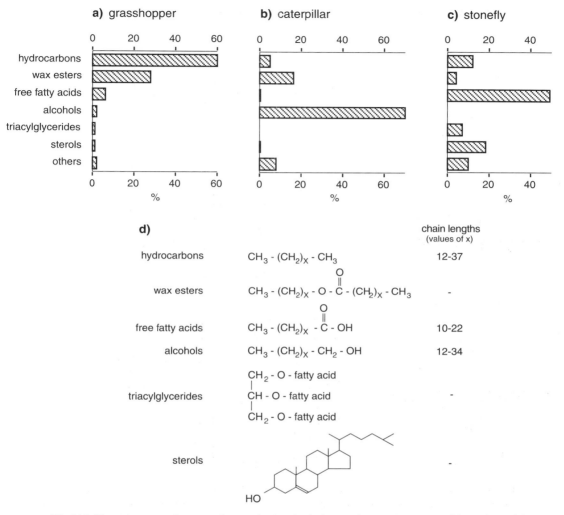

Fig. 16.5. The main groups of compounds occurring in epicuticular wax shown as percentages of the total wax: **(a)** a grasshopper – hydrocarbons predominate; **(b)** a caterpillar – alcohols predominate; **(c)** an adult stonefly – free fatty acids predominate; **(d)** the structures of the main types of compound. Chain lengths are shown in the final column, where appropriate. These values refer to X in the previous column.

enzyme in mature cuticle, producing further tanning if the epicuticle is damaged.

Outer epicuticle This is a very thin trilaminar layer. It is a highly polymerized lipid, and probably also has a protein component. Polyphenols and phenoloxidase take part in its formation. It is the first-formed layer of new cuticle produced at each molt, protecting the new procuticle from the molting enzymes. It is believed to be inextensible, setting a limit on any extension of the procuticle during growth or other activities. The material forming the outer epicuticle is often referred to as cuticulin.

Wax The epicuticular wax layer contains many different compounds (Fig. 16.5). Hydrocarbons are universally present, and may comprise over 90% of the wax, as in the cockroaches. Chain lengths range from around 12 to over 50 carbon atoms, and compounds with an odd number of carbon atoms in the chain are usually dominant. In larval Lepidoptera and Coleoptera, aliphatic alcohols are the most abundant compounds. In this case, compounds with even numbers of carbon atoms, in the range 12 to 34, are dominant. The alcohols may form esters with fatty acids, but esters are usually only minor components. In an adult stonefly, free fatty acids constitute the principal class of compounds in the wax.

It has been suggested that the molecules adjacent to the outer epicuticle are strongly oriented as a result of their polar, hydrophilic groups being adsorbed on to the surface of the cuticulin so that they form a layer only one molecule thick and hence called a monolayer. However, current evidence does not support this idea (section 18.4.1).

In some species, the wax forms a bloom on the outside of the cuticle and, in a few, very large quantities are produced. This is the case with some Fulgoroidea and scale insects, and the larva of *Calpodes*. Bees secrete large quantities of wax, which they use in the production of their larval cells. This wax is secreted by epidermal cells on the ventral surface of abdominal segments four to seven. These cells presumably contribute to the insect's epicuticular wax, but they become greatly enlarged when the bees have an adequate supply of honey which provides the basic chemicals from which the wax is produced. The secretion of the glands, which contains over 300 components, is pressed into flat scales between the overlapping sterna. The scales are removed by spines on the posterior basitarsi and are manipulated with the forelegs and mandibles to produce the cells of honeycomb (Winston, 1987).

The wax is important in waterproofing the cuticle (section 18.4.1.1) and, in some insects, is the source of chemical signals important in intra- and, perhaps, interspecific signalling. It is synthesized by the oenocytes.
Reviews: Blomquist & Dillwith, 1985; de Renobales, Nelson & Blomquist, 1991; Lockey, 1988

Cement The cement is a very thin layer outside most of the wax, perhaps consisting of mucopolysaccharide which becomes closely associated with lipids. It may serve to protect the underlying wax, although it is sometimes present as an open meshwork. It is not produced by all insects and appears to be absent from the cuticle of honeybees, for example.

Cement is the product of type 3 gland cells in the epidermis.
Review: Horwath & Riddiford, 1991

16.2.3 Pore canals and epicuticular filaments
Running through the cuticle at right angles to the surface are very fine pore canals one micron or less in diameter (Fig. 16.2b). They extend from the epidermis to the inner epicuticle and, at least early in the development of the cuticle, contain cytoplasmic extensions of the epidermal cell at densities as high as $15000\,mm^{-2}$. In adult beetles and bugs the pore canals are more abundant in the outer procuticle, and some 30 pore canals per cell in the outer region may join to a single canal lower down.

In *Rhodnius*, the pore canals are circular in cross-section, but in many species they have a flattened, ribbon-like form, the plane of flattening being parallel with the microfibers in each layer of the cuticle. As the microfibers in successive layers change direction the canal also rotates, so that it has the form of a twisted ribbon (Fig. 16.6).

It is presumed that the cytoplasmic threads within the canals are subsequently withdrawn and the canals may be filled by chitin and protein with the chitin microfibers oriented along the canal. Whether they are filled or not, the canals contain tubular filaments (called wax canal filaments, or epicuticular filaments) arising from the plasma membrane of the epidermal cell. At the epicuticle, these filaments diverge and extend to the surface of the outer epicuticle (Fig. 16.2b). The filaments range from 13 to 21 nm in external diameter in different insects, and have a lipid-filled lumen of about half this diameter. They possibly consist of tanned protein and are concerned with the transport of lipids from the epidermal cells to the surface of the cuticle (Wigglesworth, 1985).

16.3 DIFFERENT TYPES OF CUTICLE

The cuticle varies in nature in different parts of the body and between insects with different life forms. The most obvious difference is between the rigid cuticle of the sclerites and the flexible cuticle of the membranes between them.

16.3.1 Rigid and hard cuticle
Rigid cuticle is produced as a result of sclerotization in the outer part of the procuticle to form exocuticle and its stiffness increases as a greater proportion of the cuticular proteins become cross-linked (Fig. 16.7). The extent of sclerotization varies between different developmental stages and in different parts of the cuticle. In larval

a)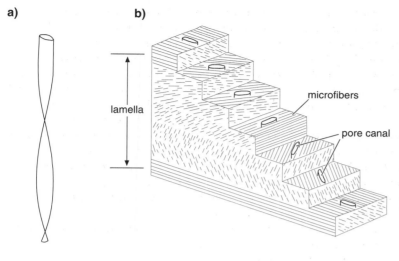

b)

microfibers

pore canal

Fig. 16.6. Pore canals in cuticle with the microfibrils arranged helicoidally (based on Neville, Thomas & Zelazny, 1969). **(a)** A single twisted ribbon-shaped canal. **(b)** A segment of cuticle with helicoidally arranged microfibrils showing the ribbon-shaped pore canals oriented parallel with the fibrils in each layer. As a consequence of this effect the complete pore canal would have the form shown in (a).

lamella

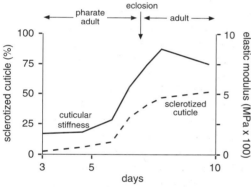

Fig. 16.7. The relationship between sclerotization and cuticular stiffness. Over the period of the molt, and principally around the time of eclosion, the degree of sclerotization of an abdominal tergite of the honeybee increases. (Sclerotization is expressed as the inverse of the percentage of protein that is extractable. Only about half the protein in the cuticle becomes bound). At the same time, the stiffness of the cuticle, measured as the elastic modulus, also increases (based on Richards, 1967).

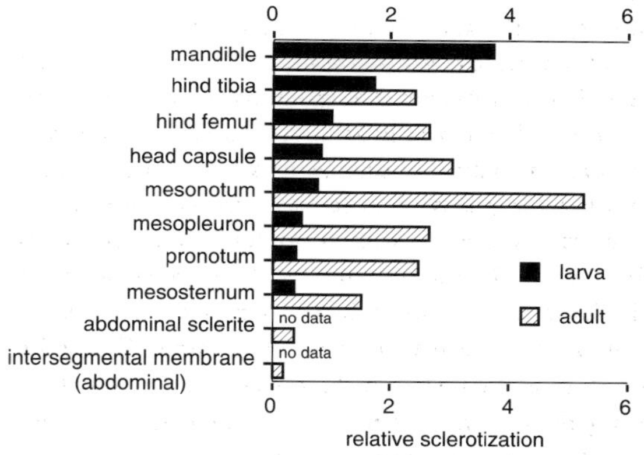

Fig. 16.8. Amounts of sclerotization of the cuticle from different parts of the body of a mid fifth-stage larva and a 10-day-old adult locust. Sclerotization is based on the amounts of ketocatechols released when the cuticle is hydrolysed (based on Andersen, 1974).

Schistocerca, for example, the exocuticle of the sclerites is sharply differentiated from the endocuticle, which never sclerotizes. In the adult, on the other hand, some changes continue for weeks after the final ecdysis and the whole of the procuticle may become differentiated. In the fully hardened larva of *Schistocerca*, the mandibles are much more heavily sclerotized than other parts of the cuticle; they are also heavily sclerotized in the adult, but to no greater degree than some other parts of the body. The

dorsal mesothorax is particularly heavily sclerotized in relation to its function in flight (Fig. 16.8).

Additional rigidity is produced in sclerotized cuticle by structural modifications. These are based on inflexions of the cuticle seen as grooves (sulci) on the outside and as ridges on the inside. Most of the grooves seen in insect cuticle and which define the areas of cuticle given specific names by morphologists are formed in this way. These inflexions provide rigidity in the same way as a T-girder.

Even greater rigidity is provided when inflexions meet internally to form an 'internal' (endophragmal) skeleton.

Fig. 16.9. Cuticular hardening. Zinc in the mandible of a caterpillar (after Fontaine *et al.*, 1991). **(a)** Scanning electron micrograph of the inner surface of a mandible. **(b)** X-ray microanalysis of the same mandible showing the presence of zinc (white areas) in the mandibular cusps.

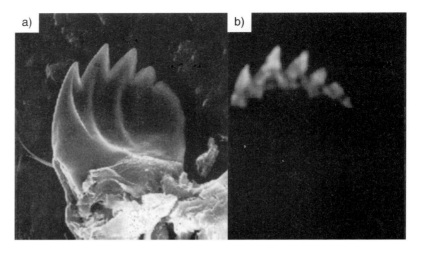

Deep, finger-like invaginations of the cuticle are called apodemes or apophyses. In the head, four apodemes join centrally to form the tentorium (see Fig. 1.5), and in the thorax of winged insects, pleural and sternal apophyses join or are held together by muscles (see Fig. 7.4). The tentorium provides rigidity to the head just above the level at which the mandibles are articulated. The junction of pleural and sternal apophyses gives lateral stiffness to the thorax which is important for the proper functioning of the flight muscles.

Hardness, as opposed to rigidity, increases relatively slowly after ecdysis and it may take several days for the cuticle to reach its final degree of hardness. During this time, the cuticle increases in thickness and this may contribute to the increased hardness (Hillerton, Reynolds & Vincent, 1982). It may also be that other changes occur in the cuticle following the initial rapid sclerotization.

Parts of the cuticle that are especially hard contain heavy metals. Zinc, manganese, or occasionally iron, are present in the mandibles of insects with biting and chewing mouthparts from several orders (Fig. 16.9). These metals are present in relatively large amounts, comprising considerably more than 1% of the dry weight of the cuticle in which they occur. These metals are also present in the claws and ovipositors of some insects. Their role in hardening is inferred from their presence in areas which are known to be especially hard, but the mechanism by which they produce hardness is not known (Chapman, 1995).

Calcium, probably as calcium carbonate, possibly contributes to cuticular hardness in some tenebrionid beetles (Leschen & Cutler, 1994), but it is not usually a significant component of insect cuticle.

16.3.2 Membranous cuticle

The sclerites are joined by flexible arthrodial membranes. In these, the procuticle remains unsclerotized (Fig. 16.10a), but it also differs qualitatively from the cuticle of sclerites in containing proteins with different amino acid composition. The extent of the membrane and the method of articulation of the two adjacent sclerites determines the degree of movement that can occur at the joint. Sometimes, as between abdominal segments, the membrane is extensive and there is no specific point of contact between adjacent sclerites, so that movement is unrestricted (Fig. 16.10a, b). More usually, the sclerites make contact with each other to form articulations and the joints are called monocondylic or dicondylic depending on whether there are one or two points of contact (see Fig. 8.1). Monocondylic articulations, such as that of the antenna with the head, permit considerable freedom of movement, whereas dicondylic articulations, which occur at many of the leg joints, give more limited, but more precise movements, and a less complex array of muscles is necessary to provide fine control of movement. The articular surfaces may lie within the membrane (intrinsic), as in most leg joints (Fig. 16.10c, d), or they may lie outside it (extrinsic), as, for example, in the mandibular articulations (Fig. 16.10e).

Apart from membranous areas, exocuticle is also absent along the ecdysial cleavage lines of larval hemimetabolous insects. The cuticle along these lines consists only of

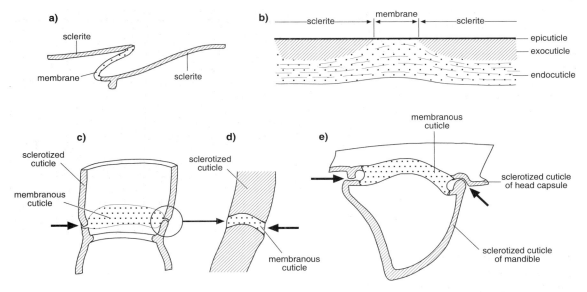

Fig. 16.10. Flexibility of the cuticle. Arrows show points of articulation at which the two sclerotized regions meet and are free to move relative to each other (partly after Snodgrass, 1935). (**a**) Intersegmental membrane. Extensive membrane with no articulation between sclerites. (**b**) Diagrammatic representation of a membranous connection similar to that in (**a**). (**c**) An intrinsic articulation where the points of articulation are within the membrane, as in most leg joints. (**d**) Detail of one of the articulations in (**c**). (**e**) Extrinsic articulations where the sclerotized parts meet outside the membrane.

epicuticle and undifferentiated procuticle (Fig. 16.11), so that they constitute lines of weakness along which the cuticle splits at ecdysis.

In many larval holometabolous insects, the greater part of the cuticle remains undifferentiated and somewhat extensible. This facilitates growth and also permits movement by changes in body form. It is also important for the conservation of materials as the bulk of the undifferentiated cuticle is digested and resorbed at molting, while sclerotized parts are lost. Hence an unsclerotized cuticle is more economic for a larva. In addition, many larval insects eat their exuviae so that they conserve as much as possible from the cuticle. Some insects are known to remove material from unsclerotized cuticle during long periods without food, presumably using it for maintenance of other tissues (Neville, 1975).

16.3.3 Elastic and extensible cuticle

Resilin Some parts of the cuticle contain a colorless, rubber-like protein called resilin in which glycine comprises 30–40% of the amino acids. The protein molecules are

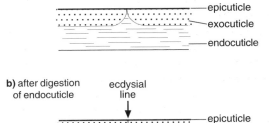

Fig. 16.11. Ecdysial cleavage line: (**a**) mature cuticle; (**b**) after digestion of the endocuticle, only the epicuticle holds the cuticle together along the cleavage line.

linked together, and the linkages are produced continuously as the resilin is laid down, unlike sclerotization in which the proteins become linked some time after they are produced. The amino acid sequence in resilin prevents other cross-links forming; these would impair its rubber-like properties.

Like rubber, resilin can be stretched under tension and stores the energy involved so that when the tension is

Fig. 16.12. Elastic cuticle (modified after Andersen & Weis-Fogh, 1964). **(a)** Transverse section through the thoracic wall and wing base of a grasshopper showing the position of the wing hinge. **(b)** The hinge enlarged showing the resilin pad.

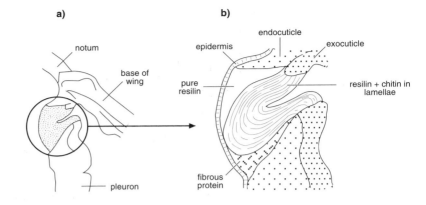

released it returns immediately to its original length. In the locust, between one-quarter and one-third of the wing's recoil energy away from its equilibrium position is due to the elasticity of the wing-hinge ligament which is between the pleural process and the second axillary sclerite (Fig. 16.12). The ligament is sharply differentiated from adjacent sclerotized cuticle and consists of a ventral region of tough, dense chitin and fibrous protein, a dorsal region containing layers of resilin separated by chitinous lamellae, and a pad of pure resilin on the inside.

Elsewhere, resilin is found in the clypeo-labral spring which keeps the labrum pressed against the mandibles, and in the food pump of reduviid bugs (Edwards, 1983). In fleas, a pad of resilin above the hind leg stores energy for the jump. In beetles, where there are no inspiratory muscles, the abdominal terga and sterna are held apart by ribs of cuticle that end, dorsally and ventrally, in pads of resilin. During expiration, these pads are compressed by the action of the dorsoventral muscles; when these muscles relax, the pads resume their original shape, pushing the terga and sterna apart so that inspiration occurs. Other examples of the functions of resilin are described by Andersen & Weis-Fogh (1964).

Plasticization of cuticle Whereas resilin is an elastic protein, undifferentiated procuticle can sometimes be plasticized in order to facilitate stretching. This probably occurs in all insects at the time of ecdysis, where plasticization facilitates expansion of the new cuticle. It also occurs in the blood-sucking bug, *Rhodnius*, during feeding, and in female grasshoppers when ovipositing.

Plasticization at ecdysis facilitates the stretching of presumptive sclerites, before the cuticle is sclerotized. In

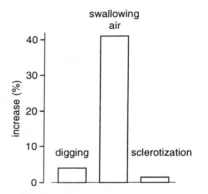

Fig. 16.13. Cuticular plasticity. The figure shows the extent to which a presumptive sclerite on the dorsal surface of the thorax of a fly (*Calliphora*) increases in length when subjected to a standard stretching force at different stages of eclosion. Having escaped from the puparium, the fly digs its way to the surface of the soil. During this period the cuticle is virtually inextensible. When the fly reaches the surface it swallows air to expand the new cuticle. At this time the cuticle is plasticized by the action of bursicon. Subsequently, bursicon induces sclerotization and the cuticle becomes hard and inextensible (data from Cottrell, 1962b).

the blowfly, *Calliphora*, plasticization of the sclerites occurs when the newly eclosed fly swallows air and pumps rhythmically to expand the cuticle. Very little extension of the cuticle can be produced before this period even though the procuticle is still undifferentiated. Once sclerotization begins, the cuticle is again inextensible (Fig. 16.13).

In *Manduca*, and probably in other insects, these changes are regulated by two hormones, eclosion hormone, which is released into the hemolymph just

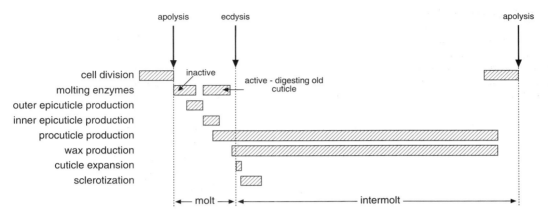

Fig. 16.14. The sequence of events involved in cuticle production. The timing of apolysis and the degree of overlap between procuticle production and cell division vary from species to species.

before ecdysis and which also switches on ecdysis behavior, and bursicon, which is released just after eclosion and which subsequently initiates sclerotization.

Rhodnius larvae normally take only one very large blood meal in each developmental stage. This results in considerable distension of the abdomen, so that its surface area increases about fourfold and the cuticle becomes considerably thinner. This distension is facilitated by plasticization of the largely unsclerotized abdominal cuticle. Plasticization is apparent within two minutes of the start of feeding, and is initiated by the pumping activity associated with food intake. Information is passed via stretch receptors to the central nervous system from which neurosecretory axons extend to the body wall where they branch extensively beneath the epidermis. They release serotonin so that a high local concentration occurs (Orchard, Lange & Barrett, 1988).

The cuticle in the plasticized regions contains only about 11% chitin, and the constituent proteins are low in aspartic and glutamic acids, but high in alanine and histidine. The low levels of amino acids with bulky side-chains may facilitate sliding of the protein molecules. As the cuticle is unsclerotized, the molecules are held together by relatively labile bonds; it is believed that a change in pH within the cuticle disrupts the bonding.

The increase in abdominal size in *Rhodnius* associated with plasticization is to a large extent reversible and the abdomen slowly shrinks to approach its original size as the blood meal is digested and excess fluid is excreted (Reynolds, 1985).

Female locusts and grasshoppers lay their eggs in the ground, often extending the abdomen to more than twice its normal length by stretching some of the intersegmental membranes (see section 13.5.2). The membranes are not extensible in males.

Review: Reynolds, 1985

16.4 MOLTING

Growth is limited by the cuticle. Sclerotized cuticle is virtually inextensible, and membranous cuticle can stretch only to the extent that folds in the outer epicuticle allow. As a result, for any marked increase in size to occur, the cuticle must be shed and replaced. Casting the cuticle is commonly known as molting, but it involves a sequence of events beginning with the separation of the old cuticle from the underlying epidermal cells and ending with the remnants of the old cuticle being shed. These processes are known as apolysis and ecdysis, respectively. After ecdysis, the new cuticle is expanded and chemically modified. In most insects, extensive deposition of new cuticle continues in the intermolt period. The timing of events is illustrated in Fig. 16.14 and changes in the integument in Fig. 16.17. The hormonal regulation of molting is discussed in section 15.4.

16.4.1 Changes in the epidermis

In most insects with well-sclerotized cuticles, such as the larvae of most hemimetabolous insects, epidermal cell density is constant through most of the intermolt period.

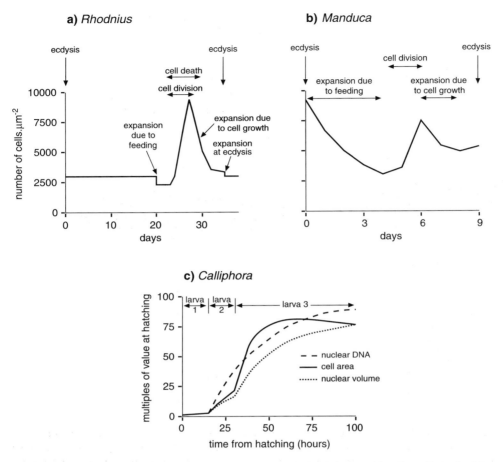

Fig. 16.15. Changes in the density or size of epidermal cells in relation to molting. Expansion and cell destruction decrease cell density; cell division increases it. (**a**) Tracheal epithelium of larval *Rhodnius*. A similar pattern probably occurs in epidermis associated with sclerites in other insects except for the expansion due to feeding (based on Locke, 1964). (**b**) Epidermis of larval *Manduca*, an example of an insect with unsclerotized cuticle (data from Wielgus & Gilbert, 1978). (**c**) In cyclorrhaphan Diptera no cell division occurs in the larval epidermis. Increase in epidermal area is entirely dependent on an increase in cell size. At the same time the DNA content of the nuclei increases due to endomitosis and the nuclei increase in volume (after Pearson, 1974).

It increases as a result of mitosis at the onset of the molting cycle and then undergoes a sharp, relatively small decrease during cuticular expansion following ecdysis (Fig. 16.15a). There is a net increase in the total number of cells so that although the number per unit area returns to its original value, the overall area of the epidermis, and hence of the cuticle which it produces, is increased. In larval forms with flexible cuticle, however, the pattern of changes is different. In the period of feeding following ecdysis, cell density decreases as the cuticle above the cells unfolds and

the cells spread laterally (Fig. 16.15b). Then, when mitosis occurs and the number of cells is increased, cell density increases sharply. At the larval–pupal molt, some increase in density probably also occurs due to rearrangement of the cells as the larva shortens. Cell growth in the period before ecdysis results in a subsequent decrease in density.

These changes in epidermal cell density are accompanied by changes in cell shape. In *Rhodnius*, the cells are columnar immediately after molting, but then become squamous until the insect feeds; after this they become

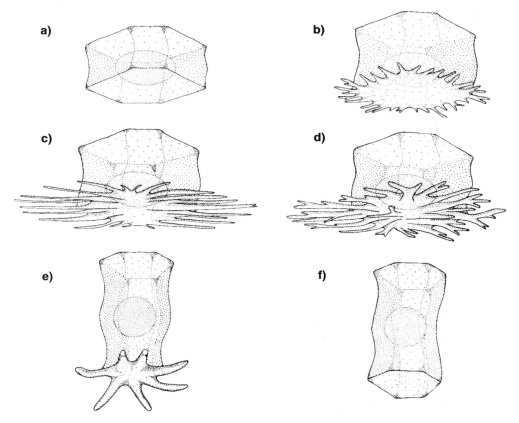

a)

b)

c)

d)

e)

f)

Fig. 16.16. Changes in the shape of epidermal cells during the final larval stage of *Calpodes* (Lepidoptera) (after Locke, 1985b): (**a**) 36 hours after ecdysis from the fourth stage; (**b**) 72 hours after ecdysis. Cell growth keeps pace with the increasing size of the larva; (**c**) the larva has reached its maximum size and the cell has extensive feet; (**d**) the feet begin to shorten as the larva shortens in the period before pupation; (**e**) cell area and the extent of feet is greatly reduced as shortening continues, but cell depth increases as the cells become crowded; (**f**) about 20 hours before ecdysis to the pupa.

deeper again. The epidermal cells of caterpillars, in contrast, become progressively deeper in the days following ecdysis, but get markedly deeper at the time of cell division and when the insect shortens prior to pupation (Fig. 16.16). Before this phase, in the final larval stage of *Calpodes* (Lepidoptera), the cells develop basal processes, or feet, which become longer as ecdysteroid titer rises, but then shorten rapidly as the larval segments also shorten prior to pupation. The feet are probably critically important in producing changes in the shape of an insect at a metamorphosis, and are probably formed in all insects.

Larval Cyclorrhapha differ from other insects because no cell division occurs in the epidermis. As a result, cell density decreases and the area of each cell increases throughout larval life as it grows (Fig. 16.15c).

16.4.2 Separation of the cuticle from the epidermis (Apolysis)

Perhaps as a result of the changes in cell shape, tension is generated at the epidermal cell surface resulting in its separation from the cuticle (Fig. 16.17b). In *Podura* (Collembola), however, the outer plasma membrane of the epidermal cells forms small outpushings that separate off as vesicles to form a foam which lifts off the cuticle. The separation of the cuticle from the underlying epidermis is known as apolysis, and the space formed between the epidermis and cuticle is called the exuvial, or subcuticular space.

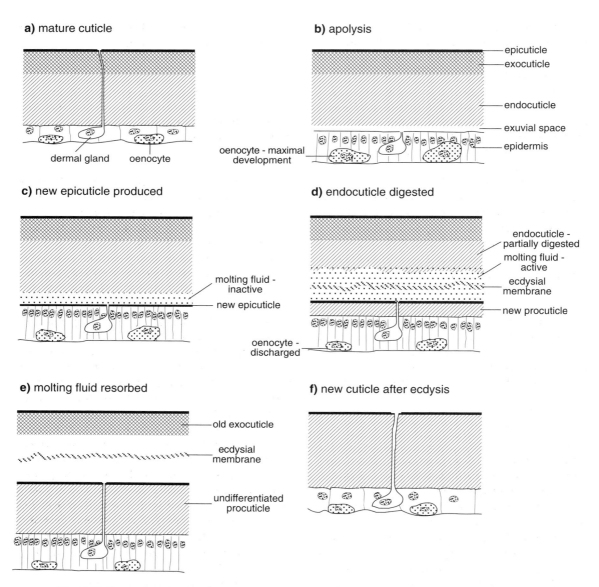

a) mature cuticle

b) apolysis

epicuticle
exocuticle
endocuticle
exuvial space
epidermis

oenocyte - maximal
development

dermal gland   oenocyte

c) new epicuticle produced

d) endocuticle digested

molting fluid -
inactive

new epicuticle

endocuticle -
partially digested
molting fluid -
active
ecdysial
membrane

new procuticle

oenocyte -
discharged

e) molting fluid resorbed

f) new cuticle after ecdysis

old exocuticle

ecdysial
membrane

undifferentiated
procuticle

Fig. 16.17. Changes in the integument over a molt. **(a)** Mature cuticle. **(b)** Apolysis follows cell division and the change in cell shape. **(c)** Molting fluid is produced. At first, the enzymes it contains are inactive. The new outer and inner epicuticles are produced. **(d)** The enzymes in the molting fluid are activated and start to digest the old endocuticle. At the same time new procuticle is laid down. **(e)** All the old endocuticle is digested and resorbed. New procuticle becomes thicker. **(f)** After ecdysis the new cuticle expands and the epidermal cells change shape. The procuticle is still undifferentiated.

16.4.3 Digestion of the old endocuticle

Before the cuticle is shed, the endocuticle is digested by enzymes secreted by the epidermal cells. A mixture of enzymes is involved, including chitinases, a trypsin-like protease and an aminopeptidase (Samuels & Paterson, 1996).

At least in some species, some of the enzymes are present in the cuticle even before apolysis. For example, in the larva of *Calpodes*, electron-dense droplets are secreted during the production of the last layers of procuticle. These exuvial droplets are probably inactive precursors of molting enzymes. In *Manduca* larva, chitinase is present in the cuticle (Bade, 1974). In many species, however, the enzymes are secreted into the exuvial space after apolysis. In *Podura*, they are produced as granules, and in various Lepidoptera they are described as forming a gel.

These enzymes do not start to digest the endocuticle until the outer epicuticle of the new cuticle is formed. They may be inactive because they are in the form of a precursor (zymogen) that requires activation, or because their substrate is not available to them. Proteolytic enzymes are probably secreted as precursors and require the secretion of a zymogen-activating enzyme before they become active. Chitinases, however, appear to have the potential to be active from the time of secretion, but depend on the prior action of proteolytic enzymes that free the chitin from its close association with proteins (Samuels & Reynolds, 1993). In *Manduca*, and probably in other insects, a chitinase attacks internal linkages in the chitin molecule, producing small oligosaccharides. These are then broken down by a β-*N*-acetylglucosaminidase. The two enzymes act synergistically.

Activation is associated with the active transport of potassium into the exuvial space accompanied by a bulk flow of water. The fluid produced is called molting fluid and its ionic composition may serve to buffer the enzyme systems against pH changes during the subsequent digestion of cuticle, as well as contributing to the solution of exuvial droplets.

The enzymes digest all the unsclerotized cuticle except the ecdysial membrane (see below), but have no effect on the exocuticle or on the muscle and nerve connections to the old cuticle. These connections persist so that the insect is still able to move and receive stimuli from the environment, but they are finally broken at ecdysis by the movements of the insect.

The products of cuticular digestion are absorbed through the mouth and anus, and possibly also directly through the integument (Cornell & Pan, 1983). As a result, up to 90% of the materials present in the cuticle may be conserved.

As a result of the activity of molting fluid, the cuticle becomes very thin and weak along the ecdysial lines (Fig. 16.11). These vary in position, but in the locust there is a Λ-shaped line on the head (see Fig. 1.3a) and a median dorsal line on the prothorax.

Review: Reynolds & Samuels, 1996

16.4.4 Ecdysis

When the molting fluid and the products of digestion of the molting fluid are resorbed, the old cuticle consists of little more than epicuticle and exocuticle and is quite separate from the new cuticle (Fig. 16.17e). Ecdysis usually follows as soon as digestion is complete (section 15.2).

A phase of preparatory activity loosens the old cuticle (pre-ecdysis behavior, Fig. 15.34). It consists of a sequence of relatively simple motor activities following each other in a more or less definite order. Carlson (1977) recognizes seven motor programs in the preparatory phase of *Teleogryllus*. Adult Lepidoptera make partial rotatory movements of the abdomen to free the themselves from the pupal cuticle just before eclosion. The sequence of activities is controlled largely endogenously by the central nervous system (section 15.4.3), but feedback from peripheral sensilla can prolong activities.

The insect splits the old cuticle by exerting pressure against it from within. An increase in blood volume, occurring before ecdysis in *Schistocerca* (Fig. 5.14) and probably in other insects, contributes to the insect's ability to do this. Then, in the preparatory phase of ecdysis, it usually swallows air or water, swelling the gut so that hemolymph pressure is increased. Blood is pumped into a particular part of the body, often the thorax, so that this expands and exerts pressure on the old cuticle causing it to split along its lines of weakness. Special muscles may be concerned in these pumping movements (section 10.2.4).

When the old cuticle is split, the insect draws itself out, usually head and thorax first, followed by the abdomen and appendages. Many insects suspend themselves freely from a support so that emergence is aided by gravity. All the cuticular parts are shed, including the intima of fore- and hind-gut, the endophragmal skeleton and the linings of the tracheae except for some delicate parts which may break off. The old cuticle is referred to as the exuviae.

Immediately after emergence, the new cuticle is unexpanded and soft, so that it provides the insect with little support. The blood probably acts as a hydrostatic skeleton, since its volume is still high, and, in adult *Calliphora*, it constitutes 30% of the body weight at this time. When expansion is complete, blood volume is reduced so that it comprises only about 10% of the body weight.

Some parts of the skeleton may be hardened before ecdysis. This pre-ecdysial hardening is usually restricted to small parts of the cuticle such as the claws, that are essential for the insect to hold on with, but it is more extensive in Cyclorrhapha and those Lepidoptera which have to escape from a pupal cell or cocoon.

16.5 CUTICLE FORMATION

16.5.1 Formation of the epicuticle
Production of the new cuticle begins with the secretion of the outer epicuticle as patches at the tips of microvilli of the epidermal cells. The patches grow at their margins and coalesce to form a continuous layer over the whole of the epidermis (Fig. 16.18). Apart from the lipids and proteins that form the structure of this layer, polyphenols and phenoloxidase are also produced so the proteins are probably stabilized soon after they are laid down. The surface pattern of the cuticle is produced as the outer epicuticle is formed, being molded by the underlying epidermal cells.

As the outer epicuticle is produced, the ecdysial membrane forms from the last few lamellae of the old procuticle that become sclerotized by the polyphenols and phenoloxidase involved in stabilizing the new outer epicuticle. Locke and Krishnan (1971) suggest that formation of the ecdysial membrane is simply a consequence of the presence of these chemicals, and that it has no functional significance. In *Calpodes*, the ecdysial membrane becomes fenestrated when the ecdysial droplets dissolve, permitting the free access of molting fluid to the more distal procuticle.

The inner epicuticle is secreted when the outer epicuticle is complete and the apical surfaces of the epidermal cells withdraw slightly. It is discharged in vesicles, which coalesce to form a discrete layer, and, in *Calpodes*, phenoloxidase is secreted at the same time. Polyphenols, however, are not present until about the time of ecdysis, so tanning of this layer does not take place until some time after its production.

16.5.2 Production of procuticle
Production of procuticle begins after the inner epicuticle is laid down. The chitin microfibrils are produced at plaques on the surface membrane of the epidermal cells. At the same

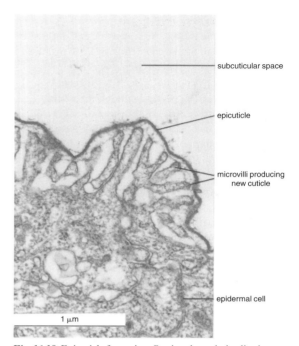

Fig. 16.18. Epicuticle formation. Section through the distal regions of epidermal cells during a molt at a stage approximately equivalent to Fig. 16.17c. The outer epicuticle is present as a complete layer. Outside (above) the epicuticle is the exuvial (subcuticular) space beneath the remains of the old cuticle, which has been removed in this preparation (courtesy of Dr. M. Locke).

time protein is laid down in the interstices between the fibrils. Some of this protein is synthesized in the epidermal cells, while some is taken up from the hemolymph, having been synthesized elsewhere. The zone of deposition of new cuticle is distinct and is sometimes called Schmidt's layer, but the precise nature of this zone is not known (Neville, 1975).

16.5.3 Hardening the cuticle
Some hardening takes place before ecdysis, but the bulk occurs soon afterwards when expansion of the new cuticle is completed. The process of hardening involves the development of cross links between protein chains and is known as sclerotization.

The proteins are laid down as the procuticle is formed, but the catecholamines producing the cross-linking are not formed until immediately prior to sclerotization. Insects cannot synthesize the ring forming the skeleton of the catecholamines, and it is usually derived from the aromatic amino acids, tyrosine and phenylalanine. Consequently, one or other of these compounds is a dietary essential for all

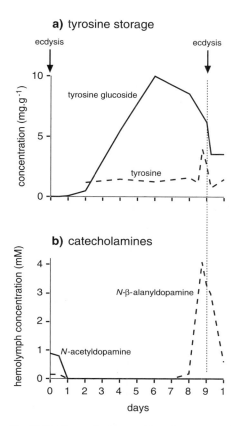

Fig. 16.19. Changes in the quantities of precursors of sclerotizing agents in the hemolymph of the last larval stage of a caterpillar (*Manduca*). (**a**) Tyrosine is stored largely as a glucoside, but before ecdysis the glucoside is hydrolysed and free tyrosine increases. It is rapidly converted to a catecholamine so the amount of free tyrosine present is never very high (after Ahmed, Hopkins & Kramer, 1983). (**b**) Catecholamines are only present in the hemolymph at the time of the molt. At the molt to the final larval stage (left) *N*-acetyldopamine is most abundant, but at the larva/pupa molt (right) only *N*-β-alanyldopamine is produced (after Hopkins, Morgan & Kramer, 1984).

insects (phenylalanine is also essential for protein synthesis).

Most insects store tyrosine over the intermolt period. Sometimes it is stored in the fat body, but often the bulk is in the hemolymph. As tyrosine is relatively insoluble it may be common for it to be conjugated with another compound so that its solubility is increased. In *Manduca*, it is stored as a glucoside (Fig. 16.19) which, in addition to increasing its solubility, may also protect it from metabolism in competing pathways.

The subsequent processing of tyrosine apparently

takes place primarily in the epidermis, although the products may be returned briefly to the hemolymph. The first steps are its conversion to dihydroxyphenylalanine (dopa) and then decarboxylation to dopamine (Fig. 16.20). From this, *N*-acetyldopamine and *N*-β-alanyldopamine, the two compounds currently known to be important in sclerotization, are produced.

phenylalanine

dihydroxyphenylalanine
(dopa)

dopamine

N - acetyldopamine

N - β - alanyldopamine

N-acetyldopamine is probably the only tanning agent in the cuticles of grasshoppers. In *Manduca*, *N*-acetyldopamine predominates at larva-to-larva molts and at the pupa-to-adult molt, but at the larva-to-pupa molt, *N*-β-alanyldopamine predominates (Fig. 16.19b). The latter has also been shown to be important in some flies.

Immediately after ecdysis, these compounds are transferred to the cuticle where they are oxidized to quinones by phenoloxidases. In the larva of *Calpodes*, a phenoloxidase is produced as an integral part of the procuticle as it is laid down, and this may be a common phenomenon. This phenoloxidase is incorporated into the cuticle only in areas destined to become sclerotized. The timing of sclerotization is regulated, in *Calpodes*, by the availability of catecholamines as the activity of the enzyme involved in cross-linking does not vary significantly. This is probably also true in other insects.

Quinones are highly reactive molecules and, once formed, their linkage to proteins is independent of enzyme activity. Covalent bonds may form directly between the quinone ring and protein molecules, or the quinones may form unstable quinone methides leading to linkages of the

Fig. 16.20. Diagram illustrating the synthesis of catecholamines from tyrosine and their roles in sclerotization. Only major pathways are shown (based on Hopkins & Kramer, 1992).

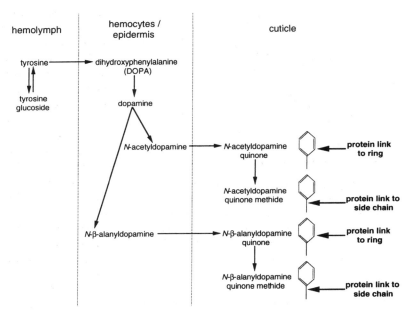

side chain of the quinone to the protein. The former is known as quinone tanning, the latter as β-sclerotization or quinone methide sclerotization.

quinone tanning

β - sclerotization

β-sclerotization produces a colorless cuticle, but quinone tanning causes the cuticle to darken.

Most of the amino groups in the proteins are already involved in the peptide linkages, -CO.NH-, between adjacent amino acids forming the protein chain. Consequently, the only amino groups available for linking to quinones are the terminal groups of the chains and those associated with dibasic acids, usually lysine. In the dibasic acids only one of the amino groups is involved in a peptide linkage, leaving the other available for cross-linking.

As the cuticle hardens, it usually also darkens. This darkening may simply result from quinone tanning, but it may also involve the polymerization of excess quinones to form melanin. It is probably generally true that, as in *Schistocerca*, only procuticle produced before ecdysis is sclerotized in larval insects. Procuticle which is produced subsequently remains undifferentiated, so that the insects come to have distinct exo- and endocuticles. In the adult locust, however, where cuticle deposition continues for more than 20 days, sclerotization continues for a similar period and so the whole thickness of the cuticle of the sclerites becomes stabilized. In the larva *N*-acetyldopamine is produced by the epidermal cells only for the first day after ecdysis, whereas in the adult it is formed continuously. This control is effected by hemocytes, which take up tyrosine from the hemolymph for a limited period at the larval molt.

Membranous regions remain unsclerotized and it is clear that the extent of sclerotization is regulated by epidermal cells responsible for producing the structural proteins and enzymes of the cuticle.

Reviews: Andersen, 1985, 1991; Hopkins & Kramer, 1992; Lipke, Sugumaran & Henzel, 1983 – Diptera; Sugumaran, 1991

16.6 EXPANSION OF THE NEW CUTICLE

In the later stages of ecdysis and immediately afterwards, the insect expands the new cuticle before it hardens. This often involves swallowing air or water. In the blowfly, air is pumped into the gut by the pharyngeal muscles, producing a steady increase in blood pressure (Fig. 16.21),

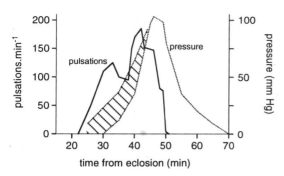

Fig. 16.21. Expansion of the new cuticle in an adult fly (*Sarcophaga*) after eclosion. Pulsations of the pharyngeal muscle show the activity of the pharyngeal pump as it pumps air into the gut. This produces an increase in blood pressure and, at the same time, contractions of ptilinal and abdominal muscles produce further, fluctuating increases in pressure (shown by the diagonally hatched area) (after Cottrell, 1962a).

while at the same time simultaneous contractions of the abdominal and ptilinal muscles produce transient increases in pressure. Expansion results partly from the opening out of deep folds in the new cuticle and partly from the pulling out of minor wrinkles in the epicuticle (Fig. 16.22). It is probably facilitated by a plasticization of the procuticle of the presumptive sclerites. The high pressure does not expand the membranous regions between the sclerites perhaps because the presumptive sclerites are held tightly together by accessory muscles (see Fig. 10.13).

As a result of these processes, the hind femur of the first instar larva of *Schistocerca* increases in length by about 35% over the period from hatching to final hardening of the cuticle, but the dimensions of the epicuticle do not change (Bernays, 1972).

As the procuticle expands it also becomes thinner, partly due to stretching and partly to dehydration. Some changes in the orientation of chitin microfibrils may occur. In the wing of the butterfly *Aglais*, the microfibrils have a helicoidal arrangement when the cuticle is first laid down, but when the wing is expanded they are oriented parallel with the veins.

16.7 CHANGES IN THE INTERMOLT CUTICLE

In many species, procuticle production continues until the next molt when cuticle digestion begins again. In *Calpodes*, and probably in other insects, the rate of cuticle deposition is greater during the intermolt period than during the molt

(Fig. 16.23a). In general, new cuticular material is added as new lamellae adjacent to the epidermis (Wolfgang & Riddiford, 1981).

In soft-bodied larvae, such as caterpillars, the surface area of the cuticle increases at the same time as the increase in thickness. The surface area of the final stage *Manduca* larva, for example, increases four-fold. This involves unfolding, but not stretching, the epicuticle. The increase in area of the procuticle results partly from addition of material in a column of cuticle above each epidermal cell (Fig. 16.24). Each column is produced by a bundle of long microvilli which extend almost to the epicuticle. They lay down cuticular microfibers at right angles to the lamellae produced by other parts of the same cell. The columns are initially about 5 μm in diameter, but increase in diameter by a factor of three or four as more cuticle is produced. At the same time, the lamellae nearest to the cuticle are stretched and reduced to about 25% of their original thickness (Wolfgang & Riddiford, 1981). Some unfolding of pleats in the original arrangement of microfibers also occurs (Carter & Locke, 1993).

It is not clear how the production of cuticle in the intermolt period is regulated. In some larval Lepidoptera it is believed that very low concentrations of ecdysone govern the process, but in *Manduca* there is evidence that ecdysone is not involved. It is suggested that a humoral factor from the brain regulates cuticle production by causing the fat body to synthesize or release precursors of cuticular proteins (Wielgus and Gilbert, 1978).

Wax is first secreted on to the outside of the cuticle at about the time of ecdysis and its production and secretion probably continues throughout much of the intermolt period. During the pupal period of *Trichoplusia*, hydrocarbon synthesis is low at first, and stops completely about half way through the period and the amount of lipid on the surface of the pupa remains constant (Fig. 16.23b). A sharp rise in synthesis occurs just before eclosion and the high rate of synthesis is then maintained for at least the first three days of adult life. A small increase in the amount of wax on the pupal cuticle occurs, but most of the hydrocarbons produced during the pupal period are stored and secreted on to the adult cuticle at the time of eclosion (de Renobales *et al.*, 1988). The secretion of wax by *Calpodes* larvae is controlled by a hormone released from the corpus allatum-corpus cardiacum complex. This hormone is only effective when it is present with ecdysone.

A layer of cement is formed over the surface of the wax soon after ecdysis in some species.

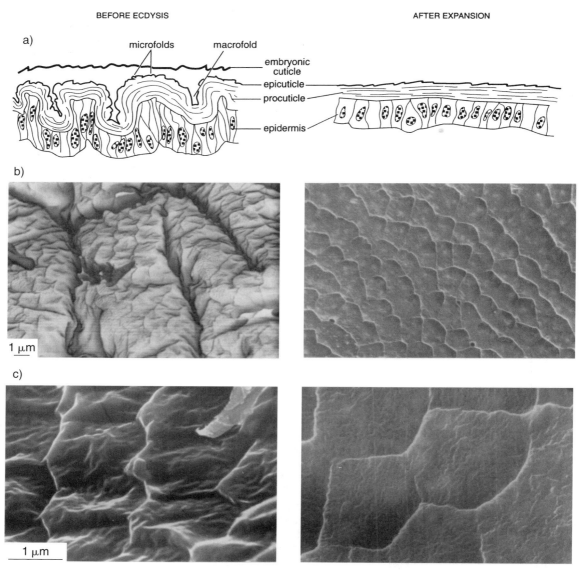

Fig. 16.22. Expansion of the new cuticle in a first stage larva of a grasshopper immediately after hatching. Similar changes occur at subsequent molts. **Left**: before ecdysis of the embryonic cuticle, the larval cuticle exhibits both macro- and microfolds; **right**: after expansion, about 30 minutes later, but before sclerotization is evident. The cuticle is flat without macrofolds and microfolds have almost disappeared. Outlines of the underlying epidermal cells producing each patch of cuticle are clearly seen (after Bernays, 1972 and courtesy Dr E.A. Bernays). (**a**) Diagrammatic sections through the integument. (**b**) Low power scanning electron micrographs of the surface of the cuticle. (**c**) Higher power scanning electron micrographs of the cuticle.

a) larval procuticle

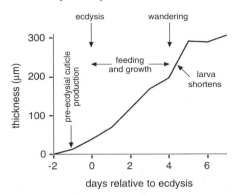

Fig. 16.23. Cuticular changes in the intermolt period. **(a)** Changes in the thickness of the cuticle during the final larval stage of a caterpillar (*Manduca*). During the period of feeding, the cuticle thickens although the caterpillar is also increasing in surface area at the same time. This is due to the continuous production of procuticle. Production stops during the wandering stage, but the cuticle becomes thicker as the larva shortens prior to pupation (based on Wolfgang & Riddiford, 1987). **(b)** Changes in the amount of cuticular hydrocarbons during the pupal and early adult stages of a moth (*Trichoplusia*). Additional hydrocarbon is deposited on the surface of the cuticle during the first half of the pupal period; subsequently it remains constant. This material is lost at ecdysis and a new layer of wax forms on the adult cuticle. The quantity of hydrocarbon increases sharply at ecdysis and continues to rise in the following hours (based on de Renobales *et al.*, 1988).

b) pupal wax

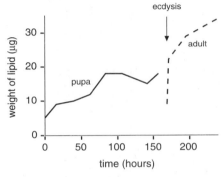

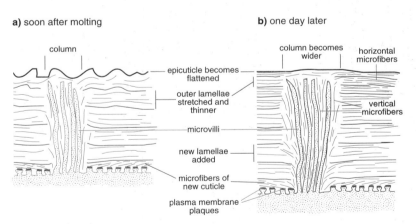

Fig. 16.24. Intermolt growth of cuticle of a caterpillar. Sections of the cuticle showing the column of microvilli above an epidermal cell (based on Wolfgang & Riddiford, 1981): **(a)** soon after ecdysis; **(b)** one day later.

16.8 FUNCTIONS OF THE CUTICLE

The cuticle is one of the features of insects which is primarily responsible for their success. It plays an important part in supporting the insect, an essential requirement in terrestrial animals. The tubular, external skeleton of the legs provides great strength and relative lightness compared with the internal skeleton of vertebrates. Further, the presence of hard, jointed appendages makes accurate movements possible with a minimum of muscle, and, by lifting the body off the ground, facilitates rapid movement. The rigidity of cuticle, forming wings, also makes flight possible.

The cuticle also provides protection. Some insects, such as adult beetles, have hardened, heavily sclerotized cuticles which make them difficult for predators to catch or parasites to parasitize. Protection from the physical environment is also afforded. Again in beetles, the upper cuticle of the abdomen, protected by the elytra, is very thin but the cuticle of the ventral surface, which is exposed and subject to abrasion by the substratum, is very thick. The cuticular lining of the fore- and hind-guts also protects the epidermis from abrasion by the food.

The cuticle also plays a major role in the success of insects as terrestrial organisms by reducing water loss. This is largely a function of the wax layer, but the whole contributes (section 18.4.1.1).

Finally, parts of the cuticle are modified to form sense organs (Chapters 23, 24) and its physical structure is also often important in the production of color (section 25.2).

REFERENCES

Ahmed, R.F., Hopkins, T.L. & Kramer, K.J. (1983). Tyrosine and tyrosine glucoside titres in whole animals and tissues during development of the tobacco hornworm *Manduca sexta* (L.). *Insect Biochemistry*, **13**, 369–74.

Andersen, S.O. (1974). Cuticular sclerotization in larval and adult locusts, *Schistocerca gregaria. Journal of Insect Physiology*, **20**, 1537–52.

Andersen, S.O. (1979). Biochemistry of insect cuticle. *Annual Review of Entomology*, **24**, 29–61.

Andersen, S.O. (1985). Sclerotization and tanning of the cuticle. In *Comprehensive Insect Physiology, Biochemistry and Pharmacology*, vol. 3, ed. G.A. Kerkut & L.I. Gilbert, pp. 59–74. Oxford: Pergamon Press.

Andersen, S.O. (1991). Sclerotisation. In *Physiology of the Insect Epidermis*, ed. K. Binnington & A. Retnakaran, pp. 123–40. Melbourne: CSIRO.

Andersen, S.O., Højrup, P. & Roepstorff, P. (1995). Insect cuticular proteins. *Insect Biochemistry and Molecular Biology*, **25**, 153–76.

Andersen, S.O. & Weis-Fogh, T. (1964). Resilin. A rubberlike protein in arthropod cuticle. *Advances in Insect Physiology*, **2**, 1–66.

Ashhurst, D.E. (1985). Connective tissues. In *Comprehensive Insect Physiology, Biochemistry and Pharmacology*, vol. 3, ed. G.A. Kerkut & L.I. Gilbert, pp. 249–87. Oxford: Pergamon Press.

Bade, M.L. (1974). Localization of moulting chitinase in insect cuticle. *Biochimica et Biophysica Acta*, **372**, 474–7.

Bernays, E.A. (1972). Changes in the first instar cuticle of *Schistocerca gregaria* before and associated with hatching. *Journal of Insect Physiology*, **20**, 281–90.

Blomquist, G.J. & Dillwith, J.W. (1985). Cuticular lipids. In *Comprehensive Insect Physiology, Biochemistry and Pharmacology*, vol. 3, ed. G.A. Kerkut & L.I. Gilbert, pp. 117–54. Oxford: Pergamon Press.

Carlson, J.R. (1977). The imaginal ecdysis of the cricket (*Teleogryllus oceanicus*) I. Organization of motor programs and roles of central and sensory control. *Journal of Comparative Physiology* A, **115**, 299–317.

Carter, D. & Locke, M. (1993). Why caterpillars do not grow short and fat. *Tissue & Cell*, **22**, 81–102.

Caveney, S. (1985). Intercellular communication. In *Comprehensive Insect Physiology, Biochemistry and Pharmacology*, vol. 2, ed. G.A. Kerkut & L.I. Gilbert, pp. 319–70. Oxford: Pergamon Press.

Caveney, S. (1991). Cell-to-cell signalling in the epidermis. In *Physiology of the Insect Epidermis*, ed. K. Binnington & A. Retnakaran, pp. 23–35. Melbourne: CSIRO.

Chapman, R.F. (1995). Mechanics of food handling by chewing insects. In *Regulatory Mechanisms in Insect Feeding*, ed. R.F. Chapman & G. de Boer, pp. 3–31. New York: Chapman & Hall.

Chino, H. (1985). Lipid transport: biochemistry of hemolymph lipophorin. In *Comprehensive Insect Physiology, Biochemistry and Pharmacology*, vol. 10, ed. G.A. Kerkut & L.I. Gilbert, pp. 115–35. Oxford: Pergamon Press.

Cornell, J.C. & Pan, M.L. (1983). The disappearance of molting fluid in the tobacco hornworm, *Manduca sexta*. *Journal of Experimental Biology*, **107**, 501–4.

Cottrell, C.B. (1962a). The imaginal ecdysis of blowflies. Observations on the hydrostatic mechanisms involved in digging and expansion. *Journal of Experimental Biology*, **39**, 431–48.

Cottrell, C.B. (1962b). The imaginal ecdysis of blowflies. Evidence for a change in the mechanical properties of the cuticle at expansion. *Journal of Experimental Biology*, **39**, 449–58.

de Renobales, M., Nelson, D.R., & Blomquist, G.J. (1991). Cuticular lipids. In *Physiology of the Insect Epidermis*, ed. K. Binnington & A. Retnakaran, pp. 240–51. Melbourne: CSIRO.

de Renobales, M., Nelson, D.R., Mackay, M.E., Zamboni, A.C. & Blomquist, G.J. (1988). Dynamics of hydrocarbon bio-synthesis and transport to the cuticle during pupal and early adult development in the cabbage looper *Trichoplusia ni* (Lepidoptera: Noctuidae). *Insect Biochemistry*, **18**, 607–13.

Edwards, H.A. (1983). Occurrence of resilin in elastic structures in the food-pump of reduviid bugs. *Journal of Experimental Biology*, **105**, 407–9.

Fontaine, A.R., Olsen, N., Ring, R.A. & Singla, C.L. (1991). Cuticular metal hardening of mouthparts and claws of some forest insects of British Columbia. *Journal of the Entomological Society of British Columbia*, **88**, 45–55.

Gu, X., Quilici, D., Juarez, P., Blomquist, G.J. & Schal, C. (1995). Biosynthesis of hydrocarbons and contact sex pheromone and their transport by lipophorin in females of the German cockroach (*Blattella germanica*). *Journal of Insect Physiology*, **41**, 257–67.

Hepburn, H.R. (1985). Structure of the integument. In *Comprehensive Insect Physiology, Biochemistry and Pharmacology*, vol. 3, ed. G.A. Kerkut & L.I. Gilbert, pp. 1–58. Oxford: Pergamon Press.

Hillerton, J.E., Reynolds, S.E. & Vincent, J.F.V. (1982). On the indentation hard-ness of insect cuticle. *Journal of Experimental Biology*, **96**, 45–52.

Hopkins, T.L. & Kramer, K.J. (1992). Insect cuticle sclerotization. *Annual Review of Entomology*, **37**, 273–302.

Hopkins, T.L., Morgan, T.D. & Kramer, K.K. (1984). Catecholamines in haemolymph and cuticle during larval, pupal and adult development of *Manduca sexta* (L.). *Insect Biochemistry*, **14**, 533–40.

Horwath, K.L. & Riddiford, L.M. (1991). Cellular differentiation of specialised epidermal cells: the dermal glands. In *Physiology of the Insect Epidermis*, ed. K. Binnington & A. Retnakaran, pp. 185–94. Melbourne: CSIRO.

Jackson, A. & Locke, M. (1989). The formation of plasma membrane reticular systems in the oenocytes of an insect. *Tissue & Cell*, **21**, 463–73.

Kramer, K.J. & Koga, D. (1986). Insect chitin. Physical state, synthesis, degradation and metabolic regulation. *Insect Biochemistry*, **16**, 851–77.

Leschen, R.A.B. & Cutler, B. (1994). Cuticular calcium in beetles (Coleoptera: Tenebrionidae: Phrenapetinae). *Annals of the Entomological Society of America*, **87**, 918–21.

Lipke, H., Sugumaran, M. & Henzel, W. (1983). Mechanisms of sclerotization in dipterans. *Advances in Insect Physiology*, **17**, 1–84.

Locke, M. (1964). The structure and formation of the integument in insects. In *The Physiology of Insecta*, vol. 3, ed. M. Rockstein, pp. 379–470. New York: Academic Press.

Locke, M. (1984). Epidermal cells. In *Biology of the Integument* vol. 1, *Invertebrates*, ed. J. Bereiter-Hahn, A.G. Matoltsy & K.S. Richards, pp. 502–22. Berlin: Springer-Verlag.

Locke, M. (1985a). A structural analysis of post-embryonic development. In *Comprehensive Insect Physiology, Biochemistry and Pharmacology*, vol. 2, ed. G.A. Kerkut & L.I. Gilbert, pp. 87–149. Oxford: Pergamon Press.

Locke, M. (1985b). The structure of epi-dermal feet during their development. *Tissue & Cell*, **17**, 901–21.

Locke, M. (1991). Insect epidermal cells. In *Physiology of the Insect Epidermis*, ed. K. Binnington & A. Retnakaran, pp. 1–22. Melbourne: CSIRO.

Locke, M. & Krishnan, N. (1971). The distribution of phenoloxidases and polyphenols during cuticle formation. *Tissue & Cell*, **3**, 103–26.

Lockey, K.H. (1988). Lipids of the insect cuticle: origin, composition and func-tion. *Comparative Biochemistry and Physiology*, **89B**, 595–645.

Neville, A.C. (1975). *Biology of the Arthropod Cuticle*. Berlin: Springer-Verlag.

Neville, A.C. (1983). Daily growth layers and the teneral stage in adult insects: a review. *Journal of Insect Physiology*, **29**, 211–9.

Neville, A.C., Thomas, M.G. & Zelazny, B. (1969). Pore canal shape related to molecular architecture of arthropod cuticle. *Tissue & Cell*, **1**, 183–200.

Noirot, C. & Quennedy, A. (1974). Fine structure of insect epidermal glands. *Annual Review of Entomology*, **19**, 61–80.

Orchard, I., Lange, A.B. & Barrett, F.M. (1988). Serotonergic supply to the epi-dermis of *Rhodnius prolixus*: evidence for serotonin as the plasticising factor. *Journal of Insect Physiology*, **34**, 873–9.

Pearson, M.J. (1974). The abdominal epi-dermis of *Calliphora erythrocephala* (Diptera) I. Polyteny and growth in the larval cells. *Journal of Cell Science*, **16**, 113–31.

Rees, H.H. (1985). Biosynthesis of ecdysone. In *Comprehensive Insect Physiology, Biochemistry and Pharmacology*, vol. 7, ed. G.A. Kerkut & L.I. Gilbert, pp. 249–93. Oxford: Pergamon Press.

Reynolds, S.E. (1985). Hormonal control of cuticle mechanical properties. In *Comprehensive Insect Physiology, Biochemistry and Pharmacology*, vol. 8, ed. G.A. Kerkut & L.I. Gilbert, pp. 335–51. Oxford: Pergamon Press.

Reynolds, S.E. & Samuels, R.I. (1996). Physiology and biochemistry of insect moulting fluid. *Advances in Insect Physiology*, **26**, 157–232.

Richards, A.G. (1967). Sclerotization and the localization of brown and black colours in insects. *Zoologische Jahrbücher. Abteilung für Anatomie und Ontogenie der Tiere*, **84**, 25–62.

Romer, F. & Bressel, H.U. (1994). Secretion and metabolism of ecdysteroids by oenocyte-fat body complexes (OEFC) in adult males of *Gryllus bimaculatus* DeG (Insecta). *Zeitschrift für Naturforschung* C **49**, 871–80.

Samuels, R.I. & Paterson, I.C. (1996). Cuticle degrading proteases from insect moulting fluid and culture filtrates of entomopathogenic fungi. *Comparative Biochemistry and Physiology*, **110B**, 661–9.

Samuels, R.I. & Reynolds, S.E. (1993). Molting fluid enzymes of the tobacco hornworm, *Manduca sexta*: timing of proteolytic and chitinolytic activity in relation to pre-ecdysial development. *Archives of Insect Biochemistry and Physiology*, **24**, 33–44.

Snodgrass, R.E. (1935). *The principles of Insect Morphology*. New York: McGraw-Hill.

Sugumaran, M. (1991). Molecular mechanisms of sclerotisation. In *Physiology of the Insect Epidermis*, ed. K. Binnington & A. Retnakaran, pp. 141–68. Melbourne: CSIRO.

Wigglesworth, V.B. (1985). The transfer of lipid in insects from the epidermal cells to the cuticle. *Tissue & Cell*, **17**, 249–65.

Wigglesworth, V.B. (1988). The source of lipids and polyphenols for the insect cuticle: the role of the fat body, oenocytes and oenocytoids. *Tissue & Cell*, **20**, 919–32.

Wielgus, J.J. & Gilbert, L.I. (1978). Epidermal cell development and control of cuticle deposition during the last larval instar of *Manduca sexta*. *Journal of Insect Physiology*, **24**, 629–37.

Willis, J.H. (1987). Cuticular proteins: the neglected component. *Archives of Insect Biochemistry and Physiology*, **6**, 203–15.

Willis, J.H. (1991). The epidermis and metamorphosis. In *Physiology of the Insect Epidermis*, ed. K. Binnington & A. Retnakaran, pp. 36–45. Melbourne: CSIRO.

Winston, M.L. (1987). *The Biology of the Honey Bee*. Cambridge, Massachusetts: Harvard University Press.

Wolfgang, W.J. & Riddiford, L.M. (1981). Cuticular morphogenesis during continuous growth of the final instar larva of a moth. *Tissue & Cell*, **13**, 757–72.

Wolfgang, W.J. & Riddiford, L.M. (1987). Cuticular mechanics during larval development of the tobacco hornworm, *Manduca sexta*. *Journal of Experimental Biology*, **128**, 19–34.

17   Gaseous exchange

Gaseous exchange in insects occurs through a system of internal tubes, the tracheal system, the finer branches of which extend to all parts of the body and may become functionally intracellular in muscle fibers. Thus oxygen is carried directly to its sites of utilization and the blood is not concerned with its transport. In terrestrial insects and some aquatic species, the tracheae open to the outside through segmental pores, the spiracles, which generally have some closing mechanism reducing water loss from the respiratory surfaces. Other aquatic species have no functional spiracles, and gaseous exchange with the water involves arrays of tracheae close beneath the surface of thin, permeable cuticle.

Reviews: Grassé, 1976 – general; Mill, 1985 – general; Miller, 1981a – cockroaches

17.1 TRACHEAL SYSTEM

17.1.1 **Tracheae**

The tracheae are the larger tubes of the tracheal system, running inward from the spiracles and usually breaking up into finer branches, the smallest of which are about 2 μm in diameter. Tracheae are formed by invaginations of the ectoderm and so are lined by a cuticular intima which is continuous with the rest of the cuticle. A spiral thickening of the intima runs along each tube, each ring of the spiral being called a taenidium (Fig. 17.1). The intima consists of outer epicuticle with a protein/chitin layer beneath it. In the taenidia the protein/chitin cuticle is differentiated as mesocuticle or exocuticle. The chitin microfibrils in the taenidia run round the trachea, while between the taenidia they are parallel with the long axis of the trachea. A layer of resilin may be present beneath the epicuticle (Whitten, 1972).

The taenidia prevent collapse of the trachea if pressure within the tube is reduced. In the wing tracheae of some insects, the taenidia are themselves twisted, giving some elasticity to the wall of the trachea (section 5.1.2.1).

In places, the tracheae are expanded to form thin-walled airsacs (see Fig. 17.4) in which the taenidia are absent or poorly developed and often irregularly arranged. Consequently, the airsacs collapse under pressure and they play a very important part in ventilation of the tracheal system as well as having other functions. Airsacs are widely distributed along the main tracheal trunks of many insects.

17.1.2 **Tracheoles**

At various points along their length, especially distally, the tracheae give rise to finer tubes, the tracheoles. There is no sharp distinction between tracheae and tracheoles, but the latter always appear to be intracellular and often retain their cuticular lining at molting, which is not usually true of tracheae. Proximally the tracheoles are about 1 μm in diameter, tapering to about 0.1 μm or less. They are formed in cells (often called tracheoblasts, but Wigglesworth, 1983, suggests the term 'tracheolar cells') which are derived from the epidermal cells lining the tracheae (Fig. 17.2). It is not certain if the tracheoles develop as truly intracellular structures or if they are really extracellular, forming in a deep fold of the plasma membrane of the tracheolar cell and so having the appearance of being intracellular (Whitten, 1972; Wigglesworth, 1983). The intima of tracheoles is some 16–20 nm thick and may consist only of outer epicuticle. It is thrown into taenidial ridges, but, unlike the taenidia of tracheae, these ridges are not filled with chitin/protein matrix (Edwards, Ruska & Harven, 1958).

The tracheoles are intimately associated with the tissues and, in some flight muscles, they indent the muscle plasma membrane and penetrate deep into the fiber (Fig. 17.1c,d), but it is probable that they never become truly intracellular within the muscle. Each tracheolar branch ends blindly; there is no good evidence that they anastomose to form a network, although this is described in some earlier literature.

Wigglesworth (1990b) describes tracheae in the abdomen of *Calliphora* in which the cuticle between the taenidia bulges outward having the appearance of a tracheole wound round the trachea in a spiral, but with the length

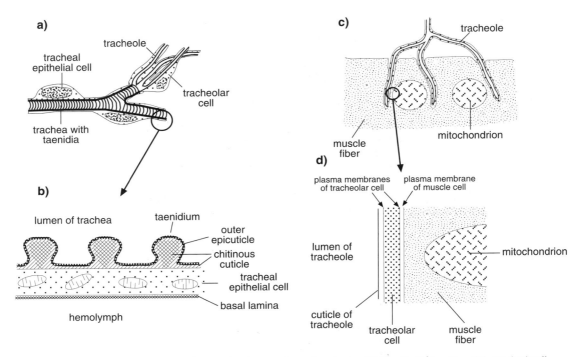

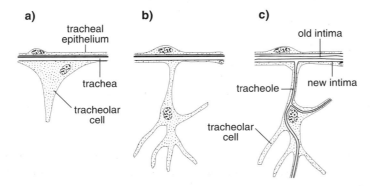

Fig. 17.1. Tracheal system. **(a)** Diagram showing a trachea and tracheoles. **(b)** Longitudinal section of the tracheal wall. Tracheal lumen above, hemolymph below. **(c)** Diagram showing tracheoles indenting a muscle fiber. **(d)** Enlargement of area circled in (c). Oxygen diffuses from the lumen of the tracheole (left) through the cuticle and the tracheolar cell to the mitochondrion in the muscle fiber.

Fig. 17.2. Development of a tracheole (after Keister, 1948). **(a)** Tracheolar cell developing from the tracheal epithelium. **(b)** Tracheolar cell with extensive cytoplasmic processes. **(c)** Tracheole within tracheolar cell and becoming connected to existing trachea at a molt.

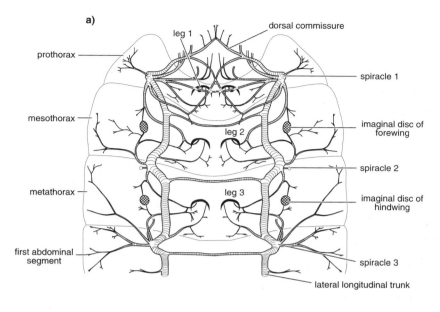

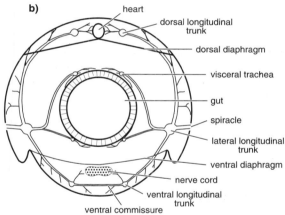

Fig. 17.3. Tracheal system (from Snodgrass, 1935). **(a)** Tracheation of the thorax and first abdominal segment of a caterpillar, dorsal view. **(b)** Diagrammatic cross-section of the abdomen of an orthopteran showing the principal tracheae and tracheal trunks.

of the lumen connecting with the lumen of the trachea. Wigglesworth calls these aeriferous tracheae and suggests that they functionally replace tracheoles as the primary supply to the ovaries.

Review: Wigglesworth, 1983

17.1.3 Distribution of the tracheae within the insect

In those Collembola that have tracheae and in the Archaeognatha, the tracheae from each spiracle form a tuft which remains separate from the tufts of other spiracles. In the majority of insects, however, the tracheae from neighboring spiracles join to form longitudinal trunks running the length of the body (Fig. 17.3). Usually there is a lateral trunk on either side of the body and these are often the largest tracheae, while, in addition, dorsal and ventral longitudinal trunks may also be present. The longitudinal tracheae are connected to those of the other side of the body by transverse commissures, while smaller branches extend to the various tissues and, in turn, give rise to the tracheoles. In general, the heart and dorsal muscles are supplied by branches from the dorsal trunks, the alimentary canal, gonads, legs and wings from the lateral trunks and the central nervous system from the ventral trunks or transverse commissures.

The arrangement of tracheae tends to follow a similar pattern in each body segment, especially in larval forms and in the abdomen of adults. This basic arrangement is modified, however, in the head, where there are no spiracles, and in the thorax. The head is supplied with air from spiracle 1 (often on the prothorax, see below) through two main tracheal branches on each side, a dorsal branch to the antennae, eyes and brain and a ventral branch to the mouthparts and their muscles.

In actively flying insects, the oxygen demands of the flight muscles during flight greatly exceed those of all the other tissues. It is essential that these demands should not deprive other important tissues, such as the central nervous system, of oxygen. In the desert locust, *Schistocerca*, this is achieved by effectively isolating the air supply to the thorax from that to the rest of the body (Fig. 17.4). In addition, the two sides are isolated from each other. Even though the head is supplied from the first thoracic spiracle, its tracheal supply is also largely separated from the thoracic supply by the small bore of some connecting tubes and the occlusion of others. This ensures a good and direct supply of air to the brain and major sense organs and, as the exhalent trunk from the head supplies the thoracic ganglia, these also have a good air supply (Miller, 1960c).

To some extent, the distribution and abundance of tracheae and especially of tracheoles reflect the demands for oxygen by different tissues. Consequently, tracheoles are most abundant in areas of high metabolic activity, whether this is associated with mechanical energy production, as in flight muscles, or with chemical synthetic activity, as in pheromone glands. They are also very abundant in the central nervous system, especially in the neuropil, perhaps suggesting that synapses have a high demand for oxygen. In the ganglia of the locust central nervous system, every point in the neuropil is within 10 μm of a tracheole, but no tracheal branching occurs in the outer parts of the ganglion where the cell bodies are situated (Fig. 17.5) (Burrows, 1980). The ovaries of several orders of insects are supplied by aeriferous tracheae (see above), but the rapidly developing terminal oocytes are individually supplied by many tracheoles (Wigglesworth, 1991).

Each flight muscle has a primary supply consisting of a large tracheal trunk or airsac running alongside or through the muscle (Fig. 17.6). If a trachea forms the primary supply it widens to an airsac beyond the muscle. From the primary supply, small, regularly spaced tracheae arise at right angles, running into the muscle. These form the secondary supply and they are often oval proximally, permitting some degree of collapse, and taper regularly to the distal end. Finer branches pass in turn from these tracheae into the muscles (Weis-Fogh, 1964a). In Odonata the terminal tracheolar branches run alongside and between the muscle fibers, but in close-packed and fibrillar flight muscle they indent the fiber membrane and the finest tracheolar branches are closely associated with the mitochondria (Fig. 17.1c,d). Thus they are functionally within the muscle fiber, although anatomically they are still extracellular. The tracheal supply to the flight muscles follows a similar pattern in all larger insects.

Major changes in tracheation occur only at a molt and, in endopterygote insects, during the pupal period. For instance, the relative volume of tracheae in the ovaries of *Schistocerca* increases 18 times at the penultimate molt and a further 16 times at the final molt. Between molts, the relative volume decreases because the ovary continues to grow; similar changes occur in the testes and male accessory glands. Thus, as a tissue grows within a larval stage it becomes relatively less well supplied with oxygen. Possibly shortage of oxygen stimulates mitosis in the tracheal cells, so that at the molt new tracheae are formed. New tracheae arise as columns of cells growing out from the existing tracheal epithelium (Fig. 17.7). A space develops between the cells and it becomes lined with cuticle and connects to the existing system at the next molt (Wigglesworth, 1954).

Small changes in the distribution of tracheoles may occur between molts. For instance, in the event of damage to the epidermis, the epidermal cells produce cytoplasmic threads which extend towards and eventually attach themselves to the nearest tracheole. These cytoplasmic threads, which may be 150 μm long, are contractile and drag the tracheole to the region of oxygen-deficient tissue (Fig. 17.8). The normal distribution of tracheoles in the epidermis might arise in a similar way (Wigglesworth, 1959). Changes in tracheole distribution occur in relation to flight muscle development in pupae of cyclorrhaphan Diptera (see Fig. 15.25).

At least partly related to altered oxygen demands, the tracheal system varies with the state of development and becomes more complex at each molt. This may involve

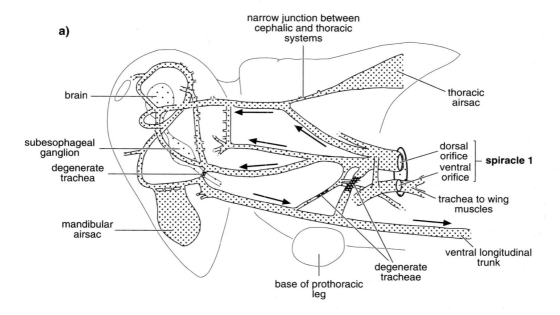

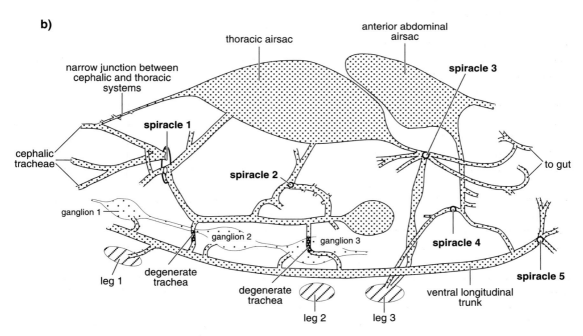

Fig. 17.4. Tracheal system of a locust (*Schistocerca*) (after Miller, 1960c). (**a**) Supply to the head from spiracle 1. Arrows indicate the probable direction of airflow resulting from abdominal ventilatory movements. (**b**) Supply to the pterothorax. Notice that the thoracic airsac is largely isolated from other parts of the tracheal system by narrow or degenerate tracheae.

Fig. 17.5. Tracheal supply to the metathoracic ganglion of the central nervous system of a locust. Two pairs of tracheae, arising from the ventral longitudinal tracheae, enter the ganglion on the ventral side. Notice that branching is largely restricted to the central region of the ganglion. The peripheral region, where the cell bodies of the neurons are present, does not have extensive branching (after Burrows, 1980).

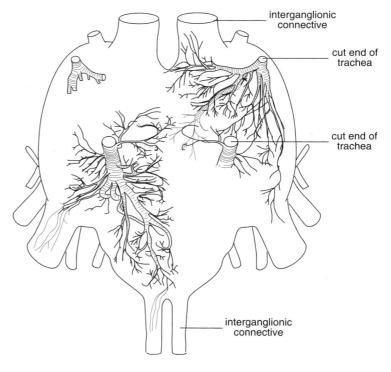

changes in the functional spiracles as well as in tracheation. For example, the first stage larva of *Sciara* (Diptera) is metapneustic, the second is propneustic and the fourth stage hemipneustic (see below). Each new system is built round that of the preceding instar, but on a larger scale and with new extensions.

17.1.4 Molting the tracheal system

The cuticular lining of the tracheae is cast at each molt, and a new, larger intima formed in its place. The longitudinal trunks break at predetermined points, the nodes, between adjacent spiracles and the old lining is drawn out through the spiracles and shed with the rest of the exuviae. Where the number of functional spiracles is reduced, the 'non-functional' spiracles persist and facilitate the shedding of the tracheal intima so that this can occur even in apneustic insects. The 'non-functional' spiracles may be visible as faint scars on the cuticle. From each scar, a strand of cuticle (formed from a collapsed trachea) connects with the longitudinal trachea and, at each molt, a tube of new cuticle is laid down round it. The old intima is withdrawn through the tube which subsequently closes and forms the cuticular strand connecting the new intima with the outer

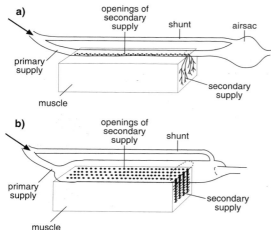

Fig. 17.6. Tracheal supply to the flight muscles. Arrows indicate the inward flow of air (after Weis-Fogh, 1964a). **(a)** Primary supply is a trachea ventilated by an airsac outside the muscle. **(b)** Primary supply is an airsac directly associated with the muscle.

a)

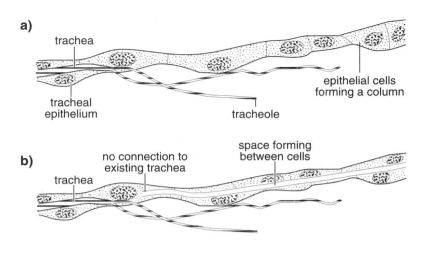

b)

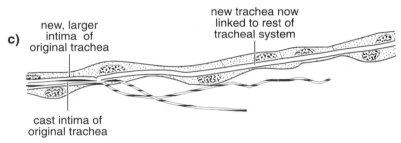

c)

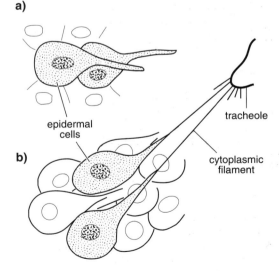

Fig. 17.7. Development of a new trachea as an extension of part of the existing system. Notice that the tracheolar cell is not shown (after Wigglesworth, 1954). (a) A column of cells extends from the tracheal epithelium. (b) A space develops between the cells of the column. (c) The space becomes lined by cuticle at a molt and connects with the new cuticle of the original trachea.

cuticle. Similar ecdysial tubes are formed next to functional spiracles when the structure of the spiracles is so complex that it does not permit the old intima to be drawn through it. This occurs, for example, in Elateridae and Scarabaeidae amongst the beetles and in some Diptera (Hinton, 1947).

The cuticular lining of the tracheoles appears to be shed in some insects and some stages, but not in others. In *Sciara*, the intima is shed at larva–larva molts, but not at the larva–pupa or pupa–adult molts, whereas, in *Rhodnius*, the linings of the tracheoles are never shed (Wigglesworth, 1954). Instead, when a new tracheal lining is formed, it pinches the old lining at the origin of the tracheoles. At the point of constriction, the old cuticle breaks and the new tracheal intima becomes continuous with the original tracheole lining, although a marked discontinuity is apparent (Fig. 17.9)

17.1.4.1 *Pneumatization*

The tracheal system is liquid-filled in a newly hatched insect, and liquid fills the space between old and new

Fig. 17.8. Change in the position of a tracheole (after Wigglesworth, 1959). (a) Two epidermal cells send out processes in the direction of a tracheole. (b) Cytoplasmic filaments from the epidermal cells attach to a tracheole and draw it towards the cells.

Fig. 17.9. Molting in the tracheal system (after Wigglesworth, 1954). **(a)** New intima forms in the trachea. The old intima is still in place. **(b)** Detail of the junction of a tracheole whose intima remains unchanged from one stage to the next.

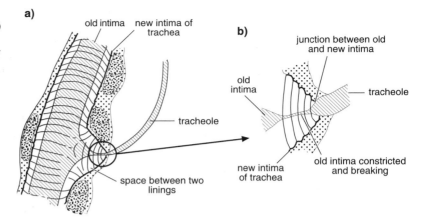

cuticles before each ecdysis. The liquid is replaced by gas in a process known as pneumatization. In terrestrial insects, gas may enter via the spiracles, but pneumatization occurs in aquatic insects, and in many terrestrial species without access to air. Gas usually appears first in a main tracheal trunk and then spreads rapidly through the system which becomes completely gas-filled in 10–30 minutes. The gas is probably forced out of solution in the liquid by physical forces resulting from the active resorption of liquid from the tracheae together with the change in the surface properties from hydrophile to hydrofuge which occurs when the cuticle is tanned. These forces lead to the rupture of the liquid column and the appearance of gas in its place.

In many insects, some liquid normally remains in the endings of the tracheoles. During periods of high energy consumption the liquid is withdrawn from the system and air is drawn further into the tracheoles. At other times, the liquid is secreted again and the air retreats. The level of liquid in the tracheoles represents the balance between the forces of capillarity and forces resulting mainly from imbibition due to colloidal substances in the cytoplasm of the tracheolar cell. These, in turn, are influenced by changes in the osmotic pressure of the tissues. When the metabolic rate is high, osmotic pressure rises and fluid is withdrawn from the tracheoles; when it is low, osmotic pressure falls and capillarity draws fluid into the tracheoles. The initial withdrawal of liquid from the tracheae after molting may similarly result from metabolic changes (Wigglesworth, 1983).

17.2 SPIRACLES

The spiracles are the external openings of the tracheal system. They are lateral in position, and in the Insecta,

there is never more than one pair of spiracles on a segment, usually on the pleuron. Often, each spiracle is contained in a small, distinct sclerite, the peritreme.

17.2.1 Number and distribution

The largest number of spiracles found in insects is ten pairs, two thoracic and eight abdominal, and the respiratory system can be classified on the basis of the number and distribution of the functional spiracles (Keilin, 1944). In numerous insects, the first spiracle is on the prothorax, but is mesothoracic in origin (Hinton, 1966). Spiracles are never present on the head.

Polypneustic – at least 8 functional spiracles on each side

Holopneustic – 10 spiracles: 1 mesothoracic, 1 metathoracic, 8 abdominal – as in dragonflies, grasshoppers, cockroaches, fleas and some hymenopteran and dipteran larvae such as the larvae of *Apis* and *Bibio*

Peripneustic – 9 spiracles: 1 mesothoracic, 8 abdominal – as in caterpillars and many other endopterygote larvae

Hemipneustic – 8 spiracles: 1 mesothoracic, 7 abdominal – as in mycetophilid larvae

Oligopneustic – 1 or 2 functional spiracles on each side

Amphipneustic – 2 spiracles: 1 mesothoracic, 1 posterior abdominal – as in many larval Diptera, especially amongst the Cyclorrhapha

Propneustic – 1 spiracle: 1 mesothoracic – as in mosquito pupae

Metapneustic – 1 spiracle: 1 posterior abdominal – as in mosquito and some aquatic beetle larvae

Apneustic – no functional spiracles – as in many aquatic larvae. Apneustic does not imply that the insect has no

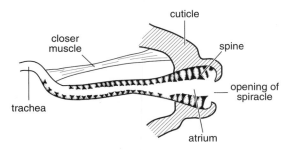

Fig. 17.10. Section through a spiracle of a louse, *Haematopinus*, showing the atrium and branched spines in which dust accumulates (after Webb, 1948).

tracheal system, but that the tracheae do not open to the outside.

Where less than ten functional spiracles are present, the tracheae may retain connections to the body wall at points where other spiracles normally occur. These 'non-functional' spiracles are open at the time of ecdysis and permit the cast intima to be shed (see above).

Amongst other hexapod groups, some Diplura, such as *Japyx*, have 11 pairs of spiracles, including four pairs on the thorax, while the sminthurids (Collembola) have only a single pair of spiracles, between the head and prothorax, from which tracheae extend, without anastomoses, to all parts of the body. Most Collembola have no tracheae at all.

17.2.2 Structure

In its simplest form, found in some Apterygota, the spiracle is a direct opening from the outside into a trachea, but generally the visible opening leads into a cavity, the atrium, from which the tracheae arise. In this case the opening and the atrium are known collectively as the spiracle. Often the walls of the atrium are lined with hairs which filter out dust (Fig. 17.10). In some Diptera, Coleoptera and Lepidoptera the spiracle is covered by a sieve plate containing large numbers of small pores. In the fifth stage larva of *Bombyx* (Lepidoptera) the pores are oval ($6 \times 3 \, \mu$m). The sieve plates perhaps serve to prevent the entry of dust or, especially in aquatic insects, water into the tracheal system.

The spiracles of most terrestrial insects have a closing mechanism which is important in the control of water loss. The closing mechanism may consist of one or two movable valves in the spiracular opening or it may be internal, closing off the atrium from the trachea by means of a constriction. In most insects, closure results from the

activity of one muscle; opening is produced by the elasticity of the cuticle associated with the spiracle, an elastic ligament, or a second muscle. Even where there is an opener muscle, its activity is often associated with an elastic component so that, in the absence of any muscular activity, the spiracle opens.

The metathoracic spiracle of grasshoppers is an example of a 'one muscle' spiracle (Fig. 17.11). It lies in the membrane between the meso- and metathorax and has two movable semi-circular valves which are unsclerotized except for the hinge and a basal pad into which a muscle is inserted. This muscle, by pulling down on the valves, causes them to rotate and so to close. The spiracle normally opens because of the elasticity of the surrounding cuticle, but in flight it opens wider as a result of the slight separation of the mesepimeron and metepisternum. These two sclerites which surround the spiracle are normally held together by an elastic bridge, but, when the basalar and subalar muscles contract, the sclerites are pulled apart. This movement is transmitted to the spiracle largely through a ligament connecting the metepisternum to the anterior valve, making the spiracle open wide (Miller, 1960b).

The mesothoracic spiracle of grasshoppers is a 'two muscle' spiracle (Fig. 17.12a). It is on the membrane between the pro- and mesothorax, and consists of a fixed anterior valve and a movable posterior valve. It is unusual in having two tracheae leading directly from the external opening. The openings of the two tracheae are called the dorsal and ventral orifices. A scelerotized rod runs along the free edge of the posterior valve, passing between the orifices and running round the ventral one. The closer muscle arises on a cuticular inflexion beneath the spiracle and is inserted into a process of the sclerotized rod, while the opener muscle, also from the cuticular inflexion, is inserted on to the posterior margin of the posterior valve. When the insect is at rest and the closer muscle relaxes, the spiracle opens to some 20–30% of its maximum as a result of the elasticity of the cuticle; the opener muscle plays no part. Contraction of the opener muscle occurs during slow, deep ventilatory movements and results in the spiracle opening fully (Miller, 1960b).

Closure of abdominal spiracles usually involves a constriction method (Figs. 17.12b, 17.13). Commonly, the atrium is pinched between two sclerotized rods, or in the bend of one rod, as the result of the contraction of a muscle. In other instances, the atrium or trachea is bent so that the lumen is occluded.

Review: Nikam & Khole, 1989

Fig. 17.11. A one muscle spiracle; second thoracic spiracle of a locust (*Schistocerca*) (after Miller, 1960b): **(a)** external view ; **(b)** internal view; **(c)** diagrammatic section through the spiracle showing how movement of the mesepimeron (arrow) causes the valves to open wide (dotted).

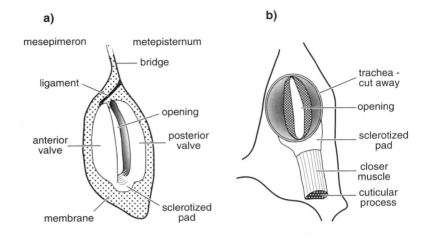

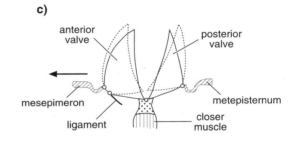

Fig. 17.12. Two–muscle spiracles of a grasshopper. **(a)** First thoracic spiracle seen from inside (based on Miller, 1960b; Snodgrass, 1935). **(b)** Abdominal spiracle (after Snodgrass, 1935).

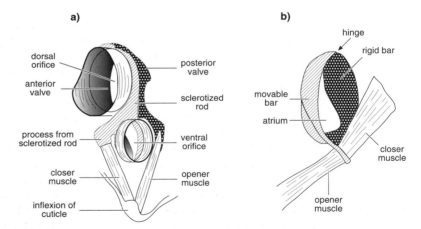

17.2.3 Control of spiracle opening

The spiracles are normally open for the shortest time necessary for efficient respiration presumably to keep water loss from the tracheal system to a minimum (but see below). Spiracle closure results from the sustained contraction of the closer muscle, while opening commonly results from the elasticity of the surrounding cuticle when the closer muscle is relaxed. The muscle is controlled by the central nervous system, but may also respond to local chemical stimuli.

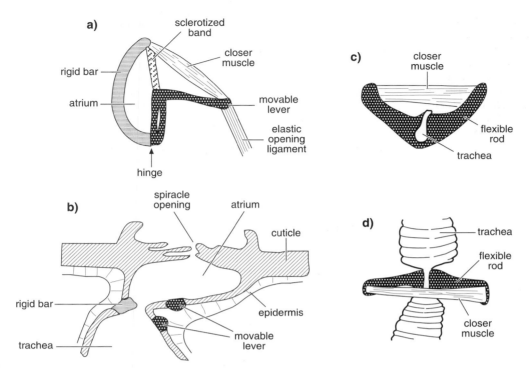

Fig. 17.13. Closing mechanisms internal to the spiracle. (**a**),(**b**) Abdominal spiracle of a caterpillar seen from inside (**a**) and in horizontal section (**b**) (from Imms, 1957). (**c**),(**d**) Constricting mechanism on a trachea of a flea seen in horizontal section (**c**) and from above (**d**) (after Wigglesworth, 1965).

The motor neurons to the spiracle muscles in each segment arise in the ganglion of the same segment or that immediately in front. The closer muscles in dragonflies, *Periplaneta* and *Schistocerca* are innervated by two motor neurons whose axons pass along the median nerve and then bifurcate, sending a branch to either side, so that the two spiracles receive the same pattern of motor impulses (see Fig. 17.21). Closer muscles may also be innervated by an inhibitory axon. The opener muscle, when it is present, is innervated from one or two neurons whose cell bodies are usually lateral in the ganglion and the spiracles of the two sides are independently innervated. In *Blaberus* and related cockroaches, a common inhibitory neuron innervates a number of abdominal spiracle opener muscles.

The closer muscle is caused to contract by the activity of its motor neurons, but the frequency of action potentials, determining the degree of contraction, may be altered by various factors acting on the central nervous system. Of particular importance are a high level of carbon dioxide and a low level of oxygen (hypoxia) in the tissues.

Both these conditions tend to arise while the spiracles are closed as carbon dioxide is produced and oxygen is utilized in respiration. Both conditions lead to a reduction in action potential frequency and so to spiracle opening. In 'two muscle' spiracles the action potential frequency to the opener muscle is increased by high carbon dioxide levels and hypoxia.

The frequency of motor impulses to the closer muscle is also affected by the water balance of the insect, possibly acting through the concentration of a particular ion. If the insect is desiccated, action potential frequency rises and the spiracles remain closed longer; with excess hydration, the converse is true and so the rate of water loss is increased.

Carbon dioxide also acts directly on the closer muscle of 'one muscle' spiracles, interfering with neuromuscular transmission so that the junction potential falls, muscle tension is reduced and the spiracle opens (Hoyle, 1960). The threshold of the peripheral response to carbon dioxide is set by the frequency of motor impulses from the central nervous system. The lower the impulse frequency, the lower the threshold to carbon dioxide. Carbon dioxide has some other, internal, effect on the muscle, not connected with neuromuscular transmission, and this also results in a reduction in tension. 'Two muscle' spiracles do not respond to peripheral stimulation by carbon dioxide and are controlled entirely by the output from the central nervous system (Miller, 1960b).

Hemolymph potassium has a direct effect on some spiracle muscles. In the case of spiracle 2 of *Schistocerca*, a 'one muscle' spiracle, concentrations of potassium above 30 mM cause the closer muscle to contract even when it is isolated from the nervous system (Hoyle, 1961). Normally the concentration in the blood is not as high as this, but it may be exceeded as a result of desiccation and also at the molt. Consequently, under these circumstances, the spiracles remain closed for most of the time, opening only when the concentration of carbon dioxide in the tissues is high. This is presumed to restrict water loss. Such sustained muscular contraction due to high concentrations of potassium does not occur in other muscles.

Review: Kaars, 1981

17.3 CUTANEOUS RESPIRATION

Some gaseous exchange takes place through the cuticle of most insects, but does not usually amount to more than a small percentage of the total gaseous movement. On the other hand, Protura and most Collembola have no tracheal system and must depend on cutaneous respiration together with transport by the hemolymph from the body surface to the tissues. Cutaneous respiration is also important in many endoparasitic and aquatic insects where it is coupled with an apneustic tracheal system. Cutaneous respiration without an associated apneustic tracheal system can only suffice for very small insects with a high surface/volume ratio.

The impermeability of most insect cuticles to oxygen arises from the epicuticle, but not from the wax layer which renders the cuticle impermeable to water (Buck, 1962). The permeability to carbon dioxide may be rather greater and the loss of this gas through the intersegmental membranes may be appreciable.

17.4 GASEOUS EXCHANGE IN TERRESTRIAL INSECTS

Oxygen passes through the tracheal system from the spiracles to the tissues and ultimately must reach the mitochondria in order to play a part in oxidative processes. There are thus two distinct phases in the transport of gases, one through the tracheal system, known as air-tube diffusion, and one through the tissues in solution in the cytoplasm, known as tissue diffusion (Weis-Fogh, 1964b).

17.4.1 Diffusion

The rate of diffusion of a gas depends on a number of factors. It is inversely proportional to the square root of the molecular weight of the gas, so that in air, oxygen, with a molecular weight of 32, diffuses 1.2 times faster than carbon dioxide, molecular weight 44. Diffusion also depends on the differences in concentration (the concentration gradient) of the gas at the two ends of the system and, in the absence of a difference in concentration, there is no net movement of gas. The change in concentration, or partial pressure (p), with distance (x) is expressed as $\delta p / \delta x$. Finally, the permeability of the substrate, in this case air or the tissues, through which the gas is diffusing affects the rate of diffusion. This factor is expressed in terms of the permeability constant, P, which is the flow of a substance through unit area per unit time when the concentration gradient is unity. Hence, the volume $(\mathcal{J})$ of a given gas transported by diffusion at normal temperature and pressure (NTP) can be represented by the equation

$$\mathcal{J} = -P \delta p / \delta x \text{ (Weis-Fogh, 1964b).}$$

The permeability constant for different substrates varies widely. That for oxygen in air at 20 °C is approximately $0.11 \text{ ml min}^{-1} \text{cm}^{-2} \text{kPa}^{-1} \text{cm}^{-1}$ whereas in water $P = 0.34 \times 10^{-6}$ and in frog muscle $0.14 \times 10^{-6} \text{ ml min}^{-1} \text{cm}^{-2} \text{kPa}^{-1} \text{cm}^{-1}$. Hence oxygen in air diffuses more than 100 000 times faster than in water or the tissues, so that although, in the insect, the path of oxygen through the tracheal system is very much longer than its path through the tissue, it will probably take longer for the gas to diffuse from the tracheole endings to the mitochondria, than from the spiracles to the tracheole endings.

Thus, the length of the tissue diffusion path is likely to be a factor limiting the size of tissues and, in particular, of flight muscles with a high requirement for oxygen. With a partial pressure difference between the tracheole and the mitochondrion of 5 kPa, a tracheole 1 μm in diameter would serve a muscle 7–15 μm in diameter if the oxygen uptake of the muscle was 1.5–3.0 ml g^{-1} min^{-1}, a level of uptake achieved in flight muscle. Hence a muscle fiber in which the tracheoles are restricted to the outside of the fibers cannot exceed about 20 μm in diameter. The fibers of the flight muscles of dragonflies are of this type and are approaching their theoretical maximum size. On the other hand, if the tracheoles indent the muscle so as to become functionally internal, as they do in flight muscles of most insects, the muscle fibers can become much larger. In *Musca* (Diptera), for instance, the indenting tracheoles are only separated by distances of 3–5 μm and lie very close to the mitochondria so that, although individual fibers may be over one millimeter in diameter, they are well within the size limits imposed by tissue diffusion (Weis-Fogh, 1964b).

Due to its greater solubility, the permeability constant of carbon dioxide in the tissues is 36 times greater than that for oxygen, so that despite its higher molecular weight, carbon dioxide travels more quickly than oxygen through the tissues for the same difference in partial pressure. Hence a system capable of bringing an adequate supply of oxygen to the tissues will also suffice to take the carbon dioxide away.

As carbon dioxide is more soluble, it is present in higher concentrations in the tissues than oxygen. Thus some carbon dioxide, instead of passing directly into the tracheal system, may be temporarily retained in the hemolymph before diffusing into the tracheae near the spiracles or passing out directly through the integument.

The exchange of gases between the tracheal system and the tissues is partly limited by the walls of the tracheae and tracheoles. The whole system may be permeable, with no marked difference between tracheoles and tracheae, but it is generally believed that the lining of the tracheoles is more permeable, especially in flight muscles (Wigglesworth, 1983, but see Wigglesworth, 1990a). In any case, as the tracheoles are more closely associated with the tissues they will, in general, be more important than the tracheae in the transfer of oxygen to the tissues.

The rate of gaseous exchange also varies with the surface area through which the gases are diffusing. In some insects, the summed cross-sectional area of the tracheal system at different distances from the spiracles is believed to remain constant. Distally this area is made up of many small tubes (tracheoles), proximally (near the spiracle) of a few large ones (tracheae), so that the summed circumference or wall area/unit length is much greater distally than proximally. Although, in *Bombyx* larva, the summed cross-sectional area of the tracheoles is less than that of the tracheae, the wall area/unit length of the tracheoles is nevertheless greater (Buck, 1962). As this is so, it again

follows that the tracheoles are generally a more important site of gaseous exchange with the tissues than are the more proximal tracheae (assuming that the permeability of their cuticular linings is similar).

It has generally been supposed that gaseous exchange in small or quiescent insects depended entirely on diffusion. However, more critical theoretical analysis of the system has led this belief to be questioned (Kestler, 1985). At the same time, experimental methods have demonstrated that in most resting adult insects, including some small ants and even phorid flies weighing less than 1 mg, gaseous exchange depends on convective movements of gas into and from the tracheal system in addition to diffusion (Miller, 1981b). The extent to which diffusion alone is important is thus an open question, although some worker ants appear not to engage in active ventilation. Presumably in aquatic insects with a closed respiratory system, diffusion normally serves to transport oxygen from the gills to the tissues.

Review: Kestler, 1985 – theoretical considerations

17.4.2 Ventilation

Ventilation of the tracheal system involves the convective movements of gases. These movements are produced primarily by changes in the volume of the tracheal system although reduced pressure within the tracheae due to the utilization of oxygen may also be important in some cases. This is known as passive suction ventilation.

Most tracheae are circular in cross-section and resist any change in form because of their taenidia, but some, such as the longitudinal trunks of larval *Dytiscus* (Coleoptera) are oval in cross-section and are subject to compression. The compression of a trachea forces air out of the tracheal system, while its subsequent expansion sucks air in again. But changes in shape of the trachea, if they occur at all, only produce small volume changes. Much larger changes, and hence better gaseous exchange, are produced by the alternating collapse and expansion of airsacs. Compression of the system, causing expiration, results indirectly as a result of hemolymph pressure. This is achieved in two ways: by reduction in the body volume, usually of the abdomen, and by displacement of hemolymph from different parts of the body while the total body volume remains constant. Expansion of the airsacs and inspiration result from the reduction of pressure due to the muscular or elastic expansion of the abdomen or the movement of hemolymph to another part of the body.

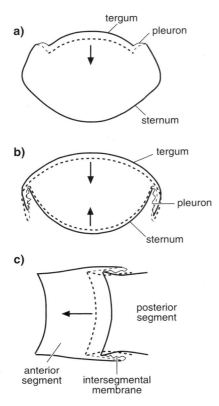

Fig. 17.14. Ventilatory movements of the abdomen. Dashed lines represent the contracted position, compressing the abdominal airsacs and causing expiration. Arrows indicate directions of movement in the compression phase (from Snodgrass, 1935). **(a)** Dorso-ventral compression involving tergal depression, transverse section. **(b)** Dorso-ventral compression involving both tergum and sternum, transverse section. **(c)** Longitudinal compression by telescoping one segment within another anterior to it, horizontal section.

Changes in abdominal volume may be produced in various ways. In Heteroptera and Coleoptera, the tergum moves up and down (Fig. 17.14a); in Odonata, Orthoptera, Hymenoptera and Diptera both tergum and sternum move (Fig. 17.14b) and this movement may be associated with telescoping movements of the abdominal segments (Fig. 17.14c); and, in adult Lepidoptera, the movement is complex and involves movements of the pleural regions as well as terga and sterna.

In adult insects in which hemolymph circulation primarily involves shunting hemolymph back and forth between thorax and abdomen (Fig. 17.15) without any marked change in body volume, these movements also produce changes in tracheal volume. For example, in the fly *Calliphora*, airsacs at the front of the abdomen occlude the hemocoel. When the heart pumps forwards, it draws hemolymph from the abdomen and transfers it to the head and thorax. The effect of this is to cause expansion of airsacs in the abdomen and to compress those in the anterior compartment. When the heart reverses, it pumps blood into the abdomen and the changes in the tracheal system are reversed. It is believed that, during forward beating of the heart, air is sucked in through the abdominal spiracles and forced out through thoracic spiracles; when the heart reverses, the converse is true. In a resting female fly, heart reversal occurs about once each minute.

Air may flow in and out of each of the spiracles. This type of air movement, which occurs in *Periplaneta*, is called a tidal flow. In many insects, however, including other species of cockroach, opening and closing of certain spiracles is synchronized with the ventilatory pumping movements of the abdomen, so that air is sucked in through some spiracles and pumped out through others and a directed flow of air through the tracheal system is produced. This is a more efficient form of ventilation than tidal flow as much of the 'dead' air, trapped in the inner parts of the system by tidal movements, is removed. In most insects the flow of air is from front to back and, in *Schistocerca*, spiracle movements are always coupled with ventilatory movements. In many insects, however, the two activities may become uncoupled so that they are not always synchronized (Miller, 1971) and the coupling may even be modified so as to produce a reversal of the airstream (Miller, 1973). In *Schistocerca*, spiracles 1, 2 and 4 are open during inspiration, but close during expiration when spiracle 10 opens (Fig. 17.16a). When the insect is more active, expiration takes place through spiracles 5 to 10. The spiracles for inspiration open immediately after the expiratory spiracles have closed and remain open for about 20% of the cycle while air is drawn in. Then, for a short time, all the spiracles are closed. The abdomen starts to contract while the spiracles are still closed, so that the air in the tracheae is under pressure; this is known as the compression phase. Then the expiratory spiracles open and air is forced out. The expiratory spiracles are only open for some 5–10% of the cycle. During activity the frequency of ventilation and spiracular movements is increased. The periods for which the spiracles are open remain the same, but the intervals between openings are

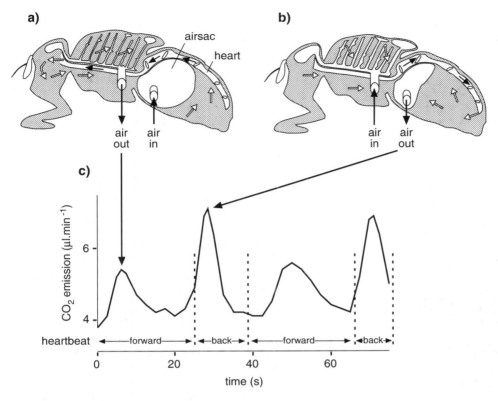

Fig. 17.15. Discontinuous gas exchange associated with reversals of the heartbeat in a fly (*Calliphora*). The tracheal system is represented diagrammatically (after Wasserthal, 1996). (**a**) Forward beating of the heart draws hemolymph from the abdomen and pushes it into the thorax and head. This expands abdominal airsacs and compresses those in the head and thorax so that air is drawn in through abdominal spiracles and forced out through thoracic spiracles. (**b**) Backwards beating of the heart draws hemolymph from the head and thorax and pushes it into the abdomen. This compresses abdominal airsacs and expands those in the head and thorax so that air is drawn in through thoracic spiracles and forced out through abdominal spiracles. (**c**) Carbon dioxide is emitted when the heart is beating in either direction, but more is forced out from the abdomen than the thorax due to the greater volume of airsacs in the former.

reduced and the compression phase eliminated (Fig. 17.16b). The net time for which the spiracles are open increases as a result (Miller, 1960b).

After periods of activity, abdominal ventilation is supplemented by other types of ventilation and in *Schistocerca* these involve protraction and retraction of the head on the prothorax (neck ventilation), and movement of the prothorax on the mesothorax (prothoracic ventilation) (Miller, 1960a). These movements primarily ventilate the head. The normal level of abdominal ventilation in *Schistocerca* pumps about $40\,\mathrm{ml\,air\,g^{-1}\,h^{-1}}$ through the body, about 5% of the air being exchanged at each movement, but this can be raised to $150\,\mathrm{ml}$, with an exchange of

about 20% of the volume at each stroke. Neck and prothoracic ventilation provide a further $50\,\mathrm{ml\,air\,g^{-1}\,h^{-1}}$. In *Dytiscus*, 60% of the air is renewed with each compression and expansion.

Reviews: Miller, 1981b; Wasserthal, 1996

17.4.3 Discontinuous gas exchange

Although ventilation is often continuous, there may be extended periods during which all the spiracles are closed. The movement of oxygen into the tracheae and carbon dioxide emission then occur in discrete bursts when the spiracles open; relatively little gas exchange occurs while they are closed. This phenomenon is known as discontinuous gas

Fig. 17.16. Spiracle activity during abdominal ventilation of a locust (after Miller, 1960b): **(a)** normal ventilation; **(b)** hyperventilation.

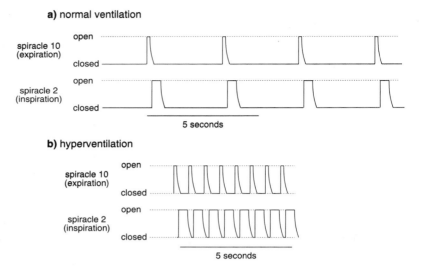

a) normal ventilation

spiracle 10 (expiration)

spiracle 2 (inspiration)

5 seconds

b) hyperventilation

spiracle 10 (expiration)

spiracle 2 (inspiration)

5 seconds

exchange or discontinuous ventilation. It is common in adult insects when they are inactive at moderate temperatures, that is, when their metabolic rates are low. The pupae of many insects also exhibit discontinuous gas exchange.

The periodicity of ventilatory bouts varies from insect to insect and with environmental conditions. In diapausing pupae of the moth, *Hyalophora*, the interval between spiracle openings is as long as eight hours. In some resting grasshoppers at 15 °C, carbon dioxide is released about once per hour, but at 30 °C the period of spiracle closure is reduced to about five minutes (Quinlan & Hadley, 1993). In the beetle, *Psammodes*, each cycle lasts about 15 minutes (Fig. 17.17), while in some ants and the honeybee it is less than five minutes (Fig. 17.18) (Lighton, 1988; Lighton & Berrigan, 1995; Wasserthal, 1996).

During the closed phase, relatively little gas exchange takes place (Fig. 17.17). As carbon dioxide is produced it is presumed to be dissolved in the hemolymph. When the spiracles open, oxygen rushes into the system and carbon dioxide and water vapor diffuse out. In some insects, such as *Psammodes*, ventilatory movements of the abdomen are made during this period so that gases are actively pumped in and out of the system until the spiracles close again (Fig. 17.17b).

Between the periods of complete closure and complete opening, some insects exhibit a period in which the spiracles repeatedly open slightly and close again, a movement called fluttering. During this period, oxygen enters the system, but relatively little carbon dioxide escapes from it (Fig. 17.17c,d). The mechanisms control-

ling the activities of the spiracle and movements of gases are not necessarily the same in all cases. In many instances, carbon dioxide release is associated with reversal of the heartbeat. This occurs even in pupal Lepidoptera where the interval between bursts is several hours, but in this case carbon dioxide release does not occur with every reversal of the heartbeat (Wasserthal, 1996).

The process is most fully understood in pupal Lepidoptera. At the end of a burst (spiracles open) in *Hyalophora* pupa, oxygen comprises about 18% of the gas in the tracheae. The spiracles close tightly and, as the oxygen is used up, the pressure in the tracheae falls. At the same time there is only a small increase in the proportion of carbon dioxide present, most of it being dissolved in the hemolymph. The low oxygen content of the tracheal air and tissues is monitored by receptors associated with the central nervous system and hypoxia reduces activity in the motor neurons to the closer muscles so that the spiracles open slightly. Because of the low pressure in the tracheal system, air rushes in. The bulk influx of air limits the outward diffusion of water vapor and of carbon dioxide, which therefore continues to accumulate in the tracheae and the tissues. Diffusion probably also contributes to the inward movement of oxygen because of the low partial pressure of oxygen in the tracheae. The influx of air temporarily increases the amount of oxygen present, so that the spiracles close again. The repeated lowering and raising of the oxygen level leads to the fluttering movement of the spiracle valves. Slowly, however, carbon dioxide

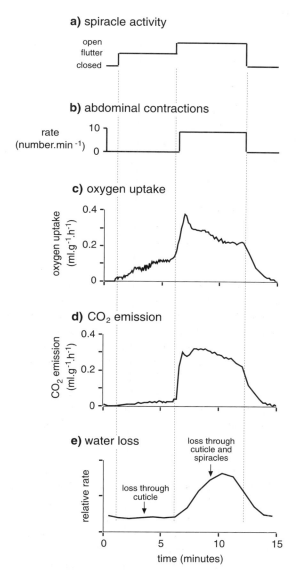

Fig. 17.17. Discontinuous gas exchange in a beetle (*Psammodes*). Changes occurring in a single bout of ventilation (after Lighton, 1988). (**a**) Activity of the spiracles. During the flutter phase, they open slightly for brief periods and then close again. (**b**) Abdominal ventilation coincides with the open phase of the spiracles. (**c**) Oxygen uptake. Some occurs during flutter, but the rate is greatly increased when the spiracles open fully. (**d**) Carbon dioxide emission is largely confined to the open phase of the spiracles. Very little occurs during spiracular flutter. (**e**) Water loss increases when the spiracles open.

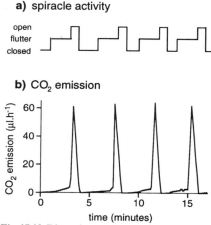

Fig. 17.18. Discontinuous gas exchange in an ant (*Cataglyphis*) (after Lighton, Fukushi & Wehner, 1993a). (**a**) Activity of the mesothoracic spiracle. During the flutter phase, it opens slightly for brief periods and then closes again. (**b**) Carbon dioxide emission occurs largely during the fully open phase of the spiracles.

accumulates in the hemolymph. Ultimately, it reaches a concentration which directly affects the activity of the closer muscle so that the spiracle opens, carbon dioxide is released from the hemolymph and leaves the system and the oxygen concentration in the tracheae is restored to normal atmospheric levels.

Spiracle fluttering in the ant, *Cataglyphis*, is also associated with reduced pressure in the tracheal system and the inward movement of air is at least partly convective. In other cases, however, it is likely that diffusion alone accounts for the inward movement of oxygen as the partial pressure in the system is continuously falling as it is used in oxidative processes (Lighton & Garrigan, 1995).

As a consequence of discontinuous gas exchange, the spiracles are closed for most of the time. As water loss through the spiracles is greatly enhanced during periods of spiracle opening, it has been assumed that the limitation of water loss is the evolutionary raison d'être for discontinuous gas exchange. However, as the amount of water lost through the spiracles is usually very small relative to that lost via the cuticle, the general application of this argument has been questioned (Hadley, 1994; Lighton, 1994; Lighton & Berrigan, 1995) although it may be valid for pupal stages and in insects with relatively high rates of spiracular water loss (Lighton *et al.*, 1993b; Williams, Rose & Bradley, 1997). **Reviews:** Hadley, 1994; Lighton, 1994, 1996; Sláma, 1988

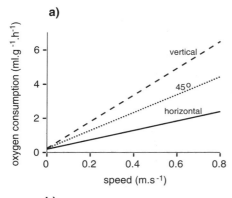

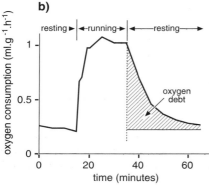

Fig. 17.19. Oxygen consumption in relation to activity. (a) Running at different speeds at different angles to the horizontal (cockroach, *Periplaneta*) (after Full & Tullis, 1990). (b) Elevated consumption continuing after a period of activity (cockroach, *Gromphadorhina* running at $33\,\mathrm{cm\,s^{-1}}$) (after Herreid, Full & Prawel, 1981).

17.4.4 Variation in gas exchange

Higher metabolic rates demand higher levels of oxygen intake. This is most obvious in flight (see Fig. 9.37). Stationary insects commonly use less than $1.0\,\mathrm{ml\,O_2\,g^{-1}\,h^{-1}}$, whereas in flight the figure rises to 15–$100\,\mathrm{ml\,O_2\,g^{-1}\,h^{-1}}$, or even more, a 30- to 100-fold increase. The demands of the flight muscles themselves may increase up to 400-fold. Walking and running also require more oxygen than just standing still. The faster an insect runs, the higher its rate of oxygen consumption, and running vertically upwards requires more energy, and so more oxygen, than running on a horizontal surface (Fig. 17.19a) (Full & Tullis, 1990). After a period of activity, oxygen consumption takes some time to return to normal, indicating that the insect has incurred an oxygen debt due

to some degree of anaerobic respiration (Fig. 17.19b) (Herreid, Full & Prawel, 1981). It is generally believed that very little anaerobic respiration occurs during flight although some data do indicate an elevated period of oxygen consumption following the cessation of flight (Krogh & Weis-Fogh, 1951).

Even in a stationary insect, and at a constant temperature, metabolic rate is not constant. After feeding, the metabolic rate of a caterpillar more than doubles. This appears to be due to increased metabolism as the food is digested and absorbed (Aidley, 1976). Oxygen consumption also varies over the course of a single larval stage (van Loon, 1988), although it is not clear how much these changes are a reflection of differences in locomotor and feeding activity. During the pupal period of holometabolous insects, oxygen uptake is highest at the beginning and end of the stage (see Fig. 15.26) and a comparable change occurs during diapause (see Fig. 15.40).

Superimposed on these general patterns, other variation may also occur. In diapausing pupae of the flesh fly, *Sarcophaga*, peaks of oxygen consumption occur at about 4-day intervals when the temperature is constant at $25\,°C$. Over a period of about 36 hours, oxygen consumption rises from a resting level of about $1.5\,\mu\mathrm{l\,h^{-1}}$ to 9 or $10\,\mu\mathrm{l\,h^{-1}}$ and then falls to the original level. The peaks in oxygen consumption coincide with peaks of protein synthesis and release. There are no parallel changes in carbon dioxide emission (Sláma & Denlinger, 1992). Similar variation in oxygen consumption occurs in the diapausing pupa of the cabbage butterfly, *Pieris*, and it is probably a common phenomenon.

The extent to which these changes in oxygen consumption are produced by variation in ventilatory activities is not clear.

17.4.4.1 *Gas exchange in flight*
The massive increase in oxygen consumption that occurs when an insect flies requires a greatly increased airflow through the tracheae to the flight muscles. During the flight of *Schistocerca*, abdominal ventilation increases in frequency and amplitude, but still only supplies about $50\,\mathrm{ml\,air\,g^{-1}\,h^{-1}}$, not sufficient to supply the needs of the flight muscles. However, the distortion of the thorax, and in particular the raising and lowering of the notal sclerites, produces large volume changes in the extra-muscular airsacs of the pterothoracic tracheal system (Figs. 17.4b, 17.6), while changes in the volumes of the muscles themselves compress the intramuscular airsacs. This

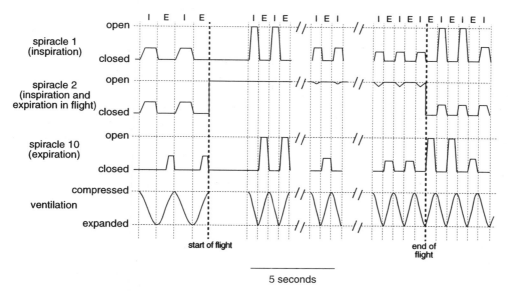

Fig. 17.20. Spiracle activity and abdominal ventilation during flight of *Schistocerca*. I, inspiration; E, expiration (after Miller, 1960c).

pterothoracic ventilation produces an airflow of about $350\,ml\,air\,g^{-1}\,h^{-1}$, which is adequate for the needs of the flight muscles.

Thoracic pumping is also important in flight in Odonata and probably in Lepidoptera and Coleoptera. In Hymenoptera and Diptera, changes in the thoracic volume during flight are not very large and abdominal pumping is of greater importance in maintaining the air supply to the flight muscles (Weis-Fogh, 1964a). In a large cerambycid beetle, *Petrognatha*, there is evidence that a stream of air, resulting from the forward movement of the insect in flight, flows in through the forwardly directed second spiracles into large tracheae running directly to the third spiracles. This stream of air ventilates the primary supply to the flight muscles while the secondary supply is probably ventilated by muscular compression (Miller, 1966a).

When *Schistocerca* starts to fly, the pattern of spiracle opening also alters (Fig. 17.20). At first, spiracles 1 and 4 to 10 close, but then they open and close rhythmically in synchrony with abdominal ventilation so that there is a flow of air via the head to the rest of the central nervous system. Increased abdominal ventilation may also improve the blood circulation and hence the fuel supply to the flight muscles. Spiracles 2 and 3 remain wide open throughout flight and although they show some incipient closures after a time these do not affect the airflow through the spiracles.

These spiracles supply the flight muscles, and, since the tracheal system of these muscles is largely isolated and the spiracles remain open all the time, there is a tidal flow of air in and out of them (Miller, 1960c).

17.4.5 Control of ventilation

As in the control of spiracles, ventilatory movements are initiated by the accumulation of carbon dioxide and, to a lesser extent, the lack of oxygen in the tracheal system (Gulinson & Harrison, 1996), acting directly on centers in the ganglia of the central nervous system. Each abdominal ganglion produces rhythmical sequences of activity controlling the movements. In *Schistocerca* and *Periplaneta*, the metathoracic ganglion acts as a pacemaker and overrides the rhythms of the other ganglia (Lewis, Miller & Mills, 1973).

In normal ventilation by *Schistocerca*, involving only dorso-ventral movements of abdominal sclerites, expiration is produced by tergosternal muscles innervated by motor neurons from the ganglion of the corresponding segment. Inspiration is produced by muscles inserted low on the tergum and innervated by axons in branches from the median ventral nerve. The perikarya of these motor neurons are in the ganglion of the preceding segment, so that the alternation of inspiratory and expiratory movements is controlled from different ganglia (Fig. 17.21a).

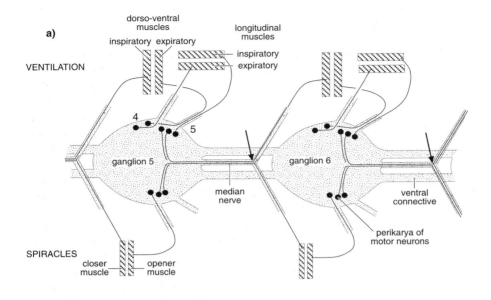

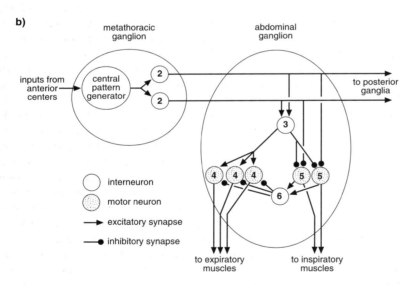

Fig. **17.21.** Control of spiracles and ventilatory movements in a locust (*Schistocerca*). (**a**) Innervation of ventilatory muscles (top) and spiracular muscles (bottom). The cell bodies of motor neurons controlling the muscles of one segment are not confined to the ganglion of that segment; some are in the ganglion of the next anterior segment. The axons in the median nerve divide (arrow) so that homologous muscles on the two sides are innervated by the same neurons. Notice that the ventilatory muscles and spiracular muscles are not controlled by the same neurons. Numbers adjacent to cell bodies in ganglion 5 relate to the cells shown in (**b**) (based on Lewis, Miller & Mills, 1973). (**b**) Diagram showing the control of ventilatory muscles. The interneurons (2) are driven by a central pattern generator and extend to all the abdominal ganglia. Motor neurons are numbered to correspond with the same numbers in (a). See text for further explanation (based on Lewis, Miller & Mills, 1973; Pearson, 1980).

Coordination is achieved by a coordinating interneuron, one in each ventral connective (cell 2 in Fig. 17.21b), which originates in the metathoracic ganglion and extends to the last abdominal ganglion. These interneurons are driven by a central pattern generator, probably in the metathoracic ganglion, whose activity is regulated by inputs from the head and other parts of the body. Action potentials in the interneuron have an excitatory effect on a local interneuron (cell 3) and probably have a weak direct inhibitory effect on the inspiratory motor neurons (cell 5 in Fig. 17.21b). The local interneuron excites the expiratory motor neurons (cell 4) and strongly inhibits the inspiratory motor neurons (cell 5). Inspiration occurs when the intersegmental interneurons are silent and the inspiratory motor neurons become spontaneously active, at the same time inhibiting the expiratory motor neurons via another local interneuron (cell 6) (Lewis *et al.*, 1973; Pearson, 1980).

The rate of ventilation is altered by reducing the interval between inspiratory bursts and is affected by sensory input from various sources. Centers sensitive to carbon dioxide are present in the head and thorax of *Schistocerca* and these modify the activity of the pacemaker, while the output is also modified by high temperature and nervous excitation generally. Proprioceptors may play some part in the maintenance of the frequency of ventilation (Miller, 1960a; Farley and Case, 1968). The coordination of the spiracles with the ventilatory movements is brought about by motor patterns derived from the ventilatory centers.

Review: Miller, 1966a

17.5 GASEOUS EXCHANGE IN AQUATIC INSECTS

Aquatic insects obtain oxygen directly from the air or from that dissolved in the water. The former necessitates some semi-permanent connection with the surface or frequent visits to the surface. Insects that obtain oxygen from water nearly always retain a tracheal system so that the oxygen must come out of solution into the gaseous phase. This is important because the rate of diffusion in the gas phase is very much greater than in solution in the hemolymph. Gaseous exchange with water takes place through thin-walled gills well supplied with tracheae, but in other cases a thin, permanent film of air is present on the outside of the body. The spiracles open into this film so

that oxygen can readily pass from the water into the tracheae.

17.5.1 **Aquatic insects obtaining oxygen from the air**
Most aquatic forms obtaining air from above the water surface must make periodic visits to the surface, but a few have semipermanent connections with the air that enable them to remain submerged indefinitely. The larva of the hover fly, *Eristalis* (Diptera), has a telescopic terminal siphon which can extend to a length of six centimeters or more in a larva only one centimeter long. By means of the siphon the larva can reach the water surface with its posterior spiracles, while the body remains on the bottom mud (Fig. 17.22).

Some other species obtain oxygen by thrusting their spiracles into the aerenchyma of aquatic plants. This habit occurs in larval *Donacia* (Coleoptera) and *Chrysogaster* (Diptera), and larvae and pupae of *Notiphila* (Diptera) and the mosquito *Mansonia*. With the exception of *Mansonia*, all of these species live in mud containing very little free oxygen (Varley, 1937). The functional spiracles are at the tip of a sharp-pointed post-abdominal siphon in larval forms (Fig. 17.23) and on the anterior thoracic horns of the pupae.

For most aquatic insects obtaining their oxygen from the air, however, periodic visits to the surface of the water are necessary to renew the gases in the tracheal system. Problems facing all insects which come to the surface are those of breaking the surface film and of preventing the entry of water into the spiracles when they submerge. The ease with which this is accomplished depends on the surface properties of the cuticle and, in particular, on its resistance to wetting. When a liquid rests on a solid or a solid dips into a liquid, the liquid/air interface meets the solid/air interface at a definite angle that is constant for the substances concerned. This angle, measured in the liquid, is known as the contact angle (Fig. 17.24). A high contact angle indicates that the surface of the solid is only wetted with difficulty and such surfaces are said to be hydrofuge. Under these conditions, the cohesion of the liquid is greater than its adhesion to the solid, and so, when an insect whose surface properties are such that the contact angle is high comes to the air/water interface, the water falls away leaving the body dry (Holdgate, 1955). The whole surface of the cuticle may possess hydrofuge properties, so that it is not readily wetted at all, or these properties may be restricted to the region around the spiracles. In dipterous larvae, for example, perispiracular glands produce an oily secretion in the

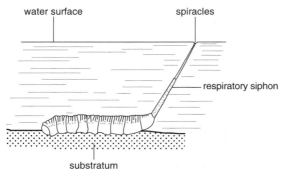

Fig. 17.22. Semi-permanent connection with the water surface. The larva (rat-tailed maggot) of the hover fly, *Eristalis*, with its respiratory siphon partly extended (after Imms, 1947).

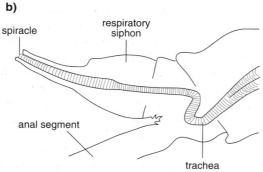

Fig. 17.23. Semi-permanent connection with the aerenchyma of a plant. The respiratory siphon of a mosquito larva (*Mansonia*) which connects with the aerenchyma of aquatic plants (after Keilin, 1944). (a) Lateral view showing saw which cuts into the plant tissue and recurved teeth which hold the siphon in place. (b) Longitudinal section showing a terminal spiracle and a trachea.

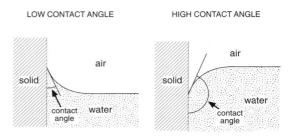

Fig. 17.24. Diagrams showing low and high contact angles.

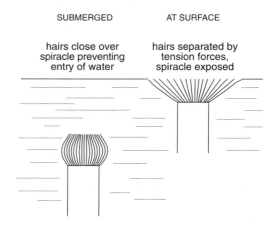

Fig. 17.25. Diagrams showing the movements of hydrofuge hairs surrounding a spiracle when an insect is submerged and at the surface. The movement of the hairs is entirely passive, depending on physical forces acting between the hairs and the water (modified after Wigglesworth, 1965).

immediate neighborhood of the spiracle. Often hydrofuge properties round the spiracle are associated with hairs, as in *Notonecta* (Heteroptera), or valves, as in mosquito larvae, which close when the insect dives, but open at the surface (Fig. 17.25), being spread out by surface tension.

In many of these insects, only the posterior spiracles are functional and they are often carried on the end of a siphon extending from the posterior body, as in larval Ephydridae and Culicidae. In these insects, only the tip of the siphon breaks the surface film. In water scorpions (Hemiptera, Nepidae), the spiracles are at the base of an air-filled tube.

An increase in the number of functional spiracles often occurs in the last stage larva since this stage is commonly less strictly aquatic than earlier stages, leaving the water in order to pupate or, in hemimetabolous insects, to facilitate

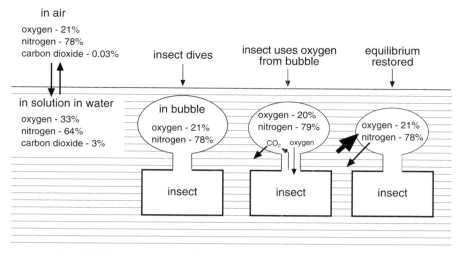

Fig. 17.26. Diagram of an air bubble acting as a physical gill. As the insect dives, the gases in the bubble are in equilibrium with those dissolved in the water. As the insect uses oxygen, the equilibrium is perturbed. It is restored by the inward movement of oxygen and the outward movement of nitrogen. This is a continuous process; it is shown as two separate steps for clarity. Note that carbon dioxide produced by the insect is immediately dissolved in the water. Oxygen comes out of solution faster than nitrogen goes in. The bubble shrinks continuously as nitrogen goes into solution.

adult emergence. The number of functional spiracles never decreases from one instar to the next (Hinton, 1947).

17.5.1.1 *Gas exchange via air bubbles*

Some insects, such as mosquito larvae, can remain submerged only as long as the supply of oxygen in the tracheae lasts, but others have an extra-tracheal air store, carrying a bubble of air down into the water when they dive. The spiracles open into this bubble, so that it provides a store of air additional to that contained in the tracheal system, enabling the insects to remain submerged for longer periods than would be possible without it.

The position of the store is characteristic for each species. In *Dytiscus*, it is beneath the elytra, and experimental removal of the hind wings increases the space beneath the elytra and enables the insect to remain submerged for longer periods. In *Notonecta*, air is held by long hydrofuge hairs on the ventral surface as well as in a store under the wings and in a thin film held by small bristles over the dorsal surface of the fore wing. The related *Anisops* (Heteroptera) has ventral and subelytral stores supplemented by oxygen loosely associated with hemoglobin in large hemoglobin cells just inside the abdominal spiracles (section 17.9).

An air store also gives the insect buoyancy, so that as

soon as the insect stops swimming or releases its hold on the vegetation, it floats to the surface. The position of the store is such that the insect breaks the surface suitably oriented to renew the air. *Dytiscus*, for instance, comes to the surface tail first and renews the subelytral air from the posterior end of the elytra.

Physical gills An air bubble provides an insect with more oxygen than that available at the time of the dive because it also acts as a temporary gill. When an insect dives, the gases in its air store are in equilibrium with the gases dissolved in the water, assuming that this is saturated with air. Normally, at the dive, the bubble would contain approximately 21% oxygen and 79% nitrogen, while the water, because of the differing solubilities of the gases, will contain 33% oxygen, 64% nitrogen and 3% carbon dioxide (Fig. 17.26). Carbon dioxide is very soluble, so there is never very much in the bubble. Within a short period of diving, the proportion of oxygen in the bubble is reduced, as the oxygen is utilized by the insect, and there is a corresponding increase in the proportion, and hence partial pressure, of nitrogen. As a result, the gases in the bubble and those in solution immediately surrounding it are no longer in equilibrium and movements of gases will occur tending to restore equilibrium. Oxygen tends to pass into the bubble from the surrounding medium, because the oxygen tension in the bubble is reduced, while nitrogen tends to pass out of the bubble into solution, because the nitrogen tension in the bubble is increased. Thus more oxygen will be made available to the insect than was originally present in the bubble, which is, in fact, acting as a gill. This effect is enhanced by the fact that the oxygen passes into the bubble about three times more readily than nitrogen passes out into solution. As a result of this, the insect is able to remain submerged for longer periods than would be possible if it depended solely on the oxygen initially available in its store.

Nitrogen, as a non-respiratory gas, is essential for the air bubble to act as a gill as, in its absence, there is no change in partial pressure as the oxygen is utilized. For this reason an insect with a bubble of pure oxygen in water saturated with oxygen does not survive for very long unless it is able to come to the surface.

With inactive insects at low temperatures, when the rate of utilization of oxygen is low, the efficiency of the air store as a gill may be adequate for the insect to remain submerged for a long time. Thus *Hydrous* (Coleoptera) can remain submerged for some months during the winter (de Ruiter *et al.*, 1952), but in more active insects, with higher oxygen requirements, the bubble lasts for only a short time and the insect must come to the surface more frequently. If the temperature is above about 15 °C, *Notonecta* uses oxygen so rapidly that the gill effect is of negligible importance and the insect soon surfaces to replenish its airstore. At 10 °C, however, it remains submerged for twice as long as would be possible with the initial oxygen supply, and below 5 °C it can survive for a very long period without access to the surface.

The efficiency of the bubble as a gill depends on the oxygen content of the water adjacent to the gill. In water devoid of oxygen, or containing only a very little, the gas tends to pass out of the bubble into solution and will be lost to the insect. Even if the oxygen tension in the water exceeds that in the bubble, the amount entering the bubble will depend on the difference in tension. Hence, the higher the oxygen tension of the outside water, the more effective will the bubble be as a gill, and this must be true of any type of gill. Consequently the frequency with which an insect visits the surface will vary inversely with the oxygen tension of the water (Fig. 17.27).

17.5.2 Insects obtaining oxygen from the water

In all insects living in water, some inward diffusion of oxygen from the water takes place through the cuticle and in many larval forms gaseous exchange takes place solely in this way.

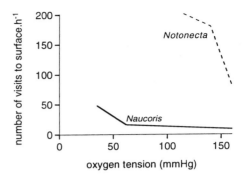

Fig. 17.27. Because an air bubble carried by an insect acts as a physical gill, the time for which an insect can remain submerged increases as the amount of oxygen dissolved in the water (oxygen tension) increases. Consequently, the insect makes fewer visits to the surface at higher oxygen tensions. Data for *Naucoris* at 20 °C; *Notonecta* at 17 °C (data from de Ruiter *et al.*, 1952).

Cutaneous diffusion depends on the permeability of the cuticle and a lower oxygen tension in the tissues as compared with the water. In many larval forms the cuticle is relatively permeable and in *Aphelocheirus* (Heteroptera), for example, the cuticle of the last larval stage is about four times as permeable as that of the adult (Thorpe and Crisp, 1947b)

In very small larvae, such as the first stage larvae of *Simulium* (Diptera) and *Chironomus* (Diptera), in which the tracheal system is filled with fluid, cutaneous diffusion into the hemolymph meets the whole oxygen requirement of the insect. In general, however, blood circulation is poor and the rate of diffusion through the blood slow, and diffusion into the blood would not suffice for most larger insects. Hence the majority of insects which obtain oxygen from water do have a tracheal system although the spiracles are non-functional. This is called a closed tracheal system. As oxygen is used by the insect, its partial pressure within the tracheae falls, creating a pressure gradient between the tracheoles just beneath the cuticle and the adjacent water. Under these conditions, oxygen from the water diffuses into the tracheal system within which it can rapidly diffuse round the body to the tissues. An incompressible tracheal system is essential for this type of gas movement to occur, as, otherwise, the tracheae would collapse under the pressure of water as oxygen was used.

17.5.2.1 *Tracheal gills*

In some insects, such as *Simulium* larvae, there is a network of tracheoles just beneath the general body cuticle, but often there are leaf-like extensions of the body forming

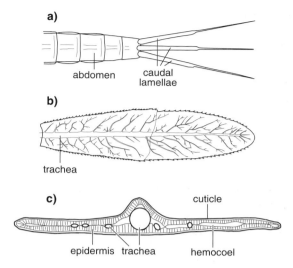

a)

abdomen   caudal lamellae

b)

trachea

c)   cuticle

epidermis trachea   hemocoel

Fig. 17.28. Tracheal gills (caudal lamellae) of a damselfly larva (Zygoptera). (a) Dorsal view of the posterior end of the abdomen. (b) One lamella showing the tracheae. Only the major branches are shown (after Gardner, 1960). (c) Transverse section of a lamella (after Tillyard, 1917).

gills. These are covered by very thin cuticle with a network of tracheoles immediately beneath and are known as tracheal gills. In most larval Zygoptera there are three caudal gills (Fig. 17.28); larval Trichoptera have filamentous abdominal gills, while, in larval Plecoptera, the position of the gills varies from species to species.

Larval Anisoptera have gills in the anterior part of the rectum, known as the branchial chamber (Fig. 17.29). Water is drawn in and out of the chamber by the activity of muscles largely unconnected with the gut. The hemocoel in the posterior part of the abdomen is isolated from the rest of the body by a muscular diaphragm across the fifth abdominal segment. Because of this, contraction of dorsoventral muscles in the posterior abdominal segments exerts pressure on the branchial chamber. Water is forced out through the partially open anus and drawn in again when the volume of the abdomen is restored, partly by the elasticity of the cuticle and partly by contraction of muscles in the diaphragm and of a transverse subintestinal muscle. During inspiration the anal valves are wide open (Fig. 17.30). A muscular prebranchial valve presumably prevents water from moving forwards into the midgut (Mill and Pickard, 1972; Pickard and Mill, 1972). Some 85% of the water in the rectum is renewed at each cycle of compression and relaxation, and the frequency of

pumping in *Aeschna* varies from about 25 to 50 cycles per minute. At the higher rates the interval between inspiration and expiration is reduced (Hughes and Mill, 1966).

In general, although a good deal of gaseous exchange probably takes place through tracheal gills (in *Agrion* (Odonata) larva 32–45% of the oxygen absorbed normally takes this route) some insects are able to survive without gills under normal oxygen tensions. Where the oxygen tension of the water is low, however, gills are of importance as they considerably increase the area available for gaseous exchange.

17.5.2.2 *Plastron respiration*

Some insects have specialized structures holding a permanent thin film of air on the outside of the body in such a way that an extensive air water interface is always present for gaseous exchange. This film of gas is called a plastron (Thorpe, 1950) and the tracheae open into it so that oxygen can pass directly to the tissues.

The volume of the plastron is constant and usually small as it does not provide a store of air, but acts solely as a gill. The constant volume is maintained by various hydrofuge devices spaced very close together so that water does not penetrate between them except under considerable pressure. Excess external pressures which may occur during the normal life of the insect are resisted. Such pressures may develop through the utilization of oxygen from the plastron so that the internal pressure is reduced, or through the insect being in deep water and therefore subjected to high hydrostatic pressure.

In adult insects, the plastron is held by a very close hair pile in which the hairs resist wetting because of their hydrofuge properties and their orientation. The most efficient resistance to wetting would be achieved by a system of hairs lying parallel with the surface of the body (Thorpe and Crisp, 1947a). *Aphelocheirus* approaches this condition in possessing hairs which are bent over at the tip (Fig. 17.31a); in other insects, such as *Elmis* (Coleoptera), the hairs are sloping (Fig. 17.31c). The ability of sloping hairs to withstand collapse depends on their being slightly thickened at the base and also on their close packing. As the air in the plastron is compressed, the closely packed hairs become pressed together so that their overall resistance to further compression increases.

In the adult *Aphelocheirus*, the plastron covers the ventral and part of the dorsal surface of the body. The hairs which hold the air are 5–6 μm high and about 0.2 μm in diameter (Fig. 17.31a). They are packed very

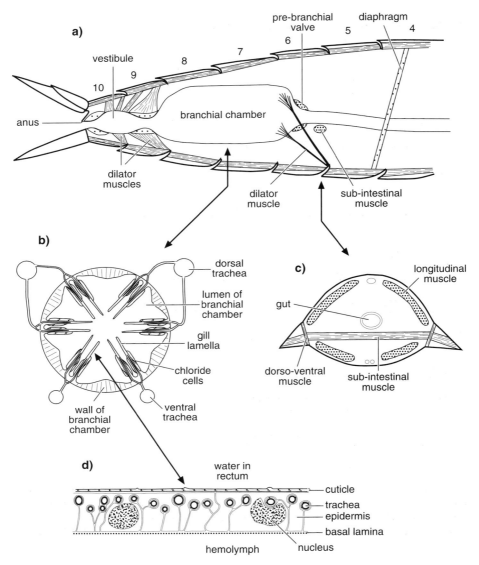

Fig. 17.29. Tracheal gills in the rectum of a dragonfly larva (after Mill & Pickard, 1972; Tillyard, 1917). (**a**) Longitudinal section through the abdomen. Numbers indicate abdominal segments. (**b**) Transverse section through the branchial chamber (**c**) Transverse section through abdominal segment 6 showing the subintestinal muscle. (**d**) Detail of gill epithelium showing the tracheae close beneath the cuticle (after Schmitz & Komnich, 1976).

close together, at about $2\,500\,000\,\text{mm}^{-2}$, and are able to withstand a pressure of about $400\,\text{kN}\,\text{m}^{-2}$ before they collapse. Hence, this is an extremely stable plastron that will only be displaced by water at excessive depths. The spiracles open into a series of radiating canals in the cuticle and these connect with the plastron via small pores. These canals are lined with hairs so that the entry of water into the tracheal system is prevented. The basal rate of oxygen utilization at $20\,^{\circ}\text{C}$ is about $6\,\mu\text{l}\,\text{h}^{-1}$ and this is readily provided by the plastron, so that, except in poorly oxygenated water, the insect need never come to the surface.

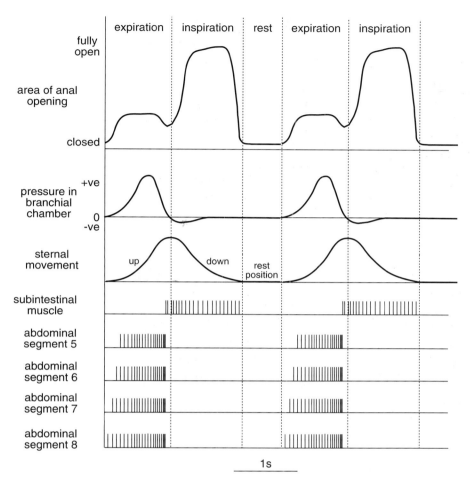

Fig. 17.30. Rectal ventilation in a dragonfly larva. Lower traces show action potential frequency in the subintestinal muscle and in the dorso-ventral muscles of successive abdominal segments. As the dorso-ventral muscles contract they raise the sterna and compress the branchial chamber and water is forced out. When the subintestinal muscle contracts the sterna are lowered, pressure in the chamber becomes negative and water flows in through the widely open anus (after Mill & Pickard, 1972).

The plastrons of other insects are generally less efficient than that of *Aphelocheirus* as they have a less dense hair pile from which the air is more readily displaced. However, a number of insects with a hair density of 3×10^4–1.5×10^5 mm^{-2} have plastrons that are usually permanent and adequate for the insect's needs. In these cases, the plastron is often supplemented by a less permanent macroplastron, consisting of a thicker layer of air outside the plastron and held by longer hairs than the plastron, as in *Hydrophilus* (Coleoptera) (Fig. 17.31b), or by

the erection of the plastron hairs, as in *Elmis* (Fig. 17.31c). The macroplastron is an air store and acts as a physical gill. It becomes reduced in size and finally eliminated when the insect is submerged, leaving only the plastron. The hairs holding the macroplastron are relatively long and flexible, so that as the gas bubble is reduced they tend to clump, leaving patches of exposed cuticle liable to wetting. To avoid this, these insects groom their hair pile and, in *Elmis*, there are brushes on the legs for this purpose. These brushes are also used to capture air bubbles and add them

a) *Aphelocheirus*

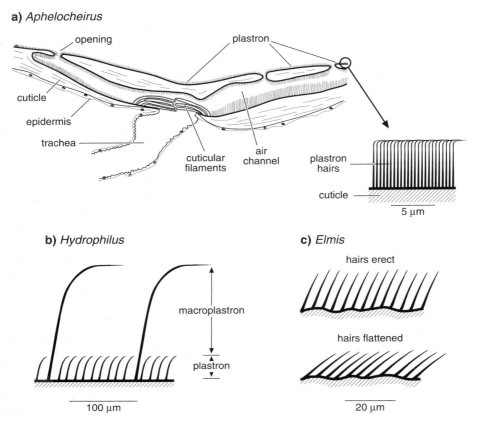

Fig. 17.31. Plastron. (**a**) Plastron of *Aphelocheirus* (Hemiptera). Section through a spiracular rosette showing the junction of a trachea with a system of channels in the cuticle connecting with the plastron. Below right, detail of the plastron (after Thorpe & Crisp, 1947a). (**b**) Macroplastron of *Hydrophilus* (Coleoptera) (after Thorpe & Crisp,1949). (**c**) Macroplastron of *Elmis* (Coleoptera) showing the hairs erect, forming a macroplastron (above) and with the hairs compressed so that only the true plastron remains (after Thorpe & Crisp, 1949).

to the macroplastron. Insects that only have a plastron do not make grooming movements of this type.
Review: Thorpe, 1950

17.5.3 Ventilation in aquatic insects

Close to a solid surface, water movement becomes laminar with little mixing between successive layers. The region of laminar flow is called a boundary layer and, in general, is greater in depth in still or slowly moving water than in fast-flowing turbulent water. This has important consequences for the availability of oxygen at any respiratory interface. In turbulent water, oxygen is moved by convection and tends to be uniformly distributed. Movement across the boundary layer, however, depends on diffusion.

As a result, water adjacent to an air bubble or a gill will tend to become depleted of oxygen because the rate of diffusion into the layer adjacent to the gill surface may not match the rate at which oxygen is taken into the gill. As movement into the gill is also a passive, diffusive process, it is obviously important to maintain the concentration of oxygen in the boundary layer at a high level. This is achieved by the insect making ventilatory movements that renew the layer of water adjacent to the gill.

In many insects, this is achieved by movements of the gills themselves. In *Ephemera* a good deal of gaseous exchange takes place through the gills, but in *Cloeon* they function almost entirely as paddles pushing water over the real respiratory surface on the abdomen. The larva of

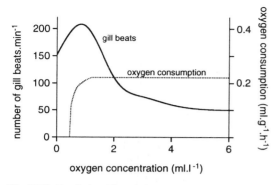

Fig. 17.32. Ventilation. The relationship between gill beat and oxygen uptake in relation to the oxygen content of the water in a mayfly larva (*Ephemera*) (after Eriksen, 1963).

Corydalis (Neuroptera) ventilates the gills using rhythmic movements of the tubercles on which the gills are mounted. Anisopteran larvae achieve the same result by pumping water in and out of the rectum. Zygopteran larvae also pump water in and out of the rectum, and, in most species, this is associated with ionic regulation (section 18.3.2.2). In *Calopteryx*, however, pumping activity is enhanced in water deficient in oxygen. The flow of water from the rectum probably stirs the water round the caudal gills, but it is possible that some gaseous exchange also occurs in the rectum (Miller, 1994).

Some insects using an air bubble as a gill or that have a plastron also ventilate. For example, the bug *Naucoris* (Heteroptera) holds on to vegetation and makes swimming movements with its back legs directing a current over its air bubble; *Phytobius relatus* (Coleoptera) ventilates its plastron with its middle legs.

By changing the frequency with which these movements are performed, the insect is able to maintain its oxygen uptake at a constant level even in water containing small amounts of dissolved oxygen. For example, in *Ephemera*, the gills beat actively in water containing little oxygen and more slowly when the oxygen tension is high (Fig. 17.32). If the oxygen content of the water is very low, however, oxygen consumption falls. Similar changes occur in the rates of ventilatory movements made by a variety of other aquatic insects (de Ruiter *et al.*, 1952; Eriksen, 1963, 1986).

Control of these ventilatory movements appears to depend on similar phenomena to those occurring in terrestrial insects. The ventilatory movements of *Corydalis* are

produced by hypoxia, while those of *Naucoris* are probably stimulated by a high concentration of carbon dioxide in the tracheal system (de Ruiter *et al.*, 1952). At least in the former, there is evidence for central pattern generators in the ganglia of abdominal segments 2 and 3, with that in segment 3 being dominant. This drives the interneurons that activate or inhibit the motor neurons (Kinnamon & Kammer, 1983).

For insects living in fast-flowing water, active ventilatory movements are not necessary, but their body forms and orientation in the current may be important in maintaining the oxygen supply. The cocoons of black flies (Simuliidae) are oriented so that their posterior, pointed ends face upstream while the spiracular gills, extending from the open end of the cocoon, are pointed downstream or upwards. The cocoons lie within the boundary layer of water flowing over the stones to which they are attached, and the flow round the cocoon produces vortices around the gills ensuring a steady supply of oxygen (Fig. 17.33a) (Eymann, 1991).

Vortices create regions of low pressure and, if the water is already saturated with air, gas will readily come out of solution, facilitating oxygen uptake by gills. In the extreme, the reduction in pressure may result in air coming out of solution as bubbles. In the pupae of Blephariceridae (Diptera), bubbles of gas appear on the gill surfaces, and this phenomenon of 'out-gassing' appears to maintain an air bubble round the adult beetle, *Potamodytes* (Fig. 17.33b) (Stride, 1955). It is possible that the facilitation of gas exchange by reduction in water pressure is a widespread phenomenon in aquatic insects living in flowing water (Pommen & Craig, 1994).

17.6 INSECTS SUBJECT TO OCCASIONAL SUBMERSION

It is relatively common for terrestrial insects to fall into water, but in general this does not occur sufficiently regularly for special respiratory adaptations to have evolved. Insects living close to the edge of water are, however, subject to more frequent alternations of submersion and emergence and so tend to be adapted for respiration in air or water. This is true for insects living intertidally, where submersion occurs regularly, and for those living at the edges of streams where submersion is much less regular.

Some intertidal beetles, such as *Bledius*, and *Dichirotrichus*, effectively avoid submergence by living in

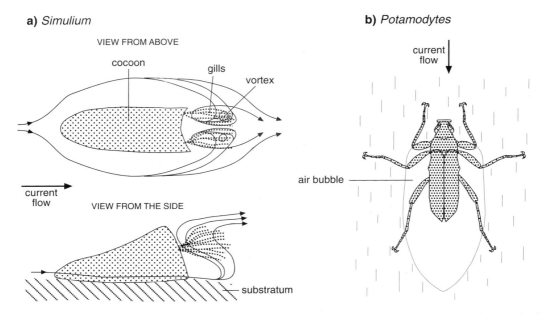

a) *Simulium*

VIEW FROM ABOVE

cocoon

gills

vortex

current
flow

VIEW FROM THE SIDE

substratum

b) *Potamodytes*

current
flow

air bubble

Fig. 17.33. Insects in fast-flowing water are oriented with respect to the current. The orientation produces vortices in the flow. These result in regions of reduced pressure so that gases tend to come out of solution. **(a)** Pupal cocoon of a blackfly (*Simulium*) from above and the side. The orientation of the cocoon in the water current results in the production of vortices round the gills (after Eymann, 1991). **(b)** The beetle, *Potamodytes*, holds on to the substratum facing into the current. The flow round the legs and body creates regions of low pressure and gases come out of solution to form an air bubble round the posterior part of the insect (after Stride, 1955).

burrows and crevices where sufficient air is trapped to enable them to survive aerobically for several hours. If they do become submerged they become quiescent as a result of anoxia. The time for recovery when returned to air is then proportional to the period of anoxia (Evans, Ruscoe & Treherne, 1971). Lice living on aquatic mammals live in the layer of air trapped in the fur, so that these, too, probably encounter no particular respiratory problems (Hinton, 1976).

The eggs of many terrestrial insects are subject to occasional flooding and those of many species have a chorionic plastron (section 14.1.2). A plastron, in the form of a spiracular gill (see below) is also present on the pupae of many species living in or close to water. By contrast, larval and adult stages living close to water generally lack any specialization of the respiratory system that would enable them to survive under water, presumably because they have the means of escaping before they become anoxic.

Review: Hinton, 1976 – marine insects

17.6.1 Spiracular gills

A spiracular gill is an extension of the cuticle surrounding a spiracle and bearing a plastron connected to the tracheal system by aeropyles. In water, the plastron provides a large gas/water interface for diffusion, while, in air, the interstices of the gill provide a direct route for the entry of oxygen, and water loss is limited because the gill opens into the atrium of the spiracle. Thus, in air, water loss through the spiracles is scarcely greater than in terrestrial insects (Hinton, 1968).

Spiracular gills occur in the pupal stages of many flies and beetles living intertidally or at the edges of streams. They also occur in the larvae of a few Coleoptera and in *Canace* (Diptera). Where they occur in pupae, the basal structure of the gill is modified to permit respiration by the pharate adult (Fig. 17.34). The spiracular gills of most dipteran pupae are prothoracic and connect with the prothoracic spiracles. In the cranefly, *Taphrophila*, for example, there is a single gill on each side with eight branches (Fig. 17.34a). The whole gill is about 1.5 mm long. Blackfly pupae

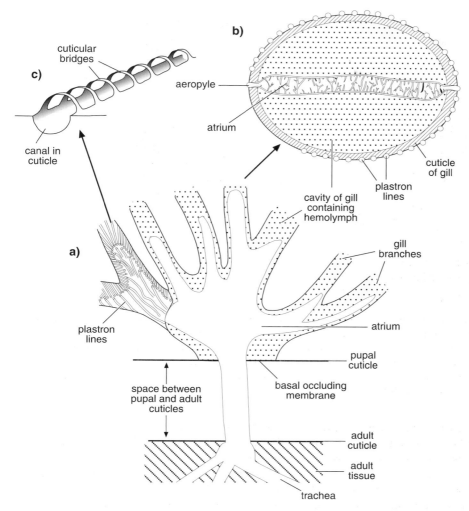

Fig. 17.34. Spiracular gills of a pharate adult cranefly (*Taphrophila*) (after Hinton, 1957). **(a)** Diagram showing basic structure and connection of gill to the tracheal system. The two left-hand branches represent the gill as seen from the outside; the remainder have the upper layer of cuticle removed to show the extent of the atrium. **(b)** Transverse section through a gill branch. **(c)** Detail of a plastron line.

(Simuliidae) also have two prothoracic gills. In this case each has at least two main branches and in some species there is abundant secondary branching. In the pupae of Psephenidae (Coleoptera), however, the gills are associated with the abdominal spiracles and in *Psephenoides volatilis* they are on abdominal segments two to seven. Each gill has between four and ten long slender branches. Abdominal spiracular gills also occur in the larvae of beetles of the genera *Torridincola*, *Sphaerius* and *Hydroscapha*.

The form of the plastron differs in different insects. In *Eutanyderus* (Diptera) and many other species, it is held by hydrofuge cuticular struts running at right angles to the surface of the gill. These struts branch at the apices and the branches of adjacent struts anastomose to form an open network. An extensive air/water interface thus exists in the interstices of the network. In these insects the plastron extends all over the gills and connects with the atrium of the spiracle at the base of each gill.

In *Taphrophila*, the spiracular atrium extends into each gill and its branches, and where the atrium meets the wall of the gill it opens to the outside through a series of small pores, the aeropyles, about 4 μm in diameter (Fig. 17.34b). The atrium is flattened in cross-section with the two walls connected by cuticular struts so that it does not collapse even if the gills dry, and the hydrofuge properties of the struts prevent the entry of water into the atrium. Running from each aeropyle, on the outside of the gill, is a shallow canal about 4 μm wide which is crossed at intervals of about 1.0 μm by cuticular bridges about 0.5 μm wide (Fig. 17.34c). The lining of the canals is strongly hydrofuge, so that in water each holds a long cylinder of air, known as a plastron line, which is not easily displaced because of the cuticular bridges. The plastron lines provide a relatively large air/water interface over which gaseous exchange occurs.

For the plastron to function efficiently it is important that the gills are turgid when the insects are submerged. In many species of tipulid and in *Psephenoides* the gills are rigid, but in other species turgor is maintained by the high osmotic content of the gills which results from the hemolymph and epidermal cells which they contain. At the pupa–adult apolysis the pupal cuticle becomes separated from that of the adult, but the gills become cut off from the rest of the tissues by a basal occluding membrane so they are still lined by an epithelium and contain hemolymph (Fig. 17.34a). The epithelium within the gills disintegrates and the cells form loose irregular clusters in the middle of the gill. This isolated tissue is important because it is able to repair damage to the surface of the gill, forming cuticular plugs in any holes which are produced, and this facility is retained even after the gills have been strongly desiccated. Gill repair is important as a damaged gill loses its high internal osmotic pressure and becomes flaccid and inefficient when in the water.

The gills of pupal *Simulium* are fully expanded in water as they have an opening at the base into which water can freely pass. A thin membrane at the base of each gill in the newly ecdysed pupa bursts due to the intake of water resulting from high internal osmotic pressure. This provides the opening for subsequent water movement into or out of the gill, ensuring that the shape of the gill is independent of the hydrostatic pressures exerted on it.

The efficiency of a plastron depends on its surface area. In spiracular gills the area of the interface varies from 1.5×10^{-4} μm^2 mg^{-1} net body weight in the pupa of *Eutanyderus* to 1.9×10^{-6} μm^2 mg^{-1} net weight in the pupa of *Simulium*. These areas are presumably large enough to permit efficient absorption of oxygen from well-aerated water. As with other plastrons, the plastron of a spiracular gill resists displacement by high hydrostatic pressures.

Review: Hinton, 1968

17.7 GAS EXCHANGE IN ENDOPARASITIC INSECTS

Endoparasitic insects may obtain their oxygen directly from the air outside the host or by diffusion through the cuticle from the surrounding host tissues. In many ichneumonid and braconid (Hymenoptera) larvae, the tracheal system of the first instar is liquid-filled and, even when it becomes gas-filled, the spiracles remain closed up to the last instar. Thus, these insects and the young larvae of most parasitic Diptera depend entirely on cutaneous diffusion. In braconid larvae the hindgut is everted through the anus to form a caudal vesicle. This is variously developed in different species, but in some, such as *Apanteles*, it is relatively thin-walled and closely associated with the heart (Fig. 17.35) so that oxygen passing in is quickly carried round the body. In these insects the vesicles are responsible for about a third of the total gaseous exchange.

When the tracheal system becomes air-filled, networks of tracheoles may develop immediately beneath the cuticle, thus facilitating the diffusion of gases away from the surface. In *Cryptochaetum iceryae* (Diptera), a parasite of scale insects, there are two caudal filaments, which in the third instar larva are ten times as long as the body and are packed with tracheae. These filaments often get entangled with the host tracheae and so provide an easy path for oxygen transfer.

Other insects, and particularly older, actively growing larvae with greater oxygen requirements, connect with the outside air by penetrating the host's body wall or respiratory system. The majority of these insects are metapneustic or amphipneustic, using the posterior spiracles to obtain their oxygen. Chalcid (Hymenoptera) larvae are connected to the outside from the first instar onwards by the hollow egg pedicel which projects through the host's body wall. The posterior spiracles of the larva open into the funnel-shaped inner end of the pedicel and so make contact with the outside air. Many tachinid (Diptera) larvae that are parasitic in other insects, tap the host's tracheal supply or pierce its body wall from within using their posterior spiracles. The host epidermis is stimulated to

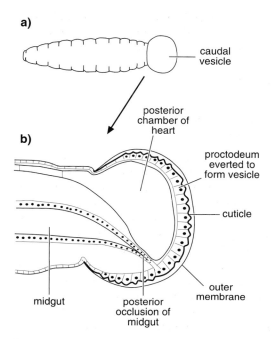

a)

caudal vesicle

posterior chamber of heart

b)

proctodeum everted to form vesicle

cuticle

outer membrane

midgut

posterior occlusion of midgut

Fig. 17.35. Caudal vesicle of a parasitic hymenopteran larva (*Apanteles*) (from Wigglesworth, 1965). (**a**) Whole larva showing the caudal vesicle. (**b**) Longitudinal section of caudal vesicle.

grow and spreads round the larva, almost completely enclosing it, and secreting a thin, cuticular membrane over its surface. The larva of *Melinda* (Diptera), parasitic in snails, respires by sticking its posterior spiracles out through the respiratory opening of the snail.

Parasites of vertebrates also often use atmospheric air. The larva of *Cordylobia* (Diptera) bores into the skin and produces a local swelling, but it always retains an opening to the outside into which the posterior spiracles are thrust. Similarly in the larva of the warble fly, *Hypoderma*, the warble opens to the outside, but here the larva bores its way out to the surface from within the host tissues (Keilin, 1944).

17.8 OTHER FUNCTIONS OF THE TRACHEAL SYSTEM

Apart from gaseous exchange, the tracheal system has a number of other functions. The whole system, and the airsacs in particular, lower the insect's specific gravity. In aquatic insects, but not in terrestrial ones, it also gives some degree of buoyancy and in the larvae of the phantom midge, *Chaoborus* (Diptera), the tracheae form hydrostatic organs enabling the buoyancy to be adjusted (Teraguchi, 1975).

The tracheal system in some insects has a major role in hemolymph circulation (section 5.1.2.1).

Airsacs, being collapsible, allow for the growth of organs within the body without any marked changes in body form. Thus at the beginning of the final stage larva the tracheal system of *Locusta* (Orthoptera) occupies 42% of the body volume, but by the end of the stage it only occupies 3.8% due to the growth of other organs causing compression of the airsacs (Clarke, 1957). The airsacs also permit changes in gut volume as a result of feeding. In *Locusta*, the increase in crop volume following a meal is accompanied by a corresponding decrease in the volume of the thoracic airsacs (Bernays and Chapman, 1973).

In some noctuids (Lepidoptera), tracheae form a reflecting tapetum beneath the eye (section 22.1.2), and tympanal membranes are usually backed by an airsac which, being open to the outside air, allows the tympanum to vibrate freely with a minimum of damping (section 23.2.3.3).

Expansion of the tracheal system may also assist when an insect inflates itself after a molt. Thus in dragonflies, spiracle closure, preventing the escape of gas from the tracheae, accompanies each muscular effort of the abdomen during expansion of the wings.

The airsacs of some insects insulate the thorax, and therefore the flight muscles, from the abdomen. This makes it possible for thorax and abdomen to have significantly different temperatures and gives the insects the capacity to regulate thoracic temperature (section 19.2).

An important general function of tracheae and tracheolar cells is in acting as connective tissue, binding other organs together.

Finally, the tracheal system may be involved in defence. In the cockroach, *Diploptera*, quinones which probably have a defensive function can be forcibly expelled from the second abdominal spiracle (Whitman, Blum & Alsop, 1990) and in *Gromphadorrhina* (Blattodea) sounds are produced by forcing air out through the spiracles.

17.9 RESPIRATORY PIGMENTS

The majority of insects have no respiratory pigments, but a few have hemoglobin in solution in the blood. The best known examples are the aquatic larvae of the midge, *Chironomus*, and related insects, the aquatic bug, *Anisops*, and the endoparasitic larvae of the bot fly, *Gasterophilus* (Diptera).

The hemoglobin of *Chironomus* has a molecular weight of 31 400, that of *Gasterophilus* is about 34 000. This is about half the molecular weight of vertebrate hemoglobin and indicates that it contains only two heme groups. It has a much higher affinity for oxygen than vertebrate hemoglobin, and is 50% saturated at partial pressures of oxygen of less than 100 Pa, compared with more than 3 kPa in vertebrates. The hemoglobin of *Anisops* is different, however, and has only a low affinity for oxygen.

Chironomus *Chironomus* larvae live in burrows in the mud under stagnant water which is commonly poor in oxygen. A flow of water is directed through the burrow by dorsoventral undulating movements of the body and the current so produced provides food and oxygen. During such bouts of irrigation, the hemoglobin in the blood is fully saturated with oxygen and apparently has no function, but during the intervals between bouts the oxygen of the surrounding medium is quickly used up. The hemoglobin has a high affinity for oxygen and only dissociates when the tension in the tissues is very low as is as soon the case during the pauses between irrigation movements. It therefore provides a small store of oxygen for these periods, but as the store only lasts for about nine minutes and the pauses often last longer than this, respiration during the rest of the time is anaerobic.

It may be that the respiratory function of hemoglobin in *Chironomus* is more important in facilitating rapid recovery from this oxygen lack when irrigation movements are resumed. The hemoglobin is able to take up oxygen and pass it to the tissues more quickly than is possible by simple solution in the hemolymph and this is especially true when the oxygen content of the water itself is low. Under these conditions, the hemoglobin takes up oxygen continuously from the water and transfers it to the tissues and never becomes fully saturated with oxygen. It thus makes muscular activity more aerobic, and hence more efficient than if physical solution of oxygen alone was involved and it also permits filter feeding, which does not occur under anaerobic conditions, to continue at low oxygen tensions.

It may be that the respiratory function of hemoglobin in *Chironomus* is secondary and that its primary function is as a protein store (for references, see Schin, Laufer & Carr, 1977).

Anisops When *Anisops* dives, it carries with it a small ventral air-store which is continuous with air under the wings. All the spiracles open into the air-store and the s-

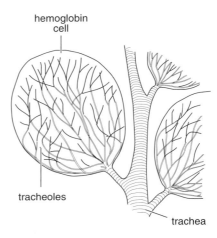

hemoglobin
cell

tracheoles

trachea

Fig. 17.36. Hemoglobin occurs in hemoglobin cells of the larva of the bot fly, *Gasterophilus* (after Keilin, 1944).

piracles of abdominal segments 5–7 are very large and covered by sieve plates. From the atria of these spiracles several tracheae arise, branching repeatedly to form 'trees', the terminal branches of which indent large cells, called hemoglobin (or tracheal) cells, that are filled with hemoglobin. The hemoglobin becomes oxygenated when the bug is at the surface and deoxygenated during a dive. The oxygen released enables the insect to remain submerged for longer than would otherwise be possible, while at the same time affecting its buoyancy. When the insect first dives, it is buoyant because of its ventral air-store, but as the store is used up this buoyancy is reduced until the density of the insect is roughly the same as water and it is able to float in mid water. This phase is maintained by the steady release of oxygen from the hemoglobin cells resulting from the reduction in partial pressure of oxygen in the store. After about five minutes, however, the hemoglobin is fully unloaded and the insect, now with a tendency to sink, swims up to the surface and renews its air-store (Miller, 1966b).

Gasterophilus The third stage larva of *Gasterophilus* is an internal parasite in the stomach of the horse. In the early instars, the larvae contain hemoglobin dissolved in the blood, but in the third instar this becomes concentrated in large hemoglobin cells. Four pairs of tracheal trunks run from the posterior spiracles and give off short branches at intervals along their lengths. Each branch breaks up into numerous tracheoles which are functionally, if not structurally, within a hemoglobin cell (Fig. 17.36).

Within the horse's stomach, the larva receives only an intermittent supply of air in gas bubbles with the food, and the hemoglobin of the hemoglobin cells enables the larva to take up more oxygen than is needed for its immediate requirements. This oxygen is used later when air bubbles are no longer available. The hemoglobin thus facilitates a more efficient use of the oxygen supply, but the store it provides is very small, lasting, at most, for four minutes (Keilin and Wang, 1946).

REFERENCES

Aidley, D.J. (1976). Increase in respiratory rate during feeding in larvae of the armyworm, *Spodoptera exempta*. *Physiological Entomology*, **1**, 73–5.

Bernays, E.A. & Chapman, R.F. (1973). The regulation of feeding in *Locusta migratoria*: internal inhibitory mechanisms. *Entomologia Experimentalis et Applicata*, **16**, 329–42.

Buck, J. (1962). Some physical aspects of insect respiration. *Annual Review of Entomology*, **7**, 27–56.

Burrows, M. (1980). The tracheal supply to the central nervous system of the locust. *Proceedings of the Royal Society of London*, B **207**, 63–78.

Clarke, K.U. (1957). On the role of the tracheal system in the post-embryonic growth of *Locusta migratoria*. *Proceedings of the Royal Entomological Society of London*, **32**, 67–79.

de Ruiter, L., Wolvekamp, H.P., Tooren, A.J. van, & Vlasblom, A. (1952). Experiments on the efficiency of the 'physical gill' (*Hydrous piceus* L., *Naucoris cimicoides* L., and *Notonecta glauca* L.). *Acta Physiologica et Pharmacologica Neerlandica*, **2**, 180–213.

Edwards, G.A., Ruska, H. & Harven, E. de (1958). The fine structure of insect tracheoblasts, tracheae and tracheoles. *Archives de Biologie*, **69**, 351–69.

Eriksen, C.H. (1963). Respiratory regulation in *Ephemera simulans* (Walker) and *Hexagenia limbata* (Serville) (Ephemeroptera). *Journal of Experimental Biology*, **40**, 455–68.

Eriksen, C.H. (1986). Respiratory roles of caudal lamellae (gills) in a lestid damselfly (Odonata: Zygoptera). *Journal of the North American Benthological Society*, **5**, 16–27.

Evans, P.D., Ruscoe, C.N.E. & Treherne, J.E. (1971). Observations on the biology and submergence behaviour of some littoral beetles. *Journal of the Marine Biological Association of the United Kingdom*, **51**, 375–86.

Eymann, M. (1991). Flow patterns around cocoons and pupae of black flies in the genus *Simulium* (Diptera: Simuliidae). *Hydrobiologia*, **215**, 223–9.

Farley, R.D. & Case, J.F. (1968). Sensory modulation of ventilative pacemaker output in the cockroach, *Periplaneta americana*. *Journal of Insect Physiology*, **14**, 591–601.

Full, R.J. & Tullis, A. (1990). Energetics of ascent: insects on inclines. *Journal of Experimental Biology*, **149**, 309–17.

Gardner, A.E. (1960). A key to the larvae of the British Odonata. In *Dragonflies*, ed. P.S. Corbet, C. Longfield, & N.W. Moore, pp. 191–225. London: Collins.

Grassé, P.-P. (1976). L'appareil respiratoire. In *Traité de Zoologie*, vol. 8, part 4, ed. P.-P.Grassé, pp. 93–204. Paris: Masson et Cie.

Gulinson, S.L. & Harrison, J.F. (1996). Control of resting ventilation rate in grasshoppers. *Journal of Experimental Biology*, **199**, 379–89.

Hadley, N.F. (1994). Ventilatory patterns and respiratory transpiration in adult terrestrial insects. *Physiological Zoology*, **67**, 175–89.

Herreid, C.F., Full, R.J. & Prawel, D.A. (1981). Energetics of cockroach locomotion. *Journal of Experimental Biology*, **94**, 189–202.

Hinton, H.E. (1947). On the reduction of functional spiracles in the aquatic larvae of the Holometabola, with notes on the moulting processes of spiracles. *Transactions of the Royal Entomological Society of London*, **98**, 449–73.

Hinton, H.E. (1957). The structure and function of the spiracular gill of the fly *Taphrophila vitripennis*. *Proceedings of The Royal Society of London* B, **147**, 90–120.

Hinton, H.E. (1966). Respiratory adaptations of the pupae of beetles of the family Psephenidae. *Philosophical Transactions of the Royal Society of London* B, **251**, 211–45.

Hinton, H.E. (1968). Spiracular gills. *Advances in Insect Physiology*, **5**, 65–162.

Hinton, H.E. (1976). Respiratory adaptations of marine insects. In *Marine Insects*, ed. L. Cheng, pp. 47–80. Amsterdam: North-Holland Publishing Co.

Holdgate, M.W. (1955). The wetting of insect cuticles by water. *Journal of Experimental Biology*, **32**, 591–617.

Hoyle, G. (1960). The action of carbon dioxide gas on an insect spiracular muscle. *Journal of Insect Physiology*, **4**, 63–79.

Hoyle, G. (1961). Functional contracture in a spiracular muscle. *Journal of Insect Physiology*, **7**, 305–14.

Hughes, G.M. & Mill, P.J. (1966). Patterns of ventilation in dragonfly larvae. *Journal of Experimental Biology*, **44**, 317–34.

Imms, A.D. (1947). *Insect Natural History*. London: Collins.

Imms, A.D. (1957). *A General Textbook of Entomology*, revised by O.W. Richards & R.G. Davies. London: Methuen.

Kaars, C. (1981). Insects – spiracle control. In *Locomotion and Energetics in Arthropods*, ed. C.F.Herreid & C.R.Fourtner, pp. 337–66. New York: Plenum.

Keilin, D. (1944). Respiratory systems and respiratory adaptations of larvae and pupae of Diptera. *Parasitology*, **36**, 1–66.

Keilin, D. & Wang, Y.L. (1946). Haemoglobin of *Gastrophilus* larvae. Purification and properties. *Biochemical Journal*, **40**, 855–66.

Keister, M.L. (1948). The morphogenesis of the tracheal system of *Sciara*. *Journal of Morphology*, **83**, 373–424.

Kestler, P. (1985). Respiration and respiratory water loss. In *Environmental Physiology and Biochemistry of Insects*, ed. K.H. Hoffmann, pp. 137–83. Berlin: Springer-Verlag.

Kinnamon, S.C. & Kammer. A. (1983). Neural control of ventilatory movements in the aquatic insect *Corydalis cornutus*: the motor pattern. *Journal of Comparative Physiology*, A **153**, 543–55.

Krogh, A. & Weis-Fogh, T. (1951). The respiratory exchange of the desert locust (*Schistocerca gregaria*) before, during and after flight. *Journal of Experimental Biology*, **28**, 342–57.

Lewis, G.W., Miller, P.L. & Mills, P.S. (1973). Neuro-muscular mechanisms of abdominal pumping in the locust. *Journal of Experimental Biology*, **59**, 149–68.

Lighton, J.R.B. (1988). Simultaneous measurement of oxygen uptake and carbon dioxide emission during discontinuous ventilation in the tok-tok beetle, *Psammodes striatus*. *Journal of Insect Physiology*, **34**, 361–7.

Lighton, J.R.B. (1994). Discontinuous ventilation in terrestrial insects. *Physiological Zoology*, **67**, 142–62.

Lighton, J.R.B. (1996). Discontinuous gas exchange in insects. *Annual review of Entomology*, **41**, 309–24.

Lighton, J.R.B. & Berrigan, D. (1995). Questioning paradigms: caste-specific ventilation in harvester ants, *Messor pergandei* and *M.julianus* (Hymenoptera: Formicidae). *Journal of Experimental Biology*, **198**, 521–30.

Lighton, J.R.B., Fukushi, T. & Wehner, R. (1993a). Ventilation in *Cataglyphis bicolor*: regulation of carbon dioxide release from the thoracic and abdominal spiracles. *Journal of Insect Physiology*, **39**, 687–99.

Lighton, J.R.B. & Garrigan, D.A. (1995). Ant breathing: testing regulation and mechanism hypotheses with hypoxia. *Journal of Experimental Biology*, **198**, 1613–20.

Lighton, J.R.B., Garrigan, D.A., Duncan, F.D. & Johnson, R.A. (1993b). Spiracular control of respiratory water loss in female alates of the harvester ant *Pogonomyrmex rugosus*. *Journal of Experimental Biology*, **179**, 233–44.

Mill, P.J. (1985). Structure and physiology of the respiratory system. In *Comprehensive Insect Physiology, Biochemistry and Pharmacology*, vol. 3, ed. G.A. Kerkut & L.I. Gilbert, pp. 517–93. Oxford: Pergamon Press.

Mill, P.J. & Pickard, R.S. (1972). Anal valve movement and normal ventilation in aeshnid dragonfly larvae. *Journal of Experimental Biology*, **56**, 537–43.

Miller, P.L. (1960a). Respiration in the desert locust. I. The control of ventilation. *Journal of Experimental Biology*, **37**, 224–36.

Miller, P.L. (1960b). Respiration in the desert locust. II. The control of the spiracles. *Journal of Experimental Biology*, **37**, 237–63.

Miller, P.L. (1960c). Respiration in the desert locust. III. Ventilation and the spiracles during flight. *Journal of Experimental Biology*, **37**, 264–78.

Miller, P.L. (1966a). The regulation of breathing in insects. *Advances in Insect Physiology*, **3**, 279–344.

Miller, P.L. (1966b). The function of haemoglobin in relation to the maintenance of neutral buoyancy in *Anisops pellucens* (Notonectidae: Hemiptera). *Journal of Experimental Biology*, **44**, 529–44.

Miller, P.L. (1971). Rhythmic activity in the insect nervous system. I. Ventilatory coupling of a mantid spiracle. *Journal of Experimental Biology*, **54**, 587–97.

Miller, P.L. (1973). Spatial and temporal changes in the coupling of cockroach spiracles to ventilation. *Journal of Experimental Biology*, **59**, 137–48.

Miller, P.L. (1981a). Respiration. In *The American Cockroach*, ed. W.J. Bell & K.D. Adiyodi, pp. 87–116. London: Chapman & Hall.

Miller, P.L. (1981b). Ventilation in active and in inactive insects. In *Locomotion and Energetics in Arthropods*, ed. C.F. Herreid & C.R. Fourtner, pp. 367–90. New York: Plenum Press.

Miller, P.L. (1994). The responses of rectal pumping in some zygopteran larvae (Odonata) to oxygen and ion availability. *Journal of Insect Physiology*, **40**, 333–9.

Nikam, T.B. & Khole, V.V. (1989). *Insect spiracular systems*. Chichester: Ellis Horwood.

Pearson, K.G. (1980). Burst generation coordinating interneurons in the ventilatory system of the locust. *Journal of Comparative Physiology*, **137**, 308–13.

Pickard, R.S. & Mill, P.J. (1972). Ventilatory muscle activity in intact preparations of aeshnid dragonfly larvae. *Journal of Experimental Biology*, **56**, 527–36.

Pommen, G.D.W. & Craig, D.A. (1994). Flow patterns around gills of pupal net-winged midges (Diptera: Blephariceridae): possible implications for respiration. *Canadian Journal of Zoology*, **73**, 373–82.

Quinlan, M.C. & Hadley, N.F. (1993). Gas exchange, ventilatory patterns, and water loss in two lubber grasshoppers: quantifying cuticular and respiratory transpiration. *Physiological Zoology*, **66**, 628–42.

Schin, K., Laufer, H. & Carr, E. (1977). Cytochemical and electrophoretic studies of haemoglobin synthesis in the fat body of a midge, *Chironomus thummi. Journal of Insect Physiology*, **23**, 1233–42.

Schmitz, M. & Komnick, H. (1976). Rectal Chloridepithelien und osmoregulatorische Salzaufnahme durch den Enddarm von Zygopteren und Anisopteren Libellenlarven. *Journal of Insect Physiology*, **22**, 875–83.

Sláma, K. (1988). A new look at insect respiration. *Biological Bulletin*, **175**, 289–300.

Sláma, K. & Denlinger, D.L. (1992). Infradian cycles of oxygen consumption in diapausing pupae of the flesh fly, *Sarcophaga crassipalpis*, monitored by a scanning microrespirographic method. *Archives of Insect Biochemistry and Physiology*, **20**, 135–43.

Snodgrass, R.E. (1935). *Principles of Insect Morphology*. New York: McGraw-Hill.

Stride, G.O. (1955). On the respiration of an aquatic African beetle, *Potamodytes tuberosus* Hinton. *Annals of the Entomological Society of America*, **48**, 344–51.

Teraguchi, S. (1975). Correction of negative buoyancy in the phantom larva, *Chaoborus americanus. Journal of Insect Physiology*, **21**, 1659–70.

Thorpe, W.H. (1950). Plastron respiration in aquatic insects. *Biological Reviews of the Cambridge Philosophical Society*, **25**, 344–90.

Thorpe, W.H. & Crisp, D.J. (1947a). Studies on plastron respiration. I. The biology of *Aphelocheirus* [Hemiptera, Aphelocheiridae (Naucoridae)] and the mechanism of plastron respiration. *Journal of Experimental Biology*, **24**, 227–69.

Thorpe, W.H. & Crisp, D.J. (1947b). Studies on plastron respiration. II. The respiratory efficiency of the plastron in *Aphelocheirus. Journal of Experimental Biology*, **24**, 270–303.

Thorpe, W.H. & Crisp, D.J. (1949). Studies on plastron respiration. IV. Plastron respiration in the Coleoptera. *Journal of Experimental Biology*, **26**, 219–60

Tillyard, R.J. (1917). *The Biology of Dragonflies*. Cambridge: Cambridge University Press.

van Loon, J.J.A. (1988). A flow-through respirometer for leaf chewing insects. *Entomologia Experimentalis et Applicata*, **49**, 265–76.

Varley, G.C. (1937). Aquatic insect larvae which obtain oxygen from the roots of plants. *Proceedings of the Royal Entomological Society of London*, A **12**, 55–60.

Wasserthal, L.T. (1996). Interaction of circulation and tracheal ventilation in holometabolous insects. *Advances in Insect Physiology*, **26**, 297–351.

Webb, J.E. (1948). The origin of the atrial spines in the spiracles of sucking lice of the Genus *Haematopinus* Leach. *Proceedings of the Zoological Society of London*, **118**, 582–7.

Weis-Fogh, T. (1964a). Functional design of the tracheal system of flying insects as compared with the avian lung. *Journal of Experimental Biology*, **41**, 207–27.

Weis-Fogh, T. (1964b). Diffusion in insect wing muscle, the most active tissue known. *Journal of Experimental Biology*, **41**, 229–56.

Whitman, D.W., Blum, M.S. & Alsop, D.W. (1990). Allomones: chemicals for defense. In *Insect Defenses*, ed. D.L. Evans & J.O. Schmidt, pp. 289–351. Albany: State University of New York Press.

Whitten, J.M. (1972). Comparative anatomy of the tracheal system. *Annual Review of Entomology*, **17**, 373–402.

Wigglesworth, V.B. (1954). Growth and regeneration of the tracheal system of an insect. *Quarterly Journal of Microscopical Science*, **95**, 115–37.

Wigglesworth, V.B. (1959). The role of the epidermal cells in the migration of tracheoles in *Rhodnius prolixus. Journal of Experimental Biology*, **36**, 632–40.

Wigglesworth, V.B. (1965). *The Principles of Insect Physiology*. London: Methuen.

Wigglesworth, V.B. (1983). The physiology of insect tracheoles. *Advances in Insect Physiology*, **17**, 85–148.

Wigglesworth, V.B. (1990a). The properties of the lining membrane in the insect tracheal system. *Tissue & Cell*, **22**, 231–8.

Wigglesworth, V.B. (1990b). The direct transport of oxygen in insects by large tracheae. *Tissue & Cell*, **22**, 239–43.

Wigglesworth, V.B. (1991). The distribution of aeriferous tracheae for the ovaries of insects. *Tissue & Cell*, **23**, 57–65.

Williams, A.E., Rose, M.R. & Bradley, T.J. (1997). CO_2 release patterns in *Drosophila melanogaster*: the effect of selection for desiccation resistance. *Journal of Experimental Biology*, **200**, 615–24.

18   Excretion and salt and water regulation

The activities of the cell are carried out most efficiently within a narrow range of conditions. It is therefore important that the environment within the cell and in the animal in general should be kept as near optimal as possible, a process known as homeostasis. This involves the maintenance of a constant level of salts and water and osmotic pressure in the hemolymph, the elimination of toxic nitrogenous wastes derived from protein and purine metabolism, and the elimination of other toxic compounds which may be absorbed from the environment. The excretory system is primarily responsible for homeostasis, often following metabolic modification of toxic compounds to chemicals more readily excreted or which can be safely stored. Some molecules entering the body from the environment may be too large or too toxic to be dealt with by the excretory system and various other tissues may be involved in metabolizing them to less toxic or more readily excretable substances. Some insects sequester toxic compounds, isolating them from the major metabolic pathways.

18.1 EXCRETORY SYSTEM

Excretion involves the production of a fluid urine that carries potentially toxic materials from the body. This process is carried out in two stages: the relatively unselective removal of substances from the hemolymph, forming the primary urine, and the selective modification of this primary urine by the resorption of useful compounds or the addition of others that may be in excess in the body. In insects, the primary urine is produced by the Malpighian tubules. Its selective modification usually occurs in the rectum, but may also take place in the Malpighian tubules themselves or in the ileum. In aquatic insects, epidermal cells outside the gut may also be involved in maintaining the insect's ionic balance.
Review: Bradley, 1985; Razet, 1976

18.1.1 **Malpighian tubules**
Malpighian tubules are long, thin, blindly ending tubes arising from the gut near the junction of midgut and hindgut (see Fig. 3.1) and lying freely in the body cavity. They range in length, in different insect species, from 2 to 100 mm and from 30 to 100 μm in diameter. In most insects, they open independently into the gut, but in some species they first come together in groups at ampullae. In crickets, all the tubules open into a single ampulla from which a cuticle-lined, tubular ureter connects with the hindgut at the posterior end of the ileum. Malpighian tubules are absent from Collembola and aphids, and represented only by papillae in Diplura, Protura and Strepsiptera, but are present in all other insects, varying in number from two in coccids to about 250 in the desert locust, *Schistocerca*. The number may increase during post-embryonic development (see Fig. 15.13). Because of the large number of tubules and the infolding of the basal plasma membrane (see below) their total outer surface area is large, facilitating an exchange of materials with the hemolymph. In adult *Periplaneta*, with more than 150 tubules, the total outer surface area of the tubules (not taking account of the infolding of the basal plasma membrane) is over 500 mm^2; the infoldings increase the surface area at least ten times to more than 5000 mm^2.

The wall of the tubule is one cell thick with one or a few cells encircling the lumen (Fig. 18.1). The principal cell type in the tubule is produced into close-packed microvilli on the lumen side, while the basal plasma membrane is deeply infolded (O'Donnell *et al.*, 1985). Laterally, the cells are held together by septate junctions near their apices. Mitochondria are abundant within and immediately beneath the microvilli and in association with the basal plasma membrane. These are the cells which produce the primary urine.

In many insects, cells that are structurally and functionally of different types may be intermingled with the principal cells throughout the tubules, or occur in discrete regions so that the tubule is divided into distinct sections. The Diptera, for example, have small stellate cells scattered amongst the principal tubule cells. They have small microvilli which do not contain mitochondria. Mixtures of cell types are also known to occur in *Periplaneta*, *Carausius* and *Tenebrio*.

a)

b)

lumen of tubule

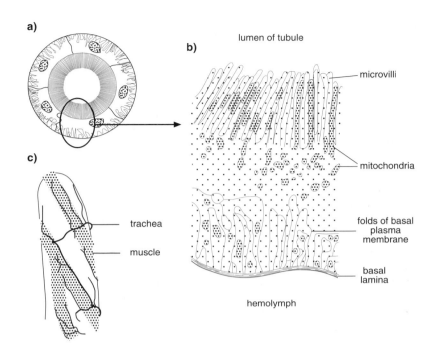

microvilli

mitochondria

c)

trachea

muscle

folds of basal
plasma
membrane

basal
lamina

hemolymph

Fig. 18.1. Malpighian tubule. (a) Cross-section of a tubule. (b) Detail of part of one cell. (c) End of a Malpighian tubule of *Apis* showing the spiral muscle strands and the tracheal supply (after Wigglesworth, 1965).

The Malpighian tubules of *Rhodnius* have different cell types in discrete regions of the tubules. The cells in the proximal part of each tubule (nearest the point of origin at the gut) differ from those in the distal part in having more widely separated microvilli which are variable in length, less complex infoldings of the basal membrane, and mitochondria concentrated in the basal region. The Malpighian tubules of *Drosophila* also have distinct regions, but, here, there are also differences between tubules. Posteriorly directed tubules have two parts while anteriorly directed tubules have a short distal section in addition to the main and lower (proximal) regions of the posterior tubules. In *Drosophila* and in *Rhodnius*, the more proximal parts of the tubules are concerned primarily with resorption, although, in *Drosophila*, the same region secretes calcium into the tubules (O'Donnell & Maddrell, 1995).

The domestic cricket, *Acheta*, has Malpighian tubules with three distinct regions: proximal, comprising about 5% of the length of the tubule; mid, 75% of the length; and distal, 20%. Each region is made up of one cell type. The cells of the middle section, forming the bulk of the tubule, resemble the principal cells described above. Those of the distal part have close-packed microvilli which almost fill the lumen of the tubule, but the basal plasma membrane is not as extensively folded as in the cells of the

middle section and most mitochondria are concentrated just below and within the microvilli (Hazelton, Parker & Spring, 1988). The distal and middle parts of the tubules are both secretory, but their rates of secretion are controlled differently. The short proximal section of tubule is probably resorptive (Kim & Spring, 1992). The water boatman, *Cenocorixa* (Hemiptera), has four discrete regions in its Malpighian tubules (Szibbo & Scudder, 1979).

The Malpighian tubules of *Rhodnius* and Lepidoptera and Diptera in general have no muscles other than a series of circular and longitudinal muscles proximally, while those of Coleoptera and Neuroptera have a continuous muscular sheath. In orthopteroids, Odonata and some Hymenoptera, strands of muscle spiral round the tubule (Fig. 18.1). These muscles produce writhing movements of the tubules in the hemocoel. This helps to maintain the concentration gradients across the tubule wall by continually bringing them into contact with fresh hemolymph. It, perhaps, also produces fluid movement within the tubules.

In some larval Coleoptera and Lepidoptera the distal parts of the Malpighian tubules are closely associated with the rectum, forming a convoluted layer over its surface. This is known as a cryptonephridial arrangement (section 18.4.1.2).

18.1.2 Hindgut of terrestrial insects

Amongst terrestrial insects, modification of the urine often occurs primarily in the rectum which is structurally modified for this function (section 3.1.3). The structure of the ileum is not well-known, but, in *Schistocerca*, the apical and basal plasma membranes are extensively folded with large numbers of associated mitochondria, and much fluid resorption occurs here (Irvine *et al.*, 1988; Lechleitner, Audsley & Phillips, 1989). In this species, the rectum also retains the structure typical of terrestrial insects and contributes to modification of the urine (see Fig. 18.7).

In fluid-feeding insects, where water conservation is of limited or no importance, the cells of the ileum resemble those just described, but the rectum does not possess the usual arrangement of enlarged cells forming rectal pads.

18.1.3 Sites of ion exchange in aquatic insects

Freshwater insects tend to lose ions to the environment; insects in saltwater tend to gain them. Thus, insects from either environment must compensate for these changes in the ionic composition of the hemolymph in addition to offsetting food-induced changes. They do this by modifying the urine as it passes along the hindgut and by moving ions directly out of, or into the surrounding medium.

In those groups where the hindgut has been examined in detail, either the ileum or the rectum appears to be associated with ionic uptake or secretion. In addition, all aquatic insects, whether they occur in fresh or salt water, possess structures which are known or inferred to be involved in ion exchange with the medium. They are present in the epidermis and may occur as isolated cells or complexes of a few cells, or as more extensive areas. Even cells in the gut may be involved in ion exchange with the medium if the insect drinks or takes in water via the anus. Because, at least in some cases, it is impossible to determine the source of fluid in the gut, structures concerned with modification of the primary urine and those concerned with ion exchange with the medium are considered together.

Cells that absorb ions from, or secrete them into, the surrounding medium are known as chloride cells, although they transport other ions in addition to chloride. They are characterized by a deeply folded plasma membrane, either at the apical or basal surface, and mitochondria are abundant, sometimes being closely associated with the folded plasma membrane (Fig. 18.2). The cuticle above the cell is often perforated. In the mayfly, *Coloburiscoides*, for example, it consists only of epicuticle and is less than 0.5 μm thick. A thin outer layer of the epicuticle is perforated by openings less than 9.5 nm in diameter; the pores in a thicker, inner layer are larger (Filshie & Campbell, 1984). Single chloride cells are scattered over the body surface of larval mayflies (Ephemeroptera) and stoneflies (Plecoptera) and of both larvae and adults of aquatic bugs (Heteroptera).

In some other aquatic insects, the chloride cells are grouped together in discrete regions. Larval Trichoptera have groups of chloride cells forming chloride epithelia on the dorsal surface of several abdominal segments. In mosquito larvae, the epithelia form discrete external structures, the anal papillae (Fig. 18.3). The cuticle covering these papillae is always very thin, 3.0 μm or less, and the tips of the short microvilli of the cells beneath are attached to it by hemidesmosomes. Folds of the basal plasma membrane extend halfway across the cell. There are many mitochondria, indicating that the cells are very active, many of them associated with the bases of the microvilli. The cells in the papilla appear to form a syncytium, both in mosquito and chironomid larvae (Edwards & Harrison, 1983). The overall size of these papillae varies inversely with the ionic concentration of the surrounding water, in at least some species (Fig. 18.3). Their ultrastructure also changes; the cells have fewer microvilli and mitochondria in insects living in water with higher concentrations of salts.

In addition to externally situated chloride cells, at least some of these insects also have epithelia concerned with ionic regulation lining parts of the hindgut. For example, in the bug, *Cenocorixa*, the cells of the ileum have basal and apical folds with associated mitochondria, but the rectum is unmodified. On the other hand, the rectum of freshwater mosquito larvae is lined by a continuous layer of cells in which both apical and basal membranes are deeply infolded and there are many mitochondria. The epithelium of the anterior part of the rectum also has this structure in saltwater mosquitoes, but the cells of the posterior rectum are deeper and have more extensive apical folding. Most mitochondria are associated with these folds. In both fresh and salt water insects, the rectum usually lacks the complex lateral scalariform junctions between cells that are such a characteristic feature of the rectum in terrestrial insects and which are associated with water uptake.

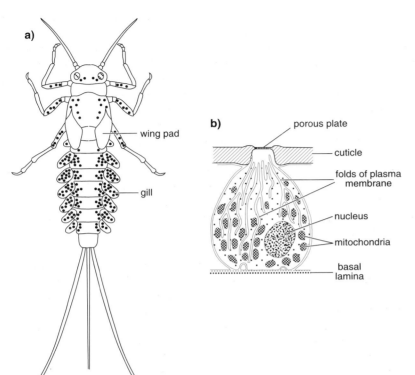

Fig. 18.2. Chloride cells in a mayfly larva (after Komnick, 1977). **(a)** Dorsal view of larva. Dots show positions of the chloride cells. Their size is greatly exaggerated. **(b)** Diagrammatic section through a chloride cell. In some chloride cells, the basal plasma membrane, rather than the apical membrane, is infolded.

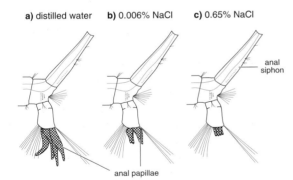

Fig. 18.3. Chloride epithelium. Posterior end of a mosquito larva showing the anal papillae (shaded), which contain the chloride epithelium. Their size is reduced in larvae reared in water containing higher salt concentrations (after Wigglesworth, 1965).

Dragonfly larvae have no external chloride cells, but comparable epithelia are associated with their rectal gills (Anisoptera) (see Fig. 17.29), and line most of the rectum in damselfly larvae (Zygoptera). In larvae of dytiscid beetles, the whole of the epithelium lining the ileum is modified, but the cells of the rectum do not have folded membranes.

Review: Komnick, 1977 – chloride cells

18.2 NITROGENOUS EXCRETION

18.2.1 Excretory products

Ammonia is the primary nitrogenous end product of protein and purine metabolism. It is highly toxic and so must either be rapidly eliminated from the body or metabolized to some less toxic compound. Because of its toxicity, its safe elimination requires that it is only present at extremely low dilutions. Consequently ammonia is excreted in quantity primarily by insects with an ample supply of water, such as those living in fresh water and others, like blowfly larvae, which live in extremely moist environments (Table 18.1, *Lucilia* larva). However, ammonia also comprises a relatively large proportion of the nitrogen excreted by *Schistocerca*, *Periplaneta* and the larva of a bruchid beetle. At least in *Schistocerca*, it is excreted as salts, probably combined with organic anions. These salts

Table 18.1. *The distribution of nitrogen in the excreta of insects expressed as a percentage of the total nitrogen in the excreta*[a]

Insect	Order	Uric acid	Allantoin	Allantoic acid	Urea	Ammonia
Terrestrial insects solid food						
Schistocerca	Orthoptera	55	–	–	+	40
Periplaneta	Blattodea	–	–	–	–	up to 90%
Melolontha (A)	Coleoptera	100	–	–	–	–
Attagenus (A)	Coleoptera	25	–	–	35	20
Pieris (L)	Lepidoptera	95	4	1	–	–
Terrestrial insects liquid food						
Rhodnius	Hemiptera	90	–	–	+	–
Dysdercus	Hemiptera	–	61	–	12	–
Pieris (A)	Lepidoptera	20	10	70	–	–
Lucilia (L)	Diptera	–	10	–	–	90
Glossina (L)	Diptera	100	–	–	+	–
Aedes (A)	Diptera	43	–	–	13	18
Freshwater insects						
Aeschna (L)	Odonata	8	–	–	–	74
Sialis (L)	Megaloptera	–	–	–	–	90

Notes:

[a] Where the total does not equal 100, other components, not shown in the table, make up the balance.

(A), adult; (L), larva; –, not present; +, present in small amounts.

are very insoluble so that ammonia is removed from solution (Harrison & Phillips, 1992). In *Periplaneta*, it is likely that bacteria in the hindgut produce at least some of the ammonia (Mullins & Cochran, 1973). These findings suggest that the production of ammonia by terrestrial insects may be more widespread than has been supposed.

For most terrestrial insects, water conservation is essential and the loss by excretion must be reduced to a minimum. Hence it has generally been argued that terrestrial insects must produce a less toxic substance than ammonia so that less water is required for its safe elimination. The substance produced by a majority of these insects is uric acid which often appears to comprise over 80% of the total nitrogenous material excreted by these insects (Table 18.1).

While there is now some question about the dominance of uric acid relative to ammonia, there is no question that its occurrence as an excretory end product of terrestrial insects is widespread. It is insoluble and relatively harmless. As a result it tends to crystallize out of solution and can be retained as a solid, non-toxic waste substance for long periods. Further, uric acid contains less hydrogen per atom of nitrogen than any other nitrogenous end product produced by animals. As hydrogen may be derived from water, this means that less water is needed for its production.

ammonia
H:N 3:1

urea
H:N 2:1

uric acid
H:N 1:1

The synthesis of uric acid occurs primarily in the fat body, although synthesis in other tissues may also occur. Its *de novo* synthesis involves the progressive addition of amino groups. It may be further metabolized to allantoin and allantoic acid.

uric acid   allantoin

allantoic acid

Allantoin is produced in small amounts by many insects, but is sometimes present in larger amounts. It is the form in which most nitrogenous waste is excreted by most species of Heteroptera that have been studied (*Rhodnius* is an exception), and by the stick insect, *Carausius*, and a beetle, *Chrysobothris*. The larvae of *Lucilia* accumulate uric acid in their tissues, but excrete it after conversion to allantoin. Allantoic acid occurs in the excreta of larval and adult Lepidoptera. It usually constitutes less than 1% of the nitrogen excreted by these insects, but may contain as much as 25% of the nitrogen in the meconium. It is also produced by larval Hymenoptera and Diptera.

The proportions of these three related compounds, uric acid, allantoin and allantoic acid may vary with the food and stage of development. For example, larvae of the grass eggar moth, *Lasiocampa*, produce mainly allantoic acid when fed on a grass, but predominantly uric acid when fed on birch. Larvae of the tortoiseshell butterfly, *Aglais*, produce mainly uric acid; this also predominates in the pupa, but, here, appreciable amounts of allantoin and allantoic acid are also formed. In the adult, allantoic acid predominates.

It is not known how uric acid is transported to the Malpighian tubules. Its insolubility probably requires that it is solubilized, perhaps by binding to a protein.

Urea, although present in the urine of a number of insects, is usually only a minor component. It is, however, the principal component of the nitrogenous excrement of a carpet beetle, *Attagenus*, and the larva of a bruchid seed beetle, *Caryedes*. The metabolic pathway by which urea is produced in insects is not known, but it is probably independent of uric acid synthesis.

Amino acids are often present in the feces of insects, but usually in small amounts that probably result largely from incomplete absorption. Under some circumstances, however, large quantities of amino acids are present. Thus in the tsetse fly, *Glossina*, most arginine and histidine from the blood of the host are excreted unchanged after absorption. These are substances with high nitrogen contents which would require a considerable expenditure of energy if they were to be metabolized along the normal pathways. *Schistocerca* actively secretes proline into the Malpighian tubules. It is subsequently reabsorbed in the rectum where it is the principal substrate for oxidative metabolism.

There is also evidence that other insects void some amino acids in large quantities when they have an abundance of nitrogen in the diet (e.g., Zanotto, Raubenheimer & Simpson, 1994), but in none of these cases is it certain that these are a result of specific elimination from the hemolymph rather than differential absorption from the midgut. Ommochromes, derived from tryptophan, are certainly excreted if high levels of tryptophan occur in the hemolymph. For example, fecal pellets of locusts are colored red by ommochromes during molting or starvation. At these times, tryptophan is probably liberated from proteins metabolized during structural rearrangement or used in energy production. Ommochromes also color the meconium of adult Lepidoptera.

Reviews: Bursell, 1967; Cochran, 1985

18.2.2 Storage excretion

Waste materials may be retained in the body in a harmless form instead of being passed out with the urine. This is known as storage or deposit excretion. Uric acid, because it is very insoluble, can be stored as a solid and is found in the fat body of many insects at some stages of development. These deposits are not always permanent. For example, in the cockroach, *Periplaneta*, uric acid accumulates in the fat body when there is ample nitrogen in the diet, but this store is depleted if the insect feeds on a diet deficient in nitrogen. It is possible that symbionts are involved in the metabolism of the uric acid. In this instance, uric acid provides a nitrogen reserve, but this is probably not common because insects without symbionts probably cannot synthesize amino acids from uric acid.

Some storage of nitrogenous end products must presumably occur in all insects during embryonic development and, in holometabolous insects, in the pupal stage when normal excretion is not possible. For example, during embryonic development of the desert locust, *Schistocerca*, the uric acid content of an egg increases progressively from about $0.5\,\mu g\,egg^{-1}$ to $25\,\mu g\,egg^{-1}$. At first, it is present in the yolk, but when the fat body develops it contains much of the uric acid (Moloo, 1973). Uric acid also builds up in the fat body of larval *Manduca* when it stops excreting it before pupation (Fig. 18.4). Then, in the days before eclosion, when morphogenesis is almost complete, the accumulation of uric acid increases sharply. Most of it is excreted with the meconium immediately following eclosion (Buckner, 1982; Levenbook, Hutchins & Bauer, 1971). Uric acid accumulates in epidermal cells of caterpillars and of *Rhodnius* during molting. In these cases it is possible that the stores simply represent the end product of metabolism of the individual cells and it is removed when the molt is completed.

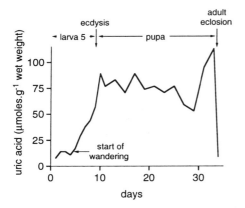

Fig. 18.4. Storage excretion. Total amounts of uric acid in the final stage larva, pupa and newly emerged adult of *Manduca*. During the feeding stage of the larva (days 1–4), uric acid is excreted. At the start of wandering it is accumulated in the body, primarily in the fat body. Immediately following eclosion, most is excreted in the meconium (based on Buckner & Caldwell, 1980; Levenbook, Hutchins & Bauer, 1971).

Sometimes uric acid and other nitrogenous compounds form permanent deposits in the epidermis and contribute to the color pattern of the insect. These can be regarded as forms of storage excretion. For example, the progressive increase in the extent of white markings in later instars of the cotton stainer bug, *Dysdercus*, results from an accumulation of uric acid (see Fig. 25.14). The white color of cabbage white butterflies, *Pieris*, is due to pterins (section 25.3.1) which represent end products of nitrogen metabolism and account for about 14% of the 'waste' nitrogen produced during the pupal period (Harmsen, 1966).

Compounds other than nitrogenous waste products may also be stored. A number of insects accumulate calcium. As with uric acid, the duration and permanence of the storage varies. Vertebrate blood contains a significant amount of calcium which *Rhodnius* stores in the cells of the upper Malpighian tubules. The accumulation increases over successive stages of development (Maddrell *et al.*, 1991). Xylem also contains relatively high concentrations of calcium and magnesium, and xylem-feeding cicadas and cercopids accumulate them in part of the tubular region of the midgut. Sometimes calcium is stored in specialized regions of the Malpighian tubules. For example, larvae of the fly, *Ephydra*, living in water containing high levels of calcium salts, store calcium in

enlarged Malpighian tubules (Herbst & Bradley, 1989) and, in the stick insect, *Carausius*, the terminal regions of the inferior tubules are expanded to store it.

Heavy metals accumulate in midgut cells of caterpillars and cercopids. Cadmium is sequestered by the posterior midgut cells of larvae of the midge, *Chironomus* (Seideman *et al.*, 1986).

18.3 URINE PRODUCTION

A fluid must be produced in the Malpighian tubules to carry excretory substances to the hindgut. This fluid is called the 'primary urine' (sometimes also tubule fluid) to differentiate it from the 'urine' that leaves the insect via the anus, having been modified on its passage through the hindgut.

Most terrestrial insects that feed on solid foods probably rarely excrete wholly liquid urine; it is usually mixed with undigested food in the rectum so the feces may be more or less fluid depending on the amount of urine they contain. Fluid feeders and aquatic insects, however, normally excrete liquid urine.

18.3.1 Formation of the primary urine

In all insects, the movement of water into the Malpighian tubules from the hemolymph depends on the active transfer of cations into the lumen of the tubule. Potassium is usually the predominant cation, but, in insects feeding on vertebrate blood which is high in sodium, sodium has a major role. An ATPase on the apical plasma membrane, activated by the mitochondria in the microvilli, pumps hydrogen ions into the lumen of the tubule. The hydrogen is then exchanged for potassium or sodium from the cell. Chloride ions follow, moving down the electrochemical gradient, and water flows through the cells (transcellularly) down the osmotic gradient created by the accumulation of ions between the microvilli. Because the wall of the tubule is highly permeable to water, this requires a difference of only a few $mosmol\,l^{-1}$ between the lumen and the hemolymph. In *Rhodnius*, and probably in other insects where the cells in the secreting regions of the tubule are all the same, all the major ions move through the principal cells, but, in *Drosophila*, it is likely that chloride passes though the stellate cells (O'Donnell *et al.*, 1996).

Once the fluid is produced within the tubule, solutes in the hemolymph also move in (Fig. 18.5). This involves passive diffusion down the concentration gradient as well

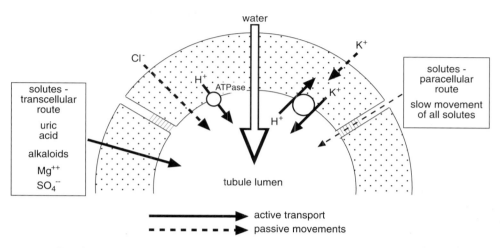

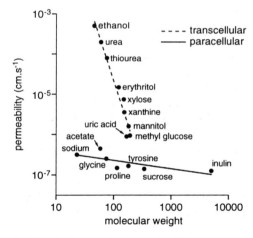

Fig. 18.5. Production of primary urine. Representation of the processes involved in the movement of water into the tubule and associated transcellular and paracellular movements of solutes (based on Maddrell & O'Donnell, 1992; Phillips, 1981).

as, in some cases, active pumping. Passive diffusion is believed to occur both between the cells (the paracellular route) and through the cells (the transcellular route). Movement between the cells is slow for two reasons: the cells are held together by septate desmosomes which greatly reduce the permeability; and the total area of the clefts between cells is only a small fraction of the total outer surface of the tubules, especially since the basal plasma membranes are deeply infolded, increasing the surface area of the cells. In *Rhodnius*, the openings of the clefts comprise less than 0.05% of the outer area of the tubule wall.

All solutes in the hemolymph will tend to diffuse into the tubules paracellularly. Large molecules, such as sucrose (molecular weight 342) and inulin (molecular weight about 5000) are unable to pass through the cell membranes unless actively transported. They will, therefore, only enter the lumen of the tubule through the intercellular gaps (Figs. 18.5, 18.6). This is also true for positively charged particles. Small uncharged molecules are able to diffuse transcellularly. Because the cell surface area is so large relative to the clefts, substances passing through the cell accumulate more rapidly in the lumen of the tubule than those passing between the cells. As an example of the effects of being positively charged, mannitol (uncharged) diffuses into the unstimulated tubule (one

Fig. 18.6. Production of primary urine. Passive movement of solutes into the unstimulated Malpighian tubules of *Rhodnius*. Charged substances, such as sodium, glycine and tyrosine, as well as larger molecules, move between the cells (paracellularly). (Note that sodium is actively transported through the cells when the Malpighian tubules are stimulated by diuretic hormone.) Uncharged substances, such as urea and mannitol, move much more quickly although they have similar molecular weights to glycine and tyrosine. They probably diffuse through the cells (transcellularly) (based on O'Donnell & Maddrell, 1983).

which is not actively secreting cations) of *Rhodnius* about 10 times faster than tyrosine (charged) which has a similar molecular weight.

Urate is probably actively pumped into the Malpighian tubules of all insects that normally excrete uric acid. In the tsetse fly, *Glossina*, the activity of the uric acid pump is proportional to the quantity of urate in the hemolymph. Subsequently uric acid may be precipitated as the urine is modified on its way to the anus. In adult *Calliphora*, uric acid is converted to allantoin by the Malpighian tubules which also actively pump it into the lumen (O'Donnell, Maddrell & Gardiner, 1983).

Inorganic ions other than potassium and sodium commonly diffuse into the urine transcellularly. In adult blowflies (*Calliphora*), the movement of most anions occurs in inverse proportion to their hydrated size. In some insects, however, some of the ions are actively transported. This is true of phosphate in *Carausius* and sulphate in *Schistocerca*.

Although the passive movement of molecules through the intercellular gaps is slow, it is important because it ensures that any foreign molecules of moderate size that enter the hemolymph are slowly removed. This provides the insect with some capacity to eliminate molecules that may be toxic and that it may never have encountered before so that specific mechanisms for their safe elimination cannot have evolved. Such a mechanism, however, also results in the loss of molecules that are important to the insect, such as hormones and nutrients. However, lipid soluble compounds, including some of the hormones, are transported through the hemolymph bound to proteins. These are too large to be lost at any significant rate. Smaller molecules of use to the insect are resorbed at some site downstream of that involved in production of the primary urine.

Some organic molecules with potential toxicity are also actively pumped into the tubules, although the data so far available are sparse. Nicotine and other alkaloids are actively transported into the urine in *Rhodnius*, and in larval *Manduca* and *Pieris*. Of these, only *Manduca* normally encounters nicotine in its diet and it must be assumed that the facility in *Rhodnius* and *Pieris* reflects the ability to excrete other compounds. The adults of *Manduca* and *Pieris* cannot excrete the alkaloids, but adults of *Calliphora* and *Musca* excrete nicotine after metabolizing it to another compound (Maddrell & Gardiner, 1976). The cardiac glycoside, ouabain, is moved passively into the Malpighian tubules of the migratory locust, *Locusta*, but active excretion is induced in the tubules of the variegated grasshopper, *Zonocerus* by previous exposure to the compound. *Locusta*, a grass feeder, would not normally encounter cardiac glycosides, but *Zonocerus* is highly polyphagous and probably does (Rafaeli-Bernstein & Mordue, 1978). Some organic anions that are the end products of metabolism of aromatic acids are also actively moved into the tubules.

As a consequence of these activities, primary urine contains all the solutes present in the hemolymph, but, except for those compounds that are actively pumped into the tubules, their concentrations are lower than in the hemolymph (Table 18.2). The fluid in the tubule is, however, isosmotic with the hemolymph (Table 18.3).

Reviews: O'Donnell & Maddrell, 1983 – transport into the tubule; Pannabecker, 1995 – physiology of Malpighian tubules

18.3.2 Modification of the primary urine
18.3.2.1 *Terrestrial insects*
The production of primary urine depends on the active movement of potassium or sodium into the Malpighian tubules followed by a passive movement of anions, primarily chloride, to restore electrical equilibrium. The insect cannot normally sustain such a high loss of ions from the hemolymph and they are recovered by resorption from the fluid. In *Schistocerca*, and perhaps in many other terrestrial insects eating solid food, this occurs in the ileum and the rectum and depends on the active resorption of chloride and sodium (Fig. 18.7). Potassium follows along the electrical potential gradient generated by the movement of chloride (see Fig. 3.26). In *Carausius*, about 95% of the sodium and 80% of the potassium in the primary urine may be resorbed. On the other hand, the rectum of adult *Pieris* shows none of the structures associated with resorption which, in this insect, occurs in the ileum (Nicolson, 1976).

In some other insects, most or all of the resorption occurs in the Malpighian tubules themselves. Absorption of potassium, chloride and water occurs in the lower, proximal regions of the tubules of *Rhodnius* although the structure of the rectum suggests that some further modification of the urine occurs there.

Useful organic compounds are also recovered from the primary urine. Glucose is actively reabsorbed from the Malpighian tubules of *Locusta* so the primary urine is modified even as it is being produced. The uptake of amino acids in the rectum may also be active, and this is certainly the case with proline in *Schistocerca*.

Table 18.2. *Composition (mM) of hemolymph and primary urine of* Schistocerca[a]

Constituent	Hemolymph	Primary urine
Na^+	103	47
K^+	12	165
Mg^{++}	12	20
Ca^{++}	9	7
Cl^-	107	88
Phosphate	6	12
Glucose	2.5	4.6
Alanine	1.0	1.0
Aspartate	0.1–0.9	0.5
Asparagine	1.0	0
Arginine	1.5	0
Glutamate	0.1–1.0	0.8
Glutamine	4	0.5
Glycine	14	4.0
Histidine	1.4	0
Isoleucine	0.4	0
Leucine	0.4	0
Lysine	1.0	0
Methionine	0.4	0
Phenylalanine	0.7	0
Proline	13	38
Serine	2–4	1.0
Threonine	0.5	0
Tyrosine	1.0	0
Valine	0.6	0

Note:

[a] Notice that potassium and proline are in much higher concentrations in the urine because they are actively secreted. Most other substances are in lower concentration in the urine; they diffuse into the tubule (after Phillips *et al.*, 1986).

The cuticle lining the rectum limits the size of molecules which can be absorbed as, in the locust, for example, it is impermeable to molecules with a radius greater than 0.6 nm. As a consequence, glucose passes readily through it, but trehalose does so only at a low rate and larger molecules are unable to do so. The effect of this is to protect the rectal cells from toxic molecules whose concentrations may increase in the rectum and to ensure that these substances are excreted.

Review: Phillips *et al.*, 1986

18.3.2.2 *Freshwater insects*

Freshwater insects tend to lose salts to the environment because most species have a highly permeable cuticle. Potassium, sodium and chloride are resorbed in the rectum, but water is not. As a result, the rectal fluid is hypotonic to the hemolymph (Table 18.3).

Salts are gained from the food, but also from the environment by the chloride cells. In the larvae of *Aedes*, *Culex* and *Chironomus*, potassium, sodium, chloride and phosphate are actively taken up (Fig. 18.8a). Larvae of *Aedes aegypti* are able to maintain their hemolymph osmotic pressure in a medium containing only $6\,\mu\mathrm{mol\,l^{-1}}$ of sodium, indicating that they are able to take up salt from extremely dilute solutions. Salts are also taken up by the rectal gills of larval Anisoptera and the rectal chloride epithelium of Zygoptera. A deficiency of inorganic ions in the environment causes larval Zygoptera to pump water in and out of the rectum, as anisopteran larvae do for respiratory purposes. The effect of this is to supply water, potentially containing ions, to the chloride epithelium in the rectum (Miller, 1994). Not all freshwater insects have the capacity to take up ions from the environment. Larval *Sialis*, for example, are unable to take up chloride.

Some freshwater insects are able to offset changes in ionic concentrations in the hemolymph with compensating changes in the non-electrolyte fraction. Probably, amino acids are produced from hemolymph proteins in sufficient quantity to maintain the osmotic pressure.

The ability of freshwater insects to regulate the composition and osmotic pressure of the hemolymph is good over the range of conditions to which they are normally subjected, but in hypertonic media regulation breaks down and the hemolymph rapidly becomes isotonic with the medium (Fig. 18.9, *Aedes aegypti*). These insects are, apparently, unable to produce a fluid in the rectum which is hypertonic to the hemolymph.

18.3.2.3 *Salt water insects*

A number of aquatic insects live in habitats in which the salinity varies widely (Fig. 18.10). *Aedes detritus*, for example, occurs in salt marshes, and the fly, *Coelopa frigida*, breeds in seaweed washed up on the sea shore. In both situations the salinity varies according to the degree of inundation and desiccation. The salinity of the salt pans in which *Ephydrella* lives varies seasonally from 0.3 to $1.3\,\mathrm{mol\,l^{-1}}$ NaCl. *Ephydra cinerea* is quite exceptional,

Table 18.3. *Comparison of osmotic pressures and ionic concentrations in hemolymph, Malpighian tubules and rectum[a]*

Species	Compartment	Osmotic pressure[b]	Na	K	Cl
Terrestrial insects					
Carausius	Hemolymph	171	11	18	87
	Tubule	171	5	145	65
	Rectum	390	8	327	—
Schistocerca	Hemolymph	214	103	12	107
	Tubule	226	47	165	88
	Rectum	433	1	22	5
Rhodnius	Hemolymph	206	174	7	155
	Tubule	228	114	104	180
	Rectum	358	161	191	—
Freshwater insect					
Aedes aegypti	Hemolymph	138	87	3	—
(larva)	Tubule	130	24	88	—
	Rectum	12	4	25	—

Notes:
[a] Notice that the osmotic pressure of the tubule fluid (primary urine) is the same as that of the hemolymph although concentrations of particular ions are different. In terrestrial insects the rectal fluid becomes hyperosmotic due to resorption of water whereas in freshwater insects it becomes hyposmotic due to resorption of salts (partly after Stobbart and Shaw, 1974; Concentrations are in $mEq l^{-1}$).

[b] Expressed as the equivalent solution of NaCl.

living in the Utah Salt Lake which has a salinity equivalent to a 20% sodium chloride solution ($3 \, mol \, l^{-1}$ NaCl).

Most of these insects regulate the ionic composition of their hemolymph so that its osmotic pressure changes very little over wide ranges of environmental salinity. This is the case in *Aedes detritus* and *Ephydra*, for example (Fig. 18.9). Others, like the caddis fly, *Limnephilus affinus*, regulate relatively poorly but can tolerate a three-fold increase in hemolymph osmotic pressure (Foster and Treherne, 1976).

Insects living in salt water gain water and salts with their diet and lose water osmotically. Ingestion of salts can be limited by selecting food of the lowest available salinity, as in *Bledius*, or by limiting drinking. *Limnephilus affinis* drinks only 3–7% of its body weight per day compared with about 50% by freshwater insects, but the rate of drinking by *Aedes taeniorhynchus* remains more or less constant irrespective of the osmotic concentration of the medium (Bradley & Phillips, 1977a).

Salt-water insects get rid of excess ions by excreting a urine which is hypertonic to the hemolymph. In salt water mosquitoes, some resorption occurs in the anterior rectum, but sodium, potassium, magnesium and chloride ions are secreted into the posterior rectum to create a hypertonic fluid (Fig. 18.8b). The osmolarity of the urine produced in the posterior rectum is proportional to that of the medium. If the insect is in a hypotonic medium, the posterior rectum is inactive and the insect produces a hypotonic urine. In this way the hemolymph osmotic pressure is regulated. At least in *Aedes campestris*, the insect adapts to the presence of different salts in the medium by excreting the appropriate ions. Thus, a larva living in water with a high concentration of sodium chloride actively secretes sodium and chloride ions into the posterior rectum; if bicarbonate is the dominant anion in the medium it is probably secreted into the rectum; if it is sulfate, this is secreted into the Malpighian tubules.

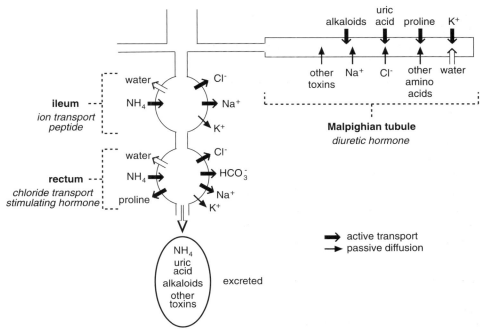

Fig. 18.7. Urine production and modification in a terrestrial insect (*Schistocerca*). Active transport of potassium into the Malpighian tubule leads to the osmotic movement of water and most other solutes follow passively. Many of the solutes are recovered as the urine moves through the hindgut, but ammonia is actively secreted into it. The hormones regulating the processes are shown in italics (partly based on Phillips & Audsley, 1995).

18.3.3 Control of diuresis

The Malpighian tubules of caterpillars appear to secrete fluid continuously, but, in most terrestrial insects, the rate of urine production varies. Many species reduce urine production and increase blood volume just before a molt and then increase urine production and rapidly reduce blood volume after cuticle expansion. Diuresis[1] is most obvious at eclosion of adult holometabolous insects. *Pieris*, for example, reduces its blood volume to less than 33% in the three hours after eclosion, and the blood volume of *Sarcophaga* is reduced to 25% in 36 hours. Insects that eat wet food also produce relatively large amounts of urine after feeding. This is most marked in blood-feeding insects. In the course of a blood meal, *Rhodnius* may ingest more than 10 times its own weight; it eliminates most of the water in the plasma, about half of the ingested volume, in the next hour.

These changes in the rates of fluid loss depend on changes in the rate of primary urine production that also necessitate changes in resorption to ensure that an increase in fluid loss does not result in excessive loss of ions or other useful compounds. An increase in the rate of primary urine production by the Malpighian tubules does not necessarily result in an increase in water loss as active recovery may occur in the lower part of the system. Individual *Schistocerca* deprived of food for one day, and so subject to some degree of water shortage, recover about 95% of the fluid from the primary urine, and, in the desert beetle, *Onymacris*, the fluid from the Malpighian tubules is directed forwards into the midgut and resorbed (Nicolson, 1991). High rates of urine flow may be maintained in these instances to remove potentially toxic materials from the hemolymph.

Production of the primary urine and subsequent resorption are regulated independently. A diuretic hormone stimulates increased fluid production by the Malpighian tubules and other hormones regulate reabsorption.

---

[1] Diuresis refers to a rapid flow of urine eliminated from the body

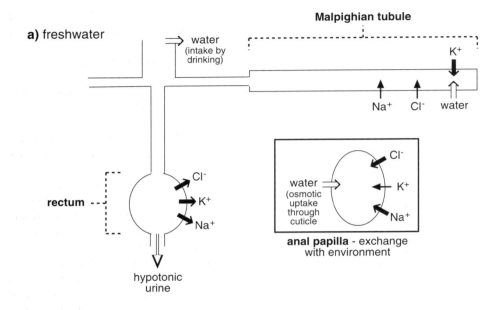

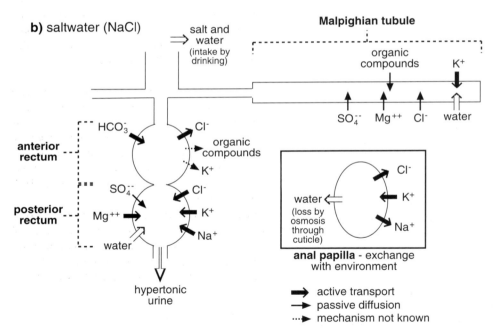

Fig. 18.8. Urine production and modification in aquatic insects (mosquito larvae). Nitrogenous excretory products not shown. (a) A freshwater insect, such as *Aedes aegypti*. The larva gains water by drinking and through the permeable cuticle of the anal papillae; excess water is removed as urine. (b) A saltwater insect, such as *Aedes campestris*. The gain of water due to drinking is greater than osmotic loss through the cuticle. Further water is lost in the urine. Note that although water is moved into the posterior rectum, the fluid produced there is hypertonic to the hemolymph and the medium (partly based on Bradley & Phillips, 1977b).

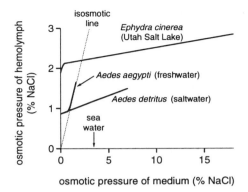

Fig. 18.9. Osmotic regulation in aquatic dipteran larvae. The freshwater species, *Aedes aegypti*, regulates only when the osmotic pressure of the medium is below its own hemolymph osmotic pressure. The two saltwater species regulate moderately well over a range of environmental osmotic pressures (after Shaw & Stobbart, 1963).

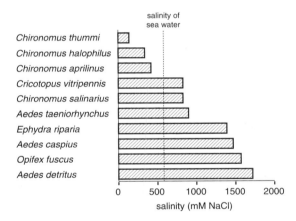

Fig. 18.10. Tolerance to salinity amongst aquatic dipterous larvae. Each bar shows the range of salinities that each species can tolerate. *Chironomus thummi* is a freshwater species included for comparison (after Foster & Treherne, 1976).

The trigger for release of the diuretic hormone in *Rhodnius* is abdominal stretching produced by the meal, and abdominal distension resulting from feeding may also be the trigger in some other insects. In *Rhodnius*, stretching results in the release of serotonin from the abdominal nerves in the hemocoel at the same time as diuretic hormone is released. The two substances act synergistically to regulate primary urine production by the Malpighian tubules (Lange, Orchard & Barrett, 1989; Maddrell *et al.*, 1991).

It is probable that secretion by the Malpighian tubules is controlled by more than one factor in other insects. In *Drosophila*, two hormones may be involved in activating the H^+-ATPase in the apical membranes of the principal cells that leads to the secretion of sodium and potassium ions into the lumen, while a third (leucokinin) increases the permeability of stellate cells to chloride (O'Donnell *et al.*, 1996). In *Acheta*, two diuretic hormones may influence secretion through different intracellular pathways in the Malpighian tubules, while an antidiuretic hormone inhibits secretion.

A variety of diuretic and antidiuretic factors have been found in the nervous system or corpora cardiaca of insects, but in most cases it is not known that they are released into the hemolymph. Until it is demonstrated that they do occur in the hemolymph at concentrations high enough to produce an effect, it cannot be concluded that they have an effect on diuresis in the intact insect (Wheeler & Coast, 1990). The role of serotonin is well-established in *Rhodnius* and larval *Aedes*, and this compound may also be important in other insects. Leucokinin-like peptides and cardioacceleratory peptides (section 21.1) are also probably diuretic hormones, while the role of vasopressin-like peptides, originally identified as diuretic hormones, has been questioned.

Resorption is regulated in most insects by one or more separate hormones. In *Schistocerca*, separate factors control resorption in the ileum and the rectum and this may also be true in other insects. An ion transport peptide from the corpora cardiaca stimulates chloride absorption in the ileum, while in the rectum this process is controlled by chloride transport stimulating hormone (Fig. 18.7) (Phillips *et al.*, 1994).

Rhodnius is unusual in that the diuretic hormone controls resorption from the lower Malpighian tubule as well as secretion into the upper part. It also regulates the movement of fluid from the blood meal into the hemocoel.

Reviews: Phillips, 1983; Phillips & Audsley, 1995; Spring, 1990

18.4 WATER REGULATION

The water content of different insects commonly ranges from about 60 to 80% of the body weight, although values as low as 40% are recorded. As these values include the cuticle, which has a relatively low water content, the water content of the living tissues is higher than these figures suggest. The degree of tolerance to water loss varies from species to species, but, ultimately, loss of water leads to death. For example, amongst the beetles, *Carabus* dies if it loses more than about 20% of its body water; the tenebrionid, *Phrynocolus* can tolerate up to 50% loss; and the rice weevil, *Sitophilus*, over 80%.

The problems faced by insects in regulating water content vary according to their habitat and so terrestrial, freshwater and salt-water insects will be considered separately.

Review: Hadley, 1994 – terrestrial insects

18.4.1 Terrestrial insects

18.4.1.1 *Water loss*

Terrestrial insects lose water by evaporation from the general body surface and the respiratory surfaces as well as in the urine. If they are to survive, these losses must be kept to a minimum and must be offset by water gained from other sources. Most water is gained from the food, but many insects drink if they are deprived of food and water is available. Metabolic water always contributes to the available water.

Water loss through the cuticle Cuticular permeability varies greatly between insects, depending on their habitat and life style. Thus the desert cicada, *Diceroprocta*, that feeds on xylem and so has ample water available to it, loses $100\,\mu g\,cm^{-2}\,mmHg^{-1}\,h^{-1}$ through the cuticle (at 30 °C in dry air), while in the beetle *Eleodes* from the same habitat the rate of cuticular loss is $17\,\mu g\,cm^{-2}\,mmHg^{-1}\,h^{-1}$, and in *Onymacris*, a beetle from the very arid Namib desert, it is $0.75\,\mu g\,cm^{-2}\,mmHg^{-1}\,h^{-1}$. Amongst grasshoppers, the rate of water loss through the cuticle varies from $15\,\mu g\,cm^{-2}\,mmHg^{-1}\,h^{-1}$ in a desert species (*Trimerotropis*) to $67\,\mu g\,cm^{-2}\,mmHg^{-1}\,h^{-1}$ in an alpine species (*Aeropedellus*). Very little water moves through the puparial cuticle of the tsetse fly, *Glossina* ($0.3\,\mu g\,cm^{-2}\,mmHg^{-1}\,h^{-1}$).

The rate of evaporation of water from insect cuticle varies with the temperature and humidity of the adjacent air (Fig. 18.11). In many species, evaporation through the cuticle shows very little increase with temperature up to a certain point but, above this, the rate of water loss increases sharply although there is probably not usually a distinct transition temperature at which the change occurs, as was once commonly believed. The increase in the rate of water loss usually occurs at temperatures well above the normal environmental temperatures that the

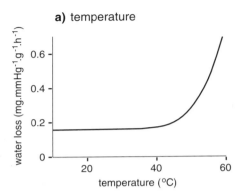

a) temperature

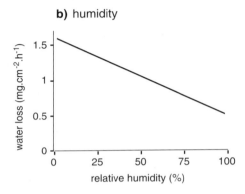

b) humidity

Fig. 18.11. Water loss through the cuticle. **(a)** Effects of temperature on water loss from *Locusta* (after Loveridge, 1968). **(b)** Effects of relative humidity on water loss from *Periplaneta* (after Appel, Reierson & Rust, 1986).

insect is likely to meet. It may be due to melting of the wax on the surface of the cuticle. The change produced by high temperature is permanent. Water loss from the cuticle is limited by high humidity of the adjacent air.

Water loss through the cuticle is restricted by the epicuticular lipids of the epicuticle and primarily by the outer layer of wax. The precise relationship between permeability and the wax is not understood. Earlier studies had suggested that a monolayer of lipid molecules was primarily responsible for waterproofing the cuticle, but there is no direct evidence for this. On the contrary, insects with more impermeable cuticles generally have greater quantities of wax per unit area than species with more permeable cuticles suggesting that the thickness of the wax is important. This is supported by evidence of changes within a species in relation to the need to resist desiccation. For example, diapause pupae of *Manduca*

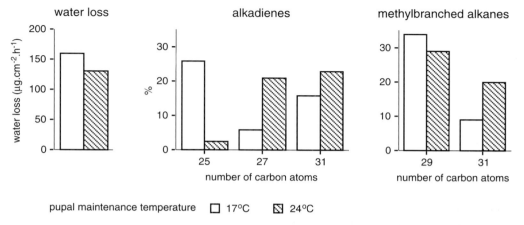

Fig. 18.12. Effects of epicuticular wax on water loss through the cuticle of adult *Drosophila*. Larvae were reared at 17 °C and pupae maintained at either 17 or 24 °C. Reduced water loss in the insects from the higher temperature is associated with higher proportions of longer chain (number of carbon atoms) components of the cuticular surface wax. Amounts of cuticular components are expressed as percentages of the total epicuticular wax. Only the major constituents are shown (data from Toolson, 1982).

have greater quantities of cuticular wax than non-diapause pupae (see Fig. 15.43) and there is three times as much wax on the surface of the desert beetle, *Eleodes* in summer, when the beetle is exposed to high temperatures, than in winter. If winter beetles are kept at high temperatures, however, they produce more wax.

The specific composition of the wax is probably also important. Lipids with longer carbon chains melt at higher temperatures than those with shorter chain lengths. In *Drosophila*, insects exposed to higher temperature during the pupal period have relatively more long chain alkadienes and methylbranched alkanes than insects from pupae kept at lower temperatures. They also lose water less rapidly (Fig. 18.12) (Toolson, 1982). Individuals of the grasshopper, *Melanoplus sanguinipes*, reared at higher temperatures have a higher proportion of *n*-alkanes to methyl-branched alkanes of similar chain length than siblings reared at lower temperatures; *n*-alkanes melt at higher temperatures. In this case, rearing temperature has no effect on the chain lengths (Gibbs & Mousseau, 1994).

The ducts of dermal glands and pore canals appear to be the main channels through which water moves through the cuticle and the rate of water loss is positively correlated with the number of dermal glands associated with the cuticle. Loss of water by this route may be hormonally regulated, perhaps by occluding and opening the channels (Machin, Smith & Lampert, 1994; Noble-Nesbitt, 1991).

Water loss from the respiratory surfaces The respiratory surfaces, being permeable, are a potential source of water loss. The loss from this source is reduced by the invagination of the respiratory surfaces as tracheae and is further limited by the spiracles. When the spiracles are open, the rate of water loss is increased (Fig. 18.18, and see Fig. 17.17). Consequently, conditions which result in more prolonged, or more frequent spiracle opening result in increased water loss. High metabolic rates due to increased temperature or activity necessitate longer overall periods of spiracle opening to permit the exchange of respiratory gases; they also lead to increased water loss (Fig. 18.13). This becomes especially important during flight when some spiracles may be continuously open. In *Schistocerca*, the rate of water loss at 30 °C and 60% relative humidity increases from $0.9 \, \text{mg} \, \text{g}^{-1} \, \text{h}^{-1}$ when the insect is at rest to $8.0 \, \text{mg} \, \text{g}^{-1} \, \text{h}^{-1}$ with the insect in tethered flight. Loss through the spiracles accounts for a large part of this increase, although not all of it. Discontinuous ventilation has been considered an adaptation to limit water loss, although not all recent studies support this thesis (section 17.4.3, and see discussion in Hadley, 1994).

Water loss in excretion Nitrogenous excretion necessitates the production of a fluid urine, and, although some resorption of water may occur as the urine moves towards the anus, there is always some water loss. The insect can

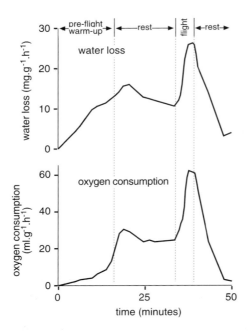

Fig. 18.13. Evaporational water loss. Changes in water loss during pre-flight warm-up and flight of a carpenter bee, *Xylocopa*. The increases are mainly due to increased loss through the spiracles as oxygen consumption increases when the insect increases its metabolic rate (after Nicolson & Louw, 1982).

control its water loss by regulating the relative rates of urine production and resorption. This is demonstrated by differences in the rates of fluid excretion occurring at different times during an insect's development (section 18.3.3). A locust feeding on wet food produces feces with a water content of up to 90%, but this is reduced to 25% when the insect feeds on dry food in a dry habitat. When experiencing some degree of water shortage, the insect probably reabsorbs as much as 95% of the water in the primary urine.

Insects with cryptonephridial systems may be able to resorb water even more effectively. Thus, the larva of *Tenebrio* can reduce the water content of its feces to 15%.

Fluid-feeding insects need to eliminate excess fluid rapidly. In discontinuous feeders, such as *Rhodnius*, this may be achieved physiologically (see above). Even here, the insect can adjust the loss to its current needs. For example, the percentage of a blood meal eliminated by *Glossina* just after feeding is reduced in insects which have previously been desiccated. In phloem- and xylem-feeding

bugs, where the intake of fluid is more or less continuous, the structure of the alimentary canal is modified so that water is shunted directly from the anterior midgut to the hindgut and does not enter the hemolymph (section 3.4.1.1).

18.4.1.2 *Gain of water*

To offset the inevitable loss by transpiration and excretion, water must be obtained from other sources. Most insects normally obtain sufficient water with their food. Locusts and some other insects adjust their food intake in relation to its water content and their own water balance (Bernays, 1990) and there is evidence that, if the moisture content of the food is very low, some insects consume more food than needed in order to extract the water from it. For example, larvae of the flour moth, *Ephestia*, and the carpet beetle, *Dermestes*, eat more food at low humidities, but it is clear that the bulk of their water is obtained as a result of the metabolism of this food rather than directly from its original water content (see Fig. 18.16). Some other insects, however, appear to maintain a relatively constant nutrient intake irrespective of its water content. Caterpillars, for example, eat more artificial diet when it is diluted with water, regulating their water content by changes in urine production (Timmins *et al.*, 1988).

Drinking Many insects drink water if they are dehydrated in the laboratory and it is assumed that they also do so under natural conditions. For example, locusts without food for 24 hours or with access only to dry food, drink if they encounter free water, but turn away from it if they have had access to moist food. Cockroaches, termites, beetles, flies, Lepidoptera and Hymenoptera are also known to drink (Edney, 1977).

Some tenebrionid beetles living in the Namib desert obtain water by drinking water that condenses from fogs. They may simply take condensate from any surface, but some species exhibit behavioral adaptations for collecting condensate. *Onymacris unguicularis* is normally diurnal, but during nocturnal fogs the insects adopt a head-down posture on the tops of sand dunes. Water condenses on the body and trickles down to the mouth, so that insects may increase their weight by as much as 12% overnight (Seely, 1976). Another species, *Lepidochora discoidalis*, constructs ridges of sand to trap moisture that it subsequently extracts (Seely and Hamilton, 1976).

Table 18.4. *Uptake of water from water vapor. Critical equivalent humidities above which some insects can absorb water*

Species	Order	Habitat	Critical equivalent humidity (%)	Temperature (°C)
Thermobia	Thysanura	Domestic	47	30
Arenivaga	Blattodea	Desert	80	30
Elipsocus	Psocoptera	Bark	76	20
Stenopsocus	Psocoptera	Leaves	76	20
Columbicola	Phthiraptera	Feathers	43	30
Xenopsylla (larva)	Siphonaptera	Rat nest	65	25
Tenebrio (larva)	Coleoptera	Stored products	88	21
Onymacris (larva)	Coleoptera	Desert	82	27

The physiological control of drinking varies in different insect species. In adult *Phormia* (Diptera) and the locust, *Locusta*, drinking is initiated by reduced blood volume, presumably measured by stretch receptors. However, in the fly, *Lucilia*, the amount of water imbibed is correlated with the concentration of chloride ions in the hemolymph. At least in *Locusta*, the termination of drinking is regulated independently of hemolymph volume, and is related to the reduction in hemolymph osmotic pressure following water intake (Bernays & Simpson, 1982).

Uptake through the cuticle All terrestrial insects passively absorb some water from the air as a result of water molecules striking the surface of the cuticle. The rate of uptake increases with humidity (reflecting the number of water molecules in the air) and the permeability of the cuticle. In most cases, the rate of uptake by this route does not balance water loss, but, in some species under some circumstances, there may be a net gain of water.

Some insects and related hexapods have special structures concerned with the absorption of water. The larva of *Epistrophe* (Diptera) can evert an anal papilla into a drop of water and absorb it. In Collembola the ventral tube (Noble-Nesbitt, 1963) and in *Campodea* (Diplura) the eversible vesicles on the abdomen have this function (Drummond, 1953).

A number of insect species actively take up water vapor from the air if they are short of water and the air humidity exceeds a critical equilibrium humidity (Table 18.4). Most of these insects are larval forms living in environments where

water is in short supply such as deserts, or environments created by vertebrates, but the phenomenon is also exhibited by a wide range of Psocoptera, including winged adults, that live in more moist environments. For most species studied, the critical equilibrium humidity is above 70% relative humidity, but Thysanura and several species of biting lice can take up water even when the humidity is less than 50%.

Water uptake in these insects either occurs in the rectum or through a modified structure on the mouthparts. In tenebrionid beetle larvae, uptake involves the close apposition of the Malpighian tubules to the rectum, a cryptonephridial arrangement, while in Thysanura and the larval rat flea, although the rectum is modified and is involved in water uptake, it is not closely associated with the Malpighian tubules. In both these cases, air enters the rectum via the open anus. Water uptake in the desert cockroach, *Arenivaga*, and in the barklice and biting lice, occurs via modifications of the hypopharynx which is extended so that the structures are exposed between the other mouthparts during periods of water uptake.

Rectal uptake is most fully understood in tenebrionid beetles. Here the distal ends of the Malpighian tubules are in close contact with the rectum enclosed by layers of flattened cells that form the perinephric sheath (Fig. 18.14). It is assumed that the sheath is impermeable, except perhaps at specific points (see below) but anteriorly it is not tightly bound to the rectal wall and hemolymph probably moves through the gap between the rectum and sheath into the perinephric space immediately surrounding the tubules. At intervals, the Malpighian tubules are swollen into flattened

Fig. 18.14. Absorption of water vapor from the rectum of a beetle larva (*Tenebrio*). (**a**) Transverse section of the cryptonephridial complex (after O'Donnell & Machin, 1991). (**b**) Diagrammatic representation of ionic and water movements. It is assumed that ions are pumped into the Malpighian tubules from the hemolymph, but not via the leptophragmata, whose function is unknown. Numbers show the osmotic pressure in Osmol kg^{-1}. Note the very high levels in the Malpighian tubule (based on Machin, 1983; O'Donnell & Machin, 1991).

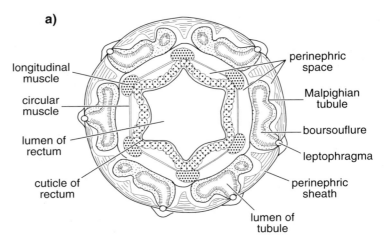

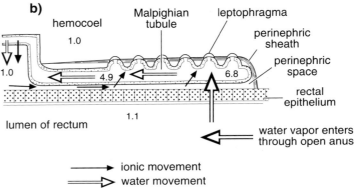

chambers, called boursouflures. These are so numerous in larval *Onymacris*, that they cover the whole surface of the rectum; in larval *Tenebrio*, they are less extensive. Each boursouflure has a cell, called a leptophragma cell, which is closely apposed to a thin region of the perinephric sheath. At these points the sheath may be permeable.

Water uptake from the rectum results from the development of a very high osmotic pressure in the Malpighian tubules in the perinephric space. Potassium, sodium and hydrogen ions are all actively transported into the tubules; chloride ions follow passively. It is believed that the solution may become supersaturated with potassium chloride. The osmotic pressure is sufficiently high that water molecules are drawn from the air in the rectum into the perinephric space and then into the tubules (Fig. 18.14b). Water and ions are presumably resorbed into the general body cavity in more proximal parts of the tubules (Machin & O'Donnell, 1991; O'Donnell & Machin, 1991).

In caterpillars, which also have a cryptonephridial system although they are not known to absorb water vapor, the Malpighian tubules pass beneath the muscle layers of the rectum and then double back on themselves to form a more convoluted outer layer. In these insects, the cryptonephridial system may be primarily important in ionic regulation although it may help to maintain the high blood volume needed for its function as a hydrostatic skeleton.

In the bark lice (Psocoptera), the hypopharynx bears a pair of bladders which are normally concealed by the labrum and labium. During periods of water uptake, however, these bladders are exposed (Fig. 18.15) and are covered with a hygroscopic fluid believed to be secreted by labial glands. At humidities above the critical equilibrium humidity, water vapor condenses on to the fluid covering the bladders and the condensate is pumped into the gut by a cibarial pump (Rudolph & Knülle, 1982). In the desert

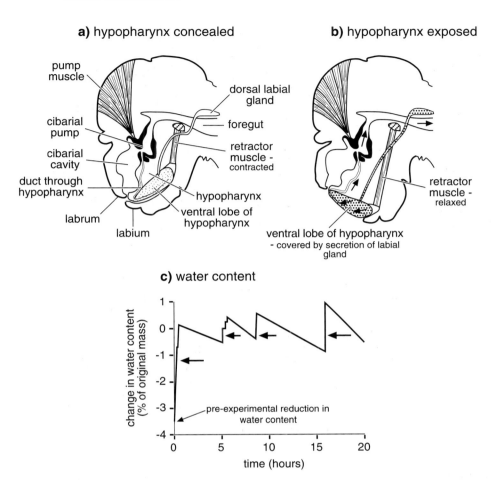

a) hypopharynx concealed

pump muscle

cibarial pump

cibarial cavity

duct through hypopharynx

labrum

labium

dorsal labial gland

foregut

retractor muscle - contracted

hypopharynx

ventral lobe of hypopharynx

b) hypopharynx exposed

retractor muscle - relaxed

ventral lobe of hypopharynx - covered by secretion of labial gland

c) water content

change in water content (% of original mass)

pre-experimental reduction in water content

time (hours)

Fig. 18.15. Absorption of water vapor via the hypopharynx. (a),(b) Diagrammatic sagittal sections through the head of a bark louse showing the hypopharynx concealed (a) and exposed (b). In the latter position the distal lobes of the hypopharynx are covered by a hygroscopic secretion (stippling in b) into which water vapor condenses. The condensed water is drawn through a duct in the hypopharynx into the gut by the cibarial pump. Arrows show the direction of water movement (after Rudolph & Knülle, 1982). (c) Regulation of body water in a biting louse. Immediately before the experiment, the louse was exposed to drying conditions which reduced its water content by 3.5%. At time 0 it was exposed to 91% relative humidity at 30 °C. The critical equilibrium relative humidity for this species is 52%. The insect absorbed water when the hypopharyngeal bladders were exposed (arrows, comparable with b above), but slowly lost water when they were concealed (comparable with a above). Notice that the insect was able to regulate within ± 1% of its normal water content (after Rudolph, 1983).

cockroach, *Arenivaga*, the cuticle of the hypopharyngeal bladders is covered by a dense mat of hydrophilic hairs. Water vapor condenses on to the hairs which swell as they take up water. Water is released from the cuticle by a solution secreted on to its surface from a gland on the epipharyngeal face of the labrum (O'Donnell, 1982).

The activities associated with water uptake, such as opening the anus, exposing the hypopharyngeal bladders,

and muscular pumping activity, only occur at humidities above the critical equilibrium humidity if the insect's water level is suboptimal. Thus the insect responds both to ambient humidity, presumably measured by humidity receptors (section 19.7), and its own water content, perhaps measured volumetrically and involving stretch receptors. As a consequence, it is able to maintain its water content within narrow limits (Fig. 18.15c).

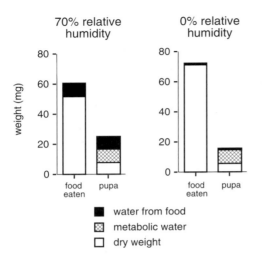

Fig. 18.16. Metabolic water provides about 50% of the water in a pupa of the flour moth, *Ephestia*, developing at 70% relative humidity. In dry air (0% relative humidity), the food contains very little water. The dry weight of the pupa is less than that of a pupa developing at 70% relative humidity despite the fact that the larva eats more. Nearly all the water in the pupa is metabolic water (data from Fraenkel & Blewett, 1944).

Metabolic water Water is an end-product of oxidative metabolism and the water so produced contributes to the water content of the insect; some species are dependent on this water for survival. The amount of metabolic water produced depends on the amount and nature of the food utilized. The complete combustion of fat leads to the production of a greater weight of water than the weight of fat from which it is derived (100 g of palmitic acid produces 112 g of water; 100 g of glycogen only 56 g of water). However, metabolism is tied to energy demands of the insect and the release of one kilocalorie leads to the production of less metabolic water if fat is oxidized to meet these demands than if carbohydrate is the substrate. Thus the insect does not gain more water from the combustion of fat unless it also expends more energy. This may be of significance during flight, but perhaps not in other circumstances.

The larvae of *Tribolium* (Coleoptera) and *Ephestia* (Lepidoptera) normally obtain much of their water from the oxidation of food, especially at low humidities. Metabolic water contributes over 90% of the body water in a pupa of *Ephestia* when the larva is reared in dry air on dry food; the food alone does not contain sufficient water to

account for most of the water in the pupa (Fig. 18.16) (Fraenkel and Blewett, 1944). In order to produce this water, these insects eat and metabolize greater quantities of food at lower humidities. Metabolic water is also of particular importance to starved insects and it enables them to survive for short periods where they would otherwise die from desiccation.

18.4.1.3 *Water balance*

Water balance is the net result of the various gains and losses experienced by the insect. A net gain is necessary for normal growth; a net loss, if sustained, will eventually lead to death. The balance will vary with the quality of food, the environmental conditions, and the physiological capabilities of the insect.

Water is normally obtained with the food or by drinking, and metabolic water is insignificant, but, when external water is not available, metabolic water may become the major source. This may occasionally be true for insects, like the locust, which are deprived of normal food (Table 18.5). In this case, however, the amount of metabolic water would not be sufficient to enable the insect to survive or develop normally. By contrast, the larva of *Ephestia* appears to metabolize excess food under dry conditions and so produces enough metabolic water for survival (Fig. 18.16). Metabolic water is also particularly important for insects that engage in long distance flights. During flight, the rate of water loss is greatly increased because of the increased airflow through the spiracles and over the surface of the body, while, at the same time, the insect is unable to replace this water from external sources. Nevertheless, the high rate of metabolism produces relatively large quantities of metabolic water and this enables the insect to sustain an appropriate water balance. For example, an aphid in flight for six hours uses 15% of its dry weight. Almost all of this will be lipid which produces more than its own weight of metabolic water. As a result, body water only declines by about 7% (Fig. 18.17) and the wet weight of the insect, expressed as a proportion of total weight, rises slightly. A consequence of this is that water loss is unlikely to be a constraint on the insect's flight; this also appears to be true for other species of insect.

In a well-hydrated insect, excretion is the major source of water loss. However, in insects that are short of water because of inadequate food, or because their habitat is very dry, transpirational losses through the cuticle and spiracles become dominant (Table 18.5).

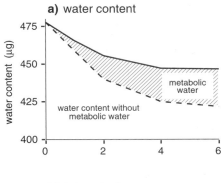

a) water content

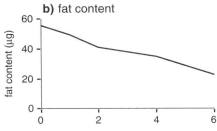

b) fat content

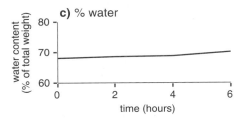

c) % water

Fig. 18.17. Metabolic water in the flight of an aphid. It is assumed that all dry weight loss is due to the combustion of fat and this provides metabolic water (data from Cockbain, 1961). **(a)** The water content falls during flight. The initial rapid fall is due to defecation. Metabolic water (hatched area) contributes significantly to the total water content after the second hour of flight when the water content is almost constant. **(b)** Decline in fat content. **(c)** The percentage water content of the insect rises slightly, despite the considerable loss of water.

The rate of water loss through the spiracles of resting insects is generally low relative to that occurring through the cuticle. Some recent studies show that over 90% of the water loss occurs through the cuticle (Fig. 18.18). However, there are exceptions. Amongst tenebrionid beetles, more than 40%, and up to 70%, of water loss may occur through the spiracles (Zachariassen, 1991). To some extent, these values vary with cuticular permeability, as, if

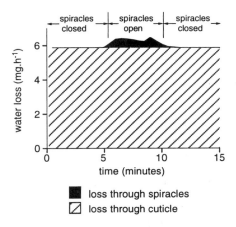

loss through spiracles

loss through cuticle

Fig. 18.18. Water loss from the cuticle and spiracles of a resting grasshopper (*Romalea*). The insect was ventilating discontinuously with the spiracles open for relatively long periods at extended intervals. At the time of opening and closing, the spiracles fluttered slightly, producing the slow rise and fall in water loss. *Romalea* lives in relatively moist environments and its cuticle may be more permeable than that of insects from drier habitats (based on Hadley & Quinlan, 1993).

the cuticle is very impermeable, a given respiratory loss must represent a greater proportion of the total water loss. The proportion of water lost through respiration also increases as the temperature rises, but still, for many insects, most water is lost through the cuticle.

Table 18.5 gives examples of the overall balance in two insects under different conditions. For the locust feeding on normal wet food, water intake greatly exceeds water loss. Balance is achieved by adjusting the rate of water loss, although, given a choice of wet and dry foods, the insect will also adjust its water intake. Metabolic water produces only an insignificant amount. However, with dry food and reduced food intake, the same amount of metabolic water provides the greater part of the water gained. Under these conditions, dry feces are produced and loss through the spiracles is minimized, but the insect still experiences a net loss of water.

Arenivaga is a desert-living cockroach. In dry air, with relatively dry food, metabolic water is the main component of water gain. However, this species has the capacity to absorb water vapor from moist air so that, at high humidities that might occur below the soil surface, most water is gained via this route. At the same time loss by transpiration through the cuticle and spiracles is reduced in moist air, so that the insect achieves a net gain of water.

Table 18.5. *Water balance sheet for a locust and* Arenivaga *under different conditions (expressed as mg water 100 mg^{-1} day^{-1}) (from Edney, 1977)*

	Locust		Arenivaga	
	Fresh food	Dry food	Dry air	88% R.H.
Gain				
Food	76.05	0.3	0.22	0.44
Metabolism	0.9	0.9	0.87	0.87
Vapor absorption	0	0	0	2.14
Total	76.95	1.2	1.09	3.45
Loss				
Feces	32.4	2.4	0.19	0.19
Cuticle	6.3	6.3	}5.43	}0.65
Spiracles	7.35	2.4		
Total	46.05	11.1	5.62	0.84
Net change	+30.9	−9.9	−4.53	+2.61

Note:

R.H., relative humidity.

Water balance in insects is also influenced by their choice of habitat. Wireworms, the larvae of the beetle, *Agriotes*, that normally live in the soil, aggregate in the wettest parts of a humidity gradient. Conversely, adult *Tenebrio* always choose the drier of two humidities even showing a slight preference for 5% over 10% relative humidities. Other species, like *Schistocerca*, show a preference for intermediate humidities, in this case between 60 and 70%.

Several species have been shown to change their responses to humidity according to their state of hydration, and this is probably a common phenomenon. For example, fully hydrated *Periplaneta* tend to turn towards a dry (40% relative humidity) airstream, but after a period of desiccation they move more readily towards a wetter airstream (Doi & Toh, 1992). Coupled with differences in the extent of locomotion, this would lead to aggregation in drier or moister regions, respectively, as has been demonstrated with the Oriental cockroach, *Blatta* and *Sitophilus* (Weston & Hoffman, 1992). The effect of such behavior will tend to minimize further water loss from insects already short of water. The response is mediated by humidity receptors on the antennae. The structure and functioning of these sensilla is described in section 19.7.

18.4.2 Aquatic insects

Freshwater insects gain water by drinking and as the hemolymph is hypertonic to the surrounding water there is a tendency for water to pass into the insect through its cuticle. Species differ in their cuticular permeability. The cuticle of adult water beetles, aquatic Heteroptera and larval *Sialis* (Megaloptera) is relatively impermeable. In these insects, the osmotic uptake of water is not excessive, about 4% of the body weight per day. The majority of aquatic larval forms, however, have highly permeable cuticles, or, at least, some areas of the cuticle are permeable. For example, osmotic uptake of water through the anal papillae of larvae of *Aedes aegypti* amounts to 30% of the body weight per day. This uptake of water is offset by the production of a copious urine. Presumably relatively little water is resorbed in the rectum (Fig. 18.8a).

Insects living in salt water are subject to osmotic loss of water because the hemolymph is often hypotonic to the medium. Water is also lost in the urine, but, as in terrestrial insects, a hypertonic urine is produced. The water lost is replaced by controlled drinking and by absorption in the midgut, which in these species is able to withstand high salt concentrations without damage, unlike that of freshwater species. The larva of *Philanisus* drinks about 25% of its own weight of sea water per day and the subsequent absorption of water in the midgut is probably linked to the salt uptake (Stobbart & Shaw, 1974). In saltwater mosquito larvae, the gain in water due to feeding and drinking is greater than that lost through the cuticle and the balance is achieved by losing water in the urine. In addition to that in the primary urine, more water is added in the posterior rectum (Fig. 18.8b).

18.5 NON-EXCRETORY FUNCTIONS OF THE MALPIGHIAN TUBULES

In a few insect species, the Malpighian tubules are modified for silk production. The tubules of larval lacewings, *Chrysopa* (Neuroptera), are thickened distally and the nuclei of the cells become branched after the second instar. These thickened regions produce the silk of the pupal cocoon. During larval life, the tubules produce a proteinaceous substance that acts as an adhesive on the anal sucker during locomotion and may, at the same time, be an excretory end product. Uric acid is stored in the fat body of these insects. Antlions (myrmeleontid larvae) also produce silk in the Malpighian tubules and store it in the rectal sac. Chrysomelid beetles produce a sticky substance in the Malpighian tubules for covering the eggs.

Amongst spittle bugs (Cercopidae, Homoptera) the proximal region of the larval tubules is enlarged, consisting of large cells without microvilli. These cells produce the material which, when mixed with air, forms the spittle within which the larvae live. Some other cercopids build rigid tubes, that of *Chaetophyes compacta* being conical and attached to the stem of the host plant. The proximal part of each Malpighian tubule in this insect is divided into two zones, one that produces the fibrils forming the basis of the tube, and another, more distal zone, that produces spittle. Fibrils pass out through the anus and are laid down by characteristic semicircular movements of the tip of the abdomen accompanied by radial pushes from the inside that push the tube into its polygonal form. Other organic materials and calcium and magnesium from the Malpighian tubules are deposited on the meshwork of fibers to form the hardened tube (Marshall, 1965).

Other species also use calcium temporarily stored in the Malpighian tubules for specific functions. For example, in the face fly, *Musca autumnalis*, and other Cyclorrhapha it contributes to hardening of the puparium (Krueger *et al.*, 1988), and, in *Carausius*, it is later taken into the hemolymph and deposited in the chorion of the eggs. In other cases, as in *Acheta* and *Calpodes*, the store probably provides a metabolic reserve (Hazelton *et al.*, 1988; Ryerse, 1979).

In the larva of the fly, *Arachnocampa luminosa*, the enlarged distal ends of the Malpighian tubules form luminous organs (section 25.7).

18.6 NEPHROCYTES

Nephrocytes are cells in the hemocoel that take up foreign chemicals of relatively high molecular weight which the Malpighian tubules may be incapable of dealing with. They take up dyes and colloidal particles but not bacteria. Some nephrocytes are usually present on the surface of the heart, when they are known as pericardial cells, others lie on the pericardial septum or the alary muscles. In larval Odonata they are scattered throughout the fat body and in the louse, *Pediculus*, form a group on either side of the esophagus, in addition. Nephrocytes of caterpillars form a sheet, one cell thick, in the hemocoel, on either side of the body. The sheet is closely associated with the fat body and extends through the thorax and abdomen (Lavenseau, Lahargue & Surleve-Bazeille, 1981). In larval Cyclorrhapha they form a conspicuous chain, sometimes called a garland, running between the salivary glands (Fig. 18.19a). In addition, larval *Calliphora* have 12–14 large pericardial cells, 140–200 μm in diameter, on each side of the posterior part of the heart, and hundreds of smaller pericardial cells, 25–60 μm in diameter, around the anterior heart.

Nephrocytes are characterized by having their plasma membrane invaginated to form a labyrinth, but, on the outside, the folds are held together by desmosome-like structures, so that the outer layer of the cell is more or less smooth (Fig. 18.19b,c) (Dallai, Riparbelli & Callaini, 1994). The cells are clothed and held together by a basal lamina. Coated vesicles arising from the invaginations become associated with an intracellular system of tubular elements. Different types of vesicles, including lysosomes lie within this zone, while a perinuclear zone contains abundant rough endoplasmic reticulum and Golgi. Nephrocytes usually have one or two nuclei.

The basal lamina and membrane junctions at the outer ends of the invaginations limit the size of molecules that can enter the labyrinth, but it is known that the nephrocytes of *Rhodnius* can absorb hemoglobin. Molecules are taken into the cell by pinocytosis and, following degradation, may be returned to the hemolymph. In addition, the

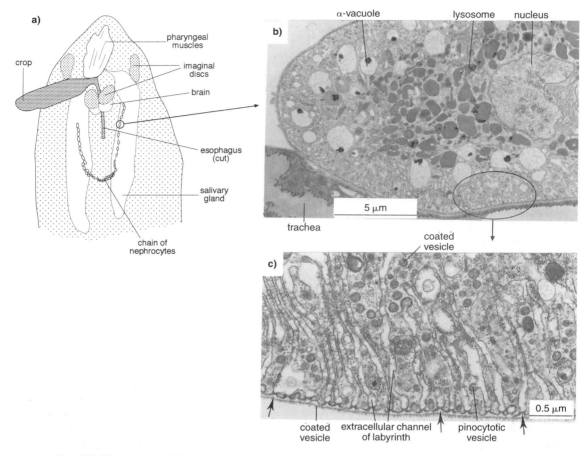

Fig. 18.19. Nephrocytes. **(a)** Dissection of the anterior part of a blowfly larva showing the chain of nephrocytes known as a garland. **(b)** Nephrocyte from the garland of *Drosophila* larva. α-vacuoles are formed from the amalgamation of pinocytotic vesicles formed in the labyrinth (after Koenig & Ikeda, 1990). **(c)** Cortical region of a nephrocyte of *Drosophila* larva showing the labyrinth. Arrows show the desmosome-like connections between adjacent projections of the cell (after Koenig & Ikeda, 1990).

nephrocytes release lysozyme which is important in resistance to disease (section 5.3.3).

Review: Crossley, 1985

18.7 DETOXIFICATION

Most plants contain chemicals that are potentially toxic to insects that feed on them. Insects usually respond to such chemicals by avoiding them or by reducing their effects metabolically. This ability is not restricted to plant-feeding insects, however. Many insects synthesize toxins which are assumed to make them unpalatable to predators. These substances are also potentially toxic to the

insect producing them and it must have the capacity to prevent autotoxicity. Parasites and predators feeding on these insects will also be exposed to the potential toxins. Moreover, as all insects have some innate capacity to metabolize insecticides, it must be supposed that insects that do not feed on food containing toxins, and do not produce their own toxins for defense (section 27.2) have acquired this ability through exposure to naturally occurring toxins which, perhaps, are able to penetrate the cuticle.

Avoidance of toxins may occur behaviorally; the insect rejects the food on the basis of its odor or taste, or as a result of rapid post-ingestive effects (Glendinning, 1996).

Table 18.6. *Induction of cytochrome P–450 by a single compound results in increased metabolism of many others*

Substrate (type of compound)	Enzyme activity		
	Without toxin in food	With toxin in food	
		Amount	% increase
Indole 3–carbinol[a,b]	0.46	1.80	391
α – Pinene (monoterpene)	3.18	16.56	521
Farnesol (sesquiterpene)	3.70	34.96	945
Phytol (diterpene)	3.70	29.97	810
β – Carotene (carotenoid)	0	1.97	>1000
Caffeine (alkaloid)	0.47	1.22	256
Sinigrin (glucosinolate)	0.30	0.70	233
Digitoxin (cardiac glycoside)	0	0.70	>500

Notes:

[a] Larvae of the fall armyworm, *Spodoptera frugiperda*, were reared on artificial diet without or with a potentially toxic compound, indole 3–carbinol. The activity of the monooxygenase system in the midguts of these insects was measured against various potential toxins. Activity is measured as mmol NADPH oxidized $min^{-1}\,mg^{-1}$ protein (data from Yu, 1987)

[b] Compound on which the insects fed.

In other insects, the food may be ingested, but the toxic chemical is not absorbed so that it never enters the insect's cells or hemolymph.

The insect's behavior can also sometimes minimize, or completely negate, the effects of ingested compounds. This is the case with some compounds, such as the furanocoumarins, that are activated by light. Some insect species that habitually feed on plants containing furanocoumarins avoid photoactivation of the compounds by constructing shelters of leaves or by feeding during periods of low radiation. Other species have heavily melanized cuticles through which light does not penetrate (Arnason, Philogene & Towers, 1992).

In a majority of insects, however, toxic chemicals that enter the tissues are excreted, often after being metabolized. The changes frequently result in the compound becoming less toxic, but the converse is sometimes true and toxicity is enhanced. Some compounds, notably those that are water soluble, are metabolized to components that are subsequently incorporated into the insect's primary metabolic pathways. This may be true of some non-protein amino acids, for example. Most lipophilic substances, however, are first converted to a water soluble product and then excreted or sequestered.

Many different enzyme systems are known to be involved in these reactions and some systems are almost certainly ubiquitous. The best known is the system of polysubstrate monooxygenases (also called mixed-function oxidases). The terminal component of this system is cytochrome P–450, so called because it absorbs light maximally at around 450 nm when complexed with carbon monoxide. Cytochrome P–450 combines with the substrate (which may be a toxin) and with molecular oxygen, catalyzing the oxidation of the substrate. This system is important because the cytochrome will combine with many different lipophilic substrates and, in addition, exists in different forms, called isoenzymes, that vary in their substrate specificity. The system is, therefore, able to metabolize a wide range of substances.

The activity of the polysubstrate monooxygenase system towards a particular substrate is usually low or may not be detectable at all if the insect has not previously encountered the substrate. Exposure of the system to an appropriate substrate, however, induces an increase in cytochrome P–450 leading to more rapid metabolism of the substrate on subsequent encounters (Fig. 18.20). Because of the broad specificity of the system, it becomes active against a range of substrates, not just that which

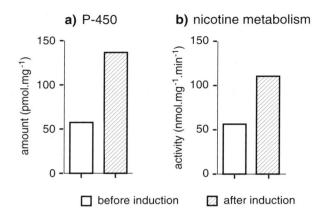

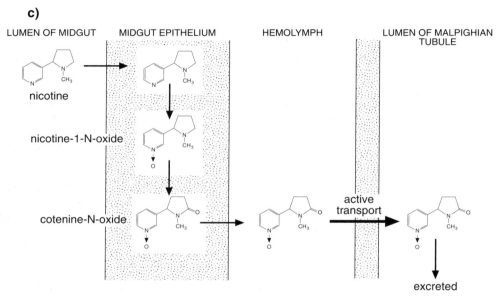

Fig. 18.20. Metabolism of a toxic compound (nicotine) by larval *Manduca* (based on Snyder, Hsu & Feyereisen, 1993; Snyder, Walding & Feyereisen, 1994). **(a)** Induction of cytochrome P-450 by feeding on artificial diet containing 0.75% nicotine. Expressed as amount of P-450 per mg of microsomal protein in the midgut. **(b)** Rate of metabolism of nicotine in relation to induction of cytochrome P-450. Expressed as nmol of product formed per mg of microsomal protein in the midgut. **(c)** Nicotine is metabolized by the P-450 enzyme to cotenine-*N*-oxide in the midgut cells. Cotenine-*N*-oxide enters the hemolymph and is actively transported into the Malpighian tubules.

caused the induction (Table 18.6). The induction is temporary, and persists only as long as an inducing agent is present. Some of the other enzyme systems involved in metabolizing toxins are also inducible.

These processes may occur in a variety of tissues as there is no organ comparable with the liver that is the focus for comparable reactions in vertebrates. Activity of the appropriate enzymes often occurs in the midgut, fat body and Malpighian tubules with the highest level of activity often in the midgut.

Different species differ widely in their ability to metabolize toxic compounds. Amongst plant-feeding insects, this variation contributes to hostplant specificity. The caterpillar of *Manduca sexta*, for example, habitually feeds on alkaloid-containing plants, including tobacco, and it is able to do this because it detoxifies the alkaloids. Nicotine from tobacco is taken into the midgut cells where most of it is metabolized to cotinine-*N*-oxide by a microsomal oxidase (Fig. 18.20c). The cotinine-*N*-oxide enters the hemolymph and passes into the urine in the Malpighian tubules which have a non-specific alkaloid pump. Any nicotine that enters the hemolymph is also rapidly oxidized (Snyder, Walding & Feyereisen, 1994).

Although the end products of these metabolic processes are commonly excreted, they are sometimes sequestered. Sequestration may also occur without any prior metabolism. In some species, the compound is stored in the cuticle, perhaps minimizing the risk to the insect itself, but other species store the defensive substances in glands or in the hemolymph. There is little understanding of how these insects avoid autotoxicity, but in the milkweed bug, *Oncopeltus*, and the monarch butterfly, *Danaus*, that sequester cardiac glycosides, the sodium/potassium pumps responsible for maintaining the ionic environment of the nervous system are relatively insensitive to ouabain (a cardiac glycoside). The pumps of insects which do not sequester cardiac glycosides are much more sensitive (section 27.2.5) (Moore & Scudder, 1986).

Reviews: Agosin, 1985 – monooxygenases; Brattsten, 1992 – metabolic defenses; Brattsten & Ahmad, 1986 – general; Hodgson, 1985 – monooxygenases; Lindroth, 1991 – detoxification systems

REFERENCES

Agosin, M. (1985). Role of microsomal oxidations in insecticide degradation. In *Comprehensive Insect Physiology, Biochemistry and Pharmacology*, vol. 12, ed. G.A. Kerkut & L.I. Gilbert, pp. 647–712. Oxford: Pergamon Press.

Appel, A.G., Reierson, D.A. & Rust, M.K. (1986). Cuticular water loss in the smokybrown cockroach, *Periplaneta fuliginosa*. *Journal of Insect Physiology*, **32**, 623–8.

Arnason, J.T., Philogene, J.R. & Towers, G.H.N. (1992). Phototoxins in plant–insect interactions. In *Herbivores. Their Interactions with Secondary Plant Metabolites*, vol. 2, ed. G.A. Rosenthal & M.R. Berenbaum, pp. 317–41. San Diego: Academic Press.

Bernays, E.A. (1990). Water regulation. In *Biology of Grasshoppers*, ed. R.F. Chapman & A. Joern, pp. 129–41. New York: Wiley.

Bernays, E.A. & Simpson, S.J. (1982). Control of food intake. *Advances in Insect Physiology*, **16**, 59–118.

Bradley, T.J. (1985). The excretory system: structure and physiology. In *Comprehensive Insect Physiology, Biochemistry and Pharmacology*, vol. 4, ed. G.A. Kerkut & L.I. Gilbert, pp. 421–65. Oxford: Pergamon Press.

Bradley, T.J. and Phillips, J.E. (1977a). The effect of external salinity on drinking rate and rectal secretion in the larvae of the saline-water mosquito *Aedes taeniorhynchus*. *Journal of Experimental Biology*, **66**, 97–110.

Bradley, T.J. and Phillips, J.E. (1977b). The location and mechanism of hyperosmotic fluid secretion in the rectum of the saline-water mosquito larva *Aedes taeniorhynchus*. *Journal of Experimental Biology*, **66**, 111–26.

Brattsten, L.B. (1992). Metabolic defenses against plant allelochemicals. In *Herbivores: Their Interactions with Secondary Plant Metabolites*, vol. 2, ed. G.A. Rosenthal & M.R. Berenbaum, pp. 176–242. San Diego: Academic Press.

Brattsten, L.B. & Ahmad, S. (eds.) (1986). *Molecular Aspects of Insect-Plant Associations*. New York: Plenum Press.

Buckner, J.S. (1982). Hormonal control of uric acid storage in the fat body during last-larval instar of *Manduca sexta*. *Journal of Insect Physiology*, **28**, 987–93.

Buckner, J.S. & Caldwell, J.M. (1980). Uric acid levels during last-larval instar of *Manduca sexta*, an abrupt transition from excretion to storage in fat body. *Journal of Insect Physiology*, **26**, 27–32.

Bursell, E. (1967). The excretion of nitrogen in insects. *Advances in Insect Physiology*, **4**, 33–67.

Cochran, D.G. (1985). Nitrogenous excretion. In *Comprehensive Insect Physiology, Biochemistry and Pharmacology*, vol. 4, ed. G.A. Kerkut & L.I. Gilbert, pp. 467–506. Oxford: Pergamon Press.

Cockbain, A.J. (1961). Water relationships of *Aphis fabae* Scop. during tethered flight. *Journal of Experimental Biology*, **38**, 175–80.

Crossley, A.C. (1985). Nephrocytes. In *Comprehensive Insect Physiology, Biochemistry and Pharmacology*, vol. 3, ed. G.A. Kerkut & L.I. Gilbert, pp. 487–515. Oxford: Pergamon Press.

Dallai, R., Riparbelli, M.G. & Callaini, G. (1994). The cytoskeleton of the ventral nephrocytes in *Ceratitis capitata* larva. *Cell & Tissue Research*, **275**, 529–36.

Doi, N. & Toh, Y. (1992). Modification of cockroach behavior to environmental humidity change by dehydration (Dictyoptera: Blattidae). *Journal of Insect Behavior*, **5**, 479–90.

Drummond, F.H. (1953). The eversible vesicles of *Campodea* (Thysanura). *Proceedings of the Royal Entomological Society of London* A, **28**, 145–8.

Edney, E.B. (1977). *Water Balance in Land Arthropods*. Berlin: Springer-Verlag.

Edwards, H.A. & Harrison, J.B. (1983). An osmoregulatory syncytium and associated cells in a freshwater mosquito. *Tissue & Cell*, 15, 271–80.

Filshie, B.K. & Campbell, I.C. (1984). Design of an insect cuticle associated with osmoregulation: the porous plates of chloride cells in a mayfly larva. *Tissue & Cell*, 16, 789–803.

Foster, W.A. & Treherne, J.E. (1976). Insects of marine saltmarshes: problems and adaptations. In *Marine Insects*, ed. L. Cheng, pp. 5–42. Amsterdam: North-Holland Publishing Co.

Fraenkel, G. & Blewett, M. (1944). The utilisation of metabolic water in insects. *Bulletin of Entomological Research*, 35, 127–39.

Gibbs, A. & Mousseau, T.A. (1994). Thermal acclimation and genetic variation in cuticular lipids of the lesser migratory grasshopper (*Melanoplus sanguinipes*): effects of lipid composition on biophysical properties. *Physiological Zoology*, 67, 1523–43.

Glendinning, J.G. (1996). Is chemosensory input essential for the rapid rejection of toxic foods? *Journal of Experimental Biology*, 199, 1523–34.

Hadley, N.F. (1994). *Water Relations of Terrestrial Arthropods*. San Diego: Academic Press.

Hadley, N.F. & Quinlan, M.C. (1993). Discontinuous carbon dioxide release in the Eastern lubber grasshopper *Romalea guttata* and its effect on respiratory transpiration. *Journal of Experimental Biology*, 177, 169–80.

Harmsen, R. (1966). The excretory role of pteridines in insects. *Journal of Experimental Biology*, 45, 1–13.

Harrison, J.F. & Phillips, J.E. (1992). Recovery from acute hemolymph acidosis in unfed locusts II. Role of ammonium and titratable acid excretion. *Journal of Experimental Biology*, 165, 97–110.

Hazelton, S.R., Parker, S.W. & Spring, J.H. (1988). Excretion in the house cricket (*Acheta domesticus*): fine structure of the Malpighian tubules. *Tissue & Cell*, 20, 443–60.

Herbst, D.B. & Bradley, T.J. (1989). A Malpighian tubule lime gland in an insect inhabiting alkaline salt lakes. *Journal of Experimental Biology*, 145, 63–78.

Hodgson, E. (1985). Microsomal monooxygenases. In *Comprehensive Insect Physiology, Biochemistry and Pharmacology*, vol. 11, ed. G.A. Kerkut & L.I. Gilbert, pp. 225–321. Oxford: Pergamon Press.

Irvine, B., Audsley, N., Lechleitner, R., Meredith, J., Thomson, B. & Phillips, J. (1988). Transport properties of locust ileum *in vitro*: effects of cyclic AMP. *Journal of Experimental Biology*, 137, 361–85.

Kim, I.S. & Spring, J.H. (1992). Excretion in the house cricket (*Acheta domesticus*): relative contribution of distal and midtubule to diuresis. *Journal of Insect Physiology*, 38, 373–81.

Koenig, J.H. & Ikeda, K. (1990). Transformational process of the endosomal compartment in nephrocytes of *Drosophila melanogaster*. *Cell and Tissue Research*, 262, 233–44.

Komnick, H. (1977). Chloride cells and chloride epithelia of aquatic insects. *International Review of Cytology*, 49, 285–329.

Krueger, R.A., Broce, A.B., Hopkins, T.L. & Kramer, K.J. (1988). Calcium transport from Malpighian tubules to puparial cuticle of *Musca autumnalis*. *Journal of Comparative Physiology*, 158B, 413–9.

Lange, A.B., Orchard, I. & Barrett, F.M. (1989). Changes in hemolymph serotonin levels associated with feeding in the blood-sucking bug, *Rhodnius prolixus*. *Journal of Insect Physiology*, 35, 393–9.

Lavenseau, L., Lahargue, J. & Surleve-Bazeille, J.-E. (1981). Ultrastructure of the 'organe rameux' of the silkworm *Bombyx mori*. *International Journal of Insect Morphology & Embryology*, 10, 235–45.

Lechleitner, R.A., Audsley, N. & Phillips, J.E. (1989). Antidiuretic action of cyclic AMP, corpus cardiacum and ventral ganglia across the locust ileum in vitro. *Canadian Journal of Zoology*, 67, 2655–61.

Levenbook, L., Hutchins, R.F.N. & Bauer, A.C. (1971). Uric acid and basic amino acids during metamorphosis of the tobacco hornworm, *Manduca sexta*, with special reference to the meconium. *Journal of Insect Physiology*, 17, 1321–31.

Lindroth, R.L. (1991). Differential toxicity of plant allelochemicals to insects: roles of enzymatic detoxication systems. In *Insect–Plant Interactions* vol. III, ed. E. Bernays, pp. 1–33. Boca Raton: CRC Press.

Loveridge, J.P. (1968). The control of water loss in *Locusta migratoria migratorioides* R. & F. I. Cuticular water loss. *Journal of Experimental Biology*, 49, 1–13.

Machin, J. (1983). Water vapor absorption in insects. *American Journal of Physiology*, 244, R187–92.

Machin, J. & O'Donnell, M.J. (1991). Rectal complex ion activities and electrochemical gradients in larvae of the desert beetle, *Onymacris*: comparisons with *Tenebrio*. *Journal of Insect Physiology*, 37, 829–38.

Machin, J., Smith, J.J.B. & Lampert, G.J. (1994). Evidence for hydration-dependent closing of pore structures in the cuticle of *Periplaneta americana*. *Journal of Experimental Biology*, 192, 83–94.

Maddrell, S.H.P. & Gardiner, B.O.C. (1976). Excretion of alkaloids by Malpighian tubules of insects. *Journal of Experimental Biology*, 64, 267–81.

Maddrell, S.H.P., Herman, W.S., Mooney, R.L. & Overton, J.A. (1991). 5-hydroxytryptamine: a second diuretic hormone in *Rhodnius prolixus*. *Journal of Experimental Biology*, 156, 557–66.

Maddrell, S.H.P. & O'Donnell, M.J. (1992). Insect Malpighian tubules: V-ATPase action in ion and fluid transport. *Journal of Experimental Biology*, 172, 417–29.

Marshall, A.T. (1965). Spittle-production and tube-building by cercopoid nymphs (Homoptera). 3. The cytology and function of the fibril zone of the Malpighian tubules of tube-dwelling nymphs. *Quarterly Journal of Microscopical Science*, **106**, 37–44.

Miller, P.L. (1994). The responses of rectal pumping in some zygopteran larvae (Odonata) to oxygen and ion availability. *Journal of Insect Physiology*, **40**, 333–9.

Moloo, S.K. (1973). Accumulation and storage sites of uric acid in the developing egg of *Schistocerca gregaria* (Forskål) (Orthoptera: Acrididae). *Journal of Entomology* A, **48**, 85–8.

Moore, L.V. & Scudder, G.G.E. (1986). Ouabain-resistant Na,K-ATPases and cardenolide tolerance of the large milkweed bug, *Oncopeltus fasciatus*. *Journal of Insect Physiology*, **32**, 27–33.

Mullins, D.E. and Cochran, D.G. (1973). Nitrogenous excretory materials from the American cockroach. *Journal of Insect Physiology*, **19**, 1007–18.

Nicolson, S.W. (1976). Diuresis in the cabbage white butterfly, *Pieris brassicae*: water and ion regulation and the role of the hindgut. *Journal of Insect Physiology*, **22**, 1623–30.

Nicolson, S.W. (1991). Diuresis or clearance: is there a physiological role for the 'diuretic hormone' of the desert beetle *Onymacris*? *Journal of Insect Physiology*, **37**, 447–52.

Nicolson, S.W. & Louw, G.N. (1982). Simultaneous measurement of evaporative water loss, oxygen consumption, and thoracic temperature during flight in a carpenter bee. *Journal of Experimental Zoology*, **222**, 287–96.

Noble-Nesbitt, J. (1963). A site of water and ionic exchange with the medium in *Podura aquatica* L. (Collembola: Isotomidae). *Journal of Experimental Biology*, **40**, 701–11.

Noble-Nesbitt, J. (1991). Cuticular permeability and its control. In *Physiology of the Insect Epidermis*, ed. K. Binnington & A. Retnakaran, pp. 252–83. Melbourne: CSIRO.

O'Donnell, M.J. (1982). Hydrophilic cuticle – the basis for water vapor absorption by the desert burrowing cockroach, *Arenivaga investigata*. *Journal of Experimental Biology*, **99**, 43–60.

O'Donnell, M.J., Dow, J.A.T., Huesmann, G.R., Tublitz, N.J. & Maddrell, S.H.P. (1996). Separate control of anion and cation transport in Malpighian tubules of *Drosophila melanogaster*. *Journal of Experimental Biology*, **199**, 1163–75.

O'Donnell, M.J. & Machin, J. (1991). Ion activities and electrochemical gradients in the mealworm rectal complex. *Journal of Experimental Biology*, **155**, 375–402.

O'Donnell, M.J. & Maddrell, S.H.P. (1983). Paracellular and transcellular routes for water and solute movements across insect epithelia. *Journal of Experimental Biology*, **106**, 231–53.

O'Donnell, M.J. & Maddrell, S.H.P. (1995). Fluid reabsorption and ion transport by the lower Malpighian tubules of adult female *Drosophila*. *Journal of Experimental Biology*, **198**, 1647–53.

O'Donnell, M.J., Maddrell, S.H.P. & Gardiner, B.O.C. (1983). Transport of uric acid by the Malpighian tubules of *Rhodnius prolixus* and other insects. *Journal of Experimental Biology*, **103**, 169–84.

O'Donnell, M.J., Maddrell, S.H.P., Skaer, H.le B. & Harrison, J.B. (1985). Elaborations of the basal surface of cells of the Malpighian tubules of an insect. *Tissue & Cell*, **17**, 865–81.

Pannabecker, T. (1995). Physiology of the Malpighian tubule. *Annual Review of Entomology*, **40**, 493–510.

Phillips, J.E. (1981). Comparative physiology of insect renal function. *American Journal of Physiology*, **241**, R241–57.

Phillips, J.E. (1983). Endocrine control of salt and water balance: excretion. In *Endocrinology of Insects*, ed. R.G.H. Downer & H. Laufer, pp. 411–25. New York: Liss.

Phillips, J.E. & Audsley, N. (1995). Neuropeptide control of ion and fluid transport across locust hindgut. *American Zoologist*, **35**, 503–14.

Phillips, J.E., Hanrahan, J., Chamberlin, M. & Thomson. B. (1986). Mechanisms and control of reabsorption in insect hindgut. *Advances in Insect Physiology*, **19**, 329–422.

Phillips, J.E., Thomson. R.B., Audsley, N., Peach, J.L. & Stagg, A.P. (1994). Mechanisms of acid-base transport and control in locust excretory system. *Physiological Zoology*, **67**, 95–119.

Rafaeli-Bernstein, A. & Mordue, W. (1978). The transport of the cardiac glycoside ouabain by the Malpighian tubules of *Zonocerus variegatus*. *Physiological Entomology*, **3**, 59–63.

Razet, P. (1976). Les organes excréteurs et l'excrétion. In *Traité de Zoologie*, Vol. 8, part 4, ed. P.–P.Grassé, pp. 517–635. Paris: Masson et Cie.

Rudolph, D. (1983). The water-vapor uptake system of the Phthiraptera. *Journal of Insect Physiology*, **29**, 15–25.

Rudolph, D. & Knülle, W. (1982). Novel uptake systems for atmospheric water vapor among insects. *Journal of Experimental Zoology*, **222**, 321–33.

Ryerse, J.S. (1979). Developmental changes in Malpighian tubule structure. *Tissue & Cell*, **11**, 533–51.

Seely, M.K. (1976). Fog basking by the Namib Desert beetle, *Onymacris unguicularis*. *Nature, London*, **262**, 284–5.

Seely, M.K. & Hamilton, W.J. (1976). Fog catchment sand trenches constructed by tenebrionid beetles, *Lepidochora*, from the Namib Desert. *Science*, **193**, 484–6.

Seideman, L.A., Bergtrom, G., Gingrich, D.J. & Remsen, C.C. (1986). Accumulation of cadmium by the fourth instar larvae of the fly *Chironomus thummi*. *Tissue & Cell*, **18**, 394–405.

Shaw, J. & Stobbart, R.H. (1963). Osmotic and ionic regulation in insects. *Advances in Insect Physiology*, **1**, 315–99.

Snyder, N.J., Hsu, E.-L. & Feyereisen, R. (1993). Induction of cytochrome P–450 activities by nicotine in the tobacco hornworm, *Manduca sexta*. *Journal of Chemical Ecology*, **19**, 2903–16.

Snyder, N.J., Walding, J.K. & Feyereisen, R. (1994). Metabolic fate of the allelochemical nicotine in the tobacco hornworm, *Manduca sexta*. *Insect Biochemistry and Molecular Biology*, **24**, 837–46.

Spring, J.H. (1990). Endocrine regulation of diuresis in insects. *Journal of Insect Physiology*, **36**, 13–22.

Stobbart, R.H. & Shaw, J. (1974). Salt and water balance; excretion. In *The Physiology of Insecta*, vol.5, ed. M. Rockstein, pp. 361–446. New York: Academic Press.

Szibbo, C.M. & Scudder, G.G.E. (1979). Secretory activity of the segmented Malpighian tubules of *Cenocorixa bifida* (Hung.) (Hemiptera, Corixidae). *Journal of Insect Physiology*, **25**, 931–7.

Timmins, W.A., Bellward, K., Stamp, A.J. & Reynolds, S.E. (1988). Food intake, conversion efficiency, and feeding behaviour of tobacco hornworm caterpillars given artificial diet of varying nutrient and water content. *Physiological Entomology*, **13**, 303–14.

Toolson, E.C. (1982). Effects of rearing temperature on cuticle permeability and epicuticular lipid composition in *Drosophila pseudoobscura*. *Journal of Experimental Zoology*, **222**, 249–53.

Weston, P.A. & Hoffman, S.A. (1992). Influence of starvation, dehydration, and humidity differential on humidity responses of *Sitophilus zeamais* (Coleoptera: Curculionidae). *Environmental Entomology*, **21**, 1345–50.

Wheeler, C.H. & Coast, G.M. (1990). Assay and characterisation of diuretic factors in insects. *Journal of Insect Physiology*, **36**, 23–34.

Wigglesworth, V.B. (1965). *The Principles of Insect Physiology*. London: Methuen.

Yu, S.J. (1987). Microsomal oxidation of allelochemicals in generalist (*Spodoptera frugiperda*) and semi-specialist (*Anticarsia gemmatalis*) insect. *Journal of Chemical Ecology*, **13**, 423–36.

Zachariassen, K.E. (1991). Routes of transpiratory water loss in a dry-habitat tenebrionid beetle. *Journal of Experimental Biology*, **157**, 425–37.

Zanotto, F.P., Raubenheimer, D. & Simpson, S.J. (1994). Selective egestion of lysine by locusts fed nutritionally unbalanced foods. *Journal of Insect Physiology*, **40**, 259–65.

19 Thermal relations

Insect systems function optimally within a limited range of temperatures. For a majority of insects, enzyme activity, tissue functioning and the behavior of the whole insect is optimal at a relatively high temperature, often in the range 30–40 °C (see Figs. 3.22, 10.17b). This chapter considers the factors that determine an insect's temperature, how body temperature may be regulated, and how insects cope with extremes of temperature.

DEFINITIONS

Ectothermal – body temperature depends on heat acquired from the environment.

Endothermal – body temperature depends on heat produced by the animal's own metabolism.

Heterothermal – body temperature is sometimes determined by the organism's metabolism (endothermy); at other times it is governed by the environment (ectothermy).

Poikilotherm – an animal whose body temperature is variable and dependent on ambient.

Homeotherm – an animal whose body temperature is constant as a result of endothermy.

19.1 BODY TEMPERATURE

19.1.1 Heat gain

19.1.1.1 *Internal heat production*

The body temperature of an insect is always a reflection of ambient conditions coupled with any heat that may be produced by metabolic activity. Because the mechanical efficiency of muscles is very low, any muscular activity produces heat, but, in insects, because of the small size of the muscles and the high rate of heat loss from the organism, the effects of muscular activity on body temperature are usually insignificant. The flight muscles, however, are relatively large and oscillate at high frequencies when generating the power needed for flight. Consequently,

their activity produces a significant amount of heat and the thoracic temperatures even of quite small insects are elevated above ambient during flight. The extent to which thoracic temperature is elevated depends on body size because smaller insects produce less heat and lose it more rapidly (Fig. 19.1a), but the relationship between size and

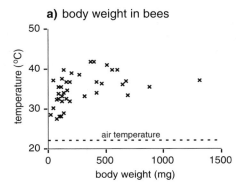

a) body weight in bees

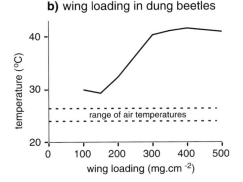

b) wing loading in dung beetles

Fig. 19.1. Elevated thoracic temperatures during flight due to heat production by the flight muscles. (a) Effects of size. Smaller insects have lower thoracic temperatures. Data for bees (Apoidea) flying at an air temperature of 22 °C. Each point relates to a different species (after Stone & Willmer, 1989). (b) Effects of wing loading. Higher wing loading results in higher thoracic temperatures. Data for five species of dung beetle flying at air temperatures of 24–26 °C (based on Bartholomew & Heinrich, 1978).

temperature is not a simple one as other factors also affect rates of heat loss (Stone & Willmer, 1989). For example, insects with high wing loading, in general, have higher wingbeat frequencies (see Fig. 9.22). This results in more heat production and higher body temperatures (Fig. 19.1b).

The flight muscles may be used to generate heat even when the insect is not flying. Because power output for flight for most species is only achieved at relatively high muscle temperatures, usually above 30 °C, flight at lower ambient temperatures is preceded by a period of warming up. Warm-up behavior is especially common in night-flying insects and those day-flying insects, such as bumble-bees, that fly early in the year in temperate climates where the ambient temperature, even during the day, is often too low for flight to occur without warm-up. Similar behavior occurs in some beetles, dragonflies and, occasionally, in locusts. Such endothermic warm-up occurs even in some quite small insects. For example it has been recorded in a bee, *Lasioglossum*, weighing only 10 mg, and a hover fly, *Syrphus*, weighing 20 mg. However, not all insects exhibit such behavior even though they may be theoretically capable of doing so. For example, horse flies appear to rely wholly on external sources of heat to produce the body temperatures necessary for flight (Schutz & Gaugler, 1992) and most Lepidoptera do not exhibit preflight warm-up.

During warm-up, the depressor and levator muscles of the wings contract together, instead of in antiphase as they do in flight. This difference results from a change in the pattern of firing of the motor neurons. In the Lepidoptera, with synchronous flight muscles, the motor neurons fire in phase instead of in antiphase (Fig. 19.2a). Diptera and Hymenoptera, however, have asynchronous flight muscles in which muscle contractions are not determined by the rate of firing of the motor neurons, but depend on stretch activation of one set of muscles by its antagonists (section 10.5.2). During warm-up, however, the muscles function as synchronous muscles. At least in the bees, *Apis* and *Bombus*, the rates of stimulation to the two sets of muscles are not exactly the same so they tend to contract irregularly; the uncoordinated twitches that result are too slow to produce stretch activation. At higher rates of stimulation the muscles contract tetanically (Esch & Goller, 1991). As the thorax warms up, the motor neurons fire more rapidly (Fig. 19.2b). This results in a progressive increase in the rate of heat production but the rate at which the thorax heats up does not match the rate of heat production

because heat is lost more quickly as the difference between thoracic and air temperatures increases.

Because the antagonistic muscles contract almost in phase with each other, wing movements are very small and this warm-up behavior is described as 'shivering'. In Hymenoptera and Diptera, the wing articulations appear to become uncoupled so that no wing movements occur at all.

In many insects that exhibit warm-up behavior, the thorax is insulated from the abdomen by airsacs or, in Hymenoptera, by the waist (Fig. 19.3a). As a result, heat is not readily transferred to the abdomen except in the hemolymph. The heart loops between the dorsal longitudinal flight muscles and opens in the head. During its passage through the muscles the hemolymph heats up so that hot hemolymph is pumped into the head and forced back into the abdomen. At the junction between thorax and abdomen (or propodeum and gaster in Hymenoptera), this backwardly flowing hemolymph surrounds the aorta in which cooler hemolymph from the abdomen is flowing forwards. This system functions as a heat exchanger, and heat from the thoracic hemolymph warms the abdominal hemolymph (Fig. 19.3b). In *Apis*, the heart forms a spiral where it passes through the waist. This increases the time and surface area for temperature exchange and so increases the efficiency of the exchanger. Heat is conserved in the thorax by its insulation and the heat exchanger, and the temperature of the abdomen (gaster in Hymenoptera) is usually much lower than that of the thorax (Fig. 19.3c).

Some insects also use the flight muscles to warm up for activities other than flight. The katydid, *Neoconocephalus*, begins singing soon after dark and will sing at air temperatures as low as 17 °C. The sound is produced by the activity of the wing muscles (section 21.1.2.1). At low temperatures, shivering raises the thoracic temperature to 30 °C or above before singing starts. During shivering the muscles contract in synchrony, but during singing they act in antiphase. Some cicadas that sing at night also warm up, presumably using the flight muscles to do so (Sanborn *et al.*, 1995). In this case, the muscles producing the heat are separate from the tymbal muscles that produce the sound (section 26.1.3).

Scarab beetles that make dung balls are able to raise their body temperatures when walking or rolling the dung. However, these activities sometimes occur without an increase in body temperature so the muscles used in walking do not produce the temperature increase. Again, it

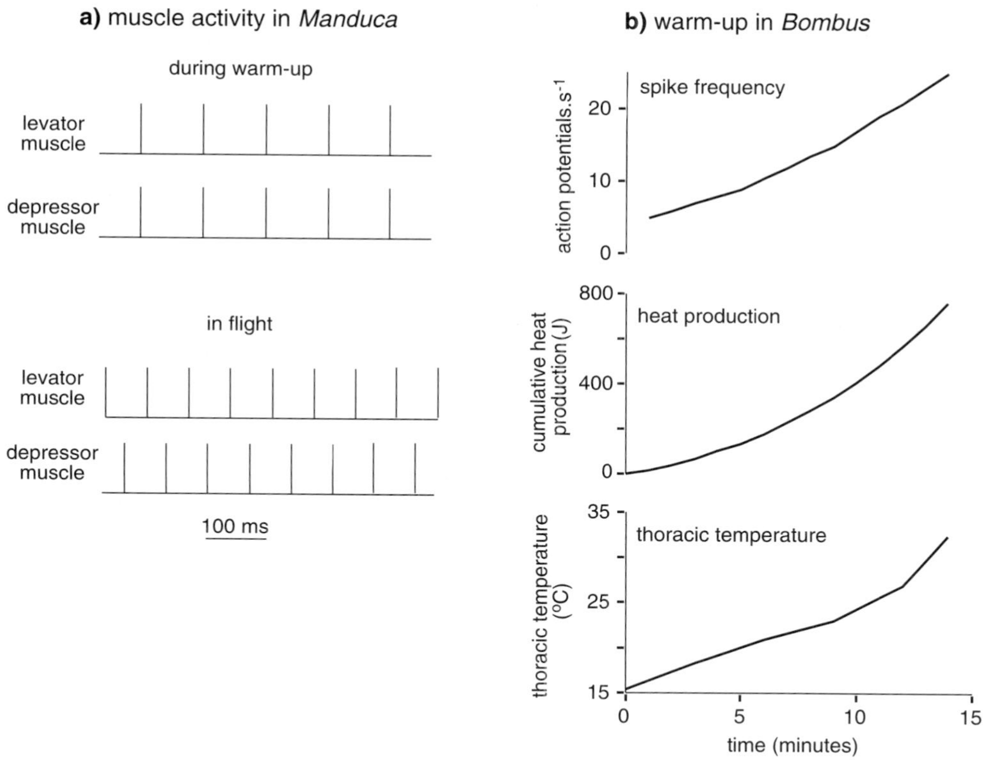

a) muscle activity in *Manduca*

during warm-up

levator
muscle

depressor
muscle

in flight

levator
muscle

depressor
muscle

100 ms

b) warm-up in *Bombus*

spike frequency

heat production

thoracic temperature

Fig. 19.2. Heat production during warm-up. (a) Activity of the flight muscles of a moth (*Manduca*). During warm-up, levator and depressor muscles contract at the same time; during flight they are in antiphase. (b) Warm-up in a bumblebee (*Bombus*). Time 0 is the start of warm-up behavior. Muscle activity (shown as spike frequency) increases over time as the thorax warms up. Notice that the total amount of heat produced, about 800 joules, is sufficient to raise the thoracic temperature by about 200 °C; it does not do so because more heat is lost as the difference between thoracic and air temperature increases. Air temperature in this experiment was 11 °C. The thorax was already hotter than this at the beginning of the experiment (after Heinrich & Kammer, 1973).

is likely that heat is produced by oscillations of the flight muscles (Bartholomew & Heinrich, 1978).

19.1.1.2 *Heat gain from the environment*
In the absence of solar radiation, the body temperature of a resting insect is close to ambient, but solar radiation will elevate body temperature above ambient. The amount by which body temperature exceeds air temperature, sometimes called the temperature excess, is directly proportional to the intensity of radiation. Small insects heat up more rapidly than large ones because of their smaller mass, but attain a lower final body temperature because they lose heat more rapidly, having a greater surface/volume ratio (Figs. 19.4, 19.5).

The extent to which body temperature is raised depends on the reflectance properties of the insect's cuticle. Dark insects, or parts of insects, reflect relatively little of the radiation that falls on them, while light colors reflect more (de Jong, Gussekloo & Brakefield, 1996). In relation to this, alpine and arctic insects are often black, and many insects of temperate regions exhibit seasonal variation in the extent of their black markings. Butterflies in the genus *Colias* illustrate these points. Species living at high elevations have more dark scales on the underside of the hindwings and absorb more radiation than those at lower elevations (Kingsolver, 1983), and *Colias eurytheme*, which occurs at lower elevations, is darker in the spring and fall than during the summer. The change is triggered

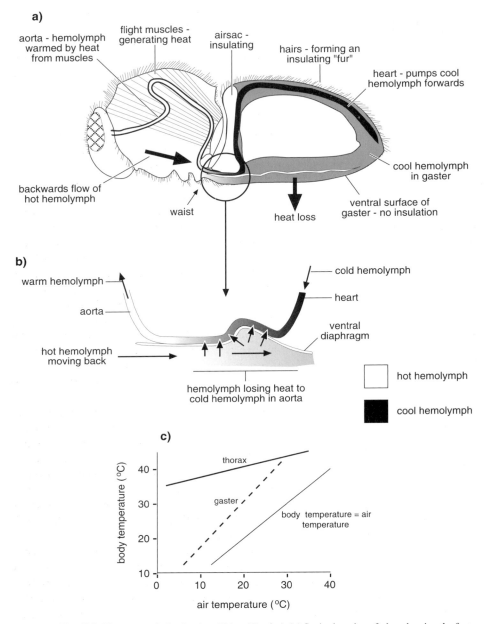

a)

aorta - hemolymph warmed by heat from muscles

flight muscles - generating heat

airsac - insulating

hairs - forming an insulating "fur"

heart - pumps cool hemolymph forwards

backwards flow of hot hemolymph

waist

heat loss

cool hemolymph in gaster

ventral surface of gaster - no insulation

b)

warm hemolymph

aorta

hot hemolymph moving back

cold hemolymph

heart

ventral diaphragm

hemolymph losing heat to cold hemolymph in aorta

hot hemolymph

cool hemolymph

c)

body temperature (°C)

air temperature (°C)

thorax

gaster

body temperature = air temperature

Fig. 19.3. Thermoregulation in a bumblebee (*Bombus*). (**a**) Sagittal section of a bee showing the features involved in regulating thoracic temperature (after Heinrich, 1976). (**b**) Detail of the heat exchange system in the waist region. As hot hemolymph flows back under the ventral diaphragm it loses heat to the cool, forwardly flowing hemolymph in the aorta (after Heinrich, 1976). (**c**) Temperature of the thorax and gaster of bumblebees in flight. As air temperature rises, the gaster gets markedly hotter. The insect cannot fly without overheating when thoracic temperature exceeds 45 °C. (after Heinrich, 1975).

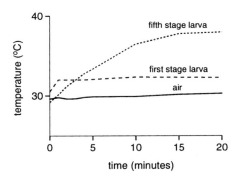

Fig. 19.4. Effect of solar radiation on body temperature of a locust (*Schistocerca*). At time 0, the insects were placed in the sun, having previously been in shade. The larger, fifth stage larva reaches a higher temperature but heats up more slowly than the smaller first stage larva (after Stower & Griffiths, 1966).

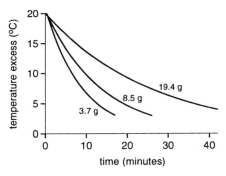

Fig. 19.5. The effects of size on heat loss from adult beetles (*Heliocopris dilloni*). Three beetles of different weights were heated until their metathoraces were 20 °C above ambient (the temperature excess) and then allowed to cool. The beetles were all of the same species so their body forms were similar. Differences in the rate of cooling result from the larger surface/volume ratio of the smaller insects (data from Bartholomew & Heinrich, 1978).

by different daylengths experienced by the later stage larvae. If the daylength is longer than 14 hours, as it is in summer, the adults have relatively few dark scales; when the daylength experienced by the larvae falls to 12 hours, as in spring and fall, the adults have more dark scales (Fig. 19.6) (Jacobs and Watt, 1994).

Developmental temperature has a direct effect on coloration in many insects. It is common for insects reared at low temperatures to be very dark, while siblings reared at high temperatures are pale, sometimes almost white.

These differences, which are well-known in grasshoppers and caterpillars, will, in turn, affect body temperature.

Insects sometimes gain heat by conduction from the substratum. This sometimes occurs in locusts, that lower their bodies so that they are in contact with the ground, and has also been recorded in flies.

19.1.2 Heat loss

Evaporation cools insects because the latent heat of vaporization is drawn from the body. In stationary insects in still air this is the most important source of heat loss. Hence factors affecting evaporation will also affect heat loss.

The rate of evaporation from a body is limited by the humidity of the immediate environment. If environmental humidity is high, little evaporation occurs, and little heat is lost by this route. In dry air, on the other hand, evaporation is much faster and an insect's body temperature in dry air may be 3–4 °C below ambient. Evaporation may be reduced by local accumulations of water vapor round the body and this effect may be enhanced by hairs and scales holding a layer of still air adjacent to the body. Restricting evaporation from the tracheal system by closure of the spiracles also tends to maintain body temperature. Air movement tends to remove local accumulations of water vapor and so to increase evaporation. Thus, at higher wind speeds, body temperature is reduced.

Evaporation is generally slight at low air temperatures, so that, even in the absence of radiation, body temperature slightly exceeds air temperature because of the heat produced by the insect's metabolism. At higher temperatures, evaporation is increased and body temperature falls below ambient. For example, in a resting grasshopper at a constant relative humidity of 60%, the temperature excess at 10 °C is 0.6 °C, at 20 °C it is 0.4 °C, while at 30 °C body temperature is 0.2 °C below ambient.

Convection of heat away from the body is also important if an insect's body temperature is above ambient and especially in moving air. Heat is lost more rapidly by smaller insects with a larger surface/volume ratio (Fig. 19.5). Convection is increased at higher wind speeds and is a major source of heat loss in a flying insect. In a flying locust, for example, 60–68% of heat loss results from convection compared with less than 10% by evaporation and 10% by long wave radiation.

Conduction is probably unimportant as a means of heat loss except in the transfer of heat between the body and the layer of air immediately adjacent to it.

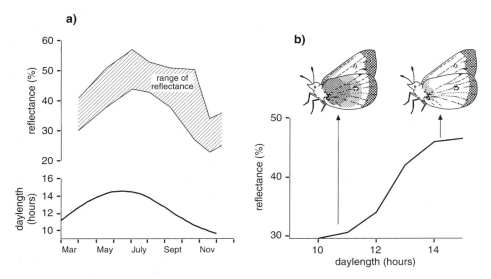

Fig. 19.6. Seasonal variation in black pigmentation of the butterfly, *Colias eurytheme* (after Hoffmann, 1973). **(a)** Above: seasonal variation in the amount of reflectance at 650 nm from the underside of the hind wing. Reflectance is inversely proportional to the abundance of black scales and so to the degree of heating. Below: seasonal changes in daylength. **(b)** Effects of experimental daylengths during the last larval stages on the dark patterning and average reflectance from the hind wing.

Heat loss from the body is reduced by insulation at the surface. Many species have the thorax clothed with hairs which hold an insulating layer of air round the body. In species of *Colias* living at high altitudes the hairs are longer than in species from low altitudes, creating a layer of fur which, in *C. meadii*, living above 3000 m, is almost 1.5 mm thick. The effectiveness of the insulation depends on the density of hairs, but, in general, the temperature excess of flying insects is increased 50–100% by their insulation, amounting to as much as 9 °C in a large hawk moth. Insulation is achieved in dragonflies by a layer of air-sacs beneath the thoracic cuticle.

Review: Prange, 1996 – evaporative cooling

19.2 THERMOREGULATION

Insects are relatively small animals; the vast majority weigh less than 500 mg. As a result, their surface/volume ratio is large compared with other, larger animals, and, as a consequence, insects are unable to maintain a constant body temperature. Many species can, however, regulate their temperature so as to approximate the optimal under some conditions. This may be done behaviorally, or, in some cases, physiologically.

19.2.1 Behavioral regulation

Extreme temperatures are avoided. At temperatures over 44 °C, approaching the upper lethal temperature, larvae of the desert locust, *Schistocerca*, become highly active. Similarly, movement into an area of low temperature promotes a brief burst of activity. This activity is undirected, but may tend to take the insect out of the immediately unfavorable area so that it is neither killed by extreme heat nor trapped at temperatures too low for its metabolism to continue efficiently.

Within the normal range of temperature in which they are active, insects have a preferred range in which, given the choice, they tend to remain for relatively long periods. This preferred temperature range is towards the upper end of the normal range of temperatures experienced by the insect and in *Schistocerca*, for example, extends from 35 to 45 °C with a peak at 40–41 °C (Fig. 19.7). The tendency to remain still in this preferred range may be regarded as a mechanism tending to keep the insects within a range of temperatures which is optimal for most metabolic processes.

In the field, many insects are known to regulate body temperature by moving into areas of sunlight when the air temperature is low, or moving to the shade when it is high. They also vary posture and orientation to the sun exposing

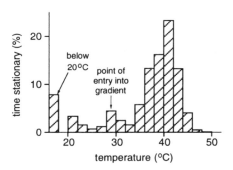

Fig. 19.7. Temperature preference of second stage larvae of the desert locust, *Schistocerca*, as shown by the amount of time for which the insects were stationary at different points in a temperature gradient. The insects had previously experienced a temperature of 35 °C. Between 32 and 43 °C, the insects often had the abdomen in contact with the floor of the apparatus. Above 42 °C, if they remained stationary, the abdomen was usually lifted clear of the floor. Above 45 °C, they were usually continuously active, often hopping. This high level of activity resulted in their moving back to a lower temperature. If they remained below 20 °C for more than a few minutes, the insects became sluggish (after Chapman, 1965).

large areas of the body if it is cool, or as little as possible when body temperature is high. Using these devices, insects are able to extend their active periods and maximize metabolic rate. Examples of such behavior are recorded in insects from many orders: grasshoppers (Chappell & Whitman, 1990), some caterpillars (Casey, 1993), butterflies (Rutowski, Demlong & Leffingwell, 1994), flies (O'Neill & Kemp, 1992) and in other groups.

19.2.2 Physiological regulation

During flight, the flight muscles raise the thoracic temperature above ambient and, for many insects, the only way to avoid overheating at high air temperature is to stop flying. Some species, however, have the ability to regulate thoracic temperature physiologically. For example, in flight, *Manduca* keeps its thoracic temperature between 38 and 42 °C over a range of ambient temperatures from 12 to 36 °C. Some other moths, bees, beetles and dragonflies are known to have a similar capacity (Fig. 19.3c).

When insects are warming up they are able to alter heat production by varying the activity of the flight muscles (see above). In flight, however, the flight muscles are dedicated to producing the aerodynamic power necessary for flight and their output can by varied only within narrow limits. For this reason, flying insects are thought, generally, to be unable to modulate heat production by altering wingbeat frequency, but exceptions are known. The large dragonfly, *Anax*, reduces its wingbeat frequency from about 35 Hz at an air temperature of 20 °C to 25 Hz at 35 °C, at the same time reducing its flight speed (May, 1995). Dragonflies also make more intermittent flights and spend more time gliding at higher temperatures. In honeybees, reduction of wingbeat frequency at high temperatures is also known to contribute to regulation of thoracic temperature (Harrison *et al.*, 1996).

In general, however, flying insects control their body temperature by regulating heat loss. Moths and bumblebees do this by varying the heat transfer to the abdomen which acts as a radiator of heat as it is less well insulated than the thorax. As the thorax warms up, the heart beats more rapidly and with greater amplitude. The effect of this is to increase the rate of circulation to the abdomen and reduce the efficiency of the heat exchanger. In bumblebees, the efficiency of the heat exchanger may be greatly reduced at high ambient temperatures by the hemolymph being pushed back into the gaster in discrete slugs. Consequently, the difference in temperature between the gaster and the air is greater at high temperatures (Fig. 19.3c). In honeybees, however, this is not true. The temperature of the gaster is always very similar to ambient probably because the efficiency of the heat exchanger is so high that little or no thoracic heat reaches the gaster.

Although some insects warm up for activities unrelated to flying (see above), physiological thermoregulation has not been described in these instances although the insects can prevent overheating by reducing their activity.

Only evaporative cooling can reduce body temperature below ambient, and most insects do not have sufficient water to use this method. Xylem-feeding insects, however, have an abundant water supply and a few cicada species are known to exhibit an increased rate of water loss at about 38 °C which can reduce body temperature to as much as 5 °C below ambient (Fig. 19.8). The water is lost by an active process through pores almost 10 μm in diameter on the dorsal surface of the thorax and abdomen. This process is only effective when the water content of the insect is high and humidity low (Hadley, Quinlan & Kennedy, 1991). Honeybees foraging for nectar also use evaporative cooling. When the head reaches a temperature of about 44 °C, a flying worker regurgitates a drop of nectar from the honey stomach and holds it on the

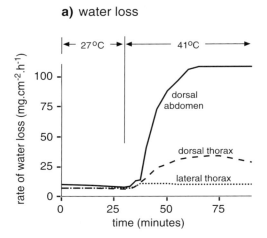

Fig. 19.8. Cooling by evaporative water loss in a desert cicada (*Diceroprocta*). **(a)** Rates of water loss from different parts of the insect. At moderate temperatures water loss is similar from all parts. When the insect is transferred to high temperature only a slight increase occurs in loss from the lateral thorax, but much bigger losses occur from the dorsal thorax and abdomen. Water loss at high temperature occurs primarily through large pores which are present at higher density on the dorsal abdomen than the dorsal thorax; there are none on the lateral thorax (after Hadley, Toolson & Quinlan, 1989). **(b)** The cooling effect of water loss from a feeding cicada (after Hadley, Quinlan & Kennedy, 1991).

mouthparts. Evaporation of the drop cools the head and as the head and thorax are in broad continuity, the thorax is also cooled (Fig. 19.9). At high ambient temperatures, honeybees switch from pollen gathering to nectar collecting presumably because pollen gatherers do not have an adequate supply of nectar for cooling (Cooper, Schaffer & Buchmann, 1985).

It is possible that some other insects use evaporative cooling very selectively. Grasshoppers have patches of thin cuticle (known as Slifer's patches) on the dorsal sides of their thoracic and abdominal segments. The patches are most extensive on the middle segments of the abdomen. The rate of evaporation through these patches is high and it has been suggested that this produces local cooling, perhaps of the gonads, when temperatures are very high (Makings, 1987). A similar function has been ascribed to patches of thin cuticle on the sternites of some Pentatomidae (Staddon & Ahmad, 1994).

19.2.3 Regulation of colony temperature by social insects

Social insects regulate the temperature of their nests so that their larvae develop under relatively constant conditions. Ants, for example, move their larvae to the most favorable situations in the nest. On warm days in

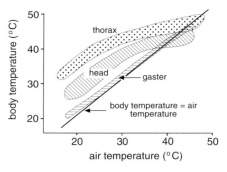

Fig. 19.9. Thermoregulation in the honeybee (*Apis*). Each shaded area represents the range of temperatures recorded from bees in free flight. Notice that at higher air temperatures, head and thoracic temperatures rise less sharply than at lower temperatures. This results from the cooling effects of evaporative water loss from nectar droplets on the mouthparts. The temperature of the gaster is always close to air temperature (body temperature = air temperature). It plays no part in cooling. (Compare Fig. 19.3c) (after Heinrich, 1980a,b).

summer, the older larvae are brought near to the surface, while in winter they may be 25 cm or more below the surface and so avoid the lowest temperatures. On hot days, *Formica* blocks its nest entrance stopping the entry of warm air.

Regulation of colony temperature is also well known in *Apis*. At high temperatures, workers stand at the entrance of the hive fanning with their wings and creating a draught through the nest. This is sufficiently effective to keep the temperature of the brood down to 36 °C when the hive is heated to 40 °C. Water may also be carried in to help cool the hive by evaporation and, at excessively high temperatures, the bees leave the combs and cluster outside so that further heating due to their metabolism is avoided. On the other hand, in winter when there is little or no brood the bees cluster together on and between a small number of combs. This behavior is seen when air temperature drops below 15 °C, and their metabolic heat maintains the inside of the cluster at 20–25 °C. By packing closer when the temperature is very low and spreading out when it is higher, the bees are able to regulate this temperature. In addition, individual bees at low temperature increase their metabolic rate, and so increase heat output, by contracting the flight muscles without moving the wings.

Queen bumblebees also warm up using the flight muscles and rest with the underside of the gaster, which is not insulated, closely pressed against brood cells. Heat is transferred to the gaster and from it to the brood. As in *Apis*, the lower the temperature, the more energy is used to generate heat and at temperatures approaching freezing the metabolic rate of a queen bumblebee is as high as during flight (Fig. 19.10).

Reviews: Heinrich, 1981, 1993

19.3 BEHAVIOR AND SURVIVAL AT LOW TEMPERATURES

Although most insects are active at moderately high temperatures, some exhibit normal behavior at temperatures close to freezing and many temperate species survive low temperatures during the winter.

19.3.1 Flight at low temperature

A number of moths fly habitually when the air temperature is close to zero. Some of these, such as *Eupsilia*, operate by raising the thoracic temperature to 30 °C or above even at an ambient temperature of 0 °C. These moths have a thick insulating hair pile on the thorax and a well-developed countercurrent heat exchange system for maintaining the thoracic temperature (Heinrich, 1987). Other species, like the winter moth, *Operophtera*, can fly without having to raise the thoracic temperature. The

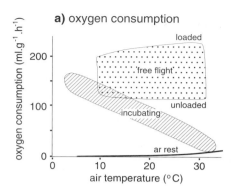

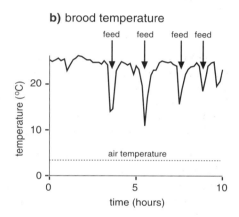

Fig. 19.10. Thermoregulation in a social insect, a bumblebee (*Bombus*). (a) Oxygen consumption of queen bees performing different activities. The shaded areas show the ranges of a series of measurements. At very low temperatures, the queen generates heat by shivering, and her metabolic rate, as indicated by oxygen consumption, is as high as in flight. At higher temperatures, muscle contractions are presumably at a lower frequency, so she produces less heat and uses less oxygen. In this way she is able to regulate the amounts of heat transferred to the brood, and so keeps the brood at a relatively constant temperature despite wide fluctuations in air temperature (after Heinrich, 1975; Kammer & Heinrich, 1974). (b) Temperature of a clump of bumblebee cells (brood) is maintained close to 25 °C although the air temperature is only about 4 °C. Whenever the queen leaves the cells in order to feed, the temperature drops sharply (after Heinrich, 1974).

flight muscles of these insects have shorter twitch durations at low temperatures than *Manduca*, for example (Fig. 19.11), but, nevertheless, even at 5 °C a cycle of contraction and relaxation may take 100 ms. Thus wingbeat frequency cannot exceed about 10 Hz and at lower

Table 19.1. *Mortality of the waxmoth,* Galleria, *at temperatures above the supercooling point*[a]

Temperature (°C)	Period of larval growth (days)	Period of survival as a larva (days)	Mortality (%)
30	4	6	2
25	7	10	2
20	12–20	60–160	20
15	30–40	100–200	100
10	0	20–50	100
5	0	3–6	100
0	0	1–3	100
−5	0	1	100

Note:

[a] Larvae were reared at 30°C and transferred to the temperatures shown on the second day of the last larval stage (after Sehnal, 1991).

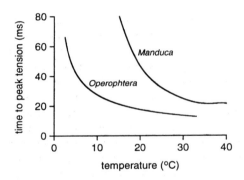

Fig. 19.11. Flight at low temperatures. The time for the flight muscles to reach peak tension is shorter in the winter moth, *Operophtera*, which flies at low thoracic temperatures, than in *Manduca* which only flies when the thoracic temperature is above 30 °C (after Marden, 1995).

temperatures must be even less. These moths are able to fly by virtue of their low wing loading. For *Operophtera*, this is about 3.2 mg cm^{-2}, ten times lower than most other moths of similar size. Thus a wingbeat frequency as low as 4 Hz produces enough power to keep the insects airborne (Heinrich & Mommsen, 1985; Marden, 1995).

Many small Diptera also fly at low air temperatures. Their small size probably precludes any significant warming so their flight muscles must be operating at low temperature, but nothing is known of their muscle physiology.

19.3.2 Survival at low temperatures (cold hardiness)

Survival at low temperatures is affected by the duration of exposure. Low temperatures above freezing are not immediately lethal for most insects, although the insects may ultimately die. For example, *Locusta* larvae move very sluggishly and do not feed if the temperature is below 20 °C but remain alive for many days even at temperatures close to 0 °C. These insects recover and develop normally if returned to a high temperature after a short interval. Similarly, last stage larvae of the wax moth, *Galleria*, grow and pupate at 25 °C and above, but at 20 °C most larvae die before pupation although some feeding does occur and a few larvae live as long as 160 days. Below 20 °C, all the larvae die, living for progressively shorter periods at lower temperatures (Table 19.1).

Insects overwintering in temperate and subarctic regions are normally exposed to temperatures below freezing for some part of the year and at certain stages of development. At temperatures below the freezing point of water, insects respond in different ways. They are usually classified as freeze-tolerant or freeze-susceptible. Freeze-tolerant insects can withstand freezing of the tissues; freeze-susceptible insects cannot. The temperature at which body fluids freeze is usually well below 0 °C, partly because the freezing point is lowered by solutes, but also because water may cool to temperatures well below its freezing point before ice crystals begin to form. This is known as supercooling and the temperature to which an insect supercools, the supercooling point, is

Table 19.2. *Supercooling points and cryoprotectants of some freeze-tolerant insects occuring in northern latitudes. The insects overwinter in the stage shown*

Species	Stage	Supercooling point (°C)	Lowest survival temperature (°C)	Cryoprotectant
Blattodea				
Cryptocercus	Adult	−6	?	Ribitol
Lepidoptera				
Gynaephora	Larva	−7	−70	Glycerol
Papilio	Pupa	−20	−30	Glycerol
Coleoptera				
Dendroides	Larva	−10	?	Glycerol, sorbitol
Pterostichus	Adult	−10	−80	Glycerol
Diptera				
Eurosta	Larva	−10	−50	Glycerol, sorbitol

influenced by many factors controlling the formation of ice crystals.

Reviews: Block, 1990 – cold tolerance; Lee, 1991 – cold tolerance; Sehnal, 1991 – morphogenesis; Storey, 1990 – biochemistry

19.3.2.1 *Freeze tolerance*

The supercooling point for most freeze-tolerant insects is in the range −5 to −10 °C, although some species supercool to lower temperatures. Some examples of freeze-tolerant insects are given in Table 19.2. Once frozen, these insects may be able to withstand very low temperatures. An extreme example is the caterpillar of *Gynaephora groenlandica* from Greenland. This insect, which freezes at −7 °C, can remain frozen for 9 months and survive temperatures as low as −70 °C.

Intracellular ice formation destroys the cells, and if many cells are affected the insect dies when it thaws. Any mechanism which reduces the rate of cooling at the surface of the cells is important in avoiding intracellular freezing. Thus, the freezing of a large amount of the extracellular body fluids beforehand is beneficial.

Ice nucleating agents in the hemolymph or cells induce freezing in a controlled way. Ice nucleating proteins and a lipoprotein are known to occur in the hemolymph of some freeze-tolerant adult beetles and larval Diptera. Larvae of the goldenrod gall fly, *Eurosta*, lack ice nucleators in the hemolymph, but a similar function is performed by crystals of calcium phosphate in the Malpighian tubules (Mugnano, Lee & Taylor, 1996). Crystals of uric acid may perform a similar function in other insects. As these compounds limit the effects of nucleation occurring round food in the gut, they may be especially important in species, such as the beetle, *Phyllodecta*, subject to regular nightly freezing, enabling them to feed during the day and retain food in the gut at night. The effects of nucleating agents are not influenced by cryoprotectants which may also be present even though in freeze-susceptible insects these same compounds act as supercooling agents.

Glycerol is the most widely occurring cryoprotectant in insects, although other polyhydric alcohols and some sugars serve similar functions. These compounds, by adding to the pool of solute molecules, affect the osmotic pressure of the hemolymph and help to regulate cell volume during extracellular ice formation and also to stabilize proteins. They are derived from glycogen stored in the fat body and, in many insects, begin to accumulate at the onset of an overwintering period (Fig. 19.12). Exposure to temperatures approaching 0 °C is often the trigger for the development of cryoprotectants and ice nucleating compounds (Fig. 19.13) (Baust, 1982).

Reviews: Duman, *et al.*, 1991 – ice nucleators; Lee, Lee & Strong-Gunderson, 1993 – ice nucleators; Storey & Storey, 1991 – cryoprotectants; Zachariassen, 1982 – ice nucleators

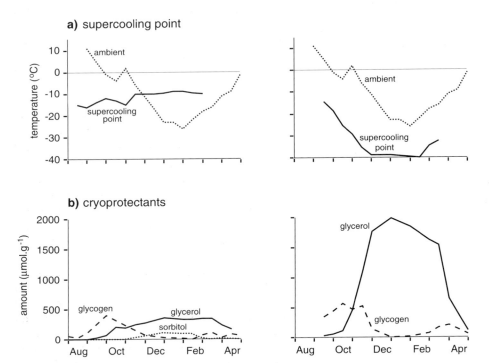

Fig. 19.12. Supercooling points and cryoprotectants of two insects that use different strategies to survive low temperatures during the winter. *Eurosta* is a gall fly that tolerates freezing. *Epiblema* is a gall moth that is freeze susceptible, but has a low supercooling point. Both species overwinter in the last larval stage in galls on the host plant, and both may occur on the same plant. Data in the figure relates primarily to insects overwintering near Ottawa, Canada. (a) Variation in supercooling points over the winter. Ambient temperatures are the minima occurring over a two-week period in one year. Notice that minimum temperatures are below the supercooling point of *Eurosta* for several months; the supercooling point of *Epiblema* is always well below the environmental minima (based on Morrissey & Baust, 1976; Rickards, Kelleher & Storey, 1987). (b) Seasonal changes in glycogen and the cryoprotectants, glycerol and sorbitol, that are derived from it. Notice that glycerol levels in the freeze susceptible insects reach levels five times higher than in the freeze tolerant insect. *Epiblema* produces very little sorbitol. Amounts are expressed in relation to the wet weights of the insects (based on Rickards, Kelleher & Storey, 1987; Storey & Storey, 1986).

19.3.2.2 *Freeze susceptibility*

Freezing of the tissues quickly kills insects in this category, but many freeze-susceptible species have very low supercooling points, often below $-20\,°C$. The supercooling point often varies seasonally, being lowest at the coldest times of year and varying inversely with the accumulation of polyhydric alcohols (Fig. 19.12). Differences also occur between different stages of a species, and eggs generally have lower supercooling points than active stages (Table 19.3). The overwintering stages of insects from temperate regions that are not normally exposed to very low temperatures nevertheless have supercooling points as low as some arctic species. For example, the supercooling point of eggs of the aphid, *Myzus persicae*, in Britain is below $-35\,°C$, similar to that of arctic aphid species (Strathdee, Howling & Bale, 1995).

It is important to recognize that, while the supercooling point is a lower lethal limit, high mortality may occur for reasons other than freezing at temperatures much higher than this. For example, in the aphid, *Myzus*, the supercooling point of first stage larvae is $-26\,°C$, but half the insects are killed if exposed to $-8\,°C$ for only one

Table 19.3. *Supercooling points and cryoprotectants of some freeze-sensitive insects occuring in northern latitudes*[a]

Species	Stage	Supercooling point (°C)	Cryoprotectant
Hemiptera			
Rhopalosiphum	Egg	−40	?
Myzus	Larva	−26	?
Lepidoptera			
Malacosoma	Egg	−41	Glycerol
Laspeyresia	Larva	−37	Glycerol
Pieris	Pupa	−26	Sorbitol
Vanessa	Adult	−21	?
Coleoptera			
Dendroctonus	Larva	−34	Glycerol
Coccinella	Adult	−24	None?
Diptera			
Hylemya	Larva	−25	?
Sarcophaga	Pupa	−23	Glycerol
Anopheles	Adult	−17	?
Hymenoptera			
Neodiprion	Egg	−32	Glycerol
Cephus	Larva	−27	None?
Megachile	Pupa	−42	Glycerol
Camponotus	Adult	−28	Glycerol

Notes:

[a] The insects overwinter in the stage shown. Values for supercooling points are the lowest values observed in mid-winter where data are available.

? not known.

minute. They do not die immediately, but do so progressively over the next three days (Bale, Harrington & Clough, 1988). On the other hand, there are some examples, such as the larva of the golden rod gall moth, *Epiblema*, where freezing at the supercooling point is the principal cause of death (Bale, 1993).

Previous exposure to moderately low temperatures often enhances an insect's ability to survive low temperatures. If *Myzus* is reared at 5 °C instead of 20 °C, half the insects survive exposure to −20 °C instead of −8 °C. Even a brief acclimation period at low temperature can protect an insect from later cold shock, the injury produced by a brief exposure to low, non-freezing temperatures. For example, most pharate adults of the flesh fly, *Sarcophaga*, reared at 25 °C are killed by a 2-hour exposure to −10 °C, but if they are first exposed to 0 °C for two hours they survive the lower temperature. Previous exposure to a moderately high temperature may have the same effect and it is possible that heat shock proteins contribute to the change in sensitivity.

Freeze-susceptible insects lack ice nucleating agents in the hemolymph and, in many species, the likelihood of ice formation is reduced by emptying the gut before the cold period as food in the gut may form nuclei for ice formation.

The reduction in supercooling point is commonly associated with a reduction in water content which increases the osmotic pressure of the hemolymph. Further reduction of the supercooling point is produced by polyhydric alcohols, especially glycerol. These are present at much higher concentrations than in freeze-tolerant species (Fig. 19.12), sometimes accounting for 20% of the wet weight of the insect. In cold-tolerant insects, temperatures close to 0 °C inhibit the conversion of glycogen phosphorylase to its inactive form so that most of the enzyme is in the active form. As a result, glycogen is phosphorylated to produce glucose-1-phosphate. At the same time, low temperature reduces the activity of glycogen synthase, just as it reduces the activity of most other enzymes, and glycogen is not resynthesized. Instead, the glucose-1-phosphate is converted in a series of steps to the sugar alcohols (Fig. 19.13). At high temperature, the enzyme balance is reversed and the alcohols are reconverted to glycogen.

In addition, some insects surviving winter conditions by supercooling are known to have compounds that stabilize the supercooling point (Duman *et al.*, 1991). These are peptides or glycopeptides that perhaps adsorb to the surface of newly formed ice crystals, and prevent new water molecules from reaching the crystals. In this way, they contain the spread of ice. These compounds are called thermal hysteresis factors or antifreeze proteins. Their production is induced by low temperatures and short photoperiods. In larval *Tenebrio*, these environmental features produce a brief elevation in hemolymph titer of juvenile hormone which may have a regulatory role in the production of the antifreeze proteins (Xu *et al.*, 1992).

Reviews: Denlinger *et al.*, 1991 – cold shock; Duman *et al.*, 1991 – antifreeze proteins; Somme, 1982 – supercooling; Zachariassen, 1991 – water relations

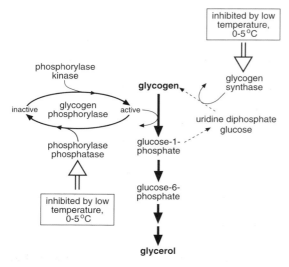

Fig. 19.13. Synthesis of glycerol from glycogen. At low temperatures, the conversion of glycogen phosphorylase from the active to the inactive form is inhibited; the production of the active form continues. As a result, the proportion of this enzyme in the active form is greatly increased so glycogen is metabolized to glucose-6-phosphate. Other enzymes (not shown) then lead to the synthesis of glycerol or sorbitol. Reconversion of glucose-1-phosphate to glycogen does not occur because the activity of glycogen synthase is inhibited by low temperature.

19.4 ACTIVITY AND SURVIVAL AT HIGH TEMPERATURES

At the upper end of the temperature range, above the preferred temperature, insects show a sharp rise in activity. At still higher temperatures this is followed by an inability to move, a phase known as heat stupor, and then by death. The temperature at which death occurs depends on the species, the duration of exposure and interaction with other factors, in particular with humidity.

Insects are cooled by evaporation, so that for short periods they can withstand higher air temperatures if the air is dry. *Periplaneta*, for instance, dies at 38 °C at high humidities, but can survive a short exposure of up to 48 °C if the air is dry. Few insects can tolerate the high rate of water loss that cooling necessitates for very long, but some desert cicadas and honeybees with access to ample water supplies do habitually reduce their body temperatures by evaporative cooling (see above).

For long exposures to high temperature, humidity has the opposite effect because at low humidities insects are adversely affected by desiccation. Thus, *Blatta* can survive for 24 hours at 37–39 °C if the air is moist, but dies as a result of similar exposure in dry air. Humidity does not affect the lethal temperature of small insects such as lice as the volume of water available for evaporation is small while the surface taking up heat is relatively large.

For many insects, the lethal temperature for short-term exposures is within the range 40–50 °C, but, for insects from particular habitats, lethal temperatures may be very different. Thus *Grylloblatta* (Grylloblattodea), living at high altitudes in the Rocky Mountains, dies when its temperature gets as high as 20 °C, while *Thermobia* (Thysanura), the fire brat, dies at 51 °C, and for chironomid larvae living in hot springs at 49–51 °C the lethal temperature must be even higher.

Some modification of the upper lethal temperature occurs, depending on the previous experience of the insect. Thus *Drosophila* reared at 15 °C and maintained at 15 °C as adults, survive for about 50 minutes in dry air at 33.5 °C, but if they are maintained at 25 °C beforehand they survive for about 130 minutes. If the larvae are also reared at the higher temperature the period of exposure that the insects can survive is still further increased to 140 minutes in adults maintained at 15 °C and to 180 minutes in adults maintained at 25 °C. Thus two types of acclimation can be recognized: long lasting acclimation due to conditions during development and short-term physiological acclimation depending on the more immediate conditions. The latter is easily reversible. The effect of physiological acclimation is more marked in dry conditions than in wet.

At least a part of the short-term acclimation to high temperatures is a result of the production of heat shock proteins. These have been demonstrated in several insect species as well as in many other organisms. *Drosophila* and *Locusta* each produce six heat shock proteins, and the larva of the gypsy moth, *Lymantria*, has seven (Yocum, Joplin & Denlinger, 1991). They belong to three different families of proteins with different molecular weights. Some expression of these proteins occurs at normal temperatures, but expression is enhanced within seconds of a sharp rise in temperature (Pauli, Arrigo & Tissières, 1992). At high temperatures, proteins become denatured and clump together as insoluble aggregates. Heat shock proteins may protect other proteins so that this does not occur or they may bind to the surface of an aggregate and promote its dissolution, at the same time causing the proteins to refold. The same proteins also give some protection against low temperatures for short periods (Burton *et al.*, 1988) and are also induced by toxic chemicals. Their production should, thus, probably be regarded as a general response to stress. High rates of expression are only sustained for periods of a few hours after the shock.

Workers of the ant, *Cataglyphis bombycina*, are active foragers in the Sahara desert at temperatures above those tolerated by other small animals. They forage for a very brief period in the middle of the day when the average ground surface temperature is above 46 °C; other species stop foraging when the surface temperature is above 45 °C. These ants have long legs that allow them to lift their bodies off the surface and they also frequently move up stalks of dry vegetation where the temperature is slightly lower than at the ground surface. In addition, they apparently have a high constitutive level of heat shock proteins so that they are already protected for the brief periods when they are exposed (Gehring & Wehner, 1995; Wehner, Marsh & Wehner, 1992).

Death at high temperatures may result from various factors. Proteins may be denatured or the balance of metabolic processes may be disturbed so that toxic products accumulate. Thus blowfly larvae kept at high temperatures accumulate organic and inorganic phosphates and adenyl pyrophosphate in the hemolymph. In some cases food reserves may be exhausted and *Pediculus* (Phthiraptera), for instance, survives better at high temperatures if it has recently fed. Sometimes, particularly over long periods, death at high temperatures may result from desiccation.

19.5 ACCLIMATION

Responses to temperature are not static, but vary according to the previous experience of the insect. Such modification is known as acclimatization or acclimation as has been discussed above in relation to survival at extreme temperatures. However, acclimation also occurs within the normal temperature range of an insect and affects both physiology and behavior. For example, the oxygen consumption of *Melasoma* (Coleoptera) adults increases with temperature, but the level of consumption depends on the temperature to which the insects were previously exposed and, for any given temperature, oxygen consumption is higher in insects acclimated to lower temperatures (Fig. 19.14). The temperature at which maximum oxygen consumption occurs is also lower in the insects preconditioned at the lower temperature. Similarly the spontaneous output from the central nervous system is related to

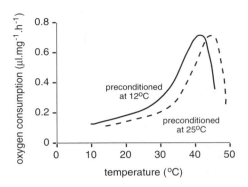

Fig. 19.14. Preconditioning. Effects of acclimation at different temperatures on oxygen consumption by adult *Melasoma* (after Wigglesworth, 1972).

preconditioning temperature (Kerkut and Taylor, 1958) as is the level of activity of the whole insect. Thus, adults of *Ptinus* previously maintained at 15 °C are less active at all temperatures than others previously maintained at 28 °C (Gunn and Hopf, 1942). Acclimation is a continuous process.

19.6 CRYPTOBIOSIS

Cryptobiosis is the term used to describe the state of an organism when it shows no visible signs of life and metabolic activity is brought reversibly to a standstill. The only insect in which this is known to occur is the larva of *Polypedilum* (Diptera), a chironomid living in pools on unshaded rocks in Nigeria. In the dry season these pools dry up and the surface temperatures of the rock probably reach 70 °C. Active larvae of *Polypedilum* die after an exposure of one hour at 43 °C, but if they are dehydrated so that their water content is less than 8% of its original value they can survive extreme temperatures for long periods. Some recovery occurs even after exposure for one minute at 102 °C or several days in liquid air at −190 °C. At room temperatures larvae can withstand total dehydration for three years and some showed a temporary recovery after ten years (Hinton, 1960).

There is some evidence that the larva of a species of *Sciara* (Diptera) and some ceratopogonid larvae also exhibit cryptobiosis.

Some other insects may possess specific tissues which exhibit the phenomenon, although the insect as a whole may not. For example, blood cells in the gills of

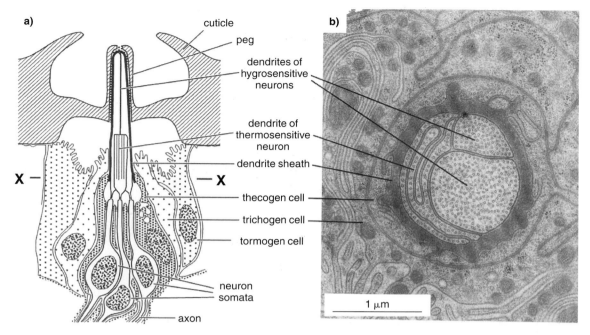

Fig. 19.15. Thermohygroreceptor. (**a**) Diagrammatic longitudinal section through a coeloconic sensillum. (**b**) Electron micrograph of a transverse section of a thermohygroreceptor at the level of X–X in (**a**). Notice the dendrites of two hygrosensitive neurons and folds of the dendrite of a thermosensitive neuron completely occupying the space in the dendrite sheath (*Bombyx*) (after Steinbrecht *et al.*, 1989).

Taphrophila (Diptera) pupae (section 17.6) and *Sialis* larvae can be desiccated for long periods, but when rehydrated show some vital activities such as clotting.

19.7 TEMPERATURE AND HUMIDITY RECEPTORS

Sense cells responsive to temperature and humidity are usually present together in a single sensillum. All insects that have been studied have such receptors on the antennae. They consist of a short peg without any pores and an immovable socket. Projecting into the peg are the distal dendrites of two neurons (Fig. 19.15). The dendrites usually occupy the whole of the space within the dendrite sheath that forms the lining of the peg and are attached to it. Microtubules within the dendrite are joined together forming a structure with some resemblance to the tubular body of mechanosensitive dendrites (section 23.1.1.2). A third dendrite ends below the base of the peg. In many insects it is produced into a series of lamellae or it may be flattened and in the form of a whorl. Proximally, the dendrites of both types have a ciliary region. On the basis of their responses, the cells with dendrites extending into the peg are believed to be hygrosensitive, while the cell which does not enter the peg is probably thermosensitive.

The sensilla may be exposed at the surface of the antenna or concealed in a pit, forming one type of coeloconic sensillum (see Fig. 24.3). When exposed at the surface, they are associated with trichoid mechanoreceptors which form a protective umbrella above them, preventing direct contact with any outside object. The number on each antenna is small. There is often a group on the terminal annulus and one or two on each of the proximal annuli. Worker honeybees have about 50 on each antenna, and *Periplaneta* has about 100. Presumably, since temperature and humidity are ubiquitous properties of the air surrounding the antenna, a large number of receptors is not necessary for the insect to obtain accurate information.

In addition to these sensilla, some multiporous olfactory sensilla on the antennae of *Periplaneta* and *Locusta* contain a cell that is thermosensitive. It is structurally indistinguishable from the chemosensitive neurons. About

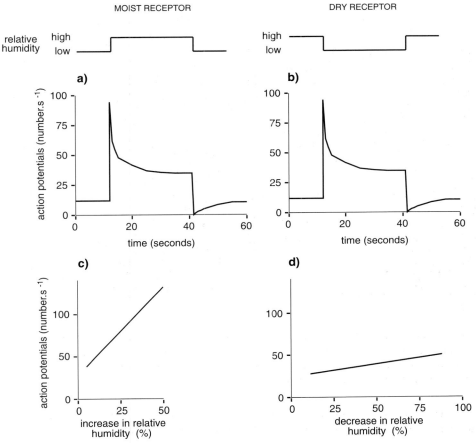

Fig. 19.16. Responses of hygrosensitive neurons to rapid changes in humidity. (**a**) Response of moist receptor to a rapid change from dry (0%) to saturated (100%) air and back again. The firing rate (action potentials s^{-1}) increases sharply immediately following the change and then maintains a higher sustained rate of firing at the higher humidity (cockroach, *Periplaneta*) (based on Yokohari & Tateda, 1976). (**b**) Response of dry receptor to a rapid change from saturated (100%) to dry (0%) air and back again. The firing rate (action potentials s^{-1}) increases sharply immediately following the change and then maintains a higher sustained rate of firing at the lower humidity (cockroach, *Periplaneta*) (based on Yokohari & Tateda, 1976). (**c**) Response of moist receptor following rapid increases in humidity (stick insect, *Carausius*). The number of action potentials is proportional to the magnitude of the increase in relative humidity (based on Tichy, 1979). (**d**) Response of dry receptor following rapid decreases in humidity. Note that the *x*-axis shows the percentage decrease in relative humidity. Larger decreases produce slightly higher firing rates (stick insect, *Carausius*) (based on Tichy, 1979).

100 multiporous sensilla with a thermosensitive neuron are also present on each maxillary palp of *Periplaneta*.

The transduction mechanisms in hygrosensitive neurons are not understood. One possibility is that changes in ambient humidity affect the water content of the cuticle associated with the peg causing a small deformation that is perceived by the dendrites connected to the inner wall of the peg; they effectively function like

mechanoreceptors. However, the presumed hygrosensitive dendrites in the silk moth, *Bombyx*, shorten when the insect is dry adapted, suggesting that there may be some direct effect of humidity on the cells (Steinbrecht & Müller, 1991).

One of the hygroreceptors responds to a rise in humidity with an increased firing rate, the other responds to a fall in humidity (Fig. 19.16), and they are called moist and dry receptors, respectively. The greater the change in humidity, the higher the firing rate of the cell responding. These receptors also fire continuously at constant humidities. In *Periplaneta*, for example, the dry receptor has a higher tonic firing rate at low humidities. It is also temperature sensitive with a higher firing rate at higher temperatures (Loftus, 1976).

The thermoreceptor responds to a fall in temperature; the greater the fall, the higher the firing rate (Fig. 19.17). It is called a cold receptor, and such cells may respond to temperature changes of less than one degree Celsius. In the larva of *Speophyes*, a cave-dwelling beetle, the firing rate of the cold receptor cell follows slow changes of less than one degree occurring over five minutes (Fig. 19.17c).

The temperature sensitive cells in cockroach olfactory sensilla differ in their responses from other thermoreceptors that have been described. They have a maximum rate of firing in the range 18–27 °C, with lower rates at lower and higher temperatures. They thus have the potential to provide the insect with an absolute measurement of air temperature (Nishikawa, Yokohari & Ishibashi, 1992).

It is unclear how this information is integrated within the nervous system. As some of the cells are both thermo- and hygrosensitive there is potential for ambiguity. Nevertheless, it is clear from their behavior that insects are able to detect changes in both temperature and humidity. Bed bugs, *Cimex*, are sensitive to changes of less than 1 °C, and honeybees can be trained to select one of two temperatures differing by only 2 °C, and may, perhaps, also detect absolute levels of temperature. This is suggested by their ability to move to, and remain in, areas of preferred temperature.

The beetle *Melanophila* responds to infrared radiation which is perceived by sensilla in thoracic pits (Schmitz, Bleckmann & Mürtz, 1997). Blood-sucking bugs may also have this capacity (Lazzari & Núñez, 1989).

Review: Altner & Loftus, 1985

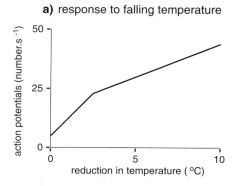

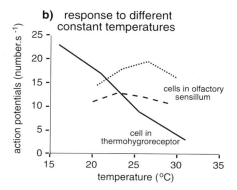

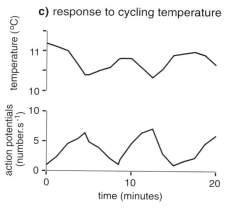

Fig. 19.17. Responses of cold receptor neurons. (a) Response to continuously falling temperature of a cell in a thermohygroreceptor of the cockroach, *Periplaneta*. Cold cells in olfactory sensilla respond in essentially the same way (based on Nishikawa, Yokohari & Ishibashi, 1992). (b) Response to a range of constant temperatures. A cell in a thermohygroreceptor of *Periplaneta* is more active at lower temperatures, but cold receptors in olfactory sensilla exhibit only slight differences with temperature, tending to have maxima at intermediate temperatures (based on Nishikawa, Yokohari & Ishibashi, 1992). (c) Response to cycling temperature of a cold receptor on the antenna of a beetle larva, *Speophyes*. The response follows changes in temperature, increasing as the temperature falls, and decreasing as it rises. Notice that the temperature changes are slow, only about 0.5 °C over five-minute periods (after Altner & Loftus, 1985).

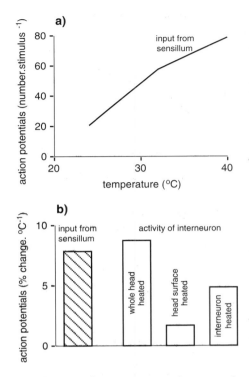

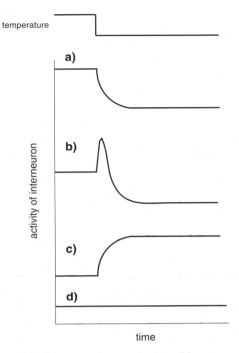

Fig. 19.18. Effect of temperature on mechanosensory input from the wind-sensitive hairs on the head and the activity of a giant interneuron in the brain of a grasshopper, *Schistocerca americana* (after Miles, 1992). (a) Sensory input (action potentials per stimulus) from a single hair deflected by 29 μm at different temperatures. (b) Changes in responses, expressed as percentage change in the frequency of action potentials for each °C change in temperature. Input from the wind-sensitive hairs increased sharply when they were heated, but the central response, as indicated by the activity of the interneuron, was dependent on the temperatures of both the sensory cells and of the interneuron. If both experienced a rise in temperature (whole head heated), the interneuron responded sharply to a change in temperature; if only the temperature of the mechanoreceptors was affected (head surface heated), the interneuron showed little change in activity; if the temperature of the interneuron increased, but that of the sensory neurons did not (interneuron heated) its activity increased, but to a lesser extent than if the whole head was heated.

19.8 TEMPERATURE-RELATED CHANGES IN THE NERVOUS SYSTEM

The rate of firing of peripheral mechanosensory neurons increases with temperature in both *Periplaneta* and

Fig. 19.19. Diagrammatic representations of the response elicited by a sharp fall in temperature on the activity of four different types of unidentified neurons in the cockroach nerve cord (after Kerkut & Taylor, 1958).

Schistocerca and this is, perhaps, a general phenomenon (French, 1985; Miles, 1985). The extent of the change differs in sensilla on different parts of the body. For example, the number of action potentials produced by a standard stimulation of the wind-sensitive hairs on the head of *Schistocerca* more than doubles over the range 25–40 °C, while changes in the input from sensilla elsewhere on the body are less marked (Fig. 19.18). Hairs on the hind tarsi exhibit a broad maximum of response over the range 30–39 °C, firing less actively at lower and higher temperatures. This corresponds approximately to the preferred temperature range (Fig. 19.7) and these sensilla possibly provide the insect with information by which it is able to determine this range when in a temperature gradient.

Changes also occur in the activities of motor neurons and interneurons. For example, the conduction velocity of the motor axon innervating the fast extensor tibiae muscle of *Schistocerca* increases with temperature, while the synaptic delay between this neuron and the flexor muscle motor neurons decreases (Burrows, 1989).

The firing rate of an interneuron receiving input from the wind sensitive hairs also increases with temperature, and its output is a resultant of its own temperature and that of the sensory neurons on the outside of the head (Miles, 1992). Within the central nervous system of *Periplaneta*, unidentified units, possible interneurons, fall into four categories with respect to their responses to temperature (Fig. 19.19). In one, the output is directly proportional to temperature; others show a similar response with the addition of a transient increase if the temperature falls or decrease if the temperature rises; in a third type the output is inversely proportional to temperature; a fourth class is not affected by temperature (Kerkut & Taylor, 1958).

Thus the response of the insect as a whole is affected by temperature in a variety of ways; by information from its peripheral temperature receptors, by effects on the input of other types of receptors, and via direct effects on the central nervous system. Changes in internal body temperature will be slower than peripheral changes, so the central nervous system will be affected primarily by relatively persistent changes in the external temperature.

REFERENCES

Altner, H. & Loftus, R. (1985). Ultrastructure and function of insect thermo- and hygroreceptors. *Annual Review of Entomology*, **30**, 273–95.

Bale, J.S. (1993). Classes of insect cold hardiness. *Functional Ecology*, **7**, 751–3.

Bale, J.S., Harrington, R. & Clough, M.S. (1988). Low temperature mortality of the peach–potato aphid *Myzus persicae*. *Ecological Entomology*, **13**, 121–9.

Bartholomew, G.A. & Heinrich, B. (1978). Endothermy in African dung beetles during flight, ball making, and ball rolling. *Journal of Experimental Biology*, **73**, 65–83.

Baust, J.G. (1982). Environmental triggers to cold hardening. *Comparative Biochemistry and Physiology*, **73A**, 563–70.

Block, W. (1990). Cold tolerance of insects and other arthropods. *Philosophical Transactions of the Royal Society of London* B, **326**, 613–33.

Burrows, M. (1989). Effects of temperature on a central synapse between identified motor neurones in the locust. *Journal of Comparative Physiology*, **165**, 687–97.

Burton, V., Mitchell, H.K., Young, P. & Petersen, N.S. (1988). Heat shock protection against cold stress of *Drosophila melanogaster*. *Molecular & Cellular Biology*, **8**, 3550–2.

Casey, T.M. (1993). Effects of temperature on foraging of caterpillars. In *Caterpillars*, ed. N.E. Stamp & T.M. Casey, pp. 5–28. New York: Chapman & Hall.

Chapman, R.F. (1965). The behaviour of nymphs of *Schistocerca gregaria* (Forskål) in a temperature gradient, with special reference to temperature preference. *Behaviour*, **24**, 283–317.

Chappell, M.A. & Whitman, D.W. (1990). Grasshopper thermoregulation. In *Biology of Grasshoppers*, ed. R.F. Chapman & A. Joern, pp. 143–72. New York: Wiley.

Cooper, P.D., Schaffer, W.M. & Buchmann, S.L. (1985). Temperature regulation of honey bees (*Apis mellifera*) foraging in the Sonoran desert. *Journal of Experimental Biology*, **114**, 1–15.

de Jong, P.W., Gussekloo, S.W.S. & Brakefield, P.M. (1996). Differences in thermal balance, body temperature and activity between non-melanic and melanic two-spot ladybird beetles (*Adalia bipunctata*) under controlled conditions. *Journal of Experimental Biology*, **199**, 2655–66.

Denlinger, D.L., Joplin, K.H. Chen, C.-P. & Lee, R.E. (1991). Cold shock and heat shock. In *Insects at Low Temperature*, ed. R.E. Lee & D.L. Denlinger, pp. 131–48. New York: Chapman & Hall.

Duman, J.G., Xu, L., Neven, L.G., Tursman, D. & Wu, D.W. (1991). Hemolymph proteins involved in insect subzero-temperature tolerance: ice nucleators and antifreeze proteins. In *Insects at Low Temperature*, ed. R.E. Lee & D.L. Denlinger, pp. 94–127. New York: Chapman & Hall.

Esch, H. & Goller, F. (1991). Neural control of fibrillar muscles in bees during shivering and flight. *Journal of Experimental Biology*, **159**, 419–31.

French, A. (1985). The effects of temperature on action potential encoding in the cockroach tactile spine. *Journal of Comparative Physiology*, **156**, 817–21.

Gehring, W.J. & Wehner, R. (1995). Heat shock protein synthesis and thermotolerance in *Cataglyphis*, an ant from the Sahara desert. *Proceedings of the National Academy of Sciences of the United States of America*, **92**, 2994–8.

Gunn, D.L. & Hopf, H.S. (1942). The biology and behaviour of *Ptinus tectus* Boie. (Coleoptera: Ptinidae), a pest of stored products. II. The amount of locomotion in relation to experimental and to previous temperatures. *Journal of Experimental Biology*, **18**, 278–89.

Hadley, N.F., Quinlan, M.C. & Kennedy, M.L. (1991). Evaporative cooling in the desert cicada: thermal efficiency and water/metabolic costs. *Journal of Experimental Biology*, **159**, 269–83.

Hadley, N.F., Toolson, E.C. & Quinlan, M.C. (1989). Regional differences in cuticular permeability in the desert cicada *Diceroprocta apache*: implications for evaporative cooling. *Journal of Experimental Biology*, **141**, 219–30.

Harrison, J.F., Fewell, J.H., Roberts, S.P. & Hall, H.G. (1996). Achievement of thermal stability by varying metabolic heat production in flying honeybees. *Science*, **274**, 88–90.

Heinrich, B. (1974). Thermoregulation in bumblebees. I. Brood incubation by *Bombus vosnesenskii* queens. *Journal of Comparative Physiology*, **88**, 129–40.

Heinrich, B. (1975). Thermoregulation in bumblebees. II. Energetics of warm-up and free flight. *Journal of Comparative Physiology*, **96**, 155–66.

Heinrich, B. (1976). Heat exchange in relation to blood flow between thorax and abdomen in bumblebees. *Journal of Experimental Biology*, **64**, 561–85.

Heinrich, B. (1980a). Mechanisms of body-temperature regulation in honeybee, *Apis mellifera* I. Regulation of head temperature. *Journal of Experimental Biology*, **85**, 61–72.

Heinrich, B. (1980b). Mechanisms of body-temperature regulation in honeybee, *Apis mellifera* II. Regulation of thoracic temperature at high air temperature. *Journal of Experimental Biology*, **85**, 73–87.

Heinrich, B. (ed.) (1981). *Insect Thermoregulation*. New York: Wiley.

Heinrich, B. (1987). Thermoregulation by winter-flying endothermic moths. *Journal of Experimental Biology*, **127**, 313–32.

Heinrich, B. (1993). *The Hot-Blooded Insects*. Cambridge, Massachusetts: Harvard University Press.

Heinrich, B. & Kammer, A.E. (1973). Activation of the fibrillar muscles in the bumblebee during warm-up, stabilization of thoracic temperature and flight. *Journal of Experimental Biology*, **58**, 677–88.

Heinrich, B. & Mommsen, T.P. (1985). Flight of winter moths near 0 °C. *Science*, **228**, 177–9.

Hoffmann, R.J. (1973). Environmental control of seasonal variation in the butterfly *Colias eurytheme*. I. Adaptive aspects of a photoperiodic response. *Evolution*, **27**, 387–97.

Hinton, H.E. (1960). Cryptobiosis in the larva of *Polypedilum vanderplanki* Hint. (Chironomidae). *Journal of Insect Physiology*, **5**, 286–300.

Jacobs, M.D. & Watt, W.B. (1994). Seasonal adaptation *vs* physiological constraint: photoperiod, thermoregulation and flight of *Colias* butterflies. *Functional Ecology*, **8**, 366–76.

Kammer, A.E. & Heinrich, B. (1974). Metabolic rates related to muscle activity in bumblebees. *Journal of Experimental Biology*, **61**, 219–27.

Kerkut, G.A. & Taylor, B.J.R. (1958). The effect of temperature changes on the activity of poikilotherms. *Behaviour*, **13**, 259–79.

Kingsolver, J.G. (1983). Thermoregulation and flight in *Colias* butterflies: elevational patterns and mechanistic limitations. *Ecology*, **64**, 534–45.

Lazzari, C.R. & Núñez, J.A. (1989). The response to radiant heat and the estimation of the temperature of distant sources in *Triatoma infestans*. *Journal of Insect Physiology*, **35**, 525–9.

Lee, R.E. (1991). Principles of insect low temperature tolerance. In *Insects at Low Temperature*, ed. R.E. Lee & D.L. Denlinger, pp. 17–46. New York: Chapman & Hall.

Lee, R.E., Lee, M.R. & Strong-Gunderson, J.M. (1993). Insect cold-hardiness and ice nucleating active microorganisms including their potential use for biological control. *Journal of Insect Physiology*, **39**, 1–12.

Loftus, R. (1976). Temperature-dependent dry receptor on antenna of *Periplaneta*. Tonic response. *Journal of Comparative Physiology*, **111**, 153–70.

Makings, P. (1987). Survival value of Slifer's patches for locusts at high temperature. *Journal of Insect Physiology*, **33**, 815–22.

Marden, J.H. (1995). Evolutionary adaptations of contractile performance in muscle of ectothermic winter-flying moths. *Journal of Experimental Biology*, **198**, 2087–94.

May, M.L. (1995). Dependence of flight behavior and heat production on air temperature in the green darner dragonfly *Anax junius* (Odonata: Aeshnidae). *Journal of Experimental Biology*, **198**, 2385–92.

Miles, C.I. (1985). The effects of behaviourally relevant temperatures on mechanosensory neurones of the grasshopper *Schistocerca americana*. *Journal of Experimental Biology*, **116**, 121–39.

Miles, C.I. (1992). Temperature compensation in the nervous system of the grasshopper. *Physiological Entomology*, **17**, 169–75.

Morrissey, R.E. & Baust, J.G. (1976). The ontogeny of cold tolerance in the gall fly, *Eurosta solidagensis*. *Journal of Insect Physiology*, **22**, 431–7.

Mugnano, J.A., Lee, R.L. & Taylor, R.T. (1996). Fat body cells and calcium phosphate spherules induce ice nucleation in the freeze-tolerant larvae of the gall fly *Eurosta solidaginis* (Diptera, Tephritidae). *Journal of Experimental Biology*, **199**, 465–71.

Nishikawa, M., Yokohari, F. & Ishibashi, T. (1992). Response characteristics of two types of cold receptors on the antennae of the cockroach, *Periplaneta americana* L. *Journal of Comparative Physiology*, A **171**, 299–307.

O'Neill, K.M. & Kemp, W.P. (1992). Behavioural thermoregulation in two species of robber flies occupying different grassland microhabitats. *Journal of Thermal Biology*, **17**, 323–31.

Pauli, D., Arrigo, A.-P. & Tissières, A. (1992). Heat shock responses in *Drosophila*. *Experientia*, **48**, 623–8.

Prange, H.D. (1996). Evaporative cooling in insects. *Journal of Insect Physiology*, **42**, 493–9.

Rickards, J., Kelleher, M.J. & Storey, K.B. (1987). Strategies of freeze avoidance in larvae of the goldenrod gall moth, *Epiblema scudderiana*: winter profiles of a natural population. *Journal of Insect Physiology*, **33**, 443–50.

Rutowski, R.L., Demlong, M.J. & Leffingwell, T. (1994). Behavioural thermoregulation at mate encounter sites by male butterflies (*Asterocampa leilia*, Nymphalidae). *Animal Behaviour*, **48**, 833–41.

Sanborn, A.F., Heath, J.E., Heath, M.S. & Noriega, F.G. (1995). Thermoregulation by endogenous heat production in two South American grass dwelling cicadas (Homoptera: Cicadidae: *Proarna*). *Florida Entomologist*, **78**, 319–28.

Schmitz, H., Bleckmann, H. & Mürtz, M. (1997). Infrared detection in a beetle. *Nature, London*, **386**, 773–4.

Schutz, S. & Gaugler, R. (1992). Thermoregulation and hovering behavior of salt marsh horse flies (Diptera: Tabanidae). *Annals of the Entomological Society of America*, **85**, 431–6.

Sehnal, F. (1991). Effects of cold on morphogenesis. In *Insect at Low Temperature*, ed. R.E. Lee & D.L. Denlinger, pp. 149–73. New York: Chapman & Hall.

Sømme, L. (1982). Supercooling and winter survival in terrestrial arthropods. *Comparative Biochemistry and Physiology*, **73A**, 519–43.

Staddon, B.W. & Ahmad, I. (1994). A further study of the sternal patches of Heteroptera – Pentatomidae with considerations of their function and value for classification. *Journal of Natural History*, **28**, 353–64.

Steinbrecht, R.A., Lee, J.-K., Altner, H. & Zimmermann, B. (1989). Volume and surface of receptor and auxiliary cells in hygro-/thermoreceptive sensilla of moths (*Bombyx mori, Antheraea pernyi* and *A. polyphemus*). *Cell and Tissue Research*, **255**, 59–67.

Steinbrecht, R.A. & Müller, B. (1991). The thermo-/hygrosensitive sensilla of the silkmoth, *Bombyx mori*: morphological changes after dry- and moist-adaptation. *Cell and Tissue Research*, **266**, 441–56.

Stone, G.N. & Willmer, P.G. (1989). Warm-up rates and body temperatures in bees: the importance of body size, thermal regime and phylogeny. *Journal of Experimental Biology*, **147**, 303–28.

Storey, J.M. & Storey, K.B. (1986). Winter survival of the gall fly larva, *Eurosta solidaginis*: profiles of fuel reserves and cryoprotectants in a natural population. *Journal of Insect Physiology*, **32**, 549–56.

Storey, K.B. (1990). Biochemical adaptation for cold hardiness in insects. *Philosophical Transactions of the Royal Society of London* B, **326**, 635–54.

Storey, K.B. & Storey, J.M. (1991). Biochemistry of cryoprotectants. In *Insects at Low Temperature*, ed. R.E. Lee & D.L. Denlinger, pp. 64–93. New York: Chapman & Hall.

Stower, W.J. & Griffiths, J.F. (1966). The body temperature of the desert locust (*Schistocerca gregaria*). *Entomologia Experimentalis et Applicata*, **9**, 127–78.

Strathdee, A.T., Howling, G.G. & Bale, J.S. (1995). Cold hardiness of overwintering aphid eggs. *Journal of Insect Physiology*, **41**, 653–7.

Tichy, H. (1979). Hygro- and thermoreceptive triad in antennal sensillum in the stick insect, *Carausius morosus*. *Journal of Comparative Physiology*, **132**, 149–52.

Wehner, R., Marsh, A.C. & Wehner, S. (1992). Desert ants on a thermal tightrope. *Nature*, **357**, 586–7.

Wigglesworth, V.B. (1972). *The Principles of Insect Physiology*. London: Methuen.

Xu, L., Duman, J.G., Wu, D.W. & Goodman, W.G. (1992). A role for juvenile hormones in the induction of antifreeze protein production by the fat body of the beetle *Tenebrio molitor*. *Comparative Biochemistry and Physiology*, **107B**, 105–9.

Yocum, G.D., Joplin, K.H. & Denlinger, D.L. (1991). Expression of heat shock proteins in response to high and low temperature extremes in diapausing pharate larvae of the gypsy moth, *Lymantria dispar*. *Archives of Insect Biochemistry and Physiology*, **18**, 239–49.

Yokohari, F. & Tateda, H. (1976). Moist and dry hygroreceptors for relative humidity of the cockroach, *Periplaneta americana*. *Journal of Comparative Physiology*, **106**, 137–52.

Zachariassen, K.E. (1982). Nucleating agents in cold-hardy insects. *Comparative Biochemistry and Physiology*, **73A**, 557–62.

Zachariassen, K.E. (1991). The water relations of overwintering insects. In *Insects at Low Temperature*, ed. R.E. Lee & D.L. Denlinger, pp. 47–63. New York: Chapman & Hall.

20 Nervous system

The nervous system is an information processing and conducting system ensuring the rapid functioning and co-ordination of effectors, producing and modifying the insect's responses to the input from peripheral sense organs. The sensory systems are considered in Chapters 22–24, and the stomodeal system, regulating gut activity, is described in section 3.1.5. This chapter discusses the basic structure and functioning of neural elements, and the anatomy and functioning of the central nervous system.

Review: Burrows, 1996 – locust central nervous system

20.1 BASIC COMPONENTS

20.1.1 Neuron

The basic element in the nervous system is the nerve cell, or neuron (Fig. 20.1). This consists of a cell body containing the nucleus from which one or more long cytoplasmic projections extend to make contact with other neurons or with effector organs, principally muscles. The cell body is called the soma or perikaryon. It contains abundant mitochondria, Golgi complexes and rough endoplasmic reticulum. Information is conducted from one cell to another along the processes of the neuron. Usually, one of the processes, called the dendrite, is specialized for the reception of information. It may arise directly from the soma or as a branch from a major projection of the cell. Input directly to the soma rarely occurs in insects. The branch of the cell which carries information to other cells is usually much longer than the input branch; it is called the axon. In the central nervous system it is common for both dendrites and axon to branch extensively; they are said to arborize. Golgi complexes and rough endoplasmic reticulum are absent from the dendrites and axon. No protein synthesis occurs in these parts so all organelles and proteins within them are manufactured in the soma. Numerous microtubules, about 20 nm in diameter, run along the branches, possibly providing pathways for transport of material from the soma.

Most insect neurons are monopolar, having only a single projection from the soma (Fig. 20.1a). This projection, or neurite, subsequently branches to form the axon and dendrite. The peripheral sense cells are bipolar with a short, and usually unbranched, distal dendrite receiving stimuli from the environment and a proximal axon extending to the central ganglia (Fig. 20.1b). Some multipolar cells, with a number of branches, (Fig. 20.1c) occur in the ganglia of the nervous system and are also associated with stretch receptors.

The points at which neurons receive information from, or convey it to other cells are known as synapses. At most synapses, where transmission from one cell to another involves a chemical, the plasma membranes of the two cells lie parallel with each and are close together, separated by a gap of 20–25 nm. These chemical synapses are characterized, when viewed in section with the electron microscope, by an area of electron-dense material adjacent to the membrane of the fiber transmitting information, the presynaptic fiber. The dense material, which possibly represents proteins involved in the release of chemical transmitter, forms a patch 150–500 nm across.

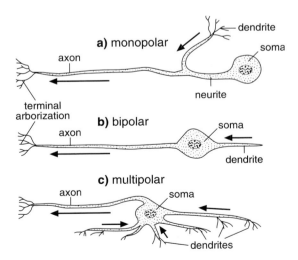

Fig. 20.1. Neurons. Diagrammatic representation of basic types of neuron. Arrow shows direction of conduction.

Fig. 20.2. Synapses. (**a**) Synapse in the optic lobe of *Drosophila*. The presynaptic cell is a retinula cell, the two postsynaptic cells are monopolar interneurons (see Fig. 20.21) (after Meinertzhagen & O'Neil, 1991). (**b**) Synapses on small sections of branches of a non-spiking neuron. Inward-pointing arrows indicate positions of input synapses from other neurons; outward-pointing arrowheads show positions of output synapses. Notice that the major branch has a mixture of outward and inward synapses; the minor branch has only input synapses. Other branches (not shown) have only output synapses. The total number of synapses on this cell would be many thousands. Hatched areas show points at which branches extend out of the plane of view (after Watson & Burrows, 1988).

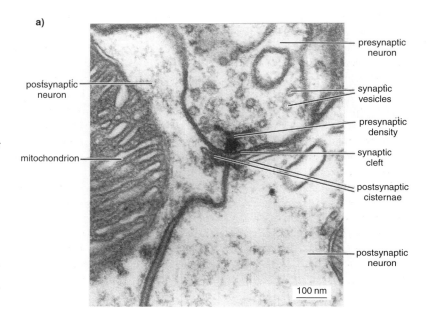

a)

postsynaptic neuron

mitochondrion

presynaptic neuron

synaptic vesicles

presynaptic density

synaptic cleft

postsynaptic cisternae

postsynaptic neuron

100 nm

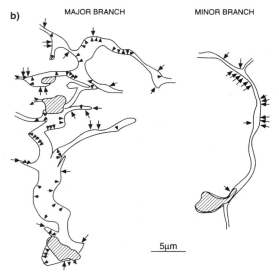

b)

MAJOR BRANCH   MINOR BRANCH

5μm

In the cytoplasm adjacent to the dense material are synaptic vesicles (Fig. 20.2a). These are often small, 50–80 nm in diameter, and electron lucent, but may be larger and electron dense. Sometimes both types are present at a single synapse. The electron-lucent vesicles contain the chemical transmitter that conveys information to the following, postsynaptic cell. They release their contents into the gap between the cells, known as the synaptic cleft, by exocytosis. The electron-dense granules contain neuropeptides with a modulatory function (see below). The postsynaptic cell has electron-dense material adjacent to the membrane opposite the presynaptic specialization and sometimes, as in Fig. 20.2a, there are cisternae.

Cells make contact with each other at many synapses. A single retinula cell in the eye of a fly has about 200 output synapses, while some interneurons of the locust may have as many as one million output synapses. Often,

Communication

A   Physiological Co-ordination Within the Insect

the output synapses of a cell occur on the branches of the axon, while input synapses are on the dendrites, but, in many neurons, input and output synapses occur on the same branches (Fig. 20.2b). Sometimes a single output synapse communicates with more than one postsynaptic cell as in Fig. 20.2a. The number of synapses in a cell is not necessarily constant; changes may occur during postembryonic development and also with the experience of the insect. A single neuron typically has input synapses connecting it to more than one presynaptic neuron and makes output to more than one postsynaptic cell.

Some neurons are in direct electrical contact with each other, instead of communicating chemically. At these electrical synapses, the membranes of the two cells are separated by a gap of only 3.5 nm, and ion channels in the two membranes link with each other across the gap. This type of connection between cells is called a gap junction.

20.1.2 Glial cells

Each neuron is, except for its finest branches, almost wholly invested by folds of one or more glial cells. Synaptic contacts between neurons occur only where glial folds are absent, although the absence of a glial fold does not necessarily indicate a synapse. Glial cells are much more numerous in the central nervous system than neurons. Some are closely associated with the somata, others are at the surface of the neuropil with processes extending inward to invest the axons and dendrites. In the case of small axons, several may be enclosed within one glial fold (Fig. 20.3a, axon B), but larger axons are usually completely enveloped. The enveloping sheath may consist of a single fold or the fold may coil round several times so that there are several layers forming the envelope (Fig. 20.3a, axon A). In general, larger axons are invested by a larger number of glial windings. The same effect may be achieved by several overlapping cells. Glial processes are often connected to each other by desmosomes, tight junctions and gap junctions.

Between the glial cells are extracellular spaces. These are relatively extensive in the outer regions of ganglia, but much more restricted within the neuropil, where they are continuous with the narrow spaces between the glial folds. These narrow spaces show periodic lacunae, which are characteristic of insect nerves (Fig. 20.3a). The fluid in the extracellular spaces bathes the neurons directly and is

therefore of great importance in determining their electrical properties (see below).

The glial cells pass nutrient materials to the neurons and, at times, contain extensive reserves of glycogen. Nutrient transfer is facilitated by finger-like inpushings into the neuronal somata. They are relatively much more abundant in neurons with large somata.

Glial cells have an important function during development of the central nervous system (Tolbert & Oland, 1990) and, perhaps, also in making repairs.

Nerve sheath Because nerve activity depends on movement of ions, it is necessary to maintain the immediate environment of the neurons as constant as possible despite the marked differences in ionic concentration that occur between it and the hemolymph. This constancy is provided by a layer of specialized glial cells, known as the perineurium, round the outside of the nervous system (Fig. 20.3a).

The cells of the perineurium are held together by tight junctions and septate desmosomes. On the inner side, the cells have relatively extensive processes which penetrate between other glial cells. The perineurium extends over the whole of the central nervous system and the larger peripheral nerves, but it is absent from the finer branches of peripheral nerves where the axons are wrapped round by individual glial cells. It forms the 'blood–brain barrier', allowing the passage only of specific substances into the neural environment (see below).

Outside the perineurium is a thick basal lamina known as the neural lamella. It is an amorphous layer of neutral mucopolysaccharide and mucoprotein in which collagen-like fibrils are present. The fibrils lie parallel with the surface, but are otherwise randomly oriented. The neural lamella is probably secreted by cells of the perineurium, although other cells may also contribute to it. It provides mechanical support for the central nervous system, holding the cells and axons together while permitting the flexibility necessitated by the movements of the insect. It offers no resistance to diffusion of material from the hemolymph into the nerve cord. The perineurium and neural lamella are collectively known as the nerve sheath.

Reviews: Carlson & Saint Marie, 1990 – glia; Edwards & Tolbert, 1998 – glia; Ito, Urban & Technau, 1995 – glial classification; Lane, 1985 – structure of neurons and glia; Strausfeld & Meinertzhagen, 1998 – neurons; Treherne, 1985 – blood-brain barrier

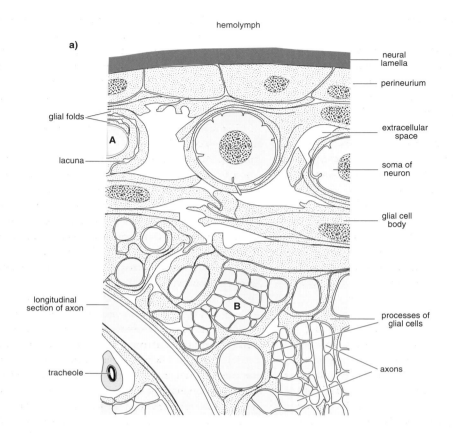

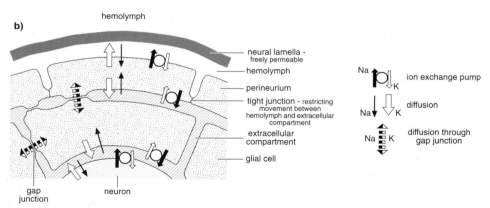

Fig. 20.3. Glia. **(a)** Diagrammatic section of part of an abdominal ganglion showing glial cells (light shading) that play different roles. Axons A and B are referred to in the text (after Smith & Treherne, 1963). **(b)** Representation of the role of glial cells in maintaining a constant ionic composition of the fluid bathing the neurons in a ganglion. The extracellular compartment is shown hatched to emphasize its isolation from the hemocoel. The pumps maintain a high concentration of sodium and low potassium in this compartment despite the diffusive movements of the ions. Broad open arrows indicate greater permeability to potassium than to sodium ions (after Treherne & Schofield, 1981).

20.2 BASIC FUNCTIONING

20.2.1 Electrical properties of the neuron

A difference in electrical potential is maintained across the plasma membrane of a neuron, the inside being negatively charged with respect to the outside. In an unstimulated cell, this resting membrane potential is often about $-70\,mV$.

The membrane potential arises from the differential distribution of positively and negatively charged ions inside and outside the cell. Several factors combine to produce this distribution: a Donnan equilibrium is established by non-diffusible organic anions within the neuron; the resting cell membrane is more permeable to potassium than to sodium ions; and membrane pumps exchange sodium ions from the cell for potassium ions outside it in the ratio 3:2. The net result is that potassium ions are at a high concentration within the cell, but low outside, while for sodium and chloride the converse is true.

Neurons differ from other types of cell in that the membrane potential can be varied considerably, and changes in potential provide the signals that enable neurons to communicate with each other. Because the resting membrane potential differs from zero, the cell is said to be polarized. A change in the potential towards zero, a decrease in its negative charge, is called depolarization; an increase in negativity is called hyperpolarization.

20.2.2 Signal transmission

Transmission of signals via the nervous system involves three different processes. First, the incoming signal, which may be visual, mechanical or chemical, is converted to electrical energy (transduction). The transduction mechanisms used by sensory cells are considered in sections 22.2.3, 23.1.2, 24.1.2 and 24.2.2. The change in membrane potential produced by the incoming signal is called the receptor potential in the case of sensory neurons, or the postsynaptic potential at synapses within the central nervous system or in muscles. Second, the electrical signal is conducted along the axon as action potentials; and, finally, the electrical signal is usually converted to a chemical signal for transmission to the following cell at the synapse.

Review: Pichon & Ashcroft, 1985

20.2.2.1 *Action potential*

Unlike the receptor or postsynaptic potential, which may vary in amplitude (see below), the action potential, or spike, is of constant amplitude. The first phase of the action potential is a small increase in permeability of the cell membrane to sodium. As a result, sodium ions flow into the axon down the concentration gradient and produce a small depolarization of the membrane. This, in turn, causes voltage-sensitive sodium channels to open, and this results in a rapid positive swing in the charge on the inside of the membrane, amounting to 80–$100\,mV$ in the cockroach. This is the rising phase of the action potential (Fig. 20.4a). The period of permeability to sodium is short-lived because the sodium channels close and are inactivated. Potassium channels activate more slowly than sodium channels, starting to open as the sodium channels close. As a result, potassium flows out of the fiber, which again becomes more negatively charged on the inside. This is the falling phase of the action potential. The total duration of the action potential is very brief, only two or three milliseconds.

A current flows in a local circuit away from the point of depolarization inside the axon, and towards it on the outside (Fig. 20.4b). Where this current reaches an area of resting membrane, it produces a slight depolarization, leading to further changes in permeability, first to sodium and then to potassium, that produce an action potential. This is a continuous process so that the action potential is propagated along the fiber. Although current flows out from the action potential in both directions, no further change occurs in the membrane potential of the area through which the action potential has just passed because the sodium channels are inactivated and the continued outflow of potassium negates any tendency for depolarization.

The number of action potentials produced is proportional to the size of the postsynaptic potential (or receptor potential in the case of sensory neurons). As all the action potentials produced by one neuron have the same amplitude, information about the magnitude of any signal can only be conveyed in their number and frequency. The frequency at which action potentials can follow each other is limited by the refractory condition of the axon immediately following the passage of an action potential. The period for which no further action potentials can be produced is known as the absolute refractory period; it lasts for two or three milliseconds (Fig. 20.4b). After this, an action potential can be produced, but only by a very strong stimulus (a large receptor potential) and, as the axon recovers, progressively weaker stimuli are able to produce action

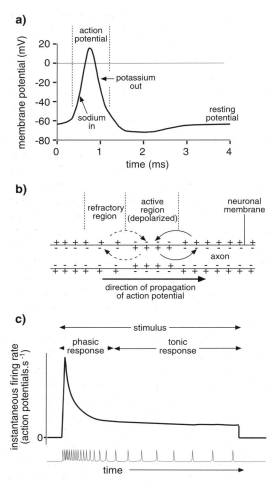

Fig. 20.4. Action potential. **(a)** Changes in the membrane potential associated the production of an action potential. **(b)** Diagram showing the electrical charge across the axonal membrane and propagation of the action potential. At the extreme ends, the membrane is polarized; the inside is negatively charged with respect to the outside. The action potential results from depolarization, the inside of the membrane becoming less negatively charged and even becoming positive. This depolarization spreads passively along the membrane (shown by arrows). In the region through which it has recently passed (the refractory region), the outward flow of potassium offsets the build up of positive charge. Consequently, the membrane repolarizes in this region, but becomes depolarized in the previously resting membrane (to the right of center in the diagram). **(c)** Below: a train of action potentials from a peripheral sensory neuron showing the phasic and tonic periods; above: the same information shown as the instantaneous firing rate.

potentials until the level of excitation returns to normal. This relative refractory period lasts for 10–15 ms.

The rate at which action potentials are produced by a neuron decreases with time after arrival of the stimulus (Fig. 20.4c). At first, action potentials are often produced in rapid sequence, but the intervals between them get longer within a few milliseconds. This type of response is called phasic. In many neurons it is followed by a more sustained response during which the rate of action potential production remains relatively constant, or declines only slowly. This is called a tonic response. The rate of action potential production is often called the firing rate. This is often expressed as the total number of action potentials produced in one second (or some other appropriate unit of time). However, because the rate of firing may vary considerably during the phasic period of a response, it is often expressed as the instantaneous firing rate, which is based on the intervals between action potentials. For example, a cell producing two action potentials separated by 10 ms is said to have an instantaneous firing rate of $100\,s^{-1}$ at that moment (the number of 10 ms intervals in one second).

The production of an action potential in a neuron normally results from stimulation of receptors, but many neurons also fire spontaneously in the absence of any such input. This spontaneous activity probably maintains the system in a highly active state so that it is more easily excited. Stimuli which would be subthreshold in an unexcited fiber may produce a response in a fiber already discharging spontaneously. The firing rate of a spontaneously active neuron may be increased, decreased or remain unaffected by the input of other neurons.

The velocity of conduction in a nerve fiber is proportional to the square root of its diameter. Giant fibers, with diameters of 8–50 μm, have conduction speeds of 3–$7\,m\,s^{-1}$; fibers of about $5\,\mu$m diameter conduct at speeds of 1.5–$2.3\,m\,s^{-1}$. The velocity of conduction also varies with temperature. Following the development of an action potential, the ionic composition within the axon is altered; the sodium concentration has increased and the potassium concentration decreased. The ionic quantities involved are very small, but, nevertheless, if the axon is to continue functioning over long periods a recovery mechanism is required to bring the ionic concentrations back to their original values. This involves a membrane pump which uses energy to exchange sodium for potassium.

Not all neurons produce action potentials. In some cells, the receptor or postsynaptic potential spreads passively throughout the cell producing changes in membrane potential at presynaptic sites that lead to the release of a transmitter substance, which then acts on a following neuron to produce action potentials. The retinula cells of the eye are of this type, as are some interneurons in the central nervous system. The latter are called non-spiking interneurons (see below). The spread of information through the cell is slow compared with that resulting from action potentials, and non-spiking neurons are present in situations where short-distance communication occurs.

20.2.2.2 *Events at the synapse*

Chemical synapses The arrival of an action potential at a presynaptic terminal of a chemical synapse results in the opening of voltage-sensitive calcium channels so that calcium ions move into the neuron. Calcium increases the probability that a synaptic vesicle, containing transmitter substance, will fuse with the neuronal membrane and release the transmitter into the synaptic cleft. The greater the number or frequency of arrival of action potentials, the more calcium enters the cell and the more synaptic vesicles release their contents.

Each vesicle contains a standard amount of transmitter so that the release of multiple vesicles increases the amount of transmitter released in a stepwise manner. The basic quantity is called a quantum. Because the release of transmitter is a probabilistic event, it occurs occasionally even in the absence of any electrical activity in the presynaptic axon.

The transmitter affects the permeability of the postsynaptic membrane either directly, causing ligand-gated ion channels to open, or indirectly, binding to membrane receptors and activating second messenger systems. The postsynaptic change may be a depolarization or a hyperpolarization of the membrane depending on the nature of the transmitter substance and the receptors present on the postsynaptic membrane. Depolarization is produced by opening cation channels in the membrane. As sodium and calcium ions flow in faster than potassium ions move out, the magnitude of the negative charge on the inside of the membrane is reduced. Hyperpolarization is produced by an inward movement of chloride ions, increasing the negative charge. A depolarization is excitatory; it increases the probability that action potentials will be produced. A hyperpolarization is inhibitory; it decreases the probability that action potentials will be produced. The changes are known as excitatory postsynaptic potentials (EPSPs) and inhibitory postsynaptic potentials (IPSPs), respectively.

The magnitude of the postsynaptic potential is variable (graded), and is proportional to the amount of transmitter substance crossing the synapse from the presynaptic terminal. The change in potential induced at the synapse spreads passively, electrotonically, along the dendrite, progressively decreasing in amplitude. Action potentials are only produced if the change in membrane potential at the action potential initiation zone, usually near the soma, exceeds a certain magnitude. This change does not depend on the activity of a single synapse, but is the sum of all the postsynaptic events occurring in the neuron; the effects are said to summate. Summation may be either spatial or temporal.

As contact between any two neurons involves multiple synapses, spatial summation is a normal feature of neuronal activity. It may also involve the inputs from more than one neuron and the inputs may be excitatory, inhibitory or both. Thus, spatial summation may result in an increase (Fig. 20.5b) or a decrease (Fig. 20.5d) in the magnitude of the postsynaptic potential, either increasing or decreasing the likelihood that an action potential will be produced by the postsynaptic neuron.

Temporal summation results from the arrival of successive action potentials at a synapse (Fig. 20.5a). The postsynaptic potential produced by the arrival of a single action potential at an excitatory synapse may not be big enough to initiate a new action potential and will slowly decay. If, however, a second action potential arrives at the synapse and causes a further depolarization of the postsynaptic fiber before the first potential has completely decayed, the total potential resulting from the two successive depolarizations is greater than would have been produced by either event by itself. It may exceed the threshold and initiate a postsynaptic action potential.

Depression of the activity of a synapse can also occur as a consequence of its previous activity (Fig. 20.5c). **Review:** Callec, 1985

Electrical synapses At an electrical synapse, there is a direct flow of current from the pre- to the post-synaptic cell. As a result, transmission is extremely rapid. Some electrical synapses in some animals are known to conduct in either direction, but it is not known if this is also true in insects.

20.2.2.3 *Ionic environment of the neurons*
As the activity of the neuron involves the movement of ions into and out of the cell it follows that neural activity is

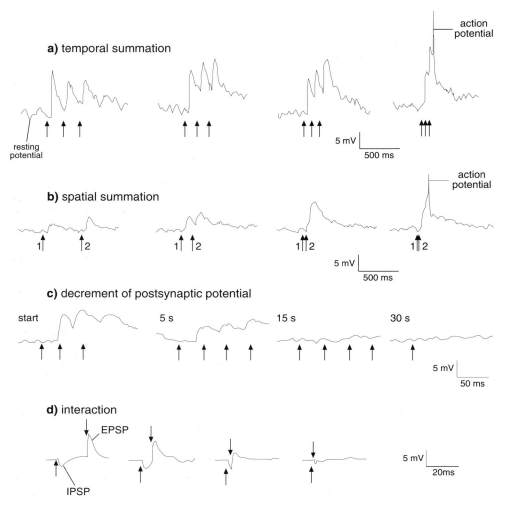

Fig. 20.5. Postsynaptic changes in membrane potential. Each trace shows changes in the membrane potential of the postsynaptic interneuron following stimulation by the presynaptic neuron. Arrows indicate the arrival of an action potential in the presynaptic axon. Notice the different time scales. **(a)** Temporal summation in an interneuron. Stimulation via a single axon results in a progressive increase in the postsynaptic potential as the intervals between stimuli are reduced. At very short intervals, an action potential is produced (extreme right) (input from locust tibial mechanoreceptor) (after Newland & Burrows, 1994). **(b)** Spatial summation in an interneuron. Stimulation via two separate axons (labelled 1 and 2) produces an action potential when the stimuli are close together (extreme right) (input from locust tibial mechanoreceptors) (after Newland & Burrows, 1994). **(c)** Depression due to previous activity. Continuous stimulation over a 30-second period results in a progressive decline in the postsynaptic potential. Figure shows the changes in membrane potential of an interneuron at various times after the start of stimulation (after Newland & Burrows, 1994). **(d)** Interaction of excitatory and inhibitory effects. Summation between an excitatory postsynaptic potential (EPSP) and an inhibitory postsynaptic potential (IPSP) is dependent on the interval between the presynaptic events shown by arrows (cockroach giant interneuron) (after Callec, 1985).

influenced by the concentration of ions in the medium bathing it. For example, if the level of potassium in the external fluid is raised, there is less tendency for potassium to flow out of the axon and hence the membrane potential is reduced. Similarly a low level of sodium in the external medium reduces the magnitude of the action potential. The external concentration of calcium is also critical because calcium is involved in the release of synaptic transmitters. Lowering the level of external calcium results in complete nerve block. In *Carausius*, magnesium is also an essential constituent of the extracellular fluid, possibly being able to replace sodium to some extent in the development of an action potential. In other species, high concentrations of magnesium block conduction.

For the nervous system to function efficiently, therefore, it should be in a constant ionic environment, but the hemolymph that bathes the tissues often differs widely in composition from that required by the nervous system. In many insects, the concentration of sodium in the hemolymph is high, while the potassium concentration is low (section 5.2.2.1), but in some, particularly in herbivorous insects such as *Carausius*, the converse is true.

The environment within the nervous tissue is kept constant by the so called 'blood–brain barrier' and the activities of other glial cells (Fig. 20.3b). The tight junctions between the perineurial cells provide an effective intercellular barrier to ionic movement, which is, therefore, largely restricted to movement through the cells. It is believed that the perineurial cells are directly connected with other glial cells by gap junctions. Ion pumps in the membranes of these other cells are presumed to regulate the ionic concentration in the fluid immediately surrounding the neuron, while the perineurial cells, exchanging ions with the hemolymph, maintain the concentrations in the system as a whole. As a consequence, the concentrations of ions in the extracellular fluid of the nervous system are different from those in the hemolymph (Table 20.1).

20.2.3 Chemical messengers of neurons

A variety of chemical messengers are produced and released by insect neurons. These substances can be grouped into four classes, based on their chemical structure: acetylcholine, biogenic amines, amino acids and peptides (Fig. 20.6, Table 20.2). Some 100 different neuropeptides have been identified in insect nervous systems, although in many cases their functions are not known. Neuropeptides are synthesized in the cell soma

Table 20.1. *Ionic concentrations (mmol kg^{-1} tissue) within the nerve sheath and in hemolymph of some insects (after Treherne, 1974)*

Species	Region	Na	K	Ca	Mg
Periplaneta	Hemolymph	156	8	4	5
	Nerve cord	76	132	—	—
Romalea	Hemolymph	56	18	—	—
	Nerve cord	69	89	—	—
Carausius	Hemolymph	15	18	7	53
	Nerve cord	64	313	30	22

and transported to release points along the axon; most other chemical messengers of neurons are synthesized in the terminal regions of the axon.

The chemical messengers have three types of functions:

Neurotransmitters are only released into the synaptic cleft and have a transient effect on the electrical potential of the postsynaptic membrane. The effect is transient due to enzymatic degradation of the transmitter molecule or its re-uptake into the presynaptic terminal.

Neuromodulators are released in the vicinity of the synapse, modifying synaptic transmission. Their effects are relatively slow and long lasting; specific degradation or re-uptake does not occur.

Neurohormones are released into the hemolymph from neurohemal release areas and function as hormones.

It is common for more than one of these chemical messengers to be produced by a single neuron, and when they occur together within one cell they are said to be co-localized. As their release from the neuron is regulated by its electrical activity, they may be co-released although the release of neuromodulators may only occur with higher levels of electrical activity. While the neurotransmitter is released only into the focal area of the synaptic cleft, neuromodulator release occurs over a broader area. It may be limited to the presynaptic membrane immediately surrounding the synapse. This is called parasynaptic release (Fig. 20.7b) and the effects of the neuromodulator occur only within a limited area. In other cases, the neuromodulator is released over a larger unspecialized region of the axon and has a more

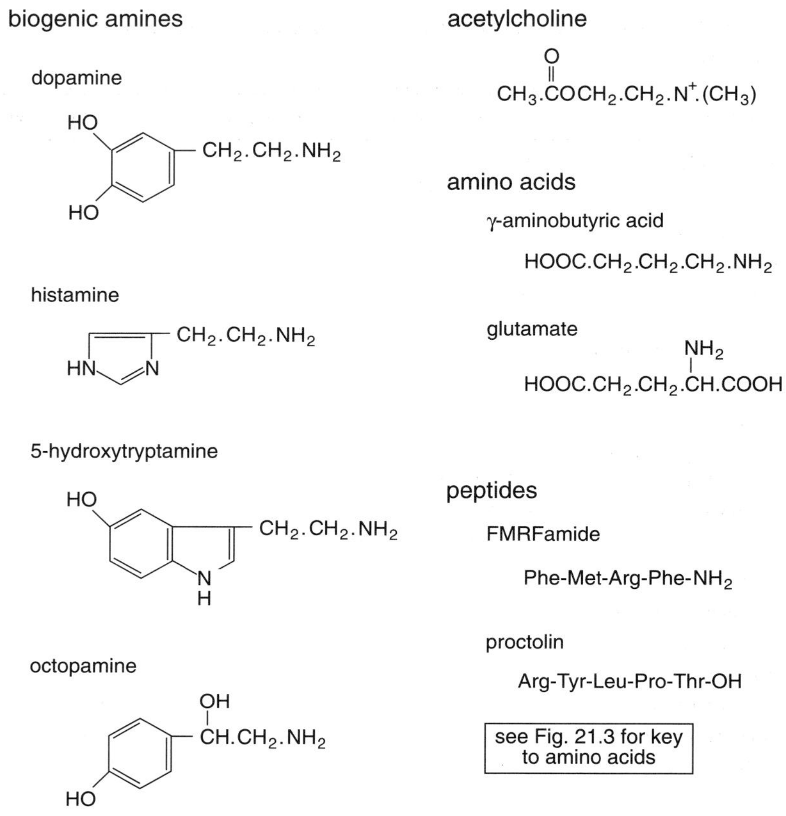

Fig. 20.6. Compounds produced by neurons that act as neurotransmitters or neuromodulators. Some may also act as neurohormones (see Fig. 21.3).

biogenic amines

dopamine

HO, HO — $CH_2.CH_2.NH_2$

histamine

$CH_2.CH_2.NH_2$

5-hydroxytryptamine

$CH_2.CH_2.NH_2$

octopamine

OH
$CH.CH_2.NH_2$

acetylcholine

$$CH_3.\overset{\overset{\displaystyle O}{\|}}{C}OCH_2.CH_2.N^+.(CH_3)$$

amino acids

γ-aminobutyric acid

$HOOC.CH_2.CH_2.CH_2.NH_2$

glutamate

$$HOOC.CH_2.CH_2.\overset{\overset{\displaystyle NH_2}{|}}{C}H.COOH$$

peptides

FMRFamide

Phe-Met-Arg-Phe-NH$_2$

proctolin

Arg-Tyr-Leu-Pro-Thr-OH

see Fig. 21.3 for key
to amino acids

extensive effect, perhaps affecting transmission at a number of synapses. This type of secretion is called paracrine release (Fig. 20.7c, d).

20.2.3.1 *Neurotransmitters*

Acetylcholine is probably the most widespread excitatory transmitter in the insect nervous system (Table 20.2). It is the major transmitter of olfactory and mechanosensory neurons and of many interneurons. Serotonin appears to be the neurotransmitter of some chordotonal and multipolar sensory neurons (Lutz and Tyrer, 1988), while histamine is the transmitter in retinula cells of the compound eyes and ocelli, and dopamine may be a transmitter in the salivary glands. Octopamine functions as a neurotransmitter in the light organs of fireflies (Christensen *et al.*, 1983).

Glutamate is the principal excitatory transmitter at insect nerve/muscle junctions, but also occurs in some interneurons. γ-aminobutyric acid is the principal

inhibitory neurotransmitter both in the central nervous system and at nerve/muscle junctions.

Other compounds amongst those listed in Table 20.2 may also sometimes function as neurotransmitters, but the distinction between neurotransmitter and neuromodulator is not always clear.

20.2.3.2 *Neuromodulators*

The synaptic interaction of one neuron with another neuron, or with a muscle, mediated by a neurotransmitter can be altered (modulated) by chemicals known as neuromodulators. Neuromodulators can act presynaptically, altering the tendency for vesicles of transmitter substance to be released, or postsynaptically, altering the response of the postsynaptic membrane to a given quantity of neurotransmitter.

They produce these effects by modifying the ionic permeability of the pre- or postsynaptic cell. Postsynaptic modulators may compete with a neurotransmitter for

Table 20.2. *Substances secreted by neurons in the central nervous system and their principal functions*

Chemical	Neurotransmitter	Neuromodulator	Neurohormone
Acetylcholine	+++ (excitatory in CNS)		
Biogenic amines			
Dopamine	+	+	
Histamine	+		
Serotonin (5-hydroxytryptamine)		++	+
Octopamine	+	+++	+
Amino acids			
γ-aminobutyric acid (GABA)	+++ (inhibitory)		
Glutamate	+++ (excitatory at muscles)		
Peptides			
Adipokinetic hormone			+++
Allatotropin		+	+
Allatostatin		+	+
Cardioactive peptide			+++
Diuretic peptide			+++
FMRFamide		+	
Leucokinin		+	
Pheromone biosynthesis (PBAN)		+	++
Proctolin	+	+++	+
Prothoracicotropic hormone			+++

Note:

+++, major function; +, occasional function.

receptor sites on the postsynaptic membrane, or they may bind to specific membrane receptors, affecting the cell's ion channels via second messenger systems.

Biogenic amines and a number of neuropeptides are widely distributed in the central nervous system suggesting that they are of considerable importance. In some cases they are known to act as neuromodulators, but their function is only inferred in a majority of instances. There are, as yet, no well established cases of neuromodulatory effects in the nervous system in the intact insect, but experimental work demonstrates the types of effects to be expected.

The overall effects of neuromodulation may be more profound than changes at single synapses might suggest. Although such effects have not yet been demonstrated in insects, in Crustacea a small set of neurons in the stomato-gastric ganglion can deliver several different output patterns depending on the relative amounts of up to 15 different neuromodulators released into the system (see Harris-Warwick *et al.*, 1992).

Homberg (1994) describes the distribution of 'neurotransmitters' (including neuromodulators) in the insect brain.

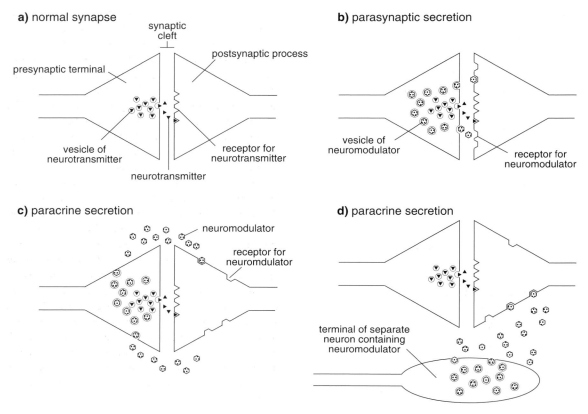

a) normal synapse

synaptic cleft

presynaptic terminal

postsynaptic process

vesicle of neurotransmitter

receptor for neurotransmitter

neurotransmitter

b) parasynaptic secretion

vesicle of neuromodulator

receptor for neuromodulator

c) paracrine secretion

neuromodulator

receptor for neuromdulator

d) paracrine secretion

terminal of separate neuron containing neuromodulator

Fig. 20.7. Neuromodulation at a synapse. Diagrammatic representation of different pathways by which neuromodulators may affect the postsynaptic element. Transmitter compound is shown as black triangles; neuromodulator as shaded hexagons. (a) A synapse with no neuromodulators. (b) Parasynaptic secretion. The neuromodulator co-occurs with the transmitter and acts close to the synapse. (c) Paracrine secretion. The neuromodulator co-occurs with the transmitter but is released round the area of the synapse. It may affect other neurons in the vicinity. (d) Paracrine secretion. The neuromodulator is produced by a separate neuron and affects the general environment around the synapse.

Octopamine Octopamine probably has a widespread modulatory effect on the activity of skeletal muscles (see section 10.3.2.4), but it also modulates the activity of some interneurons and sensory neurons. One example of its effect on interneurons concerns the generation of the pattern of activity of locust flight muscles. During flight, the interneurons produce action potentials in bursts (Fig. 20.8a), but a similar bursting pattern cannot be generated experimentally when the insect is at rest unless octopamine is applied to the preparation. It is known that the concentration of octopamine in the metathoracic ganglion increases during the early part of flight, and it is likely that this

increase is responsible for a change in the properties of the cell membrane of the interneurons (Orchard, Ramirez & Lange, 1993).

Octopamine may also act via the hemolymph as a neurohormone modulating the activity of sensory neurons. This occurs with the neuron of the locust forewing stretch receptor. Its activity is enhanced by octopamine (Fig. 20.8b) which is released into the hemo-lymph at the beginning of flight.

In locusts, the dorsal and ventral unpaired median cells in the thoracic and abdominal ganglia and some inter-neurons in the brain are octopaminergic (Stevenson *et al.*, 1992).

a) interneuron

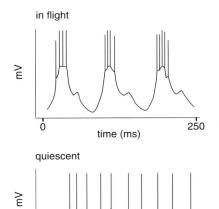

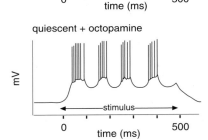

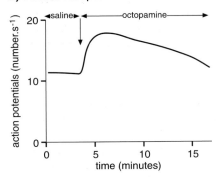

b) stretch receptor

Serotonin Relatively small numbers of serotonergic neurons, mostly interneurons, are present in the central nervous system. Adult *Calliphora*, for example, have only 154 such neurons, some occurring in each of the neuromeres. Despite the relatively small number, most regions of neuropil are invaded by the branches of serotonergic neurons. In locusts, serotonergic axons run to the gut, the reproductive organs and the salivary glands. Comparable innervation does not occur in *Calliphora*, but it is probable that, in this insect, serotonin is released as a neurohormone to act on the same organs.

An example of neuromodulation by serotonin is known from studies on the olfactory lobe of *Manduca* where a single serotonergic neuron innervates all the glomeruli (Fig. 20.23c). A low concentration of serotonin reduces the activity of olfactory-lobe neurons, while a higher concentration enhances it (Fig. 20.9) (Kloppenburg & Hildebrand, 1995). It is known that experimentally-administered serotonin alters the olfactory responses of the honeybee, *Apis*, and it may have similar effects on *Manduca* and other insects. Although the concentrations of serotonin experienced by the neurons in the glomeruli of living insects are not known, the effects produced experimentally by different concentrations are suggestive of how an insect's response might be altered completely by neuromodulatory effects.

Neuropeptides Although neuropeptides occur in cells of the central nervous system, their functions are not generally understood except for those that are neurohormones. One exception is proctolin which functions as a neuromodulator at some visceral and skeletal muscle junctions (section 10.3.2.4). Nässel & O'Shea (1987) report about 500 proctolinergic neurons, both interneurons and motor neurons, in various parts of the central nervous system of *Calliphora*.

20.2.3.3 *Neurohormones*

Neurohormones are produced by neurons in various ganglia of the central nervous system. Their distribution

Fig. 20.8. Neuromodulatory effects of octopamine in the locust. (**a**) Effect on an interneuron associated with the flight muscles. When the insect is in flight the action potentials in the interneuron occur in bursts, but stimulation of the interneuron in a quiescent insect produces a train of action potentials. A bursting pattern similar to that occurring in flight can be produced experimentally by applying octopamine to the quiescent preparation (after Orchard, Ramirez & Lange, 1993). (**b**) Output from a wing stretch receptor with the wing oscillating at 18 Hz (similar to the normal wingbeat frequency). When octopamine was added the output from the stretch receptor increased for a time (after Ramirez & Orchard, 1990).

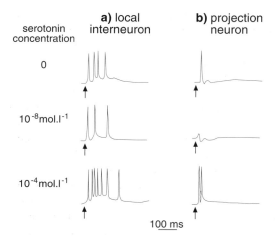

Fig. 20.9. Neuromodulatory effects of serotonin on interneurons in the olfactory lobe of *Manduca*. Notice that a low concentration ($10^{-8}\,\text{mol}\,\text{l}^{-1}$) of serotonin reduced the number of action potentials produced by both neurons when the antennal nerve was stimulated, but a higher concentration ($10^{-4}\,\text{mol}\,\text{l}^{-1}$) had the opposite effect. Arrows show the time at which the antennal nerve was stimulated (after Kloppenburg & Hildebrand, 1995). A diagrammatic representation of the two types of interneuron is shown in Fig. 20.23b. (a) Local interneuron restricted to the antennal lobe. (b) Projection neuron connecting a glomerulus to other parts of the brain.

and functions are considered in Chapter 21. The cells producing neurohormones are commonly called neurosecretory cells. With advancing knowledge it has become clear that many neurons produce comparable secretions which do not necessarily function as hormones, but rather as neuromodulators (see above). In this text, the terms neurosecretion and neurosecretory cell will be used only in cases where the secretion is known to have a hormonal function, being transported in the hemolymph.

Reviews: Golding, 1994 – release of neuromodulators; Holman, Nachman & Wright, 1990 – neuropeptides; Homberg, 1994 – neurotransmitters and modulators in the brain; Kandel, Schwartz & Jessell, 1991 – general account of neural structure and function; Nässel, 1988 – serotonin; Nässel, 1991 – neurotransmitters and neuromodulators in the visual system; Orchard, Belanger & Lange, 1989 – proctolin; Sattelle, 1990 – GABA receptors; Shephard, 1994 – general account of neural structure and function; Treherne, 1985 – blood-brain barrier; Usherwood, 1994 – glutamate receptors

20.3 ANATOMY OF THE NERVOUS SYSTEM

20.3.1 Ganglia

The somata of interneurons and motor neurons are aggregated to form the ganglia of the central nervous system. Within the ganglia, the somata are grouped peripherally. The center of each ganglion is occupied by the terminal arborizations of sensory axons, by the dendritic arborizations of motor neurons, and by axons and arborizations of interneurons. Within this mass of fibers, known as the neuropil, some fibers with a common orientation are grouped together to form distinct tracts, but, in general, the fibers lack a common orientation. Axons and dendrites are complexly interwoven and it is here that synapses occur. No synapses, apart from nerve/muscle junctions, occur outside the central nervous system.

Primitively, a pair of ganglia is presumed to have been present ventrally in each postoral segment. This arrangement is partially visible in many embryonic insects, but in postembryonic stages some segmental ganglia fuse with a presegmental ganglion to form the brain, and some degree of fusion of postcephalic ganglia also occurs. The central nervous system thus consists of the brain which is pre-oral and is positioned dorsally in the head above the foregut, followed by a series of segmental ganglia lying close above the ventral body wall. Adjacent ganglia are joined by a pair of interganglionic connectives that contain only axons and glia; there are no somata, and no synapses.

The first ganglion in the ventral chain is the subesophageal. This is a compound ganglion, lying ventrally in the head, formed by the fusion of the ganglia of the mandibular, maxillary and labial segments. In Orthoptera, the neuromeres of the separate segments are clearly visible, but the segmental origins are not apparent in Hymenoptera and Diptera. The ganglion sends nerves to the mandibles, maxillae and labium, and an additional one or two pairs to the neck and salivary glands.

There are typically three thoracic ganglia, but in some insects they fuse to form a single ganglion (Fig. 20.10). The metathoracic ganglion is often fused with one or more anterior abdominal ganglia. Each thoracic ganglion has some five or six nerves on each side which branch to innervate the muscles and sensilla of the thorax and its appendages. The arrangement of nerves varies considerably, but usually the last nerve of one segment forms a common nerve with the first nerve of the next.

The largest number of abdominal ganglia occurring in

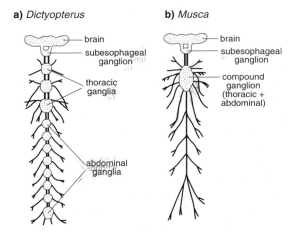

a) *Dictyopterus*

- brain
- subesophageal ganglion
- thoracic ganglia
- abdominal ganglia

b) *Musca*

- brain
- subesophageal ganglion
- compound ganglion (thoracic + abdominal)

Fig. 20.10. Numbers of ventral ganglia (from Horridge, 1965). **(a)** Most ganglia remain separate (*Dictyopterus*, Coleoptera). **(b)** All the ventral ganglia except the subesophageal are fused into a single compound ganglionic mass (*Musca*, Diptera).

larval or adult insects is eight, as in Thysanura, male *Pulex* (Siphonaptera) and many larval forms, but the last ganglion is always compound, being derived from the ganglia of the last four abdominal segments.

The abdominal ganglia are smaller than those of the thorax and, in general, fewer peripheral nerves arise from each of them than arise from the thoracic ganglia. In addition, the branching of the nerves is less diverse and variable, reflecting the relative simplicity and similar segmental organization of the abdominal musculature.

Sometimes, most or all of the ventral ganglia are fused to form a single compound ganglion as in the blood-sucking bug, *Rhodnius*, and in cyclorrhaphan flies (Fig. 20.10b). Adult holometabolous insects often have more ganglia fused than the larvae (see Fig. 15.23a), but the cyclorrhaphous Diptera are an exception.

In most cases the muscles of a segment are innervated by neurons from the ganglion of the same segment, but some innervation by axons arising in neighboring ganglia also occurs (see Fig. 10.8). Some afferent fibers may also be intersegmental (see below).

Reviews: Altmann & Kien, 1987; Weevers, 1985

20.3.2 Neurons of the central nervous system

20.3.2.1 *Motor neurons*

Each segmental ganglion contains the somata of motor neurons concerned with control of muscles, primarily in the same body segment. The number of motor neurons in each ganglion is relatively small, corresponding to the relatively small number of muscle units that insects possess; nevertheless, there are about 500 in the mesothoracic ganglion of *Periplaneta*.

The somata of motor neurons are relatively large; in the locust metathoracic ganglion they range from about 20 to 90 μm in diameter. They are roughly constant in position and homologous neurons on the two sides of a ganglion are placed more or less symmetrically. Somata of axons running within a single nerve tend to be grouped together, but even those running to the same muscle can be widely separated. For example, the somata of the fast and slow axons to the metathoracic extensor tibiae muscle are widely separated and their axons run in different nerve branches.

The neurite of each motor neuron increases in diameter on entering the neuropil, within which it gives rise to many branches (Fig. 20.11). The shape of this dendritic tree is characteristic of each motor neuron and its complexity an indication of the complexity of its synaptic connections with other neurons. The axon branches from the neurite and runs, unbranched (except for those in the median nerve), into one of the peripheral nerves.

20.3.2.2 *Interneurons*

Direct synaptic connections are known to occur between afferent (sensory) fibers and motor neurons, but these monosynaptic pathways are unlikely to occur commonly because this arrangement allows for little variability of response. In general, interneurons are interpolated between sensory and motor neurons, and most of the neurons in the central nervous system are interneurons.

Anatomically, interneurons can be divided into local and intersegmental interneurons. Local interneurons are restricted to a single ganglion; they may be spiking or non-spiking. It is estimated that there are approximately 1500 intraganglionic interneurons in the mesothoracic ganglion of *Periplaneta* compared with 200 interganglionic interneurons. Non-spiking interneurons form a significant proportion of these local interneurons, at least in the thoracic ganglia. For example, in the metathoracic ganglion of the locust about 100 non-spiking neurons are recognizable compared with about 140 spiking local interneurons, but many other neurons remain to be characterized.

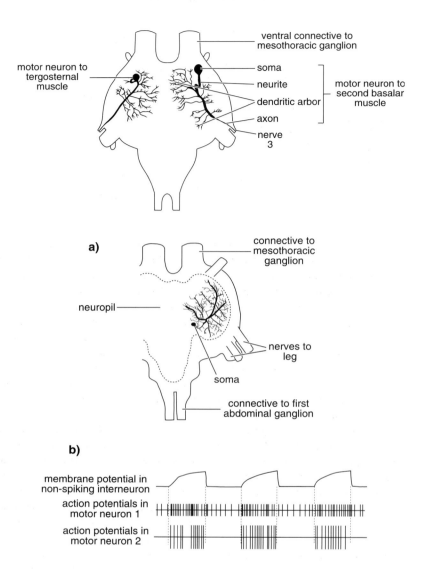

Fig. 20.11. Two motor neurons in the metathoracic ganglion of a locust showing the main dendritic branches and axons extending into lateral nerve 3. The somata are in the outer part of the ganglion while the dendrites are in the central neuropil. Left: motor neuron to a tergosternal muscle; right: motor neuron to the second basalar muscle (after Burrows, 1977).

Fig. 20.12. Non-spiking neurons. **(a)** A non-spiking neuron from the metathoracic ganglion of a locust (after Siegler & Burrows, 1979). **(b)** A non-spiking neuron affecting the activity of two motor neurons (labelled 1 and 2). An increase in the membrane potential (depolarization) causes neuron 1 to fire more actively and neuron 2 to start firing (after Pearson, 1977).

Non-spiking local interneurons Non-spiking local interneurons in the ventral ganglia often arborize extensively in the ipsilateral side of the ganglion (that is, on the same side as the soma) (Fig. 20.12a). Slow changes in membrane potential are adequate to produce synaptic transmission; spike production does not occur. Non-spiking neurons have low resting potentials, of about −30 to −50 mV, compared with spike-producing neurons (about −70 mV). Postsynaptic potentials are produced by these neurons by the release of a chemical synaptic transmitter just as in spiking neurons, and in some cases the resting membrane potential is sufficiently low to produce a tonic postsynaptic potential. This may be depolarizing, so that the postsynaptic fiber fires tonically, or hyperpolarizing, so that the activity of the postsynaptic fiber is inhibited. Changes in the membrane potential of the non-spiking neuron lead to graded changes in postsynaptic potential and hence to changes in the rate of spike production in the postsynaptic fiber (Fig. 20.12b). An experimental change in the membrane potential of as little as 2 mV may have such an effect, and, in the course of normal activity, changes of up to 15 mV can occur.

Review: Siegler, 1985

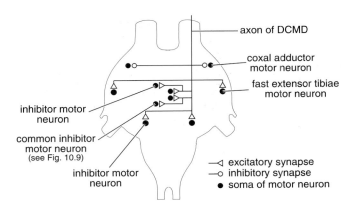

axon of DCMD

coxal adductor
motor neuron

fast extensor tibiae
motor neuron

inhibitor motor
neuron

common inhibitor
motor neuron
(see Fig. 10.9)

inhibitor motor
neuron

◁ excitatory synapse
○ inhibitory synapse
● soma of motor neuron

Fig. 20.13. Connections of a descending interneuron, the contralateral movement detector (DCMD) of the locust (*Schistocerca*). Diagrammatic representation of synaptic connections to motor neurons controlling the muscles of the hind leg. The homologous axon on the left-hand side (not shown) makes similar connections. Note that synapses occur in the neuropil, not directly with the somata as shown in this diagram (after O'Shea, Rowell & Williams, 1974).

Interganglionic interneurons Interganglionic interneurons transmit information along the nerve cord. Fibers that carry information from anterior to posterior ganglia are called descending fibers; those transmitting information to the brain or anterior ganglia are called ascending fibers. These neurons are responsible for co-ordinating the activities of different ganglia and so bring about the coordinated activity of the whole insect. The axons of these neurons may have arbors, indicating that they make synaptic connections, in several ganglia or may extend through some ganglia without branching, arborizing only in the ganglion where they end.

One of the most thoroughly studied interganglionic interneurons is the descending contralateral movement detector (DCMD) of the locust. There is one DCMD in each neural connective, running from the brain to the metathoracic ganglion. The soma lies posteriorly in the protocerebrum. Within the brain, the axon swells to form an integrating segment from which six or more major dendrites diverge so that their fine branches occupy a sphere about 200 μm in diameter. The neuron receives input from other neurons in this region. In particular, it receives information about small-scale movements in the visual field. Beyond the integrating segment, the axon crosses to the contralateral side of the brain and then passes down the ventral connective to the metathoracic ganglion. It gives rise to a single branch in the prothoracic ganglion, and to further branches in both the mesothoracic and metathoracic ganglia. These branches form less extensive arborizations than the dendritic fields of the motor neurons with which they connect. The neuron makes excitatory output synapses to a number of motor neurons, some of which are shown in Fig. 20.13. These include neurons of the extensor

tibiae muscle and the common inhibitor of the leg musculature (see Fig. 10.9). It also provides inhibitory input to the coxal adductor muscle motor neuron.

The ventral nerve cords of cockroaches, various Orthoptera, dragonfly larvae, *Drosophila* (Diptera) and possibly all insects contain a number of axons which are much bigger than the majority. They are called giant fibers. In the cockroach, there are six to eight giant fibers, 20–60 μm in diameter, in each connective as well as 10 to 12 medium-sized axons between 5 and 20 μm in diameter. (The axons of most interneurons are less than 5 μm in diameter). In the cockroach, the somata of all the giant fibers are in the terminal abdominal ganglion. In contrast, adult *Drosophila* have only two giant interneurons and they have somata in the brain.

Giant fibers run for considerable lengths of the nerve cord. In *Periplaneta* the largest fibers extend from the terminal abdominal ganglion to the subesophageal ganglion. Within the terminal abdominal ganglion each giant fiber has an extensive dendritic arborization within which synapses are made with afferent fibers from the cercal nerve (Fig. 20.14). Different giant fibers arborize in different parts of the ganglion suggesting that they receive input from different groups of sensilla.

The giant fibers allow for the rapid transmission of information over long distances because of their high conduction rates due to their large size and lack of synapses.

20.3.2.3 *Homologies*

Because the early patterning of neurogenesis is similar in different segments and even in different insects, it is sometimes possible to recognize homologous neurons in different

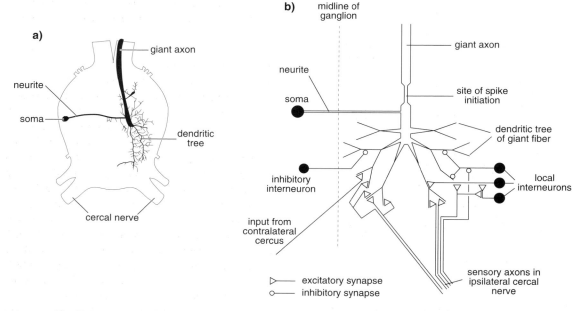

Fig. 20.14 Integration at synapses. A giant fiber in the terminal abdominal ganglion of *Periplaneta* (after Callec, 1985). **(a)** Diagram showing the dendritic arbor and contralateral position of the soma. **(b)** The neuron with the giant axon receives direct input on to its dendritic branches from sensory neurons in both the ipsilateral and contralateral cerci as well as from local interneurons. Inputs are both excitatory and inhibitory.

ganglia and even in different insect species despite the differences that develop postembryonically as a consequence of functional differentiation. For example, specific inter-neurons in the meso-, metathoracic and first abdominal neuromeres of the locust that have somewhat similar arrangements of their axons and arborizations (Fig. 20.15) develop from precisely homologous cells during early development. In other cases, where the embryonic origins are not known, possible homologies are suggested by the similarity of position and anatomy of neurons, sometimes together with their immunocytochemical characteristics.

Review: Kutsch & Breidbach, 1994

20.3.3 Peripheral nerves

The peripheral nerves radiating from the ganglia are made up of large numbers of axons. Within these nerves, the axons remain independent of each other; they do not branch or form synapses. Axons conveying information to the ganglia from the peripheral sensory neurons are said to be afferent or sensory; those that distribute information from the central nervous system are called efferent or motor fibers. Most nerves contain both afferent and

efferent axons, but the antennal and cercal nerves contain only sensory axons.

A small median nerve which runs from the posterior end of each ganglion lies between the paired connectives joining the ganglia. In the abdomen, it extends from one ganglion to the next, but, in the thorax, it extends only part of the way before branching transversely to the spiracles and some ventilatory muscles. These nerves contain small numbers of sensory and motor axons. Each motor axon divides where the median nerve branches, so that the homologous muscles on either side of the body are inner-vated by the same neuron (see Fig. 17.21).

20.4 BRAIN

The brain is the principal association center of the body, receiving sensory input from the sense organs of the head and, via ascending interneurons, from the more posterior ganglia. Motor neurons to the antennal muscles are also present in the brain, but the majority of nerve cell bodies found here belong to interneurons and the bulk of its mass is composed of their fibers. Many of the inter-

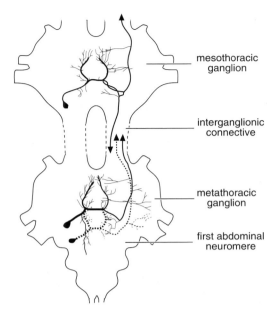

mesothoracic
ganglion

interganglionic
connective

metathoracic
ganglion

first abdominal
neuromere

Fig. 20.15. Homologous interneurons. The three interneurons shown are derived from homologous neuroblasts in the meso- and metathoracic and first abdominal ganglion in the embryo. The major branches of the neuron in the first abdominal neuromere are shown as broken lines for clarity. Arrows indicate that the axon continues in the direction shown (*Locusta*) (after Pearson *et al.*, 1985).

neurons are concerned with the integration of activities; others extend down the nerve cord to the more posterior ganglia, transmitting information that controls the insect's behavior from the brain.

Three regions are recognized in the brain, the protocerebrum, deutocerebrum and tritocerebrum (Fig. 20.16).

20.4.1 Protocerebrum

The protocerebrum is bilobed and is continuous laterally with the optic lobes. In hypognathous insects it occupies a dorsal position in the head and, as with other ganglia, the somata are largely restricted to a peripheral zone while the central region is occupied by neuropil. Anterodorsally, in a region called the pars intercerebralis, a group of somata occurs on either side of the midline (Fig. 20.17). The anterior cells of the pars contribute fibers to the ocellar nerves, while fibers from the more lateral cells enter the protocerebral bridge (pons cerebralis), a median mass of neuropil connecting with many other parts of the brain. Also within the pars intercerebralis are neurosecretory cells, the

axons of which decussate (cross over each other to the opposite side) within the brain and extend to the corpora cardiaca.

20.4.1.1 *Mushroom bodies*

At the sides of the pars intercerebralis are the mushroom bodies, or corpora pedunculata (Fig. 20.17). Each consists of a flattened cap of neuropil, the calyx, from which a stalk (peduncle) runs ventrally before dividing into two or sometimes three lobes, known as the α, β and γ lobes. In *Apis* and *Periplaneta*, the calyx is double and comprises three concentric rings of neuropil known as the lip, collar and basal ring (Fig. 20.18b). The mushroom bodies are given their form by a large number of interneurons, called Kenyon cells, that have their somata above the calyx. Each Kenyon cell has dendrites in the calyx and an axon running down the stalk and often dividing ventrally; collectively they form the α and β, and sometimes also γ, lobes (Fig. 20.18a). Morphologically, several different types of Kenyon cells are recognizable, perhaps reflecting functional differentiation.

The relative size of the corpora pedunculata is related to the complexity of behavior shown by the insects. They are small in Collembola, Heteroptera, Diptera and Odonata, of medium size in Coleoptera, Orthoptera, Blattodea, Lepidoptera and sawflies, and most highly developed in social insects (Fig. 20.19a). In termites, the volume of the peduncle and α and β lobes is large and the calyx represents only about a third of the total volume of the mushroom bodies. In bees and social wasps the calyx is large and very complex representing about 70% of the total volume. Worker honeybees, have about 170 000 Kenyon cells in each mushroom body, representing about 40% of all the neurons in the brain; the fly, *Calliphora*, has about 21 000 Kenyon cells on each side, about 12% of the total neuron population.

Changes occur in the mushroom bodies with age and experience. Over the first week of adult life, the number of fibers in the mushroom bodies of *Drosophila* increases by more than 20% (Fig. 20.19b) and is then maintained at a plateau until the flies are very old. These changes are reflected in an increase in volume of the calyx and probably result from an increase in the number of Kenyon cells (Heisenberg, Heusipp & Wanke, 1995; Technau, 1984). Changes also occur in the overall size of the mushroom bodies over the first ten days of life of worker honeybees (Fig. 20.19c).

The increase in size of the mushroom bodies is affected by the insects' experience in both *Drosophila* and *Apis*. The

Fig. 20.16. Brain. Diagrammatic representations (*Locusta*) (after Albrecht, 1953): (**a**) anterior view; (**b**) lateral view which also shows the major ganglia of the stomodeal nervous system.

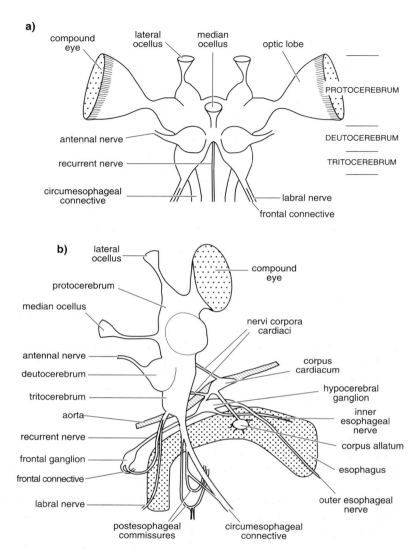

number of fibers in the peduncle increases less in flies kept in isolation compared with those in a crowd, while, amongst *Apis* workers, the lip and collar of the calyx increase more in foragers compared with nurse bees of the same age (Fig. 20.19b,c) (Durst, Eichmüller & Menzel, 1994).

In all insects, the mushroom bodies receive input from neurons carrying information from the antennal lobes (Fig. 20.18b). Within the calyx, each of these neurons communicates with a large number of Kenyon cells. In *Apis*, these connections are made in the lip of the calyx while inputs from the visual

system connect in the collar. The mushroom bodies of other insects do not have these visual connections. Another group of neurons, known as feedback neurons, arborize in both the lobes and the calyx (Fig. 20.18d). The terminals of the Kenyon cells converge on to a number of large output interneurons in the lobes. These interneurons run primarily to other parts of the protocerebrum, perhaps including the central body (Fig. 20.18c).

The mushroom bodies are involved in olfactory, and, perhaps, in some insects, visual learning.

Review: Schurmann, 1987

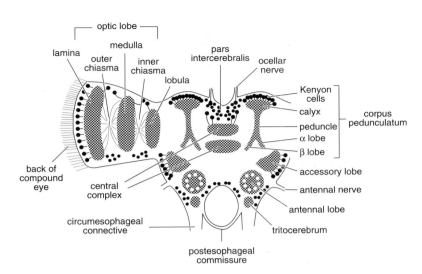

20.4.1.2 *Central complex*

The central complex is a series of interconnected neuropils in the center of the protocerebrum. The principal features are the protocerebral bridge, the central body and lateral accessory lobes (Fig. 20.20a) although different terminology has been used in different insects (see Homberg, 1987). The arrangement of neurons within the complex is well-ordered. One conspicuous feature is an array of 64 axons arising from somata in the protocerebrum and running in bundles of four through the central body. These axons arborize extensively in the protocerebral bridge and in the central body and some also extend into the lateral accessory lobes. In parallel with these fibers, although innervating different parts of the complex, is a second series which are serotonergic (Homberg, 1991). Others, arising from somata lying both dorsally and ventrally to the complex, run transversely through the central body, arborizing in its upper or lower parts (Fig. 20.20b,c). In addition to these intrinsic fibers of the central complex, many others convey information to and from the system. These enter it via the protocerebral bridge or make direct connections between the central body and the anterior and posterior protocerebrum. Some of the probable input fibers are serotonergic and have extensive arborizations in one or another part of the central complex. Most of the cells of the central body contain neuropeptides, probably used as neurotransmitters, and many of the transverse neurons innervating the lower part of the central body are GABAergic.

There are no direct connections between the central complex and the corpora pedunculata, or the optic or antennal lobes, and the functions of the central complex are not understood. As it is the only unpaired region of organized neuropil it is probably concerned with the integration of information from the right and left halves of the brain.
Review: Homberg, 1987

20.4.1.3 *Optic lobes*

The optic lobes are lateral extensions of the protocerebrum to the compound eyes. Each consists of three neuropil masses (Fig. 20.17), known as the lamina, the medulla and the lobula complex which, in Lepidoptera, Trichoptera and Diptera, is subdivided into the lobula and the lobula plate. Within these masses, the arrangements of arborizations of different sets of neurons produce a layered appearance. Between successive neuropils the fibers cross over horizontally forming the outer and inner optic chiasmata so that the neural map of the visual image is reversed and then rereversed.

In the lamina of most insects, the axons of retinula cells from one ommatidium remain together and are associated with neurons originating in the lamina and the medulla to form a cartridge. In insects with an open rhabdom (section 22.1.2), however, each cartridge contains the axons from retinula cells with the same field of view rather than from the same ommatidium. In either case, the number of cartridges in the lamina is the same as the number of ommatidia.

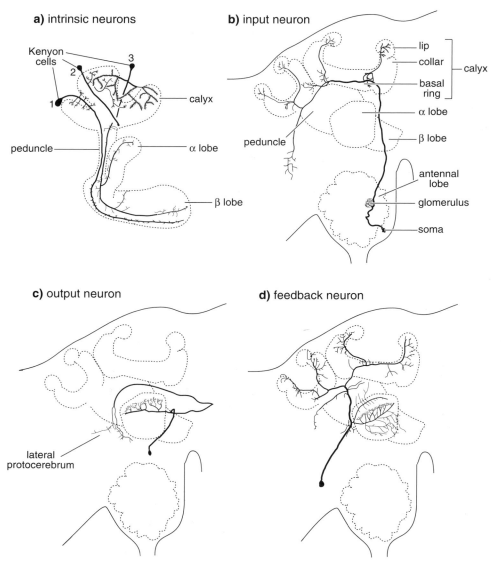

a) intrinsic neurons

Kenyon cells
2
3
1
calyx
peduncle
α lobe
β lobe

b) input neuron

lip
collar
calyx
basal ring
α lobe
peduncle
β lobe
antennal lobe
glomerulus
soma

c) output neuron

lateral protocerebrum

d) feedback neuron

Fig. 20.18. Mushroom body. Examples of neurons associated with the mushroom bodies. (a) Intrinsic neurons (Kenyon cells: labelled 1, 2 and 3) showing different types of dendritic arborization in the calyx. Notice that the axon of neuron 1 extends only into the β lobe; for neuron 2, only the proximal part is shown; the axon of neuron 3 extends to both the α and β lobes (*Apis*) (based on Mobbs, 1985). (b) Input neuron from the antennal lobe. The dendritic arbor is restricted to a single glomerulus, but this is not true of all the input neurons from the antennal lobe to the mushroom bodies. (Notice that this would be called an *output, projection* neuron from the antennal lobe, but an *input* neuron to the mushroom body). The axon has branches to both parts of the calyx (*Manduca*) (based on Homberg, 1994). (c) Output neuron with a dendritic arbor in the α lobe and an axon running to the lateral protocerebrum (*Manduca*) (based on Homberg, 1994). (d) Feedback neuron connecting different parts of the mushroom body (*Manduca*) (based on Homberg, 1994).

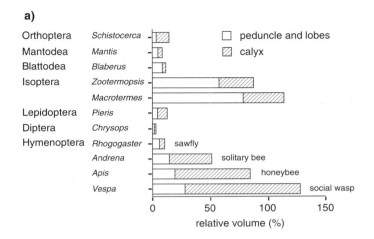

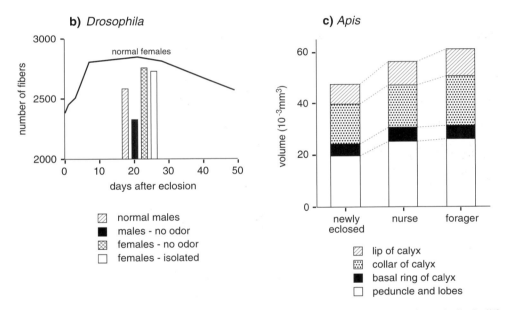

Fig. 20.19. Mushroom bodies. Variation in development. (a) The volumes of the mushroom bodies in different species of insect expressed relative to the size of the central complex. The mushroom bodies are relatively much larger in social insects than in others. Notice that in termites (Isoptera) the peduncle and lobes account for most of the volume whereas in the Hymenoptera the calyx contributes the greater part (after Howse, 1974). (b) Changes in the number of fibers (thought to reflect the number of Kenyon cells) in the peduncle of a mushroom body of *Drosophila*. The insects shown by histograms were all 21 days post-eclosion. Maintenance in environments offering less stimulation resulted in smaller numbers of fibers in both males and females (after Technau, 1984). (c) Changes in the volumes of parts of the mushroom body of *Apis*. All parts increase in volume as the insects get older. Foragers have bigger lips and collars in the calyces than nurse bees of the same age and parents (after Durst *et al.*, 1994).

Fig. 20.20. Central complex. Examples of neurons of the central complex (*Schistocerca*) (after Homberg, 1991, 1994). **(a)** A neuron linking the protocerebral bridge to the lower part of the central body. **(b)** Two neurons connecting the lateral accessory lobes with parts of the central body. Neuron 1 contributes to the system of columns in the upper central body (see text). **(c)** Two symmetrical neurons with extensive arborization in the upper central body.

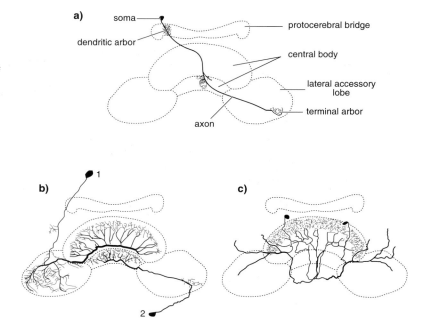

Axons from most of the retinula cells in the eye end in the cartridges, although one or two from each ommatidium pass through to the medulla (Fig. 20.21a). In flies, the six retinula cells ending in the lamina have the same range of spectral sensitivity with maximum sensitivity to green. They form part of a color-insensitive (because they all arise from cells with the same visual pigment) pathway which converges on to the giant interneurons in the lobula (see below). The two cells extending to the medulla (shown solid black in Fig. 20.21a) have different wavelength sensitivities and form part of a color-sensitive system converging on the smaller interneurons of the lobula.

The cartridges of one insect are uniform in structure. In the fly, 19 cells contribute to each one, and each group of neurons is wrapped by a glial cell which effectively isolates the cells within from extraneous inputs (Fig. 20.21a). The retinula cell axons that terminate in the lamina synapse with monopolar interneurons whose axons run to the medulla and with other fibers having cell bodies in the medulla (centrifugal fibers). The monopolar cells are of two types: some receive input only from retinula axons in a single cartridge, and they are known as small-field cells; others receive input from several cartridges, and are known as wide-field cells.

The identity of each group of fibers from the cartridges is retained in a series of columns in the medulla and it is probable that, here, as in the lamina, the pattern of neural signals is a precise representation of any image falling on the eye. This is called retinotopic mapping. Over 30 neurons contribute to each column in the medulla. The axons of the ultraviolet-sensitive retinula cells, as well as axons of lamina neurons, terminate within these columns.

In the lobula, some of the precision retained in the medulla is lost as the neural pathways converge, but the convergence involves several different neural pathways which run in parallel, each relaying information about particular aspects of the visual field such as size, shape, movement and color. The most extensive studies are on cyclorraphan flies, but similar principles probably apply to other insects.

Two types of interneuron are present in the lobula: small-field neurons which receive inputs from a small number of columns in the medulla, and wide-field neurons receiving inputs from a very large number of columns (Fig. 20.21c). Small-field neurons in the fly receive inputs from about 12 to 150 columns representing a receptive field in the eye that is 20° or less in diameter (Gilbert & Strausfeld, 1992; Strausfeld & Gilbert, 1992). Collectively, these neurons receive inputs from the whole of the retina so that a coarse-grained representation of the image is retained. They may be important in pattern recognition.

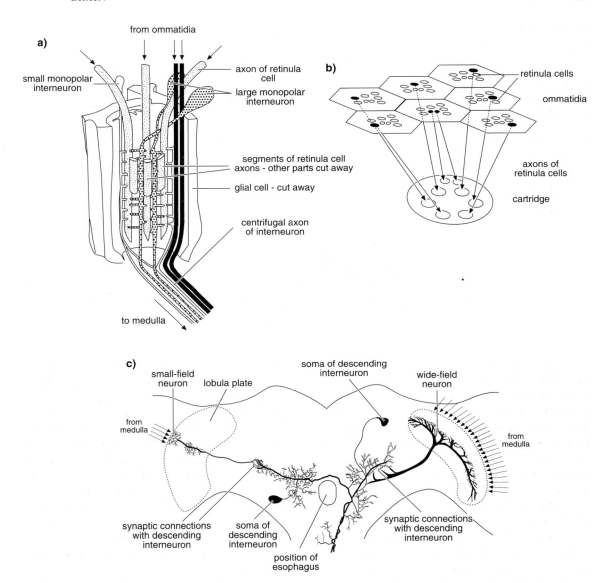

a)

from ommatidia

small monopolar
interneuron

axon of retinula
cell

large monopolar
interneuron

segments of retinula cell
axons - other parts cut away

glial cell - cut away

centrifugal axon
of interneuron

to medulla

b)

retinula cells

ommatidia

axons of
retinula cells

cartridge

c)

small-field
neuron

lobula plate

soma of descending
interneuron

wide-field
neuron

from
medulla

from
medulla

synaptic connections
with descending
interneuron

soma of
descending
interneuron

synaptic connections
with descending
interneuron

position of
esophagus

Fig. 20.21. Neurons of the optic lobe. **(a)** Diagram of an optic cartridge in the lamina of a fly. The side facing out of the page
has been cut away to show the arrangement of fibers within the cartridge (after Laughlin, 1975). **(b)** Relationship between a
cartridge in the lamina and the overlying ommatidia. In flies, which have an open rhabdom (see Fig. 22.3) the axons of retinula
cells with the same visual field (shown here in black) in adjacent ommatidia come together in one cartridge. In other insects,
each cartridge receives sensory axons only from the ommatidium immediately outside it. **(c)** Representation of wide-field and
narrow-field interneurons with arborizations in the lobula plate of a fly and their connections with descending interneurons
(based on Gronenberg & Strausfeld, 1992).

Fig.20.22. Visual perception of instability by a fly in flight based on the three horizontal neurons in the lobula plate (after Hausen, 1982). (a) Receptive fields of the three neurons on each side. Some slight contralateral effects also occur, but these are not shown. (b) Yaw to the fly's right (to the left in the diagram) produces a movement of images to the left across the eyes. The horizontal neurons of the left optic lobe would be excited while those of the right would be inhibited. (c) Pitching head down produces an upward movement of the image; the upper horizonal neurons of both sides (field 1 in a) would be excited and the lower ones inhibited. (d) Rolling does not stimulate the horizontal neurons.

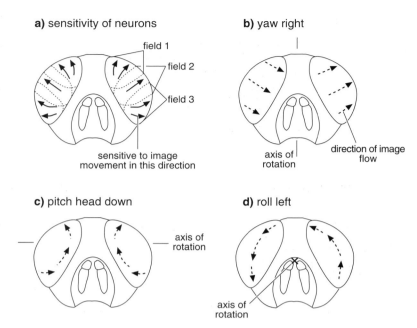

a) sensitivity of neurons

field 1
field 2
field 3

sensitive to image
movement in this direction

b) yaw right

axis of rotation

direction of image flow

c) pitch head down

axis of rotation

d) roll left

axis of rotation

In *Sarcophaga*, nine T-shaped wide-field neurons are present in the lobula plate forming the so-called vertical system, while three other cells with inputs from different parts of the eye form the 'horizontal' system (Fig. 20.21c). These cells receive inputs from large numbers of columns in the medulla and their principal function is in detection of different types of movement. The horizontal cells, for example, are excited by movement over the eye from front to back such as the insect would receive in normal forward movement, and are inhibited by movement from back to front (Fig. 20.22). They are thus well-suited to the perception of yawing movements of the insect, but they do not respond to rolling movements because image movement is vertical (Egelhauf, 1985; Hausen, 1982). Comparable wide-field neurons are present in other insects.

Both small- and wide-field interneurons have axons projecting to the deutocerebrum. Here they synapse with descending interneurons regulating the motor systems of the thorax that are involved in walking and flight (Fig. 20.21c, and see Fig. 9.32).

Reviews: Kral, 1987 – lamina; Laughlin, 1981 – lamina; Strausfeld & Nässel, 1981

20.4.2 Deutocerebrum

The deutocerebrum contains the antennal (olfactory) lobes and the antennal mechanosensory and motor center

(Figs. 20.16, 20.17). The latter is a relatively poorly defined region of neuropil containing the terminal arborizations of mechanosensory neurons from the scape and pedicel, and perhaps also from the flagellum of the antenna. It also contains dendritic arborizations of the motor neurons controlling the antennal muscles.

The antennal lobes are regions of neuropil, one in relation to each antenna, within which are discrete balls of dense synaptic neuropil which in some insects are surrounded by a layer of glial cells (Fig. 20.23a). These structures are called glomeruli. In the antennal lobe of *Locusta* there are about 1000 relatively poorly defined glomeruli, but *Periplaneta* has only 125 and *Manduca* about 64, while *Aedes* has fewer than ten. Axons from olfactory sensilla on the antenna terminate in the glomeruli, and each axon only goes to one glomerulus (Christensen *et al.*, 1995; Hansson *et al.*, 1992). Within a species, individual glomeruli appear to be constant in form and position. This is most obvious in the males of many Lepidoptera and of *Periplaneta*, insects that are attracted to females by pheromones, where two or three larger glomeruli are grouped together forming a macroglomerular complex to which the axons of sex pheromone-specific neurons converge. The axons from olfactory receptors on the maxillary and labial palps also terminate in a glomerulus which, at least in *Manduca*, does not receive inputs from the antennae.

a)

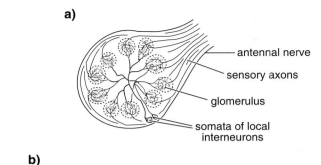

- antennal nerve
- sensory axons
- glomerulus
- somata of local interneurons

b)

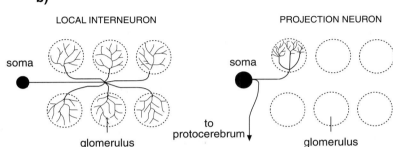

LOCAL INTERNEURON   PROJECTION NEURON

soma   soma

to protocerebrum

glomerulus   glomerulus

Fig. 20.23. Antennal lobe. **(a)** Diagram of an antennal lobe showing glomeruli. **(b)** Representation of a local interneuron and an output (projection) neuron (after Matsumoto and Hildebrand, 1981). **(c)** A serotonergic neuron with terminals in all the glomeruli and extensive branching through other parts of the brain. Notice the soma in the contralateral olfactory lobe. A similar neuron innervates the lefthand lobe (after Kent, Hoskins & Hildebrand, 1987).

c)

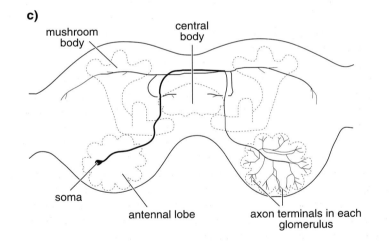

mushroom body   central body

soma

antennal lobe   axon terminals in each glomerulus

Within the antennal lobe, intrinsic (local) interneurons innervate several or even all of the glomeruli (Fig. 20.23b). These local interneurons relay information to other interneurons, called projection neurons, connecting the olfactory lobes to the mushroom bodies and other parts of the protocerebrum (Figs. 20.18b, 20.23b). Some projection neurons have uniglomerular arborizations; others are multiglomerular. It is common for the local interneurons to inhibit activity in the projection neurons (Christensen *et al.*, 1993). The physiological characteristics of the projection neurons vary. Some respond only when the antennal receptors are stimulated by a specific odor that is a regular feature in the life of the species, like a pheromone component. A few may respond when the antenna is stimulated by specific mixtures of compounds that are of particular importance to the species, such as a pheromone blend (Christensen, Mustaparta & Hildebrand, 1995). Other projection neurons, however, respond when the antenna is exposed to any one of a wide range of odors and it is probably these neurons that provide the insect with general information about odors in the environment.

Other interneurons provide input to the antennal lobes. The serotonergic cell described above is of this type and its effect is presumably to modulate the transfer of information within the glomeruli (Fig. 20.23c).

Reviews: Boeckh *et al.*, 1990; Christensen & Hildebrand, 1987; Homberg, Christensen & Hildebrand, 1989

20.4.3 **Tritocerebrum**

The tritocerebrum is a small part of the brain consisting of a pair of lobes beneath the deutocerebrum. From it, the circumesophageal connectives pass to the subesophageal ganglion, and the tritocerebral lobes of either side are connected by a commissure passing behind the esophagus (Fig. 20.17). Anteriorly, nerves containing sensory and motor elements connect with the frontal ganglion and the labrum.

20.5 CONTROLLING BEHAVIOR

The nervous system does not, in general, act as a simple relay between receptors and effectors; it integrates the activities of different parts of the body so that appropriate behavioral responses and internal regulating changes are made. Some of this integration results from the convergence of neurons on to interneurons, but integration may also involve modulation of a signal at a synapse.

20.5.1 **Integration at the synapse**

Changes in synaptic transmission may occur as a result of changes in the presynaptic or postsynaptic neuron. They may be produced either as a result of activity of the neurons themselves or by the activity of neuromodulators. Neuromodulators are considered in section 20.2.3.2.

The activity of a presynaptic fiber at a synapse may be modified by synaptic inputs from other neurons close to the output synapse. This is known to occur in some mechanosensory neurons (Watson, 1992) and in interneurons. The effect of these inputs is to increase the size of the presynaptic potential, if the input is excitatory, or decrease it, if the input is inhibitory. The net result is to change the number of transmitter vesicles releasing their contents at the synapse and thus to alter the effect on the postsynaptic neuron.

In the case of the chordotonal organ of the locust hind

femur, interaction occurs between the sensory neurons acting via an interneuron. The activity of one sensory axon activates the interneuron and causes it to inhibit activity in other sensory axons. There are about 90 sensory axons from the chordotonal organ and the effect of each one is regulated by the activity of the whole ensemble (Burrows & Matheson, 1994).

Postsynaptic integration results from temporal and spatial summation.

20.5.2 **Integration by interneurons**

Some measure of integration is achieved by individual interneurons as a result of the diversity of inputs that they receive from elsewhere, and the diversity of effector units which they supply. For instance, Fig. 20.14 shows some of the known inputs to one of the giant fibers in the terminal abdominal ganglion of the cockroach. These include the input from mechanoreceptors, some of which exhibit a purely phasic input while others are tonic or phasic/tonic. There are also inhibitory fibers from the cercus which act via inhibitory interneurons, and there is input from both ipsilateral and contralateral cerci. Any one or all of these input routes may be in action at the same time, and all this information is integrated, leading to the production of spikes at the narrow point on the axon. The number and frequency of spikes will vary with the overall input. The insect's behavior, moreover, does not depend on the activity of a single interneuron, but on that of a whole population of interneurons (Levi & Camhi, 1996).

Fig. 20.13 shows, for comparison, the motor connections of the DCMD interneuron in the metathoracic ganglion of *Schistocerca*. These include excitatory synapses with the motor neurons of the extensor tibialis muscles and inhibitory neurons of the flexor muscle. Contact is made with both ipsilateral and contralateral neurons.

20.5.3 **Neural mapping**

For the insect to make appropriate responses to external stimuli it is obviously important that the spatial pattern of incoming information is preserved within the central nervous system. In the eye, this is achieved by the system of cartridges and columns in the lamina and medulla forming a retinotopic map and feeding into the wide-field or narrow-field interneurons of the lobula (see above).

In the mechanosensory system, preservation of the spatial pattern is achieved by the spatial distribution of sensory arborizations within the ganglia. Hairs on the femur, tibia and tarsus of the locust, *Schistocerca*, arborize in progressively more posterior and lateral regions of the neuropil in the relevant ganglion (Fig. 20.24a). Dorsal, ventral, medial and lateral positions of the hairs on the legs also tend to be retained within the neuropil. Thus the sensory neuropil contains a three dimensional representation of the positions of the hairs on the legs. This is called a somatotopic map. Moreover, some of the spiking local interneurons with which the sensory axons connect, retain the pattern from a single leg (Fig. 20.24b,c), but others incorporate information from other legs losing some of the spatial content of the signal so the map becomes much more coarse-grained. Some intersegmental interneurons also retain some of the positional information (Burrows, 1992; Burrows & Newland, 1993).

This is more clearly seen in the cockroach escape response. The trichoid sensilla on the cerci of the cockroach are differentially sensitive to deflection in different directions. Their axons connect with four giant interneurons, two on each side, in a manner which retains the major components of directionality. Thus the interneurons are most responsive to deflection of the hairs backwards or forwards, to left or to right. This information is conveyed by the interneurons to the thorax where the output of motor neurons, producing a turn to left or right, depends on the balance of activity between the interneurons on the two sides (Camhi, 1993).

As the insect has contact chemoreceptors on the tarsi and other parts of the body, it is to be expected that positional information is also necessary for the gustatory system, in addition to quality coding. There is, at present, no information concerning differential distribution of the arbors of gustatory cells within the ganglia, but, in general, gustatory sensilla contain a mechanosensitive neuron (section 24.2.1). As stimulation of the gustatory neurons depends on contact with the substrate, the contact may be sufficient to stimulate the mechanosensitive neuron and it is possible that positional information about gustation is conveyed by these cells. It must also be inferred that some separation of taste quality, at least with respect to being acceptable or unacceptable, is maintained in the central nervous system, but there is no information on this.

Responses to olfactory stimuli also require processing of information about the source and the quality of the odor. Orientation with respect to an odor source is usually determined mechanically from the direction of air movement. In addition, the right and left antennae may be differentially stimulated and it is possible that the insect can also determine which part of an antenna is stimulated. This is certainly the case in male *Periplaneta* where some projection neurons from the macroglomerulus synapse with sensory axons from different regions of the antenna (Hösl, 1990).

Many insects can distinguish between different odors and this necessitates maintaining differences in the sensory input in the central nervous system. Sensilla sensitive to a specific odor appear to converge on a single glomerulus independent of their position on the antenna. Thus the array of glomeruli may contain an 'odotopic' map of the odor environment. (This is not to imply that the odors are spatially separated in the environment, but that the neural pathways responding to different odors are anatomically separated). This is most evident in the case of male perception of female sex attractant pheromones. Each part of the macroglomerular complex receives input from sensory cells responding to one of the principal components of the attractant pheromone, and all the units responding to that component terminate in the same part.

The extent of convergence of information from the olfactory sensilla to the olfactory glomeruli and subsequently to the mushroom bodies is very great. For example, in a male cockroach, *Periplaneta*, each antenna carries about 195 000 olfactory neurons responding to various environmental odors plus about 150 000 that respond to components of the female pheromone. The former converge on to 125 glomeruli, the latter to the macroglomerular complex. About 280 output neurons leave the olfactory lobe to convey information that forms the basis of behavioral responses. Similarly, the worker honeybee has over 60 000 olfactory receptors on each antenna the axons of which converge on to about 155 glomeruli.

As each glomerulus is believed to receive sensory input from receptors for a specific compound, or perhaps class of compounds, this system with its output neurons provides ample basis for the discrimination of odors. The extent to which odor discrimination is dependent on labelled lines versus across-fiber patterning is not known. A labelled line responds only to a specific compound or combination of compounds; such a system conveys information totally unambiguously. Across-fiber patterning requires the assessment of the overall input from an array of neurons with different ranges of sensitivity. The relatively large numbers of glomeruli and output interneurons provide the insect with the potential to distinguish a very large number of odors using across-fiber patterning.

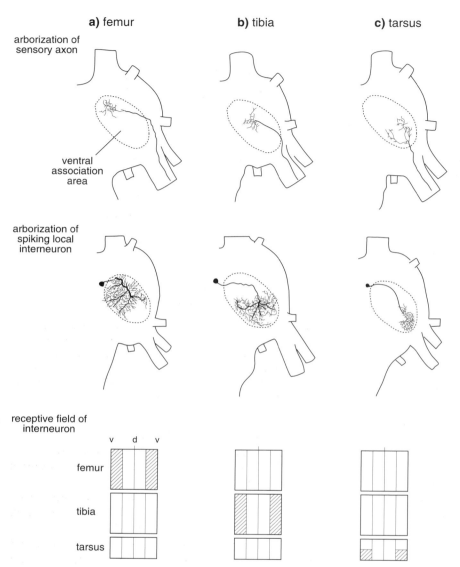

a) femur **b)** tibia **c)** tarsus

arborization of sensory axon

ventral association area

arborization of spiking local interneuron

receptive field of interneuron

v   d   v

femur

tibia

tarsus

Fig. 20.24. Somatotopic mapping of mechanosensory input from the hind leg of a locust. All the mechanosensory axons from the leg arborize in a ventral region of the neuropil known as the ventral association area (indicated by the dotted line). The diagrams at the bottom represent the leg as if slit along the ventral side and flattened so that the dorsal midline is in the center, ventral midline at the outer edges. Cross-hatching shows the area in which mechanical stimulation results in excitation of the interneuron shown in the central panel (after Burrows & Newland, 1993). (**a**) Input from the femur. Top: the arborization of a mechanosensitive neuron at the base of the femur. Center: the arborization of a spiking local interneuron which is excited by stimulation of the ventral area of the femur, but not other parts of the leg (shown at bottom). (**b**) Input from the tibia. Top: the arborization of a mechanosensitive neuron at the base of the tibia. Center: the arborization of a spiking local interneuron which is excited by stimulation of the ventral area of the tibia, but not other parts of the leg (shown at bottom). (**c**) Input from the tarsus. Top: the arborization of a mechanosensitive neuron on the basal tarsomere. Center: the arborization of a spiking local interneuron which is excited by stimulation of the ventral area of the distal tarsus, but not other parts of the leg (shown at bottom).

In the same way that the neural characteristics of different odors may be anatomically separated in the central nervous system, so may the inputs from sounds of different frequencies, at least in bush crickets.

20.5.4 Pattern generation

Some activities, such as flight, walking, ventilation and, at least in grasshoppers, oviposition, involve patterns of muscular activity which are frequently repeated and are, to some extent, stereotyped. The basic sequences of motor neuron activity in these instances are produced within the central nervous system and the essential features are produced even when the central nervous system is experimentally isolated and there is no input from peripheral receptors (see Fig. 8.19). These patterns are produced by an array of interneurons forming a central pattern generator and the cells involved appear to be, to a large extent, dedicated to a particular activity; different activities such as walking and flight have separate pattern generators (Ramirez & Pearson, 1988).

Fig. 20.25 shows how the basic oscillation of wing levator and depressor muscles may be generated in the locust central nervous system. In the absence of tarsal contact with the ground, wind acting via the aerodynamic mechanoreceptor hairs on the head stimulates interneuron 206, which in turn excites interneuron 504. This excites the levator muscle motor neurons (L), so that the wings are raised, and, at the same time, excites interneuron 301. This interneuron has a critical function; it disinhibits (removes inhibition from) the depressor muscle motor neurons (D) and also excites them, but with a time delay. Thus, it first puts the depressor motor neurons in a state of readiness to respond (since they are no longer inhibited) and then induces the response. Also acting via the time delay, it excites interneuron 501 which inhibits the activity of other cells in the system so that the levator motor neuron is inhibited as the depressor motor neuron fires. This is probably not the only rhythm-generating system of neurons involved in producing the flight pattern; it appears to be a common feature of pattern-generating systems that several such oscillators act together. Other common features are a time delay at some point in the system, and a preponderance of inhibitory connections (Robertson & Pearson, 1985). In the intact insect, the basic pattern is modified by sensory input that enables the insect to adjust its movements to differences in the environment.

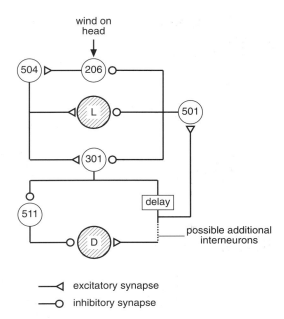

Fig. 20.25. Pattern generator. Diagrammatic representation of the interactions between a group of interneurons producing alternations in the contractions of the locust flight muscles. Interneurons represented by numbered circles; motor neurons hatched. L, motor neuron to levator muscle; D, motor neuron to depressor muscle (based on Robertson & Pearson, 1985).

20.5.5 Neural basis of learning

Although many insects have the ability to learn (Papaj & Lewis, 1993), very little is known of the neural basis of learning.

The mushroom bodies have been shown experimentally to be involved in associative learning by bees and *Drosophila* (de Belle & Heisenberg, 1994; Menzel *et al.*, 1991). Anatomical changes in the mushroom bodies are correlated with differences in experience (Fig. 20.19b,c) and even a single flight, lasting about 10 minutes, is sufficient to produce changes in the dendritic spines of Kenyon cells through which synaptic connections are made with other interneurons (Brandon & Cross, 1982). Injection of octopamine into the calyces enhances memory storage and retrieval in honeybees, whereas serotonin has the opposite effect (Menzel, Wittstock & Sugawa, 1989). Input of octopamine may normally come from ventral unpaired median neurons with cell bodies in the subesophageal ganglion and extensive arborizations in the lip and basal ring of the calyx as well as in the antennal lobes and lateral protocerebrum.

The mushroom bodies occupy relatively larger proportions of the brain in the social insects (Fig. 20.19a) than in other insects, presumably reflecting their more complex behavior.

In several animals other than insects, the development of a long-term memory is associated with protein synthesis. There is also evidence for this in *Drosophila* where memory is associated with neuropeptide synthesis regulated by a gene, *amnesiac* (Feany & Quinn, 1995). Variation in the level of a protein which regulates levels of cAMP also affects long-term memory (Yin *et al.*, 1995). In the honeybee, however, long-term memory does not involve protein synthesis (Hammer & Menzel, 1995).

Habituation to a repetitive stimulus is a common phenomenon. For example, a honeybee extends its proboscis if a drop of sucrose solution is applied to one of its antennae, but with frequent repetition of the stimulus the response wanes and finally disappears completely. As stimulation of the opposite antenna causes the response to reappear (dishabituation), the change in the nervous system that produces habituation cannot be occurring in the interneurons or motor neurons immediately responsible for proboscis extension. This leads Menzel *et al.* (1991) to suggest that changes may be occurring within the antennal lobe.

Habituation by grasshoppers to certain auditory and visual stimuli is correlated with changes in specific interneurons which convey information to other parts of the nervous system and so affect the behavioral response. For example, the locust's descending contralateral movement detector (DCMD) interneuron habituates if the insect is repeatedly presented with the same visual stimulus to the same part of the eye, and fewer action potentials are produced by each successive stimulation (Fig. 20.26). However, stimulating another part of the eye or changing the state of arousal of the insect by providing other stimuli results in a recovery to the original rate of firing. This indicates that the effect is not in the DCMD itself, but at an earlier stage in the neural pathway between the eye and this interneuron. The most probable mechanism to account for the process is depletion of synaptic transmitter of one of the neurons leading to the DCMD. However, the neuron may not fully recover its response for several days after many successive stimulations (Rowell, 1974).

Some degree of learning can occur in isolated segmental ganglia. Headless cockroaches and locusts can learn to keep a leg flexed to avoid electrical shocks. Flexion involves several muscles of which one, the coxal adductor muscle,

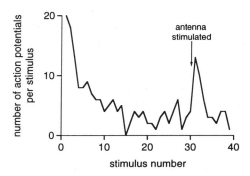

Fig. 20.26. Habituation in the central nervous system. Responses of the descending contralateral movement detector (DCMD) interneuron to repeated visual stimulation by a moving black disc. Stimuli were presented at eight-second intervals. After stimulus number 30 an antenna was stimulated electrically. This produced a short-lived dishabituation (after Rowell, 1974).

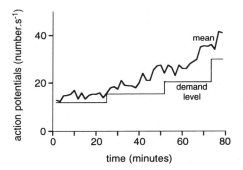

Fig. 20.27. Learning in an isolated ganglion. Each time the frequency of action potentials arriving at the coxal adductor muscle fell below an arbitrary 'demand level', the muscle was given an electric shock. After a number of shocks, the frequency of action potentials rose. By increasing the demand level, the firing rate of the motor neurons activating the muscle was increased progressively. This is the equivalent of the ganglion 'learning' to keep the leg raised in order to avoid shocks (after Hoyle, 1965).

has been most fully investigated. A reduction in the activity of the motor neuron to the adductor muscle leads to a reduction in muscle tension so that the leg drops. As a result, the leg receives a shock. This leads to increased activity in the neuron and the leg is raised (Fig. 20.27). Maintenance of the high level of activity in the motor neuron keeps the leg raised so that further shocks are avoided. Isolated ganglia only retain the information leading to leg raising for a short period. The brain and subesophageal ganglion are necessary to develop long-term memory, but, once the information is acquired, the head is no longer essential, as the segmental ganglia retain the appropriate information derived from that processed in the head.

Reviews: Eisenstein & Reep, 1985 – learning by isolated ganglia; Hammer & Menzel, 1995 – learning and memory in honeybees

20.5.6 Rhythms of behavior

Many insects, and other organisms, exhibit rhythms of behavioral or physiological activity with a 24-hour cycle. These are called circadian rhythms. These rhythms are driven by cells, known as pacemakers, in the brain and optic lobes. In *Drosophila*, the pacemakers are probably cells in which the *period* (*per*) gene is expressed. These cells are neurons or glial cells, perhaps both, with a principal focus in the central region of the brain, but also occurring elsewhere in the brain and optic lobes (Ewer *et al.*, 1992). The pacemaker cells produce a protein (PER protein) cyclically. A second gene, *timeless* (*tim*), regulates accumulation of this protein in the nucleus, and its activity also varies temporally (Vosshall *et al.*, 1994). The TIM protein is destroyed in the light, although the mechanism by which this occurs is not understood. Consequently, very little TIM is present during the day (Fig. 20.28a). In the absence of the proteins, transcription of both proteins occurs in the latter part of the day and they start to accumulate in the cytoplasm where they become physically associated. TIM protein facilitates the entry of PER protein into the nucleus where they inhibit transcription of their own RNAs and no more protein is formed. The destruction of TIM protein after dawn removes the inhibition on transcription and RNA is produced again (Lee *et al.*, 1996; Myers *et al.*, 1996). It is supposed that the PER protein acts on other genes, leading to the regulation of various physiological activities. The neural circuitry involved in the pacemaker system of insects is not known.

Adult *Drosophila* exhibit a circadian rhythmicity even if reared from the egg and maintained as adults in complete darkness, but to be effective biologically the innate rhythmicity of the *per* genes must be related to the environmental diurnal cycle. This entrainment of the rhythm usually depends on diurnal fluctuation in light or temperature (for examples see Fig. 15.2) and, in the case of light, destruction of the TIM protein during a light phase is the critical step in entrainment. If the onset of darkness is experimentally delayed, the accumulation of

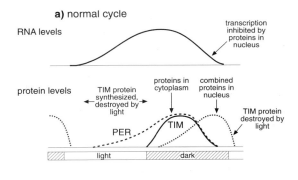

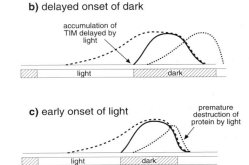

Fig. 20.28. Circadian rhythm of *Drosophila*. **(a)** In the normal cycle TIM protein is destroyed by light. The entry of PER protein into the nucleus is dependent on an association with TIM protein. Within the nucleus, the two proteins inhibit transcription of their RNAs. **(b)** If the onset of darkness is delayed, TIM protein is destroyed, but since its RNA is present it starts to accumulate at the onset of darkness. **(c)** If the light period starts early, TIM protein is destroyed. No more is synthesized immediately because no RNA is present.

proteins is also delayed and the clock is set back (Fig. 20.28b); if the photophase begins earlier than normal, the proteins are destroyed and, in the absence of RNA, no more is produced and the clock is set forward (Fig. 20.28c). Although the same genes are present in the moth, *Antheraea*, the PER protein does not enter the nucleus and the diurnal rhythm of activity is clearly produced in a different way.

In the cockroach, *Periplaneta*, the environmental light cycle is perceived via the compound eyes, but in the adult aphid, *Megoura*, and pupae of the silkmoth, *Antheraea*, light acts directly on cells in the brain.

Reviews: Page, 1985 – general account of insect circadian rhythms; Arechiga, 1993 – neural organization of rhythms

REFERENCES

Albrecht, F.O. (1953). *The Anatomy of the Migratory Locust*. London: Athlone Press.

Altman, J.S. & Kien, J. (1987). Functional organization of the subesophageal ganglion in arthropods. In *Arthropod Brain*, ed. A.P. Gupta, pp. 265–301. New York: Wiley.

Arechiga, H. (1993). Circadian rhythms. *Current opinion in Neurobiology*, **3**, 1005–10.

Boeckh, J., Distler, P., Ernst, K.D., Hösl, M. & Malun, D. (1990). Olfactory bulb and antennal lobe. In *Chemosensory Information Processing*, ed. D. Schild, pp. 201–27. Berlin: Springer-Verlag.

Brandon, J.G. & Cross, R.G. (1982). Rapid dendritic spine stem shortening during one-trial learning: the honeybee's first orientation flight. *Brain Research*, **252**, 51–61.

Burrows, M. (1977). Flight mechanisms in the locust. In *Identified Neurons and Behaviour of Arthropods*, ed. G. Hoyle. New York: Plenum Press.

Burrows, M. (1992). Local circuits for the control of leg move ments in an insect. *Trends in Neuroscience*, **15**, 226–32.

Burrows, M. (1996). *The Neurobiology of an Insect Brain*. Oxford: Oxford University Press.

Burrows, M. & Matheson, T. (1994). A presynaptic gain control mechanism among sensory neurons of a locust leg proprioceptor. *Journal of Neuroscience*, **14**, 272–82.

Burrows, M. & Newland, P.L. (1993). Correlation between the receptive fields of locust interneurons, their dendritic morphology, and the central projections of mechanosensory neurons. *Journal of Comparative Neurology*, **329**, 412–26.

Callec, J.J. (1985). Synaptic transmission in the central nervous system. In *Comprehensive Insect Physiology, Biochemistry and Pharmacology*, vol. 5, ed. G.A. Kerkut & L.I. Gilbert, pp. 139–79. Oxford: Pergamon Press.

Camhi, J.M. (1993). Neural mechanisms of behavior. *Current Opinion in Neurobiology*, **3**, 1011–19.

Carlson, S.D. & Saint Marie, R.L. (1990). Structure and function of insect glia. *Annual Review of Entomology*, **35**, 597–621.

Christensen, T.A., Harrow, I.D., Cuzzocrea, C., Randolph, P.W. & Hildebrand, J.G. (1995). Distinct projections of two populations of olfactory receptor axons in the antennal lobe of the sphinx moth *Manduca sexta*. *Chemical Senses*, **20**, 313–23.

Christensen, T.A. & Hildebrand, J.G. (1987). Functions, organization and physiology of the olfactory pathways in the lepidopteran brain. In *Arthropod Brain*, ed. A.P. Gupta, pp. 457–84. New York: Wiley.

Christensen, T.A., Mustaparta, H. & Hildebrand, J.G. (1995). Chemical communication in heliothine moths VI. Parallel pathways for information processing in the macroglomerular complex of the male tobacco budworm moth *Heliothis virescens*. *Journal of Comparative Physiology* A, **169**, 259–74.

Christensen, T.A., Sherman, T.G., McCaman, R.E. & Carlson, A.D. (1983). Presence of octopamine in firefly photomotor neurons. *Neuroscience*, **9**, 183–9.

Christensen, T.A., Waldrop, B.R., Harrow, I.D. & Hildebrand, J.G. (1993). Local interneurons and information processing in the olfactory glomeruli of the moth *Manduca sexta*. *Journal of Comparative Physiology*, A **173**, 385–99.

de Belle, J.S. & Heisenberg, M. (1994). Associative odor learning in *Drosophila* abolished by chemical ablation of mushroom bodies. *Science*, **263**, 692–5.

Durst, C., Eichmüller, S & Menzel, R. (1994). Development and experience lead to increased volume of subcompartments of the honeybee mushroom body. *Behavioral and Neural Biology*, **62**, 259–63.

Edwards, J.S. & Tolbert, L.P. (1998). Insect Neuroglia. In *Microscopic Anatomy of Invertebrates*, vol. 11B, *Insecta*, ed. F.W. Harrison & M. Locke, pp. 449–66. New York: Wiley-Liss

Egelhauf, M. (1985). On the neuronal basis of figure-ground discrimination by relative motion in the visual system of the fly II. Figure-detection cells, a new class of visual interneurons. *Biological Cybernetics*, **52**, 195–209.

Eisenstein, E.M. & Reep, R.L. (1985). Behavioral and cellular studies of learning and memory in insects. In *Comprehensive Insect Physiology, Biochemistry and Pharmacology*, vol. 9, ed. G.A. Kerkut & L.I. Gilbert, pp. 513–47. Oxford: Pergamon Press.

Ewer, J., Frisch, B., Hamblen-Coyle, M.J., Rosbash, M. & Hall, J.C. (1992). Expression of the periodic clock gene within different cell types in the brain of *Drosophila* adults and mosaic analysis of these cells' influence on circadian behavioral rhythms. *Journal of Neuroscience*, **12**, 321–49.

Feany, M.B. & Quinn, W.G. (1995). A neuropeptide gene defined by the *Drosophila* memory mutant *amnesiac*. *Science*, **268**, 869–73.

Gilbert, C. & Strausfeld, N.J. (1992). Small-field neurons associated with oculomotor control in muscoid flies: functional organization. *Journal of Comparative Neurology*, **316**, 72–86.

Golding, D.W. (1994). A pattern confirmed and refined – synaptic, nonsynaptic and parasynaptic exocytosis. *BioEssays*, **16**, 503–8.

Gronenberg, W. & Strausfeld, N.J. (1992). Premotor descending neurons responding selectively to local visual stimuli in flies. *Journal of Comparative Neurology*, **316**, 87–103.

Hammer, M. & Menzel, R. (1995). Learning and memory in the honeybee. *Journal of Neuroscience*, **15**, 1617–30.

Hansson, B.S., Ljungberg, H., Hallberg, E. & Lofstedt, C. (1992). Functional specialization of glomeruli in a moth. *Science*, **256**, 1313–5.

Harris-Warwick, R.M., Marder, E., Selverston, A.I. & Moulins, M. (eds.) (1992). *Dynamic Biological Networks: The Stomatogastric Nervous System.* Cambridge, Massachusetts: MIT Press.

Hausen, K. (1982). Motion sensitive interneurons in the optomotor system of the fly II. The horizontal cells: Receptive field organization and response characteristics. *Biological Cybernetics*, **46**, 67–79.

Heisenberg, M., Hensipp, M. & Wanke, C. (1995). Structural plasticity in the *Drosophila* brain. *Journal of Neuroscience*, **15**, 1951–60.

Holman, G.M., Nachman, R.J. & Wright, M.S. (1990). Insect neuropeptides. *Annual Review of Entomology*, **35**, 201–17.

Homberg, U. (1987). Structure and functions of the central complex in insects. In *Arthropod Brain*, ed. A.P. Gupta, pp. 347–67. New York: Wiley.

Homberg, U. (1991). Neuroarchitecture of the central complex in the brain of the locust *Schistocerca gregaria* and *S. americana* as revealed by serotonin immunocytochemistry. *Journal of Comparative Neurology*, **303**, 245–54.

Homberg, U. (1994). *Distribution of Neurotransmitters in the Insect Brain.* Stuttgart: Gustav Fischer Verlag.

Homberg, U., Christensen, T.A. & Hildebrand, J.G. (1989). Structure and function of the deutocerebrum in insects. *Annual Review of Entomology*, **34**, 477–501.

Horridge, G.A. (1965). The Arthropoda: General anatomy. In *Structure and Function of the Nervous System of Invertebrates*, ed. T.H. Bullock & G.A. Horridge, pp. 801–964. San Francisco: Freeman.

Hösl, M. (1990). Pheromone-sensitive neurons in the deutocerebrum of *Periplaneta americana*: receptive fields on the antenna. *Journal of Comparative Physiology* A, **167**, 321–7.

Howse, P.E. (1974). Design and function in the insect brain. In *Experimental Analysis of Insect Behaviour*, ed. L. Barton Browne, pp. 180–94. Berlin: Springer-Verlag.

Hoyle, G. (1965). Neurophysiological studies on 'learning' in headless insects. In *The Physiology of the Insect Nervous System*, ed. J.E. Treherne & J.W.L. Beament, pp. 203–32. London: Academic Press.

Ito, K., Urban, J. & Technau, G.M. (1995). Distribution, classification and development of *Drosophila* glial cells in the late embryonic and larval ventral nerve cord. *Roux's Archives in Developmental Biology*, **204**, 284–307.

Kandel, E.R., Schwartz, J.H. & Jessell, T.M. (1991). *Principles of Neural Science*. Norfolk, Connecticut: Appleton & Lange.

Kent, K.S., Hoskins, S.G. & Hildebrand, J.G. (1987). A novel serotonin-immunoreactive neuron in the antennal lobe of the sphinx moth *Manduca sexta* persists throughout postembryonic life. *Journal of Neurobiology*, **18**, 451–65.

Kloppenburg, P. & Hildebrand, J.G. (1995). Neuromodulation by 5-hydroxytryptamine in the antennal lobe of the sphinx moth *Manduca sexta*. *Journal of Experimental Biology*, **198**, 603–11.

Kral, K. (1987). Organization of the first optic neuropil (or lamina) in different insect species. In *Arthropod Brain*, ed. A.P.Gupta, pp. 181–201. New York: Wiley.

Kutsch, W. & Breidbach, O. (1994). Homologous structures in the nervous systems of arthropods. *Advances in Insect Physiology*, **24**, 1–113.

Lane, N.J. (1985). Structure of components of the nervous system. In *Comprehensive Insect Physiology, Biochemistry and Pharmacology*, vol. 5, ed. G.A. Kerkut & L.I. Gilbert, pp. 1–47. Oxford: Pergamon Press.

Laughlin , S.B. (1975). The function of the lamina ganglionaris. In *The Compound Eye and Vision of Insects*, ed. G.A. Horridge, pp. 341–58. Oxford: Clarendon Press.

Laughlin, S. (1981). Neural principles in the peripheral visual systems of invertebrates. In *Handbook of Sensory Physiology*, vol. 7 part 6B, ed. H.Autrum, pp. 133–280. Berlin: Springer-Verlag.

Lee, C., Parikh, V., Itsukaichi, T., Bae, K. & Edery, I. (1996). Resetting the *Drosophila* clock by photic regulation of PER and a PER–TIM complex. *Science*, **271**, 1740–4.

Levi, R. & Camhi, J.M. (1996). Producing directed behavior: muscle activity patterns of the cockroach escape response. *Journal of Experimental Biology*, **199**, 563–8.

Lutz, E.M. & Tyrer, N.M. (1988). Immunohistochemical localization of serotonin and choline acetyltransferase in sensory neurones of the locust. *Journal of Comparative Neurology*, **267**, 335–42.

Matsumoto, S.G. & Hildebrand, J.G. (1981). Olfactory mechanisms in the moth *Manduca sexta*: response characteristics and morphology of central neurons in the antennal lobes. *Proceedings of the Royal Society of London* B, **213**, 249–77.

Meinertzhagen, I.A. & O'Neil, S.D. (1991). Synaptic organization of columnar elements in the lamina of the wild type in *Drosophila melanogaster*. *Journal of Comparative Neurology*, **305**, 232–63.

Menzel, R., Hammer, M., Braun, G., Mauelshagen, J. & Sugawa, M. (1991). Neurobiology of learning and memory in honeybees. In *The Behavior and Physiology of Bees*, ed. L.J. Goodman & R.C. Fisher, pp. 323–53. Wallingford: CAB International.

Menzel, R., Wittstock, S. & Sugawa, M. (1989). Chemical codes of learning and memory in honeybees. In *The Biology of Memory*, ed. L.R. Squire & L. Lindenlaub, pp. 335–55. Stuttgart: Schattauer Verlag.

Mobbs, P.G. (1985). Brain structure. In *Comprehensive Insect Physiology, Biochemistry and Pharmacology*, vol. 5, ed. G.A. Kerkut & L.I. Gilbert, pp. 299–370. Oxford: Pergamon Press.

Myers, M.P., Wager-Smith, K., Rothenfluh-Hilfiker, A. & Young, M.W. (1996). Light-induced degradation of TIMELESS and entrainment of the *Drosophila* circadian clock. *Science*, 271, 1736–40.

Nässel, D.R. (1988). Serotonin and serotonin-immunoreactive neurons in the nervous system of insects. *Progress in Neurobiology*, 30, 1–85.

Nässel, D.R. (1991). Neurotransmitters and neuromodulators in the insect visual system. *Progress in Neurobiology*, 37, 179–254.

Nässel, D.R. & O'Shea, M. (1987). Proctolin-like immunoreactive neurons in the blowfly central nervous system. *Journal of Comparative Neurology*, 265, 437–54.

Newland, P.L. & Burrows, M. (1994). Processing of mechanosensory information from gustatory receptors on a hind leg of the locust. *Journal of Comparative Physiology* A, 174, 399–410.

Orchard, I., Belanger, J.H. & Lange, A.B. (1989). Proctolin: a review with emphasis on insects. *Journal of Neurobiology*, 20, 470–96.

Orchard, I., Ramirez, J.-M. & Lange, A.B. (1993). A multifunctional role for octopamine in locust flight. *Annual Review of Entomology*, 38, 227–49.

O'Shea, M., Rowell, C.H.F. & Williams, J.L.D. (1974). The anatomy of a locust visual interneurone; the descending contralateral movement detector. *Journal of Experimental Biology*, 60, 1–12.

Page, T.L. (1985). Clocks and circadian rhythms. In *Comprehensive Insect Physiology, Biochemistry and Pharmacology*, vol. 6, ed. G.A. Kerkut & L.I. Gilbert, pp. 577–652. Oxford: Pergamon Press.

Papaj, D.R. & Lewis, A.C. (1993). *Insect Learning*. New York: Chapman & Hall.

Pearson, K.G. (1977). Interneurons in the ventral nerve cord of insects. In *Identified Neurons and Behaviour of Arthropods*, ed. G. Hoyle, pp. 329–37. New York: Plenum Press.

Pearson, K.G., Boyan, G.S., Bastiani, M. & Goodman, C.S. (1985). Heterogeneous properties of segmentally homologous interneurons in the ventral nerve cord of locusts. *Journal of Comparative Neurology*, 233, 133–45.

Pichon, Y. & Ashcroft, F.M. (1985). Nerve and muscle: electrical activity. In *Comprehensive Insect Physiology, Biochemistry and Pharmacology*, vol. 5, ed. G.A. Kerkut & L.I. Gilbert, pp. 85–113. Oxford: Pergamon Press.

Ramirez, J.-M. & Orchard, I. (1990). Octopaminergic modulation of the forewing stretch receptor in the locust *Locusta migratoria*. *Journal of Experimental Biology*, 149, 255–79.

Ramirez, J.M. & Pearson, K.G. (1988). Generation of motor patterns for walking and flight in motoneurons supplying bifunctional muscles in the locust. *Journal of Neurobiology*, 19, 257–82.

Robertson, R.M. & Pearson, K.G. (1985). Neural circuits in the flight system of the locust. *Journal of Neurophysiology*, 53, 110–28.

Rowell, C.H.F. (1974). Boredom and attention in a cell in the locust visual system. In *Experimental Analysis of Insect Behavior*, ed. L. Barton Browne, pp. 87–113. Berlin: Springer-Verlag.

Sattelle, D.B. (1990). GABA receptors of insects. *Advances in Insect Physiology*, 22, 1–113.

Schurmann, F.-W. (1987). The architecture of the mushroom bodies and related neuropils in the insect brain. In *Arthropod Brain*, ed. A.P. Gupta, pp. 231–64. New York: Wiley.

Shephard, G.M. (1994). *Neurobiology*. New York: Oxford University Press.

Siegler, M.V.S. (1985). Nonspiking interneurons and motor control in insects. *Advances in Insect Physiology*, 18, 249–304.

Siegler, M.V.S. & Burrows, M. (1979). The morphology of local non-spiking interneurones in the metathoracic ganglion of the locust. *Journal of Comparative Neurology*, 183, 121–48.

Smith, D.S. & Treherne, J.E. (1963). Functional aspects of the organisation of the insect nervous system. *Advances in Insect Physiology*, 1, 401–84.

Stevenson, P.A., Pflüger, H.-J., Eckert, M. & Rapus, J. (1992). Octopamine immunoreactive cell populations in the locust thoracic-abdominal nervous system. *Journal of Comparative Neurology*, 315, 382–97.

Strausfeld, N.J. & Gilbert, C. (1992). Small-field neurons associated with oculomotor control in muscoid flies: cellular organization in the lobula plate. *Journal of Comparative Neurology*, 316, 56–71.

Strausfeld, N.J. & Meinertzhagen, I.A. (1998). The insect neuron: types, morphologies, fine structure, and relationship to the architectonics of the insect nervous system. In *Microscopic Anatomy of Invertebrates*, vol. 11B, *Insecta*, ed. F.W. Harrison & M. Locke, pp. 487–538. New York: Wiley-Liss.

Strausfeld, N.J. & Nässel, D.R. (1981). Neuroarchitectures serving compound eyes in Crustacea and insects. In *Handbook of Sensory Physiology*, vol. 7 part 6B, ed. H. Autrum, pp. 1–132. Berlin: Springer-Verlag.

Technau, G.M. (1984). Fiber number in the mushroom bodies of adult *Drosophila melanogaster* depends on age, sex and experience. *Journal of Neurogenetics*, 1, 113–26.

Tolbert, L.P. & Oland, L.A. (1990). Glial cells form boundaries for developing insect olfactory glomeruli. *Experimental Neurobiology*, **109**, 19–28.

Treherne, J.E. (1974). The environment and function of insect nerve cells. In *Insect Neurobiology*, ed. J.E. Treherne. Amsterdam: North-Holland Publishing Co.

Treherne, J.E. (1985). Blood–brain barrier. In *Comprehensive Insect Physiology, Biochemistry and Pharmacology*, vol. 5, ed. G.A. Kerkut & L.I. Gilbert, pp. 115–37. Oxford: Pergamon Press.

Treherne, J.E. & Schofield, P.K. (1981). Mechanisms of ionic homeostasis in the central nervous system of an insect. *Journal of Experimental Biology*, **95**, 61–73.

Usherwood, P.N.R. (1994). Insect glutamate receptors. *Advances in Insect Physiology*, **24**, 309–41.

Vosshall, L.B., Price, J.L., Sehgal, A., Saez, L. & Young, M.W. (1994). Block in nuclear localization of *period* protein by a second clock mutation, *timeless*. *Science*, **263**, 1606–9.

Watson, A.H.D. (1992). Presynaptic modulation of sensory afferents in the invertebrate and vertebrate nervous system. *Comparative Biochemistry and Physiology*, **103A**, 227–39.

Watson, A.H.D. & Burrows, M. (1988). Distribution and morphology of synapses in nonspiking local interneurones in the thoracic nervous system. *Journal of Experimental Biology*, **272**, 605–16.

Weevers, R. de G. (1985). The insect ganglia. In *Comprehensive Insect Physiology, Biochemistry and Pharmacology*, vol. 5, ed. G.A. Kerkut & L.I. Gilbert, pp. 213–97. Oxford: Pergamon Press.

Yin, J.C.P., Del Vecchio, M., Zhou, H. & Tully, T. (1995). CREB as a memory modulator: induced expression of a dCREB2 activator isoform enhances long-term memory in *Drosophila*. *Cell*, **81**, 107–15.

21 Endocrine system

Hormones are chemicals produced by an organism which circulate in the blood to regulate its long term physiological, developmental and behavioral activities. They complement the nervous system, which provides short-term coordination, and the activities of the two systems are closely linked.

General aspects of hormones are discussed in this chapter. The specific actions of hormones regulating particular functions are considered in other chapters:

molting and metamorphosis	–	Section 15.4.
yolk synthesis	–	Section 13.2.4.2.
embryonic cuticles	–	Section 14.2.10.
diuresis	–	Section 18.3.3.
mobilization of fuel for flight	–	Section 9.12.2.
polyphenism	–	Section 15.5.
diapause	–	Section 15.6.5.

General review of insect hormones: Nijhout, 1994

21.1 CHEMICAL STRUCTURE OF HORMONES

Apart from molting hormone and juvenile hormone, most known insect hormones are peptides. Some biogenic amines are also known to function as hormones.

21.1.1 Molting hormones

Molting hormones are ecdysteroids (Fig. 21.1) which, in immature insects, are produced by the prothoracic glands. In most insects these glands secrete ecdysone, but some larval Lepidoptera are also known to secrete 3-dehydroecdysone which is converted to ecdysone by enzymes in the hemolymph (Fescemeyer *et al.*, 1995). Ecdysone is a prohormone; it is converted to the active hormone, 20-hydroxyecdysone, in the fat body or epidermis by a cytochrome P-450 enzyme. In the honeybee *Apis*, and in Heteroptera, the principal ecdysteroid is makisterone A.

Insects cannot synthesize steroids. Consequently, sterols, usually cholesterol or a closely related structure, are essential dietary constituents.

21.1.2 Juvenile hormone

Juvenile hormone is a sesquiterpene (Fig. 21.2) produced by the corpora allata. Several slightly different forms, known as JH0, I, II and III containing 19, 18, 17 and 16 carbon atoms, respectively, have been isolated. JHIII is the form occurring in most insects; JHI and II are known primarily from Lepidoptera, and JH0 only from lepidopteran eggs. In the wandering stage larva of *Manduca* and in pharate adult males of *Hyalophora*, the corpora allata produce juvenile hormone acid that is subsequently converted to juvenile hormone in the imaginal discs and accessory sex glands, respectively (Sparagana, Bhaskaran & Barrera, 1985).

In addition to JHIII, the corpora allata of cyclorrhaphous Diptera secrete two related compounds, a bisepoxide of JHIII and methyl farnesoate (Yin *et al.*, 1995), and it is possible that similar compounds are also produced by other insects. The fate or function of these compounds is not known.

21.1.3 Peptide hormones

Insects have many different peptide hormones, some of which are shown in Fig. 21.3. They are produced by individual cells, mainly in the central nervous system and the midgut epithelium (see below). In some cases, as with prothoracicotropic hormone and adipokinetic hormone, their functions are well-established, but the roles of many others remain obscure even though they may have been shown, experimentally, to produce certain physiological effects.

Peptide hormones with similar functions in different insect species are often similar in size and amino acid sequences; although differences do occur, there are conserved regions in which the amino acid sequences are similar. The adipokinetic hormones illustrate this point (Fig. 21.3c). They are also closely similar to hyper-trehalosemic hormones (elevating trehalose levels in the hemolymph) and to cardiac acceleratory hormones. In other cases, peptides that are functionally the same have different structures. For example, the allatostatins in

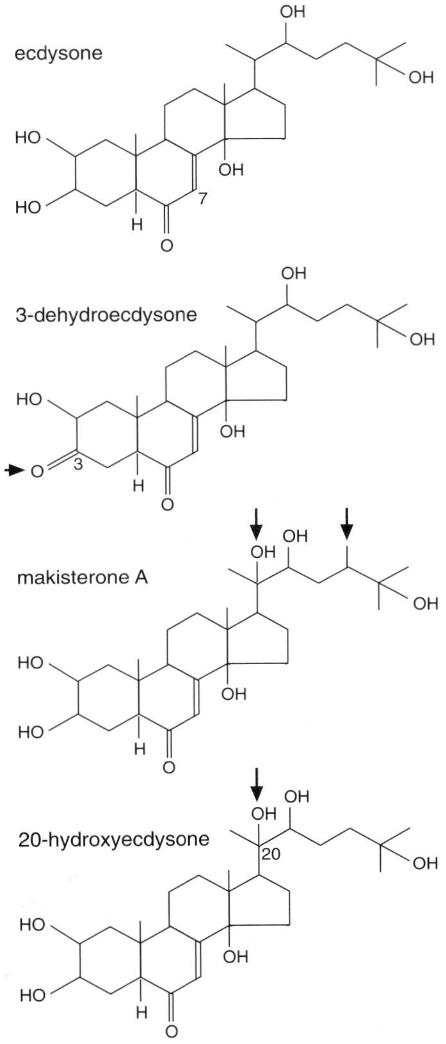

Fig. 21.1. Ecdysteroids. Structure of ecdysone and related ecdysteroids produced by insects. Arrows indicate differences from ecdysone. Numbers relate to positions of carbon atoms.

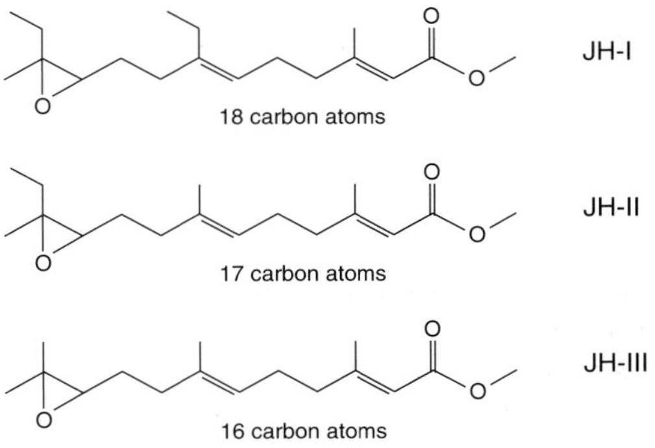

Fig. 21.2. Juvenile hormone structures.

small, consisting of only four to seven amino acids, the other large, with 22–30 amino acid residues. The larger hormone is a dimer (a molecule composed of two similar units).

Peptide hormones are derived from protein precursors which are cleaved and modified post-transcriptionally to produce the active peptide. Sometimes only a single active peptide is formed from one protein. This is the case with prothoracicotropic hormone and adipokinetic hormone. In other instances, however, several different peptides are produced from a single protein. The seven peptides with allatostatin properties in *Diploptera* are all derived from the same protein. In *Bombyx*, a single precursor protein is the source of diapause hormone and pheromone biosynthesis activating hormone, peptides with quite different functions (Yamashita, 1996).

Peptide hormones vary in the numbers of amino acid residues they contain. Some have as few as eight, but others are much larger. Pheromone biosynthesis activating neuropeptide (PBAN) has 33 amino acid residues (*Bombyx*) (Raina, 1993); eclosion hormone has 62 residues (*Manduca*) (Nijhout, 1994); and bursicon is a protein of 30 kDa (Kostron *et al.*, 1995).

Reviews: Gäde, 1990 – adipokinetic hormone; Horn & Bergamasco, 1985 – ecdysteroids; Orchard, 1987 – adipokinetic hormones; Stay, Tobe & Bendena, 1994 – allatostatins

21.2 ENDOCRINE ORGANS

The endocrine organs of insects are of two types: specialized endocrine glands, and neurosecretory cells, most of which are within the central nervous system.
Review: Joly, 1976

orthopteroid insects are structurally different from those so far identified in Lepidoptera (21.3b).

When two hormones with similar functions are present within a species, they may have similar compositions, or be different. For example, the different allatostatins produced by *Diploptera* all have the same sequence, phenylalanine-glycine-leucine-NH$_2$, at the C-terminal end although other parts of their sequences differ. Larval Lepidoptera have two prothoracicotropic hormones, one

a) peptides involved in molting

> prothoracicotropic hormone (PTTH)
> ecdysis triggering hormone (ETH)
> eclosion hormone (EH)
> crustacean cardioactive peptide (CCAP)

b) peptides regulating juvenile hormone synthesis

> allatotropin (*Manduca*)
>> Gly-Phe-Lys-Asn-Val-Glu-Met-Met-Thr-Ala-Arg-Gly-Phe-NH$_2$
>
> allatostatins
>> *Diploptera* I   Ala-Pro-Ser-Gly-Ala-Gln-Arg-Leu-Tyr-Gly-Phe-Gly-Leu-NH$_2$
>>
>> *Diploptera* III   Gly-Gly-Ser-Leu-Tyr-Ser-Phe-Gly-Leu-NH$_2$
>>
>> *Diploptera* V   Ala-Tyr-Ser-Tyr -Val- Ser-Glu-Tyr-Lys-Arg-Leu-Pro-Val-Tyr-Asn-Phe-Gly-Leu-NH$_2$
>>
>> *Manduca*   pGlu-Val-Arg-Phe-Arg-Gln-Cys-Tyr-Phe-Asn-Pro-Ile-Ser-Cys-Phe-OH

c) adipokinetic and related hormones

> *Locusta* - AKHI   pGlu-Leu-Asn-Phe-Thr-Pro-Asn-Trp-Gly-Thr-NH$_2$
> *Locusta* - AKHII   pGlu-Leu-Asn-Phe-Ser-Ala-Gly-Trp-NH$_2$
> *Manduca* - AKH   pGlu-Leu-Thr- Phe-Thr-Ser-Ser-Trp-Gly-NH$_2$
> *Carausius* - HTH   pGlu-Leu-Thr- Phe-Thr-Pro-Asn-Trp-Gly-Thr-NH$_2$
> *Gryllus* - AKH   pGlu-Val-Asn- Phe-Ser-Thr-Gly-Trp-NH$_2$

d) peptides involved in salt and water regulation

> chloride transport stimulating hormone (CTSH)
> diuretic hormone (*Locusta*)   Cys-Leu-Ile-Thr-Asn-Cys-Pro-Arg-Gly-NH$_2$
>
>   NH$_2$-Gly-Arg-Pro-Cys-Asn-Thr-Ile-Leu-Cys
>
> leucokinin   Asp-Pro-Ala-Phe-Asn-Ser-Try-Gly-NH$_2$

e) other peptide hormones

> diuretic hormone (*Bombyx*)
> pheromone biosynthesis activating neuropeptide (PBAN)

ABBREVIATIONS

Ala - alanine	Lys - lysine
Arg - arginine	Met - methionine
Asn - asparagine	Phe - phenylalanine
Cys - cystine	Pro - proline
Gln - glutamine	Ser - serine
Glu - glutamic acid	Thr - threonine
pGlu - pyroglutamate	Trp - tryptophan
Gly - glycine	Tyr - tyrosine
Ile - isoleucine	Val - valine
Leu - leucine	

Fig. 21.3. Some of the major peptide hormones, with the amino acid sequences of some. Commonly used abbreviations are given in brackets where appropriate. (**a**) Peptides involved in different stages of molting. (**b**) Peptides regulating juvenile hormone synthesis. Note the similarities of the endings (shown in dotted rectangle) of the *Diploptera* allatostatins, and the completely different structure of *Manduca* allatostatin. (**c**) Adipokinetic hormone and other hormones with similar sequences. Dotted rectangles indicate similar positions of specific residues in at least four of the compounds shown. AKH, adipokinetic hormone; HTH, hypertrehalosemic hormone. (**d**) Peptides involved in salt and water regulation. (**e**) Other peptide hormones.

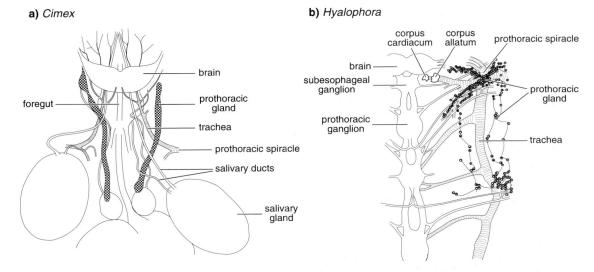

Fig. 21.4. Prothoracic glands shown by dark shading. General arrangement: **(a)** in larval bed bug, *Cimex* (after Wells, 1954); **(b)** in pupal silk moth, *Hyalophora* (after Herman & Gilbert, 1966).

21.2.1 Endocrine glands

21.2.1.1 *Glands producing ecdysteroids*

In the immature stages of all insects, molting hormones are produced by the prothoracic glands. In adult females, where the same hormones are produced to regulate embryonic development, the follicle cells in the ovary are the principal source, although there is evidence that some ecdysteroids are produced elsewhere in the abdomen of some insects, possibly by the oenocytes.

Prothoracic glands The prothoracic, or thoracic, glands are a pair of usually diffuse glands at the back of the head or in the thorax in most insects (Fig. 21.4), but in the base of the labium in Thysanura. Each gland has a rich tracheal supply and often a nerve supply. In most insects, the nervous connection is with the subesophageal ganglion, but in some there is also a connection to the prothoracic ganglion, and, in cockroaches, there is a connection to the brain. The prothoracic glands of Heteroptera have no nerve supply.

The glandular cells making up the gland in most hemimetabolous insects and in Coleoptera are small, up to 25 μm across, and they are present in large numbers and undergo cycles of mitosis. In larval *Tenebrio*, there are up to 1000 cells. However, in most other holometabolous insects and in Heteroptera, the cells are large, and they are present in relatively small numbers, ranging from about 30

in the larva of *Diatraea* to almost 250 in *Hyalophora*. The cells in these species do not divide, but undergo endomitotic division so that they become polyploid, increasing greatly in size through larval life. The glands show cycles of development associated with secretion.

The prothoracic glands degenerate in the adults of nearly all insects, but they persist in Apterygota, which continue to molt throughout life, and in the adults of solitarious locusts. Normal persistence of the glands throughout development results from the presence of juvenile hormone in the hemolymph; their breakdown after adult emergence is due to the lack of this hormone.

21.2.1.2 *Corpora allata*

The corpora allata are glandular bodies, usually one on either side of the esophagus (see Fig. 20.16) although they may be fused to form a single median organ as in higher Diptera. Each is connected with the corpus cardiacum on the same side by a nerve carrying fibers from neurosecretory cells of the brain. In addition, a fine nerve connects each corpus allatum with the subesophageal ganglion. This is a major nerve in Ephemeroptera, where the nerve from the corpus cardiacum is absent. In Thysanura, the corpora allata are in the bases of the maxillae and, in addition to a fine nerve direct from the subesophageal ganglion, they are innervated by branches from the mandibular and maxillary roots of the subesophageal ganglion.

In Thysanura and Phasmida, the corpora allata are hollow balls of cells, with gland cells forming the walls. Elsewhere they are solid organs of glandular secretory cells, often with lacunae between the cells. In hemimetabolous insects, the glands contain large numbers of small glandular cells. By contrast, in holometabolous species, the gland cells are large and few in number; there are only about 20 in *Drosophila* and some beetles.

The corpora allata produce juvenile hormone whose principal functions are regulating metamorphosis (section 15.4) and, in some species, yolk synthesis and deposition in the oocytes of adults (section 13.2.4.2).

21.2.1.3 *Corpora cardiaca*

The corpora cardiaca are a pair of organs often closely associated with the aorta, and forming part of its wall (see Fig. 20.16). In higher groups such as Lepidoptera, Coleoptera and some Diptera they become separated from the aorta. Corpora cardiaca are not known to be present in Collembola. Each organ contains the endings of axons from neurosecretory cells in the brain and other axons passing through to the corpora allata. In addition, they contain intrinsic secretory cells with long cytoplasmic projections extending towards the periphery of the organ. These cells are probably derived from neurosecretory cells, rather than being glandular cells with a separate origin (Orchard, 1987). Their projections probably facilitate the release of secretions into the hemolymph.

The corpora cardiaca store and release hormones from the neurosecretory cells of the brain, to which they are connected by one or two pairs of nerves. They serve as neurohemal organs for several different hormones. In addition, the intrinsic secretory cells produce adipokinetic hormone and some other peptides whose functions are often unknown. Sometimes storage and secretory cells are intermingled, but in *Schistocerca* and some Heteroptera one part of the corpus cardiacum is concerned with secretion and another part with storage (Cazal, 1948).

21.2.1.4 *Ring gland of cyclorrhaphous Diptera*

In the larvae of cyclorrhaphous Diptera the ring gland surrounds the aorta just above the brain (Fig. 21.5). It is formed from the corpora allata, corpora cardiaca and prothoracic glands all fused together, although the compo-

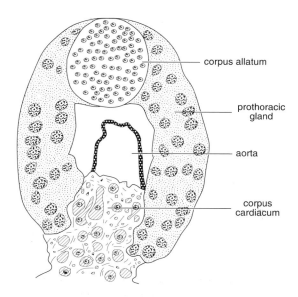

Fig. 21.5. Ring gland from early pupa of the hover fly, *Eristalis*. In this species the ring gland is fused with the hypocerebral ganglion, but this is not always the case (after Cazal, 1948).

nent elements can still be identified. The ring gland is connected to the brain by a pair of nerves and it also has a connection with the recurrent nerve.

The larvae of Nematocera have completely separate endocrine glands, but larval Orthorrhapha approach the cyclorrhaphan condition, except that the corpus allatum tends to be separated from the rest as a single median lobe.

21.2.1.5 *Endocrine cells of the midgut*

All insects have cells in the midgut epithelium which, by virtue of their ultrastructure and immunocytological properties are believed to be endocrine cells, although in no case has their endocrine function been proved (Žitnan, Sauman & Sehnal, 1993). These are isolated cells scattered amongst the principal midgut cells. They stand on the basal lamina of the midgut and are, therefore, in direct contact with the hemolymph. The basal plasma membrane is flat, not deeply infolded like that of the principal midgut cells, and they release their contents into the hemolymph by exocytosis (Andriès & Tramu, 1985). The cells are of two morphological types: those that extend to the lumen of the midgut by a slender process, and those that apparently do not. These are called open and closed cells, respectively. Where the open cells reach the lumen, they have microvilli. It is probable that the endocrine cells which extend to the gut lumen respond

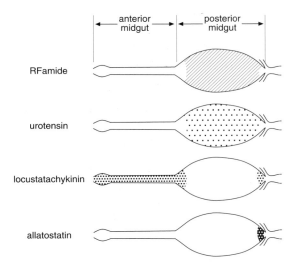

Fig. 21.6. Endocrine cells in the midgut of female mosquito, *Aedes*. Diagrams show the distribution of endocrine cells whose contents are immunoreactive to antibodies against four different peptides. The functions of these substances in the mosquito are not known (after Veenstra *et al.*, 1995).

directly to substances in the lumen. Immunocytological studies indicate that these cells produce peptides.

The adult female mosquito, *Aedes*, has about 500 endocrine cells scattered through the midgut and a ring of cells in the pyloric region (Fig. 21.6). By contrast, at least 30 000 endocrine cells are present in the midgut epithelium of *Periplaneta* (De Loof, 1987). The endocrine cells in the midgut of *Aedes* are immunoreactive to four different antisera, each cell containing only one peptide. These are perhaps the only peptides present in the midgut cells of *Aedes* (Veenstra *et al.*, 1995), but other insects are known to have as many as 10 different peptides in the midgut endocrine cells (Montuega *et al.*, 1989).

Although the functions of these peptides are not known, they are presumed to have a hormonal function relating to digestion and absorption, perhaps regulating the synthesis of digestive enzymes and post-feeding diuresis. Some of the peptides are similar to some peptide hormones secreted by endocrine cells in the vertebrate gut. For example, half of the 500 endocrine cells in the midgut of *Aedes* are recognized by an antiserum to the peptide, phenylalanine–methionine–arginine–phenylalnine–amide (FMRFamide). All the cells exhibiting this response are situated in the midgut where the blood meal is digested (equivalent to the RFamide-staining region in

Fig. 21.6) and they appear to release their contents during the six hours after a meal (Brown, Crim & Lea, 1986). It is possible that these cells are involved in the regulation of enzyme synthesis and release.

Other cells may have different functions, perhaps including the regulation of gut motility. The midgut epithelia of several caterpillars have been shown to contain peptides that induce contractions in visceral muscles (Yi, Tirry & Degheele, 1992).

21.2.1.6 *Epitracheal glands*
Epitracheal glands are known only from Lepidoptera, although it is likely that they will be found in other insects. In *Manduca*, each epitracheal gland is a group of three or four gland cells attached to a trachea near each spiracle. One of the cells increases in size prior to ecdysis, reaching a diameter up to 250 μm, and regressing again after ecdysis. These cells are the source of ecdysis triggering hormone (Žitňan *et al.*, 1996).

Reviews: Beaulaton, 1990 – prothoracic glands; Cassier, 1990 – corpora allata; Sedlac, 1985 – cell structure; Tobe & Stay, 1985 – corpora allata

21.2.2 **Neurosecretory cells**
Neurosecretory cells normally occur in the ganglia of the central nervous system. They generally resemble monopolar nerve cells, but are characterized by showing cytological evidence of secretion and, usually, by discharging their products into the hemolymph. There is no sharp distinction between neurosecretory cells, as defined here, and neurons with a neuromodulatory function. They may secrete similar compounds, but the effects of neuromodulatory cells are generally restricted to the nervous system (see section 20.2.3.2).

Each neurosecretory cell has a dendritic arbor in the neuropil of the ganglion in which the soma occurs. The axon extends through the central nervous system, but at some point, characteristic for a particular neurosecretory cell, it penetrates the blood–brain barrier, so that the cell's secretion is released into the hemocoel and is then transported to the target cells in the hemolymph. At its terminal outside the blood–brain barrier, the axon divides into fine branches which end in swellings. Within these swellings, are synapse-like structures (called synaptoids) with closely associated clusters of secretory vesicles. The secretion is presumed to be released into the hemolymph at the synaptoids.

The areas at which secretions of neurosecretory cells are released into the hemolymph are called neurohemal areas, or, if a well-defined structure is formed, neurohemal organs. The corpora cardiaca are the neurohemal organs at which many of the hormones produced in the brain enter the hemolymph (see below).

The somata of neurosecretory cells occur in all the ganglia of the central nervous system. Examples of a few of the cells in the ganglia of the caterpillar of *Manduca* for which the nature of their secretion is known are given in Fig. 21.7.

In the brain, there are often two main groups of neurosecretory cells on each side. One group is in the pars intercerebralis, near the midline. The number of apparent neurosecretory cells in this group is very different in different species. The desert locust, *Schistocerca*, for example has about 500, whereas *Aphis* has only four or five. The axons from these cells pass backwards through the brain and some or all of them cross over to the opposite side (decussate), emerging from the brain as a nerve which runs back to the corpus cardiacum. Most of the fibers end here, but a few pass through the corpus cardiacum to the corpus allatum and, in the locust, to the foregut and ingluvial ganglion. In most Apterygota, these median neurosecretory cells are contained in separate capsules of connective tissue, known as the lateral frontal organs, on the dorsal side of the brain, but in some Machilidae the cells are intercerebral as in Pterygota. *Petrobius*, also a machilid, occupies an intermediate position with some neurosecretory cells in the lateral frontal organs and others in an adjacent frontal zone of the brain (Watson, 1963).

The second group of neurosecretory cells in the brain is variable in position. It is sometimes medial to the corpora pedunculata, and sometimes between the latter and the optic lobes. In some Diptera and Hymenoptera, the cells corresponding with this group are associated with the other neurosecretory cells in the pars intercerebralis. A second axon tract passes from these cells through the brain to the corpus cardiacum and, in *Schistocerca*, some fibers also extend to the corpus allatum.

Variable numbers of neurosecretory cells occur in the ventral ganglia. Their products are released into the hemolymph at various neurohemal sites. For example, the axons of cells in the subesophageal ganglion of *Manduca* run to the corpora cardiaca via the brain or the lateral nerves (Fig. 21.7). The axons of some cells in the abdominal ganglia leave the ganglion in the median interganglionic connective and diverge laterally forming swollen structures called perivisceral or perisympathetic organs; others may reach the same structures via the lateral nerves. In some cases, the neurohemal areas occur along the lateral nerves without the development of discrete neurohemal organs.

Some neurosecretory cell bodies occur outside the central nervous system. In *Carausius* there are 22 neurosecretory cells with somata on the peripheral nerves in each abdominal segment. Unlike the neurosecretory cells in the central nervous system, they are multipolar cells. Others, called cardiac cells, are present alongside the heart. It is uncertain if these cells are connected to the central nervous system.

The secretions of neurosecretory cells are usually neuropeptides, but in some cases they are known to be biogenic amines. The products are synthesized in the soma where they are associated with a protein to form membrane-bound granules. It is these granules which are visualized in studies on neurosecretion. The granules are transported along the axon to the terminals at rates varying from 5 to 21 mm h^{-1} and released from the terminals by exocytosis. Exocytosis results in a temporary increase in the surface membrane of the axon as the membrane around the granules fuses with the plasma membrane. Membrane is subsequently retrieved by micropinocytosis and the small size of the vesicles, about 40 nm diameter, ensures that only a minimum amount of extracellular fluid is taken up for a relatively large retrieval of membrane. Presumably the small vesicles are broken down and the material re-cycled.

Neurosecretory cells resemble normal neurons in having an excitable plasma membrane. They conduct action potentials, and this electrical activity leads to release of the neurosecretory material at the axon terminals just as an action potential in a normal neuron causes the release of neurotransmitter at a synapse. The axon potentials are generally much longer lasting than those in other neurons, however, with durations up to 20 ms. The secretory activity of a neurosecretory cell is regulated by neurons making synaptic connections with its dendritic arbor. The inputs to the cell may be excitatory or inhibitory.

The endocrine activity of neurosecretory cells may take one of two forms: the cells may either produce hormones which act directly on effector organs or they may act on other endocrine organs, which, in turn, are stimulated to produce hormones. Adipokinetic hormone is an example

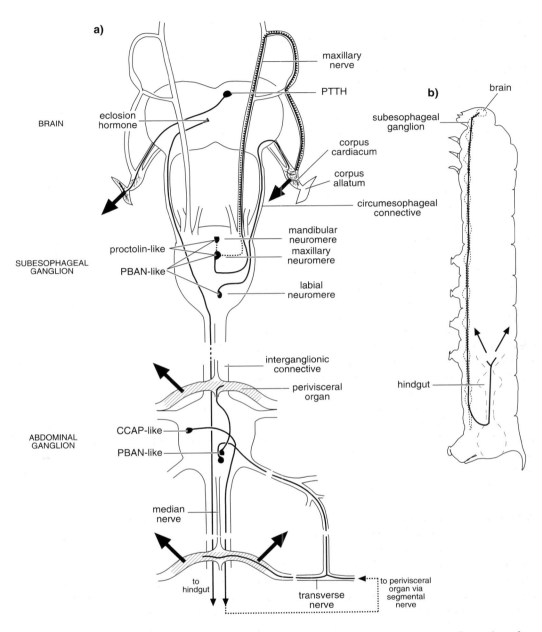

Fig. 21.7. Examples of neurosecretory cells and neurohemal organs in the larva of *Manduca*. (**a**) Connections of some neurosecretory cells with their neurohemal organs. Bold arrows indicate release of peptides into hemolymph. Only a few of the cells are shown. Note that, in most insects, the corpora cardiaca are the neurohemal organs for prothoracicotropic hormone (PTTH); the Lepidoptera are unusual in using the corpora allata for this purpose. It is important to note that the suffix 'like' on some labels indicates that the peptides in the labelled cell are similar to, but not necessarily the same as the named peptide. Their functions are generally unknown. Abbreviations: CCAP, crustacean cardioactive peptide; PBAN, pheromone biosynthesis activating neuropeptide (based largely on data of Dr. N.T.Davis). (**b**) Eclosion hormone is released from neurohemal areas on the hindgut. The position of the soma in the brain is shown in (a) (after Truman & Copenhaver, 1989).

of a peptide acting directly on an effector organ. Prothoracicotropic hormone (PTTH) and the secretions regulating corpus allatum activity, allatotropins and allatostatins, are examples of peptides regulating the synthesis of other hormones.

The contents of some neurosecretory cells react with antisera of more than one neuropeptide and it is likely that these cells secrete cocktails of peptides each producing a different, though probably related, effect (Homberg, Davis & Hildebrand, 1991).

Reviews: Orchard & Loughton, 1985; Raabe, 1975, 1983

21.3 TRANSPORT OF HORMONES

Ecdysteroids are relatively insoluble in water and are probably transported through the hemolymph bound to a protein although a binding protein has been clearly established only in *Locusta*.

Although juvenile hormone is slightly soluble in water, in the hemolymph most is bound to a protein. Calculations suggest that, in the locust, less than 0.1% of the juvenile hormone is not associated with protein, and this is probably usual for other insects. This is important because free juvenile hormone tends to bind non-specifically to surfaces. Its association with protein both solubilizes it and prevents non-specific binding.

In many insects, the binding protein is a lipophorin and although this may have a relatively low affinity for juvenile hormone, the large quantities present in the hemolymph may provide an efficient means of circulation. At least in Orthoptera and Lepidoptera, however, specific juvenile hormone binding proteins are present that are quite different from lipophorins. That in Orthoptera is a high density lipoprotein, while that of Lepidoptera is a low molecular weight protein which is not associated with significant amounts of lipid or carbohydrate. Orthopteran binding protein has a higher affinity for JHIII, the normal juvenile hormone of orthopterans, than it has for JHI or II, while the binding protein of Lepidoptera has the opposite properties.

The binding protein also gives juvenile hormone some degree of protection from enzymic degradation. In *Manduca*, it protects the hormone from non-specific esterases in the hemolymph, but not from the specific juvenile hormone esterase (Touhara *et al.*, 1995).

Peptides enter the hemolymph in neurohemal areas or organs formed by the terminals of neurosecretory cells

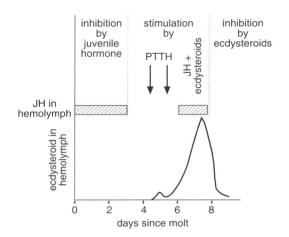

Fig. 21.8. Prothoracic glands: regulation of synthetic activity in the final stage larva of *Manduca* (based on Smith, 1995).

(see above). Once in the hemolymph, being water soluble, they require no special transport mechanisms to enable them to reach their targets.

Review: Trowell, 1992 – juvenile hormone

21.4 REGULATION OF HORMONE TITER

Hormone titers in the hemolymph vary in specific temporal patterns necessary for the proper regulation of the activities they govern (see Figs. 13.9, 13.10, 15.29, 15.30). These variations in titer are produced by changes in their synthesis, release, degradation and excretion.

Ecdysone is not stored in the prothoracic gland; its appearance in the hemolymph reflects its immediate synthesis in the gland. Synthesis is often under dual control. It is triggered by the prothoracicotropic hormone (PTTH), but the effects are modulated, at least in some insects, by an inhibitory hormone and by direct neural regulation which may be either stimulatory or inhibitory (Koolman, 1995).

The competence of the prothoracic gland to respond to PTTH varies through the developmental cycle (Fescemeyer *et al.*, 1995) and, in larval *Manduca*, ecdysteroids and juvenile hormone both affect the production of ecdysone, but their effects vary (Fig. 21.8). During the feeding period of the larva, juvenile hormone inhibits ecdysone synthesis. Subsequently, the ecdysteroids produced by the glands have a positive feedback effect which is enhanced when juvenile hormone is also present. Finally, however, when the glands are highly active, the ecdysteroids have a negative feedback and contribute to the subsequent rapid decline in hemolymph titer (Sakurai & Williams, 1989). The rate at which the prohormone ecdysone is converted to the active 20-hydroxyecdysone also varies.

Production of 20-hydroxyecdysone is offset by its degradation and excretion as well as by its conversion to conjugated forms which are inactive. As a result, the period for which the active hormone remains in the hemolymph is limited. For example, in the third stage larva of *Calliphora*, its half-life (the time within which half of the compound is converted to other compounds or excreted) is about three hours, but in the pupa its half-life is more than one day.

Conjugates of ecdysteroids are often phosphates or glucosides. In eggs, most of the ecdysteroid may be conjugated except for brief periods when active hormone is present (see Fig. 14.26). The Malpighian tubules excrete both ecdysone and 20-hydroxyecdysone, as well as various metabolites (Fig. 21.9).

Juvenile hormone is released from the corpora allata as it is produced; it is not stored. Its hemolymph titer, which varies through the course of development, is consequently a product of the rate of synthesis and the rate at which it is degraded or excreted. Estimates of its half life in various insects are usually less than two hours so that sustained high titers must reflect high rates of synthesis. Fig. 21.10a illustrates the close correspondence between synthesis and hemolymph titer in an adult cockroach.

The structure of the corpora allata varies in relation to juvenile hormone synthesis. At times of maximum juvenile hormone production, the number of cells increases, cell organelles proliferate and the enzymes necessary for the synthesis increase in abundance. All these factors decrease when the rate of synthesis is low and it is possible that their regulation involves different mechanisms, but this is not known. The overall rate of synthesis is regulated by peptides (Figs. 21.3, 21.11): allatotropins enhance synthesis; allatostatins decrease it. Five different allatostatins are known from *Diploptera*, but it is not clear which stages of the synthetic process they affect. These peptides are produced by neurosecretory cells in the brain, but their mode of action on the corpora allata is uncertain. They may act as neuromodulators as the axon terminals are present in the corpora allata and allatostatin receptors are known to be present in the corpora allata of *Diploptera*. They may also act as neurohormones, however, as they are also present in the hemolymph.

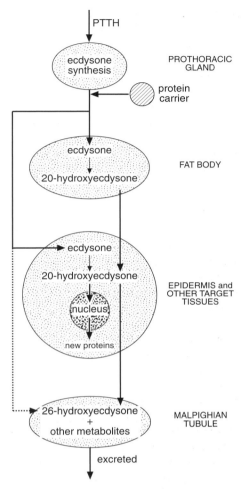

Fig. 21.9. Ecdysteroids. The principal stages in the production, activity and degradation of ecdysteroids.

These peptides are not the only regulators of corpus allatum activity. The sensitivity of the glands to the peptides varies, probably in relation to the numbers of receptors present. It is not known how this is regulated, but perhaps some independent neural input from the brain is involved. There is also evidence that octopamine affects juvenile hormone synthesis, enhancing synthesis in *Locusta* and *Apis*, but inhibiting it in adult *Diploptera* (Kaatz, Eichmuller & Kreissl, 1994).

Juvenile hormone in the hemolymph has a feedback effect on its own production, at least in adult *Diploptera*. Low titers enhance synthesis, and high titers inhibit it. Factors produced by the ovary have similar effects; an

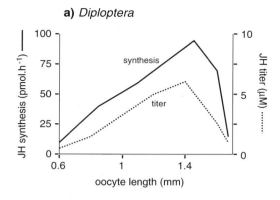

a) *Diploptera*

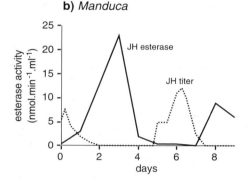

b) *Manduca*

Fig. 21.10. Juvenile hormone: regulation of hemolymph titer. **(a)** Synthetic activity of the corpora allata and hemolymph titer of juvenile hormone during an ovarian cycle in the cockroach, *Diploptera* (after Tobe & Stay, 1985). **(b)** JH esterase activity and hemolymph titer of juvenile hormone throughout the final larval stage of *Manduca* (after Jesudason, Venkatesh & Roe, 1990).

ovary with small oocytes produces a stimulating factor; one with large oocytes an inhibitory factor. The inhibitory effects are produced via allatostatins; whether allatotropins are also involved is not known (Stay *et al.*, 1994).

Once in the hemolymph, the short half-life of juvenile hormone is due to its enzymic degradation. At least in the Lepidoptera, its breakdown is catalyzed by substrate-specific enzymes: an esterase and an epoxide hydrolase. The activity of these enzymes varies through development, and peaks of activity coincide with the disappearance of juvenile hormone from the hemolymph (Fig. 21.10b). In larval *Manduca*, the esterase is usually dominant (Jesudason, Anspaugh & Roe, 1992), but in the mosquito, *Culex*, the epoxide hydrolase predominates (Lassiter, Apperson & Roe, 1995).

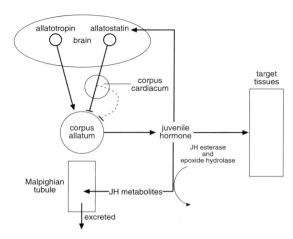

Fig. 21.11. Juvenile hormone. Regulation of hemolymph titer involves the balance between synthesis in the corpora allata and degradation and excretion by the Malpighian tubules.

The release of neurohormones is presumably controlled by other neurons in the central nervous system. **Reviews:** Feyereisen, 1985 – juvenile hormone synthesis; Goodman, 1990 – juvenile hormone; Hammock, 1985 – juvenile hormone degradation; Koolman & Karlson, 1985 – ecdysteroids; Smith, S.L., 1985 – ecdysteroids; Smith, W.A., 1995 – prothoracic glands; Stay & Woodhead, 1993 – juvenile hormone synthesis; Tobe & Stay, 1985 – corpora allata

21.5 MODE OF ACTION OF HORMONES

When a hormone reaches a target cell it initiates biochemical changes in the cell, but not all cells respond in the same way, and some do not respond at all. Specificity of response is a feature of receptor proteins in the target organs. Only those cells with the appropriate receptors will be affected by a hormone and since a receptor may exist in different forms and be expressed at different times, the hormone can have different effects in different tissues. Depending on the nature of the hormone, these receptors are in the cell membrane or within the cell.

Ecdysteroids and juvenile hormone are lipophilic and, as a consequence, pass readily through cell membranes. They bind with specific receptor proteins within the cell, directly causing the activation or inactivation of genes and the synthesis of new proteins. The pattern of expression of ecdysteroid receptors in the epidermis of *Manduca* over the period of metamorphosis differs from the pattern seen

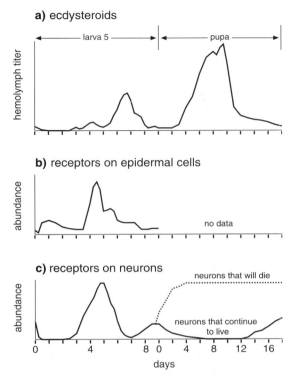

a) ecdysteroids

b) receptors on epidermal cells

c) receptors on neurons

Fig. 21.12. Ecdysteroids. Their effects on different tissues of *Manduca* are dependent on the presence of receptors in the tissues. Notice the different time scales in larva and pupa (after Riddiford & Truman, 1993). (a) Hemolymph titer of ecdysteroids in the last stage larva and pupa. (b) Changes in the abundance of ecdysteroid receptors on the epidermal cells. No data available for the pupal stage. (c) Changes in the abundance of ecdysteroid receptors on some neurons in the central nervous system.

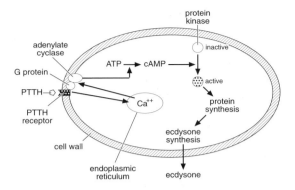

Fig. 21.13. Mode of action of PTTH. Diagram showing the probable mode of action on a cell of the prothoracic gland (after Smith, 1995).

in neurons (Fig, 21.12). Moreover, the expression of receptors on neurons that will die after eclosion differs from that on neurons that will persist, and possibly there are different types of receptor in the different cells (Riddiford & Truman, 1993).

The effect of juvenile hormone in immature insects is to modify the responses to ecdysteroid; it has no effect by itself. Nuclear receptors are known to be present in larval *Manduca*, and their presence varies temporally. In adult insects, however, where juvenile hormone affects sexual maturation and behavior, it acts independently of ecdysteroids. Here, it may function as a primer or as a regulator of the target tissue. As a primer, it acts directly on the nucleus so that the cell becomes responsive to a subsequent regulatory signal, which may be juvenile hormone itself or another hormone. For example, the fat body cells do not respond to juvenile hormone (or ecdysone in the Diptera) by synthesizing vitellogenin until they have been exposed to the juvenile hormone for some time. The process of priming may include the development of specific receptors in the cell membrane and the initiation of transcription.

As a regulator, juvenile hormone may act via membrane receptors to activate existing enzymes via a second messenger cascade, or it may act at the nucleus to promote protein synthesis. For example, in female *Rhodnius*, juvenile hormone triggers patency amongst the follicle cells, allowing hemolymph proteins to reach the oocyte. It is known that juvenile hormone stimulates a protein kinase in these cells, presumably via a second messenger cascade. The kinase then activates an ATPase in the cell membrane. By pumping sodium and potassium out of the cell this produces an osmotic gradient, drawing water out of the cells so that they shrink, and spaces appear between them (Sevala & Davey, 1989). In contrast, vitellogenin synthesis in the fat body depends on the modulation of gene expression.

Peptide hormones and biogenic amines are lipophobic, and will not pass through cell membranes. Specific receptor proteins for these hormones are present in the cell membranes. Via these receptors they activate second messenger cascades which lead to the activation of enzymes that are already in existence. The second messengers are often cAMP or cGMP, and these activate, or sometimes inactivate, specific enzymes via a protein kinase (Fig. 21.13). PTTH, adipokinetic hormone, some diuretic hormones and the pheromone biosynthesis activating neuropeptide (PBAN) act via cAMP, but the activation of glycogen phosphorylase by adipokinetic hormone appears to involve a different pathway. Eclosion hormone triggers ecdysis behavior via an increase in cGMP accompanied by an increase in inositol trisphosphate although the exact relationship between these two messengers is unclear (Morton & Simpson, 1995; Smith & Sedlmeier, 1990; Spring, 1990).

A hormone may lead to the production or activation of not just one, but a suite of enzymes necessary to produce and ensure the successful secretion of the final product. For example, prothoracicotropic hormone not only initiates ecdysone formation in the prothoracic gland, but also promotes the production of β-tubulin which may be important in the transport of ecdysteroid precursors within the cells (Rybczynski & Gilbert, 1995). In the epidermis, ecdysone initiates the *de novo* synthesis of DOPA decarboxylase and also activates the enzyme which catalyses the synthesis of a phenol oxidase from its proenzyme, making available the essential enzymes for cuticular tanning.

Reviews: Riddiford, 1994 – juvenile hormone; Riddiford, 1995 – Lepidoptera; Riddiford & Truman, 1993 – ecdysteroids; Truman, 1995 – Lepidoptera; Wyatt & Davey, 1996 – juvenile hormone in adult insects

REFERENCES

Andriès, J.C. & Tramu, G. (1985). Ultrastructural and immunohistochemical study of endocrine cells in the midgut of the cockroach *Blaberus craniifer* (Insecta, Dictyoptera). *Cell & Tissue Research*, **240**, 323–32.

Beaulaton, J. (1990). Anatomy, histology, ultrastructure, and functions of the prothoracic (or ecdysial) glands in insects. In *Morphogenetic Hormones of Arthropods*, vol. 1, part 2, *Embryonic and Postembryonic Sources*, ed. A.P. Gupta, pp. 343–435. New Brunswick: Rutgers University Press.

Brown, M.R., Crim, J.W. & Lea, A.O. (1986). FMRFamide- and pancreatic polypeptide-like immunoreactivity of endocrine cells in the midgut of a mosquito. *Tissue & Cell*, **18**, 419–28.

Cassier, P. (1990). Morphology, histology and ultrastructure of JH-producing glands in insects. In *Morphogenetic Hormones of Arthropods*, vol. 1, part 2, *Embryonic and Postembryonic Sources*, ed. A.P. Gupta, pp. 83–194. New Brunswick: Rutgers University Press.

Cazal, P. (1948). Les glandes endocrines rétro-cérébral des insectes (étude morphologique). *Bulletin Biologique de la France et de la Belgique*, supplement **32**, 227 pp.

De Loof, A. (1987). The impact of the discovery of vertebrate- type steroids and peptide hormone-like substances in insects. *Entomologia Experimentalis et Applicata*, **45**, 105–13.

Fescemeyer, H.W., Masler, E.P., Kelly, T.J. & Lusby, W.R. (1995). Influence of development and prothoracicotropic hormone on the ecdysteroids produced *in vitro* by the prothoracic glands of female gypsy moth (*Lymantria dispar*) pupae and pharate adults. *Journal of Insect Physiology*, **41**, 489–500.

Feyereisen R. (1985). Regulation of juvenile hormone titer: synthesis. In *Comprehensive Insect Physiology, Biochemistry and Pharmacology*, vol. 7, ed. G.A. Kerkut & L.I. Gilbert, pp. 391–429. Oxford: Pergamon Press.

Gäde, G. (1990). The adipokinetic hormone/red pigment-concentrating hormone peptide family: structure, interrelationships and functions. *Journal of Insect Physiology*, **36**, 1–12.

Goodman, W.G. (1990). Biosynthesis, titer regulation, and transport of juvenile hormones. In *Morphogenetic Hormones of Arthropods*, vol. 1, part 1, *Discoveries, Syntheses, Metabolism, Evolution, Modes of Action and Techniques*, ed. A.P. Gupta, pp. 83–124. New Brunswick: Rutgers University Press.

Hammock, B.D. (1985). Regulation of juvenile hormone titer: degradation. In *Comprehensive Insect Physiology, Biochemistry and Pharmacology*, vol. 7, ed. G.A. Kerkut & L.I. Gilbert, pp. 431–72. Oxford: Pergamon Press.

Herman, W.S. & Gilbert, L.I. (1966). The neuroendocrine system of *Hyalophora cecropia*. *General and Comparative Endocrinology*, **7**, 275–91.

Homberg, U., Davis, N.T. & Hildebrand, J.G. (1991). Peptide-immunocytochemistry of neurosecretory cells in the brain and retrocerebral complex of the sphinx moth *Manduca sexta*. *Journal of Comparative Neurology*, **303**, 35–52.

Horn, D.H.S. & Bergamasco, R. (1985). Chemistry of ecdysteroids. In *Comprehensive Insect Physiology, Biochemistry and Pharmacology*, vol. 7, ed. G.A. Kerkut & L.I. Gilbert, pp. 185–248. Oxford: Pergamon Press.

Jesudason, P., Anspaugh, D.D. & Roe, R.M. (1992). Juvenile hormone metabolism in the plasma, integument, midgut, fat body and brain during the last instar of the tobacco hornworm, *Manduca sexta* (L.). *Archives of Insect Biochemistry and Physiology*, **20**, 87–105.

Jesudason, P., Venkatesh, K. & Roe, R.M. (1990). Haemolymph juvenile hormone esterase during the life cycle of the tobacco hornworm, *Manduca sexta* (L.). *Insect Biochemistry and Molecular Biology*, **20**, 593–604.

Joly, P. (1976). Les organes endocrines. In *Traité de Zoologie*, vol. 8, part 4, ed. P.-P. Grassé, pp. 637–75. Paris: Masson et Cie.

Kaatz, H., Eichmüller, S. & Kreissl, S. (1994). Stimulatory effect of octopamine on juvenile hormone biosynthesis in honey bees (*Apis mellifera*): physiological and immunocytochemical evidence. *Journal of Insect Physiology*, **40**, 865–72.

Koolman, J. (1995). Control of ecdysone biosynthesis in insects. *Netherlands Journal of Zoology*, **45**, 83–8.

Koolman, J. & Karlson, P. (1985). Regulation of ecdysteroid titer: degradation. In *Comprehensive Insect Physiology, Biochemistry and Pharmacology*, vol. 7, ed. G.A. Kerkut & L.I. Gilbert, pp. 343–61. Oxford: Pergamon Press.

Kostron, B., Marquardt, K., Kaltenhauser, U. & Honegger, H.W. (1995). Bursicon, the cuticle sclerotizing hormone – comparison of its molecular mass in different insects. *Journal of Insect Physiology*, **41**, 1045–53.

Lassiter, M.T., Apperson, C.S. & Roe, R.M. (1995). Juvenile hormone metabolism during the fourth stadium and pupal stage of the southern house mosquito, *Culex quinquefasciatus* Say. *Journal of Insect Physiology*, **41**, 869–76.

Montuenga, L.M., Barrenechea, M.A., Sesma, P., López, J. & Vázquez, J.J. (1989). Ultrastructure and immunocytochemistry of endocrine cells in the midgut of the desert locust, *Schistocerca gregaria* (Forskal). *Cell and Tissue Research*, **258**, 577–83.

Morton, D.B. & Simpson, P.J. (1995). Eclosion hormone-stimulated cGMP levels in the central nervous system of *Manduca sexta*: inhibition by lipid metabolism blockers, increase in inositol(1,4,5)trisphosphate and further evidence against the involvement of nitric oxide. *Journal of Comparative Physiology* B, **165**, 417–27.

Nijhout, H.F. (1994). *Insect Hormones*. Princeton: Princeton University Press.

Orchard, I. (1987). Adipokinetic hormones – an update. *Journal of Insect Physiology*, **33**, 451–63.

Orchard, I. & Loughton, B.G. (1985). Neurosecretion. In *Comprehensive Insect Physiology, Biochemistry and Pharmacology*, vol. 7, ed. G.A. Kerkut & L.I. Gilbert, pp. 61–107. Oxford: Pergamon Press.

Raabe, M. (1975). Les organes périsympathiques. In *Traité de Zoologie*, vol. 8, part 3, ed. P.-P. Grassé, pp. 511–33. Paris: Masson et Cie.

Raabe, M. (1983). The neurosecretoryneurohaemal system of insects; anatomical, structural and physiological data. *Advances in Insect Physiology*, **17**, 205–303.

Raina, A.K. (1993). Neuroendocrine control of sex pheromone biosynthesis in Lepidoptera. *Annual Review of Entomology*, **38**, 329–49.

Riddiford, L.M. (1994). Cellular and molecular actions of juvenile hormone. I. General considerations and premetamorphic actions. *Advances in Insect Physiology*, **24**, 213–74.

Riddiford, L.M. (1995). Hormonal regulation of gene expression during lepidopteran development. In *Molecular Model Systems in the Lepidoptera*, ed. M.R. Goldsmith & A.S. Wilkins, pp. 293–322. Cambridge: Cambridge University Press.

Riddiford, L.M. & Truman, J.W. (1993). Hormone receptors and the regulation of insect metamorphosis. *American Zoologist*, **33**, 340–7.

Rybczynski, R. & Gilbert, L.I. (1995). Prothoracicotropic hormone elicits a rapid, developmentally specific synthesis of β tubulin in an insect endocrine gland. *Developmental Biology*, **169**, 15–28.

Sakurai, S. & Williams, C.M. (1989). Short-loop negative and positive feedback on ecdysone secretion by prothoracic gland in the tobacco hornworm, *Manduca sexta*. *General and Comparative Endocrinology*, **75**, 204–16.

Sedlac, B.J. (1985). Structure of endocrine glands. In *Comprehensive Insect Physiology, Biochemistry and Pharmacology*, vol. 7, ed. G.A. Kerkut & L.I. Gilbert, pp. 26–60. Oxford: Pergamon Press.

Sevala, V.L. & Davey, K.G. (1989). Action of juvenile hormone on the follicle cells of *Rhodnius prolixus*: evidence for a novel regulatory mechanism involving protein kinase C. *Experientia*, **45**, 355–6.

Smith, S.L. (1985). Regulation of ecdysteroid titer: synthesis. In *Comprehensive Insect Physiology, Biochemistry and Pharmacology*, vol. 7, ed. G.A. Kerkut & L.I. Gilbert, pp. 295–341. Oxford: Pergamon Press.

Smith, W. A.(1995). Regulation and consequences of cellular changes in the prothoracic glands of *Manduca sexta* during the last larval instar: a review. *Archives of Insect Biochemistry and Physiology*, **30**, 271–93.

Smith, W.A. & Sedlmeier, D. (1990). Neurohormonal control of ecdysone production: comparison of insects and crustaceans. *Invertebrate Reproduction and Development*, **18**, 77–89.

Sparagana, S.P., Bhaskaran, G. & Barrera, P. (1985). Juvenile hormone acid methyltransferase activity in imaginal discs of *Manduca sexta* prepupae. *Archives of Insect Biochemistry and Physiology*, **2**, 191–202.

Spring, J.H. (1990). Endocrine regulation of diuresis in insects. *Journal of Insect Physiology*, **36**, 13–22.

Stay, B., Sereg Bachmann, J.A., Stoltzman, C.A., Fairbairn, S.E., Yu, C.G. & Tobe, S.S. (1994). Factors affecting allatostatin release in a cockroach (*Diploptera punctata*): nerve section, juvenile hormone analog and ovary. *Journal of Insect Physiology*, **40**, 365–72.

Stay, B., Tobe, S.S. & Bendena, W.G. (1994). Allatostatins: identification, primary structures, functions and distribution. *Advances in Insect Physiology*, **25**, 267–337.

Stay, B. & Woodhead, A.P. (1993). Neuropeptide regulators of insect corpora allata. *American Zoologist*, **33**, 357–64.

Tobe, S.S. & Stay, B. (1985). Structure and regulation of the corpus allatum. *Advances in Insect Physiology*, **18**, 305–432.

Touhara, K., Bonning, B.C., Hammock, B.D. & Prestwich, G.D. (1995). Action of juvenile hormone (JH) esterase on the JH–JH binding protein complex. An *in vitro* model of JH metabolism in a caterpillar. *Insect Biochemistry and Molecular Biology*, **25**, 727–34.

Trowell, S.C. (1992). High affinity juvenile hormone carrier proteins in the hemolymph of insects. *Comparative Biochemistry and Physiology* **103B**, 795–807.

Truman, J.W. (1995). Lepidoptera as model systems for studies of hormone action in the central nervous system. In *Molecular Model Systems in the Lepidoptera*, ed. M.R. Goldsmith & A.S. Wilkins, pp. 323–39. Cambridge: Cambridge University Press.

Truman, J.W. & Copenhaver, P.F. (1989). The larval eclosion hormone neurons in *Manduca sexta*: identification of the brain-proctodeal neurosecretory system. *Journal of Experimental Biology*, **147**, 457–70.

Veenstra, J.A., Lau, G.W., Agricola, H.-J. & Petzel, D.H. (1995). Immunohistological localization of regulatory peptides in the midgut of the female mosquito *Aedes aegypti*. *Histochemistry and Cell Biology*, **104**, 337–47.

Watson, J.A.L. (1963). The cephalic endocrine system in the Thysanura. *Journal of Morphology*, **113**, 359–73.

Wells, M.J. (1954). The thoracic glands of Hemiptera Heteroptera. *Quarterly Journal of Microscopical Science*, **95**, 231–44.

Wyatt, G.R. & Davey, K.G. (1996). Cellular and molecular actions of juvenile hormone. II. Roles of juvenile hormone in adult insects, *Advances in Insect Physiology*, **26**, 1–155.

Yamashita, O. (1996). Diapause hormone of the silkworm, *Bombyx mori*: structure, gene expression and function. *Journal of Insect Physiology*, **42**, 669–79.

Yi, S.-X., Tirry, L. & Degheele, D. (1992). Presence of myotropins in larval midgut extracts of lepidopteran insects: *Manduca sexta*, *Agrotis segetum* and *Spodoptera exempta*. *Journal of Insect Physiology*, **38**, 1023–32.

Yin, C.-H., Zou, B.X., Jiang, M., Li, M.-F., Qin, W., Potter, T.L. & Stoffolano, J.G. (1995). Identification of juvenile hormone III bisepoxide (JHB$_3$), juvenile hormone III and methyl farnesoate secreted by the corpus allatum of *Phormia regina* (Meigen), *in vitro* and function of JHB$_3$ either applied alone or as a part of a juvenoid blend. *Journal of Insect Physiology*, **41**, 473–9.

Žitňan, D., Kingan, T.G., Hermesman, J.L. & Adams, M.E. (1996). Identification of ecdysis-triggering hormone from an epitracheal endocrine system. *Science*, **271**, 88–91.

Žitňan, D., Sauman, I. & Sehnal, F. (1993). Peptidergic innervation and endocrine cells of midgut. *Archives of Insect Biochemistry and Physiology*, **22**, 113–32.

PART V

Communication

22 Vision

Light is perceived by insects through a number of different receptors. Most adult insects and larval hemimetabolous insects have a pair of compound eyes and often three single-lens eyes, called ocelli. Larval holometabolous insects have one or more single-lens eyes, known as stemmata, on the sides of the head. Some insects are also known to possess epidermal light receptors, and, in some cases, light is known to have a direct effect on cells in the brain.

22.1 COMPOUND EYES

22.1.1 Occurrence

Compound eyes are so called because they are constructed from many similar units called ommatidia. They are present in most adult pterygote insects and the larvae of hemimetabolous insects, but are strongly reduced or absent in wingless parasitic groups, such as the Phthiraptera and Siphonaptera, and in female coccids. This is also true of cave-dwelling forms. Amongst termites, compound eyes are greatly reduced or absent from stages that are habitually subterranean, and, although present in winged reproductives, the sensory components of the eyes may degenerate. Amongst Apterygota, compound eyes are lacking in some Thysanura, but Lepismatidae have 12 ommatidia on each side. Fully-developed compound eyes are present in Archaeognatha. In the non-insect orders of Hexapoda, Collembola have up to eight widely spaced ommatidia, while Protura and Diplura have none.

Each compound eye may be composed of several thousand ommatidia. There are about 10 000 in the eyes of dragonflies and drone honeybees, 5500 in worker honeybees and 800 in *Drosophila*. At the other extreme, workers of the ant, *Ponera punctatissima*, have only a single ommatidium on each side of the head. Usually the eyes are separate on the two sides of the head, but in some insects, such as Anisoptera and male Tabanidae and Syrphidae, the eyes are contiguous along the dorsal midline, this being known as the holoptic condition.

22.1.2 **Ommatidial structure**

Each ommatidium consists of an optical, light-gathering part and a sensory part, which transforms light into electrical energy. The sensory receptor cells of most diurnal insects end close to the lens, and, because of the method of image formation, these are called apposition eyes (Fig. 22.1). Most night-flying insects, however, have eyes with a clear zone between the lenses and the sensory components; they are called superposition eyes (Fig. 22.2).

The cuticle covering the eye is transparent and colorless and usually forms a biconvex corneal lens. In surface view, the lenses are usually closely packed together forming an array of hexagonal facets. Each corneal lens is produced by two epidermal cells, the corneagen cells, which later become withdrawn to the sides of the ommatidium and form the primary pigment cells. Beneath the cornea are four cells, the Semper cells, which, in most insects, produce a second lens, the crystalline cone. This is usually a hard, clear, intracellular structure bordered laterally by the primary pigment cells.

The sensory elements are elongate neurons known as retinula cells. Generally there are eight retinula cells in each ommatidium, but some species have seven, and others nine. Each retinula cell extends basally as an axon which passes out through the basal lamina backing the eye and into the lamina of the optic lobe (see Fig. 20.17). The margin of each retinula cell nearest the ommatidial axis is differentiated into close-packed microvilli extending towards the central axis of the ommatidium at right angles to the long axis of the retinula cell. The microvilli of each retinula cell lie parallel with each other and are often aligned with those of the retinula cell opposite, but are set at an angle to those of adjacent retinula cells (Fig. 22.1d). The microvilli of each retinula cell collectively form a rhabdomere, and, as it is usual for the retinula cells to have a twist along their lengths, the orientation of the microvilli of each rhabdomere changes regularly through the depth of the eye. The visual pigment is located within the microvilli.

In most insects, the rhabdomeres abut on each other along the axis of the ommatidium forming a 'fused'

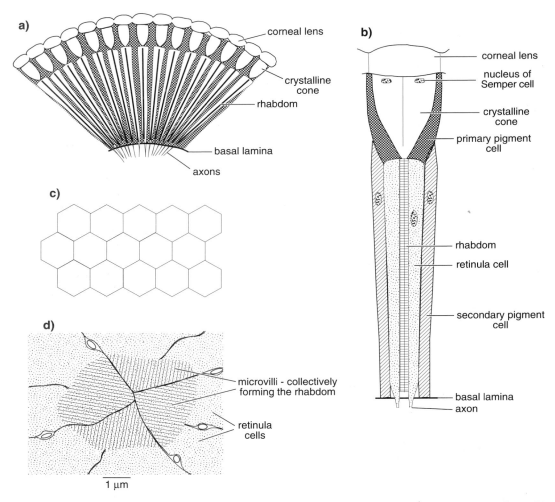

Fig. 22.1. Apposition eye: **(a)** diagrammatic section through an apposition eye showing the rhabdoms extending to the crystalline cones.; **(b)** ommatidium; **(c)** surface view of part of an eye showing the outer surfaces of some corneal lenses (facets); **(d)** cross-section through a fused rhabdom (*Apis*) (after Goldsmith, 1962).

rhabdom (although the cells are not actually fused), but Diptera, Dermaptera, some Heteroptera and some Coleoptera have widely separated rhabdomeres forming an 'open' rhabdom (Fig. 22.3). In a fused rhabdom, all the retinula cells within one ommatidium have the same field of view. In species with open rhabdoms each retinula cell within an ommatidium has a separate visual field, shared by individual cells in each of the adjacent ommatidia (see Fig. 20.21b).

The rhabdom of apposition eyes usually extends the full length of the retinula cells between the crystalline cone and the basal lamina. It is 150 μm long in the ant,

Camponotus, and, in *Drosophila*, with an open rhabdom, each rhabdomere is 60 μm long. It is usually shorter in superposition eyes, and even in apposition eyes one of the rhabdomeres may be very short (see Fig. 22.5a, cell 9).

There is much variation in the way that the clear zone in superposition eyes is bridged. In many Lepidoptera and Coleoptera, the retinula cells extend to the crystalline cone as a broad column, but the rhabdom is restricted to the basal region (Fig. 22.2c), but, in Carabidae and Dytiscidae, one of the retinula cells also has a short distal rhabdomere just below the cone. In other Lepidoptera (the Bombycoidea and Hesperioidea), the retinula cells of each

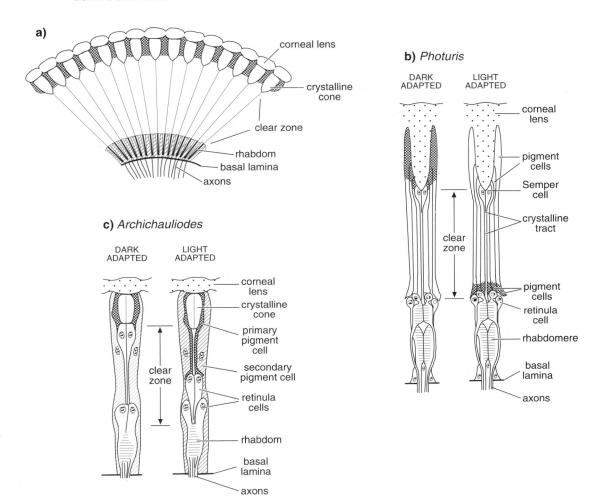

Fig. 22.2. Superposition eye. (**a**) Diagrammatic section through a superposition eye showing the clear zone between the rhabdoms and the lens systems. (**b**) Exocone eye in which the clear zone is crossed by a crystalline tract formed from the Semper cells. Note that there is no crystalline cone in exocone eyes. Left: dark-adapted; right: light-adapted (*Photuris*, Coleoptera). (**c**) Eucone eye in which the clear zone is bridged by retinula cells. Left: dark-adapted; right: light-adapted (*Archichauliodes*, Megaloptera) (after Walcott, 1975).

ommatidium form a thin strand, which may be only 5 μm across, to the lens. This strand is called a crystalline tract. Beetles with exocone eyes (see below) have a similar structure, but it is formed by the Semper cells and the retinula cells are restricted to a basal position in the ommatidium (Fig. 22.2b).

The sensory parts of each ommatidium are usually surrounded by 12–18 secondary pigment cells so that each ommatidium is isolated from its neighbors.

Tracheae pass between the ommatidia proximally in some species, and, in many Lepidoptera, closely packed

tracheae form a layer at the back of the eye. This layer which reflects light back into the eye is called a tapetum. The taenidia of the tracheae forming the tapetum are flattened and enlarged. Their cuticle consists of many successive layers each about one quarter of the wavelength of light in thickness with alternate layers having high and low refractive indices. Light is reflected at each boundary, producing interference colors (section 25.2.2). In butterflies, the reflected light from the back of the eye doubles the effective length of each rhabdom as the light passes through it twice. In moths, with superposition eyes, the

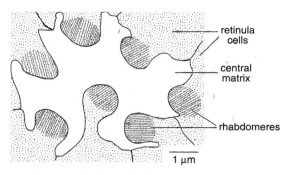

Fig. 22.3. Open rhabdom. A section through the rhabdomeres of *Drosophila* (after Wolken, Capenos & Turano, 1957).

basal region of each ommatidium is surrounded by tracheae. Here, their function is probably to reflect back light that escapes from the rhabdom.

Neural connections The axons passing back from each ommatidium in insects with fused rhabdoms are kept together in the lamina of the optic lobe, each ommatidium being represented by a separate cartridge (section 20.4.1.3). In general, six retinal axons terminate in the cartridge, while two others, which in *Apis* are from ultraviolet-sensitive cells, pass through the lamina, ending in the medulla.

The arrangement is different in the open rhabdom of Diptera. Here, well-developed cartridges are present in the lamina, but, instead of being derived from the axons of a single ommatidium, they are formed round the axons of retinula cells with the same field of view (see above). Thus they contain axons from each of seven adjacent ommatidia and bring together information about a particular area in the visual field (see Fig. 20.21b). By analogy with superposition eyes, which bring together information from a number of ommatidia (see below), these eyes are called neural superposition eyes. Axons from all the green-sensitive cells end in the cartridge; those sensitive to other wavelengths pass through the cartridge to the medulla usually without making synaptic connections (see Fig. 20.21a). The open rhabdoms of Heteroptera and Dermaptera are not known to be associated with neural superposition.

22.1.3 Variation in ommatidial structure
22.1.3.1 *Variation within a species*
The form and arrangement of ommatidia differs in different parts of the eye in many, and perhaps all, insects.

For example, in the praying mantis, *Tenodera*, the facet diameter is greatest in the forwardly directed part of the eye, and decreases all round (Fig. 22.4). Because the radius of curvature of this same part of the eye is greater (so the surface is flatter) than elsewhere, the angle between the optical axes of adjacent ommatidia (the interommatidial angle) is less than elsewhere and the rhabdoms are longer, but thinner. This area of the eye is functionally equivalent to the vertebrate fovea (see below) and similar regions are known to be present in the eyes of other insect species with a particular need for good resolution.

In males of some species, the eye is sharply differentiated into a dorsal region with relatively large facets and small interommatidial angles and a ventral region with much smaller facets and bigger interommatidial angles. This occurs in drone honeybees and in male *Bibio* (Diptera). The division is complete in the male of *Cloeon* (Ephemeroptera), where each eye is in two parts quite separate from each other. Not only are the ommatidia in these two parts different in size, they are also different in structure. Those of the dorsal part are relatively large and of the superposition type, while in the lateral part the ommatidia are smaller and of the apposition type. The eyes are also divided into two in the aquatic beetle *Gyrinus* where the dorsal eye is above the surface film when the insect is swimming and the ventral eye is below the surface. The significance of these differences is discussed below.

Many species have a band of ommatidia along the eye's dorsal rim which differ from those in the rest of the eye. Here, the retinula cells are not twisted and the microvilli of different retinula cells are at right angles to each other (see Fig. 22.15). This area of the eye is polarization-sensitive (see below).

Differences may also occur in the nature of the screening pigments in different parts of the eye. For example, the drone honeybee has transparent red screening pigment in its dorsal ommatidia which have receptors maximally sensitive to blue and ultraviolet, but a more opaque red pigment in the ventral ommatidia, which have green-sensitive receptors in addition to those responding to blue and ultraviolet. These differences relate to the wavelengths absorbed by metarhodopsin and the regeneration of rhodopsin (see below; Stavenga, 1992).

22.1.3.2 *Variation between species*
Variation in the lens system The origin and form of the crystalline cone varies in different insects. Most species have eucone eyes in which the structure is intracellular in

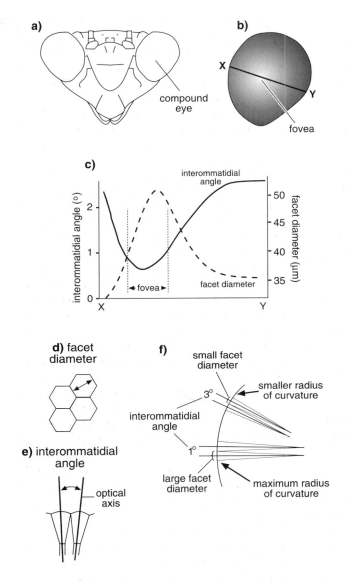

Fig. 22.4. Variation in ommatidia within an eye (*Tenodera*, Mantodea) (after Rossel, 1979). (a) Frontal view of the head showing the compound eyes. (b) Compound eye. White area is the fovea in which facet diameter is greatest and interommatidial angle smallest. Increasing density of shading shows decreasing facet size and increasing interommatidial angles. (c) variation in facet size and interommatidial angle along the transect X–Y shown in (b). (d) Facet diameter. (e) The interommatidial angle is the angle between the optical axes of adjacent ommatidia. (f) Diagrammatic section through the eye showing variation in the curvature of the surface of the eye and the associated differences in facet diameter and interommatidial angle.

the Semper cells. It is usually conical, but in some groups, notably in Collembola and Thysanura, it is more or less spherical. In a few beetles, some Odonata and most Diptera the Semper cells secrete an extracellular cone which is liquid-filled or gelatinous rather than crystalline. Ommatidia with this type of lens are called pseudocone ommatidia. The Semper cells do not produce a separate lens in some species, but their cytoplasm is clear and they occupy the position of the cone. These acone eyes are present in various families of Coleoptera, for example Coccinellidae, Staphylinidae and Tenebrionidae, in some

Diptera and in Heteroptera. In Elateridae and Lampyridae, the Semper cells do not contribute to the lens. Instead, the corneal lens forms a long cone-shaped projection on the inside (Fig. 22.2b). This is known as the exocone condition.

Variation in the retinula cells Although it is common for all the retinula cells to be of similar length and to form a rhabdomere all along the inner margin, this is not always true. The ommatidia of *Apis* have nine retinula cells. Eight of them have more or less similar rhabdomeres distally, but

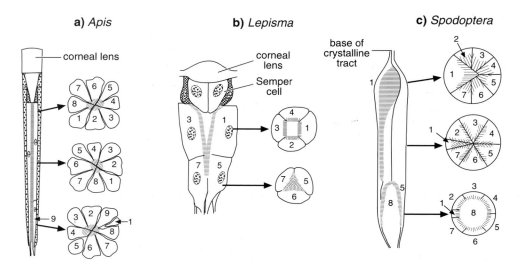

Fig. 22.5. Different rhabdoms, showing a longitudinal section of an ommatidium with transverse sections at the positions shown. Numbering of retinula cells is for clarity; the numbers have no other significance. (**a**) Twisted rhabdom. The retinula cells twist through 180° from top to bottom of the rhabdom. Each retinula cell is given the same number in cross-sections at different levels in the eye. Twisting is not shown in the longitudinal section, but note the short ninth cell which only contributes to the rhabdom proximally (*Apis*). (**b**) Tiered rhabdom. Four retinula cells contribute to the more distal rhabdom, three others to the proximal rhabdom (*Lepisma*, Thysanura) (after Paulus, 1975). (**c**) Tiered rhabdom in a superposition eye. Only the proximal part of the eye is shown. Cell 8 only contributes to the rhabdom proximally; distally, cell 1 contributes a major proportion (*Spodoptera*, Lepidoptera) (after Langer, Hamann & Meinecke, 1979).

proximally, two cells (cells 1 and 5 in Fig. 22.5a) do not contribute to the rhabdom while the rhabdomere of the short, ninth cell is present. Many other insects have retinula cells of different lengths so that they have a tiered arrangement, as in *Lepisma* (Fig. 22.5b), or as in *Spodoptera* (Fig. 22.5c). Many other arrangements are known.

22.2 FUNCTIONING OF THE EYE

22.2.1 Image formation

Image formation depends on the optical properties of the corneal lens and the crystalline cone. Refraction of light occurs at any interface with a difference in refractive index on the two sides. In most apposition eyes, the outer surface of the corneal lens is the principal or only refracting surface, although, in butterflies, further refraction occurs in the crystalline cone.

In an apposition eye, each ommatidium is separated from adjacent ommatidia by screening pigment, so each functions as an independent unit. Each lens produces a small inverted image of the object in its field of view which is in focus at the tip of the rhabdom (Fig. 22.6a). Because the retinula cells in most ommatidia are twisted, they all receive the same signal, and the image formed by the lens is not preserved. The light reaching the rhabdom in each ommatidium has an overall intensity which varies from one ommatidium to the next depending on the amount of light reflected by the object in the field of view, and collectively the rhabdoms receive a mosaic of spots of light of different intensities, which form an image of the object.

By bringing the light to a focus at the tip of the rhabdom, the insect maximizes the amount of light entering the rhabdom. In flies, with neural superposition eyes, the amount of light available from each point in space enters through seven ommatidia (see Fig. 20.21b) so the signal contains more photons than light entering through a single lens. Consequently, flies have greater sensitivity at low light intensities than insects with fused rhabdoms.

Dark-adapted superposition eyes function in a very

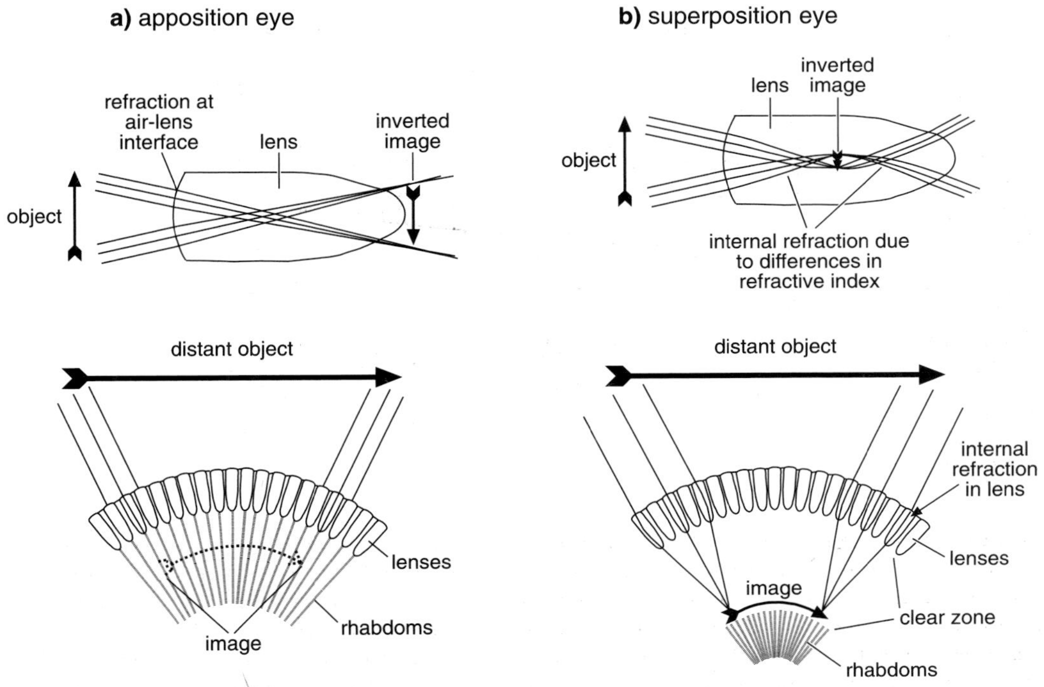

a) apposition eye

b) superposition eye

Fig. 22.6. Image formation: (**a**) in an apposition eye, each lens forms an inverted image at the tip of the rhabdom (above), but because the rhabdomeres function as a single unit the impression of this image is not retained. Consequently the output from each rhabdom is a response to the overall intensity of light that reaches it. Information concerning the object is thus represented as a series of spots differing in intensity (below; suggested by the dotted arrow); (**b**) in a superposition eye, light rays are refracted internally within the lens (above). They are unfocused as they exit the lens, but collectively form a single upright image at the tips of the rhabdoms (below).

different way. The screening pigments are withdrawn so that light leaving one lens system is not confined within a single ommatidium, but can spread to the rhabdoms of neighboring units (Fig. 22.6b). This enables the eye to function at low light intensities. Whereas, in apposition eyes, stray light is absorbed by the screening pigment, in superposition eyes much of it is utilized for image formation although with a consequent loss of resolution (see below).

The superposition eye forms a single upright image. This requires that light is refracted not just on entering each lens, but also within it, so that it follows a curved pathway. The lenses of superposition eyes are composed of concentric cylinders of material with slightly differing refractive indices (Fig. 22.7). Consequently, refraction occurs at the interfaces between these cylinders within the lens making the light path curved.

22.2.2 Resolution

Resolution refers to the degree of fineness with which an eye forms an image of an object. In compound eyes, resolution is determined by the interommatidial angle, the diameter of the lens through which light reaches the rhabdom, and the diameter of the rhabdom itself.

The interommatidial angle is the angular separation of the visual axes of adjacent ommatidia. In apposition eyes where each rhabdom functions as a unit, the fineness of the image will be greater the smaller the interommatidial angle. The interommatidial angle is often between 1 and $3°$, but is greater than $5°$ in many beetles, and as little as $0.24°$ in parts of the eye of the dragonfly, *Anax*. In many insects it varies in different parts of the eye. The greatest resolution is achieved if the visual fields of the ommatidia do not overlap and, at least in the mantis, *Tenodera*, this is approximately true. The degree of resolution, however, is

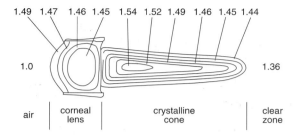

1.49 1.47 1.46 1.45 1.54 1.52 1.49 1.46 1.45 1.44

1.0

1.36

air | corneal lens | crystalline cone | clear zone

Fig. 22.7. Lens cylinder from a superposition eye. The refractive index changes within the corneal lens and crystalline cone so that refraction occurs within the lenses as well as at the air/corneal lens and crystalline cone/clear zone interfaces (after Horridge, 1975).

decreased when the eye becomes dark-adapted and light from a greater area is admitted to each unit (see below).

Diffraction within the eye results in light loss and, with very small diameter lenses, this becomes critical. High resolution requires large diameter facets, and this is achieved in some insects by a large radius of curvature of a part of the eye. In this way it is possible to maximize the lens aperture and retain small interommatidial angles. Robber flies, Asilidae, and mantids have an area at the front of the eye where the facets are larger and the interommatidial angles are smaller than elsewhere (Fig. 22.4). This gives them a high level of resolution in the part of the eye involved in prey detection and fixation. By analogy with vertebrate eyes, the zone of highest resolution is sometimes called the fovea or acute zone. Many male Diptera also have a forwardly directed acute zone which is presumed to be important in following the rapid movements of females during mating flights. Other male insects, like bibionids and mayflies, have upwardly directed acute zones in the eyes. These species form male swarms and perhaps use the eyes to detect females which fly into the swarm.

Reviews: Land, 1985, 1989, 1997; Warrant & McIntyre, 1993

22.2.3 Transduction

The conversion of light to electrical energy involves the visual pigment. This is a chromoprotein belonging to a group of chromoproteins known as rhodopsins. It consists of retinal, the aldehyde of vitamin A, conjugated with a protein (opsin) (Fig. 22.8a). It is a transmembrane protein in the microtubules forming the rhabdomeres. The 11-*cis* isomer of retinal is unstable and a photon interacting with

it causes a structural change producing, in a series of steps, metarhodopsin. This form of metarhodopsin activates G proteins with high efficiency and so initiates a second messenger cascade leading to an amplification of the energy of the reaction and, ultimately, to a change in the membrane potential of the retinula cell (Fig. 22.8b).

The metarhodopsin quickly loses its efficiency in activating G proteins as a result of phosphorylation and binding to a protein, probably arrestin. The metarhodopsin is, itself, photosensitive, absorbing light at a different wavelength from the rhodopsin giving rise to it. Photoactivation reconverts the metarhodopsin to inactive rhodopsin which, after dephosphorylation and release from arrestin, is available to transduce light again. Thus light of a different wavelength from that to which the photopigment is sensitive may contribute to the efficiency of the response and, under constant light conditions, the amounts of rhodopsin and metarhodopsin are in equilibrium. Different visual pigments produce metarhodopsins with different absorption spectra (Kiselev & Subramaniam, 1994; Stavenga, 1992).

The probability that a photon will encounter a visual pigment molecule is increased by the length of the rhabdomere and by the rhabdom acting as a lightguide or a waveguide (Fig. 22.9). In a light guide, light passes along its length as a result of reflection at the boundaries of the rhabdom with the cytoplasm of the retinula cells. Light entering the rhabdom within about 10° of its long axis (the angle varies with the diameter of the rhabdom) is totally reflected and so contained within the rhabdom, but some of the light entering at more oblique angles is lost. However, if the rhabdom, or the separate rhabdomeres in insects with an open rhabdom, is 0.5 μm or less in diameter, approaching the wavelength of light, it functions as a waveguide. This differs from a lightguide because the light waves interfere with each other producing a series of modes in which the light is not uniformly distributed across the rhabdom. These modes form stable patterns which travel along the rhabdom either within or outside it.

Because of these properties, light entering the rhabdom tends to be retained within it and a high proportion is absorbed by the visual pigment. A tapetum, when present, reflects light back into the eye and so further increases the chances that the photons and receptor molecules will interact.

Because the energy is amplified by a second messenger cascade, the interaction between a single photon and a receptor molecule is sufficient to produce a transient

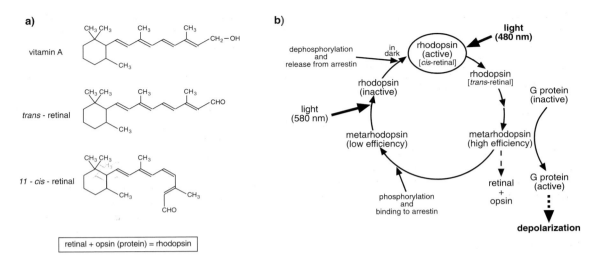

Fig. 22.8. Signal transduction: changing the energy of light to electrical energy. (a) Vitamin A and retinal. 11-*cis*-retinal is unstable and is reconverted to *trans*-retinal by the energy of light. (b) The pathway by which light at one wavelength leads to depolarization and then, at a different wavelength, to the regeneration of rhodopsin. The wavelengths shown are the absorption maxima for a blue-sensitive rhodopsin (480 nm) and the metarhodopsin (580 nm) to which it is transformed. In the active rhodopsin molecule, retinal is in the unstable *cis* form. Light with a wavelength of 480 nm produces isomerization to the *trans* form, and, in a series of steps, this is converted to a form of metarhodopsin which activates G proteins with high efficiency and leads to depolarization of the cell membrane. The metarhodopsin is unstable and is mostly transformed to a form with low efficiency for activating G proteins, but some decays to retinal and opsin, especially at high light intensities. Rhodopsin is regenerated, in an inactive form, from metarhodopsin by light of 580 nm wavelength. Active rhodopsin is slowly produced from the inactive form in the dark. Rhodopsins with different absorption maxima produce metarhodopsins absorbing maximally in different ranges from that shown (based on Kiselev & Subramaniam, 1994).

change, called a 'bump', in the membrane potential, often large enough to lead to a response in the postsynaptic cells of the lamina. The size of the receptor potential increases with the intensity of illumination as more photons interact with molecules of visual pigment, but the relationship between intensity and response is not linear and above a certain level, further increases in light intensity have no further effect on the receptor potential, which is said to be saturated. For any one insect, receptor potential increases with intensity over a range of about four orders of magnitude (Fig. 22.10). Below this range, the cells do not respond; above it, their output is maximal and shows no further increase.

Action potentials are not produced by retinula cells and depolarization is transmitted along their axons by passive conduction. In the lamina, most of the retinula cells synapse with large monopolar cells, and with other interneurons (see Fig. 20.21c). The neurotransmitter at these synapses is histamine and the signal is inverted. That is, a depolarization of the retinula cell produces a hyperpolarization of the interneuron. Any new signal is amplified at these synapses while the effects of the general level of illumination are reduced. The mechanisms producing these effects are not fully understood, but involve presynaptic interactions (Laughlin & Osorio, 1989).

22.2.4 Adaptation
The natural change from darkness to full sunlight involves a change in light intensity of some 10 log units: a white surface in bright sunlight reflects about 4×10^{20} photons $m^{-2} sr^{-1} s^{-1}$ compared with about 10^{10} photons $m^{-2} sr^{-1} s^{-1}$ on an overcast night with no moon, and 10^{14} photons $m^{-2} sr^{-1} s^{-1}$ on a moonlit night. However, the response of the retinula cells is fully saturated by an increase in intensity of about four log units (Fig. 22.10). Various devices are used to regulate the amount of light reaching the receptors so that the insect can operate over a wider range of intensities. The eye is said to adapt to the light conditions.

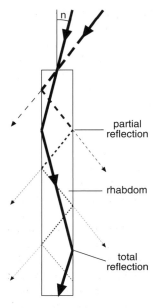

Fig. 22.9. The rhabdom acting as a lightguide. Light entering the rhabdom at an angle close to the long axis is totally reflected within the rhabdom. For insect eyes, this occurs when n is less than approximately 12°. Light entering the rhabdom more obliquely (dashed line) is only partly reflected. At more oblique angles smaller proportions are reflected internally.

This type of adaptation occurs at two different levels in the eye: 1) the amount of light reaching the photoreceptors is regulated, and 2) the receptor sensitivity can be changed. In addition, changes occur at the synapses between retinula cells and interneurons in the lamina. Although the latter does not influence the immediate response to the amount of light, it does influence the information reaching the central nervous system.

22.2.4.1 *Regulation of light reaching the receptors*
The amount of light reaching the rhabdom is regulated by movement of pigment in the screening cells, sometimes associated with anatomical changes in the ommatidium. Pigment movements are most clearly seen in superposition eyes. In the dark-adapted eye of species with a crystalline tract, pigment is withdrawn to the distal ends of the pigment cells and light leaving the lenses can move between ommatidia in the clear zone (Figs. 22.2; 22.6b). At high light intensities, however, the pigment moves so that light can now reach the rhabdom only via

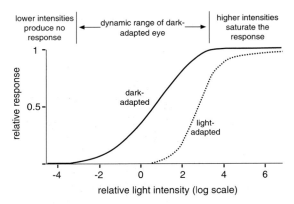

Fig. 22.10. Response of visual cells. Different light intensities can only be distinguished within the dynamic range. In the light-adapted eye, the response is shifted to the right (higher intensities). Notice that within the dynamic range of the light-adapted eye a small change in intensity results in a much bigger change in the response level than in the dark-adapted eye. In other words, contrasts are more readily differentiated by the light-adapted aye. Relative response is the receptor potential expressed as a proportion of the maximum response in dark- or light-adapted eyes (based on Matić & Laughlin, 1981).

the crystalline tract and the effective aperture of the ommatidium through which light reaches the rhabdom is very small. The eye is now functioning as an apposition eye as the ommatidia now function as separate units. In superposition eyes where the retinula cells form a broad column to the lens system, these cells undergo extensive changes in shape (Fig. 22.2c). Dark adaptation results in their extension and in compression of the crystalline cone so that the screening pigment is restricted to the most peripheral parts of the eye. In the light, the retinula cells are short and the primary pigment cells extend below the lens.

Comparable pigment movements occur in apposition eyes in some insects. For example, when the ant, *Camponotus*, is in the light, the primary pigment cells extend proximally (Fig. 22.11). At the same time, the proximal end of the crystalline cone is compressed to form a narrow crystalline tract surrounded by the pigment cells. Light must pass through the narrow opening to reach the rhabdom. At the same time, the rhabdom shortens, further increasing the likelihood that only light entering the ommatidium directly along its axis will reach the rhabdom (Menzi, 1987).

a) dark-adapted   **b) light-adapted**

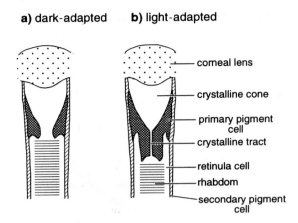

- corneal lens
- crystalline cone
- primary pigment cell
- crystalline tract
- retinula cell
- rhabdom
- secondary pigment cell

Fig. 22.11. Adaptation. Changes in the apposition eye of an ant, *Camponotus*. Only the most distal part of an ommatidium is shown (after Menzi, 1987): (a) dark-adapted; (b) light-adapted.

In the acone eyes of many Diptera and Heteroptera the retinula cells extend in the dark, carrying the distal end of the rhabdom to a more peripheral position and at the same time causing the Semper cells to become shorter and broader. This may involve a movement of 15 μm or even more by the rhabdom. At the same time, the primary pigment cells are displaced laterally so that the aperture of the optic pathway is increased.

The changes associated with dark adaptation begin within minutes of an insect's entry into darkness, but take longer to complete. In the ant *Camponotus*, the first change is already apparent within 15 minutes of an individual moving from light to dark, but completion takes about two hours. The changes are more rapid in some other insects. In addition to a direct response to environmental conditions, pigment movements are commonly entrained to the light cycle, occurring even in the absence of any environmental change. Thus the eye of the codling moth normally starts to become light-adapted about 30 minutes before sunrise and dark-adapted just before sunset, the process taking about an hour to complete.

A consequence of these changes is that the rhabdom receives light from a wide acceptance angle in dark-adapted eyes (Fig. 22.12), resulting in a reduction in image resolution.

The range of light intensities over which these pigment movements occur varies according to the insect's normal habits. For example, pigment movements in the eye of the cave dwelling tenebrionid beetle, *Zophobas*, occur at light intensities about 5 orders of magnitude lower than they do in the diurnal *Tenebrio* (Ro & Nilsson, 1993).

Pigment movements are regulated individually within each ommatidium. In the moth, *Deilephila*, a structure near the tip of the crystalline cone absorbs ultraviolet light, the principal wavelength producing pigment movements (Nilsson *et al.*, 1992), but how it effects movement is not known.

22.2.4.2 *Regulation of receptor sensitivity*
Regulation of sensitivity is achieved by structural changes in the retinula cells that limit the amount of light straying between the rhabdoms of neighboring ommatidia. In light-adapted eucone apposition eyes, granules of a screening pigment are usually present within the retinula cells (in addition to that in the screening pigment cells) close to the inner ends of the microvilli of the rhabdomeres (Fig. 22.13). This pigment has a high refractive index, so that the difference in refractive index between the rhabdom and the adjacent cytoplasm is reduced. As a result, less light is reflected internally within the rhabdom. If the rhabdom functions as a waveguide, the pigment absorbs higher order modes which are propagated outside the rhabdom. In the dark, large vesicles develop in the endoplasmic reticulum so that a clear space, known as the palisade, is formed round the rhabdom. The contrast in refractive index between the rhabdom and the surrounding cytoplasm is now enhanced and internal reflection within the rhabdom is maximized.

Sensitivity is also reduced when the eye is illuminated because rhodopsin is converted to metarhodopsin. In the light, metarhodopsin quickly becomes converted to an inactive form which accumulates together with inactive rhodopsin. Consequently, less of the active forms are available and the efficiency with which energy is transferred to the second messenger cascade is reduced (Kiselev & Subramaniam, 1994).

22.2.4.3 *Synaptic adaptation*
Adaptation also occurs at the synapses between the retinula cells and the interneurons in the lamina. The mechanisms are not understood, but probably include modulation of the quantity of transmitter released by the retinula cells. As a result of this adaptation, the input to the lamina is believed to be more or less constant despite differences in overall light intensity. Consequently, changes in stimulation of comparable magnitude are

Fig. 22.12. Dark adaptation. Effects
on the acceptance angles of ommatidia
in different parts of the eye of the
mantis, *Tenodera*. The acceptance
angle is usually measured as the angle
over which 50% of the incident light
reaches the rhabdom. Compare Fig.
22.4 (after Rossel, 1979). (**a**) An
ommatidium in the fovea where the
interommatidial angle is 0.6°. (**b**) An
ommatidium in the dorsal eye where
the interommatidial angle is 2.0°

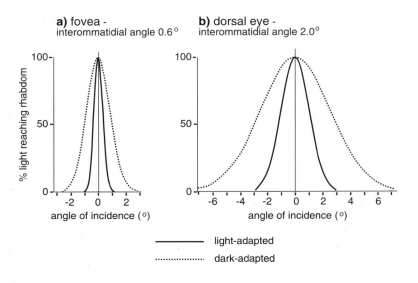

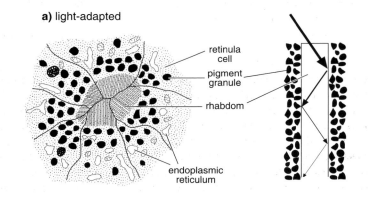

Fig. 22.13. Adaptation. Changes in
the retinula cells in an apposition eye.
Each diagram represents (left) a cross-
section of the rhabdom and associated
retinula cells, and (right) a
longitudinal section showing the
extent to which light is internally
reflected within the rhabdom (partly
after Snyder & Horridge, 1972): (**a**)
light-adapted; (**b**) dark-adapted.

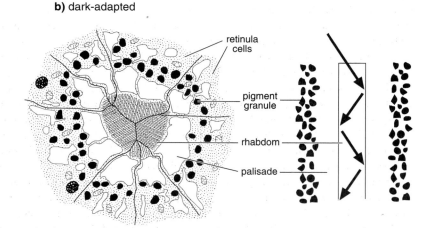

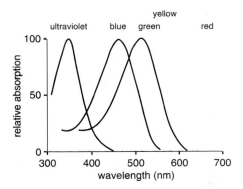

Fig. 22.14. Visual pigments: absorption of light of different wavelengths by three commonly occurring visual pigments in insect eyes with peak absorption in the ultraviolet, blue and green ranges of the spectrum. The absorption by each pigment is expressed as a percentage of the maximum for that pigment.

registered by similar activity in the medulla, irrespective of the background level of stimulation.

Review: Laughlin, 1989

22.2.5 Wavelength discrimination

Wavelength discrimination has been shown behaviorally to occur in a number of species belonging to the orders Odonata, Blattodea, Orthoptera, Hemiptera, Thysanoptera, Neuroptera, Hymenoptera, Diptera, Coleoptera and Lepidoptera. The ability to discriminate between light of different wavelengths requires the presence, in separate retinula cells, of photopigments with maximum sensitivity to light of different wavelengths.

All insects, whether or not they exhibit wavelength discrimination, have a visual pigment with maximum absorption in the green range of the spectrum (maximum 490–540 nm). Its range of absorption often extends to wavelengths below 400 nm, into the ultraviolet region, and up to about 600 nm (orange). Most insects have two additional pigments, one with maximum sensitivity to wavelengths in the ultraviolet, and another with maximum absorption in the blue region (Fig. 22.14). Thus, for a majority of insects, long wavelengths (red) do not stimulate the eyes. Dragonflies, some Lepidoptera and a few Hymenoptera, however, do have an additional pigment absorbing maximally wavelengths around 600 nm and with some absorption up to 700 nm. Five pigment types are present in a few Lepidoptera and Odonata (Bandai et al., 1992; Yang & Osoria, 1991). Blattodea have only two visual

pigments with maximum absorption at about 365 and 510 nm.

The wavelengths of light absorbed by a visual pigment are, of course, partly determined by the wavelengths reaching the receptors. In most insect eyes this is probably the same as the light reaching the eye from outside, but in some insects filtering within the eye can have a marked effect. In a few insects, the lenses are colored so that the rhabdoms of the associated ommatidia can receive only those wavelengths transmitted by the lenses. Male dolichopodid flies, for example, have alternating vertical rows of red and yellow facets. As a result, the same visual pigment in different ommatidia will be differentially activated, providing a basis for color vision even if only one type of visual pigment is present.

In other cases, the screening pigments absorb light of different wavelengths differentially. In *Calliphora* and *Drosophila*, they do not absorb light with wavelengths much above 600 nm and such light is transmitted with less attenuation than the shorter wavelengths. This has the effect of increasing the apparent sensitivity of the retinula cells to red light even though their visual pigment is maximally sensitive to green. Where the rhabdomeres are tiered (Fig. 22.5b,c), the distal rhabdomeres function as filters for those situated more proximally. For example, in flies it is probable that cell 7 filters out a high proportion of the ultraviolet and blue light, so that cell 8, which is proximal to it, receives a relatively high proportion of longer wavelengths and consequently appears to be more sensitive to them. The electrical response of the individual retinula cell depends on all these factors, but primarily on the particular photopigment which it contains.

Retinula cells with different photopigments occur within one ommatidium. In *Periplaneta*, for example, ommatidia in all parts of the eye have three ultraviolet-sensitive cells and five green-sensitive retinula cells, but in many insects the occurrence of different visual pigments varies in different parts of the eye apparently in relation to the eye's different functions. For example, in the drone honeybee, ommatidia in the dorsal part of the eye have only ultraviolet- and blue-sensitive pigments whereas, in the ventral part of the eye, green-sensitive pigment is also present. This distinction appears to occur in many flying insects, perhaps because dorsal ommatidia are skyward pointing and the ventral ones face the vegetation. Sexual differences may also occur, as in the butterfly, *Lycaena rubidus* where the dorsal ommatidia in females have a visual pigment absorbing maximally at 568 nm; this pigment is absent from the dorsal ommatidia of males (Bernard & Remington, 1991).

Sometimes retinula cells with different pigments are anatomically differentiated. For example, in *Apis* the short basal cell is one of the ultraviolet-sensitive cells, and in *Spodoptera* the basal cell (cell 8) is red-sensitive and the distal cell (cell 1) ultraviolet- or blue-sensitive (Fig. 22.5c). **Reviews:** Peitsch *et al.*, 1992; White, 1985

22.2.6 Discrimination of the plane of vibration (polarization sensitivity)

Light waves vibrate in planes at right angles to the direction in which they are travelling. These planes of vibration may be equally distributed through 360° about the direction of travel, or a higher proportion of the vibrations may occur in a particular plane. Such light is said to be polarized, and if all the vibrations are in one plane the light is plane-polarized.

Light coming from a blue sky is polarized. The degree of polarization and the plane of maximum polarization from different parts of the sky varies with the position of the sun (Wehner, 1989). It is, consequently, possible to determine the position of the sun, even when it is obscured, from the composition of polarized light from a patch of blue sky. Certain insects are able to make use of this information in navigation. It is particularly important in the homing of social Hymenoptera, and is best known in *Apis* and the ant, *Cataglyphis*. In other insects, including Odonata and some Diptera, the ability to perceive polarized light probably enables the insects to maintain a constant and steady orientation.

Detection of the plane of polarization is possible because photopigment molecules are preferentially oriented along the microvilli of the rhabdomere and maximum absorption occurs when light is vibrating in the same plane as the dipole axis of the pigment molecule. If the rhabdom is twisted, as in most ommatidia, there can be no preferred plane of absorption; polarization sensitivity depends on a uniform orientation of the visual pigment molecules within a rhabdomere. In ants and bees, and other insects responding to the plane of polarization, straight rhabdomeres are present only in a small group of ommatidia along the dorsal rim of each eye (Meyer & Labhart, 1993). These polarization sensitive ommatidia constitute only 6.6% of the total ommatidia in *Cataglyphis*, and 2.5% in *Apis*.

When light is polarized, all wavelengths are affected, but the polarization receptors of ants and bees are responsive to ultraviolet light. In *Cataglyphis*, six of the eight retinula cells in the dorsal rim ommatidia are ultraviolet-sensitive; four have a common orientation of their microvilli, those of the other two are at right angles (Fig. 22.15a). Consequently, when one set of cells is responding maximally because its microvilli are parallel with the plane of polarization, the other set is responding minimally. Each ommatidium in the dorsal rim responds maximally to polarization in one plane, and the population of receptors exhibits a range of different orientations (Fig. 22.15b). In the medulla of the optic lobe, the retinula cell axons synapse with interneurons responding maximally to polarization in a particular plane (Fig. 22.15c) (Labhart, 1996). It is believed that, by scanning the sky, the insect is able to match the inputs of its polarization receptors to the pattern in the sky. When reorienting it matches the sky pattern to the remembered neural input. **Review:** Wehner, 1989

22.2.7 Magnetic sensitivity

A number of insect species have been shown to respond to changes in magnetic field and it is possible that they use the earth's magnetic field in navigation (Schmitt & Esch, 1993; Towne & Gould, 1985). Two mechanisms have been proposed to account for this response. Some insects are known to contain particles of magnetite, an iron oxide, which might be affected by the magnetic field. In worker honeybees, the magnetite is contained in a ring of innervated oenocytes in each abdominal segment (Hsu & Li, 1994). Alternatively, it is possible that the compound eyes are involved, and that magnetism affects electron spin resonance in one of the photoreceptors. In keeping with this possibility, the orientation of male *Drosophila* in a magnetic field is affected by the wavelength of light in which they are tested. In ultraviolet light (wavelength 365 nm) they exhibited a consistent magnetic compass orientation towards a direction to which they had been trained. Tested in blue–green light (500 nm), they oriented at right angles to the training angle (Phillips & Sayeed, 1993).

22.3 VISION

Field of view Insects with well-developed compound eyes generally have an extensive field of view. For example, in the horizontal plane *Periplaneta* has vision through 360°,

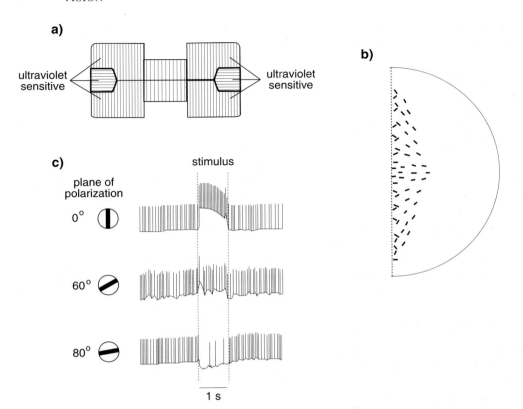

Fig. 22.15. Detection of the plane of polarization. (**a**) Arrangement of microvilli in the rhabdom in the polarization-sensitive area. Six of the eight retinula cells are ultraviolet-sensitive with orthogonally arranged microvilli (*Cataglyphis*) (after Wehner, 1982). (**b**) Orientation of the rhabdoms in the polarization-sensitive area along the dorsal margin of the right eye of *Cataglyphis* (after Wehner, 1989). (**c**) Response of an interneuron in the medulla to stimulation of the eye with light polarized in different planes. Vertically polarized light (0°) results in depolarization of the interneuron and an increase in its firing rate. Near horizontally (80°) polarized light causes a hyperpolarization and inhibition of the cell. These effects are thought to be produced by interaction of the input from retinula cells with different microvillar orientation, as in (a) (*Gryllus*) (after Labhart, 1988).

with binocular vision in front and behind the head (Fig. 22.16). In the vertical plane the visual fields of the two eyes overlap dorsally, but not ventrally (Butler, 1973). The visual fields of grasshoppers and many other insects are similar to that of the cockroach.

Distance perception Insects are able to judge distance with considerable accuracy. This is most obvious in insects, such as grasshoppers, that jump to a perch, or in visual predators like mantids, but is also true for any insect landing at the end of a flight. Two possible mechanisms exist that enable insects to estimate distance: a stereoscopic mechanism and motion parallax.

Many insects have binocular vision in front of the head and so have the potential to assess distance stereoscopically. This depends on the angle which the object subtends at the two eyes (Fig. 22.17). Errors can arise in the estimation of distance due to the size of the interommatidial angle, as this is important in determining resolution. Fig. 22.17 shows the error of estimation which might arise if the interommatidial angle was 2°; larger interommatidial angles will result in greater errors. Errors will also be larger if the distance of the insect from its prey is long relative to the distance between the eyes, and in many carnivorous insects which hunt visually, such as mantids and Zygoptera, the eyes are wide apart.

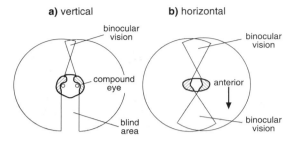

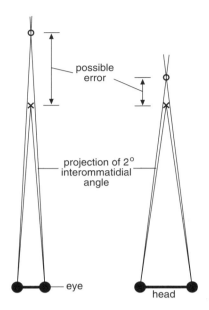

Fig. 22.16. Visual field of *Periplaneta* (after Butler, 1973). (a) In the vertical plane, the visual fields of the two eyes overlap above the head giving binocular vision. (b) In the horizontal plane, the visual fields of the two eyes overlap in front of and behind the head giving binocular vision in both directions.

Insects, such as grasshoppers, that jump to perches and need to judge distances accurately, make peering movements while looking at their proposed perch. Peering movements are side-to-side swayings of the body, keeping the feet still and with the head vertical, but moving through an arc extending 10 ° or more on either side of the body axis. Perhaps in such cases distance is estimated by motion parallax, the extent of movement over the retina as the head is swayed from side-to-side; big movements indicate that the object is close to the insect, while small movements show that it is at a greater distance. Grasshoppers are able to judge distances even when blinded in one eye, which is consistent with the idea that they are using motion parallax.

Review: Kral & Poteser, 1997

Visual tracking Visual tracking refers to an animal's ability to keep a moving target within a specific area of the retina, often when the animal itself is also moving. It occurs, for example, when a predator such as a mantis or dragonfly catches its prey, or when a male fly pursues a female. To do this, the insect must move its head or body to minimize the angle subtended by the object relative to some reference point on the retina. This point is usually in the center of a foveal region where the ommatidia have bigger facets and smaller ommatidial angles than elsewhere in the eye giving better resolution of the object.

When watching slowly moving prey against a homogeneous background, a mantis moves its head smoothly to keep the image in the foveal regions (Fig. 22.18). If the prey moves rapidly, however, or the background is heterogeneous, the insect makes rapid intermittent (saccadic)

Fig. 22.17. Distance perception. Diagrams illustrate how a wider head with greater separation of the eyes improves the estimation of distance. If the interommatidial angle is 2°, an object might lie anywhere between the cross and the open circle. Doubling the width of the head greatly reduces the possible error.

head movements. Between these movements, the head is kept still until the image has moved some distance from the center of the fovea, when another rapid change occurs.

Male flies pursuing females exhibit these same two types of behavior, but differ between species. The dolichopodid, *Poecilobothrus*, continuously reorients in flight, whereas *Musca* males make saccadic movements to keep the female in view (Land, 1993).

Review: Land, 1992

Visual flow fields During forward locomotion, objects in the environment appear to move backwards with respect to the organism. This is especially important in flight, and insects are unable to make controlled flights without visual input, requiring image movement across the eye from front to back. Like all other behaviors associated with vision, response to the visual flow field is possible because the neural circuitry within the optic lobe permits the extraction of the appropriate information from image movement over the retina. Specific interneurons within the central nervous system relay appropriate information

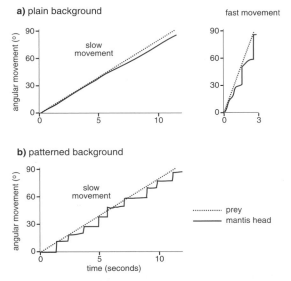

a) plain background

b) patterned background

Fig. 22.18. Visual tracking of prey by a mantis. The dotted line shows the steady movement of a potential prey item; the solid line shows the orientation of the mantid's head (after Rossell, 1980). **(a)** When the prey moves slowly against a plain background, the mantis moves its head steadily, keeping the prey within the foveal areas of the eyes (left). When the prey moves quickly (right), the mantis makes saccadic movements of the head. **(b)** With the prey against a patterned background, the mantis makes saccadic movements whatever the speed of the prey. In a saccade, the head remains still (horizontal lines) until the image of the prey has passed outside the fovea, it is then moved rapidly (vertical lines) so that the image of the prey is now in its original position. In this example, each movement of the head is through approximately 10°.

to the circuits controlling locomotion (Baader, Schäfer & Rowell, 1992; Rowell, 1988).

Form perception The eye's ability to detect the form of an object depends on its resolving power (see above). As, in diurnal insects, the interommatidial angle, and hence the angular resolution, is often 1° or less in some parts of the eye, this sets the limits of resolution. In predatory insects, like mantids (Fig. 22.4), the fovea provides a region of high quality form perception in the front of the eye where prey are detected before the strike. In other insects, ommatidia in the dorsal part of the eye have small angles of acceptance, giving better form perception than other parts of the eye. If an object subtends an angle at the eye which is less than the ommatidial angle, it will be seen only as a spot.

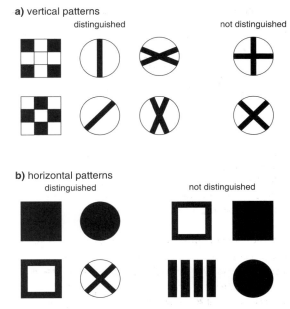

a) vertical patterns

b) horizontal patterns

Fig. 22.19. Pattern recognition by honeybees. The ability of insects to distinguish between patterns varies with the orientation of the pattern. Vertical patterns can be rotated with reference to the vertical plane because the insect normally flies dorsal side up. With horizontal patterns there is no comparable point of reference; the orientation of stripes depends on the angle at which the insect approaches them. **(a)** Vertical patterns. Left: patterns in the top row are distinguished from those immediately below; right: the insect is unable to distinguish these two patterns from each other. **(b)** Horizontal patterns. Left: patterns in the top row are distinguished from those immediately below; right: the insect is unable to distinguish those in the top row from those immediately below.

Bees provide a good model for understanding form perception because they can be trained to discriminate between different shapes. They can, for example, distinguish between solid patterns and broken patterns, and between patterns with different vertical orientations (Horridge, 1996) (Fig. 22.19).
Reviews: Land, 1997; Srinivasan, 1994

22.4 DORSAL OCELLI

Dorsal ocelli are found in adult insects and the larvae of hemimetabolous insects. Typically there are three ocelli forming an inverted triangle antero–dorsally on the head (Fig. 22.20a), although in Diptera and Hymenoptera they

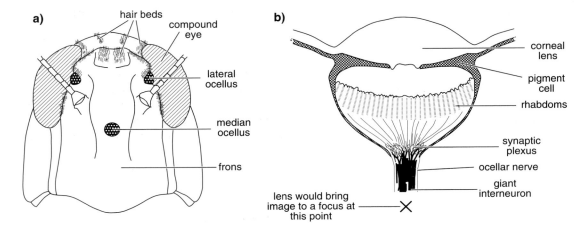

Fig. 22.20. Ocellus. **(a)** Frontal view of the head of a grasshopper showing the positions of the ocelli. **(b)** Diagrammatic longitudinal section through an ocellus of a grasshopper. The diagram shows the light-adapted condition with the pigment of the pigment cells restricting the entry of light. Notice that the image is focused below the rhabdoms (after Wilson, 1978).

occupy a more dorsal position on the vertex. The median ocellus shows evidence of a paired origin as the root of the ocellar nerve is double and the ocellus itself is bilobed in Odonata and *Bombus* (Hymenoptera). Frequently, one or all of the ocelli are lost and they are often absent in wingless forms.

A typical ocellus has a single thickened cuticular lens (Fig. 22.20b), although in some species, such as *Schistocerca* and *Lucilia*, the cuticle is transparent, but not thickened, and the space beneath it is occupied by transparent cells. Each ocellus contains a large number of retinula cells packed closely together without any regular arrangement; in the locust ocellus there are 800–1000. A rhabdomere is formed on at least one side of each retinula cell, and the rhabdomeres of between two and seven cells combine to form rhabdoms. The rhabdomeres usually occupy much of the cell boundary and, in the case of *Rhodnius*, are present all round the cells, forming a hexagonal meshwork similar to that in the stemmata of *Cicindela* (see Fig. 22.23c). The structure of the rhabdomeres in the dorsal ocelli is the same as that in the compound eye.

Pigment cells sometimes invest the whole ocellus, but in some species, like the cockroach, are lacking. A reflecting tapetum, probably consisting of urate crystals in a layer of cells, may be present at the back of the receptor cells.

Each retinula cell gives rise, proximally, to an axon which passes through the basal lamina of the ocellus and terminates in a synaptic plexus immediately behind the ocellus. Two anatomical classes of ocellar interneurons originate here. Some have giant axons, up to $20\,\mu m$ in diameter (often called large (L) fibers), others are of small diameter (S fibers). About 10 large fibers and up to 80 small ones are associated with each ocellus. In most insects studied, the large interneurons end in the brain, but in bees and flies some extend to the thoracic ganglia (Fig. 22.21). Where these descending interneurons are absent, the pathway to the thoracic motor centers is completed by second order descending interneurons. The small interneurons connect with several other centers in the brain, including the optic lobes, mushroom bodies and the central body.

The retinula cell axons synapse repeatedly and reciprocally with each other and with the interneurons which also synapse with each other. Some of the synapses between interneurons and retinula cells are input synapses to the retinula cells indicating that the interneurons may modulate the activity of the retinula cells as well as receiving information from them.

Illumination produces a sustained depolarization of the retinula cell which is proportional to light intensity. No action potentials are produced in the retinula cells, and graded receptor potentials are transmitted along the axons to the synapses. As at the first synapses behind the compound eye, the signal is amplified and the sign is reversed. Also as in the compound eye, the input signals

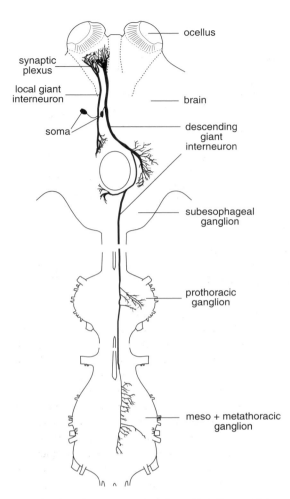

Fig. 22.21. Giant interneurons from the ocelli. A diagram of the anterior ganglia of the central nervous system of a honeybee showing examples of giant interneurons, a local neuron that does not extend beyond the brain, and a descending neuron that runs from the ocellus to the pterothoracic ganglion (based on Mobbs, 1985).

arising from contrasts in illumination are of similar amplitude even though the background level of illumination is different (Simmons, 1993). The giant interneurons transmit information to the brain, either electrotonically, or by spiking. The functions of ocelli remain uncertain. Although an image is produced by the lens it is not in focus on the retina. In addition, the extensive convergence of retinula cells on to a small number of interneurons indicates that form perception would be

extremely crude, at best. The structure and physiology of the ocelli suggest that they are adapted for the concentration of light and perception of changes in intensity, the giant interneurons providing a pathway for rapid conduction. In the locust, the ocelli are involved in detecting roll, their sensitivity to rapid changes in light intensity being well-suited for the perception of changes in the position of the horizon. In the higher Diptera, where they are on the top of the head, they are not important in the maintenance of stability.

Reviews: Goodman, 1981; Mizunami, 1994; Mobbs, 1985 – neural connections

22.5 STEMMATA

Stemmata are the only visual organs of larval holometabolous insects. They are sometimes called lateral ocelli, but this term is better avoided as it leads to confusion with the dorsal ocelli from which they are functionally and often structurally distinct. They occur laterally on the head and vary in number from one on each side in tenthredinid larvae to six on each side in larval Lepidoptera (Fig. 22.22). The larvae of fleas and most hymenopterans, other than Symphyta, have no stemmata. In larval Cyclorrhapha they are represented only by internal receptors.

Stemmata are of two types, those with a single rhabdom, and those with many rhabdoms. The former occur in Mecoptera, most Neuroptera, Lepidoptera and Trichoptera. They are also present in Diptera but, in some species, several stemmata are fused together to form a compound structure with a branching rhabdom. In Coleoptera, the stemmata of many species have a single rhabdom, but those of larval Adephaga have multiple rhabdoms. Stemmata with multiple rhabdoms also occur in larval Symphyta.

In caterpillars, each stemma has a cuticular lens beneath which is a crystalline lens (Fig. 22.22b,c). Each lens system has seven retinula cells associated with it. Commonly, three form a distal rhabdom and four form a proximal rhabdom. A thin cellular envelope lies round the outside of the sense cells and is, in turn, shrouded by the extremely enlarged corneagen cells.

All the distal cells contain a visual pigment with maximal absorbance in the green part of the spectrum, while some proximal cells contain a blue- or an ultraviolet-sensitive pigment. The rhabdomeres within a caterpillar's stemmata have different visual fields, and the acceptance

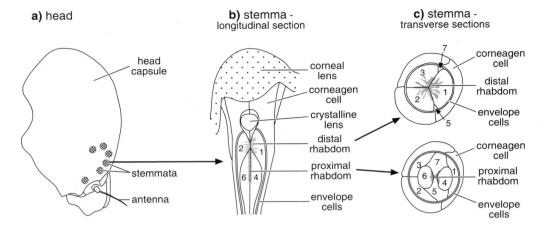

Fig. 22.22. Stemmata with a single rhabdom in a caterpillar (mainly after Ichikawa & Tateda 1980). (a) Side of the head showing the positions of the stemmata. (b) Longitudinal section of one stemma. (c) Transverse sections through the distal and proximal rhabdom.

angles of the distal rhabdomeres are close to 10° so that they have low spatial resolution. The proximal rhabdomeres have much smaller acceptance angles, of less than 2°. This, together with the fact that the focal plane of the lens is at the level of the proximal cells, gives them better spatial resolution. The visual fields of adjacent stemmata do not overlap so the caterpillar perceives an object as a very coarse mosaic, which is improved by side-to-side movements of the head enabling it to examine a larger field. It is known that caterpillars can differentiate shapes and orient towards boundaries between black and white areas (Ichikawa & Tateda, 1982).

The situation is different in larval symphytans (Hymenoptera), which have only a single stemma. These have large numbers of rhabdoms each formed by eight retinula cells and each group of cells is isolated by pigment from its neighbors. The lens produces an image on the tips of the rhabdoms which, in *Perga*, are oriented at about 5° from each other (equivalent to the interommatidial angle in compound eyes). Consequently, this type of eye is capable of moderately good form perception (Meyer-Rochow, 1975).

The larvae of the tiger beetle, *Cicindela*, have six stemmata, like caterpillars, but with a large number of retinula cells in each stemma, as in the Hymenoptera. The largest of the stemmata has about 5000 retinula cells each of which forms a rhabdomere on all sides so that the rhabdoms are in the form of a lattice (Fig. 22.23). It is possible that

spatial resolution in these eyes is limited because of optical pooling and perhaps electrical coupling (Toh & Mizutani, 1994).

The optic lobes of larval insects consist of a lamina and medulla comparable with those associated with compound eyes of adults and, at least in caterpillars, each stemma connects with its individual cartridge in the lamina.

In all these types of stemmata, the retinula cells contain screening pigment granules in addition to the visual pigment. Movement of the granules, away from the rhabdomeres in the dark and towards them in the light, provides sensitivity adjustment.

Caterpillars have three visual pigments and the neural capacity to distinguish colors and the larvae of several holometabolous species have been shown, experimentally, to respond to the plane of polarization of incident light. In neither case is the behavioral importance of these abilities understood.

Review: Gilbert, 1994

22.6 OTHER VISUAL RECEPTORS

22.6.1 Dermal light sense

A number of insects, such as *Tenebrio* larvae, still respond to light when all the known visual receptors are occluded. The epidermal cells are apparently sensitive to light. This is also suggested by the pigment movements which occur in isolated epidermal cells of some insects (section 25.5.1),

Fig. 22.23. Stemma with multiple rhabdoms in the larva of a tiger beetle (based on Toh & Mizutani, 1994): (**a**) diagrammatic longitudinal section; (**b**) diagrammatic longitudinal section through the retinula cells showing the rhabdoms; (**c**) diagrammatic transverse section through the retinula cells showing the rhabdomeres forming a continuous lattice.

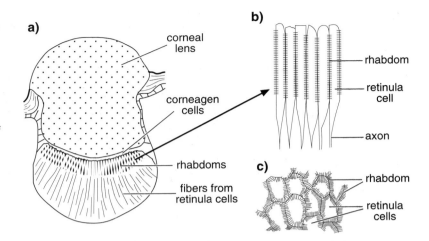

and by the production of daily growth layers in the cuticle (section 16.2.1.1).

Several families of butterflies are known to have photoreceptors on the genitalia of both sexes. In *Papilio*, there are two receptors on each side, each consisting of a single neuron lying on a nerve a short distance below the epidermis. The cuticle above the neuron is transparent. The cell extends into a number of processes and from these large numbers of microvilli arise. These are closely packed together, without any particular orientation, and appear to be contained within the cell, but are probably formed from the cell membrane as they are in similar light-sensitive structures in other invertebrates. The cells are called phaosomes and they probably monitor the positions of the genitalia during copulation (Miyako, Arikawa & Eguchi, 1993).

22.6.2 Sensitivity of the brain

In several insect species, light affects neural activity directly by acting on the brain, not via the compound eyes or ocelli. This commonly occurs in the entrainment of diurnal rhythms. In some species, daylength, regulating diapause or polyphenism, is registered directly by the brain. The light-sensitive structures are not known, although possible candidates have been described in tenebrionid beetles (Fleissner, Fleissner, & Frisch, 1993).

REFERENCES

Baader, A., Schäfer, M. & Rowell, C.H.F. (1992). The perception of the visual flow field by flying locusts: a behavioural and neuronal analysis. *Journal of Experimental Biology*, **165**, 137–60.

Bandai, K., Arikawa, K. & Eguchi, E. (1992). Localization of spectral receptors in the ommatidium of butterfly compound eye determined by polarization sensitivity. *Journal of Comparative Physiology*, **171**, 289–97.

Bernard, G.D. & Remington, C.L. (1991). Color vision in *Lycaena* butterflies: spectral tuning of receptor arrays in relation to behavioral ecology. *Proceedings of the National Academy of Sciences of the United States of America*, **88**, 2783–7.

Butler, R. (1973). The anatomy of the compound eye of *Periplaneta americana* L. I. General features. *Journal of Comparative Physiology*, **83**, 263–78.

Fleissner, G., Fleissner, G. & Frisch, B. (1993). A new type of putative nonvisual photoreceptor in the optic lobe of beetles. *Cell and Tissue Research*, **273**, 435–45.

Gilbert, C. (1994). Form and function of stemmata in larvae of holometabolous insects. *Annual Review of Entomology*, **39**, 323–49.

Goldsmith, T.H. (1962). Fine structure of the retinulae in the compound eye of the honey-bee. *Journal of Cell Biology*, **14**, 489–94.

Goodman. L.J. (1981). Organization and physiology of the insect dorsal ocellar system. In *The Handbook of Sensory Physiology*, vol. 7, part 6c, ed. H. Autrum, pp. 201–86. Berlin: Springer-Verlag.

Horridge, G.A. (1975). Optical mechanisms of clear-zone eyes. In *The Compound Eye and Vision of Insects*, ed. G.A. Horridge, pp. 255–98. Oxford: Clarendon Press.

Horridge, G.A. (1996). Pattern vision of the honeybee (*Apis mellifera*): the significance of the angle subtended by the target. *Journal of Insect Physiology*, **42**, 693–703.

Hsu, C.-Y. & Li, C.-W. (1994). Magnetoreception in honeybees. *Science*, **265**, 95–7.

Ichikawa, T. & Tateda, H. (1980). Cellular patterns and spectral sensitivity of larval ocelli in the swallowtail butterfly, *Papilio*. *Journal of Comparative Physiology*, **139**, 41–7.

Ichikawa, T. & Tateda, H. (1982). Receptive field of the stemmata in the swallowtail butterfly, *Papilio*. *Journal of Comparative Physiology*, **146**, 191–9.

Kiselev, A & Subramaniam, S. (1994). Activation and regeneration of rhodopsin in the insect visual cycle. *Science*, **266**, 1369–73.

Kral, K. & Poteser, M. (1997). Motion parallax as a source of distance information in locusts and mantids. *Journal of Insect Behavior*, **10**, 145–63.

Labhart, T. (1988). Polarization-opponent interneurons in the insect visual system. *Nature, London*, **331**, 435–7.

Labhart, T. (1996). How polarization-sensitive interneurones of crickets perform at low degrees of polarization. *Journal of Experimental Biology*, **199**, 1467–75.

Land, M.F. (1985). The eye: optics. In *Comprehensive Insect Physiology, Biochemistry and Pharmacology*, vol. 6, ed. G.A. Kerkut & L.I. Gilbert, pp. 225–75. Oxford: Pergamon Press.

Land, M.F. (1989). Variations in the structure and design of compound eyes. In *Facets of Vision*, ed. D.G. Stavenga & R.C. Hardie, pp. 90–111. Berlin: Springer Verlag.

Land, M.F. (1992). Visual tracking and pursuit: humans and arthropods compared. *Journal of Insect Physiology*, **38**, 939–51.

Land, M.F. (1993). Chasing and pursuit in the dolichopodid fly *Poecilobothrus nobilitatus*. *Journal of Comparative Physiology* A, **173**, 605–13.

Land, M.F. (1997). Visual acuity in insects. *Annual Review of Entomology*, **42**, 147–77.

Langer, H., Hamann, B. & Meinecke, C.C. (1979). Tetrachromatic visual system in the moth, *Spodoptera exempta* (Insecta: Noctuidae). *Journal of Comparative Physiology*, **129**, 235–9.

Laughlin S.B. (1989). The role of sensory adaptation in the retina. *Journal of Experimental Biology*, **146**, 39–62.

Laughlin S.B. & Osorio, D. (1989). Mechanisms for neural enhancement in the blowfly compound eye. *Journal of Experimental Biology*, **144**, 113–46.

Matić, T. & Laughlin, S.B. (1981). Changes in the intensity- response function of an insect's photoreceptors due to light adaptation. *Journal of Comparative Physiology*, **145**, 169–77.

Menzi, U. (1987). Visual adaptation in nocturnal and diurnal ants. *Journal of Comparative Physiology* A, **160**, 11–21.

Meyer, E.P. & Labhart, T. (1993). Morphological specializations of dorsal rim ommatidia in the compound eye of the dragonflies and damselflies (Odonata). *Cell and Tissue Research*, **272**, 17–22.

Meyer-Rochow, V.B. (1975). The dioptric system in beetle compound eyes. In *The Compound Eye and Vision of Insects*, ed. G.A. Horridge, pp. 299–313. Oxford: Clarendon Press.

Miyako, Y., Arikawa, K. & Eguchi, E. (1993). Ultrastructure of the extraocular photoreceptor in the genitalia of a butterfly, *Papilio xuthus*. *Journal of Comparative Neurology*, **327**, 458–68.

Mizunami, M. (1994). Information processing in the insect ocellar system: comparative approaches to the evolution of visual processing and neural circuits. *Advances in Insect Physiology*, **25**, 151–265.

Mizutani, A & Toh, Y. (1995). Optical and physiological properties of the larval visual system of the tiger beetle, *Cicindela chinensis*. *Journal of Comparative Physiology* A, **177**, 591–9.

Mobbs, P.G. (1985). Brain structure. In *Comprehensive Insect Physiology, Biochemistry and Pharmacology*, vol. 5, ed. G.A. Kerkut & L.I. Gilbert, pp. 299–370. Oxford: Pergamon Press.

Nilsson, D.-E., Hamdorf, K. & Höglund, G. (1992). Localization of the pupil trigger in insect superposition eyes. *Journal of Comparative Physiology*, **170**, 217–26.

Paulus, H.F. (1975). The compound eye of apterygote insects. In *The Compound Eye and Vision of Insects*, ed. G.A. Horridge, pp. 3–19. Oxford: Clarendon Press.

Peitsch, D., Fietz, A., Hertel, H, de Souza, D. & Menzel, R. (1992). The spectral input systems of hymenopteran insects and their receptor-based colour vision. *Journal of Comparative Physiology* A, **170**, 23–40.

Phillips, J.B. & Sayeed, O. (1993). Wavelength-dependent effects of light on magnetic compass orientation in *Drosophila melanogaster*. *Journal of Comparative Physiology* A, **172**, 303–8.

Ro, A.-I. & Nilsson, D.E. (1993). Sensitivity and dynamics of the pupil mechanism in two tenebrionid beetles. *Journal of Comparative Physiology* A, **173**, 455–62.

Rossel, S. (1979). Regional differences in photoreceptor performance in the eye of the praying mantis. *Journal of Comparative Physiology*, **131**, 95–112.

Rossel, S. (1980). Foveal fixation and tracking in the eye of the praying mantis. *Journal of Comparative Physiology*, **139**, 307–31.

Rowell, C.H.F. (1988). Mechanism of flight steering in locusts. *Experientia*, **44**, 389–95.

Schmitt, D.E. & Esch, H.E. (1993). Magnetic orientation of honeybees in the laboratory. *Naturwissenschaften*, **80**, 41–3.

Simmons, P.J. (1993). Adaptation and responses to changes in illumination by second- and third-order neurones in locust ocelli. *Journal of Comparative Physiology* A, **173**, 635–48.

Snyder, A.W. & Horridge, G.A. (1972). The optical function of changes in the medium surrounding the cockroach rhabdom. *Journal of Comparative Physiology*, **81**, 1–8.

Srinivasan, M.V. (1994). Pattern recognition in the honeybee: recent progress. *Journal of Insect Physiology*, **40**, 183–94.

Stavenga, D.G. (1992). Eye regionalization and spectral tuning of retinal pigments in insects. *Trends in Neurosciences*, **15**, 213–8.

Toh, Y. & Mizutani, A. (1994). Structure of the visual system of the larva of the tiger beetle (*Cicindela chinensis*). *Cell and Tissue Research*, **278**, 125–34.

Towne, W.F. & Gould, J.L. (1985). Magnetic field sensitivity in honeybees. In *Magnetite Biomineralization and Magnetoreception in Organisms*, ed. J.L. Kirschvink, D.S. Jones & B.J. MacFadden, pp. 385–406. New York: Plenum Press.

Walcott, B. (1975). Anatomical changes during light adaptation in insect compound eyes. In *The Compound Eye and Vision of Insects*, ed. G.A. Horridge, pp. 20–33. Oxford: Clarendon Press.

Warrant, E.J. & McIntyre, P.D. (1993). Arthropod eye design and the physical limits of spatial resolving power. *Progress in Neurobiology*, **40**, 413–61.

Wehner, R. (1982). Himmelsnavigation bei Insekten. Neurophysiologie und Vehalten. *Neujahrsblatt. Naturforschende Gesellschaft in Zuerich*, **184**, 1–132.

Wehner, R. (1989). The hymenopteran skylight compass: matched filtering and parallel coding. *Journal of Experimental Biology*, **146**, 63–85.

White, R.H. (1985). Insect visual pigments and color vision. In *Comprehensive Insect Physiology, Biochemistry and Pharmacology*, vol. 6, ed. G.A. Kerkut & L.I. Gilbert, pp. 431–93. Oxford: Pergamon Press.

Wilson, M. (1978). The functional organisation of locust ocelli. *Journal of Comparative Physiology*, **124**, 297–316.

Wolken, J.J., Capenos, J. & Turano, A. (1957). Photoreceptor structures. III. *Drosophila melanogaster. Journal of Biophysical and Biochemical Cytology*, **3**, 441–7.

Yang, E.C. & Osorio, D. (1991). Spectral sensitivities of photoreceptors and lamina monopolar cells in the dragonfly, *Hemicordulia tau. Journal of Comparative Physiology* A, **169**, 663–9.

23 Mechanoreception

Mechanoreception is the perception of any mechanical distortion of the body. This may result from touching an object or from the impact of vibrations borne through the air, water or the substratum, and thus mechanoreception includes the sense of hearing. It also includes distortions of the body which arise from the stance of the insect and from the force exerted by gravity, so some mechanoreceptors are proprioceptors. Three broad structural categories of mechanoreceptor are present in insects: cuticular structures with bipolar neurons; subcuticular structures with bipolar neurons, known as chordotonal organs; internal multipolar neurons which function as stretch receptors.

Definition: The term sensillum (plural: sensilla) refers to the basic structural and functional unit of cuticular mechanoreceptors and chemoreceptors. It includes the cuticular structure, the neuron or neurons, the associated sheath cells with the cavities they enclose and the structures they produce. Fig. 23.1 shows the basic elements of a sensillum. Sensilla often occur singly, but may also be grouped together to form functional units.

Review: McIver, 1985

23.1 CUTICULAR MECHANORECEPTORS

Cuticular mechanoreceptors fall mainly into two classes: hair-like projections from the cuticle with a basal socket, and dome-like campaniform sensilla. Both types have similar arrangements of neurons and sheath cells.

23.1.1 Structure

23.1.1.1 Cuticular components

Hair-like structures Most of the larger hairs on an insect's body are mechanoreceptors. There are no pores (apart from the molting pore, see below) in the cuticle of hairs that function solely as mechanoreceptors, and they are, therefore, known as aporous sensilla. Many contact chemoreceptors also have a mechanoreceptor function and, although they have a terminal pore, it is not relevant to the mechanoreceptive function (section 24.2.1). The

wall of the hair consists of exocuticle with an outer layer of epicuticle.

Most commonly, the hairs taper from base to tip, and they are often known as trichoid sensilla (this term also applies to chemoreceptors with a similar form). Others, on the cerci of crickets and cockroaches, for example, are very long relative to their diameter and do not taper; they are called trichobothria. Some insects, such as first stage grasshoppers, have club-shaped hairs, and the position-sensitive hairs on the cerci of some cockroaches consist of a sphere of cuticle on a short stalk. Other forms occur in relation to specific functions.

Whatever the external form, the hair-like structure arises from a socket permitting movement of the hair, and neural stimulation results from the displacement of the hair in its socket. To give this flexibility and hold the hair in its normal position, the socket usually has three components (Fig. 23.1). Externally, a thin cuticular joint membrane is continuous with the general body cuticle and that of the hair. Beneath this, bridging a gap between the body cuticle and the hair, are rings of suspension fibers. Finally, on the inside, is the socket septum. This is also fibrous, but appears more delicate than the suspension fibers. It runs across the outer part of the receptor lymph cavity to the dendrite sheath.

These suspensory structures restore the hair to its original position following movement. They often also restrict the extent to which the hair can be moved in different directions, so it is common for hairs to have directional sensitivity. Directional sensitivity is also conferred by the position at which the dendrite is inserted into the hair base (see Fig. 23.5b).

In addition to scattered mechanoreceptor hairs, all insects have groups of small hairs, known as hair plates, at some joints in the cuticle.

Campaniform sensilla Campaniform sensilla are areas of thin cuticle, domed and usually oval in shape with a long diameter usually in the range 5–30 μm. The dome of thin cuticle consists of an outer homogeneous layer with the

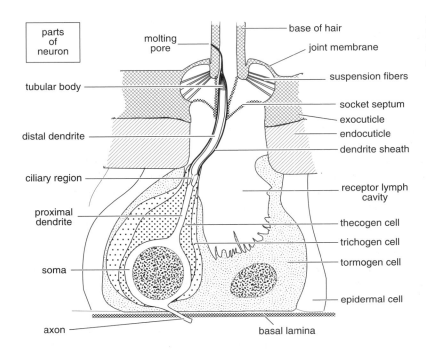

parts
of
neuron

molting
pore

base of hair

joint membrane

suspension fibers

tubular body

socket septum

exocuticle

endocuticle

distal dendrite

dendrite sheath

ciliary region

receptor lymph
cavity

proximal
dendrite

thecogen cell

trichogen cell

tormogen cell

soma

epidermal cell

axon

basal lamina

Fig. 23.1. Mechanoreceptor hair sensillum. Diagram showing the principal features of a mechanosensitive sensillum.

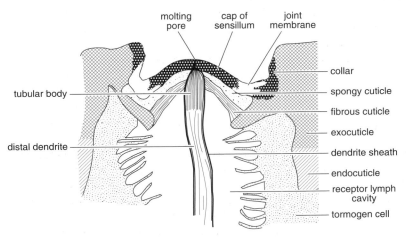

molting
pore

cap of
sensillum

joint
membrane

collar

tubular body

spongy cuticle

fibrous cuticle

exocuticle

distal dendrite

dendrite sheath

endocuticle

receptor lymph
cavity

tormogen cell

Fig. 23.2. Campaniform sensillum. Diagram showing the essential cuticular features. The arrangement of cells beneath the cuticle would be similar to that in Fig. 23.1.

appearance of exocuticle, and an inner lamellar or fibrous layer (Fig. 23.2). Sometimes these two layers are separated by a layer of transparent (in the electron microscope) cuticle referred to as spongy cuticle. The dome is connected to the surrounding cuticle by a different type of cuticle. It is called the joint membrane. The dendrite sheath, enclosing a single dendrite, is inserted into the center of the dome, and a molting pore may be visible on the outside. Details of the structure vary according to the

position of the sensillum even within a species. Grünert & Gnatzy (1987) recognize nine different types in adult *Calliphora*.

Campaniform sensilla are situated in areas of the cuticle that are subject to stress. On the appendages, where they occur most commonly, they are usually close to the joints. Fig. 8.7a shows their distribution on one leg of a cockroach. In some orthopteroid insects, there is a single campaniform sensillum at the base of each of the tibial

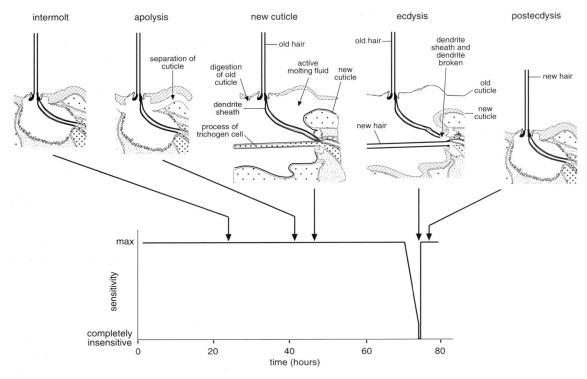

Fig. 23.3. Sensitivity of a hair sensillum in relation to molting. Diagrams in the upper panel illustrate the structural changes occurring through one larval stage from one ecdysis (time 0) to the next (compare Fig. 16.17). Lower panel shows the sensitivity of the hair to stimulation at different times. Sensitivity is completely lost only for about 30 minutes at ecdysis (partly based on Gnatzy & Tautz, 1977).

spines. Campaniform sensilla also occur on the mouth-parts, on the basal segments of the antennae, on veins close to the wing base, and on the ovipositor, while large numbers are present on the haltere of Diptera (Figs. 9.14, 9.15). An adult *Calliphora* has a total of approximately 1200 campaniform sensilla, about 36 on each leg, 140 on each wing base and 340 on each haltere (Gnatzy, Grünert & Bender, 1987). They often occur in groups, all those within a group having the same orientation and, presumably, functioning as a unit.

Dendrite sheath In all cuticular mechanoreceptors, the distal part of the dendrite is enclosed within a sheath, known as the cuticular or dendrite sheath which inserts at the hair base or in the center of a campaniform sensillum. In uniporous sensilla the sheath may continue to the tip of the hair and appears to be continuous with the epicuticle.

When an insect molts, the dendrite sheath extends across the space between old and new cuticles (Fig. 23.3), breaking off at ecdysis. The point of breakage is marked on the outside of the sensillum by a molting pore (Figs. 23.1, 23.2). During the molt the sheath presumably protects the dendrites from enzymes in the molting fluid, making it possible for the sensillum to continue functioning after apolysis until the time of ecdysis. In caterpillars of *Barathra*, for example, the sound-sensitive hairs are non-functional for only about 30 minutes before ecdysis and the new hair is functional within a few minutes of the old cuticle being shed (Fig. 23.3).

23.1.1.2 *Cellular components*
The cells of a sensillum are derived from epidermal cells and lie within the epidermal layer. One of the derivatives becomes the neuron, and two or three others are sheath cells.

Neuron In a majority of cuticular mechanosensilla there is only one neuron. Its soma is large with a large nucleus, and distally a single unbranched dendrite extends to the cuticle.

It is divided into three sections. The proximal dendrite contains mitochondria and other organelles. Some distance beneath the cuticle it narrows abruptly to the ciliary region (Fig. 23.1) which contains a ring of nine doublets of microtubules but usually lacks the central tubules typically present in motile cilia. The tubules arise from a basal body from which roots extend into the proximal region of the dendrite. After a short ciliary region the dendrite widens again. In this distal region the neurotubules increase in number, but no other organelles are present. At the extreme tip of the dendrite is a dense mass of up to 1000 microtubules known as a tubular body. The cytoplasm around the tubules of the tubular body is electron dense (when seen in section in the electron microscope) and there often appear to be cross-connections between the tubules as well as connections with the cell membrane.

Sheath cells The sheath cells wrap round the neuron. When three sheath cells are present, the innermost is called the thecogen or neurilemma cell. It closely surrounds the soma and proximal dendrite and encloses an extracellular cavity round the ciliary region. Distally it is attached to the dendrite sheath, which it produces during development of the sensillum. It often contains a complex labyrinth of extracellular spaces within its folds. Sometimes there is apparently no separate neurilemma cell and its position and roles are taken over by the trichogen cell.

Outside the thecogen cell is the trichogen cell. A fingerlike projection from this cell secretes the cuticle of the hair (Fig. 23.3, center diagram). The projection then withdraws, leaving the hair as a hollow structure. In some adult insects, the trichogen cell dies after it has secreted cuticle so the sensillum only has two remaining sheath cells.

The outermost sheath cell is the tormogen cell. It secretes the cuticle of the socket and, when it withdraws, comes to surround a cavity, known as the receptor lymph cavity, through which the dendrite sheath passes and which is continuous with the cavity of the hair. The inner margin of the tormogen cell, where it forms the receptor lymph cavity, is often produced into irregular microvilli. The outer margin is in contact with the cuticle beneath the socket, and is contiguous with unmodified epidermal cells.

All these cells are held together distally by maculae adherens desmosomes and septate desmosomes, and they are in communication via gap junctions. The thecogen cell is also connected to the neuron by desmosomes, but there are no gap junctions. The desmosomes isolate the receptor lymph cavity from the hemocoel.

23.1.2 Functioning

Two processes are involved in the functioning of cuticular mechanoreceptors. The mechanical distortions of the cuticle produced by bending a hair or compressing a campaniform sensillum must be transmitted to the dendrite, in a process known as coupling. The mechanical energy is then transformed, or transduced, to electrical energy.

Coupling Coupling in cuticular mechanoreceptors probably involves distortion of the tip of the dendrite sheath containing the tubular body, and hence of the tubular body itself. In hair sensilla, movement of the base of the hair produces this distortion, although the precise mechanism differs between hair types. The hair acts as a lever and amplifies any force exerted near its tip. Because of this, hair sensilla respond to very low forces. The position at which the tubular body is inserted also determines the effect of bending and probably contributes to the directional sensitivity commonly found in hairs.

In campaniform sensilla, coupling is effected by distortion of the cuticular dome resulting from compression. Oval-shaped sensilla are sensitive to compression along their short axis (Zill & Moran, 1981). Circular sensilla are presumed to be equally sensitive to compression in any direction, although, in some, directional sensitivity is conferred by having a fan-shaped dendrite. Sometimes, at least, compression produces an inward dimpling of the dome.

Transduction Transduction involves the deformation of the dendrite and production of a receptor potential. Although not yet demonstrated in insects, it seems certain that the plasma membrane of the neuron contains stretch-activated ion channels, that is, channels that are opened as a result of stretching the membrane so that an inward movement of ions occurs.

It also seems certain that the tubular body plays an essential role in this process, and a distortion of as little as 5 nm is sufficient to produce a receptor potential. However, some mechanoreceptor dendrites do not have a fully developed tubular body, and a receptor potential is sometimes produced even when the microtubules are destroyed experimentally.
Review: French, 1988

Neural input Hair sensilla vary in their responses to bending. In some, a response is only produced during or immediately following the movement; if the hair remains bent, no further response occurs (Fig. 23.4a). This is known

a) phasic response

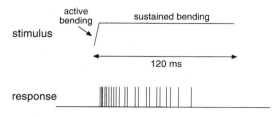

b) phasic/tonic response

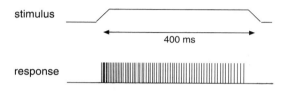

c) phasic/tonic response

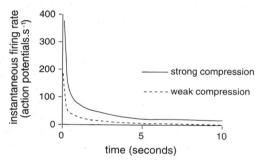

Fig. 23.4. Sensory input from cuticular mechanoreceptors. **(a)** Phasic response of a hair sensillum on the trochanter of a cockroach. The hair was bent and remained bent throughout the recording. Lower trace shows the action potentials in the axon. Notice that the sensillum stops firing even though the hair remained bent (based on Spencer, 1974). **(b)** Phasic/tonic response of a campaniform sensillum on the proximal tibia of a cockroach. Action potentials were produced as long as compression of the sensillum due to bending the leg was sustained (based on Zill & Moran, 1981). **(c)** Phasic/tonic response of the same sensillum as in **(b)** showing the instantaneous firing rate. With stronger compression, more action potentials were produced and the neuron was still active after 10 seconds of continuous stimulation (based on Zill & Moran, 1981).

as a phasic response. Other sensilla have a tonic response, action potentials continuing to be produced as long as the hair remains bent. The response often contains both components, and it is then known as a phasic/tonic response (Fig. 23.4b,c). These different response types are related to the different functions of mechanoreceptors. Hairs with a tactile function often exhibit only a phasic response, while those with a proprioceptive function and campaniform sensilla usually exhibit phasic/tonic responses.

23.1.3 Functioning in the living insect

In all insects, hair sensilla perform two types of function. Some are concerned with tactile perception of the environment (exteroception); others act as proprioceptors and this is the sole function of campaniform sensilla.

23.1.3.1 *Exteroception*

Hair sensilla responding to external stimuli are usually phasic. They are concerned with the sense of touch and sometimes with sensing air movement and with orientation with respect to gravity.

Tactile hairs Hairs responding to tactile stimulation are widely distributed over the insect's body and may be present in large numbers. On the middle leg of *Schistocerca*, for example, there are 1500–2000 mechanosensitive hairs (Mücke, 1991). These hairs usually taper from the base and are generally curved and set at an angle to the surface of the cuticle. Contact of any part of the body with an external object is perceived by these hairs and the spatial separation of axonal arborizations from the hairs within the central nervous system (see Fig. 20.24) enables the insect to make an appropriate response. Small groups of mechanoreceptor hairs are often present on the distal annulus of the antenna and on the ovipositor.

Air movement receivers Some insects have hairs that are specialized for the perception of air movements, sometimes including sounds.

Grasshoppers and some Lepidoptera have groups of small trichoid mechanoreceptors on specific regions at the front and top of the head (see Fig. 22.20a). These hairs are stimulated by the backward flow of air over the head when the insect flies. They are responsible for the maintenance of wingbeat (section 9.10.2), and are sometimes called aerodynamic sense organs. Since they are also directionally sensitive, they also contribute to the control of yaw (section 9.11.2).

Amongst Orthopteroid insects, hairs sensitive to air movement are present on the cerci and they are also responsive to sounds in some cases. Hairs responding to airborne vibration are usually long and slender, and have flexible mounting in the sockets so that they vibrate freely. Whereas tympanal organs respond to changes in air pressure associated with a sound (see below), hair sensilla are particle movement receptors, responding to movements of the air created by the sound source. Because the energy associated with particle displacement falls off rapidly with distance from the source, these receptors usually function only when the insect is close to the source, in the so-called near field. This is often no more than a few centimeters (see Ewing, 1989; Michelsen & Larsen, 1985).

Hairs responding to sounds in the range 32–1000 Hz are present in some caterpillars (Fig. 23.5) and similar hairs are present on the cerci of crickets and cockroaches. In the latter insects, the hairs are directionally sensitive and their axons make synaptic connections with giant interneurons in the terminal abdominal ganglia (see Fig. 20.14). Apart from sounds, air moving at only $4\,\mathrm{cm\,s^{-1}}$ is sufficient to stimulate the cercal hairs of the locust. In all these cases, stimulation of the hairs produces an aversive response or movement away from the source of air movement.

Orientation with respect to gravity In most insects, orientation with respect to gravity involves proprioceptors (see below), but the desert cockroach, *Arenivaga*, and a few other insects, have modified hair sensilla which contribute specifically to this function. The hairs, which are suspended beneath the cercus, are swollen at the tips. The hair's center of gravity is presumably in this swelling so that they hang vertically to the extent that their sockets permit. Consequently, any change in the insect's orientation with respect to gravity can be perceived by their movements (Hartman, Bennett & Moulton, 1987).

Cavities enclosing a statolith-like structure have been described in a few insects. While it is possible that they function as gravity receivers, this has not been clearly demonstrated.
Review: Horn, 1985

Pressure receivers Most aquatic insects are buoyant because of the air they carry beneath the water surface when they submerge, but *Aphelocheirus* (Heteroptera), living on the bottom of streams and using a plastron for gas exchange, is not buoyant. Although the plastron can with-

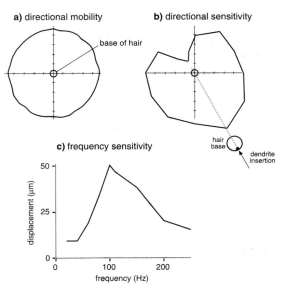

Fig. 23.5. Sound reception by a hair sensillum of the caterpillar of *Barathra* (after Tautz, 1977, 1978). (a) The hair is equally mobile in all directions. The circle shows the extent to which it can be deflected in each direction. (b) The sensory neuron associated with the hair shows marked directional sensitivity. It produces most action potentials when the hair is bent towards the side on which the dendrite is inserted, and the fewest when bent in the opposite direction. Sensitivity is shown by the distance from the intersection of vertical and horizontal axes. (c) The tip of the hair exhibits maximum displacement, that is, the hair is most sensitive, at a sound frequency close to 100 Hz. At lower and higher frequencies, displacement (sensitivity) sharply declines.

stand considerable pressures, it only functions efficiently in water with a high oxygen content, such as normally occurs in relatively shallow water. Hence some mechanism of depth perception is an advantage to *Aphelocheirus*, although it is unnecessary in the majority of other, buoyant aquatic insects.

On the ventral surface of the second abdominal segment of *Aphelocheirus* is a shallow depression containing hydrofuge hairs which are much larger than the hairs of the plastron (see Fig. 17.31). These hairs are inclined at an angle of about 30° to the surface of the cuticle and mechanosensory hairs are dispersed amongst them. The volume of air trapped by these groups of hairs depends on the balance between the pressure of air inside and the pressure of water outside. If the insect moves into deeper water, where the pressure is higher, the volume of air is decreased and the hairs are bent over, carrying with them the sensilla,

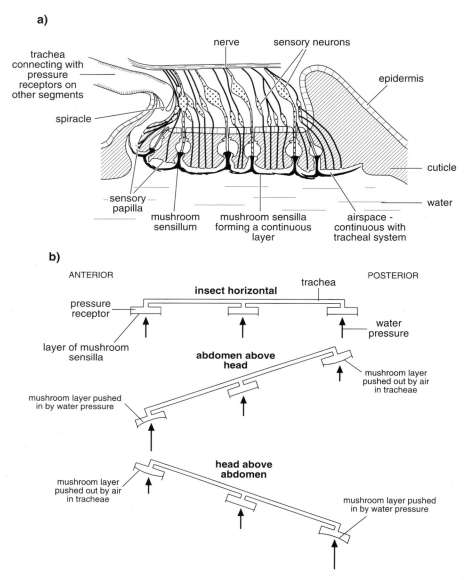

Fig. 23.6. Pressure reception by the water scorpion, *Nepa* (Hemiptera) (after Thorpe & Crisp, 1947). **(a)** Diagram of the receptor system associated with one spiracle on the ventral surface of the abdomen. The overlapping ends of the mushroom sensilla form a continuous flexible sheet holding an airspace which is continuous with air in the tracheal system. **(b)** Diagram showing how the pressure receptors of adjacent spiracles are joined through the tracheal system and the manner in which pressure differences arising when the body is tilted cause differential movements of the mushroom sensilla.

which are thus stimulated. The insect responds to such an increase in pressure by swimming up, but there is no response to a decrease in pressure (Thorpe & Crisp, 1947).

Pressure receptors are also present in *Nepa* (Heteroptera) which has a pair on the sternum of each of

abdominal segments three to five. Each consists of over 100 hair sensilla whose tips are expanded into thin plates giving them a mushroom-like appearance. The plates of adjacent sensilla overlap and enclose an airspace (Fig. 23.6a). A spiracle opens into the space and the spaces in the three organs of one side are connected through the tracheal system. The receptors give no general response to an increase in pressure, but differential stimulation of the organs of one side does produce a response. If the head is tilted upwards, the air in the system tends to rise towards the head end and the anterior mushroom-shaped plates tend to be pushed out while those on the posterior organ collapse (Fig. 23.6b). The converse is true if the head is tilted downwards. The resulting sensillar movements enable the insect to determine its orientation. The role of the sensory papillae shown in Fig. 23.6 is unknown (Thorpe and Crisp, 1947).

23.1.3.2 *Proprioceptors*

Trichoid sensilla Groups or rows of small hairs often occur at joints between some leg segments (see Fig. 8.7a), on the basal antennal segments, on the cervical sclerites and elsewhere. They monitor the position of one cuticular element relative to another. The hairs are positioned so that the flexing of one part of the cuticle with respect to another causes the hairs to bend (Fig. 23.7). Often the hairs in a hair plate (or hair bed) are graded in size, presumably giving the potential to discriminate between different degrees of flexion.

Hair plates on the first cervical sclerite and on the front of the prothorax provide the insect with information about the position of its head relative to the thorax. This is particularly important in the mantis during feeding and also aids the stability of the locust in flight when the head is oriented by a dorsal light reaction and the thorax is aligned with the head (section 9.11.2).

Hair plates are also important in orientation with respect to gravity because the extent to which the hairs are stimulated will depend on the insect's orientation. They are probably the principal gravity receptors in most terrestrial insects. Those on the legs are of particular importance, but others at the bases of the antennae also contribute. Several different hair plates probably contribute to the response at any one time, although one group may be dominant. In the stick insect, *Carausius*, hair plates on the trochantin which are stimulated by the coxa are of principal importance.

Campaniform sensilla In the insect skeleton all stresses can be expressed as shearing stresses in the plane of the surface. Such stresses produce changes in the shape of the campaniform sensilla resulting in their stimulation. The response is phasic/tonic.

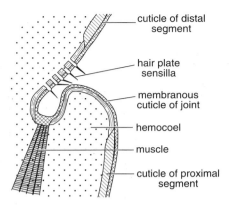

Fig. 23.7. Diagram showing how the sensilla of a hair plate are stimulated by coming into contact with adjacent cuticle.

Campaniform sensilla usually occur in groups and are oriented at right angles to the principal lines of stress (see Fig. 8.7a). For example, the campaniform sensilla on the leg are so oriented that they are stimulated when the foot is on the ground and the leg bears the insect's weight. Sensilla with different orientations are sensitive to bending in different directions as the insect's stance changes or different muscles become active.

Input from campaniform sensilla on the legs forms part of a negative feedback loop producing reflex changes in muscle tension (Schmitz, 1993). When a group of campaniform sensilla is stimulated, their input leads to a reduction in neural output to a muscle exerting stress on a leg. As a result, muscle tension falls, the stress on the leg is reduced, and the campaniform sensilla are no longer stimulated. Such effects are probably continuously in operation, enabling the insect to maintain its stance or adjust leg movements during walking.

23.2 CHORDOTONAL ORGANS

23.2.1 Structure

Chordotonal, or scolopophorous, organs consist of single units or groups of similar units called scolopidia. They are subcuticular, often with no external sign of their presence, and are attached to the cuticle at one or both ends. Each scolopidium consists basically of three cells: the neuron, an enveloping, or scolopale cell, and an attachment cell (Fig. 23.8). Some scolopidia contain more than one neuron. The dendrite narrows to a cilium-like process containing a peripheral ring of nine double filaments and with roots extending proximally within the dendrite. In

Fig. 23.8. Basic structure of a scolopidium. Diagrams on the right show the arrangement in cross-section at the points indicated. From the tympanal organ of *Locusta* (after Gray, 1960).

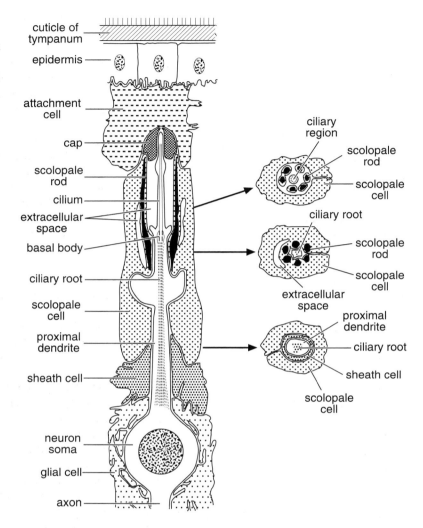

some scolopidia, and perhaps in all, the doublets are connected with the cell membrane near their origins at the basal body by a structure called the ciliary necklace because of its appearance. In some scolopidia, the tip of the cilium lies in a hollow of the extracellular scolopale cap. In others, the dendrite dilates distal to the cilium. The scolopale cell contains the scolopale which consists of fibrous material containing actin arranged in a ring or a series of rods (Wolfrum, 1991). The ciliary region of the dendrite lies in an extracellular space within the folds of the scolopale cell. The contents of the extracellular space are not known. The attachment cell connects the sensillum to the epidermis.

23.2.2 Functioning

The coupling mechanism by which external movement is transmitted to the dendrite probably varies in different chordotonal organs, but where the organ bridges the junction between two cuticular plates, movements of one relative to the other distort the scolopidium. Where the scolopidium is only connected with the cuticle at one end, other factors are responsible for the distortion (see below). Whatever the origin of the distortion, the effect is perhaps to displace the scolopale cap producing sliding of the tubule doublets which leads to bending at the base of the cilium. The actin of the scolopale may be involved in restoring the position of the cap and dendrite after bending.

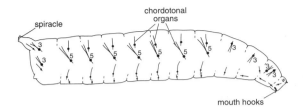

Fig. 23.9. Chordotonal organs on one side of the body of a *Drosophila* larva. The numbers show the number of scolopidia in each sense organ; those not numbered have only one (from Horridge, 1965b).

Transduction is probably dependent on stretch-activated ion channels, although this has not been demonstrated.

23.2.3 Distribution and functions in the living insect

Chordotonal organs occur throughout the peripheral regions of an insect's body. In larval *Drosophila* there are 90 such organs, each containing between one and five scolopidia, arranged in a segmental pattern and suspended between points on the body wall (Fig. 23.9). *Rhodnius* has a single scolopidium on either side of the midline in four abdominal segments. The scolopale and associated dendrite are attached to the basal lamina of the epidermis. They record pressure from within the abdomen such as that resulting from feeding (Chiang, Chiang & Davey, 1990). Female crickets, *Teleogryllus*, have chordotonal organs embedded in the epidermis of the genital chamber. They are stimulated when the wall is distended by the arrival of an oocyte in the genital chamber. Their input leads to the muscular changes that delay the backwards movement of the oocyte and move sperm down the spermathecal duct so that fertilization is effected (Sugawara, 1996).

In the thorax of many insects there are large chordotonal organs containing about 20 scolopidia that record the movements of the head on the thorax. Others in the wing bases of some insects record some of the forces which the wings exert on the body. In *Apis* there are three such organs, each with 15–30 scolopidia, at the base of the radial and subcostal veins and in the lumen of the radial vein.

Four or more chordotonal organs are usually present in each leg (see Fig. 8.7a). The first is attached proximally within the femur and is inserted distally into the knee joint. In *Machilis* this organ contains seven scolopidia; in

grasshoppers there are about 300. The second, proximally in the tibia, is the subgenual organ (see below) and the third, which in *Apis* contains about 60 scolopidia, arises in the connective tissue of the tibia and is inserted into the tibio-tarsal articulation. Finally, a small organ with only about three scolopidia extends from the tarsus to the pretarsus. In bushcrickets (Tettigoniidae) the subgenual is the most proximal of three chordotonal organs in the tibia (see Fig. 23.13b). The most distal is associated with the tympanal membranes in the forelegs (see below), but it is also present, although containing fewer scolopidia, in the middle and hind legs which have no tympanal organs (Kalmring, Rössler & Unrast, 1994).

Another chordotonal organ that occurs almost universally in insects is Johnston's organ which lies in the antennal pedicel (see below). Some insects have tympanal organs (see below).

The scolopidia of a chordotonal organ are sometimes grouped to form more than one functional unit. For example, the femoral chordotonal organ of locusts and stick insects has two functionally distinct groups (Kittmann & Schmitz, 1992) and separate groups are sometimes recognizable in Johnston's organ.

23.2.3.1 *Subgenual organs*

The subgenual organ is a chordotonal organ in the proximal part of the tibia, usually containing between 10 and 40 scolopidia. It is not associated with a joint. Processes from the attachment cells at the distal ends of the scolopidia are packed together as an attachment body, which is fixed to the cuticle at one point, while the proximal ends of the scolopidia are supported by a trachea (Fig. 23.10). The organ is often in two parts, one more proximal, called the true subgenual organ by Debaisieux (1938), and the other slightly more distal. Both are present in Odonata, Blattodea and Orthoptera. In Homoptera, Heteroptera, Neuroptera and Lepidoptera only the distal organ is present. Amongst Hymenoptera, the scolopidia enclose an extracellular space filled with acid mucopolysaccharide giving the structure an irregular spherical shape (Menzel & Tautz, 1994). Coleoptera and Diptera have no subgenual organs.

The subgenual organs respond to vibrations of the substratum. They are extremely sensitive and, in *Periplaneta*, for instance, respond to a displacement of only 0.2 nm at a frequency of 1.5 kHz (Shaw, 1994). In this species, vibrations up to 8 kHz are detected. At low frequencies, up to

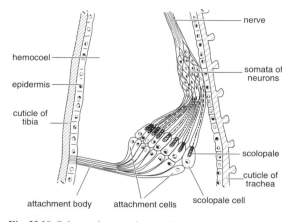

hemocoel

epidermis

cuticle of tibia

nerve

somata of neurons

scolopale

cuticle of trachea

attachment body   attachment cells   scolopale cell

Fig. 23.10. Subgenual organ of an ant (from Horridge, 1965a).

about 50 Hz, the neural response is synchronous with the stimulus, but at higher frequencies it is asynchronous. These sensilla also respond to airborne sound at high intensities.

The mechanism by which the subgenual organs are stimulated is not understood. Possibly, vibrations of the leg create vortices in the hemolymph within the leg and these move the chordotonal organ so that the neurons are stimulated. Alternatively, the leg's vibrations may cause the organ to vibrate with its own natural frequency, but, because the attachment cells are bound together and attached to the cuticle, they have a different natural frequency from the scolopale cells, which are relatively free. Hence the proximal and distal parts of the organ will vibrate at different frequencies, so that rapid and complex changes occur in the forces acting at the junction between the two parts. These rapid changes may serve to stimulate the sensory neurons.

In most insects, the detection of these substrate-borne vibrations is perhaps concerned primarily with predator avoidance, but in green lacewings, *Chrysoperla* (Neuroptera) and many Auchenorrhyncha which communicate via the substrate, the subgenual organs are probably also important in reception of intraspecific signals (section 26.3.1).

23.2.3.2 *Johnston's organ*
Johnston's organ is a chordotonal organ in the antennal pedicel with its distal insertion in the articulation between the pedicel and flagellum. It occurs in all adult insects and, in a simplified form, in many larvae. It consists of a single

mass or several groups of scolopidia which respond to movements of the flagellum with respect to the pedicel.

In *Calliphora*, most of the sensilla comprising the organ give phasic responses, action potentials only developing during and immediately after the movement, so that a single to and fro movement of the flagellum produces an 'on' and an 'off' response. The amplitude of the 'on' response increases with stimulus intensity due to different units having different thresholds and, if the stimulus is of very short duration, the 'off' response may be completely suppressed. Hence at high frequencies of stimulation small changes in stimulus pattern may produce major changes in excitation. Some of the sensilla respond to movement in any direction, others only if they are moved in a particular direction.

Movement of the flagellum relative to the pedicel may have a number of causes and Johnston's organ serves a variety of functions in any one insect. In grasshoppers, bees, Lepidoptera, some flies, and probably other insects, it acts as an air speed indicator. In *Calliphora*, for example, wind blowing on the face causes the arista to act as a lever, rotating the third antennal segment on the second. Even in a steady airflow the antenna trembles, and this is sufficient to stimulate some sensory neurons of Johnston's organ even though they are stimulated primarily by changes in the degree of rotation of the third antennal segment, rather than by a steady deflection. With different angles of rotation, more or different scolopidia are stimulated so that Johnston's organ can give a measure of the degree of static deflection of the third antennal segment as well as changes in its position. Regulation of flight speed in the locust is described in section 9.11.4.

In some insects, Johnston's organ functions as a particle movement detector, perceiving near-field sound. This occurs during courtship of *Drosophila melanogaster* when the female responds to the male's wing vibration, and during orientation dances of honeybees when workers respond to the sounds produced by an incoming dancing worker (section 26.5.1) (Dreller & Kirchner, 1993).

Males of some Diptera Nematocera detect females by their flight tone using Johnston's organ. In males of Culicidae (mosquitoes) and Chironomidae (midges), the pedicel is enlarged to house the organ. In Culicidae the base of the antennal flagellum forms a plate from which processes extend for the insertion of the scolopidia (Fig. 23.11). The latter are arranged in two rings all round the axis of the antenna and, in addition, there are three single

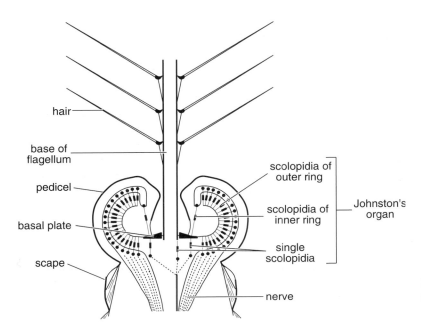

Fig. 23.11. Johnston's organ of a male mosquito (from Autrum, 1963).

scolopidia which extend from the scape to the flagellum. The males of these insects have plumose antennae with many fine, long hairs arising from each annular joint. These hairs are caused to vibrate by sound waves and their combined action produces a movement of the flagellum. The amplitude of flagellar movement is greatest near its own natural frequency which approximately corresponds to the flight tone of the mature female and stimulation at this frequency leads to the seizing and clasping response in mating. *Aedes aegypti* males are most readily induced to mate by sound frequencies between about 400 and 650 Hz, but the limits to which they respond become wider as the insect gets older and are wider in unmated than in mated males. Sounds at other frequencies and high intensities produce a variety of reactions: cleaning movements, jerking, flight or becoming motionless.

Gyrinid beetles are able to perceive ripples on the water surface, apparently due to displacements of the antennal flagellum, so that they are able to avoid collisions with other insects and, by echolocating using their own ripples, also avoid the sides of their container. In order to do this, the insect must also be able to detect the direction from which the ripples are coming.

Johnston's organ probably also contributes input relevant to an insect's orientation with respect to gravity, and, in the water boatman, *Notonecta*, it is concerned with orientation in the water. In this species, an air bubble extends between the head and the antenna so that when the insect is correctly orientated on its back the antenna is deflected away from the head. If, however, the insect is the wrong way up the antenna is drawn towards the head and Johnston's organ registers the position.

23.2.3.3 *Tympanal organs*
Structure and occurrence of tympanal organs Tympanal organs are chordotonal organs specialized for sound reception. Each consists of an area of thin cuticle, the tympanum (or tympanic membrane), which is generally backed by an air-sac so that it is free to vibrate. Attached to the inside of the tympanum or adjacent to it is a chordotonal organ which contains from one scolopidium, in *Plea* (Heteroptera), to over 1000 in some Cicadidae. Tympanal organs occur on the neck membrane of some scarab beetles (Forrest *et al.*, 1997), on the prothoracic legs of Grylloidea and Tettigonioidea, on the prothorax of at least one parasitic fly (Robert, Read & Hoy, 1994), on the mesothorax of some Hydrocorisae, such as *Corixa* and *Plea*, on the metathorax in Noctuoidea, and on the abdomen in Acrididae, Cicadidae, Pyraloidea and Geometroidea (Lepidoptera) and Cicindellidae (Coleoptera) (Spangler, 1988a). In *Chrysopa* (Neuroptera) the tympanum is on the ventral side of the radial vein of the forewing.

Acrididae (grasshoppers) have a tympanum in a recess on either side of the first abdominal segment (Fig. 23.12). It is about $2.5 \times 1.5\,mm^2$ in area in adult *Locusta*, varying in thickness from less than 1 μm to over 10 μm. Most of its cuticle is weakly sclerotized mesocuticle, but small islands of well sclerotized cuticle serve for the attachment of different groups of scolopidia. The cuticle of the airsac which backs the tympanum is only 0.2 μm thick and perhaps consists only of epicuticle. The chordotonal organ, known as Muller's organ, is complex, containing, in *Locusta*, about 80 neurons with their cell bodies aggregated into a ganglion. From the ganglion, the sensory units connect with the tympanum in four separate groups with different orientations (shown by arrows in Fig. 23.12b). The whole chordotonal organ and the auditory nerve which runs from it to the metathoracic ganglion are enclosed in folds of the airsac which backs the tympanum. Airsacs are contiguous right across the body between the two tympani.

The tympanal organs of Grylloidea (crickets) and Tettigonioidea (bushcrickets) are similar to each other in basic structure, being situated in the base of the fore tibia, which is slightly dilated and typically has a tympanum on either side, anterior and posterior with the leg at right angles to the body (Fig. 23.13a). Often the posterior tympanum is bigger than the anterior one and sometimes, as in *Gryllotalpa*, only the former is present. In most Tettigonioidea, the tympanic membranes are enclosed by forwardly projecting folds of the tibial cuticle with only a narrow gap connecting to the outside.

In both crickets and bushcrickets, the cavity of the leg between the two tympani is almost entirely occupied by a trachea divided into two by a rigid membrane, the blood space of the leg being restricted to canals anteriorly and posteriorly (Fig. 23.13b). In tettigoniids, the trachea runs proximally through the femur without branching and widens into a vestibule opening at the prothoracic spiracle. Its cross-sectional area changes from tibia to spiracle in a way that approximates the shape of an exponential horn so that sound entering the spiracle is amplified. This trachea is isolated from other tracheae and the spiracle has no closing mechanism. In crickets the trachea does not expand in this way, but the tracheae of the two sides join across the midline (see Fig. 23.15) although the airspaces are separated by a delicate septum.

The chordotonal organ associated with the tympanal organs is in the anterior hemocoelic space of the tibia (Fig. 23.13b). It is called the crista acoustica, and, in *Decticus*, it contains 33 scolopidia which lie parallel with each other in a vertical row, the sensilla becoming progressively smaller towards the distal end (see Fig. 23.17). In *Teleogryllus* they are within a tent-like membrane which extends along the tympanal organ between the wall of the trachea and the dorsal wall of the tibia. The neuron somata lie on the wall of the trachea and the dendrites project into the cavity of the tent. More proximal scolopidia are attached to large accessory cells within the tent, but, more distally, long attachment cells extend directly to the tibial epidermis. In these insects, as in the Acrididae, tympanal organs are present in all the larval stages, but they probably only become functional in the later larvae and adults.

The tympanal organs of the Noctuoidea occupy the posterior part of the metathorax (Fig. 23.14) and the tympanum faces into a space between the thorax and abdomen roofed over by the alula of the hind wing. Medial to the tympanum, and resembling it structurally, is a second membrane, but it has no sense organ. This second membrane is the counter tympanic membrane, and is probably an accessory resonating structure. The sense organ attached to the back of the tympanum contains only two scolopidia, supported by an apodemal ligament and an invagination of the tympanal frame.

In cicadas the two tympani are situated ventro-laterally on the posterior end of the first abdominal segment behind the folded membrane and beneath the operculum (see Fig. 26.10). The air-sacs by which they are backed are continuous right across the abdominal cavity. Each chordotonal organ contains about 1000 scolopidia enclosed in a cuticular tympanic capsule and attached to the posterior rim of the tympanum by an apodeme.

In *Chrysopa*, the tympanum is on the ventral side of the radial vein of the forewing. This vein has a bulbous swelling near the base that is thick walled and rigid above, but has the thin tympanum below. Unlike most other tympanal organs, the tympanum is in direct contact with the hemolymph; it is not backed by an air-sac. About 28 scolopidia are associated with the tympanum.

Functioning of the tympanal organs Sound impinging on the tympanic membrane causes it to vibrate, the amplitude of the vibrations varying with the intensity of sound and the structure of the membrane. In *Locusta*, different parts of the membrane vibrate to some extent independently of each other. Vibration is also affected by the

a) horizontal section

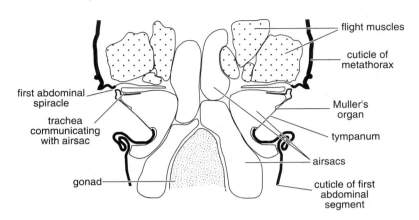

flight muscles

cuticle of metathorax

first abdominal spiracle

trachea communicating with airsac

Muller's organ

tympanum

airsacs

gonad

cuticle of first abdominal segment

b) Muller's organ

c) tympanal thickness

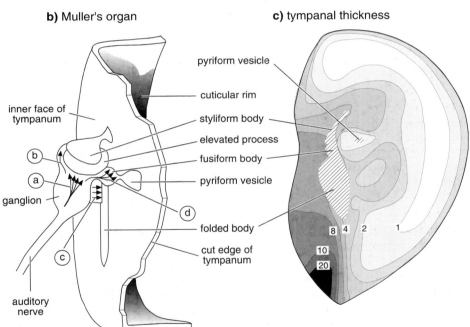

inner face of tympanum

ganglion

auditory nerve

b

a

c

d

pyriform vesicle

cuticular rim

styliform body

elevated process

fusiform body

pyriform vesicle

folded body

cut edge of tympanum

8 4 2 1

10

20

Fig. 23.12. Tympanal organ of a grasshopper. (**a**) Diagrammatic horizontal section through the metathorax and first abdominal segments showing the positions of the tympanal membranes. Notice that the airsacs form a continuum across the body (*Oedipoda*) (from Schwabe, 1906). (**b**) Diagram showing the attachment of the scolopidia to the inside of the tympanum. The cuticle of the airsac which normally covers Muller's organ and the tympanum has been removed. The orientations of scolopidia in different parts of the chordotonal organ are indicated by arrows. Letters show the positions of cells some of whose sensitivities are shown in Fig. 23.17a. The styliform body, elevated process and folded body are cuticular structures continuous with the tympanum (*Locusta*) (after Gray, 1960). (**c**) Thickness of the tympanum. Lines mark contours of thickness and numbers show the thickness in microns. Hatched areas are regions of well-sclerotized cuticle (*Schistocerca*) (after Stephen & Bennet-Clark, 1982).

Fig. 23.13. Tibial
chordotonal complex
of a bushcricket.
(a) Transverse section
through the tibia at the
level shown in (b)
(*Decticus*) (from Schwabe,
1906). (b) The complex
of chordotonal organs
exposed when the dorsal
(anterior) cuticle of the
tibia is removed (*Decticus*)
(from Schwabe, 1906).

a) transverse section of tibia

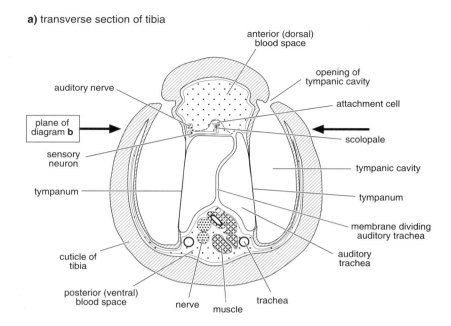

b) chordotonal organs

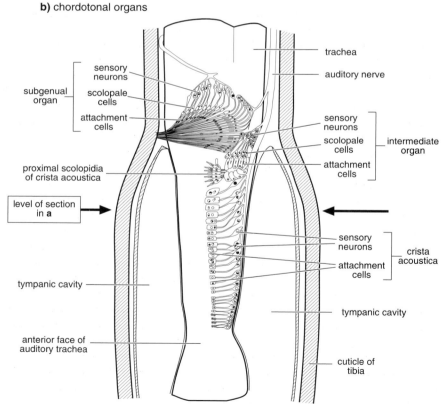

Fig. 23.13. Tibial chordotonal complex of a bushcricket. (a) Transverse section through the tibia at the level shown in (b) (*Decticus*) (from Schwabe, 1906). (b) The complex of chordotonal organs exposed when the dorsal (anterior) cuticle of the tibia is removed (*Decticus*) (from Schwabe, 1906).

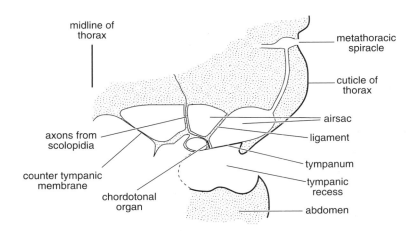

midline of
thorax

metathoracic
spiracle

cuticle of
thorax

airsac

axons from
scolopidia

ligament

counter tympanic
membrane

tympanum

tympanic
recess

chordotonal
organ

abdomen

Fig. 23.14. Tympanal organ of a noctuid moth. Diagrammatic horizontal section through the righthand side of the metathorax. The tympanum faces towards the abdomen (after Roeder & Treat, 1957).

degree of damping of the tympanum. In most insects this damping is reduced to a low level by the air-sacs backing the tympanum, but the locust tympanal membrane is damped to some degree by the mass of Muller's organ and sensitivity may be further reduced if the fat body is well developed.

Tympanal organs may be pressure receivers, in which sound impinges only on one side of the tympanum, or pressure-gradient (pressure-difference) receivers, in which sound reaches both sides of the membrane. In the latter, vibration of the tympanum depends on differences in sound pressure on its two sides resulting from differences in phase of the vibrations. The sound reaching the inner face of the tympanum is not the same as that reaching the outside because some frequencies may be differentially absorbed in transit through the system, which acts as a filter, or because the intensities of some frequencies are amplified more than others. For clear accounts of the physical principles involved see Ewing (1989) and Michelsen & Larsen (1985).

In insects, sound can reach the inner side of a tympanum via the tracheal system or through the tissues, but, because the intensity and phase of the sound reaching the inside of the tympanum depends on frequency, each tympanum can function either as a pressure receiver or as a pressure difference receiver. The change from one to the other occurs at about 10 kHz in locusts. At frequencies below this, sound reaches the tympanum from both sides and it functions as a pressure difference receiver. At frequencies above 10 kHz, most sound reaches the tympanum from the outside because sound entering the airsacs is filtered out and so the tympanum functions as a pressure receiver. In

bushcrickets, high frequency sound (above 20 kHz) entering the prothoracic spiracle (Fig. 23.15a) is amplified in the tracheal horn so that it dominates the vibrations induced in the tympanic membrane which acts as a pressure receiver. At low frequencies, however, relatively little amplification occurs and the tympanum acts as a pressure difference receiver (Michelsen, Popov & Lewis, 1994b).

In crickets, because the tympanal organs of left and right legs are connected via the transverse trachea, each tympanum may receive sound via four routes: directly from the outside, or indirectly via the ipsilateral spiracle, the contralateral spiracle, or the contralateral tympanum (Fig. 23.15b). At the frequency of the calling song, about 4.7 kHz, the last is relatively unimportant, but the input via both spiracles is essential for the insect's directional sensitivity (see below, and Michelsen et al., 1994a).

LOUDNESS. The number of action potentials produced by the neuron of a scolopidium is proportional to the intensity or loudness of the sound (Fig. 23.16). The response threshold for a single neuron varies according to the sound frequency, and different scolopidia within the same chordotonal organ may have different thresholds. When sound intensity is very high, the response saturates; no further increase is possible. Between threshold and saturation is the dynamic range in which the magnitude of the response is a measure of sound intensity. The dynamic range of a neuron is often around 30 dB, but cells with different thresholds extend the dynamic range for the insect. This is most obvious in noctuid tympanal organs in which there are only two scolopidia. These have similar frequency responses but their thresholds differ by about

a) bushcricket

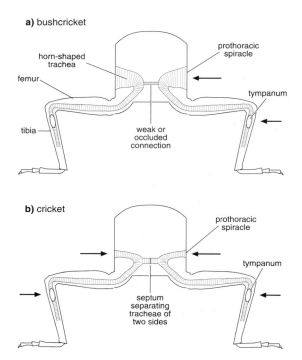

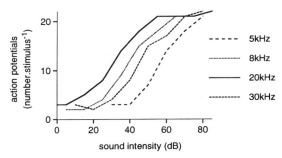

Fig. 23.16. Sensitivity. Responses of one scolopidium in the crista acoustica of a bushcricket to sound pulses of similar duration at different frequencies. At higher intensities the number of action potentials produced increases until the response becomes saturated. The dynamic range extends over about 50 dB at all frequencies, but the scolopidium is most sensitive (lowest sound intensity to produce a response) at 20 kHz (after Oldfield, 1982).

Fig. 23.15. Tympanal organs functioning as pressure difference receivers. Diagrammatic cross-section through the prothorax and the prothoracic legs. Arrows indicate how sound arriving at different points of the body may reach the tympanal organ of the righthand leg. The relative magnitudes of these effects vary with the sound frequency (see text). **(a)** In a bushcricket, the tympanal organs of the two sides operate in isolation from each other. The trachea from the prothoracic spiracle forms a horn in which some frequencies are amplified. Sound reaches the tympanum directly from the outside and via the trachea of the same side. **(b)** In a cricket the two sides are part of a continuous system through the transverse trachea. Sound reaches the tympanum directly, and indirectly through the spiracles of both sides, and possibly via the contralateral tympanum.

20 dB (Fig. 23.17c). The more sensitive cell enables the insect to hear bat sounds from greater distances, while stimulation of the less sensitive cell provides immediate information on the close proximity of the bat. Male cicadas fold their tympani when they are singing, and this greatly reduces the sensitivity of the ear and probably protects it from damage by the sound the insect is itself producing.

FREQUENCY. The tympanal organs of different groups of insects respond to different frequency ranges. Acrididae respond to sounds with frequencies from 100 Hz to 50 kHz, tettigoniids from 1 to 100 kHz, gryllids from 200 Hz to 15 kHz, and cicadas from 100 Hz to 15 kHz. Within these ranges, sensitivity is generally greatest over a more limited range, corresponding with the sound frequencies to which the insects are adapted, commonly the sound produced by conspecifics. However, in many insects the tympanal organs are sensitive to much higher frequencies and respond to the sounds emitted by bats which, during their search phase, often cover a broad band of frequencies from about 25 kHz to over 100 kHz. Noctuids, for example respond to frequencies from 1 to 140 kHz with maximal sensitivity in the range 20–40 kHz (Fig. 23.17c). Maximal responses of the tympanal organs of *Chrysopa* and of mantids are also within this range.

Frequency discrimination by tympanal organs results from differences in the mechanical parts of the system that convey vibrations to the sensory cells. Different parts of the tympanum of grasshoppers tend to vibrate maximally at different frequencies. At the point of attachment of the (a) cells in Fig. 23.12b, the membrane oscillates maximally when stimulated by sound in the range 5–10 kHz. Sound in this range is a major component of the sounds of many grasshoppers. At the point of attachment of the (d) cells in Fig. 23.12b, the membrane exhibits a maximum response in the same range, but has a second maximum between 15 and 20 kHz (Meyer & Elsner, 1996). Consequently, different scolopidia of Muller's organ tend to be stimulated by sounds of different frequencies (Fig. 23.17a) and

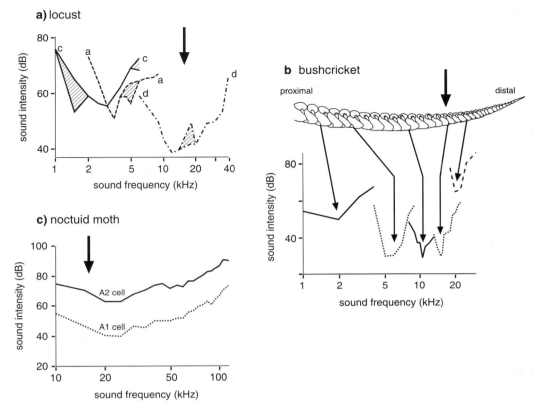

Fig. 23.17. Frequency thresholds of individual scolopidia in tympanal organs. Note that the *lowest* threshold equates with the *highest* sensitivity. Bold arrows at the top indicate the approximate upper limit of human hearing (about 15 kHz). (**a**) A locust. Letters, a, c and d, refer to the groups of scolopidia with the same lettering in Fig. 23.12b. Hatched areas show the range of sensitivities of different cells in the same group (after Michelsen & Nocke, 1974). (**b**) A bushcricket. Above: the crista acoustica. Below: responses of scolopidia at different distances along the crista acoustica. The more distal the scolopidium, the higher the frequencies to which it responds. Notice that cells responding to the highest and lowest frequencies are very insensitive, responding only at high sound intensities (after Oldfield, 1982). (**c**) A noctuid moth. The two scolopidia in the chordotonal organ respond to a wide range of frequencies, but cell A1 is much more sensitive than cell A2 (after Waters & Jones, 1996).

input in the tympanal nerve, representing the sum of membrane oscillation and the sensitivity of the sensory neurons, is maximal at 5–10 kHz, with a second maximum between 25 and 35 kHz.

In bushcrickets, the intermediate organ may be responsible for the perception of low frequency sound, while, in the crista acoustica, proximal scolopidia are most sensitive to low frequencies and more distal cells to higher frequencies (Fig. 23.17b). This tuning is not due to differences in the tympanum, and is perhaps a property of the individual scolopidium, although this has not been demonstrated.

Some of the interneurons relaying information to the brain are tuned to specific frequencies. One such interneuron in *Ancistrura* has a maximum response to stimulation of the tympanum by frequencies in the range 12–16 kHz, corresponding with the principal frequencies in the male song. The interneuron is inhibited by sound frequencies above and below this range (Stumpner, 1997).

PULSE RATE. Sound communication among insects often depends on the production and reception of discrete pulses of sound (section 26.3.1). The ability to distinguish sound pulses following each other in rapid succession depends on the characteristics of the tympanum as well as those of the sensory neurons. The chordotonal organs of the tympanal organs of cicadas and noctuids can separate sound pulses up to about 100 Hz, but above this the response becomes continuous.

DIRECTIONAL SENSITIVITY. Insects are small relative to the speed of sound, so they are unable to determine the direction from which a sound comes from differences in arrival times on the two sides of the body. Instead, directional sensitivity results from the ear's properties as a pressure difference receiver, or, at higher frequencies, as a result of sound diffraction.

The response of a pressure difference receiver may be greatly affected by the direction from which the sound reaches the body (Fig. 23.18). As a result, even a single ear may be capable of determining direction and this effect is reinforced by the integration within the central nervous system of information from ears on opposite sides of the body. In bushcrickets, the tympanic chambers resonate at the frequency of the insect's calling song and sound at this frequency is amplified if it enters the chambers directly through the outer openings (Fig. 23.18a), but other frequencies are not amplified. This gives the insects great directional sensitivity to sound of the appropriate wavelengths. In crickets, the septum across the transverse trachea in the prothorax (Fig. 23.15b) produces a large phase shift in sound reaching a tympanum from the spiracles. The effect is greatest at the frequency of the calling song. However, the difference in phase from the sound reaching the tympanum directly from outside varies with the direction of the sound source. As a result, the amplitude of vibration of the tympanum varies as the insect swings its leg forwards and back during walking. This enables the insect to determine the position of the sound source (Michelsen & Löhe, 1995).

Sound diffraction is the reflection of sound from the body or nearby objects. Low frequency sounds have wavelengths much longer than an insect's body and so are relatively undisturbed, but the body or wings may have a significant shadowing effect on high frequencies. The result is that the tympani on the two sides of the body receive signals differing in intensity, providing the basis for directional sensitivity. Tettigoniids, which produce high frequency songs (section 26.1.2.1), and moths that respond to bat sounds depend on diffraction for their

a) bushcricket

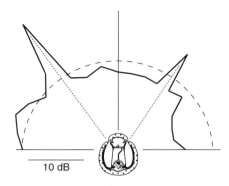

b) grasshopper

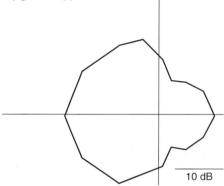

Fig. 23.18. Directional sensitivity of tympanal organs. Sensitivity is measured as decibels of sound pressure at the tympanum. The greater the distance from the point at which the vertical and horizontal axes cross, the greater the sensitivity. (a) Bushcricket, left foreleg. A cross-section through the proximal tibia containing the tympanal organ is shown at the center. The smooth arc shows the intensity of the signal (15 dB). Notice that when the sound comes from directly opposite the slit-like openings to the tympanic cavities (see Fig. 23.13) the sound is amplified. This is only true at sound frequencies close to 7.4 Hz, the principal frequency of the calling song (after Stephen & Bailey, 1982). (b) Grasshopper, left tympanum. The insect was exposed to sound of equal intensity from a series of positions all round its body. The tympanum was most strongly stimulated when the source was on the ipsilateral side (left of midline) (from Bailey, 1991).

directional sensitivity. Noctuids are able to locate a source of high frequency sound when they are flying and so they have to localize sounds in a vertical as well as a horizontal direction. The wings tend to screen the tympani, so there is a marked difference in the responses of the two organs to asymmetrical stimulation when the wings are raised, but when the wings are in the lower half of their beat there is little difference between the two sides. When the wings are down the insect is more sensitive to sounds coming from below it than to sounds from above. Thus, during flight, the auditory input from low intensity sounds of high frequency will vary cyclically and in different ways according to the position of the sound source.

The accuracy with which the insects orient to sound direction is not very great. If the sound source is within about 10° of its body axis, the grasshopper, *Chorthippus*, makes turns of up to 60° (sometimes more) to either side. If the angle of incidence is greater, the insect turns towards the correct side, but the turning angle is anywhere from 60 to 120° irrespective of the angle of incidence. This often results in the insect making a zigzag course towards the source because successive turns tend to put the sound source on the opposite side of the body. Tettigoniids appear to orient more precisely, although still with considerable variation.

Scolopidia may be stimulated by the insect's own normal activities. In *Locusta*, abdominal ventilation and flight may both modulate responses to sound, in some cases completely inhibiting the response (Hedwig, 1988, 1989).

Reviews: Ball, Oldfield & Rudolph, 1989 – crickets
Ewing, 1989; Hoy & Robert, 1996; Michelsen & Larsen, 1985; Spangler, 1988b – moths

Functions of tympanal organs Insects have the capacity to hear many different sounds, including most sounds that humans can hear (about 100 to 15 000 Hz), and, in those with high frequency sensitivity, sounds that are not audible to humans. Insects may respond behaviorally to such sounds even though they are not the primary reason for the insect's auditory sense.

Orthoptera and cicadas produce sounds for attracting mates and, in some species, for aggregation (section 26.5). Recognition of these sounds is dependent primarily on sound patterns rather than on variation in frequency, although the hearing systems of different species respond maximally to the frequencies of the sounds emitted (see

above). Directional sensitivity is obviously a critical element in these behaviors.

Sound is also important in predator avoidance and insects in several groups have tympanal organs sensitive to high frequency sounds. This is commonly assumed to relate to the avoidance of bats which emit high frequency sounds, although in few cases is there strong evidence of the importance of this behavior. Noctuid moths can detect bats about 30 m away and at such distances, where the intensity of stimulation is low, tend to turn away from the source of sound (see above). At high sound intensities, such as would occur when the bat was within about 5 m of the moth, it may close its wings and drop to the ground, or power dive to the ground, or follow an erratic, weaving course.

Most of the scolopidia in the tympanal organs of cicadas are maximally sensitive in the range 1–5 kHz, well below the principle frequencies in the calling songs of many species. It is possible that they are important in predator recognition.

Some parasitic Diptera have been shown to use their auditory sense to locate their singing hosts. This has been extensively studied in the case of *Ornia*, a parasitoid of crickets (Robert *et al.*, 1994), and it also occurs in *Emblemasoma*, a species which parasitizes cicadas (Soper, Shewell & Tyrrell, 1976).

Review: Bailey, 1991

23.3 STRETCH RECEPTORS

Stretch receptors differ from other insect mechanoreceptors in consisting of a multipolar neuron with free nerve endings, while all the others contain a bipolar neuron. Stretch receptors have a variety of forms. Sometimes they are an integral part of an oriented structure such as a muscle fiber or a strand of connective tissue, but in other cases they have no specific orientation or associated structures. In many cases, their function is inferred from light microscopy, or known from physiology without structural details.

23.3.1 Unspecialized receptors

Multipolar cells without an associated specialized structure are often associated with epithelia. The cell body, ensheathed in glial cells, is in the hemocoel, but the basal lamina around the dendrites becomes fused with that of the epithelium. In the multipolar cells of the bursa copulatrix

a)

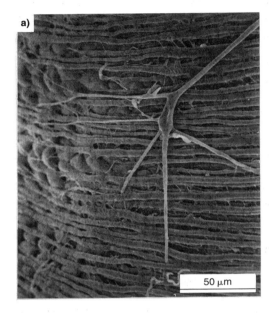

b)

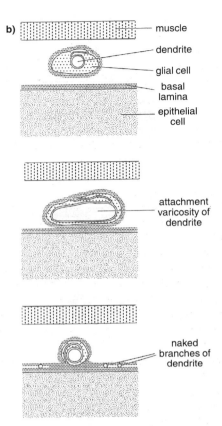

muscle

dendrite

glial cell

basal
lamina

epithelial
cell

attachment
varicosity of
dendrite

naked
branches of
dendrite

Fig. 23.19. Unspecialized stretch receptor on the outside of the bursa copulatrix of *Pieris* (after Sugawara, 1981). **(a)** The epithelium of the bursa is covered by a layer of muscles. The soma of the multipolar neuron is free in the hemolymph, although clothed by glial cells (not distinguishable in photograph). Dendritic endings of the stretch receptor pass between the muscle strands and end on the surface of the epithelium. **(b)** Diagrammatic cross-sections of a dendrite beneath the muscle layer. Top: it approaches the epithelium clothed by a glial cell. Middle: it swells to form a varicosity. Its basal lamina is confluent with that of the epithelial cell beneath. Bottom: fine branches of the dendrite extend into the basal lamina of the epithelial cell. They are no longer covered by a glial cell.

of *Pieris*, for example, each dendrite swells at intervals, the swellings, with their basal lamina, apparently contributing to the attachment. In this example, the finer dendritic branches, free of glial covering, end in the basal lamina of the epithelial cell (Fig. 23.19), but in fly larvae, the naked dendrites of stretch receptors on the epidermis end within invaginations of the epidermal cells.

In soft-bodied dipterous larvae and perhaps in all insects with largely unsclerotized cuticle, multipolar cells are attached to the inside of the epidermis. The number of these neurons in cyclorrhaphan fly larvae is constant within a segment, varying from 24 in the prothorax to 30 in each of the abdominal segments. They collectively form a

subepidermal nerve net, although they do not anastomose to form a true net. These receptors monitor changes in body shape.

Insects with a hard external cuticle are usually considered to have no subepidermal nerve net, but Knyazeva, Fudalewicz-Niemczyk & Rosciszewska (1980) report numerous multipolar neurons in the abdomen of the house cricket, *Acheta*. Blood-sucking insects have small numbers of multipolar neurons beneath the abdominal epidermis. In adult tsetse fly, *Glossina*, there are three pairs of such cells associated with the ventral body wall. They presumably monitor abdominal distension when the insect feeds. In *Schistocerca*, a few multipolar cells are present beneath the arthrodial membrane at the femoro–tibial joints. They respond to movement and position of the tibia with respect to the femur (Williamson & Burns, 1978).

Some insects are known to have stretch receptors associated with the fore and hindguts, and these perhaps are universally present. Large numbers are present on the foreguts of grasshoppers and crickets where they probably monitor the movement of food along the gut. In grasshoppers, they also monitor foregut distension, regulating meal size (Bernays & Chapman, 1973). Others monitor the amount of food present in the hindgut (Simpson, 1983) and probably regulate fecal production. A pair of multipolar stretch receptors is also present on the bursa copulatrix of the cabbage butterfly, *Pieris*. These receptors monitor distension of the bursa, which is normally caused by the placement of the spermatophore. This distension causes the female to refuse further matings. Multipolar receptors, together with chordotonal organs, at the base of the ventral gonocoxa of *Teleogryllus* monitor movement of the egg during its passage from the oviduct, and perhaps regulate the delivery of sperm from the spermatheca (Sugawara, 1993).

Multipolar stretch receptors have also been reported on or within the nerves of a variety of insects and are probably of widespread occurrence. They respond to stretching of the nerve which may arise from a variety of causes, depending on the position of the nerve.

Review: Finlayson, 1976

23.3.2 Receptors with accessory structures

Where stretch receptors are not immediately associated with an epithelium or a nerve, they are usually supported by a strand of connective tissue or a modified muscle fiber, known as strand and muscle receptors, respectively, although there appears to be a great variety of forms.

A strand receptor consists of a strand of connective tissue, extending from one structure to another, with a multipolar neuron on the outside or embedded within it (Fig. 23.20a). The neuron is sheathed in glial cells except for the tips of the dendrite which are embedded in the connective tissue. Some strand receptors of orthopteroid insects have the soma of the neuron in the central nervous system (Braunig, 1982).

Muscle receptors are associated with a modified muscle fiber which may be a component of a functional muscle, or may be a separate structure. The fiber contains typical muscle fibrils, but these are sparse in the center of the fiber, called the sarcoplasmic core, where, in the larva of *Antheraea*, a giant nucleus occurs (Fig. 23.20b). Attached to the muscle is a tube of connective tissue known as the fiber tract, which contains many fibers. These fibers are secreted by the tract cell, which lies within the connective tissue and also has a giant nucleus. The neuron gives rise to two to four main dendrites, which run along the length of the fiber tract. From these main dendrites, side branches, which are not clothed in glial cells at the tips, are embedded in dense bundles of fibers (Osborne, 1970). In adult beetles, the neurons of muscle stretch receptors associated with the mandibles indent the muscle membrane, ending near the Z-bands (Honomichl, 1978). Muscle receptors have been described in Orthoptera, Neuroptera, Coleoptera, Trichoptera and Lepidoptera.

Occurrence of strand and muscle receptors Stretch receptors of these types are probably present in the abdomen of all insects. Dorsally, a pair extends from the tergum to the intersegmental membrane or from one intersegmental membrane to the next. Orthopteroid insects and beetles also have a vertical stretch receptor on each side of each segment, and some species also have ventral stretch receptors. Comparable dorsal receptors are present in the thorax of some insects.

Elsewhere in the body, strand or muscle receptors are associated with the mouthparts, legs and wings. The wing stretch receptors of locusts play a key role in the control of wing movements (section 9.10.3).

23.3.3 Functioning of stretch receptors

Stretch receptors have none of the obvious structural features, either within the dendrites, such as the tubular body, or outside them, such as the scolopale, that occur in other mechanoreceptors. In all cases, however, large numbers of naked dendritic endings are embedded in connective tissue or in the basal lamina of the tissue with which the receptor is associated. It is inferred that transduction results from stretching the membrane of those endings. Although the somata of these cells are usually in the hemocoel, the glial cells surrounding them appear to form an effective 'blood–brain barrier' and action potentials result from an inward flow of sodium ions. However, the neurons of muscle stretch receptors of caterpillars are affected by the ionic composition of the medium in which they are bathed (Weevers, 1966a).

Fig. 23.20. Specialized stretch receptors (based on Osborne, 1970): **(a)** strand receptor; **(b)** muscle receptor. Inset below shows detail of dendrite endings in fiber bundle.

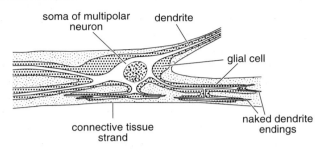

a) strand receptor

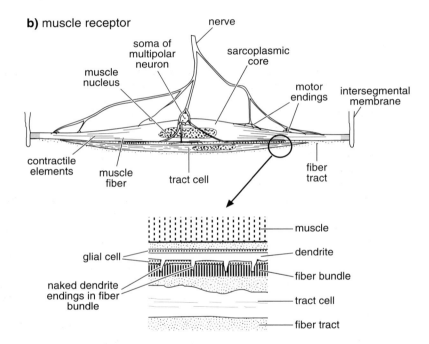

b) muscle receptor

Fig. 23.21. Instantaneous action potential frequency from a muscle receptor during stretching. The neuron fires tonically at a frequency dependent on the degree to which the receptor is already extended. As stretching begins and accelerates the response rises sharply **(a)**. Continued stretching at a constant velocity is superimposed on an increasing tonic level as the receptor lengthens **(b)**. When active stretching stops, the deceleration produces a sharp reduction in the firing rate **(c)**, which then stabilizes at the new tonic level (after Weevers, 1966b).

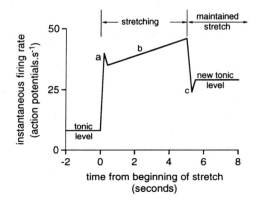

These receptors exhibit tonic activity under sustained tension that is proportional to the degree of stretching (Fig.23.21). In many cases, they also monitor the rate of stretching, firing faster the higher the rate of stretch. They may also respond to acceleration in the rate of stretching. Because of these characteristics, they have the capacity to follow low frequency oscillations in the degree of stretch.

The role of the muscle in muscle stretch receptors is not clear. In caterpillars, the activity of the motor neuron to the stretch receptor muscle is inhibited when the stretch receptor is stretched, but exhibits a transient decrease during stretching and a transient increase in activity during relaxation. It is suggested that this might protect the receptor from excessive stretch and take up the slack if stretch is suddenly reduced.

The normal role of these receptors, at least in some cases, is not simply to monitor changes in position; they may also promote negative feedback effects which tend to restore the system to its initial state. This is most clearly demonstrated in the dorsal muscle stretch receptors of caterpillars where stretching a muscle receptor organ of one segment affects the activity of at least 32 different motor units. These are on the contralateral side of the same segment and the ipsilateral side of adjacent segments (Weevers, 1966c).

REFERENCES

Autrum, H. (1963). Anatomy and physiology of sound receptors in invertebrates. In *Acoustic Behaviour of Animals*, ed. R.-G. Busnel. Amsterdam: Elsevier.

Bailey, W.J. (1991). *Acoustic Behaviour of Insects. An Evolutionary Perspective.* London: Chapman & Hall.

Ball, E.E., Oldfield, B.P. & Rudolph, K.M. (1989). Auditory organ structure, development and function. In *Cricket Behavior and Neurobiology*, ed. F. Huber, T.E. Moore & W. Loher, pp. 391–422. Ithaca: Cornell University Press.

Bernays, E.A. & Chapman, R.F. (1973). The regulation of feeding in *Locusta migratoria*. *Entomologia Experimentalis et Applicata*, **16**, 329–42.

Braunig, P. (1982). Strand receptors with central cell bodies in the proximal leg joints of orthopterous insects. *Cell and Tissue Research*, **222**, 647–54.

Chiang, R.G., Chiang, J.A. & Davey, K.G. (1990). Structure of the abdominal receptor responsive to internally applied pressure in the blood-feeding insect, *Rhodnius prolixus*. *Cell and Tissue Research*, **261**, 583–7.

Debaisieux, P. (1938). Organes scolopidiaux des pattes d'insectes. *Cellule*, **47**, 77–202.

Dreller, C. & Kirchner, W.H. (1993). Hearing in honeybees: localization of the auditory sense organ. *Journal of Comparative Physiology*, A **173**, 275–9.

Ewing, A.W. (1989). *Arthropod Bioacoustics. Neurobiology and Behaviour.* Ithaca: Cornell University Press.

Finlayson, L.H. (1976). Abdominal and thoracic receptors in insects, centipedes and scorpions. In *Structure and Function of Proprioceptors in the Invertebrates*, ed. P.J. Mill, pp. 153–211. London: Chapman & Hall.

Forrest, T.G., Read, M.P., Farris, H.E. & Hoy, R.R. (1997). A tympanal hearing organ in scarab beetles. *Journal of Experimental Biology*, **200**, 601–6.

French, A.S. (1988). Transduction mechanisms of mechanosensilla. *Annual Review of Entomology*, **633**, 39–58.

Gnatzy, W., Grünert, U & Bender, M. (1987). Campaniform sensilla of *Calliphora vicina* (Insecta, Diptera). I. Topography. *Zoomorphology*, **106**, 312–9.

Gnatzy, W. & Tautz, J. (1977). Sensitivity of an insect mechanoreceptor during moulting. *Physiological Entomology*, **2**, 279–88.

Gray, E.G. (1960). The fine structure of the insect ear. *Philosophical Transactions of the Royal Society of London* B, **243**, 75–94.

Grünert, U. & Gnatzy, W. (1987). Campaniform sensilla of *Calliphora vicina* (Insecta, Diptera). II. Typology. *Zoomorphology*, **106**, 320–8.

Hartman, H.B., Bennett, L.P. & Moulton, B.A. (1987). Anatomy of equilibrium receptors and cerci of the burrowing desert cockroach *Arenivaga* (Insecta, Blattodea). *Zoomorphology*, **107**, 81–7.

Hedwig, B. (1988). Activation and modulation of auditory receptors in *Locusta migratoria* by respiratory movements. *Journal of Comparative Physiology*, A **162**, 237–46.

Hedwig, B. (1989). Modulation of auditory information processing in tethered flying locusts. *Journal of Comparative Physiology*, A **164**, 409–22.

Honomichl, K. (1978). Feinstruktur zweier Propriorezeptoren im Kopf von *Oryzaephilus surinamensis* (L.) (Insecta, Coleoptera). *Zoomorphologie*, **90**, 213–26.

Horn, E. (1985). Gravity. In *Comprehensive Insect Physiology, Biochemistry and Pharmacology*, vol.6, ed. G.A. Kerkut & L.I. Gilbert, pp. 557–76. Oxford: Pergamon Press.

Horridge, G.A. (1965a). The Arthropoda: Receptors other than eyes. In *Structure and Function in the Nervous Systems of Invertebrates*, ed. T.H. Bullock & G.A. Horridge, pp. 1005–62. San Francisco: Freeman.

Horridge, G.A. (1965b). The Arthropoda: Details of groups. In *Structure and Function in the Nervous Systems of Invertebrates*, ed. T.H. Bullock & G.A. Horridge, pp. 1165–270. San Francisco: Freeman.

Hoy, R.R. & Robert, D. (1996). Tympanal hearing in insects. *Annual Review of Entomology*, **41**, 433–50.

Kalmring, K., Rössler, W. & Unrast, C. (1994). Complex tibial organs in the forelegs, midlegs, and hindlegs of the bushcricket *Gampsocleis gratiosa* (Tettigoniidae): comparison of the physiology of the organs. *Journal of Experimental Zoology*, **270**, 155–61.

Kittmann, R. & Schmitz, J. (1992). Functional specialization of the scoloparia of the femoral chordotonal organ in stick insects. *Journal of Experimental Biology*, **173**, 91–108.

Knyazeva, N.I., Fudalewicz-Niemczyk, W. & Rosciszewska, M. (1980). Proprioceptors in the nymph of the house cricket *Gryllus domesticus* (Orthoptera, Gryllidae). *Entomological Reviews*, **58**, 16–9.

McIver, S. B. (1985). Mechanoreception. In *Comprehensive Insect Physiology, Biochemistry and Pharmacology*, vol. 6, ed. G.A. Kerkut & L.I. Gilbert, pp. 71–132. Oxford: Pergamon Press.

Menzel, J.G. & Tautz, J. (1994). Functional morphology of the subgenual organ of the carpenter ant. *Tissue & Cell*, **26**, 735–46.

Meyer, J. & Elsner, N. (1996). How well are frequency sensitivities of grasshopper ears tuned to species-specific song spectra? *Journal of Experimental Biology*, **199**, 1631–42.

Michelsen, A., Heller, K.-G., Stumpner, A. & Rohrseitz, K. (1994a). A new biophysical method to determine the gain of the acoustic trachea in bush crickets. *Journal of Comparative Physiology*, A **175**, 145–51.

Michelsen, A. & Larsen, O.N. (1985). Hearing and Sound. In *Comprehensive Insect Physiology, Biochemistry and Pharmacology*, vol. 6, ed. G.A. Kerkut & L.I. Gilbert, pp. 495–556. Oxford: Pergamon Press.

Michelsen, A. & Löhe, G. (1995). Tuned directionality in cricket ears. *Nature*, **375**, 639.

Michelsen, A. & Nocke, H. (1974). Biophysical aspects of sound communication in insects. *Advances in Insect Physiology*, **10**, 247–96.

Michelsen, A., Popov, A.V. & Lewis, B. (1994b). Physics of directional hearing in the cricket *Gryllus bimaculatus*. *Journal of Comparative Physiology*, A **175**, 153–64.

Mücke, A. (1991). Innervation pattern and sensory supply of the midleg of *Schistocerca gregaria* (Insecta, Orthopteroidea). *Zoomorphology*, **110**, 175–87.

Oldfield, B.P. (1982). Tonotopic organisation of auditory receptors in Tettigoniidae (Orthoptera: Ensifera). *Journal of Comparative Physiology*, **147**, 461–9.

Osborne, M.P. (1970). Structure and function of neuromuscular junctions and stretch receptors. *Symposium of the Royal Entomological Society of London*, **5**, 77–100.

Robert, D., Read, M.P. & Hoy, R.R. (1994). The tympanal hearing organ of the parasitoid fly *Ornia ochracea* (Diptera, Tachinidae, Orniini). *Cell and Tissue Research*, **275**, 63–78.

Roeder, K.D. & Treat, A.E. (1957). Ultrasonic reception by the tympanic organ of noctuid moths. *Journal of Experimental Zoology*, **134**, 127–57.

Schmitz, J. (1993). Load-compensating reactions in the proximal leg joints of stick insects during standing and walking. *Journal of Experimental Biology*, **183**, 15–33.

Schwabe, J. (1906). Beitrage zur Morphologie und Histologie der tympanalen Sinnesapparate der Orthopteren. *Zoologica, Stuttgart*, no. 50, 154 pp.

Shaw, S.R. (1994). Re-evaluation of the absolute threshold and response mode of the most sensitive known 'vibration' detector, the cockroach's subgenual organ: a cochlea-like displacement threshold and a direct response to sound. *Journal of Neurobiology*, **25**, 1167–85.

Simpson, S.J. (1983). The role of volumetric feedback from the hindgut in regulation of meal size in fifth-instar *Locusta migratoria* nymphs. *Physiological Entomology*, **8**, 451–67.

Soper, R.S., Shewell, G.E. & Tyrrell, D. (1976). *Colcondamyia auditrix* nov. sp. (Diptera: Sarcophagidae), a parasite which is attracted by the mating song of its host, *Okanagana rimosa* (Homoptera: Cicadidae). *Canadian Entomologist*, **108**, 61–8.

Spangler, H.G. (1988a). Hearing in tiger beetles (Cicindelidae). *Physiological Entomology*, **13**, 447–52.

Spangler, H.G. (1988b). Moth hearing, defense, and communication. *Annual Review of Entomology*, **33**, 59–81.

Spencer, H.J. (1974). Analysis of the electrophysiological response of the trochanteral hair receptors of the cockroach. *Journal of Experimental Biology*, **60**, 223–40.

Stephen, R.O. & Bailey, W.J. (1982). Bioacoustics of the ear of the bushcricket *Hemisaga* (Saginae). *Journal of the Acoustical Society of America*, **72**, 13–25.

Stephen, R.O. & Bennet-Clark, H.C. (1982). The anatomical and mechanical basis of stimulation and frequency analysis in the locust ear. *Journal of Experimental Biology*, **99**, 279–314.

Stumpner, A. (1997). An auditory inter-neurone tuned to the male song frequency in the duetting bushcricket *Ancistrura nigrovittata* (Orthoptera, Phaneropteridae). *Journal of Experimental Biology*, **200**, 1089–101.

Sugawara, T. (1981). Fine structure of the stretch receptor in the bursa copulatrix of the butterfly, *Pieris rapae crucivora*. *Cell and Tissue Research*, **217**, 23–36.

Sugawara, T. (1993). Oviposition behaviour of the cricket *Teleogryllus commodus*: mechanosensory cells in the genital chamber and their role in the switch-over of steps. *Journal of Insect Physiology*, **39**, 335–46.

Sugawara, T. (1996). Chordotonal organs embedded in the epidermis of the soft integument of the cricket, *Teleogryllus commodus*. *Cell and Tissue Research*, **284**, 125–42.

Tautz, J. (1977). Reception of medium vibrations by thoracic hairs of caterpillars of *Barathra brassicae* L. (Lepidoptera, Noctuidae). I. Mechanical properties of receptor hairs. *Journal of Comparative Physiology*, **118**, 13–31.

Tautz, J. (1978). Reception of medium vibrations by thoracic hairs of caterpillars of *Barathra brassicae* L. (Lepidoptera, Noctuidae). II. Response characteristics of the sensory cell. *Journal of Comparative Physiology*, **125**, 67–77.

Thorpe, W.H. & Crisp, D.J. (1947). Studies on plastron respiration. III. The orientation responses of *Aphelocheirus* (Hemiptera, Aphelocheiridae (Naucoridae)) in relation to plastron respiration: together with an account of specialised pressure receptors in aquatic insects. *Journal of Experimental Biology*, **24**, 310–28.

Waters, D.A. & Jones, G. (1996). The peripheral auditory characteristics of noctuid moths: responses to the search-phase echolocation calls of bats. *Journal of Experimental Biology*, **199**, 847–56.

Weevers, R. de G. (1966a). A Lepidopteran saline: effects of inorganic cation concentration on sensory, reflex and motor responses in a herbivorous insect. *Journal of Experimental Biology*, **44**, 163–75.

Weevers, R. de G. (1966b). The physiology of a Lepidopteran muscle receptor I. The sensory response to stretching. *Journal of Experimental Biology*, **44**, 177–94.

Weevers, R. de G. (1966c). The physiology of a Lepidopteran muscle receptor III. The stretch reflex. *Journal of Experimental Biology*, **45**, 229–49.

Williamson, R. & Burns, M.D. (1978). Multiterminal receptors in the locust mesothoracic leg. *Journal of Insect Physiology*, **24**, 661–6.

Wolfrum, U. (1991). Centrin and α-actinin-like immunoreactivity in the ciliary rootlets of insect sensilla. *Cell and Tissue Research*, **266**, 231–8.

Zill, S.N. & Moran, D.T. (1981). The exoskeleton and insect proprioception. I. Responses of tibial campaniform sensilla to external and muscle-generated forces in the American cockroach, *Periplaneta americana*. *Journal of Experimental Biology*, **91**, 1–24.

24 Chemoreception

Stimulation by chemicals involves the senses of smell (olfaction) and taste (gustation). Olfaction implies the ability to detect compounds in the gaseous state. Taste, in humans, refers to detection of compounds in solution, or in a liquid state, by receptors in the oral cavity. Insects have comparable receptors on many parts of the body, often using them for purposes unrelated to feeding, and they also have the ability to detect chemicals on dry surfaces as well as in solution. For these reasons, it is usual to refer to 'contact chemoreception' in insects rather than 'taste'. The distinction between olfaction and contact chemoreception is usually clear, although olfactory receptors can respond to substances in solution and contact chemoreceptors respond to high concentrations of some odors. Processing within the central nervous system is, however, quite different. The axons from all olfactory receptors terminate in the antennal lobes in the brain (section 20.4.2) whereas axons from contact chemoreceptors terminate within the ganglion of the body segment on which the receptors occur.

Reviews: Frazier, 1992 – chemoreceptor physiology; Morita & Shiraishi, 1985 – chemoreceptor physiology; Payne, Birch & Kennedy, 1986 – olfaction

24.1 OLFACTION

24.1.1 Structure of receptors
Olfactory sensilla comprise cuticular and cellular components comparable, in a general sense, to those of cuticular mechanoreceptors and contact chemoreceptors (Fig. 24.1, and compare Figs. 23.1, 24.8).
Review: Zacharuk, 1985

Cuticular components The cuticle of olfactory receptors is characterized by the presence of numerous small pores which permit the entry of chemicals (Figs. 24.1, 24.2). These sensilla are called multiporous, and, in most cases, their function as olfactory receptors is inferred from this structure since there is relatively little experimental data. The pores vary in diameter from 10 to over 25 nm in sensilla of different insects and their density varies from about $5\,\mu m^{-2}$ on the trichoid sensilla of *Bombyx* to $125\,\mu m^{-2}$ on the plate sensilla of *Apis*. Within the pore, in many cases, is a circular cavity called the pore kettle from which fine pore tubules, 10–20 nm in diameter, pass for a short distance into the lumen of the sensillum (Fig. 24.2b). Different types of sensillum may have as many as 50 pore tubules arising from each pore kettle, or as few as five, or even none at all, although the extent to which these differences are artifacts is not certain. These tubules are the remnants of wax canals (section 16.2.3) remaining after the trichogen cell has withdrawn from the cuticle. They sometimes appear to extend inwards to make contact with the dendrite, but this is not generally true.

The gross cuticular form of the sensilla is very variable. They may be relatively long hairs with thick walls often $0.5\,\mu m$ or more in thickness (trichoid sensilla); short finger-like projections with thin cuticle less than $0.3\,\mu m$ thick (basiconic pegs); flat plates level with the general surface of the cuticle (plate or placoid sensilla); or short pegs sunk in depressions of the cuticle and opening to the exterior via a relatively restricting opening (coeloconic sensilla) (Fig. 24.3a). Amongst the latter, the wall of the peg is often hollowed out by a series of extensions of the peg lumen. These are called double wall pegs and the pores often arise from longitudinal furrows on the surface (Fig. 24.3b). Many variations on these general themes have been described. Olfactory hairs and pegs usually arise from the cuticle without any specialized socket region.

A dendrite sheath surrounds the dendrites within the receptor lymph cavity, but usually ends at the hair base, the dendrites emerging from it.

In some insects, complex olfactory organs are formed from the fusion of a number of sensilla. This is common in Homoptera where the rhinaria of aphids are formed from the fusion of up to six separate units, and the plaque organs of the fulgorid, *Pyrops*, from as many as 35. Other compound olfactory sensilla are known from some beetles and larval *Musca* (see Zacharuk, 1985).

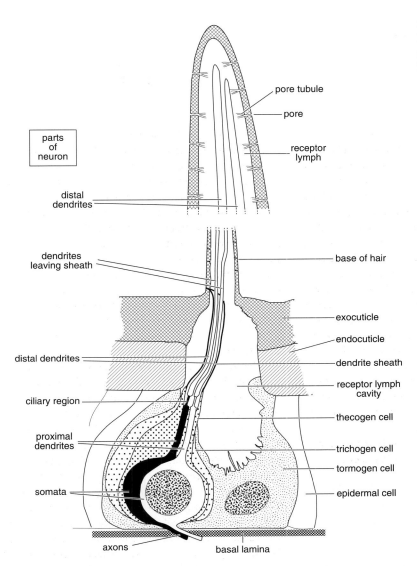

Fig. 24.1. Olfactory sensillum. Diagrammatic section through a sensillum. The terminal part of the hair is enlarged to show the pores.

Cellular components The cells of olfactory sensilla are derived from the epidermis like those of other cuticular sensilla. The number of sensory neurons in olfactory sensilla varies in different insects. Two are commonly present in the thick-walled pheromone-specific sensilla of male moths, but over 20 are present in some thin-walled receptors of grasshoppers. Coeloconic sensilla often have three neurons.

The proximal and ciliary regions of the dendrite are similar to comparable regions of mechanoreceptors, but distally the dendrite leaves the dendrite sheath and extends into the lumen of the hair, or close beneath the cuticle of pore plates (Fig. 24.1). Consequently, the distal dendrites are bathed directly by the receptor lymph. In some cases, the distal dendrites branch.

The axons from the neurons extend, without synapses, to the antennal lobes, ending in arborizations that form the olfactory glomeruli. It is generally supposed that each neuron ends in one glomerulus, and that all neurons responding to the same compound, or group of compounds, end in the same glomerulus (section 20.4.2). This is true for multiporous sensilla on the palps as well as those

Fig. 24.2. Thick-walled trichoid olfactory sensillum of *Bombyx*. **(a)** Cross-section of the whole hair showing two dendrites. In one place the section passes through the opening of a pore. Elsewhere, channels in the cuticle leading to pores are visible although the pores are not in the plane of the section. The receptor lymph is continuous with that in the receptor lymph cavity (see Fig. 24.1) (after Steinbrecht, 1980). **(b)** Detail of a pore showing the pore tubules. Notice that the outer layer of the epicuticle is continuous across the pore (after Steinbrecht, 1973)

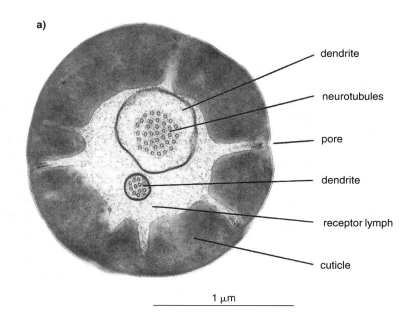

a)

dendrite

neurotubules

pore

dendrite

receptor lymph

cuticle

1 μm

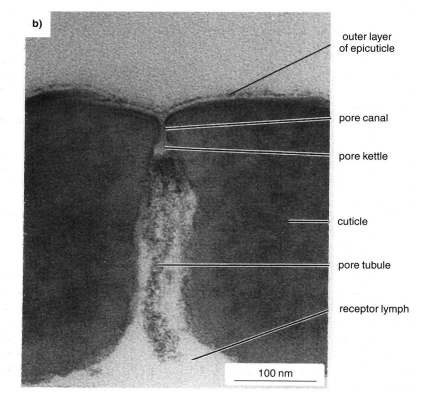

b)

outer layer of epicuticle

pore canal

pore kettle

cuticle

pore tubule

receptor lymph

100 nm

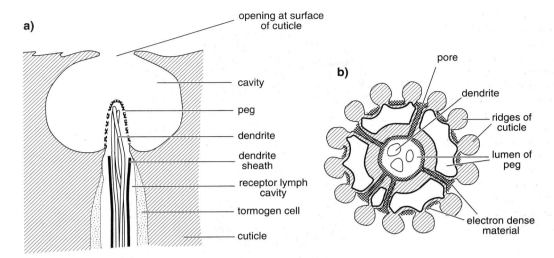

Fig. 24.3. Coeloconic sensillum. (**a**) Longitudinal section showing the peg in a cavity beneath the surface of the cuticle. (**b**) Transverse section through a double-walled peg.

on the antennae. In *Manduca*, there is a single glomerulus in each antennal lobe in which the axons from the palp sensilla end. In most cases, the inputs from olfactory receptor neurons end in the ipsilateral antennal lobe, but in Diptera many of the neurons connect with the lobes of both sides. This is also true of the palp receptors in *Manduca*.

The accessory cells of olfactory sensilla are essentially similar to those of mechanoreceptors (Fig. 24.1). They vary in number from two to four, but usually there are three: the thecogen, trichogen and tormogen cells. In addition to their development functions, these cells regulate the composition of the receptor lymph. Pheromone-binding protein (see below), is produced continuously by the trichogen and tormogen cells of antennal sensilla in *Antheraea*. The same two cells, probably together with the thecogen cell and the neuron, are also involved in the breakdown of the protein (Steinbrecht, Ozaki & Ziegelberger, 1992).

Distribution Multiporous, olfactory sensilla are probably present on the antennae of all insects, including most larval forms. In addition, some insects may have multiporous sensilla on the maxillary and labial palps. These may be relatively few in number, as in grasshoppers, or there may be large numbers as on the labial palps of Lepidoptera. Multiporous sensilla also occur on the genitalia of the sheep blowfly, *Lucilia*.

Various types of olfactory sensilla are present on the antennae of any insect. For example, three distinct types are found in grasshoppers, two types of basiconic and one of coeloconic sensilla (a second type of coeloconic sensillum responds to temperature and humidity (section 19.7)). In the midge, *Culicoides*, there are four types, two of trichoid sensilla, one of basiconic and one of coeloconic sensilla. These different types are usually associated with neurons responding to different classes of chemical. Sensilla with the same outward form may contain neurons which respond to different ranges of compounds so that the number of physiological classes may be considerably greater than the number of anatomical classes. By contrast, olfactory sensilla on the palps tend to be structurally uniform.

Reviews: Altner & Prillinger, 1980; Steinbrecht, 1987; Zacharuk, 1985; Zacharuk & Shields, 1991 – distribution on larvae

24.1.2 Functioning of the receptor system

Two processes, which can be regarded as analogous to coupling and transduction in mechanoreceptors, are involved in producing the neuronal response. The first is the capture and transport of odor molecules to the dendrite membrane, known as perireceptor events. The second is transduction, the conversion of chemical to electrical energy.

Fig. 24.4. Diagrammatic representation of the possible perireceptor events occurring during olfaction. The odor molecule is captured on the surface of the sensillum; it diffuses to the pore and enters the sensillum through a pore tubule; it becomes bound to a binding protein and transported to a receptor molecule in the dendrite membrane; here it initiates the transduction process; it is finally destroyed by an odor degrading enzyme (but see text) (based on Vogt, 1995).

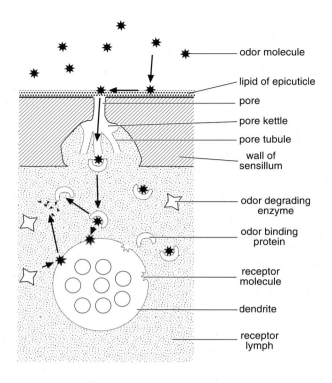

odor molecule

lipid of epicuticle

pore

pore kettle

pore tubule

wall of sensillum

odor degrading enzyme

odor binding protein

receptor molecule

dendrite

receptor lymph

Perireceptor events   Most odor molecules are to some extent lipophilic, and so they will dissolve in the epicuticular lipid forming the outer coating of a sensillum (Figs. 24.2, 24.4), but they do not readily dissolve in the insect's body fluids. A consequence of the former is that the whole outer surface of the sensillum is involved in the capture of odor molecules; it is not necessary for the molecules to strike the pores directly (Kanaujia & Kaissling, 1985). A consequence of the latter is that they must be solubilized in order to reach the dendritic membrane. This is achieved by combining with specific water-soluble proteins, called odor-binding proteins. Three classes of such proteins have been recognized: pheromone-binding proteins and two classes of general odor-binding proteins. The former usually occur only in male moths where they bind specifically with components of the female sex pheromone. General odor-binding proteins are present in the olfactory sensilla of both sexes. They are believed to bind non-specifically with many different classes of odor molecules (Steinbrecht, Laue & Ziegelberger, 1995). These proteins may be at high concentrations in the receptor lymph. For example, the pheromone-binding protein of *Antheraea* is present at a concentration of about 10 mM.

In this solubilized form, the odor molecules are carried to the membrane of the dendrite where, it is assumed, they are released from the binding proteins and become associated with receptor proteins.

Transduction   Transduction occurs at the dendrite membrane. It is thought that protein molecules in the outer surface of the cell membrane bind with odor molecules. These receptor molecules are believed to be relatively specific for the type of odor molecule with which they bind. Work on the olfactory system of some other animals suggests that there may be a large family of receptor molecules giving the organism the potential of discriminating among a wide range of compounds. A small number of possible receptor proteins have been identified in insects.

Acting via a second messenger system that probably involves inositol trisphosphate, capture of an odor molecule by a receptor molecule produces a depolarization of the cell membrane, called a receptor potential. This spreads electrotonically to the soma and leads to the production of action potentials if it is of sufficient magnitude. However, some odors produce a hyperpolarization of the

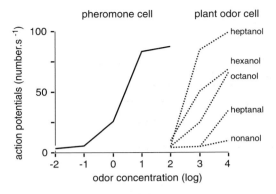

Fig. 24.5. Sensitivity of two olfactory neurons of male *Spodoptera*. The pheromone cell is sensitive to a minor pheromone component, Z9,E12-14 acetate. Notice that its dynamic range extends over odor concentrations of about two orders of magnitude. The plant odor cell responds maximally to heptanol. This cell responds to other related compounds, but less strongly. Notice that both chain length and functional group are important. Concentration is shown as the amount of each compound in μg (log scale) applied to the cartridge which acted as the source of odor. The concentration at the receptor will be proportional to this for compounds of similar volatility (after Anderson, Hansson & Löfqvist, 1995; Ljungberg, Anderson & Hansson, 1993).

cell membrane, and may act via a different second messenger system involving cAMP.

Olfactory neurons commonly exhibit a low level of spontaneous activity, producing action potentials even in the absence of any odor. Consequently an odor may enhance the level of activity, have no effect, or reduce activity. All these effects may occur in one neuron in response to different compounds.

The response to a compound that increases the rate of firing is often phasic/tonic and it may continue for a brief period after removal of the stimulus. The magnitude of the response increases with stimulus concentration until the receptor is saturated (Fig. 24.5). The dynamic range of neurons usually extends over a concentration range of two or three orders of magnitude. At lower concentrations, below threshold, the neuron does not respond; at higher concentrations it normally responds maximally. Threshold concentrations vary considerably. Pheromone-specific neurons, for example, often have much lower thresholds than neurons responding to other types of odor (Fig. 24.5).
Reviews: Vogt, 1995; Vogt, Rybczynski & Lerner, 1990

Stimulus degradation It is necessary for stimulating molecules to be removed quickly from the dendrite and the sensillum lymph in order to maintain sensitivity, and so that the neural signal is temporally related to the occurrence of the environmental stimulus. Two possible mechanisms have been suggested. One is that the odor molecules are degraded by enzymes. Two types of enzyme are known to occur in the receptor lymph of male *Antheraea*. An esterase specifically degrades the acetate component of the female sex pheromone, and an aldehyde oxidase, also present in the female, acts on a range of different compounds. The odor-binding proteins probably protect the odor molecules from the enzyme during transit to the dendrite. However, it is not certain that enzymic degradation occurs sufficiently rapidly to limit the duration of a stimulus. The alternative possibility is that the odor-binding protein serves, not only to carry the odorant molecules to the dendrite, but also to inactivate them. In a reduced form it may act as a carrier becoming oxidized at the receptor membrane so that the pheromone molecule is no longer available to stimulate (Ziegelberger, 1995). Enzymic degradation may then follow more slowly.

Because the odor molecules are first trapped at the surface of the cuticle, their transport into the sensillum is relatively slow. This could provide the insect with false information about the occurrence of a stimulus, and there is some evidence that enzymes degrading odor molecules are also present in the cuticle.

Specificity Neurons vary in the specificity of their response to odors. Many pheromone receptor neurons are highly specific, effectively responding only to a single compound and often showing marked selectivity even for a specific isomer (Fig. 24.6a). Such cells may respond to other compounds, but usually only at concentrations several orders of magnitude higher than that of the compound to which they are tuned. There is increasing evidence that some neurons responding to host-related odors are also specific. For example, in *Spodoptera littoralis* 21 classes of neurons have been identified on the basis of the compounds, other than sex pheromones, that they respond to. Except for those responding to the common green leaf compounds, most of the different classes respond only to one or two particular compounds. None of the cells responds to a wide range of compounds (Anderson, Hansson & Löfqvist, 1995; see also Blight *et al.*, 1995).

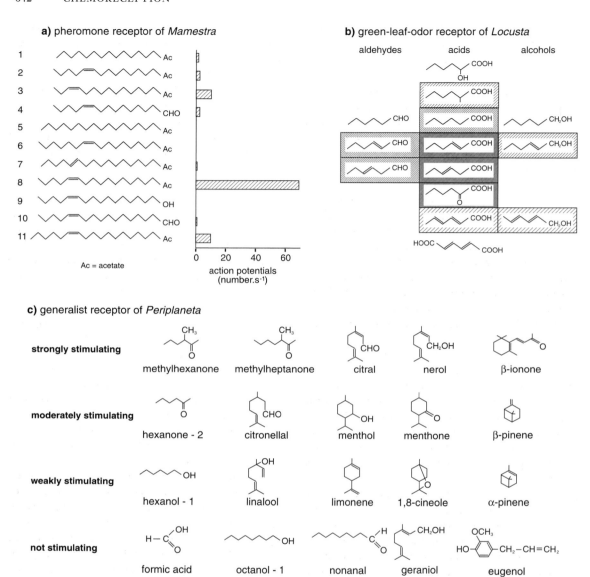

Fig. 24.6. Specificity of olfactory receptor neurons. **(a)** Specialist neuron on antenna of male *Mamestra*. This cell responds primarily to one of the components of the female sex attractant pheromone, Z11-hexadecen acetate (compound 8). Notice that it shows no response to the E11-isomer (compound 7) or to Z11-hexadecen alcohol (9) or aldehyde (10). Acetates with different chain lengths, 14 (compounds 1–3) or 17 (11) carbon atoms, produce moderate responses only if the double bond is in the Z11 position (based on data in Renou & Lucas, 1994). **(b)** Group specialist neuron in a coeloconic sensillum on the antenna of *Locusta*. The cell is sensitive to various forms of hexanoic acid and some related compounds. Degree of shading indicates intensity of response (darkest = greatest sensitivity). Compounds not in boxes did not stimulate (after Kafka, 1971). **(c)** Generalist olfactory neuron in the antenna of *Periplaneta*. This cell responds primarily to monoterpenes. Notice that nerol is strongly stimulating (top row), but geraniol, its stereoisomer, does not stimulate (bottom row). The cell responded to more compounds than those shown in the diagram. Some other cells on the antenna have similar, although not identical, response spectra, while others are quite different (data from Selzer, 1981).

Other neurons appear to be less specific, responding to a range of closely related compounds. Several leaf-feeding insects have sensory neurons that respond to a range of commonly occurring six-carbon-atom volatiles from green leaves, such as hexanoic acid (Fig. 24.6b), but not to other types of compound. Generalist receptors respond to a range of structurally different compounds, and cells of this type are present in *Periplaneta* (Fig. 24.6c) and in some bark beetles. Even here, the neurons are differentially responsive to different compounds (Mustaparta, 1975; Selzer, 1981).

A neuron's specificity is assumed to be largely a consequence of the receptor molecules present in its dendritic membrane. It is presumed that binding depends on molecular shape, and that a neuron that responds primarily to a single compound has only one type of receptor molecule with very specific binding properties. Neurons that respond to a group of related compounds possibly have a single receptor site with characteristics that enable it to bind with different molecules provided they have some characteristic structure. In this case, some compounds presumably have a less good fit with the receptor molecule and so produce lower firing rates. The plant-odor-specific neuron depicted in Fig. 24.5 is probably of this type. Neurons that respond to a number of structurally different compounds may have several different types of receptor molecules binding to structurally different odor molecules. The specificity of the odor-binding proteins may also be important in determining the types of molecule that reach the dendrite.

The sensitivity of the insect as a whole is influenced by the number and arrangement of the sensilla. The greater the number of sensilla, the more molecules will be captured and the more information sent to the brain. Thus, the males of species in which the female emits a sex attractant pheromone often have pectinate antennae with many branches providing space for a large number of sensilla. In male *Manduca*, where the flagellum is stouter than in the female, although not pectinate, there are probably 300 000 neurons responding to components of the sex attractant pheromone, and another 100 000 which respond to plant odors. Females have only this latter class of neurons with which to locate their host plants for nectar feeding and oviposition. It is believed that the difference in number is a reflection of the very low concentrations of pheromone emitted by females compared with the concentration of plant odors.

Temporal response patterns Airborne odors do not form a continuous gradient because of air turbulence; instead they occur as a discontinuous series of bursts separated by 'clean' air. The ability to distinguish successive contacts with such odor bursts is important in odor-modulated flight and the ability to do so depends on the characteristics of the sensory neurons.

Males of the moth, *Helicoverpa*, have large numbers of olfactory neurons on the antennae which respond to the principal component of the female sex attractant pheromone. Some of these exhibit a strong phasic response which rapidly falls to zero. These neurons give discrete responses to 20 ms pulses of pheromone at 1.5 Hz; at 6 Hz the response is continuous, but distinct peaks still correspond with the arrival of the stimulus. Only at 12 Hz do the responses merge into a continuum from which the insect probably cannot distinguish the pulsed nature of the stimulus from continuous stimulation. Thus, the peripheral neurons have the capacity to allow the insect to recognize a stimulus that is pulsed at quite high frequency (Almaas, Christensen & Mustaparta, 1991).

Odor discrimination It is clear from the behavior of insects that they are able to discriminate between odors. This is most obvious with respect to pheromones where closely related species respond to their own pheromones, but not to those of other species. It is also apparent from studies showing that bees and other insects can be trained to distinguish between odors.

Specific odor recognition often depends on the occurrence of sensory neurons responding only to certain compounds. This information is relayed to specific glomeruli in the antennal lobe and from here, projection neurons specific for the odor may extend to other parts of the brain (section 20.4.2). This is called a labelled line system. Labelled line systems are the product of evolved adaptations to odors of critical and usual importance to the insect and are known to exist in the case of some pheromones. An alternative way of recognizing different odors is by across-fiber patterning. Here, sensory neurons, or projection neurons from the antennal lobe, respond to a range of compounds, but cells differ in their relative sensitivity to particular chemicals. The information carried by a single projection neuron does not convey a clear message, but the patterns of response of all the neurons together may be sufficiently distinct for the insect to distinguish between odors.

Variation in response   An insect's response to an odor is dependent on its physiological state. For example, blowfly larvae are attracted by the smell of ammonia during their feeding period, but, when they are ready to pupate, their response to ammonia is reversed. The state of feeding also influences the readiness of grasshoppers and blood-sucking flies to respond to host odors.

It is probable that these behavioral changes reflect the processing of information in the central nervous system, but there is also some evidence that the responsiveness of the sensory neurons is variable. In some female mosquitoes, as in *Aedes*, host-finding is partly dependent on stimulation of a sensory neuron responding to the odor of lactic acid which is present in human sweat. The proportion of cells responding increases after adult eclosion (Fig. 24.7a), and their sensitivity falls when the insect has taken a blood meal, but rises again after oviposition when the insect is ready to feed again (Fig. 24.7b) (Davis, 1984a, b). This reduction in sensitivity is due to the presence of a hormone. Similar changes occur in another mosquito, *Culex*, in relation to diapause. The lactic acid neurons are much more excitable in post-diapause females than they are during diapause. At the same time the proportion of cells responding to ethyl propionate, a compound characteristic of the oviposition sites, increases although the sensitivity of the cells that do respond does not change (Bowen, 1990).

Such changes in sensitivity do not occur in the pheromone receptors of male moths, and perhaps they are only to be expected in insects exhibiting marked switching of behavior, as in the female mosquito which switches between blood feeding and oviposition.

Interactions between olfactory neurons   Usually, a biologically significant odor occurs within a background of other environmental odors, and important odor signals may themselves be mixtures of compounds (section 27.1.3). There is some evidence that the responses of individual sensory neurons are affected by the mixture of compounds to which they are exposed. Sometimes the effect is synergistic, as in a pheromone receptor of *Spodoptera litura* (Aihara & Shibuya, 1977), but in other cases there are negative interactions. For example, the response of cells in some placoid sensilla of *Apis* to the flower odor, geraniol, is inhibited by another flower odor, linalool (Akers & Getz, 1993).

The physiological basis for these interactions is not understood, but sometimes a compound that stimulates one cell is known to hyperpolarize another in the same sensillum. The effect of this, in a mixture, would be to sup-

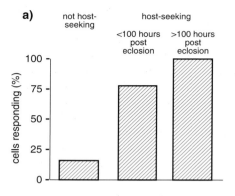

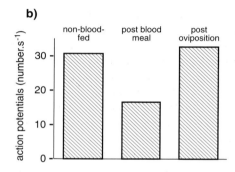

Fig. 24.7. Changes in sensitivity of olfactory neurons of female mosquitoes (*Aedes*). (a) Changes in the proportion of cells responding to lactic acid in females that had not started host-seeking and in those that were actively host-seeking but had not yet taken a meal (data from Davis, 1984b). (b) Changes in the sensitivity of neurons in relation to blood-feeding. Sensitivity declines 48 hours after taking a blood meal, but rises again when the insect has laid eggs and is ready for a further blood meal (data from Davis, 1984a).

press the effect of any compound that might stimulate the second neuron (Todd, Haynes & Baker, 1992).

24.1.3 Functions of the olfactory system

It is probably true that insects can detect, in the sense that they exhibit a sensory response to, many odors in the environment. For example, grasshoppers have a sensory response to plants of many different kinds, not only to their host plants (White & Chapman, 1990), and antennal receptors of adult Lepidoptera respond to a wide range of plant-derived compounds (Topazzini, Mazza & Pelosi, 1990). Many of these odors may produce no overt behavior; others may be critically important in locating a mate, a food source, or an oviposition site from a distance. They may also affect behavior close to the source, and responses to odors may be learned.

Because of air turbulence, continuous odor gradients cannot persist under natural conditions except for very short distances. Consequently, although odor may provide the signal that initiates upwind flight, the insect depends on a combination of mechanical and visual stimuli to achieve upwind movement. This type of behavior is known to occur in response to sex attractant pheromones, such as those produced by many female Lepidoptera. It may also occur in relation to food, or an oviposition site especially in night-flying insects where a direct visual response may only be possible close to the source.

Although single compounds can sometimes induce these behaviors, it is probably generally true that specificity of attraction, where it occurs, results from particular combinations of compounds. For example, although a single compound can attract males to a pheromone-producing female, a mixture of two or three specific compounds provides the insect with an unambiguous signal (see Chapter 27). Similarly, a specific mixture of the commonly occurring six-carbon-atom green leaf volatiles is an attractant for the Colorado potato beetle, *Leptinotarsa*, and many insects have olfactory neurons responding to these compounds. Sheep blowflies are attracted to injured sheep by the sulfur-containing breakdown products of wool. Dimethyldisulfide and ethanediol, together with hydrogen sulfide are major components of this odor, but, individually, neither dimethyldisulfide nor ethanediol has any effect on the fly's behavior (Ashworth & Wall, 1994).

Carbon dioxide is an attractant for many species that suck vertebrate blood and there are carbon dioxide receptors on the antennae of tsetse flies (Bognor, 1992), and presumably of other blood-sucking species. The carbon dioxide acts, together with acetone, octenol and other compounds commonly occurring in animal breath, to attract tsetse flies. Interestingly, many Lepidoptera have large numbers of carbon dioxide receptors in a pit on the labial palps. In *Heliothis*, these are sensitive to very small changes in the ambient level of carbon dioxide, although their role in the behavior of the insect is not known (Stange, 1992).

There are relatively few examples of odor affecting behavior at close range even though such behavior may be widespread. Flower scents on the bodies of bees returning from foraging help other workers recognize the food source. Male monarch butterflies produce an aphrodisiac pheromone which enhances the female's readiness to mate. Females of the black swallowtail butterfly, *Papilio polyxenes*

are stimulated to land on artificial leaves by the odor of carrot leaves, the normal host plant of the larvae, but this behavior is disrupted if the odor of cabbage is also present.

Examples of insects being repelled by odors are also known. For example, nepetalactone, the chemical giving a characteristic odor to catnip, repels a number of phytophagous Homoptera and Coleoptera (Eisner, 1964) and linalool is a repellent for the aphid *Cavariella*.

Reviews: Bernays & Chapman, 1994 – behavior in relation to host-plant odors; Colvin & Gibson, 1992 – host-finding by tsetse flies; Murlis, Elkinton & Cardé, 1992 – odor plumes; Payne, Birch & Kennedy, 1986 – physiology and behavior; Visser, 1986 – odor perception in phytophagous insects

24.2 CONTACT CHEMORECEPTION

24.2.1 Structure of receptors

Cuticular components Contact chemoreceptors are usually short hairs or cones characterized by the presence of a single pore at, or close to, the tip of the projection (Fig. 24.8). Sensilla having this structure, but whose function has not been experimentally investigated are often called uniporous sensilla; the assumption is that they are contact chemoreceptors. The pore may be circular or oval to slit-shaped, and is often about 0.2 μm in longest diameter. At least in some instances, a porous plug within the pore separates the dendrite endings from the environment. The plug appears to be composed of fibrils of cuticular material (Shields, 1994). The dendrite sheath extends inwards from the inner edge of the pore and appears to be continuous with the epicuticle. In the contact chemosensory hairs of grasshoppers the dendrite sheath lies in the center of the hair surrounded by the receptor lymph which extends into the hair. In flies, however, the lumen of the hair is divided into a smaller lateral compartment containing the dendrites and a larger one occupying most of the space within the hair. The dendrite sheath in this type is continuous with the walls of the lateral compartment.

Hair-like sensilla usually have sockets similar to those of mechanoreceptor hairs giving them some mobility. Cone-shaped sensilla are found primarily on the inner mouthparts. They lack an articulation, the cone being continuous with the surrounding cuticle.

Cellular components The neurons of contact chemoreceptors are essentially the same as those of olfactory receptors except that their dendrites extend to the tip of the hair or cone within the dendrite sheath (Fig. 24.8). There are commonly four chemosensitive neurons in a sensillum, but some sensilla have larger numbers. In addition, socketed sensilla have a mechanosensitive neuron which ends at the cuticle of the hair base and has a tubular body at the point of insertion.

The tips of the chemoreceptor dendrites are covered by a viscous fluid containing mucopolysaccharide which sometimes exudes through the terminal pore of the sensillum.

The axons terminate in the ganglion of the segment on which the receptor occurs. Thus, the input from sensilla on the maxillae and labium end in arborizations in the subesophageal ganglion, and those from sensilla on the legs end in the thoracic ganglia.

Two or three accessory cells are present in each sensillum, with similar functions to those of mechano- and olfactory receptors.

Distribution   Contact chemoreceptors may be present on all parts of the body, although major concentrations occur on the labrum, maxillae and labium (see Fig. 2.6). Moderate numbers occur on the antennae, especially at the tip. Chemoreceptors are not present on the mandibles. Nor do they occur within the gut[1], although many, perhaps all, insects have chemoreceptors in the wall of the cibarium. Relatively large numbers of receptors are often present on the legs, but, although contact chemoreception may be important in oviposition, only a few, and sometimes none at all, are present on the ovipositor. In addition to those associated with specific organs, grasshoppers have large numbers of small trichoid contact chemoreceptors scattered over all parts of the body. Stocker (1994) gives a complete compendium of the chemoreceptors of adult *Drosophila*.

Review: Zacharuk & Shields, 1991 – distribution in larvae

24.2.2 Functioning of contact chemoreceptors

Very little is known about the physiology of contact chemoreception. The receptors on the inner parts of the mouthparts are probably, during feeding, bathed by saliva and fluids released from the food and so respond to chemicals in solution. Sensilla of the tarsi and other parts of the body, however, probably normally respond to chemicals on

a dry surface, like the surface of a leaf or, in the case of contact pheromones, a member of the opposite sex. It is likely that compounds in the viscous fluid at the tip of the hair function as carriers for some compounds on these surfaces. In sensilla on the palps of *Periplaneta*, a sugar-binding protein is present within this material (Becker & Peters, 1989) and a protein in the same family of proteins as pheromone-binding protein in olfactory sensilla (section 24.1.2) is present in contact chemoreceptors on the proboscis and tarsi of flies (Ozaki *et al.*, 1995). The mucopolysaccharide possibly also functions as a filter as not all substances will be equally readily transported.

Transduction mechanisms probably vary depending on the type of compound to which the neuron responds, but almost nothing is known of the processes in insects. By analogy with vertebrates (see Avenet, Kinnamon & Roper, 1993), it may be that inorganic ions act directly on ion channels in the dendritic membrane, whereas there are protein receptor molecules in the membrane for other types of chemical. Two proteins which probably serve this function are present in the sugar receptors of the fly, *Phormia* (Ozaki *et al.*, 1993) (see below). Binding of the ligand to these receptors probably initiates a second messenger cascade leading to an alteration of the membrane permeability to produce the receptor potential.

The receptor potential is presumed to develop distally on the dendrite and then to spread passively to the spike (action potential) initiation zone, possibly at the origin of the axon from the soma. The magnitude of the receptor potential is proportional to the concentration of stimulating compound and as the number of action potentials produced is proportional to the size of the receptor potential, there is a direct relationship between concentration and the number of action potentials. Hyperpolarization decreases the likelihood that action potentials will be produced.

The dynamic range of contact chemoreceptor neurons extends over concentration ranges of two or three orders of magnitude, similar to that of olfactory receptors. Threshold levels differ between neurons. Cells responding to major nutrient compounds, such as sugars and amino acids, are usually sensitive only to moderately high concentrations (Fig. 24.9a, b), but cells responding to chemicals with specific ecological relevance apart from nutrition are often much more sensitive (Fig. 24.9c, d). Even within a single insect, neurons responding to the same compound may have different thresholds. For example, in *Phormia*, the neurons in the tarsal hairs that

---

[1] Receptors involved in the transport of some nutrients from the gut lumen into the midgut cells probably are present, but these have only a local effect; they are not connected with the nervous system.

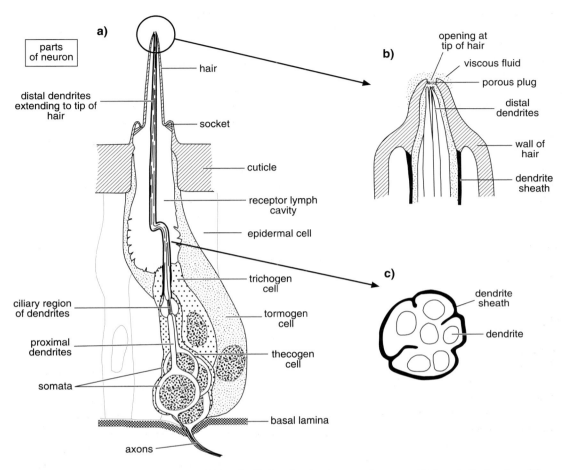

Fig. 24.8. Contact chemoreceptor. (**a**) Longitudinal section showing general structure. (**b**) Longitudinal section of tip showing detail of structure. (**c**) Transverse section through the dendrite sheath and distal dendrites in the receptor lymph cavity.

respond to sucrose have thresholds at concentrations one or two orders of magnitude lower than those in labellar sensilla (Fig. 24.10). Such differences provide the insect with a greater dynamic range of response than can occur in individual neurons.

Most contact chemoreceptors exhibit a marked phasic response often with only a short-lived tonic response during continuous stimulation. As a consequence, the input from a single neuron may fall to zero very quickly, sometimes in less than one second. Perhaps to maintain a flow of sensory information, many insects make drumming movements with the legs, palps or antennae in which they touch a surface repeatedly for very brief periods.

When a grasshopper, for example, drums on a leaf with its palps (palpates) the contact chemoreceptors on the tips of the palps touch the surface about 10 times a second, each contact lasting less than 20 ms. In the intervals between contact, the neurons disadapt and thus maintain a high level of input for a second or more.

Unlike olfactory neurons, contact chemoreceptor neurons generally fire only when stimulated, but if they do exhibit spontaneous activity, the firing rate is very low.

Specificity Each of the neurons within a sensillum responds to a different class of compound. It is common for one cell to respond primarily to inorganic salts and

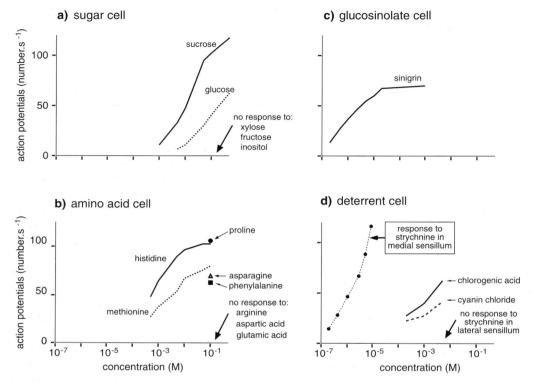

Fig. 24.9. Responses of the four cells in the lateral styloconic sensillum of the caterpillar of *Pieris brassicae*. The activity of each cell is shown during the first second of stimulation. The position of the sensillum is shown in Fig. 2.6b. **(a)** Response of sugar cell. This cell only responds to sucrose and glucose (after Ma, 1972). **(b)** The amino acid cell responds to 14 out of 22 amino acids tested (after van Loon & van Eeuwijk, 1989). **(c)** Response of the glucosinolate cell to sinigrin. This cell responds to many glucosinolates. Notice the very low threshold. The cell starts to respond at a concentration of 2×10^{-7} M sinigrin (after Blom, 1978). **(d)** Response of the deterrent cell. This cell responds to a number of phenolic acids and flavonoids. It does not respond to strychnine, but a deterrent cell in the medial styloconic sensillum is very sensitive to this compound with a very low threshold (after Blom, 1978; van Loon, 1990).

another to sugars (Fig. 24.9a). The other neurons within a sensillum are also relatively specific, although the compounds to which they respond vary from species to species and from sensillum to sensillum within a species. Amongst flies, one cell may respond to water and, in *Phormia*, to very dilute salt solutions. Cells responding to amino acids are present in some sensilla of caterpillars and chrysomelid beetles, and probably in many other insects (Fig. 24.9b). There is no evidence that insects can taste proteins.

Some species have chemoreceptor neurons that only respond to specific chemicals that are key components of the food, or to contact pheromones. For example, many larval and adult brassica-feeding insects have a cell that responds to glucosinolates (Fig. 24.9c; Städler *et al.*, 1995).

Glucosinolates are characteristic compounds of Brassicaceae. Both sexes of adult fruit flies, *Rhagoletis*, in which the female marks fruit with a pheromone after oviposition, have neurons in tarsal receptors that are specific for the compound (Städler *et al.*, 1994). In species using contact pheromones in their surface waxes, like houseflies, there are probably neurons that respond specifically to the compounds in some tarsal sensilla (Wall & Langley, 1993).

Caterpillars and grasshoppers have a cell in some sensilla that responds to chemicals from a number of different chemical classes that produce aversive behavior (see Peterson *et al.*, 1993). No cell with an obvious link to aversive behavior has been described in Coleoptera or Diptera.

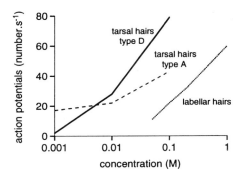

Fig. 24.10. Differences in threshold of sucrose-sensitive neurons in different sensilla of the blowfly (data from McCutchan, 1969; Omand & Dethier, 1969).

These different cell types are often called 'salt', 'sugar', 'amino acid' or 'deterrent' receptors according to the class of compound to which they respond or the type of behavioral response induced when they are stimulated, but these response ranges are not absolute. A 'sugar' receptor may also sometimes be stimulated by an inorganic salt, for example. Even within a class of cells, the specificity varies from sensillum to sensillum. For example, the 'sugar' cell in the lateral styloconic sensillum on the galea of larval *Pieris brassicae* (see Fig. 2.6b) responds only to sucrose and glucose, while that in the medial sensillum responds to a wide range of sugars including pentose sugars, and, weakly, to trisaccharides and the polyhydric alcohols inositol and sorbitol.

These different specificity ranges are assumed to be conferred by the receptor molecules present on the dendritic membrane. Highly specific neurons have a single class of receptor, whereas those responding to many different compounds probably have two or more different receptor types. It is known that the sugar-responsive neurons of flies have at least two different receptor proteins, one responding to pyranose sugars and one to furanose sugars.

As with olfactory neurons, the specificity of a single receptor molecule is believed to be a reflection of its stereochemistry (its molecular shape). It may react only with molecules having a specific shape or with a range of molecules provided these have some feature in common. It is believed that sensitivity to a particular stimulant reflects the fit of the stimulating and receptor molecules. For example, it is probable that both sucrose and glucose react with the same pyranose receptor molecule, but, in the lateral styloconic sensillum of *Pieris*, sucrose produces a

greater response than glucose at the same concentrations (Fig. 24.9a). Other pyranose sugars do not stimulate this neuron, presumably because they lack the characteristics necessary to bind with the receptor molecule.

Variation in response The responses of peripheral contact chemoreceptors are not necessarily consistent. Variation occurs in relation to feeding and, over a longer term, to the development of the insect. In locusts, and probably in flies, sensitivity declines and the neurons in some sensilla may become wholly unresponsive following a meal. These changes are due to the closing, and subsequent re-opening of the terminal pore of each sensillum. In addition, specific feedbacks can have differential effects on the sensory cells within a sensillum. The responsiveness of neurons to amino acids in the terminal sensilla of the palps in a locust fed on a high protein/low carbohydrate food is greatly reduced, but the responsiveness of cells responding to sugars is unaffected. In insects fed on a low protein/high carbohydrate food, the converse is true, the activity of the sugar cell is reduced (Fig. 24.11a). Similar changes occur in caterpillars.

These effects are controlled by blood-borne factors. Sensillum closing in locusts is a consequence of crop filling, which causes the release of a hormone from the corpora cardiaca. The more specific effects related to nutritional balance are produced directly or indirectly by the relevant compounds in the hemolymph. High levels of amino acids in the hemolymph reduce the sensitivity of amino acid-sensitive neurons, but how they affect the neurons in the sensilla is not known.

Variation in responsiveness also occurs over longer time scales in relation to the previous experience of the insect. For example, the sensitivity to salicin of the deterrent cell in *Manduca* caterpillars is reduced if the larva is forced to feed on a diet containing salicin. This effect is not confined to deterrent compounds; habituation to inositol, a sugar alcohol which causes some phagostimulation, also occurs (Fig. 24.11b).

Ontogenetic changes in the sensitivity of contact chemoreceptors have been demonstrated in a number of insects. In larval *Spodoptera littoralis*, the response to inositol is markedly higher after day 2 of the final stadium (Fig. 24.11c), while the response to sinigrin peaks in mid-stadium and then declines to a low level on the day of ecdysis. Changes over the course of a day also occur.

Review: Blaney, Schoonhoven & Simmonds, 1986

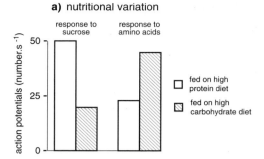

a) nutritional variation

response to sucrose | response to amino acids

☐ fed on high protein diet

◨ fed on high carbohydrate diet

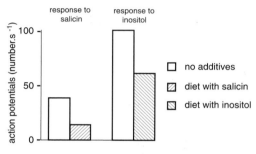

b) nutritional variation

response to salicin | response to inositol

☐ no additives

◨ diet with salicin

◨ diet with inositol

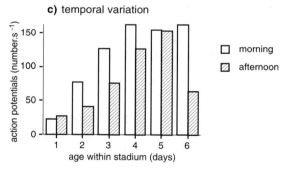

c) temporal variation

age within stadium (days)

☐ morning

◨ afternoon

Fig. 24.11. Variation in sensory response. **(a)** Effects of feeding on artificial diets containing different proportions of proteins and carbohydrates. Responses of neurons on the maxillary palp of *Locusta* after access to the diets for only four hours (data from Simpson *et al.*, 1991). **(b)** Effects of rearing on artificial diets without added chemicals or with salicin (a deterrent compound) or inositol (a phagostimulant with some nutritional value). Responses of neurons in the medial styloconic sensillum of *Manduca* were reduced after rearing on the diets for several days (after Schoonhoven, 1969). **(c)** Variation in the response of caterpillars of *Spodoptera* to inositol through the final stage. Larvae were reared on artificial diet without inositol (after Schoonhoven, Simmonds & Blaney, 1991).

Peripheral interactions resulting from mixtures of chemicals The response to a normal food, usually containing a variety of different chemicals, is not simply the sum of the inputs occurring when the various components are tested singly. Interactions occur between the chemicals or between the peripheral sensory neurons. The signals transmitted to the central nervous system are a reflection of these interactions.

Sometimes compounds have an effect on feeding even though none of the chemosensory cells responds to them. This is true of alkaloids in the food of flies. They do this by reducing the input from the cells responding to phagostimulatory compounds, such as sugars, present in the mixture. The extent of inhibition depends on the relative concentrations of the compounds (Fig. 24.12a) (Dethier & Bowdan, 1989). A similar effect is observed in insects which do have deterrent cells, such as the caterpillars and grasshoppers. Here, as the alkaloid concentration is increased, the responses of the sugar cells decrease until they are finally silenced completely. This effect is not restricted to alkaloids or to deterrent compounds; interactions involving many different compounds occur. For example, in *Stomoxys*, the activity of neurons responding to inorganic salts in the labellar hairs is inhibited by adenine nucleotides for which they appear to have purinoceptors (Ascoli-Christensen, Sutcliffe & Albert, 1990). The nucleotides are phagostimulants for these insects. Sucrose and inositol both inhibit the response to sinigrin in caterpillars of *Mamestra* and *Trichoplusia* (Fig. 24.12b). [For both these insects, sinigrin is a deterrent, although it is a phagostimulant for *Pieris* (compare Fig. 24.9c).] These interactions may be mutual so that compounds producing spike activity in different cells may each inhibit the activity in the other cell (Fig. 24.12c). Sometimes the effects are synergistic rather than inhibitory. Interactions may also occur between cells after the development of a receptor potential so that the firing rate of one neuron is affected by that of an adjacent cell.

It is likely that interactions between chemicals and between adjacent neurons is a normal phenomenon as most naturally encountered substances consist of many different chemicals.

Reviews: Chapman, 1995; Dethier, 1987; Hanson, 1987 – contact chemoreception in flies; Schoonhoven, 1987 – sensory code in caterpillars

24.2.3 Functions of contact chemoreceptors

Contact chemoreceptors are of primary importance in relation to feeding. In *Phormia*, for example, stimulation of the tarsal chemoreceptors with sugar leads to proboscis extension. This brings chemoreceptors on the labellum into contact with the food. Their stimulation causes the fly to spread its labellar lobes so the interpseudotracheal sensilla (see Fig. 2.6c) contact the food and sucking begins. A final monitoring occurs by the cibarial receptors. Continuous sensory input is necessary for continued feeding (Dethier, 1976). Essentially similar sequences occur in other insects and, for many, sugars are major phagostimulants.

Sometimes, a specific chemical is necessary to induce feeding. This is often true for insects with a relatively specific host range. It is also true, for many plant-feeding insects, that feeding is inhibited by many plant secondary compounds, especially those in non-host plants. These components stimulate deterrent cells or inhibit the activity of nutrient-specific cells.

Contact chemoreception is also important as an oviposition cue in some insects. For example, *Pieris* butterflies are induced to oviposit by glucosinolates. It is common for tarsal chemoreceptors to be important in regulating this behavior.

Insects use contact chemoreception in the perception of some pheromones. Trimethylheptatriacontane, a component of the surface wax of the female, acts as a sexual recognition pheromone for the tsetse fly, *Glossina morsitans*. Host-marking pheromones are also perceived by tarsal receptors (section 27.1.6.1).

Reviews: Chapman, 1995 – feeding; Renwick & Chew, 1994 – oviposition in Lepidoptera; Städler, 1992 – responses to plant secondary compounds; Städler, 1994 – oviposition

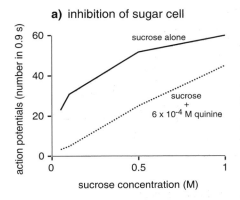

a) inhibition of sugar cell

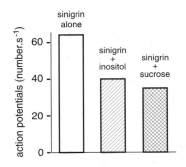

b) inhibition of deterrent cell

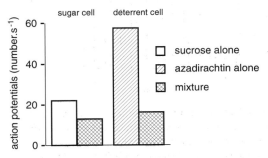

c) mutual inhibition

Fig. 24.12. Peripheral inhibition of contact chemoreceptors. **(a)** Inhibition by quinine (an alkaloid) of the activity of the cell responding to sucrose in a tarsal sensillum of *Phormia* (data from Dethier & Bowdan, 1989). **(b)** Inhibition of a deterrent cell response to sinigrin by inositol and sucrose in the medial styloconic sensillum of the caterpillar of *Trichoplusia*. Notice that, in this insect, sinigrin produces an aversive response and stimulates the deterrent cell [in *Pieris*, it is a phagostimulant and stimulates a specific glucosinolate-sensitive cell (Fig. 24.9c)](after Shields & Mitchell, 1995). **(c)** Mutual inhibition by sucrose and azadirachtin (a terpenoid which stimulates the deterrent cell). The left-hand column of each pair shows the response to the single compounds; the right-hand columns show the response of the same cells when sucrose and azadirachtin are mixed together. Each compound is at the same concentration in the mixture as when presented singly. Neurons in a tarsal sensillum of adult *Helicoverpa* (data from Blaney & Simmonds, 1990).

REFERENCES

Aihara, Y. & Shibuya, T. (1977).
Responses of single olfactory receptor
cells to sex pheromones in the tobacco
cutworm moth, *Spodoptera litura*.
Journal of Insect Physiology, 23,
779–83.

Akers, R.P. & Getz, W.M. (1993).
Response of olfactory sensory neurons
in honey bees to odorants and their
binary mixtures. *Journal of
Comparative Physiology*, 173 A, 169–85.

Almaas, T.J., Christensen, T.A. &
Mustaparta, H. (1991). Chemical
communication in heliothine moths, I.
Antennal receptor neurons encode
several features of intra- and interspecific odorants in the male corn
earworm moth *Helicoverpa zea*. *Journal
of Comparative Physiology*, 169 A,
249–58.

Altner, H. & Prillinger, L. (1980).
Ultrastructure of invertebrate chemo-,
thermo-, and hygroreceptors and its
functional significance. *International
Review of Cytology*, 67, 69–139.

Anderson, P., Hansson, B.S. & Löfqvist, J.
(1995). Plant-odour-specific receptor
neurones on the antennae of female and
male *Spodoptera littoralis*. *Physiological
Entomology*, 20, 189–98.

Ascoli-Christensen, A., Sutcliffe, J.F. &
Albert, P.J. (1990). Purinoceptors in
blood feeding behaviour in the stable
fly, *Stomoxys calcitrans*. *Physiological
Entomology*, 16, 145–52.

Ashworth, J.R. & Wall, R. (1994).
Responses of the sheep blowflies *Lucilia
sericata* and *L. cuprina* to odour and the
development of semiochemical baits.
Medical and Veterinary Entomology, 8,
303–9.

Avenet, P., Kinnamon. S.C. & Roper, S.D.
(1993). Peripheral transduction mecha-
nisms. In *Mechanisms of Taste
Transduction*, ed. S.A. Simon & S.D.
Roper, pp. 161–73. Boca Raton: CRC
Press.

Becker, A. & Peters, W. (1989).
Localization of sugar-binding sites in
contact chemosensilla of *Periplaneta
americana*. *Journal of Insect Physiology*,
35, 239–50.

Bernays, E.A. & Chapman, R.F. (1994).
*Host-Plant Selection by Phytophagous
Insects*. New York: Chapman & Hall.

Blaney, W.M. Schoonhoven, L.M. &
Simmonds, M.S.J. (1986). Sensitivity
variations in insect chemoreceptors; a
review. *Experientia*, 42, 13–9.

Blaney, W.M. & Simmonds, M.S.J.
(1990). A behavioural and electrophysiological study of the role of tarsal
chemoreceptors in feeding by adults of
Spodoptera, *Heliothis virescens* and
Helicoverpa armigera. *Journal of Insect
Physiology*, 36, 743–56.

Blight, M.M., Pickett, J.A., Wadhams,
L.J. & Woodcock, C.M. (1995).
Antennal perception of oilseed rape,
Brassica napus (Brassicaceae), volatiles
by the cabbage seed weevil
Ceutorhynchus assimilis (Coleoptera,
Curculionidae). *Journal of Chemical
Ecology*, 21, 1649–64.

Blom, F. (1978). Sensory activity and food
intake: a study of input–output relationships in two phytophagous insects.
Netherlands Journal of Zoology, 28,
277–340.

Bognor, F. (1992). Response properties of
CO_2-sensitive receptors in tsetse flies
(Diptera: *Glossina palpalis*).
Physiological Entomology, 17, 19–24.

Bowen, M.F. (1990). Post-diapause
sensory responsiveness in *Culex pipiens*.
Journal of Insect Physiology, 36, 923–9.

Chapman, R.F. (1995). Chemosensory
regulation of feeding. In *Regulatory
Mechanisms in Insect Feeding*, ed. R.F.
Chapman & G. de Boer, pp. 101–36.
New York: Chapman & Hall.

Colvin, J. & Gibson, G. (1992). Host-
seeking behavior and management of
tsetse. *Annual Review of Entomology*,
37, 21–40.

Davis, E.E. (1984a). Regulation of
sensitivity in the peripheral chemoreceptor systems for host-seeking
behaviour by a haemolymph-borne
factor in *Aedes aegypti*. *Journal of Insect
Physiology*, 36, 179–83.

Davis, E.E. (1984b). Development of
lactic acid-receptor sensitivity and
host-seeking behaviour in newly
emerged female *Aedes aegypti*. *Journal
of Insect Physiology*, 36, 211–5.

Dethier, V.G. (1976). *The Hungry Fly*.
Cambridge, Massachusetts: Harvard
University Press.

Dethier, V.G. (1987). Discriminative taste
inhibitors affecting insects. *Chemical
Senses*, 12, 251–63.

Dethier, V.G. & Bowdan, E. (1989). The
effect of alkaloids on sugar receptors
and the feeding behaviour of the
blowfly. *Physiological Entomology*, 14,
127–36.

Eisner, T. (1964). Catnip: its raison d'être.
Science, 146, 1318–20.

Frazier, J.L. (1992). How animals perceive
secondary plant compounds. In
*Herbivores. Their Interactions with
Secondary Plant Metabolites*, vol. II, ed.
G.A. Rosenthal & M.R. Berenbaum,
pp. 89–134. San Diego: Academic
Press.

Hanson, F.E. (1987). Chemoreception in
the fly: the search for the liverwurst
receptor. In *Perspectives in
Chemoreception and Behavior*, ed. R.F.
Chapman, E.A. Bernays & J.G.
Stoffolano, pp. 99–122. New York:
Springer-Verlag.

Kafka, W.A. (1971). Specificity of odor-
molecule interaction in single cells. In
Gustation and Olfaction, ed. G. Ohloff
and A.F. Thomas, pp. 61–72. London:
Academic Press.

Kanaujia, S. & Kaissling, K.E. (1985).
Interactions of pheromone with moth
antennae: adsorption, desorption and
transport. *Journal of Insect Physiology*,
31, 71–81.

Ljungberg, H., Anderson, P. & Hansson, B.S. (1993). Physiology and morphology of pheromone-specific sensilla on the antennae of male and female *Spodoptera littoralis* (Lepidoptera: Noctuidae). *Journal of Insect Physiology*, **39**, 253–60.

Ma, W.C. (1972). Dynamics of feeding responses in *Pieris brassicae* Linn as a function of chemosensory input: a behavioural, ultrastructural and electrophysiological study. *Mededelingen Landbouwhogeschool Wageningen*, **72–11**, 162 pp.

McCutchan, M.C. (1969). Responses of tarsal chemoreceptive hairs of the blowfly, *Phormia regina*. *Journal of Insect Physiology*, **15**, 2059–68.

Morita, H. & Shiraishi, A. S. (1985). Chemoreception physiology. In *Comprehensive Insect Physiology, Biochemistry and Pharmacology*, vol. **6**, ed. G.A. Kerkut & L.I. Gilbert, pp. 133–70. Oxford: Pergamon Press.

Murlis, J., Elkinton, J.S. & Cardé, R. (1992). Odor plumes and how insects use them. *Annual Review of Entomology*, **37**, 505–32.

Mustaparta, H. (1975). Responses of single olfactory cells in the pine weevil *Hylobius abietis* L. (Col.: Curculionidae). *Journal of Comparative Physiology*, **97**, 271–90.

Omand, E. & Dethier, V.G. (1969). An electrophysiological analysis of the action of carbohydrates in the sugar receptor of the blowfly. *Proceedings of the National Academy of Sciences of the United States of America*, **62**, 136–43.

Ozaki, M., Amakawa, T., Ozaki, K. & Tokunaga, F. (1993). Putative taste receptor molecules for sweetness. *Journal of General Physiology*, **102**, 201–16.

Ozaki, M., Morisaki, K., Idei, W., Ozaki, K. & Tokunaga, F. (1995). A putative lipophilic stimulant carrier protein commonly found in the taste and olfactory systems. A unique member of the pheromone-binding protein superfamily. *European Journal of Biochemistry*, **230**, 298–308.

Payne, T.L., Birch, M.C. & Kennedy, C.E.J. (1986). *Mechanisms in Insect Olfaction*. Oxford: Oxford University Press.

Petersen, S.C., Hanson, F.E. & Warthen, J.D. (1993). Deterrence coding by a larval *Manduca* chemosensory neurone mediating rejection of a non-host plant, *Canna generalis* L. *Physiological Entomology*, **18**, 285–95.

Renou, M. & Lucas, P. (1994). Sex pheromone reception in *Mamestra brassicae* L. (Lepidoptera): responses of olfactory receptor neurones to minor components of the pheromone blend. *Journal of Insect Physiology*, **40**, 75–85.

Renwick, J.A.A. & Chew, F.S. (1994). Oviposition behavior in Lepidoptera. *Advances in Insect Physiology*, **39**, 377–400.

Schoonhoven, L.M. (1969). Sensitivity changes in some insect chemoreceptors and their effect on food selection behaviour. *Koninklijke Nederlandse Akademie van Wetenschappen – Amsterdam* C, **72**, 491–8.

Schoonhoven, L.M. (1987). What makes a caterpillar eat? The sensory code underlying feeding behavior. In *Perspectives in Chemoreception and Behavior*, ed. R.F. Chapman, E.A. Bernays & J.G. Stoffolano, pp. 69–97. New York: Springer-Verlag.

Schoonhoven, L.M., Simmonds, M.S.J. & Blaney, W.M. (1991). Changes in the responsiveness of the maxillary styloconic sensilla of *Spodoptera littoralis* to inositol and sinigrin correlate with feeding behaviour during the final larval stadium. *Journal of Insect Physiology*, **37**, 261–8.

Selzer, R. (1981). The processing of complex food odor by antennal olfactory receptors of *Periplaneta americana*. *Journal of Comparative Physiology* A, **144**, 509–19.

Shields, V.D.C. (1994). Ultrastructure of the uniporous sensilla on the galea of larval *Mamestra configurata* (Walker) (Lepidoptera: Noctuidae). *Canadian Journal of Zoology*, **72**, 2016–31.

Shields, V.D.C. & Mitchell, B.K. (1995). The effect of phagostimulant mixtures on deterrent receptor(s) in two crucifer-feeding lepidopterous species. *Philosophical Transactions of the Royal Society of London* B, **347**, 459–64.

Simpson, S.J., James, S., Simmonds, M.S.J. & Blaney, W.M. (1991). Variation in chemosensitivity and the control of dietary selection behaviour in the locust. *Appetite*, **17**, 141–54.

Städler, E. (1992). Behavioral responses of insects to plant secondary compounds. In *Herbivores. Their Interactions with Secondary Plant Metabolites*, vol. II, ed. G.A. Rosenthal & M.R. Berenbaum, pp. 45–88. San Diego: Academic Press.

Städler, E. (1994). Oviposition behavior of insects influenced by chemoreceptors. In *Olfaction and Taste XI*, ed. K. Kurihara, N. Suzuki & H. Ogawa, pp. 821–6. Tokyo: Springer-Verlag.

Städler, E., Ernst, B., Hurter, J. & Boller, E. (1994). Tarsal contact chemoreceptor for the host marking pheromone of the cherry fruit fly, *Rhagoletis cerasi*: responses to natural and synthetic compounds. *Physiological Entomology*, **19**, 139–51.

Städler, E., Renwick, J.A.A., Radke, C.D. & Sachdev-Gupta, K. (1995). Tarsal contact chemoreceptor response to glucosinolates and cardenolides mediating oviposition in *Pieris rapae*. *Physiological Entomology*, **20**, 175–87.

Stange, G. (1992). High resolution measurement of atmospheric carbon dioxide concentration by changes in the labial palp organ of the moth *Heliothis armigera* (Lepidoptera: Noctuidae). *Journal of Comparative Physiology* A, **171**, 317–24.

Steinbrecht, R.A. (1973). Der Feinbau olfaktorischer Sensillen des Seidenspinners (Insecta, Lepidoptera). Rezeptorfortsätze und reizleitender Apparat. *Zeitschrift für Zellforschung und mikroskopische Anatomie*, **139**, 533–65.

Steinbrecht, R.A. (1980). Cryofixation without cryoprotectants, freeze substitution and freeze etching of an insect olfactory receptor. *Tissue & Cell*, **12**, 73–100.

Steinbrecht, R.A. (1987). Functional morphology of pheromone-sensitive sensilla. In *Pheromone Biochemistry*, ed. G.D. Prestwich & G.J. Blomquist, pp. 353–84. Orlando: Academic Press.

Steinbrecht, R.A., Laue, M. & Ziegelberger, G. (1995). Immunolocalization of pheromone-binding protein and general odorant-binding protein in olfactory sensilla of the silk moths *Antheraea* and *Bombyx*. *Cell and Tissue Research*, **282**, 203–17.

Steinbrecht, R.A., Ozaki, M. & Ziegelberger, G. (1992). Immunocytochemical localization of pheromone-binding protein in moth antennae. *Cell and Tissue Research*, **270**, 287–302.

Stocker, R.F. (1994). The organization of the chemosensory system in *Drosophila melanogaster*: a review. *Cell and Tissue Research*, **275**, 3–26.

Todd, J.L., Haynes, K.F. & Baker, T.C. (1992). Antennal neurones specific for redundant pheromone components in normal and mutant *Trichoplusia ni* males. *Physiological Entomology*, **17**, 183–92.

Topazzini, A., Mazza, M. & Pelosi, P. (1990). Electroantennogram responses of five Lepidoptera species to 26 general odourants. *Journal of Insect Physiology*, **36**, 619–24.

van Loon, J.J.A. (1990). Chemoreception of phenolic acids and flavonoids in larvae of two species of *Pieris*. *Journal of Comparative Physiology* A, **166**, 889–99.

van Loon, J.J.A. & van Eeuwijk, F.A. (1989). Chemoreception of amino acids in larvae of two species of *Pieris*. *Physiological Entomology*, **14**, 459–69.

Visser, J.H. (1986). Host odor perception in phytophagous insects. *Annual Review of Entomology*, **31**, 121–44.

Vogt, R.G. (1995). Molecular genetics of moth olfaction: a model for cellular identity and temporal assembly of the nervous system. In *Molecular Model Systems in the Lepidoptera*, ed. M.R. Goldsmith & A.S. Wilkins, pp. 341–67. Cambridge: Cambridge University Press.

Vogt, R.G., Rybczynski, R. & Lerner, M.R. (1990). The biochemistry of odorant reception and transduction. In *Chemosensory Information Processing*, ed. D. Schild, pp. 33–76. Berlin: Springer Verlag.

Wall, R. & Langley, P.A. (1993). The mating behaviour of tsetse flies (*Glossina*): a review. *Physiological Entomology*, **18**, 211–8.

White, P.R. & Chapman, R.F. (1990). Olfactory sensitivity of gomphocerine grasshoppers to the odours of host and non-host plants. *Entomologia Experimentalis et Applicata*, **55**, 205–12.

Zacharuk, R.Y. (1985). Antennae and sensilla. In *Comprehensive Insect Physiology, Biochemistry and Pharmacology*, vol. **6**, ed. G.A. Kerkut & L.J. Gilbert, pp. 1–69. Oxford: Pergamon Press.

Zacharuk, R.Y. & Shields, V.D. (1991). Sensilla of immature insects. *Annual Review of Entomology*, **36**, 331–54.

Ziegelberger, G. (1995). Redox-shift of the pheromone-binding protein in the silkmoth *Antheraea polyphemus*. *European Journal of Biochemistry*, **232**, 706–11.

25 Visual signals: color and light production

25.1 THE NATURE OF COLOR

Color is produced from white light when some of the wavelengths are eliminated, usually by absorption, and the remainder are reflected or transmitted. The wavelengths of the reflected or transmitted component determine the color observed (Fig. 25.1). If all wavelengths are reflected equally the reflecting surface appears white; if all are absorbed the appearance is black.

Differential reflection of light to produce color occurs in one of two ways: the physical nature of the surface may be such that only certain wavelengths are reflected, the remainder being transmitted, or pigments may be present which, as a result of their molecular structure, absorb certain wavelengths and reflect the remainder. Colors produced by these methods are known, respectively, as physical (or structural) and pigmentary colors. In either case, the color observed is dependent on the wavelengths of light in which the insect is viewed, usually sunlight or white light. Insects do not fluoresce or, with a few exceptions, luminesce (section 25.7).

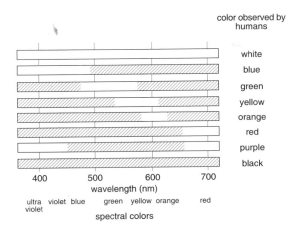

Fig. 25.1. Color production by absorption of certain wavelengths from white light. Eliminated wavelengths are hatched, reflected wavelengths clear. Note that humans do not normally see ultraviolet light, but many insects are sensitive to it (after Fox, 1953).

The colors of insects are usually produced by structures or pigments in the cuticle or epidermis, but, if the cuticle is transparent, the hemolymph, fat body, or even gut contents may contribute to the insect's color.

25.2 PHYSICAL COLORS

Surface structures are mainly responsible for the production of whites, blues and iridescent colors. Such colors may be produced by scattering, interference or diffraction.
Reviews: Hinton, 1976; Neville, 1975

25.2.1 Scattering

Light may be scattered, that is, reflected in all directions, by irregularities of a surface or by granules just beneath it. If the irregularities or granules are large relative to the wavelength of light, greater than about 700 nm, all the light is reflected and the surface appears white. Most whites in insects are produced in this way, although some white pigments also occur. Matt whites are produced by an even scattering of the light in all directions and, in some Lepidoptera, such as Pieridae, this results from deep longitudinal corrugations and fine, unordered striations on the surface of the scales. Pearly whites, such as those of *Argynnis* (Lepidoptera), are produced by scattering from a number of thin, overlapping lamellae separated by airspaces. In butterflies the lamellae are the upper and lower laminae of overlapping scales.

If the granules near the surface are small with dimensions similar to the wavelengths of blue light, the short, blue waves are reflected while the longer wavelengths are not. This type of scattering is called Tyndall scattering and produces blue or green. It depends for its effect on an absorbing layer of dark pigment beneath the fine granules. In the absence of this layer the blue is masked by light reflected from the background. Tyndall blues are uncommon in insects, but the blue of dragonflies is produced by scattering from small granules in distal regions of the epidermal cells, the dark background being provided by larger granules containing brown-violet ommochrome

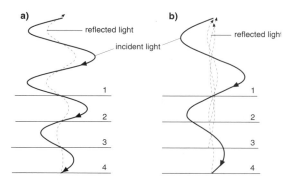

Fig. 25.2. Interference colors. Diagrammatic representation of the production of color. At each of the reflecting surfaces (numbered 1–4), some light is reflected while the remainder is transmitted at lower intensity, as indicated by the reduced amplitude. Two wavelengths are shown, differing in their relationship to the spacing between the surfaces (after Richards, 1951). (a) Spacing between reflecting surfaces is half the wavelength of the incident light. Reflections from the surfaces reinforce each other so that a strong reflection of this wavelength occurs. (b) Wavelength of the incident light bears no simple relationship to the spacing between surfaces. Light reflected at successive surfaces is out of phase; interference occurs so that reflection of this wavelength is reduced or eliminated. Since, for any given spacing between the reflecting layers, most reflected wavelengths are out of phase, the reflected light will consist only of those wavelengths depicted in (a).

which are deeper in the epidermis. Similar granules are present in the epidermis of the grasshopper, *Kosciuscola*, which is blue during the daytime (see below). In this insect, the small granules are about 170 nm in diameter, and the larger ones about 1 μm.

25.2.2 Interference

Interference colors result from the reflection of light from a series of superimposed surfaces separated by distances comparable with the wavelengths of light. As a result of this spacing, some wavelengths reflected from successive surfaces are in phase and are therefore reinforced, while others are out of phase and are cancelled out (Fig. 25.2). The net result is that only certain wavelengths are reflected and the surface appears colored.

Interference colors are common in adult Lepidoptera where the layers producing interference are formed by modifications of the scales. Each of the scales of the *Morpho* butterfly which produce its bright blue color

consists of a flat basal plate carrying a large number of vertical vanes running parallel with the length of the scale (Fig. 25.3a). Each vane is supported by a series of vertical and obliquely horizontal mullions (thickenings). Collectively, the horizontal mullions of adjacent vanes form a series of reflecting surfaces which are spaced such that a blue color is produced by interference. Similar effects are produced by a variety of scale modifications in different blue and green butterflies (Ghiradella, 1989).

Interference colors in other insects are produced by reflection at the interfaces of layers in the cuticle which differ in refractive index. The refractive indices of the alternating layers in the pupa of the danaid butterfly, *Euploea*, are 1.58 and 1.37. In jewel beetles (Buprestidae) and tiger beetles (Cicindellidae), these layers are in the exocuticle, but in tortoise beetles (Cassidinae) and some butterfly pupae they are in the endocuticle.

In some scarab beetles, the reflecting surfaces are layers of chitin/protein microfibrils with a common orientation in the transparent exocuticle. Because their orientation changes progressively in successive layers, forming a helix (section 16.2.1.1), any given orientation recurs at intervals and, if the intervals are within the range of wavelengths of light, interference colors are produced. The light reflected by microfibrils oriented in one direction is polarized, but because the orientation changes with depth in the cuticle, the plane of polarization also changes.

Whatever the nature or position of the reflecting surfaces, the wavelength of the reflected light depends on their spacing. Viewing the surface from a more oblique angle is equivalent to increasing the distance between successive surfaces so that the color changes in a definite sequence (known as Newton's series) as the angle of viewing is altered. A change in color with the angle of viewing is called iridescence and is a characteristic of interference colors.

A constant spacing between the reflecting surfaces produces relatively pure colors, as in the scales of *Morpho*. Spectral purity is enhanced when the reflecting layers are backed by a layer of black pigment which absorbs any transmitted light. By contrast, impure colors are produced if the spacing changes systematically with increasing distance from the surface of the cuticle. For example, in the pupa of *Euploea*, the thickness of the paired layers changes systematically through the depth of the endocuticle (Fig. 25.3b). This produces interference over a wide range of wavelengths and the pupae appear

a) *Morpho*

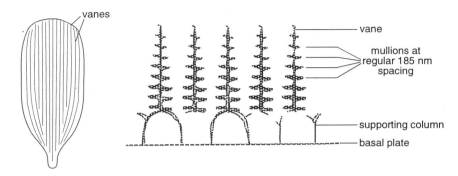

b) *Euploea*

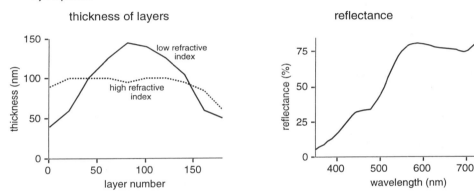

Fig. 25.3. Interference color production in different insects. (**a**) *Morpho* wing scale. Left: whole scale showing vanes running parallel with long axis; Right: cross-section of part of scale. The horizontal mullions lie parallel with each other with a regular spacing of 185 nm. The array of mullions across the scale forms a series of reflecting surfaces which reflect blue light. Notice that the supporting columns are irregularly arranged so only some are in the plane of the section (based on Smith, 1970). (**b**) *Euploea* pupal cuticle. Left: variation in the thickness of the alternating layers of high and low refractive index in the cuticle. The paired layers begin (layer 1) about 10 μm below the surface of the cuticle. Right: reflectance of light of different wavelengths from the abdominal sclerites. This produces a metallic golden color as seen by the human eye (based on Steinbrecht *et al.*, 1985).

golden to human eyes (Steinbrecht *et al.*, 1985). Bronze and silver interference colors are produced in the same way.

The wavelength produced by interference also depends on the refractive indices of the materials forming the layers. The horizontal mullions on *Morpho* wing scales are uniformly about 185 nm apart, producing a blue interference color. The paired layers in the exocuticle of the scarab beetle, *Heterorrhina*, are of similar thickness, but the interference color produced is green due to

the different refractive indices of the paired cuticular layers compared with the air/cuticle interfaces in *Morpho*.

The brightness of the reflected color increases with the number of reflecting surfaces. In *Morpho* species, there are only eight to 12 horizontal mullions, and similar numbers of layers are present in the cuticles of buprestid beetles. Only a small percentage of light, less than 40%, is reflected. By contrast, the metallic gold or silver cuticles of pupae of danaid butterflies have over 200 reflecting layers and almost 80% of the incident light over a broad band of wavelengths is reflected (Fig. 25.3b). Amongst scarab beetles, reflectivity is enhanced by large amounts of uric acid in the reflecting layers.

Apart from colors of the human visible spectrum, ultraviolet is also produced as an interference color. This occurs in the white butterflies, Pieridae. In *Eurema*, for example, certain scales possess reflecting lamellae about 55 nm thick separated by air gaps of about 80 nm. This system reflects light with a peak wavelength of 350 nm.

Interference is responsible for the iridescence of the membranous wings of many different insects.

25.2.3 Diffraction
When a ray of light strikes the edge of a groove or ridge, different wavelengths are bent to varying degrees and white light is split into its component spectral colors. If there is a series of parallel grooves or ridges, separated by about the wavelength of light, the wavelengths reflected from each line interfere with each other. In a particular direction, light of a given wavelength is reinforced, while other wavelengths are cancelled out. The conserved wavelength varies with the direction in which the light is reflected. This is known as diffraction and a series of grooves or ridges producing such an effect is called a diffraction grating.

Diffraction is responsible for the iridescence of some beetles. It occurs in at least eight beetle families most of which have diffraction gratings on the elytra, but they are on the abdomen of Staphylinidae where the elytra are short. The spacings between the lines of the gratings vary from 0.8 to 3.3 μm. Scarabaeidae have rows of microtrichia along the gratings and their effect is to make the insects iridescent when viewed from one direction, but not from the opposite direction. The rib spacing of some lepidopteran scales is appropriate to produce diffraction colors, but their irregularity produces an overlap of spectra and so white light results. Diffraction colors are not produced in dim light and beetles which iridesce due to diffraction in bright light look black or brown at low light intensities.

25.3 PIGMENTARY COLORS

Pigments appear colored because they reflect certain wavelengths of light; the remainder are absorbed and their energy is dissipated as heat. Which particular wavelengths are absorbed depends on the molecular structure of the compound. Particularly important in the production of color are the number and arrangement of double bonds, $C=C$, $C=O$, $C=N$ and $N=N$. Particular functional groups are also important. The $-NH_2$ and $-Cl$ radicals, for example, shift the absorptive region of a compound so that it tends to absorb longer wavelengths. The color-producing molecule, known as a chromophore, is often conjugated with a protein molecule, forming a chromoprotein.

Insects are able to synthesize most of their pigments, but not flavonoids or carotenoids which are, consequently, acquired in the diet. The sources of some other pigments, found only in a few insects, are unknown.
Review: Kayser, 1985

25.3.1 Pigments that are synthesized

Brown and black of cuticle   The black or brown of hardened cuticle often results from sclerotization. Quinone tanning produces dark cuticle, but β–sclerotization does not (section 16.5.3). However, cuticular hardening and darkening are not necessarily tightly linked. This is illustrated by albino strains of the locust, *Schistocerca*, which have a hard, but colorless cuticle.

Dark cuticle is also attributed to the presence of the pigment melanin, a nitrogen-containing compound which has been demonstrated in the cuticles of Blattodea, Diptera, adult and some larval Lepidoptera, and in Coleoptera. It is typically present as granules in the exocuticle. Insect melanin is a polymer of indole derivatives of tyrosine. Like sclerotization, melanization involves the production of DOPA and dopaquinone (Fig. 25.4).

Black pigments in the pupal cuticle of some Lepidoptera are not granular, and they are probably not melanin.

Pterins   Pterins are nitrogen-containing compounds, all having the same basic structure, but differing in the radicals attached to this nucleus (Fig. 25.5). They are synthesized from the purine guanosine triphosphate. Not all pterins appear colored. Some are important metabolically as cofactors of enzymes concerned in growth and differentiation and may act as controlling agents in these processes. They often occur together with pigments of another group, the ommochromes, because they are cofactors of the enzymes involved in ommochrome synthesis. The vitamin folic acid also contains a pterin.

As pigments, they are colored white (for example,

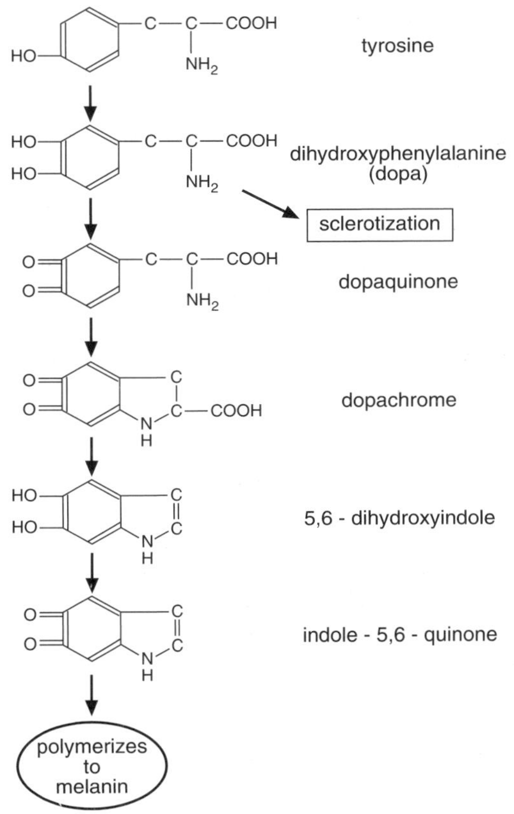

Fig. 25.4. Melanin synthesis. Note that dopa is also a key intermediate in cuticular sclerotization (see Fig. 16.20) and, because of this, melanization and sclerotization may sometimes be linked together.

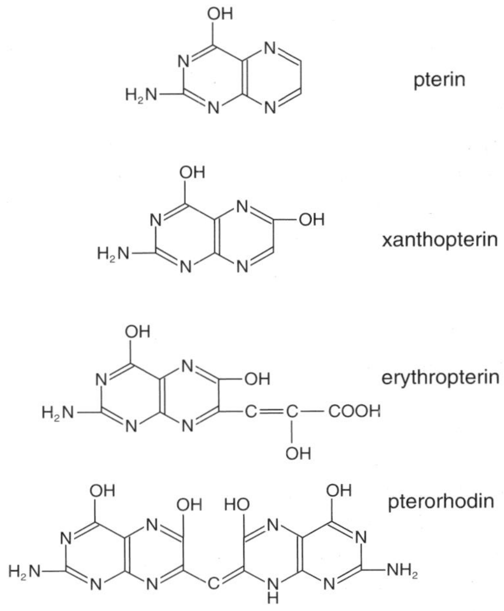

Fig. 25.5. Pterins. Pterin is the basic structure from which other pterins are derived. Xanthopterin is a yellow pigment present in many insect orders. Erythropterin is a red pigment found in Hemiptera and Lepidoptera. Pterorhodin is a red pigment occurring in some Lepidoptera. Notice that the pterorhodin molecule is essentially two xanthopterin molecules, one a mirror image of the other.

leucopterin), yellow (e.g., xanthopterin) and red (e.g., erythropterin). They are important pigments in lepidopteran scales where they have a crystalline form. Leucopterin and xanthopterin are common in the wings of Pieridae, where they supplement the structural white. The yellow of the brimstone butterfly, *Gonepteryx*, is due to chrysopterin, the brighter color of the male resulting from a higher pigment concentration than is present in the female, while the red of the orange-tip butterfly, *Anthocharis*, is due to erythropterin. The yellows of Hymenoptera are produced by granules of pterins in the epidermis.

The pterins are also abundant in compound eyes, occurring with ommochromes in the screening pigment cells of the ommatidia. In this situation, they are sometimes components of granules, but are often in solution.

Several different pterins are present, not all of them colored. Their functions in the eyes are not clear. They accumulate with age in the eyes of higher Diptera, and, as they are products of purine degradation, they may provide a means of storage excretion. The progressive accumulation provides a means of aging these insects (Wall, Langley & Morgan, 1991). However, no comparable accumulation occurs in the eyes of the moth, *Pectinophora*, although changes do occur in the pterins (Noble & Walker, 1990).

Ommochromes The ommochromes are a group of pigments derived from the amino acid tryptophan via kynurenine and 3-hydroxykynurenine. Oxidative condensation of 3-hydroxykynurenine gives rise to the ommochromes (Fig. 25.6). In larval *Drosophila*, kynurenine production takes place primarily in the fat body, but its conversion to 3-hydroxykynurenine occurs in the Malpighian tubules where it is stored. At metamorphosis, the 3-hydroxykynurenine is transported to the eyes where the

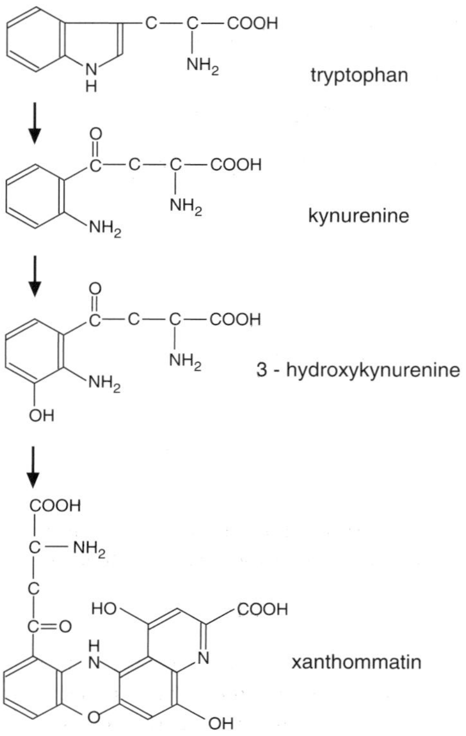

Fig. 25.6. Ommochromes. Synthesis of xanthommatin from tryptophan. Synthesis of other ommochromes involves the same pathway, differing in the final steps.

ommochromes are formed. In the adult fly, the whole process can take place in the cells of the eye although this is normally supplemented by kynurenine and 3-hydroxy-kynurenine synthesized elsewhere. The scale-forming cells on butterfly wings also have the capacity to synthesize ommochromes from tryptophan or from 3-hydroxy-kynurenine (see Fig. 25.12). Transport of ommochromes in the hemolymph is achieved by specialized binding proteins (Martel & Law, 1992).

Ommochrome production is the only way in which insects can remove tryptophan which is toxic at high concentrations such as may occur at times of high protein turnover. A transitory increase in tryptophan occurs at metamorphosis in holometabolous insects, often followed by the production of ommochromes. In Lepidoptera ommochromes are accumulated in the meconium, the accumulated waste products of the pupal period which are voided immediately following eclosion. They are responsible for its characteristic red/brown coloration.

Accumulation of ommochromes in the integument causes the larva of the puss moth, *Cerura*, to turn red just before pupation. Some of the ommochrome produced at the time of pupation contributes to the screening pigment in the eyes of the adult. Red fecal pellets containing ommochromes are produced by locusts during molting or starvation. At these times, excess tryptophan is likely to be liberated from proteins that are broken down during structural rearrangement or that are used for energy production.

Ommochromes are yellow, red and brown pigments usually occurring in granules coupled with proteins. The granules also contain accumulations of calcium. Xanthommatin is the most widely distributed ommochrome and is usually present wherever ommochromes are found.

Ommochromes, principally xanthommatin, are widely distributed in insects as screening pigments in the accessory cells of the eyes, usually associated with pterins. They are also present in the retinula cells. Yellow, red and brown body colors are produced by ommochromes in the epidermis. The pink of immature adult *Schistocerca*, for example, is due to a mixture of ommochromes. Red Odonata and probably also the reds and browns of nymphalid butterflies are due to ommochromes, while in blue Odonata a dark brown ommochrome provides the background for the production of Tyndall blue. Epidermal ommochromes sometimes directly underlie cuticular melanin, and in these cases they do not contribute to the insect's color.

Review: Summers, Howells & Pyliotis, 1982 – eye pigments

Tetrapyrroles There are two major classes of tetrapyrroles: the porphyrins, in which the pyrroles form a ring, and the bilins which have a linear arrangement of the pyrroles (Fig. 25.7).

A porphyrin having an atom of iron in its center is called a heme molecule and this forms the basis of two important classes of compounds, the cytochromes and the hemoglobins. In each case the heme molecule is linked to a protein. All insects are able to synthesize cytochromes which are essential in respiration, the different cytochromes differing in the forms of their heme groupings. Normally they are only present in small amounts and so produce no color, but where they are present in high concentrations, as in flight muscle, they produce a reddish-brown color.

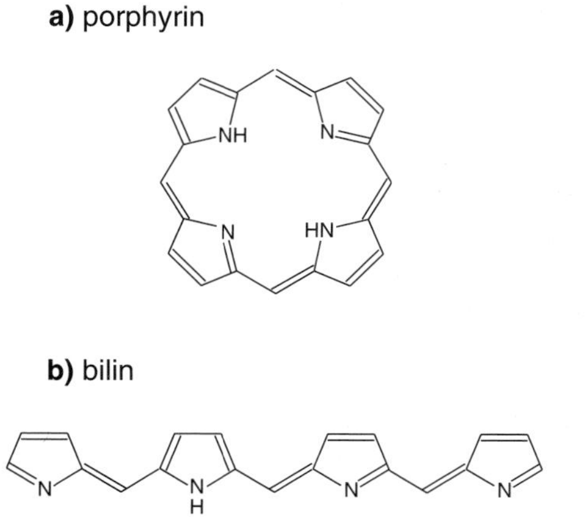

a) porphyrin

b) bilin

Fig. 25.7. Tetrapyrroles: **(a)** porphyrin skeleton; **(b)** bilin skeleton.

a) anthraquinone

b) aphin

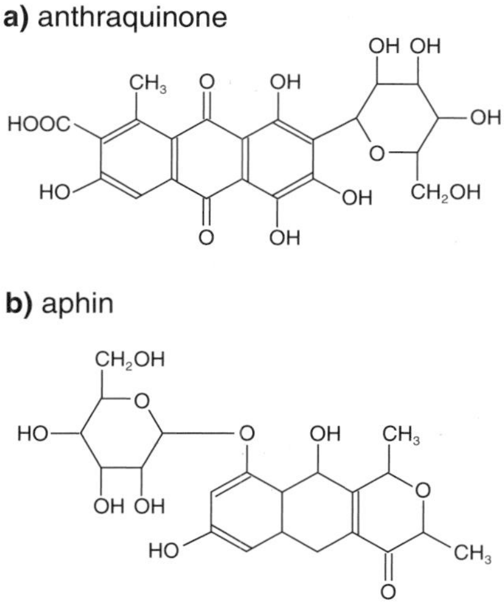

Fig. 25.8. Quinone pigments: **(a)** an anthraquinone, carminic acid, from *Dactylopius cacti*; **(b)** an aphin, neriaphin, from *Aphis nerei*.

Only a few insects, usually living in conditions subject to low oxygen tensions, contain hemoglobin and these are colored red by the pigment showing through the integument. In *Chironomus* (Diptera) larvae the hemoglobin is in solution in the blood, while in the larva of *Gasterophilus* (Diptera) it is in hemoglobin cells. Hemoglobin serves a respiratory function, but perhaps also serves as a protein store, and enables the aquatic hemipteran, *Anisops*, to regulate its buoyancy (section 17.9).

The bilins are usually associated with proteins to form blue chromoproteins. Biliverdin occurs in many hemimetabolous insects, but is also found in Neuroptera and some Lepidoptera although the latter usually contain other types of bilins. Associated with a yellow carotenoid, these pigments are responsible for the greens of many insects. Sometimes the pigments themselves are green. In *Chironomus*, bilins derived from the hemoglobin of the larva accumulate in the fat of the adult and impart a green color to the newly emerged fly. In *Rhodnius*, the pericardial cells become green due to the accumulation of bilins derived from ingested hemoglobin.

Papiliochromes Papiliochromes are yellow and red/brown pigments known only from the swallowtail butterflies, Papilionidae. Papiliochrome II is pale yellow and is formed from one molecule of kynurenine, derived from tryptophan, and one molecule of β-alanine (Ishizaki & Umebachi, 1990).

Quinone pigments The quinone pigments of insects fall into two categories: anthraquinones and aphins (Fig. 25.8). Both occur as pigments only amongst Homoptera, the former in coccids (Coccoidea), and the latter in aphids. Their origins in the insects are not known. They may be derived from plants, synthesized by the symbiotic organisms that are always present in these insects, or, perhaps, manufactured by the insects themselves. Anthraquinones are widespread in plants, and some beetles sequester them, although they do not form pigments in these insects. Aphins are known only from aphids.

Anthraquinones are formed from the condensation of three benzene rings. In the coccids they give the tissues a red, or sometimes yellow, coloration. The best known is cochineal from *Dactylopius cacti*. The purified pigment is called carminic acid (Fig. 25.8a). It is present in globules in the eggs and fat body of the female, constituting up to 50% of the body weight. The larva contains relatively little pigment.

Aphins are quinone pigments formed from the condensation of three, in the monomeric forms, or seven benzene rings, in dimeric forms. They are found in the blood of aphids, sometimes in high concentration, and impart a purple or black color to the whole insect. Neriaphin is a monomeric form (Fig. 25.8b).

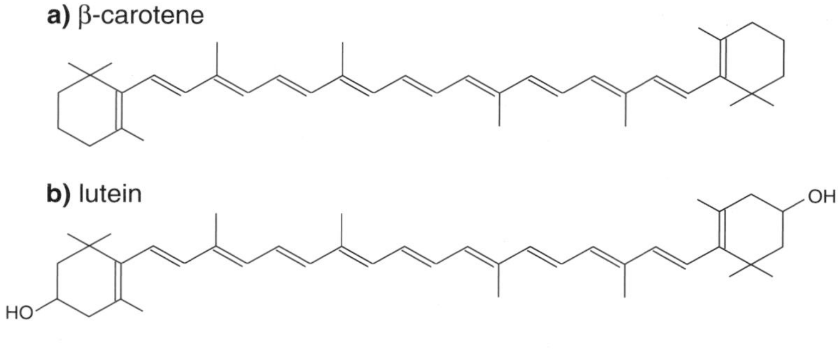

a) β-carotene

Fig. 25.9. Carotenoids: **(a)** a carotene, β–carotene; **(b)** a xanthophyll, lutein.

b) lutein

25.3.2 Pigments obtained from the food

Carotenoids   Carotenoids are a major group of pigments that are lipid soluble and contain no nitrogen. They are built up from two diterpenoid units joined tail to tail (Fig. 25.9). In nearly all insect carotenoids, the central chain contains 22 carbon atoms with nine double bonds, and each of the end groups contains nine carbon atoms. There are two major groups of carotenoids: the carotenes, and their oxidized derivatives, the xanthophylls.

Yellow, orange and red are commonly produced by carotenoids, the color depending largely on whether or not the terminal groups are closed rings and on their degree of unsaturation. If the carotenoid is bound to a protein, the color may be altered, sometimes even resulting in a blue pigment. Insects cannot synthesize carotenoids and consequently must obtain them from the diet. Uptake from the food is, at least to some extent, selective. Orthopteroids preferentially absorb carotenes, while lepidopterans favor xanthophylls. Some postingestive modification of the carotenoids may occur.

Carotenoids can occur in many different tissues and in all stages of development. A number of structurally different carotenoids may be present in one insect.

The possible metabolic functions of carotenoids in insects are not well understood. In other organisms, they protect cells from damage due to photo–oxidation by light, but their importance to insects in this regard is not known. They are also the source of retinal in the insect's visual pigments. Apart from producing the reds and yellows of many insects, in combination with a blue pigment, often a bilin, they produce green. Green produced by a carotenoid with a bilin is sometimes known as insectoverdin.

Review: Feltwell, 1978

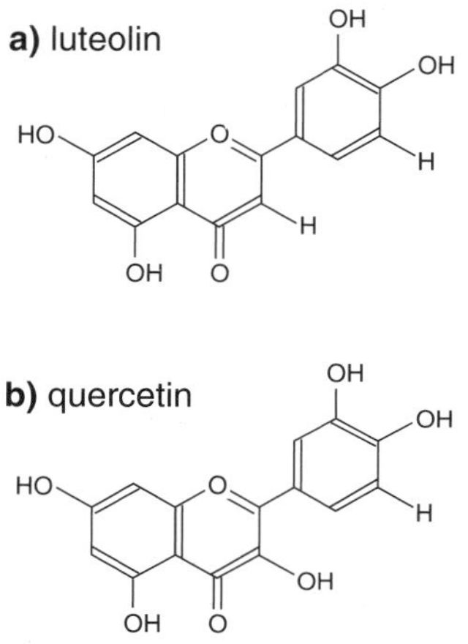

a) luteolin

b) quercetin

Fig. 25.10. Flavonoids: **(a)** luteolin; **(b)** quercetin.

Flavonoids   Flavonoids are heterocyclic compounds commonly found in plants (Fig. 25.10). In insects they are mainly found amongst the butterflies and are common in Papilionidae, Satyridae and Lycaenidae as cream or yellow pigments. At least in some species, flavonoids are present in all developmental stages, including the egg, indicating that the flavonoids are stored in internal tissues of the adult as well as in the scales. These insects acquire the flavonoids with their food, so the flavonoids present in their bodies reflect what is present in the hostplants. Some modification of structure does occur postingestively so that the flavonoids in the insect are not exactly the same as those in the host. Perhaps these changes are produced by the insects themselves, but it is possible that their gut flora is responsible.

Review: Harborne, 1991

25.4 THE COLORS OF INSECTS

Insect colors result from a variety of structures and pigments. Very often several pigments are present together, and the observed color depends on the relative abundance and positions of the pigments (see Fig. 25.14). The position of color-producing molecules relative to other structures is also important, and this may change, resulting in changes in coloration (section 25.5.1).

Table 25.1 lists the sources of color in some insect groups. In general, only the major sources of color are listed.

25.4.1 Color patterns of butterflies

In general, insects do not have a uniform coloring, but have specific and often finely detailed patterns. These are most obvious and completely studied in the Lepidoptera where the wing pattern is produced by thousands of small overlapping scales. The development of patterns is controlled by a series of organizing centers from which, it is presumed, morphogens diffuse outwards creating a concentration gradient. Such morphogens govern the development of pigment in each scale-forming cell and different concentrations induce different pigments. As diffusion extends uniformly outwards in all directions round an organization center, a symmetrical pattern is produced, but organization centers are often sufficiently close together that the patterns they produce grade into each other. In this way, a band of color rather than a series of separate rings can be produced. On the forewings of butterflies (Papilionoidea) there are three rows of organizing centers that produce three sets of paired bands (known as symmetry systems) from front to back across the wing (Fig. 25.11). Each band has a separate organizing center in each of the wing cells [the areas defined by veins (section 9.1.1.4)]. Thus the overall wing pattern consists of three rows of markings which tend to be similar within rows, but different between them. This basic arrangement is referred to as the nymphalid groundplan. The specifics of the pattern vary between species. A series of less complex groundplans occurs amongst the many families of moths.

In general, it appears that each scale contains only a single pigment, although examples of scales with two pigments are known. The pigment produced by each scale-forming cell depends on synthesis within the cell (Fig. 25.12); although some intermediates may be manufactured elsewhere, scale color is not determined by selective uptake

of presynthesized pigment (Koch, 1991). It seems, therefore, that the morphogens moving between epidermal cells regulate the synthetic machinery within each cell.
Reviews: Nijhout, 1991, 1994

25.5 COLOR CHANGE

Color changes are of two kinds: short-term reversible changes which do not involve the production of new pigment, and long-term changes which result from the formation of new pigments and are not usually reversible. Short-term reversible changes are called physiological changes. Color changes involving the metabolism of pigments are called morphological color changes.
Review: Fuzeau-Braesch, 1985

25.5.1 Physiological color change

Physiological color change is unusual in insects. It may occur where colors are produced physically as a result of changes in the spacing between reflecting layers, or, where color is produced by pigments, as a result of pigment movements.

Tortoise beetles change color when they are disturbed. Normally, the beetles are brassy yellow or green, but can change to violet and finally brown–orange in less than one minute. The change is due to a reduction in the spacing of the lamellae producing the normal interference colors of the insect, following a reduction in the state of hydration of the cuticle. Rehydration restores the original color.

The elytra of the hercules beetle, *Dynastes*, are normally yellow due to a layer of spongy, yellow-colored cuticle beneath a transparent layer of cuticle. If the spongy cuticle becomes filled with liquid, as it does at high humidities, light is no longer reflected by it but passes through and is absorbed by black cuticle underneath, making the insect black. In the field, these changes probably occur daily, so that the insect tends to be yellow in the daytime when it is feeding among leaves and dark at night, making it less conspicuous (Hinton, 1976).

Physiological color changes involving pigment movements are known to occur in the stick insect, *Carausius*, in the grasshopper, *Kosciuscola*, and in a number of blue damselflies (Zygoptera). All these insects become black at night due to the movement of dark pigment granules to a more superficial position in the epidermal cells (Fig. 25.13). *Kosciuscola* is blue during the daytime as a result of Tyndall scattering from small granules, less than 0.2 μm in

Table 25.1. *Some of the principal causes of colors in different insects*

Color	Taxon	Cause of color
Black*	Homoptera, Aphididae	Aphins
	Coleoptera	Melanin
	Diptera	Melanin
	Lepidoptera (larvae)	Melanin
Red	Odonata	Ommochromes
	Hemiptera, Heteroptera	Pterins and carotenoids
	Hemiptera, Coccoidea	Anthraquinones
	Coleoptera, Coccinellidae	Carotenoids
	Diptera, Chironomidae (larvae)	Porphyrin
	Lepidoptera	Ommochromes
Brown*	Lepidoptera	Ommochromes
	many orders – eye colors	Ommochromes + pterins
Orange	Hemiptera, Heteroptera	Pterins
Yellow	Orthoptera, Acrididae	Carotenoids, flavonoids
	Hymenoptera	Pterins
	Lepidoptera, Papilionidae	Papiliochromes
Brassy yellow	Lepidoptera	Interference color
Bronze	Coleoptera, Scarabaeidae	Interference color
Gold	Coleoptera, Cassidinae	Interference color
	Lepidoptera, Danaidae (pupae)	Interference color
Green	Orthoptera	Insectoverdin (carotenoid + bilin)
	Lepidoptera (caterpillars)	Insectoverdin
	Lepidoptera, Zygaenidae	Interference color
	Diptera, Chironomidae (adults)	Bilin
Blue	Odonata	Tyndall scattering
	Lepidoptera	Interference color
	Orthoptera, Acrididae	Carotenoids
Ultraviolet	Lepidoptera, Pieridae	Interference color
White	Hemiptera, Heteroptera	Uric acid
	Lepidoptera, Pieridae	Scattering + pterins
	Lepidoptera, Satyridae	Flavonoids
Silver	Lepidoptera, Danaidae (pupae)	Interference color
Iridescence	Coleoptera	Diffraction
	Coleoptera, Scarabaeidae	Interference colors

Note:

* Darkening may result from sclerotization.

a) organizing centers

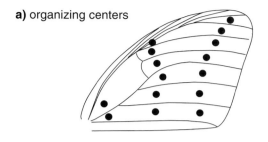

b) symmetry systems

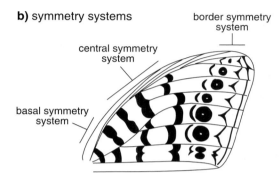

Fig. 25.11. Color patterning in the forewing of a butterfly. (a) Positions of possible organization centers shown by dots (hypothetical). (b) The nymphalid groundplan. A basic pattern of symmetry system that would arise from an arrangement of organization centers shown in (a) (after Nijhout, 1994).

diameter, composed mainly of white leucopterin and uric acid. At night, the blue is masked by the dispersal of larger pigment granules amongst the small reflecting granules which also disperse through the cell (Filshie, Day & Mercer 1975). Similar masking of Tyndall blue occurs in a number of damselflies (Veron, 1976), and, in brown individuals of the stick insect, *Carausius*, ommochrome granules occupy a superficial position at night, causing the insect to become darker, while in the daytime they occupy a proximal position in the epidermal cells making the insects paler. Green specimens of *Carausius* do not change color because they lack ommochromes. The granules move along the paths of microtubules which may be responsible for the movements (Berthold, 1980). Similar microtubules are present in the epidermal cells of *Kosciuscola*.

The color change in *Kosciuscola* and damselflies is temperature dependent. The insects are always black below a temperature which is characteristic of the species but usually about 15 °C. At higher temperatures, they tend to become blue, and, in some species, the change to blue is

enhanced by light. Here, the epidermal cells are to some extent independent effectors, responding directly to stimulation. This is true of the change from black to blue in the damselfly *Austrolestes*, but the reverse change is controlled by a secretion released from the terminal abdominal ganglion (Veron, 1973).

The significance of these changes is unknown, but they may be thermoregulatory. Dark insects absorb more radiation than pale ones so they may warm up more rapidly in the mornings and become active earlier than would be the case if they remained pale.

25.5.2 Morphological color change

Changes in the amounts of pigments can occur in response to a variety of external and internal factors.

25.5.2.1 *Ontogenetic changes*

Many insects change color in the course of development. For example, the eggs of the plant sucking bug, *Dysdercus nigrofasciatus*, are white when laid, becoming yellow as the embryo develops. The first stage nymph is a uniform yellow color when it hatches, but becomes orange and then red. In the second instar, white bands appear ventrally on some of the abdominal segments (Fig. 25.14). These become more extensive and white bands are also present dorsally in the later stages. In the final larval instar, the red becomes less intense, especially in the female, and the adults are yellow with white stripes. The yellow and red colors of this insect are produced by three pterins, the proportions of which change through development to produce the different colors. The white bands are formed from uric acid.

It is also common for caterpillars to exhibit a regular change in color during development. The larva of the puss moth, *Cerura*, turns from green to red just before pupation as a result of the production of ommochrome in the epidermis. The early larva of the swallowtail butterfly, *Papilio demodocus*, is brown with a white band at the center, whereas the late larva is green with purple markings and a white lateral stripe.

In adult hemimetabolous insects, color change is often associated with ageing and maturation. Adult male *Schistocerca* change from pink to yellow as they mature. The pink is produced by ommochromes in the epidermis which decrease in amount as the insect gets older, and the yellow is due to β-carotene which increases with age. Color changes related to sexual maturation also occur in some Odonata.

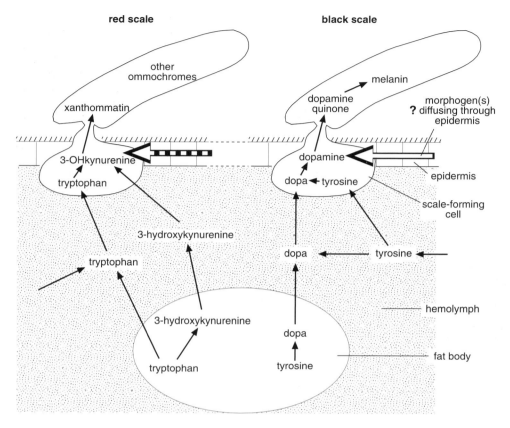

Fig. 25.12. Color patterning in Lepidoptera. Each scale-forming cell has the genetic capacity to form any of the pigments. In general, each scale only produces one pigment, probably in response to gradients of one or more morphogens. The complete process can be carried out by the cells, although some precursors, synthesized in the fat body, may also be taken up. The large horizontal arrows in the epidermis represent different morphogens (based on Koch, 1991).

These changes are controlled by changes in hormone levels associated with molting and sexual maturation, but the effects may be very different in different species. The changes in larval *Cerura* are initiated by a low level of ecdysteroid in the hemolymph leading to the metabolism of 3-hydroxykynurenine to an ommochrome. Perhaps the commitment peak which occurs before pupation is normally responsible for the change (section 15.4.2). Juvenile hormone leads to the accumulation of xanthommatin in *Bombyx* caterpillars, but prevents it in *Manduca* (Sawada *et al.*, 1990). Juvenile hormone, concerned in sexual maturation in adult *Schistocerca* (section 13.2.4.2), also regulates the accumulation of carotenoids which make the insects yellow as they become sexually mature. The hormone bursicon, which controls

sclerotization, probably also regulates darkening occurring immediately after ecdysis.

25.5.2.2 *Homochromy*

The colors of many insects change to match the predominant color of the background. This phenomenon is called homochromy. The changes may involve the basic color of the insect or may involve a general darkening. For example, larvae of the grasshopper, *Gastrimargus*, tend to assume a color more or less matching the background when reared throughout their lives on that background (Fig. 25.15). Insects reared on black become black, those reared on white are pale grey. On a green background, however, most insects develop a yellowish coloration. The differences are produced by different amounts of black

a) blue

b) black

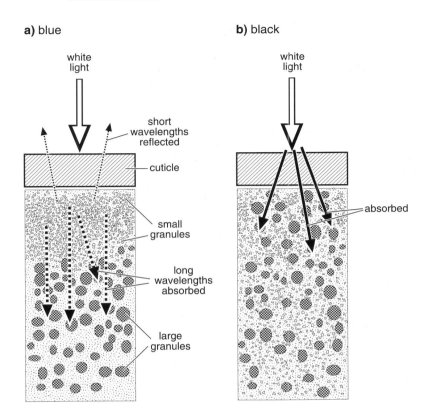

Fig. 25.13. Physiological color change in *Kosciuscola*. The two diagrams have the same numbers of large and small granules (based on Filshie, Day & Mercer, 1975). **(a)** At moderately high temperatures, the larger granules are restricted to the proximal parts of the epidermal cells. Short wavelengths of light are scattered by the small granules in the more distal parts of the epidermal cell, longer wavelengths travel further into the cell and are absorbed. The insects appear blue. **(b)** At lower temperatures, the pigments are generally dispersed through the cell. All the light is absorbed by the larger granules and the insects appear black.

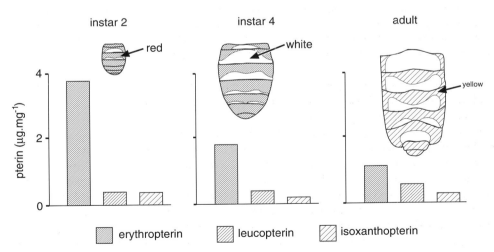

Fig. 25.14. Ontogenetic changes in pigmentation in *Dysdercus nigrofasciatus*. Underside of the abdomen of second and fourth instar larvae and adult male. The change from red to yellow is accompanied by a reduction in the relative amount of erythropterin present (expressed as μg of pterin per mg body weight). The white areas are due to uric acid in the epidermal cells (data from Melber & Schmidt, 1994).

Fig. 25.15. Homochromy in the grasshopper, *Gastrimargus*. Each graph shows the ground color of insects in the final larval stage after rearing from the first stage in containers with different colored backgrounds (after Rowell, 1970).

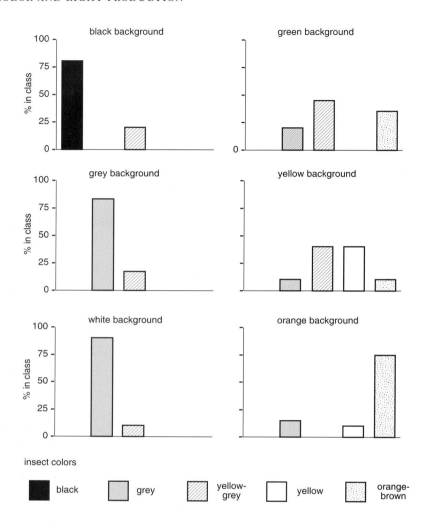

pigment, possibly melanin, in the cuticle and a dark ommochrome in the epidermis, together with yellow and orange pigments in the epidermis. The darkening of the stick insect, *Carausius*, which occurs on a dark background is also due to the accumulation of two ommochromes, xanthommatin and ommin (Bückmann, 1977). In this insect, changes occur at any time during the course of a stadium, but in most other examples the change is first seen at a molt. Ommochrome deposition begins after about five days on the new background, and the changes in other insects generally require a similar period of exposure. Caterpillars, too, may exhibit some degree of homochromy. Larvae of the popular hawk moth, *Laothoe populi*, may be almost white or yellow–green depending on

the plants on which they are reared (Grayson & Edmunds, 1989), and the pupae of some butterflies that are not concealed in a cocoon or in the soil, may be green, or dark or pale brown according to their surroundings.

An extreme example of homochromy occurs in African grasshoppers after bush fires. Many species become black within a few days, the change occurring in adults as well as larvae, but only in bright sunlight. In diffuse light the change is much less marked.

All these changes depend on vision, although the details may vary from species to species. In *Carausius*, darkening results from weak stimulation, or the absence of stimulation of the lower parts of the compound eye such as would occur when the insect is on a dark surface. In

Gastrimargus, darkening occurs if a high incident light intensity is associated with a low reflectance background, and changes in the yellow and orange pigments are dependent on the wavelength of light. Homochromy in caterpillars and lepidopteran pupae is also determined by the wavelength of light and the contrasts in intensity reaching the different stemmata.

The visual information affects the activity of neurosecretory cells in the central nervous system. Homochromy in pupae of the peacock butterfly, *Inachis io*, is produced by changes in the relative amounts of melanin and lutein (a carotenoid) in the cuticle. The accumulation of both pigments is controlled by a single neuropeptide which is widely distributed in the central nervous system before being released into the hemolymph. A low hormone titer immediately before pupation stimulates melanization, a higher titer results in increased incorporation of lutein into the pupal cuticle (Starnecker, 1997).

25.5.2.3 *Other factors affecting color*
Temperature is important in pigment development and there is a general tendency for insects reared at very high temperatures to be pale, and those developing at low temperatures to be dark. Humidity affects the color of many Orthoptera. Green forms are more likely to occur under humid conditions, and brown forms under dry conditions.

Crowding influences color in some insects, the most extreme examples being locusts. Locust larvae reared in isolation are green or fawn, while rearing in crowds produces yellow and black individuals. The colors and patterns change as the degree of crowding alters (Stower, 1959). The larvae of some Lepidoptera, such as *Plusia* and the armyworms, undergo comparable changes, some occurring in the course of a stadium, but the most marked alterations occur only at molting (Long, 1953). In caterpillars, these changes are known to be regulated by a neuropeptide.

25.5.3 Color polymorphism
Insects of many orders exhibit a green/brown polymorphism, tending to be green at the wetter times of year and brown when the vegetation is dry. The two forms are genetically determined, but homochromy that develops when the insect moves to a new background may be superimposed on this. In green morphs the production of ommochromes in the epidermis is largely or completely inhibited; if it occurred, the green would be obscured.

In some insects there are marked differences in color between successive generations correlated with seasonal changes in the environment. Such seasonal polyphenism and its physiological regulation is discussed in section 15.5. Extreme cases of color polymorphism occur in some butterflies in relation to mimicry (see below).

25.6 SIGNIFICANCE OF COLOR

Insect pigments are often the end products of metabolic processes and may have evolved originally as forms of storage excretion. Pterins, for example, may be derived from purines, such as uric acid. Similarly melanin production might be a method of disposing of toxic phenols ingested or arising from metabolism and it may be significant that melanin in the cuticle is often produced above metabolically active tissue such as muscle. Tryptophan in high concentrations reduces the rate of development of *Drosophila* and *Oryzaephilus* and it is noteworthy that ommochrome production follows the appearance of unusually high levels of tryptophan in the tissues (Linzen, 1974). However, it is clear that storage excretion is not the sole, or even primary function of pigments in most insects. For example, most insects excrete most of the end product of the biologically active hydrogenated pterins and do not store them. *Pieris*, on the other hand, synthesizes much larger amounts than would result from normal metabolism and accumulates them in the wings, where they contribute to the color (Ziegler and Harmsen, 1969).

In most cases, the colors of present-day insects are ecologically important. Color is most frequently involved as a defence against vertebrate predators and it may also be important in intraspecific recognition. It also has important consequences for body temperature, and color changes may contribute to thermoregulation (see above).

25.6.1 Predator avoidance
In many species, color and color patterns have evolved as part of a strategy to avoid predation. The color patterns may function in a variety of ways in different insects: for concealment (crypsis), to deflect attack from the most vulnerable parts of the body, to startle a predator, or to advertise distastefulness.
Review: Edmunds, 1974

Crypsis   Color often helps to conceal insects from predators. Many insects are known to select backgrounds on which they are least conspicuous. Often, homochromy is associated with some appropriate body form and behaviour as in stick insects and many mantids and grasshoppers which may be leaf-like or twig-shaped according to the backgrounds on which they normally rest. Protection may also be afforded by obliterative shading. Objects are made conspicuous by the different light intensities which they reflect as a result of their form. A solid object usually looks lighter on the upper side and darker beneath because of the effect of shadows, but by appropriate coloring this effect can be eliminated. If an object is shaded in such a way that when viewed in normal lighting conditions all parts of the body reflect the same amount of light, it loses its solid appearance. Such countershading is well-known in caterpillars where the surface towards the light is most heavily pigmented and the side normally in shadow has least pigment. To be successful in concealing the insect, this type of pigmentation must be combined with appropriate behavior patterns; if the larva were to rest with the heavily pigmented side away from the light it would become more, not less, conspicuous. Color may also afford protection if the arrangement of colors breaks up the body form. Such disruptive coloration is most efficient when some of the color components match the background and others contrast strongly with it. Disruptive coloration occurs in some moths which rest on tree trunks.

Review: Edmunds, 1990

Deimatic behavior   Some insect species have colored wings, or other parts of the body, which are normally concealed, but are suddenly displayed when a potential predator approaches. This behavior, sometimes associated with the production of a sound, has been shown, in a few cases, to startle the predator. It is known as deimatic behavior.

In many insects exhibiting this behavior, the hindwing is deep red or black. It is normally concealed beneath the forewing, but is revealed by a sudden partial opening of the wings. This occurs in a number of stick insects and mantids, and in moths that normally rest on vegetation. Moths in the family Arctiidae have bright red or yellow abdomens, often with black markings. When disturbed, the abdomen is displayed by opening the wings. The deimatic display of some mantids involves the front legs which have conspicuous marks on the inside. *Galepsus*, for example, displays the insides of the fore femora and coxae which are orange, at the same time exposing a dark mark in the ventral surface of the prothorax.

Some Lepidoptera have a pattern of scales on the hind wing forming an eyespot which is displayed when the insect is threatened. One of the best known examples is the peacock butterfly, *Inachis io*. It has one eyespot on the upper surface of each wing. These eyespots are primarily black, yellow and blue, surrounded by dark red. The butterfly rests with its wings held up over its back, the upper surfaces of the forewings juxtaposed so that the eyespots are concealed. If the insect is disturbed by visual or tactile stimuli it lowers the wings so that the eyespots on the forewings are displayed and then protracts the forewings to expose the hindwing eyespots. At the same time the insect makes a hissing sound with a tymbal on the forewing (section 26.1.3.2). The forewings are then retracted and partly raised and the sequence of movements repeated, sometimes for several minutes. While displaying, the body is tilted so that the wings are fully exposed to the source of stimulation and at the same time the insect turns so as to put the stimulus behind it. This eyespot display causes flight behavior in at least some birds. Some mantids also have a large eyespot on the hind wing.

Deflection marks   Small eyespots, often present on the underside of the hindwings of butterflies, appear to deflect the attention of birds away from the head of the insect. There is no sharp distinction between eyespots used for deimatic behavior and those concerned with deflection. In general, deflecting spots are probably smaller than those used in intimidation, but it is possible that some may serve either function depending on the nature and experience of the predator.

Some lycaenid butterflies have tails on the hind wings and a color pattern that, at least to humans, has the illusion of the insect's head. It is probable that this serves as a deflection device to some bird predators.

Aposematic coloration   Many insects are distasteful by virtue of chemicals they produce themselves, or that are sequestered from their food; other insects sting (section 27.2.7). Such insects are commonly brightly colored, and are usually red or yellow combined with black. Such coloration is a signal to predators that the potential prey is distasteful and should be avoided. It is called aposematic coloration. For such coloration to be effective, the predator must exhibit an innate or a learned avoidance response. Aposematic species occur in many orders of insects (Table 25.2).

Reviews: Guilford, 1990; Rothschild, 1972

Table 25.2. *Aposematic insects. Examples from different orders*

Order and species	Stage	Color	Basis of unpalatability
Orthoptera			
Zonocerus variegatus	Adult	B/Y and R markings	Various chemicals
Romalea guttata	Adult	B/R hind wings	Various chemicals
Hemiptera			
Aphis nerii	All	Bright Y	Cardiac glycosides
Oncopeltus fasciatus	Adult	Y/B spots	Cardiac glycosides
Coleoptera			
Coccinella septemfasciata	Adult	R/B spots	Alkaloids
Tetraopes oregonensis	Adult	R	Cardiac glycosides
Hymenoptera			
Vespula vulgaris	Adult	Y/B stripes	Sting
Lepidoptera			
Tyria jacobaeae	Larva	B/Y stripes	Pyrrolizidine alkaloids
Tyria jacobaeae	Adult	B/R marks	Pyrrolizidine alkaloids
Battus philenor	Larva	R/B spots	Aristolochic acids
Danaus plexippus	Larva	W/B and Y stripes	Cardiac glycosides
Zygaena filipendula	Adult	B/R spots	Cyanogenic glycosides

Note:

B, black; Y, yellow; R, red; W, white.

Mimicry Predators learn to avoid distasteful insects with distinctive colors, but theoretically a predator must learn to avoid each individual species separately. If, however, the color patterns of some species are similar to each other, learning to avoid one species because it is distasteful also produces an avoidance of the other. Resemblance of one species to another is called mimicry.

Mimicry takes two forms, Mullerian and Batesian. Species exhibiting Mullerian mimicry are all distasteful. Here, the advantage to the insects is that predation on any one species is reduced. For example, many social wasp species, such as *Vespa* and *Vespula* which all have a sting, have the same basic black and yellow pattern; if a predator learns to avoid one species, it is likely to avoid others with a similar appearance. Mullerian mimicry is also common amongst Lepidoptera, with the genus *Heliconius* having been especially well studied. Sometimes the mimics have different life-styles. Cotton stainer bugs of the genus *Dysdercus* usually have similar red and black coloration and all the species are distasteful. The predaceous reduviid

bug, *Phonoctonus*, lives with *Dysdercus* and preys on them. It, too is unpalatable to predators, and is a color mimic of *Dysdercus* so that avoidance of one leads to avoidance of both species.

In Batesian mimicry, only one of a pair of species is distasteful, the other is not. They are called the 'model' and 'mimic', respectively. Here, the palatable species gains some advantage from a resemblance to a distasteful species. In this type of mimicry it is essential that the mimic is uncommon relative to the model. If this were not the case a predator might learn to associate a particular pattern with palatability rather than distastefulness. This limits the numbers or distribution of a mimetic form, but the limit may be circumvented by the mimic becoming polymorphic with each of its morphs resembling a different distasteful species. The best-known example of such polymorphism is that of the female *Papilio dardanus* (Lepidoptera), which has a large number of mimetic forms mimicking a series of quite different-looking butterflies.

Because the unpalatability of an individual may be affected by the nature of the food that it eats, mimicry may vary temporally and spatially. For example, the viceroy and queen butterflies of north America are both distasteful and normally exhibit Mullerian mimicry (Ritland, 1991). However, sometimes larvae of the queen feed on plants that are so low in the cardenolides sequestered by the insect that the resulting adults are not distasteful. It may be supposed that the palatable queen butterflies now depend for protection on their resemblance to other members of the species that have sequestered cardenolides, and on unpalatable viceroy butterflies (Ritland, 1994).

Mimicry is common in Hemiptera, Hymenoptera, Diptera and Coleoptera as well as Lepidoptera. The model and mimic are often in different orders.
Review: Turner, 1984 – butterflies

25.6.2 Intraspecific recognition
Color is important in intraspecific recognition in diurnally active insects, and its role is most fully understood in damselflies and dragonflies (Odonata), and in butterflies. In both groups it has two principal functions in male behavior: the recognition of females, and the recognition of conspecific males.

In some dragonflies, male color is important in defending a territory against other males. For example, the dorsal side of the abdomen of male *Plathemis lydia* is blue. If a second male enters a territory, the resident male faces the intruder and displays the color by raising the abdomen towards the vertical. This has an inhibiting effect on the intruding male. In the presence of a female, the abdomen is depressed so that the blue is not visible as the male approaches her. The females of many species change color as they become sexually mature and this is associated with a change in male behavior towards them. Amongst the damselflies, the females of some species are dimorphic with one of the morphs resembling the male, and there are distinct differences in the behavior of males towards these two morphs (Robertson, 1985).

In butterflies, the general color of a female may be more important than the details of pattern in attracting males, although size and movement are also important. For example, males of the African butterfly, *Hypolimnas misippus*, are attracted by the female's red-brown wing color; the black and white markings sometimes present on the forewings are unimportant, although white on the hindwing may have an inhibitory effect (Smith, 1984; Stride,

1957). Males of *Colias eurytheme* are attracted by the yellow underside of the female hindwing which is exposed when the female is at rest. In this species, the male reflects ultraviolet from the upper side of its wings. This signal is used by the female in interspecific mate recognition, and it also inhibits attraction by other males, reducing the likelihood of further male intervention when a male is copulating (Silberglied & Taylor, 1978). It may be generally true that the colors of males are more important in male-male interactions than for recognition by females.
Review: Silberglied, 1984 – butterflies

25.7 LIGHT PRODUCTION

A number of insects appear to luminesce, but in many cases this luminescence is due to bacteria. Self-luminescence, not involving bacteria, is only known to occur in a few Collembola, such as *Onychiurus armatus*, in the homopteran *Fulgora lanternaria*, in a few larval Diptera belonging to the families Platyuridae and Bolitophilidae, and in a relatively large number of Coleoptera, primarily in the families Lampyridae (fireflies), Elateridae and Phengodidae (railroad worms). In these families luminescence may occur in both sexes or be restricted to the female; it also occurs in some larval forms.

The light-producing organs occur in various parts of the body. *Onychiurus* emits a general glow from the whole body, but in most beetles the light organs are relatively compact, and are often on the ventral surface of the abdomen. In male *Photuris* (Coleoptera), there is a pair of light organs in the ventral region of each of the sixth and seventh abdominal segments. In the female the organs are smaller and often only occur in one segment. The larvae have a pair of small light organs in segment eight, but these disappear at metamorphosis when the adult structures form. Larvae and females of railroad worms (Phengodidae) have 11 pairs of dorso-lateral light organs on the thorax and abdomen and another on the head. In *Fulgora* the light organ is in the head.

The light organs are generally derived from the fat body, but in *Arachnocampa* (Diptera) they are formed from the enlarged distal ends of the Malpighian tubules.

25.7.1 Structure of light-producing organs in Coleoptera
Each light organ of an adult firefly consists of a number of large cells, the photocytes, which lie just beneath the

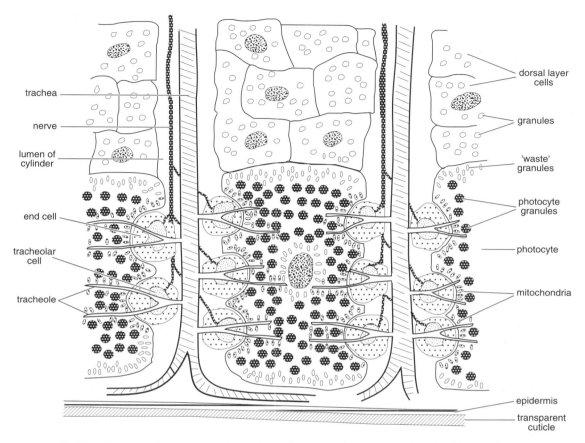

Fig. 25.16. Diagrammatic section through part of the light organ on the ventral side of an adult *Photuris*. The tracheoles pass between the photocytes, but do not penetrate into the cells (based on Smith, 1963).

epidermis and are backed by several layers of cells called the dorsal layer cells (Fig. 25.16). The cuticle overlying the light organ is transparent. The photocytes form a series of cylinders at right angles to the cuticle with tracheae and nerves running through the core of each cylinder. Each trachea gives off branches at right angles and as these enter the region of the photocytes they break up into tracheoles, which run between the photocytes parallel with the cuticle. The tracheoles are spaced 10 to 15 μm apart and as the photocytes are only about 10 μm thick the diffusion path for oxygen is short. The origin of the tracheoles is enclosed within a large tracheal end cell, the inner membrane of which is complexly folded where it bounds the tracheolar cell (section 17.1.2). In some species, the end cells are only poorly developed. The neurons entering the photocyte cylinder end as spatulate terminal processes between the plasma membranes of the end cell and the tracheolar cell within which the tracheoles arise. In adult *Pteroptyx* (Coleoptera) and some other genera, nerve endings occur on the photocytes as well as on the tracheal end cells.

The photocytes are packed with photocyte granules, each containing a cavity connecting with the outside cytoplasm via a neck. It is presumed that the reactants involved in light production are housed in these granules. Smaller granules also occur dorsally and ventrally. Mitochondria are sparsely distributed except where the cell adjoins the end cells and tracheoles. The dorsal layer cells also contain granules, generally regarded as urate granules, and it has been supposed that the cells form a reflecting layer. There is, however, no direct evidence for this and it has also been suggested that the oxyluciferin irreversibly produced in light production (see below) is stored in them.

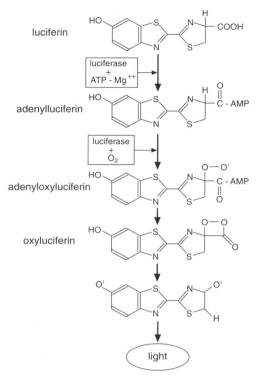

Fig. 25.17. Basic reactions involved in light production from luciferin.

It is estimated that the two lanterns of *Photinus* together contain about 15 000 photocytes forming some 6000 cylinders, each with 80–100 end cells.

The lanterns of larval fireflies contain the same elements, but their organization is simpler. The tracheal system is diffuse and there are no tracheal end cells. Nerve endings occur on the photocytes, and are not separated from them by the tracheal end cells as in adult *Photuris*.

25.7.2 Mechanism of light production

Basically, light is produced by the oxidation of luciferin, in the presence of the enzyme, luciferase (Fig. 25.17). Luciferin is first activated by ATP in the presence of magnesium and luciferase to produce adenylluciferin. This remains tightly bound to the enzyme and is oxidized to form excited oxyluciferin, which decays spontaneously with the production of light. The reaction is very efficient, some 98% of the energy involved being released as light.

In many insects the light produced by the light organs

is yellow-green in color, extending over a relatively narrow band of wavelengths, 520–650 nm in *Photinus* and *Lampyris* (Coleoptera). In larval and adult female railroad worms such as *Phrixothrix*, the light organs on the thorax and abdomen produce green to orange light, depending on the species, in the range 530–590 nm. That on the head produces red light with peak emission at 620 nm, but extending from about 580 to over 700 nm. The light produced by *Arachnocampa* is blue-green, that of *Fulgora* white.

At least in the beetles, the differences in wavelengths of light produced are due to differences in the enzyme, luciferase, the luciferin is identical in the different species. **Review:** McElroy & DeLuca, 1985

25.7.3 Control of light production

The light organs of adult *Photuris* are innervated by three and four dorsal unpaired median (DUM) cells in the last two abdominal ganglia, respectively. Like other DUM cells (section 10.3.2.4), the axons from these cells divide to send symmetrical branches to the lanterns on each side. In most adult fireflies, the axons terminate on the tracheal end cells, but in larvae, where there are no end cells, they innervate the photocytes directly. This difference indicates a different method of controlling light production although in both larvae and adults the neurotransmitter appears to be octopamine.

In adult *Photuris*, light production appears to be regulated by the availability of oxygen. As the DUM neurons terminate on the tracheal end cells it is presumed that neural activity causes a change in these cells facilitating the flow of oxygen to the photocytes. Flash duration in adult fireflies is generally very brief, a few hundred milliseconds, with flashes following each other at regular intervals. This implies that the oxygen supply to the photocytes is closely regulated. Larval fireflies produce a prolonged glow and here the octopamine acts directly on the photocytes (Christensen *et al.*, 1983).

Species of fireflies of the genus *Pteroptyx*, notably in South-east Asia, form male groups in which the individual insects flash in synchrony. The males of these species, in isolation, can flash regularly with almost constant intervals between flashes. Buck and Buck (1968) observed one individual of *Pteroptyx* with a mean flash-cycle length of 557.3 ms and 95% of all flash cycles fell within 5 ms of this mean. If an individual detects a flash within a critical period of having produced its own flash, it immediately flashes

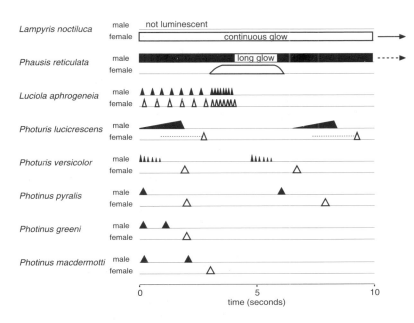

Fig. 25.18. Flash patterns of different fireflies (Lampyridae). The height of each symbol represents the intensity of the flash; the shape shows intensity rising to a maximum and then declining. The dotted line in female *Photuris lucicrescens* shows the interval during which the female flash may occur (based on Case, 1984).

again. As the individuals produce flashes at regular intervals, this resetting rapidly results in synchronous flashing by all the insects.

25.7.4 Significance of light production

In most luminous insects light production has sexual significance. Light signals are used in two basic ways in Lampyridae (Lloyd, 1971). In some species, such as *Lampyris*, the female is sedentary and attracts the male to herself; in other species, such as *Photuris* and *Photinus*, one sex, usually the male, flies around flashing in a specific manner. Flash duration and the interval between flashes is often characteristic for the species (Fig. 25.18) and flashing is associated with distinct flight patterns. For instance, male *Photinus pyralis* produce a flash lasting about 500 ms at six-second intervals. During the flash the male climbs steeply and then hovers for about two seconds. If a female flashes 1.5 to 2.5 s after the male flash, he flies towards her and flashes again three seconds later. He does not respond to flashes occurring after different time intervals. Repeated flashing sequences bring the male to the female. Precise timing requires a very well-defined time marker and flashes begin or end (sometimes both) sharply; these transients probably provide the temporal signals.

Females of *Pteroptyx* are attracted by the flashes of the male group.

After mating, the females of some *Photuris* species change their behavior so their patterning of flashing comes to resemble that of females of other species (Vencl, Blasko & Carlson, 1994). In this way they lure males of those species on which they then feed. The luminescence of *Arachnocampa* (Diptera) larvae also serves as a lure, attracting small insects into networks of glutinous silk threads on which they become trapped. The larvae then eat them.

In railroad worms, the lateral light-producing organs may be suddenly illuminated if the insect is attacked, and they possibly have a defensive function. It is suggested that the red head light provides these insects with illumination, presumably when they are searching for the millipedes on which they prey. The relatively long wavelengths emitted by this organ may not be visible to most other insect predators, but the eyes of the railroad worms probably can detect them (Viviani & Bechara, 1997).

The sensitivity of the dark adapted compound eyes of fireflies varies diurnally, as it does in at least some other insects. It increases rapidly by about four orders of magnitude at a time approximately corresponding with the time at which the insects normally flash (Lall, 1993).

Review: Case, 1984

REFERENCES

Berthold, G. (1980). Microtubules in the epidermal cells of *Carausius morosus*. Their pattern and relation to pigment migration. *Journal of Insect Physiology*, **26**, 421–5.

Buck, J.B. & Buck, E.M. (1968). Mechanism of rhythmic synchronous flashing of fireflies. *Science*, **159**, 1319–27.

Bückmann, D. (1977). Morphological colour change: stage independent, optically induced ommochrome synthesis in larvae of the stick insect, *Carausius morosus* Br. *Journal of Comparative Physiology*, **115**, 185–93.

Case, J.F. (1984). Vision in mating behavior of fireflies. *Symposium of the Royal Entomological Society of London*, **12**, 195–222.

Christensen, T.A., Sherman, T.G., McCaman, R.E. & Carlson, A.D. (1983). Presence of octopamine in firefly photomotor neurons. *Neuroscience*, **9**, 183–9.

Edmunds, M. (1974). *Defence in Animals*. Harlow: Longman.

Edmunds, M. (1990). The evolution of cryptic coloration. In *Insect Defenses*, ed. D.L. Evans & J.O. Schmidt, pp. 3–21. Albany: State University of New York Press.

Feltwell, J. (1978). The distribution of carotenoids in insects. In *Biochemical Aspects of Plant and Animal Coevolution*, ed. J.B. Harborne, pp. 277–307. London: Academic Press.

Filshie, B.K., Day, M.F. & Mercer, E.H. (1975). Colour and colour change in the grasshopper, *Kosciuscola tristis*. *Journal of Insect Physiology*, **231**, 1763–70.

Fox, D.L. (1953). *Animal Biochromes and Colours*. Cambridge: Cambridge University Press.

Fuzeau-Braesch, S. (1985). Color changes. In *Comprehensive Insect Physiology, Biochemistry and Pharmacology* vol. 9, ed. G.A. Kerkut & L.I. Gilbert, pp. 549–89. Oxford: Pergamon Press.

Ghiradella, H. (1989). Structure and development of iridescent butterfly scales: lattices and laminae. *Journal of Morphology*, **202**, 69–88.

Grayson. J. & Edmunds, M. (1989). The causes of colour and colour change in caterpillars of the popular and eyed hawkmoths (*Laothoe populi* and *Smerinthus ocellata*). *Biological Journal of the Linnean Society*, **37**, 263–79.

Guilford, T. (1990). The evolution of aposematism. In *Insect Defenses*, ed. D.L. Evans & J.O. Schmidt, pp. 23–61. Albany: State University of New York Press.

Harborne, J.B. (1991). Flavonoid pigments. In *Herbivores: Their Interactions with Secondary Plant Metabolites*, vol. 1, ed. G.A. Rosenthal & M.R. Berenbaum, pp. 389–429. San Diego: Academic Press.

Hinton, H.E. (1976). Recent work on physical colours of insect cuticle. In *The Insect Integument*, ed. H.R. Hepburn, pp. 475–96. Amsterdam: Elsevier.

Ishizaki, Y. & Umebachi, Y. (1990). Further studies on dopamine and n-acetyldopamine during the pupal stage of *Papilio xuthus*. *Comparative Biochemistry and Physiology*, **97B**, 563–7.

Kayser, H. (1985). Pigments. In *Comprehensive Insect Physiology, Biochemistry and Pharmacology*, vol. 10, ed. G.A. Kerkut & L.I. Gilbert, pp. 367–415. Oxford: Pergamon Press.

Koch, P.B. (1991). Precursors of pattern specific ommatin in red wing scales of the polyphenic butterfly *Araschnia levana* L.: hemolymph tryptophan and 3-hydroxykynurenine. *Insect Biochemistry*, **21**, 785–94.

Lall, A.B. (1993). Nightly increase in visual sensitivity correlated with bioluminescent flashing activity in the firefly *Photuris versicolor* (Coleoptera: Lampyridae). *Journal of Experimental Zoology*, **265**, 609–12.

Linzen, B. (1974). The tryptophan–ommochrome pathway in insects. *Advances in Insect Physiology*, **10**, 117–246.

Lloyd, J.L. (1971). Bioluminescent communication in insects. *Annual Review of Entomology*, **16**, 97–122.

Long, D.B. (1953). Effects of population density on larvae of Lepidoptera. *Transactions of the Royal Entomological Society of London*, **104**, 543–84.

Martel, R.B. & Law, J.H. (1992). Hemolymph titers, chromophore association and immunological cross-reactivity of an ommochrome-binding protein from the hemolymph of the tobacco hornworm, *Manduca sexta*. *Insect Biochemistry and Molecular Biology*, **22**, 561–9.

McElroy, W.D. & DeLuca, M. (1985). Biochemistry of insect luminescence. In *Comprehensive Insect Physiology, Biochemistry and Pharmacology* vol. 4, ed. G.A. Kerkut & L.I. Gilbert, pp. 553–63. Oxford: Pergamon Press.

Melber, C. & Schmidt, G.H. (1994). Quantitative variations in the pteridines during postembryonic development of *Dysdercus* species (Heteroptera: Pyrrhocoridae). *Comparative Biochemistry and Physiology*, **108B**, 79–94.

Neville, A.C. (1975). *Biology of the Arthropod Cuticle*. Berlin: Springer-Verlag.

Nijhout, H.F. (1991). *The Development and Evolution of Butterfly Wing Patterns*. Washington: Smithsonian Institution Press.

Nijhout, H.F. (1994). Symmetry systems and compartments in Lepidopteran wings: the evolution of a patterning mechanism. *Development*, 1994 supplement, 225–33.

Noble, R.M. & Walker, P.W. (1990). Pteridine compounds in adults of the pink-spotted bollworm, *Pectinophora scutigera*. *Entomologia Experimentalis et Applicata*, **57**, 77–83.

Richards, A.G. (1951). *The Integument of Arthropods*. Minneapolis: University of Minnesota Press.

Ritland, D.B. (1991). Revising a classic butterfly mimicry scenario: demonstration of Müllerian mimicry between Florida viceroys (*Limenitis archippus floridensis*) and queens (*Danaus gilippus berenice*). *Evolution*, **45**, 918–34.

Ritland, D.B. (1994). Variation in palatability of queen butterflies (*Danaus gilippus*) and implications regarding mimicry. *Ecology*, **75**, 732–46.

Rothschild, M. (1972). Secondary plant substances and warning colouration in insects. *Symposium of the Royal Entomological Society of London*, **6**, 59–83.

Robertson, H.M. (1985). Female dimorphism and mating behaviour in a damselfly, *Ischnura ramburi*: females mimicking males. *Animal Behaviour*, **33**, 805–9.

Rowell, C.H.F. (1970). Environmental control of coloration in an acridid, *Gastrimargus africanus* (Saussure). *Anti-Locust Bulletin*, no. 47, 48 pp.

Sawada, H., Tsusué, M., Yamamoto, T. & Sakurai, S. (1990). Occurrence of xanthommatin containing pigment granules in the epidermal cells of the silkworm, *Bombyx mori*. *Insect Biochemistry*, **20**, 785–92.

Silbergleid, R.E. (1984). Visual communication and sexual selection among butterflies. *Symposium of the Royal Entomological Society of London*, **11**, 207–23.

Silbergleid, R.E. & Taylor, O.R. (1978). Ultraviolet reflection and its behavioural role in the courtship of the sulfur butterflies, *Colias eurytheme* and *C.philodice*. *Behavioral Ecology and Sociobiology*, **3**, 203–43.

Smith, D.A.S. (1984). Mate selection in butterflies: competition, coyness, choice and chauvinism. *Symposium of the Royal Entomological Society of London*, **11**, 225–44.

Smith, D.S. (1963). The organization and innervation of the luminescent organ in a firefly, *Photuris pennsylvanica* (Coleoptera). *Journal of Cell Biology*, **16**, 323–59.

Smith, D.S. (1970). Links between cellular structure and function. *Symposium of the Royal Entomological Society of London*, **5**, 1–15.

Starnecker, G. (1997). Hormonal control of lutein incorporation into pupal cuticle of the butterfly *Inachis io* and the pupal melanization reducing factor. *Physiological Entomology*, **22**, 65–72.

Steinbrecht, R.A., Mohren, W., Pulker, H.K. & Schneider, D. (1985). Cuticular interference reflectors in the golden pupae of danaine butterflies. *Proceedings of the Royal Society of London B*, **226**, 367–90.

Stower, W.J. (1959). The colour patterns of hoppers of the desert locust (*Schistocerca gregaria* Forskål). *Anti-Locust Bulletin*, no. 32, 75 pp.

Stride, G.O. (1957). Investigations into the courtship behavior of the male of *Hypolimnas misippus* L. (Lepidoptera, Nymphalidae), with special reference to the role of visual stimuli. *British Journal of Animal Behavior*, **5**, 153–67.

Summers, K.M., Howells, A.J. & Pyliotis, N.A. (1982). Biology of eye pigmentation in insects. *Advances in Insect Physiology*, **16**, 119–66.

Turner, J.R.G. (1984). Mimicry: the palatability spectrum and its consequences. *Symposium of the Royal Entomological Society of London*, **11**, 141–61.

Vencl, F.V., Blasko, B.J. & Carlson, A.D. (1994). Flash behavior of female *Photuris versicolor* fireflies (Coleoptera: Lampyridae) in simulated courtship and predatory dialogues. *Journal of Insect Behavior*, **7**, 843–58.

Veron, J.E.N. (1973). The physiological control of the chromatophores of *Austrolestes annulosus* (Odonata). *Journal of Insect Physiology*, **19**, 1689–703.

Veron, J.E.N. (1976). Responses of Odonata chromatophores to environmental stimuli. *Journal of Insect Physiology*, **22**, 19–30.

Viviani, V.R. & Bechara, E.J.H. (1997). Bioluminescence and biological aspects of Brazilian railroad-worms (Coleoptera: Phengodidae). *Annals of the Entomological Society of America*, **90**, 389–98.

Wall, R., Langley, P.A. & Morgan, K.L. (1991). Ovarian development and pteridine accumulation for age determination in the blowfly *Lucilia sericata*. *Journal of Insect Physiology*, **37**, 863–8.

Ziegler, I. & Harmsen, R. (1969). The biology of pteridines in insects. *Advances in Insect Physiology*, **6**, 139–203.

26 Mechanical communication: producing sound and substrate vibrations

Many insects communicate using vibrational signals. These may be air- or water-borne sounds, which are sometimes audible to humans, or they may be transmitted through the substrate on which the insect is resting. Sometimes the vibrations are purely adventitious, resulting from the insect's normal activities, such as the sounds produced by wing vibration in flight or by the mandibles of an insect chewing, but they often have some special significance for the insect and, in most cases, are produced independently of other activities.

Vibrations may be produced in water and solids with little or no anatomical specialization. By contrast, insects using airborne vibrations to communicate use structures and behaviors that enhance sound radiation. The structure transmitting vibrations to the air is known as the sound radiator; it is often a part of the wing which is suspended from the surrounding cuticle in a way that facilitates its free vibration.

A sound radiator functions most efficiently in producing sounds with wavelengths similar to or less than its own dimensions. Consequently, because of their small size, insects are constrained towards producing sounds with short wavelengths (and high frequencies). However, high frequency sounds attenuate more rapidly than lower frequency, longer wavelength sounds so they are less good for distance communication. In practice, airborne sounds produced by different insects cover a very wide range of frequencies and are often ultrasonic to humans (Fig. 26.1).

Sound is produced on both sides of a vibrating membrane. If the membrane is small relative to the wavelength of the sound produced, sound waves from the two sides tend to interfere destructively because they are in antiphase. Such interference can be reduced or eliminated by a baffle separating the air on the two sides of the membrane. In many insects using part of the wing to produce sound, the non-vibrating parts of the wing surrounding the vibrating membrane may be used as a baffle. Some species extend the size of the baffle by employing structures in the

environment (see below). If the sounds from the two sides of the vibrating membrane are brought into phase with each other, constructive interference occurs, increasing the intensity of the sound produced. This is achieved by the cricket, *Anurogryllus* (see below).

Reviews: Bailey, 1991; Ewing, 1989

26.1 MECHANISMS PRODUCING VIBRATIONS

26.1.1 Percussion

Percussion refers to vibrations produced by the impact of part of the body against the substrate or by bringing two parts of the body sharply into contact with each other.

Many insects signal by striking the substrate with some part of the body, usually without any related structural modifications. Some termites and beetles use their heads. Soldiers of *Zootermopsis*, for example, make vertical oscillating movements using the middle legs as a fulcrum so that the head rocks up and down banging the tips of the

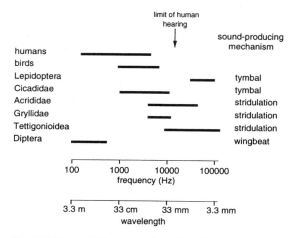

Fig. 26.1. Range of airborne sounds produced by some insect groups, birds and humans. Within a group, not all species produce the full range of frequencies.

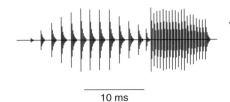

10 ms

Fig. 26.2. Drumming signals produced by male *Isoperla* (Plecoptera). The figure shows airborne sounds recorded close to the substratum in an experiment. The pattern resembles that occurring in the substratum. The insects respond to substrate vibrations (after Szczytko & Stewart, 1979).

mandibles on the floor of their gallery and, less frequently, the top of the head against the roof. Usually, two or three taps are produced successfully followed by an interval of about 500 ms before being repeated (Kirchner, Broeker & Tautz, 1994). Workers and larvae produce a lower intensity sound by hitting their heads on the roof in similar vertical oscillating movements. The death watch beetle, *Xestobium*, produces tapping sounds by banging its frons against the wood in which it is burrowing. This is repeated seven or eight times in a bout lasting almost a second. Both sexes perform this behavior (Birch & Keelyside, 1991).

Plantlice (Psocoptera) and some stoneflies of both sexes (Plecoptera) drum with the abdomen (Fig. 26.2) and some species have a small knob on the ventral surface of the abdomen which they use as a hammer to tap the ground. The stonefly, *Siphonoperla*, makes similar drumming movements with its abdomen, but the abdomen does not touch the substrate. Instead, the vibrations are transmitted to the substrate through the legs. Some lacewings, *Chrysoperla*, also transmit abdominal vibrations to the substrate via the legs (Henry, 1980). In many other insects, vibrations of the body produced by different mechanisms are also transmitted via the substrate (see below).

Some Orthoptera drum with the hindlegs. Males of the grasshopper, *Oedipoda*, drum on the ground with their hind tibiae at a rate of about 12 beats per second, while females drum more slowly. Males of the bush cricket, *Meconema*, also drum, but only use one leg. The frequency of impacts varies with temperature, but is usually in the range 30–60 Hz. Usually, a series of short impact trains, in which the tarsus strikes the leaf five or six times, is followed by several long trains comprising 30 or more impacts (Sismondo, 1980).

A single tap on a solid surface produces a complex wave pattern that varies according to the nature of the solid. In addition, it will produce airborne sounds that are sometimes audible to humans. For most insects using percussion, communication via the substrate is probably more important than sound communication.

A few insects are known to produce airborne sounds by striking one part of the body against another. When noctuid moths are flying, their wingtips may meet above the body and, in two species, this is known to produce a brief burst of sound at about 50 kHz (Agee, 1971). This may be a common phenomenon, but there is no evidence to indicate that the sound production is more than accidental, although it can be heard by other moths (Agee, 1971). However, the males of whistling moths, *Hecatesia*, produce sound by banging together specialized areas of the forewings known as castanets. Each castanet consists of a series of cuticular knobs on the costal area of the forewing separated from the rest of the forewing by a crescent of corrugated cuticle. *H. thyridion* produces the sounds while in flight. Each pulse of sound lasts about 0.2 ms and pulses are produced at a rate of about $80\,\mathrm{s}^{-1}$, corresponding with the wingbeat frequency. The sound emitted covers a broad range of frequencies, peaking between 17 and 20 kHz. Another species, *H. exultans*, produces sound by vibrating its wings while at rest, and only small wing movements are necessary to produce the impacts. This species produces a much purer tone than *H. thyridion* probably because the space between the wings forms a resonator. The peak frequency is about 30 kHz.

Oedipodine grasshoppers often produce clicking or crackling sounds in flight known as crepitation. These sounds may result from the impact of the wings against the legs, but is possibly produced by changes in tension in parts of the wing membrane when the wings are in certain positions (Otte, 1970).

26.1.2 Stridulation

Stridulation refers to the production of vibrations by moving a cuticular ridge (called the scraper or plectrum) on one part of the body over a toothed ridge (known as the file or strigil) on another. Repeated contacts of the scraper against the teeth of the file cause part of the body to vibrate. Vibration of the insect's body, acting via the legs, may give rise to substrate vibrations. If the vibrating membrane is of appropriate dimensions it will also produce airborne sounds. This is the case in many Orthoptera where substrate-borne vibrations are generally assumed to be unimportant.

This method of producing vibrations is employed by insects in many different orders, but is particularly associated with Orthoptera, Heteroptera and Coleoptera. In one insect or another almost every part of the body has become modified to produce vibrations in this way.

26.1.2.1 *Stridulation in Orthoptera*
The Orthoptera employ two main methods of stridulation: tegminal stridulation in crickets (Grylloidea) and bush crickets (Tettigonioidea) and femoro-tegminal stridulation in grasshoppers (Acridoidea).

Crickets   In male Grylloidea, each tegmen has a file on the underside of the second cubital vein near its base, while the ridge forming the scraper is on the edge of the opposite tegmen (Fig. 26.3a,b). Mole crickets are able to use the scraper on either wing by varying the relative positions of the wings, but in field crickets the right tegmen always overlaps the left so that only the right field and left scraper are functional. In producing the sound, the tegmina are raised at an angle of 15–40° to the body and then opened and closed so that the scraper rasps on the file. Sound is produced by both tegmina, each wing closure producing a single syllable of sound.

The principal sound radiator on the tegmen is an area of thick sclerotized wing membrane with no cross veins, although sometimes with oblique veins. It is called the harp and is enclosed by the first and second cubital veins (Fig. 26.3a). It is separated from the rigid subcostal, radial and medial veins, by a band of flexible cuticle so that it is free to vibrate as a plate. In *Gryllus*, a more distal area of thin cuticle, called the mirror, appears not to be specifically involved in sound production, although a comparable area in Tettigoniidae is very important (see below).

The resonant frequency of the harp is approximately the same as the tooth impact frequency, and this contributes to the pure tones of most cricket songs (Fig. 26.3c) although the impact frequency may vary slightly during a single wingstroke and from stroke to stroke. However, the airspace between the body and the raised tegmina also forms a resonator which both filters and amplifies the sound produced. The insect may thus be able to control the frequency of its sound output by adjusting the volume of this space (Stephen & Hartley, 1995).

Some gryllids use environmental features to improve sound radiation. Species of oecanthine crickets use leaves as baffles. They may simply choose a singing position between two leaves which then function as baffles, but *Oecanthus burmeisteri* cuts a small hole in a leaf and, when singing, sits with its head projecting through the hole and its tegmina tightly pressed against the leaf surface. It emits a sound with a frequency of about 2 kHz and a wavelength of 170 mm. The sound-producing membrane on the wings is only about 3 mm in diameter, but it uses leaves ranging in size from 70×80 mm to 170×300 mm and even the smallest of these is an efficient baffle (Prozesky-Schulze *et al.*, 1975).

Anurogryllus produces complementary interference of its song by standing in a saucer-shaped depression and singing with the tegmina horizontal at a height equal to one quarter of the wavelength of its principal frequency. The sound waves from the underside of the wings are reflected upwards from the saucer. These reflected waves are in antiphase with the waves coming down and cancel them out, but they are in phase with the sounds produced on the upper side of the wing and so complement them.

The mole cricket, *Gryllotalpa vineae*, amplifies its song by singing in a burrow shaped like an exponential horn with two openings (Fig. 26.4). The dimensions of the horn are such that when the forewings are raised in the singing position they form a diaphragm across it. The sound produced has a fundamental frequency of 3–5 kHz, and is amplified so that muscular energy is converted to acoustic power with very great efficiency. As a result, the sound carries over great distances, up to 600 m, and because of the tuning of the burrow the song is purer than when the insect is removed from the burrow. The sound is beamed in an arc by the arrangement of the openings of the burrow.

Female gryllids do not usually possess stridulatory apparatus, but male larvae of the later instars have the apparatus and may stridulate.
Review: Bennet-Clark, 1989

Bush   crickets   The   stridulatory   apparatus   of Tettigonioidea is similar to that of gryllids, but the left tegmen overlaps the right and, in most fully winged forms, only the left file and the right scraper are present. In some species, however, in which the hindwings are absent and the tegmina are short and rounded (Fig. 26.5), being retained only for the production of sound, a file and scraper are present on each tegmen, although only the left file is functional. The file of *Platycleis* has an average of 77

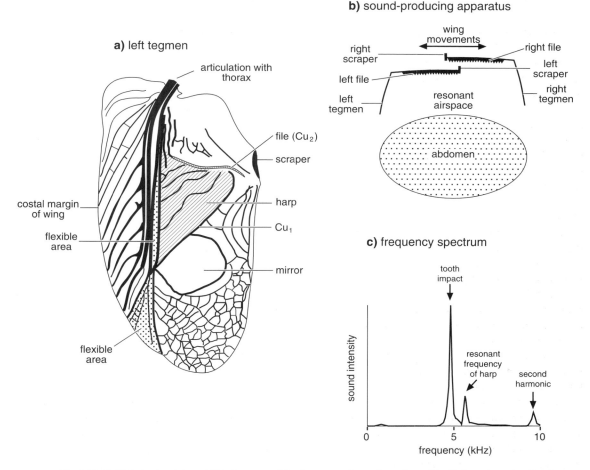

Fig. 26.3. Stridulation in crickets. (**a**) Left tegmen of *Gryllus campestris* seen from above. The file is on the underside of Cu_2 and so would not be seen in this view (after Bennet-Clark, 1989). (**b**) Diagrammatic cross section through an anterior abdominal segment of a stridulating cricket. The sizes of the files and scrapers are exaggerated. The size of the airspace can be varied so that it resonates at the frequency of the harp (not shown in this diagram). (**c**) Frequency spectrum of the sound produced by *Gryllus bimaculatus* (after Stephen & Hartley, 1995).

teeth, and similar numbers of teeth are present in other species. Close to the plectrum on the right tegmen is an area of thin, clear cuticle known as the mirror supported by a frame of thickened veins. As the scraper passes over each tooth of the file, the frame is first distorted and then returns to its original form, vibrating at its natural frequency as it does so.

In most tettigoniids the natural frequency of the mirror and its frame is higher than the tooth impact frequency and each impact produces a short burst of sound waves (an impulse, Fig. 26.5a) which is rapidly damped before the

next tooth impact occurs. The sound produced extends over a much wider range of frequencies than that of crickets (Fig. 26.5b) perhaps due to vibrations produced by different parts of the wing. The range often extends well into the human ultrasonic range and may be entirely inaudible to humans. The songs of *Conocephalus* species, for example, begin at about 20 kHz and extend above 100 kHz.

As in gryllids, sound is generally produced on wing closure, but sometimes, as in some individuals of *Platycleis intermedia* and in *Ephippiger* (Fig. 26.5), a syllable of sound is also produced on wing opening.

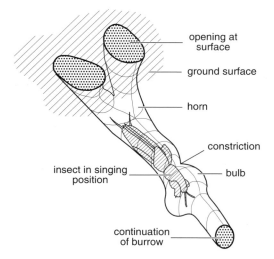

opening at
surface

ground surface

horn

constriction

insect in singing
position

bulb

continuation
of burrow

Fig. 26.4. Diagram of a singing burrow of *Gryllotalpa vineae* (after Bennet-Clark, 1970).

The wing movements are produced by the flight muscles, opening by the direct flight muscles and closure by the indirect muscles. In some species, as in *Neoconocephalus robustus*, the frequency of wing movements is as high as 200 Hz, much higher than the normal wingbeat frequency of these insects in flight. The muscles producing these movements are synchronous muscles, although synchronous muscles usually do not exhibit such high frequencies of contraction.

Some female tettigoniids have stridulatory apparatus, but this is usually much less well developed than in males.

In addition to tegminal stridulation, some bush crickets stridulate by rubbing ribs on the anal veins against a series of ridges on some abdominal tergites. In *Panteophylus*, this activity produces sound over a broad band of frequencies up to 160 kHz with a peak at 10 kHz (Heller, 1996).

Short-horned grasshoppers Many Acridoidea produce sounds by rubbing the hind femora against the tegmina. In Acridinae, a ridge on the inside of the hind femur rasps against an irregular intercalary vein, while in Gomphocerinae a row of pegs on the femur is rubbed against ridged veins on the tegmen (Fig. 26.6). A single tooth impact produces a highly damped vibration of the tegmen which has no specialized sound radiating area. The sound produced by one leg movement consequently

has a principal frequency at the tooth impact frequency. Sounds may be produced on both the up and down strokes. Both hind legs are used in stridulation, and, in most of the Gomphocerinae which have been studied, the legs are slightly out of phase and produce slightly different song patterns.

The stridulatory leg movements are produced by coxal promoter and remotor muscles. These are bifunctional muscles. In flight, where they contribute to wing depression, both sets of muscles contract simultaneously, but during leg movement they alternate (section 26.4).

Stridulatory apparatus is often present in female acridids as well as in males. Many other stridulatory mechanisms occur in other Acridoidea. They are reviewed by Kevan (1955).

26.1.2.2 *Stridulation in other insects*

Hemiptera Stridulation occurs widely in the Heteroptera, and many different parts of the body are involved, varying with the species. In Pentatomomorpha, the most common mechanisms involve a file on the ventral surface rubbed by a scraper on the leg, or a file on the wing rubbed against a scraper on the dorsal surface of the body. For example, both sexes of *Kleidocerys resedae* have a vein-like ridge on the underside of the hind wing (Fig. 26.7a). This ridge bears transverse striations about 1.7 μm apart and is rubbed on a scraper projecting from the lateral edge of the metapostnotum. In general, the Cimicomorpha do not stridulate, but the Reduvioidea nearly all have a file between the front legs which is rasped by the tip of the rostrum (Fig. 26.7b). This apparatus is present in males, females and larvae.

Many aquatic Heteroptera stridulate under water using a variety of parts of the body. *Ranatra*, for example, has a file on the fore femora with a scraper on the coxa; *Buenoa* has a similar arrangement in addition to a file on the fore tibia which it scrapes with the rostrum. In these aquatic bugs, the air bubble normally carried beneath the surface for respiration acts as a resonator. In corixids, the bubble is caused to vibrate by the movements of the head as it is rubbed by the femoral file. Because the bubble gets smaller the longer the insects remains submerged, its resonant frequency increases. When the insect dives after surfacing the bubble is large with a lower resonance. Thus, the sound frequency produced by these insects varies in relation to the size of the air bubble.

a) stridulatory mechanism

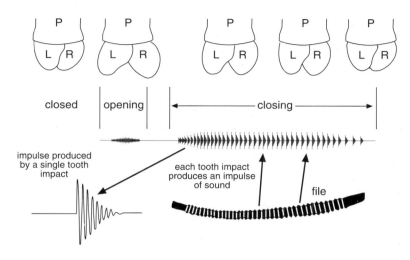

Fig. 26.5. Stridulation in bush
crickets (*Ephippiger*). (a) Stridulatory
mechanism. Top: opening and closing
of the forewings (tegmina). In this
species the wings are very short and
not used for flying. P, pronotum; L,
left tegmen; R, right tegmen. Center:
the file showing how each tooth
impact produces a sound impulse.
Detail of a single impulse to the left
(based on Pasquinelly & Busnel,
1955). (b) Frequency spectrum of the
sound produced (based on Dumortier,
1963b).

b) frequency spectrum

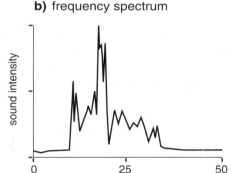

Amongst Homoptera, frictional methods of stridulation occur in the aphid, *Toxoptera*, and most Psyllidae.

Coleoptera Stridulation occurs in many beetles, especially amongst Carabidae, Scarabaeidae, Tenebrionidae and Curculionidae, and in the aquatic Dytiscidae and Hydrophilidae. Different species use different parts of the body to produce sounds, but most commonly the elytra are involved. In *Oxycheila*, for instance, there is a striated ridge along the edge of the elytron which is rubbed by a ridged area on the hind femur.

Larval Lucanidae, Passalidae and Geotrupidae also stridulate, rubbing a series of ridges on the coxa of the middle legs with a scraper on the trochanter of the hind leg. In larval passalids the hind leg is greatly reduced to

function as a scraper and is no longer used in locomotion (Fig. 26.8).

Review: Lyal & King, 1996 – Curculionidae

Lepidoptera The larvae and pupae of some Lepidoptera are able to stridulate. The caterpillars of lycaenid and riodinid butterflies which are dependent on ants for their development stridulate using a file on an abdominal segment. In *Arhoplala*, for example, a file on segment six is rubbed by segment five. Caterpillars of species not associated with ants are not known to stridulate (DeVries, 1990).

Three main types of stridulation occur in pupal Lepidoptera (Hinton, 1948). In ten families, notably the Hesperiidae, Papilionidae, Lymantriidae and Saturniidae there are coarse transverse ridges on the anterior edges of

a) stridulatory mechanism

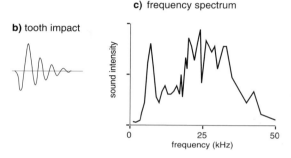

c) frequency spectrum

b) tooth impact

Fig. 26.6. Stridulation in grasshoppers. **(a)** Stridulatory mechanism. The inner side of a hind femur showing the position of the row of stridulatory pegs (the file) with detail of a few pegs (*Stenobothrus*) (after Roscow, 1963). **(b)** Sound produced by a single tooth impact. **(c)** Frequency spectrum of the sound produced by *Chorthippus biguttulus* (based on Dumortier, 1963b).

certain abdominal segments which are rubbed against fine tubercles on the posterior edges of the preceding segments. In Noctuidae, the pupa may have rough areas on the head, thorax and abdomen which are rubbed against the inside of the cocoon or the inside of the cocoon itself may be ridged so that the pupa's wriggling movements produce a scraping sound. The pupa of *Gangara thyrsis*, a hesperiid, has a pair of transverse ridges on either side of the ventral midline of the fifth abdominal segment. The long proboscis extends between and beyond these ridges and is itself transversely striated, so that when the abdomen contracts the proboscis rubs against the ridges and produces a hiss.

Stridulation appears to be rare in adult Lepidoptera, but male *Urania* have scales with scalloped edges in a shallow concavity of the anterior femur. These are rubbed by long scales on the coxa producing an irregular sound extending over a wide frequency range with a peak at 25–30 kHz (Lees, 1992).

Other groups Relatively isolated instances of frictional stridulation are widespread in other groups of insects. A few examples are given here.

The larva of *Epiophleha* (Odonata) has lateral ridged areas on abdominal segments three to seven. These are rubbed by the ridged inner side of the hind femur to produce a sound. Similarly, larval Hydropsychidae (Trichoptera) have ridges on the side of the head and a scraper on the front femur.

Amongst ants, stridulation occurs in Ponerinae, Dorylinae and primitive Myrmicinae, striations at the base of the gaster being rubbed by a scraper on the petiole (Fig. 26.9). Finally, in the Tephritidae (Diptera) stridulation is probably widespread. For example, in the male of *Dacus tryoni* the cubito-anal area of each wing vibrates dorso-ventrally across two rows of 20–24 bristles on the third abdominal segment, thus producing a noise.

Review: Aiken, 1985 – aquatic insects

26.1.3 Tymbal mechanisms

A tymbal is an area of thin cuticle surrounded by a rigid frame. Vibrations are produced when the tymbal buckles, usually as a result of the activity of a muscle attached to its inner surface. Tymbals are found in Hemiptera, especially in the Homoptera, and in some adult Lepidoptera, primarily in Arctiidae.

26.1.3.1 *Tymbals in Hemiptera*

The tymbal mechanism is most fully studied in Cicadidae where it is normally restricted to the males. Dorso–laterally on each side of the first abdominal segment is an irregular area of thin cuticle called the tymbal. It consists of a membrane of resilin supported by a sclerotized rim. Set in the membrane is a series of dorso-ventral sclerotized ribs and, posteriorly, a more extensive sclerotized area called the tymbal plate. Dorsally, the membrane consists of a thickened pad of resilin (Fig. 26.10). The tymbal is protected by a forward extension of the abdominal cuticle forming the tymbal cover. Internally, a cuticular compression strut runs from the ventral body wall to the posterior edge of the supporting rim and a tymbal muscle, running parallel with the compression strut, arises ventrally and is inserted into an apodeme attached to the tymbal plate. Running from a backwardly projecting knob on the metathorax to the anterior rim of the tymbal is the tensor muscle. When it contracts it pulls the rim of the tymbal so that the curvature of the latter is increased.

The tymbal is backed by an airsac which surrounds the

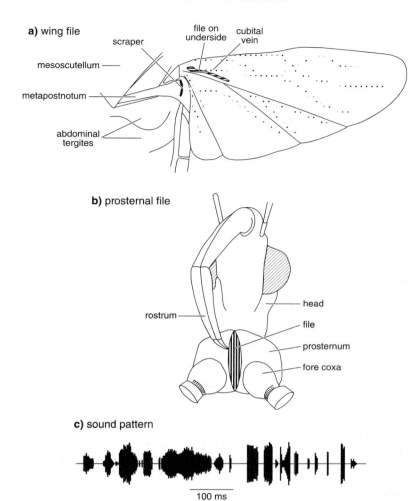

a) wing file

scraper

mesoscutellum ——

metapostnotum ——

abdominal tergites

file on underside

cubital vein

b) prosternal file

rostrum ——

head

file

prosternum

fore coxa

c) sound pattern

100 ms

Fig. 26.7. Stridulation in Heteroptera. (a) File on hind wing, scraper on metathorax. Note that the file is on the underside of the wing (*Kleidocerys*) (modified after Leston, 1957). (b) File on prosternum, tip of rostrum forms scraper (*Coranus*) (after Dumortier, 1963a). (c) Irregular sound pattern produced by *Coranus* (after Haskell, 1961).

muscle and communicates with the outside via the metathoracic spiracle. The presence of the airsac leaves the tymbal free to vibrate with a minimum of damping. Projecting back from the thorax on the ventral surface is the operculum, which encloses a cavity containing the tympanum (used in sound reception) and an area of thin, corrugated cuticle, the folded membrane, which separates the airsacs from the cavity beneath the operculum. When the abdomen is raised the membrane is stretched.

Sound is produced when the tymbal muscle contracts, causing the tymbal to buckle inwards. Depending on the arrangement of the vertical ribs on the tymbal and on the tension exerted by the tensor muscle, the buckling may occur in a single movement or in a series of movements as the ribs give way. Each component of the buckling may produce a pulse of sound, so one or several pulses may be

produced by a single contraction of the muscle, depending on the arrangement of the ribs (Fig. 26.10b, c). When the muscle relaxes, the tymbal is returned to its original position by the dorsal resilin pad. This outward movement of the tymbal also produces sound. In some species sound produced on the outward movement is of such low intensity that it probably does not contribute to the biological signal, but in other cases it does appear to form part of the song.

Buckling of the tymbal causes it to vibrate at its resonant frequency and this vibration is transferred to the airsacs backing the tymbal. If their resonant frequency is close to that of the tymbal, the sound is greatly amplified. In the bladder cicada, *Cystosoma*, the airsacs are fused and fill the whole abdomen except for a small space dorsally containing the viscera. By varying the volume of the airsacs, other cicadas can bring them in and out of tune so

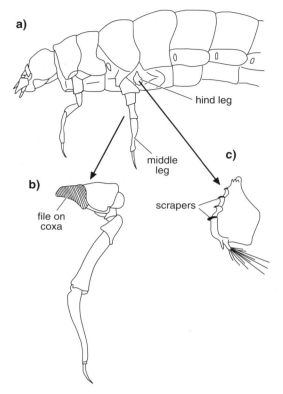

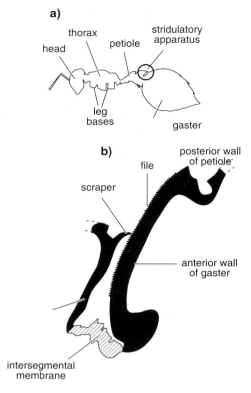

Fig. 26.8. Stridulation in a beetle larva (*Passalus*) (after Haskell, 1961). (a) Outline drawing of the anterior end of the larva showing the position of the stridulatory apparatus. (b) File on posterior face of the coxa of the middle leg. (c) Scrapers on reduced hind leg.

Fig. 26.9. Stridulation in ants (*Myrmica*) (after Dumortier, 1963a). (a) Outline drawing showing the position of the stridulatory apparatus. (b) Diagrammatic sagittal section through the cuticle of the stridulatory apparatus. The size of the file teeth is exaggerated for clarity.

that sound intensity is modulated. In some cicadas, such as *Cyclochila*, sound radiates from the tympani which also function as hearing organs; the tymbals themselves are unimportant, but in other species both tymbals and tympani, as well as other parts of the abdomen wall, act as radiators (Young & Bennet-Clark, 1995).

Repeated oscillations of the tymbal muscle produce a sustained sound. In most species, including *Psaltoda* where they oscillate at 225 Hz, the tymbal muscles are driven neurogenically (section 10.3.2.2), but in *Platypleura*, with an oscillation frequency of 390 Hz, the tymbal muscles are myogenic fibrillar muscles. In some species, the muscles of the two sides contract in synchrony, while in others they contract alternately thus doubling the pulse repetition frequency. Some species are known to vary the degree of synchrony between the two sides.

The frequency of the sound produced is determined by

the natural frequency of the tymbal. In *Cystosoma* this is 850 Hz, while in *Platypleura* it is about 4500 Hz. The sound may be a relatively pure tone with few harmonics, but in other species a broad range of frequencies is produced.

Amongst other Homoptera, tymbal mechanisms are present in the males of all the Auchenorrhyncha examined and in both sexes of many families. In these insects the tymbals are not backed by airsacs and so damping of the vibrations of the tymbals is very high. Tymbals are also present in some pentatomids, cydnids and reduviids where they are on the dorso-lateral surfaces of the fused first and second abdominal terga with an airsac beneath. In these insects, tymbals usually occur in both sexes, although often only those of the male are functional. The vibrations are transmitted to the substrate, presumably through the legs.

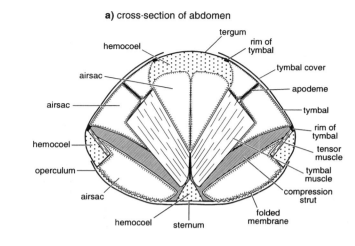

a) cross-section of abdomen

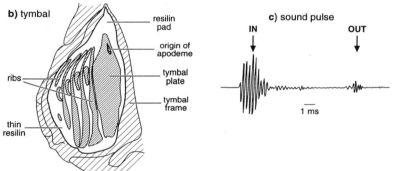

b) tymbal

c) sound pulse

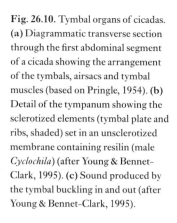

Fig. 26.10. Tymbal organs of cicadas. (a) Diagrammatic transverse section through the first abdominal segment of a cicada showing the arrangement of the tymbals, airsacs and tymbal muscles (based on Pringle, 1954). (b) Detail of the tympanum showing the sclerotized elements (tymbal plate and ribs, shaded) set in an unsclerotized membrane containing resilin (male *Cyclochila*) (after Young & Bennet–Clark, 1995). (c) Sound produced by the tymbal buckling in and out (after Young & Bennet–Clark, 1995).

26.1.3.2 *Tymbals of Lepidoptera*

Some Arctiidae have a tymbal on each side of the meta-thorax (Fig. 26.11). It is covered by scales posteriorly, but anteriorly has a band of parallel horizontal striations, comparable with the vertical ribs on the tymbals of cicadas. These vary in number in different species from 15 to as many as 60.

Buckling of the tymbal is not produced by a muscle directly attached to it, as it is in the cicadas, but by the coxo–basalar muscle attached above and below it. When this muscle contracts, the tymbal buckles inwards, starting dorsally and proceeding down the striated band, each rib being stressed to the point of buckling and then suddenly giving way and producing a pulse of sound. When the muscles relax, the tymbal springs out due to elasticity and a further series of pulses is produced. In *Melese*, the sounds produced on the IN movement show a progressive fall in frequency, those on the OUT movement a progressive rise (Fig. 26.11b). Most of the sound produced lies within a range of frequencies from 30–90 kHz, but the overall spectrum extends from 11 to 160 kHz.

Tymbals are also present in some Pyralidae. In the wax moth, *Galleria*, the tegula at the base of the forewing is modified to form a tymbal. No muscles are directly associated with it, but it is actuated by twisting the wing base presumably by the direct flight muscles. In males of another pyralid, *Symmoracma*, sound is produced by a tymbal on the underside of the terminal abdominal segment (Heller & Krahe, 1994).

Some butterflies, of which *Inachis io* (the peacock butterfly) is the best known example, produce sounds as they open their wings due to the buckling of a small area of the forewing between the costal and subcostal veins.

26.1.4 Vibrations produced by the flight muscles

In addition to the stridulatory movements of some Orthoptera that involve the flight muscles (see above), oscillation of the flight muscles produces thoracic and

a) tymbal

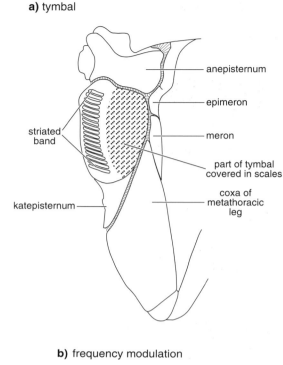

b) frequency modulation

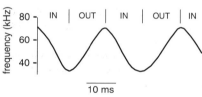

Fig. 26.11. Tymbal organ of an arctiid moth (*Melese*) (after Blest, Collett & Pye, 1963). (a) Diagram of the left side of the metathorax showing the position of the tymbal relative to the surrounding sclerites. (b) Modulation of the principal frequency component of the sound produced. Each cycle is the result of one in–out movement of the tymbal.

wing vibrations which may be used in communication by a number of insect species.

Vibration of the wings in flight produces a sound with a fundamental frequency the same as the frequency of the wingbeat. However, other components may be added to this fundamental frequency as a result of the varied structure of different parts of the wing and the vibration of the thorax, so the overall sound produced is complex consisting primarily of harmonics of the wingbeat frequency. The flight noise of a locust, *Schistocerca*, with a wingbeat

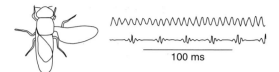

Fig. 26.12. Sound produced by wing muscles. Wing vibration by male *Drosophila melanogaster*. Left: diagram showing the position of the wing during sound production. Right: two different songs. The sounds are of very low intensity and are perceived in the near field (after Cowling & Burnet, 1981).

frequency of about 25 Hz, is a complex sound with frequencies extending from 60 to 6400 Hz although mainly falling between 3200 and 5000 Hz. Pulses of sound are produced at the rate of 17–20 per second, corresponding to the wingbeat frequency.

Insects such as Lepidoptera with very low wingbeat frequencies, of the order of 20 Hz, produce a flight tone that is inaudible to humans, but insects with higher wingbeat frequencies produce clearly audible sounds and these may sometimes provide relevant signals to the insects. The flight tone of *Apis* is about 250 Hz and that of culicine mosquitoes from about 200 to over 500 Hz. In general, smaller species have a higher wingbeat frequency and flight tone than larger species (Belton, 1986) (see Fig. 9.22). Hard-bodied insects usually produce a higher intensity of sound than soft-bodied insects.

Many *Drosophila* species and some trypetid flies produce sounds when they are not flying by vibrating the partially opened wings. Male *D. melanogaster* produces sound with one wing extended sideways at 90° to the body; the sound frequency is only about 100 Hz (Fig. 26.12). *Drosophila erecta*, which sings with its wings only partially open, at 20–40° from the body axis, produces a high frequency sound, about 450 Hz. These sounds are of very low intensity and they are perceived as particle movements by the antennae of the female. They may be produced as a continuous tone, or as a series of sound pulses.

Honeybees use their flight muscles to generate vibrations important in social communication (Fig. 26.13). A worker bee returning from foraging performs a waggle dance during which airborne vibrations are produced by the wings which are held horizontally over her back. These vibrations have a frequency of 250–300 Hz, close to the wingbeat frequency, and are produced in 30-second pulses. They are detected as near field sounds (particle

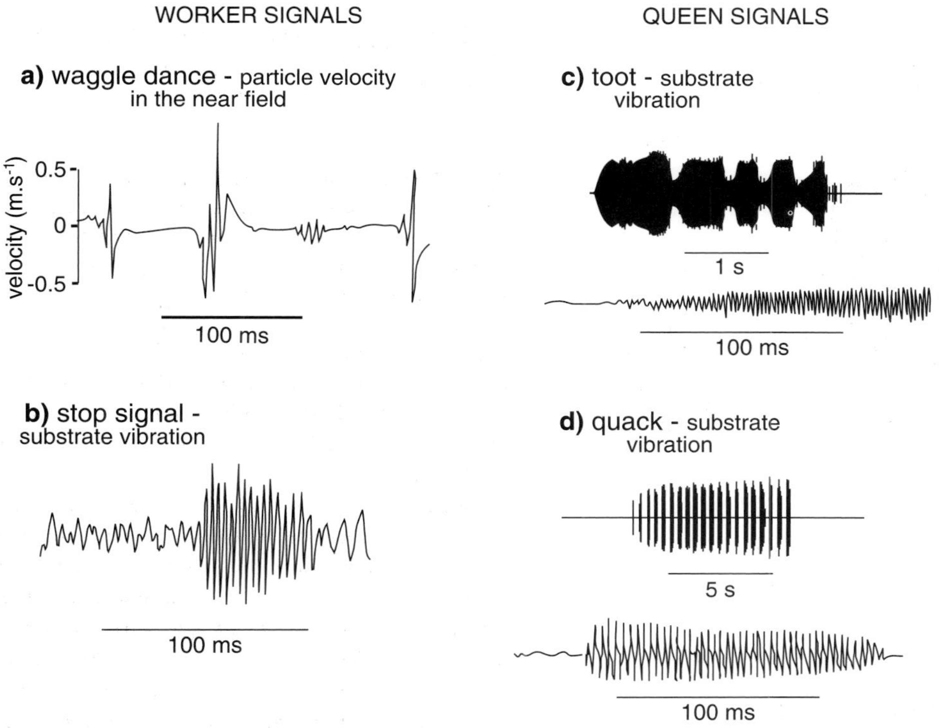

WORKER SIGNALS   QUEEN SIGNALS

a) waggle dance - particle velocity in the near field

c) toot - substrate vibration

b) stop signal - substrate vibration

d) quack - substrate vibration

Fig. 26.13. Vibrations produced by wing muscles of worker and queen honeybees (*Apis*). (**a**) Particle displacement in the near field produced by workers performing the waggle dance. The variation in amplitude of alternate pulses results from the side-to-side movement of the gaster (after Michelsen *et al.*, 1987). (**b**) Substrate vibration produced as a stop signal ('begging') by a bee following a dancer (after Michelsen, Kirchner & Lindauer, 1986b). (**c**) Tooting. A substrate borne vibration produced by a young queen. Detail of the vibrations below (after Michelsen *et al.*, 1986b). (**d**) Quacking. A substrate borne vibration produced by a young queen. Detail of the vibrations below (after Michelsen *et al.*, 1986a).

displacement) by the antennae of following workers who, at the same time, may produce stop signals (previously known as begging calls). These, too, are produced by vibration of the flight muscles, but are transmitted through the insects' legs to the comb of the hive which transmits the vibrations.

Young virgin queen bees produce piping sounds by thoracic vibration. The thorax is pressed to the comb so that vibrations are transferred directly to the comb. Queens that have escaped from the cells produce a 'toot'; those still in the cells produce a 'quack' which is a rapid train of short periods of vibration (Fig. 26.13).

26.1.5 Air expulsion

The death's head hawk moth, *Acherontia*, sucks air in through the mouth by dilating the pharynx. This produces a sound with a range of frequencies up to about 14 kHz and a maximum at 7–8 kHz. Vibrations of the epipharynx produce a series of sound pulses (Fig. 26.14). Contraction of the pharynx with the epipharynx held erect expels the air producing a whistle. These sequences are repeated rapidly.

The hissing cockroach, *Gromphadorhina*, makes sounds by expelling air through the enlarged fourth abdominal spiracle. The trachea leading from the spiracle tapers to a narrow connection with the longitudinal tracheal trunk. Air is forced through this connection when the expiratory muscles contract and all other spiracles are closed, producing a hissing noise. The frequency spectrum of the sound has a maximum at about 10 kHz but extends up to about 40 kHz.

Fig. 26.14. Sound produced by a vibrating column of air. Diagrammatic sagittal sections through the head of the death's head hawk moth, *Acherontia atropos*, showing the method of sound production (after Dumortier, 1963a). **(a)** On inspiration the epipharynx vibrates producing a pulsed air stream and intermittent sounds (below). **(b)** On exhalation the airstream is continuous, producing a continuous whistle (below).

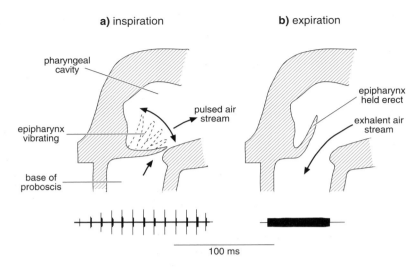

a) inspiration   **b)** expiration

pharyngeal cavity

pulsed air stream

epipharynx vibrating

base of proboscis

epipharynx held erect

exhalent air stream

100 ms

26.2 SIGNAL TRANSMISSION

The signals produced in these various ways may be transmitted through the air as sounds or as substrate vibrations. Substrate vibration may be produced by percussion, stridulation, a tymbal or by contractions of the flight muscles. Except in the case of percussion, the vibrations are usually transmitted to the substrate via the insect's legs, although queen bees bring the ventral surface of the body into contact with the comb. Mechanisms producing substrate vibration may also produce airborne sounds, but is it probable that, for most insects, the relevant signal is conveyed through the substrate.

In soil, vibrations are transmitted for only a few centimeters due to heavy damping, but, on plants, signals may be transmitted for two meters. They are propagated as bending waves at velocities ranging from less than 0.5 to over $1.0\,\mathrm{m\,s^{-1}}$, depending on characteristics of the plant stem and on the wavelengths of the vibrations. Higher frequencies are propagated at higher rates. The vibrations produced in a plant by an insect extend over a relatively broad frequency range from below 200 Hz to over 1500 Hz. Their magnitudes (amplitudes) also vary, but may be as much as $4\,\mu\mathrm{m}$ peak to peak with accelerations measuring as much as $8\,\mathrm{m\,s^{-2}}$. This is well within the sensitivity ranges of the subgenual organs of many insects (section 23.2.3.1).

In air and water, vibrations (sounds) occur as both pressure waves and particle movements. Close to the source, most of the energy of the signal involves particle displacement; at greater distances pressure waves are more important. The region in which particle movement is dominant is called the near field; beyond this, where pressure waves carry most of the energy, is the far field. These distinctions are important. Very low intensity sounds with insufficient energy to stimulate a pressure receiver may have sufficient energy in the near field to stimulate particle displacement receptors (sections 23.1.3.1, 23.2.3.2).

The near field usually extends less than one wavelength of sound from the source. Consequently, it is important for insects only when they are very close together, one or two centimeters apart. In water, the near field is about four times greater than in air for a given wavelength.

Reviews: Markl, 1983 – physics of substrate vibration; Michelsen & Larsen, 1983 – environmental effects on sound propagation; Michelsen *et al.*, 1982 – vibration of plants

26.3 PATTERNS OF VIBRATIONAL SIGNALS

26.3.1 Organized patterns.

The vibrational signals produced by insects are often modulated to produce discrete patterns characteristic of the species, and some species also vary their signal in relation to particular functions. In the case of sounds produced by stridulation different patterns are produced by varying the tooth impact frequency, the number of file teeth that impact the scraper on each complete wing or leg movement, and the frequency of the movements.

Various classifications have been used to describe the hierarchies of sounds produced by orthopterans, but the system currently most widely used was proposed by Elsner

(1974) (Fig. 26.15). The sound of a single file tooth striking the scraper is called an impulse; the sound produced by a single up or down movement of a leg or by opening or closing a wing is called a syllable; and the sound produced by one whole sequence of leg movements, as shown in Fig. 26.16, is a chirp. Chirps may also be organized into higher order sequences.

The grasshopper, *Chorthippus mollis*, provides an example of the mechanisms involved in producing a distinctive song pattern. At the start of stridulation it produces a sharp, high amplitude syllable of sound by moving one leg sharply downwards (Fig. 26.16). It then makes oscillating up and down movements with the femur which gradually increase and then decrease in amplitude. These oscillations are superimposed on a slow raising and lowering of the general position of the femur. As a result, different sections of the file strike the scraper and produce sound syllables that vary in intensity. Towards the end of the chirp the rapidity with which each leg movement is made becomes reduced and the syllables occur at greater intervals. The hindleg on the other side makes a similar sequence of movements, but is slightly out of phase. The number of syllables in each chirp also varies. At the beginning of stridulation each chirp consists of 10–12 pairs of syllables; towards the end of a sequence, there may be 26–28 pairs.

The small movements at the beginning of a song sequence by *Chorthippus mollis* obviously only employ a small proportion of the file teeth and it is usual that not all the teeth are used in any one leg or wing movement. The bush cricket, *Platycleis intermedia*, has, on average, 77 teeth on the file and produces a chirp of three syllables, one syllable on opening the wings and two on subsequent closures. These pulses employ, respectively, only 31%, 41% and 67% of the file teeth.

In tettigoniids, the frequency of muscle oscillations producing wing closure varies considerably even in closely related species. For example, while the wing muscles of *Neoconocephalus robustus* oscillate at about 200 Hz during singing, those of *N. ensiger* oscillate at only 10–15 Hz. These differences result in similar differences in the frequencies of syllable production.

Because the rate of muscular contraction is temperature dependent, the movements involved in stridulation are also affected by temperature. Thus the frequency with which sound impulses or syllables are produced by some crickets and bush crickets varies linearly with temperature

a) impulse - one tooth impact

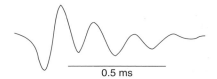

0.5 ms

b) syllable - one down or up movement of leg (opening or closing wing)

tooth impacts

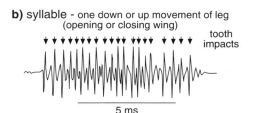

5 ms

c) chirp - one complete cycle of leg (wing) movements

down   up   down   up   down   up

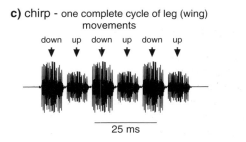

25 ms

d) first order sequence - continuous series of chirps

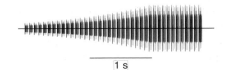

1 s

Fig. 26.15. Terminology used to describe orthopteran sounds (grasshopper, *Chorthippus biguttulus*) (after Elsner, 1974). Similar terminology is applied to sounds produced by wing movements in crickets and bush crickets. (a) Impulse. Sound produced by the impact of a single file tooth against a scraper. (b) Syllable. Sound produced by a single up or down movement of the leg (or opening or closing the wings in crickets). (c) Chirp. Sound produced by a complete cycle of leg movements. (Fig. 26.16 shows the leg movements in a cycle, but note that the species is different). (d) Sequence. Sound produced by a series of chirps following in rapid succession with only brief, less than 100 ms, intervals between them.

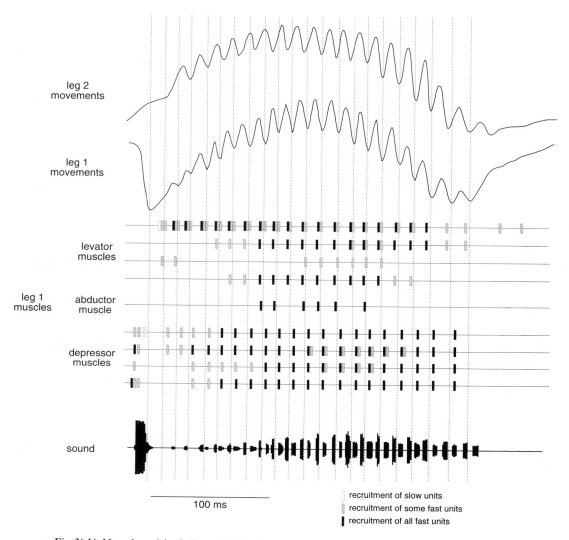

leg 2
movements

leg 1
movements

leg 1
muscles

levator
muscles

abductor
muscle

depressor
muscles

sound

100 ms

▯ recruitment of slow units
▨ recruitment of some fast units
▮ recruitment of all fast units

Fig. 26.16. Muscular activity during stridulation by a grasshopper (*Chorthippus mollis*). Top: the activity of the two hind legs during a single chirp which consists of a sequence of up and down movements of each leg superimposed on a slower upward and downward movement. The effect of the slow upward movement is to bring progressively more proximal teeth of the femoral file into contact with the scraper. In this species, the two legs are slightly out of phase. Center: the pattern of activity in muscles moving leg 1. Each horizontal line represents a separate muscle. Each muscle is innervated by fast and slow axons whose activity in recruiting the muscle units is shown by rectangles. Note that the greatest number of muscles and muscle units are activated by the fast axons during the middle of the chirp when the amplitude of leg movements is greatest (above) and the amplitude of sound greatest (below). The muscles are activated shortly before leg movement is observed. Bottom: sounds produced by the movements of leg 1. The vertical axis shows the amplitude of sound. Syllables of sound are produced on both the up- and downstrokes, but sound intensity is greatest on the downstroke (after Elsner, 1975).

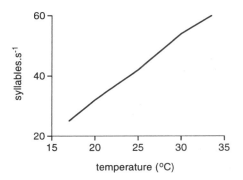

Fig. 26.17. The effects of temperature on sound production. The frequency of syllable production (wing movements) produced by the cricket, *Oecanthus* (after Walker, 1962).

(Fig. 26.17). The increase in pulse repetition frequency with temperature may result solely from a more rapid opening of the tegmina so that more sound-producing closures are possible in unit time as in *Oecanthus*, or may also involve a reduction in the number of file teeth employed so that short, rapid closing strokes of the tegmina occur as in *Gryllus rubens*.

Thus, although different species within a group of orthopterans have basically the same stridulatory apparatus, they are able to produce distinctive songs by using species-specific motor patterns (Fig. 26.18). An individual may also be able to produce different songs in different circumstances which may differ in intensity and sound frequency as well as in pattern (Fig. 26.19, 26.22a).

Similar principles apply to the cicadas where the mechanism of sound production is similar in all species (Fig. 26.20a). Repetition frequency is varied between species by differences in the frequencies of muscle oscillation. For example in *Cystosoma* each tymbal muscle oscillates about 40 times per second; in *Abricta* the frequency is about 75 Hz, while in *Psaltoda* it is 225 Hz. Repetition rate is also varied by changes in the tension of the tymbal, which can be varied even within a species, by differences in the activity of the tensor muscle. The sequential buckling of the ribs on the tymbal may also produce discrete pulses of sound differing from between species. Finally, the tymbal muscles of the two sides may be in or out of phase; in the latter case the repetition rate is doubled.

Amplitude modulation is produced in cicadas by changes in the activity of the tensor muscle which distorts the tymbal frame and so changes the elastic properties of the tymbal (Hennig *et al.*, 1994). Raising or lowering the

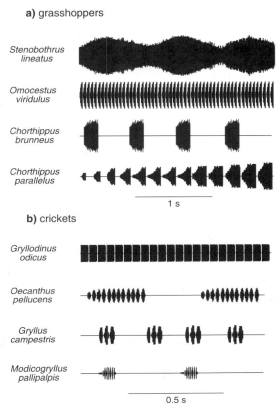

Fig. 26.18. Patterns of sounds produced by stridulatory mechanisms in Orthoptera. **(a)** Femoro-tegminal stridulation in grasshoppers. All the species are members of the Gomphocerinae and employ a similar method of sound production (after Haskell, 1957). **(b)** Tegminal stridulation in crickets. All the species employ a basically similar mechanism (after Popov *et al.*, 1974).

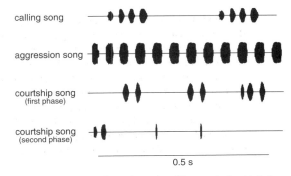

Fig. 26.19. Patterns of sounds produced by tegminal stridulation by a male of *Gryllus campestris* in different situations (after Huber, 1962).

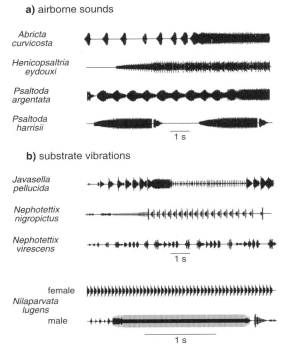

a) airborne sounds

Abricta curvicosta

Henicopsaltria eydouxi

Psaltoda argentata

Psaltoda harrisii

1 s

b) substrate vibrations

Javasella pellucida

Nephotettix nigropictus

Nephotettix virescens

1 s

female
Nilaparvata lugens
male

1 s

Fig. 26.20. Vibration patterns produced by tymbal organs. All the insects have a basically similar mechanism for producing vibrations. (**a**) Airborne vibrations produced by different species of cicada (after Young, 1973). (**b**) Substrate vibrations produced by planthoppers and leafhoppers. *Javasella* (Delphacidae), *Nephotettix* (Cicadellidae) and *Nilaparvata* (Delphacidae) (after de Vrijer, 1982; Claridge, 1983; Claridge, den Hollander & Morgan, 1985).

abdomen increases the space between the abdomen and the operculum, altering the resonant frequency of the airsacs so that they can be roughly tuned to the tymbal to increase the intensity of sound. Each species has a characteristic pattern of activity of these different muscles.

Organized patterns of vibrations are also produced by some insects that communicate via the substrate. This occurs in Plecoptera and Neuroptera, in many Homoptera, and in honeybees. These insects generate vibrations using different mechanisms – abdominal oscillations in Plecoptera and Neuroptera, tymbals in Homoptera (Fig. 26.20), and the flight muscles in *Apis*, but in most cases they are transmitted to the substrate via the legs.

In these cases a principal variant is the frequency with which discrete pulses of oscillation occur, each pulse being produced by a single impact or bout of muscle activity. The signal strength can also be varied, presumably by changing the force exerted by the muscles (Fig. 26.16). Fig. 26.2 shows the pattern produced when the male stonefly *Isoperla* strikes the substrate with its abdomen. The female responds with a different pattern and other species also have distinct sequences.

More examples of the patterns of vibration produced by species in different groups of insects using different mechanisms to produce vibrations are illustrated in the following works:

Crickets – tegminal stridulation   Bennet-Clark, 1989

Grasshoppers – femur/tegmen stridulation   Elsner, 1983

Cicadas – tymbal, airborne sound   Claridge, 1985

Planthoppers – tymbal, substrate vibration   Claridge, 1985

Stoneflies – percussion, substrate vibration   Szczytko & Stewart, 1979

Lacewings – abdominal vibration, via substrate   Henry, 1980

Drosophila – wing vibration, near field   Ewing, 1989

26.3.2 Unorganized vibrations

Many insect species produce vibrations, which may be airborne or substrate-borne, that are not organized in precise temporal patterns (Fig. 26.7c). In these cases, the vibrations themselves, in context, are sufficient to convey information to conspecifics or other species. Such sounds may be produced by any of the mechanisms described above.

26.4 NEURAL REGULATION OF SOUND PRODUCTION

The patterns of muscular activity resulting in sound production are generated by pattern generators in the central nervous system. In crickets, where the sound is produced by movements of the forewings, the pattern generator is in the mesothoracic ganglion; in grasshoppers, where stridulation involves movements of the hind legs, the generator is in the metathorax. The pattern generators are presumed to be networks of interneurons whose output regulates the activity of the motor neurons controlling the muscles.

However, each of the muscles is made up of several units innervated by fast and slow axons. As a result, the

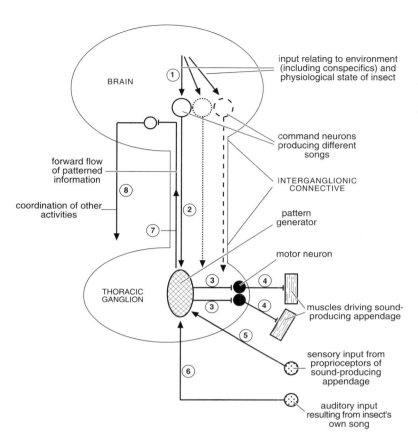

Fig. 26.21. Neural control of sound production in Orthoptera. The diagram is based on information from crickets and grasshoppers; all the steps shown have not necessarily been demonstrated in any one species. Numbers indicate the sequence of neural events.

force exerted by a muscle can be varied, and this affects sound production. When a grasshopper produces a chirp, the number of muscle units involved in the oscillating movements of the femur, and their activation via fast axons, first increases and then decreases (Fig. 26.16). These changes, with slight changes in the timing, produce the alterations in up and down movements during the chirp. Variation in the activity of other muscles, the coxal adductor and abductor muscles, produces changes in sound intensity by pulling the coxa towards or away from the body and varying the pressure of the femur against the tegmen.

Although the basic movements are controlled by a pattern generator, they are modulated by sensory feedback from proprioceptors. In crickets, a hair plate and campaniform sensilla on the tegmina are essential for the production of normal song. In addition, at least in some species, the sound of the insect's own song is important. *Gryllus bimaculatus*, for example, is less well able to regulate its

tooth–impact frequency when it cannot hear its own song. By adjusting this frequency and altering its wing position it is able to tune its song (Stephen & Hartley, 1995). In the cicada, *Cystosoma*, the frequency of oscillation of the tymbal, maintained by a central pattern generator in the metathoracic ganglion, is modulated by feedback from a chordotonal organ inserted on the rim of the tymbal.

Singing is initiated by descending interneurons from the brain, known as command neurons (Fig. 26.21). In a male cricket, different command neurons initiate different songs. In addition, at least in grasshoppers, it is probable that output from the metathoracic pattern generator is fed forwards to the subesophageal ganglion and then redistributed to other parts of the body. This information may provide positive feedback resulting in increased amplitude of leg movements and may also be involved in coordination of the left and right legs. It probably also permits coordination of other activities with those directly involved in stridulation (Lins & Elsner, 1995).

Both external and internal factors are important in regulating song production. Males only sing when they are sexually mature and, in grasshoppers, there is a brief interruption of singing after copulation. Males of *Gryllus campestris* only sing when they are carrying a spermatophore. After copulation, when the spermatophore is transferred to the female, singing stops. Female grasshoppers only sing when they are in a sexually responsive state and, like other aspects of sexual maturation in these insects, it is possible that singing is regulated by juvenile hormone.

Review: Kutsch & Huber, 1989 – crickets

26.5 SIGNIFICANCE OF VIBRATIONAL SIGNALS

The vibrations produced by insects can be classified according to whether they are intraspecific, used to signal to other members of the same species, or interspecific, giving signals to other species.

26.5.1 Vibrations having intraspecific significance

Vibrations having intraspecific significance are usually organized sounds with a regular pulse repetition frequency. The regularity of these patterns confers species specificity and contributes to reproductive isolation between species. Often the vibration patterns are concerned with attraction of the opposite sex from a distance; in other cases they are produced during courtship when the two sexes are close together. In many of these species, only the males stridulate.

Sometimes an individual insect produces different sounds in different contexts. Thus many Orthoptera produce a calling song to attract the opposite sex, a courtship song when the sexes are close together, and sometimes also an aggression song sung by one male in the presence of another. These songs are characterized by differences in their patterns (Fig. 26.19).

Reproductive isolation   Patterns of vibration probably contribute to sexual isolation in many insects, but in a few species they are known to have a critical role. For example, the two grasshoppers, *Chorthippus biguttulus* and *Chorthippus brunneus* are sympatric and very similar morphologically. In the laboratory they interbreed and produce fertile offspring, but this does not occur in the field. Here the differences in song play a major role in sexual isolation. The same is probably true with two

species of *Drosophila*, *D.pseudoobscura* and *D.persimilis*. Some populations of the brown planthopper of rice, *Nilaparvata lugens*, exhibit little interbreeding, even in the laboratory, and it appears that differences in the patterns of substrate borne vibrations produced by the different populations are primary isolating mechanisms (Claridge, den Hollander & Morgan, 1985).

These three examples demonstrate that airborne sounds, in both the far field (*Chorthippus*) and near field (*Drosophila*), and substrate-borne vibrations can contribute to sexual isolation.

Attraction from a distance   Attraction over the greatest distances involves airborne sounds in the far field, perceived by pressure or pressure-difference receivers. These occur in the Orthoptera and the cicadas. Attraction to a source of sound is called phonotaxis. The distances over which attraction occurs are probably usually no more than a few meters. For example, the maximum distance at which the grasshopper, *Ligurotettix coquilletti*, can hear a calling male is about 14 m so attraction can only occur within this distance (Bailey, Greenfield & Shelly, 1993). The sound of the mole cricket, *Gryllotalpa*, however, is audible to humans over several hundred meters from the source and presumably attracts females from some distance. Frequently, as in crickets and most cicadas, only the males sing and the female moves towards the source of sound. In gomphocerine grasshoppers, however, both sexes stridulate. If a female *Chorthippus brunneus* in the responsive state hears the song of a male, she sings in reply, with a similar song to the male's calling song. The two insects orient and move towards each other, stopping to sing at intervals and thus carrying out a mutual search eventually leading to visual contact with each other.

In the mosquito, *Aedes*, the male recognizes a female from the sound produced by her wings in flight. In this case, attraction occurs from only a short distance, with the antennae acting as displacement receivers in the near field.

Substrate vibrations are also used to attract members of the opposite sex. This occurs in Delphacidae and Cicadellidae amongst the Homoptera, and in Plecoptera and Chrysopidae. In many of these species, the females also produce vibrations, returning the calls of the males. In these insects, communication probably occurs over only one or two meters because transmission is limited by the size and characteristics of the plant on which the insects are situated.

Sometimes the sounds produced by male insects attract other males as well as females. This occurs in some crickets and a few cicadas. The males in such groups then sing together (chorusing), although factors other than sound may also contribute to the formation of the aggregations.

Chorusing occurs in a few species of bush crickets, grasshoppers and cicadas. Sometimes the singing activities of the individuals are not precisely timed relative to each other although bouts of singing are separated by periods of silence. In general, the insects only sing when they hear a conspecific singing and stop when there is no feedback. This is called unsynchronized chorusing.

Some other species exhibit synchronized chorusing in which the calls of individuals are precisely timed with respect to others. In these species, a follower male starts its song at a precise interval after the start of another insect's song, and it adjusts the timing by altering the duration of its chirps and silent intervals.

In some other Orthoptera, chorusing involves the alternation of the songs of individual insects. This occurs in the grasshopper, *Ligurotettix planum*, for example, where an individual alternates with other nearby individuals, but ignores the song of more distant insects. The alternation involves a set time delay following the onset of a neighbor's call (Minckley, Greenfield & Tourtellot, 1995).

Courtship When the two sexes are close together, many insect species exhibit some courtship behavior before mounting and copulation. The precise functions of courtship are unclear. It is sometimes suggested that some 'sexual threshold' (perhaps relating to contact with another insect) in the female must be lowered and it may provide the opportunity for sexual selection. In some insects, courtship involves sound.

In crickets and grasshoppers using a calling song to attract a mate, song pattern changes to the courtship song when the sexes are within view of each other. The courtship song is often of lower intensity (that is, quieter) than the calling song and, in many crickets, the principal component of the courtship song is also at a higher frequency than that of the calling song, attenuating more rapidly with distance (Fig. 26.22). It may also be less regularly organized and species-specific than the calling song.

Other insects may only produce sounds during courtship; it is not involved in bringing the sexes together. This

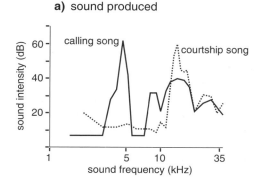

a) sound produced

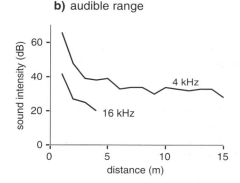

b) audible range

Fig. 26.22. Calling and courtship songs of *Gryllus campestris*. (a) Frequency spectrum of the two songs. The sound intensity of the calling song is greatest at about 4 kHz, with higher but less intense harmonics. The courtship song has its highest intensity at about 16 kHz (after Nocke, 1972). (b) Sound intensity at different distances from a male singing the calling song. High frequency sounds attenuate much more rapidly with distance. As a result, the courtship song (principal frequency about 16 kHz) would be audible to another cricket with an auditory threshold around 30 dB only within two or three meters of the source. The lower frequencies of the calling song would be above threshold for a much greater distance (after Popov *et al.*, 1974).

is the case amongst the species of *Drosophila* which produce sound by wing vibration.

Territorial behavior and aggressive stridulation Aggressive stridulation nearly always occurs in relation to reproductive behavior. It has the effect of spacing males in the environment, and is sometimes associated with territorial behavior. The calling song, while attracting females, may, at the same time, serve as an aggressive signal to other males, but many species have a separate aggression song.

Aggressive stridulation is well illustrated by crickets. Each male of *Oecanthus* has a territory of some $50 \, \text{cm}^2$ in which he sings his normal calling song. If another male cricket enters the territory, the resident male sings his aggression song and the intruding male replies. Fighting may occur, the males lashing each other with their antennae, sparring and biting until one male retires. The dominant males in a colony stridulate more in aggression than others and, at the end of an encounter, the dominating male may continue to stridulate.

Stridulation by the grasshopper, *Ligurotettix coquilletti*, is also involved in territory defense as well as female attraction. In this example, the territory is also a food resource, a creosote bush. Dominant males defend bushes with the highest nutritional quality against other males (Greenfield & Shelly, 1990).

The whistling sounds produced by the moth, *Hecatesia*, function in territory defense as well as in mate attraction (Alcock & Bailey, 1995).

Larval caddis flies in the family Hydropsychidae live in silken retreats attached to stones in the water. A larva may attempt to occupy the retreat of another individual, leading to fights. If one of the larvae stridulates, it wins the contest.

Communication in social insects Both worker and queen honeybees use vibrations for signalling within a colony (Fig. 26.13). Workers produce a pulsed sound during the straight run of their dance and the number of pulses and total period of sound production is proportional to the distance from the food (Fig. 26.23). These sounds are modulated by the side to side movements of the gaster and may be used to indicate the distance of the food to other members of the colony. They are perceived by the antennae of following workers acting as particle receivers. In addition, the side-to-side swaying of the bee may produce displacements of the cell walls in the plane of the comb that can be perceived by other workers (Sandeman, Tautz & Lindauer, 1996). The following workers also produce stop signals which are transmitted via the comb and cause the dancing worker to stop and regurgitate nectar to the follower. The piping noises of young queen bees are produced in relation to swarming.

Bees in the genera *Melipona* and *Trigona* may also use vibrational signals. The signals of returning foragers of

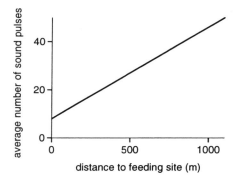

Fig. 26.23. Communication of distance by worker bees in the hive. Pulses of sound are produced as the abdomen swings from side to side (see Fig. 26.13). The number of pulses produced during the straight run of the dance is proportional to the distance to the food (after Wenner, 1962).

Trigona excite other workers in the colony and if, at the same time, the workers are offered nectar they will also begin to forage.

The stridulations of leafcutter ants, *Atta*, produced while cutting leaves, serve several functions. The vibrations are transmitted to the leaf via the head. The substrate vibrations attract other workers to the site. Other large workers are induced to start cutting, while minim workers climb on to the leaf fragments being cut. They are carried back to the nest on the leaf, protecting the major workers from parasitic phorid flies. The vibrations of the mandibles also produce a stiffening effect on tender leaves making them easier to cut (Roces & Hölldobler, 1995; Tautz, Roces & Hölldobler, 1995). Stridulation by the ant, *Messor*, is also substrate-borne. It leads to recruitment of other workers to prey (Baroni-Urbani, Buser & Schilliger, 1988).

The percussion sounds produced by *Zootermopsis* are important as a warning to other members of the colony, and are initiated by any disturbance, especially by vibration of the substrate. The sound leads to the retreat of other colony members to the remoter parts of the nest.

26.5.2 Sounds having interspecific significance

Sounds having interspecific significance are usually unorganized having no regular pulse repetition frequency and covering a broad spectrum of frequencies with sharp transients. They are usually produced by both males and females and sometimes also by larvae. Sounds of this type, which include the stridulation of reduviids, beetles and lepidopterous pupae, are presumed to be concerned with defence against predators, inducing a startle response.

This type of sound production sometimes accompanies visual displays employed against predators (section 25.6.1). In Arctiidae, sound production by the tymbal organs may be associated with a display of warning colors. These are distasteful species and it is generally true that sounds are more readily elicited from the less distasteful species amongst them. Sounds associated with visual display are also produced by some mantids and grasshoppers.

There is also the suggestion that sound mimicry occurs. For example, not only do certain syrphids look like bees, they also sound like them, and the sounds produced by *Nicrophorus* when it is disturbed are similar to those of a torpid bumble bee, to which the beetle also bears a superficial visual resemblance. Whether these resemblances are due to selection or to chance similarities is not known.

The sound frequencies produced by the tymbals of some arctiid moths are within the range of the signals produced by echolocating bats. The moths respond to a bat's signals with clicks of sound causing the bat to veer away. The mechanism by which the bat's behavior is altered is not fully understood.

Although interspecific vibrational signals usually have a defensive role, those produced by some caterpillars are exceptional. The larvae of Riodinidae and Lycaenidae that develop in association with ants use substrate vibrations to attract ants. Species not tended by ants do not stridulate (DeVries, 1990). Bumblebees use vibrations of the body produced by oscillation of the flight muscles to displace pollen from the anthers of some flowers. This process, known as buzz pollination, requires the bee to make bodily contact with the anthers either by gathering them to its body with the legs or by grasping an anther with the mandibles (King, 1993).

REFERENCES

Agee, H.R. (1971). Ultrasound produced by wings of adults of *Heliothis zea*. *Journal of Insect Physiology*, **17**, 1267–73.

Aiken, R.B. (1985). Sound production by aquatic insects. *Biological Reviews of the Cambridge Philosophical Society*, **60**, 163–211.

Alcock, J. & Bailey, W.J. (1995). Acoustical communication and the mating system of the Australian whistling moth *Hecatesia exultans* (Noctuidae: Agaristidae). *Journal of Zoology, London*, **237**, 337–52.

Bailey, W.J. (1991). *Acoustic Behaviour of Insects. An Evolutionary Perspective*. London: Chapman and Hall.

Bailey, W.J., Greenfield, M.D. & Shelly, T.E. (1993). Transmission and perception of acoustic signals in the desert clicker, *Ligurotettix coquilletti* (Orthoptera: Acrididae). *Journal of Insect Behavior*, **6**, 141–54.

Baroni-Urbani, C., Buser, M.W. & Schilliger, E. (1988). Substrate vibration during recruitment in ant social organization. *Insectes Sociaux*, **35**, 241–50.

Belton, P. (1986). Sounds of insects in flight. In *Insect Flight: Dispersal and Migration*, ed. W. Danthanarayana, pp. 60–70. Berlin: Springer-Verlag.

Bennet-Clark, H.C. (1970). The mechanism and efficiency of sound production in mole crickets. *Journal of Experimental Biology*, **52**, 619–52.

Bennet-Clark, H.C. (1989). Songs and the physics of sound production. In *Cricket Behavior and Neurobiology*, ed. F. Huber, T.E. Moore & W. Loher, pp. 227–61. Ithaca: Comstock Publishing Associates.

Birch, M.C. & Keelyside, J.J. (1991). Tapping behavior is a rhythmic communication in the death-watch beetle, *Xestobium rufovillosum* (Coleoptera: Anobiidae). *Journal of Insect Behavior*, **4**, 257–63.

Blest, A.D., Collett, T.S. & Pye, J.D. (1963). The generation of ultrasonic signals by New World arctiid moth. *Proceedings of the Royal Society of London* B, **158**, 196–207.

Claridge, M.F. (1983). Acoustic signals and species problems in the Auchenorrhyncha. In *Proceedings of the First International Workshop on Leafhoppers and Planthoppers of Economic Importance*, ed. W.J. Knight, N.C. Pant, T.S. Robertson & M.R. Wilson, pp. 111–20. London: Commonwealth Institute of Entomology.

Claridge, M.F. (1985). Acoustic signals in the Hemiptera: behavior, taxonomy, and evolution. *Annual Review of Entomology*, **30**, 297–317

Claridge, M.F., den Hollander, J. & Morgan, J.C. (1985). Variation in courtship signals and hybridization between geographically definable populations of the rice brown planthopper, *Nilaparvata lugens* (Stål). *Biological Journal of the Linnean Society*, **24**, 35–49.

Cowling, D.E. & Burnet, B. (1981). Courtship songs and genetic control of their acoustic characteristics in sibling species of the *Drosophila melanogaster* subgroup. *Animal Behaviour*, **29**, 924–35.

DeVries, P.J. (1990). Enhancement of symbioses between butterfly caterpillars and ants by vibrational communication. *Science*, **248**, 1104–6.

Dumortier, B. (1963a). Morphology of sound emission apparatus in Arthropods. In *Acoustic Behaviour of Animals*, ed. R.-G. Busnel, pp. 277–345. Amsterdam: Elsevier.

Dumortier, B. (1963b). The physical characteristics of sound emissions in Arthropods. In *Acoustic Behaviour of Animals*, ed. R.-G. Busnel, pp. 346–73. Amsterdam: Elsevier.

Elsner, N. (1974). Neuroethology of sound production in gomphocerine grasshoppers (Orthoptera: Acrididae) I. Song patterns and stridulatory movements. *Journal of Comparative Physiology*, **88**, 67–102.

Elsner, N. (1975). Neuroethology of sound production in gomphocerine grasshoppers (Orthoptera: Acrididae) II. Neuromuscular activity underlying stridulation. *Journal of Comparative Physiology*, **97**, 291–322.

Elsner, N. (1983). A neuroethological approach to the phylogeny of leg stridulation in gomphocerine grasshoppers. In *Neuroethology and Behavioral Physiology*, ed. F. Huber & H. Markl, pp. 54–68. Berlin: Springer-Verlag.

Ewing, A.W. (1989). *Arthropod Bioacoustics: Neurobiology and Behaviour*. Ithaca: Comstock Publishing Associates.

Greenfield, M.D. & Shelly, T.E. (1990). Territory-based mating systems in desert grasshoppers: effects of host plant distribution and variation. In *Biology of Grasshoppers*, ed. R.F. Chapman & A. Joern, pp. 315–35. New York: Wiley & Sons.

Haskell, P.T. (1957). Stridulation and associated behaviour in certain Orthoptera. I. Analysis of the stridulation of, and behaviour between males. *Animal Behaviour*, **5**, 139–48.

Haskell, P.T. (1961). *Insect Sounds*. London: Witherby.

Heller, K.-G. (1996). Unusual abdomino-alary, defensive stridulatory mechanism in the bushcricket *Panteophylus cerambycinus* (Orthoptera, Tettigonioidae, Pseudophyllidae). *Journal of Morphology*, **227**, 81–6.

Heller, K.-G. & Krahe, R. (1994). Sound production and hearing in the pyralid moth *Symmoracma minoralis*. *Journal of Experimental Biology*, **187**, 101–11.

Hennig, R.M., Weber, T., Moore, T.E., Huber, F., Kleindienst, H.-U. & Popov, A.V. (1994). Function of the tensor muscle in the cicada *Tibicen linnei*. *Journal of Experimental Biology*, **187**, 33–44.

Henry, C.S. (1980). The importance of low-frequency, substrate-borne sounds in lacewing communication (Neuroptera: Chrysopidae). *Annals of the Entomological Society of America*, **73**, 617–21.

Hinton, H.E. (1948). Sound production in lepidopterous pupae. *Entomologist*, **81**, 254–69.

Huber, F. (1962). Central nervous control of sound production in crickets and some speculations on its evolution. *Evolution*, **16**, 429–42.

Kevan, D.K.McE. (1955). Méthodes inhabituelles de production de son chez les Orthoptères. In *Colloques sur l'acoustique des Orthoptères*, ed. R.-G. Busnel, pp. 103–41. *Annales des Épiphytes, Fascicule hors série*.

King, M.J. (1993). Buzz foraging mechanism of bumblebees. *Journal of Apicultural Research*, **32**, 41–9.

Kirchner, W.H., Broeker, I. & Tautz, J. (1994). Vibrational alarm communication in the damp-wood termite *Zootermopsis nevadensis*. *Physiological Entomology*, **19**, 187–90.

Kutsch, W. & Huber, F. (1989). Neural basis of song production. In *Cricket Behavior and Neurobiology*, ed. F. Huber, T.E. Moore & W. Loher, pp. 262–309. Ithaca: Cornell University Press.

Lees, D.C. (1992). Foreleg stridulation in male *Urania* moths (Lepidoptera: Uraniidae). *Zoological Journal of the Linnean Society*, **106**, 163–70.

Leston, D. (1957). The stridulatory mechanisms in terrestrial species of Heteroptera. *Proceedings of the Zoological Society of London*, **128**, 381–400.

Lins, F. & Elsner, N. (1995). Descending stridulatory interneurons in the suboesophageal ganglia of two grasshopper species II. Influence upon the stridulatory patterns. *Journal of Comparative Physiology* A, **176**, 823–33.

Lyal, C.H.C. & King, T. (1996). Elytro-tergal stridulation in weevils (Insecta: Coleoptera: Circulionoidea). *Journal of Natural History*, **30**, 703–73.

Markl, H. (1983). Vibrational communication. In *Neuroethology and Behavioral Physiology*, ed. F. Huber & H. Markl, pp. 332–53. Berlin: Springer-Verlag.

Michelsen, A., Fink, F., Gogala, M. & Traue, D. (1982). Plants as transmission channels for insect vibrational songs. *Behavioral Ecology and Sociobiology*, **11**, 269–81.

Michelsen, A., Kirchner, W.H. Andersen, B.B. & Lindauer, M. (1986a). The tooting and quacking vibration signals of honeybee queens: a quantitative analysis. *Journal of Comparative Physiology*, **158**, 605–11.

Michelsen, A., Kirchner, W.H. & Lindauer, M. (1986b). Sound and vibrational signals in the dance language of the honey bee, *Apis mellifera*. *Behavioral Ecology and Sociobiology*, **18**, 207–12.

Michelsen, A., & Larsen, O.N. (1983). Strategies for acoustic communication in complex environments. In *Neuroethology and Behavioral Physiology*, ed. F. Huber & H. Markl, pp. 321–31. Berlin: Springer-Verlag.

Michelsen, A., Towne, W.F., Kirchner, W.H. & Kryger, P. (1987). The acoustic near field of a dancing honeybee. *Journal of Comparative Physiology* A, **161**, 633–43.

Minckley, R. L., Greenfield, M.D. & Tourtellot, M.K. (1995). Chorus structure in tarbush grasshoppers: inhibition, selective phonoresponse and signal competition. *Animal Behaviour*, **50**, 579–94.

Nocke, H. (1972). Physiological aspects of sound communication in crickets (*Gryllus campestris* L.). *Journal of Comparative Physiology*, **80**, 141–62.

Otte, D. (1970). A comparative study of communicative behavior in grasshoppers. *Museum of Zoology, University of Michigan, Miscellaneous Publications*, no. 141, 168 pp.

Pasquinelly, F. & Busnel, M.-C. (1955). Études preliminaires sur les mécanismes de la production des sons par les Orthoptères. In *Colloques sur l'acoustique des Orthoptères*, ed. R.-G. Busnel, pp. 145–53, *Annales des Épiphytes, Fascicule hors Séries*.

Popov. A.V., Shuvalov, V.F., Svetlogorskaya, I.D. & Markovich, A.M. (1974). Acoustic behavior and auditory system in insects. In *Mechanoreception*, ed. J. Schwartzkopff, pp. 281–306. Opladen: Westdeutscher Verlag.

Pringle, J.W.S. (1954). A physiological analysis of cicada song. *Journal of Experimental Biology*, **31**, 525–60.

Prozesky-Schulze, L., Prozesky, O.P.M., Anderson, F. & van der Merwe, G.J.J. (1975). Use of a self-made sound baffle by a tree cricket. *Nature*, **255**, 142–3.

Roces, F. & Hölldobler, B. (1995). Vibrational communication between hitchhikers and foragers in leaf-cutting ants (*Atta cephalotes*). *Behavioral Ecology and Sociobiology*, **37**, 297–302.

Roscow, J.M. (1963). The structure, development and variation of the stridulatory file of *Stenobothrus lineatus* (Panzer) (Orthoptera: Acrididae). *Proceedings of the Royal Entomological Society of London* A, **38**, 194–9.

Sandeman, D.C., Tautz, J. & Lindauer, M. (1996). Transmission of vibration across honeycombs and its detection by bee leg receptors. *Journal of Experimental Biology*, **199**, 2585–94.

Sismondo, E. (1980). Physical characteristics of the drumming of *Meconema thalassinum*. *Journal of Insect Physiology*, **26**, 209–12.

Stephen, R.O. & Hartley, J.C. (1995). Sound production in crickets. *Journal of Experimental Biology*, **198**, 2139–52.

Szczytko, S.W. & Stewart, K.W. (1979). Drumming behavior of four western Nearctic *Isoperla* (Plecoptera) species. *Annals of the Entomological Society of America*, **72**, 781–6.

Tautz, J., Roces, F. & Hölldobler, B. (1995). Use of a sound-based vibratome by leaf-cutting ants. *Science*, **267**, 84–7.

Vrijer, P.W.F. de (1982). Reproductive isolation in the genus *Javasella* Fenn. *Acta Entomologica Fennica*, **38**, 50–1.

Wenner, A.M. (1962). Sound production during the waggle dance of the honey bee. *Animal Behaviour*, **10**, 79–95.

Walker, T.J. (1962). Factors responsible for intraspecific variation in the calling songs of crickets. *Evolution*, **16**, 407–28.

Young, D. (1973). Sound production in cicadas. *Australian Natural History*, **1973**, 375–80.

Young, D. & Bennet-Clark, H.C. (1995). The role of the tymbal in cicada sound production. *Journal of Experimental Biology*, **198**, 1001–19.

27 Chemical communication: pheromones and chemicals with interspecific significance

Hormones are concerned with regulation within the organism. Other chemical substances are concerned with communication between organisms, both intra- and interspecifically. Chemicals with an intraspecific function are called pheromones; those with an interspecific function are often called allelochemicals. Some chemicals can serve both functions, even in one insect (Blum, 1996). Most pheromones and allelochemicals are produced by ectodermal glands secreting to the outside of the body. The basic structure of these exocrine glands is similar irrespective of whether they produce pheromones or allelochemicals, but their positions on the body and the chemicals they produce exhibit extraordinary diversity. Glands producing chemicals for interspecific communications usually have reservoirs while most pheromone glands do not. Pheromones used for marking, however, are usually stored in a reservoir. Many chemicals used for defense are stored within the body and are not secreted.

27.1 PHEROMONES

A pheromone is a secretion which, if passed to another individual of the same species, causes it to respond, physiologically or behaviorally, in a particular manner. Pheromones are thus concerned with signalling between individuals of the same species and are therefore often important in sexual behavior and in regulating the behavior and physiology of social and subsocial species. It is probably true that pheromones are used in communication by at least some species in all the insect orders.

Many pheromones are perceived as scents by olfactory receptors and affect the recipient via the central nervous system. In other cases, pheromones are ingested by the recipient. These may be perceived by the sense of taste, exerting their effects via the central nervous system, or the pheromone, once ingested, may be absorbed and influence biochemical reactions within the recipient. Pheromones which affect behavior directly through the nervous system

are called releaser pheromones; those that affect metabolism are known as primer pheromones.

27.1.1 Structure and distribution of pheromone glands

In most insects, pheromones are produced by glandular epidermal cells concentrated in discrete areas beneath the cuticle, but, in some species, gland cells are scattered through the epidermis of different parts of the body. In male desert locusts (*Schistocerca gregaria*), for example, class 3 gland cells (section 16.1.1) are scattered over the head, thorax and abdomen. In an immature insect, they are small and restricted to the basal part of the epidermis, but as the insect matures they enlarge and extend distally towards the cuticle. The cytoplasm in a mature insect, viewed with the electron microscope, contains large numbers of electron-dense granules with numerous clear vesicles close to, and probably discharging into, a terminal cavity. The contents of the cavity are discharged on the surface of the cuticle via a ductule. Similar scattered glands are present in some Heteroptera. In the higher Diptera, the cuticular hydrocarbons forming the sex pheromone are produced by epidermal cells, primarily in the abdomen.

Where the secretory cells are aggregated to form a gland, they are usually class 1 gland cells (section 16.1.1) without ducts to the exterior and abutting directly on to the cuticle. The glands are often concealed beneath a fold of cuticle, such as an intersegmental membrane between abdominal segments. They have no reservoir and the pheromone is released directly following its synthesis. In other cases, the glands open into an epidermal invagination which may then serve as a reservoir in which the pheromone accumulates. This is the case with the marking pheromone of social Hymenoptera where larger quantities of pheromone are produced.

In many female Lepidoptera, the glands producing the male sex attractant pheromone lie beneath an intersegmental membrane of the posterior abdominal segments,

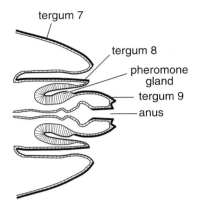

tergum 7

tergum 8

pheromone gland

tergum 9

anus

Fig. 27.1. Pheromone gland. Horizontal section of the tip of the abdomen of a female moth (*Plodia*) in which the glands producing the sex attractant pheromone are in a lateral position. The gland is exposed when the female extends her abdomen (from Wigglesworth, 1972).

usually between segments eight and nine (Fig. 27.1). It is often ventral in position, but in some species is dorsal and in others occurs as a ring all round the body. In some moths, *Heliothis* and *Manduca*, for example, the gland is innervated from the terminal abdominal ganglion, but in others, such as *Ostrinia*, this is not the case (Christensen, Lashbrook & Hildebrand, 1994; Ma & Roeloffs, 1995). Some arctiid moths are unusual in having a pair of branched, tubular glands invaginated from the dorsal surface of the interseg-mental membrane between segments eight and nine. In this case the lumen of the tubes forms a reservoir.

Many male Lepidoptera have pheromone-producing glands which are often associated with structures used for pheromone dispersal. Male Noctuidae produce an aphro-disiac pheromone in special glands, called Stobbe's glands, in the second abdominal segment (see Fig. 27.11). Each gland consists of a number of glandular cells which are greatly enlarged in the pharate adult, each enclosing a large central cavity continuous with the base of a hair. After eclosion, the fluid in the cavity is discharged into the hair, which is tubular and ends at the base of a brush of scales. Many male Arctiidae have a pair of eversible sacs, called coremata, in the posterior abdomen. Male danaids have a pheromone-producing pocket on each hind wing and other male Lepidoptera produce pheromones from glands associated with scales.

Other insects have glands which underlie abdominal sternites or tergites. Termites, for example, have glands

beneath sternites IV or V in which the trail pheromone is produced. Except in Rhinotermitidae these glands have no ducts. Male cockroaches have glands producing an aphro-disiac pheromone beneath the tergites of certain abdomi-nal segments. The individual gland cells open at the surface of the cuticle through small ducts, sometimes in association with tufts of long setae which possibly assist in dispersal of the pheromone.

In aphids, alarm pheromone is produced by glands associated with the cornicles, while the sex pheromone of female *Schizaphis* is released from glands on the hind tibiae.

The hindgut produces pheromones associated with aggregation in bark beetles, and glands in the genital atrium, called atrial glands, produce the sex attractant pheromones of *Periplaneta* (Abed *et al.*, 1993). The oviposition marking pheromone of the fruit fly, *Rhagoletis*, is produced in the posterior midgut. This is one of the few examples of a pheromone not produced by glands derived from ectoderm.

Exocrine glands of social Hymenoptera Social Hymenoptera have a variety of exocrine glands, some of which produce pheromones, others defensive compounds, while others are involved in digestion and feeding the brood. In different species, the secretions of these glands can perform different functions. Fig. 27.2a shows the prin-cipal exocrine glands of ants (ants: Hölldobler & Wilson, 1990; honeybee: Winston, 1987).

Many Hymenoptera have mandibular glands in the head which open via a duct at the base of each mandible. From the duct, a groove runs into a depression on the inner face of the mandible (Fig. 27.3a). The gland is a sac-like invagination of the epidermis with an epithelium of class 3 secretory cells lined by a thin cuticular intima. These glands are well-developed in queen and worker honeybees, but are greatly reduced in drones.

In ants, other sources of pheromones are the poison gland and Dufour's gland, both of which are absent in males as they are associated with the sting, and Pavan's gland, which opens on the ventral surface of the abdomen above the sixth abdominal sternite. The poison gland con-sists of a pair of glandular tubules which unite to form a duct opening into a reservoir (Fig. 27.2c). The gland cells of the poison gland are class 3 cells. In contrast, the cells forming Dufour's gland are class 1. The gland opens into the poison duct near the base of the sting. It is a small,

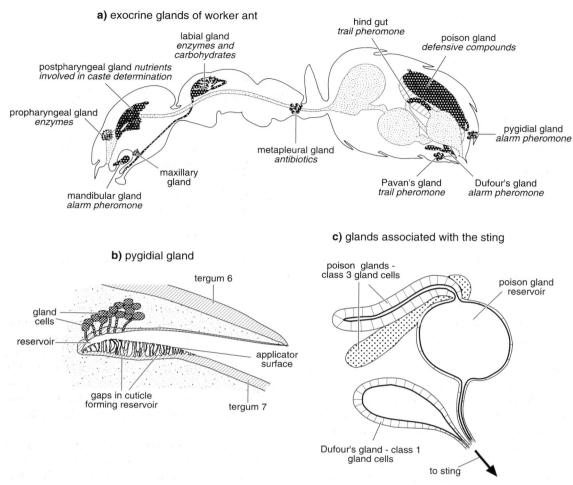

Fig. 27.2. Exocrine glands of worker ants. **(a)** Longitudinal section of a worker ant showing the principal glands with dark shading. All the glands shown may not be found in one species. Examples of the functions of secretions from the glands are shown in italics, but note that the secretion from a particular gland may serve different functions in different species (based on Hölldobler & Wilson, 1990). **(b)** Pygidial gland of a worker ant, *Pachycondyla* (based on Hölldobler & Traniello, 1980). **(c)** Poison gland and Dufour's gland of a worker ant (*Myrmica*). The organization of the gland is simpler than that depicted in **(a)** (based on Billen, 1986).

simple sac with thin glandular walls and a delicate muscular sheath.

Worker honeybees have a gland, Nasonov's gland, beneath abdominal tergite six which is exposed by depressing the tip of the abdomen. It is made up of about 500 large class 3 gland cells each with a duct to the surface of the cuticle (Fig. 27.3b). Oenocytes and fat body cells are closely associated with the gland cells, but the gland is not innervated (Cassier & Lensky, 1994). Most ants have a

pygidial gland with a similar structure, the ducts from the cells opening into a reservoir between abdominal tergites six and seven (Fig. 27.2b).

Review: Percy-Cunningham & MacDonald, 1987 – sex pheromone glands

27.1.2 Chemical characteristics of pheromones

There is great variety in the chemicals used as pheromones depending on species, function and mode of action. The

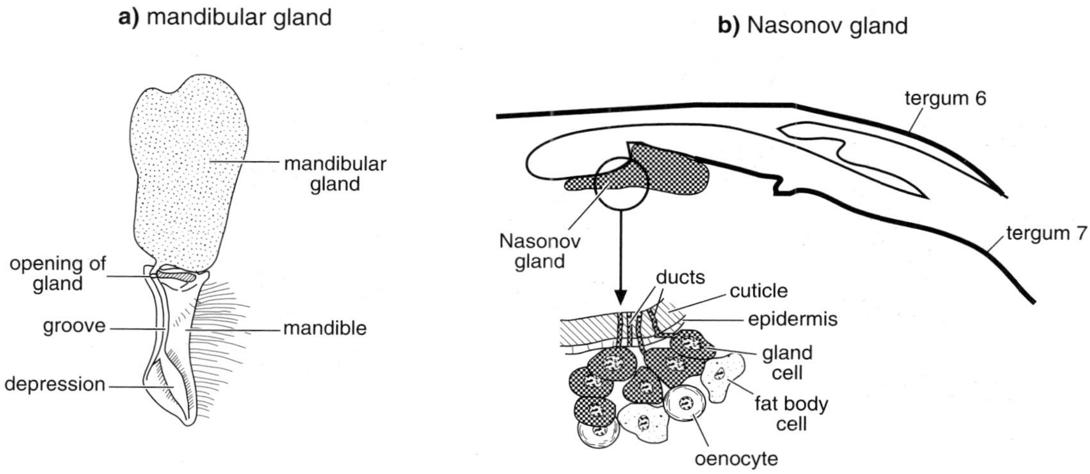

a) mandibular gland

b) Nasonov gland

Fig. 27.3. Two exocrine glands of worker honeybees: (a) mandibular gland and mandible (after Snodgrass, 1956); (b) Nasonov gland, with part of the gland enlarged (below).

chemical nature of volatile pheromones is, to some extent, constrained by the conflicting requirements of volatility and specificity. Small molecules are more volatile, but the number of structural variations which they can form is limited. The appropriate degree of specificity can only be achieved with larger molecules which allow more variation. Hence the molecular size of many volatile pheromones represents a balance between the two opposing requirements of volatility and specificity. The degree to which these features are required depends on the particular functions of the pheromone in question. Sex attractant pheromones are, of necessity, both specific and volatile. Alarm pheromones, by contrast, should be highly volatile in order to alert other nearby insects to a source of danger, but also to minimize the period of disruption. They do not need, therefore, to be highly specific and consequently tend to be made up from small, highly volatile compounds. Marking pheromones, such as oviposition deterrents and trail pheromones, on the other hand might be expected to be specific, but less volatile as their persistence is a requisite. In some cases, however, persistence is achieved through continual remarking. Contact pheromones are not constrained by the need for volatility (see Fig. 27.4).

The chemical components of pheromones are, in most cases, not unique and the same compound may be used by a variety of species. Aliphatic hydrocarbons and terpenoids are most commonly employed, but a variety of other chemical classes are used by different insect species.

Most pheromones are mixtures of compounds, sometimes only two, but sometimes as many as ten. The components may all be similar and, perhaps, derived through a single synthetic pathway (see Table 27.1; Fig. 27.7), but chemicals of more than one class may also be present. The different components of a pheromone are sometimes known to act synergistically, and in other cases evoke different responses from the recipient (see Fig. 27.13).

Aliphatic hydrocarbons Aliphatic hydrocarbons are used as pheromone components by many insects. The sex attractant pheromones of many female Lepidoptera are straight chain hydrocarbons, most commonly with chain lengths of 12, 14 and 16 carbon atoms (Fig. 27.4). Most moth pheromones are acetates. Hydrocarbons of similar size are the sex pheromones of some Coleoptera, but, in these insects, the compounds are often methylated. Short-chain molecules also form the pheromones of scale insects (Coccidae, Homoptera), sawflies and the honeybee where the principal sex attractant is 9-oxodecanoic acid. In most bumblebees, the main components of the marking pheromone are hydrocarbons with chain lengths of 16, 18 or 19 carbon atoms. The scent scales of some male pierid butterflies contain alkanes with chain lengths of 10 to 13 carbon atoms. In Tenebrionidae (Coleoptera) and Pentatomidae (Hemiptera) short-chain aldehydes function as aggregation pheromones.

Shorter chain length hydrocarbons are often used as alarm pheromones. Amongst the Heteroptera, hexenal

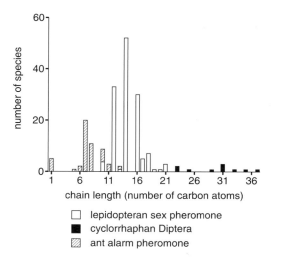

Fig. 27.4. Chain lengths of aliphatic hydrocarbons used as alarm pheromones by ants, long-range sex attractants by moths, and contact pheromones by cyclorrhaphan Diptera.

and octenal have been identified as components of alarm pheromones; in bees there are short chain (5–9 carbon atoms) acetates, alcohols and ketones; and myrmicine ants produce ketones with chain lengths of 6–9 carbon atoms. Formic acid is a component of the alarm pheromone of formicine ants; it is also used as a defensive chemical.

At the other extreme, the contact or short-range sex pheromones of some calypterate Diptera are alkanes or alkenes with odd-numbered chain lengths of more than 20 carbon atoms, and, in many species, more than 30.

Aromatic hydrocarbons Aromatic hydrocarbons have not been widely reported as pheromones, but they do occur in the male scent brushes of a number of noctuid moths. Phenyl ethanol, benzyl alcohol and benzaldehyde are examples.

Terpenoids Mono- and sesquiterpenes are commonly used as components of alarm pheromones. This is true for aphids, termites and formicine ants. Some ants also use terpenes as trail pheromones, and they are the marking compounds used by a few bumblebees. Bark beetles use a variety of monoterpenoid compounds as attractants and aggregation pheromones, and this is also true of the cotton boll weevil, *Anthonomus*.

Nitrogen-containing cyclic compounds A number of ants use heterocyclic nitrogen compounds as trail pheromones,

and species of *Odontomachus* use them as alarm pheromones. Some male arctiid moths and danaid butterflies produce pheromones from pyrrolizidine alkaloids.

27.1.3 Specificity

Communication within a species requires specificity of the signals. With pheromones, this specificity is achieved in two ways: on the basis of molecular structure, and by specific mixtures (blends) of compounds. This demands not only that the sender should produce specific chemicals in appropriate proportions, but also that the neural system of the receiver is tuned to the appropriate blend.

Specificity has been studied most in the sex attractant pheromones of Lepidoptera. Four characteristics of the molecule are of particular importance: chain length, the nature of the functional groups, the number and position of double bonds, and the configuration of the bonds. For the families Noctuidae, Pyralidae and Tortricidae, the major pheromone component has 10, 12, 14, 16 or 18 carbon atoms in a straight chain (Fig. 27.5). The functional group is usually a primary alcohol, an aldehyde, or acetate, and there may be one, two or three double bonds in a variety of positions. This combination of characters provides an enormous range of different compounds, and neural receptors in the receiver may be highly specific (see Fig. 24.6).

Specificity also results from the use of particular blends of compounds. Closely related, sympatric species may produce pheromones with the same suite of chemicals but differing in the proportions of the components. Table 27.1 illustrates some of these points in relation to the moth genus, *Spodoptera*. The chain lengths for the major components of most species are the same, 14 carbon atoms, but the North American *S.frugiperda* is distinct as its primary component has only 12 carbon atoms. In all species, acetate is the functional group on the major compounds and the principal differences lie in the number of double bonds, their positions, and the proportions of the different compounds. Except for *S.littoralis* and *S.litura*, all the species are distinguishable from these characteristics and, presumably, have the capacity to distinguish conspecific from heterospecific insects. The geographic ranges of the species do not all overlap, so that some species never encounter others. For example, *S.litura* occurs in eastern Asia, while *S.littoralis* is west Asian and African, and perhaps this geographical separation makes it possible for the two species to have such similar pheromones. *Spodoptera descoinsi* and *S. latifascia*, however, are sympatric

Table 27.1. *Components of sex attractant pheromones produced by females of different species of* Spodoptera[a]

Chain length	12	14	14	14	14	14	14	14	14	14	14	16
Double bond	Z9	—	Z9	Z9	Z9	Z9 E11	Z9 E12	Z9 Z12	E9 E12	Z11	E11	Z11
Functional group[b]	ac	ac	ac	ald	ol	ac	ac	ac	ac	ac	ac	ac
descoinsi			41	2		1	42		3	<1		11
eridania			80				20					
exempta			95				5					
exigua			48		4		40	6				2
frugiperda	90		10									
latifascia			78	8		1	7		1	<1		8
littoralis		+	+			95	+			+	+	
litura						91	9					

Notes:
[a] Numbers give percentage of each component. For most species they represent the amounts present in the gland.
[b] ac, acetate; ald, aldehyde; ol, alcohol.
+, reported in small amounts, differing between populations.

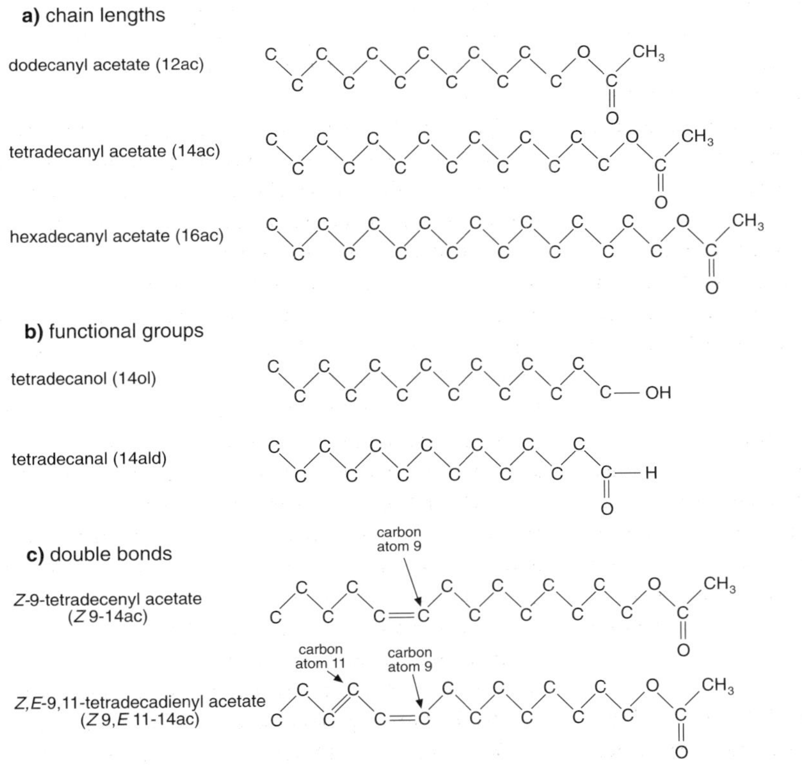
a) chain lengths
dodecanyl acetate (12ac)
tetradecanyl acetate (14ac)
hexadecanyl acetate (16ac)
b) functional groups
tetradecanol (14ol)
tetradecanal (14ald)
c) double bonds
Z-9-tetradecenyl acetate (Z9-14ac)
carbon atom 9
Z,E-9,11-tetradecadienyl acetate (Z9,E11-14ac)
carbon atom 11
carbon atom 9

Fig. 27.5. Pheromone specificity. Factors contributing to the specificity of a pheromone based on aliphatic hydrocarbons. Commonly used abbreviations are given in brackets. **(a)** Differences in chain length. **(b)** Differences in functional groups. **(c)** Differences in the number and position of double bonds.

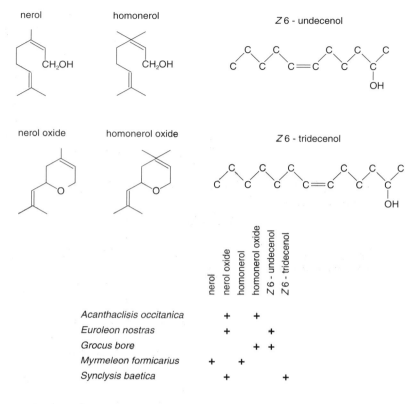

Fig. 27.6. Pheromone specificity. Male antlions use a two-component pheromone. One component is nerol or a derivative, the second may be an aliphatic alcohol (based on Bergström *et al.*, 1992).

and they will interbreed in captivity. They are clearly distinguishable from the relative proportions of Z9-tetradecyl acetate and Z9,E14-tetradecyl acetate and presumably this difference contributes to sexual isolation in the field.

Sometimes the blend includes structurally different chemicals. For example, the male attractant pheromones of antlions contain two compounds, which, in some species, are an aliphatic alcohol and a monoterpene. Specificity is conferred by small differences in the structure of these molecules (Fig. 27.6).

27.1.4 Origins of pheromone chemicals

Pheromones may be synthesized *de novo* in the glandular cells, or derived from precursors present in the food. Many pheromones with ring structures are known to be derived from such precursors, but there is also evidence, in some cases, that the insects can synthesize them. In other cases, symbiotic microorganisms may contribute to synthesis.

Pheromone synthesis Most pheromones of most insects are synthesized *de novo* from smaller molecules. The hydrocarbon pheromones of many Lepidoptera and Diptera, for example, are synthesized from fatty acids in a series of steps. These involve chain shortening or elongation, desaturation and, finally, the modification of the functional group by reduction, acetylation or, sometimes, oxidation (Fig. 27.7).

Honeybees produce mandibular gland pheromones from stearic acid. In workers this is hydroxylated at the terminal (C18) position, while, in queens, C17 becomes hydroxylated. The products become shortened to chains of 10 carbon atoms and desaturated at the 2 position to produce 10-hydroxydecenoic acid in the workers and 9-hydroxydecenoic acid in queens. Oxidation produces other related chemicals (Plettner *et al.*, 1996).

Ants synthesize the terpenoids that often contribute to their alarm pheromones.

Pheromone precursors from food Some insects are known to require specific chemicals from their food which are then modified to form pheromones.

Many of the compounds in the pheromones of male

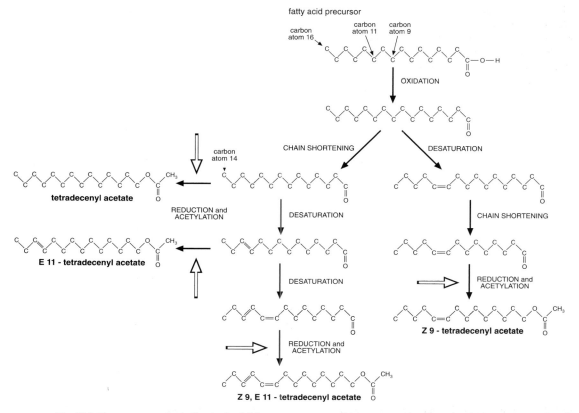

Fig. 27.7. Pheromone synthesis. Synthesis of different components of the sex attractant pheromone of *Spodoptera littoralis* from a fatty acid. The processes occurring at each step are shown in capitals, and names of the final products are in bold type. Open arrows show the stage which is probably regulated by the pheromone biosynthesis activating neuropeptide (based on Fabriás, Marco & Camps, 1994).

Lepidoptera that have androconial brushes are derived from chemicals in the larval food. These may be commonly occurring compounds, such as benzaldehyde, benzyl alcohol and butyric acid used by a number of Noctuidae, but, in other cases, the pheromones are derived from compounds characteristic of the host on which the larvae feed. For example, some arctiid moths use pyrrolizidine alkaloids from their host plants as pheromone precursors. The males of many danaid butterflies also synthesize their pheromones from pyrrolizidine alkaloids (Fig. 27.8), but in these species the larvae do not feed on plants containing the alkaloids. Instead, they are acquired by the adult males, which lick alkaloid-containing plants.

Bark beetles and the boll weevil, *Anthonomus grandis*, also produce pheromones from resins of their host plants (Fig. 27.8). *Ips paraconfusus*, for example, derives ipsenol

and ipsdienol from myrcene, and *cis*-verbenol from α-pinene, but *Ips duplicatus* synthesizes similar pheromones *de novo* (Ivarsson & Birgersson, 1995).

In *Reticulitermes* an unsaturated alcohol, *cis*–3, *cis*–6, *trans*–8-dodecatrienol, which occurs in wood infected with the fungus *Lenzites*, is used as a trail pheromone.

Synthesis by microorganisms Bacterial symbionts in the hindgut of some bark beetles and the beetle, *Costelytra*, are able to synthesize some of the compounds used by their host insects as pheromones. It seems to be generally true, however, that the host insects are also capable of synthesis and the relative importance of the symbionts is unknown.

Reviews: Prestwich & Blomquist, 1987; Schneider, 1987 – Lepidoptera, alkaloids; Wood, 1982 – bark beetles

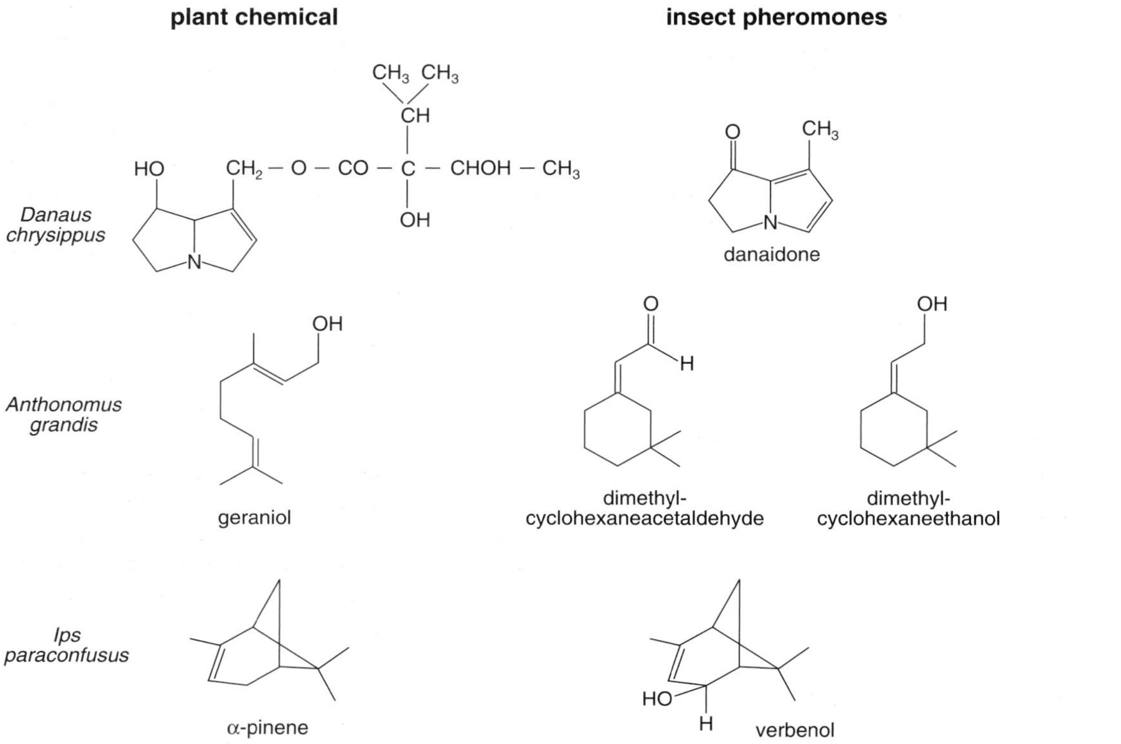

plant chemical   **insect pheromones**

Fig. 27.8. Pheromones derived from chemicals in plants. Top: pyrrolizidine alkaloids obtained from a plant by adult male *Danaus chrysippus*. Center: geraniol, a monoterpene, in cotton is a source of pheromone components of the boll weevil, *Anthonomus grandis*. Bottom: α-pinene, a monoterpene, in pine bark is a source of pheromone components in the bark beetle, *Ips paraconfusus*.

27.1.5 Control of synthesis and release

Release of pheromones into the environment involves two separate processes: synthesis and dispersal. Where the pheromone gland is not associated with a reservoir, synthesis often leads directly to dispersal, but, where a reservoir is present, the pheromone may be stored, and synthesis and dispersal are temporally separated.

27.1.5.1 *Pheromone synthesis*

In some female moths, such as *Trichoplusia ni*, pheromone synthesis is continuous and the amount present in the gland at any time represents the balance between synthesis and release. Consequently, pheromone accumulates during the daytime, when none is released, and then decreases during darkness as a result of pheromone release (Fig. 27.9a). In other species, such as *Helicoverpa zea*, synthesis only occurs at particular times of day (Fig. 27.9b), which vary with the insect's age and mated status. Virgin

female *H. zea* produce increasing amounts of pheromone each night for the first three nights after eclosion, but mating almost completely inhibits further production.

The enzymes necessary for pheromone synthesis are produced late during development of the pharate adult, and the development of competency to produce pheromone is regulated by the falling titer of ecdysteroids. The system is subsequently controlled by a peptide, pheromone biosynthesis activating neurohormone (PBAN) which is produced by cells in the subesophageal ganglion. The method by which it affects the abdominal pheromone gland is unclear; it may be released into the hemolymph via the corpora cardiaca, or, in those instances where the pheromone glands are innervated, may be transmitted neuronally directly to the glands.

PBAN regulates the enzymes involved in pheromone synthesis. In some species, such as *Helicoverpa*, it appears to govern the production of intermediates from fatty acids,

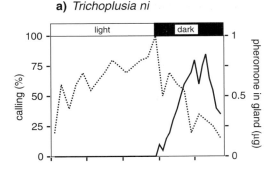

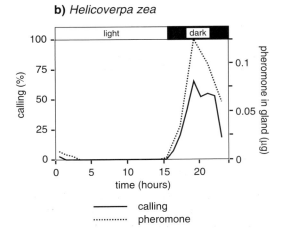

Fig. 27.9. Temporal variation in calling behavior and amounts of pheromone in the glands of two nocturnal noctuid moths.
(a) The quantity of pheromone in the gland increases during the photophase, but declines during the scotophase when the females are calling. Pheromone synthesis is continuous (*Trichoplusia ni*) (data from Hunt & Haynes, 1990) . (b) Pheromone synthesis is synchronous with calling behavior. None accumulates in the gland during the photophase (*Helicoverpa zea*) (data from Raina, Davis & Stadelbacher, 1991).

but in others it acts at the stage of reduction and acetylation shown in Fig. 27.7 (Fang, Teal & Tumlinson, 1995).

The release of PBAN is regulated by environmental factors, especially daylength. Host-plant odors may also stimulate its release.

Other factors may be involved in pheromone synthesis and release. The amount of octopamine, which is known to stimulate pheromone production, reaches a peak in the pheromone gland of *Helicoverpa* just before the start of the scotophase (Christensen *et al.*, 1992). In *Manduca*, experimental neural stimulation of the gland also leads to the

immediate release of pheromone, and pulses of pheromone are released in response to successive stimulations (Christensen, Lashbrook & Hildebrand, 1994). How these effects are related to that of PBAN is not understood.

Pheromone synthesis is depressed following mating (see Fig. 12.19) apparently by a factor passed from the male to the female during sperm transfer. In *Helicoverpa*, this factor is a peptide which inhibits the release of PBAN from the subesophageal ganglion; in *Bombyx*, however, the inhibitory signal acts via the nervous system, perhaps arising from mechanical stimulation during copulation (Ando *et al.*, 1996).

There is some evidence for a comparable control mechanism in the Hessian fly, *Mayetiola*. In this species, pheromone synthesis is controlled by a factor produced in the head, and is inhibited following mating by a substance transferred from the male (Foster *et al.*, 1991).

In contrast, pheromone synthesis in bark beetles (Scolytinae) is regulated by the rising titer of juvenile hormone associated with oogenesis (Ivarsson & Birgersson, 1995).

Review: Raina, 1993

27.1.5.2 *Pheromone dispersal*

Dispersal of pheromones into the environment involves specific behavior and is often enhanced by cuticular structures.

Where the gland cells are of class 3, with a duct extending to the surface of the cuticle, the pheromone can pass freely on to the surface of the cuticle. Where, as in many glands, the gland cells are class 1 and have no ducts, the cuticle contains large numbers of wax canal filaments or, as in the glands producing the honeybee alarm pheromone, it may be perforated by enlarged pore canals (Cassier, Tel-Zur & Lensky, 1994). In all these cases, the pheromone passes through the cuticle very quickly after it is synthesized.

Dispersal of volatile pheromones from the surface of the cuticle is believed to be facilitated by microtrichia. These may be on the cuticle immediately above the gland, as in the abdominal glands of some female Lepidoptera, or they may be on an adjacent part of the cuticle, as is the case with the alarm pheromone of the honeybee (Lensky, Cassier & Tel-Zur, 1995). The peak rate of release of the sex attractant pheromone by female *Trichoplusia ni* into the air is about $10 \, \mathrm{ng \, min^{-1}}$, but it is considerably lower in some other moths.

Many male Lepidoptera have modified scales which are known or (more frequently) inferred to be involved in

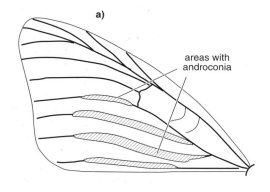

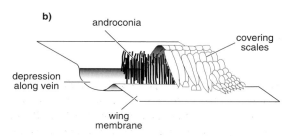

Fig. 27.10. Androconia on the forewing of a male butterfly (*Argynnis*) (based on Barth, 1944). (**a**) Forewing showing the positions of the androconia along the veins. (**b**) Diagram showing part of a vein. The androconia arise from a depression along the vein. They are normally covered by a roof of scales.

dispersal of aphrodisiac pheromones. These may be single scales directly associated with gland cells or highly complex structures which are sometimes separate from the gland cells. They are called androconial organs (Boppré, 1984) although other terms are used to refer to specific structures. Androconial organs are often present on the wings of male butterflies where they may be isolated elongate scales, ending in a brush-like row of processes, or they may be well defined structures along the veins (Fig. 27.10). Male Noctuidae and Danainae have brush-like tufts of scales mounted on a hinged cuticular lever (Fig. 27.11). Frequently, the brushes are housed in longitudinal pockets on each side of the abdomen; in other species they occur at the tip of the abdomen. During courtship the brushes are everted by muscular action so that pheromone is dispersed from the hairs. The scales forming the brushes are sometimes open lattices of cuticle providing a large surface area for evaporation. In Danainae and some other Lepidoptera, androconial scales fragment to form large numbers of small cuticular particles to which the pheromone adheres.

They are called pheromone transfer particles. The coremata of male Arctiidae are everted by hemolymph pressure and are covered by hairs from which pheromone is presumed to evaporate. In *Creatonotus*, the size of the coremata is dependent on the amount of pyrrolizidine alkaloid, the pheromone precursor, ingested by the larva.

Pheromone release commonly involves some specific behavior. Marking pheromones are often applied by direct contact and may be associated with specific cuticular structures functioning as applicators. This is the case in the ant, *Pachycondyla*, where the cuticle associated with the pygidial gland is deeply folded, the folds coming together to form a smooth surface with narrow spaces opening at the surface (Fig. 27.2b). The pheromone is secreted into the space above the applicator and is believed to be stored temporarily in the cuticular spaces. When marking, the ant curls its abdomen so that the dorsal surface of the terminal segments touches the substrate and the cuticle of the applicator is rubbed on the surface. In other ant species, and in flies producing an oviposition marking pheromone in the gut, the insect drags the tip of the abdomen over the surface as it runs. In some ants, the sting has lost its stinging function and serves solely to apply chemicals produced in the poison gland and Dufour's gland to the substrate. Pheromones produced in labial glands of bumblebees are transferred to the vegetation by biting.

Some male butterflies, such as *Hipparchia* and *Argynnis*, which have androconia on the forewings, land facing a female and close the forewings around the female's antennae causing them to touch the scent scales. Species with brush-like androconia extend them when close to the female, either when hovering or on the ground, and the species producing pheromone transfer particles dust them on to the female's antennae.

Release of the sex attractant pheromone by female moths involves a specific behavior pattern, known as calling, in which the pheromone gland is exposed by depressing the tip of the abdomen, if the gland is in a dorsal intersegmental membrane, or by extension of the abdomen. In some species, the gland is everted by hemolymph pressure. Female arctiid moths that have a tubular gland, forcibly eject the pheromone as an aerosol by compressing the gland (Yin, Schal & Cardé, 1990). Calling behavior is exhibited only at certain times of day (Fig. 27.9).

Exposure of the gland is sometimes accompanied by wing vibration which facilitates pheromone dispersal. Similar behavior occurs in worker honeybees when they expose the Nasonov gland at the nest entrance or a foraging site.

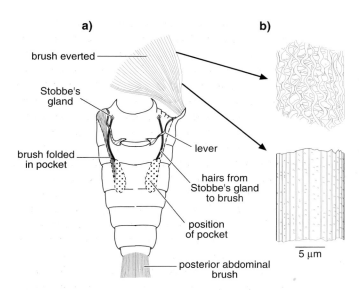

a)

brush everted

Stobbe's
gland

brush folded
in pocket

lever

hairs from
Stobbe's gland
to brush

position
of pocket

posterior abdominal
brush

b)

5 μm

Fig. 27.11. Pheromone brushes of a male moth (*Phlogophora*). Pheromone from Stobbe's gland is transferred to the brushes through tubular hairs (after Birch, 1970). **(a)** Ventral surface of the abdomen showing one brush folded in its cuticular pocket (left), and the other partially everted (right). **(b)** Details of a hair from the brush. Above: the distal region forms an open lattice from which the pheromone readily evaporates. Below: the proximal part of the hair has a solid wall.

Short-chain, unsaturated compounds in general have higher volatility than longer chain, saturated molecules. This affects the rate of evaporation of different compounds from the surface of a pheromone gland. The rate of emission also varies with temperature, but the effect of temperature differs between compounds so that small changes in the ratios of the components of a blend occur (Liu & Haynes, 1994).

Reviews: Birch, Poppy & Baker, 1990 – moth androconia; Boppré, 1984 – butterfly androconia

27.1.6 Functions of pheromones

27.1.6.1 *Pheromones affecting behavior*

Releaser pheromones are commonly involved in various aspects of sexual behavior, but also occur in relation to other aspects of behavior necessitating communication between insects of the same species.

Long-distance attractants Pheromones are employed by a large number of insects to bring the sexes together for mating. These include sex attractant pheromones, produced by one sex and attracting the other, and aggregation pheromones, commonly produced by one sex, but attracting individuals of both sexes.

Sex attractant pheromones are produced by the females of some Homoptera, such as aphids and scale insects, by many Lepidoptera and Coleoptera, and by some Hymenoptera. Male ant lions (Neuroptera) are also known to produce sex attractant pheromones. Amongst the cockroaches (Blattodea), a sex attractant pheromone is produced

by the females of some species and the males of others (Farine *et al.*, 1994). Aliphatic hydrocarbons and terpenoids are commonly used as sex attractants (Table 27.2), the insects within a particular group tending to use similar compounds although in the Lepidoptera and Coleoptera different families may use different compounds.

These pheromones are only released when the insect is ready to mate and at a time of day appropriate for the species. Virgin female *Aphis fabae*, for example, do not exhibit calling behavior for the first three or four days of their adult life, but after seven days call regularly by raising the abdomen and hind tibiae, (Thieme & Dixon, 1996). They call mainly in the morning, although there are differences between subspecies. Most moths call only at night (Fig. 27.9). At least in some species, calling behavior occurs independently of the amount of pheromone in the gland indicating that regulation of calling and pheromone production are independent processes.

Sex attractant pheromones stimulate upwind flight by members of the opposite sex bringing individuals of the two sexes into proximity. In some cases, the sex attractant odor appears sufficient to produce mating, but frequently vision or an odor produced by the opposite sex are important in the final stages.

Aggregation pheromones attract both sexes and, in some cases also larvae. They are commonly associated with feeding behavior but also facilitate the meeting of the sexes. Bark beetles (Scolytidae) aggregate in large numbers on host trees selected for attack and many species are known to produce attractant pheromones. When a

Table 27.2. *Sex attractant pheromones. Examples of chemicals which are commonly major constituents of the pheromone in different families of insects. The term hydrocarbon is used here to refer to compounds with an unbranched chain of carbon atoms.*

Order/family	principal compounds	example	sex producing	position of gland
Blattodea				
Blattidae cockroaches	periplanones	periplanone - B	female	atrial glands
Hemiptera				
Aphididae aphids	terpenes	nepetalactone	female	hind tibia
Diaspididae scale insects	branched hydrocarbons	3 - methylene, 7 - methyl, 7 - octen propanoate	female	?
Miridae mirid bugs	hydrocarbons	butyl butyrate	female	?
Neuroptera				
Myrmeleontidae antlions	monoterpene + hydrocarbon	(see Fig. 27.6)	male	thorax
Hymenoptera				
Tenthredinidae pine sawflies	branched hydrocarbons	3,7 - dimethyl - 2 -tridecanol	male	?
Apidae honeybees	hydrocarbons	*E* 2, 9 - oxodecanoic acid	female	mandible
Trichoptera				
Rhyacophilidae caddis flies	hydrocarbons	nonan-2-one	female	abdomen sternum 5
Lepidoptera				
many families of moths	hydrocarbons	(see Fig. 27.7)	female	abdomen intersegment 8-9
Coleoptera				
Scarabaeidae scarab beetles	hydrocarbons	*Z* - 7 - tetradecen - 2 - one	female	?
Scolytidae bark beetles	terpenes	see Figs. 27.8 and 27.13	both sexes	?

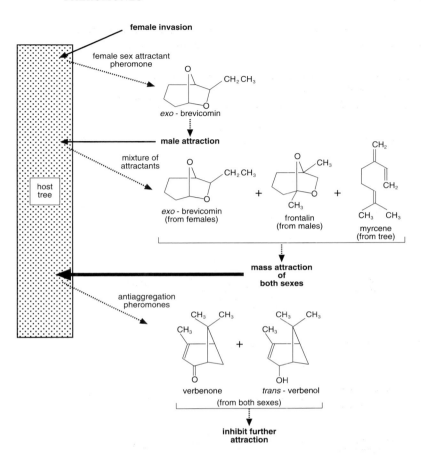

Fig. 27.12. Aggregation pheromones. Compounds involved in the invasion of a tree by the bark beetle, *Ips paraconfusus*.

female *Dendroctonus* bores into the phloem of a host tree she produces a pheromone, exo-brevicomin (Fig. 27.12) in the feces. It is attractive to both sexes and so more beetles invade the tree. The males, when they arrive, produce a closely related pheromone, frontalin, which, together with the volatile myrcene from the tree itself, enhances the effect of the exo-brevicomin so that more and more beetles arrive at the tree. As male beetles enter the female galleries they stridulate, leading to the release of pheromones which reduce the attractant response. These 'antiaggregative' pheromones are trans–verbenol, verbenone and others.

Aggregation pheromones are also produced by males of some scarab and nitidulid beetles, by some male weevils, and by male Heteroptera from a number of families. In many cases they are methyl branched hydrocarbons (Aldrich *et al.*, 1994; Bartelt, Weaver & Arbogast, 1995; Hallett *et al.*, 1995; James, Moore & Aldrich, 1994; Oehlschlager *et al.*, 1995). Both sexes of the caddis fly,

Hydropsyche, which forms male swarms, also produce an aggregation pheromone with nonan–2-one as a major component. Males of a number of *Drosophila* species produce an aggregation pheromone consisting of a mixture of aliphatic hydrocarbons. This, acting together with odors from the habitat, attracts both sexes. It becomes transferred to females during mating and the females then transfer it to oviposition sites so that more insects are attracted (Hedlund *et al.*, 1996).

Gregarious locusts also produce aggregation pheromones containing a number of aromatic hydrocarbons. Adults of both sexes as well as larval stages produce the pheromone, apparently from the gut, although there are some differences in the compounds produced by adults and larvae (Torto *et al.*, 1994).

Sex pheromones active at close quarters Many insects employ pheromones when the two sexes are in close proximity. These may serve to identify the opposite sex with greater certainty and may also have the effect of increasing the willingness of a female to mate. They are, therefore, sometimes called aphrodisiac pheromones. They come into operation when the sexes have already been brought together by some other signal, which may be an attractant pheromone, a visual or mechanical stimulus, or a common response to features of the habitat.

Aphrodisiac pheromones are believed to be produced by males of many Lepidoptera, although their presence is often inferred from the insects' behavior rather than from the identification of the chemicals. In Danaidae, the males are visually attracted to the females and fly in pursuit. When a male *Danaus gilippus* overtakes a female he extrudes his hair pencils and dusts her with pheromone transfer particles. These carry the pheromone which stimulates the olfactory sensilla of the antennae and induce the female to land. The male may then hover over the female, continuing to dust her with pheromone particles, until finally he lands and copulates. The brushes of male Noctuidae are used in an essentially similar way, being everted just before copulation when courtship is already in progress. The associated release of pheromone is essential for copulation to occur.

In Danaidae, the pheromones are derived from pyrrolizidine alkaloids taken up by the males (see Fig. 27.8) and similar compounds are used by some male arctiid moths. Many noctuid moths employ mixtures of aromatic hydrocarbons as aphrodisiac pheromones, but some other families use aliphatic hydrocarbons or terpenoids. These compounds are produced in larger quantities than those forming the attractant pheromones produced by females.

Males of some butterflies produce aphrodisiac pheromones from androconia on the wings. For example, in *Eumenis semele* the male follows the female visually, and, ultimately, if she is virgin, the female lands. The male has scent scales in a patch on the upper side of the fore wing (similar to those shown in Fig. 27.10) and courtship is completed by the male standing in front of the female and bowing towards her with the wings partly open so that the female's antennae come into contact with the scent areas. The female then allows the male to move round and copulate.

Glands which may produce aphrodisiac scents also occur in various Neuroptera, Trichoptera, Diptera and Hymenoptera, sometimes in the male, sometimes in the female. Some male cockroaches produce a substance from a dorsal abdominal or thoracic gland which is fed on by the

female and induces her to mount on the male's back. Mechanoreceptors near the gland are stimulated when the female feeds indicating to the male that she is in the right position for mating (Farine *et al.*, 1996).

Many cyclorrhaphan Diptera use cuticular hydrocarbons for sex recognition. These usually have chain lengths of more than 25 carbon atoms and are often methylated. They function as contact pheromones and, perhaps, as volatile attractants when the insects are already very close together. After mating, the composition of the female's cuticular hydrocarbons changes, in *Drosophila* due to a change in synthesis, but in *Glossina* due to transfer of cuticular components from the male. The effect of these changes is to inhibit further courtship (Blomquist & Dillwith, 1985; Howard, 1993).

Marking pheromones These pheromones are used by social insects to define feeding or nest sites and to mark trails. Amongst non-social insects they serve to mark positions occupied by eggs, or high densities of insects. They might therefore be expected to be more persistent than most other pheromonal signals and so to be less volatile. While this is sometimes the case, it is not always true; a lack of persistence may sometimes be overcome by frequent marking.

Pheromones which cause the members of a population to become more widely spaced are sometimes called epideictic pheromones. Females of many parasitic Hymenoptera and some phytophagous species of Hymenoptera, Diptera and Lepidoptera, as well as some beetles, mark their oviposition sites with a pheromone that often inhibits further oviposition in the same host or part of a plant (Table 27.3). These pheromones are called oviposition marking pheromones. In parasitic Hymenoptera they are produced in Dufour's gland, while in the fruit fly, *Rhagoletis*, they are synthesized in the posterior midgut. The pheromones are deposited after oviposition when the female drags the tip of her abdomen over and round the site. Few of these pheromones have been identified. The principal component of that of the ichneumonid wasp, *Nemeritis*, is heneicosane, the codling moth, *Cydia*, uses a range of fatty acids with chain lengths of 14 to 18 carbon atoms, with linoleic acid (C18) being the most abundant (Thiéry *et al.*, 1995), while the apple maggot fly, *Rhagoletis*, produces taurine.

Epideictic pheromones are also produced by the larvae of some species. For example, the larvae of the flour moth,

Table 27.3. *Oviposition marking pheromones.*

Order/family/genus	example	origin
Diptera Tephritidae *Rhagoletis* fruit fly	$CH_3CH(CH_2)_6CH(CH_2)_6CONHCH_2CH_2SO_3H$ *N* - taurine	midgut
Hymenoptera Ichneumonidae *Nemeritis* parasitic wasp	heneicosane	Dufour's gland
Lepidoptera Tortricidae *Cydia* codling moth	linoleic acid	?
Pieridae *Pieris* cabbage butterfly	miriamide	?
Coleoptera Anobiidae *Lasioderma* cigarette beetle	serricorone	?

Ephestia, produce a pheromone from their mandibular glands when they encounter other larvae of the same species. This pheromone causes increased wandering, later pupation and ultimately results in smaller pupae. The moths emerging are consequently also smaller and, as fecundity is proportional to size, fewer eggs are laid and population density in the next generation is reduced.

Marking pheromones are also used by honeybees. The secretion of the Nasonov gland is made up of seven different terpenoids, including citral and geranic acid. The bees do not deposit the pheromone on a surface, but worker bees expose the gland and disperse the scent by fanning with their wings. They exhibit this behavior at the nest entrance, guiding returning workers, and during clustering when a colony swarms. They may also occasionally use this pheromone to guide other workers to sources of food or water.

Trail pheromones are another category of marking pheromones. They are produced primarily by termites and ants, social insects that forage on the ground, but a few nonsocial insects, such as tent caterpillars, also produce them. The persistence of such trails depends on the number of individuals depositing the pheromone, the nature of the pheromone, the substrate on which it is laid, and the prevailing environmental conditions. The trail laid by a single ant is often relatively short-lived and it is only by constant reinforcement that more persistent trails are produced. Tent caterpillars lay a silk trail at the same time as their scent trail and the silk contributes to the ability of the caterpillars to follow the trails.

Termites produce a trail pheromone from a gland on the ventral surface of the abdomen. In most species the pheromone is produced continuously and a drop is deposited each time the abdomen touches the surface. In a number of instances a constituent of the food forms one component of the pheromone system. In *Coptotermes formosanus* the major component of the trail pheromone is

Table 27.4. *Trail pheromones. Examples of chemicals which are commonly major constituents of the pheromone in different insects. Notice that different ant species (Formicidae) produce the pheromone in different glands.*

Order/family/genus	principal compounds	example	origin
Isoptera Termitidae *Coptotermes*	aliphatic hydrocarbons	 *Z,Z,E* - 3,6,8 - dodecatrienol	ventral abdominal gland
Lepidoptera Lasiocampidae *Malacosoma* (larva)	steroids	 5β - cholestane - 3,24 - dione	ventral abdominal gland
Hymenoptera Formicidae Ponerinae *Megaponera*	heterocyclic nitrogen compounds	 *N,N* - dimethyluracil	poison gland
Myrmicinae *Atta*	heterocyclic nitrogen compounds	 methyl 4 - methylpyrrole - 2 - carboxylate	poison gland
Solenopsis	terpenoids	 *Z,E* - α - farnesene	Dufour's gland
Dolichoderinae *Iridomyrmex*	aliphatic hydrocarbons	 *Z* - 9 - hexadecenal	sternal gland
Formicinae *Lasius*	aliphatic hydrocarbons	 hexanoic acid	hindgut

Z,Z,E–3,6,8–dodecatrienol with *Z,E,E*–3,6,8–dodecatrienol as a minor component (Table 27.4).

Ants lay scent trails which they use to mark foraging paths. The scent is produced in different glands in different subfamilies of ants and the products of one gland often differ between species. In many cases the pheromone is a mixture of chemicals, perhaps giving specificity to the trail, and, in some species, the different chemicals may be contributed by different glands. For example, workers of *Daceton* can lay a trail using chemicals from the poison gland or from the sternal gland. The trail produced by the former is much more persistent, perhaps related to the higher molecular weight compounds commonly produced by the poison gland.

Tent caterpillars lay trails by pressing the ventral surface of the terminal abdominal segment against the substratum. If they find food, they overmark these exploratory trails as they return to the nest and these trails are then followed preferentially by other caterpillars. The major components of the trail pheromone are two steroids, and one, a monoketone, is more effective than the other, a diketone, in inducing trail following. It is possible that the caterpillars modify the balance between these two chemicals when laying different types of trail (Fitzgerald, 1993).

Chemical trails are also utilized by some bees, though here the trail is followed in flight. In the stingless bees of the genus *Trigona* a pheromone is used to mark the route to a food source. A worker bee who has located a suitable food source marks it by biting and depositing the pheromone from the mandibular glands. Further odor spots are made at intervals of about 15 m back to the nest and recruited workers follow the succession of odor spots to the food source. Citral constitutes 95 % of the pheromone from the mandibular glands of *T. subterranea*, but, in some other species, alcohols and ketones are amongst the major constituents.

Male bumblebees, *Bombus*, use marking pheromones from the labial glands to mark out routes which they then patrol. When it lands to mark, the bee grasps the object to be marked in the mandibles and gnaws at its surface. The pheromone comprises a mixture of short-chain hydrocarbons and, in some species, monoterpenes (Bergman & Bergström, 1997). The mixture is species-specific and this specificity is enhanced by the tendency of different species to mark different objects and to fly at different heights (Appelgren *et al.*, 1991). Each time a male comes to a marked point he hovers for a short period before flying on to the next point. The function of these routes is not clear although it is assumed that they facilitate mate finding. Comparable scent routes are also made by males of some other bees such as *Psithyrus* and *Anthophora*.

The importance of chemical mixtures in conferring specificity to the signals provided by pheromones is illustrated by members of the genus *Trigona*. *Trigona postica* is unable to follow the trails of either *T. spinipes* or *T. xanthotricha*, but *T. xanthotricha* can follow the trail of *T. postica*. All these species produce 2-heptanol as a major component of the trail pheromone, but in *T. spinipes* there are only three known components compared with at least ten in the other two.

Alarm pheromones Alarm pheromones are produced as a response to disturbance by some Thysanoptera and Hemiptera that tend to feed in groups and also by social insects. They are usually highly volatile compounds, monoterpenes, sesquiterpenes or short chain aliphatic hydrocarbons, although other classes of compound are produced by some ants (Table 27.5). These pheromones, although commonly comprising mixtures of several compounds, tend to be less specific than other types of pheromone. Alarm pheromones are commonly produced in the mandibular glands of social insects, although some species employ chemicals from other glands.

The effect of alarm pheromone in non-social insects is to cause dispersal. Aphids, for example, may fall from their hostplant when they perceive the alarm pheromone. Amongst social insects, the response varies according to the species, but commonly involves attraction of other workers or soldiers and the adoption of aggressive postures with the head raised and jaws wide apart.

Different components of an alarm pheromone may elicit different behavior in recipients and, because of their different volatilities, and perhaps due to differences in the sensitivity of the receiving insects, the areas over which they produce active responses by the ants vary (Fig. 27.13). For example, hexanal is the most volatile of the components produced by *Atta* so the area over which it spreads rapidly increases in size. It alerts the soldier ants, who raise their heads and open their jaws. Hexanol, the other major component, is rather less volatile and spreads more slowly; at low concentrations it stimulates movement towards the site of release. Other compounds are present in small quantities and are effective only close to the point of deposition. This is true of both 3-undecanone and 3-butyl–2-octenal which both induce biting.

The alarm pheromone of the bee *Trigona* is also produced by the mandibular glands. At least some of the chemicals which are used in trail-laying also function in producing alarm, the nature of the response depending on the situation. An alarm pheromone is produced in the mandibular glands of *Apis*, and another is released via the sting chamber. Guard bees at the entrance to a hive use this to mark intruders and to stimulate aggressive activity by other bees.

Reviews: Aldrich, 1995 – heteropteran secretions; Birch, Poppy & Baker, 1990 – male moth pheromones; Blum, 1985 – alarm pheromones; Boppré, 1984 – male moth pheromones; Borden, 1985 – aggregation pheromones; Haynes & Birch, 1985 – trail pheromones; Pickett *et al.*, 1992 – aphid pheromones; Prokopy, 1981 – spacing pheromones; Tamaki, 1985 -sex pheromones; Traniello & Robson, 1995 – trail pheromones of social insects

Table 27.5. *Alarm pheromones. Examples of chemicals which are commonly major constituents of the pheromone in different families of insects. The term hydrocarbon is used here to refer to compounds with an unbranched chain of carbon atoms.*

Order/family	principal compounds	example	origin
Isoptera Termitidae	terpenes	limonene	cephalic gland
Thysanoptera Thripidae	hydrocarbons	decyl acetate	?
Hemiptera Aphididae	terpenes	*E* - β - farnesene	cornicles
Tingidae	hydrocarbons + monoterpenes	*E* 2 - hexenal / geraniol	dorsal abdominal gland
Coreidae	hydrocarbons	hexanal	metathoracic gland
Hymenoptera Apidae	hydrocarbons	2 - heptanone	mandibular glands + poison gland
Formicidae Ponerinae	heterocyclic nitrogen compounds	2,5 - dimethyl - 3 - isopentylpyrazine	mandibular glands
Myrmicinae	branched hydrocarbons	4 - methyl - 3 - heptanone	mandibular glands

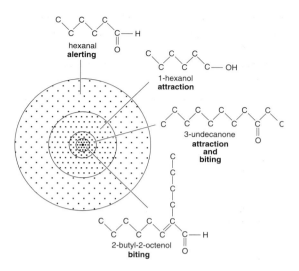

Fig. 27.13. Differential effects of pheromone components. The alarm pheromone of the leaf-cutting ant, *Atta*, has hexanal and 1-hexanol as major components. The area over which they are perceived by other workers depends on the volatility of the compounds and the insects' sensitivity. Consequently hexanal is effective over a greater area than hexanol, and the two compounds elicit different behaviors from recipient ants. 3-undecanone and 2-butyl-2-octenol are minor components with much smaller effective areas (based on Bradshaw, Baker & Howse, 1979).

27.1.6.2 *Pheromones affecting physiology*

Most instances of pheromones affecting physiology are in social insects, but some examples are also known in non-social insects. Mature male desert locusts (*Schistocerca gregaria*), for example, accelerate the maturation of other, less mature locusts of either sex. The pheromone responsible for this is produced in scattered epidermal glands. Receiving individuals may perceive the pheromone either as a scent or by bodily contact, the latter being more effective. The mode of action of the pheromone is not understood, but its effect is to stimulate the activity of the corpora allata. This leads to maturation of the gonads and to some synchronization of maturation throughout the population.

Both sexes of *Tenebrio* produce a pheromone which accelerates oogenesis in females and also causes them to produce and release more attractant pheromone. Males are more effective than females in inducing these responses (Happ, Schroeder & Wang, 1970).

Review: Loher, 1990 – locust pheromones

27.1.6.3 *Pheromones of social insects*

Social insects possess pheromones comparable with those of other insects, but, in addition, pheromones have particularly important roles in effecting communication between workers and in the maintenance of colony structure.

The differentiation of castes in termites appears to be regulated by a series of pheromones, although the pheromones have yet to be isolated. In the dry wood termite, *Kalotermes*, the numbers of a particular caste are regulated by adjusting the balance between production and elimination of members of the caste concerned. For example, in the absence of the king and queen, replacement reproductives are produced. Usually an excess of replacements is produced and these are eaten by the pseudergates so as to leave only one pair. This is believed to be controlled by pheromones produced by the reproductives. Normally, the queen produces a substance which inhibits the further development of female pseudergates, and in the absence of this substance any female pseudergates that are competent to do so become replacement reproductives. The male is believed to produce a comparable pheromone which inhibits the development of male pseudergates. The male produces a further substance which, in the absence of the female inhibitor, stimulates the production of females. Finally, the behavior of the pseudergates is affected by other pheromones which leads them to eat any excess replacement reproductives which have been produced. The male produces a substance leading to the elimination of excess males, the female produces one leading to the elimination of excess females. The production of soldiers is probably controlled in a similar way. In large colonies of *Kalotermes* there are, on the average, three soldiers for every 100 individuals in the colony. If soldiers are removed, others develop in their place. Conversely, when a colony is producing alates, the number of individuals and hence the number of soldiers increases, but when the alates swarm the population of the colony suddenly decreases leaving an excess of soldiers. The excess soldiers are eliminated, being eaten by the pseudergates.

Colony structure in honeybees is regulated by a pheromone produced in the mandibular glands of the queen. This has two components, 9-oxodecanoic acid (Table 27.2) and 9-hydroxydecenoic acid. The former is called queen substance. From the mandibular glands the pheromone becomes distributed over the whole body of the queen. It attracts workers who touch the queen with

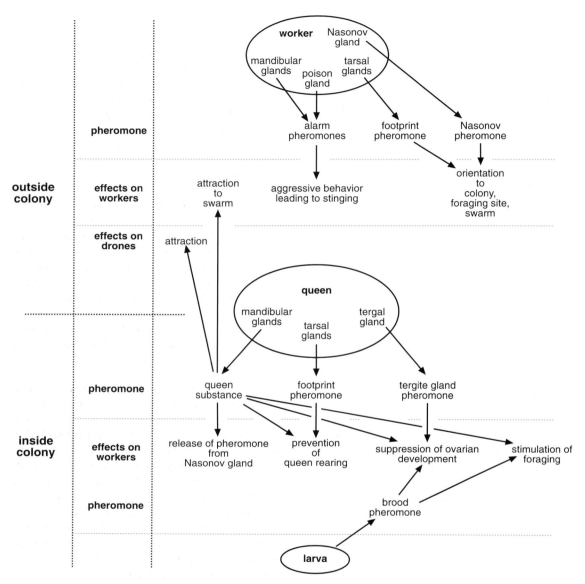

Fig. 27.14. Pheromones of the honeybee, *Apis.*

their antennae and tongues resulting in the transfer of the pheromone to their bodies. Transfer through the colony results from the frequent contacts made by these workers with others. One major function of the pheromone is to inhibit the development of the workers' ovaries (Fig. 27.14), but it also inhibits queen rearing by the workers.

Different castes produce different pheromones even from the same glands. For example, while queen honey-bees produce queen substance in the mandibular glands, workers produce 2-heptanone, an alarm pheromone. In the leaf-cutting ant, *Atta sexdens*, minor workers whose principal duties are within the nest, produce the same chemicals in their mandibular glands as sexual males and females. Major workers, engaged in foraging, and soldiers, with a defensive role, have completely different major components (Fig. 27.15).

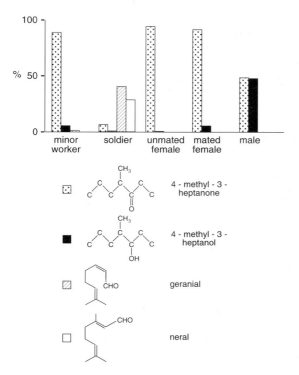

Fig. 27.15. Caste differences in mandibular gland secretions of the leaf-cutting ant, *Atta* (based on Nascimento *et al.*, 1993).

Kin recognition is an important aspect of social insect behavior both with respect to recognizing members of other colonies and in recognizing brood. Colony recognition probably often involves innate characteristics together with acquired environmental characters. Both are learned by the individual insects when they eclose. Differences between colonies in the composition of cuticular waxes have been demonstrated in some social wasps and ants, providing a basis for colony recognition (Layton, Camann & Espelie, 1994). In *Apis* colonies, small differences in cuticular hydrocarbons also exist between individuals that are full sisters, having the same father, and half sisters, having different fathers; individual bees can discriminate between the two.

Recognition of brood by adult social insects depends largely on the presence of pheromones on the larval cuticle.

Reviews: Hölldobler & Wilson, 1990 – ant pheromones; Smith & Breed, 1995 – social recognition; Winston, 1987 – honeybee pheromones; Vander Meer *et al.*, 1998 – social insects

27.2 SECRETIONS WITH INTERSPECIFIC SIGNIFICANCE

Many insects produce chemicals which affect the behavior or physiology of other organisms, often having some defensive function. Like pheromones, these chemicals are often produced in exocrine glands, but they are sometimes stored within the insect's body and only become evident if a potential predator makes contact.

Reviews: Blum, 1981; Whitman, Blum & Alsop, 1990

27.2.1 Structure and distribution of glands

Glands involved in producing chemicals for interspecific signalling in general differ from pheromone glands in having a reservoir in which the chemicals are stored; the quantities of chemical involved are much greater than those employed as pheromones. Most defense glands develop as invaginations of the epidermis and are lined by cuticle, the obvious exception being those salivary glands which produce defensive substances. The glands can be grouped into those which are not everted, but from which the chemicals are expelled, sometimes forcibly, and those which can be everted and from which chemicals diffuse away or are effective only on contact.

Non-eversible glands In non-eversible glands, the gland cells may line a reservoir, or the secretory region and reservoir may be separate and joined by a duct. The gland cells usually have ducts leading through the cuticle (class 3 cells), but in the prothoracic glands of notodontid larvae they do not (class 1 cells). The reservoir is lined by cuticle and its opening to the exterior is normally closed by the elasticity of the surrounding cuticle and opened by specific muscles.

Exocrine glands producing chemicals that are known or presumed to have an interspecific role are known from species in almost all orders of insects except for the Phthiraptera and Siphonaptera, which are parasitic on vertebrates, and for Thysanura and Mecoptera. Their positions and form are very variable, but within a particular insect group they are more uniform. A few examples are given to illustrate the variety of structures. The soldiers of Rhinotermitidae and Termitidae have a gland called the frontal gland which discharges its secretion through a median anterior pore in the head. In some rhinotermitids the reservoir is extremely large and occupies more than half the abdominal cavity, but in termitids it is restricted to the head capsule. Nasute termites have the front of the head modified to form a rostrum with the gland opening at the

a) reservoir associated with gland

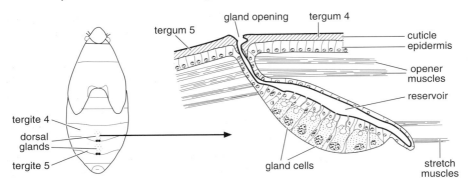

b) separate reservoir

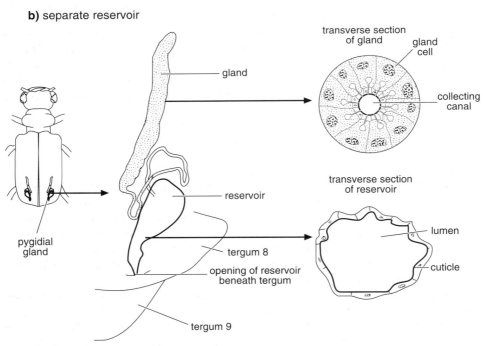

Fig. 27.16. Non-eversible exocrine glands. **(a)** Gland cells directly associated with reservoir. Abdominal glands of larval Heteroptera, *Pyrrhocoris* (after Staddon, 1979). **(b)** Gland cells separate from reservoir. Coleopteran pygidial gland, *Cicindela* (based on Forsyth, 1970).

tip. The soldiers of some Termitidae have the salivary glands modified for defense, the reservoir occupying the front half of the abdomen, with the secretion being ejected from the mouth as a result of abdominal contraction.

The romaleid grasshoppers and a cockroach, *Diploptera*, have glandular epithelium lining a trachea inside the metathoracic and second abdominal spiracles, respectively. In the grasshoppers, the secretion is stored in the trachea which is not normally used in respiration.

Heteropteran larvae typically have glands in the dorsal abdomen opening behind the tergites, usually by a single median pore (Fig. 27.16a). Sometimes they occur in

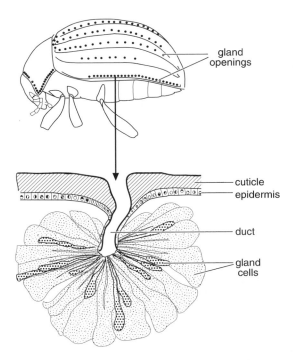

Fig. 27.17. Non-eversible exocrine gland with no reservoir. Coleopteran pronotal and elytral glands, *Leptinotarsa* (after Deroe & Pasteels, 1977).

segments 3 to 6, but often they are reduced in numbers. They sometimes persist in adults, which also typically have a ventral metathoracic gland. This is often complex with lateral and median reservoirs which open via lateral ducts between the meso- and metathoracic legs. It exhibits great variation between species and is often reduced in species that sequester chemicals from plants.

Nearly all adephagan beetles have a pair of pygidial glands, in which the secretory lobes are distinct from the reservoirs, opening behind abdominal tergite 8 (Fig. 27.16b). Adult chrysomeline beetles have glands opening on the pronotum and the elytra (Fig. 27.17). In this group, there are many separate glands which have no reservoirs. Instead, the secretions are stored in vacuoles and in extracellular spaces. Many other groups of beetles also have exocrine glands with a probable defensive function (Dettner, 1987).

Amongst Lepidoptera, larval Notodontidae have a gland opening ventrally in the neck membrane. The gland is single-chambered in some species, but has two chambers in others (Attygalle *et al.*, 1993).

Hymenoptera produce chemicals having interspecific significance from many of the same glands used to produce pheromones (Fig. 27.2). The mandibular and poison glands are particularly important in this respect.

Eversible glands Eversible defensive glands have been described in relatively few species. They are normally completely invaginated within the body, but when the insect is disturbed they are everted by hemolymph pressure. These glands have no separate reservoir, although secretion presumably accumulates within the folds of the invaginated structure. They are pulled back within the body by muscles attached directly to the cuticle of the gland.

A few species of grasshoppers have such a gland between the pronotum and mesonotum. In larval swallowtail butterflies (Papilionidae), a bifurcate tubular structure known as the osmeterium can be everted from the dorsal neck membrane. The larvae of many chrysomelid beetles have pairs of small eversible glands on the meso- and metathorax and on the first seven abdominal segments, while the adult staphilinid beetle, *Stenus*, has an eversible vesicle on each side of the anus.

Reviews: Grassé, 1975; Staddon, 1979 – Heteroptera; Whitman, Blum & Alsop, 1990

27.2.2 Storage within the body

Chemicals used in interspecific communication are often stored within the body, as well as, or instead of, in glands. Fig. 27.18 shows the distribution of ouabain, a cardiac glycoside, in the adult of the milkweed butterfly, *Danaus plexippus*. About 80% is in the integument of the wings and the rest of the body, but an appreciable amount is in the fat body.

The integument is a storage site for cardiac glycosides in the milkweed bug, *Oncopeltus*, and in *Danaus plexippus*; for cyanogenic glycosides in the larva and adult of the burnet moth, *Zygaena*; and quinolizidine alkaloids in the caterpillar of *Uresiphita*. In larval *Zygaena*, there are large cavities in the cuticle of the meso- and metathorax and of the first eight abdominal segments in which the glycosides are stored (Fig. 27.19a). Smaller cavities are present in the rest of the cuticle of the dorsal and lateral regions of the body. In *Oncopeltus*, there is a layer of vacuolated cells immediately beneath the epidermis over much of the thorax and abdomen (Fig. 27.19b).

Storage in the hemolymph occurs most frequently

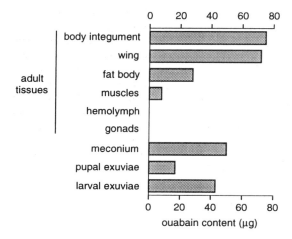

Fig. 27.18. Sequestration in different parts of the body. Distribution of ouabain, a cardiac glycoside, in an adult *Danaus plexippus* after feeding by the larva on food containing ouabain. An appreciable amount of ouabain is also present in larval and pupal exuviae and in the meconium. Ouabain is not present in milkweeds (the normal food of this insect), but is structurally very similar to cardiac glycosides that are normally present (after Frick & Wink, 1995).

amongst the Coleoptera. The blister beetles (Meloidae) store cantharidin at concentrations above 10^{-3} M, coccinellids store alkaloids, and the rove beetle, *Paederus*, an amide, pederin (Kellner & Dettner, 1995).

Potentially toxic compounds are often present in insect eggs, often being derived from those present in the female, but sometimes coming from the male (section 12.4.4). For example, among the beetles, the eggs of the staphylinid, *Paederus*, contain the amide, pederin, those of the chrysomelid, *Gastrophysa*, contain oleic acid, and those of *Chrysolina*, cardiac glycosides.

27.2.3 Chemicals used in interspecific communication

Many types of chemical are employed by insects for interspecific communication, usually for defensive purposes. In most instances the chemicals are produced and stored in their active form, but in some cases a chemical reaction occurs at the time of release to produce the active principle. Sometimes the chemicals are volatile, producing their effect at a distance; in other cases, their volatility is very low and they act on contact. In nearly every case, the secretion comprises a mixture of compounds. The

chemicals are often widely occurring compounds, but, in some cases, as with the alkaloids of coccinellid beetles, they are unique to the insects producing them. The following account gives a few examples of the major types of compounds occurring in insects.

Aliphatic hydrocarbons are present in the frontal-gland secretions of some soldier termites. The reservoirs of these species are often very large and the dry weight of secretion may constitute almost 50% of the insect's dry weight. The fungus-growing termites of the genus *Macrotermes* produce alkanes and alkenes with chain lengths of 25 to 35 carbon atoms, and some Rhinotermitidae produce vinyl ketones and ketoaldehydes.

Formic acid is produced by numerous insects. It is present in the prothoracic gland secretion of some notodontid caterpillars and the pygidial gland secretions of some carabid beetles, as well as in the poison glands of many ants. *Polistes*, a vespid wasp, produces carboxylic acids, with octodecanoic acid the most abundant, in van der Vecht's gland.

Phenolic compounds are common in insect exocrine secretions. They are produced by the pygidial glands of water beetles. For example, *Platambus* produce *p*-hydroxybenzaldehyde, benzoic acid, *p*-hydroxybenzoic acid methylester and hydroquinone. The scent glands of some water bugs contain *p*-hydroxybenzaldehyde and methyl *p*-hydroxybenzoate.

Quinones are employed by other insects. Soldier *Odontotermes* employ a secretion which includes both benzoquinone and protein from the salivary gland reservoir. Quinones are also produced by many beetles. They are present in the pygidial gland secretions of Carabidae and the abdominal gland secretions of some Staphylinidae and many Tenebrionidae where toluquinone and ethyl quinone are produced by all the species examined. Bombardier beetles from various subfamilies generate a quinonoid spray from hydroquinones.

Volatile terpenoids are produced by many insects. Soldier termites of many species produce terpenes in the frontal glands. In major soldiers of *Trinervitermes*, the secretion of the frontal gland constitutes 8% of the body weight with α-pinene as the most abundant monoterpene. The phasmid, *Anisomorpha*, produces the terpene dialdehyde anisomorphal as a defensive spray and many Heteroptera produce terpenoids in the abdominal or metasternal glands. The secretions from the prothoracic

a) storage in cuticular cavities

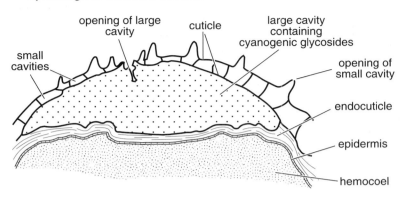

Fig. 27.19. Sequestration in the integument. (a) Diagrammatic section through the integument of the larval burnet moth, *Zygaena*, showing the cavities in which cyanogenic glycosides are stored (based on Franzl & Naumann, 1985). (b) Accumulation of cardiac glycosides in a layer of cells immediately inside the epidermis (Heteroptera: *Oncopeltus*) (after Scudder, Moore & Isman, 1986).

b) storage in subepidermal layer

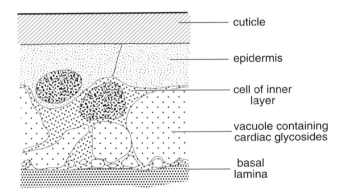

exocrine glands of some water beetles contain a range of sesquiterpenes, while some staphylinid beetles produce citral in their abdominal glands.

Steroidal compounds, with a presumed defensive function, are present in the bodies of insects from several orders. The grasshopper, *Poekilocerus*, has cardiac glycosides in the secretion of its dorsal thoracic gland. Cardiac glycosides are also sequestered in the integument of the milkweed bug and the milkweed butterfly and are synthesized by some chrysomelid beetles where they are present in the hemolymph.

Alkaloids are synthesized or sequestered by a variety of insects. For example, unique alkaloids, produced by the insects, are present in the hemolymph of coccinellid beetles and the venom of fire ants, *Solenopsis*, and Pharaoh's ant, *Monomorium*. On the other hand, some

chrysomelid beetles, several lepidopteran species, and a grasshopper, *Zonocerus*, sequester alkaloids from their host plants.

A number of Lepidoptera contain cyanogenic compounds including the glycosides linamarin and lotaustralin from which hydrogen cyanide may be produced. A few beetles also contain cyanogenic glycosides.

Biogenic amines and proteins are also secreted by some insects. An amide, pederin, is produced by some staphylinid beetles (Kellner & Dettner, 1995). Pompilid and sphecid wasps produce histamine and 5-hydroxy-tryptamine, and sometimes also acetylcholine, in their poison glands. Peptides known as melittins constitute about 50% of the dry weight of bee venom, and other peptides are present in smaller amounts. The venom of bees and some ants also contains several enzymes:

hyaluronidase, phospholipase and phosphatase. Tent caterpillars and adults of some moths, like *Euproctis* have specific proteins associated with urticating hairs.

In addition to these major classes of substances which are commonly identified in known or supposed defensive secretions, insects produce a variety of chemicals belonging to other chemical classes.

Reviews: Aldrich, 1988 – Heteroptera; Brower, 1984 – butterflies; Davis & Nahrstedt, 1985 – cyanogenesis; Dettner, 1987 – Coleoptera; Piek, 1985 – Hymenoptera; Prestwich, 1984 – Isoptera

27.2.4 **Origins of chemicals**

Many chemicals are synthesized *de novo* by insects, but many plant-feeding species sequester chemicals from their hosts. Some insects are able to synthesize as well as sequester some chemicals.

Synthesis Synthesis of interspecific signalling chemicals is of widespread occurrence in insects. This applies not only to relatively simple structures like aliphatic hydrocarbons and monoterpenes, but to some much more complicated structures. For example, cardiac glycosides are synthesized by some chrysomelid beetles, and alkaloids by coccinellid beetles and fire ants. Insects with this capacity produce the same suite of compounds irrespective of their food.

Sequestration Many plant-feeding insects sequester plant toxins which often are not secreted to the outside, but are retained within the body so they are effective only when the insect is tasted by a predator. The species which do this are often specialists on plants containing specific classes of chemical, presumably reflecting the ability of the insects to tolerate high concentrations of the particular chemicals. An exception to this generalization is the lubber grasshopper, *Romalea*, which feeds on many different plants and sequesters a range of different chemicals.

Compounds commonly sequestered are cardiac and iridoid glycosides, cucurbitacins and pyrrolizidine alkaloids (Table 27.6). Sequestration is selective and compounds may be stored unchanged or after some reconstruction. For example, *Euphydryas* sequesters some iridoid glycosides unchanged, others are metabolized before storage, and others are excreted (Fig. 27.20). The larvae of *Danaus plexippus* selectively store polar cardiac glycosides, but excrete those that are apolar. Some chrysomelid beetles synthesize salicaldehyde from salicin in their food.

The amounts and particular chemicals sequestered are affected by differences in the food (Fig. 27.21), and the quantities may be very large. For example, the iridoid glycosides sequestered by the checkerspot butterfly, *Euphydryas*, may constitute 10% of the insect's dry weight.

In a few cases, chemicals are separated from the food during ingestion. Larvae of the sawfly, *Perga*, feed on *Eucalyptus*. Oils from the food are sequestered in a diverticulum on the ventral side of the foregut; very little oil reaches the midgut and in the last stage larva the full gland may make up 20% of the body weight. Similar adaptations occur in larvae of *Neodiprion* feeding on *Pinus* and in larvae of *Myrascia* (Lepidoptera) feeding on Myrtaceae (section 3.1.1).

Reviews: Blum, 1987 – synthesis; Bowers, 1990 – sequestration; Duffey, 1980 – sequestration

27.2.5 **Avoidance of autotoxicity**

A critical problem for insects that produce or sequester chemicals that are potential toxins is the avoidance of autotoxicity. Two processes contribute to this: first, the insects may exhibit some degree of tolerance to the toxins, and second, the enzymes and substrate producing a toxin may be in separate compartments so that synthesis only occurs outside the body.

Cardiac glycosides are toxic to animals because they inhibit the enzyme Na^+/K^+-ATPase, but in insects that sequester these compounds, this enzyme is 100 times less sensitive to inhibition than in some other, susceptible, insects. In *Danaus plexippus* this insensitivity results from a change in a single amino acid on a binding site of the enzyme (Holzinger & Wink, 1996).

Insects that store cyanogenic glycosides in the cuticle are also less sensitive to cyanide poisoning than most other organisms, although the mechanism of resistance is not known. Zygaenid larvae have a β-glucosidase, with high specificity for the glycosides, in the hemolymph. Cyanide is produced only if the cuticle is punctured, so that hemolymph and glycosides come together.

Quinones are often produced outside the body. For example, the quinones produced by tenebrionids are probably derived from phenol glucosides by hydrolysis followed by oxidation. These processes may take place in the end-apparatus of the secretory cells, which is extracellular.

An extreme example of the extracellular production of the active component occurs in bombardier beetles. Quinones are generated explosively by the oxidation of

Table 27.6. *Sequestration. Examples of chemicals sequestered by different insects.*

class of compounds	example	plants containing	insects sequestering
cardiac glycosides	calotropin	Asclepiadaceae	*Poekilocerus bufonius* (Orthoptera) *Aphis nerii* (Hemiptera) *Oncopeltus fasciatus* (Hemiptera) *Danaus plexippus* (Lepidoptera)
iridoid glycosides	catalpol	Scrophulariaceae	*Euphydryas phaeton* (Lepidoptera)
pyrrolizidine alkaloids		Asteraceae	*Zonocerus variegatus* (Orthoptera) *Tyria jacobaeae* (Lepidoptera) *Oreina cacaliae* (Coleoptera)
isoquinoline alkaloids	aristolochic acid	Aristolochiaceae	*Battus philenor* (Lepidoptera)
triterpenoids	cucurbitacin B	Cucurbitaceae	*Diabrotica undecimpunctata* (Coleoptera)

hydroquinones so that they are ejected as a spray through the openings of the pygidial glands. The hydroquinones are stored with hydrogen peroxide in the glands, but at the moment of discharge these chemicals are mixed with catalases and peroxidases from an accessory chamber. This causes the sudden decomposition of the hydrogen peroxide and oxidation of the hydroquinones so that the quinones are forcibly ejected with an audible explosion and the generation of temperatures up to 100 °C.

27.2.6 Dispersal of chemicals

Dispersal of the chemicals from the glands depends on the opening of the ducts leading to the exterior which are held closed by the elasticity of the cuticle. This is achieved by opener muscles. Where the gland is surrounded by a muscle layer, as in the stick insect, *Extatosoma*, and the pygidial glands of adephagan beetles, muscle contraction forces the secretion through the ducts to the outside. In other cases, contraction of body wall muscles acting on the

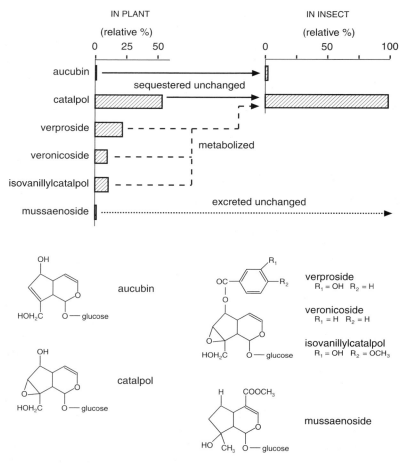

Fig. 27.20. Sequestration. Differential utilization of iridoid glycosides from a host plant, *Besseya plantaginea*, by *Euphydryas anicia*. Aucubin and catalpol are sequestered unchanged, others are metabolized, and mussaenoside is not used. A proportion of all the compounds is excreted without being absorbed (based on Gardner & Stermitz, 1988; L'Empereur & Stermitz, 1990).

gland via the hemolymph leads to compression of the gland. Eversible glands are also inflated by hemolymph pressure.

In some cases, compression of the reservoir is so rapid that fluid is forcibly ejected, in several cases travelling up to 20 cm from the insect. Sometimes the ejection is accompanied by behavior that directs the spray towards the aggressor. Notodontid larvae, for instance, raise the front half of the body and turn to face the potential danger before they spray. *Parachartergus colobopterus*, a small social wasp, directs a spray of venom from its poison gland at a source of disturbance by twisting its gaster so that it points forwards (Jeanne & Keeping, 1995). A few grasshoppers with dorsal glands also eject a spray.

Romaleid grasshoppers that have glands associated with a trachea emit a frothy bubble of chemicals from the metathoracic spiracles by closing all the other spiracles and compressing the abdomen to force air through the secretion in the metathoracic trachea.

A few species have special applicators with which they apply chemicals to surfaces. For example, some vespid wasps have a gland, known as van der Vecht's gland, with ducts opening along the anterior edge of the sixth abdominal sternite. Hairs surrounding the openings are used as a brush to apply the repellent contents of the gland to the pedicel supporting the nest (Dani, Morgan & Turillazzi, 1995).

The hairs of some urticating caterpillars and moths can penetrate the skin of vertebrates and so deliver their toxins subcutaneously. Tent caterpillars have eight tufts of barbed hollow hairs associated with the glands producing urticating substances, while some moths have a dense abdominal brush of hairs with holes at the tips through which, it is presumed, their urticating protein is delivered.

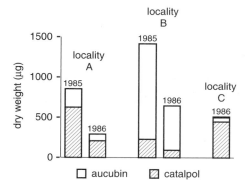

Fig. 27.21. Sequestration. Variation in total amounts and relative proportions of different iridoid glycosides sequestered by male *Euphydryas anicia* from different localities and different years. Insects at locality C also sequestered a small proportion of a third glycoside, macfadineoside, that was not present in insects from the other populations (indicated by the thicker black line at the top of the bar for 1986) (based on Gardner & Stermitz, 1988; L'Empereur & Stermitz, 1990).

Where the chemicals are volatile, the cuticle outside the gland opening may be highly sculptured, perhaps to facilitate evaporation. Over most of the osmeterium of papilionid caterpillars, the cuticle is produced into numerous papillae which hold the secretion when the gland is invaginated and facilitate evaporation when it is expanded.

Where chemicals are stored in the cuticle, they may be present on the surface so that they directly affect potential predators. A larval zygaenid compresses its body when attacked so that bubbles of fluid containing cyanogenic glycosides are forced out from the reservoirs, rupturing the cuticle at points of weakness at the bases of the exit pores. When the insect relaxes, the bubbles are withdrawn. In *Oncopeltus*, cardiac glycosides from the subepidermal reservoirs are also forced through weak points where the cuticle is much thinner than elsewhere. Insects with compounds stored in the hemolymph exhibit reflex bleeding. They have points of weakness in the membranous cuticle, often at joints in the legs, through which they are able to force drops of hemolymph.

27.2.7 Interspecific functions of chemicals

The main functions of these chemicals are the incapacitation or repulsion of potential parasites or predators, prey capture, and protection against micro-organisms. In a few cases, the chemicals are essential components of

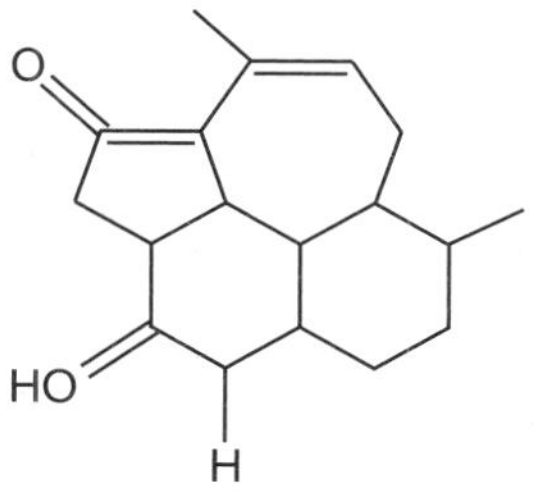

A tetracyclic diterpene that contributes to the glue secreted by soldier *Nasutitermes*.

Fig. 27.22. A tetracyclic diterpene that contributes to the glue secreted by soldier *Nasutitermes*.

mutualistic relationships. Compounds may have more than one function.

Defense against predators Enemies may be incapacitated by the physical nature of some secretions. The secretion of the frontal glands of nasute termites hardens to a sticky thread in air due to the presence of oxygenated diterpenes (Fig. 27.22). It is ejected on contact with predators such as ants, and its irritant action, which results from the presence of monoterpenes, causes the ant to preen, spreading the entangling material even more. The ant may die if it is unable to free itself of the sticky substance. Some compounds of the secretion also act as an alarm pheromone attracting other termite soldiers from up to 30 cm away. In *Macrotermes subhyalinus*, on the other hand, the frontal gland secretion of the soldiers apparently kills ants by preventing wound healing following a bite. The oily secretion oozes from the gland and coats the surface of the termite's head and is transferred to the enemy during combat.

Repellent effects of sprays are recorded in a number of insects. *Anisomorpha* directs the spray from its thoracic glands at enemies. Predatory ants and beetles elicit this response if they bite any part of the stick insect; but blue jays are sprayed before they touch the insect and a single experience is sufficient for a bird to learn to avoid the insect subsequently. An adult female *Anisomorpha* can produce up to five discharges in a short time, but it then takes up to 15 days for the store of deterrent to be replenished; when its store is depleted it is susceptible to attack. The secretions of *Poekilocerus*, which can be ejected for up to 40 cm, are also effective in repelling vertebrate predators. In this insect, too, the store of repellent is exhausted after about five expulsions.

Many insects produce secretions in response to an immediate attack, but the volatility of the chemicals may then result in an effect over a bigger area. Terpenoids are common components of these secretions and they often also function as alarm pheromones for conspecifics (Table 27.5). In the Hymenoptera, larvae of the sawfly, *Neodiprion*, emit a drop of fluid from the foregut reservoir when attacked. The fluid remains on the mouthparts and repels ants and birds from a distance. Similarly, the cynipid hyperparasite of aphids, *Alloxysta brevis*, releases a secretion from its mandibular glands if attacked by an ant. This repels the ant and the effect lasts for about five minutes, giving the wasp time to oviposit (Völkl, Hübner & Dettner, 1994). Many of the secretions produced by Heteroptera are in this category.

A few social insects are known to apply secretions to act as a barrier to potential predators. For example, the vespid wasps that coat the pedicels of their nest with secretion from van der Vecht's gland perform this behavior in response to the odor of formic acid, produced by many ants, or just before they leave the nest. The secretion consists of carboxylic acids, octodecanoic acid being most common in *Polistes*.

The function of some components of glandular secretions may be to enhance the action of other chemicals. For example, hydrocarbons produced by some beetles may act as solvents for the quinones that are the primary signalling chemicals, and 2-tridecanone may act as a wetting agent in the sprays of notodontid caterpillars, causing the formic acid to spread when it strikes a surface.

Compounds that are present in the cuticle or sequestered elsewhere in the body are usually of low volatility and will, in general, only be effective if a potential predator makes direct contact with the insect. They include alkaloids, steroids and tricyclic alcohols (Table 27.6). Their presence is commonly associated with aposematic coloration (section 25.6.1). The effectiveness of such compounds is suggested by the readiness with which the coccinellid, *Epilachna*, is eaten by spiders immediately after eclosion, before it has synthesized its alkaloid, compared with seven days later when alkaloid is present in the hemolymph (Fig. 27.23).

A few Lepidoptera produce proteins in subcuticular glands associated with hairs. These hairs are able to penetrate the skin of vertebrates and so introduce the protein which subsequently causes skin lesions. Larval tent caterpillars, *Thaumatopoea*, produce such proteins, as do adults of the gold tail moth, *Euproctis*.

In bees and social wasps, the venom is used principally for defense, although this is not true for non-social species

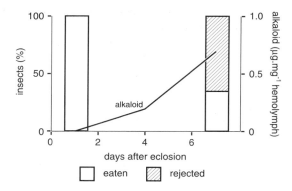

Fig. 27.23. Defence against predators. Relationship between alkaloid content of the hemolymph of adult beetles, *Epilachna varivestis*, and palatability to jumping spiders. By day 7 after eclosion, the beetles had a high concentration of alkaloid in the hemolymph and most were rejected by spiders (based on Eisner *et al.*, 1986).

(see below).

Defense against pathogens Many insects produce phenolic compounds with antibiotic properties. The secretion of the metapleural gland in some ants, which contains phenolics, is produced continuously in small amounts, and presumably has this function (Table 27.7). In aquatic insects such compounds are probably important in preventing micro-organisms from becoming established on the cuticle. Water beetles produce these compounds in their pygidial glands, spreading them over the body surface with the legs when they leave the water. Phenolic compounds probably perform this function in Dytiscidae, while sesquiterpenes have a similar role in Gyrinidae. Phenolics from the metathoracic gland of *Ilyocoris* also have an antibiotic function.

Prey capture Amongst the Hymenopteran superfamilies, the Bethyloidea, Scolioidea, Pompiloidea and Sphecoidea collect other insects and spiders with which they feed their larvae. These insects paralyse their prey with a chemical injected by the sting, which is a modified ovipositor. The paralysis may be temporary or last for many days or even months, but eventually the prey dies even if it is not eaten by the parasitic larvae. The injected venom does not keep the prey 'fresh' for the larvae to feed on; the prey survives only as long as its food reserves last. The venom is usually injected into or close to one of the ganglia of the ventral nerve cord of the prey, often near to one of the thoracic

Table 27.7. *Antibiotic effects. Phenolic compounds known to have antibiotic effects that occur in the secretions of insects.*

Order/family/genus	example	origin
Hemiptera Notonectidae *Notonecta*	 *p* - hydroxybenzaldehyde	metathoracic gland
Hymenoptera Formicidae *Crematogaster*	 3 - pentylphenol	metapleural gland
Atta	 phenylacetic acid	metapleural gland
Coleoptera Dytscidae *Ilybius*	 methyl 8-hydroxyquinoline-2-carboxylate	pygidial gland
Dytiscus	 benzoic acid	pygidial gland

ganglia. *Ampulex compressa* (Sphecoidea) makes two injections, the first into the prothorax and the second into, or near, the subesophageal ganglion of a cockroach. The first injection has a short-lived paralysing effect on its prey; the second is irreversible. The prey is still able to walk and the venom does not affect either sensory input or motor output, but has a neuromodulatory effect at specific synapses within the thoracic ganglia (Fouad, Libersat & Rathmayer, 1996).

The principal constituents of the venom and their effects on the prey differ from species to species. Histamine, acetylcholine and serotonin (5-hydroxytryptamine) are present in the venoms of some species. The venom of *Philanthus triangulin* (Sphecoidea), which preys exclusively on bees, contains a specific compound known as philanthotoxin, which causes a neuromuscular block by blocking glutamate receptors (Fig. 27.24). The venoms of some species affect a wide variety of prey species, but in other cases they appear to be relatively specific, so that injection into an inappropriate species has no effect.

Many parasitic Hymenoptera, such as Braconidae, Ichneumonidae and Chalcididae, also use a venom which serves to immobilize the host. This is true even of species which lay their eggs on the outsides of their hosts. The effect is often to cause a partial and temporary paralysis of the host while the insect is ovipositing. The venom is followed, in some species where the larvae are internal parasites, by fluid from the calyces of the oviducts which, in some species, contains a polyDNA virus (section 5.3.4).

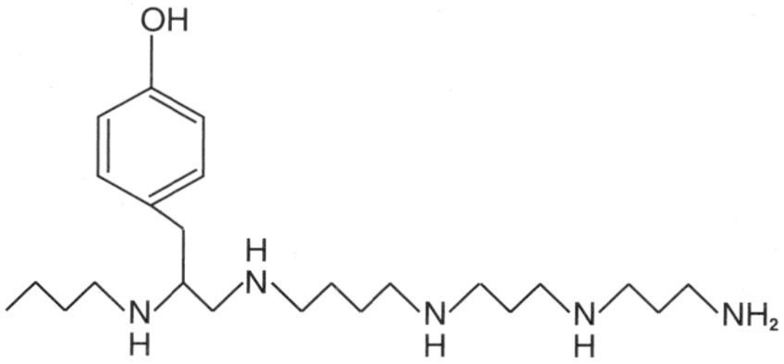

Fig. 27.24. Venom of predaceous wasp. Philanthotoxin 433 from the venom of *Philanthus triangulum*. The venom blocks glutamate receptors.

Amongst predaceous Heteroptera, the prey is rapidly subdued, apparently by fast-acting lytic processes caused by enzymes rather than by specific toxins.

Mutualisms, chemical mimicry and camouflage

Exocrine secretions are important in the mutualistic relationships of some insects with other organisms. The larvae of many lycaenid and riodinid butterflies are dependent on ants for their development. Lycaenid larvae produce secretions from glands on the dorsal surface of the seventh abdominal segment as well as from single-celled glands scattered in the epidermis. The secretions contain high concentrations of sugars, glucose, sucrose and fructose, and, in some cases, also some amino acids. They are attractive to certain ants and prevent the larvae being attacked by the ants, which, in turn, protect the larvae from parasites and other predators.

Some non-social insects and other arthropods habitually live in colonies of social insects. These inquilines are accepted in the colonies because their epicuticular waxes resemble those of the host species. Sometimes the inquilines produce wax which is qualitatively similar to that of the host. This is true, for example, of staphylinid beetles in the genera *Trichopsenius* and *Philotermes* that are inquilines of the termite, *Reticulitermes*, and of the larvae of a syrphid, *Microdon*, which are predaceous on the brood of ants in several genera, including *Camponotus*. This type of concealment is called chemical mimicry by Howard & Akre (1995). In other cases, the inquilines camouflage themselves by acquiring wax components from their hosts. This occurs with the scarab beetle, *Myrmecaphodius* that lives in the nests of *Solenopsis*.

Reviews: Baylis & Pierce, 1993 – caterpillar/ant mutualisms; Bowers, 1993 – caterpillar defenses; Howard & Akre, 1995 – chemical mimicry & camouflage; Piek, 1985 – insect venoms; Whitman, 1990 – grasshopper defenses

REFERENCES

Abed, D., Cheviet, P., Farine, J.-P., Bonnard, O., le Quéré, J.-L., & Brossut, R. (1993). Calling behavior of female *Periplaneta americana*: behavioural analysis and identification of the pheromone source. *Journal of Insect Physiology*, **39**, 709–20.

Aldrich, J.R. (1988). Chemical ecology of the Heteroptera. *Annual Review of Entomology*, **33**, 211–38.

Aldrich, J.R. (1995). Chemical communication in the true bugs and parasitoid exploitation. In *Chemical Ecology of Insects 2*, ed. R.T. Cardé & W.J. Bell, pp. 318–63. New York: Chapman & Hall.

Aldrich, J.R., Oliver, J.E., Lusby, W.R., Kochansky, J.P. & Borges, M. (1994). Identification of male-specific volatiles from Nearctic and Neotropical stink bugs (Heteroptera: Pentatomidae). *Journal of Chemical Ecology*, **20**, 1103–11.

Ando, T., Kasuga, K., Yajima, Y., Kataoka, H. & Suzuki, A. (1996). Termination of sex pheromone production in mated females of the silkworm moth. *Archives of Insect Biochemistry and Physiology*, **31**, 207–18.

Appelgren, M., Bergström, G., Svensson, B.G. & Cederberg, B. (1991). Marking pheromones of *Megabombus* bumble bee males. *Acta Chemica Scandinavica*, **45**, 972–4.

Attygalle, A.B., Smedley, S.R., Meinwold, J. & Eisner, T. (1993). Defensive secretion of two notodontid caterpillars (*Schizura unicornis, S. badia*). *Journal of Chemical Ecology*, **19**, 2089–104.

Bartelt, R.J., Weaver, D.K. & Arbogast, R.T. (1995). Aggregation pheromone of *Carpophilus dimidiatus* (F.) (Coleoptera: Nitidulidae) and responses to *Carpophilus* pheromones in South Carolina. *Journal of Chemical Ecology*, **21**, 1763–79.

Barth, R. (1944). Die männlichen Duftorgane einiger *Argynnis*-Arten. *Zoologische Jahrbücher (Anatomie)*, **68**, 331–62.

Baylis, M. & Pierce, N.E. (1993). The effects of ant mutualism on the foraging and diet of lycaenid caterpillars. In *Caterpillars: Ecological and Evolutionary Constraints on Foraging*, ed. N.E. Stamp & T.M. Casey, pp. 404–21. New York: Chapman & Hall.

Bergman, P. & Bergström, G. (1997). Scent marking, scent origin, and species specificity in male premating behavior of two Scandinavian bumble-bees. *Journal of Chemical Ecology*, **23**, 1235–51.

Bergström, G., Wassgren, A.-B., Högberg, H.-E., Hedenström, E., Hefetz, A., Simon, D., Ohlsson, T. & Löfqvist, J. (1992). Species-specific, two-component volatile signals in two sympatric ant-lion species: *Synclysis baetica* and *Acanthaclisis occitanica* (Neuroptera, Myrmeleontidae). *Journal of Chemical Ecology*, **18**, 1177–88.

Billen, J.P.J. (1986). Morphology and ultrastructure of Dufour's and venom gland in the ant, *Myrmica rubra* (L.) (Hymenoptera: Formicidae). *International Journal of Insect Morphology & Embryology*, **15**, 13–25.

Birch, M.C. (1970). Structure and function of the pheromone-producing brush-organs in males of *Phlogophora meticulosa* (L.) (Lepidoptera: Noctuidae). *Transactions of the Royal Entomological Society of London*, **122**, 277–92.

Birch, M.C., Poppy, G.M. & Baker, T.C. (1990). Scents and eversible scent structures of male moths. *Annual Review of Entomology*, **35**, 25–58.

Blomquist, G.J. & Dillwith, J.W. (1985). Cuticular lipids. In *Comprehensive Insect Physiology, Biochemistry and Pharmacology*, vol. 3, ed. G.A.Kerkut & L.I.Gilbert, pp. 117–54. Oxford: Pergamon Press.

Blum, M.S. (1981). *Chemical Defenses of Arthropods*. New York: Academic Press.

Blum, M. (1985). Alarm pheromones. In *Comprehensive Insect Physiology, Biochemistry and Pharmacology*, vol. 9, ed. G.A. Kerkut & L.I. Gilbert, pp. 193–224. Oxford: Pergamon Press.

Blum, M.S. (1987). Biosynthesis of arthropod exocrine compounds. *Annual Review of Entomology*, **32**, 381–413.

Blum, M.S. (1996). Semiochemical parsimony in the Arthropoda. *Annual Review of Entomology*, **41**, 353–74.

Boppré, M. (1984). Chemically mediated interactions between butterflies. *Symposium of the Royal Entomological Society of London*, **11**, 259–75.

Borden, J.H. (1985). Aggregation pheromones. In *Comprehensive Insect Physiology, Biochemistry and Pharmacology*, vol. 9, ed. G.A. Kerkut & L.I. Gilbert, pp. 257–85. Oxford: Pergamon Press.

Bowers, M.D. (1990). Recycling plant natural products for insect defense. In *Insect Defenses. Adaptive Mechanisms and Strategies of Prey and Predators*, ed. D.L. Evans & J.O. Schmidt, pp. 353–86. Albany: State University of New York Press.

Bowers, M.D. (1993). Aposematic caterpillars: life-styles of the warningly colored and unpalatable. In *Caterpillars: Ecological and Evolutionary Constraints on Foraging*, ed. N.E. Stamp & T.M. Casey, pp. 331–71. New York: Chapman & Hall.

Bradshaw, J.W.S., Baker, R. & Howse, P.E. (1979). Multicomponent alarm pheromones in the mandibular glands of major workers of the African weaver ant, *Oecophylla longinoda*. *Physiological Entomology*, **4**, 15–25.

Brower, L.P. (1984). Chemical defence in butterflies. *Symposium of the Royal Entomological Society of London*, **11**, 109–39.

Cassier, P. & Lensky, Y. (1994). The Nassanov gland of the workers of the honeybee (*Apis mellifera* L.): ultra-structure and behavioural function of the terpenoid and protein components. *Journal of Insect Physiology*, **40**, 577–84.

Cassier, P., Tel-Zur, D. & Lensky, Y. (1994). The sting sheaths of honey bee workers (*Apis mellifera* L.): structure and alarm pheromone secretion. *Journal of Insect Physiology*, **40**, 23–32.

Christensen, T.A., Lashbrook, J.M. & Hildebrand, J.G. (1994). Neural activation of the sex-pheromone gland in the moth *Manduca sexta*: real-time measurement of pheromone release. *Physiological Entomology*, **19**, 265–70.

Christensen, T.A., Lehman, H.K., Teal, P.E.A., Itagaki, H., Tumlinson, J.H. & Hildebrand, J.G. (1992). Diel changes in the presence and physiological actions of octopamine in the female sex-pheromone glands of heliothine moths. *Insect Biochemistry and Molecular Biology*, **22**, 841–9.

Dani, F.R., Morgan, E.D. & Turillazzi, S. (1995). Chemical analysis of sternal gland secretion of paper wasp *Polistes dominulus* (Christ) and its social parasite *Polistes sulcifer* (Zimmermann) (Hymenoptera: Vespidae). *Journal of Chemical Ecology*, **21**, 1709–18.

Davis, R.H. & Nahrstedt, A. (1985). Cyanogenesis in insects. In *Comprehensive Insect Physiology, Biochemistry and Pharmacology*, vol. 11, ed. G.A. Kerkut & L.I. Gilbert, pp. 635–54. Oxford: Pergamon Press.

Deroe, C. & Pasteels, J.M. (1977). Defensive mechanisms against predation in the Colorado beetle (*Leptinotarsa decemlineata* Say). *Archives de Biologie, Bruxelles*, **88**, 289–304.

Dettner, K. (1987). Chemosystematics and evolution of beetle chemical defenses. *Annual Review of Entomology*, **32**, 17–48.

Duffey, S.S. (1980). Sequestration of plant natural products by insects. *Annual Review of Entomology*, **25**, 447–77.

Eisner, T., Goetz, M., Aneshansley, D., Ferstandig-Arnold, G. & Meinwold, J. (1986). Defensive alkaloid in blood of Mexican bean beetle (*Epilachna varivestis*). *Experientia*, **42**, 204–7.

Fabriás, G., Marco, M.P. & Camps, T. (1994). Effect of the pheromone biosynthesis activating peptide on sex pheromone biosynthesis in *Spodoptera littoralis* isolated glands. *Archives of Insect Biochemistry and Physiology*, 27, 77–87.

Farine, J.-P., Everaerts, C., Abed, D., Ntari, M. & Brossut, R. (1996). Pheromonal emission during the mating behavior of *Eurycotis floridana* (Walker) (Dictyoptera: Blattidae). *Journal of Insect Behavior*, 9, 197–213.

Farine, J.-P., le Quere, J.-L., Duffy, J., Everaerts, C. & Brossut, R. (1994). Male sex pheromone of cockroach *Eurycotis floridana* (Walker) (Blattidae, Polyzosteriinae): role and composition of tergites 2 and 8 secretions. *Journal of Chemical Ecology*, 20, 2291–306.

Fang, N., Teal, P.E.A. & Tumlinson, J.H. (1995). PBAN regulation of pheromone biosynthesis in female tobacco hornworm moth, *Manduca sexta* (L.). *Archives of Insect Biochemistry and Physiology*, 29, 35–44.

Fitzgerald, T.D. (1993). Trail following and recruitment: response of eastern tent caterpillar *Malacosoma americanum* to 5β–cholestane–3,24–dione and 5β–cholestan–3-one. *Journal of Chemical Ecology*, 19, 449–57.

Forsyth, D.J. (1970). The structure of the defence glands of the Cicindellidae, Amphizoidae, and Hygrobiidae (Insecta: Coleoptera). *Journal of Zoology*, 160, 51–69.

Foster, S.P., Bergh, J.C., Rose, S. & Harris, M.O. (1991). Aspects of pheromone biosynthesis in the Hessian fly, *Mayetiola destructor* (Say). *Journal of Insect Physiology*, 37, 899–906.

Fouad, K., Libersat, F. & Rathmayer, W. (1996). Neuromodulation of the escape behavior of the cockroach *Periplaneta americana* by the venom of the parasitic wasp *Ampulex compressa*. *Journal of Comparative Physiology* A, 178, 91–100.

Franzl, S. & Naumann, C.M. (1985). Cuticular cavities: storage chambers for cyanoglucoside–containing defensive secretions in larvae of a zygaenid moth. *Tissue & Cell*, 17, 267–78.

Frick, C. & Wink, M. (1995). Uptake and sequestration of ouabain and other cardiac glycosides in *Danaus plexippus* (Lepidoptera: Danaidae): evidence for a carrier-mediated process. *Journal of Chemical Ecology*, 21, 557–75.

Gardner, D.R. & Stermitz, F.R. (1988). Host plant utilization and iridoid glycoside sequestration by *Euphydryas anicia* (Lepidoptera: Nymphalidae). *Journal of Chemical Ecology*, 14, 2147–68.

Grassé, P.P. (1975). Les glandes tégumentaires des insects. In *Traité de Zoologie*, vol. 8, part 3, ed. P.P. Grassé, pp. 199–320. Paris: Masson et Cie.

Hallett, R.H. & others. (1995). Aggregation pheromone of coconut rhinoceros beetle, *Oryctes rhinoceros* (L.) (Coleoptera: Scarabaeidae). *Journal of Chemical Ecology*, 21, 1549–70.

Happ, G.M. Schroeder, M.E. & Wang, J.C.H. (1970). Effects of male and female scent on reproductive maturation in young female *Tenebrio molitor*. *Journal of Insect Physiology*, 16, 1543–8.

Haynes, K.F. & Birch, M.C. (1985). The role of other pheromones, allomones and kairomones in the behavioral responses of insects. In *Comprehensive Insect Physiology, Biochemistry and Pharmacology*, vol. 9, ed. G.A. Kerkut & L.I. Gilbert, pp. 225–55. Oxford: Pergamon Press.

Hedlund, K., Bartelt, R.J., Dicke, M. & Vet, L.E.M. (1996). Aggregation pheromones of *Drosophila immigrans*, *D.phalerata*, and *D.subobscura*. *Journal of Chemical Ecology*, 22, 1835–44.

Hölldobler, B. & Traniello, J.F.A. (1980). The pygidial gland and chemical recruitment communication in *Pachycondyla (=Termitopone) laevigata*. *Journal of Chemical Ecology*, 6, 883–93.

Hölldobler, B. & Wilson, E.O. (1990). *The Ants*. Cambridge, Massachusetts: Harvard University Press.

Holzinger, F. & Wink, M. (1996). Mediation of cardiac glycoside insensitivity in the monarch butterfly (*Danaus plexippus*): role of an amino acid substitution in the ouabain binding site of Na$^+$,K$^+$-ATPase. *Journal of Chemical Ecology*, 22, 1921–37.

Howard, R.W. (1993). Cuticular hydrocarbons and chemical communication. In *Insect Lipids: Chemistry, Biochemistry and Biology*, ed. D.W. Stanley-Samuelson & F.R. Nelson, pp. 178–226. Lincoln: University of Nebraska Press.

Howard, R.W. & Akre, R.D. (1995). Propaganda, crypsis and slave-making. In *Chemical Ecology of Insects 2*, ed. R.T. Cardé & W.J. Bell, pp. 364–424. New York: Chapman & Hall.

Hunt, R.E. & Haynes, K.F. (1990). Periodicity in the quantity and blend ratios of pheromone components in glands and volatile emissions of mutant and normal cabbage looper moths, *Trichoplusia ni*. *Journal of Insect Physiology*, 36, 769–74.

Ivarsson, P. & Birgersson, G. (1995). Regulation and biosynthesis of pheromone components in the double spined bark beetle *Ips duplicatus* (Coleoptera: Scolytidae). *Journal of Insect Physiology*, 41, 843–9.

James, D.G., Moore, C.J. & Aldrich, J.R. (1994). Identification, synthesis and bioactivity of a male-produced aggregation pheromone in assassin bug, *Pristhesancus plagipennis* (Hemiptera: Reduviidae). *Journal of Chemical Ecology*, 20, 3281–95.

Jeanne, R.L. & Keeping, M.G. (1995). Venom spraying in *Parachartergus colobopterus*: a novel defensive behavior in a social wasp (Hymenoptera: Vespidae). *Journal of Insect Behavior*, 8, 433–42.

Kellner, R.L.L. & Dettner, K. (1995). Allocation of pederin during lifetime of *Paederus* rove beetles (Coleoptera: Staphylinidae): evidence for poly-morphism of hemolymph toxin. *Journal of Chemical Ecology*, **21**, 1719–33.

Layton, J.M., Camann, M.A. & Espelie, K.E. (1994). Cuticular lipid profiles of queens, workers, and males of social wasp *Polistes metricus* Say are colony-specific. *Journal of Chemical Ecology*, **20**, 2307–21.

L'Empereur, K.M. & Stermitz, F.R. (1990). Iridoid glycoside content of *Euphydryas anicia* (Lepidoptera: Nymphalidae) and its major hostplant, *Besseya plantaginea* (Scrophulariaceae), at a high plains Colorado site. *Journal of Chemical Ecology*, **16**, 187–97.

Lensky, Y., Cassier, P. & Tel-Zur, D. (1995). The setaceous membrane of honey bee (*Apis mellifera* L.) workers' sting apparatus: structure and alarm pheromone distribution. *Journal of Insect Physiology*, **41**, 589–95.

Liu, Y.-B. & Haynes, K.F. (1994). Temporal and temperature-induced changes in emission rates and blend ratios of sex pheromone components in *Trichoplusia ni*. *Journal of Insect Physiology*, **40**, 341–6.

Loher, W. (1990). Pheromones and phase transformation in locusts. In *Biology of Grasshoppers*, ed. R.F. Chapman & A. Joern, pp. 337–55. New York: Wiley.

Ma, P.W.K. & Roeloffs, W.L. (1995). Sites of synthesis and release of PBAN-like factor in the female European corn borer, *Ostrinia nubilalis*. *Journal of Insect Physiology*, **41**, 339–50.

Nascimento, R.R.do, Morgan, E.D., Billen, J., Schoeters, E., Lucia, T.M.C.della & Bento, J.M.S. (1993). Variation with caste of the mandibular gland secretion in the leaf-cutting ant *Atta sexdens rubropilosa*. *Journal of Chemical Ecology*, **19**, 907–18.

Oehlschlager, A.C., Prior, R.N.B., Perez, A.L., Gries, R., Gries, G., Pierce, H.D. & Laup, S. (1995). Structure, chirality, and field testing of a male-produced aggregation pheromone of Asian palm weevil *Rhynchophorus bilineatus* (Montr.) (Coleoptera: Curculionidae). *Journal of Chemical Ecology*, **21**, 1619–29.

Percy-Cunningham, J.E. & MacDonald, J.A. (1987). Biology and ultrastructure of sex-pheromone-producing glands. In *Pheromone Biochemistry*, ed. G.D. Prestwich, & G.J. Blomquist, pp. 27–75. Orlando: Academic Press.

Pickett, J.A., Wadhams, L.J., Woodcock, C.M. & Hardie, J. (1992). The chemical ecology of aphids. *Annual Review of Entomology*, **37**, 67–90.

Piek, T. (1985). Insect venoms and toxins. In *Comprehensive Insect Physiology, Biochemistry and Pharmacology*, vol. 11, ed. G.A. Kerkut & L.I. Gilbert, pp. 595–633. Oxford: Pergamon Press.

Plettner, E., Slessor, K.N., Winston, M.L. & Oliver, J.E. (1996). Caste-selective pheromone biosynthesis in honeybees. *Science*, **271**, 1851–3.

Prestwich, G.D. (1984). Defense mecha-nisms of termites. *Annual Review of Entomology*, **29**, 201–32.

Prestwich, G.D. & Blomquist, G.J. (eds.) (1987). *Pheromone Biochemistry*. Orlando: Academic Press.

Prokopy, R.J. (1981). Epideictic pheromones that influence spacing pat-terns of phytophagous insects. In *Semiochemicals: Their Role in Pest Control*, ed. D.A. Nordlund, R.L. Jones & W.J.Lewis, pp. 181–213. New York: Wiley.

Raina, A.K. (1993). Neuroendocrine control of sex pheromone biosynthesis in Lepidoptera. *Annual Review of Entomology*, **38**, 329–49.

Raina, A.K., Davis, J.C. & Stadelbacher, E.A. (1991). Sex pheromone produc-tion and calling in *Helicoverpa zea* (Lepidoptera: Noctuidae): effect of temperature and light. *Environmental Entomology*, **20**, 1451–6.

Schneider, D. (1987). The strange fate of pyrrolizidine alkaloids. In *Perspectives in Chemoreception and Behavior*, ed. R.F. Chapman, E.A. Bernays & J.G. Stoffolano, pp. 123–42. New York: Springer-Verlag.

Scudder, G.G.E., Moore, L.V. & Isman, M.B. (1986). Sequestration of cardeno-lides in *Oncopeltus fasciatus*: morpho-logical and physiological adaptations. *Journal of Chemical Ecology*, **12**, 1171–87.

Smith, B.H. & Breed, M.D. (1995). The chemical basis for nestmate recognition and mate discrimination in social insects. In *Chemical Ecology of Insects 2*, ed. R.T.Cardé & W.J.Bell, pp. 287–317. New York: Chapman & Hall.

Snodgrass, R.E. (1956). *Anatomy of the Honey Bee*. London: Constable.

Staddon, B.W. (1979). The scent glands of Heteroptera. *Advances in Insect Physiology*, **14**, 351–418.

Tamaki, Y. (1985). Sex pheromones. In *Comprehensive Insect Physiology, Biochemistry and Pharmacology*, vol. 9, ed. G.A. Kerkut & L.I. Gilbert, pp. 145–91. Oxford: Pergamon Press.

Thieme, T. & Dixon, A.F.G. (1996). Mate recognition in the *Aphis fabae* complex: daily rhythm of release and specificity of sex pheromones. *Entomologia Experimentalis et Applicata*, **79**, 85–9.

Thiéry, D., Gabel, B., Farkas, P. & Jarry, M. (1995). Egg dispersion in codling moth: influence of egg extract and of its fatty acid constituents. *Journal of Chemical Ecology*, **21**, 2015–26.

Torto, B., Obieg-Fori, D., Njagi, P.G.N., Hassanali, A. & Amiani, H. (1994). Aggregation pheromone system of adult gregarious desert locust *Schistocerca gregaria* (Forskål). *Journal of Chemical Ecology*, **20**, 1749–62.

Traniello, J.F.A. & Robson, S.K. (1995). Trail and territorial communication in social insects. In *Chemical Ecology of Insects 2*, ed. R.T. Cardé & W.J. Bell, pp. 241–86. New York: Chapman & Hall.

Vander Meer, R.K., Breed, M.D., Winston, M.L. & Espelie, K.E. eds. (1998). *Pheromone Communication in Social Insects*. Boulder: Westview Press.

Völkl, W., Hübner, G. & Dettner, K. (1994). Interactions between *Alloxystia brevis* (Hymenoptera, Cynipoidea, Alloxystidae) and honeydew-collecting ants: how an aphid parasitoid overcomes ant aggression by chemical defense. *Journal of Chemical Ecology*, **20**, 2901–15.

Whitman, D.W. (1990). Grasshopper chemical communication. In *Biology of Grasshoppers*, ed. R.F. Chapman & A. Joern, pp. 357–91. New York: Wiley.

Whitman, D.W., Blum, M.S. & Alsop, D.W. (1990). Allomones: chemicals for defense. In *Insect Defenses. Adaptive Mechanisms and Strategies of Prey and Predators*, ed. D.L. Evans & J.O. Schmidt, pp. 289–351. Albany: State University of New York Press.

Wigglesworth, V.B. (1972). *The Principles of Insect Physiology*. London: Methuen.

Winston, M.L. (1987). *The Biology of the Honey Bee*. Cambridge, Massachusetts: Harvard University Press.

Wood, D.L. (1982). The role of pheromones, kairomones, and allomones in the host selection and colonization behavior of bark beetles. *Annual Review of Entomology*, **27**, 411–46.

Yin, L.R.S., Schal, C. & Cardé, R.T. (1990). Sex pheromone gland of the female tiger moth *Holomelina lamae* (Lepidoptera: Arctiidae). *Canadian Journal of Zoology*, **69**, 1916–21.

Erratum

All page numbers higher than 415 are incorrect in these indexes. To obtain the correct page number subtract 2 from each page number higher than 415.

Taxonomic Index

Page numbers in bold denote illustrations. The major orders are omitted.

Subject Index

Page numbers in bold refer to a major reference or an illustration